T0181684

Book of Abstracts of the 69th Annual Meeting of the European Federation of Animal Science

The European Federation of Animal Science wishes to express its appreciation to the
Ministero delle Politiche Agricole Alimentari e Forestali (Italy) and the
Associazione Italiana Allevatori (Italy)
for their valuable support of its activities.

Book of Abstracts of the 69th Annual Meeting of the European Federation of Animal Science

Dubrovnik, Croatia, 27th – 31st August, 2018

EAN: 9789086863235
e-EAN: 9789086868711
ISBN: 978-90-8686-323-5
e-ISBN: 978-90-8686-871-1
DOI: 10.3920/978-90-8686-871-1

ISSN 1382-6077

First published, 2018

Wageningen Academic
Publishers

Welcome to Dubrovnik and Croatia

On behalf of the Croatian Organising Committee, we are pleased to invite you to attend the 69th Annual Meeting of the European Federation of Animal Science (EAAP). The meeting will be held in the Valamar Resort in Dubrovnik, one of the most prominent tourist destinations in the Mediterranean, from 27th to 31st August 2018.

For decades, the Annual Meeting has hosted scientists and experts from the field of animal science, not only from Europe but also from other countries around the globe. The EAAP Congress provides insights into the latest research results from many areas of animal science. It is a unique opportunity for industry and scientists to meet and acquire new knowledge as well as to exchange experience. Carried out through many sessions, presentations and discussions about scientific achievements in the European and world livestock production are also an opportunity for the application of new ideas in practice. Furthermore, there will be a focus on international research collaboration and knowledge exchange towards innovation. All these preferences make the EAAP one of the largest animal science congresses in the world – we expect approximately 1000 participants from more than 50 countries.

The main topic of the congress is 'Conventional and traditional livestock production systems – new challenges' and it includes sustainability, animal welfare, agroecology and product quality. The programme contains various disciplines and the latest findings regarding farm animals such as genetics, nutrition, management, health, welfare and physiology of cattle, sheep, goats, pigs, horses, poultry and fur animals, as well as the use of insects for feed and precision livestock farming..

We are delegated to invite you to participate in the 69th Annual Meeting of EAAP which focuses on translating research into animal production practice.

Assist. Prof. Zdravko Barać,
Chairman of the Organising Committee,
Croatian Agricultural Agency

Mr. Tomislav Tolušić,
Minister of Agriculture,
Patron of the 69th Annual Meeting

National Organisers of the 69th EAAP Annual Meeting

National Organising Committee:

President

- **Assist. Prof. Zdravko Barać,** Croatian Agricultural Agency

Secretary General

- **Dr. Marija Špehar,** Croatian Agricultural Agency

Members

- **Prof. Boro Mioč,** University of Zagreb – Faculty of Agriculture
- **Prof. Zlatko Janječić,** University of Zagreb – Faculty of Agriculture
- **Prof. Josip Leto,** University of Zagreb – Faculty of Agriculture
- **Assist. Prof. Nikica Šprem,** University of Zagreb – Faculty of Agriculture
- **Prof. Vesna Gantner,** University J. J. Strossmayer in Osijek – Faculty of Agriculture
- **Prof. Pero Mijić,** University J. J. Strossmayer in Osijek – Faculty of Agriculture
- **Prof. Anđelko Opačak,** University J. J. Strossmayer in Osijek – Faculty of Agriculture
- **Prof. Zvonko Antunović,** University J. J. Strossmayer in Osijek – Faculty of Agriculture
- **Prof. Velimir Sušić,** University of Zagreb – Faculty of Veterinary Medicine
- **Prof. Anamaria Ekert Kabalin,** University of Zagreb – Faculty of Veterinary Medicine
- **Assoc. Prof. Slaven Zjalić,** University of Zadar
- **Assoc. Prof. Bosiljka Mustać,** University of Zadar
- **Mr. Zdravko Tušek,** Agricultural Advisory Service
- **Dr. Danko Dežđek,** Croatian Chamber of Agriculture
- **Dr. Maja Dražić,** Croatian Agricultural Agency

European Federation of Animal Science (EAAP)

President:	M. Gauly
Secretary General:	A. Rosati
Address:	Via G. Tomassetti 3, A/I I-00161 Rome, Italy
Phone/Fax:	+39 06 4420 2639
E-mail:	eaap@eaap.org
Web:	www.eaap.org

Council Members

European Federation of Animal Science has close established links with the sister organisations American Dairy Science Association, American Society of Animal Science, Canadian Society of Animal Science and Asociación Latinoamericana de Producción Animal.

Friends of EAAP

By creating the 'Friends of EAAP', EAAP offers the opportunity to industries to receive services from EAAP in change of a fixed sponsoring amount of support every year.

- The group of supporting industries are layered in three categories: 'silver', 'gold' and 'diamond' level.
- It is offered an important discount (one year free of charge) if the sponsoring industry will agree for a four years period.
- EAAP will offer the service to create a scientific network (with Research Institutes and Scientists) around Europe.
- Creation of a permanent Board of Industries within EAAP with the objective to inform, influence the scientific and organizational actions of EAAP, like proposing choices of the scientific sessions and invited speakers and to propose industry representatives for the Study Commissions.
- Organization of targeted workshops, proposed by industries.
- EAAP can represent and facilitate activities of the supporting industries toward international legislative and regulatory organizations.
- EAAP can facilitate the supporting industries to enter in consortia dealing with internationally supported research projects.

Furthermore EAAP offers, depending to the level of support (details on our website: www.eaap.org):

- Free entrances to the EAAP annual meeting and Gala dinner invitation.
- Free registration to journal *animal*.
- Inclusion of industry advertisement in the EAAP Newsletter, in the banner of the EAAP website, in the Book of Abstract and in the Programme Booklet of the EAAP annual meeting.
- Inclusion of industry leaflets in the annual meeting package.
- Presence of industry advertisements on the slides between presentations at selected standard sessions.
- Presence of industry logos and advertisements on the slides between presentations at the Plenary Sessions.
- Public Recognition by the EAAP President at the Plenary Opening Session of the annual meeting.
- Discounted stands at the EAAP annual meeting.
- Invitation to meetings (at every annual meeting) to discuss joint strategy EAAP/Industries with the EAAP President, Vice-President for Scientific affair, Secretary General and other selected members of the Council and of the Scientific Committee.

Contact and further information

If the industry you represent is interested to become 'Friend of EAAP' or want to have further information please contact jean-marc.perez0000@orange.fr or EAAP secretariat (eaap@eaap.org, phone : +39 06 44202639).

The Association

EAAP (The European Federation of Animal Science) organises every year an international meeting which attracts between 900 and 1500 people. The main aims of EAAP are to promote, by means of active co-operation between its members and other relevant international and national organisations, the advancement of scientific research, sustainable development and systems of production; experimentation, application and extension; to improve the technical and economic conditions of the livestock sector; to promote the welfare of farm animals and the conservation of the rural environment; to control and optimise the use of natural resources in general and animal genetic resources in particular; to encourage the involvement of young scientists and technicians. More information on the organisation and its activities can be found at www.eaap.org

What is the YoungEAAP?

YoungEAAP is a group of young scientists organized under the EAAP umbrella. It aims to create a platform where scientists during their early career get the opportunity to meet and share their experiences, expectations and aspirations. This is done through activities at the Annual EAAP Meetings and social media. The large constituency and diversity of the EAAP member countries, commissions and delegates create a very important platform to stay up-to-date, close the gap between our training and the future employer expectations, while fine-tuning our skills and providing young scientists applied and industry-relevant research ideas.

YoungEAAP promotes Young and Early Career Scientists to:

- Stay up-to-date (i.e. EAAP activities, social media);
- Close the gap between our training and the future employer expectations;
- Fine-tune our skills through EAAP meetings, expand the special young scientists' sessions, and/or start online webinars/trainings with industry and academic leaders;
- Meet to network and share our graduate school or early employment experiences;
- Develop research ideas, projects and proposals.

Committee Members at a glace

- Dr. Christian Lambertz (President)
- Dr. Anamarija Smetko (Secretary)
- MSc Torun Wallgren (Secretary)

Who can be a Member of Young EAAP?

All individual members of EAAP can join the YoungEAAP if they meet one of the following criteria: Researchers under 35 years of age OR within 10 years after PhD-graduation

Just request your membership form (christian.lambertz@unibz.it) and become member of this network!!!

70th Annual Meeting of the European Federation of Animal Science

Ghent, Belgium, 26 – 30 August 2019

Contact
Conference organizer
Semico Ghent Office
Korte Meer 16-20
9000 Ghent
Belgium
T +32 9 233 86 60 – F +32 9 233 85 97
EAAP2019@semico.be

Organizing Committees

The 70th EAAP annual meeting is organized by the Flanders Research Institute for Agriculture, Fisheries and Food (ILVO) under the patronage of the Department of Agriculture and Fisheries (DAF) from the Flemish Government

Belgian Steering Committee

President:	Sam De Campeneere (ILVO)
Executive secretaries:	Bart Sonck (ILVO), Sam Millet (ILVO), Johan De Boever (ILVO)
Members:	Hanne Geenen (DAF), Ivan Ryckaert (DAF), Jurgen Vangeyte (ILVO), Stefaan De Smet (UGent), Nadia Everaert (ULG)
Conference Organizer:	Lieve Ectors (Semico)

Belgian Scientific Committee

President:	Johan De Boever (ILVO)
Vice-president:	Sam De Campeneere (ILVO)
Executive secretaries:	Sam Millet (ILVO), Bart Sonck (ILVO)
Animal Genetics:	Nadine Buys (KUL), Stefaan De Smet (UGent), Nicolas Gengler (ULG)
Animal Health and Welfare:	Ben Aernouts (KUL), Dominiek Maes (UGent), Frank Tuyttens (ILVO)
Animal Nutrition:	Yves Beckers (ULG), Eric Froidmont (CRA-W), Veerle Fievez (UGent), Yvan Larondelle (UCL)
Animal Physiology:	Nadia Everaert (ULG), Jo Leroy (UA), Geert Janssens (UGent)
Livestock Farming Systems:	Jérôme Bindelle (ULG), Didier Stilmant (CRA-W), Daan Delbare (ILVO)
Cattle Production:	Frédéric Dehareng (CRA-W), Geert Opsomer (UGent), Leen Vandaele (ILVO)
Horse Production:	Steven Janssens (KUL), Charlotte Sandersen (ULG), Myriam Hesta (UGent)
Pig Production:	Joris Michiels (UGent), Marijke Aluwé (ILVO), Jeroen Dewulf (UGent)
Sheep and Goat:	Pierre Rondia (CRA-W), Bert Driessen (KUL)
Insects:	Mik Van Der Borgt (KUL), Veerle Van linden (ILVO)
Precision Livestock Farming:	Jurgen Vangeyte (ILVO), Tomas Norton (KUL), Hélène Soyeurt (ULG)

Conference website: www.eaap2019.org

Commission on Animal Genetics

Name	Position / Country	Affiliation / Email
Erling Strandberg	President Sweden	Swedish University of Agricultural Sciences erling.strandberg@slu.se
Eileen Wall	Vice-President United Kingdom	Scotland's Rural College eileen.wall@sruc.ac.uk
Jan Lassen	Vice-President Denmark	Aarhus University jan.lassen@mbg.au.dk
Han Mulder	Vice-President The Netherlands	Wageningen University han.mulder@wur.nl
Alessio Cecchinato	Secretary Italy	Padova University alessio.cecchinato@unipd.it
Birgit Gredler	Industry rep. Switzerland	Qualitas AG birgit.gredler@qualitasag.ch

Commission on Animal Nutrition

Name	Position / Country	Affiliation / Email
Giovanni Savoini	President Italy	University of Milan giovanni.savoini@unimi.it
Eleni Tsiplakou	Vice-President Greece	Agricultural University of Athens eltsiplakou@aua.gr
Geert Bruggeman	Secretary Belgium	Nuscience Group geert.bruggeman@nusciencegroup.com
Roselinde Goselink	Secretary The Netherlands	Wageningen Livestock Research roselinde.goselink@wur.nl

Commission on Health and Welfare

Name	Position / Country	Affiliation / Email
Hans Spoolder	President The Netherlands	Wageningen Livestock Research hans.spoolder@wur.nl
Gurbuz Das	Vice-President Germany	Leibniz Institute for Farm Animal Biology gdas@fbn-dummerstorf.de
Evangelia Sossidou	Vice-President Greece	Hellenic Agricultural Organization sossidou.arig@nagref.gr
Stefano Messori	Secretary France	STAR-IDAZ IRC Secretariat stefano.messori@yahoo.it
Valentina Ferrante	Secretary Italy	Università degli Studi di Milano valentina.ferrante@unimi.it

Commission on Animal Physiology

Name	Position / Country	Affiliation / Email
Helga Sauerwein	President Germany	University of Bonn sauerwein@uni-bonn.de
Isabelle Louveau	Vice-President France	INRA isabelle.louveau@inra.fr
Chris Knight	Secretary Denmark	University of Copenhagen chkn@sund.ku.dk
Arnulf Troescher	Industry rep. Germany	BASF arnulf.troescher@basf.com

Commission on Livestock Farming Systems

Muriel Tichit	President France	INRA muriel.tichit@agroparistech.fr
Michael Lee	Vice-President United Kingdom	University of Bristol michael.lee@rothamsted.ac.uk
Raimon Ripoll Bosch	Secretary The Netherlands	Wageningen University raimon.ripollbosch@wur.nl
Monika Zehetmeier	Secretary Germany	Institute Agricultural Economics and Farm Management monika.zehetmeier@lfl.bayern.de
Thomas Turini	Industry rep. France	Terre-ECOS t.turini@terre-ecos.com

Commission on Cattle Production

Marija Klopčič	President Slovenia	University of Ljubljana marija.klopcic@bf.uni-lj.si
Sven König	Vice-President Germany	Justus-Liebig Universität Giessen sven.koenig@agrar.uni-giessen.de
Yuri Montanholi	Vice-President United Kingdom	Harper Adams University ymontanholi@harper-adams.ac.uk
Birgit Fürst-Waltl	Secretary Austria	University of Natural Resources and Life Sciences birgit.fuerst-waltl@boku.ac.at
Massimo De Marchi	Secretary Italy	Padova University massimo.demarchi@unipd.it
Ray Keatinge	Industry rep. United Kingdom	Agriculture & Horticulture Development Board ray.keatinge@ahdb.org.uk

Commission on Sheep and Goat Production

Joanne Conington	President United Kingdom	Scotland's Rural College joanne.conington@sac.ac.uk
Nóirín McHugh	Vice-President Ireland	Teagasc noirin.mchugh@teagasc.ie
Ouranios Tzamaloukas	Secretary Cyprus	Cyprus University of Technology ouranios.tzamaloukas@cut.ac.cy
Vasco Augusto Pilão Cadavez	Secretary Portugal	CIMO - Mountain Research Centre vcadavez@ipb.pt
John Yates	Industry rep. United Kingdom	Texel johnyates@texel.co.uk

Commission on Pig Production

Egbert Knol	President The Netherlands	TOPIGS egbert.knol@topigs.com
Sam Millet	Vice-President Belgium	ILVO sam.millet@ilvo.vlaanderen.be
Giuseppe Bee	Secretary Switzerland	Agroscope Liebefeld-Posieux ALP giuseppe.bee@alp.admin.ch
Antonio Velarde	Secretary Spain	IRTA antonio.velarde@irta.es
Paolo Trevisi	Secretary Italy	Bologna University paolo.trevisi@unibo.it

Commission on Horse Production

Ana Sofia Santos	President Portugal	CECAV - UTAD - EUVG assantos@utad.pt
Rhys Evans	Vice-President Norway	Norwegian University College of Agriculture and Rural Development rhys@hlb.no
Katharina Stock	Vice-President Germany	VIT-Vereinigte Informationssysteme Tierhaltung w.V. friederike.katharina.stock@vit.de
Klemen Potočnik	Vice-President Slovenija	University of Ljubljana klemen.potocnik@bf.uni-lj.si
Isabel Cervantes Navarro	Secretary Spain	Complutense University of Madrid icervantes@vet.ucm.es
Melissa Cox	Industry rep. Germany	CAG GmbH – Center for Animal Genetics melissa.cox@centerforanimalgenetics.de

Commission on Insects

Teun Veldkamp	President The Netherlands	Wageningen Livestock Research teun.veldkamp@wur.nl
Michelle Epstein	Vice-President Austria	Medical University of Vienna michelle.epstein@meduniwien.ac.at
Jørgen Eilenberg	Secretary Denmark	University of Copenhagen jei@plen.ku.dk
Alessandro Agazzi	Secretary Italy	University of Milan alessandro.agazzi@unimi.it
Marian Peters	Industry rep. The Netherlands	IIC (International Insect Centre) marianpeters@ngn.co.nl
Alexis Angot	Industry rep. France	Ynsect aan@ynsect.com
Roel Boersma	Industry rep. The Netherlands	Protix roel.boersma@protix.eu

Commission on Precision Livestock Farming (PLF)

Ilan Halachmi	President Israel	Agriculture research organization (ARO) halachmi@volcani.agri.gov.il
Jarissa Maselyne	Vice-President Belgium	ILVO jarissa.maselyne@ilvo.vlaanderen.be
Matti Pastell	Vice-President Finland	Natural Resources Institute Finland (Luke) matti.pastell@luke.fi
Shelly Druyan	Secretary Israel	ARO, The Volcani Center shelly.druyan@mail.huji.ac.il
Radovan Kasarda	Secretary Slovakia	Slovak University of Agriculture in Nitra radovan.kasarda@uniag.sk
Malcolm Mitchell	Industry rep. United Kingdom	SRUC malcolm.mitchell@sruc.ac.uk

Sponsors

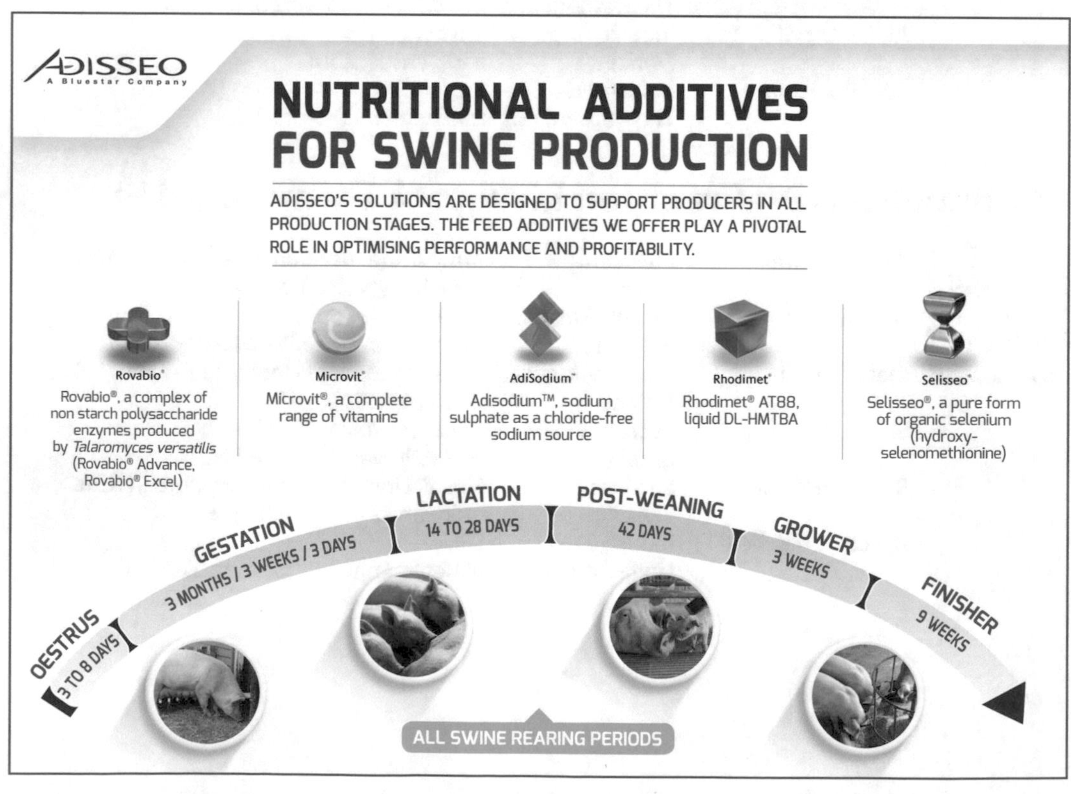

ADISSEO
A Bluestar Company
NUTRITIONAL ADDITIVES
FOR SWINE PRODUCTION
ADISSEO'S SOLUTIONS ARE DESIGNED TO SUPPORT PRODUCERS IN ALL PRODUCTION STAGES. THE FEED ADDITIVES WE OFFER PLAY A PIVOTAL ROLE IN OPTIMISING PERFORMANCE AND PROFITABILITY.
Rovabio®
Rovabio®, a complex of non starch polysaccharide enzymes produced by Talaromyces versatilis (Rovabio® Advance, Rovabio® Excel)
Microvit®
Microvit®, a complete range of vitamins
AdiSodium™
Adisodium™, sodium sulphate as a chloride-free sodium source
Rhodimet®
Rhodimet® AT88, liquid DL-HMTBA
Selisseo®
Selisseo®, a pure form of organic selenium (hydroxy-selenomethionine)
OESTRUS
3 TO 8 DAYS
GESTATION
3 MONTHS / 3 WEEKS / 3 DAYS
LACTATION
14 TO 28 DAYS
POST-WEANING
42 DAYS
GROWER
3 WEEKS
FINISHER
9 WEEKS
ALL SWINE REARING PERIODS

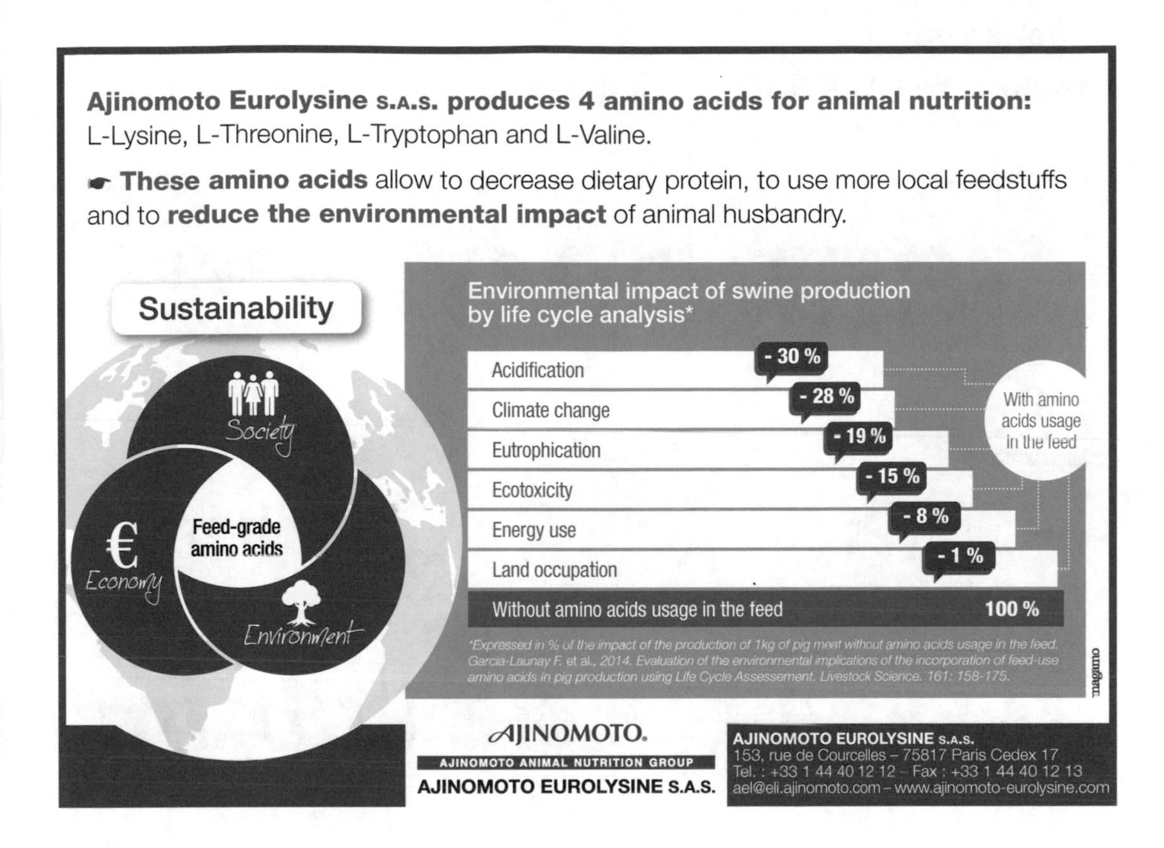
Ajinomoto Eurolysine S.A.S. produces 4 amino acids for animal nutrition:
L-Lysine, L-Threonine, L-Tryptophan and L-Valine.
These amino acids allow to decrease dietary protein, to use more local feedstuffs and to reduce the environmental impact of animal husbandry.
Sustainability
Society
Economy
Environment
Feed-grade amino acids
Environmental impact of swine production by life cycle analysis*
Acidification
- 30 %
Climate change
- 28 %
Eutrophication
- 19 %
Ecotoxicity
- 15 %
Energy use
- 8 %
Land occupation
- 1 %
With amino acids usage in the feed
Without amino acids usage in the feed
100 %
*Expressed in % of the impact of the production of 1kg of pig meat without amino acids usage in the feed. Garcia-Launay F. et al., 2014. Evaluation of the environmental implications of the incorporation of feed-use amino acids in pig production using Life Cycle Assessement. Livestock Science. 161: 158-175.
AJINOMOTO.
AJINOMOTO ANIMAL NUTRITION GROUP
AJINOMOTO EUROLYSINE S.A.S.
AJINOMOTO EUROLYSINE S.A.S.
153, rue de Courcelles – 75817 Paris Cedex 17
Tel. : +33 1 44 40 12 12 – Fax : +33 1 44 40 12 13
ael@eli.ajinomoto.com – www.ajinomoto-eurolysine.com

Tereos

SUGAR BEET AND CEREAL-BASED PRODUCTS
FOR ANIMAL NUTRITION
• Proteins
• Fibre-rich ingredients
• Energy-rich ingredients
• Moist and liquid feed ingredients
• Starches and derivatives
• Fructo-oligosaccharides (scFOS)
www.tereos.com - Tel. 03 28 38 79 30

Thank you

to the 69th EAAP Annual Congress Sponsors and Friends

Bronze sponsor

ICEROBOTICS

Sponsors

Supporters

BARTEC BENKE

Industry exhibition

Acknowledgements

an integrated infrastructure for increased research capability and innovation in the European cattle sector

WORLD ASSOCIATION for ANIMAL PRODUCTION

Scientific programme EAAP 2018

Monday 27 August 8.30 – 12.30	Monday 27 August 14.00 – 18.00	Tuesday 28 August 8.30 – 12.30
ANIMAL GENETIC RESOURCES SYMPOSIUM		
Session 01 Added value of local breeds Chair: R.B. Baumung	**Session 09** Genetic diversity and conservation Chair: J. Bennewitz	
PHYSIOLOGY DAY		
Session 02 Effects of nutrition and risk factors in early life on later performance and health Chair: H. Sauerwein	**Session 10** Hepatic, intestinal and mammary adaptive reactions: (patho-) physiology and biomarkers Chair: J.J. Gross	
	Session 11 Young Train: Dairy innovative research and extension (in cooperation with Young Scientists Club; includes NOVUS award) Chair: C. Lambertz / A. Kuipers	
ONE-DAY SEMINAR WITH INTERBULL		
Session 03 Optimization of a genomic breeding program for small sized cattle populations Chair: S. König / R. Reents	**Session 12** Cow and large population genotyping for genomic selection, management and marketing in cattle Chair: S. Mattalia / H. Benhajali	**Session 20** **Plenary session** **Opening** **and** **Leroy Lecture** Chair: M. Gauly
Session 04 Farm to fork influences on product quality in sheep and goat meat and milk (with H2020 Project iSAGE) Chair: V.A.P. Cadavez **Session 05** Free communications in animal nutrition: pigs Chair: G. Savoini **Session 06** Introducing the SMARTCOW infrastructure project Chair: J.F. Hocquette	**Session 13** Microbiability and heritability: how microbiome, nutrition, and genetics are related Chair: P. Trevisi / E.F. Knol **Session 14** Postpartum key factors associated with better lamb and kid survival Chair: J.M. Gautier **Session 15** Nutrition, inflammation and oxidative stress Chair: E. Tsiplakou **Session 16** Balancing production / consumption – animal farming in Europe vs human and planetary health? Chair: J.L. Peyraud	
COST ACTION IPEMA SYMPOSIUM		
Session 07 Entire male pigs and immunocastration as alternatives to surgical castration of male piglets: opportunities and drawbacks Chair: G. Bee / M. Čandek-Potokar	**Session 17**	
Session 08 Pig and rabbit welfare Chair: V. Ferrante	**Session 18** Ruminant health Chair: G. Das **Session 19** Options for sustainable livestock in Europe Chair: M. Zehetmeier	
11.30 – 12.30 **Commission meeting** • Physiology	12:30 – 14:00 and 20.00 – 21.00 **Poster session**	

Tuesday 28 August 14.00 – 18.00	Wednesday 29 August 8.30 – 12.00	Wednesday 29 August 14.00 – 18.00
Session 21 What the hell is resilience and efficiency? Chair: N.C. Friggens **Session 22** Physiological diversity between individuals: do we need 'personalized farming'? Chair: J. Maselyne / E. Hessel / I. Halachmi **Session 23** Dairy and meat products: from traditional to novel quality traits Chair: M. De Marchi **Session 24** Novel traits (health/quality related traits) based on images or sensors (in cooperation with ICAR) Chair: B. Fuerst-Waltl / G. Thaller **Session 26** Genetics & Genomics – connectedness and application of breeding schemes for sheep & goats Chair: N. Mc Hugh **Session 27** Free communications in animal nutrition Chair: G. Zervas **Session 28** Welfare and behaviour in animal training Chair: M.L. Cox **Session 29** Variability in the pig production chain - problems and opportunities Chair: S. Millet **Session 30** Role of livestock in sustainable landuse and resource uses Chair: M.R.F. Lee	**Session 31** Unravelling genetic architecture of complex traits using mulit-omics approaches Chair: S. Pegolo **Session 32** Precision Livestock Farming (PLF) in nutrition, genetics, and in physiology Chair: R. Kasarda / J. Maselyne / R. Mateescu **Session 33** Improving carcass and meat quality (in cooperation with Cattle Network WG, InterBeef & International Meat Secretariat) Chair: J.F. Hocquette / A. Cromie **Session 34** Innovation in cattle husbandry - research, techniques and dissemination methods, regional customisation Chair: A. Kuipers / Y. Montanholi **Session 35** How can animal breeding contribute to sustainable and global animal production Chair: J. Van Arendonk	**Session 42** Development of methods to analyse complex data Chair: N. Ibañez-Escriche **Session 43** Pig behaviour and/or machine learning Chair: E.F. Knol **Session 44** Resilient Dairy Farming - economic, environmental and social sustainability of (European) dairy farming (in cooperation with EuroDairy project) Chair: R. Keatinge / M. Klopčič **Session 45** Improving management practices and feed efficiency in cattle Chair: A. De Vries / A. Kuipers **Session 46** Free communications: animal genetics Chair: J. Lassen
	ONE-DAY SYMPOSIUM ON INSECTS FOR FEED	
	Session 36 Cost effective insect production and high quality insects for feed Chair: A. Agazzi	**Session 47** Safety, regulatory issues and effects of insects in animal feed Chair: T. Arsiwalla
	Session 37 How to design your career for the livestock industry? Chair: C. Lambertz / A. Smetko / T. Wallgren **Session 38** Sustainable control of production diseases in pigs and poultry, with emphasis on early survival Chair: G. Bee **Session 39** Opportunities (or agendas) for international collaboration in equine research Chair: C. Vial **Session 40** Interactive workshop on behavioural measurements Chair: L.A. Boyle **Session 41** New insights into efficiency Chair: M. Tichit	**Session 48** New technologies to improve feed efficiency Chair: A. Baldi **SYMPOSIUM ON LOCAL BREEDS IN PIG PRODUCTION** **Session 49** High quality pig production with local breeds (with H2020 project TREASURE) Chair: S. Millet / M. Čandek-Potokar **Session 50** Importance and stakes of the horse industry at the European level Chair: A.S. Santos **Session 51** Sheep & Goat breeding and free communications Chair: O. Tzamaloukas **Session 52** The contribution of livestock farming to the provision of ecosystem services Chair: R. Ripoll Bosch / E. Sturaro
12:30 – 14:00 **Poster session**	11.30 – 12.30 **Commission meeting** • Insect	12:30 – 14:00 **Poster session**

Thursday 30 August 8.30 – 11.30	Thursday 30 August 14.00 – 18.00
Session 53 Breeding programs Chair: E. Sell-Kubiak **Session 54** Towards an integrated system from (PLF/farm/lengthwise the chain/big) data to a solution or decision Chair: M.A. Mitchell / S. Druyan **Session 55** Heat stress / Functional traits in cattle Chair: M. Klopčič / B. Fuerst-Waltl	**Session 63** Genotype by environment interaction, non-additivity and robustness Chair: H.A. Mulder **Session 64** Precision Livestock Farming (PLF) and communication in poultry, pigs, red-deer and rabbits Chair: M. Pastell / R. Kasarda **Session 65** Practices and prospects for adapting to a challenging Mediterranean environment Chair: G. Hadjipavlou **Session 66** Free communications: animal genetics Chair: E. Wall
ONE-DAY SYMPOSIUM ON MULTIDISCIPLINARY APPROACHES	
Session 56 Multidisciplinary approaches for improving sustainable livestock production: research needs, opportunities and difficulties Chair: J. Van Milgen / M.H. Pinard-Van der Laan	**Session 67** Multidisciplinary approaches for improving sustainable livestock production: user application needs Chair: M.H. Pinard-Van Der Laan / J. Van Milgen
Session 57 Overcoming technological barriers in sheep and goat production and breeding Chair: J. Yates **Session 58** Non-invasive biomarkers in nutritional studies Chair: R.M.A. Goselink **- - - - - SYMPOSIUM ON LOCAL BREEDS IN PIG PRODUCTION** **Session 59** Genetics of local pig breeds (with H2020 project TREASURE) Chair: S. Millet / M. Čandek-Potokar **Session 60** Ethical questions in equine research/ training Chair: R. Evans / A.S. Santos **Session 61** Diseases control in poultry and pig farming: current strategies and future alternatives Chair: S. Messori **Session 62** Free communications LFS Chair: M. Tichit	**Session 68** Free communications in cattle Chair: Y. Montanholi **Session 69** Nutritional and feeding strategies to face consequences of climate change Chair: G. Bruggeman **Session 70** Large litters discussion and free communications pigs Chair: E.F. Knol **Session 71** Genetics and genomics in horse production Chair: K.F. Stock **Session 25** Perspectives in sustainable aquaculture Chair: I. Župan

11.30 – 12.30
Commission meeting

- Cattle
- Genetics
- Health and Welfare
- Horse
- LFS
- Nutrition
- Pig
- PLF
- Sheep and Goat

Scientific programme

Session 01. Added value of local breeds

Date: Monday 27 August 2018; 8.30 – 12.30
Chair: R.B. Baumung

Theatre Session 01

Poster Session 01

Session 02. Effects of nutrition and risk factors in early life on later performance and health

Date: Monday 27 August 2018; 8.30 – 11.30
Chair: H. Sauerwein

Theatre Session 02

Poster Session 02

Session 03. Optimization of a genomic breeding program for small sized cattle populations (with Interbull)

Date: Monday 27 August 2018; 8.30 – 12.30
Chair: S. König / R. Reents

Theatre Session 03

Poster Session 03

Session 04. Farm to fork influences on product quality in sheep and goat meat and milk (with H2020 Project iSAGE)

Date: Monday 27 August 2018; 8.30 – 12.30
Chair: V.A.P. Cadavez

Theatre Session 04

Poster Session 04

Session 05. Free communications in animal nutrition: pigs

Date: Monday 27 August 2018; 8.30 – 12.30
Chair: G. Savoini

Theatre Session 05

Poster Session 05

Session 06. Introducing the SMARTCOW infrastructure project

Date: Monday 27 August 2018; 8.30 – 12.30
Chair: J.F. Hocquette

Theatre Session 06

Session 07. Entire male pigs and immunocastration as alternatives to surgical castration of male piglets: opportunities and drawbacks (part 1) (with COST Action IPEMA)

Date: Monday 27 August 2018; 8.30 – 12.30
Chair: G. Bee / M. Čandek-Potokar

Theatre Session 07

Poster Session 07

Session 08. Pig and rabbit welfare

Date: Monday 27 August 2018; 8.30 – 12.30
Chair: V. Ferrante

Theatre Session 08

Poster Session 08

Session 09. Genetic diversity and conservation

Date: Monday 27 August 2018; 14.00 – 18.00
Chair: J. Bennewitz

Theatre Session 09

Poster Session 09

Session 10. Hepatic, intestinal and mammary adaptive reactions: (patho-) physiology and biomarkers

Date: Monday 27 August 2018; 14.00 – 18.00
Chair: J.J. Gross

Theatre Session 10

Poster Session 10

Session 11. Young Train: Dairy innovative research and extension (in cooperation with Young Scientists Club; includes NOVUS award)

Date: Monday 27 August 2018; 14.00 – 18.00
Chair: C. Lambertz / A. Kuipers

Theatre Session 11

Poster Session 11

Session 12. Cow and large population genotyping for genomic selection, management and marketing in cattle (in cooperation with INTERBULL)

Date: Monday 27 August 2018; 14.00 – 18.00
Chair: S. Mattalia / H. Benhajali

Theatre Session 12

Poster Session 12

Session 13. Microbiability and heritability: how microbiome, nutrition, and genetics are related

Date: Monday 27 August 2018; 14.00 – 18.00
Chair: P. Trevisi / E.F. Knol

Theatre Session 13

Poster Session 13

Session 14. Postpartum key factors associated with better lamb and kid survival

Date: Monday 27 August 2018; 14.00 – 18.00
Chair: J.M. Gautier

Theatre Session 14

Poster Session 14

Session 15. Nutrition, inflammation and oxidative stress

Date: Monday 27 August 2018; 14.00 – 18.00
Chair: E. Tsiplakou

Theatre Session 15

Poster Session 15

Session 16. Balancing production / consumption – animal farming in Europe vs human and planetary health?

Date: Monday 27 August 2018; 14.00 – 18.00
Chair: J.L. Peyraud

Theatre Session 16

Session 17. Entire male pigs and immunocastration as alternatives to surgical castration of male piglets: opportunities and drawbacks (part 2) (with COST Action IPEMA)

Date: Monday 27 August 2018; 14.00 – 18.00
Chair: G. Bee / M. Čandek-Potokar

Theatre Session 17

Poster Session 17

Session 18. Ruminant health

Date: Monday 27 August 2018; 14.00 – 18.00
Chair: G. Das

Theatre Session 18

Poster Session 18

Session 19. Options for sustainable livestock in Europe

Date: Monday 27 August 2018; 14.00 – 18.00
Chair: M. Zehetmeier

Theatre Session 19

Poster Session 19

Session 20. Plenary Session and Leroy lecture

Date: Tuesday 28 August 2018; 8.30 – 12.30
Chair: M. Gauly

Theatre Session 20

Session 21. What the hell is resilience and efficiency?

Date: Tuesday 28 August 2018; 14.00 – 18.00
Chair: N.C. Friggens

Theatre Session 21

Poster Session 21

Session 22. Physiological diversity between individuals: do we need 'personalized farming'?

Date: Tuesday 28 August 2018; 14.00 – 18.00
Chair: J. Maselyne / E. Hessel / I. Halachmi

Theatre Session 22

Session 23. Dairy and meat products: from traditional to novel quality traits

Date: Tuesday 28 August 2018; 14.00 – 18.00
Chair: M. De Marchi

Theatre Session 23

Poster Session 23

Session 24. Novel traits (health/quality related traits) based on images or sensors (in cooperation with ICAR)

Date: Tuesday 28 August 2018; 14.00 – 18.00
Chair: B. Fuerst-Waltl / G. Thaller

Theatre Session 24

Poster Session 24

Session 25. Perspectives in sustainable aquaculture

Date: Thursday 30 August 2018; 14.00 – 18.00
Chair: I. Župan

Theatre Session 25

Poster Session 25

Session 26. Genetics & Genomics – connectedness and application of breeding schemes for sheep & goats

Date: Tuesday 28 August 2018; 14.00 – 18.00
Chair: N. Mc Hugh

Theatre Session 26

Poster Session 26

Session 27. Free communications in animal nutrition

Date: Tuesday 28 August 2018; 14.00 – 18.00
Chair: G. Zervas

Theatre Session 27

Poster Session 27

Session 28. Welfare and behaviour in animal training

Date: Tuesday 28 August 2018; 14.00 – 18.00
Chair: M.L. Cox

Theatre Session 28

Poster Session 28

Session 29. Variability in the pig production chain – problems and opportunities

Date: Tuesday 28 August 2018; 14.00 – 18.00
Chair: S. Millet

Theatre Session 29

Poster Session 29

Session 30. Role of livestock in sustainable landuse and resource uses

Date: Tuesday 28 August 2018; 14.00 – 18.00
Chair: M.R.F. Lee

Theatre Session 30

Poster Session 30

Session 31. Unravelling genetic architecture of complex traits using multi-omics approaches

Date: Wednesday 29 August 2018; 8.30 – 12.30
Chair: S. Pegolo

Theatre Session 31

Poster Session 31

Session 32. Precision Livestock Farming (PLF) in nutrition, genetics, and in physiology

Date: Wednesday 29 August 2018; 8.30 – 12.30
Chair: R. Kasarda / J. Maselyne / R. Mateescu

Theatre Session 32

Poster Session 32

Session 33. Improving carcass and meat quality (in cooperation with Cattle Network WG, InterBeef & International Meat Secretariat)

Date: Wednesday 29 August 2018; 8.30 – 12.30
Chair: J.F. Hocquette / A. Cromie

Theatre Session 33

Poster Session 33

Session 34. Innovation in cattle husbandry – research, techniques and dissemination methods, regional customisation

Date: Wednesday 29 August 2018; 8.30 – 12.30
Chair: A. Kuipers / Y. Montanholi

Theatre Session 34

Poster Session 34

Session 35. How can animal breeding contribute to sustainable and global animal production

Date: Wednesday 29 August 2018; 8.30 – 12.30
Chair: J. Van Arendonk

Theatre Session 35

Session 36. Cost effective insect production and high quality insects for feed

Date: Wednesday 29 August 2018; 8.30 – 11.30
Chair: A. Agazzi

Theatre Session 36

Session 37. How to design your career for the livestock industry?

Date: Wednesday 29 August 2018; 8.30 – 12.30
Chair: C. Lambertz / A. Smetko / T. Wallgren

Theatre Session 37

Session 38. Sustainable control of production diseases in pigs and poultry, with emphasis on early survival

Date: Wednesday 29 August 2018; 8.30 – 12.30
Chair: G. Bee

Theatre Session 38

Poster Session 38

Session 39. Opportunities (or agendas) for international collaboration in equine research

Date: Wednesday 29 August 2018; 8.30 – 12.30
Chair: C. Vial

Theatre Session 39

Poster Session 39

Session 40. Interactive workshop on behavioural measurements

Date: Wednesday 29 August 2018; 8.30 – 12.30
Chair: L.A. Boyle

Theatre Session 40

Poster Session 40

Session 41. New insights into efficiency

Date: Wednesday 29 August 2018; 8.30 – 12.30
Chair: M. Tichit

Theatre Session 41

Poster Session 41

Session 42. Development of methods to analyse complex data

Date: Wednesday 29 August 2018; 14.00 – 18.00
Chair: N. Ibañez-Escriche

Theatre Session 42

Poster Session 42

Session 43. Pig behaviour and/or machine learning

Date: Wednesday 29 August 2018; 14.00 – 18.00
Chair: E.F. Knol

Theatre Session 43

Session 44. Resilient Dairy Farming – economic, environmental and social sustainability of (European) dairy farming (in cooperation with EuroDairy project)

Date: Wednesday 29 August 2018; 14.00 – 18.00
Chair: R. Keatinge / M. Klopčič

Theatre Session 44

Poster Session 44

Session 45. Improving management practices and feed efficiency in cattle

Date: Wednesday 29 August 2018; 14.00 – 18.00
Chair: A. De Vries / A. Kuipers

Theatre Session 45

Session 46. Free communications: animal genetics

Date: Wednesday 29 August 2018; 14.00 – 18.00
Chair: J. Lassen

Theatre Session 46

Session 47. Safety, regulatory issues and effects of insects in animal feed

Date: Wednesday 29 August 2018; 14.00 – 18.00
Chair: T. Arsiwalla

Theatre Session 47

Poster Session 47

Session 48. New technologies to improve feed efficiency

Date: Wednesday 29 August 2018; 14.00 – 18.00
Chair: A. Baldi

Theatre Session 48

Poster Session 48

Session 49. High quality pig production with local breeds (with H2020 project TREASURE)

Date: Wednesday 29 August 2018; 14.00 – 18.00
Chair: S. Millet / M. Čandek-Potokar

Theatre Session 49

Poster Session 49

Session 50. Importance and stakes of the horse industry at the European level

Date: Wednesday 29 August 2018; 14.00 – 18.00
Chair: A.S. Santos

Theatre Session 50

Poster Session 50

Session 51. Sheep & Goat breeding and free communications

Date: Wednesday 29 August 2018; 14.00 – 18.00
Chair: O. Tzamaloukas

Theatre Session 51

Poster Session 51

Session 52. The contribution of livestock farming to the provision of ecosystem services

Date: Wednesday 29 August 2018; 14.00 – 18.00
Chair: R. Ripoll Bosch / E. Stuaro

Theatre Session 52

Session 53. Breeding programs

Date: Thursday 30 August 2018; 8.30 – 12.30
Chair: E. Sell-Kubiak

Theatre Session 53

Poster Session 53

Session 54. Towards an integrated system from (PLF/farm/lengthwise the chain/big) data to a solution or decision

Date: Thursday 30 August 2018; 8.30 – 11.30
Chair: M.A. Mitchell / S. Druyan

Theatre Session 54

Session 55. Heat stress / Functional traits in cattle

Date: Thursday 30 August 2018; 8.30 – 11.30
Chair: M. Klopčič / B. Fuerst-Waltl

Theatre Session 55

Poster Session 55

Session 56. Multidisciplinary approaches for improving sustainable livestock production: research needs, opportunities and difficulties

Date: Thursday 30 August 2018; 8.30 – 12.30
Chair: J. Van Milgen / M.H. Pinard-Van der Laan

Theatre Session 56

Session 57. Overcoming technological barriers in sheep and goat production and breeding

Date: Thursday 30 August 2018; 8.30 – 11.30
Chair: J. Yates

Theatre Session 57

Poster Session 57

Session 58. Non-invasive biomarkers in nutritional studies

Date: Thursday 30 August 2018; 8.30 – 11.30
Chair: R.M.A. Goselink

Theatre Session 58

Poster Session 58

Session 59. Genetics of local pig breeds (with H2020 project TREASURE)

Date: Thursday 30 August 2018; 8.30 – 11.30
Chair: S. Millet / M. Čandek-Potokar

Theatre Session 59

Poster Session 59

Session 60. Ethical questions in equine research/training

Date: Thursday 30 August 2018; 8.30 – 11.30
Chair: R. Evans / A.S. Santos

Theatre Session 60

Session 61. Diseases control in poultry and pig farming: current strategies and future alternatives

Date: Thursday 30 August 2018; 8.30 – 11.30
Chair: S. Messori

Theatre Session 61

Poster Session 61

Session 62. Free communications LFS

Date: Thursday 30 August 2018; 8.30 – 11.30
Chair: M. Tichit

Theatre Session 62

Poster Session 62

Session 63. Genotype by environment interaction, non-additivity and robustness

Date: Thursday 30 August 2018; 14.00 – 18.00
Chair: H.A. Mulder

Theatre Session 63

Poster Session 63

Session 64. Precision Livestock Farming (PLF) and communication in poultry, pigs, red-deer and rabbits

Date: Thursday 30 August 2018; 14.00 – 18.00
Chair: M. Pastell / R. Kasarda

Theatre Session 64

Session 65. Practices and prospects for adapting to a challenging Mediterranean environment

Date: Thursday 30 August 2018; 14.00 – 18.00
Chair: G. Hadjipavlou

Theatre Session 65

Poster Session 65

Session 66. Free communications: animal genetics

Date: Thursday 30 August 2018; 14.00 – 18.00
Chair: E. Wall

Theatre Session 66

Poster Session 66

Session 67. Multidisciplinary approaches for improving sustainable livestock production: user application needs

Date: Thursday 30 August 2018; 14.00 – 18.00
Chair: M.H. Pinard-Van Der Laan / J. Van Milgen

Theatre Session 67

Poster Session 67

Session 68. Free communications in cattle

Date: Thursday 30 August 2018; 14.00 – 18.00
Chair: Y. Montanholi

Theatre Session 68

Session 69. Nutritional and feeding strategies to face consequences of climate change

Date: Thursday 30 August 2018; 14.00 – 18.00
Chair: G. Bruggeman

Theatre Session 69

Poster Session 69

Session 70. Large litters discussion and free communications pigs

Date: Thursday 30 August 2018; 14.00 – 18.00
Chair: E.F. Knol

Theatre Session 70

Poster Session 70

Session 71. Genetics and genomics in horse production

Date: Thursday 30 August 2018; 14.00 – 18.00
Chair: K.F. Stock

Theatre Session 71

Poster Session 71

Specific products with added value for local breeds: lessons from success and non-success stories

E. Verrier[1], L. Markey[2] and A. Lauvie[3]

[1]AgroParisTech, INRA, 16 rue Claude Bernard, 75005 Paris, France, [2]Institut de l'Elevage, 19 bis Rue Alexandre Dumas, 80096 Amiens, France, [3]INRA, Campus international de Baillarguet, 34398 Montpellier 5, France; etienne.verrier@agroparistech.fr

Preserving or developing local breeds usually requires added value, through their products or other goods and services, which keeps them viable. The aim of this paper is to provide with some lessons, drawn from concrete experiences, on how to simultaneously manage the choice of breeding goals, the preservation of genetic variability and the search for added value through food products. Different case studies have been analysed, mainly from France, dealing with dairy or meat products from different livestock species. From these experiences, it highlights five main points. First, the large diversity of ways to find added value: products with or without official seal of quality or origin, resulting from farmers' individual initiatives, collective organizations involving farmers and processors or initiatives of small companies. Second, the need to build and maintain a clientele and to co-adapt production and demand, through quantity and distribution during the year. Third, the challenge of maintaining a close control by farmers, so that the added value will benefit them. Fourth, the importance of cohesion between the different actors and of a balanced attention paid to product development and breed management. Fifth, the issue of the appropriation, by the breed or by the product, of the name that carries the public favourable image.

Quantifying ecosystem services to add value in pasture-based livestock systems

A. Bernués

Centro de Investigación y Tecnología Agroalimentaria de Aragón (CITA), Avda. Montañana 930, 50059, Spain; abernues@gmail.com

Adding value to livestock products is mostly done through intrinsic quality attributes of meat and milk, i.e. those attributes that depend on their chemical and physical characteristics. However, quality perceptions and demands for food products are changing and consumers, and the public at large, increasingly hold 'ethical' concerns about the model of agriculture and the food chain. In this context, the so called extrinsic quality attributes, those that depend on the production process rather than on the product itself (e.g. animal feeding, origin, animal welfare, environmental implications), can constitute an opportunity for market segmentation and consumer-led adding value strategies. The multifunctional landscapes in which local breeds are integrated are characterized by delivering multiple ecosystem services to society (provisioning, regulating, supporting and cultural), however many of these ecosystem services constitute public goods that do not have market price, and therefore farmers do not have incentives to produce them. We first describe the most important ecosystem services linked to grazing agroecosystems in different European regions: (1) the preservation of agricultural landscape (cultural ecosystem services); (2) the conservation of biodiversity (supporting); (3) the prevention of forest fires in Mediterranean conditions, the maintenance of soil fertility in Atlantic conditions and the preservation of water quality in Alpine conditions (regulating); and (4) the provision of specific quality food products linked to the agroecosystems (provisioning). Next, we quantify these ecosystem services with an integrated approach that combines socio-cultural and economic valuation techniques. Finally, we present a conceptual framework for innovative food value chains that aims at linking multifunctional landscapes (and the breeds they hold), the farmers, and the consumers and citizens, using the ecosystem service concept. We operationalize the framework with empirical data of sheep farming systems in Mediterranean Spain.

A critical perspective on the current paradigm of high-quality products marketing strategy

D. Martin-Collado[1], C. Díaz[2], M. Serrano[2], R. Zanoli[3] and S. Naspeti[3]
[1]CITA, Avenida Montañana, 930, 50059 Zaragoza, Spain, [2]INIA, Crta. de la Coruña, km. 7,5, 28040 Madrid, Spain, [3]UNIVPM, Piazza Roma 22, 60121 Ancona, Italy; dmartin@cita-aragon.es

The Spanish long sheep farming tradition is reflected in the large variety of native breeds, production systems and products which coexist nowadays. However, despite its historic and cultural importance, the Spanish sheep sector has suffered an almost two-decade long crisis. This crisis caused a drop of sheep numbers and a decrease of meat consumption from 2.7 kg per capita (2006) to 1.7 kg (in 2014). Conversely, cheese production has increased slowly but steadily during the last decade. These trends respond to changes in consumption habits and in consumers preferences which are influenced by urban lifestyles. One of the current paradigms to overcome this crisis is by creating breed-product specific labels and link them with high-quality and environmental and cultural values. This way, labelled products would increase profitability either by increasing the market share of the product or by entering into added-value markets. As part of iSAGE H2020 EU project, four consumers' focus groups, of 10-12 participants each, were conducted in Madrid. These focus groups explored urban consumers' awareness, attitudes and preferences towards meat and dairy sheep products to understand how to develop new marketing strategies for the sector. Focus group showed that despite their high consumption of cheese, participants lacked knowledge on cheese products, sometimes not even knowing whether they are purchasing cow, sheep or goat cheese. They also showed a wide lack of knowledge about of sheep meat products and sheep production systems. Participants also perceived environmental and animal welfare issues as important, however acknowledged that price continue to be the main driver of sheep products consumption. We argue the paradigmatic strategy explained above might not be directed toward the most common consumer type but to a less frequent consumer profile, and therefore oriented towards the development of niche market and not towards mainstream markets. We conclude that the paradigm explained above should be followed with care as it success depends on the consideration of some critic issues. These issues will be analysed and discussed during the presentation.

Valorization of locally produced dairy and meat products from native livestock breeds

S.J. Hiemstra[1], N. Remijn[2], A.H. Hoving-Bolink[1], L. Amat[3] and D. Traon[3]
[1]Wageningen University & Research, Centre for Genetic Resources, the Netherlands (CGN), P.O. Box 338, 6700 AH, Wageningen, the Netherlands, [2]Stichting Zeldzame Huisdierrassen (SZH), Dreijenlaan 2, 6703 HA Wageningen, the Netherlands, [3]Arcadia International, 1220 A4.3 Chaussée de Louvain, 1200 Woluwe Saint Lambert/Brussels, Belgium; sipkejoost.hiemstra@wur.nl

A compendium of 56 projects has been made available for the purpose of exchange of experiences and good practices, contributing to valorisation of underutilized livestock breeds and plant varieties (AGRI-2015-EVAL-09). All 56 examples in the compendium aim to valorise the use of underutilized or neglected breeds and varieties in an economically viable way. A specific fiche dedicated to each project includes a description of the particular genetic resource, the objectives of the project, actors involved and their roles, steps and activities undertaken, results to date, next steps, lessons learned and good practices, and contact information of the responsible person/coordinator of the project. One fiche titled 'Zeldzaam Lekker (Rare and Tasty) – Adding value to recognised products and services from rare domestic animal breeds in the Netherlands' was submitted by the Dutch Rare Breeds Foundation (SZH), in collaboration with CGN. In this case study, the enthusiasm and leadership of the breeders was listed as a key success factor. Locally produced dairy and meat products, which can be connected to quality labels (e.g. Slowfood, Organic, Zeldzaam Lekker), are often marketed through short supply chains and direct sales to consumers. A dedicated regional communication and marketing approach to consumers and the wider society was mentioned as an important factor to become successful. Variability of the quality of the product, lower profitability per working hour, legislation, and the use of different labels for the same product, were identified as bottlenecks. Besides sharing the experiences and good practices of individual compendium projects, identification of similarities and differences between projects can also contribute to further strengthening of ongoing projects, or to start successful new initiatives.

Promote breed conservation by implementing breed specific traits for a local sheep breed

J. Schäler[1], S. Addo[1], G. Thaller[1] and D. Hinrichs[2]
[1]Christian-Albrechts-University of Kiel, Institute of Animal Breeding and Husbandry, Hermann-Rodewald-Straße 6, 24098 Kiel, Ghana, [2]University of Kassel, Department of Animal Science, Nordbahnhofstraße 1a, 37213 Witzenhausen, Germany; jschaeler@tierzucht.uni-kiel.de

In recent decades, local breeds were replaced by high yielding breeds due to profitability. Many local breeds are threatened by extinction and loss of native genetic diversity. The need to conserve these breeds and its genetic diversity has a major importance due to its necessity for genetic change within and between populations. In order to achieve this genetic diversity, novel approaches have to be explored and extended. The aim of this study was the implementation of breed specific estimated breeding values for a small local sheep breed in northern Germany in order to promote breed conservation. The data comprised pedigree information, breeding values of several traits, and phenotypic information from a field experiment for two novel traits: (1) extensive average daily gain and (2) ultrasonic measurements of muscle and fat depth. The experimental design included a dataset of 47 progeny from 14 pure bred rams of the White headed mutton sheep. The methodical approach can be divided into three parts: (1) Analysis of the breeding program, (2) Identification of breed specific traits, and (3) Estimation and implementation of a novel breeding value(s). Genetic parameters and correlations were conducted by applying linear mixed models. The estimates for the heritability (repeatability) are between 0.60 and 0.69 (0.19 and 0.14). The difference between repeatability and heritability points to a huge impact of small data size of individuals. The genetic correlation were positive (0.30) and in accordance with the phenotypic correlation (0.09). Breed conservation can be possible by implementing and accelerating genetic gain for breed specific traits of this small local sheep breed. This analysis can be extended for several local breeds in order to keep their native genetic variance, native genetic diversity, and competitiveness towards conventional livestock breeds.

Session 01 Theatre 6

Added value to local breeds through goods and services: a diversity of ways of valorisation

A. Lauvie[1], G. Alexandre[2], V. Angeon[2], N. Couix[3], O. Fontaine[1], C. Gaillard[4], M. Meuret[1], C. Mougenot[5], C.H. Moulin[1], M. Naves[2], M.O. Nozières-Petit[1], J.C. Paoli[6], L. Perucho[6], J.M. Sorba[6], E. Tillard[1] and E. Verrier[7]
[1]INRA-CIRAD-MontpellierSupagro-Univ Montpellier UMR SELMET, Montpellier, 34060, France, [2]INRA URZ, Petit Bourg, 97170, France, [3]INRA-INPT UMR AGIR, Castanet Tolosan, 31326, France, [4]AgroSupDijon, Dijon, 21079, France, [5]SEED-Ulg, Arlon, 6700, Belgium, [6]INRA LRDE, Corte, 20250, France, [7]AgroParisTech-INRA, Univ. Paris-Saclay UMR Gabi, Paris, 75005, France; anne.lauvie@inra.fr

Beside valorising food products, production of services is often mentioned to add value to local breeds. However, such added value can modify breeds' management, for instance by introducing new stakeholders or selection criteria. We addressed this issue by analysing, through a common grid, the available data on 9 French breeds' management (cattle, sheep or goat) and raised in contrasted environments. The aim of this study was to identify: (1) the goods and services produced in systems using those breeds; (2) the role played by the breeds to produce these goods and services; and (3) the questions raised concerning underlying processes. We first described the diversity of products and other services from these systems. We identified their contribution to ecosystem management through grazing, and its consequences on landscape. We pointed to associated services provided to society like fire prevention, the uses for cultural and worship concerns, the contribution to social local dynamics, etc. For these productions the breed can play multiple roles, through its biological characteristics (abilities of the animals), through attributes not directly linked to biological characteristics, as a core element of the search of innovations in agroecosystems and/or of a collective action. We second illustrated how taking the service productions into account contributes to a better understanding of the stakes underlying the management of the breeds, and of their management dynamics through time. Finally, we highlighted several research questions that need to be addressed like: (1) how to identify the stakeholders concerned by services? (2) how to report and account for the diversity of value and valorisation processes, including the non-market ones? (3) how to better comprehend the interactions between services?

Session 01 Theatre 7

Developing sustainable value chains for small-scale livestock producers and locally adapted breeds

G. Leroy, R. Baumung, P. Boettcher and B. Besbes
Food and Agriculture Organization of the United Nations, Viale delle Terme di Caracalla, 00153, Italy; gregoire.leroy@fao.org

In developing countries, small-scale livestock producers are major stakeholders in food production and, with it, in human health and landscape management, relying largely on locally adapted breeds to produce food as well as to provide a wide diversity of services. They face however many challenges, including growing environmental constraints, poor access to markets and related services, and limited capacities, which may hamper their productivity and competitiveness vis-à-vis their larger counterparts. The sustainable food value chain development (SFVCD) framework may appear as an interesting approach to address these challenge. SFVCD is driven by several principles which can be summarized as follows: (1) measuring performance: this considers the three dimensions of sustainability, namely economic, social and environmental ones, including synergies and trade-offs; (2) understanding performance: taking a holistic perspective, the approach views the systems considered as interconnected and dynamic, governance-centred, and market-driven; (3) improving performance: implementing effective interventions requires a clear vision and upgrading strategy, but the process must also be scalable and multilateral. The specific characteristics of small-scale livestock production systems (i.e. multi-functionality, input and output provision, health and quality, social equity, etc.) underline the need to analyse value chains in a holistic manner, considering not only the supply chain itself, but also its broad environment, its dynamics and its connections to other systems. In that extent, the particular contribution of locally adapted breeds toward non-food ecosystem services (especially cultural ones), as well as their capacity to thrive in low-input and harsh production environment, should be fully considered when designing interventions.

Session 01 Theatre 8

Performance, behaviour and health of dual-purpose chicken under mountain farming conditions

C. Lambertz, K. Wuthijaree and M. Gauly
Free University of Bozen-Bolzano, Faculty of Science and Technology, Universitätsplatz 5, 39100 Bolzano, Italy; christian.lambertz@unibz.it

Production systems for dual-purpose chicken where both sexes are raised together during the first weeks, before males are separated during their final fattening stage, while females are kept for an entire laying period, might be an economic alternative for small-scale mountain farmers in South Tyrol, Northern Italy. However, systems (husbandry, feeding, slaughter age) have to be adjusted. The aim was to compare two genotypes in such a system, namely purebred (P) Les Bleues (n=300) and crossbred (C) New Hampshire × Les Bleues (n=300). One-day-old mixed-sex chicken were raised for 12 weeks under floor husbandry conditions and fed a standard broiler diet. Thereafter, males were moved to a mobile chicken house with free-range access. Males were slaughtered at weekly intervals from 12th to 16th week of age. Hens were kept for one laying period in the mobile house. Live weight development, feed consumption, health parameters, carcass and meat quality were measured in males and laying performance, egg quality, feed consumption, health parameters, behaviour and slaughter traits in females. At 12 weeks of age, males of P reached a live weight of 2,075 g and C of 1,865 g ($P<0.05$), while at 16 weeks both had more than 2,500 g ($P>0.05$). Dressing percentage was about 1% higher in C than P ($P>0.05$). Proportion of legs, breast and wings was 34.3, 16.0 and 11.0% in P and 34.7, 15.5 and 12.1% in C ($P>0.05$). Laying performance was 54.5% in P and 54.2% in C ($P>0.05$). Egg breaking strength decreased during the laying period, but remained above 30 N. Feed conversion was 3.4 kg feed/kg egg. On average, 25% of the animals stayed outdoors ($P>0.05$). Keel bone deformations were observed in 10% and breast blisters in 20% of the hens at the end of laying. Slaughter weight of hens was 1,850 g with a leg proportion of 31% ($P>0.05$). Under the specific conditions of marketing products as high-value under mountain farming conditions, performances resulted in an overall economic benefit, which was higher for P than C. The use of dual-purpose chicken to avoid the killing of one-day-old chicken and mobile housing may be ways to substantiate premium prices in such a system.

Session 01 Theatre 9

Genome-wide scan for selection signatures within two local cattle breeds in Slovakia

N. Moravčíková, O. Kadlečík, A. Trakovická and R. Kasarda
Slovak University of Agriculture in Nitra, Department of Animal Genetics and Breeding Biology, Tr. A. Hlinku 2, 94976 Nitra, Slovak Republic; nina.moravcikova1@gmail.com

The aim of this study was to detect selection signatures within and between Slovak Spotted and Slovak Pinzgau cattle, identify genomic regions subjected to differential selection histories of those local populations and characterize significantly affected regions located directly or close to QTLs associated with economically important traits. In total 236 animals, representing the nucleus of both local populations, were included in this study. The sample of Slovak Pinzgau cattle, consisting of 37 AI sires, 35 sire dams and 79 dam of dams, were genotyped by using Illumina BovineSNP50v2 BeadChip. The sampled population of Slovak Spotted cattle (37 AI sires, 48 sire dams) were genotyped by using ICBF International Dairy and Beef v3. Initially, the consensus map file were constructed (40,033 SNPs). Autosomal loci with call rate lower than 90% and minor allele frequency lower than 0.01% were filtered out. After quality control of genotyping data the final dataset consisted from 39,261 SNPs that covered overall 2,497,077 Mbp of the bovine genome. Three different approaches were used to identify signatures of selection. The first approach was based on the detection of very different allele frequencies between populations based on genome-wide F_{ST} scan. The second approach analysed the ancestral states in connection with extended haplotype homozygosity of derived alleles (iHS score) and the third approach identified the regions under selection pressure based on variation in linkage disequilibrium. Each of applied approach indicated strong genetic differentiation between analysed populations. The common signal of selection was found on BTA6 directly in genomic regions of *KIT* gene that is responsible for level of coat spotting in cattle. In addition, on chromosome 6 was found the strongest signal near genomic sequence of *ABCG2* gene that was associated with milk yield, fatty acid profile and somatic cell score. Identification of mutations important for the genetic control of economically important traits in local cattle populations in Slovakia are part of the overall breeding strategy and future genetic progress.

Session 01 Theatre 10

Milk phenotypic characterization and heterogeneity of variance across herds in Italian Jersey breed

G. Niero[1], C. Roveglia[1], R. Finocchiaro[2], M. Marusi[2], G. Visentin[2] and M. Cassandro[1]
[1]University of Padova, Department of Agronomy, Food, Natural Resources, Animals and Environment (DAFNAE), Viale Dell'Università 16, 35020 Legnaro (Padova), Italy, [2]Associazione Nazionale Allevatori Frisona Italiana (ANAFI), Via Bergamo 292, 26100 (Cremona), Italy; giovanni.niero@studenti.unipd.it

Jersey dairy cattle breed is characterized by relatively low milk yield (MY), but high milk solids content. The low productivity and its udder conformation make Jersey a suitable breed for once-a day-milking, resulting in less stress load for animals and higher milk quality. For these reasons Jersey is appreciated by the dairy industry. The present study aimed to assess the phenotypic variation of MY and milk quality traits in Italian Jersey (IJ), and to evaluate the heterogeneity of variance across IJ herds. Sources of variation were investigated through a mixed linear model, including the fixed effects of days in milk, parity, calving season, milking frequency, recording type, the interaction between days in milk and parity, and the random effects of herd-test-day (HTD) and cow. Heterogeneity of variance was assessed across herds classified as low (L), medium (M) or high (H) according to their average MY and milk quality traits. Results highlighted the high contents of milk fat (5.18%), protein (4.08%) and casein (3.16%) of IJ cows. Somatic cell score (SCS) averaged 3.35 units, which is quite high and suggest that SCS should be monitored carefully in IJ herds, especially through an improvement of management practices. Milk yield showed moderate negative correlations with milk composition and SCS. Milk with greater fat content had also greater protein and casein content, whereas SCS resulted moderately and negatively correlated with lactose content. Phenotypic variances of MY and milk quality traits were mainly related to animal effect; on the other hand phenotypic variance of milk urea nitrogen content was mainly affected by HTD effect. Finally, heterogeneity of variance was observed across L, M and H herds, in a way that highly ranked herds showed greater standard deviation. Future research should investigate the impact of accounting or not for heterogeneity of variance in genetic evaluation of IJ breed.

Dimensional reduction for breed assignment models with a high number of strongly linked markers

J.A. Baro[1], F. Bueno[1], F.J. Cañón[2], C. Díaz[3], J. Piedrafita[4], J. Altarriba[5] and L. Varona[5]
[1]Universidad de Valladolid, CC. Agroforestales, ETSIIAA, 34047 Palencia, Spain, [2]Universidad Complutense de Madrid, Genética, Facultad de Veterinaria, Ciudad Universitaria, Madrid, Spain, [3]INIA, Crtra. de la Coruna, 28040 Madrid, Spain, [4]Universidad Autónoma de Barcelona, Facultad de Veterinaria, Barcelona, Spain, [5]Universidad de Zaragoza, Facultad de Veterinaria, Zaragoza, Spain; baro@agro.uva.es

Breed assignment is required as a forensic tool for detection of fraud on meat products for designated breed branding. Tests achieve only low levels of assignment power for sets of autochthonous cattle breeds that have been differentiating over a period of just a few centuries. Previous work pointed out that a classical approach with a few hundred multi-allelic (microsatellite) markers attained moderate true-classification rates with either logistic regressions, data mining, or machine learning methods, while maximum likelihood methods gave the best results. However, these arrangements were based on modelling allele segregation and their computing costs are not ready for massive SNP based genotypes. Besides, closely linked markers result in largely over-parameterized models, and policies for radically culling the number of markers are required. We had investigated the use of ranking markers on Fisher exact tests for logistic regressors, but now turned to the partial least squares (PLS) method, which picks the best sets of latent components to be used with a cross-validation approach. Out of 778,000 SNP markers, LD (R2 of 0.09), HWE & MAF filters resulted in almost 29,000 prospective regressors, and 55 of them were enough to achieve AUC (area under the ROC curve) of 1. When markers are chosen at random, more than 100 are needed to achieve the same AUC. Further, when more markers are rejected on lower linkage rates (R2 of 0), valuable markers for breed assignment are lost and almost 70 are required by PLS for full AUC.

An alternative beef production system in mountain areas of northern Spain: genotype effect

L.R. Beldarrain[1], N. Aldai[1], P. Lavín[2], R. Jaroso[2] and A.R. Mantecón[2]
[1]University of the Basque Country (UPV/EHU), Department of Pharmacy & Food Sciences, Paseo de la Universidad 7, 01006 Vitoria-Gasteiz, Spain, [2]Mountain Livestock Institute (CSIC-ULE), Finca Marzanas, 24346 Grulleros, León, Spain; loretx_@hotmail.com

In mountain areas of northern Spain (Nansa Valley, Cantabria), the most common beef production system utilizes primarily local breeds and resources. This system is centred on calf production which farmers sell (5-6 months, mo) to bigger feedlots for fattening and commercialization(12-14 mo of age at slaughter). This production system provides limited income to local small farmers and, therefore, an alternative option was tested where farmers could produce a high quality veal (8-12 mo at slaughter) avoiding intermediate feedlots. They were thus able to simultaneously meet sustainability and multi-functionality expectations in the rural areas. Under this unconventional system three genotypes were compared: 'Asturiana de los Valles' (AV; n=7), 'Tudanca' (T; n=6), and 'Tudanca × Limousin' (T×L; n=6). Calves were allowed to graze freely on mountain pastures while suckling their mothers and with free access to concentrate until 6-7 months of age. Then, after weaning, calves were finished indoors with *ad libitum* grass hay and concentrate. Cold carcass weight and parameters (conformation, fatness) were measured and meat fatty acid (FA) composition was determined by GC. No differences were observed in carcass weight (mean of 205±6.6 kg) although AV animals were younger at slaughter (274d) compared to other genotypes (mean of 346d; $P<0.001$). Moreover, AV carcasses showed better conformation ($P<0.01$) but lower fatness ($P<0.05$) compared to other more rustic genotypes. Related to fat, meat from AV animals had the lowest total FA content (1.76%), T meat had the highest (2.90%), while T×L meat showed intermediate values (2.38%; $P<0.05$). In terms of FA profile (%), no significant differences were observed for saturated, monounsaturated and n-6 polyunsaturated FAs. However, higher level of n-3 was observed in AV meat compared to other meats (1.7% vs mean of 0.85%; $P<0.01$) but healthier *trans*-18:1 profile was observed in meat from T and T×L calves due to a higher 11*t*-18:1 ($P<0.001$) and lower 10*t*-18:1 content ($P<0.001$) compared to AV genotype.

Session 01 Poster 13

A decade of restoration of the Carpathian goat population in Poland

J. Sikora, A. Kawęcka and I. Radkowska
National Biobank of National Research Institute of Animal Production, Krakowska 1, 32-083 Balice, Poland; jacek.sikora@izoo.krakow.pl

As part of the global action aimed to conserve farm animal biodiversity, an attempt was made to restore an old local breed of goats, which was kept in the Polish Carpathian Mountains in the 19th century. It has been more than a decade since the process of Carpathian goat restoration was began in the early 21st century by the National Research Institute of Animal Production. Carpathian goats are multipurpose animals of medium size. Height at withers averages 55-65 cm for adult females and 65-75 cm for males. They have a well-proportioned body and well-formed udders. Goats have a well-shaped head, a long neck, a beard, often wattles under the neck, and are horned. Bucks have large spreading, twisted horns, a thick beard. Many males and females come with a distinct fringe over the eyes. The hair is white, semi-long, with occasional down underfur. The coat splits on the back, falling evenly on both sides of the trunk. The average hair length is 17 cm in females and 24 cm in males. Carpathian goats display features typical for autochthonous populations, such as high resistance to disease, longevity, and undemanding feeding habits. The breeding work was conducted alongside research work aimed to describe the breed being restored. A provisional characteristics of the genetic makeup of the Carpathian goat was provided in relation to other breeds found in Poland, using selected microsatellite markers recommended by FAO for evaluation of biodiversity in goats. High genetic distance was found between the Carpathian goat and the White Improved and Pygmy goats. Reproductive performance parameters were estimated: mean fertility was 87% and prolificacy was 168%. Milk performance averaged 354 kg. The highest yield was 526 kg milk in the third lactation. The milk contains 3.1% protein and 3.4% fat on average. The current population of Carpathian goats in Poland is 134 mother goats, 21 first kidders and 14 breeding bucks. Breeders who maintain Carpathian goats and implement the conservation programme for this breed, receive subsidies for breeding.

Session 01 Poster 14

Use of Polish Konik horses for avifauna conservation in selected ecosystems of Poland

M. Pasternak[1], A. Kawęcka[1], T. Gruszecki[2] and A. Junkuszew[2]
[1]National Research Institute of Animal Production, Krakowska 1, 32-083 Balice, Poland, [2]University of Life Sciences, Akademicka 13, 20-950 Lublin, Poland; marta.pasternak@izoo.krakow.pl

The Polish Konik horse is a conserved native breed descended from wild tarpans. It is undemanding in terms of environmental and feed requirements, long lived, and resistant to disease. Konik horses may remain out throughout the year. Their additional advantage is the hard hoof horn, which allows them to do well in overgrown wetlands, making them suitable for grazing in inaccessible, environmentally valuable areas. The aim of the study is to present the main efforts undertaken in recent years in Poland which combine the grazing of Polish Koniks with active conservation of biodiversity and bird habitats. Grazing of Polish Koniks is a key element shaping the habitats of water and marsh birds, for which overgrowth of wetlands is the main threat. Polish Koniks eat shrubs and bushes, making the area more accessible to protected species and harder to reach for predators. By eating sedges, reeds and tree seedlings on wet meadows of the Biebrza National Park, Polish Koniks can prevent migration of rare species of birds: *P. pugnax*, *A. paludicola*, *Charadriiformes* and predatory birds such as *C. clanga*. The main initiative was taken by the Polish Society for the Protection of Birds, which introduced Polish Koniks into the Narew Valley and also into Biebrza Marshes, the Białowieża Forest and the Knyszyn Forest, where, as part of the Natura 2000 project, the active conservation of *T. tetrix* and *C. pomarina* feeding grounds was conducted. Grazing is also performed at the Szczecin Lagoon, for the purpose of conserving the habitats of *V. vanellus*, *L. limosa*, *G. gallinago* and *T. totanus*. In addition to nesting, it facilitates foraging because the consumption of plants by the Polish Koniks allows small invertebrates, which form a forage base for many bird species, to reach the surface. In summary, Polish Koniks are highly useful in the efforts to protect endangered bird habitats in Poland, combining the use of native breeds and conservation of biodiversity. Project 'The uses and the conservation of farm animal genetic resources under sustainable development' – BIOSTRATEG2/297267/14/NCBR/2016.

CA.RA.VA.N project: toward implementation of a modern camel selection system in Northern Africa?

E. Ciani[1], V. Landi[2], G. Fernandez De Sierra[3], M. Ould Ahmed[4], S. Bedhiaf-Romdhani[5], W. Bensalem[5,6], I. Boujenane[7], P.A. Burger[8] and S.B.S. Gaouar[9]

[1]Università degli Studi di Bari 'Aldo Moro', Dipartimento di Bioscienze, Biotecnologie e Biofarmaceut, bari, bari, Italy, [2]University of Cordoba, Cordba, Spain, University of Cordoba, Cordba, Spain, University of Cordoba, Cordba, Spain, University of Cordoba, Cordba, Spain, Spain, [3]Asociación de Criadores del Camello Canario, Lanzarote, Spain, Lanzarote, Lanzarote, Spain, [4]Institut Supérieur d'Enseignement Technologique, Département de Production et Sante Animales, Rosso, Rosso, Rosso, Mauritania, [5]National Agricultural Research Institute of Tunisia, Tunis, Ariana, Tunisia, Tunisia, [6]OEP, OEP, OEP, Tunisia, [7]Institut Agronomique et Vétérinaire Hassan II, Rabat, Morocco, Rabat, Rabat, Morocco, [8]Research Institute of Wildlife Ecology, Vetmeduni Vienna, Vienna, Austria, Vienna, Vienna, Austria, [9]Abou Bakr Belkaid University, Tlemcen, Tlemcen, Algeria; elena.ciani@uniba.it

A project entitled 'CA.RA.VA.N., Toward a CAmel tRAnsnational VAlue chain' (http://www.arimnet2.net/index.php/researchprojects/projects-2nd-call-2/ca-ra-va-n), involving research institutions from Spain, Italy, France, Algeria, Morocco, and Tunisia, has been recently financed by EU through the Arimnet2 instrument. Additional partners from eligible (Tunisia) and non-eligible areas (Mauritania and Austria) have been associated to the project due to relevant interest in, and expected contribution to, the topic of the project. General goals and activities planned within the CA.RA.VA.N. project, together with an overview of the already launched initiatives, known constraints and emerging opportunities for the development of the dromedary sector in Northern African countries will be discussed.

Session 01 Poster 16

Role of horses and other aspects of wild nature in the image of the Camargue touristic natural area

C. Vial[1] and C. Costa[2]

[1]Ifce, MOISA, INRA, CIHEAM-IAMM, CIRAD, Montpellier Supagro, Univ Montpellier, 2 place Pierre Viala, 34060 Montpellier, France, [2]MOISA, INRA, CIHEAM-IAMM, CIRAD, Montpellier Supagro, Univ Montpellier, 2 place Pierre Viala, 34060 Montpellier, France; celine.vial@inra.fr

The natural park of the Camargue in France is Western Europe's largest river delta, with exceptional biological diversity, and home to unique breeds of horses, cattle, and 400 species of birds, including Pink Flamingos. It is also a popular touristic area with around one million tourists coming each year. Investigations have often focused on motivations and behaviours of tourists towards natural environments, but have rarely explored the relationship between the presence of animals or wild nature and tourism development of these areas. In this context, we propose to explore the role of horses (but also other aspects of wild nature) in the touristic image and attractiveness of the Camargue. First, 28 qualitative interviews enabled to identify the meanings tourists ascribed to this place. Then, a quantitative survey among 205 visitors of the park aimed to analyse the influence of place meanings on place attachment and destination loyalty. Confirmatory factor analysis and structural equation modelling were used to test a model. Results support that horses are an important meaning of the Camargue but among others mainly linked to animals, wild nature, and local culture. As some meanings become more salient to the individual, the individual develops a greater degree of attachment and loyalty. Moreover, place attachment influences loyalty significantly. The findings also highlight the effect of animal attachment and connectedness to nature on place attachment through a moderation of the influence of the meanings related to the presence of animals and wild nature. In the case of the Camargue, this research demonstrates the importance of adapting touristic marketing strategies according to the type of clients: focusing on animals (including horses) and wild nature will be useful for individuals attached to animals and connected to nature, whereas others will be more sensitive to freedom, unicity or local culture.

Session 02 — Theatre 1

Early-life programming effects on long-term productivity of dairy calves

M. Hosseini Ghaffari
University of Bonn, Institute for Animal Science Physiology & Hygiene, Katzenburgweg 7-9, 53115 Bonn, Germany; morteza1@uni-bonn.de

A new understanding of the relationship between early-life programming and later productivity has emerged as a new frontier of research for dairy production. Metabolic programming begins during foetal life and continues well into postnatal life. Adverse prenatal experiences such as maternal heat stress or maternal undernutrition may impact immune function and metabolism of the offspring as well as future lactational performance. Research in several species indicates that delivery of milk-borne bioactive factors from mother to offspring in early life plays a pivotal role in the programming of later life performance by affecting cellular signalling mechanisms, maturation of the immune system, and growth and differentiation of the digestive tract. However, the exact mechanisms are still not well understood. In addition to early postnatal life, the time of preweaning and weaning is another important period for growth performance. Nutritional management during this stage can have long-term effects on lactation performance of dairy cows. Finally, epigenetics studies may provide a biological mechanism to explain the effect of early-life programming on later growth, health, and performance outcomes.

Session 02 — Theatre 2

Foetal programming of mammary development in ruminants

S.A. McCoard[1] and Q. Sciascia[2]
[1]AgResearch Limited, Nutrition & Physiology Team, Private Bag 11008, Palmerston North 4474, New Zealand, [2]Leibniz Institute for Farm Animal Biology, Institute for Nutritional Physiology, Wilhelm-Stahl-Allee 2, 18196, Germany; sue.mccoard@agresearch.co.nz

The structure and development of the mammary gland is critical for future milk production. Foetal and pre-pubertal mammary development has been linked to future lactation performance, highlighting the importance of these periods of development. Development of the mammary gland during early life is dependent upon complex interactions between the mammary mesenchyme and epithelium mediated by cell signalling networks driven by nutrients, hormones and mitogens. A key pathway that mediates the effects of nutrients, hormones and growth factors such as IGF-1 is the mitogen-activated protein kinase/mammalian target of rapamycin (mTOR) pathway. This pathways play a key role in the regulation of a wide array of cellular processes, including hypertrophy, hyperplasia and differentiation in mammalian species. In foetal sheep, the development of the mammary fat pad is reduced following a reduction in the plane of maternal nutrition in mid-late gestation, likely mediated via reduced hyperplasia which provides insights into why female lambs born to nutrient-restricted ewes have improved first lactation performance. The reduction in fat pad hyperplasia was associated with an increased abundance of MAPK and fat-pad specific mTOR pathway signalling proteins that regulate the growth promoting pathways and protein synthetic capacity in response to potential paracrine IGF-1 and 2 signalling. In pre-weaning calves, mammary development is not only enhanced by increasing milk allowance which is associated with elevated IGF-1, but further growth is observed following supplementation with arginine and glutamine, which are key regulators for amino-acid controlled cell growth through the mTOR signalling pathways. Collectively, these results highlight the potential importance of nutritional programming of the mammary fat pad development during early life and insights into the potential molecular mechanism mediating the effects which may have important implications for future lactation performance.

Effect of fatty acids and epigenetic modifications on immune cells functions

P. Lacasse, C. Ster, N. Vanacker and A. Roostaee
AAFC-Sherbrooke R&D Centre, 2000 College, J1M 0C8, Sherbrooke, QC, Canada; pierre.lacasse@agr.gc.ca

Cows undergoing energy deficit have a weakened immune system and are at greater risks of infection. Negative energy balance usually occurs during the transition period in high-yielding dairy cows, but also occurs in cows under acute stress that are unable to eat the required amount of feeds. Our recent researches have indicated that high blood non-esterified fatty acid (NEFA) concentrations impair lymphocyte and neutrophil functions. This research program improves our understanding of the mechanism by which NEFA affects cow immune cell functions. A better understanding of the mechanisms involved in the NEFA-induced immunosuppression, may enable the development of approaches to reduce immunosuppressive effects of NEFA. Beyond changes in immunity related to the nutritional status, there are variations between individuals. There are a growing number of evidences that suggest that environment, including intra-uterine environment, can affect animal phenotype several years later. This phenomenon may be related to the regulation of transcription by epigenetic modifications of the genome, such as DNA methylation and histone acetylation. Our work shows that immune functions are particularly affected by epigenetic modulations. This open-up an exciting new field of research.

Study of the MaSC/progenitors committed to the development of the bovine mammary gland

L. Finot, E. Chanat and F. Dessauge
INRA, UMR 1348 PEGASE, 16 Le Clos, 35590 Saint-Gilles, France; laurence.finot@inra.fr

Milk production is highly dependent on the extensive expansion of the mammary epithelium which largely occurs during puberty. Within the mammary epithelium, the Mammary Stem Cells (MaSC) and their progeny play a key role in driving the development of the future functional mammary gland. Later, these cells contribute to regenerate the secretory epithelium in order to ensure future lactations. To elucidate the epithelial cell lineage involved in the establishment of the functional gland, we investigated the MaSC/progenitors populations in the bovine mammary tissue at various physiological stages. We used flow cytometry to phenotype epithelial cells on the basis of CD49f and CD24 expression in animals at three contrasted physiologic stages: puberty (heifers of 17 months of age), lactation (multiparous cows producing 35 kg of milk) and drying off (cows at 6 years after drying off). Comparison of the cytometric profiles highlighted the presence of a distinct epithelial population of CD49flowCD24pos cells at puberty (23.7%±3.6) which substantially decreased in the others stages (4.1%±1 at lactation and 6%±1 at drying off). These CD49flowCD24pos cells had features of a dual lineage, with luminal and basal characteristics (CD10, ALDH1 and keratin 7 expression) and were considered to be early common progenitors. The MaSC fraction was recovered in the sub-population of CD49fhighCD24pos cells. When cultured *in vitro* (in a 3D culture system with matrigel), these cells formed mammospheres, a functional property of MaSC. At puberty, total mammary cells contained 2.8%±0.3 of MaSC. This proportion of MaSC was significantly reduced at lactation (1.1±0.3%) and at drying off (0.4±0.3%). Further phenotypic characterization of the CD49fhighCD24pos cells showed that they expressed CD10 and keratin 14 at any stages. This in-depth characterization of the epithelial sub-populations at various physiological stages provide new insights into the epithelial cell hierarchy in the bovine mammary gland from puberty to drying off.

Passive transfer of dam immunoglobulins to calf in two beef breeds undernourished in early pregnancy

A. Noya[1], I. Casasús[1], J.L. Alabart[1], B. Serrano-Pérez[2], D. Villalba[2], J.A. Rodríguez-Sánchez[1], J. Ferrer[1] and A. Sanz[1]
[1]CITA de Aragón, Animal Production and Health, Avda. Montañana 930, 50059 Zaragoza, Spain, [2]Universitat de Lleida, Animal Science, Av. Alcalde Rovira Roure 191, 25198 Lleida, Spain; anoya@cita-aragon.es

Poor nutrient diet in early pregnancy, a common scenario in extensive systems, could interfere with the correct foetal programming and the colostrum yield in the dam. The accurate colostrum intake plays an essential role to acquire passive immunity by the agammaglobulinemic newborn. The aim of this study was to evaluate the effects of maternal undernutrition in first third of gestation on immunoglobulin G and M (IgG and IgM) concentration of dams (plasma and colostrum) and their offspring (plasma) in two beef breeds. Thirty one Parda de Montaña (PA) and 21 Pirenaica (PI) multiparous cows were artificially inseminated and randomly allocated to a control (Control, n=19) or nutrient-restricted (Subnut, n=33) group, which were fed at 100 or 65% of their estimated energy requirements during the first 82 days of pregnancy, and thereafter received a control 100% diet until parturition. Plasma IgG concentration in dams decreased from 8th month of gestation to parturition in all groups, but not significantly in PI-Control. All groups showed similar plasma IgM levels, regardless breed or maternal subnutrition, both at 8th month of gestation and at parturition. Both colostrum IgG and IgM values strongly fell down from period 1 (0-12 hours postcalving) to period 2 (12-24 hours postcalving). Although no Ig concentration differences among groups were registered in colostrum samples in period 1, PI-Control presented higher IgG concentrations than their counterparts in period 2. Neither breed nor maternal nutritional treatment caused differences in calf Ig concentrations. These results would confirm the correct passive transfer of dam immunoglobulins throughout colostrum regardless maternal undernutrition in early gestation, although a breed-nutritional treatment effect could diminish the physiological IgG depletion in late gestation and maintain high colostrum IgG values over time in PI breed.

Performance of calves fed either pelleted concentrate and TMR or only TMR from 10 to 18 weeks of age

M. Vestergaard, A. Jensen, N. Drake and M.B. Jensen
Aarhus University, Department of Animal Science, Foulum, 8830 Tjele, Denmark; mogens.vestergaard@anis.au.dk

Some rosé veal farmers feed calves a high-energy total mixed ration (TMR) already from weaning at 8 weeks, while others feed pelleted concentrate or a combination of both in this post-weaning period. The objective was to test how weaned dairy bull calves perform from 10 to 18 wk of age when offered either free choice between pelleted concentrate and TMR (FREE) or only TMR (TMR). A total of 32 calves were purchased from 4 dairy herds at 11±1 days of age and 47±1 kg LW (mean ± SE). Until weaning at 8 wk of age, calves were offered a total of 224 l milk replacer (23% CP, 19% fat). From 2 to 10 wk of age, calves had free access to a pelleted concentrate (20% CP) and artificially-dried chopped hay (14% CP). At wk 10 (97±2 kg LW) calves were assigned to either FREE or TMR. The pelleted concentrate and the TMR had (kg DM basis) similar NE (8.8 MJ), CP (213 g), digestible cell walls (195 g), and starch contents (328 g), but varied in DM (88 vs 70%), feed composition, physical form and chewing time (4 vs 12 min). 31 calves completed the experiment (16 FREE and 15 TMR). Hay comprised 2.5% of NEI for both FREE and TMR. TMR intake comprised 61% and concentrate pellet intake 36.5% of NEI for FREE calves. LW at 18 wk of age was 178 and 173 kg (n.s.) and ADG from wk 10 to 18 was 1.46 and 1.36 kg/d ($P<0.05$) for FREE and TMR, respectively. The long-term consequences for growth and carcass value were studied from 18 wk until slaughter at 42 wk of age in which period all calves were fed the same TMR diet (15% crude protein). The TMR calves compensated in this period with higher ADG than FREE (1.44 vs 1.36 kg/d) leading to similar ADG from birth to slaughter, carcass weight and EUROP classification. In conclusion, both post-weaning feedings can be applied without any log-term consequences for overall productivity.

Session 02 Theatre 7

Nutritional programming of piglet growth and development in the early neonatal period

Q. Sciascia and C.C. Metges
Leibniz Institute for Farm Animal Biology, Institute of Nutritional Physiology, Wilhelm-Stahl-Allee 2, 18196 Dummerstorf, Germany; sciascia@fbn-dummerstorf.de

The early neonatal period (~first 7 days of life) is one of the most stressful periods in the life of a piglet as they abruptly transition from primarily placenta supplied nutrients to enteral supplied substances from colostrum and milk. This is particularly true for low birth weight (LBW) piglets which are known to have lower energy reserves, delayed access to colostrum, lower milk intake and impaired nutrient absorption and metabolism. During this period colostrum and milk play critical roles in the stimulation and regulation of organ growth and development in suckling piglets, however studies with lactating sows suggest that specific dietary components provided by milk are insufficient for tissue protein accretion in piglets. This is important as the early neonatal period is a time of rapid organ growth which requires high amounts of substrates required for energy generation and protein turnover. Although the early neonatal period has been known to be crucial to piglet survival, with a significant percentage dying within the first seven days of life, there is a paucity of studies investigating the effect of supplementing functional dietary components during this important developmental stage – in comparison to other developmental stages (late gestation, weaning, post-weaning). To date, studies have tended to focus on amino acids (AA) such as arginine and glutamine as the concentration of each AA is not sufficient in sow's milk to support piglet protein synthesis requirements. Additionally, they are functional AA in that, as well as their role as protein building blocks they can stimulate various metabolic pathways that support organ growth and development. However, as our understanding of this important developmental stage deepens, other potential nutritional intervention avenues may have opened up to support the growth and development of LBW piglets.

Session 02 Theatre 8

Birth interval or duration of parturition: which is relevant to risk of stillbirth and intervention?

P. Langendijk, M. Fleuren and T.A. Van Kempen
Trouw Nutrition R&D, Stationstraat 77, 3811 MH Amersfoort, the Netherlands; pieter.langendijk@trouwnutrition.com

Multiparous Large White × Landrace sows (Hypor, Hendrix Genetics, n=256) were monitored continuously around parturition to record time of birth of each piglet, and whether a piglet was born alive, was a fresh stillborn, or was a non-fresh stillborn. Non-fresh stillborn were piglets that had obviously died before parturition, based on discoloration, or abnormal developmental conformity, were not included in the data. Only litters with at least 12 piglets were included. Risk of stillbirth for various birth intervals (0 to 30 min, etc.), and for various times from onset of parturition (0 to 2 h, etc.), was tested using a Chi-square test. Sows had 15.3±0.5 total born piglets on average. Duration of parturition was 280 min on average, and average stillbirth rate was 6.6%. The following data were based on 3,924 piglets in total. Piglets born within 0 to 30 min, 30 to 60 min, 60 to 90 min or more than 90 min after the previous piglet had a risk of being stillborn of 5.6, 6.8, 7.4 and 18.0%, respectively. Only piglets born more than 90 min after the previous had a significantly ($P<0.01$) higher stillbirth risk compared to shorter birth intervals. Piglets born within 2 h, 2 to 4 h, 4 to 6 h, 6 to 8 h, or more than 8 h after onset of parturition had a risk of being stillborn of 2.7, 6.9, 10.7, 13.4 and 27.3%, with the risk of stillbirth increasing significantly ($P<0.01$) with each time window that elapsed since onset of parturition. Risk of stillbirth was clearly related to the cumulative duration of parturition, and exceeded the average risk of stillbirth when 4 h had elapsed since the onset of parturition. Birth intervals between successive piglets had far less impact on the stillbirth rate, and only when the interval exceeded 90 min, was the risk of stillbirth increased compared to shorter intervals. In terms of management, this implicates that interventions will not have much benefit if they are applied based on birth intervals shorter than 90 min, and that they rather be initiated based on total duration of farrowing exceeding 4 h.

Session 02 Poster 9

Effects of adrenocorticotropin stimulation on hormones and growth factors in milk of lactating sows

E. Kanitz, A. Tuchscherer and M. Tuchscherer
Leibniz Institute for Farm Animal Biology, Wilhelm-Stahl-Allee 2, 18196 Dummerstorf, Germany; ellen.kanitz@fbn-dummerstorf.de

The maternal environment exerts important influences on offspring growth, metabolism, neurobiology, immune function and behaviour. The hypothalamus-pituitary-adrenal (HPA) axis, which regulates the secretion of glucocorticoids, is one of the systems highly susceptible to early-life programming. During lactation, mother's milk is an important physiological pathway for nutrient transfer and hormone signalling that potentially influences offspring phenotype. The aim of the study was to investigate whether maternal stress in pigs may activate the HPA axis and alter bioactive factors in milk. As a model for maternal stress, German Landrace sows were given adrenocorticotropin (ACTH; Synacthen Depot, 100 IU per animal) twice a day from lactation day 2 until 14. During the treatment period, saliva and milk samples were taken and maternal behaviour was observed. Saliva was examined for cortisol levels and milk was analysed for cortisol, oxytocin, interleukin-6 (IL-6), tumour necrosis factor-α (TNF-α) and transforming growth factor-β (TGF-β) concentrations by commercially available enzyme immunoassays either directly or after extraction. Physiological and behavioural data were evaluated by MIXED or GLIMMIX procedures in SAS/STAT software. Administration of ACTH increased the cortisol concentrations in saliva ($P<0.01$) and milk ($P<0.001$) compared to control animals during the treatment period. This increase of cortisol levels was associated with a decrease of TGF-β ($P<0.01$) and oxytocin ($P=0.06$) concentrations in milk. There were no significant effects of ACTH administration on IL-6 and TNF-α concentration. Furthermore, the ACTH treatment affected the maternal behaviour, which resulted in a more restless behaviour of the sow and in a lower suckling stability of their piglets (both $P<0.01$). In conclusion, the present results indicate that an enhanced HPA axis activity may be reflected in milk cortisol levels, and also may alter behaviour, hormones and cytokines with potential adverse effects on the offspring`s development in the context of health and welfare.

Session 02 Poster 10

Effects of maternal factors during pregnancy on the birth weight of lambs in dairy sheep

J. Pesántez[1,2], L. Torres[2], M. Sanz[2], A. Molina[3], N. Pérez[4], M. Vazquez[3], C. Garcia[2], P. Feyjoo[3], F. Cáceres[3], M. Frias[3], S. Mateos[3], F. Fernandez[5], J. Gonzalez[3,4], A. Gonzalez[2,3] and S. Astiz[2]
[1]University of Cuenca, Av. 12 de Octobre, 010220, Ecuador, [2]INIA, Av. Pta. de Hierro s/n, 28040, Spain, [3]UCM, Av. Pta de Hierro s/n, 28040, Spain, [4]TRIALVET, Encina 22, 28721, Spain, [5]Granja Cerromonte, Avila, 05358, Spain; congarcon@gmail.com

The aim of the present study was to evaluate the influence of physiological and productive factors of ewes (age and type of pregnancy and level of milk production) during pregnancy on birth weight of lambs. The study was carried out in a single commercial farm on 334 Lacaune dairy sheep, classified by age (187 mature and 147 maiden ewes), pregnancy type (155 single vs 179 multiple pregnancies), milk production average milk yield per day (YDIM), from conception to drying off; 45 Low yielding, LYDIM, <1.37 l/d; 70 Average yielding, AYDIM, 1.37 to <1.8 l/d; 72 High yielding, HYDIM, >1.8 l/d; and average milk yield per day during month of conception (Yc); 45 Low yielding, LYc, <0.91 l/d; 70 Average yielding, AYc, 0.91 to <1.3 l/d; and 72 High yielding, HYc, >1.3 l/d). Lambs body weight was recorded in 576 lambs (253 males, 323 females) at birth and at 18 days old. Differences among groups were evaluated with ANOVA and Kruskal-Wallis test when non-normal distributed. The lambs born to mature ewes were heavier than those from maiden ewes both at birth (4.2±0.8 vs 3.5±0.8 kg; $P<0.0001$) and at 18 d-old (8.8±2.0 vs 6.8±1.6 kg; $P<0.0001$). Newborns from single pregnancies were heavier at birth than those born from multiple pregnancies (4.3±0.9 vs 3.8±0.8 kg; $P<0.0001$), but no difference was found at 18 d-old (8.0±2.1 vs 7.9±2.1 kg; $P>0.05$). The birth weight of lambs was similar in the groups HYDIM and AYDIM and heavier in these both groups than lambs from the groups LYDIM (4.3±0.8; 4.3±0.8; 4.0±0.8 kg; $P<0.05$); there were no differences at 18 d-old. Concomitantly, lambs of the groups HYc and AYc were heavier than lambs in the group LYc at birth (4.4±0.8; 4.2±0.7; 4.1±0.8 kg; $P>0.05$), but values were similar among the three groups at 18 d-old. The present study suggests that, age, type of pregnancy and level of milk production during gestation in ewes are determinants of lamb birth weight.

RNA-seq reveals persistent differences in lipid metabolism of bulls with divergent birth weight

K. Komolka[1], N. Trakooljul[2], E. Albrecht[1], K. Wimmers[2], C. Kühn[2] and S. Maak[1]
[1]Leibniz Institute for Farm Animal Biology (FBN), Institute of Muscle Biology and Growth, Wilhelm-Stahl-Allee 2, 18196 Dummerstorf, Germany, [2]Leibniz Institute for Farm Animal Biology (FBN), Institute of Genome Biology, Wilhelm-Stahl-Allee 2, 18196 Dummerstorf, Germany; maak@fbn-dummerstorf.de

Birth weight and body composition in later life are associated in pigs and other species. Data on effects of birth weight on carcass traits in bulls with a longer, intensive fattening period are rare, however. Most reports recorded either effects until puberty or included nutritional interventions. We have selected two groups of bulls from an experimental F2 cross (Charolais × Holstein) with high (HBW: 59.7±1.0 kg, n=17) and low birth weight (LBW: 36.4±0.4 kg, n=15) for RNA sequencing of the M. longissimus. The animals were fattened from day 121 on a standardized, concentrate based diet *ad libitum* and were slaughtered at 18 months of age. The higher birth weight led to significantly higher hot carcass weights (420±8.4 vs 377±6.2 kg) despite similar live weights in both groups between 4 and 14 months of age. Higher feed intake in the HBW group was not accompanied by a better feed conversion compared to LBW bulls. However, carcass composition differed significantly with higher protein and lower fat accretion in HBW bulls. This was reflected in lower absolute and relative weights of the different fat depots in this group. RNA-sequencing of the muscle samples resulted in a total of 78 differently expressed genes (DEGs) at a false discovery rate (FDR)<0.05. Ingenuity Pathway analyses revealed three networks comprising the term 'Lipid Metabolism' containing 11-16 DEGs in each network. This is in accordance with the phenotypic data. Besides well-described adipokines (e.g. insulin induced gene 1, *INSIG1*) also potential candidates with uncharacterized function in cattle (e.g. apolipoprotein L domain containing 1, *APOLD1*) were identified. Our data indicate long-term metabolic effects of different birth weights in cattle and provide a basis for further investigations.

Milk yield and gene expression in the udder of beef heifers depending on pre and post- weaning diets

E. Dervishi[1], M. Blanco[2], J.A. Rodríguez-Sánchez[2], A. Sanz[2], J.H. Calvo[2,3] and I. Casasus[2]
[1]University of Alberta, 116 St & 85 Ave, Edmonton AB T6G 2R3, Canada, [2]CITA-IA2, Av. Montañana 930, 50059 Zaragoza, Spain, [3]ARAID, Av. de Ranillas 1-D, 50018 Zaragoza, Spain; jhcalvo@aragon.es

Raising female calves and heifers on high energy planes of nutrition during the pre-pubertal period has been proposed to lower the age at first calving, reducing their 'unproductive' period. The objectives of the present study were to evaluate nutrition-induced changes on the performance of the primiparous cow and its calf, milk yield and composition, and in gene expression of the mammary gland in the first lactation. Parda de Montaña female calves (n=16) were used in a 2×2 factorial experiment. In the pre-weaning period (PRE-W, 0-6 months), calves were either creep-fed or fed only on their dam's milk (Creep vs Control). In the post-weaning period (POST-W, 6-15 months), heifers received either a high (91.7 MJ/d) or a moderate energy diet (79.3 MJ/d) (High vs Moderate). After breeding (15 months), all the heifers received the same diet to the end of their first lactation (32 months). Primiparous cow and calf weights and gains during the first lactation were not affected by the PRE- or POST-W feeding management; however, creep feeding during PRE-W period reduced milk yield, milk fat, protein, casein, lactose throughout their first lactation and increased somatic cell count at 3rd and 4th months of lactation, and affected the gene expression patterns in their mammary gland at the end of the first lactation. The POST-W energy level had no impact on milk production and composition. Gene expression in mammary gland was affected by both PRE-W (307 genes; FDR<0.005) and POST-W treatments (7 genes; FDR<0.005), with PRE-W diet having the greatest impact. Overall, creep feeding during PRE-W period resulted in up-regulation of genes related with immune response and chemokine activity and down-regulation of ribosome and spliceosome genes. The data confirmed the lack of clinical mastitis, however, the possibility that the animals might be at greater risk to develop subclinical mastitis cannot be excluded. Therefore, increasing the energy level during the POST-W period would be advisable to reduce the age at first calving of heifers, without impairing milk yield or immune status.

Gestational immaturity has a bilateral dysregulatory effect on the HPA axis in adult horses

J. Clothier[1,2], A. Small[2], G. Hinch[1] and W. Brown[1]
[1]University of New England, School of Environmental & Rural Science, Armidale, NSW 2351, Australia, [2]CSIRO, Agriculture, Armidale, NSW 2350, Australia; jane.clothier@csiro.au

Human medical research shows that gestational immaturity can have a developmental impact on health, with studies involving adults reporting a link between extremely low birthweight and ongoing HPA axis dysregulation. A bilateral effect has been observed amongst affected groups, with both elevated and depressed basal cortisol levels reported, as well as altered pituitary-adrenal or adrenocortical responses to stressors amongst adults. Similar studies involving pigs, sheep and cattle have mostly reported elevated cortisol responses, while limited equine studies have reported an elevated pituitary-adrenal response due to neonatal stress. We speculated that an adapted endocrine stress response would be present in mature horses with a history of gestational immaturity, and that this would be identifiable through an ACTH stimulation and measurement of free cortisol in saliva. We assembled 10 horses aged 3-12 y known to have been small for gestational age at birth due to prematurity or dysmaturity, plus 7 control horses of similar age living in the same herds. All horses received low-dose exogenous synthetic ACTH by intramuscular injection (0.1 µg/kg bwt; Tetracosactrin 250 mg/ml, Synacthen®). Five saliva samples from each horse were collected between 0 and 2.5 h post-stimulus. These were later assayed using a salivary cortisol enzyme immuno-assay kit (Salimetrics). Data were analysed using R Studio. Mean basal salivary cortisol concentration (SCC) for all horses was within normal ranges (1.4±0.18 nmol/l). Mean peak SCC values (corrected for baseline) for affected horses fell into 2 distinct groups, comprising low cortisol responders (mean 5.18±1.42 nmol/l) and high cortisol responders (13.99±1.61 nmol/l), both significantly different to controls (9.89+1.51 nmol/l; ANOVA P<0.001). These results suggest that although basal cortisol levels remain unaffected, gestational immaturity affects the adrenocortical stress response, leading to a blunted or elevated cortisol response to stress in adult horses. It appears that, like other species, gestationally immature horses can experience adrenocortical dysregulation in later life.

Prenatal heat stress on emotional reactivity and behavioural reaction of female goats kids

W. Coloma-Garcia, N. Mehaba, A.A.K. Salama, X. Such and G. Caja
Group of Research in Ruminants (G2R). Universitat Autònoma de Barcelona, Department of Animal and Food Sciences, Edifici V, Travessera dels Turons, 08193 Bellaterra, Spain; wncg_100583@hotmail.com

Consequences of stress during the pregnancy period can affect the normal development of the offspring. In this study, the effects of heat stress were studied in Murciano-Granadina goats (n=30; 41.8±5.70 kg) which were exposed to 2 treatments: thermal-neutral (TN; n=15; 15 to 20 °C.); and heat stress (HS; n=15; 30 to 37 °C), and in their offspring. The TN and HS goats were maintained under the ambient treatment conditions from 15 d before mating until 45 d of pregnancy. Female kids born in TN group (n=16) and HS group (n=10) 30±15 day-old were subjected to a novel arena test (NAT) and a novel object test (NOT). In both tests, goat kids were entered a 4×4 m^2 arena. For the NAT, distance travel, number of squares entered, number of jumps and numbers of sniffs of the arena were recorded. The same parameters were recorded in NOT also including the latency to sniff the object and the number of object sniffs. For NAT and NOT count data, repeated measures and a simple linear model were used under a Poisson or negative binomial distribution. Compared to TN, HS goats had 3-d shorter pregnancy duration (P<0.01) and their offspring tended to show a reduction of 7% of the birth weight (P<0.13). In NAT, HS kids displayed a lower number of sniffs (P<0.01) and vocalizations (P<0.10). In NOT, HS kids also tended to show a lower number of sniffs (P<0.10). In conclusion, heat stress during the first third of pregnancy reduced the duration of gestation with probable effects on the weight of the offspring. In addition, behavioural tests suggested an altered emotional reactivity during the postnatal life of the goat kids after the heat stress suffered in utero.

Session 02 Poster 15

Feeding the transition sow *ad libitum*: a healthy start for suckling piglets

P. Langendijk and M. Fleuren
Trouw Nutrition R&D, Stationstraat 77, 3811 MH Amersfoort, the Netherlands; pieter.langendijk@trouwnutrition.com

Multiparous sows (n=48) and gilts (n=33) of a commercial line (Hypor, Hendrix Genetics, Netherlands) were allocated to a traditional feeding curve or to *ad libitum* feeding from a week before farrowing through to weaning, distributing parities equally across treatments. Sows and gilts on the traditional feeding curve were restrict fed (2.5 to 3.5 kg, depending on parity) before parturition, and subsequently feed allowance was stepped up gradually to a maximum of around 8.5 kg. Similar to the *ad libitum* fed sows, restrict fed sows were free to choose their time of feeding, the difference being only the limit in allowance. The electronic feeders allowed portioned delivery of feed, and feeder access and intake were recorded real time. Prior to farrowing, feed intake was higher in ad lib fed sows (4.7 vs 3.3 kg/d) and gilts (3.5 vs 2.6 kg/d). Ad lib fed sows consumed more than traditionally fed sows (5.0 vs 4.3 kg/d; P<0.03) during the first week of lactation. During the remainder of lactation, feed intake of the two groups was similar at around 8 kg/d. Over the whole lactation period, average intake was 7.1 kg/d for ad lib fed sows and 6.6 kg for restricted sows (P<0.08). Feed intake for gilts was similar between *ad libitum* and the traditional feeding curve. No adverse effects of *ad libitum* feeding, such as overeating with subsequent drops in feed intake, constipation, or oedema were observed. Stillbirth rate (7.8% in gilts and 8.5% in sows) was not affected by treatments. Litters with ad lib fed sows gained 7 kg more during lactation and their piglets were 0.5 kg heavier at weaning compared to traditionally fed sows (P<0.05). Average daily gain was significantly higher (247±6 vs 231±4 g/d; P<0.05) over the whole lactation, and numerically higher in the first 14 d. In gilts there was no significant difference in litter performance. Ad lib feeding from one week before farrowing clearly increased feed intake in multiparous sows, and the extra intake apparently went to benefit milk production, based on the increased litter gain.

Session 03 Theatre 1

Identification of bovine copy number variants in two genomic information sources

A.M. Butty[1], F. Miglior[1,2], K. Krivushin[3], J. Grant[3], A. Kommadath[3], F.S. Schenkel[1], P. Stothard[3] and C.F. Baes[1]
[1]Centre for the Genetic Improvement of Livestock, University of Guelph, Guelph, ON, Canada, [2]Canadian Dairy Network, Guelph, ON, Canada, [3]Dept of Agricultural, Food, and Nutritional Science, University of Alberta, Edmonton, AB, Canada; cbaes@uoguelph.ca

Copy number variants (CNV) are the most commonly found structural variants in the genome. Their identification and characterization are in early stages, but studies have shown the importance of CNV for economically relevant traits in dairy cattle. Low false discovery rates of CNV are needed to enable accurate downstream analysis. In this study, we identified CNV in 38 Holstein bulls based on two sources of genomic information. Furthermore, we compared the resulting sets to each other and to previously reported CNV in the publicly available Genomic Variant archive database of EMBL-EBI. We identified CNV using: (1) a multi-sample approach on the read depth of their whole-genome sequences using the default parameters of the software cn.MOPS (WGS_CNV); and (2) a single sample approach on the signal intensities and B-allele frequencies of their high-density array genotypes using the default parameters of the software PennCNV (HD_CNV). Identification of CNV following the WGS_CNV approach was expected to have higher power and lower FDR than the HD_CNV approach, as all samples were considered simultaneously to disentangle biological and technical sources of read depth variation. Less CNV were found in the WGS_CNV set (n=1,054) compared to the HD_CNV set (n=1,098). Significant differences between methods were observed in the average number of samples supporting the presence of a CNV ($P<10^{-7}$). Approximately 70% of the WGS_CNV and 80% of the HD_CNV were private (i.e. found in only one individual). Of the 60 CNV commonly found in both sets, 48 were not previously described in the public database. Compared to the HD_CNV set, a smaller proportion of WGS_CNV were private and a higher percentage of these were already known, suggesting a better accuracy of the identification based on sequence. Moreover, comparison of two genomic information sources allowed for pre-validation of 48 novel CNV. The identified CNV will be further described and annotated before inclusion in association analyses.

The use of runs of homozygosity in managing long term diversity in livestock populations

C. Maltecca[1], G. Gebregiwergi[2], C. Baes[3] and F. Tiezzi[1]
[1]NCSU, Box 7621 NCSU, 27695, USA, [2]NMBU, Universitetstunet 3, 1433 Ås, Norway, [3]University of Guelph, Guelph, Ontario, Canada; f_tiezzi@ncsu.edu

Population diversity management is accomplished in most cases through pedigree information and various systems aimed at constraining the accumulation of inbreeding in mating populations. Optimal contribution selection (OCS) has been a popular method to guarantee long term gains without compromising variability. Little information is available on the impact of using ROH genomic information in conjunction with OCS. In this research, we investigated the use of alternative metrics of ancestry in OCS in simulated scenarios using genomic information. One production trait and one fitness trait were generated with the GenoDiver software following a typical dairy population structure. For the production trait, a polygenic trait (h2 0.5, 1000 QTL) was simulated. For the fitness trait, partial dominance was simulated with varying proportions of lethal and sub-lethal fitness trait loci (FTL) ranging from 0 to 5% of the total FTL number. OCS was simulated for 30 generations. At each generation, genomic information was used to obtain breeding values of individuals, while different relationship measures (pedigree, genomic, and ROH (10 and 20 Mb)) were employed for optimal contribution. Selection was performed only on the production trait. Genetic progress for both production and fitness in all scenarios was measured at generation 30. Diversity measures analysed included homozygosity, lethal equivalents, fitness, segregating (sub-)lethals, QTL and FTL lost. All methods were compared to a baseline of no OCS. In all cases, OCS maintained greater genomic variability in respect to the baseline. ROH-based methods achieved larger genetic gains compared to other methods. Pedigree-based methods maintained the largest variability with the lowest genetic gain. Genomic-based OCS was best at constraining homozygosity at lethal loci, while ROH-based methods were more effective for constraining sub-lethal loci. Both ROH- and genomic-based OCS were effective, with the ROH resulting in a good compromise between short-term genetic response and long-term fitness management.

A 2-step-stratey to infer genome-wide associations for endoparasite traits in local DSN cattle

K. May[1,2], C. Scheper[2], K. Brügemann[2], T. Yin[2], C. Strube[1], P. Korkuc[3], G.A. Brockmann[3] and S. König[2]
[1]Institute for Parasitology, University of Veterinary Medicine Hannover, Buenteweg 17, 30559, Germany, [2]Institute of Animal Breeding and Genetics, Justus-Liebig-University of Gießen, Ludwigstraße 21B, 35390 Gießen, Germany, [3]Faculty of Live Science, Humboldt-Universität of Berlin, Animal Breeding Biology and Molecular Genetics, Invalidenstraße 42, 10115 Berlin, Germany; ka.may@gmx.de

A 2-step approach was applied in order to detect genome wide associations (GWAS) and physiological pathways in local DSN cattle with a limited no. of phenotypes for endoparasite traits. In step 1 (correction for environmental effects), we used the full phenotypic dataset for endoparasite traits from cows from all genetic black and white selection lines (DSN, HF from Germany, HF from New Zealand, HF-crosses) in mixed model analyses. The total dataset included 2,006 records for the endoparasite traits gastrointestinal nematodes (GIN), the bovine lungworm (*Dictyocaulus viviparus*) and the liver fluke (*Fasciola hepatica*) from 1,166 cows. In step 2, residuals from mixed models for the DSN cows (i.e. rFEC-GIN, rFLC-DV, rFEC-FH) were dependent traits in GWAS. 148 DSN cows were genotyped with the BovineSNP50 Bead Chip and subsequently imputed to Illumina HD Bead Chip level (i.e. 700,000 SNPs) using a multi-breed reference panel of 2,188 animals. The final dataset contained 423,654 SNP. For rFEC-GIN, GWAS detected 68 SNP markers above the candidate threshold (pCand=1×10^{-4}). For rFEC-FH, 3 SNP on BTA 7 were significant for the genome-wide significance threshold (pBonf=4.47×10^{-7}), and 53 variants for pCand. For rFLC-DV, 41 and 311 SNP associations surpassed pBonf and pCand, respectively. Identified associated SNP markers were related to potential candidate genes using the current gene annotations from the ENSEMBL and NCBI databases. The DAVID database was used to infer pathways of candidate genes. Marker associations included annotations to 53 candidate genes for all three endoparasite traits in the DSN population. Pathway analyses identified seven pathways being related to immune response mechanisms, or involved in host-parasite interactions. SNP effect correlations were favourable between milk yield and endoparasite traits within important chromosomal segments for endoparasite traits.

Session 03 — Theatre 4

Genetic parameters of novel mid-infrared predicted milk traits in three dual-purpose cattle breeds

S. Vanderick, F.G. Colinet, A. Mineur, R.R. Mota, N. Gengler and H. Hammami
ULiege-GxABT, TERRA Teaching and Research Centre, Passage des Deportes 2, 5030 Gembloux, Belgium; sylvie.vanderick@uliege.be

The objective of this study was to estimate genetic parameters of 39 novel mid-infrared predicted milk traits (e.g. nutritional quality, technological properties, metabolic status, environmental fingerprint) for three dual purpose cattle breeds (i.e. Dual-Purpose Belgian Blue (dpBB), Montbéliarde (MON) and Normande (NOR)), which are also used in organic farming in the Walloon Region of Belgium, as part of the 2-Org-Cows project. Edited data included 21,287, 10,062 and 4,637 first-lactation test-day records collected in the Walloon region of Belgium from 2,988, 1,330 and 621 dpBB, MON and NOR cows, respectively. Genetic parameters were estimated using REML applied to single-trait random regression test-day models for six conventional traits (yields, contents and somatic cell score) and the 39 novel mid-infrared predicted milk traits. Results for conventional traits allowed comparison to literature showing values that were close to the expected ones. For novel traits, comparison with available literature values for Holstein breed showed generally similar estimated heritabilities. Reported average daily heritabilities estimated for the 39 novel traits tended to be higher for dpBB (0.13-0.64) than MON and NOR (0.03-0.60) breeds. Few novel traits showed large differences between breeds except between dpBB and NOR for milk composition traits. However, results for NOR breed have to be taken very carefully given the low number of animals. Even if the used methane prediction equation was not yet validated for these breeds, estimated average daily heritability was moderately high for dpBB (0.41) and MON (0.36) and moderate for NOR (0.23) indicating that this prediction might also be useful in these dual purpose breeds.

Session 03 — Theatre 5

RekomBre – a tool to simulate and compare large scale breeding programs

T. Pook[1], M. Schlather[2] and H. Simianer[1]
[1]University of Goettingen, Animal Sciences, Albrecht-Thaer-Weg 3, 37075 Goettingen, Germany, [2]University of Mannheim, Stochastics and Its Applications, A5, 6 B 117, 68161 Mannheim, Germany; torsten.pook@uni-goettingen.de

When setting up a breeding program proper design is needed to generate high genetic progress. A possible aid in finding a good allocation of resources is the comparison of possible breeding programs via simulation. Over the years a lot of tools have been developed for this task (e.g. ZPLAN, QMSim) – many of these applications are not user-friendly or not versatile enough to be applied to the different scenarios in the study of interest. Therefor most simulation studies are performed with custom-made solutions that focus on the current topic of interest but thereby lack accuracy in other aspects of the underlying genetics. Here, we propose the use of the R-package 'RekomBre' which supports the simulation of complex and large scale breeding programs with focus on livestock and crop populations. User-input can be inserted in a web-based application including a tool to visualize the mating scheme between different breeding groups. Possible features include gender (or individual) specific recombination rates, flexible mutation rates and different options for gene editing. In regard to breeding values, different genetic architectures ranging from additive single marker QTL to complex epistatic effects can be simulated. Multiple breeding values with correlated phenotypes, adjustable heritabilities and different environments are supported. Critical parts of RekomBre concerning memory requirements and computing times are written in C. By using SSE2 operations and bit-wise storing computation speed can be massively increased leading to about 10 times faster matrix multiplications than the regular R implementation while needing only 1/16 of the regularly needed memory. Additionally haplotypes are calculated on-the-fly via storing of mutations, recombination and founders. Overall this leads to a massive reduction of computation requirement enabling the simulation of large breeding problems on small machines. A lot of standard procedures applied in breeding like GBLUP and OGC are implemented but the use of own methods in the selection process are possible and can be accelerated by the usage of computationally efficient methods implemented in RekomBre.

Testing different genomic selection scenarios in a small cattle population by simulation

J. Obsteter[1], J. Jenko[2], J.M. Hickey[2] and G. Gorjanc[2]
[1]Agricultural Institute of Slovenia, Hacquetova ulica 17, 1000 Ljubljana, Slovenia, [2]The Roslin Institute and Royal (Dick) School of Veterinary Studies, The University of Edinburgh, Easter Bush, Midlothian, EH259RG, Edinburgh, United Kingdom; jana.obsteter@kis.si

Genomic selection increases genetic gain by reducing generation interval and increasing accuracy of selection. However, success of selection depends also on the extent and strategy of using genomic information in the breeding program. The questions relate to how many and which animals to genotype and in what extent to use the genomic information. We developed a simulator of realistic cattle population with overlapping generations and selection. We tested five sire selection scenarios including conventional and four genomic selection scenarios that differed in the criterion for sire selection (progeny / genomic testing). All five scenarios were tested within three sire use strategies differing in the number of sires selected per year and years kept in use. Scenarios were compared by genetic gain and efficiency of selection. The latter was defined as conversion of genetic variance into genetic gain. In all sire use strategies the increasing use of genomics increased the genetic gain. Using genomic information for pre-selection of calves for progeny testing increased genetic gain between 35 and 46% and using genomic information for the immediate selection of sires increased the gain up to 123% compared to the conventional scenario. These results were in line with the reduced generation interval. In the original sire use strategy (five bulls selected each year and used for five years) the conventional scenario was the least efficient and the complete genomic selection scenario was not the most efficient. Despite higher genetic gain, increasing intensity of selection (only one bull selected each year and used for five years) decreased the efficiency of selection due to larger loss in genetic variance. Reducing the generation interval even further (replacing all five bulls every year) remedied the low efficiencies – but they were still decreased compared to the original strategy due to a lower accuracy of prediction. The developed simulator will be used as a modelling tool for answering further questions regarding genomic selection in specific populations.

Breed specific reference genomes in cattle

B. Czech[1], M. Mielczarek[1,2], M. Frąszczak[1] and J. Szyda[1,2]
[1]Biostatistics group, Department of Genetics, Wroclaw University of Environmental and Life Sciences, Kozuchowska 7, 51-631 Wroclaw, Poland, [2]National Research Institute of Animal Production, Krakowska 1, 32-083 Balice, Poland; bwczech@wp.pl

The UMD3.1 reference genome for *Bos taurus* was created based on two related individuals representing the Hereford breed. Genetic analyses directed to other breeds may be therefore erroneous or inaccurate. This constructing breed-specific reference genomes will allow a more accurate genomic analyses. Rapidly developing methods of Next Generation Sequencing (NGS) provide informative data sets, such as whole-genome sequences of 936 bulls used in this study, which allowed us to construct specific reference genomes of seven breeds: Angus, Brown Swiss, Fleckvieh, Hereford, Jersey, Limousin and Simmental. Based on seven sets of SNPs identified for bulls representing each breed we defined variants unique for each breed. The selection criteria comprised variants whose number of missing genotypes was not more than 7%, and the frequency of the alternative allele was equal to unity. The highest number of breed-specific SNPs was identified for Jersey (130,070) and the lowest – for the Simmental breed (197). Furthermore, the Variant Effect Predictor (VEP) software was used for genomic annotation of those SNPs. VEP showed that breed-specific variants were located in coding sequences of 78 genes in Angus, 140 genes in Brown Swiss, 132 genes in Fleckvieh, 100 genes in Hereford, 643 genes in Jersey, and 10 genes in Limousin. In Simmental breed-specific SNPs did not overlap with coding sequences. Genes harbouring breed-specific SNPs were further related to OMIA phenotypes and were identified for Angus, Brown Swiss, Jersey and Limousin. Moreover, genes harbouring breed-specific SNPs overlapped with QTL found in the QTLdb has been checked. Identified variants represent potential characteristic features of the compared breeds. Created reference genomes form the basis for further research on the breeds. In the case of cattle, the phenotypic differences between breeds are much more pronounced than in humans, which is due to the different purpose of each breed. Because of quick developing of NGS constructing more and more specific reference genomes allow for more accurate genomic inferences.

Opportunities and challenges for small populations of dairy cattle

H. Jorjani
Interbull Centre, Department of Animal Breeding & Genetics, Swedish University of Agricultural Sciences, Box 7023, 75007 Uppsala, Sweden; hossein.jorjani@slu.se

In the beginning of the genomic era, the usefulness of genomic evaluations was judged mainly by the gain in reliability (REL) of the EBV compared to the conventional genetic evaluations. With this in mind, higher density of SNP arrays (number of SNPs included in the genomic evaluation), and larger reference population (genotyped animals with 'enough' phenotypic data) were considered to be the two main contributing factors to higher gain in REL. The SNP densities higher than about 50,000 were proved to have marginal effects. Therefore, the emphasis on larger reference populations became stronger, and to the consolidation of consortia who advocated genotype sharing in dairy cattle populations. At the same time, it could be shown that the predicted REL of EBV was higher (and sometimes, much higher) than the realized REL values. The obvious conclusion was that the gain in REL values must be calculated from validation tests, which lead to even more emphasis on larger reference population size. Given this background, the advice for the smaller populations has been to form common reference populations by pooling reference animals from different populations (countries). One question, with uncertain answer, for the smaller dairy cattle populations is what is the minimum size of a reference population that would make the genomic evaluation worthwhile? Here, based on the genomic evaluations of small Brown Swiss and Holstein populations, with as low as 171 reference bull for a single population (country), it is claimed that: (1) correlation of the DGV and GEBV with EBV is invariably different from unity, which is helpful in selection decisions, i.e. distinguishing between bulls of equal EBV and REL; (2) any non-zero addition of the genomic information to a genetic evaluation improves the breeding program; (3) using MACE EBV leads to highly consistent estimates of SNP effects, and therefore a common reference population of any size is useful; and finally (4) the three points mentioned above leads to the conclusion that the validation test of the genomic evaluation systems (GEBV-test) is not the only way to judge the usefulness of a genomic program.

Application of ssGBLUP using random regression models in the Ayrshire and Jersey breeds

H.R. Oliveira[1,2], L.F. Brito[1], D.A.L. Lourenco[3], Y. Masuda[3], I. Misztal[3], S. Tsuruta[3], J. Jamrozik[1,4], F.F. Silva[2] and F.S. Schenkel[1]
[1]University of Guelph, Department of Animal Biosciences, 50 Stone Rd. East, N1G 2W1, Guelph, Canada, [2]Universidade Federal de Viçosa, Department of Animal Science, Av. P.H. Rolfs, 36570-900, Viçosa, Brazil, [3]University of Georgia, Department of Animal and Dairy Science, 30602, Athens, USA, [4]Canadian Dairy Network, 660 Speedvale Av. West, N1K 1E5, Guelph, Canada; holivier@uoguelph.ca

The feasibility of using random regression models (RRMs) in single-step genomic BLUP (ssGBLUP) evaluation was investigated in the Ayrshire and Jersey breeds. A total of 204,429 and 157,718 animals, in which 1,812 and 1,005 were genotyped, were used for the respective breeds. Individual additive genomic random regression coefficients for each trait were predicted using a multiple-trait RRM for milk, fat and protein yields in the first, second, and third lactations. Further, the additive coefficients were used to derive genomic estimated breeding values (GEBVs) for each day in milk, which were compared to parent average (PA). To investigate the effect of scaling the genomic relationship information to be compatible to the pedigree information, different weights for the inverse of the genomic (G) and pedigree (A_{22}) relationships matrices were tested (i.e. 1.0, 1.5 and 2.0 for G^{-1}; and 0.6, 0.7, 0.8, 0.9 and 1.0 for A_{22}^{-1}). The ssGBLUP method yielded considerably higher validation reliabilities compared to the method disregarding genomic information (PA). Average reliability for PA and GEBV were 0.42 and 0.46 for the Ayrshire, and 0.44 and 0.57 for the Jersey breed, respectively, when using no weights for G^{-1} and A_{22}^{-1}. Weights used to combine G^{-1} and A_{22}^{-1} had, in general, small influence in the validation reliabilities; however, greater effect was observed on the validation regression coefficients. Less biased regression coefficients were obtained by the ssGBLUP method when compared to PA when weights were applied to G^{-1} and A_{22}^{-1}. The weights that produced the least biased predictions and the highest validation reliabilities were 2.0 and 0.6 for the Ayrshire; and 1.5 and 0.9 for the Jersey breed, for G^{-1} and A_{22}^{-1}, respectively. The use of ssGBLUP based on RRMs is a feasible alternative to implement genomic evaluation of production traits in Ayrshire and Jersey breeds.

Improving genomic prediction in numerically small Red dairy cattle populations

J. Marjanovic[1], B. Hulsegge[1,2], A. Schurink[2] and M.P.L. Calus[1]
[1]Wageningen University & Research, Animal Breeding and Genomics, Droevendaalsesteeg 1, 6700 AH Wageningen, the Netherlands, [2]Wageningen University & Research, Center for Genetic Resources, the Netherlands (CGN), Droevendaalsesteeg 1, 6700 AH Wageningen, the Netherlands; jovana.marjanovic@wur.nl

Red dairy breeds (RDB) represent valuable cultural asset and are often a source of unique genetic diversity. These breeds, however, have difficulties to compete with other, highly productive, dairy breeds, which endangers their existence. One of the strategies to conserve RDB is to increase their competitiveness, by accelerating genetic improvement of production traits using genomic selection. Recently, new genotype data on five Dutch RDB (MRY, Groningen White Headed (GWH), Dutch Belted (DB), Dutch Friesian (DF), and Deep Red (DR)) has been generated. Many of these RDB are numerically small and obtaining a sufficiently large breed-specific reference population is challenging. Adding individuals from other breeds to the reference populations, may help overcome this issue. Effective number of chromosome segments (Me) can be used as an indicator of relatedness between individuals from different breeds and to directly predict expected accuracy of genomic prediction that relies on use of information from such breeds. The aim of this study was to estimate Me for each of the available Dutch RDB and between each pairwise combination of breeds. The Me are used to predict the accuracy of GP within each of the breeds and accuracy of any possible combination of different breeds in the reference population. This study is part of the ERA-NET SusAn project ReDiverse, that aims to increase resilient and competitive use of European RDB. The analysis involved BovineSNP50 data on at least 41 animals for each of the RDB involved. The Me were computed as the reciprocal of the variance of the difference between genomic and pedigree relationships using calc_grm software. Based on the results, MRY and DR are most closely related. For these two breeds, at least ten individuals from the other breed are needed to contribute as much information as an individual from the breed itself. DF was most closely related to DB, and vice versa, while for GWH, DF was the closest breed. The most distant relationships were observed between DR and DB, DR and GWH, and GWH and MRY.

An equation to predict accuracy of multi-breed genomic prediction model with multiple random effects

B. Raymond[1,2], A.C. Bouwman[2], Y.C.J. Wientjes[2], C. Schrooten[3], J. Houwing-Duistermaat[4] and R.F. Veerkamp[2]
[1]Biometris, Wageningen University and Research, Droevendaalsesteeg 1, 6700 AA Wageningen, the Netherlands, [2]Animal Breeding and Genomics, Wageningen University and Research, Box 338, 6700 AH Wageningen, the Netherlands, [3]CRV BV, P.O. Box 454, 6800 AL Arnhem, the Netherlands, [4]Department of Medical Statistics and Bioinformatics, Leiden University Medical Centre, Einthovenweg 20, 2333 ZC Leiden, the Netherlands; biaty.raymond@wur.nl

The accuracy of genomic prediction (GP) strongly depends on the size of the reference population. This limits the application of GP in numerically small breeds or for difficult to measure traits, in which case only a few individuals have phenotypes. In our recent study, we developed a new multi-breed GP model that utilises information on previously identified QTLs i.e. separate the effects of known QTLs from those of other markers by fitting them in different genomic relationship matrices. This way the model captures both QTL variance and the polygenic variance that is missed by QTLs. In addition, the contribution of breeds in the reference population is appropriately weighted by the genetic correlation between breeds in the new model. Our results using both real and simulated data clearly demonstrate the superiority of the new model in terms of predictive ability, in comparison with traditional within, across or multi-breed prediction models. Given the parameters of the new model, the objective of this study is to use selection index theory to develop a deterministic equation for predicting the accuracy of multi-breed GP model with more than a single random genetic effect. The new equation can serve as a decision tool for optimal GP scenarios for small populations and can also be used to assess how much benefit there is in putting more efforts on identifying QTLs. We will validate the performance of the equation using simulated data.

Integration of external information into the national multitrait evaluation model

T.J. Pitkänen[1], M. Koivula[1], I. Strandén[1], G.P. Aamand[2] and E.A. Mäntysaari[1]
[1]Natural Resources Institute Finland, Myllytie 1, 31600 Jokioinen, Finland, [2]NAV Nordic Cattle Genetic Evaluation, Nordre Ringgade 1, 8000 Aarhus, Denmark; timo.j.pitkanen@luke.fi

Integration of external information, e.g. international estimated breeding values (EBVs) from Interbull or Interbeef, into domestic evaluation model, is presented. The method requires EBVs and reliabilities for internationally evaluated animals (hereinafter external animals), and domestic EBVs and reliabilities for their relatives, i.e. typically their offspring. The method has three steps. First, the amount of extra information in international evaluation for each animal is extracted using reversed reliability calculation where International reliability is used as input for the external animals, and national reliability for their offspring. This is a reverse to the reliability calculation, i.e. output is effective record contributions (ERC). For an external animal, the obtained ERC represents extra information from international evaluation due to the domestic evaluation. In the second step, pseudo-observations for the external animals are calculated by deregression. For this, international EBVs are used as observations for the external animals, and domestic EBVs for their offspring. ERCs calculated in the first step are used as weights. In the third step, modified domestic evaluation model is fitted where each external animal has deregressed genetic predictions (DRPs) as pseudo-observation and ERC as weight. The method was tested on an artificial setup where EBVs from joint evaluation model for Nordic countries were blended to domestic evaluation model for Denmark. Because the model was multi-lactation model, a multi-trait reverse reliability calculation was implemented and DRPs were solved using three trait multi-trait deregression. For 'imported' sires the correlation of first lactation breeding values from joint and from blended national evaluations were 0.994, and the regression of blended EBV on joint evaluation was 1.03.

New approach to calculate inbreeding effective population size from runs of homozygosity

G. Gorjanc[1], J.M. Hickey[1] and I. Curik[2]
[1]The Roslin Institute of the University of Edinburgh, Easter Bush, Midlothian, EH25 9RG, United Kingdom, [2]University of Zagreb, Faculty of Agriculture, Department of Animal Science, Svetošimuska 25, 10000 Zagreb, Croatia; icurik@agr.hr

Inbreeding effective population size (N_{eI}) is among the most important parameters in conservation genetics and breeding as it is functionally linked to population extinction risk and decreased fitness. In livestock populations with overlapping generations, NeI is often estimated by regressing individual inbreeding coefficients, pedigree (F_{PED}) or genomic (F_{ROH}), on year of birth and accounting for generation interval. We present a simple new approach that enables estimation of inbreeding effective population size, here called N_{eFROH}, from single generation samples. We estimate N_{eFROH} from the accumulation of genomic inbreeding over past generations, approximated by the runs of homozygosity (ROH) of different length when calculating realized inbreeding. While calculation of N_{eFROH} is possible even from a single F_{ROH} value, its estimation accuracy depends on the sample size. The new approach, with slight modifications, also enables calculation of recent historical changes in inbreeding population size. Properties of N_{eFROH} are further analysed and tested with simulations. We also present N_{eFROH} obtained in some livestock populations. Our method is simple and when SNP array information is available it enables 'fast' warning on the potential risk caused by recent inbreeding increase in a population.

Inbreeding and effective population size of Holstein cattle in Brazil

V.B. Pedrosa[1], M.F. Sieklicki[1], H.A. Mulim[1], A.A. Valloto[2] and L.F.B. Pinto[3]
[1]Ponta Grossa State University, Department of Animal Science, Av Gen Carlos Cavalcanti 4748, 84030-900 Ponta Grossa PR, Brazil, [2]Paraná Holstein Breeders Association – APCBRH, Rua Professor Francisco Dranka 608, 81200-404 Curitiba PR, Brazil, [3]Federal University of Bahia, Av. Adhemar de Barros s/n, 40170-115 Salvador BA, Brazil; vbpedrosa@uepg.br

The objective of the present study was to evaluate the inbreeding coefficient (F), the realized effective population size (Ne) and the average relatedness. The data used in this research were provided by the Paraná Holstein Breeders Association (APCBRH), with a pedigree file of 206,796 Holstein animals born between 1970 and 2014. The coefficient of inbreeding was estimated using the methodology: $F_t=1-(1-\Delta F)^t$, where ΔF was the inbreeding rate from one generation (t) to the next one. The realized effective population size was calculated from (ΔF), computed by averaging the ΔFi of 'n' individuals included in a given reference subpopulation, as: Ne= 1/((2ΔF)). This method of computing effective population size was chosen for not being dependent on the whole reference population mating policy, but on the matings carried out throughout the pedigree of each individual. The average relatedness was calculated by the algorithm: C' = (1/n)1'A, where A was the relationship matrix of n × n size; and 1' was a vector of 1 × n. Inbreeding parameters were calculated using the software ENDOG v.4.5. The population average inbreeding was 4.99% that could be divided into three classes, being the progeny with 4.55%, followed by 7.25% of the cows and 0.22% for the bulls. The realized effective population size in the studied period ranged from 22 in the initial period of evaluation to 114 nowadays. The average relatedness was 0.71%. According to the level of inbreeding presented, it was observed that the dissemination of the genetic material of the main breeders between the generations, resulted in good indicators of genetic diversity, allowing a suitable genetic structuring of the population, facilitating the conduction of breeding programs in the evaluated herds. Funded by Fundação Araucária.

Session 03 Poster 15

Genomic inbreeding estimation and effective population size of three SA dairy cattle breeds

J. Limper, C. Visser and E. Van Marle Koster
University of Pretoria, Animal and Wildlife Sciences, Private Bag x 20, 0028, South Africa; carina.visser@up.ac.za

Small effective population sizes and inbreeding are major challenges faced by dairy cattle populations worldwide. Estimates of pedigree-based inbreeding (Fped) is unreliable due to a lack of pedigree data and pedigree errors and that may lead to an underestimation of inbreeding rates. No genomic estimates for inbreeding in South African (SA) dairy breeds are currently available. The aim of the study was to estimate genomic inbreeding and effective population sizes for three SA dairy breeds. The FSNP and FROH inbreeding levels as well as the effective population sizes for each breed was estimated. Genotypes representing the SA Holstein (n=404), SA Jersey (n=414) and SA Ayrshire (n=84), obtained from the Dairy Genomics Program (DGP) of South Africa were analysed. Genotypes were generated with the GeneSeek Genomic Profiler (GGP) 50K bovine bead chip. Samples and autosomal, mapped SNPs were quality filtered with PLINK v1.09. The average MAF across autosomes was 0.269±0.14, and the observed heterozygosity values ranged between 0.339 (Jersey) and 0.359 (Holstein). Levels of inbreeding was investigated through both FIS and the identification of runs of homozygosity (ROH) in PLINK. Effective population size (Ne) was estimated using SNeP version 1.1. FIS values ranged between -0.007 (Holstein) and -0.03 (Ayrshire). ROH was estimated at various lengths: ROH=0.5 Mb, ROH1=1 Mb, ROH2=2 Mb, ROH3=4 Mb, ROH4=8 Mb, ROH5=16 Mb. FROH16 was estimated as 0.252, 0.247 and 0.209 compared to FROH1 0.097, 0.084 and 0.095 for the Holstein, Jersey and Ayrshire respectively. This indicates an increase in inbreeding in recent years and is supported by the effective population size which shows a decrease from 577, 430 and 528 (120 generations ago) to 133, 120 and 110 (13 generations ago) for the Holstein, Jersey and Ayrshire respectively. The increase in inbreeding need to be addressed as inbreeding leads to reduction in animal performance and thus major economic losses. The monitoring and control of inbreeding using effective population size as a parameter, is essential for implementation of genetic improvement programs based on genomic information.

Effect of including only genotype of animals with accurate proofs in ssGBLUP using random regression

H.R. Oliveira[1,2], L.F. Brito[1], D.A.L. Lourenco[3], Y. Masuda[3], I. Misztal[3], S. Tsuruta[3], J. Jamrozik[1,4], F.F. Silva[2] and F.S. Schenkel[1]

[1]University of Guelph, Department of Animal Biosciences, 50 Stone Rd. East, N1G 2W1, Guelph, Canada, [2]Universidade Federal de Viçosa, Department of Animal Science, Av. Peter Henry Rolfs, 36570-900, Viçosa, Brazil, [3]University of Georgia, Department of Animal and Dairy Science, 30602, Athens, USA, [4]Canadian Dairy Network, 660 Speedvale Av. West, N1K 1E5, Guelph, Canada; holivier@uoguelph.ca

In the single-step genomic evaluations (ssGBLUP), phenotypic records are simultaneously combined with pedigree and genomic information, in order to generate Genomic Estimated Breeding Values (GEBVs). However, the effect of including only genotypes of animals with accurate proofs instead of all available genotypes in the training population when performing ssGBLUP is still unknown, especially when evaluating small populations and using complex models such as random regression (RRM). In this context, GEBVs for milk, fat and protein yields for the first three lactations of Jersey cows were predicted based only on genotypes of animals with accurate proofs (i.e. animals with reliability for the estimated breeding values (REL_{EBV}) higher than 0.50) n=917; or considering genotypes available for all animals, n=1,427. The GEBVs were estimated using ssGBLUP based on a multiple-trait RRM. The pedigree relationship matrix contained 157,718 animals. Prediction reliability for each trait was evaluated in the validation population (88 bulls with REL_{EBV} greater than 0.65 based on complete EBV), as a Pearson correlation between GEBVs and EBVs. Similar validation reliabilities and regression coefficients (bias) were estimated when considering genotypes available for only animals with accurate proofs or all animals in the reference population. Average reliability considering only accurate and all animals were 0.50 and 0.52; 0.59 and 0.59; and 0.61 and 0.62; for milk, fat and protein yields, respectively. Average regression coefficients considering only accurate and all animals were 0.83 and 0.84; 0.96 and 0.94; and 0.85 and 0.84; for milk, fat and protein yields, respectively. Therefore, our findings indicate that including all available genotypes or only for accurate animals yields similar genomic predictions in terms of validation reliability and bias.

Effect of phenotypic selection on the genomic inbreeding in Nelore cattle

M.E.Z. Mercadante[1], D.F. Cardoso[2], D.J.A. Santos[2], S.F.M. Bonilha[1], J.N.S.G. Cyrillo[1], H. Tonhati[2] and L.G. Albuquerque[2]

[1]Institute of Animal Science, Sertãozinho, SP, 14.174-000, Brazil, [2]São Paulo State University, Jaboticabal, SP, 14884-900, Brazil; mezmercadante@gmail.com

Runs of homozygosity (ROH) are recognized as useful proxy for genomic inbreeding measure in diverse species. This study aimed to compare inbreeding estimates in three experimental lines of Nelore cattle that have independently undergone selection for growth in Brazil. Two of the three lines referred to as Nelore Selection (NeS, n=120), and Traditional (NeT, n=170) have been kept under directional selection for increased yearling body weight (YW), being NeS a closed line while NeT eventually experienced gene flow. An additional control line (NeC, n=60) has been kept as a closed line under stabilizing selection for YW. A subsample of 763 animals (89 NeC, 189 NeS, and 485 NeT) born from 2004 to 2012, and genotyped for 716,581 autosomal SNPs, were used. All samples and markers presented call-rate higher than 0.90. ROHs were defined in PLINK software, as genomic segments higher than 1 kb, with at least 30 consecutive homozygous SNPs spaced for no more than 500 kb. Genomic inbreeding coefficients (F_{ROH}) were estimated by individual proportion of genome size covered by ROHs. Pedigree inbreeding (F_{PED}) were estimated to this subsample through of diagonal elements in the relationship matrix including all animals of experimental population since the lines establishment (n=9,551). Estimated averages of F_{ROH} were equal to 0.13±0.03, 0.14±0.03 and 0.14±0.02 to NeC, NeS and NeT lines, respectively, while the corresponding values to F_{PED} were 0.05±0.02, 0.03±0.01 and 0.03±0.01. F_{ROH} estimates were higher than F_{PED}, likely due better suitability of genomic information to trace old inbreeding, however, the two estimates presented correlation of 0.54. NeC's average of F_{ROH} were slightly higher than NeS and NeT's, on the contrary to observed with F_{PED}. A potential explain to this result is that directional selection do increase inbreeding while stabilizing selection potentially maintain heterozygosity and F_{PED} estimates do not account to specific autozygosity caused by selection pressure. This hypothesis was reinforced face observation of ROH hotspots in genomic regions markedly to the presence of QTL for growth traits in NeS and NeT, but not in the NeC line.

Genomic predictions of principal components for growth traits and visual scores in Nellore cattle

G. Vargas[1], F.S. Schenkel[2], L.F. Brito[2], L.G. Albuquerque[1], H.H.R. Neves[3], D.P. Munari[1] and R. Carvalheiro[1]
[1]Sao Paulo State University (UNESP), Via de Acesso Prof. Paulo Donato Castelane, 14884-900, Jaboticabal, SP, Brazil, [2]University of Guelph, 50 Stone Road East, N1G2W1, Guelph, ON, Canada, [3]GenSys Associated Consultants, Rua Guilherme Alves, 170, 90680-000, Porto Alegre, RS, Brazil; gi.vargas@hotmail.com

Brazil is among the top beef producers in the world and its high international competitiveness is due in most part to the adaptation and productive efficiency of the country's most predominant breed, the Nellore cattle. However, to remain competitive, increasing genetic progress for various economically important traits is crucial and investigating alternative selection strategies, such as combining economically important traits through a Principal Components (PC) approach. This study was carried out to investigate the feasibility of genomic predictions of PCs for growth traits (weaning and post-weaning weight gain), visual scores (conformation, finishing precocity and muscling), and a reproductive trait (scrotal circumference) in Nellore cattle. The dataset used for this study included phenotypic and pedigree information from 355,524 animals and genotypes from 3,519 animals (genotyped either with the 777K (HD) or the 76K SNP chip, and, then, accurately imputed to HD). Covariance components and estimated breeding values were obtained from a multi-trait analysis using a mixed linear animal model. Then an eigen-decomposition of the additive genetic covariance matrix (A) was used to calculate the EBVs for the first three PCs, which were used as pseudo-phenotypes in the genomic analyses. A forward-prediction validation scheme was implemented, in which the dataset was split into training (birth years: 1990 to 2008) and validation (birth years: 2009 to 2010) populations. Prediction accuracy was measured as the Pearson's correlation between pseudo-phenotypes and Genomic Estimated Breeding Values (GEBV) and were calculated to be equal to 0.59, 0.50 and 0.43 for the first, second and third PC, respectively. These results indicate the possibility of genomically select PC for growth, visual score and reproductive traits in Nellore cattle.

Estimation of inbreeding and effective population size in Simmental cattle using genomic information

N. Karapandža[1], I. Curik[2], M. Špehar[1] and M. Ferenčaković[2]
[1]Croatian Agricultural Agency, Ilica 101, 10000 Zagreb, Croatia, [2]Faculty of Agriculture, Department of Livestock Sciences, Svetosimunska 25, 10000 Zagreb, Croatia; nkarapandza@hpa.hr

Precise estimates of effective population size and a degree of inbreeding are of great importance for population and conservation genetics. Experience has shown that estimates based on pedigree data are often unreliable and imprecise. High resolute genomic information, as one of the most polymorphic types of information, was highlighted as much more precise for this kind of calculations. In this research, we used genotypes of 107 Simmental young animals. Dual purpose Simmental breed represents dominant part of Croatian cattle population (62.8%) under selection scheme. Effective population size (Ne) was estimated based on a functional relationship of Ne and correlation r2 and recombination rate (c). To estimate Ne through generations we used a multithreaded tool SNeP. Regression analysis was used to estimate current effective population size, and a value of Ne=110 (95%CI: 101.98-118.02) was obtained. Historical estimates of the Ne during the last 100 generations showed linear decrease of 29.24 per generation. We also estimated inbreeding levels based on runs of homozygosity (ROH) for every animal. Averaged values were as follows: FROH1=0.026 (0.015-0.045), FROH2=0.01 (0.004-0.019), FROH4=0.002 (0-0.006) while we did not detect any inbreeding originating from up to 6.25 generations ago. ROH and LD calculations were done by PLINK while data manipulation, regression analysis and visualizations were done using SAS 9.4. For comparison, we calculated Ne based on pedigree information using ENDOG software, which computes Ne via individual increase in inbreeding. For a population of 107 genotyped animals we obtained a value of 229.18, and for the population of 682 bulls that are of great importance for artificial insemination value of 264.92. Mean inbreeding was 0.42 and 0.27%. We can conclude that genomic information will be very useful to understand genetic diversity of the Simmental population with respect to genetic selection. In addition, it will help to identify genomic regions which have been preferentially selected.

Session 04 Theatre 1

Improvement of grasslands in dry areas of the Mediterranean region

C. Porqueddu and R.A.M. Melis
CNR-ISPAAM, Via Traversa la Crucca, 3, 07100 SASSARI, Italy; c.porqueddu@cspm.ss.cnr.it

As general rule, grassland-based products show best nutritional and sensory properties than conserved-forage/ concentrate based products. Unfortunately, in Europe, grassland-based farming systems are currently threatened by agricultural intensification and abandonment under a climate change scenario. There are some compulsory goals to pursue to reverse the trend of grasslands abandonment, namely increasing the resilience of grasslands, improving forage production and rehabilitating permanent grasslands. In Mediterranean environments, the desired goals could be achieved acting on some key factors: (1) Sowing annual and perennial species with high summer drought survival. In annual legumes, earlier maturity, delayed softening of hard seeds and greater hardseededness are the main traits needed. In perennial species, desired characteristics include dormancy or low growth during the drought period, survival across drought periods, and high water use efficiency during the growing season. (2) Increasing legume utilization. Among perennial legumes, lucerne is the most appreciated species but some limitations to its use arise under rainfed conditions. Other perennial legumes, e.g. sulla, are summer-dormant and have high forage quality also in relation to their content of condensed tannins. (3) Promoting the use of grassland mixtures. Mixtures belonging to different functional groups yield higher dry matter per area unit, show a better seasonal forage distribution, a better weed control and higher forage quality than pure stands of each species. (4) Benchmarking pastoral resources. The knowledge of grassland typology is needed to adopt the best management practices; in fact, differences in vegetation and phytosociological associations are relevant in Mediterranean grasslands. Agronomic typologies based on the forage value of dominant or reference species, or synthetic indexes were designed in different countries. (5) Extending forage availability. The use of wide diversity of meadow and species, including forage shrubs, may extend forage availability and increases the self-sufficiency. Moreover, short distance transhumance has potential to be exploited, and could generate ecosystem services.

Session 04 Theatre 2

Can the use of novel forages increase omega-3 in lamb?

N.L. Howes
AbacusBio International Ltd, Roslin Innovation Centre, Easter Bush, EH259RG Edinburgh, United Kingdom; nhowes@abacusbio.co.uk

This research investigated the potential of *Cichorium intybus* (chicory) to increase omega-3 in lamb to meet the World Health Organisation's (WHO) recommended ratios for fat and achieve international food labelling requirements. Due to the complexity of rumen interactions, achievement of these standards in the red meat sector has been limited to production systems with access to rumen-protected lipid supplements, regardless of pasture being a source of omega-3 fatty acids. The 8-week experiment involved 740, 5-month old lambs randomly assigned to 0, 2, 4, 6, or 8 weeks grazing on chicory prior to slaughter, with remaining weeks spent grazing mixed swards of *Lolium perenne* (perennial ryegrass) and *Trifolium repens* (white clover). Grazing treatments contained similar numbers of ewe, ram, and wether lambs that were managed under a commercial operation with animal performance and herbage routinely measured. Linear models were used to fit fatty acids, carcass data, and herbage measures, and linear mixed models for animal growth data. Fatty acid data from *longissimus lumborum* were modelled with and without inclusion of intramuscular fat (IMF) and carcass weight. Animal performance and muscle lipids differed across treatments, irrespective of similar ($P>0.05$) IMF. Fatty acid differences due to gender were not observed ($P>0.05$). WHO recommendations were met for omega-6:omega-3 by all treatments but PUFA:SFA reached half the recommended standard. Alpha-linolenic acid (ALA) estimated per cooked serve was highest ($P<0.05$) in lambs that grazed chicory for 8 weeks (115±3.3 mg/100 g) and lowest for lambs that grazed the grass sward (71.3±3.4 mg/100 g). Long-chain omega-3 concentrations were similar ($P>0.05$) for lambs that spent 8 weeks on chicory or ryegrass but were approximately 5 mg/100 g lower in remaining groups, putting intermediary groups below the threshold for European 'source' claims. In conclusion, chicory showed potential to increase ALA whilst long-chain omega-3 were less susceptible to dietary changes, due to diverse biological roles, but showed potential to meet labelling requirements. Further work is required to establish the role of feed intake and the effect of dietary profiles on muscle composition in lamb.

The use of seemingly unrelated regression to predict *in vivo* the carcass composition of lambs

V.A.P. Cadavez
CIMO Mountain Research Center, School of Agriculture, Polytechnic Institute of Braganza, Animal Science, Campus de Santa Apolónia, 5300-253 Braganza, Portugal; vcadavez@ipb.pt

The aim of this study was to develop and evaluate models for predicting *in vivo* the carcass composition of lambs. One hundred and twenty five lambs (82 males and 43 females) of Churra Galega Bragançana Portuguese local breed were scanned, using an ALOKA SSD-500V ultrasound machine equipped with a probe of 7.5 MHz, at lumbar region. The images were analysed in order to measure subcutaneous fat thickness between the 12th and 13th thoracic vertebrae (C12) and between the 1st and 2nd lumbar vertebrae (C1). Lambs were slaughtered after 24-h fasting and carcasses were cooled at 4 °C for 24 hours. Left side of carcasses was dissected and the proportions of lean meat (LMP), subcutaneous fat (SFP), intermuscular fat (IFP), kidney and knob channel fat (KCFP), and bone plus remainder (BP) were obtained. Models were fitted using the seemingly unrelated regression (SUR) estimator, and compared to ordinary least squares (OLS) estimates. Models were validated using the PRESS statistic. Our results showed that SUR estimator performed better in predicting LMP () and IFP than the OLS estimator. The results suggest that prediction *in vivo* of carcass composition could be improved by using the SUR estimator.

Casein milk proteins: novel genctic variation and haplotype structure

S.A. Rahmatalla[1,2], D. Arends[2], M. Reissmann[2], S. Krebs[3] and G.A. Brockmann[2]
[1]University of Khartoum, Faculty of Animal Production, Shambat P.O. Box 32, 13314 Khartoum North, Sudan, [2]Humboldt-Universität zu Berlin, Albrecht Daniel Thaer-Institut für Agrar- und Gartenbauwissenschaften, Invalidenstraße 42, 10115 Berlin, Germany, [3]University of Munich, Laboratory of Functional Genome Analysis, Gene Center, Feodor-Lynen-Strasse 25, 81377 Muenchen, Germany; rahmatas@hu-berlin.de

Caseins are the main protein component of goat milk. Knowledge of casein protein variation at the haplotype level has been a useful tool in biodiversity studies and in improving breeding strategies. In goats, four casein genes (*CSN1S1*, *CSN2*, *CSN1S2* and *CSN3*) are located as a cluster on chromosome 6. Our study aimed to find novel non-synonymous single nucleotide polymorphisms (SNPs) in different goat breeds using high density capture sequencing: Nubian, Desert, Taggar, Nilotic, Saanen, Nubian ibex, Alpine ibex, and Bezoar ibex. Six non-synonymous variants were identified within the *CSN1S1* gene; two novels: one segregating in Saanen goat, the other in alpine ibex. Six haplotypes were detected of which two were not known before. In the CSN2 gene, five non-synonymous variants were detected; three of which are novel and were only detected in the Nubian and alpine ibex. In total, we detected seven haplotypes for *CSN2* gene. Six SNPs were identified in the *CSN1S2* gene; four out of six were novel; one novel SNP was detected in Taggar and Saanen goats, whereas the other three novel SNPs were found only in Nubian ibex. Five haplotypes were found for *CSN1S2* gene. In the *CSN3* gene, two out of five non-synonymous SNPs were novel and segregated in alpine ibex. Four haplotypes were detected for the *CSN3* gene. In conclusion, 22 non-synonymous SNPs were identified within the casein genes of the seven goat breeds of which 11 were novel. We investigated haplotype structure and observed eight new haplotypes not seen before in these casein genes. Furthermore, we would like to stress that most of the novel non-synonymous SNPs were found only in the critically endangered Nubian ibex; highlighting the importance of preservation and studying rare and endangered breeds.

Fatty acid and mineral composition of Italian local goat breeds

S. Currò[1], C.L. Manuelian[1], G. Neglia[2], P. De Palo[3] and M. De Marchi[1]
[1]University of Padova, Department of Agronomy, Food, Natural Resources, Animals and Environment, Viale dell'Università 16, 35020 Legnaro (Padova), Italy, [2]University of Napoli Federico II, Department of Veterinary Medicine and Animal Production, Via Delpino 1, 80137 Napoli, Italy, [3]University of Bari Aldo Moro, Department of Veterinary Medicine, S.P. per Casamassima km 3, 70010 Valenzano (BA), Italy; sarah.curro@phd.unipd.it

This study investigated chemical, fatty acid (FA) and mineral composition of milk from 5 Italian goat breeds: Garganica (GA), Girgentana (GI), Jonica (JO), Maltese (MA) and Mediterranean-red (MR), and 1 cosmopolitan breed (Saanen, SA). A total of 39 multiparous dairy goats reared in the same herd were sampled monthly during lactation. Milk chemical composition, FA and minerals (n=237) were determined by Fourier Transform infrared, gas-chromatography and inductively coupled plasma atomic emission spectroscopy, respectively. Data were analysed using a mixed linear model that accounted for breed, week of lactation and parity as fixed effects, and animal and residual as random terms. Local breeds (LB) produced less (P<0.05) milk (1.40, 1.40, 1.20, 1.06 and 1.25 kg/d for JO, MA, RM, GI and GA, respectively) than SA (1.82 kg/d). Milk fat (%) did not differ (P>0.05) among breeds. Saturated FA and C16:0 were the most abundant FA in milk (75.5 and 23.5% of identified FA, respectively). Mainly differences between SA and GI where observed (P<0.05) for some individual and group of FA. Milk of SA had greater (P<0.05) C14:0 and C16:0, and lower (P<0.05) C18:0, C18:1n9, Unsaturated FA (UFA) and Monounsaturated FA (MUFA) than GI milk. On average, SA milk had greater (P<0.05) conjugated linoleic acid content than milk of LB (1.10 vs 0.88% of identified FA, respectively). Concerning mineral composition (ppm), goat milk was richer in Ca (1,073), P (794) and Na (364). Differences among breeds where observed (P<0.05) only for Na, P, Mg and Zn. Week of lactation affected (P<0.05) all the studied traits, whereas parity affected (P<0.05) the UFA and MUFA content. This study is a first characterization of milk FA and mineral composition of Italian goat breeds reared in the same herd. Results might be useful for biodiversity issues and to valorise the dairy products of those local breeds.

Session 04 Theatre 6

Alleviation of climatic stress of dairy goats in Mediterranean climate

N. Koluman Darcan[1], J. Agossou[2] and A. Koluman[3]
[1] Cukurova University, Animal science, Faculty of Agriculture, 01330 Adana, Turkey, [2]Redrock Group, Mevlana Bulvarı, Ege Plaza 182, Kat. 7 no. 33 Çankaya, Ankara, Turkey, [3]Pamukkale University, Faculty of Technology, Department of Biomedical Engineering, Denizli, Turkey; nazankoluman@gmail.com

High ambient temperatures and solar radiation cause thermal heat stress in dairy goats leading to decline in performance. A variety of methods such as spraying and/or ventilation can eliminate thermal stress in animal husbandry. The aim of this study was to improve the performance of dairy goats by minimizing the adverse effects of heat stress in hot summer days of Eastern Mediterranean part of Turkey. Twenty four Alpine dairy goats were classified into 2 groups. The experimental group (EG) was sprayed and ventilated one hour a day (from 12:00 to 13:00) while the control group (CG) was neither sprayed nor ventilated. The rectal temperature, skin temperature, pulse and respiration rate, anatomical, morphological adaptation mechanisms, and daily feed and water consumption of each goat in both groups were recorded four times a day. The quantity of milk detected twice a day. Daily feed and water consumption were recorded once individually in morning time. At the end of this experiment, 21% more milk yield was obtained from EG than was from CG (P=0.012). Physiological features of the two groups were observed to be significantly different from each other. Our data suggest that minimizing goats' heat stress through spray and ventilation improves their feed conversion rate as well as increasing their milk yield.

How is grazing perceived by stakeholders of the dairy goat chain in Western France?

A.L. Jacquot, P.G. Marnet, J. Flament and C. Disenhaus
Pegase, INRA, Agrocampus Ouest, 65 rue Saint-Brieuc, 35042 Rennes, France; catherine.disenhaus@agrocampus-ouest.fr

Goat milk chains of Bretagne and Pays de la Loire regions (France) are on a positive dynamic. Due to a low feed self-sufficiency rate (55%) dairy goat systems are sensitive to an increase in feed costs and volatility of agricultural commodities. Pasture is a potential way to improve feeding self-sufficiency. As 90% of goats do not pasture, the present study aims to investigate farmer's and stakeholder's perception of grazing in dairy goat. In October 2017, a semi-quantitative survey was performed using 19 farms, 21 chain stakeholders (food supply, advisers, cheese makers and banks), 24 consumers and 3 cheese retailers. The farming systems differed on the size of the flock (from 25 to 1000 goats), the type of farming (13 conventional and 6 organic farms), the main forage used (maize silage (2), hay (9), pasture (6)). All participants agreed with the positive effect of grazing on feed self-sufficiency in farming and being consistent with social expectations by reducing the gap between consumers who believe in farming system based on small grazing flocks and reality. All farmers agreed with the negative effects of grazing on goat health due to parasitism which appears as the main technical issue. However, farmers with a grazing system manage to deal with parasitism, are satisfied with the application of grazing. Other farmers and agricultural advisors consider grazing as technically difficult and it requires high technical skills (parasite and grass offer). Most of milk factories and some farmers think that goats are not suitable for grazing. Grass production is also highly dependent to weather conditions implying day-to-day and seasonal variations of milk production. Only organic milk factories agree on handling those variations due to grazing systems. Nonetheless, some conventional milk factories point out an improvement of organoleptic qualities of cheese made with milk from pasture. In conclusion, surprisingly, despite the high expectations from consumers and common interest for feed self-sufficiency through grazing, most of stakeholders remain reluctant to developing grazing in goat systems due to the lack of knowledge and technical advices.

Seasonal changes on milk quality from sheep farms using hay or silage

R. Bodas[1], D. Delgado[1], C. Asensio-Vegas[1], F. Bueno[1], J.J. García-García[1], A. Garzón[2] and L.A. Rodríguez[1]
[1]Agrarian Tecnological Institute of Casille and León (ITACyL). Dept. of Agriculture and Livestock, Subdirectorate of Research and Technology, Ctra. Burgos, km 119, 47071 Valladolid, Spain, [2]University of Córdoba, Department of Animal Production, Campus de Rabanales. Ctra. Nacional IV, km 396, 14014 Córdoba, Spain; bodrodra@itacyl.es

The technological quality of sheep milk intended for making of dairy products is determined by their physicochemical and hygienic characteristics as well as the presence of cross contaminations from the milking parlour, animal housings, water and feeds. The use of silage to feed sheep is becoming widespread, since it provides energy and fibre, to reduce the use of other forages and concentrates and to drive down the cost of rations, with potential positive effects on milk yield and quality. However, their use is not devoid of controversy due to the development of major defects in the cheese. An experiment was conducted to study the effects of forage (hay or silage) and season (winter, spring, summer, autumn) on sheep milk quality. During 2017, bulk tank milk samples were taken from 10 farms using vetch silage and 10 farms using vetch or alfalfa hay as forage basis for lactating animals. Milk physicochemical, hygienic and technological quality was assessed. Milk collected in winter and autumn showed higher ($P<0.05$) content of protein, fat and total solids, as well as lower ($P<0.05$) conductivity. The urea content was higher ($P<0.05$) in autumn than in winter and spring. Moulds counts were higher ($P<0.05$) in winter. No effects of season were observed in the somatic cell counts. Curd yield and its total solids content were higher ($P<0.05$) in winter and autumn, with the lowest values in spring. Farms using silage had greater ($P<0.05$) milk moulds, yeasts and somatic cell counts. Farms using silage as a base diet showed higher ($P<0.05$) urea and lower ($P<0.05$) lactose content. The season and the forage type did not affect ($P<0.05$) milk colour, total bacterial and butyric spores counts. From a practical point of view, sheep milk quality is greatly influenced by season, whereas regardless this seasonal effect, the use of silage may impair milk physicochemical and hygienic characteristics.

Session 04 Theatre 9

Effect of different dietary protein sources on carcass and sex organs of Egyptians growing lambs

H. Metwally[1] and S. El Mashed[2]

[1]Faculty of Agriculture, Animal Nutrition, Ain Shams University, Shubra, Cairo, Egypt, [2]Animal Production Research Institute, Animal Production Department, Ministry of Agriculture, Dokki, Giza, 11241, Egypt; sh.elmashed@hotmail.com

In a one way classification experimental design, five different dietary sources were used to study their effect on carcass characteristics, male sex organ development and some blood parameters. Fifty male lambs weighing 25.5±0.92 kg were divided into five groups (10 lambs each) for experimental feeding period of 100 days. Five iso-nitrogenous, iso-energetic diets contained either cotton seed meal (CSM), bean beans (BB), alfalfa hay (AH), sunflower seed meal (SSM) or linseed meal (LSM) as protein sources. Protein sources were evaluated for its solubility in three different solvents, for *in vitro* nitrogen disappearance and for its electrophoresis analysis. Results indicated that protein sources had significant effects on total weight gain and final body weight; where (SSM) caused the highest total weight gain (17.5±0.51 kg) while bean beans had the lowest value (15.9±0.53 kg). Empty body weight was the highest ($P<0.05$) with both alfalfa hay and LSM and the lowest with bean beans. Alfalfa hay (AH) caused the highest hot carcass weight and yield 19.135 kg and 44.5%, respectively. Significant differences were detected between groups concerning scrotal circumferences (cm), paired testes weight and mean testes volumes. The largest scrotal circumference was recorded for alfalfa hay group (38.0 cm) while the smallest was recorded with (SSM) group (34.8 cm). Group fed LSM had the highest testes weight (360.2 g) and the (CSM) group had the lowest (310.8 g). Testes volumes was the highest for (LSM) group (18.0 ml) and the lowest for (CSM) group (160.0 ml). It was concluded that dietary protein source may be a key factor for the development of male sex organs.

Session 04 Theatre 10

Holistic approach to improving European sheep and goat sectors – iSAGE

C. Thomas[1], K. Zaralis[2], R. Zanoli[3], A. Del Prado[4], G. Rose[5], G. Banos[6], A. Pompozzi[1] and G. Arsenos[5]

[1]European Federation of Animal Science (EAAP), Via G Tomassetti 3A/1, 00161 Rome, Italy, [2]Organic Research Centre, Elm Farm, RG20 0HR Newbury, United Kingdom, [3]Universita Politecnica delle Marche, Piazza Roma 22, 60121 Ancona, Italy, [4]Basque Centre for Climate Change, Almeda de Urquijo 4, 48008 Bilbao, Spain, [5]Aristotle University of Thessaloniki, University Campus, 54124 Thessaloniki, Greece, [6]SRUC, West Mains Rd, EH93JG, United Kingdom; cledwyn.thomas@gmail.com

Innovation for sustainable sheep and goat production in Europe (iSAGE) is an EU H2020 funded project (no. 679302). The consortium undertaking the work is composed of 33 partners from 6 EU countries and Turkey of which 18 are industry partners representing ~16,000 farmers and ~5.5 million sheep and goats supported by 15 diverse research partners. It is coordinated by Aristotle University of Thessaloniki, Greece. The project aims to improve the overall sustainability and innovative capacity of sheep and goat sectors in Europe. This will be achieved by: enhancing the efficiency and profitability of the sector, meeting the needs of consumers and increasing its societal acceptance and improving the delivery of ecosystem services. iSAGE will develop new socio-economic, animal welfare and sustainability assessment tools for the whole supply chain to understand barriers to innovation and development, define future opportunities for a competitive edge, develop farm management tools and innovative breeding strategies and derive solutions for social, welfare and consumer issues. The project will provide solutions to re-design sheep and goat systems that will result in improved sustainability and innovative capacity of the sector together with enhanced consumer and societal acceptance. New efficiency traits will be developed to enable selection of animals more adapted to environmental changes and increased animal welfare. The multi-actor approach with industry and research organisations working together will enable the research to be more relevant to practice and improve impact. For further information go to www.isage.eu.

Identification of meat marker in Carpatina kids for improving carcass quality evaluation by PCR-RFLP

C. Lazăr, A.M. Gras, R.S. Pelmuș, E. Ghiță and M.C. Rotar
National Research Development Institute for Animal Biology and Nutrition, Animal Genetic Resources Management, Calea Bucuresti, 077015, Romania; cristina_lazar17@yahoo.com

In Romania goat meat quality was evaluated using classical method with measurements after slaughtering and we consider appropriate to use molecular markers correlated with carcass meat quality. Objective of this study was to identify insulin growth factor *(IGF-1)* gene polymorphism by *PCR RFLP* technique in order to improve goat meat evaluation. Insulin like growth factor -1 play an important stimulator role in skeletal growth, cell differentiation and metabolism. Also it has an important role in control of hair cycles and it is involved in development of wool fibre. Therefore, it is considered as an appropriate candidate gene for meat quality evaluation. Blood samples were collected from 12 Carpatina kids and *DNA* was extracted using, Wizard Genomic DNA Purification Kit. Polymorphism of *IGF-1* was determined by *PCR* amplification followed by *RFLP* method using restriction enzyme *Hae III*. In the present study homozygous individuals were identified by RFLP with enzyme *Hae III* which cuts the amplicon in two places obtaining genotype *BC* with three migration bands at 363, 264 and 99 bp. Restriction enzyme doesn't cut the amplicon of the homozygous individuals for the genotype *BB*, and there is no restriction site, so amplified DNA fragment migration can be visualized in one band of 363 bp. Based on results, two genotypes *BB* and *BC* were identified. It was identified homozygous genotype *BB* with 75% and heterozygous genotype *BC* with 25%. *B* allele frequency was 87% and for allele *C* was 13%. Observed and expected values of *IGF-1* genotypes were found in Hardy Weinberg equilibrium, after $\chi 2$ test was calculated. That is why further investigation are necessary to be carried out on a high number of animals in order to discover other mutations correlated with meat quality in Carpatina goat breed.

Effect of suckling on composition and fatty acid profile of milk in Damascus goat pre-weaning

C. Constantinou, S. Symeou, D. Miltiadou and O. Tzamaloukas
Cyprus University of Technology, Department of Agricultural Sciences, Biotechnology and Food Science, P.O. Box 50329, Lemesos, Cyprus; despoina.miltiadou@cut.ac.cy

Previous work on lactating ewes has shown that partial suckling of lambs for 8 h per day during the first 5 weeks of lactation, affected both the fat content and the fatty acid (FA) profile of milk obtained by machine milking. The objective of the present work was to investigate the effect of continuous suckling of kids on composition of surplus commercial milk of their dams during the pre-weaning period. Thirty purebred Damascus goats in their first lactation, were allocated as they kidded, to the following two homogenous groups of 15 animals each: (1) no-kid group (NK), where kids were weaned immediately after birth and their dams were machine milked twice a day, and (2) mixed-rearing group with goats suckling their kids (1.4 kids/goat) and machine milked twice daily (SK group). Suckling kids were weaned at 49±3 days, and goats continued to be machined milked twice daily post weaning. Milk composition and FA profile were determined weekly throughout the experiment. In both groups, the feeding regime was the same covering does requirements. The results showed that fat percentage of commercial milk collected during the pre-weaning period was higher ($P<0.05$) for the SK group compared to NK in every weekly measurement (mean fat % was 4.2 and 3.6, respectively). No such difference was observed post-weaning. FA profile of milk during the pre-weaning period was also affected by the suckling. Milk of the SK group contained less total saturated FA ($P<0.05$) compared with milk of the NK group. This difference was more profound for C14:0 and C16:0 acids during the pre-weaning period, but it was not observed post-weaning. The unsaturated FA or other milk constituents (protein, lactose, solids non-fat) were not affected by the rearing method. In conclusion, commercial milk of machine milked Damascus does during suckling, had increased fat content and reduced saturated lipids compared to control non-suckling animals.

Session 04 Poster 13

Quality of meat from Carpathian goat kids depending on duration of fattening

A. Kawęcka, J. Sikora and A. Miksza-Cybulska
National Research Institute of Animal Production, Krakowska 1, 32-083 Balice, Poland; aldona.kawecka@izoo.krakow.pl

Goat meat is considered a delicatessen product and gaining more and more supporters. The Carpathian goat used in this study is an old native Polish breed which is perfectly suited to local, often harsh environmental conditions. The aim of the study was to evaluate the quality of meat from Carpathian goat kids depending on the duration of fattening. All the kids were 7 months of age and their initial body weight was 16.5 kg. The animals were maintained in a semi-intensive system: they received meadow hay and straw *ad libitum* as well as about 0.4 kg of concentrate per animal. Slaughter was performed at 9 and 12 months of age. The evaluation of meat quality included the assessment of physicochemical traits and the analysis of meat composition. No differences (P>0.05) were found in the content of individual components depending on the date of slaughter. Kid meat contained 23.5% DM, 2.3% fat, 20.8% protein and 0.02 mg/100 g cholesterol. The meat colour analysis showed differences in lightness (L*): the meat of older kids was darker (P<0.05) than that of younger kids. The yellowness (b*) was higher in younger animals. No differences (P>0.05) were found between the groups in the drip loss, cooking loss, and shear force of the heat-treated meat from the Carpathian goat kids. The organoleptic evaluation of roasted meat revealed differences (P<0.05) in colour, aroma, juiciness, and taste. The sensory assessors gave higher (P<0.05) scores to the meat of older animals. Their meat was darker, more aromatic, more tender and more juicy. The intramuscular fat of the leg muscle from both age groups contained more unsaturated (over 60%) than saturated fatty acids. No differences (P>0.05) were found between the groups in SFA and UFA content. The meat of younger kids contained higher (P<0.05) MUFA, and that of older kids more PUFA, including n-6 PUFA. This translated into a higher PUFA/SFA ratio for this group. Project 'The uses and the conservation of farm animal genetic resources under sustainable development' co-financed by the National Centre for Research and Development within the framework of the strategic R&D programme 'Environment, agriculture and forestry' – BIOSTRATEG, contract number: BIOSTRATEG2/297267/14/NCBR/2016.

Session 04 Poster 14

Perception of climate change impact by smallholders in agropastoral systems of Mediterranean region

N. Koluman Darcan[1], J.D. Agossou[2] and M. Ledsome[2]
[1] Cukurova University, Animal science, Faculty of Agriculture, 01330 Adana, Turkey, [2]Redrock Group, Mevlana Bulvarı, Ege Plaza 182, Kat. 7 no. 33 Çankaya, Ankara, Turkey; nazankoluman@gmail.com

This study investigates the perception of historic changes in climate and associated impact on local agriculture among smallholders in agropastoral systems of East Mediterranean Mountainous Region. The Mediterranean is one of the most vulnerable European regions to climate change, e.g. in terms of future water shortages, losses of agricultural potential and biome shifts. In fact, climatology characteristics such as ambient temperature and rainfall patterns have great influence on pasture and food resources availability cycle throughout the year, and types of disease and parasite outbreaks among animal populations. Also, the ability of livestock to breed, grow, and lactate to their maximal genetic potential, and their capacity to maintain health is strongly affected by climatic features. The performance, health, and wellbeing of livestock animals are strongly affected, directly or/and indirectly, by climate change and variability. These impacts are more severe in developing countries and also due to poor access to technologies and financial supports. In order to increase the profitability and ensure sustainability of livestock production, innovative political, specific managerial strategies and practices have to develop. This study will help to develop strategies to reduce waste from animal husbandry activities and the negative impact of animal husbandry on environment. We will draw on empirical data obtained from farm household surveys that will be conducted in 2 districts (Adana and Gaziantep). It will involve agropastoral associations and 100 farm households, all located in East Mediterranean Mountainous Region. Based on a questionnaire, the study will focus and review the impact of climate change on farm households over last decades as public awareness started to grow.

Consumers' preferences for functional lamb meat

M.C. Agúndez[1], P. Gaspar[1], M. Escribano[1], F.J. Mesías[1], A. Horrillo[1], A. Elghannam[2] and A. Eldesouky[3]
[1]University of Extremadura, Faculty of Agriculture, Avda. Adolfo Suarez, s/n, 06007 Badajoz, Spain, [2]Damanhour University, Faculty of Agriculture, Damanhour, Elbeheira, Egypt, [3]Zagazig University, Faculty of Agriculture, Zagazig, Sharkia, Egypt; pgaspar@unex.es

In recent years the lamb market in Spain has been decreasing due to several reasons. Since 2002 the production of lamb meat has been reduced by approximately 50% and its consumption by 30%. Among the causes of this reduction in consumption are the consumers' perception of lamb as an expensive meat, the seasonal and occasional consumption and, especially, the fact that it is closely associated with the age of the consumer, being more consumed by the elderly and little consumed in households with children and young people. All this generates unfavourable prospects for commercialization, which leads producers to look for new alternatives to boost their sales. The incorporation of functional characteristics to lamb meat might be one option. The market for functional foods has increased in recent years due to a growing consumer interest in their health and the concern about food. In recent years, different tests have been carried out in Extremadura region (SW Spain) to feed lambs with by-products from the tomato processing industry. The contribution of antioxidants (lycopene) of this by-product increases the oxidative stability of fresh meat, as well as the concentration of α-tocopherol in the muscle, which could allow the commercialization of meat as functional or enriched. The objectives of this work are (1) to know the preferences of the consumers of lamb meat particularly to know the given value to the attribute 'enriched with antioxidants and omega 3', being this one of the main characteristics shown by functional foods and (2) a segmentation of consumers based on their attitudes towards functional lamb meat. A online survey of 312 form using Google Forms has been conducted. The results have shown that consumers' perception towards functional foods is unequal. Consumers value positively the fact that the meat is enriched with omega3 or antioxidants, although this attribute is less relevant than others. The segmentation according to preferences has allowed us to identify three segments: Anti-functional consumers, local consumers, pro-functional consumers.

Ricotta fatty acid profile and sensory qualities as a reflection of dairy ewes feeding strategies

I. Fusaro[1], M. Giammarco[1], M. Odintsov[1], M. Chincarini[1], G. Mazzone[1], A. Formigoni[2] and G. Vignola[1]
[1]Università degli Studi di Teramo, Facoltà di Medicina Veterinaria, Località Piano D'Accio, 64100 Teramo, Italy, [2]Università degli Studi di Bologna, Dipartimento di Scienze Mediche Veterinarie, Via Tolara di Sopra 50, 40126 Ozzano dell'Emilia, Italy; modintsovvaintrub@unite.it

Ricotta is a traditional Italian dairy product particularly appreciated for its freshness and its low-fat content, generally considered as healthy food. The aim of this study was to evaluate the effect of different feeding treatments: pasture (P), standard TMR feeding (F) and TMR with 0.190 kg/h/d linseed supplementation (L) on chemical and textural characteristics of ricotta cheese. 54 Comisana ewes were divided into three groups and fed accordingly for 80 days. After which, pooled milk was collected to make Ricotta. The data of the chemical and fatty acid compositions of ricotta were analysed using ANOVA in a GLM procedure with SAS, dietary treatment being the only fixed factor in the model. Sensory data were normalized, standardizing each assessor by his standard deviation to reduce effect of the different use of scale. The normalized data was subjected to analysis of variance for repeated measures, with diet as a sole factor. Duncan's test was used to determine the groups significantly different from each other. Chemical composition of ricotta showed a higher ($P<0.05$) fat value in L treatment compared with the two others, while having reduced ($P<0.05$) saturated FA (SFA) and increased ($P<0.05$) monounsaturated FA. Ricotta cheese from F group had a highest ($P<0.05$) SFA/UFA ratio among the groups. Regarding total PUFA concentration, L group receiving linseed supplementation resulted in having an increase ($P<0.05$) in n-3 FAS, specifically of alpha-linolenic acid. P group resulted having an intermediate level of PUFA ($P<0.01$) but with the highest ($P<0.05$) level of CLA. Sensory evaluation showed that ricotta of L treatment, was greasier ($P<0.05$) and more spreadable ($P<0.05$) than products made from P and F milk. Ricotta obtained from P group had higher ($P<0.05$) granulosity score which could be linked to a higher level of protein ($P<0.05$), also displaying a lower ($P<0.05$) whiteness score compared to others ($P<0.001$). In conclusion, extruded linseed had a beneficial effect on ricotta quality and sensory properties.

Meat quality of lambs finished on a permanent sward or a plantain-chicory mixture

R. Rodriguez[1], D. Alomar[1] and R. Morales[2]
[1]Universidad Austral de Chile, Instituto de Producción Animal, Independecia 631, valdivia, 5090000 valdivia, Chile, [2]Instituto de Investigaciones Agropecuarias, INIA Remehue, Carretera Panamericana Sur Km. 8 Norte, osorno, 5290000 osorno, Chile; rominarodriguezmv@gmail.com

In Chile, lamb production systems are based on grazing of temperate swards, which can be affected by droughts in late spring and early summer. *Plantago lanceolata* and *Cichorium intybus* can maintain growth and quality in these environmental conditions. In this work we evaluated final weight (LW), meat quality characteristics and fatty acid (FA) composition of lambs finished on a plantain-chicory mix or a grass-based permanent sward. Forty-eight Austral weaned male lambs (LW: 31.27±2.46 kg) were finished on a ryegrass-based permanent sward (GBS) or on plantain-chicory mixture (PCH) during a six week period from October to December of 2016. Lambs were weighed weekly until slaughtering. Samples of longissimus muscle were obtained and analysed for pH, colour, tenderness and FA composition using a gas chromatography. Statistical analysis for LW, meat quality and FA was performed by the SAS software, using ANOVA in general linear model. No differences (P>0.05) were found between treatments for final LW (PCH: 41.4 and GBS: 39.3 kg), meat pH, colour and tenderness. However, differences in polyunsaturated fatty acids (PUFA) were detected (PCH: 120.5 and GBS: 73.8 mg/100 g of meat), as well as in omega-6 (n-6) content (PCH: 65.9 and GBS: 40.8 mg/100 g of meat) and omega-3 (n-3) content (PCH: 58.9 and GBS: 36.8 mg/100 g of meat) with higher (P<0.05) values observed in lambs that grazed PCH The ratio n-6:n-3 was 1.12 and 1.13 for PCH and GBS treatment, respectively; which is below 4.0 as recommended by the World Health Organisation. The ratio PUFA:SFA (PCH: 0.31 and GBS: 0.28) was also below 0.4 for both treatments. Lambs grazing PCH presented similar meat quality characteristics and a higher content of PUFA than those on the GBS. It is concluded that a plantain-chicory mixture is an attractive alternative to a permanent grass-based sward for finishing lambs.

Ultrasound measurements of LD muscle properties and subcutaneous fat thickness in Tsurcana lambs

E. Ghita, R. Pelmus, C. Lazar, C. Rotar, M. Gras and M. Ropota
National Research-Development Institute for Animal Biology and Nutrition, Animal Genetic Resources Management, 1 Calea Bucuresti, 077015 Balotesti, Ilfov, Romania; elena.ghita@ibna.ro

The local Tsurcana sheep is the breed with the largest stock in Romania, accounting for 56% of the total 10.5 million sheep. Adapted to the harsh mountain conditions, with high resistance to diseases, and with a high capacity to make use of roughages, the Tsurcana sheep have a high potential for improved meat production. The purpose of the present work was the evaluation of Tsurcana lamb's carcass quality *in vivo* as compared the Teleorman Black Head breed. Ultrasound technology was used to measure Longissimus Dorsi muscle depth (LDD), area (LDA) and perimeter (LDP), and subcutaneous fat thickness (SFT) on a total of 70 lambs (46 male and 24 females from which breed) aged 165 days, with an average body weight of 35.4±0.43 kg. The ultrasound measurements were performed with an Echo blaster 64 using LV 7.5 65/64 probe, supplied by TELEMED ultrasound medical systems. The ultrasound images were recorded using Echo Wave II software version 1.32/2009. The first measurement was taken 5 cm from the spine, at the 12th rib, while the second was taken between 3rd and 4th lumbar vertebrae. The average values at the two measurement points were 2.05/2.20 mm for SFT, 22.03/21.51 mm for LDD, 10.87/10.61 cm^2 for LDA and 136.98/136.22 mm for LDP. Comparing the average values for Tsurcana lambs with those from the local Teleorman Black head lambs, with similar body weight, showed that there were no differences (P>0.05) for LDP and LDA. On the other hand, highly significant differences (<0.001) were found for the SFT. Tsurcana lambs produced leaner (P<0.05) carcasses than Teleorman Black head lambs. The phenotypical correlations between live weight and LD muscle properties were within the range of 0.14 to 0.69 (P<0.05), and the strongest correlations were observed with longissimus muscle eye area. As a general conclusion, we can say that the quality of the carcasses obtained from lambs of the Tsurcana breed was similar to that of lambs of the Teleorman Black Head breed.

The lamb fattening results of different origin Latvia dark head sheep

D. Barzdina, D. Kairisa and J. Vecvagars
Latvia University of Life Science, Institute of Agrobiotechnology, Liela iela 2, Jelgava, 3002, Latvia; dace.barzdina@llu.lv

The Latvian Dark Head (LD) sheep breed was created at the beginning of the 20th century. In LD breed selection was allowed use of rams of related breeds (German blackhead (GB) and Oxforddown). The aim of the study was to explain the long term influence of used GB rams on LD breed lamb fast growth. The study was carried out in year 2017 at the breeding ram testing station 'Klimpas'. Fodder and hay were unlimited for fattened lambs. In addition were given minerals and drinking water. For study purposes were created 2 lamb groups: control group – purebred LD lambs (n=40), research group – lambs of the LD breed with GB blood from 6.25 to 12.69% (n=17). During the study no significant differences were observed in any of the studied traits. The average age in both lamb groups and the live weight at the start of fattening was similar – 98±1.5 days in control and 97±1.8 days in the research group. Live weight was respectively 27.2±0.51 kg and 27.3±0.53 kg. The average fattening time of the control lambs was 58±1.6 days, while the average daily live weight gain was 391.4±10.1 g. Research group lambs were fattened 2 days longer, but they reached in average 401.9±15.6 g live weight gain per day. The results of the study indicated that the lambs with a small amount of GB blood did not surpass fattening results of purebred lambs.

The results of Latvian dark head and Charolais purebred and crossbred lamb fattening

D. Kairisa, D. Barzdina and J. Vecvagars
Latvia University of Life Science, Institute of Agrobiotechnology, Liela iela 2, Jelgava, 3002, Latvia; daina.kairisa@llu.lv

Interest about in Europe popular meat-type sheep breeding in Latvia has increased after joining the European Union. One of the lesser known and less commonly used in the Latvia is Charolais sheep breed. To explain the Charolais sheep breed use possibilities, within the framework of Latvian Ministry of Agriculture funds, was carried out research on fattening was results of pure-bred Latvian dark head (LD), Charolais (CA) and their crosses (rams). Lambs were fattened in the control station 'Klimpas'. During the lamb fattening phase, fodder produced in Latvia was used (which was dry matter 227.3±3.8 g protein /kg). On the station's territory collected hay with average quality (1 kg of dry matter 96.6±10.4 g of crude protein) was used. In addition, lambs were supplemented with minerals and had access to drinking water at all times. All livestock feeds were provided *ad libitum*. The study used a total of 43 lambs; 30 purebred LD, 6 purebred CA and 7 crossbreed LD × CA lambs. At the start of fattening lamb live weight in different breed groups was not significantly different and ranged from 25.1±0.70 kg (CA) to 26.8±0.56 kg (LD), in comparison with CA bred lambs significantly older were LD bred lambs, respectively, 77.2±1.78 days and 93.5±29.1 days (<0.05). Purebred LD and crossbred LD × CA group lambs were fattened 60 days in average, but the purebred CA group lambs – 64 days, obtaining respectively 22.9±0.79, 26.0±2.54 and 28.3±2.76 kg live weight gain. The results of the research indicate on the possibility of using the CA breed for the obtainment of fast growing fattening lambs.

Preference of porcine mucosa products and plasma in newly weaned pigs

R. Davin[1], M. Bouwhuis[1], L. Heres[2], C. Van Vuure[2] and F. Molist[1]
[1]Schothorst Feed Research, Meerkoetenweg 26, 8200 AM, the Netherlands, [2]Sonac/Darling Ingredients International, Kanaaldijk Noord 21, 5690 AA Son, the Netherlands; rdavin@schothorst.nl

Highly digestible ingredients are commonly added in post-weaning piglet diets to enhance feed palatability and to control post-weaning diarrhoea. A study was conducted to investigate the palatability of three hydrolysed porcine mucosa products (MucoPro®; Darling Ingredients Inc.) and a spray-dried plasma product (Proglobulin®; Darling Ingredients Inc.) by a double-choice feeding trial. One-hundred sixty-eight 26-day-old piglets were weaned and allocated in 28 pens. Pens were distributed to 7 different treatments at random: 2.5 and 5.0% of MucoPro80, MucoPro90 (both powder products) and Proglobulin, and 2.5% DM basis of MucoPro30 (liquid product), and rotated in three 4-day periods. The seven treatments were compared with a reference diet (RefD) that contained 10% skimmed milk powder. Diet formulations were adjusted to contain the same amount of lactose. Feed intake (FI) from the two feeders containing one of the experimental diets or RefD was used to calculate preference %. FI and preference data were analysed with a mixed model (REML procedure of Genstat) with treatment, round and their interaction as main factors. Moreover, preference value from each experimental treatment was compared to the neutral value (50%) using a pairwise t-test. Results showed that there were no interaction effects between treatment and round on FI and preference (P>0.10). As expected FI intake was affected by round (P<0.01) due to piglets age, however preference was not (P=0.28). There was a treatment effect on experimental diet FI (P<0.01) that consequently affected (P<0.01) preference results. Diets containing 2.5% of either MucoPro80 or MucoPro90 were preferred over RefD (61.3 and 59.8%, respectively), to a similar extend than diets containing spray-dried plasma at both 2.5 and 5% inclusion rate (63.6 and 63.9%, respectively). Diets containing 5% of either MucoPro80 or MucoPro90 were not preferred over RefD diet (38.4 and 35.8%, respectively). MucoPro30 was neither preferred nor rejected (41.4%) by piglets. In conclusion, moderate inclusion rate of MucoPro80 and MucoPro90 may serve as a strategy to stimulate feed intake just after piglets weaning.

Effect of crude protein levels with L-valine supplementation in weaning diet

Y.G. Han, T.W. Goh, I.Y. Kwon, J.S. Hong and Y.Y. Kim
Seoul National University, 1 Gwanak-ro, Gwanak-gu, Seoul, 08826, Korea, South; hanyounggur@naver.com

This study was conducted to evaluate optimal crude protein (CP) levels with L-valine supplementation in weaning pig diet about growth performance, blood profiles, carcass characteristics of weaning to finishing pigs. A total of 160 weaning pigs ([Yorkshire×Landrace]×Duroc) with an average body weight of 7.86±1.05 kg were used for 22 weeks feeding trial. Pigs were allotted into one of five treatments by body weight and sex in 4 replicates with 8 pigs per pen in a randomized complete block design. The treatments were included (1) Corn-SBM based diet with L-valine supplementation + protein content suggested by NRC (1998) (HP), (2) Basal diet + protein content suggested by NRC (1998) – 1% (MHP), (3) Basal diet + protein content suggested by NRC (1998) – 1.5%, (4) Basal diet + protein content suggested by NRC (2012) and (5) Basal diet without L-valine supplementation + protein content suggested by NRC (2012). The experimental data were analysed as a randomized complete block design using the general linear model procedure of SAS. Reducing CP level had negative effect on body weight (BW) (P<0.01) and average daily gain (ADG) (P<0.01) during early growing phase. However, there was no significant effect of reducing CP level on BW and ADG at the end of experiment. Also, average daily feed intake (ADFI) decreased as dietary protein level decreased in 7-10 week (linear, P<0.05). On the other hand, L-valine supplementation in weaning diet improved BW in 22 week, ADG in 7-14 week and ADFI in 11-14 week compared LP group and LPV- group (P<0.01, P<0.01 and P<0.05, respectively). In blood analysis, there was a positive effect of reducing CP level and L-valine supplementation in weaning diet on blood urea nitrogen concentration in both weaning and growing-finishing period (P<0.01; linear, P<0.05). In carcass characteristics, pH and water holding capacity decreased linearly as dietary protein level decreased (linear, P<0.05). However, crude fat, Hunter L* and b* value showed increasing linear response as dietary protein level decreased (linear, P<0.05). Consequently, lowering dietary CP level down to NRC (2012) had no detrimental effect in weaning to finishing pigs only when L-valine was supplementation in weaning diet.

Effects of former food products as cereal substitute on growth performance in post-weaning pig

M. Tretola, M. Ottoboni, A. Luciano, L. Rossi, A. Baldi and L. Pinotti
Department of Health, Animal Science and Food Safety, University of Milan, Milan, 20133, Italy; luciano.pinotti@unimi.it

Ex-food or former food products (FFPs) represent a way by which convert losses from the food industry into ingredients for the feed industry, thereby keeping food losses in the food chain. Typical ex-food used in animal nutrition are biscuits, bread, breakfast cereals, chocolate bars, pasta, savoury snacks and sweets, all because of their high energy content in the form of sugars, starch and fat. Accordingly, the aim of the study was to evaluate the effects of substituting 30% conventional cereals for 30% FFP in post-weaning piglet's diets. The diets were iso-energetic (15.27 MJ/kg DM) and iso-protein (19.86% DM), and contained all essential amino acids in the recommended amounts. After an adaptation period (7d), pigs (n=12, 28d old) were housed for 12d in individual pen and assigned to two experimental groups: CRT, receiving a standard diet (n=6; 9.20±1 kg of BW) and FFP (n=6; 8.60±1 kg of BW), receiving a diet in which 30% of conventional cereals (wheat, barley, corn) were substituted for 30% ex-food Both diets were in grounded forms and piglets had *ad libitum* access to the feed and fresh water throughout the whole trial period. Individual feed intake was recorded daily, piglet's bodyweight (BW, kg) was recorded on d1, 5, 9 and 12 of the experiment. Average daily gain (ADG kg/day), average daily feed intake (ADFI kg/day) and Feed Conversion Ratio (FCR kg/kg) have been calculated. Raw data means were analysed by IBM SPSS Statistics version 25 software (SPSS Inc.). At the end of the experiment no differences in BW have been observed between groups (14.1 and 13.6 kg for CTR and FFP respectively; P=0.32). Similarly, ADG (0.29 and 0.31 kg/d respectively, P=0.71) and ADFI (0.45 and 0.44 kg/d respectively, did not differ (P=0.65) between groups. Conversely, piglets on the FFP diet showed a better FCR (CTR 1.55 vs FFP 1.39 kg/kg, P<0.05). Overall these results suggest that substituting 30% conventional cereals for 30% FFP in post-weaning piglet's diets has not detrimental effects on growth performance in post-weaning piglets. However, present study is based on a limited number of piglets, thus FFP potential in pig nutrition and feeding deserve further investigation.

Session 05 Theatre 4

Predicted glycemic index and hydrolysis index in in former food products intended for pig nutrition

M. Ottoboni, M. Tretola, D. Cattaneo, A. Luciano, C. Giromini, E. Fusi, R. Rebucci and L. Pinotti
Department of Health, Animal Science and Food Safety, University of Milan, Via Celoria 10, 20133 Milano, Italy; matteo.ottoboni@unimi.it

'Ex-food' or 'Former Food Products' (FFPs) means foodstuffs, other than catering reflux, which were manufactured for human consumption in full compliance with the EU food law but which are no longer intended for human consumption. Typical former foodstuffs used by are biscuits, bread, breakfast cereals, chocolate bars, pasta, savoury snacks and sweets, all because of their high energy content in the form of sugars, oils and starch. Based on nutritional fact reported for humans, in general FFPs are extremely rich in carbohydrate, free sugars, and depending on their origin, also in fat, even though their nutritional potential is not yet fully exploited. Thus, the aim of the present study was to evaluate both the predicted glycaemic index (pGI) and hydrolysis index (HI) in FFPs intended for pig nutrition. Six samples of bakery/confectionary FFPs, and two conventional feed ingredients (corn meal and flaked wheat), were analysed using an *in vitro* method based on the Englyst-assay simulating gastric and small intestinal digestion for starch, and that has been proposed for predicting the glycaemic index of starchy foods. White bread was used as reference material. Hydrolysis indices (HI) were calculated as a summary of the temporal glucose pattern, defined in terms of the area under the digested glucose curve for each sample. To white bread was assigned a HI of 100, while the HIs for the other samples were calculated as the area under the curve for each test material. Results have been expressed as a percentage of the mean area under the curve relative to white bread. From the HI, predicted glycaemic index (pGI) was calculated for each sample using the equation reported by Giuberti. Considering conventional feed ingredients, the lowest HI value was observed in corn meal (81.05) the same value was 93.59 in flacked wheat. In FFPs HI ranged from 86.01 to 113.45 with an overall mean of 104. Predicted GI values, were 82.1 in corn meal and 94.81 in flacked wheat. When FFPs were considered pGI ranged from 87.1 to 114.9, with five samples above the reference values of white bread. This indicates a higher glycaemic index potential in FFPs compared to common cereals.

Session 05 Theatre 5

Betaine increases net portal flux of volatile fatty acids in Iberian pigs

M. Lachica[1], J. García-Rodríguez[2], M.J. Ranilla[2] and I. Fernandez-Figares[1]
[1]Estacion Experimental Zaidin. CSIC, Fisiologia y Bioquimica de la Nutricion Animal, Profesor Albareda 1, 18008 Granada, Spain, [2]Universidad de Leon. Instituto de Ganadería de Montaña. CSIC-ULE, Produccion Animal, Campus de Vegazana, s/n, 24071 León, Spain; ifigares@eez.csic.es

Betaine and conjugated linoleic acid (CLA) have the potential to alter growth and body composition in swine, but their effects on net portal absorption of nutrients is not known. Sixteen Iberian barrows (19 kg BW) were randomly assigned to one of four isoenergetic and isonitrogenous dietary treatments: control, 5 g/kg betaine, 10 g/kg CLA, or 5 g/kg betaine + 10 g/kg CLA. Pigs were surgically fitted with 3 chronic indwelling catheters: in the portal vein and the carotid artery for blood sampling, and the ileal vein for marker (para-aminohippuric acid; PAH) infusion. Blood samples were simultaneously taken from the carotid artery and portal vein every 30 min for 4 h and hourly until 6 h after feeding 1,200 g of the diet and centrifuged, and the plasma was stored at -20 °C. Portal blood flow was determined by the PAH dilution method and net flux of volatile fatty acids (VFA) were determined according to the Fick principle. Net portal drained viscera (PDV) flux (mmol/h) of acetate, propionate, butyrate, isobutyrate, valerate and caproate were determined. Betaine increased (125% $P<0.001$) PDV flux of acetate, relative proportion of acetate NPA/total FVA NPA (34%; $P<0.001$), caproate (114%; $P<0.001$) and sum of total VFA (97%; $P<0.001$) and decreased PDV flux of isobutyrate (57%; $P<0.01$). CLA decreased PDV flux of isobutyrate (53%; $P<0.01$) and caproate (78%; $P<0.001$). Pigs fed diets supplemented with both betaine and CLA decreased PDV flux of butyrate (89%; $P<0.001$). In conclusion, betaine increased VFA PDV absorption, which may be useful in Iberian pigs fed with diets rich in non-digestible carbohydrates. Funded by Ministerio de Economía y Competitividad (AGL2016-80231), Spain.

Session 05 Theatre 6

Effect of two different feeding strategies on growing-finishing pigs in a rotational pasture system

L. Juul[1], T. Kristensen[1], M. Therkildsen[2] and A.G. Kongsted[1]
[1]Aarhus University, Department of Agroecology, Blichers Allé 20, 8830 Tjele, Denmark, [2]Aarhus University, Department of Food Science, Blichers Allé 20, 8830 Tjele, Denmark; loju@agro.au.dk

Outdoor rearing of growing pigs allows the animals to perform foraging behaviour. However, rooting destroying the grass sward, high concentrate feed inputs and high stocking densities combined with the characteristic pig excretory behaviour may cause serious nitrogen hotspots. Restrictive feeding throughout the growing-finishing period can be a strategy to reduce concentrate feed inputs and motivate direct foraging, but is known to reduce growth and impair meat quality. It has been demonstrated that restrictive feeding during the growing phase followed by a realimentation period in the finishing phase can have positive effects on the feed conversion ratio (FCR), average daily gain (ADG) and meat quality. Further, rotational paddock systems have been suggested as a strategy to maintain vegetation cover, improve manure distribution and thereby reduce the risk of nutrient leaching. The objective was to investigate the effects of restrictive feeding (RES) during the growing phase and *ad libitum* feeding (AL) of concentrate during finishing phase on ADG, FCR, foraging behaviour and meat quality in a rotational pasture system. In total, 64 pigs were randomly assigned to the treatment AL or RES-AL (2 kg concentrate from 30 to 70 kg live weight (LW) and AL from 70 to 110 kg LW) in four replicates. Paddocks were expanded twice a week to ensure continuous access to pasture. Preliminary results show that RES-AL pigs had a higher ADG during the finishing phase (1,418 g) compared with AL pigs (1,347 g), but RES-AL pigs had a lower overall ADG. No significant effect of treatment was found on FCR. Meat quality and behavioural data is yet to be analysed. Results show the potential for compensatory growth in outdoor organic pig production, and a high FCR for both groups indicate that the pigs were able to supplement daily concentrate feeding with direct foraging.

Growth, health and body condition of piglets fed 100% organic diets

N. Quander, B. Früh and F. Leiber
Research Institute of Organic Agriculture (FiBL), P.O. Box, 5070 Frick, Switzerland; nele.quander@fibl.org

The organic regulation in the EU will implement the 100% organic feeding for monogastric animals in Organic Agriculture from 2019, and high quality potato and corn gluten from conventional sources will be banned. In preparation, it is necessary to generate optimized feed rations especially for young animals, which will not compromise animal health and growth. The aim of the current study was to conduct feeding experiments with piglets on-farm, investigating three different optimized 100% organic diets formulated by a feed company. From a usual commercial piglet diet (A) with 95% organic components consisting mainly of barley, oat flakes, horsebean, protein peas, soy cake and conventional potato protein (4%), an optimized 100% organic diet (B) with changed ratios of the same components was created excluding potato protein. Two further 100% organic diets, one (C) with milk powder (3%) and another one (D) including fermentatively produced lysine (0.3%), both based on diet B, were investigated. Instead of rapeseed cake diet D contained sunflower cake. In 5 subsequent sample periods a total of 400 piglets, 100 for each of the 4 diets, were studied from birth until an age of 9 weeks. The respective diets were offered *ad libitum* to piglets during this period. Each litter was stabled with its sow in single pens until weaning at 6 weeks, thereafter piglets were separated. Piglets were weighed after birth and in week 3, 6 and 9 with simultaneous assessment of body condition score (BCS) and health status. One-way ANOVA was employed to compare means of daily weight gain for 2 different periods (DWG1 between birth and 6 weeks and DWG2 between 6 and 9 weeks) and showed no significant difference between treatments (P=0.782; P=0.057). Regarding the final weight at 9 weeks no significant difference was found (P=0.503). Also BCS showed no significant difference between treatments (P=0.391). Diet C tended to result in an improved health status of piglets was observed. The lowest mortality was found for treatment B (3.6%), the highest mortality was observed for treatment D (16.5%) while treatment A and C resulted in 11.8 and 10.9% deceased animals. In conclusion, 100% organic diets for piglets with and without milk powder are feasible without impacts on performance and health.

Effects of various feed additives in growing pigs

W.G. Kwak, W. Yun, J.H. Lee, C.H. Lee, H.J. Oh, J.S. An, S.D. Liu, S.H. Lee and J.H. Cho
Chungbuk National University, Department of Animal Science, 344, s21-5, 1, Chungdae-ro, Seowon-gu, Cheongju-si, Chungcheongbuk-do, Republic of Korea, 28644, Korea, South; kwakwooki@naver.com

Experiment 1 was a feeding trial in which 75 (Landrace × Yorkshire)× Duroc pigs with average initial body weight (BW) of 26±1 kg were used. It was assigned to three pigs/pen and five pens / treatment. Experiment 2 was a metabolic trial in which 25 (Landrace × Yorkshire) ×Duroc pigs with average initial body weight (BW) of 36±1 kg were used. The basal diet consisted of maize (57.71%), soybean meal (32.45%), and wheat bran (5%). Treatments consisted of (1) CON (control diet); (2) OE (organic acid + essential oils), that is, CON + 0.05% OE; (3) OC (organic charcoal), that is, CON + 0.1% OC; (4) AE (anise extracts), that is, CON + 0.015% AE; and (5) PB (probiotics), that is, CON + 0.1% PB. Average daily gain and gain/feed ratio were significantly higher in the OE and AE groups than in the others (P<0.05). Average daily feed intake for OE, OC, and AE treatments was significantly higher than that for CON and PB treatments (P<0.05). Dry matter digestibility was significantly higher in AE than CON (P<0.05). Crude protein was significantly higher in OE, AE, and PB than CON ({P<0.05). Ammonia emissions were significantly lower in all treated groups than in CON (P<0.05). Among the treatments, ammonia emissions were lowest in OE and AE. hydrogen sulphide emissions were significantly lower in OE, AE, and PB than in CON (P<0.05). Among the treatments, OE and AE were most effective at reducing hydrogen sulphide emissions. In conclusion, this experiment was shown to be more effective on growth performance, malodour emission, blood profiles, and nutrient digestibility in OE and AE than other additives.

The effect of mixed or single sex rearing and mineral supplementation on gestating gilts hoof health

P. Hartnett[1], L. Boyle[1], B. Younge[2] and K. O'Driscoll[1]
[1]Teagasc, Pig Development Department, Moorepark, Co.Cork, P61 P302, Ireland, [2]University of Limerick, Limerick, V94 T9PX, Ireland; phoebe.hartnett@hotmail.com

In Ireland, replacement gilts are often reared with pigs destined for slaughter, including entire males. Finisher diets are not formulated to meet the needs of developing gilts, and may not supply the necessary minerals to promote hoof health. Moreover, gilts reared with males are exposed to higher levels of mounting and aggression which may increase the risk of injury and stress. This experiment investigated the effect of group composition (GC; single (SS) or mixed sex (MS)), with or without supplementary minerals (MIN- = control diet; MIN+= control + Cu, Zn and Mn) during rearing on salivary cortisol levels, hoof health and locomotory ability of replacement gilts during pregnancy using a 2×2 factorial design. Saliva samples were collected on day 63, 74, 96, 108 and 110 of gestation and locomotory ability was scored (0 – 5) on d 10, 32, 53, 73, 95 and 109 of gestation. Hooves were scored for 8 disorders (heel erosion (HE), sole heel separation/joining (JOIN), white line disease (WLD), vertical cracks (VERT), horizontal cracks (HORIZ), dew claw length (DEW) and dew claw cracks (DCC)) on d118 and d153 of age, d73 of gestation, and at weaning (28 days postpartum). Data were analysed using SAS v 9.4. There was no effect of MIN or GC on locomotion score, or of MIN on salivary cortisol level. However, gilts reared in MS groups had higher cortisol levels (0.667±0.077 ug/Dl) than those in SS groups (0.471±0.057 ug/Dl; P<0.05). Both SS rearing (P<0.001) and a MIN+ diet improved hoof health scores,(P<0.05). MIN+ pigs had reduced HE (P<0.05) and HORIZ (P<0.01) scores. The SS pigs had reduced WLD (P<0.001) and VERT (P<0.05) scores. Overall, rearing replacement gilts in single sex groups and with the addition of a mineral supplement had benefits for hoof health and the welfare of the gilts during pregnancy.

Session 05 Poster 10

The influence of faeces drying method on nutrient digestibility of growing pigs

A.D.B. Melo[1,2], B. Villca[2], E. Esteve-García[2] and R. Lizardo[2]
[1]PUCPR, Rua Imaculada Conceição, 1155, 80215-901 Prado Velho, Curitiba, PR, Brazil, [2]IRTA, Centre Mas de Bover, Crta Reus-El Morell Km. 3.8, 43120 Constantí, Spain; rosil.lizardo@irta.es

Elimination of moisture from faeces samples is crucial to preserve them before laboratory analyses, and it might affect estimation of nutrient digestibility. Freeze-drying is the standard method but is most expensive and time-consuming when compared with oven-drying methods. The aim of this study is to evaluate the influence of drying methods on nutrient digestibility in growing pigs. One hundred forty-four growing pigs were housed by two in 72 pens, and *ad libitum* fed for 6 weeks. Titanium dioxide (0.5%) was included into diets as a tracer. Faeces samples were collected during the 5th week of experiment and aliquoted to oven-dry at 60 °C for 72 h or freeze-dry (condensation at -60 °C, freezing at -35 °C, lyophilisation 25 °C) for 72 h. Samples were analysed for dry matter (DM), gross energy (GE), crude protein (CP), ether extract (EE), crude fibre (CF), ash, organic matter (OM), N-free extracts (NFE) and titanium dioxide contents. Calculations were performed and data analysed using the GLM procedure of SAS. In general, oven-dry method showed a lower coefficient of variation, which could explain slightly higher significant DM (83.2 vs 84.2%), GE (83.8 vs 85.3%), CP (82.79 vs 80.24%), EE (81.09 vs 80.04%), CF (40.34 vs 32.44%), ash (46.30 vs 44.15%) and OM (85.5 vs 86.4%) digestibility than freeze-drying. According to these results, it can be concluded that the oven-dry method (60 °C, 72 h) can replace freeze-drying to determine major nutrient digestibility for growing pigs.

Session 05 Poster 11

Influence of linseed oil sediment in the diets of pigs on fatty acid composition of muscle tissue

V. Juskiene, R. Leikus, R. Juska and R. Juodka
LUHS Institute of Animal Science, Department of Ecology, R. Zebenkos 12, Baisogala, 82317, Baisogala, Lithuania; violeta.juskiene@lsmuni.lt

The objective of this study was to determine the influence of dietary linseed oil sediment on fatty acid composition in the muscle tissue of pigs. 68 crossbred Swedish Yorkshire × Norwegian Landrace pigs were allocated to two trials with two different levels of linseed oil sediment. 24 pigs in Trial 1 were allotted into control C1 and experimental E1 groups of 12 animals each and 44 pigs in Trial 2 were allotted into control C2 and experimental E2 groups of 22 animals each. In both treatments control and experimental groups were formed from animals analogous by origin, gender, weight and condition score. Control pigs were fed identic diet *ad libitum*. The treated pigs were fed the same diet as control pigs, but vegetable oil was replaced by linseed oil sediment at a rate of 2.5% (experimental group E1) in Trial 1 and 5% (experimental group E2) in Trial 2. Analogous pigs during slaughtering were selected from each control and treated groups to collect samples of longissimus lumborum muscle for evaluation of the composition of intramuscular fat. Fatty acid composition in intramuscular fat, was determined by gas chromatograph after extraction by the method of Folch and using transmethylation. Atherogenic, thrombogenic and hypo/hypercholesterolemic indices were calculated on the basis of fatty acid analysis. The results indicated that in linseed oil sediment addition in the diets increased the content of n-3 ά-linolenic (C18:3n-3), eicosatrienoic (C20:3n-3), eicosapentaenoic (C20:5n-3) acids, total n-3 PUFA and decreased the ratio of C18:2n-6/C18:3n-3 and n-6/n-3 ratio also the thrombogenic index of meat. Moreover, the addition of 5% linseed oil sediment resulted in higher content of docosapentaenoic (C22:5n-3) fatty acid, total PUFA and PUFA/SFA ratio. The atherogenic index of meat was by 0.03 times lower in experimental groups E1 and E2 in comparison with control groups, but the differences were insignificant.

Session 05 Poster 12

ADF and NDF apparent digestibility of cull chickpeas and peanut meal in growing pigs

J.M. Uriarte, H.R. Guemez, J.A. Romo, R. Barajas, J.M. Romo, N.A. Lopez and D. Jimenez
Universidad Autonoma de Sinaloa, Facultad de Medicina Veterinaria y Zootecnia, Blvd. Miguel Tamayo Espinoza De Los MONTEROS, 2358, 80020, Mexico; jumanul@uas.edu.mx

To determine the effect of the substitution of soybean meal and sorghum for cull chickpeas and peanut meal on apparent digestibility of ADF and NDF in growing diets for pigs; six crossbred pigs (BW=39.14±1.74) were used in a replicated Latin Square Design. The pigs were assigned to consume one of the three diets with similar levels of energy and protein: (1) Diet with 17.8% CP and 3.27 Mcal ME/kg, containing sorghum 69.5%, soybean meal 28%, and premix 2.5% (CONT); (2) Diet with 17.7% CP and 3.28 Mcal ME/kg with sorghum 42.5%, cull chickpeas 40%, soybean meal 12.0%, vegetable oil 3%, and premix 2.5% (CHP), and (3) Diet with 17.8% CP and 3.26 Mcal ME/kg with sorghum 51.4%, cull chickpeas 30%, peanut meal 14%, vegetable oil 2%, and premix 2.5% (CHPN). Pigs were individually placed in metabolic crates (0.6×1.2 m).The adaptation period was 6 days and sample collection period was 4 days. From each diet and period, one kg of diet was taken as a sample and the total faecal production was collected. Apparent digestibility of DM with values of 82.04, 82.89 and 83.36%, for CONT, CHP, and CHPN, was affected among treatments ($P<0.05$); where apparent digestibility of ADF ($P<0.05$) was altered by treatments (17.67, 34.76 and 25.88%, respectively). Apparent digestibility of NDF was not altered ($P=0.54$) by CHP and CHPN inclusion, and apparent digestibility of OM was not altered ($P=0.35$) by CHP and CHPN inclusion. It's concluded that cull chickpeas and cull chickpeas-peanut meal can be used in growing pig improving ADF digestibility.

Session 06 Theatre 1

SmartCow: integrating research infrastructures to foster innovation in the European cattle sector

R. Baumont[1], R. Dewhurst[2], B. Kuhla[3], C. Martin[1], L. Munksgaard[4], C. Reynolds[5], M. O'Donovan[6] and A. Rosati[7]
[1]INRA, UMR Herbivores, 63122 Saint-Genès-Champanelle, France, [2]SRUC, West Mains Road, EH9 3JG Edinburgh, United Kingdom, [3]FBN Leibniz, Wilhelm-Stahl-Allee 2, 18196 Dummerstorf, Germany, [4]Aarhus Universitet, Nordre Ringgade 1, 8000 Aarhus, Denmark, [5]University of Reading, Whiteknights Campus, RG6 6AH Reading, United Kingdom, [6]Teagasc, Moorepark, Fermoy, Cork, Ireland, [7]EAAP, Via Giuseppe Tomassetti 3 A/1, 00161 Roma, Italy; rene.baumont@inra.fr

SmartCow is an H2020 Research Infrastructure European project which integrates key European cattle research infrastructures to promote their coordinated use and development, and thereby help the European cattle sector face the challenge of sustainable production. SmartCow has been launched on 1st February. This session shall just introduce the project (which has not achieved any results yet). Covering all the relevant scientific fields and the diversity of cattle types and production systems, SmartCow will provide the academic and private research communities with easy access to 11 major research infrastructures from 7 countries of high quality services and resources. These are needed to develop innovative and ethical solutions for efficient use of animal and feed resources that reduce greenhouse gas emissions and promote animal welfare and healthy livestock. SmartCow combines strong scientific and technical skills in animal nutrition (*in vivo* methods for nutrient utilization and emissions measurements), genetics (genotyped animals, phenotyping capabilities), health and welfare (sensors and automatic recordings of physiological and behavioural traits) and ethics in animal experimentation. A central promotion and management of transnational access to the research infrastructures will provide access to around 10,000 cow-weeks and facilitate up to 30 research projects. Networking activities will harmonize and standardize procedures in animal care and measurements, design of experiments, data recording and analysis thanks to a cloud-based data-platform. Joint research activities will produce refined methods and proxies to evaluate feed efficiency and emissions, develop new protocols to reduce the use of animals and produce new methods to exploit sensor data for cattle husbandry. Promotion of transnational access and dissemination of SmartCow outcomes will be supported by a Stakeholder Platform composed of pre- and post-farm gate industry, farmer organizations, NGOs, policy makers at national and EU levels.

Session 06 Theatre 2

Transnational access to leading European cattle research facilities in the EU project SmartCow

B. Esmein[1], R.J. Dewhurst[2], L. Munksgaard[3] and R. Baumont[4]
[1]EAAP, Theix, 63122, Italy, [2]SRUC, West Mains Road, EH9 3JG, United Kingdom, [3]Aarhus Universitet, Nordre Ringgade 1, 800 Aarhus, Denmark, [4]INRA, UMR Herbivores, Theix, 63122 Saint-Genès Champanelle, France; bernard.esmein@orange.fr

Newly-established EU Infrastructures project, SmartCow integrates leading European cattle research. Covering all the relevant scientific fields and the diversity of cattle types and production systems. These are needed to develop innovative and ethical solutions for efficient use of animal and feed resources that promote animal welfare and healthy livestock, as well as sustainable competitiveness. At the networking level, SmartCow will allow better use of existing research infrastructures (RIs), and stimulate collaboration across research fields, production systems, and national borders. It will provide a system for harmonized and optimized transnational access (TNA) to state-of-the-art cattle RIs for any relevant scientific field and implement a TNA program that will address key European research challenges in a coordinated way. TNA funding will facilitate access to high-quality animal experimental facilities in other EU regions by funding costs for access to facilities and this will operate through managed grant funding calls. The objective will be to support high-quality research projects that address the most important challenges in European animal science. The TNA process will catalogue administrative and legal prerequisites at each partner institute, define and run funding calls, evaluate proposals, manage successful projects and review and refine operation of procedures. This second intervention of this session will present the TNA calls and processes for the potential users of the TNA.

Session 06 Theatre 3

SmartCow's training and capacity building activities

P. Dumonthier[1], P. Esmein[2] and R. Baumont[3]
[1]Institut de l'Elevage, Theix, 63122, France, [2]EAAP, Via Giuseppe Tomassetti 3 A/1, 00161 Roma, Italy, [3]INRA, UMR Herbivores, Theix, 63122 Saint-Genès Champanelle, France; philippe.dumonthier@idele.fr

The aim of the SmartCow training program is to transfer and ensure sustainability of the knowledge generated by the project, and to foster innovation, by the means of web-conferences, face-to-face courses and study tours. The courses target researchers and/or technicians from academia, advisory bodies and industries. They will cover the different fields of SmartCow, such as: Ethics in animal experimentation; Ontologies (ATOL and EOL) and their utilization; Digestibility, N balance, feed efficiency, CH_4 emissions measurements and proxies; Validation and user of sensor outputs, etc. 12 training courses (6 two-day face-to-face sessions, and 6 two-hour e-learning sessions) will be set up, targeting around 120 trainees. Special focus will be given to multi-, inter- and transdisciplinary approaches where possible. In addition, study tours will be organized to reinforce interaction between RIs and European stakeholders, i.e. to increase information and appropriation of improved methodologies and best practices, and to promote TNA activities. SmartCow will organize up to 4 tailor made, 1-day thematic visits of 4 project infrastructures, for groups of around 15 participants. These will include visits of other experimental and commercial farms, run by research institutions. A catalogue featuring furthering details on the topics and schedule of this program will be disseminated and updated as the project advances.

Session 06 Theatre 4

Stakeholders engagement in SmartCow

F. Macherez[1], B. Eismein[2] and R. Baumont[3]
[1]Institut de l'Elevage, 149 Rue de Bercy, 75595 Paris Cedex 12, France, [2]EAAP, Via Giuseppe Tomassetti 3 A/1, 00161 Roma, Italy, [3]INRA, UMR Herbivores, Theix, 63122 Saint-Genès Champanelle, France; florence.macherez@idele.fr

The interactions with the stakeholders is essential for SmartCow. It will help to define cattle research priorities, technology developments and the dissemination of project outputs throughout the cattle production chain. For that purpose SmartCow has set up a Stakeholder Platform composed of representatives of pre- and postfarm gate industry, farmers' organizations, policy makers/regulators at national and EU level, including European stakeholders in animal production (FEFANA, FEFAC, FABRE-TP, EFFAB, UECBV, ATF, ICAR, DG Agri, and EuroGenomics). The mission of the Stakeholder Platform is to: (1) ensure relevance of project activities for the pre- and post-farm gate industries, advisory bodies, farmers' organizations, notably by expressing their needs in terms of technology development and services through questionnaires or through the participation to workshops organized by the SmartCow project; (2) steer the calls for transnational access (link with WP2) by defining the research priorities with key industry stakeholders and academic partners; (3) support the project dissemination, notably by (a) providing advices on the main topics of interest for trainings and visits of the project research infrastructures (task 4.3), and by (b) promoting access to the RIs, to harmonized methodologies and best practices among European stakeholders, industrial and technical researchers. Special attention will be given to the identification and mobilization of local stakeholders in each partner's country. All partners will organize one national workshop targeting their local stakeholders to present project achievements and promote transnational facilities in their local language.

Session 07 Theatre 1

Pork production with entire males and immunocastrates in Australia

D.N. D'Souza, R.J.E. Hewitt and R.J. Van Barneveld
SunPork Group, 29 Smallwood Place, 4172, Murarrie, Australia; darryl.dsouza@sunporkfarms.com.au

Australian pork producers ceased surgical castration of entire male pigs in the late 1970's, primarily due to the growth performance inefficiencies and low slaughter weights (<90 kg liveweight). During this period most pork produced was predominantly used for the Australian market. The mid-1990s saw Australia kickstart its export market development program with chilled carcases to Singapore. With Singapore only accepting female pigs, a disproportionate number of entire male pigs were channelled onto the Australian fresh pork market. This increase, coupled with a gradual increase in slaughter weights (>105 kg liveweight) brought with it a greater risk of Australian consumers being exposed to boar tainted pork. The use of entire male pigs also comprised other eating quality parameters, including tenderness and flavour, exacerbated by very lean pork with little to no intramuscular fat. Over the last 10 years, eating quality systems have been a major focus of industry to improve the eating experience and consistency of Australia pork. Immunocastration has enabled Australian pork producers to continue to grow entire male pigs in the most efficient manner, whilst eliminating boar taint and some of the negative behaviours associate with entire male pigs, and to improve the quality and consistency of pork. Given the impact of higher feed and non-feed costs, the Australian pork industry is keen to increase carcase weights in an effort to further improve production efficiency and reduce cost of pork production. The immunocastrate vaccine allows pork producers flexibility in achieving these carcase weight increases in the most efficient manner whilst ensuring a high-quality pork product for consumers in domestic and export markets. Much has also changed in the last 5 years, with all Australian retailers allowing the use of immunocastrates (to eliminate boar taint and improve welfare). Immunocastration has been pivotal in allowing individual supply chains to develop their eating quality assurance pathways to differentiate their brands in the market place.

Session 07 Theatre 2

Pork production with entire males and immunocastrates in South America

J.V. Peloso
Private consultant, Industry researcher, Rua Vereador Luiz Soares 60, 88302-584 Itajaí SC, Brazil; pelosojv@gmail.com

Brazil is the largest pork producer in Latin America with an output of 42,000,000 harvested pigs in 2017. Approximately 75% of the total production comes from integrated system, either fully verticalized or cooperatives and nearly all pork is marketed with a brand. The local pork industry comprises a wide variety of product options at the retail case, ranging from unprocessed chilled or frozen cuts to ready-to-eat processed products. However, the majority of pork is still consumed processed, such as cooked hams, sausages, bacon and salamis. There has never been the intentional production of entire males in the Brazilian pork market. The physical castration of male piglets at young age was widespread in the local industry where the few exceptions were non-selected breeding boars from purebred farms. In the old days, when the male pig reached the market well after the onset of puberty, physical castration was mandatory, both by processors and the Federal Meat Inspectors. The fear of tainted pork products being associated with their brands halted any initiative to stop physical castration of male piglets. Surprisingly, the industry does not know how sensitive to androstenone or skatole consumers are. As pig production became more professional and integrated, with old and small farms being replaced by modern industrial facilities, environmental and welfare issues have arisen. In fully integrated pork production systems, the introduction of new technologies are somehow more practical or effective. From the breeding sows up to the sliced bacon, everything is under the same roof and everybody within the pig meat industry knows that entire males are the most cost effective gender to produce. With that in mind, immunocastration became a way to go for many pig meat processors in Brazil. At the beginning, ten years ago, few suspicious carcass from immunocastred pigs were tested for taint using fat samples. Today 60% of the male pigs are immunocastrated in the Brazilian industry. The other alternative to physical castration would be to raise entire male pigs and mix their meat and fat with that from females at the processing level. This approach places a limit at the number of boars one particular plant could use and it is not suitable to all processed products.

Whole chain approach for moving from surgical castrate to entire male production

G.B.C. Backus
Connecting Agri and Food, Oostwijk 5, 5400 AM, Uden, the Netherlands; g.backus@connectingagriandfood.nl

This study presents a unified analysis of measures to reduce boar taint prevalence and detection methods for boar taint. Effectiveness of measures to reduce boar taint was determined based on literature review and using data collected in observational and experimental studies. Benefits and costs associated with raising boars are calculated for Dutch conditions. These benefits and costs are not equally distributed across chain segments. Boars have a better feed conversion ratio and carcass quality, compared to barrows, generating a calculated net return of €7.71 per pig. At slaughterhouse level, plants are confronted with costs of testing for tainted carcasses and a price reduction for identified tainted carcasses. Our findings show that the cost-effectiveness of preventive measures to reduce boar taint prevalence varies greatly. Changes in breeding programs become attractive only when they assure considerable decreases in tainted carcasses, otherwise the advantage due to reduction of tainted carcasses is outweighed by the decline in genetic progress in other important traits. Changes in feeding and housing conditions are cost-effective, especially to reduce skatole. At slaughter plant level, companies are confronted with costs of testing for tainted carcasses and price reductions for tainted carcasses. In the Netherlands, the cost of boar taint tests based on human nose scoring is €1-2 per carcass. Valuable parts of tainted carcasses are not sold in fresh meat markets. Tainted carcasses are used for processed cold meat products. Given that an average price difference between meat used for sale in the fresh meat market and for production of processed meat products is €0.28 per kilogram, the return on tainted carcasses is ca. €25 lower than on not tainted ones. In conclusion, combinations of preventive measures to reduce boar taint that include breeding and appropriate feeding and housing conditions result in cost-effective solutions. In a non-integrated chain preventive and costly measures that reduce boar taint will only be implemented when incentives are imposed to induce farmers to prevent developing boar taint. The optimal combination of preventive measures and investments in boar taint testing can vary greatly with the boar taint threshold level.

Genetic approaches for rearing entire males

C. Larzul[1], L. Fontanesi[2], E. Tholen[3] and M. Van Son[4]
[1]GenPhyse, Université de Toulouse, INRA, ENVT, 24 chemin de Borde Rouge, 31320 Castanet-Tolosan, France, [2]University of Bologna, Department of Agricultural and Food Sciences, Viale Fanin 46, 40127 Bologna, Italy, [3]Institute of Animal Sciences, Animal Genetics, Endenicher Allee 15, 53115 Bonn, Germany, [4]Topigs Norsvin, Storhamargata 44, 2317 Hamar, Norway; catherine.larzul@inra.fr

Boar taint in the pork meat from entire male pigs results from the presence of two molecules: androstenone and skatole. The genetic determinism of androstenone and skatole has been studied for a few decades. It is well known that the heritability values are moderate to high and the genetic correlation between both levels is moderately positive. Nevertheless, application of selecting against boar taint has been limited in breeding schemes to obtain low boar taint lines. An important topic is the relationships between boar taint risk and other trait of economic interest. The genetic correlations with production traits such as growth rate, feed efficiency or carcass quality are mostly favourable, and selection applied in sire lines should tend to decrease boar taint. In dam lines, the genetic relationships with reproductive traits still need to be further explored. Some studies showed low correlations between boar taint risk and litter traits or fertility, but other evidences led to the conclusion that selection against boar taint would unfavourably impact female reproductive traits. Several issues have been addressed regarding potential criteria to predict boar taint at slaughter. One of the most promising has been human nose scoring at slaughter house, but relevant predictors are still needed that can directly be measured on live animals. New instrumental methods have to be developed for breeding purposes. In that respect, genomic information and a better understanding of genes involved in boar taint metabolism will provide useful tools for breeding entire males. The review was prepared within the framework of the Cost Action CA 15215 IPEMA.

Effect of divergent selection for cortisol level on boar taint

C. Larzul[1], E. Terenina[1], Y. Billon[2], L. Gress[1], R. Comte[3], A. Prunier[3] and P. Mormede[1]
[1]GenPhyse, Université de Toulouse, INRA, ENVT, 24 chemin de Borde Rouge, 31326 Castanet-Tolosan, France, [2]GenESI, INRA, Le Magneraud, 17700 Surgères, France, [3]PEGASE, INRA, Agrocampus Ouest, Chemin de la Prise, 35590 St-gilles, France; catherine.larzul@inra.fr

Plasma cortisol levels measured one hour after injection of ACTH reflects the hypothalamic-pituitary-adrenocortical (HPA) axis activity. The test was performed on piglets at 6 weeks of age. From a base population of Large White pigs, two divergent lines were selected for 3 generations, for high and low cortisol levels respectively. At the 3rd generation of selection, the divergence between the two lines was about 5 genetic standard deviations. At 158 days of age, thirty-two entire males were slaughtered with or without mixing before slaughtering. A blood sample was collected at exsanguination. Urine was collected at carcass evisceration directly from the bladder. A fat sample was taken at the neck level 24 hours post mortem. Skin lesions were counted on the fore, middle and rear parts of the carcass. At slaughter, the high cortisol line showed a significantly higher level of plasma cortisol (81.7 vs 31.8 ng/ml, $P<0.001$) and urinary deconjugated cortisol (106.3 vs 72.3 ng/ml, $P=0.009$), cortisone and dopamine, as well as a higher level of estradiol in plasma (3.85 vs 3.39 on a logscale, $P=0.038$) and androstenone in fat (0.189 vs -0.366 on a logscale, $P=0.019$), with a tendency for higher levels of testosterone in plasma and skatole in fat. There was no difference between lines for the number of skin lesions and urinary levels of noradrenaline and adrenaline. Stress had a significant effect on cortisol which increased in urine from mixed animals but not on fat androstenone and skatole. Mixing also increased lesions on the fore and middle parts of the body, as an index for an increased number of fights. This experiment confirmed the effect of mixing on cortisol levels and skin lesion scores. It also provided evidence that selection for cortisol level influences steroidogenesis in entire male pigs.

Genome-wide DNA methylation analysis reveals candidate epigenetic biomarkers of boar taint in pigs

X. Wang[1], M. Drag[2] and H.N. Kadarmideen[1]
[1]Technical University of Denmark, Department of Bio and Health Informatics, Kemitorvet Building 208, 2800, Denmark, [2]University of Copenhagen, Department of Veterinary and Animal Sciences, Grønnegårdsvej 7, 1870, Denmark; xiwa@bioinformatics.dtu.dk

Breeding pigs with genetic or epigenetic factors in order to reduce levels of boar taint (BT) compounds such as androstenone and skatol can avoid surgical castration – an animal welfare concern. This study conducted Reduced Representation Bisulfite Sequencing (RRBS) of DNA methylation in high versus low BT pigs and identified candidate epigenetic biomarkers associated with BT that could be used in selective breeding. Nine samples of three different BT levels were analysed using RRBS data. Filtering and quality control by Trimmomatic software removed RRBS adapters and transcripts with low read counts. Clean reads were mapped to pig reference genome (Sscrofa11.1/susScr11) by Bismark Bisulfite Mapper. Methylation levels of cytosine were analysed by R package methylKit. Differentially methylated cytosine (DMC) was defined with regard to CpG islands, CpG islands shores and the proximity to nearest transcription start site (TSS) using R package genomation. Gene ontology (GO) enrichment and pathway terms were analysed by GenCLiP 2.0 software. Interactive gene networks were visualized in Cytoscape Web. The rate of uniquely mapped clean read pairs was 48.7%. The mean distribution of cytosine methylation rate in CpG, CHG and CHH sites were 49.0, 0.9 and 0.7%, respectively. The distribution of DMC annotation within CpG islands, CpG islands shores and other regions were 57.2, 14.7 and 28.1%, as well as 5.30, 1.22, 3.79 and 89.7% in promoter, exon, intron and intergenic regions, respectively. Co-analysis of differentially expressed (DE) genes and significant DMCs found 32 significant co-identified genes. Joint analysis of GO terms, pathways and gene networks revealed that *DMAP1*, *EGFR* and *PEMT* were very important in regulating gene expression underlying BT, especially *PEMT* might be positive to oestrogen and BT status. To our knowledge, we are the first to report epigenetic mechanisms and epigenetic markers using genome-wide DNA methylation profiles for BT in pigs. These results could be used in biotechnology and breeding industries.

Mining whole genome resequencing data to identify functional mutations in boar taint-candidate genes

G. Schiavo[1], S. Bovo[1], S. Cheloni[1], A. Ribani[1], C. Geraci[1], M. Gallo[2], G. Etherington[3], F. Di Palma[3] and L. Fontanesi[1]

[1]University of Bologna, Department of Agricultural and Food Sciences, Viale Fanin 46, 40127 Bologna, Italy, [2]ANAS, via Nizza 53, 00198 Roma, Italy, [3]Earlham Institute, Research Park Innovation Centre Colney Ln, NR47UZ Norwich, United Kingdom; luca.fontanesi@unibo.it

Next generation sequencing applied in pigs have recently produced re-sequenced pig genome data from different individuals belonging to a large variety of breeds. The availability of these large datasets is opening new opportunities to mine public nucleotide archives and identify mutations that could putatively affect economic relevant traits. Moreover, resequencing data from pooled pig DNA could provide cost-effective whole genome information from a large number of animals. In this study, we mined 110 individual pig genomes retrieved from the European Nucleotide Archive and from proprietary datasets generated from pigs of 28 different breeds. This dataset was integrated from 8 pooled whole genome resequencing datasets generated from 35 individuals each from 8 distinct commercial or autochthonous breeds (Italian Large White, Italian Duroc, Italian Landrace, Apulo Calabrese, Cinta Senese, Casertana, Mora Romagnola, Nero Siciliano), respectively. Individual and pooled pig genome datasets were searched for polymorphisms in 135 annotated candidate genes, including 25 genes involved in androsterone and skatole biochemical related pathways. Short reads from these genomes were aligned using bowtie to a customized reference sequence generated from the reference pig genome, including sequence of selected genes (with depth ranging from 4 to 40X for each genome). A total of 100k variants were identified (2.3% in coding regions with about 500 missense mutations and a few other potential functional mutations). About 15% of these numbers refers to genes encoding enzymes involved in the androsterone and skatole biochemical pathways. This study provided an overview of the variability in targeted gene regions potentially involved in determining boar taint in pigs. Partially funded by European Union's H2020 RIA program (grant agreement no. 634476). Abstract reflects the authors' view. European Union Agency is not responsible for any use that may be made of the information it contains.

The lysine requirement of growing entire male pigs (10-15 weeks of age): a dose-response study

S. Millet[1], M. Aluwé[1], J. De Sutter[2], W. Lambert[3], B. Ampe[1] and S. De Campeneere[1]

[1]Flanders research institute for agriculture, fisheries and food (ILVO), Scheldeweg 68, 9090 Melle, Belgium, [2]Orffa Belgium NV, Rijksweg 10G, 2880 Bornem, Belgium, [3]Ajinomoto Eurolysine S.A.S., 153, rue de Courcelles, 75817 Paris Cedex 17, France; sam.millet@ilvo.vlaanderen.be

With the production of entire male pigs and immunocastrates, there is renewed interest in the nutrient requirements of male pigs. In the present study, we evaluated the effect of increasing dietary lysine levels (6 levels: 7.5; 8.5; 9.5; 10.5; 11.5; 12.5 g/kg) on growth performance of entire male pigs from 10 to 15 weeks of age. The CP level was fixed at 170 g/kg and all diets were formulated to meet or exceed the ideal amino acid (AA) profile for pigs. Glutamic acid was used to replace essential amino acids and keep diets isonitrogenous. Because of the high energy content assigned to glutamic acid, net energy level decreased slightly form the lowest to highest lysine level (from 9.9 to 9.6 MJ/kg). Amino acid analysis revealed slightly higher lysine levels than anticipated and requirements may be slightly underestimated. Both for daily gain and feed conversion ratio, animals on the diet with 10.5 g SID LYS showed significantly better performance than the 3 groups with lower SID lysine levels. Daily gain increased from 674±51 g/day at the lowest level to 833±90 g/day at 10.5 g/kg. Similarly, feed conversion ratio improved from 2.11±0.05 to 1.77±0.06 g/g with dietary SID LYS level between 7.5 g and 10.5 g/kg. For daily gain, it was not possible to fit a linear plateau model, but a quadratic plateau model gave an optimal SID lysine level at 12.95 g/kg. As this breakpoint is outside the measured range, it should be interpreted with caution. For feed conversion ratio, optimal SID lysine levels were 10.65 and 12.11 g/kg (1.09 or 1.25 g SID LYS/MJ NE), for a linear and quadratic plateau model, respectively. The slope of the linear plateau model indicates that an improvement in FCR of 0.11 can be expected with each g increase of the dietary SID lysine level, between 7.5 and 10.6 g/kg.

Development of long-term, pre-finishing immunocastration protocols for male Iberian pigs: 1. Efficacy

F.I. Hernández-García, M. Izquierdo, M.A. Pérez, A.I. Del Rosario, N. Garrido and A. Montero
Center of Scientific & Technological Research of Extremadura (CICYTEX), Animal Production, Autovía A-5, Km 372, 06187-Guadajira (Badajoz), Spain; francisco.hernandez@juntaex.es

Male immunocastration (IC) should be adapted to the long cycle of Iberian (IB) pigs before the expected end of male pig castration in the EU. We previously developed long-term, 3-dose, pre-finishing IC protocols for male Iberian pigs whose efficacy seemed to be influenced by nutritional level. Study 1 aimed to improve the efficacy of these protocols by short-term increase in feeding intake. Pigs (IC males; ICM; n=47) were fed concentrate in an extensive system and immunized against GnRH at 11, 12 and 14 months (m) of age. Entire males (n=5) were used as general controls. Pigs were slaughtered at 16 m. Treated subgroup (23 ICM) was submitted to a 15-day *ad libitum* feeding period starting at the 3rd vaccination. The remaining ICM were the Control subgroup. A 100% efficacy was reached by the Treated ICM, which all had <150-g testes (which was the threshold for blood testosterone presence in our earlier studies). In contrast, 4/24 Control ICM had >150-g testes. Study 2 aimed to further adjust the protocol to suit the chronology of the acorn-feeding free-ranging period (*montanera*; MT). We tested whether improving homogeneity of body condition at the start of MT would enhance and homogenize testicular atrophy. Control pigs (C; n=18 IB males) were immunized at 10.5, 12 and 13.5 m. Treated pigs (T; n=17 IB males) were immunized at 10.5, 11.5 and 13 m, with a 15-day *ad libitum* (AL) feeding period starting at the 3rd dose. *Montanera* started at 13.5 m, and slaughter took place at 16 m. The AdLib group (n=15 IB × Duroc males) were fed AL with concentrate during the growth and finishing phases in a regular extensive system, with the vaccinations taking place at 8, 9 and 10 m and the slaughter at 13 m. The Adlib and T treatments showed a 100% efficacy. Testicular and epidydimal weights were significantly smaller for AdLib and T than for C. Bulbourethral gland weight was significantly smaller for T than for C. Blood testosterone data is not yet available from both studies. In conclusion, nutritional level can be used to improve IC efficacy, and treatment is compatible with *montanera* system.

Increasing levels of condensed tannins from Sainfoin may reduce the environmental impact of pigs

E. Seoni[1,2], G. Battacone[1], F. Dohme-Meier[2] and G. Bee[2]
[1]University of Sassari, Department of Agricultural Sciences, Via Enrico de Nicola 7, 07100 Sassari, Italy, [2]Agroscope, Route de la Tioleyre 4, 1725 Posieux, Switzerland; eleonora.seoni@agroscope.admin.ch

The effects of increasing level of condensed tannins (CT) from Sainfoin on nitrogen (N) excretion of grower-finisher pigs was investigated. A total of 48 Swiss Large White entire males (EM; BW 24.8±5.1 kg) were assigned within litter to 1 of 4 isoenergetic and isonitrogenous grower (25-60 kg BW) and finisher (60-105 kg BW) diets supplemented with 0 (T0), 5 (T5), 10 (T10) and 15% (T15) of Sainfoin, respectively. Individual feed intake and total amount of urine and faeces produced over 7 d per pigs in the grower and finisher period were measured. Data were analysed with a one-way ANOVA using litter and experimental groups as fixed effects and mean differences tested using the Tukey post-hoc test. Despite similar N intake, T15 pigs had 31.5% lower ($P<0.01$) urinary N excretion compared to T0 and T5 pigs in the grower period and 32.3% lower ($P<0.001$) urinary N excretion compared to the others groups in the finisher period. Faecal N excretion was 78.8% greater ($P<0.001$) in T15 and T10 pigs compared to T0 with intermediate values for T5 pigs in the finisher period. When expressed as percentage of total N intake, urinary N excretion tended ($P=0.08$) to be lower by 26.8% and was lower ($P<0.01$) by 30.5% in T15 pigs compared to T0 pigs with intermediate values for T5 and T10 pigs in the grower and finisher period, respectively. Faecal excretion linearly increased ($P<0.001$) by 55.6% compared to T0 pigs in the grower period and was 64% greater ($P<0.001$) in T15 pigs compared to T0 and T5 pigs in the finisher period. Body N retention was greater ($P<0.05$) in T0 pigs compared to T5 pigs, with intermediate values for T10 and T15 pigs in the grower period whereas no treatment differences were observed in the finisher period. Moreover, urinary urea levels were numerically lower in T15 pigs compared to control in both periods. In conclusion, with increasing dietary CT inclusion a distinct shift in N excretion from urine to faeces was observed. This could be beneficial in reducing urinary ammonia emission from pig production.

Testicular development, sex hormones and boar taint in pig lines divergent for residual feed intake

A. Prunier[1], Y. Billon[2], J. Ruesche[3], S. Ferchaud[4] and H. Gilbert[3]
[1]INRA, UMR PEGASE, 35590 Saint-Gilles, France, [2]INRA, GenESI, 17700 Surgères, France, [3]INRA, GenPhyse, 31320 Castanet-Tolosan, France, [4]INRA, GenESI, 86480 Rouillé, France; armelle.prunier@inra.fr

Improving feed efficiency and rearing entire male pigs are relevant strategies to reduce feed cost and environmental waste in pig production. The major constraint for rearing entire male pigs being boar taint, an experiment was performed to determine the consequences of a divergent selection on residual feed intake (RFI: low RFI = LRFI; high RFI = HRFI) on pubertal development and boar taint. Purebred French Large White male pigs from two divergent lines for RFI (9th generation of selection, n=45 or 43 pigs/line from 33 litters) were reared in two batches (n=19 to 24 pigs/line/batch). Blood samples were drawn at 15 and 166±1 days of age (mean ± SD) and pigs were weighed. After slaughter at 167±1 days of age, a backfat sample was collected in the neck and the genital tract was removed for testis and epididymis weighing after tissue trimming. Percentages of testis and epididymis weight relatively to liveweight were calculated for statistical analyses. All data were analysed by ANOVA using R, including line and batch as fixed effects and litter as a random effect. When necessary, a log transformation was applied before analysis and adjusted means were back calculated. Before slaughter, LRFI pigs were lighter than HRFI pigs (100±2 vs 106±2 kg liveweight, $P<0.02$). Plasma testosterone at both ages and plasma estradiol-17b at 15 days were similar in both lines ($P>0.1$). However, plasma estradiol-17b before slaughter (9.0 vs 17.5 pg/ml), fat androstenone (0.22 vs 0.34 µg/g pure fat), testis (3.5 vs 4.4‰) and epidydimis (1.18 vs 1.62‰) relative weights were lower in HRFI than LRFI pigs ($P<0.001$). Fat skatole was lower in HRFI than in LRFI (0.10 vs 0.28 µg/g pure fat, $P<0.001$) in the first batch with slaughter performed on the 17th July 2016 during a summer heat wave. Fat skatole was low in both lines (0.08 µg/g) in the second batch with slaughter performed on the 29th August 2016. Overall, these data indicate a lower testicular activity in the HRFI than in the LRFI line, which suggests the existence of genetic links between feed efficiency and reproduction.

Genetic determinism of boar taint and relationship with meat traits

C. Dugué[1], A. Prunier[2], M.J. Mercat[3], M. Monziols[3], B. Blanchet[4] and C. Larzul[1]
[1]GenPhySE, université de Toulouse, INRA, ENVT, 24 Chemin de Borde Rouge, 31326 Castanet-Tolosan, France, [2]PEGASE, INRA, Agrocampus Ouest, Domaine de la Prise, 35590 St-Gilles, France, [3]IFIP, La Motte au Vicomte, 35650 Le Rheu, France, [4]UEPR – INRA, Domaine de la Motte, 35653 Le Rheu, France; claire.dugue@inra.fr

Entire male meat can have a major quality defect called boar taint, partly due to the presence of androstenone in fat. This study evaluates the feasibility of a selection to directly decrease back fat androstenone level or indirectly by a selection on the plasma estradiol level and estimate the consequences on meat traits in purebred or crossbred pigs. Pure Pietrain (P) and Pietrain Large White crossbred pigs (X) were measured for hormone levels: estradiol (Est) and testosterone (Tes), growth traits: average daily gain, feed conversion ratio (FCR), average daily feed intake (ADFI), carcass composition: carcass yield (CY), lean percentage (L%) and quality traits: pH in Ld and ham, drip loss, intramuscular fat and back fat androstenone level (Andr). The number of skin lesions (SL) was measured at three stages. Carcass additional measures were obtained by computerized tomography: loin eye area (LEA) and density, femur density, ham muscle/bone length ratio (HFR). The number of measured animals varied from 553 to 712 for P and from 556 to 736 and for X. Heritabilities were of medium values for estradiol level and high values for androstenone level. A selection to decrease P Andr level would increase HFR and pH in ham and decrease FCR and Tes in P pigs. On X it would increase CY, LEA, L% and HFR and decrease SL at fattening entrance, FCR, drip loss, ADFI and femur density. A selection to decrease P Est level would decrease Andr, FCR, ADFI and Tes in P pigs and Andr, SL at fattening entrance and Tes in X pigs. Heritabilities and genetic correlations indicate that a selection to decrease estradiol level would have overall favourable effects on meat traits. The authors are extremely grateful to the UEPR personnel, PEGASE technicians and IRSTEA. This study has been granted by ANR (ANR-10-GENOM_BTV-015, ANR-15-CE20-0008), Alliance R&D, InaPorc and FranceAgrimer.

Genetic analysis of boar taint and fertility traits including hormone profiles in dam lines

I. Brinke[1], K. Roth[1], M.J. Pröll[1], C. Große-Brinkhaus[1,2], I. Schiefler[2], K. Schellander[1] and E. Tholen[1]
[1]University of Bonn, Institute of Animal Science, Endenicher Allee 15, 53115 Bonn, Germany, [2]Förderverein Bioökonomieforschung e.V., Adenauerallee 174, 53113 Bonn, Germany; ibri@itw.uni-bonn.de

The issue of animal welfare has become an important factor in pig production that determines consumer acceptance. Thus, surgical castration of piglets without anaesthesia will be banned by law in Germany starting from 2019. Fattening of entire males is considered as a feasible option with respect to the metabolic benefits concerning lean muscle content and feed conversion. However, the antagonistic relationships between boar taint components and reproduction traits can be expected as described by Mathur *et al*., so that selection against boar taint in female lines is of relevant risk. The on-going G-I-FER project was designed to detect the antagonistic relationships between boar taint components and fertility/robustness in Landrace (LR) and Large White (LW) nucleus populations in German pig breeding organizations. Concentrations of the main boar taint components androstenone (AND) and skatole (SKA) are quantified in back fat samples from 2,000 performance tested entire males by SIDA-DI-SPME-GC/MS. Serum progesterone, luteinizing hormone, follicle stimulating hormone, cortisol and testosterone (only tested in boars) will be measured in blood samples from entire males and their full-sisters at standardized live weight to generate a hormone profile for each animal. All animals are being genotyped to perform a genome-wide association study (GWAS) for all boar taint components. Based on uni- and multivariate approaches heritabilities and genetic correlations of boar taint components and reproduction traits were estimated for each line. First results show that the estimated heritability for AND and SKA was 0.41. Within both lines, genetic correlations differ but indicate unfavourable relations between boar taint components and specific reproduction traits exceeding a threshold set of |0.3|. Results will be used in further studies to identify SNP-markers that are associated with boar taint and reproduction traits. In addition, biomarkers will be developed to support animal welfare and to help producers to breed less odorous animals without declines in reproduction traits.

Boar taint and hormonal development in immunocastrated pigs of 3 breeds

E. Heyrman[1], L. Vanhaecke[2], S. Millet[1], F. Tuyttens[1], S. Janssens[3], N. Buys[3], J. Wauters[2] and M. Aluwé[1]
[1]ILVO (Flanders Research Institute for Agriculture, Fisheries and Food), Animal Sciences Unit, Scheldeweg 68, 9090 Melle, Belgium, [2]Ghent University, Faculty of Veterinary Medicine, Department of Veterinary Public Health and Food Safety, Salisburylaan 133, 9820 Merelbeke, Belgium, [3]KU Leuven, Livestock Genetics, Department of biosystems, Kasteelpark Arenberg 30, 3001 Heverlee, Belgium; marijke.aluwe@ilvo.vlaanderen.be

Boar taint is an unpleasant odour in meat and fat that has been mainly associated with uncastrated male pigs (entire males). The main compounds responsible for boar taint are androstenone (AND), skatole (SKA), and to a lesser extent indole (IND). Immunocastration has been put forward as a method to eliminate boar taint. Uncastrated male pigs (n=109) of 3 male lines – Belgian Piétrain (BP), French Piétrain (FP), and Canadian Duroc (CD) – were immunocastrated at approximately 4 weeks before slaughter. Blood samples were taken from these pigs at 4 occasions: at first vaccination (V1 – 8 weeks before slaughter), at second vaccination (V2 – 4 weeks before slaughter), 2 weeks after second vaccination (V2+2) and 1 day before slaughter (S) to determine serum AND, SKA, and IND concentrations and testosterone (TES), estradiol (EST), and progesterone (PRO) levels. From 47 pigs, fat biopsy samples were taken from neckfat at V2 and V2+2, and neckfat samples were collected at the slaughterline to evaluate AND and SKA. TES and EST showed a decline after V2 (from 4.0 to 0.4 nmol/l, and from 20.2 to 7.7 ng/l), while for PRO no time effect was noted. AND in serum showed a rise up to V2 for CD and BP (3.3 and 3.4 ppb) and declined at V2+2 (2.5 and 2.6 ppb) up to S (2.5 and 2.6 ppb) while it remained low at all occasions for FP.. SKA in blood showed a rise up to V2 (8.8 ppb) for all breeds and declined to V2+2 (6.7 ppb) and to S (5.6 ppb). IND in blood rose from V2 until S for CD (4.2 ppb, 9.6 ppb) and FP (5.0, 9.9 ppb) but remained low for BP. AND in fat rose from V2+2 (82 ppb) to S (203 ppb), while SKA in fat declined from V2 (176 ppb) to S (87 ppb). The results generally confirm the effectiveness of immunocastration to inhibit sexual development and reduce boar taint. Further investigation of the results is needed to clarify the different patterns found for AND in plasma and fat.

Influence of terminal sire on fattening traits of pigs of four genders

G. Kušec[1], I. Djurkin Kušec[1], M. Škrlep[2], K. Gvozdanović[1] and V. Margeta[1]
[1]Faculty of Agriculture in Osijek, Department of Applied Anima Science, Vladimira Preloga 1, 31000 Osijek, Croatia, [2]Agricultural Institute of Slovenia, Hacquetova ulica 17, 1000 Ljubljana, Slovenia; gkusec@pfos.hr

The investigation was carried out on 240 fatteners reared on a commercial farm in Croatia. Sows were inseminated using the semen from three terminal sires: line A (Pietrain × Large White), line B (Pietrain NN genotype) and C (Pietrain × Duroc × Large White). Their offspring was assigned in groups according to the experimental treatment: surgically castrated males (SC, 60 animals), immuno-castrated males (IC, n=60), intact males (EM, n=60) and gilts (G, n=60). Immunocastration was performed by the application of Improvac™ vaccine at the age of 72 and 147 days. Live weight (LW) was measured at weaning (W, age=25 days), at the time of first vaccine application (V1, age=72 days), at the beginning of early fattening period (EF, age=116 days), at the time of second vaccine application (V2, age=147 days), and before slaughter (FW, age=168 days). Based on those measures average daily gains (ADG) were calculated for different periods of fattening: from W to V1 (FP; feeder period), from V1 to EF (PP – porker period), from EF to V2 (EFP – early fattening period), from V2 to FW (LFP – late fattening period). At the age of 168 days, the pigs were slaughtered. Terminal sire line significantly influenced LW of the fatteners at V1 (P<0.001) and at EF (P<0.001). Sex had significant influence on the LW of fatteners at EF (P<0.001), V2 (P<0.001) and FW (P<0.001), while sire line × sex interaction was observed for LW at EF (P=0.047), V2 (P=0.004), and FW (P=0.041). Sex significantly affected ADG in following periods: PP (P<0.001), EFP (P=0.002), LFP (P<0.001) and the total ADG. Terminal sire line influenced ADG only in the FP and EFP periods (P<0.001), while their interaction was observed for FP (P=0.007) and total ADG (P=0.044). Until V2 SC pigs had the highest ADG, thereafter the highest growth rate was observed in IC followed by EM. IC male pigs originating from the line C terminal sires had the highest total ADG and the highest final LW.

Does the background matter: people's perception of pictures of pigs in different farm settings

S. Gauly[1], M. Von Meyer-Höfer[1], A. Spiller[1] and G. Busch[2]
[1]Georg-August-University Göttingen, Department of agricultural economics and rural development, Platz der Göttinger Sieben 5, 37073 Göttingen, Germany, [2]Free University of Bolzano, Faculty of Science and Technology, Universitätsplatz 5, 39100 Bozen-Bolzano, Italy; gesa.busch@unibz.it

Public discussions about animal welfare in livestock farming involve the increased use of pictures in the media. Such pictures can be composed very differently ranging from airbrushed pictures for marketing purposes up to shocking pictures about grievances in farming. Thereby, the animals' appearance and the environment in which the animal is depicted may influence picture perception. The aim of this study is to analyse the effects of pig and pen composition on peoples' perception of pig welfare and pen evaluation. To achieve this aim, 1,019 German residents took part in an online survey in June/July 2016. Using a 2×2 factorial design, four modified pictures showing a pig ('happy' or 'unhappy'-looking) in a pen (slatted floor or straw bedding) were shown to respondents. They evaluated the pictures regarding perceived pig welfare and pen characteristics using 5-point semantic differential scales from -2 to 0 to +2. A General Linear Model and t-tests were calculated to analyse effects on picture evaluation. The effect of the pen was the largest on picture evaluation ($\eta2=0.392$, $P\leq0.001$) followed by the effect of the pig ($\eta2=0.099$, $P\leq0.001$). The welfare of both, the 'happy' and 'unhappy' pig, was always rated lower in the slatted floor setting than in the straw setting and the 'unhappy pig' on straw achieved better welfare-values compared to the 'happy pig' on slatted floor (e.g. evaluation of health: 'happy' pig μ-straw=0.87, μ-slatted=0.22; 'unhappy' pig: μ-straw=0.49, μ-slatted=-0.16; $P\leq0.001$). The straw pen is rated more positive compared to the slatted floor pen and the evaluation only slightly differed depending on pigs' expression (e.g. evaluation of species-appropriateness: straw pen μ-happy=0.56, μ-unhappy=0.48, slatted floor pen: μ-happy=-1.08, μ-unhappy=1.16, $P\leq0.001$). The results show that public perceptions of a contentious husbandry system (pen with slatted floor) is not altered by showing positive appearing animals whereas a more positive perceived system (pen with straw) enhances the welfare perception of the animal on pictures.

Farmers' point of view towards the applicability of a guideline to assess animal welfare of pigs

M. Pfeifer, A. Koch and E.F. Hessel
Federal Research Institute for Rural Areas, Forestry and Fisheries, Thuenen-Institute of Agricultural Technology, Bundesallee 47, 38116 Braunschweig, Germany; alex.k.g@t-online.de

In Germany the Animal Welfare Act commits all livestock owners to record animal welfare indicators. The publication `Animal Welfare Indicators: A Guideline for on-farm self-assessment – pigs´ is meant as a supporting tool for the recording. In this guideline the indicators *daily weight gains*, *animal losses*, *treatment incidence with antibiotics*, *slaughter checks* and *water supply* on herd level as well as the indicators *tail length*, *runts*, *manure on body*, *skin lesions*, *ear lesions*, *tail lesions*, *lameness* and *evidence of ectoparasites* on animal individual level are suggested. In order to assess this publication from farmers' point of view, guideline-based interviews have been performed with 20 fattening pig breeders. It was asserted that, with the exception of *tail length* and *manure on body*, at least 65% of the respondents attributed a statement of animal welfare to the other suggested indicators. *Tail lesions* and *runts* were identified as the most representative indicators regarding animal welfare. The suggestion of the guideline to assess the animal individual indicators for a sample size of 150 growing pigs of the livestock is only evaluated as the right sample size by 40% of the farmers. The feasibility of the recording of the suggested indicators was assessed on a five-point scale from very easy to very hard. At least 95% of the farmers classified the recording of the indicators *water supply*, *ear lesions* and *tail lesions* into the range of very easy and easy. 20% of the farmers stated that the recording of the indicator *manure on body* is challenging. The overall applicability of the guideline for on-farm self-assessment of animal welfare was assessed with 2.48 on average on a five-point scale (from 1 = very good to 5 = very bad). The farmers named some possible improvements. For example, the sample size of the examined pigs should be determined in relation to the overall number of fattening pigs per farm.

Test-retest reliability of the Welfare Quality® protocol applied to sows and piglets

L. Friedrich[1], I. Czycholl[1], N. Kemper[2] and J. Krieter[1]
[1]Institute of Animal Breeding and Husbandry, Christian-Albrechts-University, Olshausenstr. 40, 24098 Kiel, Germany, [2]Institute for Animal Hygiene, Animal Welfare and Farm Animal Behaviour, University of Veterinary Medicine Hannover, Foundation, Bischofsholer Damm 15, 30173 Hannover, Germany; iczycholl@tierzucht.uni-kiel.de

Reliability is important for assessment protocols of animal welfare. The present study aimed at testing the 'Welfare Quality® animal welfare assessment protocol for sows and piglets' for its test-retest reliability (TRR). The study was performed on 13 farms in Northern Germany, which were visited by the same observer 5 times within 10 months. The entire protocol was executed including a Qualitative Behaviour Assessment (QBA), direct behavioural observations (BO), a human-animal relationship test (HAR), scans for stereotypies (ST) and individual indicators (IN). TRR was calculated using Spearman's rank correlation coefficient (RS), intraclass correlation coefficient (ICC), smallest detectable change (SDC) and limits of agreement (LoA). Farm visit 1 (F1; day 0) as reference was compared to farm visits 2 to 5 (F2-5; day 3, week 7, month 5, month 10). Acceptable TRR was assigned when RS and ICC, respectively, ≥0.40, SDC≤0.10 and LoA≤∈(-0.10;0.10). Direct comparison of the adjectives of the QBA indicated poor TRR, e.g. 'playful' (e.g. F1-F4: RS 0.24 ICC 0.36 SDC 0.38 LoA ∈(-0.51;0.17)). Acceptable TRR could be found for BO, e.g. pen investigation, and ST, e.g. floorlicking (e.g. F1-F4: RS 0.63-0.72 ICC 0.45-0.52 SDC 0.04-0.06 ∈(-0.03;0.06)-(0.08;0.04). Alike, acceptable TRR was assigned to HAR due to acceptable correlation among farm visits (e.g. F1-F4: RS 0.41 ICC 0.51 SDC 0.52 ∈(-0.67;0.25)). In terms of IN, mostly acceptable TRR were achieved. Poor TRR was detected for instance for IN body condition score, metritis, bursitis and panting in sows and scouring, lameness and huddling in piglets (e.g. F1-F4: RS -0.49-0.39 ICC 0.00-0.30 SDC 0.13-1.01 ∈(-0.12;0.19)-(0.98;1.00)). Differences in the assessment can be explained by seasonal effects, movement, rare prevalences and different conditioned sow groups. Still, the Welfare Quality® protocol for sows and piglets can be a reliable approach regarding its TRR to assess welfare in sows and piglets. Due to seasonal effects, a half-yearly assessment interval is advised.

Reliability of different behavioural tests on growing pigs on-farm

I. Czycholl, S. Menke, C. Straßburg, K. Büttner and J. Krieter
Institute of Animal Breeding and Husbandrs, Christian-Albrechts-University Kiel, Olshausenstraße 40, 24098 Kiel, Germany; iczycholl@tierzucht.uni-kiel.de

Behavioural tests might have potential for the evaluation of certain aspects of welfare such as the emotional state or the human-animal relationship. However, reliability assessments of these tests are rare. Therefore, in this study, different behavioural tests on growing pigs were assessed regarding their reliability on-farm. 11 growing pig farms were visited 3 times each by 2 experienced observers. The farm visits took place with a time interval of 4 weeks in between the first and the second farm visit and a time interval of 1 week in between the second and the third farm visit. The observers carried out a Novel Object Test (NOT), a Voluntary Human Approach Test (VHAT), a Forced Human Approach Test and a Human-Animal Relationship Test (HAR) in the home pen of the pigs. The results of the different farm visits and observers, respectively, were compared by a combination of statistical parameters: Spearman's Rank Correlation Coefficient (RS), Intraclass Correlation Coefficient (ICC), Limits of Agreement (LoA) and Smallest Detectable Change (SDC). For all tests, the interobserver reliability was acceptable (e.g. NOT: RS: 0.70, ICC: 0.56, SDC: 0.10, LoA: -0.09-0.10) to good (e.g. FHAT: RS: 0.98, ICC: 0.98, SDC: 0.01, LoA: -0.01-0.01). However, the test-retest reliability was rather insufficient: For the FHAT and HAR, only in the comparison of the second to the third farm visit, RS and ICC reached values between 0.45 and 0.56, however, SDC and LoA were >0.10. For the other two comparisons as well as for all comparisons of the NOT and VHAT, no reliability was detected (e.g. VHAT, first to second farm visit: RS: 0.07, ICC: 0.00, SDC: 0.29, LoA: -0.25-0.48). Thus, it can be concluded that despite a sufficient interobserver reliability, there is insufficient consistency over time to make these behavioural tests useful as welfare assessment indicators at present. The specific time effects will be analysed in the proceeding of this study.

Behavioural tests: suitable indicators for measuring the affective state of growing pigs?

K.L. Krugmann, F.J. Warnken, I. Czycholl and J. Krieter
Institute of Animal Breeding and Husbandry Kiel University, Olshausenstraße 40, 24098 Kiel, Germany; kkrugmann@tierzucht.uni-kiel.de

Animal welfare consists of various parts. Relating to the interests of politics and consumers, there is a need to investigate affective states of livestock. This study examined whether a Human Approach Test (HAT) and a Novel Object Test (NOT) are suitable indicators for measuring emotions in growing pigs. Therefore, two batches in three farms (F1, F2 and F3) with different housing systems were tested three times during their fattening resulting in 297 individually tested pigs. To assess the pig's behaviour and thereby the affective state different variables were used: the number of contacts (NC), the duration of contacts (DC) and latencies (LA) to the human or the novel object. A linear mixed model containing the fixed effects farm, sex, point of test (start, middle and end of the fattening) and batch and a random effect for each individual was applied. The model showed a significant influence of farm in the HAT ($P\leq0.001$) resulting in higher DC on F1 than on F2 and F3 (81.5±2.5 vs 15.3±3.1 resp. 30.3±6.9). This farm effect can be explained due to differences in housing systems with barren or enriched habitat. There was also a significant effect of sex in the HAT ($P\leq0.001$) occurring in lower LA in all sows than in the boars (85.3±3.8 vs 102.5±4.4). Supposedly, the sex effect results of the castration of the boars. Habituation can be the reason for the significant point of test effect ($P\leq0.0001$) as the LA in the HAT and NOT declined over time (130.5±4.4 vs 71.8±4.4 resp. 61.1±3.6 vs 33.96±3.6). As expected, the correlations showed negative results between the NC and LA in HAT and NOT (rp=-0.67 resp. rp=-0.52) and within the DC and LA in HAT and NOT (rp=-0.60 resp. rp=-0.46). Further analysis is needed to answer the question whether HAT and NOT are suitable to measure the affective state of growing pigs objectively. Therefore, the play behaviour, tail- and ear postures and physiological parameters (e.g. the salivary protein diversity) of the pigs are further assessed.

Measuring the affective state in pigs: the role of immunoglobulin A

F.J. Warnken[1], K.L. Krugmann[1], I. Czycholl[1], R. Lucius[2], A. Tholey[3] and J. Krieter[1]
[1]Institute of Animal Breeding and Husbandry, Kiel University, Olshausenstr. 40, 24098 Kiel, Germany, [2]Institute of Anatomy, Kiel University, Olshausenstr. 40, 24098 Kiel, Germany, [3]Institute for Experimental Medicine, Kiel University, Niemannsweg 11, 24105 Kiel, Germany; fwarnken@tierzucht.uni-kiel.de

The affective state is an important conception of animal welfare, but reliable indicators to measure this part are still missing. Based on the consumer's perception of the affective state as the most important part of animal welfare, there is a need for objective indicators to measure the affective state of livestock. Referring to human studies, this project aimed at evaluating the suitability of IgA as an indicator for a positive affective state in pigs. The study assessed 288 fattening pigs held in three different housing systems (F1, F2 and F3) in Northern Germany. At the beginning, in the middle and at the end of the fattening period, the animals' behaviour was evaluated using a Human Approach Test (HAT) and a Novel Object Test (NOT), assessing the variables: number of contacts, duration of contacts und latency. Saliva samples were taken shortly before slaughtering and brains and adrenal glands were removed at abattoir. The salivary IgA levels were detected by the use of an ELISA and general salivary protein content (GPC) was determined. A linear mixed model including the fixed effects farm, sex and batch was applied to the salivary IgA levels. F2 showed significant higher IgA values than F1 and F3 (106.2±1.1 vs 48.6±1.1, and 65.8±1.1; P≤0.001) and males had significant higher values then females (77.5±1.1, vs 62.9±1.1; P=0.02). There was no significant influence of batch (1: 69.4±1.1, 2: 70.2±1.1; P=0.89). The sex effect can be explained by hormonal influences. Further, there are different reasons for the farm effect, e.g. emotions or illnesses. There were no correlations within the other parameters supposed to be connected with positive affective states (HAT, NOT, GPC, adrenal glands), therefore further research on the saliva protein composition, the number of astroglia cells and the hippocampal size of the brains is in progress to draw a definitive conclusion.

Session 08 Theatre 7

Effects of an intensified animal care on tail biting during the rearing period

K. Büttner, I. Czycholl, H. Basler and J. Krieter
Institute of Animal Breeding and Husbandry, Christian-Albrechts-University, Olshausenstr. 40, 24098 Kiel, Germany; kbuettner@tierzucht.uni-kiel.de

Tail biting in pigs is a serious welfare problem with multifactorial causes. In this study, the effect of an intensified animal care consisting of positive interactions between humans and animals such as stroking and talking calmly was analysed in depth. Data from 662 piglets in 4 batches separated in two treatment groups were analysed. Tails were not docked and males were not castrated. The trial group differed only in the intensified animal care from the control group which was carried out three times a week by one person for the duration of 15 min in each pen. Once a week the tails were scored regarding tail lesions (no lesions, superficial lesions, small / large lesions), tail losses (original length, partial loss <1/3, partial loss <2/3, partial loss ≥2/3) and tail postures (inconspicuous: curled, lifted; conspicuous: wagging, hanging, jammed). Additionally, a human approach test was performed once a week. For all traits generalized linear mixed models were implemented checking for the fixed effects treatment group, batch, gender and week of age. For tail lesions, tail losses and tail posture, a significant interaction between treatment group and batch could be obtained. The trial group showed in all batches (B) lower estimated frequencies (%) of tail lesions (B1: 20, B2: 22, B3: 30, B4: 44), tail losses (B1: 62, B2: 70, B3: 76, B4: 91) and conspicuous tail postures (B1: 47, B2: 46, B3: 56, B4: 68) compared to the control group (tail lesions: B1: 46, B2: 26, B3: 57, B4: 50; tail losses: B1: 84, B2: 76, B3: 98, B4: 98; conspicuous tail posture: B1: 56, B2: 50, B3: 71, B4: 72). Furthermore, the trial group showed a lower latency in the human approach test compared to the control group with a clear decrease with increasing age. Although the intensified animal care carried out was not able to prevent tail biting completely it was able to reduce it significantly compared to the control group.

Sow behaviour towards humans – an important trait in loose farrowing systems

J. Neu[1], N. Göres[1], J. Kecman[2], H. Swalve[2], B. Voß[3] and N. Kemper[1]
[1]University of Veterinary Medicine Hannover, Foundation, Institute for Animal Hygiene, Animal Welfare and Farm Animal Behaviour, Bischofsholer Damm 15, 30173 Hannover, Germany, [2]Martin-Luther-University Halle-Wittenberg, Institute of Agricultural and Nutritional Sciences, Theodor-Lieser-Str. 11, 06120 Halle, Germany, [3]BHZP GmbH, An der Wassermühle 8, 21368 Dahlenburg-Ellringen, Germany; nicole.kemper@tiho-hannover.de

Due to animal welfare, free-movement-pens for lactating sows gain importance. However, the sows' protective instinct during the lactation period can represent a serious risk for humans handling the animals. Therefore, the aim of the study was to develop tests to characterise sows' behaviour towards humans as a first step for the potential consideration of these traits in breeding goals. The study was carried out on a farm (BHZP GmbH) with purebred landrace sows (db.01), kept in single housing free-movement pens. The period of data collection ranged from 10/2016 -12/2017. Three tests were conducted. Nervousness and fear of novel objects were evaluated with the 'Towel Test' (TT, 531 sows, 1642 observations) by throwing a towel in the direction of the sow's head during a resting period and scoring her reaction in a score from 1 to 4. Aggression towards human interaction was evaluated by the 'Dummy Arm Test' (DAT, 525 sows, 824 observations) with a plastic hand imitation, located near the sow's head while a piglet was animated to squeak, scoring the sows behaviour from 1 to 4. Moreover, a routine situation was simulated with the 'Trough Cleaning Test' (TC, 530 sows, 1602 observations), scoring the reaction of the sow from 1 to 3. TT and TC were carried out approximately on days 4 and 10 post partum with closed and open pens, respectively. All data was analysed using SAS and ASReml 3.0. In DAT, 90.42% of the sows showed a calm reaction (score 1 and 2) towards humans, while only 2.2% bit the dummy arm (score 4). In TT and TC, the sows kept calm, too (percent of calm scores between 62.4 and 75.1). The number of attacking sows was quite low but not zero (ranged from 4.1 to 6.0% in TC and TT). First variance components analyses showed heritabilities of 0.12 (DAT), 0.17 (TT), and 0.17 (TC). These results suggest that a measurable response of sows towards humans exist and a further use of these traits in breeding programs is possible.

Modification of piglet behaviour and welfare by dietary antibiotic alternatives

S.Y.P. Parois[1], J.S. Johnson[2], B.T. Richert[3], S.D. Eicher[2] and J.N. Marchant-Forde[2]
[1]INRA, PEGASE, Agrocampus Ouest, 35590 Saint-Gilles, France, [2]USDA-ARS, LBRU, West Lafayette IN 47907, USA, [3]Purdue University, Animal Sciences, West Lafayette IN 47907, USA; jeremy.marchant-forde@ars.usda.gov

Society is demanding a decrease in prophylactic use of antibiotics in animal production. This could impact animal welfare unless alternatives can be found that confer similar benefits without risk of AMR. The objectives of these studies were to determine whether two alternatives – a probiotic and an amino acid supplement – would impact piglet behaviour and welfare post-weaning. In Experiment 1, 240 weaned piglets were assigned to 3 diets for a 2-wk period post-transport: A – an antibiotic diet including Chlortetracycline + Tiamulin, NA – a control diet, and GLN – a diet including L-glutamine. After the 2-week period, all piglets were fed the same control diet. At weaning, piglets were transported for 12 hours. Tear staining and skin lesions were recorded pre- and post-transport. Novel object tests were done in groups in the pigs' home pen 4 times post-weaning. In Experiment 2, 36 female piglets were assigned to 2 supplement treatments from 24-h to 28 d of age: SYN – a synbioticcontaining *Lactobacillus*, fructo-oligosaccharide and β-glucan in chocolate milk, and CTL – chocolate milk only. Piglets were subject to episodic-like (Object Recognition), working (Barrier Solving) and long-term (T-maze) memory tests. In Expt 1, NA pigs had larger tear stains than A and GLN pigs. NA pigs had more skin lesions post-mixing than A and GLN pigs. In the first novel object test, A pigs avoided the object more than NA pigs. In later tests, NA pigs spent less time exploring the object and took longer to interact with the object than GLN and A pigs. In Expt 2, in the object recognition test, SYN piglets interacted more quickly with the novel object. In the barrier solving test, SYN piglets had shorter distances to finish the test. In the T-maze test, SYN piglets were quicker to learn the task. Overall, the results demonstrate that short-term feeding strategy can have both short- and long-term effects on behaviour and welfare. Supplementation with L-glutamine appears to confer similar benefits to dietary antibiotics and the synbiotic supplement improved piglet cognitive performance.

Session 08 — Theatre 10

The behaviour of low-risk and high-risk crushing sows in free-farrowing pens

C.G.E. Grimberg-Henrici[1], K. Büttner[1], O. Burfeind[2] and J. Krieter[1]
[1]Insitute of Animal Breeding and Husbandry, Christian-Albrechts-University, Olshausenstr. 40, 24098 Kiel, Germany, [2]Chamber of Agriculture of Schleswig Holstein, Gutshof 1, 24327 Blekendorf, Germany; cgrimberg@tierzucht.uni-kiel.de

Pre-weaning mortality of piglets due to crushing is a multifactorial problem and remains an economical and ethical issue. The aim of the present study was to investigate low-risk crushing sows (LRC, ≤20% crushed piglets; n=10) and high-risk crushing sows (HRC, ≥35% crushed piglets; n=10) in free-farrowing pens regarding their lying down and rolling behaviour and their piglets' positions during postural changes in the first 72 hours post partum. Video observations (lying down and rolling behaviour, piglets' positions) were analysed to obtain information about critical situations of piglets being crushed. A generalised linear model was used with the fixed effects group (LRC, HRC), parity class (1: 1; 2: 2-4; 3: ≥5) and the interaction between group and parity class. Relationships were tested with the Spearman Correlation Coefficient. HRC and LRC sows did not differ in their frequency of lying down movements. However, HRC sows performed more lying down movements without using the pen walls (LRC: 7.29 vs HCR: 12.0; P<0.05). Furthermore, piglets of LRC sows were more active during lying down, which was negatively correlated (r=-0.56) with the number of crushed piglets (P<0.05). Moreover, HRC sows rolled more (LRC: 24.6 vs HRC: 47.6; P<0.05) especially with more rolling movements from one side to the other (LRC: 1.18 vs HCR: 8.41; P<0.05). These were highly correlated (r=0.81) with the number of crushed piglets (P<0.05). Furthermore, piglets of LRC sows were more active and synchronous in their behaviour during rolling movements (P<0.05). The activity of piglets of LRC sows during lying down and rolling movements may indicate more interaction of LRC sows with their piglets during postural changes. The safest place for piglets is the nest, however, piglets of HRC sows were more frequent in the nest during postural changes (P<0.05), it did not decrease the incidence of crushing. In conclusion, the detailed observation of HRC and LRC sows in free-farrowing pens showed high variation in their maternal behaviour and in their postural changes in free-farrowing pens.

Session 08 — Theatre 11

Individual feed intake during lactation as a trait to improve animal welfare

N. Göres[1], J. Neu[1], B. Voß[2] and N. Kemper[1]
[1]University of Veterinary Medicine Hannover, Foundation, Institute for Animal Hygiene, Animal Welfare and Farm Animal Behaviour, Bischofsholer Damm 15 (Building 116), 30173 Hannover, Germany, [2]BHZP GmbH, An der Wassermühle 8, 21368 Dahlenburg-Ellringen, Germany; nina.goeres@tiho-hannover.de

During lactation sows' body reserves need to be mobilized if the energy needed is not covered by feed intake Subsequent weight losses can negatively affect animal health, wellbeing and biological performance. The aim of this study was to investigate individual feed intake during lactation in relation to changes in body condition as a trait to evaluate and improve animal welfare and performance. The study was carried out at a basic breeding operation (BHZP GmbH) with purebred landrace db.01 sows, kept in single housing free-movement pens. From October 2016 until January 2018 n=776 litters in 21 batches were observed. Sows' bodyweight (BW), backfat thickness (BFT) and Body Condition Score (BCS, Scores 1-5) were documented when entering the farrowing system, 12-36 hours after farrowing, and when leaving the farrowing unit. All piglets, including stillborn, were counted and weighed after birth. The individual feed intake per sows was recorded daily. The influence of batch and parity as fixed effects on feed-intake and changes in body condition were analysed using PROC MIXED of SAS. On average the sows were fed 4.95 (±0.57 SD) kg feed per day and lost 24.04 (±13.70) kg BW, 0.46 (±0.66) points in BCS and 2.53 (±0.17) mm in BFT between farrowing and weaning. They had 15.76 (±3.61) piglets with a total litter-weight of 19.75 (±4.50) kg. Parity showed a significant effect on daily feed-intake, weight-loss and BCS-change as well as on litter size and total litter weight (P<0.0001). Correlations between post-partum (pp.) weight-loss and feed-intake as well as changes in BCS and BFT were moderate (r_p=0.35, 0.34, 0.36, respectively) and highly significant (P<0.0001). Due to cross-fostering, correlations between litter size and weight-loss pp. were low (r_p=0.13, P=0.0002). The results indicate that breeding for improved feed intake, or rather intake of metabolizable energy, is an important step to optimize animal health and performance during lactation period.

Performance of rabbit does and litters kept collectively with different management or individually

M. Birolo[1], C. Zomeño[2], F. Gratta[1], A. Trocino[2], A. Zuffellato[3] and G. Xiccato[1]
[1]University of Padova, Department of Agronomy, Food, Natural resources, Animals and Environment, Viale dell'Università 16, 35020 Legnaro, Padova, Italy, [2]University of Padova, Department of Comparative Biomedicine and Food Science, Viale dell'Università 16, 35020 Legnaro, Padova, Italy, [3]A.I.A. Agricola Italiana Alimentare S.p.A, Piazzale Apollinare Veronesi, 1, 37036, San Martino Buon Albergo, Verona, Italy; cristina.zomenosegado@unipd.it

The European Parliament asks for housing rabbits collectively to comply with current societal demands for animal welfare, but first results from group-housed reproducing females show several weaknesses. This study assessed the effect of housing and group management on doe and kit performance throughout a reproductive cycle. Sixty pregnant multiparous rabbit does were housed in collective pens (C) (2.0 m^2; 4 connected individual pens) (48 does) or individual pens (I) (0.5 m^2) (12 does). All C does were kept in groups from 9 d to 2 d before kindling and then individually until re-grouping (2 d or 12 d after kindling, 24 does per treatment). Within re-grouping time, half of the pens maintained a stable group; the other half changed one doe every week. Litters were standardized to nine kits. Data were analysed by two models: (1) with housing system as fixed effect and pen as random effect (all data); (2) with grouping time, group composition, and their interaction as fixed effects, and pen as random effect (collective data). Doe milk production (244 g/d vs 262 g/d, P=0.10) and litter growth (131 g/d vs 142 g/d, P=0.05) from 3 d to 19 d were lower in C than in I pens. Thus, C litters were lighter at 19 d (2,741 g vs 2,914 g; P<0.10) and at weaning (7,281 g vs 7,916 g, P<0.05). Does re-grouped 2 d after kindling and with variable group composition showed the lowest milk production from 3 d to 12 d (192 vs 214 g/d on average; re-grouping time × group composition interaction, P<0.10). Their litters had the lowest weights at 19 d (2,584 vs 2,794 g on average; P<0.10), but differences disappeared at weaning (33 d after kindling). In conclusion, under our conditions, collective housing impaired litter performance; grouping time and composition had weak effects on doe and litter performance.

The relationship between the status of pigs detained ante-mortem and their meat inspection outcome

D. Lemos Teixeira[1], J. Gibbons[2], A. Hanlon[2] and L. Boyle[3]
[1]Pontificia Universidad Católica de Chile, Departamento de Ciencias Animales, Santiago, Chile, [2]University College Dublin, School of Veterinary Medicine, Dublin, Ireland, [3]TEAGASC, Pig Development Department, Fermoy, Ireland; laura.boyle@teagasc.ie

Slaughter pigs are detained ante-mortem (AM) for closer inspection in the lairage if there are concerns about their health and welfare. The aim of this study was to investigate the relationship between the AM status of detained pigs and the outcome of their post-mortem inspection (PMi). Data on all 5,055 pigs detained in an Irish abattoir during 240 slaughter days in 2013 was compiled from handwritten AM (reasons for detaining pigs) and PMi reports. The association between reasons for AM detention and the outcome of the PMi was analysed using univariate binomial logistic regression models. 99.9% pigs were passed as fit for slaughter. Lameness (45.2%), stress (11.8%), recumbency (10.4%), hernias (9.8%) and tail lesions (7.8%) were the main reasons for detaining pigs. 57.4% of pigs were fully passed as fit for human consumption at PMi, 15.1% of carcasses were fully and 27.1% were partially condemned. Of the 2,130 carcasses that were condemned 47.7% had reasons recorded. Abscess was the most common reason for CC. Significant relationships were found between reasons for AM detention and the likelihood of CC. Pigs detained for abdominal distension, external abscesses and tail lesions were 9.4, 5.0 and 1.7 times more likely to be fully condemned than to pass PMi, respectively. Similarly, pigs detained for recumbency, external abscess, lameness and limb swellings were respectively 6.9, 3.0, 2.2 and 2.0 times more likely to be partially condemned than to pass PMi. Significant relationships were also found between reasons for AM detention and reasons for CC. There was a high pass rate of pigs detained AM as fit for slaughter in spite of a high likelihood of CC associated with conditions recorded AM. This offers producers an incentive to send such pigs for slaughter instead of euthanising them on farm. Standardised recording and thereafter informing producers on reasons for AM detentions has, together with PMi information, potential for use as a tool to improve pig health and welfare and losses associated with CC.

Pastured organic rabbit farming: growth of rabbits under different herbage allowance and quality

H. Legendre[1], G. Martin[2], J. Le Stum[3], H. Hoste[4], J.P. Goby[3] and T. Gidenne[1]
[1]INRA, GenPHySE, Phase, BP 52627, 31326 Castanet-Tolosan, France, [2]INRA, AGIR, EA, BP 52627, 31326 Castanet-Tolosan, France, [3]Université de Perpignan, IUT, chemin Passo Vella, 66962 Perpignan, France, [4]INRA, ENVT, IHAP, 23 Ch. des Capelles, 31076 Toulouse, France; h.hoste@envt.fr

Organic rabbit farming is developing in France, but growth performances and herbage intake are still slightly documented. Our study aimed to describe rabbit herbage intake under a wide range of grazing conditions and to characterise the factors that control herbage intake and growth. Three trials were performed in winter, summer and spring, using growing rabbits reared in moving cages (0.4 m^2 of grazing area per rabbit, 2 groups of 5 cages per season with 3 rab. per cage), to compare two 'types' of pasture dominated by legumes (LEG) or grass/forbs (GRF). Each trial began at weaning (45 d old) when rabbits were transferred to moving cages, and ended at slaughter (100 d old), and they received 60 g/d/rabbit of a complete pelleted feed. Mean herbage allowance was 27% higher in LEG (62.3 g DM/kg metabolic weight (MW), equal to kg0.75) than in GRF (49.2 g DM/kg MW). For both pasture types, herbage intake was logarithmically related to herbage allowance and plateaued around 75 g DM/kg MW. Mean total intake was 70.3±19.5 g DM/kg MW, of which half was pelleted feed. Crude protein (CP) and digestible energy (DE) and intake differed by pasture type and season. Mean CP intake was 50% higher in LEG (15.0 g/kg MW) than in GRF (10.7 g/kg MW). In summer, mean DE intake was 27% higher in LEG than in GRF but no significant differences in DE intake were found between LEG and GRF in winter and spring. Maximum DE intake plateaued near 1000 kJ/kg MW. Daily weight gain was always higher for rabbits grazing LEG (mean=22.6 g) than GRF (mean=16.0 g). CP intake was significantly related to weight gain, while DE intake had no significant relations. Meeting the objective of mean daily weight gain of 20 g requires herbage intake of 32 and 50 g DM/kg MW in LEG and GRF, respectively. Therefore, according to the herbage use efficiency observed in our experiments, herbage allowance must reach 41 and 76 g DM/kg MW in LEG and GRF, respectively. When herbage allowance is lower, rabbits cannot meet the CP intake (13 g/kg MW) required for the weight gain objective.

Effect of fattening management on animal welfare of Iberian pigs

J. García-Gudiño[1], A. Velarde-Calvo[1], M. Font-I-Furnols[1] and I. Blanco-Penedo[2]
[1]IRTA, Animal Welfare and Product Quality Programs, Finca Camps i Armet, 17121 Monells, Spain, [2]SLU, Department of Clinical Sciences, Almas Allé 8, 75007 Uppsala, Sweden; javier.garciag@irta.es

Iberian pig is a local breed located in the southwest of the Iberian Peninsula. The fattening management varies from extensive to intensive conditions. The aim of the present work was to evaluate the effect of both fattening managements on animal welfare. Fourteen farms (seven of each fattening management) were assessed with an adaptation of the Welfare Quality® protocol during the finish period. The welfare assessment was based on three welfare principles (good feeding, good housing and good health). At the slaughterhouse, skin and tail lesions of the pigs from nine farms (four intensive and five extensive) were scored. Statistical analysis was performed using SAS software. Preliminary results indicated that extensive pigs presented a significantly ($P<0.05$) lower frequency of bursitis, lameness and respiratory disorders. Moreover, mortality rate was higher on intensive farms ($P<0.05$). No significant differences were found in skin and tail lesions in slaughtered. However, the stocking density was a determinant factor because the intensive farm with lower space allowance showed the highest percentages of lesions. Generally, extensive systems had better results than intensive systems at on-farm assessment (absence de manure on the body, huddling, tail biting, scouring, skin condition and hernias). In the conditions of this experiment, it can be concluded that animal welfare in Iberian extensive farms is greater than in Iberian intensive farms during the fattening period. Acknowledgements: INIA and Inga Food S.A.

Session 08 Poster 16

Effects of a novel housing system for fattening rabbits on skin injuries, daily gain and hygiene

S.L. Rauterberg, J. Bill, N. Kemper and M. Fels
University of Veterinary Medicine Hannover, Foundation, Institute for Animal Hygiene, Animal Welfare and Farm Animal Behaviour, Bischofsholer Damm 15, 30173 Hannover, Germany; sally.rauterberg@tiho-hannover.de

Increased demands on animal welfare also require adaptations of conventional rabbit housing. In the present study, an innovative housing system for fattening rabbits was developed which was expected to have positive impact on animal welfare. The study was conducted on a commercial rabbit farm, where rabbits were kept under innovative housing conditions (IC) or under conventional conditions (CC). IC housing was characterized by large groups with up to 65 animals, slatted plastic floor (812 cm^2 per animal), and environmental enrichment such as elevated platforms, plastic tubes and different gnawing materials. IC rabbits were born in this system and were weaned at the age of 31 days. At weaning, the does were removed and up to six litters were mixed and remained in the system until slaughtering at the age of 78 days. Rabbits from CC housing were born in conventional wire cages. After weaning, they were moved to new cages and mixed in groups of eight. These cages were equipped with wire mesh floor (428 cm^2 per animal), an elevated platform and one piece of wood. The occurrence of skin lesions and pododermatitis as well as daily weight gain of 324 fattening rabbits were analysed from weaning to slaughter in a total of three batches. Furthermore, cleanliness of feet and floor was investigated using scoring systems. IC rabbits showed higher daily weight gains than CC rabbits (46.61 vs 43.17 g, $P<0.001$), while the skin lesion score was lower in rabbits from IC than from CC ($P<0.05$) at all observation times. The cleanliness of feet and floor differed between the housing systems at any time with CC being cleaner than IC ($P<0.001$). In both housing systems, pododermatitis was not observed. The number of animal losses was not different between IC and CC system (8.4 vs 8.0%), except for batch 2 where mortality was higher in IC than in CC because of diarrhoea (19.7 vs 6.8%). A lower incidence of injuries and higher daily weight gain in IC rabbits may indicate increased welfare in rabbits from innovative housing conditions. However, this housing system poses some hygienic challenges and further adjustment is necessary.

Session 08 Poster 17

Early detection of tail biting from behavioural changes in finisher pigs

M.L.V. Larsen[1], H.M.-L. Andersen[2] and L.J. Pedersen[1]
[1]Aarhus University, Department of Animal Science, Blichers Allé 20, 8830 Tjele, Denmark, [2]Aarhus University, Department of Agroecology, Blichers Allé 20, 8830 Tjele, Denmark; lene.juulpedersen@anis.au.dk

Tail biting is an animal welfare problem. One strategy to prevent tail biting from developing into serious tail damage is to detect tail biting at an early stage through for example behavioural changes observed prior to tail damage. This could inform the farmer of pens in high risk of developing tail damage and make timely intervention possible. The aim of the current study was to investigate the development in pigs' activity level, object manipulation and tail posture prior to an event of tail damage to possible later use these three behaviour types in machine learning algorithms implemented in a Precision Livestock Farming tool. The study included 112 finisher pens out of which 55 developed an event of tail damage (day0; at least one pig in the pen with a bleeding tail wound). Each tail damage pen was matched with control pens that was never scored with an event of tail damage. Pigs' activity level and object manipulation (directed towards two hard wooden sticks) was observed from video recordings the last seven days prior to day0 from 06:00-08:00 h, 16:00-18:00 h and 22:00-24:00 h (only activity level). Pigs' tail posture was observed from video recordings the last three days prior to day0 from 08:00-11:00 h and 15:30-18:30 h. All three behavioural types was analysed using generalised linear mixed models. Activity level developed differently for the tail damage and control pens, resulting in a higher activity level in the tail damage pens the last five days prior to day0. Object manipulation and tail posture did neither increase nor decrease prior to day0, but a general higher manipulation rate and a higher probability of having a hanging tail, respectively, was seen in the tail damage pens. Thus, all three behaviour types seem related to on-going tail biting, but not sufficiently to be conclusive on their abilities to be early detectors of tail biting. Instead, it is suggested to use the tail damage pens as their own controls and focus on changes from the normal behaviour pattern of that specific pen. This strategy demands that the normal behaviour pattern is known which in turn demands automatic monitoring methods for the behaviour of interest.

Session 08 Poster 18

Livestock animal welfare status and certification standard in Korea

J.Y. Lee[1], F. Leenstra[2], S.E. Woo[1], D.H. Lee[1], K.S. Kwon[1], J.H. Jeon[1], H.C. Choi[1] and K.Y. Yang[1]
[1]National Institute of Animal Science, Animal Environment Division, Wanju-gun, 55365, Korea, South, [2]Wageningen UR, Wageningen Livestock Research, De Elst 1, P.O Box 338, the Netherlands; andrerwlee@korea.kr

In Korea, the livestock industry accounts for about 45% of the agricultural sector. The size of livestock industry has been developed in quantitative terms, but it is still lower than that of advanced countries in terms of quality such as productivity. In recent years, consumer confidence in animal products has been decreasing steadily due to outbreak of disease such as FMD and AI and animal product safety issues such as insecticide eggs. In recent years, there has been constantly demand for improvement of livestock production systems. Therefore, in 2011, the Korean government enacted the Animal Welfare Livestock Farm Certification System in Animal Protection Act in the Asian countries for the first time in order to increase animal welfare level of livestock. By this Act, animal welfare certification system was introduced from 2012 to 2015 for laying hens, pigs, broilers, dairy and beef cattle, and goats. Currently, the proportion of certified animal welfare farms is 10% for laying hens, 2% for broilers and 0.3% for pigs based on the number of livestock farms, respectively. The main reason for low proportion of certified farms is that farmers prefer highly profitable traditional farming system. In the animal welfare standard, use of battery cages for laying hens are prohibited and only alternative non-cage system such as floor housing w/ or w/o aviary and free-range are allowed. Enriched cage system allowed in EU is not allowed in Korea. The requirement for drinker, feeder, perches, nests and floor surface for laying hen in Korean standard is similar to those in EU legislation. The stock density must not exceed 9 laying hens (17 hens in aviary) per m^2 usable area. Beak trimming and forced molting are prohibited. The stall is prohibited and sows and gilts must kept in groups with space allowance 3.0 and 2.3 m^2 respectively (2.25 and 1.64 m^2 in EU). After 5 days of farrowing, the sow should be able to move freely. The tail docking and tooth clipping are prohibited but castration is allowed.

Session 08 Poster 19

Effect of the different parity on the lying behaviour of farrowing sows

K.Y. Yang, S.E. Woo, D.H. Lee, K.S. Kwon, H.C. Choi, J.H. Jeon and J.Y. Lee
National Institute of Animal Science, Animal Environment Division, Wanju-gun, 55365, Korea, South; andrewlee@korea.kr

The time spent lying and the lying position are significant behavioural indicators of sow comfort and welfare in conventional pig husbandry. It was known that lower parity sows change posture in the pre-partum period 4-fold more frequently often than that in postpartum. Different parity are one of the important factors for studying the lying behaviour and reproductive performance of sows. This study was conducted to compare the effect of different parities on the lying behaviours before and after farrowing with 8 crossbred sows. Sows were housed in farrowing pens (W 2.2 × D 1.8 × H 1.2 m) on the partially slatted plastic floor. The farrowing crates were equipped with a nipple drinker, a feed trough, and a heater on the right side of sows. The sows used in this study were between the first and sixth parity in parity1-2(P1), parity 3-5(P3), and parity 6-7(P6), respectively. Lying behaviours were classified into lying on the left side (LL), lying on the right side (LR), and lying on the abdomen (LA) behaviours. Behavioural images were recorded for 24 h before and after farrowing using IR camera. The overall lying behaviour of P1, P3, and P6 was 65% (24% on the LL, 34% on the LR, and 7% on the LA), 76% (38% on the LL, 24% on the LR, and 14% on the LA), and 70% (36% on the LL, 16% on the LR, and 18% on the LA) respectively. On the other hand, P1, P3, and P6 were on the after farrowing lying behaviour of 96% (40% on the LL, 46% on the LR, and 10% on the LA), 97% (42% on the LL, 32% on the LR, and 24% on the LA), and 98% (41% on the LL, 29% on the LR, and 28% on the LA). Therefore, the LL behaviour before and after farrowing of P1 was higher than P3 and P6 ($P<0.05$). The LR behaviour before and after farrowing of P3 and P6 were higher than P1 ($P<0.05$). The number of piglets born alive were 15.5±1.5 in P1, 14±1.0 in P3, and 12.7±0.9 in P6 ($P<0.05$). In conclusion, the data showed that lying behaviour pattern of sow in farrowing crate before and after farrowing could be affected by different parities. As increase parity, in the spent lying on the right side facing the udder to the piglets was observed after farrowing, which could be explained by maternal behaviour for piglets.

Welfare of highly productive and native breeds of pigs kept in overstocked conditions

J. Walczak[1], W. Krawczyk[1], M. Pompa-Roborzyński[1] and M. Sabady[2]
[1]National Research Institute of Animal Production, Sarego 2, 31-047 Kraków, Poland, [2]Veterinary Inspectorate, Wilsona 21, 97-500 Radomsko, Poland; jacek.walczak@izoo.krakow.pl

The experiment used 84 sows and 870 fatteners of the Puławska (Pu) breed as well as Polish Landrace or Polish Large White (Pl) commercial crossbreds. The sows and fatteners of each breed were overstocked at 130% of the directive recommendations (2008/120/EC). In terms of the production results, the native breed responded positively to the modification by increasing initial piglet weight (1.31 kg) in relation to comparable animals kept under standard conditions (1.16 kg). Unfortunately, no such differences were observed for piglet weaning weights, body weight gains, and mortality. Statistically significant differences for each breed were obtained for the number of live born piglets, to the advantage of those kept in larger space (Pu 10.71 vs 10.92 kg; Pl 10.81 vs 11.1 kg)). The additional living area of the sows contributed to an increase in moving activity (Pu 16.5 vs 21.2%; Pl 13.6 vs 18.3%) and a decrease in lying time in the daily behavioural time budget of all the breeds. Significant differences were also found concerning decreased stereotypies in each breed (PU 4.3 vs 0.9%; Pl 3.1 vs 1.1%). All sows with greater space allowance showed significantly lower levels of stress hormones (Pu cortisol 90.4 vs 81.4 nmol/l; ACTH 53.2 vs 48.3 pg/ml) and elevated T4 concentration (PU 3.31 vs 3.7 µg/dl). The lower stress level in these animals also improved morphotic blood elements. In the case of fatteners, significantly higher weight gains were only noted for Pu in larger space. Mortality was also lower. The behavioural response of these animals was identical to that of sows. Stress hormone levels were also beneficial and reached significantly lower values in animals kept in larger space (Pu cortisol 36.8 vs 32.7 nmol/l; ACTH 21.1 vs 17.3 pg/ml). Such a beneficial effect was only observed for erythrocyte count in all breeds. In addition, the native breed had lower lymphocyte percentage under these conditions (Pu 32 vs 30%) The results of the present study clearly show that the modified housing conditions had a positive effect on improving the welfare of pigs. This response was strongest for the native breed.

Welfare indicators and gross pathological findings in euthanised pigs from hospital or home pens

J.A. Calderon Diaz[1], P. O'Kelly[2,3], A. Diana[1,3], M. MacElroy[2], E.G. Manzanilla[1], J. Moriarty[2], S. McGettrick[2] and L.A. Boyle[1]
[1]Teagasc Moorepark, Pig Development Dpt., Fermoy, P61 C996, Ireland, [2]DAFM, Central Veterinary Research Laboratory, Backweston, W23 X3PH, Ireland, [3]UCD, School of Veterinary Medicine, Dublin, D04 V1W8, Ireland; julia.calderondiaz@teagasc.ie

The aim of this study was to characterise welfare and pathologies in growing pigs selectively euthanized from home (HOME) and associated hospital (HOSP) pens on a 1,500 sow farrow-to-finish farm from August to November 2015. Pigs were selected for euthanasia from HOME (n=29) pens on the basis of clinical abnormalities (e.g. poor body condition [PBC], lameness, hernias) and from HOSP (n=27) pens on signs of severe clinical illness. Pigs were euthanised by an overdose of sodium pentobarbital by a veterinarian. The presence of lesions to the tail (TL), ear (EL) and body (BL) and PBC was recorded at euthanasia. Within 24 h a pathologist weighed the carcasses and carried out necropsies to classify gross pathologies. Data were analysed in SAS. HOSP pigs were lighter (6.6±3.49 vs 21.8±14.06 kg; $P<0.05$) and more tended to be in PBC (52 vs 21%; $P=0.07$) than HOME pigs. There was no difference ($P>0.05$) between HOME and HOSP pigs in the prevalence of EL (58.5%) and TL (48%). A higher ($P<0.05$) % of pigs with BL were observed in HOME (72%) compared to HOSP (30%) pens. There was no difference ($P>0.05$) between HOME and HOSP pigs in the prevalence of respiratory (53.5%), gastrointestinal (39%) and heart (17.5%) pathologies. The high prevalence of EL and TL in both cohorts raises welfare concerns and indicates that pigs are not hospitalised for such conditions. It is not surprising that fewer HOSP pigs were affected by BL as these are associated with aggression which sick pigs do not engage in. Poorer weight and body condition of these pigs was consistent with their severe clinical appearance so the lack of differences in gross pathological findings is surprising. Results suggest that on this farm, compromised grower pigs were allowed to remain in the home pens. These findings raise concerns for the management of health and welfare compromised pigs on large commercial farms.

Removal of haplotype segments originating from foreign breeds using optimum contribution selection

R. Wellmann, Y. Wang and J. Bennewitz
University of Hohenheim, Institute of Animal Science, Farm Animal Genetics and Breeding, Garbenstraße 17, 70599 Stuttgart, Germany; r.wellmann@uni-hohenheim.de

High performance livestock breeds have often been used to upgrade local breeds. This displacement crossing can progress to the point where the original genetic background of the local breed vanishes. Hence, besides making the breed profitable, breeding programs for local breeds often have the additional objective to de-extinct these breeds by the enrichment of native haplotype segments. Unfortunately, the native genetic contributions are often negatively correlated with total merit. Moreover, the native effective size, i.e. the effective size at native alleles, is typically smaller than the effective size, meaning that the homozygosity at native segments increases at a faster rate. Advanced optimum contribution selection methods have been developed to account for these issues. They are implemented in the R package optiSel. Simulation studies using genotype data of Angler cattle reveal that maximizing the genetic contribution from native ancestors while restricting the rate of inbreeding is not the optimal approach because this may lead to the loss of other rare alleles and thus decreases the genetic diversity across breeds. Instead, constraining the kinship across breeds and the kinship at native alleles can be recommended. In this case, the objective is either to maximize the genetic contribution from native ancestors or to maximize a selection index in which an appropriate weight is given to the native genetic contribution. However, genetic progress can be substantially lower in breeding programs that aim to recover the original genetic background and removal of the introgressed genetic material requires many generations of selection.

Phylogenetic analysis of mitochondrial DNA in the East Adriatic goats

I. Drzaic, D. Novosel, I. Curik and V. Cubric-Curik
University of Zagreb, Faculty of Agriculture, Department of Animal Science, Svetosimunska 25, 10000 Zagreb, Croatia; ikovac@agr.hr

In human history, domestic goat has been an important source of milk, meat, skin and fibre. Goats have spread into Europe from the Fertile Crescent through the Danubian and Mediterranean corridors during the Neolithic agricultural revolution. Croatian Spotted goat is an 'old' indigenous breed representing goat populations from the East Adriatic. Here, we sequenced a 660-bp fragment from the D-loop of mitochondrial DNA in 25 Croatian Spotted goats(CSG) randomly sampled from Eastern Adriatic part of Croatia. In addition, we retrieved goat, ancient goat and sheep sequences from GenBank, thus, providing three data sets used in the calibration of the molecular clock and phylogenetic analysis. All CSG individuals have clustered in haplogroup A while high haplotype diversity (0.967) and several unique haplotypes were observed. Principal component analysis based on Reynolds distance matrix revealed CSG to be similar, with exception of Swiss goats, to neighbouring Mediterranean goat populations (Albania, Italy and Greece). This pattern was also confirmed by the evolution analysis using the most common ancestor tree that was calculated in BEAST 2.4.8 software package. Phylogenetic analysis of CSG confirmed heterogenic origin and suggested that at least one population expansion happened in the breed history. Our future analysis will be based on Caprine remains from numerous archeological sites along East Adriatic.

Population structure and genetic diversity of Drežnica goat from Slovenia: preliminary results

M. Simčič[1], I. Medugorac[2], D. Bojkovski[1] and S. Horvat[1]
[1]University of Ljubljana, Biotechnical Faculty, Jamnikarjeva 101, 1000 Ljubljana, Slovenia, [2]LMU Munich, Faculty of Veterinary Medicine, Veterinaerstr. 13, 80539 Munich, Germany; mojca.simcic@bf.uni-lj.si

The aim of the study was to obtain unbiased estimates of the genetic diversity parameters, population structure, inbreeding level and possible admixture in the autochthonous Drežnica goat in Slovenia. This breed suffered strong decrease after the Second World War due to demographic aspects, prohibitions of goat grazing, increased demand for food and consequently governmental policies to support cattle breeding. Drežnica goat are multi-coloured, hardy, adaptable animals reared in the Alpine climate and mountainous pastures while maintaining good health and efficient production – dairy does produce 350 kg of milk in 200 days of lactation with 4.3% fat and 3.4% proteins. Today, a small population of 629 breeding animals is at high risk of extinction also due to a small area with a radius of less than 30 km in Slovenian western Alps where they are reared. Genetic analyses were performed on the genome-wide Single Nucleotide Polymorphism (SNP) Illumina Caprine SNP50 array data of 96 animals of Drežnica goat and 577 animals from 13 reference populations representing ten Alpine goat breeds from Switzerland and three Angora breeds as outgroups. To obtain unbiased estimates we used short haplotypes spanning four markers instead of single SNPs to avoid an ascertainment bias of the array. The average population inbreeding coefficient based on unified additive relationship was 0.114. Phylogenetic analyses demonstrated distinct genetic identity of Drežnica goat with a very high number of private alleles. Genetic distance matrix presented by *Neighbour-Net* revealed an independent origin of the breed also confirmed by the *ThreeMix* analysis. *Multivariate outlier test* showed only a small number of outliers in Drežnica breed. Animals identified as the most purebred represent an important genetic nucleus for the conservation of the autochthonous genetic background. We conclude that phenotypic selection and environmental-natural selection pressure acting over several centuries resulted in the well-adapted dual-purpose breed able to produce exclusive products based on grazing in Alpine pastures.

Genetic diversity of Tibetan Terrier

M. Janes[1], M. Zorc[2], V. Cubric-Curik[1], I. Curik[1] and P. Dovč[2]
[1]University of Zagreb, Faculty of Agriculture, Department of Animal Science, Svetosimunska 25, 10000, Croatia, [2]University of Ljubljana, Faculty of Biotechnology, Department of Animal Science, Groblje 3, 1230 Domžale, Slovenia; mateja.janes86@gmail.com

The Tibetan Terrier (TT) is a breed believed to be more than 2,000 years old and to descend from one of the oldest breed of dogs. In European TT population two lines were established – Lamleh and Luneville. The European population represents a rather narrow gene pool, which has been enriched by several outbreeding episodes with dogs of similar appearance during the history of the breed. We were recently able to collect an exclusive sample of native TT from Tibet, European samples of Lamleh and Luneville lineage as well as F1, F2 and F3 generations of crosses between native population and Lamleh lineage. We analysed 24 Tibetan Terrers by CanineHD Illumina BeadChip and 48 Tibetan Terriers by 18 microsatellite loci and mtDNA D-loop region. The aim of this study was to provide an assessment of the genetic structure of TT, their relationship with other breeds from the Terrier and Companion and Toy Dogs groups and to trace back their origin and history using HD density SNP chips, control region of mtDNA and microsatellites. We confirmed the assumption that European TT population not only represent a small part of genetic variation present in the native population of TT, but that native TT are genetically more similar to other Tibetan breeds from FCI Companion and Toy Dog Group than to other breeds from FCI Terrier Group.

Session 09 Theatre 5

Genomic diversity using copy number variations in worldwide chicken populations

E. Gorla[1], F. Bertolini[2], M.G. Strillacci[1], M.C. Cozzi[1], S.I. Roman-Ponce[3], F.J. Ruiz[3], V.V. Vega[3], C.M.B. Dematawewa[4], D. Kugonza[5], A. Elbeltagy[6], C.J. Schmidt[7], S.J. Lamont[2], A. Bagnato[1] and M.F. Rothschild[2]
[1]University of Milan, Department of Veterinary Medicine, Via Celoria 10, 20133 Milano, Italy, [2]Iowa State University, Department of Animal Science, 1221 Kildee Hall, 50011 Ames, USA, [3]Instituto Nacional de Investigaciones Forestales, Agrícolas y Pecuarias, Progreso 5,Barrio de Santa Caterina, 04010 Delegación Coyocan, Mexico, [4]University of Peradeniya, Labuduwa Sri D. Mawatha, 20400 Peradeniya, Sri Lanka, [5]Makerere University, np, Kampala, Uganda, [6]Animal Production Research Institute, Nadi El Said, Dokki, Egypt, [7]University of Delaware, 531 S College Ave, 19716 Newark, DE, USA; erica.gorla@unimi.it

Recently, many studies in livestock have focused on the identification of Copy Number Variants (CNVs) using high-density Single Nucleotide Polymorphism (SNP) arrays, but few have focused on studying chicken ecotypes. CNVs are polymorphisms, which may influence phenotype and are an important source of genetic variation in populations. The aim of this study was to explore the genetic structure and peculiarities, using a high density SNP chip in 936 individuals from seven different countries (Brazil, Italy, Egypt, Mexico, Rwanda, Sri Lanka and Uganda). The DNA was genotyped with the Affymetrix Axiom®600k Chicken Genotyping Array and then processed with stringent quality controls to obtain 559,201 SNPs in 915 individuals. The Log R Ratio and the B Allele Frequency of SNPs were used to perform the CNV calling with PennCNV software based on a Hidden Markov Model analysis. A total of 15,787 CNVs were detected of which 9,081 were duplications and 6,706 deletions. A total of 5,679 CNVRs (Copy Number Variant Regions) were obtained using the software BedTool (-merge command) that allowed us to merge CNVs which overlap for at least 1 bp. The intersection analysis performed between the chicken gene database (Gallus_gallus-5.0) and the 5,679 CNVRs have allowed the identification (within or partial overlap) of 8,849 Ensembl gene IDs, corresponding to 5,945 genes with an official gene ID. The CNVRs identified here represent the first comprehensive mapping in a large worldwide populations, using high-density SNP chip.

Session 09 Theatre 6

Analysis of a large cattle data suggests ZFAT has pleiotropic effects on cattle growth and lethality

J. Jenko[1], M.C. McClure[2], D. Matthews[2], J. McClure[2], G. Gorjanc[1] and J.M. Hickey[1]
[1]The Roslin Institute and Royal (Dick) School of Veterinary Studies, The University of Edinburgh, Easter Bush, Midlothian, EH25 9RG, United Kingdom, [2]Irish Cattle Breeding Federation, Bandon, Co. Cork, P72 X050, Ireland; janez.jenko@roslin.ed.ac.uk

Infertility has a major effect on profitability in cattle farming across the globe. Due to the strong selection on productive traits and increased frequency of recessive lethal alleles, fertility deteriorated in many countries over the last decades for both dairy and beef production systems. Recessive lethal alleles cause early embryonic death or reduced viability when an individual is homozygous. To search for putative recessive lethal haplotype alleles, genotypes from five main Irish beef cattle breeds (Aberdeen Angus, Charolais, Hereford, Limousin, and Simmental) were extracted from the Irish Cattle Breeding Federation database that contains more than one million genotypes of different densities. The obtained genotypes were imputed and phased using a sliding window approach. Seven different methods were used for the detection of putative recessive lethal haplotype alleles. Their lethality was further on tested on phenotype and breeding records from the Irish Cattle Breeding Federation, specifically artificial insemination, calving, and postnatal survival data. We found four statistically significant putative recessive lethal haplotype alleles that were supported by phenotypic data. Three of them were located between the 7,514,610 and 9,345,140 base pair on chromosome 14 of Aberdeen Angus and one between the 51,738,339 and 53,354,164 base pair on chromosome 16 of Simmental cattle breed. All putative recessive lethal haplotype alleles showed pleiotropic effects on economically important traits in beef production. Genome annotation suggests ZFAT as a causal gene for lethality in Aberdeen Angus. Several candidate genes were identified for the Simmental putative recessive lethal haplotype allele potentially causing prenatal or perinatal lethality. Efforts to identify a causal variant in the genome are ongoing. Finally, implementation of these haplotype alleles into breeding program will improve the breeding success and the mean fitness of the population.

The CRB-Anim web portal: access to biological resources for animal sciences

S. Marthey[1], A. Delavaud[2], N. Marthey[1] and M. Tixier-Boichard[1]
[1]INRA, AgroParisTech, Paris-Saclay University, GABI, CRJ, 78352 Jouy-en-Josas Cedex, France, [2]FRB, 195 rue St Jacques, 75005 Paris, France; michele.tixier-boichard@inra.fr

Many biological samples are collected for research projects as well as for the preservation of genetic resources. These samples are an asset for future research and should be stored safely and re-used to better benefit from them. Sharing require a good documentation of samples and procedures to trace their origin and their use. CRB-Anim is a network of French Biological Resource Centers (BRC) funded by ANR, which preserve and share biological material for research & development on domestic animals, including both reproductive and samples suitable for -omics (blood, DNA, milk, tissues, faeces). A web portal https://crb-anim.fr/access-to-collection has been developed in order to centralize access to 500,000 samples and associated data, from five BRCs for 20 domestic animal species. A list of 70 descriptors has been established, 25 being mandatory, regarding the population, the animal, the sample, the conditions for access and use, the legal status, the BRC identity. International standards (BioSamples EBI, Uberon or Brenda ontologies) have been used to document the sampled tissues. A minimum data set has been established by concertation with eight users' groups representing the different species, considering also the pre-requisite of Biosamples. Tailored connectors have been developed for each BRC to update its data on the portal from its local data management system. The portal makes possible to search for information and to monitor requests for storing or obtaining samples. At the entry into BRC, an agreement is signed with the researcher or breeder providing samples, it includes the conditions for further sample distribution, and most providers require to be informed of each request on the samples they have stored. A veto right may be provided, although not encouraged. The portal operates as a market place where each step from entry to distribution of samples is traced, including communication between the BRC and the user. The workflow involves a mandatory step to sign a material transfer agreement with document related to the Nagoya protocol. Finally, the portal provides indicators on the use of the collections in order to help the management.

Conservation genomic analyses of Croatian autochthonous pig breeds

M. Ferenčaković[1], B. Lukić[2], D. Šalamon[1], V. Orehovački[3], M. Čačić[3], I. Curik[1], L. Iacolina[4,5] and V. Cubric-Curik[1]
[1]Faculty of Agriculture, University of Zagreb, Department of Animal Science, Svetošimunska cesta 25, 10000 Zagreb, Croatia, [2]Faculty of Agriculture, University of J.J. Strossmayer, Department of Special Zootechnics, Trg Sv. Trojstva 3, 31000 Osijek, Croatia, [3]Croatian Agricultural Agency, Ilica 101, 10000 Zagreb, Croatia, [4]Aalborg Zoo, Mølleparkvej 63, 9000 Aalborg, Denmark, [5]Aalborg University, Department of Chemistry and Bioscience, Frederik Bajers Vej 7H, 9000 Aalborg, Denmark; mferencakovic@agr.hr

Turopolje (TUP) and Black Slavonian (BSP) pigs are the only two recognized Croatian autochthonous pig breeds. Here we analysed 16 animals from each breed using Illumina Infinium PorcineSNP60 v2 BeadChip and estimated genomic inbreeding levels (F_{ROH}) and linkage disequilibrium effective population size (Ne_{LD}). We also defined genomic position of TUP and BSP with respect to 29 domestic world-wide pig breeds and to six wild boar populations. The inbreeding level based on runs of homozygosity (F_{ROH}) was much higher in TUS then in BSP for both $F_{ROH>2Mb}$, 0.43 vs 0.12, and $F_{ROH>8Mb}$, 0.33 vs 0.07, respectively. Estimated Ne_{LD} for TUP was very small, 18 (95%CI: 14-22) while higher Ne_{LD} estimates, 48 animals (95%CI: 37-59), were obtained for BSP. Principal Components Analysis of the overall dataset (666 animals) positioned TUP close to the Mangalica and some Mediterranean autochthonous breeds (MED cluster), not so far from the European wild boar populations. BSP was remote from MED cluster in direction of UK and USA breeds (Berkshire, Large black, Tamworth). To our knowledge, this is the first genomic analysis of TUP and BSP providing insight into their conservation status. Although BSP population is genetically small, an organised management program has been established in the last decade. In contrast, extremely high inbreeding level and small effective population size in TUP, ring alarm bells for an urgent conservation management plan.

Genome-wide diversity and admixture of Angler and Red-and-White dual purpose cattle

S. Addo[1], M. Morszeck[1], D. Hinrichs[2] and G. Thaller[1]
[1]Institute of Animal Breeding and Husbandry, CAU, Kiel University, Olshausenstr. 40, 24098 Kiel, Germany, [2]Department of Animal Breeding, University of Kassel, Nordbahnhofstr. 1a, 37213 Witzenhausen, Germany; saddo@tierzucht.uni-kiel.de

Pedigree analyses in previous studies show some level of common ancestry of the Angler (RVA) and Red-and-White dual purpose (RDN) cattle breeds, but it is still unclear if this is the result of a recent ancestry or gene flow among them. The aim of the present study was to investigate admixture events between RVA and RDN cattle, and to determine molecular estimates of inbreeding within and across the breeds. Data on 147 RVA and 68 RDN bulls including 38,042 SNPs after quality control, were used in an admixture and a high definition network-visualization (NetView) analyses. In addition, the observed (H_o) and expected (H_e) heterozygosities as well as estimates of inbreeding were assessed. We observed similar estimates of observed and expected heterozygosities, with the average H_o value being slightly higher for the RVA breed (0.37 vs 0.35). These estimates are an indication of a remarkably high genetic diversity within the studied breeds. A multidimensional scaling plot analysis based on genetic distance matrix revealed a separation of the breeds into two clusters. At K=2 of the admixture analysis however, we observed common ancestry of the two breed, particularly, some individuals of the RDN breed having over 40% of genetic background of the RVA breed. The K value that showed the lowest cross-validation error was 11. At this value, the RDN breed showed a more distinct genetic background while the RVA had a highly mixed ancestry. The admixture results were confirmed by NetView, which provided fine-tuned graphics of two separate clusters, each consisting of several subpopulations within which individuals (nodes) are interconnected. The RVA had more subpopulations (7 vs 4) but within both clusters, different proportions of genetic background of individuals were evident. Therefore, RVA and RDN breeds have a common ancestry which could have arisen from cross breeding schemes in the past. Further studies would focus on including breeds that have been historically crossed with either the RVA or RDN in order to correctly map out unknown ancestor groups.

Session 09 Theatre 10

The conservation status of Dalmatian pramenka sheep using high-throughput molecular information

D. Šalamon[1], M. Ferenčaković[1], I. Drzaic[1], E. Ciani[2], J.A. Lenstra[3], I. Curik[1] and V. Cubric-Curik[1]
[1]Faculty of Agriculture, University of Zagreb, Department of Animal Science, Svetosimunska 25, 10000 Zagreb, Croatia, [2]University of Bari, Aldo Moro, Department of Biosciences, Biotechnologies and Biopharmaceutics, Bari, Bari, Italy, [3]Utrecht University, Faculty of Veterinary Medicine, Utrecht, Utrecht, the Netherlands; dsalamon@agr.hr

The Dalmatian Pramenka (DAL) is the largest autochthonous population of sheep in Croatia that belongs to the long tailed Pramenka type of sheep, which are widespread over South-East Europe. We genotyped 32 DAL individuals on the Illumina Ovine SNP50 K BeadChip, and calculated parameters that are important for the conservation status: observed genome-wide heterozygosity (oHet), effective population size estimated via linkage (Ne_{LD}), and gametic (Ne_{GD}), disequilibrium and the ROH-based inbreeding level ($F_{ROH\leq50g}$ and $F_{ROH\leq10g}$). A comparison with 24 Appenninica, 24 Arawapa, 24 Australian Merino, 24 Churra, 24 Finnsheep, 21 Merino Landschaf, 24 Massese and 24 Sardinian White revealed that DAL has the highest observed heterozygosity (oHet=0.375) and the second highest Ne_{GD} (124), as well as the lowest ROH-based inbreeding levels ($F_{ROH\leq50g}$=0.021 and $F_{ROH\leq10g}$=0.009). Interestingly, DAL has the highest estimated decline in Ne_{LD} (ΔNe_{LD}=13), resulting in an extremely low Ne_{LD} of 15-31 (95% CI). Most likely, this notable difference is the consequence of the population growth of DAL in last generations, in contrast to the population decline in the period considered in historical Ne_{LD} estimates (12 to 100 generations back).

Genomic data as a prerequisite for efficient conservation programme of the Czech Red cattle

K. Novák, J. Kyselová, V. Czerneková and V. Mátlová
Institute of Animal Science, Přátelství 815, 104 00 Prague, Uhříněves, Czech Republic; novak.karel@vuzv.cz

The Czech Red cattle is a historical local breed of the combined type originating from the mountain regions of Bohemia and Moravia. Like the Harz Red Cattle, it is sometimes being associated with the ancient Celtic Red cattle. The first conservation efforts can be dated back to the beginning of the 20th century. Whole genome sequencing was performed in individuals representing animals with the highest proportion of the original gene pool (the highest value 38%). In parallel, pooled population samples were re-sequenced in order to obtain a survey of the total variability. The Illumina HiSeq technology for 2×150 reads was applied with 10× coverage for individual animals and 30× coverage for the populations of 80 animals. Paired sequences were mapped to the UMD3.1.1 reference genome with Geneious (Biomatters) algorithm. In total, 12,447,000 of structural variations were found in comparison to the reference genome of the Hereford breed, confirming a significant genetic distinctness. Twenty four polymorphisms in the mitochondrial genome determine haplogroups, however, no indications of the aurochs admixture were detected. Consistently, a higher diversity was found in the population sample. Selected genes coding for the innate immunity components are genotyped in individual animals with designed panels of the primer extension reactions. The remnants of the original gene pool of this breed are being identified by comparison to the representatives of the main admixture. There is a belief that a significant proportion of the original gene pool has been preserved in spite of the repeated bottleneck stages and subsequent breed reconstruction in the twenties and seventies of the 20th century.

Conservation genomics in the management of the Mouflon population from the hunting area Kalifront

V. Cubric-Curik[1], M. Oršanić[2], D. Ugarković[2], M. Ferenčaković[1] and I. Curik[1]
[1]University of Zagreb Faculty og Agriculture, Department of Animal Science, Svetosimunska 25, 10000 Zagreb, Croatia, [2]Universty of Zagreb Faculty of Forestry, Department of Forest Ecology and Silviculture, Svetosimunska, 10000 Zagreb, Croatia; vcubric@agr.hr

Mouflon population (MKA) has been founded in 1998 by a 33 female and 8 male individuals in the open state hunting area Kalifront (Rab Island, Croatia). Since its foundation this population has been closed and currently there are >200 individual. However, decline in trophy values and body weight has been observed during last 10 years. We genotyped 32 mouflon individuals by OvineSNP50 BeadChip array and estimated runs of homozygosity based inbreeding levels with respect to different base populations ($F_{ROH>4Mb}$, $F_{ROH4\text{-}8Mb}$, $F_{ROH>8Mb}$ and $F_{ROH>16Mb}$). The same calculations were also performed on a dataset taken from the public digital repository (http://dx.doi.org/10.5061/dryad.2p0qf) for Sardinian mouflon (SAM), European mouflon (EMO) and Soay sheep (SOA) population as well as for the several domestic sheep populations. Extreme inbreeding level was observed for MKA at all levels ($F_{ROH>4Mb}$=0,205, $F_{ROH4\text{-}8Mb}$=0,086, $F_{ROH>8Mb}$=0,120 and $F_{ROH>16Mb}$=0,050). Obtained inbreeding levels were lower in comparison to EMO, but much higher in comparison to SOA, SAM and other domestic sheep populations. We further analysed future actions and options for introgression of less related mouflons.

Genomic homo- and heterozygosity in a commercial turkey population

C.F. Baes[1], B.J. Wood[1,2], F. Malchiodi[1], K. Peeters[3], P. Van As[3] and G. Marras[1]
[1]University of Guelph, Animal Biosciences, 50 Stone Road East, N1E 2W1, Canada, [2]Hybrid Turkeys, Genetics, 650 Riverbend Drive, Suite C, N2K 3S2, Kitchener, Canada, [3]Hendrix Genetics Ltd., Spoorstraat 69, 5831 CK Boxmeer, the Netherlands; cbaes@uoguelph.ca

Runs of homozygosity are increasingly being used for genomic characterization and inbreeding estimation, and to detect selective signatures in the genome of mammal and avian species. As observed in dairy, inbreeding coefficients estimated using genomic information better account for realized inbreeding rather than relying on probabilities. Here we characterize runs of homozygosity and heterozygosity in the turkey genome (Meleagris gallopavo) using commercial lines. Samples were collected between 2010 and 2017 and were genotyped using a proprietary 60k array; pedigree records were also available for genotyped birds and their ancestors. Runs of both homozygosity and heterozygosity were detected. Relatively long and abundant runs of homozygosity were detected, with comparatively few runs of heterozygosity. Inbreeding coefficients calculated using ROH were higher than those calculated using pedigree. These results provide a preliminary characterization of the turkey genome in terms of runs of homozygosity and heterozygosity-rich genomic regions. Inbreeding calculated from ROHs was higher than that calculated from the pedigree. These results provide initial information on characterization of the turkey genome in terms of runs of homozygosity and heterozygosity-rich genomic regions.

Quality of biological material stored at the National Biobank of the NRIAP

J. Sikora[1], A. Kawęcka[1], M. Puchała[1], P. Majchrowski[2] and A. Szul[2]
[1]National Biobank of National Research Institute of Animal Production, Krakowska 1, 32-083 Balice, Poland, [2]Malopolskie Center of Biotechnik, Krasne 32, 36-007 Krasne, Poland; jacek.sikora@izoo.krakow.pl

Ex situ conservation of genetic resources is carried out through the collection of genetic material (semen, embryos, oocytes) subjected to cryopreservation. The material is stored in accordance with veterinary hygiene standards to allow its use in conservation programmes and for breeding purposes. These activities fulfil the obligations assumed by Poland as part of the Convention on Biological Diversity and the Global Plan of Action for Animal Genetic Resources. The National Biobank of National Research Institute of Animal Production (NRIAP), launched in 2014, is comprised of 4 species banks for farm animals included in the genetic resources conservation programme as well as for chosen breeding and productive animals. These banks are for cattle, horses, pigs, and sheep and goats. The biological material within the collection was divided into the historical part, which has not left the bank; the active part, which can be used in active breeding; and the part intended for tests and quality control of the collected material. The cattle bank stores the semen of cows covered by the genetic resources conservation programme, such as Polish Red (RP), Polish Red-and-White (ZR), Polish Black-and-White (ZB), and Polish Simmental (SM). The highest mean ejaculate volume was noted in RP bulls (5.9 ml). SM and ZB breeds had similar volumes of around 5 ml. The smallest ejaculates were obtained from ZR bulls (4.1 ml). The highest mean ejaculate concentration was determined for the RP and SM breeds (more than $1,300\times10^6$/ml), followed by ZB semen (around $1,200\times10^6$/ml) and ZR semen (1000×10^6/ml). The highest individual motility before freezing was observed for the spermatozoa of SM bulls (80%) and the lowest for those of ZB bulls (69%). Progressive motility after freezing for all the breeds ranged from 50 to 55% and met the requirements of the semen collection centres for commercial semen. Project 'The uses and the conservation of farm animal genetic resources under sustainable development' co-financed by the National Centre for Research and Development, contract number: BIOSTRATEG2/297267/14/NCBR/2016.

The genetic resources preservation of Romanian Grey Steppe cattle by using cytogenetic screening

I. Nicolae and D. Gavojdian
Research and Development Institute for Bovine Balotesti, sos. Bucuresti-Ploiesti, km 21, Balotesti, Ilfov, 077015, Romania; ioana_nicolae2002@yahoo.com

In Europe over 40% of the livestock breeds are at risk of becoming extinct over the next two decades. Intensive production systems have significantly contributed to the threats facing the European cattle breeds, with production being focused only on a few high yielding breeds in the detriment of rare or minority breeds, which are likely to represent important genetic resources because of their environmental adaptability, disease resistance, fertility and unique product qualities. The indigenous cattle breeds in Romania are being preserved for scientific purposes as sources of genes used in animal breeding. In the frame of a national project (MCI Project 2PS/2.11.2017), a special interest has been dedicated to preserve the genetic resources of the endangered Romanian Grey Steppe cattle breed. One critical tool to preserve and protect the genetic heritage and diversity of Gray Steppe cattle is the cytogenetic screening. For this reason, we developed a cytogenetic investigation on a small nucleus herd of 25 Grey Steppe cattle, (23 cows and 2 bulls) reared in Eastern Romania. The chromosomal investigations revealed normal karyotype for 14 cattle (13 cows and 1 bull). The other 11 remaining animals presented a higher percentage of abnormal cells (gaps, chromatid breaks, chromosome breaks and fragments) compared with the normal animals. Additional investigations have been performed by using sister chromatid exchanges test (SCEs) to highlight that cytogenetic screening could be used to check the genomic stability of the Romanian Gray Steppe cattle breed, in order to aid farm animal genetic resources (FAnGR) preservation efforts.

Biodiversity of the Carniolan honeybee in Croatia

Z. Puskadija, M. Kovacic, K. Tucak, N. Raguz and B. Lukic
Faculty of Agriculture in Osijek, J.J.Strossmayer University, Vladimira Preloga 1, Osijek, 31000, Croatia; blukic@pfos.hr

The Carniolan honeybee (Apis mellifera carnica), the only honeybee subspecies in Croatia, is faced with the negative impacts on its biodiversity, alongside other bee populations. Contrary to the world trend, which shows steady growth of bee colonies, the numbers in Europe have been on the decline for the last 40 years. The primary cause could be various diseases (Varroa destructor, viruses, Nosema ceranae, etc.), monotonous nutrition (monoculture) or significant use of pesticides in plant production. Although numerous studies have been carried out, the actual cause of this decline is not completely clear. In Croatia, detailed analyses of the Carniolan honeybee biodiversity have not been conducted so far. Moreover, the phenotypic variation according to geographic distribution is also poorly investigated. According to the present studies, there are three ecotypes of Carniolan honeybee in Croatia: Panonian, Subalpine and Mediterranean (Dalmatian). In order to investigate the genetic structure and morphological characteristics of the Carniolan honeybee population in the frame of BioBeeCro Project, our aim is to collect and analyse bees from 300 apiaries from the entire Croatian territory. To encompass all local ecotypes, bee samples will be collected from the beekeepeers who keep bees on the stationary apiaries and do not buy queens on the market. The research will include modern genomic methodologies with sufficient coverage of single nucleotide polymorphisms (200 loci) along the genome. The parameters of the genetic structure i.e. the allele frequencies, the degree of heterozygosity and the fixation index will be determined. The analysis of ecotypes will include discriminant analyses. In order to determine the phenotypic and genetic correlations, morphometric measurements of the wings will be performed. The results will therefore provide the deeper insight into the genetic and morphologic variability together with the presence and the distribution of ecotypes. Moreover, the results will create the basis for the improvement of breeding program for honeybees in Croatia. The BioBeeCro project is funded by the Paying Agency for Agriculture, Fisheries and Rural Development (PAAFRD) in Croatia.

The genetic diversity and origin of the Belgian Milksheep using pedigree and genomic information

R. Meyermans, K. Wijnrocx, N. Buys and S. Janssens
KU Leuven, Livestock Genetics, Department of Biosystems, Kasteelpark Arenberg 30 bus 2456, 3001 Heverlee, Belgium; roel.meyermans@kuleuven.be

The Belgian Milksheep is a dairy type breed and is currently considered to be rare in Flanders (Belgium). It is known for its excellent milk production with a creamy taste. Moreover, it has good mothering abilities and a high fertility, but the current size of the active population is low (approx. 500 animals). In order to support the conservation efforts, a study was performed using pedigree data (n=8287) and genotypes of 192 sheep (144 Belgian Milksheep, 22 Friesian sheep and 22 Flemish sheep). From those 192 sheep, 144 were genotyped on the Illumina OvineSNP50 Chip and 48 on the 15K SNP Chip of the International Sheep Genomics Consortium. Furthermore, the Sheep HapMap project of the International Sheep Genomics Consortium provided genotypes of 74 other sheep breeds. The pedigree analysis indicated that during the period of 2011-2016, annually 128 to 201 litters were registered. The average rate of inbreeding, based on pedigree records, over the same period was between 10.0 and 13.5%. The rate of inbreeding per generation (3.5 years) fluctuated between 2.09 and 3.50%. Based on the genomic analysis, the average rate of inbreeding of the population was estimated to be 11.94% (runs of homozygosity analysis). The effective population size was estimated at 24, based on pedigree information, and at 23, based on genotype information. These values are far below the FAO-guideline of 100 animals, indicating that the risk of losing genetic diversity is high. In a principal component analysis, the Belgian Milksheep population clustered closely together with the Flemish and Friesian sheep populations (the Friesian sheep genotyped in this study, and the East-Friesian White and Brown breeds from the Sheep Hapmap project). This group of breeds was then closest related to the Scandinavian Spaelsau and the Finnsheep. This study contributes to the conservation of the Belgian Milksheep breed and helps the breed association in setting up their long term goals for the conservation of the breed. Moreover, it creates an insight in the genetic origin of these local sheep breeds.

Economic impact of inbreeding over weaning weight, yearling weight and cow productivity in Nellore

G.C. Mamani[1], B.F. Santana[1], G.L. Sartorello[1], B.A. Silva[1], E.C. Mattos[1], J.P. Eler[1], G. Morota[2] and J.B.S. Ferraz[1]
[1]University of Sao Paulo, Av. Duque de Caxias Norte, 225, 13635-900 Pirassununga, SP, Brazil, [2]University of Nebraska-Lincoln, Animal Science, 3940 Fair St, P.O. Box 830908, USA; gerardo.mamani@usp.br

Inbreeding depression is well known in livestock production, but few studies are available about economic impact on cattle traits. The objective of this work was to calculate the economic impact of inbreeding on productive traits in Nellore cattle over 10 years. We analysed pedigree of 594,123 animals in one farm in Brazil with 487,350, 138,327, and 37,093 records for weaning weight, yearling weight, and cow productivity, respectively. The pedigree inbreeding coefficient was calculated using the CFC program. For each trait, we evaluated the extent of the inbreeding depression by fitting an animal mixed model. The fixed effects were contemporary groups, linear and quadratic dam age at calving, age at the measurement, and the inbreeding coefficients. The analysis was performed using the software BLUPF90. In total, 12.94% of animals showed inbreeding and the mean and maximum values of inbreeding were 0.16 and 33%, respectively. A 1% increase of inbreeding coefficient was associated with -0.38, -0.39 and -1.12 kg change of weaning weight, yearling weight, and cow productivity, respectively, compared to non-inbred animals. Subsequently, inbreeding coefficients were multiplied by the estimated inbreeding depression to derive loss in kg for each trait. These values were multiplied by the market prices deflated to calculate the yearly loss in kg. The nominal marketed prices (R$/kg) were obtained by the General Price Index – Internal Supply of Brazil for 2007 to 2016. For example, the prices deflated in Reais (Exchange rate for January 2018, according to The Central Bank of Brazil: USD: 1.00=R$ 3.23) were 3.6 and 7.74 for calf (240 days of age) and 3.21 and 5.75 for fat beef cattle (455 days of age) for 2007 and 2016, respectively. The economic impact was on average 8,848.7, 20,297.6 and 9,314.4 Reais/year for weaning weight, yearling weight, and cow productivity in 2007 and 2016, respectively.

Genetics of coat colour variation in South African Nguni cattle

L.M. Kunene[1], E.F. Dzomba[1], F.C. Muchadeyi[2], G. Mészáros[3] and J. Sölkner[3]
[1]University of KwaZulu-Natal, Discipline of Genetics, School of Life Sciences, Private Bag X01, 3209 Scottsville, South Africa, [2]Agricultural Research Council, Biotechnology Platform, Private Bag X5, 0110 Onderstepoort, South Africa, [3]University of Natural Resources and Life Sciences, Vienna, Division of Livestock Sciences, Gregor-Mendel Str.33, 1180 Vienna, Austria; gabor.meszaros@boku.ac.at

The Nguni breed of cattle is a Sanga type breed with mixed *Bos taurus* and *Bos Indicus* ancestry and proven resistance to ticks, diseases and other harsh conditions of the African geographical landscape. The coat colour of the breed is not uniform but is based on at least four colours and a myriad of spotting patterns. In the current study, we examined the genomic architecture of colour variation based on a sample of 128 South African Nguni cattle. The genome-wide association study was done using a univariate linear mixed model based on 669,009 autosomal SNPs remaining after standard quality control. The phenotype data included various characteristics such as spotting (yes/no), muzzle colour (white or black), and prevailing coat colour (white, black, red, brown pairwise combinations and white compared to all others). The SVS GenomeBrowse BTAU5.0.1 reference genome and the Bovine QTL database, were used to investigate genes and QTLs associated with the significant SNPs. The most significant results with significant peaks above the Bonferroni threshold revealed three regions at BTA16 associated with white versus black main coat colour. The white vs brown comparison reported significant SNPs on chromosomes BTA6, BTA20 and BTA25. The white-red coat colour comparison yielded 6 significant SNPs on chromosomes BTA1, BTA10 and BTA24. Interestingly, the 150 Mb region on BTA1 was coming up whenever the white colour was included in the analysis. None of the regions contained any genes with an obvious connection to colour related traits. Several significant or indicative SNPs, however, fell into regions with identified, but functionally uncharacterized genes. The study presented a preliminary analysis on the genetics of coat and other phenotypic diversity of the Nguni cattle. Previously unknown genetic components could be involved in the coat colour genomic architecture of cattle, with direct or regulatory functions.

Estimation of effective population size based on genealogical and molecular data by different method

L. Vostry[1], H. Vostra-Vydrova[2], B. Hofmanova[1], Z. Vesela[2], A. Novotna[2] and I. Majzlik[1]
[1]Czech University of Life Science Prague, Kamycka 129, 16500 Prague, Czech Republic, [2]Institut of Animal Science, Pratelstvi 815, 10400 Prague, Czech Republic; vostry@af.czu.cz

There are three important Czech draft horse breeds: Silesian Noriker, Noriker and Czech-Moravian Belgian. Silesian Noriker and Czech-Moravian Belgian are included among the endangered breeds. These populations are currently closed to outside breeding. Effective population size is an important parameter used to assess genetic diversity for conservation program in these endangered breeds. Effective population size was assessed by pedigree and microsatellite data. Pedigree records for reference population, which included animals born in a 14-year period (1996-2010) were used for calculation. The genotyping data based on 13 microsatellites have been collected from a total of 1,298 individuals. Realized effective population size with respect to the breed based on pedigree data was estimated from the individual rate in coancestry (N_{eC}), from the individual increase in inbreeding (N_{eF}) and by the LDN_e method based on molecular data with respect to the breed. The effective population size based on molecular data was estimated by different value of minor allele frequency (MAF=0.0, 0.1, 0.2 and 0.4). The values of effective population sizes were found low for all three analysed breeds – N_{eF}=79.6, N_{eC}=95.2 NS LDN_e=97.6 for Silesian Noriker, N_{eF}=97.6, N_{eC}=195.1, LDN_e=100.6 for Noriker and N_{eF}=87.4, N_{eC}=101.4 and LDN_e=146.6 for Czech-Moravian Belgian breed, respectively. The most suitable value of MAF was selected as 0.0. Obtained low estimates of Ne suggested small founder populations or losses of genetic variation through random genetic drift. The analysis shows that the estimates of effective populations size obtained using genealogical data correspond to those obtained using molecular genetic data. These results point to a quality pedigree records of analysed breeds. These statistics suggest that the genetic variability has decreased, and without changes in breeding strategy the genetic variability might continue to decline. The study provides data and information utilizable in the management of conservation programs focused on reducing inbreeding level and minimization of genetic variability loss.

Session 09 Poster 21

Conservation of livestock (including horses) biodiversity in Poland

A. Chełmińska and I. Tomczyk-Wrona
National Research Institute of Animal Production, Department of Horse Breeding, Sarego 2, 31-047 Kraków, Poland; agnieszka.chelminska@izoo.krakow.pl

Conservation of biodiversity is extremely important because the number of breeds used in agri-production is decreasing. Therefore, various forms of activity are undertaken internationally, including in Poland, to minimize threats. Over the last 20 years, when formalized activities to conserve animal genetic resources in agriculture were begun in Poland, most of them concentrated on *in situ* conservation. As a result, the number of conserved populations increased to 83, which is an increase of 48% in relation to the year 2000. Currently 4 breeds of cattle, 3 breeds of pigs, 15 breeds of sheep, 1 breed of goat, 7 breeds of horses, 35 breeds of poultry, 13 breeds of fur animals, and 5 breeds of bees are conserved. Over 103,000 animals kept in 3,000 farms are under i*n situ* conservation. Compared to the year 2000, this is a 6-fold increase in the number of animals and a 27-fold increase in the number of herds. This also shows that the majority of native breed animals are kept in small family farms. In 2000, the conservation programmes covered 430 horses representing only 2 breeds (Hucul, Polish Konik), whereas today a total of 6,800 animals representing 7 breeds are conserved; this demonstrates that the conserved population increased almost 16-fold. This success is partly attributed to payments in support of the local breeds, as part of consecutive agri-environmental programmes. Currently the most important goal that breeders are facing is to promote alternative forms of using the native breeds and their products. The main task of the current and future agri-environmental programmes is to reduce the negative environmental impact of agriculture and to maximize its positive impact on biodiversity and rural landscape. Considering the EU headline target for 2020, namely to halt biodiversity loss, it is necessary to ensure continuity in activities for the conservation of native breeds and to secure adequate funding for this goal.

Session 09 Poster 22

The maintenance of genetic diversity in local Rendena cattle through optimal contribution selection

C. Sartori[1], N. Guzzo[2] and R. Mantovani[1]
[1]Dept. of Agronomy Food Natural resources Animals and Environment, Viale dell'Universita', 16, 35020 Legnaro (PD), Italy, [2]Dept. of Comparative Biomedicine and Food Science, Viale dell'Universita', 16, 35020 Legnaro (PD), Italy; roberto.mantovani@unipd.it

The Rendena is an Italian small dual purpose cattle breed used both in the mountain area of origin (Rendena valley, Trento province) and in the plains (Padova and Vicenza provinces). Most of milk produced by this breed is converted into typical Spressa and Grana Padano cheeses. The selection process (both for milk and meat), has been always carried out considering the risk of inbreeding. Recently, Optimal Contribution Selection (OCS) has been developed to assess the optimal balance between inbreeding (F) and genetic gain. This study aimed to evaluate the effects of OCS on candidate bull dams and bull sires in Rendena mating programs for years 2014-2017, comparing it with a full conservation policy or the traditional selection method based on the assignment of a given mating percentage to each bull sire. OCS was applied considering different penalties to average relationships (AR) to determine a scenario combining AR and predicted breeding value in offspring (EBV). Considering the mating of 2014, OCS allowed to obtain a predicted offspring with lower EBV than in traditional selection, but also with a lower F, similar to that one obtained under conservation. Subsequent years showed the actual effects of the OCS program: in 2015, before the effective introduction of OCS in routinely breeding plans, both EBV and AR were increased. In 2016 and 2017, when bull sires from OCS mating were used, a reduction in EBV occurred but the AR increase was almost null, meaning that OCS was effective to counteract the F increase naturally occurring under selection. Moving from 2015 to 2016 the predicted F rate even decreased, and increased slightly in 2017. The study confirmed OCS as an effective tool for long-term conservation of local breeds under genetic improvement, which is important for biodiversity and sustainable use of the genetic resources.

Resources and characteristics of gene pool of native breed Latvian Blue cow in 2017

I. Sematovica[1], M. Lidaks[2] and I. Kanska[2]
[1]Latvia University of Life Sciences and Technologies, Faculty of Veterinary Medicine, Helmana 8, 3004, Latvia, [2]Animal Breeders Association of Latvia, Republikas laukums 2, Rīga, 1010, Latvia; isem@inbox.lv

The Latvian Blue cow (LZ) breed is one of the primitive cattle breeds (*Bos primigenius taurus*). It is a very rare breed, characteristic only for Latvia and it is actually in the FAO extinction category. Measures have been taken to protect LZ breed with support of the Latvian government. Now due to the ERAF project No. 1.1.1.1/16/A/025, *BioReproLV* it is become possible to save LZ breed by using multiple ovulation and embryo transfer (MOET). The LZ variety is mainly used for milk production. The aim of this study was to evaluate resources of LZ cow breed and to analyse results of cow's linear parameters. Data were taken of Agricultural Data Centre Republic of Latvia in 2017 and evaluation *in vivo*. In 2017, 339 LZ cows were genebank (GB) animals and 140 of them were culled by different reasons in this year. There were 544 (190 of which were primiparous) standard lactations registered in 2009 and only 314 (56 of which primiparous) in 2017. Productivity of all registered LZ cows was 5,320.6±2,060.51 kg/lactation with milk protein (MP) 3.4±1.36% and milk fat (MF) 4.24±1.73%, but LZ GB cow average milk yield was 5,381.2±1,857.08 kg/lactation, MP 3.4±1.22%, MF 4.23±1.56%. At the moment the oldest LZ GB cow is 16.3 years old and 20% of LZ GB cows are more than 10 years old, and only 10% are younger than 5 years. The overall exterior rating was established more than 80 points for 64.4% of evaluated LZ GB cows. Some LZ GB linear parameters did not change significantly (P>0.05) with age – legs and hoofs, udder, dairy form, rump angle, foot angle, fore udder attachment, rear teat position, milking speed and cow temperament. No statistically significant differences (P>0.05) regarding all parameters of linear evaluation among LZ and LZ GB dairy cows except tendency regarding to body depth (P=0.054). No 100% purebred LZ cow exists. To become one of LZ GB cows, an animal must have certain properties characteristic to LZ breed and appropriate origin at least 60% of LZ. It is clear that the measures to save LZ breed was not effective enough so far. MOET will be one of the promoter instrument to fulfil Pedigree Law regarding to LZ GB cow.

Hepatic molecular changes induced by a high-fat high-fibre diet in growing pigs

F. Gondret, A. Vincent, S. Daré and I. Louveau
INRA, Pegase, Rennes, 35042, France; florence.gondret@inra.fr

The introduction of more fibre to cereal-based diets in pigs gained interest due to new economic considerations and to potential health and welfare benefits. The addition of fat to a fibre-rich diet is required to maintain dietary energy value for performance. Feeding pigs a high-fat high-fibre diet, however, changes the energy source and nutrients as compared to a low-fat high-starch diet. The liver plays a central role in energy metabolism. This study was undertaken to investigate hepatic molecular pathways in pigs fed diets with contrasting sources of energy and nutrients. From 74 d of age onwards, 48 Large White castrated male pigs were fed a high-fat high-fibre diet (HF, n=24) or a low-fat high-starch diet (LF, n=24). Diets were formulated to be isoenergetic and isoproteic. Starch derived from cereal grains (wheat and barley) in the LF diet was partially replaced by rapeseed and soybean oils in the HF diet and crushed wheat straw (insoluble fibre) was included as a diluent of dietary energy in this diet. At 132 d of age, the liver was excised, weighed and processed for biochemical and molecular analyses. Transcriptomics analysis was performed using porcine microarrays (Agilent, GPL16524, 8 × 60K). Functional pathways were deduced from genes declared as differentially expressed (P<0.01) using DAVID Bioinformatics Resources and Ingenuity Pathway Analysis. Compared with LF pigs, HF pigs had a lower ADG and ADFI (P<0.01) during the test period. At slaughter, the liver was lighter (-7%, P=0.03) in HF than in LF pigs. In liver, the protein content was unaffected but the glycogen content and glucokinase activity were reduced in HF pigs compared with LF pigs (P<0.05). A total of 802 annotated genes were differentially expressed between the two diets. In HF pigs, genes involved in glycogen and hexose metabolism and genes participating to oxidative phosphorylation and ATP synthesis were down-regulated. Conversely, genes contributing to cell growth, cell cycle phase, cell death and cell adhesion were up-regulated. Liver hyper-proliferation, hepatic fibrosis, and liver necrosis were suggested as top-toxicity functions responding to diet. In conclusion, pig liver functions can be affected by dietary components such as fibres and lipids.

Session 10 Theatre 2

Liver oxygen uptake, triiodothyronine and mitochondrial function vary with feed efficiency in cattle

Y.R. Montanholi[1], J.E. Martell[1] and S.P. Miller[2,3]

[1]Harper Adams University, Animal Production, Welfare and Veterinary Sciences, Edgmond, TF10 8NB, Newport, United Kingdom, [2]University of Guelph, Department of Animal Biosciences, 50 Stone Road East, N1G 2W1 Guelph, Canada, [3]Angus Genetics Inc., 3201 Frederick Avenue, MO 64506, Saint Joseph, USA; ymontanholi@harper-adams.ac.uk

Feed costs are a major expense to beef cattle operations, which justifies efforts to improve the efficiency of feed utilization. The evaluation of liver function is an avenue to identify potential proxies for productive efficiency, given the remarkable impact of hepatic metabolism on body energetics. Our objective was to assess the relationships of liver oxygen uptake (O_2), citrate synthase (CS) and triiodothyronine (T3) with feed efficiency, measured via residual feed intake (RFI; DM kg/d). Crossbred beef bulls (n=63) were performance tested during 112 days. Daily assessments of individual feed intake of a corn-based ration, and every 28-day measures of body weight and composition were computed to determine RFI prior to processing at the age of 12.9±1.0 months. During slaughter, liver samples were collected for microcalorimetry analysis to determine O_2 (μmol/min/g DM) and CS activity (μmol/min/g DM). Blood samples were also collected for T3 (ng/dl) determination. A categorical statistical analysis was conducted with bulls ranked according to RFI and divided into two RFI differing (P<0.05) groups, Efficient=32 and Inefficient=31(RFI=-0.72 vs 0.90 kg/d). The O_2 was increased in the Inefficient in comparison to the Efficient bulls (1.48 vs 1.27, P<0.05). Results also indicate an increased concentration of blood plasma T3 in the Inefficient in comparison to the Efficient bulls (94.73 vs 88.02, P<0.05). The CS activity was lower in the Inefficient in comparison to the Efficient bulls, (2.30 vs 2.59, P<0.05). These results suggest that Efficient bulls are speculated to have a more efficient production of adenosine triphosphate (ATP), by having a greater number of mitochondria in the hepatocytes. This phenotypic evidence contributes to further our understanding of the biological basis underlying feed efficiency. Such evidence should also be combined with other findings within the subject of feed efficiency, as an effort to determine a robust multi-trait assessment of feed efficiency.

Session 10 Theatre 3

Exploring the effect of dietary L-carnitine inclusion on the performance of hyper prolific sows

H.B. Rooney[1,2], K. O'Driscoll[1], J.V. O'Doherty[2] and P.G. Lawlor[1]

[1]Teagasc, Pig Development Department, AGRIC, Moorepark, Co. Cork, P61 C996, Ireland, [2]University College Dublin, Belfield, D04 V1W8, Ireland; hazel.rooney@teagasc.ie

Genetic selection for hyper-prolificacy in sows has resulted in reduced piglet birth-weight and increased piglet mortality. More precise nutritional strategies to optimize piglet development *in-utero* may help ameliorate these problems. Supplementing gestating sows with L-carnitine (L-car) has previously been found to increase litter size and piglet birth-weight, and milk quality is improved when L-car is supplemented during lactation. The objective here was to investigate whether supplementation with L-car during gestation and/or lactation can increase piglet survival and pre-weaning growth. At d1 of gestation, sows (n=66) were blocked by parity, weight and P2 back-fat, and randomly assigned to one of the four L-car treatments; Control (0 g/d), Gest (0.125 g/d throughout gestation until farrowing), Lact (0.125 g/d throughout lactation), and Both (0.125 g/d throughout gestation and lactation). Sow weight and P2 back-fat were recorded on d71 and d108 of gestation and at weaning (~d26 post-partum). Total number born (TNB) and number of piglets born alive (NBA) were recorded at birth. Piglets were individually weighed at birth, d2, d14 and at weaning. The data was analysed as a 2×2 factorial arrangement using the mixed models procedure in SAS (v.9.4). Sow weight was unaffected by treatment. However, the Control sows lost less weight during lactation compared to all other treatments (P<0.05). The Lact sows had reduced P2 back-fat at weaning compared to all other treatments (P<0.05). The TNB was higher for Gest sows (17.58±0.53) compared to Control sows (15.98±0.54; P<0.05). The NBA also tended to be higher for Gest sows (15.63±0.50) compared to Control sows (14.71±0.51; P<0.10). Piglet birth-weight was unaffected by treatment. Piglets born to Control sows were heavier at d14 compared to all other treatments (P<0.05), while piglet weight was not influenced by treatment at weaning. In conclusion, L-carnitine supplementation during gestation increased total number born and tended to increase number born alive without negatively affecting piglet birth-weight. Sow P2 back-fat was reduced at weaning when L-carnitine was supplemented during lactation.

Insights into the microbiota composition and metatranscriptome at the gut-body interface

M. Wagner[1,2], S.U. Wetzels[1,2,3], Q. Zebeli[1,3], B. Metzler-Zebeli[1,3], S. Schmitz-Esser[1,4] and E. Mann[1,2]
[1]Research Cluster Animal Gut Health, Veterinaerplatz 1, 1210 Vienna, Austria, [2]Institute for Milk Hygiene, University for Veterinary Medicine, Veterinärplatz 1, 1210 Vienna, Austria, [3]Institute for Animal Nutrition and Functional Plant Compounds, Veterinaerplatz 1, 1210 Vienna, Austria, [4]Department of Animal Science, Iowa State University, 1029 North University Boulevard, 50011 Ames, USA; martin.wagner@vetmeduni.ac.at

Although microbiome studies of farm animals have tremendously increased since 2008, comparatively little is known about the composition and function of the epimural bacterial microbiota. We find the epimural microbiome and its function particularly interesting in health and disease as we hypothesize that a number of metabolic functions is more likely associated with the bacteria attached to the gut wall than with transient bacteria acting in the lumen. In a series of studies, we investigated the composition of the mucosa-associated or epimural microbiota in pigs and cows and found it being essentially different from the luminal microbiome. In pigs, we could demonstrate that the mucosa-associated microbiome was relatively little affected by feeding antibiotics, neither was the microbiome of ileal lymphnodes affected. In cows, we studied the shifts of ruminal epimural microbiome after a subacute rumen acidosis (SARA) challenge. In addition, we investigated differences in the gene expression of the epimural bacterial microbiota before and after a long-term high-concentrate diet challenge. Our results showed that urea and starch degradation are important functions of the rumen wall microbiota. Furthermore, our results indicate that rumen wall bacteria are exposed to oxidative stress possibly caused by oxygen diffusion from the rumen wall tissue. In conclusion, we found that a high-concentrate diet induced SARA challenge largely affects the composition of the ruminal epimural bacterial microbiota without strong effects on the functional metatranscriptome.

The role of ceramide in the dairy cow: an overview of current understanding

J.W. McFadden and J.E. Rico
Cornell University, Department of Animal Science, 264 Morrison Hall, Ithaka, New York 14853, USA; jer358@cornell.edu

Dairy cows have developed distinct metabolic adaptations to ensure appropriate nutrient delivery to the fetus and the newborn calf. These homeorhetic adaptations including insulin antagonism and adipose tissue lipolysis which support nutrient partitioning. If uncontrolled, fatty liver may develop and trigger an associated metabolic disease, infertility, and compromise lactation in dairy cows. The application of mass spectrometry-based lipidomics has the potential to transform our understanding of the complex physiological and pathological processes that develop during gestation and lactation, and serve as a tool for discovering new interventions aimed at controlling nutrient partitioning in dairy cattle. To work towards this goal, we have focused our attention on the potential role of ceramide. The sphingolipid ceramide deserves attention because it is a potent inhibitor of insulin signalling in non-ruminants. Our investigations have revealed that the peripartal development of hyperlipidemia and hepatic steatosis develops with the accumulation of ceramide in plasma, low-density lipoproteins, liver, and skeletal muscle. Early evidence suggests that these metabolic events develop with reduced peripheral insulin sensitivity in all cows; albeit, ceramide accrual is exacerbated in early lactation animals with heightened adiposity. Our data also suggests that the bovine liver is a central organ for the synthesis of ceramides, actively exporting these sphingolipids within very low-density lipoproteins. In support, the induction of negative energy balance via nutrient-restriction as well as the intravenous infusion of triglycerides increases hepatic and circulating ceramide. Similarly, feeding palmitic acid, a required substrate for ceramide synthesis, promotes ceramide accumulation, relative to non-added fat controls and other fatty acids including stearic acid. Finally, we have shown that ceramide impairs insulin responsiveness in bovine primary adipocytes by inactivating protein kinase B-mediated insulin signaling. Collectively, our lipidomic studies reveal ceramide as a novel lipid mediator of nutrient partitioning in dairy cattle. On-going efforts focus at developing nutritional therapies that target ceramide to prevent metabolic disease and enhance performance in dairy cattle.

Session 10 — Theatre 6

Plasma cholesterol and adaptation of metabolism and milk production in feed restricted cows

J.J. Gross[1], A.-C. Schwinn[1], E. Müller[2], A. Münger[2], F. Dohme-Meier[2] and R.M. Bruckmaier[1]
[1]Veterinary Physiology, Vetsuisse Faculty, University of Bern, Bremgartenstrasse 109a, 3012 Bern, Switzerland, [2]Agroscope, Tioleyre 4, 1725 Posieux, Switzerland; josef.gross@vetsuisse.unibe.ch

Concomitantly to the negative energy balance in early lactation, low plasma concentrations of total cholesterol (TC) concentrations are observed. We have investigated if the level of TC in early lactation is related to short-term adaptations of metabolism and milk production, when cows are exposed to a transient concentrate withdrawal of one week that further aggravates energy deficiency. Multiparous Holstein cows (n=15) were investigated during a period of 21 days beginning at 24+-7 days in milk (mean+-SD). Cows were kept on pasture and received additional concentrate in experimental weeks 1 and 3, while in week 2 concentrate was withdrawn. Blood was sampled once and milk samples twice a day. Based on their average TC concentration during week 1 (prior to concentrate withdrawal), cows were retrospectively grouped into a high (HC; n=8, TC≥3.36 mmol/l) and a low TC group (LC; n=7, TC <3.36 mmol/l). A mixed model with group, time as fixed and cow as random effect was used to evaluate metabolic and performance responses in terms of concentrate withdrawal. Plasma concentrations of phospholipids and lipoproteins were higher in HC than in LC throughout the study (P<0.05). During concentrate withdrawal in week 2, milk yield and plasma concentrations of glucose and insulin decreased similarly in both groups, while milk fat and acetone content, and plasma concentration of BHB increased (P<0.05). Compared to initial values, plasma NEFA, TG and VLDL increased in both groups within 2 days after concentrate withdrawal (P<0.05), but declined again thereafter. Changes in insulin and glucose occurred within 1 day. Despite re-introduction of concentrate in week 3, milk yield in HC was lower compared with week 1 (P<0.05) and not different between weeks 2 and 3 (P>0.05), whereas milk yield in LC fully recovered. Activity of aspartate aminotransferase was higher in HC than in LC cows in week 2 (P<0.05). In conclusion, circulating cholesterol concentrations are associated with the extent of short-term adaptation responses to energy availability in early lactation.

Session 10 — Theatre 7

Liver damage and learning ability in female chicks

L. Bona[1], N. Van Staaveren[1], B. Pokharel[1], M. Van Krimpen[2] and A. Harlander-Matauschek[1]
[1]University of Guelph, Animal Biosciences, N1G 2W1, Guelph, Canada, [2]Wageningen Livestock Research, P.O. Box 338, 6700 AH Wageningen, the Netherlands; aharland@uoguelph.ca

In mammals, consumption of low protein energy-rich (LPER) diets increases susceptibility to liver damage and can have adverse effects on cognition. However, the effects of LPER diet on physical and mental welfare in broilers are unknown. Forty female Ross broiler chicks (1d old) were housed in pens fed a standard commercial diet. At 18-20d, half of the birds were gradually introduced to a control (19% CP, 3,200 kcal/kg ME) or LPER (17% CP, 3,300 kcal/kg ME) diet, which was the main diet from 21-51d. Visual discrimination training (1-10d) was done in a Y-maze until 80% of birds reached the learning criterion of at least 5/6 correct trials for 2 consecutive days. Reversal of the task occurred at 38-46d with the same criteria. Blood samples were collected (17-18d, 46d) and plasma levels of ammonia (NH4), alanine aminotransferase (ALT), aspartate aminotransferase (AST) and gamma-glutamyl transferase (GGT) analysed. Birds were euthanized at 52d and liver haemorrhage (0-5) and colour (1-5) assessed. The effect of LPER diet on indicators of liver damage and ability to learn a visual discrimination reversal task was assessed using generalized linear mixed models. All chicks had signs of liver haemorrhage, liver colour scores, and AST levels >230 U/l indicative of liver damage. LPER birds that successfully completed the reversal discrimination task tended to have lower haemorrhagic score (0.98±0.09) and had lower AST:ALT ratio (214.8±18.95) than birds fed a control diet (1.29±0.09, P=0.05 haemorrhagic score; 297.4±20.88, P<0.05 AST:ALT ratio). No difference was observed between LPER (6.2±0.47) and control chicks (5.6±0.49) in the number of sessions needed to successfully complete the reversal discrimination learning task, however LPER birds (3.2±0.04 kg) had lower body weights than control birds (3.5±0.04 kg, P<0.01) which could have influenced their ability to complete the task. This suggests that a LPER diet does not increase susceptibility to liver damage and does not impact reversal discrimination learning in broiler chicks. However, the high prevalence of liver damage in broiler chicks as shown by haemorrhage are of concern and highlights a need for further research on liver health.

Systems biology of amino acid use by mammary gland: milk protein synthesis and beyond

J. Loor[1] and Z. Zhou[2]
[1]University of Illinois, Animal and Nutritional Sciences, 1207 West Gregory Drive, 61801, Urbana, Illinois, USA, [2]Clemson University, Animal and Veterinary Sciences, 130 McGinty Ct., 29634, Clemson, South Carolina, USA; jloor@illinois.edu

The continued demand for producing high-quality milk in a more efficient way is a major driver for ongoing efforts to increase our understanding of mammary nutrient use. Although the role of essential amino acid (EAA) supply and energy metabolites such as glucose and acetate in the control of milk protein synthesis is well-established, less is known about the gene networks, signalling proteins, and their interactions. Knowledge accumulated in the last 10-15 years on the transcriptional and posttranscriptional regulation of genes coding for proteins involved in the synthesis of protein underscores their relevance in determining EAA uptake by the mammary gland. The interactions between insulin and mechanistic target of rapamycin kinase (mTOR) signalling pathways explain some of the nutritional effects on bovine milk protein synthesis. Besides the well-known role of STAT5, the fact that bovine casein genes have a number of transcription factor binding sites (conserved across mammalian genomes) underscores the complexity of transcriptional regulation. Methylation status of casein genes and certain microRNA (epigenetics) exert control on rate of transcription. Amino acid taste 1 receptors can sense extracellular EAA and activate mTOR in mammary cells via calcium signalling. *In vitro*, these pathways respond differently to profile and availability of EAA. Recent *in vivo* data point at interactions between the methionine cycle and polyamine synthesis pathway in the regulation of milk protein synthesis. Furthermore, some evidence suggests that EAA signalling through these pathways can also impact lipogenesis and oxidative stress status in mammary cells. Integrating the various components regulating milk protein synthesis within mammary gland using systems biology approaches should allow for continued progress in understanding the multi-faceted role of EAA supply.

Session 10 Theatre 9

Unraveling the genetic background of αs1- and αs2-casein phosphorylation in Dutch Holstein Friesian

Z.H. Fang[1,2,3], H. Bovenhuis[3], P. Martin[2] and M.H.P.W. Visker[3]
[1]Institute of Agricultural Sciences, Animal Genomics, ETH Zurich, 8092 Zurich, Switzerland, [2]INRA, AgroParisTech, Université Paris-Saclay, Génétique Animale et Biologie Intégrative, Domaine de vilvert, 78350 Jouy-en-Josas, France, [3]Wageningen University and Research, Animal Breeding and Genomics, P.O. Box 338, 6700 AH Wageningen, the Netherlands; zih-hua.fang@usys.ethz.ch

Phosphorylation of caseins (CN) is a crucial post-translational modification that allows caseins to form colloid particles known as casein micelles. The α_{s1}- and α_{s2}-CN show varying degrees of phosphorylation (isoforms) in cow's milk and are suggested to be more relevant for stabilizing internal micellar structure than β- and κ-CN. Differences between cows in α_{s1}- and α_{s2}-CN phosphorylation affect the milk technological properties and are, therefore, highly relevant for dairy product manufacturing. Previous studies suggest that variation in the degree of phosphorylation of α_{s1}- and α_{s2}-CN is to a great extend determined by genetic factors. In this study, we aimed to identify genomic regions associated with individual α_{s2}-CN phosphorylation isoforms and the relative abundance of α_{s1}- and α_{s2}-CN isoforms with higher degrees of phosphorylation. Genome wide association was studied using 50K SNP genotypes for 1,857 Dutch Holstein-Friesian cows. Phenotypes comprised milk protein composition determined by capillary zone electrophoresis, milk production traits, and content and yield of phosphorus in milk. A total of 10 QTL regions were identified for all studied traits on 10 *Bos taurus* autosomes (BTA 1, 2, 6, 9, 11, 14, 15, 18, 24 and 28). Regions associated with multiple traits were found on BTA 1, 6, 11, and 14. Our results show that the effects of detected QTL regions are divided into two groups: one is associated with milk protein synthesis including regions on BTA 1 and BTA 6 (casein gene cluster), and the other one with the relative abundance of α_{s1}- and α_{s2}-CN isoforms with higher degrees of phosphorylation including regions on BTA 1 (*SLC37A1*), BTA 11 (*PAEP*) and BTA 14 (*DGAT1*). Furthermore, the effects of the detected QTL specifically on phosphorylation might be related to their effects on the change of milk synthesis and regulation of phosphorus output in milk.

Session 10 Theatre 10

Transcriptional regulations of milk protein and lactose synthesis by the diet in dairy cows

M. Boutinaud, J. Guinard-Flament, S. Lemosquet, V. Lollivier, F. Dessauge, L. Maubuchon, F. Barley, L. Herve and H. Quesnel
INRA, UMR PEGASE, 16, le Clos, 35590 Saint Gilles, France; marion.boutinaud@inra.fr

Diet is one of the main factors affecting milk yield and lactose and protein synthesis in dairy cows. It is well known that the synthesis of milk lactose and protein results from the differentiation of mammary epithelial cells (MEC) which are able to express genes coding for alpha-lactalbumin (LALBA), several caseins and specific glucose transporters. While the effects of mammary development and galactopoietic hormones on the expression of these genes are known, the impact of nutrition has been less investigated. We and other carried out studies to investigate the expression of genes involved in lactose and protein synthesis in response to feed restriction (4 trials) or increasing metabolizable protein supply in dairy cows (2 trials). The gene expression was evaluated by qPCR in mammary tissue samples or MEC purified from milk. Feed restriction induced reductions in milk lactose and protein synthesis in all trials but its effects on gene expression differs between studies. In two studies, moderate feed restriction (70% of allowance) induced a decrease in the transcript levels of SLC2A1, one of the main transmembrane transporters of glucose in MEC. Moderate feed restriction increased milk casein and LALBA transcripts in early lactation (in 1 trial) but not on a later stage (in 4 trials). In contrast, a severe feed restriction (- 50%) in early lactation inducing a 38% milk yield loss was associated with a decreases in LALBA and K casein gene expression without modifying alphaS1 casein gene expression. Decreasing protein supply in 2 trials (from 125 to 70% or from 120 to 100% of allowance) decreased milk and lactose yields and surprisingly increased the expression of LALBA transcript. Thus, the effects of feed restriction or dietary protein levels on mammary transcripts involved in milk lactose and protein synthesis likely depends on the severity of the restriction and the stage of lactation. In some cases, manipulating the diet induced a variations in transcript expression opposite to that observed for milk lactose and protein yields suggesting that other mechanisms than modification of gene expression may modulate protein and lactose synthesis.

Session 10 Theatre 11

Impact of breeding factors on milk spontaneous lipolysis in dairy cows

E. Vanbergue[1,2], J.L. Peyraud[2] and C. Hurtaud[2]
[1]Idele, Monvoisin, 35650 Le Rheu, France, [2]PEGASE, INRA, Agrocampus Ouest, Le clos, 35590 Saint Gilles, France; elise.vanbergue@idele.fr

Lipolysis is an enzymatic reaction which leads to off-flavour in milk and impairs technological properties of milk. Lipolysis corresponds to the hydrolysis of milk fat located in milk fat globules (MFG) by the lipoprotein lipase (LPL) and its cofactors, leading to free fatty acid accumulation. Variability of spontaneous lipolysis (SL) levels depend on individuals and on breeding systems. Five trials were set up to understand SL variations. The effect of breed, parity, physiological stage, milking frequency, milking time, feeding systems and lipid supplementation on SL were evaluated by continuous designs. The effects of milking time, feeding restriction and nature of forage on all enzymatic system were evaluated by two cross-over studies. Data were analysed using linear models. Cows from the five trials could be classified in two groups according to their phenotype: 'susceptible' and 'non-susceptible' to SL, confirming the strong impact of the individual effect. Among 'susceptible' cows, we confirmed the effects of breed, parity, physiological stage, milking time, milking frequency and feeding systems. SL was higher in Holstein cows' milk than in Normande cows' milk ($P<0.001$). For high merit multiparous cows, SL was higher in early lactation than in mid lactation probably due to negative energy balance. SL was also higher in late lactation than in mid and early lactation ($P<0.001$). SL was higher in evening milks than in morning milks ($0.03<P<0.001$) due to 10 h/14 h milking interval or circadian rhythms of cows. SL increased with the increase in milking frequency ($P<0.001$), with feeding restriction ($P<0.01$), when fresh grass and conserved grass based diets were replaced by maize silage based diets ($P<0.05$) and with some lipid supplementation ($P<0.001$ for microalgae). The mechanism of action might involve proteose peptone 5, a degradation product of casein which acts as a SL inhibitor. The MFG membrane might play an important role on MFG integrity, LPL and MFG interactions, and cofactors balance. LPL activity did not impact SL variations. More investigation should be done to rank the relative impact of zootechnical and biochemical factors on SL.

Acute phase protein and cathelicidin gene expressions in milk and blood of SRLV-infected dairy goats

D. Reczynska[1], M. Czopowicz[2], M. Zalewska[1], J. Kaba[2], E. Kawecka[1], P. Brodowska[1] and E. Bagnicka[1]
[1]Institute of Genetics and Animal Breeding, Postepu str. 36A, 05-552 Jastrzebiec, Poland, [2]Veterinary Medicine Faculty, Warsaw University of Life Sciences, Nowoursynowska 159C, 02-787 Warsaw, Poland; d.reczynska@ighz.pl

Since the health status of animal affects production, its constant control is essential. One of the first defence lines of vertebrates are acute phase proteins (APPs). Their levels change during many diseases. One of the goat pathogens is small ruminant lentivirus (SRLV), which causes caprine arthritis-encephalitis leading to premature culling. The aim of the study was to estimate the expressions of APPs and cathelicidins mRNA and protein levels in milk somatic cells (MSC) and blood leukocytes (BL), and in milk and serum of SRLV-infected and non-infected goats. In this study, dairy goats (n=24) were divided into two groups: non-infected (n=12) and naturally SRLV-infected (n=12); confirmed infection at least two years before the start of observations) but without any clinical signs. Parity was identical in both groups. Milk and blood samples were collected five times during lactation (1st, 30th, 60th, 120th, 180th day). The expression of genes was measured using qPCR with cyclophilin A as a reference gene. The protein level was established using ELISA. The only difference between the groups was higher concentrations of Fb (APP) and MAP28 (cathelicidin), and lower level of Cp (APP) in serum of infected goats (P<0.05). On the other hand, in the infected group, transcripts of *MAP28* were higher in MSC than in BL (P<0.05), whereas the CRP (APP) concentration was lower (P<0.05) in milk than in serum. The changes were tissue-specific but the gene expression pattern was almost the same in blood and milk in both groups. It may mean that there are small differences in the immune system between uninfected and SRLV-infected goats without clinical symptoms of caprine arthritis-encephalitis. The differences between mRNA and protein levels can be explained by epigenetic regulations. Financed by NSC Poland, 2016/21/N/NZ9/01508, 2013/09/B/NZ6/0351.

Acetylation of mitochondrial proteins during negative energy balance

M. Garcia Roche[1,2], A. Casal[1], M. Carriquiry[1], A. Cassina[2] and C. Quijano[2]
[1]Departamento de Producción Animal y Pasturas, Facultad de Agronomía, Universidad de la República, Av. Gral Garzón 809, 11200 Montevideo, Uruguay, [2]CEINBIO, Facultad de Medicina, Universidad de la República, Av. Gral Flores 2125, 11200 Montevideo, Uruguay; garciaroche.m@gmail.com

Negative energy balance (NEB) in dairy cows is a critical period that may lead to excessive lipid mobilization and fatty liver. Mitochondrial dysfunction is involved in fatty liver pathogenesis. We demonstrated that mitochondrial function was impaired in grazing cows compared to TMR-fed cows during early lactation NEB. Thus, our objective was to compare mitochondrial function and lysine-acetylation (AcK), a post-translation modification that regulates mitochondrial energy metabolism, in this period. Twenty-four Holstein-Friesian multiparous cows (664±65 kg BW, 3.0±0.4 BCS, spring calving) were assigned in a randomized block design to G0: total mixed ration (TMR) fed *ad libitum* (70% forage: 30% concentrate) or G1: grazing plus supplementation from 0 to 180 days postpartum (DPP). G1 cows grazed on Festuca arundinacea or *Medicago sativa* in one or two sessions depending on heat stress conditions (20-30 kg DM/cow/day allowance) and were supplemented with 5.4 kg DM of a commercial concentrate or 50% *ad libitum* TMR. From 180 to 250 DPP, all cows grazed Festuca arundinacea (10 h; 30 kg DM/cow/day allowance) and were supplemented with 50% *ad libitum* TMR. Liver biopsies and blood samples were collected at 35 (NEB) and 250 DPP (positive energy balance: PEB). Data were analysed as repeated measures in a mixed model including feeding strategy, DPP and their interaction. Plasma B-hydroxybutyrate (BHB) and liver triglyceride (TAG) were measured to assess NEB while protein AcK was studied in isolated mitochondria. Plasma BHB and liver TAG were two-fold greater (P<0.001) in 35 DPP in G1 cows. Similarly, mitochondrial AcK was 76% greater in 35 DPP (1.76±0.15 vs 1.04±1.5 relative intensity, P<0.01) in G1 cows. Neither plasma BHB, liver TAG or mitochondrial AcK changed in G0 cows. NEB markers and acetylation were positively correlated (r>0.5 and P<0.001), while maximum respiratory capacity and acetylation were negatively correlated (r=-0.6 and P<0.001). These results indicate that acetylation occurs during NEB and may be one of the underlying mechanisms in mitochondrial dysfunction.

Session 10 Poster 14

Genetic evaluation of hepatic markers related to metabolic adapation and robustness in dairy cows

N.-T. Ha[1], J.J. Gross[2], A.R. Sharifi[1], M. Schlather[3], C. Drögemüller[4], U. Schnyder[5], F. Schmitz-Hsu[6], R.M. Bruckmaier[2] and H. Simianer[1]

[1]Animal Breeding and Genetics, University of Goettingen, Albrecht-Thaer-Weg 3, 37075 Göttingen, Germany, [2]Veterinary Physiology, Vetsuisse Faculty, University of Bern, Bremgartenstrasse 109a, 3012 Bern, Switzerland, [3]School of Business Informatics and Mathematics, Stochastics and Its Applications; Univ. Mannheim, A5, 6 B 117, 68161 Mannheim, Germany, [4]Institute of Genetics, University of Bern, Bremgartenstrasse 109a, 3012 Bern, Switzerland, [5]Qualitas AG, Chamerstrasse 56, 6300 Zug, Switzerland, [6]Swissgenetics, Meielenfeldweg 12, 3052 Zollikofen, Switzerland; josef.gross@vetsuisse.unibe.ch

The early lactation period in dairy cows is marked by severe metabolic stress resulting from a discrepancy between the high energy demand for rapidly increasing milk production and the concomitantly limited feed intake. The need to mobilise body reserves to compensate energy and nutrient deficiency is associated with a high risk to develop production diseases. Insufficient metabolic adaptation results in increased susceptibility to health problems. However, even under the same environmental factors and production level, the variability of how each cow deals with metabolic load is substantial, leading to the hypothesis that there might be an underlying genetic basis. In particular, the liver is crucially involved in the control of metabolic stress. Our objective was to elucidate the genetic basis of metabolic adaptation (with focus on hepatic metabolism) from a genomic, transcriptomic and breeding point of view. With a genome-wide association study and gene expression analyses, we identified several genetic factors associated with metabolic adaptation during the transition period of dairy cows. We further developed a robustness measure in form of breeding values for each bull describing how sensitive his daughters are to the metabolic load during transition. Our results strongly support the hypothesis of the genetic basis of metabolic robustness and provide an effective tool for the dairy industry to breed for metabolically robust dairy cows.

Session 10 Poster 15

Liver proteomics in response to dietary lipid sources in Nile tilapia (*Oreochromis niloticus*)

S. Boonanuntanasarn[1], P. Rodrigues[2], D. Schrama[2] and R. Duangkaew[1]

[1]Suranaree University of Technology, Institute of Agricultural Technology, 111 University Avenue, Suranaree, Muang, Nakhon Ratchasima, 30000, Thailand, [2]Universidade do Algarve, Campus de Gambelas, CCMAR, FCT, Edifício 7, 8005-139, Faro, 8005-139, Portugal; surinton@sut.ac.th

Different dietary lipid sources may reflect on not only fatty acid contents in fish body but also lipid metabolism. Recently, proteomics which refers to the large-scale experimental analysis of protein expression has been used as a tool to study biological mechanism in physiological pathways of various living organisms. Therefore, this research aimed to investigate differentially metabolic and physiological pathways which response to dietary lipid sources in Nile tilapia liver. Two-dimension polyacrylamide gel electrophoresis (2D-PAGE) was used to employ the proteomics in responses to various dietary lipid sources including palm oil (source of saturated fatty acid), linseed oil (source of n3-poly unsaturated fatty acid (n3-PUFA)) and soybean oil (source of n6-PUFA). Tilapia (initial weight ~250 g) were fed with three different dietary oil for 90 days. Liver (4 replicates/treatment diet) were collected and performed proteomic study. Total 183 protein spots were differentially expressed, and twenty-four protein spots were identified. The differential abundance of hepatic proteins was involved in lipid, carbohydrate and protein metabolism. In addition, several proteins that are involved in anti-oxidation, cytoskeletal characterization and immune processes were detectable to be different responding to dietary lipid sources. Therefore, different dietary lipid sources induce the differences in a number of metabolic and physiologic pathways in liver which would provide information how tilapia utilize different dietary lipids.

Session 10 Poster 16

Effect of over-conditioning on mTOR and ubiquitin proteasome gene expression in muscle of dairy cows

M.Hosseini Ghaffari[1], H. Sadri[1,2], K. Schuh[1,3], L. Webb[1], G. Dusel[3], C. Koch[4] and H. Sauerwein[1]
[1]University of Bonn, Institute of Animal Science, Physiology and Hygiene Unit, Katzenburgweg 7, 53115 Bonn, Germany, [2]University of Tabriz, Department of Clinical Science, Faculty of Veterinary Medicine, 5166616471 Tabriz, Iran, [3]University of Applied Sciences Bingen, Department of Life Sciences and Engineering, Animal Nutrition and Hygiene Unit, 55411 Bingen am Rhein, Germany, [4]Educational and Research Center for Animal Husbandry, Hofgut Neumuehle, 67728 Muenchweiler an der Alsenz, Germany; sadri.ha@gmail.com

Mobilisation of body reserves during the negative energy balance in early lactation concerns mainly adipose tissue and skeletal muscle. So far, research was mainly focusing on adipose tissue, whereas the regulation of muscle protein turnover has hardly been investigated. Our objective was to investigate the effects of body condition on the mRNA abundance of key genes of the mTOR pathway and the ubiquitin-proteasome-system (UPS) in skeletal muscle of dairy cows in late pregnancy and early lactation. Multiparous Holstein cows were initially pre-selected 15 weeks ante partum from the entire herd based on their previous course of body condition score (BCS) and backfat thickness (BFT) and were classified as either normal-conditioned (NBCS, n=18, <3.5 BCS or <1.2 cm BFT) or over-conditioned cows (HBCS, n=18, >3.75 BCS or >1.4 cm BFT). Muscle (M. semitendinosus) biopsies were collected at d -49, +3, +21, +84 relative to calving, RNA was extracted and mRNA abundance determined via qPCR. The mRNA abundance of mTOR and 4E-BP1 in muscle at d +21 was greater in HBCS compared to NBCS cows (P<0.05), whereas S6K1-mRNA was not different between groups. The mRNA abundance of UBA1, UBE2G1, atrogin-1 at d +21, and MuRF-1 at d 3 in muscle was greater in HBCS compared to NBCS cows (P<0.05), whereas UBE2G2-mRNA was not different. The results suggest that muscle protein turnover may be more stimulated in over-conditioned cows as indicated by the changes in the mRNA abundance of key components of mTOR signaling and UPS.

Session 10 Poster 17

Performance and gut morphology of Ross 308 broiler chickens fed low tannin sorghum based diets

M. Mabelebele[1], T.G. Manyelo[2], J.W. Ngambi[2] and D. Norris[3]
[1]University of South Africa, Agricuture and Animal Health, 100 Christiaan de Wet, Florida Park, Roodepoort, Johannesburg, Gauteng, South Africa, 1724, Gauteng, South Africa, [2]University of Limpopo, Agricultural Economics and Animal Production, Private Bag X1, 1106, Sovenga, South Africa, [3]Botswana International University of Science and Technology, Science and Technology, Palapye, 0005, Palapye, Botswana; mabelebelem@gmail.com

An experiment was conducted to evaluate the influence of sorghum as replacement of maize on the performance and carcass traits of Ross 308 broiler chickens aged 1-42 days. One hundred and sixty Ross 308 broiler chickens weighing 41±6 g were assigned to a complete randomized design with five dietary treatments replicated 4 times with eight birds per replicate. Five diets were formulated to contain low tannin white sorghum replacement levels of 0% (M100S0), 25% (M75S25), 50% (M50S50), 75% (M25S75) and 100% (M0S100). Body weight and feed intake were measured on weekly basis and used to calculate feed conversion ratio (FCR). A total of 80 birds were slaughtered at 21 and 42 days of age, intestinal morphology and carcass characteristics were evaluated. The general linear model procedure of the statistical analysis software (SAS) was used to analyse the data. When maize was replaced by low tannin sorghum meal feed intake, body weight and feed conversion ratio were not affected (P>0.05) at ages 1-42 days. Replacing maize with low tannin sorghum diet had no effect (P>0.05) on crop, gizzard, caecum, large intestine weights and intestinal morphology of Ross 308 broiler chickens aged 1-21 days of age. Drumstick, thigh and wing weights, drumstick, thigh and wing colour, meat sensory evaluation and meat pH of male Ross 308 broiler chickens were not affected (P>0.05) by replacement of maize with sorghum-based diet. It can therefore, be concluded that low tannin sorghum can replace maize at 25, 50, 75 and 100% levels without causing adverse effects on overall productivity of broiler chickens aged 1-42 days.

Session 10 Poster 18

Metabolic and endocrine profile of dairy cows with or without grazing pastures

M. Carriquiry, M. Garcia-Roche, A. Casal, A. Jasinsky, M. Ceriani and D.A. Mattiauda
Facultad de Agronomía, Universidad de la República, Departamento de Producción Animal y Pasturas, Ave. Garzón 780, 12400 Montevideo, Uruguay; mariana.carriquiry@gmail.com

The aim of the study was to evaluate the impact of two feeding strategies (with or without pasture grazing) on metabolic and endocrine profiles of dairy cows during the first 180 days of lactation (DOL). Twenty-four multiparous Holstein cows (664±65 kg BW, 3.0±0.4 BCS, spring-calving) were assigned in a randomized block design to either (G0) total mixed ration (TMR) fed *ad libitum* (70% forage: 30% concentrate) or (G1) grazing plus supplementation from 0 to 180 DOL. The G1 cows grazed on *Festuca arundinacea* or *Medicago sativa* (forage allowance = 20-30 kg DM/cow/d) in two (18 h) or one (6 h) grazing sessions depending on heat stress conditions and were supplemented with 5.4 kg DM/d of a grain-based ration or 50% of the *ad libitum* TMR (16 kg DM/d), respectively. Cows were milked twice a day and milk yield was determined daily. Cow BCS and blood samples were collected every 14 days from -14 to 180 DOL. Data were analysed as repeated measures in a mixed model including treatment, DPP and their interaction as fixed effects, block and cow as random effects and calving date as a covariate. Average milk yield did not differ between treatments (32.6 vs 31.4±1.3 kg/d for G0 and G1, respectively) but cow BCS was 0.25 units greater ($P<0.01$) for G0 than G1 cows from 70 to 180 DOL Concentrations of NEFA and BHB peaked in early lactation and plasma NEFA at 21 DOL was greater ($P<0.01$) but plasma BHB between 35 and 49 DOL was less ($P<0.01$) for G0 than G1 cows. Plasma glucose was greater ($P<0.01$) for G0 than G1 cows between 35 and 49 DOL and at 120 DOL. Similarly, insulin concentrations were greater ($P<0.01$) for G0 than G1 between 60 to 180 DOL. However, the revised quantitative insulin sensitivity check index was reduced ($P<0.01$) for G0 than G1 cows from 21 to 120 DPP. Concentrations of adiponectin and leptin did not differ between treatments but adiponectin to leptin ratio tended to be less ($P<0.06$) for G0 than G1 cows. Although milk production did not differ between cow groups, the metabolic and endocrine profiles reflected a better metabolic status in G0 cows which could impact in health and reproduction responses.

Session 10 Poster 19

Feeding ensiled olive cake affected fat and fatty acid composition of cow milk

M.C. Neofytou[1], D. Sparaggis[2], C. Constantinou[1], S. Symeou[1], D. Miltiadou[1] and O. Tzamaloukas[1]
[1]Cyprus University of Technology, Department of Agricultural Sciences, Biotechnology and Food Science, P.O. Box 50329, Lemesos, Cyprus, [2]Agricultural Research Institute, Animal Production Section, P.O. Box 22016, Nicosia, Cyprus; ouranios.tzamaloukas@cut.ac.cy

The objective of the present study was to investigate the addition of ensiled olive cake (OC) in the diet of lactating cows regarding yield, composition and fatty acid (FA) profile of milk. Twenty four Holstein Friesian lactating cows were allocated into two groups of 12 animals (homogenous for age, milk yield, period of lactation and body weight) and given the following iso-energetic and iso-nitrogenous feeding regimes: (1) no inclusion of ensiled olive cake (control group), and (2) inclusion of 5 kg/day/cow (OC group), while the other ingredients of the diets remained the same covering the animal requirements. Measurements of milk yield, milk content (fat, protein, lactose) and FA profile were taken. There was no significant differences in milk yield between the two groups. Regarding milk composition, only fat was affected by the treatment. Milk obtained from the OC group was higher ($P<0.05$) in fat % compared with the control group (average values of 3.5 and 3.2, for the OC and control group, respectively). The FA profile was also affected by the treatment and lipid unsaturation was increased by the OC addition in the cow diets. The major differences between groups were observed for saturated FA, particularly C10:0, C12:0, C14:0 and C16:0, which were decreased ($P<0.05$), and mono-unsaturated FA, such as oleic acid, which were increased ($P<0.05$) in milk of OC cows (expressed in g/100 g of fat, average values for oleic acid was 23.4 and 20.1 for milk of OC and control cows, respectively). In contrast, the amount of poly-unsaturated FA was not affected by the feeding treatment. Overall, the results support the use of ensiled OC in diets of high yielding dairy cows as a means to reduce saturated and increase beneficial monounsaturated FAs without adverse effects on milk yield.

Fresh herbage as winter TMR forage basis: effects on dairy cows' milk yield and fatty acid profile

D. Enriquez-Hidalgo, C. Sanchez, S. Escobar, M. De Azevedo, D.L. Teixeira and E. Vargas-Bello-Pérez
Departamento de Ciencias Animales, Pontificia Universidad Católica de Chile, Santiago, Chile; daniel.enriquez@uc.cl

Alfalfa hay/maize silage mixture is the forage basis used for winter TMR in dairy systems. Fresh forages can improve milk quality and systems sustainability. Berseem clover (BC) is a productive winter forage commonly sown with annual ryegrass. Winter growth of alfalfa is poor and, contrary to BC, can cause bloat in cows. The objective of this study was to evaluate the effect of using fresh alfalfa (ALF), fresh mixed annual ryegrass/BC (MIX) or alfalfa hay/maize silage (CON) as different forage basis for TMR during the winter period on dairy cows' milk yield and fatty acid (FA) profile. Lactating dairy cows (n=21) were allocated to each TMR according to a 3 diets × 3 periods Latin Square design. Cows were individually stalled and received the TMR *ad libitum*. TMRs were isoenergetics and formulated as a 50:50 forage-to-concentrate ratio. Each period consisted of 14d adaptation and 7d for data collection. TMR samples (3/wk) were analysed for DM, crude protein (CP) contents and DM digestibility (DMD). Feed intake and milk yield were estimated daily. Individual milk composition and FA were estimated once per period. Data were analysed using linear models including TMR and period in the model. TMRs differed (<0.001) in DM (ALF: 392; CON: 456; MIX: 402; 8.7 g/kg), DMD (ALF: 784; CON: 815; MIX: 779; 2.41 g/kg) and CP (ALF: 165; CON: 164; MIX: 159; 1.8 g/kg DM) contents. Cows had similar intakes (24.1±1.06 kg DM/d), milk yield (31.5±1.51 kg/d) and composition (3.82±0.216 fat% and 3.61±0.07 protein%). Milks differed in rumenic FA content (ALF: 3.76; CON: 2.40 and MIX: 3.30; 0.163 g/100 g FA; P<0.05). CON milk had greater mono-unsaturated FA (27.1 vs 24.7; 0.59 g/100 g FA; P<0.01) and vaccenic FA (0.63 vs 0.38; 0.075 g/100 g FA; P<0.05) contents than the others. ALF milk had greater poly-unsaturated FA contents than the others (5.95 vs 4.90; 0.207 g/100 g FA; P<0.001). MIX milk had greater saturated FA content than ALF and CON milk (70.3 vs 67.2; 0.78 g/100 g FA; P<0.05). The use of fresh forages as forage basis for TMR during the winter period did not affect milk yield or composition, but improved FA profile from human health standpoint. The study was supported by project FONDECYT 11160697.

Bioactive properties of organic milk in Poland

T. Sakowski[1], K.P. Puppel[2] and G. Grodkowski[1]
[1]Institute of Genetics and Animal Breeding, Jastrzębiec, Postępu 36A, 05-552, Poland, [2]Warsaw University of Life Sciences, Animal Breeding & Production Department, Ciszewskiego 8, 02-786, Poland; g.grodkowski@ighz.pl

The aim of this study was to determine the nutritional value of organic milk in Poland, investigate the influence of diet on antioxidant capacity, and to examine the effect of season on the bioactive properties of milk from organic farms. From April to September (at monthly intervals), 120 milk samples were collected from organic farms during indoor feeding season (IDS) and pasture feeding season (PS). Milk obtained during PS was found to have a higher fat content, slight but significantly lower protein content compared with milk from IDS. The study showed that the content of monounsaturated fatty acids (MUFA) in milk fat was strongly linked to the concentration of polyunsaturated fatty acids (PUFA) and, to a lesser extent, on the supply of MUFA. The IDS data in May (concentration of CLA and PUFA n-3) showed the highest values compared with the June, July, August and September. In turn, the highest concentration of MUFA was demonstrated in August. This research was supported by the National Science Center and executed within the project CoreOrganicPlus/2-ORG-COWS/125/IGHZ/2015.

Session 10 Poster 22

Comparison of the efficacy of a 'homemade' versus commercial indwelling jugular catheter for sheep

S. Turner, B. Malaweera, M. Payne, L. Carroll, C. Harvey-Clark and D. Barrett
Dalhousie University, 58 Sipu Awti, B2N 5E3, Truro, NS, Canada; david.barrett@dal.ca

Scientists who use animals strive to use the best tools available to optimize animal health/welfare and data/sample collection. Indwelling jugular vein catheters are one of the tools used for blood collection and injection of solutions. The objective of this study was to compare the efficacy of a 'homemade' vs commercial indwelling jugular catheter for blood collection from sheep. One catheter (Access™ Technologies polyurethane Hydrocoat™ (AC; commercial; 0.6 mm inner diameter (i.d.); 1.1 mm o.d.; USA) or TE Connectivity Ltd. PVC tubing (TE; 'homemade'; 1.0 mm i.d.; 1.5 mm o.d.; Australia)) was inserted into each ewe's (n=5 per catheter) left or right jugular (random; balanced) the day before blood sampling. The largest i.d./o.d. AC and TE catheters made had to fit through a 14G catheter. Blood was collected every 3 h starting at 02:00 h (0 h), ending 66 h later. At 0 h, a 10 ml sample was collected, followed by 5 ml, and then alternating volumes were collected for each successive sample. The collection time for each sample was noted. Personnel could see the blunt needle (two colours) and tubing (two colours/textures) that had to be exposed for collection. Catheters were flushed with saline and filled with heparinized saline after each collection. Data was analysed by ANOVA and Tukey Test. There was a time effect ($P<0.005$), but no catheter or interaction effect on the sampling time ($P>0.05$). Sampling was slower at 6 h (8.0±0.7 min) than 63 h (4.2±0.7 min) and slower at 9 h (8.1±0.7 min) than both 45 h (4.5±0.7 min) and 63 h ($P<0.05$). There was a tendency for a catheter effect ($P=0.050$), but no time or interaction effect on the difference from the catheter group mean for sampling time ($P>0.05$). The sampling time tended to be more variable for the AC (1.8±0.2 min) than TE ewes (1.6±0.1 min; $P=0.050$). The min. and max. sampling time per ewe were similar for each catheter type ($P>0.05$). Two AC ewes had a slow or no draw catheter for 18 h or 48 h. One TE ewe had a kinked catheter. The blunt needle came out once during sampling for two AC ewes. The TE catheters appear to be better than the AC catheters for collecting samples quickly, probably because of the larger i.d.

Session 11 Theatre 1

The relationship between milk composition associated with ketosis using artificial neural networks

E.A. Bauer, E. Ptak and W. Jagusiak
University of Agriculture in Krakow, Department of Genetic and Animal Breeding, Al.A. Mickiewicza 21, 31-120, Poland; e.bauer@ur.krakow.pl

Clinical ketosis is one of the most frequent metabolic diseases in high-producing dairy cattle during lactations. The procedure currently used for detection of cows at risk of ketosis is not effective enough. Many cows are diagnosed too late, causing serious economic losses. The aim of the study was to estimate relationship between milk composition and β-hydroxybutyric acid level in blood (BHBA), an indicator of ketosis with using artificial neural networks (ANN). An artificial neural network is a powerful computational technique for correlating data by using a large number of simple processing elements. The ANN simulator implemented in the STATISTICA12® package was used in the study. ANN was investigated with the use of one type of artificial neural networks: Multi-Layer Perceptron (MLP). The multilayer feed forward was used, which is the most popular of the many architectures currently available. The analyses included data set on population of 1,085 polish Holstein-Friesian cows made available by the Polish Cattle Breeders and Milk Producers. The cows were recorded over a period from 01.2012 till 03.2012. The calculations were made using the MLP and milk data composition as: milk yield(kg), protein(%), fat(%), fat-to-protein ratio(%), lactose(%), somatic cell count(cell/ml) and content of urea(mmol/l), β-hydroxybutyric acid (BHBA) ≥1.2 mmol/l, and acetone as input variables. The content of BHBA in blood was as target variable in all prediction models and was used as an output variable. Each of four functions (linear, logistic, exponential and hyperbolic tangent) was used as activation function in the hidden layer. Our results show that the benefit of milk composition information to predictive ability of the prediction model is high. By reducing the number of variables, preliminary results were obtained: at high parameters of sensitivity analysis for urea, acetone, fat to protein ratio. The study allows concluding that neural network model has met the expectations in terms of its predictive abilities in this matter.
Acknowledgement: The Research was financed by the National Science Center, Poland no.2017/25/N/NZ5/00793.

Session 11 Theatre 2

Temporal relationship between milk MIR predicted metabolic disorders and lameness events

A. Mineur[1], C. Egger-Danner[2], J. Sölkner[3], S. Vanderick[1], H. Hammami[1] and N. Gengler[1]

[1]ULiege-GxABT, Unité de Zootechnie, Passage des Déportés 2, 5030 Gembloux, Belgium, [2]ZuchtData, Dresdner Str. 89, 1200 Wien, Austria, [3]BOKU – Institut für Nutztierwissenschaften (NUWI), Gregor-Mendel-Straße 33, 1180 Wien, Austria; axelle.mineur@uliege.be

Lameness is an often occurring consequence of various metabolic disorders, such as sub-acute ruminal acidosis (SARA) or ketosis. Recent research showed that these metabolic disorders can be predicted with reasonable accuracy with mid-infrared (MIR) spectral data. In order to study the potentially complex temporal relationship between MIR predicted metabolic disorders and lameness events over the course of the lactation, data from 3,895 cows on 122 farms, representing the Simmental, Brown-Swiss and Holstein breeds. A total of 38,316 lameness and 11,419 MIR records were collected over a period from July to December 2014 through the Efficient Cow Project. Lactations were subdivided into 30 days lactation stage classes. Milk MIR predicted metabolites such as ketone bodies, acetone, citrates and fatty acids (C18:1cis9), and lameness scores were averaged over animals and these classes. In order to assess the temporal link between occurrences of metabolic disorder and lameness events, correlations were computed between averaged metabolites and lameness scores across the lactation stage classes. Correlations tended to be higher when comparing predicted metabolites with lameness in the three following months, rather than the same one. Results showed differences between breeds, Simmentals showing lower correlations than Holsteins or Brown-Swiss. Especially very early values for milk MIR predicted metabolites (first month), and therefore suspected metabolic disorders, were correlated more strongly to later occurring lameness events in Brown Swiss. In Holsteins, higher correlation between metabolites and lameness were observed during later lactation. In general, given the use of classes, the correlations tended to be unstable. Alternative methods, such as covariance functions, might therefore be useful to get a clearer picture. However these first results seem to support the idea of temporal relationships between metabolic disorders and later lameness events during the lactation.

Session 11 Theatre 3

Genetic correlations between methane production and traits from Polish national evaluation

M. Pszczola[1], M.P.L. Calus[2] and T. Strabel[1]

[1]Poznan University of Life Sciences, Department of Genetics and Animal Breeding, Wolynska 33, 60-637 Poznan, Poland, [2]Wageningen University & Research Animal Breeding and Genomics, P.O. Box 338, 6700 AH Wageningen, the Netherlands; mbee@up.poznan.pl

Methane (CH_4) emission is an environmentally and potentially economically important trait in dairy cattle breeding. So far genetic correlations between CH_4 production and other traits in dairy cattle remain largely unknown. We aimed to gain an insight into direction and strength of genetic correlations between CH_4 production and traits included in the breeding program for Polish Holstein-Friesian cows. Daily CH_4 production on 484 cows collected on two farms in western Poland was available. Methane concentration was measured during milking in Automated Milking System using the FTIR analyser. Daily CH_4 production was quantified in litres based on a ratio of measured CH_4 and CO_2 concentrations multiplied by an expected daily CO_2 production predicted based on productivity, live weight, and physiological status. Out of 484 cows with phenotypes, 281 cows were genotyped with the 50k SNP chip. Breeding values for 305d CH_4 production were estimated using single-step REML using a random regression test-day model. Breeding values for the traits included in Polish national evaluation were obtained from results of the National Genomic Evaluation. Genetic correlations were calculated as correlations between breeding values scaled by their reliabilities. The estimated genetic correlations were low to intermediate, reaching highest values for Chest Width (0.48). Only some type traits were negatively correlated to 305d CH_4 production with the lowest value of -0.16 for Rear Leg Set Side View. Methane 305d production had a low to moderate genetic correlations with traits currently included in the Polish breeding goal. Therefore, further selection in accordance with the current Polish breeding objective will lead to an absolute increase in CH_4 production per cow. Further studies are needed to verify the impact of the current Polish breeding goal on CH_4 production per kg of milk.

Usability of bacteriological milk samples for genetic improvement of udder health in Austrian cattle

M. Suntinger[1,2], B. Fuerst-Waltl[1], W. Obritzhauser[3], C. Firth[3], A. Koeck[2] and C. Egger-Danner[2]
[1]University of Natural Resources and Life Sciences, Department of Sustainable Agricultural Systems, Division of Livestock Sciences, Gregor-Mendel-Str. 33, 1180 Vienna, Austria, [2]ZuchtData EDV-Dienstleistungen GmbH, Dresdnerstraße 89/19, 1200 Vienna, Austria, [3]University of Veterinary Medicine, Institute of Veterinary Public Health, Veterinärplatz 1, 1210 Vienna, Austria; suntinger@zuchtdata.at

For the estimation of routine mastitis breeding values the state of infection is currently not considered. Through the integration of results of bacteriological milk samples into the central Austrian cattle database, developed in a pilot study, the first standardised data set including milk culture results was available. Therefore the objectives of this study were to derive traits from this data source for potential use in breeding for improved mastitis resistance and to investigate genetic associations to udder health traits routinely recorded in Austrian Fleckvieh. Traits were defined as binary, apart from SCS where measures were available continuously. Multivariate analysis using a linear animal model was applied for estimating heritabilities and genetic correlations. Three traits based on culture results were defined: presence of first occurrence of infection by (1) Gram-positive (GRAM+), (2) Gram-negative bacteria (GRAM-) and (3) presence of any bacteriological infection (BACI). Diagnoses of acute (AcM) and chronic mastitis (ChM) as well as culling due to udder health problems (CULL) were considered as direct udder health traits. SCS_305 was defined as lactation mean somatic cell score considering test-day records. The heritabilities for BACI, GRAM+ and GRAM- were 0.01. For AcM, ChM and CULL, heritabilities were 0.04, 0.03 and 0.02, respectively. As expected, the highest heritability was estimated for SCS (h2=0.19). Genetic correlations between BACI and GRAM+ towards mastitis and SCS traits were positive and high (0.6 to 0.9). The genetic relationship of GRAM+ seemed to be particularly high with traits that are indicators for persistent, poorly healing udder diseases (ChM, CULL). Genetic correlation between GRAM+ and GRAM- was close to zero, indicating differences in genetic control. Using pathogen data as additional direct information may enable more efficient breeding for improved mastitis resistance.

Genetic analyses of cow-specific diet digestibility based on spectral analysis of faecal samples

T. Mehtiö[1], P. Mäntysaari[1], T. Kokkonen[2], S. Kajava[3], E. Prestløkken[4], A. Kidane[4], S. Wallén[4], L. Nyholm[5], E. Negussie[1], E.A. Mäntysaari[1] and M.H. Lidauer[1]
[1]Natural Resources Institute Finland (Luke), Humppilantie 14, 31600 Jokioinen, Finland, [2]University of Helsinki, P.O. Box 28, 00014 Helsinki, Finland, [3]Natural Resources Institute Finland (Luke), Halolantie 31A, 71750 Maaninka, Finland, [4]Norwegian University of Life Sciences, Arboretveien 6, 1432 Aas, Norway, [5]Valio Ltd, P.O. Box 10, 00039 Valio, Finland; terhi.mehtio@luke.fi

The objective of this study was to estimate genetic parameters of dry matter digestibility (DMD) and indigestible neutral detergent fibre (iNDF) concentration in faeces ($iNDF_{fc}$). Faeces and feed samples were collected from three research farms in Finland and one research farm in Norway during 2012 to 2016. Faecal samples were collected during different stages of lactation and spot samples from three to five consecutive days were combined to one composite sample. The iNDF contents in faeces and feed samples were predicted by near-infrared reflectance spectroscopy (NIRS) and predictions were used to calculate DMD = 1-($iNDF_{feed}$/$iNDF_{fc}$). The final data consisted of 819 DMD and $iNDF_{fc}$ observations from 311 cows. Variance components were estimated by REML method using the DMU software package. A repeatability animal model was fitted including fixed effects of herd, breed (Nordic Red, Holstein), parity (primiparous, multiparous), lactation stage (3 classes) and year × season (8 classes), and random effects of permanent environment, animal genetic effect and random residual. The pedigree was pruned to 8 generations and had a total of 6,567 animals. The heritability estimate for DMD was 0.15 whereas it was only 0.01 for $iNDF_{fc}$. Several reasons may have contributed to the low heritability estimate for $iNDF_{fc}$. The data used in this study was collected from different farms, over time there were changes in the sampling protocols, genetic ties between some herds were weak and accuracy of NIRS predictions differed across herds. Modelling of this data shall be improved to better evaluate whether $iNDF_{fc}$ could be a potential indicator for DMD. Nevertheless, the estimate of heritability for DMD indicates that there is genetic variation among cows in the ability to digest feed.

Session 11 Theatre 6

A new methodology to estimate protein feed value using the milk protein biological response

F. Dufreneix[1,2], J.L. Peyraud[2] and P. Faverdin[2]

[1]Agrial, 14 rue des Roquemonts, 14000 Caen, France, [2]PEGASE, Agrocampus Ouest, INRA, 35590 Saint Gilles, France; florence.dufreneix@inra.fr

New processes to increase the proportion of by-pass protein from feeds for ruminant are emerging (essential oils, natural tannins). However, the traditional methodologies used to predict nutritional value (degradation *in vitro* or *in sacco*) are not suitable with these new techniques that claim more systemic effects in rumen. The objective of this trial was to test the biological milk protein response to the supply of metabolizable protein as a new methodology for assessing the protein value of feed. This biological response is not linear because the efficiency of metabolizable protein (milk protein/metabolizable protein intake) decreases when the protein supply increases. We assumed a linear decrease of protein efficiency to be able to predict the protein value of a diet from the observed milk production with only two reference extreme diets of known feed value. Four isoenergetic treatments, defined by their protein supply, were applied in a Latin square design to 24 cows divided in 6 homogenous groups (based on potential production of cows and parity) over four 3-wk periods. The two extreme treatments contained either 3.4 kg DM of energetic concentrate (C++S0, 101 g PDIE/kg MS) or 3.4 kg DM of soybean meal (C0S++, 242 g PDIE/kg MS). Two intermediary diets with mixes of these two feeds (C++S+, 152 g PDIE/kg MS and C+S++, 199 g PDIE/kg MS) where used to assess their metabolizable protein from their milk protein responses. Increasing protein supply has a significant positive effect on milk protein yield (895 and 964 g protein.d-1 for C++S0 and C0S++ respectively). Estimation of the protein value of the two intermediate diets according to the milk protein response were 17 and 15 g PDIE.kg MS-1 higher for C++S+ and C+S++ respectively than the calculated one (i.e. an error of 11 and 7% respectively). This error could be partly explained by small differences in energy intake. In spite of this slight overestimation, milk protein response seems a relevant approach, independent of the protection process, to estimate the protein value of feed, especially for the new feed protections.

Session 11 Theatre 7

Role of milk protein fractions on coagulation, curd firming and syneresis

N. Amalfitano[1], C. Cipolat-Gotet[2], A. Cecchinato[1], M. Malacarne[2], A. Summer[2] and G. Bittante[1]

[1]University of Padova, Department of Agronomy, Food, Natural resources, Animals and Environment, Viale dell'Università 16, 35020 Legnaro (PD), Italy, [2]University of Parma, Department of Veterinary Science, Via del Taglio 10, 43126 Parma, Italy; nicolo.amalfitano@studenti.unipd.it

The aim of the present work was to assess the role of milk single protein fractions (α_{s1}-CN, α_{s2}-CN, β-CN, κ-CN, β-LG, α-LA), which are expressed as content in milk (g/l) or percentage of total casein content (%cas), on milk rennet coagulation, curd firming and syneresis. Individual milk samples were collected from 1,271 Brown Swiss cows reared in 85 herds in the Trento Province (North-East Italy). The herds were divided in 4 types according to their farming system (from traditional to modern systems). The coagulation process was assessed on each milk sample using computerized lactodynamographs and extending the test to 60 min. This analysis provided the traditional milk coagulation properties (MCP): rennet clotting time (RCT), curd firming time (k_{20}), curd firmness at 30 (a_{30}) and 45 min (a_{45}). The curd firmness recorded over time every 15 sec (CF_t) was processed using a four-parameters nonlinear model, which yielded new parameters of coagulation, curd firming and syneresis: RCT estimated from the equation (RCT_{eq}), the asymptotic potential CF (CF_P), the curd firming and syneresis instant rates constant (k_{CF} and k_{SR}), the maximum CF value (CF_{max}) and the time of attainment (t_{max}). All these traits were analysed fitting two linear mixed models according to the expression of milk protein fractions included: M-g/l and M-%cas. The other effects included were herd (random effect), daily milk production (only for M-g/l), casein content (only for M-%cas), farming system, parity, days in milk, pendula of the instruments and CSN2, CSN3, BLG genotypes. The results showed a positive role of α_{s1}-CN and β-CN on the CF_{max}, with the former having an effect almost double than the latter. The κ-CN had a favourable effect on the milk coagulation ability, reducing RCT, increasing the k_{CF} and k_{SR} and allowing to reach a higher CF_{max}. In contrast, the results showed an unfavourable effect of α_{s2}-CN on both RCTs and of β-LG on curd firming. These results could be useful for the improvement of cheese production traits in cattle breeding programs.

Mid-infrared spectroscopy prediction of total antioxidant activity of bovine milk

M. Franzoi, G. Niero, M. Cassandro and M. De Marchi
University of Padova, Department of Agronomy, Food, Natural Resources, Animals and Environment (DAFNAE), Viale dell'Università 16, 35020 Legnaro (PD), Italy; marco.franzoi@studenti.unipd.it

Milk contains several antioxidant compounds that are able to inhibit the oxidation of other molecules. Such antioxidants belong to different molecular families such as vitamins (mainly A, C and E), phenols and low molecular weight thiols. At the same time, macromolecules such as casein-derived peptides and whey proteins, with particular regard to lactoferrin, have been studied for their antioxidant properties. Food antioxidants are essential in preventing human pathologies and inactivating free radicals, and they are responsible for lipids peroxidation, oxidative alteration of proteins and DNA damages. The characterisation and quantification of single antioxidant species is expensive and time-consuming. For this reason, the measurement of total antioxidant activity (TAA) is often preferred. The present study aimed to develop a mid-infrared spectroscopy calibration model to predict TAA of bovine milk. Total antioxidant activity assays were performed on 1,249 individual milk samples of Holstein Friesian cows and reference data were expressed as Trolox equivalents. Mid-infrared spectroscopy prediction model was developed using uninformative variable elimination procedure and partial least squares regression analysis. The resulting calibration reports a coefficient of determination in cross validation (R^2CV) of 0.47, a standard error in cross validation of 0.47 and a ratio of performance to deviation of 2.01. The obtained prediction models might be used to provide large-scale phenotyping while for the enhancement of TAA through breeding programs and for the study of factors affecting TAA in dairy herd management, further investigations are required.

Effect of milk yield and milk content curve shapes on first lactation survival in large herds

M. Grayaa[1,2,3], A. Ben Gara[3], S. Grayaa[2], R.R. Mota[4], H. Hammami[4], S. Vanderick[4], C. Hanzen[1] and N. Gengler[4]
[1]Faculté de Médecine Vétérinaire, Boulevard de Colonster 20, 4000, Belgium, [2]Insitut National d'Agronomie de Tunisie, 43 Avenue Charles Nicolle, 1082, Tunisia, [3]Ecole Supérieure d'Agriculture de Mateur, Route Tabarka, 7030, Tunisia, [4]Gembloux AgroBiotech, Passage de Déportés 2, 5030, Belgium; marwa.grayaa@doct.uliege.be

Genetic parameters of first lactation survival and curve shape traits of milk yield, fat and protein percentages were estimated using information of 25,981 primiparous Tunisian Holsteins belonging to large herds. For each trait lactation peak, apparent persistency, real persistency and level of production adjusted to 305 days in milk were defined. Variance components were estimated under three bivariate animal models with a linear random regression model. Milk yield as well as fat and protein percentages were modelled by fixed herd × test day interaction effects, fixed classes of 25 days in milk × age of calving × season of calving interaction effects, random environment effects, and random additive genetic effects. Survival was modelled by fixed herd × year of calving interaction effects, age of calving × season of calving interaction effects, random environment permanent effects, and random additive genetic effects. Heritability estimates were 0.03 for survival, 0.23, 0.29 and 0.30 for average milk yield, fat and protein percentages adjusted to 305 days in milk, respectively. Genetic correlations between survival and average milk yield, fat and protein percentages adjusted to 305 days in milk were 0.33, -0.33 and -0.14, respectively. Genetic correlations between survival and real persistency for fat and protein percentages were -0.24 and -0.15, respectively. Cows that had higher persistencies for fat and protein percentages, and therefore flatter fat and protein percentages curves, were more likely not to survive. This was due to higher fat percentages at the end of the lactation leading to the hypothesis that cows producing higher fat percentage dispose of less energy available for gestation and were therefore less likely to be or remain pregnant and, therefore, to survive.

Estimation of genetic parameters for young stock survival in Danish beef × dairy crossbred calves

R.B. Davis[1,2], E. Norberg[1,3] and A. Fogh[2]
[1]Aarhus University, QGG, Blichers alle 20, 8830 Tjele, Denmark, [2]SEGES, Agro Food Park 15, 8200, Denmark, [3]NMBU, Universitetstunet 3, 1430 Ås, Norway; rubd@seges.dk

Calf mortality leads to economic losses for the farmer and is an animal welfare issue. Currently, only calf mortality within the first 24 hours is accounted for in the Danish breeding goal for beef × dairy calves. However, survival throughout the rearing period is also of upmost importance. Therefore, the aim of this study was to estimate genetic parameters for young stock survival, to evaluate if it is feasible to implement such a trait. Data on 90,926 crossbred calves was extracted from the Danish Cattle Database and was provided by the Danish research centre, SEGES. Two traits were defined, young stock survival from 1-30 days and 31-200 days after birth. The traits were analysed with a univariate animal model using the AI-REML algorithm in the DMU package. The model contained a fixed effect of year × month of birth, herd, sex, breed combination, parity of the dam, transfer and a random effect of the calf and herd × year. The pedigree was traced back 5 generations, for both the sires and dams. Results showed low but significant heritabilities (0.045-0.075) for both survival traits. Breed combinations with Danish Blue cattle sires outperformed all other sire breeds. The lowest survival rates were found for breed combinations with Jersey dams or Blonde d'Aquitaine sires. Breeding values were calculated using DMU4. Sufficient genetic variation between sires for young stock survival was found. The breeding values of the sires had an effect on young stock survival that ranged from -2.5 to 3.5% and -5.4 to 4.7% for survival from 1-30 days and 31-200 respectively. It is therefore feasible to implement both young stock survival traits in a genetic evaluation for beef × dairy crossbred calves. This will increase the survival rate of the calves and hereby increase animal welfare and decrease economic loss for the veal producers.

Closing the gap between research and extension: mathematical tools for sustainable dairy farming

A.D. Soteriades[1], K. Rowland[2], D.J. Roberts[3] and A.W. Stott[4]
[1]Bangor University, SENRGy, Deiniol Road, LL57 2UW, Bangor, United Kingdom, [2]Kingshay Farming & Conservation Ltd, Bridge Farm, West Bradley, BA6 8LU, Somerset, United Kingdom, [3]SRUC, Barony Campus, Parkgate, DG1 3NE, Dumfries, United Kingdom, [4]SRUC, Peter Wilson Building, Kings Buildings, EH9 3JG, Edinburgh, United Kingdom; a.d.soteriades@bangor.ac.uk

The need for improving farm management for a sustainable and resource use efficient dairy production has never been greater. Farm management is typically guided by efficiency ratios and other metrics that benchmark farms and identify best practice. Such indicators fail to holistically capture the multifaceted nature of dairy farming. This can be overcome with the use of novel mathematical models designed to benchmark farms against each other using a range of output metrics including resource use, productivity, profitability, environmental and other (e.g. animal health or welfare) performance criteria. An example is Data Envelopment Analysis (DEA), which ranks farms by combining a range of said criteria rather than relying on a simple ratio (e.g. kg milk/cow) that may favour perverse responses/outcomes not captured by the ratio (e.g. soil erosion). It also identifies the best ways for each farm to improve its ranking, thus providing a means to add value to farm business advisory services while supporting public good outcomes. Nonetheless, DEA has largely remained an academic method and little effort has been placed on tailoring it to the needs of the farm advisor. The aim of this study was to close this gap by (1) demonstrating DEA's potential for identifying cost-reducing and profit-making opportunities for farmers with a series of examples drawn from commercial UK dairy farm data; and (2) discussing ideas for extending DEA's applicability in the agricultural industry, such as the use of carbon footprints and other farm sustainability indicators in DEA analyses. As farm advisors are sourcing an increasing volume of data, the use of DEA in extension/advising could offer invaluable insights into the wider picture of farm performance.

Expression of immune system genes in mammary gland tissues of dairy cows infected with staphylococci

M. Zalewska[1], E. Kawecka[1,2], D. Reczyńska[1], P. Brodowska[1], D. Słoniewska[1], S. Marczak[1], S. Petrykowski[1] and E. Bagnicka[1]
[1]Institute of Genetics and Animal Breeding PAS, Postępu St. 36A, Jastrzębiec, 05-552 Magdalenka, Poland, [2]Warsaw University of Life Sciences, Faculty of Veterinary Medicine, Nowoursynowska St. 159, 02-776 Warsaw, Poland; m.zalewska@ighz.pl

Mastitis is still considered as one of the main problems of the dairy husbandry and industry. Cisternal lining epithelial cells (CLEC) are the first line of udder defence and serve as a mechanical barrier. However, there are premises suggesting that this tissue provide more functions. The aim of the study was to analyse serum amyloid A (*SAA*), haptoglobin (*Hp*), ceruloplasmin (*Cp*), lactoferrin (*Lf*), lipopolysaccharide-binding protein (*LBP*) gene expressions in mammary gland parenchyma and CLEC samples. The samples were collected from 40 Polish Holstein-Friesian cows of Black-and-White variety suffering from chronic and recurrent mastitis. Altogether, 51 quarter samples were collected: infected with coagulase-positive (CoPS) (n=25), coagulase-negative staphylococci (CoNS) (n=13), and non-infected (n=13) ones. The gene expression was measured using qPCR method (LightCycler480, Roche) with *GAPDH* as a reference. There were no differences in the gene expressions between both types of tissue in the animals with the same health state. The expression level of *Hp* was higher for CoPS- and CoNS-infected ($P<0.05$) udder quarters in comparison to infection-free ones in the parenchyma, while only between CoPS and infection-free ($P<0.05$) quarters in CLEC. Moreover, higher expression of *SAA* in parenchyma and *Cp* in CLEC in CoPS-infected cows than in infection-free group was noted, which may indicate a slightly different response of different udder tissue to chronic inflammation. No difference was noted in expression levels of other tested genes. However, the strong and positive correlation between *Cp* and *SAA* in parenchyma and CLEC in CoPS- and CoNS- infected groups was stated. Presented results may suggest that CLEC plays not only mechanical but also enzymatic role in the host defence system. The research was funded by the National Science Center (Grant no. 2015/17/B/NZ9/01561).

Derivation of the metabolic status in dairy cows – prediction of daily mean ruminal pH

A. Mensching[1], M. Zschiesche[2], J. Hummel[2] and A.R. Sharifi[1]
[1]University of Goettingen, Department of Animal Sciences, Animal Breeding and Genetics Group, Albrecht-Thaer-Weg 3, 37075, Goettingen, Germany, [2]University of Goettingen, Department of Animal Sciences, Ruminant Nutrition Group, Kellnerweg 6, 37077, Goettingen, Germany; andre.mensching@uni-goettingen.de

The metabolism of dairy cows is a complex system, which is exposed to particular challenges, especially during the transition phase and early lactation. Regarding to the actual high performances, the adequate ruminant feeding in consideration of the occurrence of subacute ruminal acidosis and subclinical ketosis is an extremely relevant research field with respect to animal welfare. The objective of this study was to apply meta-analytical methods to investigate the association between the daily mean pH in the ventral rumen sack and diet-, milk ingredient- and cow- specific parameters in an explorative approach. The resulting model should be able to derive the metabolic status of dairy cows depending on parameters that are available in agricultural practice. In total, data from 52 studies were collected, whereof 39 studies used continuous pH measurement. Due to the fact that 25% of data from potential predictors for the daily mean ruminal pH were missing, an algorithm for Multiple Imputation by Chain Equations as described by Azur *et al.* was applied. After imputation, a stepwise variable selection taking into account linear and quadratic effects as well as interactions was used to identify an appropriate statistical model, which best describes the daily mean ruminal pH. The final model ($R^2=0.705$) includes four diet parameters (starch, crude protein, ether extract, and physically effective neutral detergent fibre >8 mm), three milk based parameters (protein, lactose, and the fat to protein ratio) and the daily dry matter intake of cows. Within a sensitivity analysis, whereby the diet parameters were considered as fixed based on three different example diets, it could be shown that the model identified known and new associations for prediction of daily mean ruminal pH.

Session 11 — Theatre 14

Prediction of milk mid-infrared spectrum using mixed test-day models

P. Delhez[1,2], S. Vanderick[2], F.G. Colinet[2], N. Gengler[2] and H. Soyeurt[2]
[1]National Fund for Scientific Research, Egmont 5, 1000 Bruxelles, Belgium, [2]University of Liège, Gembloux Agro-Bio Tech, Passage des Déportés 2, 5030 Gembloux, Belgium; pauline.delhez@doct.uliege.be

Mid-infrared (MIR) analysis of milk currently allows the measurement of many variables of interest for the dairy sector related to milk nutritional quality, milk technological properties, cow's status or environmental fingerprint. The aim of this study was to explore the ability of a test-day model to predict milk MIR spectra, and therefore all the resultant variables, for a future test day of a known cow or for a new cow based on easily known characteristics of cows. This is useful for instance for herd management (e.g. detecting problems, predicting potential of heifers) or to predict future environmental impacts of a dairy herd. A total of 467,496 milk MIR spectra from 53,781 Holstein cows in first lactation were used for the calibration data set. First, 323 wavelengths out of the 1,060 wavelengths of the milk spectra were conserved. This spectral information was reduced by using principal component analysis (PCA). A total of 8 principal components (PC) were kept, representing 99% of the spectral information. Then 8 univariate test-day models including the day in milk, herd×year and herd×month as fixed effects and herd×test date, permanent environment and genetics as random effects were applied for each PC. From the solutions of the models and by using a back reversing operation using eigenvectors of the PCA, the predicted 323 wavelengths of the spectra were re-obtained. The calibration correlations between observed and predicted spectral data ranged from 0.76 to 0.93. Correlations between observed and predicted milk fat and protein contents obtained from the modelled spectra were 0.83 and 0.89, respectively. These findings demonstrate the moderate ability of a test-day model to predict milk MIR spectra.

Session 11 — Poster 15

Influence of bacteria on technological parameters of dairy cow milk

P. Brodowska[1], D. Reczynska[1], M. Zalewska[1], E. Kawecka[1,2], D. Sloniewska[1], S. Marczak[1], S. Petrykowski[1] and E. Bagnicka[1]
[1]Institute of Genetics and Animal Breeding, Postepu str. 36A, 05-552 Jastrzebiec, Poland, [2]Veterinary Medicine Faculty, Warsaw University of Life Sciences, Ciszewskiego 8 str, 02-786 Warsaw, Poland; d.reczynska@ighz.pl

Mastitis concerns the population of dairy cattle around the world causing economic losses and reducing milk quality. The most common cause of udder inflammation are pathogenic bacteria as well as the environmental and opportunistic one. The aim of the study was to analyse the impact of the presence of the environmental or contagious bacteria in the cow mammary gland on technological milk parameters. The study was carried out on 154 Polish HF cows of Black-and-White variety. The dataset comprised the 211 records on daily milk yield and its components, clotting time, whey and cheese quality, and cheese yield. The whey and cheese quality was assessed arbitrarily in six-point scale, with 1 as the highest quality. The milk component contents (12 parameters) were established using MilkoScan FT2, while somatic cell count using IBCm apparatus. The microbiological analysis was done using VITEK2 equipment. The milk samples were divided into three groups: (1) bacteria-free samples (n=22), (2) the samples with environmental and opportunistic bacteria (mainly coagulase-negative staphylococci) (n=150), and (3) samples with major mastitis pathogens such as *Staphylococcus aureus* or *Escherichia coli* (n=39). The GLM procedure of SAS package was used for analyses of variance method. There were no differences in cheese yield, clotting time, pH and component contents between groups of samples. As expected, the highest milk yield was found for cows which milk was free from any bacteria, while the lowest milk yield was obtained by cows with confirmed presence of *E. coli* or *S. aureus* in milk ($P<0.01$). However, there was no difference between 1st and 2nd groups. The same pattern of differences for the cheese and whey quality as well as somatic cell score between groups were also noted. We stated that presence of major mastitis pathogenic bacteria negatively influences milk yield and milk technological parameters, but no impact of environmental bacteria on all studied parameters was noted.

New quantification method for soluble and micellar minerals in milk and application in Holstein cows

M. Franzoi, G. Niero, M. Cassandro and M. De Marchi
University of Padova, Department of Agronomy, Food, Natural Resources, Animals and Environment (DAFNAE), Viale dell'Università 16, 35020 Legnaro (PD), Italy; marco.franzoi@studenti.unipd.it

Milk has been extensively investigated as source of macro- and micro-nutrients such as essential minerals which are relevant both from technological (e.g. cheese-making) and nutritional (e.g. infant formula, fluid milk) points of view. Calcium and magnesium are essential for the stabilization of casein structure and the addition of calcium salts to milk improve milk technological traits, such as paracasein reticulum firmness. Minerals in milk can be soluble or associated to other milk components such as casein micelles and both contributes to the cheese making process. For this reason, the distinction between micellar and soluble minerals is a relevant issue in the dairy industry. Current methods for quantification of soluble and micellar minerals in milk have some limitations; for example, the rennet coagulation method needs correction factors, due to exclusion effect of major constituents, irrespective of single sample variability. The aim of this study was to develop a method for quantification of soluble and micellar minerals consistent with single sample variability. We proposed a new rennet coagulation based method, introducing a whey dilution step to exclude quantification biases from whey trapped in curd and excluded volume. The contents of Ca, Mg and K in milk, whey and diluted whey were quantified by acid digestion and inductively coupled plasma optical emission spectrometry. The repeatability of the method for micellar Ca, Mg and K was between 2.07 and 8.96%, whereas reproducibility ranged from 4.01 to 9.44%. Recovery of total milk minerals over 3 spiking levels ranged from 92 to 97%. The method has been applied to quantify micellar Ca, Mg and K to study variation of these minerals across lactation and parity in Italian Holstein cows. The results will be presented.

Control of chlorate and trichloromethane residue levels in bulk tank milk

L.F. Paludetti[1,2], A.L. Kelly[1], B. O'Brien[2] and D. Gleeson[2]
[1]University College Cork, School of Food and Nutritional Sciences, Cork, T12 K8AF, Ireland, [2]Teagasc Moorepark, Animal & Grassland Research and Innovation Centre, Fermoy, Cork, P61 C996, Ireland; lizandra.paludetti@teagasc.ie

Chlorate (CHLO) and trichloromethane (TCM) residues are formed when chlorine-based sanitation products are used inappropriately for milking equipment cleaning. Those residues are a concern in milk powder and butter manufacture, respectively. The aim of this study was to measure CHLO and TCM in 67 farm bulk milk tanks during mid- and late-lactation, and to investigate the main factors that influence these residue levels in milk. The CHLO and TCM levels were assessed using gas chromatography and high performance liquid chromatography with tandem mass spectroscopy, respectively. Questionnaires regarding sanitation practices were completed on farms. Differences between lactation periods with regard to residue levels, and the effect of equipment cleaning practices on TCM and CHLO in milk, were calculated using the MIXED procedure in SAS 9.3. The median TCM levels in milk during mid- and late-lactation were significantly different (0.0005 mg/kg [CI: 0.0004-0.0006 mg/kg] and 0.0011 mg/kg [CI: 0.0009-0.0014 mg/kg], respectively; $P<0.0001$), and were lower than the limit applied by butter manufacturers (0.0015 mg/kg). However, 6 and 21 farms in mid- and late-lactation, respectively, had TCM levels greater than that limit. In mid-lactation, less milk samples had CHLO detected (14 samples [range: 0.0010-0.0070 mg/kg]) compared to late-lactation (32 samples [range: 0.0010-0.6500 mg/kg]). In mid-lactation, all of the samples had CHLO levels lower than the European default limit for milk (0.010 mg/kg); however, in late-lactation 5 samples had CHLO levels greater than that limit. The higher levels of TCM and CHLO in late-lactation were associated with incorrect sanitation practices on-farm: insufficient rinse water used after milking ($P=0.03$) and after the detergent wash ($P=0.01$) and detergent type used ($P=0.010$). Generally, less milk is stored in bulk tanks during late-lactation, due to the advancing stage of lactation; consequently, residue concentrations may be higher in late-lactation milk. This study highlighted production conditions that could be targeted to minimize TCM and CHLO levels in farm milk.

Session 11 Poster 18

Cytokine gene expression in the mammary gland parenchyma of dairy cows infected with staphylococci

E. Kawecka[1,2], M. Zalewska[2], D. Reczyńska[2], P. Brodowska[2], D. Słoniewska[2], S. Petrykowski[2], S. Marczak[2] and E. Bagnicka[2]

[1]Warsaw University of Life Sciences, Faculty of Veterinary Medicine, Nowoursynowska St. 159, 02-776 Warsaw, Poland, [2]Institute of Genetics and Animal Breeding PAS, Postępu St. 36A, Jastrzębiec, 05-552 Magdalenka, Poland; m.zalewska@ighz.pl

Coagulase-positive (CoPS) and coagulase-negative staphylococci (CoNS) are common cause of the bovine mastitis. Cytokines are glycoproteins participating in both anti-inflammatory and pro-inflammatory processes. The aim of the study was to analyse the expression of cytokine genes (interleukin 8 –*IL8*, interleukin 18 –*IL18*, interleukin1β –*IL1-β*, C-X-C motif ligand 5 – *CXCL5*, Tumour Necrosis Factor – *TNFα* and C-C Motif Chemokine Ligand – *CCL2*) in the mammary gland parenchyma infected with coagulase-positive or negative staphylococci. The 51 quarter samples were obtained from 40 Polish Holstein-Friesian cows of Black-and-White variety. Three groups of samples were distinguished: infected with CoPS (n=25), infected with CoNS (n=13), and non-infected ones (n=13). The gene expression was analysed using qPCR techniques (LightCycler480, Roche). The *GAPDH* was used as a reference. There were no differences in *IL1-β*, *TNFα* and *CCL2* expressions. The differences in expression of *IL8* and *CXCL5* were stated. However, the differences were observed only between CoPS vs non-infected (P<0.01) and CoNS vs non-infected (P<0.05) samples. There was no difference between CoNS and CoPS infected samples. Moreover, the difference in *IL18* expression between non-infected and CoPS infected samples (P<0.05) was noted. The expression of all three genes was the lowest in non-infected samples. The studied interleukins (*IL8*, *IL18*) are pro-inflammatory cytokines and the level of their expression increase when the inflammatory state occurs. Moreover, *CXCL5* is a chemokine that is also released during the onset of inflammation in the organism. The coagulase-negative staphylococci are considered as environmental bacteria but our results indicate that both types of staphylococci cause the inflammation in the mammary gland. The study was financed by a grant from the National Scientific Center, Poland, No. 2015/17/B/NZ9/01561.

Session 11 Poster 19

Plasma pregnancy specific protein B (PSPB) in days 25, 26 and 28 in two beef cattle breeds

A. Noya[1], I. Casasús[1], D. Villalba[2], J.L. Alabart[1], B. Serrano-Pérez[2], J.A. Rodríguez-Sánchez[1], J. Ferrer[1] and A. Sanz[1]

[1]CITA de Aragón, Animal Production and Health, Avda. Montañana 930, 50059 Zaragoza, Spain, [2]Universitat de Lleida, Animal Science, Avda. Alcalde Rovira Roure 191, 25198 Lleida, Spain; anoya@cita-aragon.es

Extensive beef cattle farming systems are progressively implementing new methods to make more technical the production cycle. Detection of pregnancy specific protein B (PSPB) could be an accurate and early pregnancy diagnosis to reduce the calving interval. The aim of this study was to determine, based on PSPB concentrations, the earliest day to accurately diagnose pregnancy in beef cows. Seventy-four lactating Parda de Montaña and 40 Pirenaica multiparous cows were synchronized and inseminated on day 75.8±13.5 postpartum. Plasma EDTA samples were obtained on days 25, 26 and 28 post insemination and PSPB ELISA assays were performed. Pregnancy diagnosis was confirmed by transrectal ultrasonography on day 37 post insemination. No differences in PSPB concentrations were found in non-pregnant cows among different days (0.67, 0.41 and 0.48 ng/ml on days 25, 26 and 28, respectively). In pregnant cows, PSPB concentrations were similar on days 25 and 26, but on day 28 the highest values were recorded (1.15, 1.22 and 1.82 ng/ml on days 25, 26 and 28, respectively). For pregnancy diagnosis at day 25, the area under the ROC curve was 0.755, but no cut-off value was proposed because of the overlap between pregnant and non-pregnant PSPB values. On days 26 and 28, the area under the curve was 0.880 and 0.930 respectively, but no significant differences were found between the logistic models. The optimum cut-off value for pregnancy discrimination was 0.57 ng/ml (94.3% of sensitivity and 78.9% of specificity) and 0.91 ng/ml (94.3% of sensitivity and 80.8% of specificity) on days 26 and 28, respectively. In conclusion, plasma PSPB analysis on day 26 was a reliable tool for early pregnancy diagnosis in beef cows, with a similar accuracy to that obtained on day 28 and avoiding the lack of precision obtained on day 25.

Meta-analysis of genome wide association studies to estimate SNP effects and breeding values

M.E. Goddard
University of Melbourne and Agriculture Victoria, Parkville, Melbourne, Australia; meg@unimelb.edu.au

In recent years, the power of genetics has been greatly increased by the availability of high throughput technologies or 'omics'. In human genetics, it is common to combine results from many genome wide associations studies (GWASs) and from public omics resources. This is necessary because it is impossible to combine the raw data for privacy and other reasons. However, the use of meta-analysis has proven to be very successful because it has generated the equivalent of very large sample sizes, which are necessary because the individual SNP effects are so small. In livestock, we have a need to obtain larger sample sizes and an opportunity to do this both by combining raw data and by meta-analysis of the results from individual studies. Examples will be given from human and cattle genetics. Interbull traditionally does a meta-analysis to estimate breeding values of dairy bulls from different countries. I will present a proposed method of meta-analysis of SNP effects and hence calculation of EBVs based on SNP data that Interbull could use in future. The combination of omics data and very large sample size ('big data') could allow us in a few years time to have annotations of livestock genomes in which the phenotypic effects of every polymorphism are known. This will be an enormous advance in our understanding of livestock biology.

Possible implications of limited dimensionality of genomic information

I. Misztal, I. Pocrnic and D.A.L. Lourenco
University of Georgia, Animal and Dairy Science, 30602, Georgia; ignacy@uga.edu

The purpose of this study was finding possible explanation on peculiarities of dimensionality (M) of genomic information. The gene content matrix derived from 35-60k SNP chips has a limited M as determined by singular value decomposition; identical results are obtained with eigenvalues of genomic relationship matrix. Even with a very large number of animals, M ranges from about 4,000 for commercial pigs and broiler chicken, to about 15,000 in Holsteins. This number is normally attributed to the expected number of chromosome junctions as derived by Stam: M=4NeL, where Ne is effective population size and L is genome size. However, approximation of realized accuracies assuming M for animals with same information is not accurate. Accuracies of genomic prediction assuming M/4 animals in genomic recursions and the APY algorithm are >90% of those assuming full dimensionality. These recursions also suggest that predictions based on M animals with very high reliability should be both very accurate and persistent, and predictions from large national evaluations in Holsteins could converge. However, the real accuracies seem lower than expected. The genome in a population can be visualized in two ways. First, as Ne haplotypes within each 1/4 Morgan segment. Second, as 4NeL sequential segments. Eigenvalues analyses of the genomic information shows that popular segments cluster along the genome. Subsequently the number of segments can be higher than determined by singular values. In particular, M/4 clusters could account for 90% of segments. SNP selection decreases the dimensionality; the minimum is the number of causative SNP. SNP selection can eliminate clusters without substantial variation but point to clusters with high variation, potentially creating high GWAS signals not related to QTLs. Some ideas in this study were derived from simulated populations assuming complete genome coverage and an additive model. It remains to be seen whether accuracy predictions in real populations are affected by additional factors such as incomplete genome coverage and non-additive effects. Singular value analysis of gene content (or eigenvalue analysis of genomic relationship matrix) helps understand the complexity of genomic selection.

Session 12 Theatre 3

Implementing genomic prediction models in generic evaluation of large populations

J. Ten Napel[1], G.C.B. Schopen[1], J. Vandenplas[1], A.R. Cromie[2], E.M. Van Grevenhof[1] and R.F. Veerkamp[1]
[1]Wageningen University & Research, P.O. Box 338, 6700 AH Wageningen, the Netherlands, [2]Irish Cattle Breeding Federation, Highfield House, Shinagh, Bandon, Co. Cork, Ireland; jan.tennapel@wur.nl

Breeding value estimation requires genetic similarity between animals and it should use all available pedigree and genomic information. Utilising a very large number of genotyped animals in combination with non-genotyped animals is cumbersome. In this study we compare different single-step methods that have been proposed over time to utilise all pedigree and genomic information in a large routine evaluation. For comparison, we used data of existing routine breeding value estimation (beef cattle, Ireland) of over 3.5 million records for finisher carcass weight, 12.0 million animals in pedigree of which over 613,000 animals have been genotyped for 50,240 SNP. Analyses were performed with MiXBLUP and calc_grm. Three methods were compared: pedigree BLUP for which the relationships are modelled using only pedigree information, ssGBLUP for which relationships are modelled by combining pedigree and genomic relationship matrices, and ssRRBLUP for which SNP genotypes are imputed for non-genotyped animals and are considered as random effects. For ssGBLUP, the inverse of the genomic relationship matrix was approximated using the APY algorithm with 40,000 randomly chosen core animals. Results are expressed in computational demands and convergence. Convergence of ssRRBLUP was much slower than ssGBLUP., suggesting that ssRRBLUP requires an alternative preconditioner or a different solver for faster convergence. The size of the APY inverse of G (ssGBLUP) increases linearly with 290 Gb per million genotyped animals added, whereas the SNP covariate matrix (ssRRBLUP) increases with 50 Gb per million data records added or with 10 Gb per 10,000 SNP added. Number of residual polygenic effects to be estimated (ssRRBLUP) decreases as number of animals genotyped increases. Utilising a very large number of genotyped animals is possible, but as yet very slow. Increasing the number of genotyped animals slows down ssGBLUP further and is likely to benefit ssRRBLUP. ssRRBLUP appears to be a more suitable method than ssGBLUP for very large numbers of genotyped animals, once convergence issues have been resolved.

Session 12 Theatre 4

Ideas for continuous genomic evaluation for newly genotyped Walloon Holstein females and males

S. Naderi, R.R. Mota, S. Vanderick and N. Gengler
TERRA Teaching and Research Centre, Gembloux Agro-Bio Tech, University of Liège, Passage des Déportés, 2, 5030 Gembloux, Belgium; s.naderi@uliege.be

Crucial for large scale use in dairy cattle of genotyping for females is that any newly genotyped animal (calves, cows and heifers but also bulls) receives very quickly genomic breeding values (GEBV) even outside the official schedule for routine evaluations. In this study, a system was developed to estimate initial GEBV for newly genotyped animals before their inclusion in the official routine release of genomic evaluations. The system was setup to be run on request, featuring the setup of a 'continuous' evaluation, also being quick and simple enough to be used at least on a weekly base. For animals without own records or descendants, official GEBV were approximated using selection-index like method by combining direct genomic values (DGV) of newly genotyped animals and their parent average (PA). DGV for new genotyped animals were calculated based on SNP effects from the previous official routine evaluations (April and August, 2017). Depending on GEBV accessibility from parents of a given animal, PA was calculated based on conventional phenotypic information (cPA), and parent GEBV (gPA). To expand the system for animals with progeny, a subset of genotyped animals was selected, and conventional estimated breeding values (cEBV) and cPA of selected animals were combined with DGV and gPA in order to obtain GEBV for animals with progeny. The weights were calculated based on the covariance between DGV and gPA for animals without progeny, and between DGV, gPA, cEBV and cPA for animals with progeny. Correlations between initial and April official evaluations for 60 new genotyped animals without progeny varied from 0.87 to 0.95 for conformation, fertility and production traits, whereas correlations between initial and August official evaluations varied 0.84 to 0.92 (n=25 new genotyped animals). On the other hand, correlations between initial and August official evaluations for 120 genotyped animals with progeny varied from 0.95 to 0.97 for production traits. Study showed potential to use simple selection index based methods in continuous genomic evaluations, a way to support genotyping of females for genomic selection but also for management and marketing.

World survey of dairy cattle; breeding, genotyping and subpopulations

D. Matthews[1,2], B.W. Wickham[3], J.F. Kearney[1,2] and P.R. Amer[4]
[1]GplusE consortium, UCD, Dublin, Ireland, [2]Irish Cattle Breeding Federation, Bandon, Cork, Ireland, [3]ICAR, Via Savoia, Roma, Italy, [4]AbacusBio Ltd, P.O. Box 5585, Dunedin, New Zealand; dmatthews@icbf.com

Rates of uptake of technological advances in the field of dairy breeding differ between countries. Of specific interest here is that these technological advances can be used at higher intensities in subpopulations of dairy cattle within countries. Focusing genetic gain technology into subpopulations allows significant proportions of potential benefits to be captured, without inflicting both cost and inconvenience across subsets of commercial farmers whose interest in genetic gain is much more passive. The frequency of subpopulations of dairy cattle within countries' breeding programs has not previously been formally quantified. In order to determine this, a survey was developed in conjunction with ICAR and Interbull. In total 36 evaluation centres were sent the survey and 17 centres, covering 19 countries, provided a response. Based on cow population statistics an aggregate of over 31 million dairy cows have been covered by the survey. Information on the levels of genotyping performed in various subpopulations were sought. Nearly all organisations have very high levels of genotyping in AI bulls. However, all countries covered by the survey have small proportions of commercial cows genotyped with no region having greater than 10% of commercial cows providing phenotypic and genotypic data for evaluations. Ten organisations responded that research herds contribute phenotype data to genomic predictions used in evaluations. Other countries have research herds but these cows are not included in the reference population for genomic predictions. Elite herds of dairy cows are often used in breeding programs by AI companies in an attempt to breed future generations of AI bulls. Seven organisations have responded that nucleus herds are utilised in their country's breeding program. The survey evaluated the frequency and use of subpopulations in different regions' breeding plans. It also establishes the levels of genotyping within subpopulations of dairy cattle in participating countries or regions. On the basis of the results, it is clear that breeding organisations follow different paths in their drive to increase genetic trends.

Effects of different groups of cows in the reference population on genomic breeding values

L. Plieschke, C. Edel, E.C.G. Pimentel, R. Emmerling and K.-U. Götz
Bavarian State Research Center for Agriculture, Institute of Animal Breeding, Prof.-Dürrwaechter-Platz 1, 85586 Poing-Grub, Germany; laura.plieschke@lfl.bayern.de

At present, the question of integrating cows in the reference population arises for many breeds as the number of genotyped cows has increased significantly in recent years. Genotyping of cows includes both genotyping initiated by farmers for better on-farm selection (hereafter called routine-cows), as well as genotyping in the frame of scientific projects or structured sampling (hereafter called project-cows). It can be assumed that farmers preselect elite cows for genotyping, whereas project-cows will in most cases represent an unselected random sample. In a recent simulation study we found that integrating preselected cows into the reference population may have negative effects on validation parameters. Studies of Wiggans *et al.* and Dassonneville *et al.* dealt with the consequences of preferential treatment and provide additional evidence of potentially biasing effects. The study presented here investigates the question whether the two groups of cows show differences in their impact on genomic predictions. A minus 4 year validation study was conducted to investigate the effect of including these two different groups of cows into the reference population. We studied effects on regression slope and validation reliability of the GEBV-test. A reference population consisting of bulls (n=6,313) was expanded: (1) with all cow genotypes (n=3,559), (2) with routine-cows only (n=366), or (3) project-cows only (n=3,193). Three traits were considered: milk yield, fat yield and protein yield. If all cows were integrated into the reference population, validation reliability increased whereas regression slope remained constant. Analysing the results of the groups of cows in separate runs suggested that beneficial effects found were caused by the project-cows only. And even though the number of routine-cows was small compared to the number of project-cows or bulls, they seemed to have a rather negative impact on the validation parameters. However, the number of routine-cows was still limited in the present data, because genotyping of cows was not very common four years ago. Therefore, the effect of routine-cows must be re-examined in the future.

Genotyping and phenotyping of new health traits in Spanish dairy cattle herds: I-SA project

N. Charfeddine[1], J. Blanco[2] and M.A. Peréz-Cabal[3]
[1]CONAFE, Spanish Holstein Association, Valdemoro, 28340, Spain, [2]Complutense University of Madrid, Department of Medicine and Animal Surgery, Madrid, 28040, Spain, [3]Complutense University of Madrid, Department of Animal Production, Madrid, 28040, Spain; nouredine.charfeddine@conafe.com

Genomic selection requires a genomic evaluation for new phenotypes in order to justify and to encourage the genotype of young stock. Female genotyping in Spanish dairy cattle is making progress but slowly, in comparison with countries of our environment. In 2015, the Spanish Holstein Association started a new project called I-SA, which aims to record new health traits and to genotype all cows in selected herds in order to provide new selection tools on health traits for breeders using the Spanish genotype program CONAFE GENOMICS. Herd selection was made using the three following criteria: availability of extra information such as milking flow and claw health data, and interest shown in genotyping. Health traits recorded are ketosis, milk fever, metritis, mastitis, ovarian disorder, displaced abomasum, and retained placenta. It is expected to build a cow reference population with 20,000 genotyped and phenotyped cows. The expected cost of setting the project will be paid over a period of time of 5 years.

Feed intake breeding value estimation in German HF cows using single-step genomic evaluation

I. Harder[1], E. Stamer[2], N. Krattenmacher[1], W. Junge[1] and G. Thaller[1]
[1]Institute of Animal Breeding and Husbandry, Kiel University, Olshausenstraße 40, 24098 Kiel, Germany, [2]TiDa Tier und Daten GmbH, Bosseer Str. 4C, 24259 Westensee/Brux, Germany; iharder@tierzucht.uni-kiel.de

At the beginning of lactation, high performing dairy cows often experience a severe energy deficit which in turn is strongly associated with metabolic diseases. Increasing feed intake in this period could improve the metabolic stability and thus the health of the animals. Genomic selection enables for the first time the inclusion of this hard-to-measure trait in breeding programs. For this purpose in the project optiKuh 1,374 Holstein-Friesian dairy cows from twelve German research farms were phenotyped; feed intake data recording was standardized across farms. After data editing phenotypic data comprised a total of 40,012 average weekly dry matter intake records with a mean of 21.8±4.3 kg/d. 1,128 of these phenotyped cows were also genotyped with the Illumina BovineSNP50 BeadChip and 35 animals were genotyped but not phenotyped. Pedigree information contained sires and dams four generations back. Variance components and breeding values were estimated using both pedigree relationships and single-step genomic evaluation each carried out with the DMU software package. The underlying random regression animal model included the fixed effects of herd test week alternatively herd group test, parity and days in milk from 5 to 350. Lactation stage was modelled by the function of Ali and Schaeffer, and for both the random permanent environmental effect and the random additive genetic effect, third-order Legendre polynomials were chosen. Repeatability was high, ranging between 0.6 and 0.8. Heritability estimates ranged between 0.21 and 0.47 and increased towards the end of lactation. For the genotyped cows without phenotypes the inclusion of genomic relationship improves the average reliability of the breeding value for feed intake by nearly 10%.

Exploiting genome data from bovine hospital cases to improve animal welfare on cattle farms

O. Distl[1], S. Reinartz[1], M. Braun[1], K. Doll[2], S. Lehner[1], A. Beineke[3] and J. Rehage[4]
[1]University of Veterinary Medicine Hannover, Institute for Animal Breeding and Genetics, Buenteweg 17p, 30559 Hannover, Germany, [2]Justus-Liebig-University Giessen, Clinic for Ruminants and Swine, Faculty for Veterinary Medicine, Frankfurter Straße 110, 35392 Giessen, Germany, [3]University of Veterinary Medicine Hannover, Department for Patholgy, Buenteweg 17, 30559 Hannover, Germany, [4]University of Veterinary Medicine Hannover, Clinic for Cattle, Bischofsholer Damm 15, 30173 Hannover, Germany; ottmar.distl@tiho-hannover.de

In veterinary university clinics, a large number of cases with a wide spectrum of diagnoses are collected. We started a joint project with veterinary cattle clinics to exploit data and samples of cattle for genotyping and whole genome sequencing to improve animal welfare on farms. Since more than 10 years, all clinical findings and diagnoses are stored in a database and for each case tissue samples are asserved. Pedigree data, findings from necropsy and patho-histological examinations as well as controls are supplemented. Genotyping is done for animals with specific disease entities, a proven heritability or Mendelian inheritance when a sufficient power of the study design is given. We perform whole genome sequencing to identify causal mutations. In Holstein cows, left sided displacement of abomasum (LDA) is a frequent disease with moderate heritability. We genotyped 126 cases and 280 population-based controls with the bovine Illumina HD beadchip. We identified six genomic regions significantly associated with LDA. A validation study on 1,554 Holstein cows confirmed these loci. We are employing a Bayesian mixture model (BayesR) modelling the underlying genetic architecture to estimate genomic breeding values (GEBVs). These GEBVs are useful for breeding and management decisions on the cow, their relatives and the use of sires. Congenital malformations are estimated at frequencies of 0.02-0.08%. We offer farmers a cost-free diagnosing system for deciphering involvement of genetics and the malformation causing genetic variant as well as detection of founder animals. Examples include bulldog-syndrome, polydactyly and growth retardation. In summary, our system employed ensures a quick feedback to farmers, AI and breeding organizations for an appropriate response.

Characterization of copy number variants in a large multi-breed population of cattle using SNP data

P. Rafter[1,2], D.P. Berry[1], A.C. Parnell[2], C. Gormley[2], F. Kearney[3], M.P. Coffey[4], T.R. Carthy[1] and D. Purfield[1]
[1]Teagasc, Animal & Grassland Research, Moorepark, Fermoy, Co.Cork, P61 C996, Ireland, [2]University College Dublin, School of Mathematics and Statistics, Belfield, Dublin 4, D04 V1W8, Ireland, [3]ICBF, Highfield House, Shinagh, Bandon, Co.Cork, P72 X050, Ireland, [4]SRUC, Animal & Veterinary Sciences, Roslin Institure Building, Easter Bush, Midlothian, EH25 9RG, United Kingdom; pierce.rafter@teagasc.ie

Copy number variants (CNVs) are a form of genomic variation that changes the structure of the genome through deletion or duplication of stretches of DNA. The objective of the present study was to characterize CNVs in a large multi-breed population of beef and dairy cattle. The CNVs were called in 5,551 cattle from 22 beef and dairy breeds. The CNVs were called on the autosomes of all animals individually, using two freely available software suites, QuantiSNP and PennCNV. The CNVs were classified in two categories, either deletions or duplications. The distributions of the number of deletions and duplications per animal were both positively skewed. The first quartile, median, and third quartile of the number of deletions per animal was 16, 29, and 117, respectively. The first quartile, median, and third quartile of the number of duplications per animal was 8, 12, and 23, respectively. Per animal, there tended to be twice as many deletions as duplications. The distribution of the length of deletions was positively skewed, as was the distribution of the length of duplications. The first quartile, the median, and the third quartile for the distribution of the length of deletions were 25 kb, 52 kb, and 101 kb, for the first quartile, median, and third quartile, respectively. The first quartile, median, and third quartile for the distribution of the length of duplications were 46 kb, 109 kb, and 235 kb, respectively. Per animal, duplications tended to be twice as long as deletions. In 58.5% of the population, less than 0.1% of their autosomes were composed of CNVs. Most of the CNVs detected in the population were rare; 80.2% of duplications and 69.1% of deletions were present in only one animal in the population. Only 0.154% of all CNVs identified were present in more than 50 animals in the population.

Session 12 Theatre 11

The impact of genomic selection on genetic diversity and genetic gain in French dairy cattle breeds

A.-C. Doublet[1,2], P. Croiseau[2], S. Fritz[1,2], C. Hozé[1,2], A. Michenet[1,2], D. Laloë[2] and G. Restoux[2]
[1]ALLICE, 149 rue de Bercy, Paris, France, [2]GABI, INRA, AgroParisTech, Université Paris-Saclay, Domaine de Vilvert, 78350 Jouy-en-Josas, France; anna-charlotte.doublet@inra.fr

Since the beginning of genomic selection in French dairy cattle in 2010, genetic gain has substantially increased. However, if not controlled, it can be associated with a quick loss of genetic diversity. This loss has a detrimental economic effect because of the decrease in additive genetic variance, inbreeding depression, and loss of adaptive potential in these populations. In this context, we assessed the impact of genomic selection and of the different subsequent breeding schemes on the evolution of genetic diversity and genetic merit in 7 different French dairy cattle breeds, between 2000 and 2018. Pedigree data and 50k chip data were available for 7 French dairy cattle breeds (Brune, Montbeliarde, Normande, Prim'Holstein, Abondance, Tarentaise and Vosgienne). Genetic diversity was estimated for these breeds, by computing pedigree- and molecular-based estimates of kinship and inbreeding coefficients and the levels of heterozygosity. The annual evolution of these estimates before and after the establishment of genomic selection was assessed by considering cohorts based on birth year. The effect of genomic selection on Runs Of Homozygosity (ROH) patterns was also analysed. The evolution of these estimates was then paralleled with that of annual genetic gain. This allowed to put into perspective a possibly faster annual loss of genetic diversity in regards to higher annual genetic gain resulting from genomic selection, possibly due to a shorter turn-over of bulls in these breeding schemes. Furthermore, genetic diversity measured at the whole-genome scale might not reflect well its variations along the genome. Indeed, genomic regions purged of mutation load or target of selection could undergo higher local autozygosity without any deleterious consequences on genetic gain or fitness. Further work will highlight this issue to develop new and sustainable methods to manage inbreeding.

Session 12 Theatre 12

Effect of mating strategies on genetic and economic outcomes in a Montbéliarde dairy herd

M. Berodier[1,2], M. Brochard[1,3], C. Dezetter[4], N. Bareille[5] and V. Ducrocq[2]
[1]MO3, 259 route des Soudanières, 01250 Ceyzériat, France, [2]UMR GABI, AgroParisTech, INRA, Université Paris-Saclay, Domaine de Vilvert, 78350 Jouy-en-Josas, France, [3]Umotest, 259 route des Soudanières, 01250 Ceyzériat, France, [4]Unité de Recherche sur les Systèmes d'Elevage (URSE), Ecole Supérieure d'Agricultures (ESA), Université Bretagne Loire, 55 rue Rabelais, 49007 Angers, France, [5]BIOEPAR, INRA, Oniris, Université Bretagne Loire, La Chantrerie, 44307 Nantes, France; marie.berodier@inra.fr

This study compared the genetic and economic evolution of a 77-cow Montbéliarde dairy cattle herd after 15 years of simulation including 10 years with 8 different mating strategies: with or without genotyping of all female dairy calves combined with or without use of sexed semen and combined with or without use of beef semen. A mechanistic, stochastic and dynamic model was used to mimic the farmer's decisions and individual cow's biology. Females true breeding values for milk yield, fat content, protein content, fertility, longevity and udder health traits influenced production, reproduction, health and culling of the animal. For scenarios with sex semen, the best heifers and 1st lactation cows on breeding objective (calculated as a linear combination of the 6 traits modelled) were inseminated with sexed semen to guarantee replacement with genetically better females. For scenarios allowing beef cross, the worst cows on breeding objective were inseminated with beef breed semen to obtain higher economic value from calves. After 10 years of alternative mating strategies, variations in genetic gain (+12% to +19%) and gross margin (+40% to +49%) show a clear advantage for the scenarios using sexed semen and no beef semen compared to the scenario without female genotyping and with conventional semen only. However, those scenarios require much larger total expenses and strongly increase the number of heifers to be reared. Therefore, it is highly dependent on economic assumptions, especially on the price of 'ready-to-calve' heifers. The scenario using sexed semen, beef semen and female genotyping performed better than the scenario with conventional semen only and without female genotyping. It allows important genetic gains at herd level and a diversification of the sales when market conditions fluctuate.

Session 12 Poster 13

Genomic selection in Pinzgauer cattle

H. Schwarzenbacher and C. Fuerst
ZuchtData GmbH, Dresdner Straße 89/19, 1200 Vienna, Austria; schwarzenbacher@zuchtdata.at

A stochastic simulation study was carried out to assess the potential of single trait genomic selection in Pinzgauer cattle based on a female training population over several generations. For this purpose a simulation software was developed that combined the software 'QMSim' with own routines to allow more flexibility in the design. An evolution process was emulated to obtain an LD structure similar to that observed in the Pinzgauer population. After several generations of selection based on conventional ebvs, ten generations of genomic selection was applied to assess the performance of single step genomic selection relative to conventional blup selection. The single step system produced significantly higher reliabilities in all selection groups and consequently a larger genetic gain. The results suggest that genomic selection based on a female training population could be interesting even in a small cattle population if the genotyping costs are low.

Session 12 Poster 14

Assessing genetic architecture and signatures of selection of dual purpose Gir cattle populations us

J.A.I.I.V. Silva, A.M. Maiorano, D.L. Loureça, A.M.T. Ospina, R.A.S. Faria, L.E.C.S. Correia, A. Vercesi Filho and M.E.Z. Mercadante
Unesp/Faculdade de Ciencias Agrarias e Veterinaria, Via de Acesso Prof.Paulo Donato Castellane s/n, 14884-900, Brazil; jaugusto@fmvz.unesp.br

Studies of population structure and the signatures of selection in divergent Gir populations are scarce and need more attention. 173 and 273 animals selected for growth traits and for milk production, respectively, were genotyped on the Illumina GGP Bovine LDv4. Principal Component Analysis (PCA) and Discriminant Analysis of Principal Components (DAPC) were performed using the adegenet R package. The fixation index (Fst) was computed using the HierFstat R package and was used for verifying genetic divergence and selection signatures between the populations. The threshold to call a SNP signature of selection (outlier) was defined as three standard deviations above the mean. The percentages of variance explained by the first and second principal components were 6.79 and 1.98, respectively. For DAPC, the lowest BIC value showed two clusters should be considered. According to the PCA and DAPC analyses, there is a clear genetic differentiation between the beef and dairy populations, and this division reflects the variation existent between the populations. The average value for Fst was 0.033. The Fst approach enables the detection of 488 SNPs as selection signatures. In total, 157 out of the 488 significant Fst values were located in 151 gene regions. Fst results indicate selection for different criteria led to genetic divergence and difference in the allele frequencies between the two cattle populations. These findings are supported by the population structure results. The different breeding objectives, the populations were exposed to, were expected to imprint a degree of genomic differentiation even between close populations.

Session 12 Poster 15

Characterization of Korean cattle breeds through a high-density SNP Chip

S.C. Kim, K.W. Kim, J.W. Lee, H.J. Roh, D.Y. Jeon, S.S. Lee and C.Y. Cho
Animal Genetic Resources Research Center, National Institute of Animal Science, RDA, Namwon, 55717, Korea, South; kisc@korea.kr

The aim of this study was to compare the genetic diversity and divergence of Korean cattle breeds using the BovineHD 777K chip. Linkage disequilibrium (LD) decay could be attributed to population history events, including selection pressure, effective population sizes and admixture with wild-type ancestors. And, the effective population size (Ne) is number of breeding individuals in population and is particularly important as it determines the rate at which genetic variation is lost. So, Effective population size is strongly associated with genetic variability and adaptation. The genotype data comprised a total of 216 samples. A total of 226,694 single nucleotide polymorphisms (SNPs) were used for genetic diversity analysis. We used the squared correlation coefficient between two loci (r^2) as a measure of LD. For pairwise comparisons of all SNPs separated by a maximum distance of 100 Kb, the r^2 value based on haplotype frequencies estimated via the Expectation-Maximization algorithm was predicted using SNP & Variation Suite(SVS) 8 software. The r^2 levels start at 0.568, 0.551 and 0.557 for Brown Hanwoo (BH), Brindle Hanwoo (BRH) and Jeju Balck (JB), respectively, when using the 2Kb bin of SNPs. The Ne of each breed was estimated through SNP-based LD analysis with SNeP program. Among the breeds, BH and BRH had the high Ne of 260 and 202 respectively, until 13 generations ago while JB had the lowest Ne of 55. In this study, we observed a sharp decline in the effective population size of all the cattle breeds. The sharp decline was observed at ~50 generations ago. This was the time of formation of the current breeds. This was the time when selection and development of breeding programs had just begun. The modern-day cattle genetic structure, LD and Ne is a result of the various historic events and extensive artificial and natural selection. An analysis of differentiation based on high-density SNP chip showed the various differences between Korean cattle breeds and the closeness of breeds corresponding to the geographic region in which they are evolving.

Session 12 Poster 16

Association of BoLA-DRB3 alleles with mastitis in Romanian cattle breeds

D.E. Ilie[1], D. Gavojdian[2], R.I. Neamt[1], F.C. Neciu[1] and L.T. Cziszter[3]
[1]Research and Development Station for Bovine Arad, Calea Bodrogului 32, 310059 Arad, Romania, [2]Research and Development Institute for Bovine Balotesti, Sos. Bucuresti-Ploiesti, km 21, 077015 Balotesti, Romania, [3]Banat's University of Agricultural Sciences and Veterinary Medicine 'King Michael I of Romania', Calea Aradului 119, 300645 Timisoara, Romania; danailie@animalsci-tm.ro

Mastitis is the most common and costly production disease in dairy species, with large impacts on farming efficiency and animal welfare. The incidence of mastitis is associated with both cows' exposure to bacteria and the cows' genetic make-up for resistance to pathogens. Thereby, to improve animals' ability for mastitis resistance in succeeding generations and to reduce the disease prevalence rate one feasible approach could be the genetic selection of cows for the most potentially protective immune response. The objective of our study was to evaluate the association of 27 SNPs in exons 1-4 of BoLA-DRB3 gene with mastitis in 298 cows from Romanian Spotted (n=250) and Romanian Brown (n=48) breeds, using the Kompetitive Allele Specific PCR (KASP™) assay. The study was carried out at the Research and Development Station for Bovine Arad (46°10′36″N 21°18′4″E) where animals were managed under identical conditions. Animals were monitored for signs of clinical mastitis during one lactation, and the prevalence was determined as the proportion of animals affected. The clinical and sub-clinical mastitis incidence rates were 4.12±0.72 and 24.63±0.15% in Romanian Spotted cows and 2.08±0.82 and 27.08±0.64% in Romanian Brown cows, respectively (P>0.05). Polymorphic SNPs were detected for BoLA-DRB3 g.25472281G>A (rs42309897), BoLA-DRB3 g.25475692C>T (rs208816121) and BoLA-DRB3 g.25476219A>G (rs110124025) loci with a MAF >10%. The single SNPs and their genetic effects on mastitis were evaluated and a significant association was found in rs110124025 for the Romanian Spotted ($P \leq 0.001$) and the Romanian Brown ($P \leq 0.05$) breeds, respectively. The current results are relevant for future cattle genomic studies and suggest that polymorphisms in BoLA-DRB3 g.25476219A>G locus influence the ability of cows to resist to mastitis infections.

Rapid and reliable assays for inherited disease detection in Russian Holstein cattle population

E.E. Davydova, E.V. Krylova, O.O. Golovko, E.S. Riabova and I.V. Soltynskaya
The All-Russian State Center for Quality and Standardization of Veterinary Drugs and Feed (VGNKI), Molecular Biology, Zvenigorodskoe shosse, 5, 123022, Moscow, Russian Federation; e.davydova@vgnki.ru

Bovine leucocyte adhesion deficiency (BLAD), deficiency of uridine monophosphate synthase enzyme (DUMPS), complex vertebral malformation (CVM), factor XI deficiency (FXID), bovine citrullinemia (BC), brachyspina syndrome (BY), cholesterol deficiency (HCD) are the most common inherited diseases in Holstein population of cattle. We developed sensitive, rapid and robust PCR and PCR-pyrosequencing assays for detection of the above disorders. The assays are able to discriminate a wild-type and defective alleles, so that carriers and affected animals can be easily distinguished. PCR assay has good diagnostic performance for chromosomal rearrangements. The pairs of primers flanking target regions of bovine genome have been designed in such a way as to allow implementation of PCR with electrophoretic detection to identify OMIA 000151-9913 (Chr21: 3.3 kbp deletion, 25-27 exons, FANCI gene, BY); OMIA 001965-9913 (Chr11: 1.3 kbp insertion, 24-25 exons, APOB gene, HCD); OMIA 000363-9913 (Chr27: 76 bp insertion, 12 exon, FXI gene, FXID) disorders. Pyrosequencing was used for detection of single polymorphisms OMIA 001340-9913 (Chr3: SLC35A3_559G>T, CVM), OMIA 000595-9913 (Chr1:ITGB2_383A>G, BLAD), OMIA 000262-9913(Chr1:UMPS_1213C>T, DUMPS), OMIA 000194-9913 (Chr11:ASS1_256C>T, BC). A total of 256 imported and local Holstein bulls were tested. Fifteen HCD, seven CVM, three BY, three FXID and two BLAD-carriers were identified, corresponding to heterozygote frequencies of 6; 3; 1, 1 and 0.8% respectively. No BC and DUMPS-carriers were identified. Relatively high prevalence of HCD-carriers in the Russian livestock are explained by the fact that the HCD-associated mutation was reported only in 2016 while DUMPS, BLAD, CVM, BC mutations have been known for about twenty years. Last decade routine testing of sperm and heifers allowed to gradually eradicate the deleterious alleles from the Holstein population. The high frequency of the HCD-allele in cattle shows that implementation of HCD-testing of bull sperm is necessary.

Pyrosequencing technology for inherited disease detection in Russian cattle population

E.V. Krylova, I.V. Soltynskaya, B.U. Vetoshnikova, M.A. Pleskacheva and E.E. Davydova
The Russian State Center for Animal Feed and Drug Standardization and Quality (FGBU VGNKI), 5, Zvenigorodskoe shosse, Moscow, 1230022, Russian Federation; e.krylova@vgnki.ru

Cattle inherited diseases are usually caused by recessive alleles proceed from increased inbreeding in population. Pyrosequencing is the one of the most appropriate techniques to discriminate a wild-type and defective alleles associated with single-nucleotide polymorphisms (SNP). Pyrosequencing is a real time method of DNA sequencing based on the detection of released pyrophosphate during a nucleotide incorporation into a growing DNA chain. The released pyrophosphate undergoes a series of enzymatic transformations, as a result of which a chemiluminescent signal is detected. The set of signals corresponds to the nucleotide sequence of the target genetic locus. Pyrosequencing facilitates short-read sequencing, rendering it easier and more rapid than Sanger sequencing does. Sensitive, rapid and robust pyrosequencing based tests have been developed for the detection of cattle inherited diseases such as spinal muscular atrophy (SMA), spinal dysmyelination (SDM), Wiwer (W), brown swiss haplotype 2 (BH2), mannosidosis alpha (MA), developmental duplications (DD), dwarfism (DW). These diseases are the most common inherited diseases in Brown Swiss and Aberdeen Angus cattle populations and are associated with single polymorphisms – OMIA 000939-9913 (Chr24:KDSR_490C>T, SMA), OMIA 001247-9913 (Chr11:SPAST_560G>A, SDM), OMIA 000827-9913 (Chr4:PNPLA8_1703G>A, W), OMIA 001939-9913 (Chr19:TUBD1_rs383232842, BH2), OMIA 000625-9913 (Chr7:MAN2B1_560G>A, MA), OMIA 001465-9913 (Chr26:NHLRC2_932T>C, DD), OMIA 000299-9913 (Chr6:PRKG2_2032C>T, DW). Carriers and affected animals can be easily distinguished using the developed techniques. A total of 62 Brown Swiss and 49 Angus bulls were tested. Three SMA and four DD -carriers were identified, corresponding to heterozygote frequencies of 5 and 8% respectively. Exclusion of sperm carriers defective alleles from breeding work will reduce the prevalence of these anomalies in the cattle population.

On host-microbiota interactions and livestock phenotypes: an overview of recent results in pigs

J. Estellé
INRA, UMR 1313 GABI, AgroParisTech, Université Paris-Saclay, Jouy-en-Josas, France; jordi.estelle@inra.fr

Holobionts, defined as individual hosts and their associated symbiotic microbial communities, are the ultimate selection units into which evolution and selection act. Indeed, the microbiome of the intestinal tract may be considered as a new host organ playing major roles in livestock health and efficiency phenotypes. In the current research scenario, the high-throughput genomic technologies allow an unprecedented precision for the analysis of genomes and metagenomes. This is particularly true for the pig as it is the first livestock species, and only the third species after humans and mice, for which the genome reference sequence and the gut metagenome catalogue are both available. This oral communication will perform an overview of the recent results obtained in pigs when combining host genetics and gut microbiome analyses, which have confirmed that the gut microbiota composition is partially under host genetic control. In addition, the results linking pig gut microbiota composition with health and growth traits will be summarized by highlighting the relevance of this ecosystem for the porcine production. Overall, it is foreseen that the study of animals as holobiont entities by using explicit hologenetics approaches will emerge as a cutting edge discipline for deciphering the determinism of complex traits in livestock. The studies being currently performed in pigs present valuable opportunities to obtain relevant conclusions not only for the porcine production but also for the understanding of host-microbiota interactions in biomedicine.

Session 13 Theatre 2

Genetic approach of rumen metagenome: state of the art in small ruminant and perspectives

C. Marie-Etancelin[1], S.J. Rowe[2], A. Jonker[3] and A. Meynadier[1]
[1]GenPhySE, INRA-ENVT-INPT, Université de Toulouse, Castanet-Tolosan, France, [2]AgResearch Ltd, Invermay, Mosgiel, New Zealand, [3]AgResearch Ltd, Grasslands, Palmerston North, New Zealand; christel.marie-etancelin@inra.fr

The ruminal microbiome plays a central role in the nutrition of the ruminant host, directly affecting production and undesirable by-products such as methane. Studies on ruminal microbiota highlight a significant effect of the host, but few publications have reported results concerning the impact of host genetics on microbial community composition. Rumen microbiota can be described using targeted or whole genome sequencing: rRNA is primarily used for the determination of taxonomic abundance of bacteria/archaea/fungi/protozoa, and a shotgun approach for abundance of genes in the rumen. An alternative high-throughput genotyping by sequencing technique was recently proposed by Hess *et al.* for describing microbial community composition in large numbers of animals. In sheep, a comparison of extreme animals (phenotypically or genetically) showed differences in their microbiota. Ellison *et al.* reported differences in bacteria and archea abundances in sheep having low or high residual feed intake, with some interactions with the diet. There was little evidence of links between feeding rate in sheep and ruminal bacteria abundances. From divergent lines, Kittelmann *et al.* showed 3 bacterial community types linked with genetic level of methane emission and de Barbieri *et al.* reported that selection for fleece weight is associated with differences in the diversity of ruminal bacteria. The first estimates of microbiota heritabilities in sheep were provided by Rowe *et al.* who reported genetic control of rumen microbial communities and genetic links to methane emissions. For this conference, Marie-Etancelin *et al.* estimated that ¼ of the taxa genera abundances in the rumen have a heritability greater than 0.1. It would be useful if these preliminary results on the targeted metagenome approach were followed up by whole gene quantification to characterize the functionality of microbiota. As in cattle, microbiota metagenome and host's genome contributions to the variability of traits should be considered simultaneously in sheep experiments.

Effect of gut microbiota on production traits, interaction with genetics

H. Gilbert[1], S. Lagarrigue[2], L.M.G. Verschuren[3], O. Zemb[1], M. Velasco[4], J.L. Gourdine[5], R. Bergsma[3], D. Renaudeau[2], J.P. Sanchez[4] and H. Garreau[1]
[1]INRA, UMR GenPhySE, 31320 Castanet Tolosan, France, [2]INRA, AgroCampusOuest, UMR PEGASE, 35042 Rennes, France, [3]Topigs Norsvin Research Center B.V., Schoenaker 6, 6641 Beuningen, the Netherlands, [4]Institute for Food and Agriculture Research and Technology, IRTA, 08140, Torre Marimon, Spain, [5]INRA, URZ, 97170 Petit-Bourg, France; helene.gilbert@inra.fr

Gut microbiota is a key contributor to feed use in monogastric species, in particular via the digestion dietary fibres. Molecular techniques are now available to run large studies and decipher the potential of gut microbiota to improve livestock. Studies on human and mice are more advanced: different factors have been demonstrated to influence the gut microbiota composition and functions, including maternal transmission, environment (diet composition and quantity, humidity and heat), age and physiological status. Studies also evaluated if the host controls its gut microbiota. In pigs, chicken and rabbits, microbiota differences between animals of extreme phenotypes within populations, and between lines divergently selected for specific traits, are reported. Other studies reported that some microbiota abundancies are heritable. Linear mixed models have been used to evaluate its contribution to trait variability, or microbiability, with different data (full vs 16S sequencing of gut or faecal contents) and different variance matrices. They showed significant contribution to production traits, reaching more than 30%. However, confounding effects exist, such as the maternal inheritance of the microbiota and the genetic determinism by the host: many studies show reduction of the genetic additive variance, in addition to reduction of the residual variance with microbiability. Specific datasets, testing a given genetic in different environments or using cross-fostering, are used to better understand the relative contribution of each effect (host genetics and microbiota) to the trait variance, and propose solutions to livestock. This work is part of the European Union's H2020 Feed-a-Gene Project (grant 633531).

Gut microbiome provides a new source of information to improve growth efficiency in swine

D. Lu, F. Tiezzi and C. Maltecca
North Carolina State University, Box 7621 NCSU, 27695, USA; f_tiezzi@ncsu.edu

Gut microbiome has been proven to affect pork production *via* nutritional, physiological, and immunological processes. We studied gut bacteria of the pig from host genetics gut microbiome perspectives, seeking to incorporate such relationship in genetic improvement of pigs. There were 1,205, 1,295, and 1,283 rectal samples collected from pigs at weaning (18.6±1.09 d), 15 weeks post weaning (118.2±1.18 d), and end of feeding trial (196.4±7.86 d). Of these 1,039 animals had samples collected at all 3 time points. The microbiome data was analysed at operational taxonomic unit (OTU) level, including 1,755 OTUs. The animals were also genotyped with the Illumina PorcineSNP60 Beadchip. From our association analyses, 131 OTUs were identified as large contributors to the variance of backfat thickness (BF), live weight (WT), and loin depth (LD), at 3 time points, week 14, 18, and 22, for each phenotypic record. Three OTUs, including OTU17, OTU758, and OTU1163, had the largest contribution to the total variance of the traits. Heritabilities of the 3 OTUs varied between 0.13±0.05 and 0.40±0.06 for OTU17, 0.02±0.03 and 0.20±0.06 for OTU758, 0.02±0.03 and 0.21±0.06 for OTU1163 for the 3 time points. Single nucleotide polymorphisms (SNP) that had consistently large effects on OTU17 and OTU758, at week 15 and end of the test, were also identified on chromosomes 3, 6, and 7.We further included microbiome data in estimating breeding values (BV) for BF and average daily weight gain (ADG) at 22 weeks post-weaning. We found that providing microbiome information, under the form of relatedness among individuals based on similarity of microbial communities, significantly improved the model fit for both BF and ADG, as well as reduced standard error of predictions for BF and ADG breeding values. This analysis was a preliminary attempt to effectively include gut microbiome data to improve the accuracy of BVs in the pork industry. We have plans to incorporate the results from our association analyses in forming microbiome-based and genotype-based relationship matrices to be used in estimating BVs. We have proven interaction between host genetics and its gut microbiome in regulating the host phenotypic records. Such interaction can be used to improve the accuracy of genetic evaluation of the pig.

Session 13 Theatre 5

Faecal microbiome profiles can predict complex traits in pigs

D. Schokker[1], L.M.G. Verschuren[1,2], R. Bergsma[2], F. Molist[3] and M.P.L. Calus[1]

[1]Wageningen Livestock Research, Genomics, Droevendaalsesteeg 1, 6708 PB, Wageningen, the Netherlands, [2]Topigs Norsvin Research Center, Schoenaker 6, 6641 SZ, Beuningen, the Netherlands, [3]Schothorst Feed Research, Meerkoetenweg 26, 8218 NA, Lelystad, the Netherlands; dirkjan.schokker@gmail.com

Evidence accumulates that (early life) intestinal microbiota drive health and metabolic phenotypes in livestock. It has also been shown that gut microbiota profiles can be regarded as complex polygenic traits, that are influenced by both the host and the environment. The objective of this study was to identify faecal microbiota profiles associated to a complex trait, here focusing on average daily gain (ADG) and feed intake (FI), which are important economical traits in pigs. In order to investigate this, phenotypic data was collected of 160 three-breed cross pigs, 80 males and 80 females, coming from 20 litters. Pigs were divided in two different groups based on typical market diets, a diet based on corn/soybean meal and a diet based on wheat/barley/by-products. The day before slaughter 142 faecal samples were collected (74 female and 67 male) and subsequently the faecal microbiome was profiled by sequencing the 16S hypervariable region of bacteria. With these microbiota data, we estimated the 'microbiability', $m^2 = \sigma^2_m / (\sigma^2_m+\sigma^2_e)$, which was 49% (±14%) for ADG and 46% (±16%) for FI. The accuracy of these metagenomic predictions of the phenotypic ADG was 0.197 (R^2=0.039) and FI was 0.20 (R^2=0.04), and were both significantly greater than zero. In conclusion, we have shown that the faecal microbiome profiles can be used to predict complex phenotypes of the host, in this case ADG and FI.

Session 13 Theatre 6

Changes in rumen microbiome interaction explain the methane emissions differences in beef cattle

M.D. Auffret[1], M. Martinez-Alvaro[1], R.J. Dewhurst[1], C.-A. Duthie[1], J.A. Rooke[1], R.J. Wallace[2], T.C. Freeman[3], R. Stewart[3], M. Watson[3] and R. Roehe[1]

[1]Scotland Rural College (SRUC), The Roslin Institute, EH25 9RG Edinburgh, United Kingdom, [2]Rowett Institute, University of Aberdeen, Rowett Institute, AB25 2ZD, United Kingdom, [3]The Roslin Institute, University of Edinburgh, The Roslin Institute, EH25 9RG Edinburgh, USA; marc.auffret@sruc.ac.uk

Methane is produced from anaerobic microbial fermentation in the rumen. CH_4 emissions from cattle have negative impacts on animal production. Methanogens are able to use different methanogenic pathways, and interact with other members of the rumen microbial community. The degradation of dietary intake produces particularly H2, lactate and volatile fatty acids (VFA), impacting differently on CH_4 synthesis. The aim of our research was to use network analysis of rumen metagenomes to predict species-species interactions to explain differences between low- (LME) and high- (HME) CH_4 emitting beef cattle. Total DNA was extracted from post-mortem ruminal digesta samples taken from 66 beef cattle balanced between Aberdeen Angus and Limousin and basal diet (forage or concentrate). Taxonomic community analysis and microbial gene abundances were based on Kraken and KEGG genes databases, respectively. A network analysis using Miru defined individual clusters considered as different ecological niches and helped to study the changes and interactions within the microbiome between LME and HME. A comparison of the network between LME and HME showed one cluster containing *Methanobrevibacter* species interacting with a small number of populations in HME (8/131) and almost all the genes involved in CH_4 emissions except for genes encoding for formate dehydrogenase. In a second cluster, bacterial species within Firmicutes or Proteobacteria were significantly higher ($P<0.05$) in HME or LME respectively. Competition for H2 between microorganisms including methanogens as shown by cluster reorganization between HME and LME or different populations associated with VFA metabolism explained change in CH_4 emissions. This study improves our understanding of rumen microbial interactions associated with CH_4 production. This knowledge can be used to develop new CH_4 mitigation strategies targeting microbial populations significantly interacting with the key methanogenic groups.

Subacute ruminal acidosis and the global profile of ruminal and faecal microbiota of dairy cows

J.L. Martinez[1], E. Sandri[2], Y. Couture[3], R. Gervais[4], J. Levesque[2], D. Roy[1] and D.E. Rico[2]
[1]INAF, Université Laval, 2440 Boulevard Hochelaga, G1V 0A6, Quebec, QC, Canada, [2]CRSAD, 120-A chemin du Roy, G0A 1S0, Deschambault, QC, Canada, [3]Department of Veterinary Medicine, Université de Montreal, 3200 Rue Sicotte, J2S 2M2, Saint-Hyacinthe, QC, Canada, [4]Animal Science Department, Université Laval, 2425 Rue de l'agriculture, G1V 0A6, Quebec, QC, Canada; daniel.rico@crsad.qc.ca

Subacute ruminal acidosis (SARA) could alter the microbiote composition of the intestinal tract and productivity of dairy cows. Twelve ruminally cannulated cows (120±52 DIM; 35.5±8.9 kg of milk/d; mean ± SD) were randomly assigned to either (1) induction of SARA (2) recovery from SARA, and (3) control in a 3×3 Latin square design with 21-d periods. SARA was induced by feeding a diet containing 29% starch and 24% NDF, whereas the recovery and control diets contained 20% starch and 31% NDF. Whole ruminal and faecal samples were taken on d 0, 7 and 21 of each period and subjected to high-throughput Illumina sequencing of the V3-V4 regions of bacterial 16S rRNA gene. On average, 25,279 high-quality sequences were generated per sample. *Firmicutes* (59.8%), *Bacteroidetes* (27.4%), and *Proteobacteria* (7.4%) were the predominant ruminal phyla in rumen samples, whereas *Firmicutes* (86.5%) and *Bacteroidetes* (8.5%) were predominant in the faeces. The Chao 1index was lower in faeces compared with rumen ($P<0.05$), but no differences were detected between treatments. Principal component analysis of weighted UniFrac distances revealed a distinct clustering for SARA-associated microbiota compared with control on d 21, whereas recovery was intermediate. Linear discriminant analysis effect size (LEfSe) indicated that the ruminal proportions of the genus *Roseburia* and the order *Rickettsiales* were increased in response to the SARA diet relative to control ($P<0.05$). A pair-wise correlation analysis indicated that the class *Gammaproteobacteria* was positively associated with the rumen acidosis index (AI; area under pH 5.8/DMI) and the *trans*-10 to *trans*-11 18:1 ratio (0.55 and 0.70, respectively; $P<0.001$). Induction of SARA impacted particular bacterial clades which could be used as markers of this condition in dairy cows.

Can rumen microbes improve prediction of subclinical ketosis in dairy cows?

G.F. Difford[1,2], G. Gebreyesus[1,2], P. Løvendahl[2], A.J. Buitenhuis[2], J. Lassen[2], B. Guldbrandtsen[2] and G. Sahana[2]
[1]Wageningen University & Research Animal Breeding and Genomics, Wageningen, P.O. Box 338, 6700 AH, the Netherlands, [2]Center For Quantitative Genetics and Genomics, Aarhus University, Department of Molecular Biology and Genetics, Blichers Alle, 8830, Denmark; gareth.difford@mbg.au.dk

Sub-clinical metabolic disorders such as ketosis cause substantial economic losses in dairy cattle. Typically, these disorders are lowly heritable and difficult to predict in dairy cows. To remedy this, the use of biomarkers such as β-hydroxybutyric acid (BHBA) and acetone (ACE) concentrations in milk for early detection has gained impetus in recent years. A total of 484 lactating Danish Holstein cows from 4 robotic milking herds were sampled for rumen contents by oesophageal insertion of a rumen flora scoop. These cows were genotyped using Illumina BovineSNP50 BeadChip and imputed to Illumina BovineHD markers genotype. Milk samples were drawn and profiled using liquid chromatography/electrospray ionization-mass spectrometry (LC/ESI-MS) to determine BHBA and ACE levels. Rumen microbial 16S rRNA gene libraries were constructed for rumen bacteria and archaea and sequenced. Taxon counts were converted to an array of operational taxonomic units (OTUs). Preliminary findings show that microbial composition explains 4-6 times more variance in BHBA and ACE than additive genetic effects do. The relative predictive abilities of GBLUP (SNPS) and MBLUP (based on 16S rRNA OTUs) are examined.

Multivariate and network analysis identified microbial biomarkers linked to methane emission

Y. Ramayo-Caldas[1,2], A. Bernard[3], L. Zingareti[4], M. Popova[3], N. Mach[2], J. Estelle[2], E. Rebours[2], R. Muñoz-Tamayo[5], A. Rau[2], M. Mariadassou[6], M. Perez-Enciso[4], D. Morgavi[3] and G. Renand[2]

[1]IRTA, Torre Marimon, 08140, Caldes de Montbui, Spain, [2]INRA, AgroParisTech, Génétique Animale et Biologie Intégrative, Jouy-en-Josas, 78350, France, [3]INRA, Herbivore Research Unit, Saint Genès-Champanelle, 63122, France, [4]CRAG-UAB, Department of Animal Genetics, Bellaterra, 08193, Spain, [5]Mosar, UMR 791 AgroParisTech, Inra, Paris, 75231, France, [6]INRA, MIG, Jouy en Josas, 8350, France; yuliaxis.ramayo@inra.fr

The rumen microbiota from 65 Holstein cows was characterized through sequencing of the 16S rRNA (bacteria) gene. Methane yields ($CH_4y = CH_4/DMI$) and dry matter intake (DMI) were individually measured during 3 weeks with the GreenFeed and Calan gate systems, respectively. Twenty of these cows were measured twice in two successive months. A combination of multivariate, clustering and microbial co-abundance network analysis was implemented to (1) identify a set of OTUs jointly associated with CH_4y and (2) discriminate cows based on the structure of ruminal bacterial communities. In addition, repeated measures were used to provide an estimation of the stability of ruminal bacteria within a month, which was R=0.64. Three ruminotype-like clusters (R1, R2 and R3) were identified; and R2 was associated with higher CH_4y emission. Sparse partial least squares discriminant analysis (sPLS-DA) was done based on sample classification following two criteria: ruminotype-like cluster assignation and CH_4y phenotype classification (low, high and intermediate). A sPLS regression model using the phenotype variation of CH_4y emission was also implemented. By combining these approaches, 28 OTUs explaining 52% of the phenotype variation in CH_4y were identified. The taxonomic classification of these OTUs included families linked to CH_4 emission such as *Christensenellaceae*, *Lachnospiraceae*, and *Ruminococcaceae*. According to the interaction patterns, 83% (25/28) of these OTUs belong to the same sub-network module. In summary, our results suggest a relative stability of the rumen microbiota as well as a common set of interacting OTUs simultaneously linked to CH_4y emission and to the microbial community structure of the rumen.

Genetic determinism of dairy sheep ruminal microbiota

C. Marie-Etancelin[1], B. Gabinaud[1], G. Pascal[1], R. Tomas[2], J.M. Menras[2], F. Enjalbert[1], C. Allain[1], H. Larroque[1], R. Rupp[1] and A. Meynadier[1]

[1]INRA-GenPhySE, Animal Genetics, Chemin de Borde Rouge, 31326 Castanet Tolosan, France, [2]INRA La Fage, Animal Genetics, Experimental farm, 12250 Saint Jean et Saint Paul, France; christel.marie-etancelin@inra.fr

The microbiota of herbivorous animals plays a central role in the nutrition of its host, directly affecting his health and his ability to produce. Very few publications reported results concerning the impact of host genetics on the composition of ruminal microbiota. Thus, we proposed to study the genetic determinism of bacterial relative abundances of sheep rumen microbiota. 369 dairy Lacaune ewes raised indoor at the INRA Experimental Farm of La Fage, had a sampling of their rumen fluid done. These ewes were adult animals, fed with a 93% hay-silage based diet and belonged to 4 different lines (lines divergently selected on somatic cells count or on milk persistency). Ruminal metagenome were sequenced using 16s rRNA gene with Illumina Miseq technology. Bioinformatics analysis of the microbiota sequences were implemented with FROGS pipeline to obtain relative abundances of bacteria and R Phyloseq package to estimate biodiversity indices. Heritability estimates of the square root of relative abundances were computed in single trait using the VCE 6.0 software. FROGS pipeline allowed clustering the 4,944,307 informative sequences into 2,135 OTUs, which represented 247 bacteria taxas (140 genera, 50 families, 31 orders, 17 classes and 9 phyla). Significant differences between lines were observed: 4 bacteria generas have abundancies differences according to CCS lines (*Olsenella*, *Prevotella* 1, *Prevotellaceaea* Ga6a1, *Syntrophococcus* with always higher values for CCS+) and 4 others according to PERS lines (*Coprococcus* 1, *Olsenella*, *Succonivibrionaceae* U2, *Syntrophococcus*). At the genera scale, heritabilities estimates ranged from 0.00 to 0.49 with a standard error of 0.11 on average: 22% of genera had heritabilities higher than 0.1, which is lower than Estellé *et al.* on Large White (50% genera with h2>0.1). The most heritable genera (h2>0.25) were *Ruminococcaceae* UCG002, *Lachnospira*, *Atopobium* and *Oscillospira* which also were taxa with low abundances.

Session 13 Theatre 11

Effect of *in ovo* microbiome stimulation on immune responses in different chicken breeds

A. Slawinska[1], A. Dunislawska[1], M. Siwek[1], A. Kowalczyk[2] and E. Lukaszewicz[2]
[1]UTP University of Science and Technology, Mazowiecka 28, 85-084 Bydgoszcz, Poland, [2]Wroclaw University of Environmental and Life Sciences, Chelmonskiego 38c, 51-630 Wroclaw, Poland; slawinska@utp.edu.pl

Intestinal microbiota in animals are responsible for immune system maturation. The beneficial profile of the microbiome should be developed in perinatal period, which is critical for building overall immunocompetence but also gaining oral tolerance by gut-associated lymphoid tissue (GALT). In chickens it is possible to enhance the microbiome development during embryonic phase with use of *in ovo* technology. The goals of this paper were to (1) determine the peripheral and systemic immune responses in chickens to pro-inflammatory antigens and (2) estimate the effects of host genetics and *in ovo* stimulation with prebiotic, probiotic or synbiotic on the strength of the immune responses mounted upon the challenge. The two trials have been conducted based on full-factorial design. The animals used in the trials were broiler chicken for Trial 1 and native chicken for Trial 2. Both trials started with *in ovo* injection of prebiotic (GOS, galactooligosaccharides), probiotic (*L. lactis* subps. *cremoris*) or synbiotic (GOS + *L. lactis* subps. *cremoris*), carried out on day 12 of egg incubation. The birds were housed in litter pens (4 replicates/group, 8 animals each) for 42 days. At slaughter day chickens were injected intraperitoneally with lipopolysaccharide (LPS), lipoteichoic acid (LTA) or mock-injected with saline. Upon injection, LPS and LTA trigger transient immune responses. Animals were sacrificed two hours post-injection and gene expression of the major immune mediators was performed in caecal tonsil and spleen. Gene panel included cytokines responsible for inflammation (*IL-1B*, and *IL-12*), Th1/Th2 polarization (*IL-6*, *IL-10* and *IL-12*), Th2 (*IL-2* and *IL-4*) and Th17 (*IL17*) immune responses. We discuss G×E effects of the chicken breed (broiler vs native chicken) and *in ovo* treatment (prebiotic vs probiotic vs synbiotic) on the immune responses against environmental antigens. Acknowledgements: the research was supported by a project UMO-2013/11/B/NZ9/00783 (NSC, Cracow, Poland).

Session 13 Poster 12

Effect of successive fibre diets on nutrient digestibility and faecal microbiota composition in pigs

M. Le Sciellour[1], E. Labussière[1], O. Zemb[2] and D. Renaudeau[1]
[1]PEGASE, INRA, Agrocampus-Ouest, 35042 Rennes, France, [2]GenPhySE, Université de Toulouse, INRA, INPT, INP-ENVT, 31320 Castanet Tolosan, France; mathilde.lesciellour@inra.fr

Gut microbial population acts in complement with its host through nutrient digestion and health of the gastrointestinal tract. Changes in microbiota composition may then lead to changes in nutrient digestibility. The present study aimed at determining the effects of dietary fibre content on gut microbiota composition and apparent faecal nutrient digestibility in pigs. Furthermore, the relationships between microbiota and digestibility coefficients were investigated. Growing-finishing pigs (from 35 to 74 kg mean body weight) were fed alternatively a low-fibre (LF) and a high-fibre (HF) diet during 4 successive 3-week periods. Data collection for digestibility measurements was achieved during the last week of each period and faecal microbiota was collected at the end of each period for 16S rRNA gene sequencing. The two diets fed by the pigs could be discriminated using 31 predicting OTUs in a sparse partial least square discriminant analysis (mean classification error-rate 3.9%). Furthermore, microbiota was resilient to diet effect. Pearson correlations between microbiota composition and apparent digestibility coefficients of energy, protein, cellulose and hemicellulose emphasized the fact that in LF group, *Clostridiaceae* and *Turicibacter* were negatively correlated with protein and energy digestibility coefficients whereas *Lactobacillus* was positively correlated. In addition, *Lachnospiraceae* and *Prevotella* were negatively correlated with cell wall components digestibility. In HF diet, no significant correlation between microbiota and digestibility was found. The present study demonstrates that 3 weeks of adaptation to a new diet seem to be sufficient to observe resilience in growing pigs gut microbiota. In addition, faecal microbiota can be used to classify pigs according to their diet. Because some bacterial family and genera are favourable to digestibility, this study suggests that manipulations of bacterial populations can improve digestibility and feed efficiency. This study is part of the Feed-a-Gene Project, funded from the European Union's H2020 Programme under grant agreement no. 633531.

Session 13 Poster 13

Resistant starch type 4 modulates key metabolic pathways in the caecal bacterial metagenome of pigs

B.U. Metzler-Zebeli, M.A. Newman and Q. Zebeli
University of Veterinary Medicine Vienna, Veterinaerplatz, 1210 Vienna, Austria; barbara.metzler@vetmeduni.ac.at

Both phylogeny and functional capabilities within the gut microbiota populations are of great importance as they both impact host physiology and health. Whereas progress has been made to improve our knowledge on resistant starch (RS)-related phylogenetic shifts, little is known regarding the functional adaptations in the porcine gut microbiome due to dietary RS consumption. The present study investigated the functional alterations in the caecal bacterial metagenome of growing pigs fed RS type 4 (RS4). DNA for shotgun metagenomic sequencing was extracted from caecal samples of pigs which were either fed RS4 or control starch diets (CON) for 10 days. Samples from RS4-fed pigs were enriched in genes mapped to phenylpropanoid biosynthesis, lysozyme, several genes related to the carbohydrate metabolism, such as starch and sucrose metabolism and citrate cycle, as well as genes mapped to metabolism of cofactors and vitamins, such as ubiquinone biosynthesis, porphyrin and thiamin metabolism, and lipopolysaccharide biosynthesis (P<0.05). Functional analysis further showed that TGS-fed pigs had fewer reads of pathways related to 'folding, sorting and degradation', 'cell motility' and 'signal transduction' in their caecal metagenomes than CON-fed pigs. Changes in key metabolic genes were the result of changes in taxa associated with each type of starch. Accordingly, reads mapped to lipopolysaccharide biosynthesis were contributed mostly by the RS4-related increase in *Prevotella* and *Veillonella* (P<0.05). By contrast, the fewer reads of 'flagellar assembly' were associated with the decreased abundance of *Aeromonas*, *Spirochaeta*, *Treponema* and *Vibrio* in RS4- compared to CON-fed pigs (P<0.05). With respect to the hit counts mapped to 'starch and sucrose metabolism', the contribution of *Acidaminococcus* and *Prevotella* increased, whereas less hits were provided by *Vibrio*, *Treponema* and *Aeromonas* in RS4- compared to CON-fed pigs (P<0.05). In conclusion, metagenomic sequencing showed distinct caecal bacterial metagenome profiles in CON- and RS4-fed pigs, with functional capacities clearly linked to the shifts in bacterial taxa abundances caused by the RS4.

Session 13 Poster 14

On the influence of host genetics on gut microbiota composition in pigs

J. Estellé[1], N. Mach[1], Y. Ramayo-Caldas[1], F. Levenez[2], G. Lemonnier[1], C. Denis[1], M. Berri[3], M.J. Mercat[4], Y. Billon[5], J. Doré[2], C. Larzul[6], P. Lepage[2] and C. Rogel-Gaillard[1]
[1]GABI, INRA, AgroParisTech, Université Paris-Saclay, Vilvert, Jouy-en-Josas, France, [2]MICALIS, INRA, AgroParisTech, Université Paris-Saclay, Vilvert, Jouy-en-Josas, France, [3]ISP, INRA, Université de Tours, Nouzilly, Tours, France, [4]IFIP-BIOPORC, Pôle génétique, Le Rheu, France, [5]GENESI, INRA, Le Magneraud, Surgères, France, [6]GenPhySe, INRA, INP, ENSAT, Université de Toulouse, Castanet-Tolosan, Toulouse, France; jordi.estelle@inra.fr

Microbiomes and their effects on hosts have emerged as outstanding factors to take into account in livestock production. Despite the well-acknowledged impact of maternal colonization and environmental factors for driving the gut microbiota composition, the genetics of the host is also playing a role. In this study, we aimed at studying the influence of host genetics on the variation of gut microbiota composition in pigs. A cohort of over 500 Large White 60-day-old piglets was scored for faecal microbiota composition by sequencing the 16S rRNA bacterial gene. In parallel, these animals were genotyped with the Illumina PorcineSNP60 DNA chip. The relative abundances of operational taxonomic units (OTUs) and bacterial genera were characterized by using the Qiime package. Genetic parameters were estimated for a set of 63 bacterial genera present in the gut microbiota of pigs included in the study. Results showed that heritability was low (0.10.4) for eight genera. Positive and negative genetic correlations were found between the relative abundances of various bacterial genera, with *Prevotella*, *Oribacterium*, *Selenomonas*, *Dialister* and *Megasphaera* genera being positively correlated. Genome-wide association studies (GWAS) revealed significant associations between genomic regions and relative abundances of *Flexispira*, *Megasphaera*, *Mitsuokella* or *Streptococcus* genera. GWAS uncovered also additional shared genomic regions associated with variations in OTU abundances for several genera. Overall, our results provide new evidences that the gut microbiota composition is influenced by host genetics.

Session 13 Poster 15

Rumen bacterial populations of dairy cows fed molasses or corn with varying rumen degradable protein

E. Gunal[1], M. Hall[2], G.I. Zanton[2], P.J. Weimer[2], G. Suen[3] and K.A. Weigel[1]

[1]University of Wisconsin, Department of Dairy Science, 1675 Observatory Drive, Madison, WI 53706, USA, [2]USDA-ARS, Dairy Forage Research Center, 1920 Linden Drive, Madison, WI 53706, USA, [3]University of Wisconsin, Department of Bacteriology, 1550 Linden Drive, Madison, WI 53706, USA; kweigel@wisc.edu

This study sought to evaluate changes in the ruminal bacterial populations of Holstein cows that were fed molasses or corn grain while varying the level of rumen degradable protein (RDP). Twelve ruminally cannulated multiparous Holstein cows (185±56 days postpartum; 41.3±6.3 kg milk/day) were assigned to high or low RDP diets randomly for the duration of the study; within each RDP level, 6 cows received diets with molasses substituted for corn grain at 0, 5.25, or 10.5% of dry matter in a Latin square with 3 periods. Samples of the liquid and solid fractions of rumen contents were taken on 3 consecutive days at the end of each 28-d period and pooled within cow-period. Sequencing of the bacterial 16S rRNA gene was used to identify operational taxonomic units (OTUs, a proxy for species), which were summed at the genus level. The 30 genera with greatest relative abundance were analysed using a mixed linear model with a random cow effect and fixed effects of fraction, molasses level, RDP level, and molasses by RDP interaction. Least-squares means for 26 of 30 genera differed (P<0.05) by fraction, with genera in the *Bacteroidetes* at greater abundance in the solid fraction and genera in *Firmicutes* at greater abundance in the liquid fraction. Relative abundances of 4 genera were affected (P<0.05) by dietary molasses, with 2 increasing (*YRC22* and *Anaerostipes*) and 2 decreasing (*Succiniclasticum* and *Anaeroplasma*) as molasses level increased. Similarly, relative abundance of 6 genera differed (P<0.05) by RDP level, with 2 increasing (*Prevotella* and *Succiniclasticum*) and 4 decreasing (*YRC22*, *Anaerostipes*, *Pseudobutyrivibrio*, and unclassified) as RDP level increased. Only one interaction (*YRC22*) was observed (P<0.05) between RDP and molasses. This research suggests that changing the source and/or level of dietary carbohydrates and protein can alter the rumen microbiota of lactating dairy cows.

Session 13 Poster 16

Effects of dietary medium chain fatty acids on production and gut microbiota of laying hens

J.E. Rico[1], J.L. Martinez[2], R. Gervais[3], J. Rhonholm[4] and D.E. Rico[5]

[1]Department of Animal Science, Cornell Univeristy, 507 Tower Road, 14853, Ithaca, NY, USA, [2]INAF, Université Laval, 2440 Boulevard Hochelaga, G1V 0A6, Canada, [3]Department of Animal Science, Université Laval, 2425 Rue de l'Agriculture, G1V 0A6, Quebec, QC, Canada, [4]McGill University, Department of Food Science and Agricultural Chemistry, 111 Lakeshore Road, H9X 3V9, Ste Anne de Bellevue, QC, Canada, [5]CRSAD, 120 A Chemin du Roy, G0A 1S0, Deschamabult, QC, Canada; jer358@cornell.edu

Medium chain fatty acids (MCFA) have been shown to reduce body weight in mammals and to have antibacterial properties. This study evaluated the effects of feeding MCFA from coconut oil (Coco; rich in 12:0 and 14:0) or medium chain triglycerides (MCT; rich in 8:0 and 10:0) on animal performance, egg yolk fatty acid (FA) profile, and faecal microbiota. Twelve laying hens at 30 weeks of age were used in a crossover design (2 periods of 12 d). Coconut and MCT oils were added at 5% of DM during treatment periods, and canola oil was used during the pre-trial and washout periods (10 d in length). Dry matter intake was higher in Coco than in MCT on d 1 (P<0.05), but was not different thereafter. Body weight did not differ between treatments and decreased by 4% from d 0 to d 12 (P<0.001). No treatment effects were observed for egg production, egg weight and egg yolk weight, but the Coco treatment resulted in lower values of red intensity in the green red colour scale of egg yolks (P<0.01). Concentration of total FA<16 C in egg yolk increased by 300% from d 0 to d 12 in both treatments (P<0.01). Faecal samples taken on d 0 and 12 of each period were subjected to high-throughput Illumina sequencing of the V4 region of the bacterial 16S rRNA gene. Linear discriminant analysis effect size (LEfSe) showed LDA scores >2.5 (P<0.05) for the genera *Turicibacter* and *Staphylococcus* in Coco, whereas the family *Veillonellaceae* was increased by MCT. Compared with Canola feeding (d 0), Coco increased the genera *Stackia, Eggerthella* and *Bifidobacterium* and the families *Ruminococcacea* and *Coriobacteriaceae* (P<0.05). The source of MCFA had little effect on egg quality, but modified the faecal microbiota composition of laying hens.

Session 13 Poster 17

Microbiota as a tool to promote spontaneous liver steatosis in Greylag geese: promising results

C. Knudsen[1], M. Even[2], J. Arroyo[3], S. Combes[1], L. Cauquil[1], G. Pascal[1], X. Fernandez[1], F. Lavigne[3], S. Davail[2] and K. Ricaud[2]

[1]Université de Toulouse, INRA, INPT, ENVT, GenPhySE, 24 Chemin de Borde Rouge, 31326 Castanet Tolosan, France, [2]INRA, UPPA, NuMeA, 371 Rue du Ruisseau, 40002 Mont de Marsan, France, [3]ASSELDOR, La Tour de Glane, 24420 Coulaures, France; karine.brugirardricaud@univ-pau.fr

Greylag geese are used for the production of fatty liver, also known as 'foie gras' but the conventional production system based on overfeeding is questioned today for ethical reasons. In previous studies we demonstrated that Greylag geese are able to develop variable levels of spontaneous liver steatosis when submitted to a feed restriction period, associated with a reduction of day length, followed by 12 weeks of *ad libitum* corn feeding during winter season. To optimize liver fattening without overfeeding, we evaluated in the present study the correlations between hepatic steatosis and intestinal microbiota assessed by V3 V4 16S miSeq sequencing in Greylag geese. Geese fed an identical corn mixture, were either slaughter before steatosis stimulation (C, n=12) or submitted to overfeeding (OF, n=16) or spontaneous fattening stimulation (SFS, n=20). SFS geese were separated into two subgroups according to their liver weight (LW): NegSFS (LW<100 g) and PosSFS (LW>200 g). Bacterial community diversity index (Shannon) did not differ (P>0.5) with the type of stimulation (OF vs SFS) or level of steatosis (PosSFS vs NegSFS). Bacterial community was strongly modulated by fattening stimulation (OF or SFS vs C according to nMDS analyses). However, likely due to the low levels of steatosis obtained in this experiment compared to previous studies (269 g for the PosSFS group), no strong modulations of the bacterial community were observed between NegSFS and PosSFS. Using sPLS-DA analysis NegSFS geese could be distinguished by low abundance (<0.1%) bacterial taxa belonging the *Lactobacillus*, *Staphylococcus* and *Moraxella* genera. Regardless, this study enabled us to evidence correlations between certain bacterial taxa and the steatosis level and thus opens the way to new studies on the causal link between steatosis and intestinal microbiota in geese.

Session 14 Theatre 1

What is needed to improve sheep productivity in EU and Turkey?

I. Beltrán De Heredia[1], R.J. Ruiz[1], C. Morgan-Davies[2], C.M. Dwyer[2], T.W.J. Keady[3], A. Carta[4], D. Gavojdian[5], S. Ocak[6], F. Corbière[7] and J.M. Gautier[8]

[1]NEIKER-Tecnalia, Instituto Vasco de Investigación y Desarrollo Agrario, Agrifood Campus of Arkaute, 01080 Arkaute, Spain, [2]Scotland's Rural College (SRUC), Kirkton, FK20 8RU, Crianlarich, Scotland, United Kingdom, [3]Animal & Grassland Research & Innovation Centre, Teagasc, Grassland Science, Co Galway, H65 R718, Athenry, Ireland, [4]Agris Sardinia, Department for Research on Livestock Production, 07040 Olmedo, Italy, [5]Universitatea de Ştiinţe Agricole şi Medicină Veterinară a Banatului, Calea Aradului 119, 300645 Timişoara, Romania, [6]Togen R&D, 01170, Adana, Turkey, [7]INRA-ENVT, UMR IHAP, 1225, France, [8]Institut de l'Elevage, BP 42118, 31321 Castanet Tolosan Cedex, France; ibeltran@neiker.eus

Within the SheepNet network, a survey was undertaken among stakeholders in the sheep industry in the 6 main European sheep-producing countries (FR, IR, IT, RO, SP, UK) and Turkey, to assess the main challenges to improve sheep productivity (efficient reproduction and gestation, reduced lamb mortality). A total of 794 respondents completed the survey. The respondents were farmers/shepherds/farm workers (60%), advisors/consultants (16%), veterinarians (9%) and scientists (9%). Regarding farmers, 310 were in meat and 140 in dairy sheep systems. The respondents ranked, in order of importance, up to a maximum of 5 challenges to enhance pregnancy rate (16 options) and success (13 options), and the main management (11 options) and animal factors (11 options) involved in low lamb mortality. The main challenges identified were: (1) to improve pregnancy rate: BCS, nutrition and grassland management, and flock health status; (2) to enhance pregnancy success: nutrition, control and prevention of abortion, and pregnancy diagnosis; (3) to reduce lamb mortality (management): sheep shed conditions, advanced preparation for lambing, and nutrition/grassland management; (4) to reduce lamb mortality (animal factors): colostrum, lamb vigour, weight and health at birth. Slight differences were observed between countries, between farmers and technicians, and between meat or dairy sheep farmers, regarding the importance assigned to each issue. The SheepNet network will try to provide solutions to face these challenges within each region and production system across Europe.

Reproductive indicators in sheep farming systems in Europe and Turkey

R.J. Ruiz[1], I. Beltran De Heredia[1], C. Morgan-Davies[2], C. Dwyer[2], P. Frater[2], T.W.J. Keady[3], A. Carta[4], D. Gavojdian[5], S. Ocak[6], F. Corbière[7] and J.M. Gautier[8]
[1]NEIKER-Tecnalia, Instituto Vasco de Investigación y Desarrollo Agrario, Agrifood Campus of Arkaute, 01080 Arkaute, Spain, [2]Scotland's Rural College (SRUC), Kirkton, FK20 8RU, Crianlarich, Scotland, United Kingdom, [3]Animal & Grassland Research & Innovation Centre, Teagasc, Grassland Science, Co Galway, H65 R718, Athenry, Ireland, [4]Agris Sardinia, Department for Research on Livestock Production, 07040, Olmedo, Italy, [5]Universitatea de Ştiinţe Agricole şi Medicină Veterinară a Banatului, Calea Aradului, 300645 Timisoara, Romania, [6]Togen R&D, 01170, Adana, Turkey, [7]INRA-ENVT, UMR, IHAP, 1225, France, [8]Institut de l'Elevage, BP 42118, 31321 Castanet Tolosan Cedex, France; rruiz@neiker.eus

The efficiency of reproductive management is crucial to the profitability of sheep production. There is a wide diversity in the sheep systems due to ewe genotype, climate, environmental conditions (mountain areas, lowlands, etc.), productive aptitude (meat, dairy or dual purpose), level of intensification, nutrition management, etc. Ewe productivity and lamb mortality data was collated for 22 systems of sheep production from the 7 countries (FR, IR, IT, RO, SP, UK and Turkey) involved in SheepNet. Average fertility values in sheep systems managed with a one-lambing-season strategy range from 83 to 95%, and in accelerated systems from 88 to 116%. There is a significant number of non-productive sheep in flocks (from 5 to 17%), which may represent up to 50% of the flock in the less efficient farms. Litter size tended to be higher in lowlands systems (1.40) in comparison to those in hills or mountain areas (1.33). Despite the higher complexity of management for accelerated reproductive strategies, the average litter size achieved (1.48) did not differ much from that obtained in systems following a 1-lambing-season-per-year pattern (1.36). There is a lack of valid and reliable data for abortion and lamb mortality risks, and low utilisation of technologies available (oestrus synchronization, artificial insemination and scanning). As a result, the number of lambs produced per ewe joined to the ram is in general low (<1.5). The SheepNet network will try to propose solutions to increase sheep productivity.

Genetic parameters for lamb mortality, birth coat score and growth in divergently selected Merinos

S.W.P. Cloete[1,2], J.B. Van Wyk[3] and J.J. Olivier[1]
[1]Directorate Animal Sciences: Elsenburg, Private Bag X1, Elsenburg 7607, South Africa, Private Bag X1, 7607 Elsenburg, South Africa, [2]University of Stellenbosch, Department of Animal Sciences, Private Bag X1, 7602 Matieland, South Africa, [3]University of the Free State, Department of Animal, Wildlife and Grassland Sciences, P.O. Box 339, 9300 Bloemfontein, South Africa; schalkc@elsenburg.com

Lamb mortality (LM) has obvious economic impacts in livestock operations. However, it has also been scrutinised from an ethical perspective, as it has welfare implications. It is important to consider methods to reduce LM. Between 4,769 (birth coat score – BCS) and 7,021 (LM) records from a Merino flock divergently selected for number of lambs weaned per joining (NLW) over a 30-year period (from 1986 to 2015) were thus analysed. Selection was initially based on maternal records for NLW but was augmented with breeding values for NLW from a single-trait repeatability model since 2002. Overall, the line selected for an increased NLW had a lower BCS and LM but a higher BW and WW than the line selected in the downward direction. Least-squares means derived from single-trait analyses for birth years indicated divergence between lines for BCS, LM and WW up to 2008. At this stage, migrant sires were introduced to the flock for linkage to industry. Heritability estimates amounted 0.13 for birth weight (BW), 0.54 for BCS, 0.08 for LM and 0.13 for weaning weight (WW). Dam permanent environment (PE) estimates were respectively 0.10, 0.03, 0.05 and 0.12. Maternal genetic estimates were 0.19 for BW and 0.07 for WW. LM was not genetically correlated to any trait. However, LM was significantly correlated to BW on the PE and phenotypic levels (-0.36 and -0.16 respectively). BCS was not correlated with weight at any level. The weight traits, however, were correlated at all levels. The initial divergence in annual least-squares means for BCS, LM and WW were supported by genetic trends. These traits suggested that lambs in the line selected in the upward direction became more hairy, heavier at weaning with a lower LM. These results suggest that selection for a composite trait like NLW could benefit LM, both at the phenotypic and genetic levels. Selection will also result in lambs becoming heavier at weaning but could lead to a more hairy BCS in South African Merinos.

Session 14 Theatre 4

Keeping watch on ewes: an approach to lambing difficulties in sheep

S. Schmoelzl[1], J. McNally[1], H. Brewer[1] and A. Ingham[2]
[1]CSIRO, Agriculture and Food, New England Highway, Armidale NSW2350, Australia, [2]CSIRO, Agriculture and Food, 306 Carmody Road, St Lucia QLD 4067, Australia; sabine.schmoelzl@csiro.au

Lamb survival is a key welfare and productivity issue for the Australian sheep industry. Multi-year studies across several sites using genetically linked flocks of sheep, representing Merino, maternal and terminal sheep breeds have established through standardised lamb autopsies of all lamb mortalities that hypoxia-related birth injuries cause at least around 40% of lamb losses. This figure does not figure in transient effects from which the lamb eventually recovers yet impair its ability to stand, suckle and follow, leading to starvation and/or mismothering losses. In the present study, we aimed to derive objective measurements of parturition in sheep. We observed a total of 133 lambing events of Merino ewes after mating with either Merino or Border Leicester rams, and determined by observation: time of first signs of impending parturition; time of parturition; birth type; total duration of parturition. A qualitative descriptor of 'normal' or 'difficult' was assigned to each lambing by one human observer. Across all observed lambing events, the average time from first sign of parturition to time of birth was 162.7±21.6 min (95% CI=119.9, 205.6). Parturition duration corresponded well with observer classification in most cases; however, 12 events with durations of >253 min were classified as 'normal', while 11 events with durations <129 min were classified as 'difficult'. These discrepancies highlight the need to identify better objectives measures of parturition duration. For 55 lambing events, lamb fate was recorded (mean duration 271.6±39.3 min; 95% CI=192.7, 350.6). A total of 12 lamb losses were observed, and of those, 7 occurred during birth and 5 after birth. All 7 perinatal lamb losses occurred after parturition events of more than 281 min (mean 350.2±116.7 min) and they had been classified as 'difficult'. All 5 postnatal lamb losses occurred after lambing events which had been classified as 'normal' and only one of those had a longer than average parturition duration (mean 135.2±58.2 min). This observation underpins the notion that peri- and postnatal lamb losses point to different underlying mechanisms.

Session 14 Theatre 5

Better knowledge for colostrum production and transfer of passive immunity in sheep

F. Corbiere[1] and J.M. Gautier[2]
[1]INRA, UMR 1225 Interactions Hotes Agents Pathogènes, Ecole Nationale Vétérinaire de Toulouse, 23 chemin des Capelles, 31076 Toulouse, France, [2]IDELE, BP 42118, 31321 Castanet Tolosan, France; f.corbiere@envt.fr

Two studies have been carry out in order to have a better knowledge on the quality of colostrum, the total IgG mass produced with relation to the passive immune transfer (PIT) in the lamb. The first study aims (1) to obtain references on the sheep colostrum concentrations of Immunoglobulin G (IgG), Fat and Protein, (2) to evaluate the influence of ewe age, body condition score and litter size on colostrum composition (3) to evaluate the relationship between the IgG colostrum concentration and the quality of PIT in lamb and (4) to assess the relationship between PIT and survival of Lambs at 30 days. The second study objective was to describe the evolution of the IgG colostrum concentration and the total IgG mass produced during the first 12 hours after lambing. 90 ewes from the Vendéen breed and their 163 lambs were included in the first study and 48 ewes (Blanche du Massif Central, Noire du Velay and Lacaune breeds) in the second one. The average IgG colostrum concentration at lambing was 89.3 g/l, with high variability (from 28.2 to 180.3 g/l). Similar variability was found for Fat and Protein concentrations. A rapid decrease (reduction by 2 in the 9 postpartum hours) of the IgG colostrum concentration was also observed. The average production of IgG by udder half in the first twelve hours after lambing was 33.5 g (from 6.3 g to 71.1 g), with 25% of ewes producing less than 24 g witch is the need for a lamb of 4 kg at birth. The total IgG1 mass produced in the first 12 hours after lambing was significantly influenced by the ewe breed and body condition score and by the udder health status. Conversely, no difference in peripartum serum progesterone, prolactin and cortisol concentration profiles was fund between ewes producing either high or low IgG mass in colostrum during the first twelve hours after lambing. Failure of transfer of passive immunity (IgG <10 g/l) was evidenced in 12.7% of lambs. No correlation between lamb plasma and colostrum IgG concentrations was evidenced. Finally, morbidity and mortality before 30 days of age was significantly greater in lambs with low birth weight and with failure of transfer of passive immunity.

Session 14 — Theatre 6

Lamb rearing options for New Zealand dairy sheep systems

S. McCoard[1], T. MacDonald[2], P. Gatley[3], M. King[4], J. Ryrie[2] and D. Stevens[5]
[1]AgResearch Limited, Animal Nutrition & Physiology Team, Private Bag 11008, Palmerston North 4474, New Zealand, [2]Spring Sheep Dairy, Broadlands Road, Taupo, New Zealand, [3]Maui Milk, Central Plateau, Turangi, New Zealand, [4]Kingsmeade Artisan Cheese, First Street, Landsdown, Masterton 5810, New Zealand, [5]AgResearch Limited, Farm Systems Team, Puddle Alley, Mosgiel, New Zealand; sue.mccoard@agresearch.co.nz

Sheep dairying is an emerging industry in New Zealand and is undergoing rapid growth. Dairy ewes produce around 25% of their total milk yield in the first 30 days of lactation, corresponding to the time when lambs would naturally be suckling. Therefore, systems that allow milk to be harvested for commercial sale, as well as rearing of the lamb crop are required. Rearing of lambs from dairy ewes, either as replacements or for meat production, can be a major cost in sheep-milking enterprises. Optimising lamb survival, growth and health is also important to generate good quality ewe replacements, enable cost-effective utilisation of surplus lambs to generate additional revenue as well as meeting market-driven expectation on animal welfare. As a result, a variety of lamb-rearing options are employed worldwide ranging from lamb removal soon after birth and artificial-rearing, mixed systems that enable suckling and milking to traditional natural rearing of the lamb by the ewe coupled with early weaning. In contrast to most dairy-sheep production systems globally, the New Zealand dairy sheep industry is a pasture-based production system which can pose challenges associated with feed supply and management. Furthermore, as an emerging industry, cost-effective lamb rearing systems that also deliver good quality ewe replacements that meet growth targets for hogget mating, as well as good quality lambs for meat production, coupled with ethically exemplary farming systems are required. Scale of the farming operation is also a key consideration with New Zealand dairy sheep farming operations, highlighting the need for a variety of system options. Our recent research in both experimental and commercial systems will be described with a focus on lamb growth and health within a range of artificial lamb rearing systems that differ in feeding management and weaning practices.

Session 14 — Theatre 7

Poster session

J.M. Gautier
Institut de l'Elevage, BP 42118, 31321, France; jean-marc.gautier@idele.fr

Three posters related to the topic of the session will be briefly presented in order to identify their author: (1) Maternal behaviour of ewes of Chios and Florina (Pelagonia) breeds – differences and utilization (M.A. Karatzia); (2) Lamb survival in highly fertile Finnsheep and best practice in Finland (M.L. Sevon-Aimonen); (3) Effects of farm management practices on reproduction efficiency and lamb survival (D. Gavojdian).

SheepNet – increasing ewe productivity in the EU and Turkey

T.W.J. Keady[1] and J.M. Gautier[2]
[1]Teagasc, Athenry, Co Galway, Ireland, [2]Idele, Castanet Tolosan, 31321, France; tim.keady@teagasc.ie

Sheep meat and milk production are very important farm enterprises in Europe and neighbouring countries. However, since 2000 the number of producers has declined in the EU by 50%. In order to reinforce the attractiveness of the sheep sector, it is fundamental to increase, in a sustainable way, ewe productivity of meat sheep (the number of lambs reared per ewe joined) and of milk sheep (the number of milking ewes per ewe joined). Ewe productivity is a combination of reproduction success, embryonic and lamb survival and litter size. SheepNet (www.sheepnet.network) is an EU funded thematic network on sheep productivity and was initiated in November 2016. SheepNet involves the 6 main EU sheep producing countries (Ireland, France, United Kingdom, Romania, Spain and Italy) and Turkey but is open to all EU countries, stakeholders, sheep producers. SheepNet is a network about practice-driven innovation and practical knowledge among stakeholders. Through multi-actor and codesign approaches SheepNet promotes and establishes durable exchange of scientific and practical knowledge among researchers, farmers and advisors across Europe to stimulate knowledge exchange and promote the implementation and dissemination of innovative and best technologies and practices for the improvement of sheep productivity. Mid-term into the project, SheepNet has identified the sources of communications used by all stakeholders in sheep production to obtain information. SheepNet has defined the systems of sheep production in member countries. SheepNet has identified the needs of each country to increase ewe productivity. SheepNet has presented 55 practical solutions that match with end-user's needs in member countries, showing that solutions exist to many issues in other regions and are easily transferable. SheepNet has organised many national and transnational workshops and made publications. The aim of this session is to share the SheepNet results to date and facilitate audience interaction to provide feedback and suggestions for improving ewe productivity that have not yet identified by SheepNet. Be part of SheepNet and its objective to improve ewe productivity and the sustainability of the EU sheep sector.

Effects of farm management practices on reproduction efficiency and lamb survival

D. Gavojdian[1,2], L.T. Cziszter[1] and I. Padeanu[1]
[1]Banat's University of Agricultural Sciences and Veterinary Medicine 'King Michael I of Romania', Timisoara, Calea Aradului 119, 300645, Romania, [2]Research and Development Institute for Bovine Balotesti, Balotesti, sos. Bucuresti-Ploiesti, km 21, Ilfov, 077015, Romania; gavojdian_dinu@animalsci-tm.ro

Sheep reproduction efficiency decidedly constrains the overall farm returns in both dairy and meat production systems. Lamb mortality has a significant impact on financial margins and represents a major animal welfare concern. However, significant between-flock variation is known to exist for both flock fertility and lamb survival, with estimates ranging from 3 to 50% for mortality rates in lambs under a wide range of production systems and environments. Aim of the current research was to evaluate the effects that farm management practices have on sheep reproduction outputs and lamb survival under extensive production systems. Data was collected from 20 commercial sheep farms located in Timis County Romania (average flock size of 912.6 ewes), which reared the dual-purpose Turcana breed during a whole production year (September 2015 – August 2016). All farms were included in the performance recording schemes and practiced extensive production rearing, with one lambing per year and a stoking rate of 5 to 7 ewes per hectare. A comparative study was made between top 5% better performing farms and the average. Flock fertility was on average 94.3%, with limits ranging from 88 to 98%, while the average for the best farms was 97%. Abortion rate was on average 2.4%, with limits between 1.1 and 5.3%, while the best farms had an abortion incidence of 1.4%. Lamb mortality from birth to weaning (75±10 days) was 2.7%, with limits ranging between 1.2 and 4.9%, with the best farms losing 1.3% of the lambs born alive. Lamb mortality from weaning until finishing on pasture up to the age of 7 months was on average 2.4%, with limits between 1.7 and 4.1%, while for the best farms the average was 1.1%. The average number of lambs produced per ewe put to ram was 1.06, with limits ranging from 0.84 to 1.27, while the best farms produced 1.19 lambs per breeding ewe. Current results highlight that within the same production system and similar environment, there are significant farm effects when flock fertility and lamb survival are concerned.

Maternal behaviour of ewes of Chios and Florina (Pelagonia) breeds – differences and utilization

M.A. Karatzia, D. Tsiokos, B. Kotsampasi and V. Christodoulou
Research Institute of Animal Science, HAO-Demeter, Paralimni-Giannitsa, 58100, Greece; karatzia@rias.gr

The integral prerequisite to the survival and growth of lambs is the rapid development of a close and exclusive bond between the ewe and her offsprings. Maternal behaviour plays an important role in the establishment of this bond, and is influenced by numerous factors, such as genotype, previous maternal experience, temperament and others. The aim of the present study was to detect differences in maternal behaviour in ewes of two Greek indigenous breeds. Thus, 33 ewes and their offsprings (50 lambs) of the Florina (Pelagonia) breed and 30 ewes and their offsprings (50 lambs) of the Chios breed were used. All ewes were clinically healthy, were managed and fed identically and housed indoors in the experimental sheep farm of the Research Institute of Animal Science. Lambs were delivered naturally and remained in lambing pens with their mothers for 3 days. Maternal behaviour was measured by using a 5-point scale scoring system (MBS), based on the proximity and vocalizing of the ewe to her lamb as it is handled (ear tagged and weighted) for the first time (12-24 h after birth). MBS were grouped into 3 categories for each breed, poor (MBS:1-2), good (MBS:3) and excellent (MBS:4-5). Lactation, litter size, lamb weight at birth and weaning and survival rate were recorded and all data were analysed using SPSS© v.24. Poor MBS cases were significantly lower in Florina ewes ($P \leq 0.001$), while excellent MBS was awarded to 50% of them and only to 30% of Chios ewes. Florina ewes appear to exhibit excellent MBS when lambs' weight at birth is significantly lower than Chios lambs' (F:3.35±0.105 kg and C:3.85±0.127 kg, $P<0.05$), while the opposite was observed for poor MBS (F:3.95±0.143 kg and C:3.30±0.150 kg, $P<0.05$). An increase of 10.43% in weaning weight from poor to excellent MBS in Chios ewes was observed. Significantly younger Florina ewes achieved excellent MBS (2.26±0.345 lactations) in comparison to Chios ewes (3.12±0.217 lactations) ($P<0.05$). It can be concluded that maternal behaviour differs between the two breeds, with Florina ewes exhibiting desirable characteristics in a more pronounced way. These results indicate that MBS can be utilized as an additional management tool in selection of breeding stock.

Session 14 Poster 11

Lamb survival in highly fertile Finnsheep and best practice in Finland

M.L. Sevon-Aimonen[1], S. Eklund[2] and S. Alamikkotervo[3]
[1]Natural Resources Institute Finland (Luke), Alimentum, Myllytie 1, 31600 Jokioinen, Finland, [2]ProAgria Southern Finland, P.O. Box 97, 33101 Tampere, Finland, [3]University of Helsinki, Department of Agricultural Sciences, P.O. Box 27, 00014 Helsinki, Finland; sanna.eklund@proagria.fi

Finnsheep ewes can commonly breed out of season/all year round but the highest litter sizes are reached in spring time. Good management and the correct feeding of ewes before and during the pregnancy are very important factors when farmers want to gain a good lamb survival. Also, the lambing pens are important to use during the lambing to improve lamb survival. ProAgria has got a pedigree and performance recording software for sheep in Finland and Finnsheep are also recorded in the software. The average lambing percentage for Finnsheep is 230% for older ewes (over two years old) and 170% for young ewes (under two years old). The average mortality rate for the lambs (died before the age of 14 days) is 9.6-11.8% for the older ewes and 8.0-9.7% for the younger ewes. ProAgria is participating with Luke to develop new breeding indexes. When the number of lambs born alive per ewes was one or two, only 1% of the lambs needed artificial feeding. When the numbers of lambs born alive increased to over three lambs per ewe, the artificial feeding was needed for more than 20% of the lambs. Farmers are sharing the ideas of good farming practices to each other to maintain high lamb survival rate.

Session 15 Theatre 1

Nutritional strategies to counteract oxidative stress: benefits and challenges

A. Baldi, L. Pinotti, C. Giromini, G. Invernizzi and G. Savoini
University of Milan, Department of Health, Animal Science and Food Safety, Via Celoria 10, 20133 Milano, Italy, 20133 Milano, Italy; antonella.baldi@unimi.it

Farm animals can experience oxidative stress (OS) during their life, such as in the periparturient period, at weaning, transport or heat exposure. The loss of the redox homeostasis can be originated from increased exposure to /or production of oxidants, from decreased dietary intake, de novo synthesis or increased turnover of antioxidants. OS can be implicated in the development and progression of several metabolic and infectious diseases. In dairy cow, particularly, the increased lipid mobilization, associated with the onset of lactation, can increase the severity of OS, and a consequent relationship with negative energy balance can be observed. Supplementation of dietary antioxidants or boosting endogenous antioxidant defences of the body have been proposed as possible strategies in maintaining the steady-state redox balance. Among dietary antioxidants, vitamin E and selenium are the most recognised, even though this category includes different classes of other compounds, as vitamin C, carotenoids and polyphenols, all of them largely advised as feed additives or ingredients in farm animal nutrition. More recently, this list has been elongated with other nutrients, such as methylated compounds, that historically were not known as antioxidants. Despite the clear benefits of dietary antioxidants, difficulties in defining nutritional recommendations exist, also because of the lack of a whole evaluation of antioxidant status in farm animals. Recently, research made significant advances in the definition of adequate and standardized biomarkers of oxidative stress, making it possible to better define the subsequent effect of antioxidant supplementation, and also address some regulatory discrepancies about these substances.

Session 15 Theatre 2

Amino acid supplementation to mitigate stress responses of weaner pigs exposed to acute stressors

S.O. Sterndale, D.W. Miller, J.P. Mansfield, J.C. Kim and J.R. Pluske
Murdoch University, School of Veterinary and Life Science, 90 South St, 6150 Murdoch, Australia; samantha@sterndale.com

This study tested the hypothesis that gamma-aminobutyric acid (GABA) supplementation in water, and glutamine (Gln), glutamate (Glu) and tryptophan (Trp) supplementation in the diet, will improve growth performance and reduce indices of the stress response in weaned pigs exposed to a number of production stressors. At weaning (d0), 72 male pigs were allocated to pens with their litter mates (3 per pen) and allowed to acclimate for 14 days. At d14 of the study, pigs were allocated to a 2 by 2 factorial design with the factors being (1) without/with feed deprivation for 12 h, and (2) without/with supplementation of GABA and the amino acids. All pigs were mixed at 06:00 h on d14 to create a mixing challenge, and at this time pigs were either taken off feed for 12 h or were not deprived of feed. At 18:00 h on the same day, one half of the pigs were supplemented with approximately 50 mg/day GABA in water (from d14-18) and 0.3% L-Glu, 0.1% L-Gln and 0.34% L-Trp added to a commercially diet (from d14-22). The study finished on d28. Weekly live weights and feed intakes were measured, and blood was collected at d11, 14 (18:00 h) and 18. Feed conversion ratio (FCR) from d15-21 was lower in the supplemented pigs compared to non-supplemented pigs regardless of feed deprivation (main effect; 1.27 vs 1.42, P=0.002). At d14, plasma cortisol and nonesterified fatty acid levels were higher in pigs deprived of feed for 12 h compared to pigs not deprived of feed (main effect; 12.40 vs 7.15 ng/ml, P=0.09, and 0.34 vs 0.09 mmol/l, P<0.001, respectively). Plasma glucose on d14 was higher in pigs not deprived of feed compared to their feed-deprived counterparts (main effect; 6.34 vs 5.69 mmol/l, P<0.001), however this difference was reversed by d18 (6.62 vs 7.09 mmol/l, P<0.01). Pigs fed the non-supplemented diet had lower plasma glucose at d18 compared to supplemented pigs (main effect; 6.71 vs 7.01 mmol/l, P<0.1). Collectively, these results indicate that Gln, Glu, Trp and GABA supplementation can improve FCR and increase plasma glucose in pigs exposed to an acute stressor, in this case feed deprivation for 12 h.

Provision of enrichment to piglets attenuated the immune response to weaning

C. Ralph[1], S. Barnes[1], S. Kitessa[1], M. Hebart[2] and G. Cronin[3]
[1]Animal Welfare Science Centre, SARDI, Roseworthy 5371, Australia, [2]The University of Adelaide, Roseworthy, 5371, Australia, [3]The University of Sydney, Camden, 2570, Australia; cameron.ralph@sa.gov.au

Providing enrichment to animals can decrease antagonistic behaviours, enhance learning ability and affect immune system function. We investigated the effect of providing sucker and weaner pigs enrichment in the form of lick-blocks (Ridleys Corporation Ltd) and assessed immune response around weaning. We hypothesised that the provision of enrichment blocks would attenuate the inflammatory response to weaning. Piglets (Large White × Landrace) were housed in conventional farrowing crates for 21 days during lactation and then in group pens until 11 wks of age. Pigs were raised in an enriched or barren pen in the sucker phase and then weaned to an enriched or barren pen. There were four treatments in a 2×2 factorial design: enriched in sucker and weaner phases (n=10), enriched in the sucker phase and barren in the weaner phase (n=12), barren in the sucker phase and enriched the weaner phase (n=11) and barren in the sucker and weaner phases (n=12). Food and water were provided *ad libitum* in the weaner phase. Blood samples (3 ml) were collected via jugular venepuncture 24 h before weaning, 24 h after weaning, and 21 and 65 days after weaning. Samples were assayed for interleukin-10 (IL-10), tumour necrosis factor-α (TNF-α), interferon gamma (IFN-ɣ) and interleukin-6 (IL-6). Data were analysed using a general linear model and a repeated measures analysis of variance. The mean (±SEM) concentration of IL-10 and TNF-α was greater 24 h after weaning for pigs housed in barren pens during the sucker phase than pigs provided with enrichment in the sucker phase ($P<0.05$). The mean (±SEM) concentration of IFN-ɣ and IL-6 was greater 24 h after weaning for pigs provided with enrichment than pigs housed in barren pens in the weaner phase ($P<0.05$). These data suggest that pigs provided with enrichment in the form of enrichment blocks had an attenuated inflammatory response to weaning and an overall attenuated inflammatory status. This supports the hypothesis and is evidence that provision of enrichment blocks can influence the immune function of sucker and weaner pigs. The long-term welfare implications for sucker and weaner pigs requires further investigation.

Effects of the probiotic *Enterococcus faecium* on primary cultured porcine immune cells

S. Kreuzer-Redmer, F. Larsberg, P. Korkuc, N. Wöltje, K. Hildebrandt and G.A. Brockmann
Humboldt Universität, Thaer-Institut, Unter den Linden 6, 10099 Berlin, Germany; filip.larsberg@gmx.de

Feeding the lactic acid-producing *Enterococcus faecium* NCIMB 10415, a licensed probiotic, has been described to promote growth performance and health in pigs. However, the underlying mechanism of action is still elusive. *E. faecium* may either directly influence immune cells or the composition of the intestinal milieu. We hypothesize, that *E. faecium* is capable of directly interfering with cells of the adaptive immune system. To this end, we established a porcine *in vitro* cell culture model to explore direct interactions of lymphocytes (1×10^6/ml) with vital or UV-inactivated *E. faecium* in a ratio of 2:1, or 10:1 (immune cell:bacterium) for 1, 5 or 20 hours. We followed changes in the composition and activation status of lymphocytes via flow cytometry and qPCR. Data was analysed using R and statistical significance between treatment and control was tested using pairwise Mann-Whitney-U-tests. We detected higher relative cell counts of CD8b+ cytotoxic T-cells in the treatment group with vital *E. faecium* compared to untreated controls ($P<0.05$). Additionally, the relative cell count of activated cytotoxic T-cells (CD8+CD27-) was higher when treated with vital *E. faecium* bacteria compared to the control without *E. faecium* ($P<0.1$). Treatments with UV-inactivated bacteria did not affect the T-cell status. We observed a different pattern for B-cells. The relative cell counts of CD21+ B-cells ($P<0.05$) and equally of CD79+ B-cells ($P<0.1$) were higher in treatments with UV-inactivated *E. faecium*. In contrast, the relative cell counts of B-cells did not change significantly when treated with vital *E. faecium*. Furthermore, we observed a trend towards lower expression of B-cell regulatory (IGLC, IGKC) and activation marker genes (CD40, CD2) in treatments with vital *E. faecium* on magnetically sorted CD21+ B-cells. We suggest that treatment with vital *E. faecium* presumably influences the direction of immune response towards an enhanced cytotoxic T-cell answer at the expense of the B-cell response. Hence, this study could provide evidence of a direct immunomodulatory effect of *E. faecium* on adaptive immune cells *in vitro*.

Effect of olive bioactive extracts on immune response in lipopolysaccharide challenged heifers

L. Rostoll-Cangiano[1], M.F. Kweh[1], M.G. Zenobi[1], I.R. Ipharraguerre[2], C.D. Nelson[1] and N. Dilorenzo[1]
[1]University of Florida, Department of Animal Sciences, 2250 Shealy Dr, 32608, Gainesville, FL, USA, [2]Institute of Human Nutrition and Food Science Christian-Albrechts-University, Herrmann Rodewald Str. 6, 24118, Kiel, Germany; lautaro.rostoll@ufl.edu

During infection, peripheral tissues spare fuels to support the immune response (IR). An excessive IR can hinder normal physiological function. Olive bioactive extracts (OBE) have anti-inflammatory activity, and it could modulate the IR. A randomized block design was used to evaluate the effects of feeding OBE on newly weaned Angus and Brangus heifers (210±19 kg of BW) that where challenged i.v. with lipopolysaccharide (LPS; 0.10 µg/kg of BW). Animals were divided into 3 periods, and randomly assigned to 1 of 4 treatments (Trt): Negative control (receiving saline, CTL-); Positive control (receiving LPS, CTL+); low and high dose of OBE on the diet (receiving LPS, and OBE at 0.04 and 0.16% for OBE L and OBE H respectively).Animals were adapted for 21 days prior to the LPS challenge. Blood samples were collected on hour (h) 0, 1, 2, 4, and 8. Vaginal temperature was recorded using I buttons. Peripheral blood leukocyte counts and expression of cluster of differentiation CD11b, CD14 and CD62L proteins were measured by flow cytometry. Data were analysed for fixed effect of Trt, and Trt × time. Orthogonal contrasts were performed comparing CTL+ vs OBE, OBE L vs OBE H, and CTL+ vs CTL-. Significance declared at *P*<0.05, and tendency at *P*<0.10. Compared to CTL+, OBE H group tended to have lower average temperature (*P*=0.06; 39.5 vs 39.9 °C) at h 2. CTL+ had decreased lymphocytes, monocytes, and neutrophils counts compared to CTL- (*P*=<0.02). Monocyte and lymphocyte counts were greater for OBE compared to CTL+ (*P*=0.03). OBE group tended (*P*=0.06) to have less CD11b, CD14 and CD62L expressed on monocytes and less CD62L expressed on neutrophils compared with CTL+. In conclusion, feeding OBE modulated the change in peripheral blood leukocyte counts in response to endotoxin potentially by altering expression of leukocyte adhesion proteins (CD11b and CD62L). As CD14 is a co-receptor for LPS, a reduction in monocyte CD14 expression could have decreased the response to LPS.

Impact of nutrition on the immune system of cattle

S. Dänicke
Federal Research Institute for Animal Health, Institute of AAnimal Nutrition, Bundesallee 37, 38116, Germany; sven.daenicke@fli.de

To understand the complexity and the assumed plasticity of the homeorhetic regulation of different physiological states of the transition dairy cow it seems to be helpful to simplify the main interrelationships contributing to the health of the animal. The aim of the immune system is to protect the animal from environmental factors potentially disturbing its current physiological steady state. In order to stay responsive, the energy and nutrient requirement of the immune system needs to be met. On the other hand it is clear that the nutritional status and therefore the functionality of the immune system is determined by the diet composition and feed intake. Diet composition itself might largely influence the level of feed intake and is therefore related not only to qualitative but also to quantitative aspects of maintaining the steady state between metabolism and immune system and the environment in general. Moreover, the diet not only contains nutrients but also unwanted compounds such as mycotoxins and anti-nutritive substances which not only might adversely affect the metabolism but also the immune system. A stimulated immune system might induce pro-inflammatory cytokines which in turn decrease feed and consequently nutrient intake. Based on these facts it becomes clear that nutrition and feeding play a decisive role amongst the environmental factors challenging the steady state between metabolism, nutritional status and the immune system. In order to understand how the steady state between metabolism, nutritional status and the immune system is maintained under challenging conditions, such as an infection or vaccination, various animal models can be used which specifically manipulate the nutritional status. Such models aim at exploring the immunological response to a challenge under largely varying nutritional states such as negative energy balance (NEB) typically occurring in cows after parturition. Amongst others, NEB has been linked to an impairment of the immune system and an increased incidence of infectious diseases such as metritis, mastitis, and laminitis. Thus, counteracting the NEB by nutrition might support the defense system of the cow and prevent production disorders.

Sainfoin pellets for preventive parasite control and improved protein efficiency in dairy goats

S. Werne[1], N. Arnold[2], E. Perler[1] and F. Leiber[1]
[1]Research Institute of Organic Agriculture, Department of Livestock Sciences, Ackerstrasse 113, 5070 Frick, Switzerland, [2]ZHAW Life Sciences und Facility Management, Gruentalstrasse 14, 8820 Waedenswil, Switzerland; steffen.werne@fibl.org

The legume sainfoin (*Onobrychis viciifolia*) containing condensed tannins is often associated with beneficial effects when used in animal husbandry. We tested two potentially beneficial effects by feeding 700 g sainfoin or 700 g alfalfa pellets to 10 dairy goats each. Target parameters were nematode egg excretion per gram faeces (EPG) and daily milk protein and urea yields. Our hypothesis was that the diet with sainfoin pellets would reduce the EPG of the goats in order to reduce pasture infectivity and subsequently the severity of the infection level of the goats. Beyond that, we hypothesised that the condensed tannins will improve ruminal protein-efficiency indicated by lower urea and higher protein yields with the goat milk. During a trial period of 7 weeks, EPG, milk protein and urea yield per animal and day was measured regularly. Except for milking and pellet feeding, all animals were kept in one group and had access to pasture for approx. 5 hours daily. Beyond the experimental feeds and pasture, all goats had *ad libitum* access to non-tanniferous hay when stabled. Intake from pasture or hay was not determined. Concentration of condensed tannins in sainfoin and alfalfa was 4 and 0.3%, respectively. Crude protein content corrected for 100% dry matter was 18.2% for sainfoin and 20.1% for alfalfa. Even though arithmetic average of EPG in the sainfoin group was 18% lower compared to the control group, a repeated measurement analysis could not reveal significant differences (P=0.148). Likewise, total daily milk protein (P=0.700) and total daily urea (P=0.410) per animal did not differ between groups. In many cases, other studies making use *ad libitum* sainfoin diets, resulted in reduced EPG. As a dose dependent effect for condensed tannins is assumed, and the 700 g sainfoin roughly corresponds to one third of the total daily dry matter intake, the total amount of condensed tannins in our trial might have been too low to provoke an effect. There were no indications of a sainfoin-effect on protein efficiency at all.

More methionine in sows and piglets diets for better growth and immune response of weaned piglets

B.Y. Xu[1], L. Zhao[1], D.I. Batonon-Alavo[2], Y. Mercier[2], D.S. Qi[1] and L.H. Sun[1]
[1]Huazhong Agricultural University, Department of Animal nutrition and Feed Science, Wuhan, Hubei, 430070, China, P.R., [2]ADISSEO France SAS, CERN, 6 Route Noire, 03630, Malicorne, France; dolores.batonon-alavo@adisseo.com

The objective of this study was to determine the effects of an increased consumption of methionine either as DL-Met (DLM) or HMTBA (OH-Met) by sows and piglets on the ability of piglets to cope with an inflammatory challenge during the post-weaning period. Sows received during the last month of gestation and the lactating period three treatments: a control diet at the requirement in TSAA and two treatments supplemented with either DLM or OH-Met at 25% above the requirement. During the lactating period, protein and lactose contents in the milk were significantly increased with OH-Met in comparison to DLM and the Control. At postnatal d14, piglets fed OH-Met or DLM had higher body weights than control-fed piglets. Piglets were weaned at 21d and received three diets consistently with sows' diets: a control diet at the requirement in TSAA (CON-P) and two treatments DLM (DLM-P) or OH-Met (OH-Met -P) supplemented 25% above TSAA requirement. After 2 weeks of feeding, 20 male piglets from each treatment were submitted to a 2×3 factorial design that included the dietary treatments and immunological challenge (saline and LPS) at d35. Growth performance of piglets were measured at d35, d49 and d63 respectively. At d35, piglets were significantly heavier in OH-Met-P group (8.47±0.38 kg) than in CON-P group (7.66±0.24 kg), DLM-P group (8.25±0.27 kg) was intermediate. The LPS significantly affected piglet performance in all groups. However, OH-Met-P and DLM-P fed piglets showed the highest body weights following the LPS challenge at d49 and d63, compared to the CON-P. Moreover, body weight gain and feed to gain ratio were improved in the OH-Met-P treatment during and after LPS stress. OH-Met is known to be better trans-sulfurated than DLM, thus leading to more glutathione in comparison to DLM. This improved antioxidant status might explain the better growth performance observed with OH-Met under LPS stress.

Methionine alone or combined with choline and betaine affects ewes' milk and antioxidant capacity

E. Tsiplakou[1], A. Mavrommatis[1], K. Sotirakoglou[2], N. Labrou[3] and G. Zervas[1]
[1]Agricultural University of Athens, Nutritional Physiology and Feeding, Iera Odos 75, GR-11855, Greece, [2]Agricultural University of Athens, Plant Breeding and Biometry, Iera Odos 75, GR-11855, Greece, [3]Agricultural University of Athens, Enzyme Technology, Iera Odos 75, GR-11855, Greece; eltsiplakou@aua.gr

This study investigated the impact of dietary supplementation with rumen protected methionine alone or in combination with rumen protected choline and betaine on: (1) milk chemical composition and fatty acids (FA) profile, and (2) blood plasma glutathione transferase (GST) activity of dairy ewes. Additionally, the oxidative stress indicators for total antioxidant and free radical scavenging activity [ferric reducing ability of plasma (FRAP) and 2,2'-azino-bis(3-ethylbenzothiazoline-6-sulphonic acid) (ABTS) assays] were also determined in plasma and milk of ewes. Thus, forty-five ewes were divided into three equal groups. Each animal of the control group fed daily with a basal diet. The same diet was offered also in each animal of the other two groups. However, the concentrate fed to M group was supplemented with 2.5 g/kg rumen protected methionine, while the concentrate fed to MCB group with 5 g/kg of a commercial product which contained a combination of methionine, choline and betaine, all three in rumen protected form. The results showed that the M diet, compared with the control, increased significantly the ewe's milk fat and the total solids content. Both M and MCB diets had not noticeable impact on ewes' milk FA profile. Significantly higher FRAP values in the blood plasma of ewes fed the MCB, and in the milk of ewes fed with the M diet, compared with the control, were found. Additionally, significantly higher GST activity in the blood plasma of ewes fed the M diet, compared with the control, was observed. Moreover, a significant increase (by 20%) in the growth rate of lambs nursing ewes fed with M diet, compared to controls, was found.

Effects of alkaloids in the total mixed rations on milk yield and metabolic status of dairy cows

H. Scholz[1], D. Weber[1] and A. Ahrens[2]
[1]Anhalt University of Appled Scienes, Faculty LOEL, Strenzfelder Allee 28, 06406 Bernburg, Germany, [2]Thuringia Animal Health Fund, Victor-Goerttler-Straße 4, 07745 Jena, Germany; heiko.scholz@hs-anhalt.de

It has demonstrated that alkaloids to have anti-inflammatory, antimicrobial and immunomodulatory effects in various species. The main objective of this study was to estimate the effect of supplementation with alkaloids (SANGROVIT) on milk production and milk quality in German Holstein cows in the first 100 days of lactation. 93 Cows were randomly divided into two groups: control group (CG; 49 dairy cows) and experimental group (EG; 44 dairy cows) with 8 g SANGROVIT from 3 weeks ante partum to 15 weeks post-partum. Milk recording data and blood samples for plasma metabolites were taken. In order to assess the effect of SANGROVIT on inflammation and related liver metabolism, a spectrum of diverse parameters in plasma was determined. As parameters of general metabolic condition, concentrations of BHB and NEFA were determined. Concentration of CRP, representing an acute phase protein, was determined as a measure of inflammation. Concentrations of albumin, retinol, TAG, cholesterol, bilirubin and activities of ASAT and GLDH were determined as indicators of liver metabolism. Concentrations of tocopherols, β-carotene, TEAC and TBARS were determined as measures of the antioxidant system. For analysis, the statistical software SPSS and R were used for analysis of variance between both groups. 100-day-milk yield of cows were 195 kg higher in experimental group than by cows of control group, but there was no significant differences. Parameters of metabolism, with the only exception of CRP, were influenced by time of sampling ($P<0.05$). Several of those parameters (BHBA, retinol, TAG, ASAT, γ-tocopherol, TEAC, TBARS) were also influenced cow parity ($P<0.05$). Supplementation of SANGROVIT (EG) has less effect on the plasma variables determined. The only significant change due to SANGROVIT (EG) supplementation over all the sampling times was an increase of the albumin concentration in plasma. The present study indicates that use of alkaloids in dairy cows during the transition period can be positively influenced the milk yield during the first 100 days of lactation. No effects on metabolic status were found.

Foetal supplementation with vitamins C and E via maternal intake improves antioxidant status

V.H. Parraguez[1], F. Sales[2], O.A. Peralta[1], C. Serendero[1], G. Peralta[1], S. McCoard[3] and A. González-Bulnes[4]
[1]University of Chile, Fac. of Veterinary Sciences and Fac. of Agrarian Sciences, Santa Rosa 11735, Santiago, Chile, [2]INIA-Kampenaike, Angamos 1056, Punta Arenas, Chile, [3]AgResearch Grasslands, Tennent Drive, Palmerston North, New Zealand, [4]INIA-Madrid and Universidad Complutense de Madrid, Av. Puerta de Hierro s/n, Madrid, Spain; vparragu@uchile.cl

In sheep, undernutrition and/or twining results in intrauterine growth restriction (IUGR), associated with foetal hypoxemia and oxidative stress. A possible strategy for counteracting oxidative stress of the fetoplacental unit and thus IUGR is the maternal supplementation with antioxidant vitamins C and E. However, it is necessary to determine if such vitamins are adequately transferred to the foetus and to evaluate their effect on foetal antioxidant capacity. The present study was undertaken using 64 Corriedale sheep bearing single or twin foetuses that were allocated to natural prairie grasslands supplying ~70% of NRC requirements from 30 days of pregnancy. Half of the animals in each gestation rank were maintained at ~70% of NRC requirement and the remaining half received daily concentrate supplementation to satisfy ~100% of NRC requirements. In addition, half of the ewes from each group received daily oral administration of 10 mg/kg vitamin C and 9 IU vitamin E during the treatment period. At 140 days of pregnancy, blood samples from the maternal jugular vein and the foetal umbilical vein (obtained after caesarean section under spinal anaesthesia) were collected to measure vitamins C and E concentrations and foetal total antioxidant capacity (TAC). Vitamins supplementation significantly increased (P<0.05) vitamins C (24%) and E (25%) in maternal blood, and vitamin E (14%) and TAC (40%) in umbilical vein blood, but the increase in vitamin C (35%) was not significant. A rank × nutritional plane interaction for cord blood vitamin E concentration was observed (P<0.01). Twining resulted in higher vitamin E and lower TAC in blood cord (P<0.05), while undernutrition showed the inverse results. It is concluded that maternal supplementation with vitamins C and E increased the vitamins delivery to the foetuses and improved their redox status. This research contributes to potential future strategies to prevent IUGR. Support: FONDECYT 1160892.

Session 15 Poster 12

Differentially expressed genes (DEGs) in the backfat of pigs fed with cDDGS

M. Oczkowicz, M. Świątkiewicz and T. Szmatoła
National Research Institute of Animal Production, ul Krakowska 1, 32-083 balice, Poland; maria.oczkowicz@izoo.krakow.pl

cDDGS is relatively cheap and easy available source of protein, fat, energy and micro-nutrients in animal`s feeding. We performed the transcriptome analysis of the backfat of pigs fed with cDDGS and without cDDGS in order to evaluate its impact on gene expression. The animals were kept in uniform conditions and were divided into four feeding groups differing in the presence of cDDGS and source of fat in the diet (no cDDGS+rapeseed oil – group I n=6), (+cDDGS + rapeseed oil – group II n=6) (+cDDGS+beef tallow – group II n=5), (+cDDGS+coconut oil group – group IV n=5). After the slaughter, fragments of backfat were collected and immediately frozen at -80 °C. After RNA isolation, the cDNA libraries were prepared with SMARTer® Stranded Total RNA Sample Prep Kit – HI Mammalian (Takara, Clonetech), than the libraries were sequenced on the HiScanSQ System (Illumina). We have performed a two step statistical analysis since there were two dietary factors in our experiment (cDDGS and source of fat). Using DeSeq2 we performed pairwise comparison between all groups which revealed that there are 93 DEGs between group I and II, 126 between group I and IV and 13 between group I and III. Next, we combined all groups with cDDGS and compared them with group without cDDGS. We identified 172 DEGs (e.g *MOGAT2, UCP2, FAS, ACE, VSIG4, ARG1, APOA4*) Genes downregulated in +cDDGS group were highly significantly overrepresented in several Reactome Pathways (e.g. Metabolism FDR<2.97E-07, The citric acid (TCA) cycle and respiratory electron transport FDR<7.98E-03) and Biological Processes (coenzyme metabolic processes, P<0.003). Moreover several genes witch are potential therapeutic targets for obesity *(e.g.MOGAT2)* were downregulated in a group obtaining cDDGS in the diet. Our results suggest that cDDGS could be considered as a potential therapeutic factor for obesity in animal (for example breeding or companion ones) and human diet, but further research is necessary. This work was supported by NCN Poland grant no. 2014/13/B/NZ9/02134.

Session 15 Poster 13

Effect of quinoa and/or linseed on immune response and quality of meat from lambs

R. Marino[1], A. Della Malva[1], G. Annicchiarico[2], F. Ciampi[1], M.G. Ciliberti[1] and M. Albenzio[1]
[1]University of Foggia, Department of Agricultural Food and Environmental Sciences, via Napoli, 25, 71122 Foggia, Italy, [2]Agricultural and Economic Research Council (CREA), ss 7 Via Appia, 85051, Bella Muro (PZ), Italy; rosaria.marino@unifg.it

The aim of this study was to evaluate the effects of supplementation based on linseed, quinoa seeds and their combination on immune response, productivity and organoleptic quality of meat from merinos derived lambs. 32 Altamurana weaned lambs were divided in 4 groups with different diet supplementation: control (CO), quinoa (QS), linseed (LS) and combination of quinoa and linseed (QS+LS). The humoral and cell-mediated immune response and glucocorticoid secretion after a loading test were evaluated. After 50 days of supplementation, animals were slaughtered and pH, colour and Warner Bratzler shear force (WBSF) were estimated All data were subjected to analysis of variance using the GLM procedure including fixed effect due to diet, time of sampling and their interactions as fixed effects for immunological parameters and cortisol and fixed effect due to dietary supplementation for organoleptic parameters. The LS+Q group registered the highest increase of anti-KLH IgG at 25, 35 and 45 days. When the cell-mediated immune responses of lambs was investigated, the LS and Q groups at 15 day, and the LS group at 35 day of the experiment showed higher skinfold thickness than C and LS+Q. The cortisol secretion increased in all lambs 10 min after handling, loading and transport; in particular, the increase in LS and LS +Q lambs, expressed as % with respect to pre-test levels, was significantly higher than in C and Q lambs. LS lambs displayed the lowest concentration on average during the loading test. Muscle pH was affected by diet supplementation showing higher values at 1, 3 and 6 hours post-mortem in LS, Q and LS+Q lambs compared to control group. Lambs from LS and Q groups showed lower WBSF values than meat from control group. Dietary supplementation affected the meat colour parameters, in particular higher values of lightness, redness and yellowness were detected in meat from lambs of all supplemented groups. Results indicate that linseed and quinoa seeds supplementation may help the animal to cope stress events improving meat quality.

Session 15 Poster 14

Dietary energy, vitamin C, vitamin E and selenium improve antioxidant activity in the hen

S. Khempaka[1], P. Pasri[1], S. Okrathok[1], P. Maliwan[1], M. Sirisopapong[1], W. Molee[1], N. Gerard[2] and P. Mermillod[2]
[1]School of Animal Production Technology, Institute of Agricultural Technology, Suranaree University of Technology, 111 Unversity Avenue, 30000, Thailand, [2]Physiologie de la Reproduction et des Comportements, UMR085, INRA, CNRS, Université de Tours, IFCE, Nouzilly, 37380, France; khampaka@sut.ac.th

Oxidative stress in poultry result from the imbalance between reactive oxygen species (ROS) and the antioxidants that neutralize ROS that can damage cell membranes in different tissues, affecting important functions, such as reproduction, and leading to production loss and leads to poor in growth rate and production. Antioxidants such as vitamin C, vitamin E and selenium (Se) are the scavengers of ROS and have been used to reverse the adverse impact of free radicals. This study aimed to examine the effects of the dietary metabolizable energy (ME) level, vitamin C, vitamin E and Se on antioxidant activity in egg yolk and uterine fluid of laying hens. A total of 128 ISA Brown female laying hens aged 33 weeks were randomly allotted to 4 groups of 32 birds each in a Completely Randomized Design (CRD). All birds were fed diets of 109 g/day and free access to water, under 17 hours of light per day. The experimental diets of the 4 groups were: (1) low energy (2,650 ME kcal/kg) diet; (2) low energy diet supplemented with 200 mg/kg of vitamin C, 100 mg/kg of vitamin E, and 0.3 mg/kg of Se; (3) normal energy diet (2,900 ME kcal/kg); and (4) normal energy diet supplemented with 200 mg/kg of vitamin C, 100 mg/kg of vitamin E, and 0.3 mg/kg of Se. The results showed that the supplementation of vitamin C, vitamin E, and Se can enhance glutathione peroxidase activity in egg yolk and uterine fluid under both energy levels. The antioxidant capacity (DPPH) and low TBARS of egg yolk in the normal energy diet with supplementation was improved compared to the other groups ($P<0.05$). In conclusion, it was shown that dietary energy, vitamin C, vitamin E, and Se can enhance antioxidant activity in both egg yolk and uterine fluid in laying hens.

Influence of yeast culture on blood metabolites and acute phase response in beef heifers

Y.Z. Shen[1,2], H.R. Wang[2], T. Ran[1], I. Yoon[3], A.M. Saleem[1] and W.Z. Yang[1]
[1]Lethbridge Research and Development Centre, 5403, 1 Ave S, Lethbridge T1J4B1, Canada, [2]Yangzhou University, East of Wenhui Road, Yangzhou, 225009, China, P.R., [3]Diamond V, 2525 60th Ave SW, Cedar Rapids, 52404, USA; wenzhu.yang@agr.gc.ca

The objectives of this study were (1) to investigate the effects of sites of delivering Saccharomyces cerevisiae fermentation product (SCFP) on faecal IgA, blood metabolites and acute phase response in beef heifers fed high-grain diets, and (2) to examine the use of SCFP as a potential alternative for current industry standard antibiotics used in beef cattle feeds. Five beef heifers (initial BW=561±11.7 kg) equipped with ruminal and duodenal cannulas were used in a 5×5 Latin square design with 28-d periods, including 21 d of adaption and 7 d of data collection. Five treatments were: (1) control diet that contained 10% barley silage and 90% barley concentrate (DM basis); (2) control diet supplemented with antibiotics (ANT; 330 mg monensin/d and 110 mg tylosin/d); (3) ruminal (top dress) delivery of SCFP (rSCFP; NaturSafe®, Diamond V, 18 g SCFP/d); (4) duodenal delivery of SCFP (dSCFP; 18 g SCFP/d, via duodenal cannula); and (5) a combination of rSCFP and dSCFP (rdSCFP; 18 g rSCFP and 18 g dSCFP). Data were analysed using the MIXED procedure of SAS with model including treatment as fixed effect and the random effects of heifer and period. The PDIFF option was included in the LSMEANS statement to account for multiple comparisons among treatments. Intake of DM (kg/d) tended ($P<0.10$) to be greater for heifers fed rdSCFP (13.0) than control (12.2), ANT (11.8) and rSCFP (11.8) diets. Faecal IgA concentration (µg/g) was highest with ANT (85.5), intermediate with dSCFP (72.6) and rdSCFP (79.4), and lowest with control (59.7) and rSCFP (45.1). No treatment effects on blood concentrations of glucose (73.4 mg/dl), urea N (15.4 mg/dl), and NEFA (49.4 µM) were observed. Acute phase response measured with blood serum amyloid A (33 µg/ml) and LPS-binding protein (196 µg/ml) did not differ among treatments. These results suggest an improvement of intestinal mucosal immunity by SCFP with delivery of SCFP to the duodenum. The SCFP performed better or at least equal to antibiotics currently used in beef cattle rations and could be a natural alternative for beef cattle production.

Differentially expressed genes (DEGs) in the backfat of pigs fed with different fats

M. Oczkowicz, M. Świątkiewicz and T. Szmatoła
National Research Institute of Animal Production, ul Krakowska 1, 32-083 balice, Poland; maria.oczkowicz@izoo.krakow.pl

Appropriate ratio of saturated and unsaturated fatty acids in the diet is very important for health of animals and humans. We aimed to analyse by RNA-seq the transcriptome changes in the backfat of pigs fed with different sources of fat in the diet (rapeseed oil – group I n=6), (beef tallow – group II n=5), (coconut oil group – group III=5) to investigate genes responsiveness for diet treatment. The animals were kept in uniform conditions and obtained isoenergetic and isonitrogenous diet which differed only in source of fat. After the slaughter, fragments of backfat were collected and immediately frozen at -80 °C. After RNA isolation, the cDNA libraries were prepared with SMARTer Stranded Total RNA Sample Prep Kit – HI Mammalian (Takara, Clonetech). Then the libraries were sequenced on the apparatus and HiScanSQ System (Illumina, San Diego, USA). DeSeq2 analysis revealed that there are 29 DEGs between repassed oil and beef tallow groups. Analysis of these genes with String software showed that genes engaged in complement and coagulation cascades are overrepresented FDR<0.03 (PLAU, F3, THBD, SPP1). Interestingly, these genes have been suggested as therapeutic targets for hypertension and arteriosclerosis, while diet rich in animal fats is connected with increased risk of these diseases. Only, two DEGs (CD209 and FAM101A) were identified between rapeseed oil and coconut oil. No DEGs were identified between coconut oil group and beef tallow group. In conclusions, our results suggest that the well-established negative effect of animal fats in the diet on CVD (cardiovascular diseases) is associated with changes in expression of genes engaged in the regulation of blood pressure and blood coagulation. This work was supported by NCN Poland grant no. 2014/13/B/NZ9/02134.

Session 15 — Poster 17

Reconstituted alfalfa hay in starter feed improves health status of dairy calves during pre-weaning

S. Kargar[1], K. Kanani[1], M.G. Ciliberti[2], M. Albenzio[2], A. Della Malva[2], A. Santillo[2] and M. Caroprese[2]
[1]Shiraz University, Department of Animal Sciences, Fars Province, 65, 71441-65186, Shiraz, Iran, [2] University of Foggia, Department of the Sciences of Agriculture, Food and Environment (SAFE), Via Napoli, 25, 71122, Foggia, Italy; maria.ciliberti@unifg.it

Twenty 3-d-old Holstein dairy calves were used to investigate the effects of feeding starter feed containing dry (AH) vs reconstituted (RAH) alfalfa hay at 10% of dietary dry matter on health status during the pre-weaning period (d 1 to 49; n=10 calves per treatment). Dietary treatment of calves were represented by 6 l/d of milk from d 1 to d 43, by 4 l/d of milk from d 44 to d 46, and by 2 l/d of milk from d 47 to d 49 of study and weaned on d 50. Health status of calves was monitored several times daily, and health scores were recorded once daily. Faecal scores in a scale from 1 (normal and well-formed) to 5 (watery, mucous, and bloody) were assigned. Scale of respiratory scores was from 1 (normal) to 5 (wet cough). Calves were considered to be affected by diarrhoea when the faecal score was ≥3 and to be affected by respiratory illness when the respiratory score was >1. Models for occurrence of diarrhoea, respiratory illness and needs for medication by logistic regression using a binomial distribution in the GLIMMIX procedure in SAS were evaluated. Frequency and duration of diarrhoea, respiratory illness and administration of medication with a Poisson distribution using the GENMOD procedure of SAS were tested. The occurrence of diarrhoea (odds ratio=2.02; *P*=0.007) and respiratory illness (odds ratio=4.74; *P*=0.01) was increased in calves fed AH vs RAH. Feeding AH vs RAH decreased the chance of administration of medication when calves experienced diarrhoea (odds ratio=0.49; *P*=0.007) and respiratory illness (odds ratio=0.21; *P*=0.01). Calves fed RAH vs AH showed a reduction of days with diarrhoea (2.4 vs 4.7 d; *P*=0.01), a decreasing of frequency (0.1 vs 0.4; *P*=0.001), duration (0.3 vs 1.4 d; *P*=0.001) and medication days (0.3 vs 1.4 d; *P*=0.001) of calves experienced respiratory illness. Overall, feeding RAH improved health status through decreasing the occurrence of diarrhea and respiratory illness.

Session 15 — Poster 18

Role of the dietary grape seeds meal given to fattening pigs on the oxidative status of meat

M. Saracila, T.D. Panaite, M. Cornescu, M. Olteanu and A. Bercaru
National Research-Development Institute for Animal Biology and Nutrition (IBNA), Laboratory of Chemistry and Nutrition Physiology, Calea Bucuresti, Balotesti,1, 077015, Romania; mihaela.saracila@yahoo.com

A feeding trial was performed on 12 Topigs growing- finishing pigs (Large White × Pietrain) × (Talent), with an initial bodyweight of 66.42 kg±10. The pigs were assigned to 2 groups (C, E), housed in the same hall, in two distinct pens, with 6 pigs/pen. The conventional diet (C) had corn, wheat, soybean meal and rapeseeds meal as basic ingredients (3,110 kcal/kg ME and 17.60% CP). Compared to C diet, the experimental diet (E) included 7.5% flax meal as source of polyunsaturated fatty acids and 1% grape seeds meal as natural antioxidant. In the end of experiment, in agreement with the experimental protocol approved by the Ethics Commission of IBNA, all pigs were slaughtered and meat (shoulder, neck, loin) samples were collected. At 0 and 7 days after refrigeration (4°C), the malonaldehyde concentration (ppm MDA) and the antioxidant capacity (mM equivalent ascorbic acid) of the collected samples was evaluated. Both at 0 and at 7 days of refrigeration, the shoulder, neck and loin samples from group E had significantly (P≤0.05) lower malonaldehyde concentrations than those from group C. After 7 days of refrigeration, the antioxidant activity of the samples from group E (shoulder: 14.099±2.78; neck: 14.148±0.20; loin: 10.473±2.39; mM ascorbic acid equivalent) was significantly (P≤0.05) higher than in the samples from group C (shoulder: 12.279±0.72; neck: 11.043±1.56; loin: 9.470±0.11 mM ascorbic acid equivalent). The conclusion of the study was that the grape seeds meal added to high polyunsaturated fatty acids diets for fattening pigs inhibited the propagation of the oxidative reactions within the muscle tissues, thus proving its antioxidant activity.

Effect of dietary herbal supplementation on the health of organically raised Rhode Island Red hens

E. Sosnówka-Czajka, I. Skomorucha and E. Herbut
National Research Institute of Animal Production, Department of Poultry Breeding, 1, Krakowska Street, 32-083 Balice near Krakow, Poland; ewa.sosnowka@izoo.krakow.pl

The objective of the study was to determine the effect of dietary herb supplementation on the health of native Rhode Island Red (RIR) hens. The experimental material consisted of RIR pullets, which were assigned to four experimental groups: group I (control), and groups II, III and IV, in which pullets from 20 wk of age were fed diets supplemented with *Thymus vulgaris* L., *Echinacea purpurea* L. and a mixture of *T. vulgaris* and *E. purpurea*, respectively. All the groups were reared according to organic farming recommendations. At 40 wk of rearing, blood was collected from 10 birds per group as were lymphoid organs, which were weighed and their proportion in body weight was calculated. The blood samples were analysed for haemoglobin concentration (using the Epoll 20 photometer), the immunoglobulin complex, cholesterol and glucose (using the biochemistry analyser Mindray BS-120). Haematocrit value was determined using haematocrit capillary tubes by centrifuging blood in an MPW-52 centrifuge. A Nikon YS 100 microscope was used to calculate the erythrocyte and leukocyte count as well as the heterophil to lymphocyte ratio (H:L). The results were statistically analysed using one-way analysis of variance, and significant differences were estimated with Duncan's test. Hens receiving the feed supplemented with *T. vulgaris* and *E. purpurea* were characterized by a higher percentage of lymphoid organs, higher H:L ratio, and higher glucose levels compared to the other groups. Birds from the experimental groups showed lower blood haematocrit in relation to the control group. The blood of experimental birds had more erythrocytes and a greater amount of immunoglobulin complex compared to the control. The experimental factors had no effect on the other blood parameters under analysis. In summary, the dietary herbal supplementation of RIR hens improved general immunity by increasing the proportion of lymphoid organs and by increasing the amount of immunoglobulin complex in the blood of these birds.

Introduction ATF-EAAP Special Session

J.L. Peyraud
Animal Task Force, 149 rue de Bercy, 75012 Paris, France; mail@animaltaskforce.eu

The EAAP & ATF Special Session during the EAAP Annual Meeting aims to bring together animal science with practice of animal production and connect researchers, policy-makers, industry representatives and societal organisations. Every year, a different topic is addressed during this half-day session. Background It is often communicated in the media and among the general public that European citizens should reduce their consumption of animal products/proteins. Very often, the justification is both based on human health and sustainability point of views (use of resources, impact on climate, environmental footprint, AMR, animal welfare…). Can we find a consensus on recommended shares of animal products in our diets, at the junction of human health and planetary health? What share of animal-derived food in our diet is sustainable on the economic, environmental, social point of views? Format of the EAAP & ATF Special Session The session would like to engage discussion with farmers, food processors and industries, retailers, nutritionists, scientists, but also with the society. Most important findings will be discussed with a panel. The outcomes of the session will be discussed in more details during the ATF seminar, in Brussels, on Nov. 7th 2018, where a large panel of European stakeholders will be invited. The Special Session aims to contribute to: (1) engage a dialogue with various stakeholders; (2) support knowledge development and innovation, foster appropriation by farmers and industries; (3) address how research and innovation can help the livestock sector; and (4) provide input to European research and innovation agendas and to public policies to secure Europe's role as a leading global provider for safe and healthy animal derived products.

Narrative ATF-EAAP special session

J.L. Peyraud
Animal Task Force, 149 rue de Bercy, 75012 Paris, France; mail@animaltaskforce.eu

Narrative of the session Human consumption and nutritional issues – From the scientific/medicine point of view: What is the state of the art in human health? It is relevant to recommend a general decrease in livestock consumption in the EU? Can we be more specific? We should acknowledge the risks associated to an over-consumption of animal products, but to what extent? Are there parts of the population who risk deficiencies in decreasing their consumption (children, young women, adults, old age)? Is there a consensus on the recommended share of livestock products in our diets from a human health perspective? – From the industrial technology point of view: What are the good practices implemented by the livestock industry to address nutritional issues? What are the substitutes to meat and dairy products (plant proteins, insects, algae, *in vitro* meat…), can they reach animal products quality? What are the risks associated to plant proteins (phyto-hormones…), to highly processed food meat substitutes…? Production and environmental health issues – What are the global impacts of livestock farming on planetary health regarding feed efficiency, GHG emissions, AMR, resource use, etc.)? What are the associated benefits (ecosystem services, employment, gender issues, etc.)? – Can we find an optimal balance of European animal productions regarding emissions and resource use? – What are the most efficient livestock systems? Can we find/showcase good examples in the EU (optimisation of impacts through GHG mitigation measures, methane production, circularity, etc.)? Human & planetary health drivers to economic prospects in production – How to compare the sustainability of various diets, e.g. vegan vs vegetarian vs conventional diets? – What would be the impacts of a downsizing of livestock farming in Europe on human health, agriculture, territories, economy, trade balance, ethics, standards of production…? What are the current and projected trends in consumption and production in the EU and globally?

Session 16 Theatre 3

Closing ATF-EAAP special session

J.L. Peyraud
Animal Task Force, 149 rue de Bercy, 75012 Paris, France; mail@animaltaskforce.eu

The outcomes of the session will be discussed in more details during the ATF seminar, in Brussels, on Nov. 7th 2018, where a large panel of European stakeholders will be invited.

Session 17 Theatre 1

Weight of sex glands as an on line tool to discriminate entire males from immunocastrates

N. Batorek Lukač[1], G. Fazarinc[2], M. Prevolnik-Povše[1], M. Škrlep[1] and M. Čandek-Potokar[1]
[1]Agricultural Institute of Slovenia, Hacquetova ulica 17, 1000 Ljubljana, Slovenia, [2]University of Ljubljana, Veterinary Faculty, Gerbičeva 60, 1000 Ljubljana, Slovenia; nina.batorek@kis.si

Active immunisation against the hypothalamic GnRH referred to as immunocastration (IM) seems a suitable alternative to surgical castration of pigs, as this method is effective in prevention of boar taint. However, some pigs may escape the effective IM (so called non-responders) thus it is important to find a simple and reliable indicator of IM effectiveness at slaughter line. Namely, laboratory analysis of androstenone and skatole, substances responsible for boar taint, are time consuming and expensive. The weight of reproductive organs could serve as an indicator of successful IM. In the present study data from 55 entire males (EM) and 76 immunocastrates (IC) varying in body weight (103.6±13.5 kg) and delay between IM and slaughter (2 to 9 weeks) were used. Testes and accessory sex glands (vesicular and bulbourethral) were dissected, weighed and used to distinguish between IC and EM by discriminant analysis. Better discrimination was achieved for IC than EM. To distinguish between EM and IC, testes weight or bulbourethral gland alone were less reliable (92 and 93%, respectively) than using the weight of all three sexual glands (96% success rate). Actually the success rate of correct prediction of IC was 100%, because the one misclassified IC was likely a non-responder (i.e. EM). On the other hand correct recognition of EM was lower (91%). The results of the present study indicate that taking into consideration the weight of all three sex glands could be a reliable tool to test the effectiveness of IM. The authors acknowledge the financial support of the Slovenian Agency of Research (grant P4-0133) and the COST action CA12215 IPEMA.

Session 17 Theatre 2

Health and welfare issues regarding surgical castration of male piglets and its alternatives

E. Von Borell[1], A. Prunier[2] and U. Weiler[3]
[1]Martin-Luther-University Halle-Wittenberg, Theodor-Lieser-Str. 11, 06120 Halle, Germany, [2]INRA, Domaine de la Prise, 35590 Saint-Gilles, France, [3]University of Hohenheim, Garbenstrasse 17, 70599 Stuttgart, Germany; eberhard.vonborell@landw.uni-halle.de

Surgical castration without anaesthesia is increasingly recognized as a painful and welfare relevant procedure, although the systematic use of anaesthesia with or without additional analgesia for pain relief during surgical castration is currently only practiced in some countries where this was mandated or part of their national quality assurance scheme. From scientific studies it is evident that the effectiveness of pain intervention during and after surgical castration is only given when anaesthesia is combined with preemptive analgesia, although it has to be considered that additional stress and associated health and welfare risks may be imposed by the intervention itself. Previous surveys among stakeholders evidenced that the acceptance and likelihood of anaesthesia implementation will depend on authorisation of farmers to do the pain interventions after special training. However, such protocols are difficult to apply taking into account legal and economic constraints. Immunocastration is another already licensed and practiced alternative performed by farmers. Although this intervention was proven to be effective with less impact on health and welfare of pigs, this method has not achieved a broad application due to the resistance of some chain actors that fear low acceptance of this method by consumers. The widely favoured option of fattening entire males poses a risk for injuries due to unwanted behaviours under current housing and management conditions, unless pigs are slaughtered before puberty in order to reduce critical behaviours and to limit boar taint. Raising entire young males and immunocastration seem to be the preferred short term and widely usable options, whereas fattening of matured heavy entire males as a sustainable and effective long-term goal would necessitate the implementation of new genetic schemes as well as new housing and management systems and a significant effort in research. The paper was prepared within the framework of the COST Action 15215 IPEMA.

Session 17 — Theatre 3

Finishing heavy boars for lower taint, suitable welfare and optimal performance

L. Martin[1], A. Frias[1], R.P.R. Da Costa[1], M.A.P. Conceição[1], R. Cordeiro[2] and A. Ramos[1]
[1]Inst Politec Coimbra, ESAC, DCZ, Bencanta, 3040-316 Coimbra, Portugal, [2]Uzaldo Lda., R. Balastreira 8, 3090-649 Figueira da Foz, Portugal; luisam@esac.pt

Entire male pig production is considered economically and environmentally advantageous due to superior feed efficiency and growth rates compared to barrows. Nevertheless, finishing heavy boars is challenging not only because of aggressive behaviour at puberty, but also due to boar taint. Inulin has been indicated as an additive to address this problem, though dosage needs to be refined. This study aims to assess boar taint and welfare in finishing entire males subjected to a diet with inulin, lower stocking density and an enriched environment. A model will be developed to support pig farmers decision on finish heavier male pigs without castration and better profit. A total of 60, 3-cross Pietrain × F1 (Landrace × Large White) boars (114±9.5 kg) were selected and randomly assigned to 6 pens where a 3×2 factorial design experiment took place. 3 isoproteic (15.5% CP) and isoenergetic (2.3 Mcal NE) diets of 0, 3 and 6% inulin, balanced for essential amino acids were used. Different stocking density (1 or 1.9 m^2/pig) pens were provided, for 10 boars each. The lower density pens had 2 nipple drinkers instead of 1 and had 2 extra anti-bite toys. A seven-week trial took place and measurements were taken through the trial and at slaughter. The average final weight was 154±11.7 kg with an average daily gain of 0.83±0.17 kg. Mean carcass yield was 73.5±3.8% and lean-meat 58±2.6%. For a confidence level of 95%, inulin and environment had no significant differences on growth, carcass yield or lean-meat. Skin lesions score (Welfare Quality® method) was numerically lower on pigs fed 6% of inulin compared with pigs fed 0 and 3% inulin. Animal dirtiness (WQ®) was higher on pens with higher stocking rates and higher intake of inulin (which coincided with more liquid faeces). Plasma skatole, indole (HPLC-FLD), androstenone and cortisol (ELISA) levels will be presented. A model to optimize slaughter weight for heavier boars, reflecting welfare and maximising profit will be set. Nevertheless, as inulin extracted from chicory is an expensive additive, its use is only beneficial if board taint is removed or reduced.

Session 17 — Theatre 4

Pork production with immunocastration: welfare and environment

V. Stefanski[1], E. Labussière[2], S. Millet[3] and U. Weiler[1]
[1]University of Hohenheim, Institute of Animal Science, Garbenstrasse 17, 70599 Stuttgart, Germany, [2]PEGASE, INRA, Agrocampus-Ouest, 35042 Rennes, France, [3]Flanders Research Institute for Agriculture, Fisheries and Food (ILVO), Scheldeweg 68, 9090 Melle, Belgium; volker.stefanski@uni-hohenheim.de

Societal pressure to end surgical castration of piglets without anaesthesia and pain relief is increasing. Two alternatives to surgical castration currently exist: raising entire males (EM) or immunocastrated males (IC), both with advantages and disadvantages. Considering animal integrity, EM may be the preferred choice. However, research has shown that IC may solve some of the welfare issues associated with EM production, especially with regard to aggression (i.e. penile injuries). Compared to barrows, EM and IC grow more efficiently. Despite increased nutrient requirements for the animals, both probably imply lower nutrient excretion and a lower carbon footprint per kg pork. More research is nevertheless required on the feed of IC after the second vaccination to further enhance the economic and environmental sustainability because of the large increase in their voluntary feed intake. Increased knowledge is necessary to further improve the sustainaibility of pork production with immunocastration, which is also the purpose of the ERA-Net project 'SuSI' (Sustainability in pork production with immunocastration). This presentation will provide a brief overview of the impact of IC on animal behaviour, welfare and on nutritional efficiency to minimize the environmental footprint.

Immunocastration, avoiding teeth clipping and tail docking improve piglets' production and welfare

L. Morgan[1], E. Klement[1], L. Koren[2], J. Meyer[3], D. Matas[2], S. Novak[1], L. Golda[1], Y. Cohen[1], W. Abu Ahmad[1] and T. Raz[1]
[1]Koret School of Veterinary Medicine, Hebrew University of Jerusalem, 7610001 Rehovot, Israel, [2]Bar-Ilan University, Ramat-Gan, 52900002, Israel, [3]University of Massachusetts, MA, 01003, USA; liat.morgan@mail.huji.ac.il

Piglets routinely undergo a set of invasive procedures during the first days of their lives, which commonly include surgical castration, tail docking and teeth clipping. Those procedures are a top welfare concern since it potentially resulting in pain and stress, which may have long term effects on the animals' health and production. Our objective was to examine production and welfare parameters of pigs from birth to slaughter under different managements, by avoiding these procedures, while providing alternatives such as anti-GnRH vaccine and environmental enrichment. Litters (n=32 sows; 329 piglets; 3 days after farrowing) were allocated randomly into one of 4 groups; G1: Surgical castration, tail docking and teeth clipping, without environmental enrichment; G2: same as G1, but meaningful environmental enrichment was provided; G3: Non-surgical sterilization with anti-GnRH vaccine (Improvac®), tail docking, teeth clipping, with environmental enrichment; G4: none of the invasive procedures were performed, piglet were vaccinated (Improvac®), and environmental enrichment was provided. Mixed-effects Linear Regression model revealed that slaughter weight significantly increased when invasive procedures were avoided and environmental enrichment was provided. Weight interval from birth to slaughter was higher in G4 (G1: 99.2±1.07, G2: 99.9±1.4, G3: 103.6±.1.58, G4: 106.5±1.6 kg; $P<0.05$). The odds ratio to be weak, dead or injured in the conventional, non-enriched G1, was 89% higher than in G4 ($P<0.05$). Hair cortisol at weaning, as a marker for chronic stress during lactation, decreased gradually as management becomes welfare friendlier (-3.66 pg/mg for each step group, G1→G2→G3→G4; $P<0.05$). Anti-GnRH vaccine was effective in reducing serum and hair testosterone, similar to surgical castration. In conclusion, replacing surgical castration by anti-GnRH vaccine, avoiding teeth clipping and tail docking, and providing environmental enrichment, are better alternatives that would substantially benefit both the animals and farmers.

Behavioural response to an intermittent stressor is higher in entire compared to castrated male pigs

M. Holinger[1,2], B. Früh[2], P. Stoll[3], M. Kreuzer[1], R. Graage[4], A. Prunier[5] and E. Hillmann[1,6]
[1]ETH Zurich, Universitätstsstrasse 2, 8092 Zürich, Switzerland, [2]FiBL, Ackerstrasse 113, 5070 Frick, Switzerland, [3]Agroscope, Tioleyre 4, 1725 Posieux, Switzerland, [4]University of Zurich, Winterthurerstrasse 260, 8057 Zürich, Switzerland, [5]INRA, Domaine de la Prise, 35590 Saint-Gilles, France, [6]Humboldt Universität zu Berlin, Philippstrasse 13, 10115 Berlin, Germany; mirjam.holinger@bluewin.ch

Entire male pigs show more agonistic and sexual behaviour, which might negatively affect welfare in group-housed entire male fattening pigs. In order to assess chronic stress in entire and castrated male pigs, we conducted an experiment with a 2×2×2 factorial design, comprising the factors castration (entire / castrated), chronic intermittent social stress (yes/no) and provision of grass silage (yes/no). Effects of grass silage will not be discussed here. The stress treatment consisted of repeated short-term confrontations (30 min each) and separations (20 min each). We observed behaviour in the home pen several times throughout the fattening period, measured circadian rhythm of salivary cortisol and scored stomachs after slaughtering with respect to gastric ulceration. Data analysis with general linear mixed effect models revealed that the stress treatment affected behaviour: Stress-exposed pigs displayed a reduction in posture changes and in agonistic behaviour. Entire male pigs responded more pronouncedly to the stress treatment in terms of posture changes and play behaviour, implying an increased stress response. The frequency of pathological changes in the gastric mucosa and of gastric ulcers was larger in stress-exposed pigs but did not differ between entire and castrated male pigs. Thus, gastric ulceration showed potential as gender-independent reference indicator of chronic stress. Salivary cortisol was considerably higher in castrated compared to entire males, but we did not find a change in circadian rhythm neither with respect to stress nor to castration. Based on the parameters, which responded to the stress treatment, we can conclude that chronic stress level is not higher in entire than castrated male pigs. However, their increased behavioural stress response might have implications for pig management and handling.

Session 17 Theatre 7

Meat quality issues in entire male and immunocastrated pigs

M. Škrlep and M. Čandek-Potokar
Agricultural Institute of Slovenia (KIS), Hacquetova 17, 1000 Ljubljana, Slovenia; martin.skrlep@kis.si

Introduction of alternatives to surgical castration like rearing of entire males (EM) and immunocastrates (IC) is becoming a reality, thus the European pig sector is confronting a considerable challenge. In addition to solving the issues related to economics, nutrition, welfare, etc., there is also a problem of meat technological quality, potentially affecting processing aptitude and consumer acceptability. Apart from the problem of boar taint, EM were shown to deposit less fat which is more unsaturated, have less intramuscular fat (IMF) and tougher meat, along with indications for inferior water holding capacity (WHC) compared to surgical castrates (SC). With regard to the effect on meat colour and pH the literature is inconsistent. Lower amount of IMF or higher collagen content in EM could directly affect tenderness. Indirectly, higher fat unsaturation and oxidative metabolic profile could be related to higher protein oxidation, leading to inferior WHC and tougher meat. In contrast, indications of higher proteolytic potential of EM muscle also exist. Regarding the IC, they are mostly similar to SC but superior to EM in terms of higher IMF levels, more tender meat, although some indications of lower WHC and lighter meat colour exist. Their resemblance to either EM or SC depends on the length of the interval between the effective immunisation and slaughter. As intensive metabolic changes occur during this period (increased appetite, growth rate and fat deposition) this could induce changes of proteolytic and lipolytic properties of IC tissues. Yet, similarly as in the case of EM, the differences are not fully confirmed and aetiology poorly explained which supports the need for further research, either on fresh or processed meat. The authors acknowledge the financial support of the Slovenian Agency of Research (grants P4-0133, L4-5521). The review was prepared within the framework of the COST Action CA12215 IPEMA.

Session 17 Theatre 8

Comparison of muscle proteome profile between entire males and surgically castrated pigs

K. Poklukar, M. Škrlep, U. Tomažin, N. Batorek Lukač and M. Čandek-Potokar
Agricultural institute of Slovenia, Hacquetova ulica 17, 1000, Slovenia; klavdija.poklukar@kis.si

Literature indicates that entire males (EM) and surgically castrated pigs (SC) differ in meat quality traits (e.g. water holding capacity, intramuscular fat content), but the underlying mechanisms are unclear. The objective of the present study was to explain differences in meat quality between EM and SC using proteomic approach. We have compared protein expression of the *longissimus lumborum* muscle between EM (n=12) and SC (n=12) of Landrace × Large white crossbred pigs raised under the same environmental conditions and slaughtered at the age of 198±4 days. Muscle samples were collected 24 h post-mortem. Proteins were extracted and separated by two-dimensional electrophoresis, gel images digitalised and data on the relative abundance of protein spots were analysed with SAS statistical software. Spots of interest were identified by mass spectrometry (MALDI-TOF-TOF). A total of 124 protein spots were differentially expressed ($P<0.05$) between EM and SC, 104 of them were more abundant in EM. Among the identified protein spots (n=32), higher expression of malate dehydrogenase, Na/K transferring ATP-ase α-subunit and blood plasma proteins indicates higher oxidative metabolism in EM than SC, while higher abundance of identified protein fragments (mainly of actin and myosin heavy chain molecules) are indicative of higher level of proteolysis in EM. The authors acknowledge the financial support of the Slovenian Agency of Research (grants P4-0133 and L4-5521) and the COST action CA12215 IPEMA.

Quality control in entire male pig production with particular emphasis on boar taint detection

M. Font-I-Furnols[1], M. Čandek-Potokar[2], N. Panella-Riera[1], J.-E. Haugen[3] and I. Bahelka[4]
[1]IRTA, Finca Camps i Armet, 17121 Monells, Catalonia, Spain, [2]Agricultural Institute of Slovenia, Hacquetova ulica 17, 1000 Ljubljana, Slovenia, [3]Nofima AS, Osloveien 1, 1430 Ås, Norway, [4]NAFC, RIAP, Hlohovska 2, 94992 Nitra, Slovak Republic; maria.font@irta.cat

On line determination of meat quality in the slaughter plant, together with carcass grading, would allow an optimization of its processing. Due to the declaration of intentions to voluntary end with surgical castration it is important that pork quality is not reduced due to the increased share of entire and immunocastrated males in the meat chain. An issue to consider is rapid detection of boar tainted carcasses to use them for specific processing and to guarantee taint free meat. This is a challenge due to the lack of rapid boar taint detection methods and definition of cut-off levels to guarantee consumer acceptance. Within Cost Action IPEMA, WG4 deals with the innovation of grading and meat quality control systems advancing present classification on lean meat content by including important traits for processing and eating quality (in particular detection of boar taint). Several technologies such as near infrared/raman spectroscopy, ultrasounds and sensors have been envisaged to determine meat quality traits (intramuscular fat, fatty acids, water holding capacity) at/on line. Various instrumental methods/measurement principles are still in the research and development stage, and a few may have the potential for future boar taint detection (androstenone/skatole) at industrial level, after performance (sensitivity/selectivity) validation to meet industrial requirements, capacity and costs. There are still gaps to fill and aspects to improve to have a complete on-line integrative quality control system covering the requested meat quality grading parameters. Regarding boar taint, instrumental methods should also be compared to sensory analysis to draw quality control conclusions. Currently the olfactory human nose method is mostly used for on-line detection of boar tainted carcasses in the slaughter plants. A colorimetric method for at-line determination of skatole/indole has also been developed. Acknowledgments: Review prepared within the framework of the Cost Action CA 15215 IPEMA.

Fully automated and rapid at-line method for measuring boar taint related compounds in back fat

C. Borggaard, R.I.D. Birkler, S. Støier and B. Lund
Danish Technological Institute, Danish Meat Research Institute, Gregersensvej 9, 2630 Høje Taastrup, Denmark; rub@teknologisk.dk

A rapid mass spectroscopic method, capable of measuring the malodorous boar taint compounds androstenone and skatole in fat samples from male pig carcasses was developed. The method is well suited for use in commercial abattoirs as an at-line method to detect the presence of these compounds in pig carcasses. The developed chemical assay is based on salt assisted liquid-liquid extraction followed by direct measurement with Laser Diode Thermal Desorption-MS/MS. The method, capable of giving a result for both androstenone and skatole every 10 seconds, will when implemented in an abattoir as an automated at-line method give a single MS-MS instrument a measuring capacity of up to 2,880 male pig carcasses per 8-hour workday. The LOQ for the rapid method is 0.05 μg/g and 0.10 μg/g for skatole and androstenone respectively. The LOQ is to be compared with expected sorting thresholds at the abattoir of 0.25 μg/g for skatol and at least 1-2 μg/g for androstenone. Coefficient of variation is 5% for skatol and 3% for androstenone. The method will be implemented at a large Danish abattoir during 2018. From each uncastrated male carcass on the slaughter line, fat biopsies (0.3-0.5 g) are automatically extracted, placed in a deep well plate and weighed. Deep well plates containing 24 samples are conveyed to an in-house laboratory for extraction and MS-MS analysis. Total time from a sample is acquired on the slaughter line until the analysis result is in the abattoir data base is less than 40 minutes. The pure running costs of analysis per carcass, including consumables, is below 1€.

Raising entire male pigs: a framework for sensory quality control

D. Moerlein
University of Goettingen, Department of Animal Sciences, Albrecht-Thaer-Weg 3, 37075 Göttingen, Germany; daniel.moerlein@agr.uni-goettingen.de

A framework for establishing and monitoring sensory quality control schemes is provided. Standardized procedures for selection of sensory assessors including reference materials as well as the thorough use of statistical performance parameters are suggested. The application of standardized smell tests is demonstrated thus facilitating objective characterization of potential assessors olfactory acuity. Both the capability to correctly discriminate androstenone and skatole at relevant concentration levels and the ability to correctly identify these substances need to be assured. A statistical simulation showed the importance of large enough sample sizes for performance evaluations. The quantitative relationship of key boar taint substances with the olfactory perception of deviant smell was modelled based on a large sample set (1000+ carcasses). A nonlinearity of perception was shown as was the relatively higher importance of skatole. We suggest a curved approach instead of the so called 'safe box' when comparing sensory with chemical boar taint evaluation as it better reflects the sensory perception. As the performance of a single assessor usually is inferior to a group of assessors, preventive measures should be taken for slaughter house quality control given the risk of consumer complaints. The quantitative model can be used as reference / calibration for sorting schemes based on technical measurement of androstenone and or skatole. A strictly consumer-driven approach is recommended. That is, both the assessor performance criteria and the sorting limits should be aligned with consumer acceptance data. Further research is needed with regard to the long-term performance of slaughter house panels as boredom and adaptation may be distractors. Acknowledgement: this review was prepared within the framework of the Cost Action CA 15215 IPEMA.

Citizen attitudes and consumer acceptability towards meat from boars and immunocastrates in Europe

M. Aluwé[1], L. Tudoreanu[2], L. Lin[3], A.R.H. Fischer[3] and M. Font I Furnols[4]
[1]ILVO, Animal sciences unit, Scheldeweg 68, 9090 Melle, Belgium, [2]USAMV, Faculty of Veterinary Medicine, 59 Mărăşti Boulevard, 011464 Bucharest, Romania, [3]Wageningen University, Marketing and Consumer Behaviour, Hollandseweg 1, 6706 Wageningen, the Netherlands, [4]IRTA, Finca Camps i Armet, 17121 Monells, Spain; marijke.aluwe@ilvo.vlaanderen.be

Alternatives for surgical castration are production of boars and immunocastrates. Each of these face advantages and disadvantages which may result in different citizen attitudes (influenced by information about process characteristics, beliefs and feelings) and consumers acceptability (influenced by price, taste, health and convenience). For boars, it should be possible to reduce and detect boar taint at the slaughterline. Besides, poorer results for intramuscular fat and water-holding capacity may also result in a lower sensory quality. However, consumer preference may differ across countries. Application of immunocastration can solve a major part of the quality issues related to boar production, but market acceptance is low as consumer acceptance of this practice is questioned. While some consumers are in favor, others question the practice from food safety aspect. Therefore understanding when immunocastration is acceptable to whom is needed. Communication and a harmonization towards production systems with the alternatives are crucial, but this needs to be tailored to the local and individual context as citizen attitudes and consumer acceptability may strongly differ between countries and consumers. It is assumed that consumers consider a trade-off between price, animal welfare, food quality, and food safety when purchasing products. However, the lack of general knowledge of citizens on the pig production chain makes it hard to predict. Research linking sensory evaluation and attitudes may result in better insights in consumer perception towards the alternatives and will help to set up information strategies. Paper prepared within the framework of the Cost Action CA 15215 IPEMA.

Session 17 Theatre 13

Consumer's opinion on animal welfare and pig castration in Croatia

I. Djurkin Kusec[1], G. Kusec[1], L. Guerrero[2], M. Font-I-Furnols[2] and I. Tomasevic[3]
[1]Faculty of Agriculture in Osijek, Josip Juraj Strossmayer University of Osijek, Vladimira Preloga 1, 31000 Osijek, Croatia, [2]IRTA Food Industries, Finca Camps i Armet, 17121, Monells, Girona, Catalonia, Spain, [3]Faculty of Agriculture, University of Belgrade, Nemanjina 6, 11080 Belgrade, Serbia; idurkin@pfos.hr

The surgical castration without anaesthesia is still a routine zootechnical practice in Croatia. However, due to the growing concern on animal welfare this practice is being seriously questioned in most European countries. Subsequently, a declaration of intention to abandon it has been signed by many different stakeholders of the porkchain (farmers' representatives, meat industry, retailers, scientists, veterinarians and animal welfare NGOs). A number of alternatives to surgical castration have been proposed, immunocastration and entire male production being feasible ones at the moment. However, besides its advantage on animal welfare, the consumer's opinion should also be considered. Having this in mind, we carried out a survey in order to understand consumer beliefs and attitudes towards pig castration and animal welfare, as well as to identify their willingness to pay more for meat of animals that were not surgically castrated and that were treated according to animal welfare legislation. A total of 301 consumers participated in the survey through self-administered questionnaires during 2017 in Croatia. They were selected by gender and age to follow the National distribution and no previous information was given to them about production practices and meat quality. The results of the questionnaire show that Croatian consumers are concerned about animal welfare and strongly believe that animals raised for human consumption should be treated with dignity. The consumers are also very much concerned about the transport of animals, as well as the slaughter systems used in Croatian abattoirs. Generally, Croatian consumers prefer the meat from castrated pigs, however they accept immunocastration as an alternative to surgical castration. The results of this study indicate that immunocastration is a feasible alternative in Croatia and that abandon of surgical castration should be supported.
Acknowledgments: The authors would like to acknowledge the contribution of the COST Action IPEMA CA15215.

Session 17 Theatre 14

Consumer expectations towards meat from castrated and immunocastrated pigs: a segmentation approach

A. Claret[1], L. Guerrero[1], M. Font-I-Furnols[1] and A. Dalmau[2]
[1]IRTA, Food Industries, Finca Camps i Armet, 17121 Monells, Spain, [2]IRTA, Animal Welfare, Veïnat de Sies, 17121 Monells, Spain; maria.font@irta.cat

The aim of this work is to identify segments of consumers according to their expectations and acceptability of Iberian pork from five sex types. Iberian entire females (EF), castrated females (CF), immunocastrated females (IF), castrated males (CM) and immunocastrated males (IM) pigs (n=83, live weight=155.7±8.4 kg) were used. Consumer studies were carried out in Barcelona (n=150) and Madrid (n=100). Consumers evaluated one sample of each sex type at three information conditions (1) liking in blind condition, (2) expectations knowing sex type and (3) liking in informed condition. Cluster analysis was performed to identify segments and mixed model with repeated measures was applied to each cluster. Three clusters were identified. Cluster 1 (n=79) was named as 'No preference regarding expectations' and it was composed by consumers with similar expectation toward all the products. In blind condition they preferred meat from CM. In informed condition too but not significantly different from those of EF, CF and IF. Assimilation effect occurred in meat from EF and IF and negative disconfirmation in all sexes. Consumers of cluster 2 (n=79) were labelled as 'Against castration and immunocastration' since they expected better quality for EF meat. Nevertheless, meat from CM was the most preferred in blind conditions (not different from CF) and in informed conditions (not different from EF and IM). Negative disconfirmation occurred in EF loins and positive disconfirmation in CF, CM and IM. Consumers from cluster 3 (n=94) expected lower quality for immunocastrated pork and were named as 'Against immunocastration'. In fact, when they tasted loins in blind conditions no significant differences were found between sex types while in informed conditions IF and IM were less preferred. Negative disconfirmation was detected in CF and EF and positive disconfirmation in IF and IM. Assimilation was detected in EF, IF and IM. In conclusion, segments of consumers with different expectations towards meat from castrated and immunocastrated pigs were detected, and they have disconfirmation expectations in most of the types of meat.
Acknowledgements: Project funded by INIA (RTA2010-00062-CO2).

Session 17 Poster 15

Quality of pig carcasses of surgical- and immunocastrated males slaughtered at different live weight

M. Povod[1], O. Kravchenko[2], A. Getya[3] and O. Kodak[2]

[1]Sumy State National University, G.Kondratyeva str., 160, 40021, Sumy, Ukraine, [2]Poltava State Agrarian Academy, Skovorody str., 1/3, 36003, Poltava, Ukraine, [3]National University of Life and Environmental Sciences of Ukraine, Gen. Rodimtseva str., 19, 03041, Kyiv, Ukraine; getya@ukr.net

Nowadays in the world there is trend to slaughter the pigs at more heavy live weight, because of better economic effectiveness of fattening. Traditionally Ukrainian pork producers slaughter the animals with live weight 100-110 kg but would like to raise the weight before slaughter, working with hybrid pigs. According to reports of some researchers the increasing of weight before slaughter causes worsening of carcass quality, especially in surgically castrated and entire males. As a solution the immunological castration can be taken into account especially for rising heavy weight pigs. This assumption has been investigated in presented research. The trial has been organized under the condition of commercial farm on hybrid animals (Yorkshire×Langrass×Maxgro). Two groups of pigs with 30 heads in each were formed (surgical and immunological castrated males). Surgical castration was done at the age 4 of days, immunological – by application of Improvac vaccine in accordance to the instructions for use. All animals were kept together and slaughtered at the same time but with different weights: from 100 till 120 kg. Carcass evaluation was made using Fat-o-Meater S71. The thickness of the backfat was measured at the level of 6-7 thoracic vertebra and at sacrum. Carcass grading has been done using EUROP and Ukrainian scale. Sensory evaluation of carcasses after slaughter was performed. The results suggest that increasing of live weight before slaughter of immunocastrated males till 120 kg does not influence negative their carcass quality. Dressing percentage of carcasses did not depend on live weight before slaughter in both groups. Carcasses of immunocastrated males had lower backfat thickness measured at 6-7 thoracic vertebra: slaughter at 100 kg – by 2.8 mm; slaughtered at 120 kg – by 3.3 mm. After slaughter at weight 120 kg 70% of carcasses of immunocastrated males were graded E + U, while in the group of surgical castrated males such carcasses were only 50%. It was not detected critical level of boar odour in all slaughter weights.

Session 17 Poster 16

Influence of housing conditions on antibody formation and testosterone after Improvac vaccinations

K. Kress, U. Weiler and V. Stefanski

University of Hohenheim, Behavioral Physiology of Livestock (460f), Garbenstr. 17, 70599, Germany; volker.stefanki@uni-hohenheim.de

Immunocastration of entire male pigs is a valuable tool to avoid surgical castration in pork production and to reduce both, problems in welfare due to aggressive and sexual behaviour and product quality as well. The amount of 'non-responders' to a twofold vaccination with Improvac is discussed controversially. In physiological studies a pronounced variation in the formation of Anti-GnRH antibodies may be observed, pointing to the fact, that the reaction is not an all-or-nothing phenomenon, but systematical studies on relevant environmental factors are missing. Thus in a current study the effect of standard housing conditions is compared to enriched conditions (organic production) and stressful housing (mixing of groups) on the formation of GnRH-antibodies after two Improvac vaccinations. Additionally the consequences for testosterone levels and aggressive behaviour are evaluated. First results of a larger study are presented. The study aims on the identification of factors reducing the effectiveness of immunocastration to improve the reliability of immunocastration under practical conditions.

Session 17 Poster 17

Effect of immunocastration on performance and fresh ham qualty of heavy gilts

D. Martin[1], C. Carrasco[1], M. Hebrero[2], M. Nieto[2], A. Fuentetaja[2] and J. Peinado[1]
[1]Imasde Agroalimentaria, S.L., C/ Nápoles, 3, 28224 Madrid, Spain, [2]Comercial Pecuaria Segoviana, S.L. (Copese), Camino Moraleja, 5, 40480 Segovia, Spain; dmartin@e-imasde.com

Immunocastration can be used as an alternative to surgical castration of male piglets to avoid boar taint. However, it can also be interesting in heavy gilts destined to dry-cured products industry to increase fat deposition and provide primal cuts with a higher added value. A total of 110 crossbred gilts (Duroc × Landrace × Large White) of 31.5±1.0 kg of initial BW were randomly allotted into two treatments: entire gilts (E-GI) and immunocastrate gilts (IM-GI). The experimental unit was the box with 11 gilts housed together (5 replicates per treatment). Feeding program was common for all the gilts and consisted of five diets offered *ad libitum* (2.44, 2.36, 2.46, 2.44 and 2.58 Mcal NE/kg and 1.28, 1.14, 1.02, 0.86 and 0.67% lys from 20 to 30, 30 to 35, 35 to 60, 60 to 90 and 90 kg BW to slaughter, respectively). Gilts BW was measured at nine moments (85, 93, 114, 128, 149, 166, 187, 201 and 219 days of age) and feed intake was recorded daily. The IM-GI group received two injections (114 and 149 days of age) of Improvac® (Zoetis-Pfizer; GnRH analogue protein conjugate) subcutaneously behind the ear. Gilts were slaughtered at 130.0±2.0 kg in a commercial slaughterhouse. Carcass weight, carcass yield and fat thickness, pH and T 24 h *post-mortem* of hams were measured in all the carcasses. Data were analysed as a completely randomised design by GLM of SPSS, including treatment as main effect. Carcass weight was used as covariable. Although no differences were found for feed conversion ratio between groups, IM-GI tend to eat more (2.124 vs 2.310 kg/d; P=0.08) and reached higher average daily gain for all the experimental period (0.696 vs 0.742 kg/d; P<0.05). This greater growth could be motivated by the higher consumption observed in IM-GI from the application of the second vaccine to slaughter. It resulted in a shorter fattening period and higher fat thickness of ham (22.5 vs 26.5 mm; P<0.05) for IM-GI. However, E-GI had higher carcass yield (80.1 vs 79.4%; P<0.08). It is concluded that IM-GI grew faster and showed hams with higher fat content; therefore they could be preferred for dry-cured products industry.

Session 17 Poster 18

Consumer's attitudes towards surgical castration of pigs in three Western Balkan countries

I. Tomasevic[1], S. Novakovic[1], I. Djekic[1], D. Nakov[2], L. Guerrero[3] and M. Font-I-Furnols[3]
[1]University of Belgrade, Faculty of Agriculture, Nemanjina 6, 11080 Belgrade, Serbia, [2]University St.Cyril and Methodius, Faculty of Agricultural Sciences and Food, blvd. Goce Delcev 9, 1000 Skopje, Macedonia, [3]IRTA-Food Industries, Finca Camps I Armet, 17121 Monells, Girona, Spain; tbigor@agrif.bg.ac.rs

Surgical castration of male piglets without anaesthesia is performed routinely by farmers or veterinarians in Western Balkan countries to eliminate the risk of boar taint in pork meat. The aim of the study was to investigate consumers' attitudes towards surgical castration of piglets in Serbia, Bosnia and Herzegovina (B&H) and Macedonia. A representative consumer survey was carried out in these three countries in 2017. Over twelve hundred (1,287) questionnaires were answered by pork eaters. Likert scale data were considered as ordinal values and non-parametric statistical tests have been used since data were not normally distributed. Mann-Whitney U test has been performed to compare the statements between genders and age and Kruskal-Wallis H test between countries (P=0.05). Western Balkan consumers agree (5.2±1.2) that surgical castration produces pain to the animal. They are significantly more likely to agree in Macedonia that meat from castrated pigs is of better quality (4.8±1.5) than in Serbia (4.6±1.3) and B&H (4.5±1.1). Macedonians agree the least (3.9±1.3) followed by more ambiguous consumers from Serbia (4.4±1.1) and B&H (4.5±1.0) that meat from castrated pigs is more expensive. Macedonians are also significantly more likely to disagree (3.6±1.6) that castration is not necessary, compared to Serbian (4.7±1.1) and B&H (4.9±0.9) consumers. On average, Western Balkan consumers, slightly disagree that meat from castrated pigs is leaner (3.7±1.1) and they neither agree nor disagree that they prefer to eat meat from castrated pigs (4.0±1.3) or that pig castration with vaccines improves pork quality (4.0±1.5). There are differences in consumer's attitudes towards surgical castration by country of origin, but further work is needed to find segments of consumers according to their attitudes.
The authors would like to acknowledge networking support by the COST Action IPEMA CA 15215 'Innovative approaches in pork production with entire males'.

Session 17 Poster 19

Quality and sensory evaluation of meat from entire males after adding hydrolysable tannins to diet

I. Bahelka[1], O. Bučko[2] and E. Hanusová[1]
[1]National Agricultural and Food Centre, Research Institute for Animal Production, Hlohovecká 2, 951 41 Lužianky, Slovak Republic, [2]Slovak Agricultural University, Trieda A. Hlinku 2, 949 01 Nitra, Slovak Republic; bahelka@vuzv.sk

The aim of study was to evaluate pork quality parameters as well as sensory properties of meat from entire male pigs after supplementation of diet with hydrolysable tannins. Thirty pigs – progeny of Landrace sows and Yorkshire × Pietrain boars – from 6 litters were assigned to 3 experimental groups (10 per group). They were housed in pairs in the pen and fed a commercial diet (12.8 MJ ME, 17.2% crude proteins). The trial started at 30 kg live weight and finished at 122±4 kg. The diets were supplemented with 0, 1 or 2% (T0, T1 and T2) of chestnut wood extract (Farmatan, Slovenia). A day after slaughter, pork quality parameters were measured and samples of *longissimus dorsi* muscle (100 g) for sensory evaluation as well as backfat for androstenone and skatole analyses were taken. Sensory properties of pork were evaluated 4 days after slaughter in a consumer test. Significant differences between T0 and T1, resp. T2 were found in electrical conductivity measured 24 hours in *musculus semimembranosus* (3.71 vs 2.27, 2.55 μS). Other pork quality traits were not influenced by tannin supplementation. Effect of the tannin addition on odour, flavour, juiciness and tenderness of pork was not significant when compared to the T0. However, tannins supplementation significantly reduced skatole concentration in fat tissue at 2% addition compared to control (0.011 vs 0.34 μg/g, $P<0.05$).

Session 17 Poster 20

About analysis of boar taint compounds in meat compared to subcutaneous fat

S. Ampuero Kragten and G. Bee
Agroscope, Animal Production Systems and Animal Health, Rte de la Tioleyre 4, 1725 Posieux, Switzerland; silvia.ampuero@agroscope.admin.ch

Most of the studies dealing with boar taint, whether from genetic, nutrition or management point of view, but especially for sensory and consumer tests, including meat products, are heavily based on quantification of boar taint compounds in subcutaneous fat. Indeed, boar taint compounds being strongly lipophilic tend to concentrate in fat hence facilitating their analysis. However, the fat content of meat products hardly goes beyond 30%, whereas that of any fresh meat cutlet can barely approach 20%, except for bacon. Hence the analysis of boar taint compounds in meat seems necessary. At the best of our knowledge, few attempts of performing the analysis of boar taint compounds in meat have been published. Among them, Rius and Garcia-Regueiro found Pearson correlation coefficients (r) between concentration in back fat and Longissimus Dorsi lean muscle (LD) of 0.63 and 0.91 for skatole and indole respectively, with however significantly reduced skatole and indole levels in LD. A similar observation was made by Wauters *et al.*, r ranging from 0.47 to 0.99, from 0.76 to 0.98 and from 0.82 to 0.95 for androstenone, skatole and indole respectively, again with considerably lower levels in meat than in back fat. 48 entire male pigs (Swiss large white breed) were reared at Agroscope and slaughter at 105 kg live weight and 170±5 d of age. LD muscle samples and subcutaneous fat from the neck region were analysed for androstenone, skatole and indole with HPLC reversed phase column; the method was adapted for the analysis of meat. Although good correlations were found between boar taint compounds in subcutaneous fat and meat ($r>0.8$), the concentration of these compounds in meat seems to be related to phospholipid content. These results raise further questions: The need for new threshold values and a better comprehension of sensory perception with regard to boar taint compounds measured in meat?

Sex neutralization of heavy pigs from Iberian Peninsula breeds: solutions and limitations

R. Charneca[1], F.I. Hernández-García[2] and M. Izquierdo[2]
[1]Instituto de Ciências Agrárias e Ambientais Mediterrânicas, Universidade de Évora, Apartado 94, 7000-803 Évora, Portugal, [2]Center of Scientific & Technological Research of Extremadura (CICYTEX), Autovía A-5, Km 372, 06187-Guadajira, Badajoz, Spain; rmcc@uevora.pt

This report gives an overview of the production system with Alentejano and Iberian swine breeds, the need for sexual neutralization of these pigs and the possible limitations related to the use of immunocastration (IC). The Alentejano (AL, in Portugal) and Iberian (IB, in Spain) pig breeds are genetically related and are produced under the same free-range system. For the certification of high grade dry-cured products like the 'Pata Negra' ham, the animals are slaughtered with an age of at least 14 months and a body weight range of 145-210 kg. During the fattening period pigs have access to acorns and grass from the Mediterranean forest. Until now the gonadectomy of both males (avoid boar taint and aggressive and sexual behaviour) and females (avoid mating by wild boars) is a common practice. The foreseen voluntary end of surgical castration (SC) without pain relief in the EU requires the use of alternatives in these swine breeds management. Taking into account that age and weight are key factors for the final products, the only options are SC with pain relief or IC. However, IC in these systems entails various difficulties and raise questions regarding effectiveness, practicability and effects. Studies in IB pig have shown that for females a 3 dose protocol starting before puberty is effective until the usual slaughter age, suppressing ovarian cyclicity. For males a 3 dose protocol is also needed but in this case the immunisation efficacy has been variable, although a 100% efficacy was recently reached with a protocol in which the 3rd dose was administrated before the acorn-feeding. The effects of IC on male meat quality seem limited, but no information is available regarding cured products. Also, no scientific studies on IC vaccine are available for AL breed. Further studies for protocol optimization and impact of IC on final high grade products from AL and IB pigs are needed.

Effect of essential oil addition on masking boar taint in fresh pork sausage

B. Šojić[1], V. Tomović[1], P. Ikonić[2], B. Pavlić[1], N. Džinić[1], N. Batorek-Lukač[3] and I. Tomašević[4]
[1]University of Novi Sad, Faculty of Technology, Bulevar cara Lazara 1, 21000 Novi Sad, Serbia, [2]Institute of Food Technology, Bulevar cara Lazara 1, 21000 Novi Sad, Serbia, [3]Agricultural Institute of Slovenia, Hacquetova ulica 17, 1000 Ljubljana, Slovenia, [4]University of Belgrade, Faculty of Agriculture, Nemanjina 6, 11080 Belgrade, Serbia; tbigor@agrif.bg.ac.rs

Boar taint is an off-flavour in the meat of entire male pigs, mainly related to the presence of androstenone and skatole. Approximately 99% of consumers are sensitive to skatole. In order to reduce boar taint in meat and meat products, different spices and plant extracts were used. The aim of this study was to evaluate the effect of Satureja montana essential oil addition on masking boar taint in fresh pork sausages. Fresh pork sausages were produced using meat and fat from castrates. Skatole (0.2 ppm and 0.4 ppm) and Satureja montana essential oil (0.150 µl/g) were added. The consumers (students and staff members of the Faculty of Technology Novi Sad) evaluated the samples on a nine point hedonic scale from dislike very much (1) to like very much (9). Control sample had significantly ($P<0.05$) higher overall liking score compared to all samples containing skatole. These results indicated that consumers can detect skatole odour and flavour at concentrations above 0.2 ppm. Satureja montana essential oil had a positive effect to mask boar taint in sausages produced with 0.2 and 0.4 ppm of skatole. The data from this study showed that addition of essential oils and plant extracts in fresh pork sausages may be an alternative solution for commercializing meat of entire male pigs. The authors acknowledge support from the cost action CA15215 IPEMA.

Session 17 Poster 23

Characterisation of eQTLs associated with androstenone, skatole and indole in porcine testis

M. Drag[1], L.J.A. Kogelman[2], H. Maribo[3], L. Meinert[4], P.D. Thomsen[1] and H.N. Kadarmideen[5]
[1]University of Copenhagen, Department of Veterinary and Animal Sciences, Grønnegårdvej 7, L106, 1870, Denmark, [2]University of Copenhagen, Department of Neurology, Danish Headache Center, Valdemar Hansens Vej 1-23, 2600 Glostrup, Denmark, [3]SEGES, Danish Pig Research Center, Axeltorv 3, 1609 København V, Denmark, [4]Danish Technological Institute, Danish Meat Research Institute, Gregersensvej 9, 2630 Taastrup, Denmark, [5]Technical University of Denmark, Department of Bio and Health Informatics, Kemitorvet building 208, 2800 Kgs. Lyngby, Denmark; hajak@bioinformatics.dtu.dk

Boar taint is an offensive odour/flavour from a proportion of non-castrated pigs (boars), which affects market price of the carcass. Castration is effective but under debate due to animal welfare concerns. As concentrations of the main compounds responsible for boar taint (androstenone, skatole and indole [ASI]) exhibit heritability, biomarker-assisted selection for low boar taint has been proposed but candidate biomarkers are lacking. The aims of this study were to identify expression quantitative trait loci (eQTLs) associated with ASI for candidate biomarker development and to generate more insights into underlying biological / metabolic pathways. RNA-Seq and genotypes were obtained from testis and loin. A high correlation was found between estimated breeding values (EBVs) of androstenone and androstenone concentrations (r^2=0.48) and a moderate correlation was found between carcass weight and ASI concentrations (r^2=0.32). A total of 213 eQTLs were associated with ASI concentrations, covering *MVP*, *COX1*, *MAP2K4* and *RDH16*, which encode proteins with functions within membrane, extracellular exosomes, ATP binding, MAPK cascade and oxidative stress. The highest densities of the eQTLs were on pig chromosomes (SSC) SSC18, SSC12, SSC2, SSC7 and SSC14. The genomic coordinates of 170 eQTLs enriched 42 known traits from PigQTLdb. The traits included health, meat quality and reproduction traits. A total of 33 candidate eQTLs enriched genomic coordinates for androstenone and smell intensity in fat. Our findings showed that biomarker discovery for boar taint is promising but careful monitoring of important productions parameters is necessary to avoid negative consequences of selection.

Session 17 Poster 24

Boar taint and meat quality characteristics of entire male and surgically castrated male pigs

I.G. Penchev[1], S. Ribarski[1] and T. Stoyanchev[2]
[1]Trakia University, Meat and meat products, Student's Campus, Trakia University, Faculty of Agriculture, 6000 Stara Zagora, Bulgaria, [2]Trakia University, Food safety and control, Student's Campus, Trakia University, Faculty of Veterinary medicine, 6000 Stara Zagora, Bulgaria; igpenchev@uni-sz.bg

The aim of the study was to evaluate the effect of castration on meat quality and boar taint with meat from entire male pigs. Forty-six (Landrace × Danube White) crossbred pigs were assigned to two experimental groups: entire males (EM) and surgically castrated males (CM). Pigs were reared in two pens per sex and slaughtered at an average of 150-160 days of age. The values of pH_1 (45 min post mortem) and boar taint were evaluated at slaughterhouse. *M. Longissimus thoracis et lumborum* and *M. Semimembranosus* were selected for sampling and measurement of meat quality. The values of pH were measured by a digital pH-meter 'Testo 205' (Testo Inc., United States). Boar taint detection was estimated on the slaughter line, by the trained person and by the 'hot iron' technique described by Aluwé *et al.*. The water-holding capacity was determined by the compression method. A Minolta Chroma Meter (model CR-400, Minolta Co., Ltd., Osaka, Japan) with 8 mm measuring head at D65 lighting was used for Minolta L*(lightness), a* (redness), and b* (yellowness) measurements. Cooking losses were assayed by cooking at 150 °C for 20 minutes. Marbling was scored on a 1 to 10 scale based on the standards set by the National Pork Producers Council. There wasn't significant difference in chemical composition on meat between EM and CM. EM showed higher water and lipid content in comparison with CM. The values of pH measured in the two sex are minimal and not significant. 51,8% of EM showed boar taint in comparison with CM respectively 31.6%. There were no significant effects of animal type on colour of meat. Assay of meat quality showed cooking losses from 36.2 to 36.5% and WHC from 20 to 23%, but with no significant differences between EM and CM. Intramuscular fat was higher in EM (1.6%) with respect to CM (1.0%) (P<0.001). Regarding ours results, we can conclude that from the point of view of meat quality breeding of entire male pigs could be a good decision to the surgical castration.

Session 17 Poster 25

Retrospective by sex of the new EU lean meat content of pig carcasses

G. Daumas
IFIP-Institut du Porc, BP 35104, 35601 Le Rheu Cedex, France; gerard.daumas@ifip.asso.fr

A new EU regulation on carcass classification applies from July 2018. This regulation changes the definition of the reference lean meat percentage (LMP) which becomes the LMP in the carcass (from total dissection). Nevertheless, each Member State can choose when he will update its grading methods. The aim of this work is to simulate this change on the last 5 years in France for each sex. Conversely to most of European countries sex is registered online in France during pig classification. Statistics per sex are regularly published by the regional classification organisations. The production of entire males started in France in 2013 (13% of the males), increased quickly and is now stable (27% of the males in 2017), around 2.5 million a year. More than 95% of entire males are classified with the classification method CSB Image-Meater® (IM) approved by the EU in 2013. The present LMP prediction equation contains two fat depths (G3 and G4) and two muscle depths (M3 and M4). A sample of 180 pigs was uniformly stratified on sex: 60 entire males, 60 females and 60 castrated males. All cuts were CT scanned allowing to calculate the LMP. Prediction equation was calculated by a general linear model including the 4 depths as well as the interactions with sex. The stepwise procedure by using BIC selected a model with 1 fat depth (G3) and 2 muscle depths (M3 and M4), G3 and M3 coefficients depending on sex. The RMSE was 2.15. Removal of M4 only decreased the RMSE of 0.01. The fat coefficient of entire males was more than twice that of females, while that of castrates was in the middle. Differences on M3 coefficient had lower impact on LMP. The equation was applied on the annual classification averages from 2013 to 2017. In 4 years LMP increased, but differently according to sex: +0.11 for entire males, +0.25 for females and +0.34 for castrated males. In the same period the difference between entire and castrated males decreased from 4.02 to 3.79, while the difference between entire males and females decreased from 1.37 to 1.24. When the Image-Meater method will be updated, it seems worthwhile to consider separate slopes for entire males, females and castrated males.

Session 17 Poster 26

Sensory analysis of cooked ham from entire pigs raised with different feeding and housing conditions

R. Pinto, N. Reis, C. Barbosa, R. Pinheiro and M. Vaz Velho
Instituto Politécnico de Viana do Castelo, Av. Atlântico, Viana do Castelo, Portugal; rpinto@ipvc.pt

This study aims at evaluating sensory quality of cooked ham produced with entire male pigs raised under specific conditions in order to reduce or eliminate the boar taint. Meat from entire male pigs raised under six different conditions (normal housing versus improved housing, fed with different levels of added inulin: 0, 3 and 6%) was used to produce six batches of cooked ham, processed under the same conditions. Samples were coded as N0%, N3%, N6%, C0%, C3% and C6%, where 'N' means normal housing, 'C' improved housing and 0, 3 and 6 correspond to percentage of added inulin in animal feed. A quantitative descriptive analysis was performed by 11 trained panellists, assessing odour and flavour of skatole and androstenone, texture and appearance, on a 1 to 10 scale. Perception of defects on overall appreciation and appearance was evaluated on a 1 to 5 scale, where scores below 3 meant there was a perceived problem. A total of eight coded samples, with the six conditions and two replicates to evaluate repeatability. ANOVA with a *post hoc* Fisher's LSD test was used to investigate significance of observed differences and Canonical Variates Analysis (CVA) with judges as replicates, was carried out to discriminate groups of samples. Results show significant differences ($P<0.05$) in the evaluation of attributes such as flavour and odour of androstenone and skatole, being N0%, N3% and N6% the samples with highest intensity scores meaning they had stronger boar taint odour and flavour. CVA separates N0% from the other for almost every conditions. Ham from animals with improved housing conditions had better appearance scores than normal housing. No differences were found in texture and overall appreciation. It can be concluded that the effect of inulin addition in pig diets was not clear, but it was found that its addition could have positive effects when combined with better housing conditions. As expected, the presence of boar taint was perceived in meat samples where no addition of inulin and no improved conditions were applied. Acknowledgments POCI-01-0247-FEDER-017626 – PIGS+CARE, co-financed by FEDER through COMPETE 2020 – Operational Program for Competitiveness and Internationalization.

Session 17 Poster 27

Effect of the method of castration on growth performance and boar taint

K. Zadinova, R. Stupka, J. Citek and N. Lebedova
Czech University of Life Sciences Prague, Department of Animal Husbandary, Kamycka 129, 165 00 Prague, Czech Republic; zadinova@af.czu.cz

The method of castration can affect growth performance and the quality of pork meat. Compared with barrows and gilts, fattening entire male pigs appears to be more profitable, mainly due to better growth efficiency due to improved feed conversion and improved carcass leanness. However, the occurrence of 'boar taint' in pork of boars represents a significant problem. Boar taint can be eliminated by castration or immunocastration. The latter prevents the occurrence of boar taint while preserving the positive effects of testicular steroids and anabolic hormones occurring in males before second vaccination. For this study a total of 52 pigs of the D × (LW × L) crossbreed were assigned into 3 groups: boars (n=18); immunocastrates (n=16); barrows (n=18). Barrows were surgically castrated on day 3 of age. The immunocastrates were injected with Improvac® when they were 94 and 115 days old. All pigs were fed *ad libitum* a commercial diets. The pigs were slaughtered at 154 days of age. Androstenone and skatole levels were measured according to Okrouhlá *et al*. The following traits were monitored and calculated weekly: body weight (kg), daily feed intake (kg/d) and average daily gain (g/d), feed conversion ratio (kg/kg). In addition, carcass weight (kg), lean meat (%) and backfat thickness (mm) were measured at slaughter. The data were analysed using the GLM procedure in SAS. Overall growth performance did not differ between immunocastrates and boars. On the other hand, significant differences between barrows and the other groups were observed. The greatest backfat thickness, the lowest lean meat percentage and the greatest daily feed intake ($P<0.05$) were observed in barrows. Acknowledgments This study was supported by the Internal Grant Agency of the CULS (CIGA) (Project No. 20172005) and the EU Framework Programme for Research and Innovation Horizon 2020 (COST action IPEMA, CA15215).

Session 18 Theatre 1

Guidelines for recording, validation and use of claw health data

N. Charfeddine[1], B. Heringstad[2], K.F. Stock[3], M. Alsaaod[4], M. Holzhauer[5], G. Cramer[6], J. Kofler[7], N. Bell[8], G. De Jong[9] and C. Egger-Danner[10]
[1]CONAFE, Spanish Holstein Association, Valdemoro, 28340, Spain, [2]Norwegian University of Life Science, As, 1430, Norway, [3]IT Solutions for Animal Production, Verden, 27283, Germany, [4]Vetsuisse-Faculty, University of Bern, Bern, 3012, Switzerland, [5]GD Animal Health, Deventer, 7418 EZ, the Netherlands, [6]College of Veterinary Medicine, University of Minnesota, St. Paul, MN 55108, USA, [7]University of Veterinary Medicine Vienna, Vienna, 1210, Austria, [8]Herd Health and Production Medicine, University of Nottingham, Nottingham, NG7 2RD, United Kingdom, [9]CRV, Aenhem, 6843 NW, the Netherlands, [10]ZuchtData EDV-Dienstleistungen GmbH, Vienna, 1200, Austria; nouredine.charfeddine@conafe.com

Since 2015 ICAR working group of functional traits and claw health experts have been working in setting guidelines for claw health data. The aim of this contribution is to present the guidelines which will be available on ICAR website soon. The purpose of these ICAR guidelines is to provide recommendations on recording, data validation and use of claw health information. We focus mainly on claw trimming data, even though data could be recorded by farmers or be veterinary diagnoses. Data validation consists in two main steps, data screening and data verification. The exhaustiveness and the completeness of each process depend on the purpose of use and data sources. Editing criteria have been suggested at trimmer/vet, herd, animal and record level. When it comes to the use of claw data, health status of each cow provides an important insight into the health status of the entire herd which makes claw health monitoring feasible. Herd management reports should answer whether or not claw health status has changed and whether the herds trimming goal was met. The use of claw data for benchmarking might be a useful tool in order to know the current performances, to motivate producers to adopt preventive practices and to foster claw health documentation. Several indicators and parameters have been defined in this guideline as how to calculate incidence and prevalence rates, how to define cows at risk and time period at risk, how to identify new lesion and chronic cow, which are essential concepts for working with claw health data.

Session 18 Theatre 2

Associating types of hoof disorders with mobility score of dairy cows in pasture-based systems

A. O Connor[1,2], E.A.M. Bokkers[1], I.J.M. De Boer[1], H. Hogeveen[3], R. Sayers[2], N. Byrne[2], E. Ruelle[2] and L. Shalloo[2]
[1]Department of Animal Sciences, Animal Production systems, Wageningen University & Research, the Netherlands, [2]Teagasc, Animal & Grassland Research & Innovation Centre, Moorepark, Fermoy, Co. Cork, Ireland, [3]Department of Social Sciences, Business Economics, Wageningen University & Research, the Netherlands; aisling.oconnor@teagasc.ie

Sub-optimal mobility can be broadly defined as abnormal gait which causes a deviation from the optimal walking pattern of a cow. It is widely accepted that hoof disorders (HD) have an impact on mobility, however their association to mobility score (MS) has not yet been extensively researched. This study aims to characterise sub-optimal mobility, by using prevalence by type of HD to predict MS. A dataset with 7,602 cows from 52 spring-calving pasture-based dairy farms in Ireland was used. MS was recorded for all cows and HD data included two types; (1) non-infectious HD, i.e. overgrown claw, bruised claw, whiteline disease and ulcer and (2) infectious HD, i.e. digital dermatitis and interdigital phlegmon. For this study the DairyCo mobility scoring method was used. The effects of the different HD on MS were assessed statistically using a forward stepwise regression approach. Output variables were analysed with logistic regressions as multinomial (MS 0, 1, 2, or 3, where 0 is good and 3 is bad mobility). The presence/absence of each HD type was included in the model to predict cow MS. Analyses were performed using R statistical software (functions multinomial logistic regressions). Non-infectious type HD affected 59.10% of all the cows. Non-infectious type HD and had an impact on MS wherein a cow with one or more non-infectious type HD was more likely to have a MS>0 versus MS=0 compared to a cow with no non-infectious type HD at all levels of MS (P<0.001). Infectious type HD only affected 3.03% of the cows, yet had greater impact on MS (in terms of odds ratios) at all levels of MS (P<0.001). This analysis confirms an association between MS and HD. The results indicate that any form of HD presence is a relevant predictor of MS. From the results, it is clear however that infectious type HD, (although they are far less prevalent in cows in pasture-based system) have a significantly greater impact on MS.

Session 18 Theatre 3

Are sole ulcer and white line disease causes of lameness in finishing beef cattle?

L. Magrin, M. Brscic, G. Cozzi, L. Armato and F. Gottardo
University of Padova, Department of Animal Medicine, Production and Health, Viale dell'Università 16, 35020 Legnaro (PD), Italy; luisa.magrin@studenti.unipd.it

Sole ulcer (SU) and white line disease (WLD) are 2 of the most debilitating and painful claw lesions in dairy cows related to almost 30% of lameness events and largely consequences of metabolic disorders and mechanical loading. The aim of this study was to assess *post-mortem* the prevalence of SU and WLD on hind feet of finishing beef cattle and to test their variation among breed, sex and time point. The study was carried out in 2 abattoirs in Northern Italy at 3 time points: first (April-June 2016), second (September-October 2016), and third (February-March 2017). A total of 4,292 hind feet were inspected belonging to 2,748 bulls and 1,544 heifers from 153 batches (animals/batch 14.0±2.8). Feet were collected after slaughter. A vet trimmed the claws with a grinder, then identified claw disorders using the ICAR Claw Atlas classification and classified them for the affected site (lateral/medial) and sole zone (1 to 6). Effects of breed, sex and time point for all lesions (expressed as % at batch level) were assessed using the non-parametric Kruskall-Wallis test. The 52.3% of batches showed at least one WLD in the lateral claws and the prevalence of affected feet within batch ranged from 0 to 50%. Prevalence of feet having at least one WLD in the lateral claws differed among breeds, being lower in Limousine than in other breeds (Charolaise, French and Italian crossbreds and other Italian breeds). In the medial claws, WLD was detected in 13.7% of batches, with a prevalence of affected feet within batch ranging from 0 to 17%, similar among breeds. Zone 3 of the sole resulted the most frequently affected by WLD in both claws. Sole ulcer was detected in 2% of batches in zone 3 and 4 of the lateral claws and in zone 4 of the medial ones. Prevalence of feet affected by SU within batch varied from 0 to 3% in both claws resulting similar among breeds. Neither sex nor time point affected the prevalence of feet showing WLD and SU in both claws. Results point out similar prevalence of SU and WLD on apparently healthy beef cattle to those on dairy cows suggesting that also in finishing cattle these disorders might lead to potential painful conditions and performance losses.

Pathogen-specific production losses in bovine mastitis

A.-M. Heikkilä[1], E. Liski[1], S. Pyörälä[2] and S. Taponen[2]
[1]Natural Resources Institute Finland (Luke), P.O. Box 2, 00791 Helsinki, Finland, [2]University of Helsinki, Faculty of Veterinary Medicine, P.O. Box 57, 00014 Helsinki, Finland; anna-maija.heikkila@luke.fi

Prevention and treatment of mastitis are the main reasons for antimicrobial drug use for dairy cows. Efficient and responsible measures to control mastitis require pathogen-specific information on the disease. Generally, reduction in long- term milk yields represents a notable share of the economic losses caused by bovine mastitis. The aim of this study was to determine the production loss caused by the six most common udder pathogens in Finnish dairy herds: non-*aureus* staphylococci (NAS), *Staphylococcus (Staph.) aureus, Streptococcus (Strep.) uberis, Strep. dysgalactiae, Corynebacterium (C.) bovis,* and *Escherichia (E.) coli*. The materials consisted of milk and health recordings and microbiological diagnoses of mastitic quarter milk samples of 20,580 dairy cows in 2010-2012. Only one pathogen was detected in the milk samples from a cow that were accepted into the data. We used a multilevel model to estimate the lactation curves for lactations with and without mastitis for comparison. Mastitis caused by each pathogen resulted in milk production loss. The extent of the reduction depended on the pathogen, timing of mastitis during lactation, and the type of mastitis (clinical (CM) vs subclinical (SCM)). *E. coli* CM, diagnosed before peak lactation, caused the largest loss, 10.6% of the 305-d milk yield (3.5 kg/d). The corresponding loss for *Staph. aureus* CM was 7.1% (2.3 kg/d). The loss was almost equal in CM and SCM caused by *Staph. aureus*. Mastitis caused by *Strep. uberis* and *Strep. dysgalactiae* resulted in a loss ranging from 3.7% (1.2 kg/d) to 6.6% (2.0 kg/d) according to type and timing of mastitis. CM diagnosed before peak lactation and caused by the minor pathogens, *C. bovis* and NAS, also had a negative impact on milk production: 7.4% (2.4 kg/d) and 5.7% (1.8 kg/d), respectively. In conclusion, the minor pathogens should not be underrated as a cause of milk yield reduction. On single dairy farms, control of *E. coli* mastitis would bring about a significant increase in milk production. Reducing *Staph. aureus* mastitis is the greatest challenge for the Finnish dairy sector.

Reducing antibiotic use: essential oils as potential alternative treatments for ovine footrot

A. Westland and A. Smith
University Centre Myerscough, Agriculture, St Michaels Road, Bilsborrow, Preston, Lancashire, PR3 0RY, United Kingdom; awestland@myerscough.ac.uk

Ovine footrot is a major cause of lameness in the UK sheep flock accounting for 80% of lameness cases costing between £24-80 million p.a. and considered one of the top five globally economic important diseases. Recent studies have shown that 'best practice' for controlling this infectious bacterial disease is prompt treatment with parenteral and topical antibiotic but with the challenge of antibiotic resistance there is a need to find alternative treatments. Organic farmers, pharmaceutical and health industries have all noted strong anti-bacterial and anti-microbial properties of certain oils such as Tea Tree, Citronella and Bergamot, so the aim of this trial was to investigate the potential of these oils to control the growth of *Dichelobacter nodosus* which causes clinical foot rot. Isolating *D. nodosus* from clinical cases is difficult as it cannot colonise without prior infection by *Fusobacterium necrophorum*. An *in vitro* investigation was therefore carried out to determine the efficacy of Tea Tree (TT), Citronella (Ci) and Bergamot (Bg) oils to combat the growth of *F. necrophorum* grown on trypticase soy agar enriched with 5% sterilized horse blood. The essential oils were diluted with a carrier mineral oil to achieve 1 and 10% concentration, with Bergamot being additionally diluted to 0.1% concentration. Each dilution was applied at 1 ml per plate, 21 plates per treatment, sealed in a gas pack incubator at 37 °C for 4 days under anaerobic conditions to allow bacterial growth. Colony forming units (cfu) were counted and compared to 19 control plates of a pure culture *F. necrophorum* diluted at 1001. All three oils had a significant effect (P<0.0001) in reducing the number of colonies of *F. necrophorum* when applied at 10% (TT=33 cfu, Ci=47 cfu, Bg=5 cfu) and at 1% (TT=45 cfu, Ci=200 cfu, Bg 70=cfu) when compared to the control at 797 cfu. Bergamot oil applied at 0.1% concentration was still significantly effective at reducing colony count (Bg 90=cfu, P<0.0001). In conclusion, Tea Tree, Citronella and Bergamot oils were all considered effective treatments against *F. necrophorum* cultured *in vitro*.

Effects of protein sources in calf starter on health-related parameters in plasma

S.H. Rasmussen[1], T. Larsen[1], C. Berthelsen[1], C. Brøkner[2] and M. Vestergaard[1]
[1]Aarhus University, Department of Animal Science, Foulum, 8830 Tjele, Denmark, [2]Hamlet Protein A/S, Horsens, 8700, Denmark; zqw841@alumni.ku.dk

Unprocessed soybeans are not well-digested in unweaned calves so toasting is used to reduce the level of anti-nutritional factors, e.g. trypsin inhibitors and lectins, and to improve digestibility of soybean meal used in calf starters. However, further processing of soybeans by use of aqueous ethanol extraction, fermentation, or enzyme treatment to reduce the level of heat stable anti-nutritional factors, e.g. oligosaccharides, glycinin, and β-conglycinin, can further improve digestibility. We tested a commercial soy protein product in the concentrate offered to milk-fed dairy calves to study the effects on metabolic and health-related parameters in plasma. A total of 32 calves were purchased at 12.8±0.6 d of age and 49.6±1.1 kg of BW (means ± SE) from 4 commercial herds. Calves were divided into 2 blocks of 16 based on herd, BW and age. Within block and herd, calves were randomly allocated on 2 treatment groups; control (CON) and test (TEST). From 2 wk of age CON was offered a traditional calf starter with 30% soybean meal and TEST was offered a calf starter with 23% HP Rumen Start (Hamlet Protein A/S). Both starters had (on a DM basis) the same CP (24%), starch (27-30%), aNDF (14%) and Net Energy (7.8 MJ) content. Calves were offered skim-milk-based milk replacer (up to 8 l/d), grass hay and calf starter, were weaned at 8 wk of age, and followed until 10 wk of age. Blood samples were obtained at 2 (before start), 4, 6 and 10 wk of age and plasma was analysed. There were no differences between CON and TEST at 2 wk of age, except for a 23% lower haptoglobin in TEST ($P<0.05$). There was an effect of wk of age ($P<0.05$ to 0.001) but no effect of treatment ($P>0.05$) on albumin, total protein, IgG, IgA, IgM, NEFA, glucose, BOHB, alfa-2 macro-globulin and haptoglobin across the 4, 6 and 10 wk samples. TEST had 23% lower TNF-alfa than CON ($P<0.05$). For urea, there was a wk × treatment interaction ($P<0.05$), showing highest urea on CON at 10 wk in agreement with a higher concentrate intake at this stage (2.6 vs 2.4 kg/d). The results suggest that soy-protein source in these two starter concentrates only marginally affect metabolic and health-related parameters in dairy calves.

Bovine appeasing pheromone to decrease stress and BRD in beef cattle

B. Mounaix[1], M. Guiadeur[1], L. Michel[2], J. Boullier[3] and S. Assie[3]
[1]Institut de l'Elevage, Santé et bien-être des ruminants, 149 rue de Bercy, 75595 PARIS Cedex, France, [2]TERRENA, Innovation, La Noëlle, 44150 Ancenis, France, [3]ONIRIS-INRA, UMR 1300, La Chantrerie, 44307 Nantes, France; beatrice.mounaix@idele.fr

Bovine respiratory disease (BRD) is the most frequent disease encountered by beef calves when entering feedlots and accounts for about 3% of mortality. It is responsible for the main use of antibiotic molecules in the fattening industry: 20% of fattening beef calves receive antibiotics for the cure of clinical bronchopneumonia infections (BRD). Stress is a major factor in the apparition and the development of respiratory disease in fattening cattle by impeding the immune system efficiency. An on-farm experiment was conducted in 2017 to investigate effects of bovine appeasing pheromone (BAP) on BRD prevalence in commercial conditions. A total of 265 Charolais male young cattle (BW=366 kg) were treated with BAP (pour-on) or placebo (Transcutol[Ò], pour-on) before entering fattening lots in 4 farms in the North-West of France in winter with higher BRD risk conditions. The effect of BAP on stress and BRD was evaluated trough behavioural and clinical observations. Permutation statistics were necessary to account for farm, lot, number of cattle/lot and day effects and crossed effects in behaviour data. Health data were processed with mixed regression and hierarchical classification. BAP treated cattle showed higher activity (exploratory motion behaviour) during the first week of fattening ($P=0.001$), but no significant difference in eating and agonistic behaviour was observed. This may indicate that when the level of stress is low, cattle adapt better. However, effects on growth performance were low. Our results also showed 10% less BPI clinical signs (veterinary examination + treatment) at 30 days of fattening ($P=0.001$), whereas no significant difference was observed at 8 days. This suggests that decreasing stress is a promising way to reduce the incidence of respiratory diseases, but BAP only is probably not efficient enough for farmers to diminish first-line antibiotic treatment.

Climate change and animal disease: Vectors and vector borne pathogens in Croatia

R. Beck[1], T. Šarić[2], S. Bosnić[1] and R. Brezak[1]
[1]Croatian Veterinary Institute, Savska cesta 143, 10 000 Zagreb, Croatia, [2]University of Zadar, Trg kneza Višeslava 9, 23 000, Croatia; relja.beck@gmail.com

Globalization and climate change have an unprecedented worldwide impact on emergence and re-emergence of animal diseases, especially vector borne diseases. Climate change is transforming natural ecosystems and providing more suitable environments for infectious diseases allowing the movement of bacteria, viruses, parasites and fungi into new areas where they can infect wildlife and domestic species, as well as humans. Diseases that used to be limited only to tropical areas are now increasingly spreading to other previously unaffected region as is happening in region and Croatia. The aim of this manuscript is to present new insights into vectors and emergent/re-emergent vector borne pathogens in Croatia. Bluetongue virus (BTV) serotypes 1 and 16 were introduced in Oryx antelope from the Sultanate of Oman during the quarantine period on the Island of Veliki Brijun (Croatia) in 2010. In 2002 first outbreak was caused with serotypes 9 followed by serotype 16 in 2004. Serotypes 4 and 1 were responsible for recent outbreak in 2014. Except predominant vectors from *Obsoletus* complex and *Pulicaris* complex several 'new' species were molecularly confirmed but their vector capacity is unknown. In several wild ungulates *A. bovis* and *A. centrale* have been detected in the Croatian costal region. Both pathogens cause bovine anaplasmosis, economically one of the most important diseases. *Anaplasma ovis* was recently confirmed in sheep flock and questing *Rhipicephalus bursa* and ticks collected from sheep and goats from same region. Presence of Babesia cf. crassa in *Rhipiephalus sanguineus* from Pelješac and questing *Hamaphisalis parva* from Slavonia represent one of first findings in Europe. Theileria cf. buffeti and Babesia sp. Angola Isolate were detected in *I. ricinus* tick from Cres and *Rhipicephalus turanicus* from Pelješac. In *R. turanicus* and *R. bursa* ticks as well in sheep *Theileria* ovis and *Babesia ovis* have been confirmed from Southern Croatia. For the end vaccination of cattle against Lumpy skin disease was mandatory for last two years due to outbreaks in neighbouring countries. The improvement of epizootic understanding together with better programs for diagnose, prevention and trace pathogens, their origin and routes of infection, play a key role in future prevention of vector-borne diseases in Croatia as well in worldwide.

Case report: First description and evidence of *Anaplasma ovis* infection in the sheep in Croatia

T. Šarić[1] and R. Beck[2]
[1]University of Zadar, Trg kneza Višeslava 9, 23000 Zadar, Croatia, [2]Croatian Veterinary Institute, Savska cesta 143, 10000 Zagreb, Croatia; tosaric@unizd.hr

Anaplasma ovis is gram-negative rickettsial bacteria transmitted by different tick species. Usually it causes only mild clinical symptoms in small ruminants, but in immunosuppressed animals bacteria can cause severe clinical symptoms even with lethal outcome. Infections occur in certain areas in southern Europe (Italy, Greece, France, Portugal) and so far it has not been proven on the Eastern coast of the Adriatic Sea. Because of clinical signs of piroplasmosis blood sample from a ram was analysed with PCR for *Babesia/Theileria* species as well as for *Anaplasma sp.* Frequent occurrence of clinical signs that correspond to the sheep piroplasmosis, in the last years in Zadar County, induced us to conducted field study. After collecting anamnestic data from veterinary practitioners, we have found out that clinical signs usually include fever, apathy and anaemia as well as lack of imidocarb-diproprionate efficiency in sick animals. Blood samples were collected from another ram presenting identical clinical signs such as fever, apathy, anaemia and icterus, as well as samples from all 13 sheep in the herd. Rams were treated with long-acting oxytetracycline and after few days they fully recovered. Polymerase Chain Reaction and subsequent sequencing of 16 S rRNA and groEL showed 100% similarity with *A. ovis* in another ram with clinical signs disease. Also infection with *A. ovis* was found in 92% of healthy sheep from the same herd. Repeated blood sampling and molecular analysis from same animals after 7 months showed persistent infection of *A. ovis* in all examined sheep. Identical sequences from *A. ovis* were also detected in *Rhipicephalus bursa* ticks from same region. Despite the fact that sheep anaplasmosis is considered as 'tropical' disease and global climatic changes could affected the distribution of vectors and possible spread of this pathogen we believe that *A. ovis* present an 'autochthonous' species. The presented data in this research support this theory, as usually rams exhausted during the breeding season develop disease. This study presents the first evidence of asymptomatic infection in sheep and clinical signs in rams from Croatia.

Insights in the genetics of paratuberculosis in the goat

G. Minozzi[1,2], F. Palazzo[1,3], E. Pieragostini[4], F. Petazzi[4], S. Biffani[5], V. Ferrante[2], G. Gandini[2], G. Pagnacco[2] and J.L. Williams[6]

[1]Parco Tecnologico Padano Srl, Via Einstein, 26900 Lodi, Italy, [2]Università degli Studi di Milano, DIMEVET, Via Celoria 10, 20133 Milano, Italy, [3]Università di Teramo, Faculty of Bioscience and Technology for Food, Agriculture and Environment, via R. Balzarini 1, 64100 Teramo, Italy, [4]Università degli Studi di Bari Aldo Moro, Piazza Umberto I, 70121 Bari, Italy, [5]IBBA-CNR Milano, Via Bassini 15, 20133 Milano, Italy, [6]University of Adelaide, School of Animal and Veterinary Sciences, Roseworthy, SA 5371, Australia; giulietta.minozzi@unimi.it

Mycobacterium avium ssp. *paratuberculosis* (MAP) causes chronic enteritis in different ruminant species. In goats, MAP causes a chronic disease called Johne's disease, or paratuberculosis, characterised by ileal lesions that limit nutrients absorption, leading to weight loss, rough hair coat, and skin peeling. Several genome wide association studies, over the last few years, have identified genetic loci putatively associated with MAP susceptibility in cattle and sheep. The aim of this work was to perform a case-control study in the caprine species, using the 50K SNP panel (Illumina CaprineSNP50 BeadChip), to identify genes underlying the mechanisms of susceptibility to Johne's disease. During 2011- 2015 in Italy more than 9,000 goats have been tested for antibodies against MAP. In goat herds with high occurrence of Johne's diseases, 352 animals have been selected for the association study. All samples were classified based on the serum antibodies produced in response to MAP, 163 MAP ELISA positive and 163 negative. Whole genome association analysis was performed using the R package GenABEL with the Grammar-CG approach. The analysis identified two chromosomal regions associated with the ELISA status on chromosomes 16 and 27 with high significance ($P<5\times10^{-6}$). These results provide evidence for genetic loci involved in the antibody response to MAP in goats.

Vaccination: still practical roadblocks to overcome in beef cattle farms

B. Mounaix

Institut de l'Elevage, Santé et bien-être des ruminants, 149 rue de Bercy, 75595 Paris Cedex, France; beatrice.mounaix@idele.fr

Bovine Bronchopneumonia Infections (BPI) is the most frequent disease encountered by beef calves when entering feedlots. Known primary infections by viruses predisposing to BPI include: BRSV, PI-3, bovine coronavirus and BVDV. *Mannheimia haemolytica* and *Pasteurella multocida* are the most often encountered bacterial agents of BPI. Bacteria are usually commensal in bovine upper respiratory system and opportunistically spreads up to the lung after viral infection leading to lung tissue destruction and inflammation. Antibiotics are then need to stop the replication of bacteria into the lung, to restrict the damages of lung tissues and to prevent the animal to die, but collective treatments lead to growing risks of antibio-resistance. Vaccination for viruses, and to a lesser extent bacterial pathogens is one solution, but it is most often implemented too late in the French beef cattle industry. A survey was conducted in 2015 in 32 French beef cattle farms and their veterinaries to explore points of view, obstacles and incentives associated with vaccination in beef cattle. It showed that 90% of farmers had a good perception of vaccination (generally speaking) and almost 70% found it worth the cost. They acknowledged the preventive role of vaccination, but still implemented it after first sick animals show up. Although they considered it efficient (60%), a majority of them wouldn't vaccinate beef cattle at weaning because of no expected positive cost/benefit balance for them. But text-analysis of answers showed that perceived obstacles to vaccination at weaning were mainly practical: difficulty of vaccination at pasture and uncertainty of young cattle marketing dates preventing to complete vaccination. Veterinarians also expressed difficulties in advising farmers on that specific matter and occasional reluctance of farmers to change working habits to implement vaccination. All surveyed stakeholders shared strong expectations towards a new chain of quality between nursing farms and fattening lots. New management systems to make vaccination possible look like a promising solution for the cattle fattening production chain. One is currently being experimented in France.

Session 18 Poster 12

Effects of age and physiological state of cows on selected haematological parameters

W. Krawczyk, P. Wójcik, I. Radkowska and A. Szewczyk
National Research Institute of Animal Production, Sarego 2, 31-047 Kraków, Poland; wojciech.krawczyk@izoo.krakow.pl

The aim of this study was to determine the effects of age and physiological state on some haematological and biochemical parameters of dairy cows. Blood samples for analysis were collected from 300 cows, in the period at around parturition and then on the 200 day of lactation. The results differed between the groups, but all were within the scope of national veterinary standards. The values of red blood cells (RBC), haemoglobin and haematocrit were usually higher in animals on day 200 of lactation, while in perinatal cows the RBC parameters decreased and the number of the white blood cells (WBC) increased. It was found that the number of neutrophils decreased with the age of dairy cows, but increased when the level of cortisol and adrenaline increased. In healthy cows in a state of equilibrium, the number of lymphocytes increased with age. Similarly, the average cell volume (MCV) changed with the effort and age of the cow, reaching higher values in older cows ($P \leq 0.01$). The level of stress hormones indicated the cow's ability to adapt to living conditions. Changes in levels of ACTH, cortisol and TSH depended on the physiological states of the cows. At around parturition their level tended to increase regardless of the age of the animals. It was found that the level of these hormones was lower for longevous cows, which shows that they were much more resistant to stress resulting both from innate psychophysical constitution and individual experience acquired throughout life ($P \leq 0.01$). In conclusion, longevous cows were characterized by a much better adaptation to environmental conditions and better resistance compared to other short-lived ones, which is confirmed by the haematological and biochemical parameters. The features of functional longevity of cows should be promoted, and the genetic material of these cows used to create maternal lines.

Session 18 Poster 13

Post-mortem rumen mucosa, lung and liver alterations in intensively finished beef cattle

M. Brscic, L. Magrin, I. Lora, G. Cozzi and F. Gottardo
University of Padova, Department of Animal Medicine, Production and Health, Viale dell'Università 16, 35020 Legnaro (PD), Italy; luisa.magrin@studenti.unipd.it

This study aimed at assessing *post-mortem* the prevalence of rumen mucosa, lung and liver alterations at slaughter as a welfare monitoring tool for intensively finished beef cattle and to study their variation among breeds and sex. It was carried out in 2 abattoirs in Northern Italy. Two trained veterinarians inspected 15 animals/batch directly in the slaughter-line and recorded presence of redness of mucosa, signs of ruminitis, hyperkeratosis, plaque, star scar, ulcer, adherence on rumen wall and parasitosis on rumen; signs of pneumonia, presence of fibrin filaments and of pleuritis on lungs; and signs of steatosis, parasitosis and adherence or abscess on livers as binary. Data were expressed as percentages at batch level. A total of 2,161 animals belonging to 153 batches (97 bulls; 56 heifers) were observed during 30 slaughter days. Percentages of rumens with plaque, star scar and adherence; of lungs with severe pneumonia; and of livers with signs of steatosis, parasitosis and adherence or abscess had a null prevalence representing over 80% of the sample. Rumens were involved (mean ± SD) by redness of mucosa (47.6±31.5%), signs of ruminitis (29.9±24.1%), hyperkeratosis (57.5±24.4%), and parasitosis (36.6±22.9%). Lungs had at least a sign of pneumonia in 39.6±20.1%, fibrin filaments in 23.1±14.0%, and signs of pleuritis in 17.8±12.1% of cases. The effect of the breed was significant only for rumen parasitosis; the lowest prevalence recorded for Italian compared to imported French beef can be linked to the fact that Italian cattle are always reared indoors while those imported spend a period on pasture where they might be exposed. Lungs with signs of pleuritis had a greater percentage in bulls compared to heifers (24.3±2.1 vs 22.6±2.1%, LS-mean ± SEM; $P=0.001$). This study gives an overview of the alterations occurring on rumens, lungs and livers of finishing beef cattle that were considered as healthy on arrival at the slaughterhouse. The wide variability observed for given alterations points out criticisms and needs for further investigation of potential predisposing factors related to management quality and feeding systems in some farms.

Efficacy of a new subunit vaccine against bovine mastitis in the field

L. Saenz, V. Bugeño, J. Quiroga, M. Molina, M. Duchens and M. Maino
Universidad de Chile, Laboratory of Veterinary Vaccine, Santa Rosa 11735, La Pintana, 8820808, Chile; leosaenz@uchile.cl

Mastitis is one of the most important problems that affects milk production and can cause up to a 50% loss in economic margins. The search for new tools to prevent this pathology is essential to improve the competitiveness of the dairy sector in the national and international economy. Vaccines against mastitis are a viable and safe alternative. In this study a subunit vaccine was developed using a biotechnological process that incorporates fragments of bacterial membranes in the form of nanovesicles. The objective of this study was to validate the safety of the vaccine against mastitis caused by *Escherichia coli* and compare its efficacy against commercial bacterin vaccine based on J5 strain. We assessed the state of mammary health by quantifying the incidence of mastitis and severity, as well as duration of the infection. The protocol, identical for both groups, included two doses: first at 21-27 days before calving and the second 7-10 days postpartum. A total of 226 primiparous cows belonging to one intensive dairy herd of the central zone of Chile were used. To compare the incidence of mastitis a Mann-Whitney test was used. Severity assessment was evaluated with a Kruskal-Wallis test and the duration of mastitis was evaluated based on the analysis of variance using one factor. The results showed that during the evaluation period (14 days post immunization), animals immunized with a dose three times higher than the recommended dose showed no local or systemic signs, except for one slight local reaction. The incidence of mastitis was 16.8% (19 cases) in the treated group and 18.6% (21) in the control group (commercial vaccine), with no significant differences between the two groups. The average duration of clinical mastitis was 6 and 5.6 days for the treated and the control groups, respectively ($P>0.05$). These results indicate that the vaccine developed shows at least the same performance as the vaccine currently in use.

Effects of homeopathic blend on serum metabolites and cortisol of weaned Holstein calves

T.H. Silva[1], I.C.S.B. Guimarães[2], R. Melotti[3] and A. Saran Netto[1]
[1]School of Animal Science and Food Engineering, University of Sao Paulo, Av. D. Caxias, 225, 13635-900, Pirassununga, SP, Brazil, [2]School of Veterinary and Animal Science, University of Sao Paulo, Av. D. Caxias, 225, 13635-900, Pirassununga, SP, Brazil, [3]Real H, Technical manager, Av. Zilá C. Machado, 7904, 12068, Campo Grande, MS, Brazil; saranetto@usp.br

Although in recent years farmers have shown interest in using homeopathy, the reliability of its products has still to be demonstrated on health parameters of the animals on the farm environment. The aim of the study was evaluate the effect of homeopathic blend supplementation on serum metabolites and cortisol of weaned Holstein calves in a double-blind placebo-controlled trial. One hundred and eighty-four weaned Holstein calves (83.01±7.9 days old; 112.5±11.7 kg) were allocated to 8 paddocks in a completely randomized design experiment. During a 112 days period, animals received a total mixed ration with the following treatments: control (basal diet + calcium carbonate, top-dressed at 30 g/animal) and homeopathic blend (basal diet + TopVita™-Real H, top-dressed at 30 g/animal – *Sulphur*:10^{-60}+*Viola tricolor*:10^{-14}+*Caladium seguinum*:10^{-30}+*Zincum oxydatum*:10^{-30}+*Phosphorus*:10^{-60}+*Cardus marianus*:10^{-60}+*Colibacillinum*:10^{-30}+*Podophyllum*:10^{-30}+Vehicle: calcium carbonate – q.s. 1 kg). Blood samples were collected each 28 days until d112. Data were analysed by a MIXED procedure for repeated measurements of SAS. No differences in serum AST and urea were found among treatments. Serum GGT level was higher in control group (22.6 vs 20.2, $P<0.05$) compared to homeopathic group. At d28 and d56 serum albumin was higher in homeopathic treatment ($P<0.05$), but at d84 the control group demonstrated higher concentration ($P<0.05$). Serum creatinine concentration was higher in control treatment at d28 and, at d84 and d112 higher levels were detected to the homeopathic treatment ($P<0.05$). Serum total protein was higher in control treatment at d84 ($P<0.05$). Glucose and cortisol levels were elevated in control group at d28 ($P<0.05$). Despite all parameters were within the reference limits, the homeopathic blend altered some blood components mainly at d28 of study, which can demonstrate that the animal's organism reacted better to environmental challenges, diseases for example, compared to control group.

Session 18 Poster 16

Effects of lameness on milk yield and reproduction efficiency in Simmental cows

D. Gavojdian[1], R. Neamt[2], L.T. Cziszter[3] and D.E. Ilie[2]
[1]Research and Development Institute for Bovine Balotesti, Balotesti, sos. Bucuresti-Ploiesti, km 21, Ilfov, 077015, Romania, [2]Research and Development Station for Bovine Arad, Calea Bodrogului 32, Arad, 310059, Romania, [3]Banat's University of Agricultural Sciences and Veterinary Medicine 'King Michael I of Romania', Timisoara, Calea Aradului 119, 300645, Romania; gavojdian_dinu@animalsci-tm.ro

The economic relevance of lameness is attributed to the costs of treatment and control, impaired reproductive performance, decreased milk yield and the increased number of culled cows. Moreover, because of the acute pain, discomfort and high prevalence of lameness in cattle, this disorder represents a major animal welfare concern. Aim of the current research was to evaluate the effects of lameness on milk yield and reproduction efficiency in dual-purpose Simmental cows. The study was carried out at the Research and Development Station for Bovine Arad (46°10'N 21°19'E) on a herd consisting out of 751 Simmental cows. Animals were monitored for clinical signs of lameness during one lactation starting year 2016, and were classified as: healthy, with no signs of lameness; one clinical lameness diagnostic per lactation (n=10); two clinical lameness diagnostics per lactation (n=6). Lameness incidence within the herd was 2.89±0.82%, significantly lower compared to data reported in the literature. The number of artificial inseminations per gestation was on average 1.21±0.12 in healthy cows, 1.78±0.07 in cows that had one lameness diagnostic and 1.90±0.13 in cows that were affected twice (P>0.05). The number of days open was on average 94.12±10.73 in healthy cows, 128.56±6.50 and 133.00±10.92 in cows affected by a single or two lameness episodes, respectively (P>0.05). Milk yield was on average 5,888.7±239.78 kg in healthy animals, while the cows with one lameness episode produced 5,469.2±92.08 kg and 5,180.9±205.81 kg produced the group with two lameness diagnostics (P>0.05). Although statistical differences between healthy and lame cows for all studied traits were not found to be significant, there are noticeable production disparities between the lame and non-lame animals. Current results suggest that lameness, particularly when the prevalence is high, leads to a lower reproduction efficiency at herd level and a decreased milk production.

Session 18 Poster 17

Influence of mastitis on reproduction and milk production in Romanian Simmental cows

L.T. Cziszter[1], D. Gavojdian[2], R.I. Neamt[3] and D.E. Ilie[3]
[1]Banat's University of Agricultural Sciences and Veterinary Medicine 'King Michael I of Romania', from Timisoara, Calea Aradului 119, 300645, Romania, [2]Research and Development Institute for Bovine Balotesti, Şos. Bucuresti-Ploiesti, km 21, 077015, Romania, [3]Research and Development Station for Bovine Arad, Calea Bodrogului 32, 310059, Romania; cziszterl@animalsci-tm.ro

Mastitis is one of the most frequently occurring and costly diseases in the dairy industry with an incidence ranging from 25 to 40% and represents one of the major reasons for involuntary culling and productive loss. Mastitis is a food safety as well as an animal welfare problem. The aim of the study was to evaluate the influence of mastitis on reproduction and milk production in dual-purpose Romanian Simmental cows. The study was carried out at the Research and Development Station for Bovine Arad, on 751 lactations. Clinical signs of mastitis were observed during lactation, classifying animals as healthy with no signs of mastitis (H), one mastitis episode (M1) and two or more mastitis episodes (M2+) during one lactation. Clinical mastitis was very low at farm level, while subclinical mastitis was six times higher (4.12±0.72 and 24.63±0.15%, respectively). The number of artificial inseminations per gestation was significantly (P<0.05) higher when more mastitis episodes were observed during one lactation, 1.73±0.05 in healthy, 2.05±0.21 in M1 and 3.12±0.87 in M2+ cows. Calving interval and days dry were not significantly influenced (P>0.05) by the mastitis episodes. The number of days open increased in mastitis affected cows compared to healthy cows (P<0.05), from 116.21±10.74 in healthy cows to 128.01±5.72 and 129.50±41.50 in M1 and M2+ cows, respectively. Milk production was 22-25% higher in healthy cows (P<0.05) compared to those affected by clinical mastitis once or several times during a lactation. Healthy cows produced 6,561.5±295.50 kg milk per lactation, while milk yield was 5,442.1±80.55 for M1 cows and 5,238.9±312.08 for M2+ cows. Current findings show that mastitis has a significant influence not only on milk production but also on selected reproduction indices, such as the number of artificial inseminations per gestation and days open.

Session 19 Theatre 1

Livestock sustainability is more than controlling environmental footprint

M. Tichit

INRA, UMR SADAPT, AgroParisTech, Université Paris-Saclay, 75005 Paris, France; muriel.tichit@agroparistech.fr

Improving livestock sustainability requires more than just controlling its environmental footprint. Solutions must be built on a comprehensive consideration of environmental, social, and economic benefits and costs along with their synergies and trade-offs. Current decision-making regarding sustainable development of the livestock production sector is hindered by a lack of evidence-based knowledge about the multi-dimensional consequences of innovations. Multi-dimensional consequences include impacts on the three pillars of sustainability (economic, environmental, and social) and impacts at multiple levels (i.e. farm, region, nation, EU). A single innovation might positively affect one or several sustainability issues but negatively affect others. Similarly, an innovation might be effective at a given level (e.g. the farm) but ineffective at EU level. We postulate that livestock farming can be made more sustainable by developing solutions that combine innovations at multiple levels to exploit synergies among benefits and mitigate trade-offs between benefits and costs.

Session 19 Theatre 2

Benefits and costs of livestock production across scales: a case study of Dutch egg production

E.M. De Olde, A. Van Der Linden and I.J.M. De Boer

Wageningen University & Research, Animal Production Systems group, P.O. Box 338, 6700 AH Wageningen, the Netherlands; evelien.deolde@wur.nl

Livestock production provides benefits to society, but is also associated with considerable costs. These benefits and costs emerge at interlinked scales. Mapping these benefits and costs across scales is key to identify options for sustainable livestock production. It requires the involvement of stakeholders active at different scale levels and in different segments of the food chain. To identify options for sustainable egg production in the Netherlands, we interviewed 24 stakeholders related to egg production in the Netherlands. Interviewees included farmers' organizations, innovative farmers, suppliers, processing industry, researchers and (non-)governmental organizations. Based on these interviews, we identified benefits and costs, as well as innovations that could reduce costs related to egg production in the Netherlands. Results show that farm profitability, for example, is a benefit at the farm level, whereas employment is relevant at region level and product quality is determined by the chain. Examples of costs are animal welfare, which emerges at farm level, particulate matter, which is relevant at regional level and greenhouse gas emissions, which is a global issue. Whereas the egg sector is generally appreciated for its relatively low carbon footprint and high protein efficiency, the associated import of feed, feed-food competition and the processing of manure can be considered as (externalized) costs. Moreover, results demonstrate how changes at farm level (a shift towards alternative housing systems as a response to changing consumer demand and the ban on battery cages), can result in costs at regional level through the increased emission of particulate matter and disease risks (i.e. avian influenza). Recognizing the impact of such environmental, economic and social trade-offs and synergies within and across scales is critical to present sustainable options for livestock production. Correspondingly, interviewees expressed the need for integrative approaches and innovations in feed (use of left-over streams), housing systems (covered outdoor area), technology (sensors to measure emissions) as well as regulatory changes.

Session 19 — Theatre 3

SusSheP – how to increase sustainability and profitability of European sheep production?

C. Morgan-Davies[1], P. Creighton[2], I.A. Boman[3], T. Blichfeldt[3], A. Krogenaes[4], N. Lambe[1], E. Wall[5], T. Padiou[5], M.G. Diskin[2], K.G. Meade[6], N. McHugh[6], X. Druart[7] and S. Fair[8]

[1]SRUC, Kirkton, Crianlarich, United Kingdom, [2]Teagasc, Athenry, Co Galway, Ireland, [3]NSG, Moerveien, 104, Aas, Norway, [4]Norwegian University of Life Science, Ullevålsveien 72, Oslo, Norway, [5]SheepIreland, Shinagh, Bandon, Ireland, [6]Teagasc, Moorepark Fermoy, Co Cork, Ireland, [7]INRA, Val de Loire, 37380 Nouzilly, France, [8]University of Limerick, Castletroy, Limerick, Ireland; claire.morgan-davies@sruc.ac.uk

European Sheep Production must be underpinned by environmentally sustainable and welfare friendly practices which are profitable and labour efficient for farmers, thus encompassing all 3 pillars of the sustainability triangle: economic competitiveness, environment and society. Sheep are unproductive (but carbon productive) until they produce their 1st lamb crop, and often only produce 4 crops of lambs in the lifetime. Despite its importance both from an economic and environmental perspective, ewe longevity is not included in sheep breeding indexes across Europe. Parallel to this, European sheep production systems are varied in production types (meat vs milk), breeds (prolific vs non-prolific) and management approaches (use of Electronic Identification technology, breeding indexes and/or artificial insemination). It is paramount to identify the most carbon and labour efficient production systems to enable the development of strategies to reduce the labour input and carbon hoofprint per kg of output. Moreover, in order to breed more efficient sheep, developing a more socially acceptable sheep artificial insemination method, which farmers could use themselves, is essential. The paper presents 'SusSheP', a project funded under European Research Area Network Sustainable Animal Production Framework (ERA NET SusAn), which investigates these three topics between 4 countries (Ireland, UK, Norway and France). This paper focuses on the effects on labour (measured on sample days in each country during the sheep year using GoPro cameras) and carbon hoofprint (using AgreCalc (c)) when changing from one sheep production system to another. SusSheP will give farmers better indications on how to make their own system more sustainable, thus ensuring uptake across the farming industry in Europe.

Session 19 — Theatre 4

Pasture and manure management for sustainable dairy farming: a life cycle assessment

A.D. Soteriades[1], A.M. Gonzalez-Mejia[1], D. Styles[1], A. Foskolos[2], J.M. Moorby[2] and J.M. Gibbons[1]

[1]Bangor University, SENRGy, Deiniol Road, LL57 2UW, Bangor, United Kingdom, [2]Aberystwyth University, IBERS, Ceredigion, SY23 3EB, Aberystwyth, United Kingdom; a.d.soteriades@bangor.ac.uk

Pasture-based dairy farming can have positive impacts on farm biodiversity, milk quality and farming costs. It is also the most common farming method in several regions across the globe. However, nitrogen (N) excretion from grazing animals is a major source of N leakage, which is associated with major environmental impacts. Numerous studies show that feeding cows with 'high-sugar grasses' (HSG), i.e. grasses that have been bred to express elevated concentrations of water soluble carbohydrates, can significantly reduce the proportion of N lost in urine. There is a gap in existing knowledge on the potential of HSG to make pasture-based farms sustainable from a whole-farm system, life cycle assessment (LCA) perspective. We evaluated the LCA performance of the typical, six-month grazing – six-month indoors UK dairy farm when conventional perennial ryegrass (scenario Sc-CTR) was replaced by HSG (scenario Sc-HSG). The LCA impacts studied were global warming, eutrophication, acidification and resource depletion potentials. To obtain a more holistic representation of farm management, for each scenario we also considered a range of manure spreading and storage choices that are prevalent in UK dairy farms. Estimated impacts were 1-9% lower for Sc-HSG relative to Sc-CTR under identical manure management. The largest reductions (5-40%) occurred when switching from the Sc-CTR scenario with inefficient manure technologies to the Sc-HSG scenario with the most efficient manure management technologies. Simple land engineering combined and improved farm technology can work synergistically to deliver environmental gains from pasture-based dairy farming.

Relationships between dairy farming and mountain agro-ecosystems: a focus group approach

G. Faccioni[1], D. Bogner[2], R. Da Re[3], L. Gallo[1], P. Gatto[3], M. Ramanzin[1] and E. Sturaro[1]
[1]University of Padova, DAFNAE, Viale dell'Università 16, 35020 Legnaro, Italy, [2]Umweltbüro GmbH, Bahnhofstraße 39, 9020 Klagenfurt, Austria, [3]University of Padova, TESAF, Viale dell'Università 16, 35020 Legnaro, Italy; enrico.sturaro@unipd.it

The multifunctionality of mountain livestock systems can be addressed using the ecosystem services (ESs) framework, to ascribe values to services and goods that contribute to human well-being and the attractiveness of mountain areas. This study is part of a project (IR VA Italia-Österreich 'TOPValue') aiming at: (1) implementing the optional quality term 'mountain product' (MP) as defined by EU Reg. 665/2014; (2) empowering the optional MP quality term by identifying and quantifying ESs delivered by local food chains, with a particular focus on mountain dairy cattle farming. Specifically, this study aims at understanding stakeholders' opinion on the opportunities to generate added value given by the Mountain Product EU regulation. First, we performed a stakeholder analysis using the snowball sampling method. Second, we developed an online survey to discover stakeholders' background knowledge on Ecosystem Services (ESs). The results obtained during the survey have been used to design a focus group, the last step of the study. This procedure was performed for each one of the study areas. We involved 19 participants (8 in Austria and 11 in Italy), including the following categories: producers (farmers) and dairy cooperatives, regional policymakers, local communities, and tourist operators. Stakeholders have a positive perception on the effects of the livestock production chains on the mountain areas and vice versa, with maintenance of landscape, tourist attractiveness and production of high quality foods indicated as the most relevant ESs. There was some discordance on possible null or negative impacts (mainly GHG emissions and water quality) among the Austrian and Italian stakeholders. Based on our results, collaboration among stakeholders is highly suggested to improve a targeted communication of added value and of positive externalities generated by the dairy production chains in mountain areas.

Behaviour patterns to the intensification vary differently within dairy producers

A.-C. Dalcq[1], Y. Beckers[1,2], B. Wyzen[3], E. Reding[3], P. Delhez[1,2,4] and H. Soyeurt[1,2]
[1]ULiège-GXABT, AGROBIOCHEM, Passage des Déportés, 2, 5030 Gembloux, Belgium, [2]ULiège-GXABT, Terra Research and Teaching Centre, Passage des Déportés, 2, 5030 Gembloux, Belgium, [3]AWE, Rue des Champs Elysées, 4, 5590 Ciney, Belgium, [4]National Fund for Scientific Research, Egmont 5, 1000 Bruxelles, Belgium; anne-catherine.dalcq@uliege.ac.be

Intensification (Int) of dairy production is a great issue and will maybe be a component of the future dairy management. This phenomenon has occurred in various parts of world and in most European countries. This research aimed to study the different kinds of evolution of the level of Int of dairy producers in the Walloon Region in Belgium. A total of 144 farmers' accounts provided Int variables as milk and milking cows per hectare of forage area, composition of the forage area, etc. An index of Int was created by using a principal component analysis carried out on 15 Int variables. The index was modelled by year and year^2 as fixed effects, which provided 144 intercepts, linear and quadratic regression coefficients. In function of the level of significance and the sign of these parameters equation, several patterns of evolution of Int over time were defined. The principal patterns represented in the population were constant (27%), linear (7.6%), quadratic (24.3%), quadratic before 2012 (7%) and quadratic after 2012 (7%) relationships between Int and time. The producers with a quadratic pattern showed an average peak of the relationship in 2012, which leads to assume that this milk price and input price crisis impacted the level of Int of a large number of dairy producers. Extensive and intensive producers, defined by using the intercept of their equation, were distributed equally in all the principal patterns. Only the quadratic after 2012 pattern counted more intensive producers ($P<0.05$). It seems that this kind of individuals can be more disturbed by a dairy crisis. In conclusion, dairy producers have followed different evolutions of Int. The highest proportion of them presented a constant evolution and so an ability or a will of constant level of Int, notwithstanding the varying economic and climatic context.

Session 19 Theatre 7

How should we measure the sustainability of livestock?

J.M. Gibbons
Bangor University, SENRGy, Thoday building, Deiniol Road, Bangor, LL57 2UW, United Kingdom; j.gibbons@bangor.ac.uk

Milk and beef production can contribute significantly to ecosystem damage via greenhouse gas (GHG) emissions, nutrient losses to water, ammonia (NH_3) emissions to air, biodiversity impacts and depletion of finite resources. However, milk and beef provide calories and important nutrients to diet and livestock production contributes substantially to the rural economy. In addition, livestock can be produced in areas too marginal for crop production for human consumption. Sustainability can be measured as a footprint of production, relating units of output to inputs and environmental burdens. Previous studies at the farm and sector level have generally shown that intensification reduces the footprint of production by increasing animal productivity and feed conversion efficiency. These results suggest that there is potential for land sparing by concentrating production in smaller areas. However, footprints for the same farm will differ depending on where the system boundary is drawn. Accurate footprints must recognise that farms and farm processes are intricately linked in a complex web of local and global interactions. Dairy and beef systems are linked through calf production and livestock production is linked to arable production through feed consumption. This means that an apparent reduction in footprint in e.g. dairy may simply displace burdens to beef production or that changing feed may reduce in country burdens but displace burdens overseas. In these circumstances any potential for land sparing is illusory. Using case studies from the UK dairy and beef sector I show how accurate evaluation of these trade-offs can be undertaken using detailed farm models that differentiate across farm management practices and a broad measurement of burdens that includes multiple environmental impact categories and considers upstream and indirect effects. I also show that accurate measures of sustainability require accurate representation of the complexity and diversity of farm systems. I conclude that assessment of livestock sustainability requires this broad system wide approach.

Session 19 Theatre 8

Benefits and costs of livestock systems in ten European case studies

D. Neumeister[1], M. Zehetmeier[2], E. De Olde[3], T. Valada[4], M. Tichit[5], T. Rodriguezo[6], C. Morgan-Davis[7], I. De Boer[3], C. Perrot[1] and A.C. Dockes[1]
[1]Institut de l'Elevage, Paris, France, [2]LFL, Muenchen, Germany, [3]Wageningen University, Wageningen, the Netherlands, [4]IST-ID, Lisboa, Portugal, [5]INRA, Paris, France, [6]CITA-IA2, Zaragoza, Spain, [7]SRUC, AYR, United Kingdom; delphine.neumeister@idele.fr

The ERANET Animal Future project focuses on an analysis of livestock farming benefits and costs, in order to improve livestock systems sustainability, by fostering the benefits that livestock offers to the territories and by reducing its costs. Benefits are defined as activities with positive impact on three dimensions of sustainability (economic, environmental and social issues). Costs are defined as activities with negative impact on these issues. To measure the importance of territorial context, ten livestock regions (dairy cows, cattle, sheep pig, poultry) distributed in six European countries (France, Germany, the Netherlands, Portugal, Scotland, Spain) have been used as case studies. Seven workshops with local and regional influencing stakeholders (farmers, farmer organizations, advisers, processors, governments, etc.) have been organized, using participatory approaches (SWOT analysis, prioritization…) to identify main issues in livestock production in the territory. The main portfolios of benefits and costs at farm and regional levels were identified in small groups, to gain insight into innovations that can maximize the benefits and minimize (if not overcome) the costs, but also into the relationships, trade-off and synergies between the different benefits and costs. The workshops' synthesis shows that the perception of benefits and costs largely depends on territorial issues and illustrates why some systems are identified as sustainable only in a defined area. Moreover, increasing societal expectations play an important role in the livestock perception: agricultural and social stakeholders sometimes express conflictual views on benefits and costs resulting from livestock production. Finally, the results of the workshops demonstrate the importance of taking a multi-dimensional approach as livestock systems balance depends on economic, environmental and social benefits and costs; fostering one dimension of sustainability may lead to trade-offs in another.

Session 19 Theatre 9

Effect of low and high concentrate supplementation on health and welfare in mountain dairy farms

L. Flach, S. Kühl, E. De Monte, C. Lambertz and M. Gauly
University of Bolzano, Faculty of Science and Technology, Universitätsplatz 5, 39100 Bolzano, Italy; laurafranziska.flach@natec.unibz.it

The optimal level of concentrate supplementation in dairy production in general and in particular under mountain farming conditions, where concentrates largely have to be imported, is critically discussed. Therefore, the objective was to identify relationships between concentrate supplementation and breed with health and fertility traits, animal welfare and husbandry conditions. A total of 49 farms (ø 12.5 dairy cows/farm) raising either Brown Swiss (BS) or Tyrolean grey (TG) were classified based on concentrate supplementation into low-input (LI; n=5 BS farms; n=11 TG farms) and high-input (HI; n=21 BS farms; n=12 TG farms). Farms were visited during the barn period and 614 animals were assessed for body condition score (BCS), cleanliness, injuries and hairless patches. Milk yield per cow differed significantly between breeds (BS 7,151 kg, TG 5,533 kg; $P<0.05$) and level of intensification (HI 7,366 kg, LI 5,317 kg, $P<0.05$). The breed had an effect on the number of lactations ($P<0.05$), which ranged between 3.42 in TG-LI and 2.44 in BS-HI. Number of lactations were positively correlated with grazing days ($r=0.307$; $P<0.05$), which differed between breeds and intensity levels ($P<0.05$) and were highest for TG-LI (84.54) and lowest for BS-HI (16.4). Number of grazing days were negatively correlated with milk yield ($r=-0.492$; $P<0.05$) emphasizing that cows in intensive farms less often have access to pasture. Breed and system had an effect ($P<0.05$) on the insemination index, which was lower in TG-LI (1.69) than in TG-HI (2.24), BS-LI (2.1) and BS-HI (2.2). The highest incidence of mastitis treatments, which was affected by breed and level of intensification ($P<0.05$), was found in BS-HI, while the lowest was found in TG-LI (0.86 and 0.29 treatments/cow). Cleanliness was better in TG than in BS ($P<0.05$). On average, cows showed 0.18 injuries and 1.55 hairless patches, which were correlated with one another ($r=0.41$; $P<0.05$). Breeds and intensification level did not significantly affect hairless patches, injuries, BCS, somatic cell count and number of veterinary treatments. The high variation within the groups of both breeds already show that results should be interpreted with caution.

Session 19 Theatre 10

Regional diversity in bundles of impacts and services points to options for livestock future

B. Dumont[1], J. Ryschawy[2], M. Duru[2], M. Benoit[1], V. Chatellier[3], L. Delaby[4], C. Delfosse[5], J.Y. Dourmad[4], P. Dupraz[3], N. Hostiou[6], B. Méda[7], D. Vollet[6] and R. Sabatier[8]
[1]INRA, UMRH, 63122 St-Genès-Champanelle, France, [2]INRA, INPT, UMR AGIR, 31326 Castanet-Tolosan, France, [3]INRA, Smart-Lereco, 35000 Rennes, France, [4]INRA, Pegase, 35590 St-Gilles, France, [5]Univ. Lyon, Etudes Rurales, 69363 Lyon, France, [6]Irstea, INRA, UMRT, 63000 Clermont-Fd, France, [7]INRA, UMR BOA, 37380 Nouzilly, France, [8]INRA, Ecodéveloppement, 84914 Montfavet, France; bertrand.dumont@inra.fr

Livestock farming (LF) is increasingly constrained by competition for natural resources, climate change and changing sociocultural values. Analysing interactions between the various services and impacts provided by LF makes it possible to identify the origin of potential trade-offs. Here, we first use two simple criteria, livestock density (in livestock unit/ha Utilized Agricultural Area; UAA) and the proportion of permanent grassland (including rangelands) within the UAA to identify and map five types of LF areas, based on Eurostat data for 2010 processed at NUTS3 level. In each area, we then use 'the barn', a representation derived from a social-ecological approach, to visualize how LF systems interact with their physical, economic and social environment. This reveals the classical distinction between production-oriented areas and areas that provide more environmental and cultural services. In production-oriented areas, LF systems get local-level acceptance based on the territorial vitality services they provide, such as on-farm jobs and social bonds linked to livestock. The barn also allows a multilevel analysis to account for niche-systems hidden by the dominant socio-technical regime. For instance, Label Rouge chickens (LRc) in Pays de la Loire represent 41% of French LRc production, and provide a very different bundle of services compared to the dominant LF regime of this region. Finally, the barn points to LF areas where bundles of services are more balanced: (1) some PDO cheese production areas where territorial governance forms create more value by enhancing the use of local resources and regulating production by dedicated professional organisations, (2) agricultural and recreational landscapes in peri-urban areas that benefit from consumer trust in local food systems.

Description and validation of the Teagasc pig production model

J.A. Calderón Díaz[1], L. Shalloo[2], J. Niemi[3], M. McKeon[1], G. McCutcheon[1] and E.G. Manzanilla[1]
[1]Teagasc Moorepark, Pig Development Dpt, Fermoy, P61C996, Ireland, [2]Teagasc Moorepark, Livestock Systems Dept, Fermoy, P61C996, Ireland, [3]Natural Resources Institute Finland, Seinäjoki, 60320, Finland; julia.calderondiaz@teagasc.ie

This paper describes the development and validation of the Teagasc Pig Production Model (TPPM), a bio-economic simulation model for a farrow-to-finish Irish pig farm. It is expected that the TPPM will be used as a decision tool on different aspects of production such as investments, nutrition, welfare and health. The TPPM was built using real Irish data obtained from multiple sources including the Teagasc e-profit monitor (ePM), Teagasc research data and input from members of the Teagasc pig advisory team. The model simulates, on a weekly basis, the annual production of a farm. Biological inputs include herd size, conception and farrowing rate, No. of litters/sow/year, No. of piglets born alive per litter and mortality rate for each production stage. These inputs are used to calculate physical (e.g. feed usage and number of pigs slaughtered) and financial (e.g. annual cashflow, profit and loss account and a balance sheet) outputs. Net profit is calculated on a total farm basis, as well as per pig produced and per kg of carcass sold (DW). The TPPM was validated using the Delphi method where a group of experts (i.e. pig advisors and researchers) revised the methodology and values used for the model. Once the experts agreed, a second validation was carried out by comparing TPPM outputs with real farm data from the ePM. The model was parameterised to simulate the biological performance of 16 Irish pig farms and the simulated results were compared to the average performance of such farms. Results from the validation show that the TPPM closely simulates the ePM farms with similar total income (€129.16 vs €126.03 per pig and €1.54 and €1.52 per kg/DW, respectively) and total farm costs (€119.32 vs €112.12 per pig and €1.44 and €1.34 per kg/DW, respectively). However, the TPPM under-estimated net profit per pig (€6.71 vs €17.05) as well as per kg of DW (€0.08 vs €0.20) compared to ePM farms. This is attributed to the lower variable costs associated with the ePM farms, most notably labour, replacement female costs and depreciation as some farms do not record them.

Coping with citizen demands: a field study of of suckling processes in dairy herds

C. Disenhaus[1], A. Michaud[2], N. Genot[1], L. Valenzisi[1], D. Pomies[2], B. Martin[2], C. Chassaing[2] and Y. Le Cozler[1]
[1]PEGASE, INRA, Agrocampus-Ouest, Rennes, 35042, RENNES, France, [2]Université Clermont Auvergne, INRA, VetAgro Sup, UMR Herbivores, Theix, 63122 Saint-Genès-Champanelle, France; catherine.disenhaus@agrocapus-ouest.fr

In dairy farms calves and cows are usually separated shortly after birth but this practice raises the question of animal welfare. This social concern may lead to bring back dairy calves to their mother. Even though suckling is quite seldom in dairy production, some farmers are using such a practice for many years. The present study aims to better know farmer's motivations, practical implementation and farmer's perception of the impact of suckling on performance and animal behaviour. In January 2018 a semi-quantitative survey was performed on 44 farms where calves suckle at least 24 hours their mother or a nurse cow. The farming systems differed on the size (20 to 140 cows), the type (19 conventional, 25 organic farms), the suckling (28 by the mothers, 16 by nurses), and the breed. Results indicate that main farmers' motivations are calves' health (52%), better working conditions (41%) and saving time (34%), with little consideration on animal welfare (7%). Numerous practices were found in terms of allowance of daily cow-calf contact and suckling duration with a difference between male and female calves. According to farmers, these practices were efficient to improved calves' health (70%), save time (75%) and improved working conditions (52%). Weaning was considered as a stressful situation, especially after long suckling periods. Cow mooing usually stopped 2.5 d after weaning. At least 34 farmers are fully satisfied with this practice. The main challenge appears to be the management of the wildness of future heifers, by investing time to manipulate calves before weaning. The results of this survey could help to propose solutions to farmers who want to cope with the societal demand of keeping calves with dams in dairy herds.

Session 19 Poster 13

Which sustainability indicators are addressed by livestock models applicable in Europe?

A. Van Der Linden, E.M. De Olde, P.F. Mostert and I.J.M. De Boer
Wageningen University & Research, Animal Production Systems group, De Elst 1, 6708 WD Wageningen, the Netherlands; aart.vanderlinden@wur.nl

Models are widely used as tools to assess the sustainability performance of farms with livestock, and to evaluate effects of innovations on sustainability. Such models usually calculate the sustainability performance with a specific set of environmental, economic, and/or social indicators, depending on model objectives. Our aim is to create a library of livestock models and to evaluate their indicators. This library is developed to ultimately explore synergies and trade-offs among sustainability impacts of innovations in livestock farming systems in Europe. A review of the scientific literature is currently conducted to list models simulating livestock production at the farm, herd, flock, or animal level. Experts from various European countries were asked to propose models too. After de-duplication, records describing model applications and models not applicable to European livestock farming were excluded. Sustainability indicators and model characteristics were listed and analysed. Model characteristics that were investigated include: the livestock species and purpose, applicability domain, country of origin, programming language, and availability. Preliminary results indicate that environmental and economic indicators are included in many models, whereas the social dimension of sustainability is hardly represented. Livestock models for dairy and beef cattle are most abundant. The majority of the models has been developed in Western Europe. Models are usually applicable to a limited number of regions and countries in Europe. A diverse range of programming languages is used, and few models are accessible and available on the internet. Our preliminary conclusion is that the model library is useful to determine which (combination of) models can be used to assess sustainability based on a broad set of sustainability indicators. This model library can be used to get *a priori* insight in the synergies and trade-offs among environmental and economic indicators that are likely to emerge when introducing specific agricultural innovations in different livestock systems in Europe.

Session 19 Poster 14

Economic sustainability of bull fattening operations in the Czech Republic

J. Syrůček[1], L. Bartoň[1], J. Kvapilík[1], M. Vacek[2] and L. Stádník[2]
[1]Institute of Animal Science, Cattle breeding, Přátelství 815, 104 00 Praha Uhříněves, Czech Republic, [2]Czech University of Life Sciences Prague, Department of Animal Husbandry, Kamýcká 129, 165 00 Praha 6 Suchdol, Czech Republic; syrucek.jan@vuzv.cz

Beef production is one of the most important segments of the cattle industry. In most European countries, the main source of beef is represented by intensively fattened bulls of beef and dairy breeds, or their crossbreds. They contribute approximately 50% to the overall beef production in European Union as well as in the Czech Republic (CR). Profitability of fattening operations is a fundamental prerequisite for sustainability and further growth of domestic beef production. Therefore, the objectives of this study were to evaluate the economic efficiency of fattening bull operations in the CR, to determine minimum profitability requirements, and to perform a sensitivity analysis. From 2013 to 2016, input data were annually collected from 16 to 18 farm operations located in different regions of the CR using a questionnaire. The average total cost including the purchase price of a weaned calf was 37,429 CZK per bull. Total costs per feeding day according to the herd size tended to decrease with the increasing number of fattened bulls. The average selling price was 32,316 CZK per bull and thus, the fattening operations suffered a loss (negative profitability on average -7%). To compensate for these losses, a higher average live weight gain by 151 g/day would have been required. Break-event points as minimum requirements to achieve zero profitability were calculated for average initial bull´s price (10,132 CZK), selling price (95 CZK/kg of carcass) and total costs (34,864 CZK/bull). The results of the sensitivity analysis indicated that the economic efficiency of the fattening operation can be substantially altered by only a small change in prices, e.g. in carcass price, initial price of bulls at the beginning of fattening, or feed prices. Additional supporting measures aimed directly at fattening operations should also contribute to improving the overall profitability and thus to ensure the sustainability of fattening.

Session 19 Poster 15

Scandinavian beef production models explored on efficiency on producing human edible protein

C. Swensson and A. Herlin
Swedish University of Agricultural Sciences, Biosystems and Technology, P.O. Box 103, 23053 Alnarp, Sweden; anders.herlin@slu.se

Ruminants are superior at converting feed into the nutrient-rich foods such as milk and meat. But how effective are they? And are humans and ruminants competing for the same feed and or food. This study analysed the production of beef from two different systems, beef from beef breed cattle and six systems with beef from bull calves deriving from dairy production system. Three different systems from beef breed production were compared, a Danish extensive system based on high roughage intake from natural and permanent pastures and two more intensive systems from Denmark and Sweden. Two production models for steers were analysed. Four different production models using bull calves as the starting animal were compared with two veal calf models and two production scenarios for young bulls All models and feeding scenarios were obtained from Mogensen *et al*. Protein efficiency was calculated as the total protein efficiency and for the cattle's ability to turn feed protein to human edible animal protein using factors for different feeds derived from Wilkinson. Results show that more human edible protein is obtained than put into the production model from the extensive beef breed system (factor 2) and from the two steer production system (about factor 1.2) than from the other production models. These model are based on large pasture intake and large amounts of silage which results in high efficiency in converting feed to human edible protein. Mogensen *et al.* found that intensive beef production systems, especially veal and to some extent young bulls, had a much less carbon foot print than the extensive systems based on pasture and silage (steers and beef breed cattle). The management of the conflict of objectives between climate impact and protein efficiency is one of the key challenges for ruminant-based meat production.

Session 19 Poster 16

What are the limitations for organic conversion of extensive livestock systems in dehesas?

A. Horrillo[1], P. Gaspar[1], M. Escribano[1], F.J. Mesías[1] and A. Elghannam[2]
[1]University of Extremadura, Faculty of Agriculture, Avda. Adolfo Suarez, s/n, 06007 Badajoz, Spain, [2]Damanhour University, Faculty of Agriculture, Damanhour, Elbeheira, Egypt; pgaspar@unex.es

The dehesa is a unique agroforestry system within the European territory; it provides an important set of environmental, cultural, aesthetic, and economic values. Extensive livestock farms raising cattle, sheep and Iberian pigs are the main economic activity in these systems, unfortunately, its profitability is decreasing. Extremadura is a Spanish Autonomous Community with the largest extension of dehesas (2.2 million has) which are considered the basis of socio-economic activities in the region. A regional research project is ongoing with the objective of fostering the conversion of livestock farms located in dehesa ecosystems into a sustainable organic livestock production system. This way marketing of products certified as organic could add value to all products increasing its profitability. The extensive livestock systems located in dehesas are very close to the organic production models. Although its conversion is possible from a technical point of view, its practical applicability is rather limited. Livestock farms are located in the same environmental, social and legal context. Therefore, the problems and possible solutions are shared, and in many cases the policy makers at regional level are those who will have to adopt policies and to take decisions. At the first stage of the project, a participatory approach with the aim to identify what are the main limitations that dehesa farmers find for their conversion to organic models has been conducted. Four focus group sessions along the region have been carried out with a total of 33 stakeholders (farmers, advisors, researchers and regional government employees) participating. The first findings have identified a set of problems grouped into categories. Insufficient training for local advisors, difficulties in the use of common land pastures for organic farmers, measures applied against tuberculosis by the regional government, excess of bureaucracy and lack of processing facilities in the region a have been the main items specifically identified in Extremadura.

Session 19 Poster 17

Improving extensive livestock farming in Karrantza Valley concerning the risks of wolf predation

M. Sáenz De Buruaga[1], R.J. Ruiz[2], M.Á. Campos[1], J. Gómez[3], J. Batiz[4], G. Calvete[1], K. Mas[4], I. Bilbao[2], N. Mandaluniz[2], R. Atxaerandio[2], F. Canales[1] and N. Navamuel[1]

[1]Consultora de Recursos Naturales, Castillo de Quejana 11, 01007 Vitoria-Gasteiz, Spain, [2]NEIKER-Tecnalia, Instituto Vasco de Investigación y Desarrollo Agrario, Agrifood Campus of Arkaute, 01080 Arkaute, Spain, [3]Enkarterrialde, Barrio San Miguel de Linares 1, 48879 Artzentales, Spain, [4]LORRA S Coop, Barrio Garaioltza 23, 48196 Lezama, Spain; rruiz@neiker.eus

In 2016, 4 entities, under the 'Livestock-Wolf' Operative Group, submitted a proposal within the framework of the funding programme for cooperation corresponding to Regulation (EU) No. 1305/2013 of the European Parliament and the Council of 17 December, and the Rural Development Program of the Basque Country 2015-2020. The initiative aims to establish preventive mechanisms to reduce the vulnerability of extensive livestock to wolf predation, seeking for a better cohabitation between livestock and this species in the valley of Karrantza and surroundings (Bizkaia, Basque Country). This area includes 2 Natura 2000 Network sites where there have been several attacks of wolfs on cattle in recent years: SAC Ordunte (ES2130002) and SAC Armañón (ES2130001)/Armañón Natural Park. The following systems of prevention of damages are being tested: (1) joint and professional surveillance of extensive cattle, and (2) introduction of autochthonous livestock guarding dogs (LGD). In addition, GPS devices for cattle and LGD are tested to better assess the work of the dogs. We will also analyse the technical and budgetary possibilities of relating the presence of wolves with extensive livestock farming in the food products marketed within the area, so trying to add value to the presence of the wolf in these places. The project started in February 2017 and is scheduled for completion in December 2018. There are already 16 extensive livestock farms attached, and 10 high quality LGD puppies have been delivered and already impregnated with the type of livestock to be protected. A first phase of surveillance was undertaken (summer-autumn 2017) by a team of four people, and GPS devices were placed on 15 cattle, 1 sheep and 2 goats to geolocalise the different herds kept on the grasslands of these mountains, and at least 7 LGD will also be equipped with GPS devices.

Session 19 Poster 18

Preliminary results of the BEEF CARBON initiative in innovative beef farms in Italy

S. Carè, G. Pirlo and L. Migliorati

CREA, Research Centre for Animal Production and Aquaculture, Via A.Lombardo n.11, 26900, Italy; sara.care@crea.gov.it

Beef production is considered one of the most contributors of GHG. Public concern imposes to study mitigation strategies and to demonstrate that reduction of GHG emissions is possible. For this reason, the EU has founded the 'BEEF CARBON ACTION PLAN' project, which aims to reduce the carbon footprint of beef production by 15% over 10 years in France, Ireland, Italy and Spain. It also promotes innovative livestock farming systems, ensuring the technical, economic, environmental and social sustainability of beef farms. In Italy, the project involves 100 demonstrative farms in Piemonte and Veneto and considers three different production systems: suckler to store producing weaners; suckler to beef producing bulls/heifers and specialized fattening. Nineteen fattening innovative farms have been identified where different GHG mitigation strategies will be applied and their effect on GHG emissions will be evaluated. These strategies will refer to animal, soil, waste management and energy consumption. The functional unit is 1 kg of live weight gain and the environmental categories are: global warming, eutrophication, acidification, water consumption and contribution of beef farms to the rural landscape. The environmental impact is estimated using CAP2ER® niveau 2 developed at the French Institut de l'Elevage. Initial (before the introduction of mitigation strategies) average carbon footprint of the young bulls is 8.49±1.42 kg CO_2eq/LWG; acidification is 0.05±0.02 kg SO_2eq/LWG and eutrophication 0.06±0.03 kg PO_4 eq/LWG. A linear regression analysis (PROC REG, SAS) has been made between carbon footprint and average daily gain, initial and final live weight, length of the fattening period, age at purchase and age at sale. Regression results significant only for age at sale with an intercept and slope equal to -2.86 and 0.64 ($P<0.01$). Regression between average daily gain and carbon footprint shows an intercept and slope of 12.63 and -3.45 ($P=0.07$). The relationship between carbon footprint and age of sale can be explained because feed efficiency decreases with age. This work was carried out in collaboration with ASPROCARNE, UNIVARVE and CRPA, it was funded by European Commission area Environment (LIFE BEEF CARBON project).

Session 19 Poster 19

Integral use of rapeseed as a sustainable alternative for Idiazabal PDO cheese-making

N. Mandaluniz[1], A. García-Rodríguez[1], I. Goiri[1], J. Arranz[1], F. Ajuria[2] and R.J. Ruiz[1]
[1]NEIKER, Animal Production, P.O. Box 46, E-01080, Spain, [2]Kerexara s.l.l., Barrio Kerixara s/n, Aramaio (Spain), E-01169, Spain; igoiri@neiker.eus

Dairy sheep production in the Basque Country has been traditionally based on a pasture-based farming system with a local dairy breed. However, feeding management practices have changed during the past few decades as a result of the intensification of these traditional systems, with a higher dependency from inputs, which means direct consequences on environmental impacts. Therefore, the utilization of local energy and protein sources can contribute to enhance the competitiveness, efficiency and sustainability of livestock farming. Under this scenario, TURTLIO (www.turtolio.com) Operative Group submitted a proposal within the framework of the funding programme for cooperation corresponding to Regulation (EU) No. 1305/2013 of the European Parliament and the Council of 17 December, and the Rural Development Program of the Basque Country 2015-2020. The objective of TURTOLIO was to perform an integral use of rapeseed in an on-farm assay, to assess the potential impact of rapeseed's by-products after cold pressed: cake for dairy sheep feeding and oil to heat milk within the on-farm Idiazabal PDO cheese-making process. During 2017 an on-farm assay was carried out in Kerexara S.L.L. sheep were fed including 20% of cold-pressed rapeseed cake in the concentrate during the milking season (February-August) and oil was used in the boiler for cheese-making. During this period sheep consumed 6,500 kg of rapeseed cake and 8,000 kg of Idiazabal cheese were produced. According to previous results, cold-pressed rapeseed cake reduces about 15% feeding costs and 3% enteric methane emissions, while improves the fatty acid profile of milk by increasing the omega-3 concentration. In addition to technical-economic and environmental benefits, the integral use of rapeseed increases self-sufficiency on farms and promotes the circular economy and energy efficiency, increasing the sustainability and competitiveness of the flocks.

Session 19 Poster 20

Alternative to eCG- treatment for fix time AI in piglet production with group farrowing system

M. Waehner[1] and H.P. Knoeppel[2]
[1]Anhalt University of Applied Sciences, Agriculture, Strenzfelder Allee 28, 06406 Bernburg, Germany, [2]MSD Tiergesundheit Deutschland GmbH, Feldstraße 1a, 85716 Unterschleißheim, Germany; martin.waehner@hs-anhalt.de

The following requirements are imposed on pig production in present time: (1) Timed organization of important processes for piglet production, (2) Securing continuously high reproductive and growth performance, (3) Maintaining high animal health stable management according to the 'all in - all out' system, (4) Ensuring the required qualities in the animal and the product. (5) Keep animal health and welfare. The piglet production is organized mainly according to the group farrowing system. Fix time insemination of all sows in a group, which are weaned in the same time, requires the simultaneous entry of estrus in all sows of the group. In practical farms several hormonal treatments are used for securing the simultaneous entry of oestrus in all sows. Following hormones are available: (1) Gonadotrophins (eCG, hCG) and their combinations (PG600; MSD) (2) Releasinghormones (GnRH, Porceptal®/MSD) (3) Synthetic lGnRH-III (Peforelin®/ Veyx). In present time the production of PMSG with the help of pregnant mares is viewed critically. Therefore, it is necessary to look for alternatives that spare the use of eCG. Especially synthetic Releasinghormones (GnRH) are favoured for use in commercial piglet production with fix time insemination and group farrowing system. Compared to eCG-treatment for simultaneous onset of estrus in sows and gilts equal fertility results are achieved. The proven production management can be maintained. Reproduction performances of sows after different biotechnical treatment for fix time insemination. Study was performed in practical farm with 1.350 sows with 4 week lactation and 7-day batch farrowing. Two methods were used: (1) eCG/ Heat-Orientated (d-hAI) (n=29) (2) GnRH (Buserilin)/Double Fixed AI (B-dFTI) (n=26) d-hAI: farrowing rate: 93,1%; 13,7±4.1 live born/litter, 1.275 live born/100 first AI B-dFTI: farrowing rate: 92.3%; 15.0±2.8 live born/ litter, 1.389 live born/100 first AI Generally, the use of GnRH is a biotechnical alternative to the eCH treatment for fix time inseminations in commercial piglet production with group farrowing system.

Session 19 Poster 21

Can we combat anthelmintic resistance in ruminants?

J. Charlier[1], S. Sotiraki[2], L. Rinaldi[3], E. Claerebout[4], E. Morgan[5], G. Von Samson-Himmelstjerna[6], F. Kenyon[7] and H. Hoste[8]
[1]KREAVET, Hendrik Mertensstraat 17, 9150, Kruibeke, Belgium, [2]Vet Res Institute HAO-Demeter, NAGREF Campus Thermi, 57001, Greece, [3]University of Naples Federico II, Dept of Veterinary Medicine and Animal Productions, Via Delpino 1, 80137 Napoli, Italy, [4]Faculty of Veterinary Medicine Ghent University, Laboratory of Parasitology, Salisburylaan 133, B-9820 Merelbeke, Belgium, [5]Queen's University Belfast, Institute for Global Food Security, Lisburn Road, BT9 26 7BL Belfast, United Kingdom, [6]Freie Universität, Institute for Parasitology and Tropical Veterinary Medicine, 30 Robert-von-Ostertag-Str. 7-13, 14168 Berlin, Germany, [7]Moredun Research Institute, Pentlands Science Park, Edinburgh EH26 0PZ, United Kingdom, [8]UMR 1225 IHAP INRA/ENVT, 23 Chemin des Capelles, 31076 Toulouse, France; smaro_sotiraki@yahoo.gr

Helminth parasitic pathogens cause severe disease and are amongst the most important production-limiting diseases of grazing ruminants. Frequent anthelmintic use to control these infections has resulted in the selection of drug resistant helminth populations. Anthelmintic resistance (AR) is today found in numerous major helminth species across the EU and globally. COMBAR (COMBatting Anthelmintic Resistance in Ruminants) is a recently launched COST Action (2017) which aims to advance research on the prevention of AR in helminth parasites of ruminants and disseminate current knowledge among all relevant stakeholders. COMBAR aims to integrate, evaluate and assess the economic trade-off of the novel developments in the field mainly by networking and has already attracted scientists from 27 countries. The Network has been organised around three Working Groups (WG): WG1 'Improving Diagnosis' which aims to prioritise, evaluate and implement cost-effective methods for the diagnosis of helminth infections and AR; WG2 'Understanding the socio-economic aspects' which aims to develop, disseminate and apply methods to study the economics and human behaviour in the field of helminth control in ruminants and WG3 'Innovative, sustainable control methods' which aims to develop practical and sustainable helminth control strategies that integrate current insights from diagnostics, Targeted (Selective) Treatment approaches, epidemiology, anti-parasitic forages, vaccinology, farm economics and human behaviour.

Session 19 Poster 22

Angus and Hereford breed bulls suitability for fattening with grass forage

I. Muizniece and D. Kairisa
Institute of Agrobiotechnology, Latvia University of Agriculture, Liela Street 2, 3001 Jelgava, Latvia; muiznieceinga@inbox.lv

Beef cattle fattening with grass forage is an environment friendly fattening method, which prevent direct competition between animals and human for food products (feed no food) and ensure the sustainable use of resources. Cattle fattening with grass should be inexpensive and profitable to the farmer, therefore the most suitable beef cattle breeds should be find out, as well as the pasture seasonal benefits should be used, because the grass forage in Latvian weather conditions is the cheapest feeding material during the summer period. In the study was used Angus (n=21) and Hereford (n=26) breed bulls which fattened and slaughtered in year 2015, 2016 and 2017 within framework of project 'Baltic Grassland Beef'. Bulls were born in the spring season (March, April), after weaning for fattening was used grass forage: in winter period – silage and hay, in summer period – pasture grass. For the study, bulls with live weight before slaughter from 459 kg to 594 kg were selected. The average age before slaughter of Hereford breed bulls was 614 days, but Angus bulls age was – 558 days ($P \leq 0.05$). The biggest slaughter weight (281.8 kg) was obtained from Angus breed bulls, which was 11.6 kg more than the Hereford breed bulls. The biggest dressing percentage also was obtained from Angus breed bulls – 52.0% (the Hereford breed bulls – 50.9%). Results of EUROP classification showed that the best carcass conformation score was for Angus group bulls – average 2.8 points (76% R, 24% O), but Hereford breed bulls got in average 2.5 points (50% R, 50% O) ($P \leq 0.05$). Carcasses from both groups were evaluated as 2nd and 3rd fat class, on average in groups receiving 2.1 (Angus) and 2.2 (Hereford) points. Obtained results showed that the bulls of both breeds are suitable for fattening with grass forage but Angus breed bulls can be realized younger with better slaughtering results.

Future expectations of producers and consumers from poultry genetics

R. Preisinger
EW GROUP GmbH, Hogenbögen 1, 49429 Visbek, Germany; rudolf.preisinger@ew-group.de

In commercial poultry breeding, large closed gene pools are extensively tested and selected for market requirements which must be anticipated at least 5 years ahead. Animal welfare and cage-free housing dominate the future needs of the global market. Retailers are setting individual standards for feeding and welfare criteria which are often a clear contrast to efficiency and sustainability. Federal animal welfare labels can be used to set national or even global standards and harmonise market needs. The ban on mutilations and new housing systems have shifted selection priorities. Stronger shells for longer production cycles have to be combined with better bones, minimal tendency to develop feather-pecking or cannibalism, and perfect nesting behaviour. Selection for adjusted beak shape, skeletal integrity, shell stability and better feed efficiency, are combined in a balanced selection approach for performance, quality and welfare traits. No big gene effects can be expected to control multifactorial problems like feather-pecking and cannibalism. Adjusting the shape of the beak by accurate data recording and selection with a heritability of 0.10-0.25, can contribute to reduce the risk of feather damage and severe cannibalism. For better skeletal integrity, the assessment of bone quality in pedigree birds housed in enriched cages is done by palpation of the keel bone or ultrasound measurement of the humerus. The combination of pure line, cross-line testing and genome-wide marker analysis, are established tools used to generate more progress for a balanced performance and behaviour profile. Line-specific SNP chips with 50-70,000 SNPs are used to select young males before sexual maturity, i.e. those which do not have own performance for sex-limited traits. Dual-purpose breeds are no alternatives to avoid culling day-old males. Sex determination in the egg after 4 days of incubation will be a global solution for this major topic in egg production where semen sexing is not an alternative.

Cope with climate change through knowledge

M. Pasqui
Istituto di Biometereologia, Centro Nazionale delle Ricerche, Via Giovanni Caproni 8, 50145 Firenze, Italy; m.pasqui@ibimet.cnr.it

Air temperature warming in the Central Mediterranean basin represents one of the observed climate change clearest footprints mainly related to the anthropic activities. Present knowledge of long-term temperature variability along with the heat waves occurrence in the last decades are key feature to cope impacts of climate change in the agricultural sector. Recent results highlight how heat waves frequency increased over last decades and that exists a robust functional dependence between heat waves in the Central Mediterranean basin and large-scale circulation over Northern Africa. These high impact weather events represent a critical challenge for adapting agriculture to climate change. Thus, new interdisciplinary modelling approaches are needed to provide climate change impacts integrated assessment and new climate knowledge paradigms, like climate services, are now available and ready to be used in the decision-making chains. Such ready-to-use climate knowledge will add new options to cope climate change impacts on a wide range of agricultural activities.

Session 20 Theatre 3

Food 2030: the role of livestock sector and animal science in achieving the Sustainable Development Goals

B. Besbes, A. Mottet, R. Baumung, P. Boettcher, G. Leroy, G. de'Besi and A. Acosta
Animal Production and Health Division, Food and Agriculture Organization of the United Nations (FAO), Viale delle Terme di Caracalla, 00153 Rome, Italy; badi.besbes@fao.org

In January 2016, the United Nations officially launched the 2030 Agenda for Sustainable Development with its 17 Sustainable Development Goals (SDGs) and 169 targets. The SDGs seek to address, in a sustainable manner, the root causes of poverty and hunger and the universal need for development. The livestock sector can contribute directly or indirectly to each of the SDGs, and in particular to SDGs 1, 2, 3, 5, 8, 10, 13 and 15. For decades, the livestock debate has focused on how to produce more with less. With the adoption of the UN 2030 Agenda, the focus has shifted from fostering sustainable livestock production per se, to enhancing the sector's contribution to the achievement of the SDGs. Because of the complex interactions among the SDGs, optimizing livestock's contribution to the UN 2030 Agenda requires careful planning, implementation and monitoring. There are synergies to be optimized and trade-offs to be managed. For example, increasing efficiency in livestock production reduces poverty and hunger as well as environmental impact, and can reduce disease burden. At the same time, increasing production by degrading and overexploiting resources reduces long-term productivity and sustainability. Maximizing livestock's contribution to the 2030 Agenda not only involves steering sector development along a sustainable pathway, it also requires policies and investments beyond the livestock sector. Strategies and reforms are needed to enhance small-scale producers' access to productive assets and rural services, make markets more transparent and efficient, have better access to financial resources and technological innovation, strengthen institutions to enable small-scale producers to act collectively and thereby be more competitive. Livestock policy frameworks that help governments and stakeholders to enhance the contribution of the sector to the 2030 Agenda are required. This calls for a change from a one-dimensional approach that addresses components of sustainability independently, towards a system approach that integrates its different components. Animal science and research will play a critical role in enhancing contributions of the sector to the SDGs by providing the necessary objectivity in the debate on livestock sustainability, which has become sometimes passionate. Policy and scientific actions required to address the future challenges and opportunities for the sector are discussed according to four major themes related to sustainability: food security and nutrition, livelihoods and economic development, animal and public health, and climate change and the environment.

Session 20 Theatre 4

A changing role for the animal scientist: understanding and responding to societal concerns regarding animal production practices

D.M. Weary
Animal Welfare Program, Faculty of Land and Food Systems, University of British Columbia, 2357 Main Mall, Vancouver, BC, V6T 1Z4, Canada; danweary@mail.ubc.ca

Over of the past century, much of the focus of animal scientists and those working in the livestock industries has been on increasing production and improved economic efficiency. In this talk I will argue that the greatest challenge now facing livestock production is that farming practices are often out of step with societal values. Public criticism of industry practices may be temporarily avoided by avoiding public scrutiny (e.g. using 'ag-gag' legislation to prevent the release of videos exposés), but I will review evidence showing how secrecy diminishes public trust. Some criticisms may also be blunted by attempts to 'educate' the public about farm practices, but I will review evidence showing how these attempts can also backfire. Instead I argue that our challenge is to find platforms for meaningful engagement with concerned citizens. I will review research showing how, as academics, we can use social science methods to better understand the attitudes of farmers, citizens and others, and thus identify shared solutions and barriers to change. I will also review examples of how new biological research can address important gaps in understanding, for example, in developing rearing systems that address public concerns around freedom of movement and social contact without putting animals at increased risk of disease.

Session 20 Theatre 5

Do animals have a role in future food systems?

I.J.M. De Boer
Wageningen University, Animal Production Systems Group, P.O. Box 338, 6700 AH Wageningen, the Netherlands; imke.deboer@wur.nl

It is widely recognized that the food system generates a broad range of environmental impacts and that the contribution of livestock is significant. For these and other reasons, the future role of animals in the food system is heavily debated. A central question is: what role, if any, could animals play in an environmentally sustainable food system? We demonstrate that animals raised under the circular economy concept could provide a significant, non-negligible part of our daily nutrient needs. We suggest that the role of animals in the food system should be centred on converting biomass that we cannot or do not want to eat into valuable products, such as nutrient-dense food (meat, milk, and eggs) and manure. By converting these leftover streams, livestock recycle nutrients back into the food system that otherwise would have been lost in food production. The availability of these biomass streams for livestock then determines a boundary for livestock production and consumption. We argue, therefore, that we should no longer focus on improving life-time productivity of animals, but on improving the efficiency with which animals recycle biomass unsuited for direct human consumption back into the food system. Future research, therefore, should be directed at questions, such as: which combination of animal systems (including insects and fish) convert available biomass streams most efficiently? Or, which technological or biological treatments can improve the safety, digestibility and nutrient availability of biomass streams available for animals? Acknowledging this recycling role of animals in the food system also offers potential to account for other functions livestock provide to humans and to produce with respect for the animal. It however also implies that developed countries have to significantly reduce their consumption of animal-source food.

Session 21 Theatre 1

What the hell do resilience and efficiency mean in the real world and what's the underlying biology

N.C. Friggens[1] and J.M.E. Statham[2]
[1]INRA, MoSAR, AgroParisTech, 16 Rue Claude Bernard, 75005 Paris, France, [2]RAFT Solutions, Mill Bank, Ripon, HG4 2QR, United Kingdom; nicolas.friggens@agroparistech.fr

Resilience and efficiency are increasingly sought after characteristics in livestock systems. As these systems move towards more sustainable livestock production there is a shift from more intensive management that seeks to fully control the animals' proximal environment to more open systems in which the animal may need to cope with greater environmental fluctuations (even if farm management in such systems will seek to buffer extremes). Thus, there is a growing need to select animals that have increased resilience as well as good efficiency. This is easier said than done; resilience and efficiency are complex traits for which multiple definitions exist. Although it may be relatively easy to subjectively agree on what is meant by an efficient or and resilient animal, when it comes to characterizing and quantifying these traits it becomes much more difficult to pin them down. This presentation aims to highlight the key elements of resilience and efficiency by considering the underlying biology of animal functioning alongside the practical perspective of the livestock manager, i.e. the frontline in development and management of the bespoke mix of animals offering the optimal balance of efficiency and resilience to best match the local production environment, and thereby contribute to improving the farms durability. Common elements that emerge from this exploration are: (1) the need for resilience definitions that consider not just effects of environmental perturbations on production but also effects on health, reproduction and welfare, (2) the timespan over which efficiency is measured needs to be long enough to account for trade-off effects of selection on performance, (3) on-farm technologies have a huge potential to contribute to phenotypic quantification of resilience and efficiency that is currently underused, and (4) farmers and vets have an important role to play in contextualising efficiency and resilience information and approaches.

What efficiency and resilience gains have we actually achieved in the past century or decade, and at what cost

J. Lassen[1] and Y. De Haas[2]
[1]Vikinggenetics, Ebeltoftvej 16, 8960 Randers, Denmark, [2]Wageningen University and Research, P.O. Box 338, 6700 AH Wageningen, the Netherlands; jalas@vikinggenetics.com

For the last two decades most countries have changed their breeding goal from main focussing on yield and conformation to a broader focus on health, fertility and longevity in a total merit index. This have been implemented due to the recognition of a phenotypic decline in the cost reducing traits. The reason why this has become possible is a much stronger focus and investment in new phenotypes and methods to register these phenotypes. Also, the development of national databases where the farmers can get use of the registrations is crucial in order to get good data. At the same time there is still much that can be achieved and improved. For fertility most countries use days from calving to first service and days from first to last service as the phenotype, which are clearly not good fertility traits. They are management inflated and not necessarily telling anything about the cows ability to show heat or get pregnant. For health traits treatments are often used. Again these traits are dependent on the farmers management decision and therefore not necessarily saying much about the cows disease resistance. With the introduction of total merit indices and use of these most countries show positive genetic trends for fertility, health and longevity and at the same time strong positive genetic trends for production. These trends though do not always show phenotypically. For feed efficiency very little data exists. The ongoing strong selection for production also gives a genetic response in feed intake, but the correlation is not one, which means that negative energy balance will get more and more severe problem unless proper genetic evaluations for feed intake and efficiency will be available. Over the last decades we have achieved genetic gains for efficiency and resilience traits because the focus on balanced breeding has been implemented in most countries and breeds. Over the future decades this will be improved even more through better phenotypes and genomic herd management tools. The GENTORE project will contribute to this development.

Session 21 Theatre 3

How fast can we change resilience and efficiency through breeding and management?

N. Gengler[1], M. Hostens[2] and Genotype Plus Environment Consortium (www.gpluse.eu)[3]
[1]ULiège-GxABT, Passage des Déportés 2, 5030 Gembloux, Belgium, [2]UGent, Salisburylaan 133, 9820 Merelbeke, Belgium, [3]Lead Partner UCD, Belfield, Dublin 4, Ireland; nicolas.gengler@uliege.be

The efficiency of dairy cattle has to be balanced against their resilience to disease challenges but also their individual responses to internal and external environmental stressors. A holistic approach is required as both efficiency and resilience have to be defined in a broad sense. Efficiency is more then only productive efficiency, the dilution of maintenance and indirect cost effects have to be considered through reduced feed, rearing, health and replacement costs; environmental costs should also not be forgotten. With this broader definition of efficiency, resilience as a major factor to reduce health costs contributes directly to a holistic view on efficiency. An underlying issue is here the old question if we change environments to address the needs of animals, or do we change animals to adapt to the environment. Previously, it was common understanding to prioritize first approach. However changing profoundly and suddenly environments and therefore management practices is difficult, disruptive and costly. However the EU FP7 project GplusE develops Hazard Analysis Critical Control Point (HACCP) and Evolutionary Operation (EVOP) based methods to optimize dairy cow management in a given production circumstance. The important advantage of this approach is that it is not disruptive but allows a slow but continued process of optimization. It is therefore not very different from the continuous process of cumulative optimization of the animals achieved by genomic selection, another important research topic of the GplusE project. Both approaches are complementary optimization opportunities for resilience and efficiency. However they are not possible without the development of appropriate response variables describing efficiency and resilience. The GplusE project has made the choice to develop novel milk based bio-markers and proxies for, often difficult to obtain; traits describing efficiency and resilience. Changes through breeding or management have to be continuous and well balanced considering the whole system. Changing environments or animals too fast should be avoided as this may lead to unforeseen consequences.

Session 21 Theatre 4

Interaction between the concepts of resilience and efficiency: how they apply to agriculture

V. Eory
Scotland's Rural College (SRUC), Peter Wilson Building, Kings Buildings, West Mains Road, Edinburgh, EH9 3JG, United Kingdom; vera.eory@sruc.ac.uk

Changing climate, environmental problems, inequality, increasing global demand for food: a multifaceted problem agriculture is facing all over the world. Improving efficiency and increasing the resilience of our food production systems entail changes which can help tackling those major problems. However, both resilience and efficiency can be considered in different systems, at various temporal and spatial scales and in relation to a range of outcomes. Since resilience emerged over forty years ago as a useful concept for describing the robustness of ecological systems, its use has been extended to cover economic and social systems in relation to a diverging range of perturbances. Similarly, efficiency – originally an engineering concept – has been used to assess biological, environmental, economic and social processes. This paper analyses the connection between the multiple resilience concepts, the relationship between the various efficiency aspects and the interplay of resilience and efficiency – in the context of agricultural systems. A literature review maps resilience and efficiency concepts in the joint domains of environmental, economic and social sustainability and presents operational frameworks used for quantitative assessment of agricultural systems. The aim of the paper is to provide an overview and clarification of the concepts for research modelling agricultural production and to suggest methods of defining and quantifying resilience and efficiency in relation to particular research questions and stakeholders.

Session 21 Theatre 5

Panel discussion on resilience and efficiency

N. Friggens
INRA, MoSAR, AgroParisTech, 16 Rue Claude Bernard, 75005 Paris, France; nicolas.friggens@agroparistech.fr

Panel discussion with stakeholders: (1) What should be the real-world goals of improving Resilience and Efficiency (2) Where will we find future gains in Resilience and Efficiency? (3) If you had €10m funding to improve Resilience and Efficiency: what would you blow it all on?

Session 21 Theatre 6

Operational measures of efficiency: make them measureable on large scale

H. Gilbert[1] and E.F. Knol[2]
[1]INRA, UMR GenPhySE, 31320 Castanet Tolosan, France, [2]Topigs Norsvin Research Center, bv, Schoenaker 6, 6641 Beuningen, the Netherlands; helene.gilbert@inra.fr

Finisher pigs tend to be 3-way crossbreds kept in pens of 8-15 in Europe and 20-30 in the Americas, with many exceptions. In theory it is possible to feed finisher pigs individually, but costs and management requirements of automatic feeding systems, and availability of easy to handle decision making software limit their use on commercial farms. Management unit is, therefore, the pen, even though there is still a lot of variation among individual animals (among others, mendelian sampling). In addition, there is phenotypic variation, since piglets differ in birth weight (which is a maternal trait), in colostrum intake and in establishment of gut microbiome coming from their (foster) dam, and later depend on their pen mates influence (activity, hierarchy) and farm ambient conditions. Measuring feed efficiency, and predicting it to feed animals according to requirements given a certain environmental condition, thus remains a challenge, but yet animal sorting options are being put into practice to optimize feeding on a pen level. Alternatives to individual electronic feeder measurements are also tested to produce individual measurements, including on-farm identification of drivers of the biological basis of feed efficiency (genomic and bio-markers) and measurements of components of feed efficiency (body composition, activity, gut microbiome composition). Feed efficiency is (still) mainly energy efficiency, and it has been the main driver of selection until now. However, energy requirements are now most often properly covered, protein efficiency is coming up and other efficiencies (vitamins, minerals) are being explored that might become of interest. In addition, by feeding animals high quality easy to digest feed, feed efficiency is essentially metabolic efficiency, but the increasing diversity of feed resources, including industry byproducts with higher dietary fibre contents, questions the opportunity to now focus on the genetic variability of digestive efficiency.

Session 21 Theatre 7

Developing resilience indicator traits based on longitudinal data: opportunities and challenges

H.A. Mulder, H.W.M. Poppe and T.V.L. Berghof
Wageningen University and Research Animal Breeding and Genomics, P.O. Box 338, 6700 AH, the Netherlands; han.mulder@wur.nl

A high general resilience is the ability to withstand environmental disturbances with minimal loss of performance, health and welfare. Disease resilience is a special form of general resilience related to how animals deal with diseases in terms of resistance and tolerance to infections. The aim of this contribution is to discuss recent developments to define resilience indicator traits for genetic studies and routine breeding value estimation based on longitudinal data and secondly to define the economic value of resilience. The two most obvious statistical measures of resilience are the residual variance and the slope of the reaction norm. Animals with a larger variance or a steeper slope (= decline) are hypothesised to be more sensitive to the environment than animals with a lower variance and a flatter slope. The utility of both concepts depends on whether the environmental disturbance affects the whole herd all at once, such as heat stress or a disease epidemic, or that only part of the animals are affected, such as mastitis or ketosis in cows. Simulations show that the residual variance is more suitable for animal-specific disturbances, while the reaction norm is more suitable for whole-herd disturbances. One of the main difficulties with defining resilience indicators is the validation (of these indicators), because often there are no clear validation traits available, such as disease records. On the other hand, resilience has an economic value in its own right. Resilient animals need less labour than less-resilient animals due to the difference in alerts they may get from management systems. We show that resilience is increasingly important for farms that grow and have limited labour available. Developing breeding tools for resilience will help to breed trouble-free animals and availability of data from sensor technology opens that avenue.

Session 21 Theatre 8

Genetics and economics of a feed efficiency breeding value for New Zealand dairy cattle

M.D. Camara[1], J.R. Bryant[1], K.A. MacDonald[1], M. Olayemi[1,2] and P. Amer[3]

[1]DairyNZ, New Zealand Animal Evaluation Limited, Private Bag 3221, Hamilton 3240, New Zealand, [2]Sugar Research Australia, 50 Meiers Road, Indooroopilly, Queensland,4068, Australia, [3]AbacusBio Ltd., 442 Moray Place, Dunedin 9016, New Zealand; mark.camara@dairynz.co.nz

Since 1996, the New Zealand dairy National Breeding Objective (NBO) has been to 'identify animals whose progeny will be the most efficient converters of feed into farmer profit.' To achieve this, New Zealand Animal Evaluation Ltd. (NZAEL) produces and publishes estimates of Breeding Worth (BW), an economically weighted index of overall genetic merit comprised of eight production and robustness traits (milk volume, fat, protein, liveweight, fertility, somatic cell score, body condition score, and residual survival). BW explicitly targets high gross lifetime efficiency, but currently does not include a direct measure of feed efficiency. The only available breeding values for feed efficiency in New Zealand are genomic estimates of residual feed intake (RFI) produced by Livestock Improvement Corporation (LIC) for Holstein-Friesian dairy sires. This breeding value is based on a small reference population and consequently has low reliability and low uptake from farmers. In 2011, as part of seven-year Primary Growth Partnership (PGP) called 'Transforming the Dairy Value Chain,' the NZ Ministry for Primary Industries, LIC, Fonterra, and DairyNZ initiated research on the feasibility and potential benefits of developing a routine animal evaluation for RFI. Research to-date has focused on young bulls and heifers using both traditional and genomic approaches to estimate the heritability of RFI, its genetic correlations with other components of BW, the genetic correlation between the sexes, comparing Holstein-Friesian and Jersey heifers, the economic value of RFI during the growth phase, and the utility of less expensive, higher-throughput predictor traits such as thermal imaging and mid-infrared analyses of blood. This presentation will summarize the results of this long-term research effort and evaluate potential strategies for routine national-scale evaluation and implementation of a feed efficiency breeding value in New Zealand.

Session 21 Theatre 9

Towards the quantitative characterization of piglet robustness to weaning: a modelling approach

M. Revilla[1,2], N.C. Friggens[1], L.P. Broudiscou[1], G. Lemonnier[2], F. Blanc[2], L. Ravon[3], M.J. Mercat[4], Y. Billon[3], C. Rogel-Gaillard[2], N. Le Floch[5], J. Estellé[2] and R. Muñoz-Tamayo[1]

[1]MoSAR, INRA, AgroParisTech, Université Paris-Saclay, 75231 Paris, France, [2]GABI, INRA, AgroParisTech, Université Paris-Saclay, 78352 Jouy-en-Josas, France, [3]GenESI, INRA, 17700 Surgères, France, [4]IFIP, Institut du porc-BIOPORC, 35651 Le Rheu, France, [5]PEGASE, INRA, 35000 Rennes, France; manuel.revilla-sanchez@inra.fr

Weaning is a critical phase in swine production conditions because piglets cope with different stressors that impact its health. During this period, the prophylactic use of antibiotics is still frequent to limit piglet morbidity, which raises economic and public health concerns such as the growing number of antimicrobial-resistant agents. With the interest of developing tools for assisting health and management decisions around weaning, it is key to provide robustness indexes that inform on the animal resilience to weaning. This task is hampered by the multiple-component nature of robustness. This work aimed at developing a modelling approach for facilitating the quantification of piglet resilience to weaning. We monitored 414 Large White pigs housed and fed conventionally during the post-weaning period without antibiotic administration. Body weight and diarrhoea scores were recorded before and after weaning. We constructed a dynamic model based on the Gompertz-Makeham law to describe live weight during the first 75 days after weaning following the rationale that the animal response is partitioned in two time windows (a perturbation and a recovery window). The transition time between the two windows is individual specific as well as model calibration, performed for each animal. The model captured the weight dynamics of animals at different degrees of perturbation. The power of the model is that it provides biological parameters that inform on the amplitude and length of perturbation, and the rate of animal recovery. We are currently investigating how to combine these proxy parameters to provide a robustness/resilience index to weaning. Next step is to correlate this index with individual diarrhoea scores and other health status measurements such as blood biomarkers and faecal microbiota diversity. We foresee that this study will provide a step forward in the quantitative characterisation of robustness.

Development of resilience indicators using deviations in milk yield from the lactation curve

H.W.M. Poppe, H.A. Mulder and R.F. Veerkamp
Wageningen University & Research, Animal Breeding and Genomics, P.O. Box 338, 6700 AH Wageningen, the Netherlands; marieke.poppe@wur.nl

Cows encounter various environmental challenges during their lives, such as infectious agents and heat waves. The extent to which an animal is affected in its functioning by environmental challenges is called resilience. Intensive recording of animal performance provides the opportunity to detect short term fluctuations in functioning and possibly define resilience. The objective of this study is to define resilience indicators using milk yield records of automatic milking systems, and to perform a genetic analysis. The data will consist of approximately 210,000 lactations finished between 1997 and 2017 from first, second and third parity cows from 500 Dutch herds. Individual lactation curves will be fitted using smoothing and random regression. Three resilience indicators will be defined: the variance, first-order autocorrelation and skewness of the residuals from the fitted curve. The genetic variance and heritability of the resilience indicators and the genetic correlations among them will be estimated using ASREML. To test the usefulness of the resilience indicators, genetic correlations between the resilience indicators and mastitis, ketosis, claw health and heat tolerance will be estimated. Moreover, validation will be done by investigating differences between resilient and non-resilient cows in incidence of health disorders and the response in milk yield during periods with high ambient temperatures. It is expected that the resilience indicators will show sufficient genetic variation and will be heritable. Furthermore, it is expected that cows that are identified as resilient, will be more resistant to diseases, and will show lower declines in milk yield during heat waves than less resilient cows. It is expected that the best resilience indicators found in this study can also be used on other sensor data and other species.

Indirect traits for feed efficiency

C. Egger-Danner[1], A. Koeck[1], C. Fuerst[1], M. Ledinek[2], L. Gruber[3], F. Steininger[1], K. Zottl[4] and B. Fuerst-Waltl[2]
[1]ZuchtData, Dresdner Straße 89/19, 1200 Vienna, Austria, [2]University of Natural Resources and Life Sciences Vienna (BOKU), Gregor-Mendel-Strasse 33, 1800 Vienna, Austria, [3]Agricultural Research and Education Centre, Raumberg 38, 8952 Irdning, Austria, [4]LKV Niederösterreich, Pater Werner Deiblstr. 4, 3910 Zwettl, Austria; egger-danner@zuchtdata.at

To meet the demand for sufficient food supply to provide for the world's growing population (around 10 billion in 2050), feed efficiency traits are gaining importance. Even in the age of genomic selection, enough phenotypes need to be available for the target traits. Feed efficiency traits are usually recorded at research stations. Smaller breeds do face the limitation of available phenotypes. In the framework of the Austrian project 'Efficient Cow' various indirect traits, which can be recorded on commercial farms, were analysed. The focus was on feeding information, body weight, possibilities of body weight prediction, mobilization and health. For predicted dry matter intake based on feeding information a heritability of 0.18 was estimated for Fleckvieh. Body weight as an indirect predictor for feed efficiency was assessed. Body weight prediction models were developed. The correlation between body weight (on scale) and predicted body weight varied depending on the information used for prediction. If only linear traits, which are routinely recorded, were used, the correlation between breeding values was up to 0.80. If muscularity or preferably heart girth were used, the correlation increased to 0.88. Various analyses within the project in the context of feed efficiency and fertility and health showed that e.g. more efficient cows mobilize more body reserves and have more fertility problems. If body weight is used as an indirect trait for feed efficiency the positive correlation to body condition score has to be taken into account. A selection for reduction of body weight will decrease body condition score. A lower body condition score results in increased health problems and higher culling rates. Therefore, it is recommended to include information on mobilization for prediction of feed efficiency. To fully exploit the potential of feed efficiency, animals have to be healthy as analyses within the project show.

Survival analysis for prediction of productive herd life in Nguni cows

M. Ngayo[1], V. Ducrocq[2,3], M.D. Fair[3], F.W.C. Neser[3], M.M. Scholtz[1,3] and J.B. Van Wyk[3]
[1]ARC, Animal Production Institute, Private Bag X2, 0062 Irene, South Africa, [2]GABI, INRA, AgroParisTech, Université Paris-Saclay, 78350 Jouy-en-Josas, France, [3]University of the Free State, Animal, Wildlife and Grassland Sciences, P.O. Box 339, 9301 Bloemfontein, South Africa; fairmd@ufs.ac.za

The productive herd life of Nguni cows was analysed using survival analysis. A semi-parametric Cox model, featured in the Survival was assumed. The data consisted of 1,245 calving records of cows, starting from 1996 to 2015 with parities from one up to 9. Animal records for cows in the productive herd alive on the 1st of January 1984 were truncated at that date. Records of cows still alive on 31 December 1995 were censored because of lack of information on the exact culling or death date. The cow herd was stratified into 3 different age groups at first calving (AFC) (18, 24 and 36 months). The likelihood risk ratio results indicated that year, AFC and parity all had a significant effect on risk of culling ($P<0.05$). The model yielded a heritability estimate of 0.02 and a standard error of 0.003 for productive herd life. Risk ratios for culling differed for different ages at first calving, over the years (from 1996 to 2015) and at different parities (from the 1st to the 9th parity). For AFC at 24 months, the culling risk ratio was the least compared to the other age groups. Heifers that had their first calf at 18 months of age were at a 0.061 risk of being culled, whereas the ones that had their first calf at 24 months were at a 0.042 probability of being culled in comparison to 1.000 for heifers that calved for the first time at 36 months of age. In 2015 about 250 cows were culled (more than any other year). This could have been due to a severe drought. More animals were culled in their 9th parity (120 cows) while the least number were culled in their first parity (4). Culling risk is high for cows calving for the first time at a very young age or at a very old age. Based on these results, it is recommended that heifers calve at 24 months of age for the first time in order to have the lowest risk of being culled from the herd.

Farmers' perceptions on parameters defining suckler cow efficiency

I. Casasús, S. Lobón and A. Bernués
CITA Aragón-IA2, Avda. Montañana 930, 50059 Zaragoza, Spain; icasasus@aragon.es

Most beef cattle breeding programs focus on traits related to calving ease and calf growth during lactation and fattening, selected because of their economic importance, easy measurement and adequate heritability to allow for genetic improvement via classical breeding programs. Other traits can also play a major role on cow lifetime productivity, such as number of weaned calves or cumulative weaning weight, but they have low heritability and long generation intervals. In the context of a survey that analysed the efficiency and resilience of suckler cattle farming systems in Spanish mountain areas (GenTORE H2020), farmers were asked to score the relative importance (1-not important to 5-very important) of several traits in order to define the efficiency of their cows: age at first calving, calving ease, fertility, cumulative number of weaned calves, calf weight at birth, at 90 days and at weaning, calf carcass conformation, cow size, cow udder conformation, feet and legs morphology, docility and use of low quality feedstuffs. We also asked if they actually registered these traits, and if they provided the information to any breeder association. Preliminary results indicate that despite 67% of the farmers belonged to breeder associations, only 14% of them delivered data for their breeding programmes. In fact, data were registered by relatively few farmers (age at first calving by 48%, fertility and calf birth weight by 29%, calving ease by 24%, calf weaning weight by 19%), mainly in large farms (>100 cows) and those that fattened their calves. However, most of these traits were considered important or very important to determine cow efficiency, with the highest scores given to calving ease (4.9), fertility (4.8), and docility and cumulative number of weaned calves (4.3). Calf conformation (4.1) and adult udder (4.1) and leg conformation (4.0) were also considered important. Surprisingly, calf weight traits were scored lower (3.9 at birth, 3.2 at 90 days, 3.7 at weaning), and the less important trait was cow size (2.8). At this stage, some trends were found between scores given by farmers with different farm size (below vs over 100 cows), predominant cow breed (authoctonous vs specialized beef breed) and type of marketed product (weaned vs fattened calf).

Session 21 Poster 14

A survey on sensors availability on Italian dairy farms: potential tools for innovative selection

I. Lora[1], A. Zidi[2], M. Cassandro[2], F. Gottardo[1] and G. Cozzi[1]
[1]University of Padova, Department of Animal Medicine, Production and Health, Viale dell'Università 16, 35020 Legnaro (PD), Italy, [2]University of Padova, Department of Agronomy, Food, Natural resources, Animals and Environment, Viale dell'Università 16, 35020 Legnaro (PD), Italy; flaviana.gottardo@unipd.it

The aim of the GenTORE project (Genomic Management Tools to Optimise Resilience and Efficiency across the Bovine Sector) is to develop new models for cow selection for improving animal Resilience and Efficiency towards environmental challenges, increasing also the sustainability of the bovine sector. Farm sensors could provide precious phenotypical information on individual cows. However, there is a lack of knowledge on the actual spread of sensor systems in Italian dairy farms. In this survey, type of sensor installed (pedometer, collar, or eartag) and parameters recorded (activity, rumination, eating, resting, and localization in the barn) were investigated on 993 dairy farms of the northeastern Italy through questionnaires submitted to the farmers by the breeders' association (ARAV) technicians. The average herd size was 68±66 (±SD) cows (min=2; max=668), for a total of 66,779 cows reared (72% Holstein, 10% crossbred, 7% Brown, 6% Simmental, and 5% local breeds). Overall, 15% of the farms had a sensor system, of which 87% were farms with more than 50 cows. The most frequent type of sensor was the collar (54%), followed by the pedometer (44%), and the eartag (only 1%). Among breeds, sensors were used mainly on Holstein cows (79%), followed by Browns and crossbreds (8% each), local breeds (3%), and Simmentals (2%). Both collar and pedometer were used mainly for heat detection (100%), and only 39% of the collars were able to record also eating and rumination. The most sophisticated eartags gave even information on resting behaviour and localization of the cow in the barn. Refined parameters such as eating and rumination were available only on 3% of the overall dairy farms, covering 5% of the farms with more than 50 cows and 6% of the overall cows reared. Therefore, the use of sensors in Italian dairy farms is not widespread yet, but the potentiality of those systems are enormous particularly if sensor recordings will be compared with individual cow productive, reproductive, and health data.

Session 21 Poster 15

Changes on female fertility aggregate index in Italian Holstein dairy cattle

G. Visentin, M. Marusi, R. Finocchiaro, J.B.C.H.M. Van Kaam and G. Civati
Associazione Nazionale Allevatori Frisona Italiana (ANAFI), via Bergamo 292, 26100 Cremona (CR), Italy; giuliovisentin@anafi.it

The aggregate index for female fertility was introduced into the national Holstein cattle breeding objectives in 2005, and included five selection criteria, namely days from calving to first service (DTFS), non-return rate at 56 d from first insemination (NR56), calving interval (CI), angularity (ANG), and equivalent-mature milk yield (MY). Conception rate at first service (CR) was the only breeding goal. The objective of the present study was to revise this aggregate index by including new information from both linear scoring (body condition score, BCS), services and pregnancy testing (interval from first to last insemination, IFL). After edits, information was available on 5,466,546 primiparae calving since 1994; to reduce computational time, six subsets of approximately 15,000 cows were extracted. (Co) variance components of DTFS, NR56, IFL, MY, BCS, and CR were estimated simultaneously with a multi-trait animal model, which included the fixed effects of herd-year-season of calving (DTFS, NR56, IFL, MY, and CR), age-year of calving (DTFS, NR56, IFL, CR), month of calving (DTFS, IFL, CR), month of insemination (NR56), herd-year-season of linear scoring (BCS), age-stage of lactation at linear scoring (BCS), and year of calving (BCS). Random terms were animal additive genetic and the residual. At a national level, means of DTFS, NR56, IFL, BCS, MY, and CR were 88 d, 59%, 71 d, 2.98, 10,483 kg, and 35%, respectively. (Co)variances estimated in each subset were averaged to derive G and R matrixes. Heritability estimates of fertility traits ranged from 0.013 (NR56) to 0.076 (DTFS); heritability estimates of BCS and MY were 0.18 and 0.29, respectively. Selection index methodology was employed to derive appropriate index weights from (co)variances between breeding goal (CR) and (among) selection criteria. Results from the present study indicated that the genetic response of the breeding goal using information on BCS and IFL, substituting ANG and CI, respectively, increases by 5.6% after one selection round.

Session 21 Poster 16

The value of commercial farm-management data to evaluate Pietrain boars for vitality and robustness

W. Gorssen, S. Janssens and N. Buys
KU Leuven Livestock Genetics, Biosystems, Kasteelpark Arenberg 30, bus 2456, 3001 Heverlee, Belgium; wim.gorssen@kuleuven.be

Breeding more viable and robust pigs is becoming a priority. Farmers desire easy to manage – labour efficient – animals which are economically efficient despite environmental disturbances. Drivers for this trend are the increasing number of animals per farm, the worldwide distribution of genetic material in combination with climate change and an increasing public concern considering animal welfare and health. This study evaluates the usefulness of data collected by commercial farmers for evaluating the genetic merit of Pietrain boars for piglet vitality/robustness. Litters where scored by farmers for vitality within 24 hours after birth. Scores were given on a scale from 0 (low vitality) to 5 (high vitality). Pietrain boars (n=400) where mated with commercial crossbred sows (n=2,112) to produce 3,609 litters at six different farms spread across Flanders (Belgium). Genetic parameters and EBVs of terminal Pietrain boars were calculated using a sire model via the BreedR package in R. A two-trait model was used including the traits vitality and total number born alive (TNBA). Fixed effects where parity of the sow, farm, time period (month) and gestation length. Sow was included as a random effect. Results showed a low heritability for TNBA (7.2±3.4%) and a low to intermediate heritability for vitality (14.4±4.9%). For TNBA, the variance explained by the genetic sire-effect was about 10 times smaller compared to the sow-effect (0.22 vs 2.28), indicating that little genetic progress can be made in TNBA by selecting on the paternal side. For vitality however, the estimated variance of the genetic sire-effect was considerable compared to the sow-effect (9.2 vs 17.4), indicating that substantial genetic progress can be made by selecting on terminal Pietrain boars. Furthermore, genetic correlations between TNBA and vitality were close to zero (-0.01), pointing out that selecting for an increased vitality does not negatively affect TNBA. These results suggest that it is possible to use data collected on commercial farms for genetic evaluation of terminal boars and to improve litter vitality by selection.

Session 22 Theatre 1

Evaluation of models to predict feed intake in dairy cows

V. Ambriz-Vilchis[1,2], M. Webster[1], J. Flockhart[2], D. Shaw[3] and J. Rooke[2]
[1]BioSimetrics Ltd., Kings Buildings, EH9 3JG Edinburgh, United Kingdom, [2]SRUC, Future Farming Systems, Kings Buildings, EH9 3JG Edinburgh, United Kingdom, [3]Royal (Dick) School of Veterinary Studies and The Roslin Institute, The University of Edinburgh, EH25 9RG Roslin, United Kingdom; virgilio.ambriz@sruc.ac.uk

Mathematical models in the form of regression equations or dynamic mechanistic models have been used to describe animal systems. Feed intake (FI) is paramount in the performance of livestock and has been of interest when creating such models. The aim of the present study was to evaluate four models in their prediction of FI in dairy cows fed total mixed rations (TMR): BSM-Milk (BioSimetrics Ltd.) a dynamic mechanistic whole cow model, the FI equation part of the CNCPS, the model and that by Vadiveloo and Holmes. A trial was carried out at SRUC's Dairy Research Centre, Scotland UK. Two contrasting TMR were offered to forty Holstein Friesian cows. The diets were: forage (g/kg/DM Grass silage 0.40, maize silage 0.23, crimped wheat 0.11, beans 0.25 and min 0.01) and concentrate based (Wholecrop 0.40, Megalac 0.02, whey 0.08, min 0.01 and a concentrate 0.50). Using electronic feeders (HOKO, Insentec, The Netherlands) individual FI were recorded daily. Details of the animals and the TMR were used as inputs. FI predictions were compared to those obtained on-farm. To evaluate the predictions regression analysis, the limits of agreement (LoA) method and the concordance correlation coefficient (CCC) were used. All statistical analyses were carried out using R. The evaluated models predicted FI with different levels of success. Obtained R2 values were: 0.78 BSM-Milk, 0.48 CNCPS, 0.42 NRC and 0.48 VH. The CCC were 0.88 BSM-Milk, 0.58 CNCPS, 0.61 NRC and 0.34 VH. The LoA showed that BSM-Milk predicted FI in average 0.19 higher than observed (limits -3.80 to 4.19), similarly CNCPS predicted FI 0.98 higher than observed (limits -5.06 to 7.03), NRC predictions were -0.41 lower than observed (limits -6.80 to 5.98) and lastly VH predictions were 5.31 higher than observed (limits -0.70 to 11.31). BSM-Milk was the model with the best performance when compared with the rest of the evaluated models. Future research should compare BSM-Milk predictions to those obtained with other dynamic mechanistic models.

A new method of monitoring body temperature in horses with a microchip

S. Benoist[1], L. Wimel[1] and P. Chavatte-Palmer[2]
[1]The French Horse and Riding Institute (IFCE), Experimentation research center of Chamberet, Domaine de la Valade, 19370 Chamberet, France, [2]UMR BDR, INRA, ENVA, Université Paris Saclay, 78350, Jouy en Josas, France; stephanie.benoist@ifce.fr

In mammals, body temperature is regulated by thermoregulation and a circadian rhythm, which is an endogenous process. In veterinary medicine and in physiology, temperature represents a reference measurement, indicator of health of horse. The rectal temperature is the most common and usual method to measure temperature of a horse. The objective of study is to automatically analyse body temperature in foals using an identification microchip implanted in the neckline. The temperature is recorded each time the foal is in close proximity to specific antenna located near the drinking trough. Body temperatures recorded by the microchip in 62 foals (25 males and 37 females) from 6 to 12 months of age, over 3 generations (2015 to 2017), during the winter period, totalling 141,613 temperature readings. Foals were housed in 3 stalls within the same building (two groups of females and one group of males of the same generation) with a straw bedded and an exercise area. The stall occupied by the males was always the same. Food (concentrated feed and forage) was provided twice daily, with free access to water. Preliminary analyses with ANOVA for repeated measures indicate a circadian rhythm, a difference in body temperature between males and females (males having a higher temperature accounting for 3.6% of total variance, $P<0.0001$, with significant year effect accounting for 5.13% of total variance, $P<0.0001$ and sex × year interaction, accounting for 0.56% of total variance, $P<0.0001$). There was also a significant effect of the time of the day ($P<0.0001$), with lowest temperatures in the night and daily highest observed at 11:00 and 18:00, corresponding to 1-2 hours after food distribution. Now that we have developed this non-invasive tool to explore the role of extrinsic and intrinsic factors on body temperature in yearlings, the aim of our work is to explore intrinsic and extrinsic causes to body temperature changes.

Real-time animal response to climate changes

H. Levit[1], S. Goldshtein[1], S. Pinto[2], A. Kleinjan Elazari[3], V. Bloch[1], Y. Ben Meir[3], E. Gershon[3], J. Miron[3] and I. Halachmi[1]
[1]Agricultural Research Organization-The Volcani Center (A.R.O), Institute of Agricultural Engineering, (PLF) Lab, HaMaccabim Road 68, Rishon LeTsiyon, P.O. Box 15159, 7505101, Israel, [2]Leibniz Institute for Agricultural Engineering and Bioeconomy (ATB), Max-Eyth-Allee 100, Potsdam 14469, Germany, [3]Agricultural Research Organization-The Volcani Center (A.R.O), Institute of Animal science, HaMaccabim Road 68, Rishon LeTsiyon, P.O. Box 15159, 7505101, Israel; harelle@volcani.agri.gov.il

The immense metabolic demand of high yielding cows is constantly requires disposing (by panting, sweating and vasodilatation) large amount of metabolic heat. Inability to do so causes heat stress that results animal suffering and deteriorates production and reproduction. In many arid and semi-arid zones, fans and water sprinklers are used in barn, in cooling yards and/or in the feeding lane to help the cows to get rid of excessive heat. It is common to activate the cooling management by surrounding temperature-humidity index (THI) measurements and not by real time body temperature measurements of the cow that might be different (due to production level, body condition, gynaecologic status, herd genetic variance) in response to heat stress, etc. Real-time temperature can now be measured using reticulorumen bolus (SmaXtec). In the current study, we developed a model that calibrate the cow temperature from the reticulorumen to the vaginal temperature loggers – the vaginal temperature was our gold standard. A total of 30 lactating cows were randomly assigned to one of two treatment groups (evenly by lactation, energy corrected milk-ECM and days in milking), fed the same TMR: Treatment 1, which is the common cooling methods used in farms (time based cooling), was compare to sensor based cooling regime – treatment 2. The sensor based cooling group showed higher milk fat (3.65 vs 3.43%), milk protein (3.23 vs 3.13%), ECM (42.84 vs 41.48), FCM 4% (fat corrected milk; 42.76 vs 41.34) and lower body temperatures (38.6 °C). The preferred cooling regime, after carrying out a series of tests on cooling time and duration along the trial, result with a stabilize animal reaction. The sensor base cooling found to be an effective tool to detect and ease heat stress in intensive dairy cows in arid and semi-arid zones.

Session 22 Theatre 4

Physiological diversity between individuals: when do we need 'personalized livestock farming (PLF)'?

I. Halachmi, N. Barchilon, V. Bloch, A. Godo, Y. Lepar, H. Levit, E. Vilenski, M. Kaganovich, R. Bezen, O. Geffen, T. Glasser and S. Druyan
PLF Lab., Institute of Agricultural Engineering; Animal Sci Inst. Ramat Hanadiv, Agricultural Research Organization (A.R.O.), The Volcani Centre, P.O. Box 15159, HaMaccabim Road 68, Rishon LeZion, Israel; halachmi@volcani.agri.gov.il

Precision livestock farming (PLF) can be defined as real-time monitoring technologies aimed at managing the smallest manageable production unit's temporal variability, known as 'the per animal approach'. This lecture will go through five successful PLF case studies, based on five EU projects run in the Israeli PLF lab at the ARO: Attenuating heat stress (Bolus temperature, OptiBarn), Automatic body condition scoring (health, hunger, OptiScore), A sensor for early detection of cow lameness (attenuating pain, BioBuseness), Early detection of calving diseases (health); (5) Sensing and caring thermoregulation of broiler, PLF in small ruminants, Monitoring the cow individual feed deficiency (EU-PLF).

Session 22 Theatre 5

Automatic lameness detection in sows using the sow stance information system (SowSIS): a pilot study

P. Briene, O. Szczodry, P. De Geest, A. Van Nuffel, J. Vangeyte, B. Ampe, S. Millet, F. Tuyttens and J. Maselyne
ILVO (Flanders research institute for agriculture, fisheries and food), Burg. Van Gansberghelaan 92 bus 1, 9820 Merelbeke, Belgium; jarissa.maselyne@ilvo.vlaanderen.be

Lameness is a very prevalent problem in breeding sows, which often goes undetected for long periods of time. This can have severe consequences for animal welfare and has impact on the productive performance of sows. Lameness in sows is often hard to detect as sows are not observed while walking on a daily basis. Automatic detection of lameness in sows could help pig farmers to recognize the problem sooner, allowing them to act and thus possibly preventing the problem from getting worse. The SowSIS consists of 4 force plates, providing output for each leg separately. It is built into a Nedap electronic sow feeder (ESF) to allow continuous data collection without interfering with the sows. Several stance information variables can be extracted from the data. In a previous study comparing single measures of 6 sound and lame sows, weight asymmetry, number of weight shifts and number of kicks were the most promising variables to detect lameness extracted from the SowSIS data. The SowSIS has since been improved by using multiple load cell-mounting for higher accuracy per plate. A pilot test study is currently ongoing where data is continuously collected by 4 separate SowSIS in group-housing for a period of 74 days during gestation. Feeding visits are extracted from the data and linked to the ESF data to allow identification of individual sows. All sows are visually scored for gait twice a week using a 150 mm continuous tagged visual analogue scale (tVAS) to determine whether they are lame or not. Case studies of lame and non-lame sows will be selected from the data of 2 production groups containing 16 and 18 sows. These data will be analysed and compared to the visual gait scores to determine the ability of the SowSIS to correctly detect lame sows and to test which of the pre-determined variables are most useful to detect lameness in different cases. The dataset is unique as it provides insight into the development and the course of lameness over a longer period of time in individual breeding sows. The results from several case studies will be presented at the conference.

Advantages of individual feed consumption and weight monitoring in growing-finishing pigs

A. Peña Fernández[1], T. Norton[1], A. Youssef[1], C. Bahr[2], E. Vranken[1,3] and D. Berckmans[1]
[1]KU Leuven, Animal and Human Health Engineering division, M3-BIORES Laboratory, Kasteelpark Arenberg 30, 30001 Heverlee, Belgium, [2]AGRIFIRM Belgium N.V., Industrieweg 18, 2280 Grobbendonk, Belgium, [3]FANCOM B.V., Research Department, P.O. Box 7131, 5980 AC Panningen, the Netherlands; alberto.penafernandez@kuleuven.be

Feed is the largest item cost in pig production, around 70% of the total cost of growing a pig. Today's feeding strategies at commercial farms are set at group level according with the needs of the most demanding pig. Precision Livestock Farming (PLF) technologies gives the opportunity to monitor each pig individually and provide them individualized feeding strategies which meets their nutritional needs, improving their health and welfare. The aim of this work is to show the potential of applying real-time modelling and forecasting to monitor the weight development of grower-finisher pigs at individual level, according to their feed consumption. Three different experiments, with 80 growing-finishing pigs in each one of them, have been performed in order to check pig´s response to sudden changes in features of the feed composition, such as energy and/or lysine content, and restriction periods along the fattening period. The experiments were carried out in the facilities of an experimental farm in The Netherlands, following typical Dutch pig husbandry conditions. Pigs were housed in 4 pens of 16 m^2 and each one equipped with a feeding station. Applying both, Transfer Function (TF) and Dynamic Linear Regression (DLR) approaches to the time-series for the amount of feed consumption and weight, it is possible to monitor and forecast the response of an individual pig over the growing-finishing period to its feed consumption. On average, a group level approach leads to an average performance of the models of 67% fitting agreement, meanwhile the individualized TF model approach reaches a fitting agreement of 94% on average. Moreover, testing the forecasting properties of the individualized approach, the Mean Relative Prediction Error (MRPE) of applying the DLR approach with a window size of 5 days and a forecasting horizon of 1 day is 2%. These properties allow to monitor individual pig's growth and provide the farmer useful information to manage the pigs individually.

Technologies for identification, localisation and tracking of indoor housed livestock

E.F. Hessel
Thuenen-Institute of Agricultural Technology, Federal Research Institute for Rural Areas, Forestry and Fisheries, Bundesallee 47, 38116 Braunschweig, Germany; engel.hessel@thuenen.de

Sizes of livestock housing systems has increased in the last recent years. In order to ensure animal welfare even in large-scale housing systems technologies have been developed to monitor livestock behaviour and to assist farmers in their decisions. Among other, technologies for animal identification, localisation and tracking are key elements for monitoring livestock on an individual level. Passive Radio Frequency Identification (RFID) systems have been successfully used in animal facilities for many years and research has identified potential applications in behaviour monitoring for automated illness detection. However, by using passive RFID animals are only identified in the range of antennas. Active livestock localization systems can detect positions, movements and velocities of farm animals, as well as record travelled paths within the whole stable. By tracking the routes also activities like social interaction will be detectable. Besides the simplest function of localization systems for finding animals in the barn, these systems especially offer new monitoring instruments for practical farmers. Individual localisation of animals realises individual monitoring. Different activities like walking, standing, feeding, etc. are detectable. If routes and the circadian activities can be measured, farmers can be informed about suspicious changes at an early stage in order to monitor health and to treat animals in case of illnesses at an individual level. Localisation in combination with other sensors (e.g. rumination sensors) can provide an even more precise information about individual animals. In this contribution, techniques like active RFID and ultra-wide band (UWB) for localisation and tracking will be described. Advantages, disadvantages of different technologies will be discussed so that the limitations of systems for indoor localisation and tracking of livestock will be comprehensible.

Session 22 Theatre 8

Debating physiological diversity between individuals: do we need 'personalized farming'?

J. Maselyne[1] and E. Hessel[2]
[1]ILVO, Belgium, [2]Thuenen, Germany; jarissa.maselyne@ilvo.vlaanderen.be

Debating physiological diversity between individuals: do we need 'personalized farming'? The debate will discuss common issues that will be raised during the session. The session speakers and posters' authors will form a panel-of-experts that will answer questions from the audience and identify common research challenges and potential engineering solutions. 15 minutes just before coffee/diner.

Session 23 Theatre 1

Fatty acid profile of conventional, organic, and free-range retail milk in the United Kingdom

S. Stergiadis[1], C.B. Berlitz[1,2], B. Hunt[1], S. Garg[1], K.E. Kliem[1] and I. Givens[1]
[1]University of Reading, School of Agriculture, Policy, and Development, Animal, Dairy and Food Chain Sciences, P.O. Box 237, Earley Gate, RG6 6AR, Reading, United Kingdom, [2]Federal University of Rio Grande do Sul, Department of Animal Science, Av Bento Gonçalves, 7712, 91540-000, Porto Alegre, RS, Brazil; s.stergiadis@reading.ac.uk

Dairy production system is known to affect UK retail milk fatty acid (FA) profile, but the extent of the differences across the year and the FA profile of free-range retail milk, is unknown. This study assessed the FA profiles of retail milk from three UK dairy production systems (conventional, CON; organic, ORG; free-range, FR), over 12 months. Milk samples (n=120) from four CON, four ORG and two FR brands were collected monthly. ANOVA was derived from linear mixed effect models using production system, month, and their interaction, as fixed factors and milk ID as random factor. Overall, ORG milk contained more of the nutritionally beneficial polyunsaturated FA (+12.5%), omega-3 FA (+47.6%), vaccenic acid (+40.0%) rumenic acid (RA; +34.5%), α-linolenic acid (ALNA; +53.0%), eicosapentaenoic acid (+39.3%) and docosapentaenoic acid (+28.8%), and less of the nutritionally undesirable palmitic acid (C16:0; -5.1%), when compared with conventional milk. Differences were more pronounced in the outdoor/grazing than the indoor months, also including lower concentrations of saturated FA (-2.5%) in ORG milk in April-May. These findings may be due to the higher forage:concentrate ratio and increased contribution of pasture and clover in the ORG than CON cow diets (differences which are less distinct in the indoor months). CON and FR milk had an overall similar FA profile, but FR milk contained less C16:0 (-6.6%) and more ALNA (+23.2%) and RA (+24.1%), in specific months in the outdoor/grazing months. Consuming ORG milk may lead to increased intakes of beneficial FA and reduced intakes of undesirable C16:0, but potential health effects of the long-term exposure of consumers to these differences is unknown.

Session 23 Theatre 2

Cheese-making properties and cheese yield of milk from Holsteins and 3-way rotational crossbred cows

S. Saha, N. Amalfitanò, G. Bittante and L. Gallo
University of Padova, DAFNAE, Via dell'Università 16, 35020 Legnaro, Padova, Italy; sudeb.saha@studenti.unipd.it

Crossbreeding is regarded as a viable strategy to counterbalance the declining fertility, resilience and longevity of purebred Holstein (HO) cows, but knowledge about cheese yield properties of milk from crossbred (CROSS) cows is still scarce. The aim of this study was to investigate the effects of a 3-way rotational crossbreeding (ProCROSS) of Holstein (HO) cows with Viking Red in first (F1, VR×HO), Montbéliarde in second [F2, MO×(VR×HO)], and again HO sires in third generation {F3: HO×[MB×(VR×HO)]} on milk coagulation (MCP), curd firming and curd yield (CY) traits. Milk samples (n=333) were collected once during the morning milking from HO (n=194) and CROSS (n=139: 37 F1, 44 F2, 58 F3) cows reared in a commercial farm located in Northern Italy. Individual milk samples were analysed for assessing milk composition, single-point MCP and modelling curd-firming over time (CF_t) traits, and CY measured through a laboratory cheese-making procedure. Data were analysed using a linear model including the fixed effects of parity, DIM class and breed combinations. No difference between HO and CROSS was found for milk yield (34.8 kg/d), fat (3.91%), lactose (5.09%), urea (20.6 mg/dl) and pH (6.59). Milk from CROSS cows showed a greater content of protein (3.71 vs 3.60%) and casein (2.89 vs 2.79%) and a smaller SCS (2.81 vs 3.34), without differences among generations. Milk from CROSS cows was also characterized by a more favourable firming time (8.3 vs 9.8 min) and curd firmness at 30 min (20.2 vs 14.8 mm) than HO, whereas some differences in favor of F2 and F3 generations respect to F1 were found for coagulation time and curd firmness after 30 min. The recovery of milk protein in the curd was greater for CROSS cows (80.4 vs 79.4%), whereas recovery of fat and total solids, and CY were not different. Also the daily production of fresh cheese per cow was not different between HO (6.2 kg/d) and CROSS (6.1 kg/d) cows. In conclusion, the 3-way rotational crossbreeding scheme considered allows to obtain daily milk and cheese yield similar to purebred HO breeding and some advantage for milk quality and technological properties. The trial is currently in progress to increase cow sample size and number of herds involved.

Session 23 Theatre 3

Comparing cheese making properties of milk from lowland and highland farming systems

G. Niero[1], M. Koczura[2], M. De Marchi[1], S. Currò[1], M. Kreuzer[2], G. Turille[3] and J. Berard[2]
[1]University of Padova, Department of Agronomy, Food, Natural Resources, Animals and Environment (DAFNAE), Viale Dell'Università 16, 35020 Legnaro (Padova), Italy, [2]ETH Zurich, Institute of Agricultural Sciences, Universitaetstrasse 2, 8092 Zurich, Switzerland, [3]Institut Agricole Régional d'Aoste (IAR), Regione La Rochère 1/A, 11100 Aosta, Italy; giovanni.niero@studenti.unipd.it

In various European mountainous regions with transhumance established since centuries, dairy cows are moved from lowland (LO) to highland (HI) pastures. Highland sojourn lasts from late spring to late summer, and this activity is important for the preservation of typical regional dairy products, marginal areas and biodiversity. Previous studies indicated that cheese-making properties of HI milk is lower than that of LO milk, but this was mainly tested with high-genetic merit cow types. The present study aimed to investigate the effect of HI sojourn on cheese making properties of Aosta Red pied (ARP) cows, a dual purpose cattle breed farmed in Aosta Valley. Fontina PDO cheese, is manufactured from unpasteurized ARP milk, within 2 h after every single morning and evening milking. Milk coagulation properties were measured through a lactodynamographic tool in the milk of 47 cows, sampled in May before transhumance, and in June and July after transhumance. Sources of variation were investigated using linear mixed models, including parity, site, milking time, the interaction parity × site, milking time × site, and milking time × parity. Cow was nested within site, and used as subject for repetition, and sampling date was included as repeated factor. Morning and evening milk were similar in coagulation properties. Curd firming time and firmness did not differ between LO and HI. Rennet coagulation time was prolonged in HI compared to LO both in primiparous (16.4 vs 18.5 min) and multiparous animals (17.5 vs 21.1 min, respectively), indicating that experience of cows was not effective in this respect. The percentage of non-coagulating samples was greater in HI (15.0%) compared to LO (8.5%). The lower milk reactivity to rennet addition in HI seems to be mostly related to the simultaneously increasing somatic cell score rather than to the milk composition in terms of protein and casein, which were not affected by transhumance.

Chemical composition, hygiene characteristics and coagulation aptitude of milk for Parmigiano Reggia

F. Righi[1,2], P. Franceschi[2], M. Malacarne[1,2], P. Formaggioni[2], C. Cipolat-Gotet[2], M. Simoni[2] and A. Summer[1,2]
[1]University of Parma, MILC Center, Parco Area delle Scienze, 59/A, 43124 Parma, Italy, [2]University of Parma, Department of Veterinary Science, Via del Taglio, 10, 40126 Parma, Italy; federico.righi@unipr.it

The aim of the research was to compare milk quality parameters among herd characterized by different levels of milk production. The trial involved 1,080 bulk milk samples collected from 30 dairy herds producing milk for Parmigiano Reggiano cheese. Herds were grouped in 3 classes (10 herds per class) according to their production level (kg/cow/lactation):from 6,000 to 7,999 (L), from 8,000 to 9,999 (M) and from 10,000 to 12,000 (H). The average herd size were 64, 69 and 64 in the L, M and H, respectively. Samples were collected monthly in each herd during 3 years period and the following parameters were assessed on each sample: fat and crude protein (CP), titratable acidity (TA), total bacterial count (TBC), somatic cells (SCC), coliforms bacteria (CB) and Clostridia spores. Least mean values were obtained by ANOVA univariate using the herd class (L, M or H) as fixed factor. Fat was higher in L milk than in the M and H ones (3.60 vs 3.28 vs 3.30 g/100 g; $P<0.001$ respectively). CP resulted lower in L milk compared to M and H milks (3.17 vs 3.37 vs 3.43 g/100 g; $P<0.001$ respectively). The TA was higher in M milk and lower in L and H ones (3.25 vs 3.22 vs 3.23 °SH/50 ml; $P<0.01$ respectively). These differences, although statistically significant, probably were not significant from a cheese-making perspective. The parameters of SCC, TBC and CB resulted higher in L milk and lower in H milk. Milk produced in H herds showed a better microbial quality, with less SCC, TBC and CB values. No differences were observed for Clostridia spores in the 3 types of milks.

Milk and cheese authentication using FTIR, NIR spectra, fatty acid, and volatile organic compounds

M. Bergamaschi, A. Cecchinato and G. Bittante
University of Padova, Department of Agronomy, Food, Natural Resources, Animals and Environment (DAFNAE), Viale dell'Università, 16, 35020 Legnaro, Padova, Italy; matteo.bergamaschi@unipd.it

The aim of this work was to compare different sources of information for discriminating milk and cheese derived from different dairy systems. Data were from 1,264 cows farmed in 85 traditional or modern herds. Gas chromatography measured fatty acid (FA) concentration and Fourier-Transform-Infrared (FTIR) spectral data were available for all sampled cows. Milk samples were processed in individual model cheeses that, after 60 d of ripening, were analysed to obtain the Near-Infrared spectra (NIR), and volatile organic compounds (VOCs, PTR-ToF-MS) of cheeses. Using a linear discriminant analysis, the milk and cheese samples were assigned to different groups according to either 3 or 5 dairy systems differing in their management (traditional or modern), available facilities (automatic feeder, total mixed rations, milking parlor), diets (concentrates, hay, and/or silage). The dataset was randomly divided into 2 sub-sets: a calibration set (approximately 75% of the group) and a validation set, comprising the remaining data. The process was repeated to perform a 10-fold cross-validation. Using FTIR spectra, a correct classification was observed for 73.5 and 65.0% of milk samples according to 3 and 5 dairy systems, respectively. The FA results (77.3 and 65.1% for 3 and 5 dairy systems) were comparable to those observed with FTIR, while their combination was slightly more accurate (77.9 and 70.3% for 3 and 5 dairy systems). The correct classification obtained with the NIR spectra (67.1 and 52.1%) and VOCs (66.9 and 48.2%) of cheese was considerably lower than those of FTIR and GC. We conclude that FTIR spectra, FA profile, NIR spectra, and Volatile Organic Compounds of milk and cheese contain valuable information useful to classify samples from different dairy systems. Moreover, it seems that the classification of milk is more effective than that of cheese.

Session 23 Theatre 6

Urea-PAGE patterns of PDO Évora cheese manufactured with *Cynara cardunculus* L. ecotypes

C.M.S.C. Pinheiro[1], A.L. Garrido[2], S. Freitas[2], E. Lamy[3], L. Rodrigues[3], N.B. Alvarenga[4], J. Dias[5], A.P.L. Martins[6] and F. Duarte[7]

[1]University of Évora/ICAAM/ECT, Animal Science Department, Apartado 94, 7002-554 Évora, Portugal, [2]University of Évora, Apartado 94, 7002-554 Évora, Portugal, [3]University of Évora/ICAAM, Apartado 94, 7002-554 Évora, Portugal, [4]INIAV/LEAF/ ESA-IPBeja, UTI, Qt. Marques, 2780-157 Oeiras, Portugal, [5]ESA-IPBeja/GeoBioTecRes.Inst., Apartado 6155, 7800 295 Beja, Portugal, [6]INIAV/LEAF, UTI, Qt. Marques, 2780-157 Oeiras, Portugal, [7]CEBAL/ICAAM, R.Pedro Soares, 7800 295 Beja, Portugal; ccp@uevora.pt

Some Portuguese and Spanish ewe's cheese are made with the aqueous extracts of *Cynara cardunculus* L. dried flowers. Évora cheese is a Portuguese PDO cheese which is characterized with a semi-hard or hard consistency, a light yellowed colour, by few or no hole and by an intense flavour. It is obtained by draining slowly the curd, after the coagulation of raw ewe's milk with the action of *C. cardunculus* vegetable coagulant. The renewed interest in the enzymes of this extracts prompted to the investigation of its proteolytic effect in Évora cheese. Urea-PAGE of the casein fraction was done to enable the evaluation of the extent of proteolysis of Évora cheese made with three *C. cardunculus* ecotypes (Cynara 1, Cynara 2, Cynara 3), compared with a commercial animal rennet (Animal) after 1, 7, 14, 21, 35, 49 and 60 days of ripening and therefore to establish the pattern of the casein fractions degradation. The results showed a higher protein degradation of cheeses made with vegetable coagulant than cheeses made with animal coagulant. Up to 35 days of ripening there was an increase of the casein degradation rate, remaining relatively constant until the end of maturation. After 63 days of ripening, αS-caseins (54.90%) were more degraded than β-caseins (37.27%) in cheeses made with *C. cardunculus* ecotypes.

Session 23 Theatre 7

Quality attributes of PDO dry-cured hams as affected by low-protein diets and genetic line of pigs

G. Carcò, M. Dalla Bona, L. Carraro, L. Gallo and S. Schiavon

University of Padova, Department of Agronomy, Food, Natural Resources, Animals and Environment, Viale dell'Università, 16, 35020 Legnaro (PD), Italy; giuseppe.carco@phd.unipd.it

This study aimed to evaluate the effects of a decrease in the crude protein (CP) of diets on several quality attributes of traditional dry-cured hams provided by pigs of two genetic groups (GG) characterized by different potential for lean growth. Hams data from 96 pigs were used; the half fed a conventional diet (CONV) representative of those commonly distributed to heavy pigs in Italy, the others a low-protein (LP) diet, with a 20% lower CP content with respect to CONV. Pigs were slaughtered at 165 kg average BW and all trimmed hams obtained were processed to produce typical San Daniele dry-cured hams. At the end of seasoning, 10 left dry-cured hams for each diet × GG combination were selected and analysed for chemical (fat and salt content, proteolysis index, water activity) and texture traits and fatty acid composition, and evaluated for some sensorial attributes through a consumer test. The effects of diets, GG and their interaction were included in the statistical model. Pigs fed LP diets provided hams with lower protein content ($P=0.003$), whereas a greater fat content was found only in the low lean growth potential GG. Conversely, texture traits and fatty acid composition were unaffected by diet. Hams from high lean growth potential GG showed greater losses at processing ($P<0.001$) and deboning ($P=0.015$), an only nominal greater salt content, greater protein content ($P=0.039$) and a greater proportion of total polyunsaturated fatty acids ($P<0.001$) and ω6 ($P<0.001$) than the other GG. In conclusion, lowering the dietary CP has only minimal effects on dry-cured ham quality. Due to its positive effects on sustainability of dry-cured ham chain by decreasing pig farm nitrogen excretion and feeding costs, lowering the dietary CP content seems an advisable strategy for the feeding of traditional PDO heavy pigs. Genetic groups of different lean growth potential exerted only minor effects on the dry-cured ham quality.

Tenderstretching and ageing time are key parameters for premium beef

I. Legrand[1], J.F. Hocquette[2], C. Denoyelle[3], R. Polkinghorne[4] and P. Bru[5]
[1]Institut de l'Elevage, Service Qualité des Viandes, MRAL, 87000 Limoges, France, [2]INRA, UMRH, Theix, 63122 Saint-Genès Champanelle, France, [3]Institut de l'Elevage, 149 Rue de Bercy, 75012 Paris, France, [4]Birkenwood Pty Ltd, 8431 Timor Road, Murrurundi, NSW 2338, Australia, [5]Beauvallet, CV Plainemaison, 18 rue des abattoirs, 87000 Limoges, France; jean-francois.hocquette@inra.fr

Australia has developed the Meat Standards Australia (MSA) grading scheme to predict eating quality of beef. Through large-scale sensory testing of meat by untrained consumers, MSA has identified animal and carcass factors along the supply chain that impact on consumer palatability. This experiment was designed to study the effects of tenderstretching and ageing time on beef eating quality. Nine Limousin cows from to 5 to 14 years of age were slaughtered in a commercial slaughterhouse. Carcass weights ranged from 400 to 455 kg. For each cow, two hanging methods were used during 48 h post-slaughter for the two half carcasses: AT (Achilles tendon) or TS (Tenderstretch from the ligament). For each hanging method, four cuts of high value (striploin, cube roll, eye of rump, topside) were sampled, aged 10 or 20 days and then scored by 240 untrained consumers for tenderness, flavour liking, juiciness and overall liking after grilling using the MSA protocol. The statistical analysis included animal number ($P<0.05$) and consumer number ($P<0.001$) as covariates plus fixed effects: side of the carcass (NS), cut, ageing time and hanging method ($P<0.05$ to $P<0.001$ depending on the score). Across cuts, TS increased tenderness by 23 and 14% for beef aged 10 or 20 days respectively and overall liking by 18 and 8%. Increasing aged time from 10 to 20 days increased tenderness by 14 and 5% for AT and TS respectively, and overall liking by 13 and 3%. In conclusion, both TS and a longer ageing time induced a better beef palatability, the effect of TS being more important. The effect of TS was also higher at 10 days of ageing than at 20 days, but both effects are partially additive. This experiment allowed the private company Beauvallet to launch a new premium beef brand 'Or Rouge' which is in development in France.

Consumers' awareness of PDO and PGI labelled food products in Slovenia

M. Klopčič[1], M. Kos Skubic[2] and K. Erjavec[3]
[1]University of Ljubljana, Biotechnical Faculty, Dept. of Animal Science, Groblje 3, 1230 Domžale, Slovenia, [2]Ministry of Agriculture, Forestry and Food, The Administration of the Republic of Slovenia for Food Safety, Veterinary Sector and Plant Protection, Dunajska 22, 1000 Ljubljana, Slovenia, [3]University of Novo mesto, Faculty of Economy and Informatics, Na Loko 2, 8000 Novo mesto, Slovenia; marija.klopcic@bf.uni-lj.si

Food labelling is an important communication means between consumers and food industry. This study was aimed at determining the consumers' importance, awareness and determinants of use of the Protected Designation of Origin (PDO) and Protected Geographical Indication (PGI) in Slovenia. A structured questionnaire was administrated to a representative sample of Slovenian consumers. In *total,* 333 *questionnaires* were fully completed. Results showed a low awareness of PDO or PGI labelled products, as 64.53% of the respondents declared they had no knowledge of food products with of food products with a PDO and PGI labels. Consumers tend to pay greater attention to the taste of a food product, its positive impact on their health, and ingredients in the product. The presence of a PDO label on the package of a food product was ranked, together with labels of a special quality (prize, award, certificate of control), brand name, and package attractiveness, as not being too important for the respondents. Consumers' use of food products with a PDO and PGI labels was triggered by their interests and quality perceptions of food products with a PDO and PGI labels. Apart from building general awareness of the quality schemes, these results highlight the need to extend and intensify promotional and communication activities to increase interest in origin and getting information about product quality through PDO and PGI labels.

Native breeds and their products in consumers' opinion

P. Radomski, M. Moskała and J. Krupiński
National Research Institute of Animal Production, Krakowska 1, 32-083 Balice, Poland; pawel.radomski@izoo.krakow.pl

Poland is a country that pays close attention to the conservation of farm animal genetic resources. Also of importance is the promotion of high processing suitability, nutritional value, health-promoting raw materials, and products obtained from local breeds of cattle, pigs and sheep. The emergence of the market of traditional and regional foods in Poland is a direct effect of the integration with the European Union, where Organic Agriculture, Protected Designation of Origin, Protected Geographical Indication and Traditional Speciality Guaranteed are conferred as a tool of the food quality policy. The production of such food improves the situation of semi-subsistence farms and local producers while increasing the competitiveness and tourist attractiveness of the regions. The aim of the research was to analyse the consumer requirements for the quality and attractiveness of the local products made from raw materials obtained from conservation breeds and small populations. The survey was carried out by means of a questionnaire with 35 questions at agricultural exhibitions and during organized conferences and trainings. More than 87% of the respondents are familiar with the term native breed and more than 75% are able to mention at least one such breed. In the opinion of those surveyed, which concerns qualitative traits of the products from native breeds, more than 60% regard them as healthy and delicious; without artificial additives (52%); and of high quality, made with traditional methods (43%). It is also worth noticing that as many as 71% of those surveyed would accept a higher price of such products compared to the price of the analogous ones coming from mass production.
The research was realised within the project 'The uses and the conservation of farm animal genetic resources under sustainable development' co-financed by the National Centre for Research and Development within the framework of the strategic R&D programme 'Environment, agriculture and forestry' – BIOSTRATEG, agreement number: BIOSTRATEG2/297267/14/NCBR/2016.

Are views towards egg farming associated with egg consumers' purchasing habits?

D. Lemos Teixeira[1,2], R. Larraín[2] and M.J. Hötzel[1]
[1]Universidade Federal de Santa Catarina, Laboratório de Etologia Aplicada, Rod. Admar Gonzaga 1346, 88034-000, Brazil, [2]Pontificia Universidad Católica de Chile, Departamento de Ciencias Animales, Av. Vicuña Mackenna 4860, Chile; dadaylt@hotmail.com

In many industrialised countries, public rejection of intensive animal production systems has led to the development of legislation and industry actions that have resulted in significant changes in animal care at the farm level. However, little is known about the views of citizens from emerging countries regarding animal production. The aims of this study were to explore the views of Brazilian and Chilean consumers towards egg farming, and to investigate if these views are associated with participants' eggs purchasing habits and reported willingness to pay (WTP) more for eggs produced in the conditions they perceive as important. In an open question, participants (n=716) were asked to describe an ideal egg production farm and explain their subjacent reasons. This was followed by closed questions asking egg purchasing habits, WTP for eggs of perceived higher quality and demographic information. Participants main concerns were with animal welfare, naturalness, hygiene, production, and ethical aspects, which many associated with improved health, sensory, and nutritional quality of the eggs. The views of participants towards an ideal egg production farm were associated, to some extent, with type of egg purchasing habits and WTP a premium for organic or free-range eggs. Our results suggest a demand for more natural, animal friendly egg production systems; furthermore, they indicate a disconnect between lay citizens' expectations and industry practices, given that intensive confined systems typically fail to supply many of the expected characteristics.

The effect of Virginia fanpetals (*Sida hermaphrodita*) silage on the mineral content of cow's milk

Z. Nogalski[1], C. Purwin[2], M. Fijałkowska[2] and M. Momot[1]
[1]University of Warmia and Mazury in Olsztyn, Department of Cattle Breeding and Milk Evaluation, Oczapowskiego 5, 10-179 Olsztyn, Poland, [2]University of Warmia and Mazury in Olsztyn, Department of Animal Nutrition and Feed Science, Oczapowskiego 5, 10-179 Olsztyn, Poland; zena@uwm.edu.pl

Milk is a rich source of minerals such as calcium, phosphorus, potassium, magnesium, sodium, iodine, chlorine, iron and zinc. The mineral composition of milk varies depending on genetic, physiological and environmental factors, including the feeding regime. Virginia fanpetals (*Sida hermaphrodita*) can play a role in milk production. This perennial plant species has low soil requirements and provides 10 to 20 t DM per year. The aim of this study was to determine the effect of Virginia fanpetals silage on the mineral content of cow's milk. Virginia fanpetals silage and alfalfa silage were made from first-cut herbage harvested in the budding stage, and maize silage was made from first-cut herbage harvested in the dough stage. The experiment was performed on 3 groups of dairy cows (of 9 animals each), characterized by a similar calving date, age and productivity. Over a 4-month period, the cows received a roughage ration composed of maize silage and: alfalfa silage – group I, alfalfa silage and Virginia fanpetals silage in a ratio of 1:1 – group II, and Virginia fanpetals silage – group III. The ration was supplemented with a concentrate. Milk samples were collected once a week over 3 months. Mineralized milk samples were assayed for the content of K, Ca and Na by atomic emission spectroscopy (AES), and Mg and Zn by atomic absorption spectrometry (AAS). The inclusion of Virginia fanpetals silage to the diet of dairy cows, as a substitute for alfalfa silage, did not decrease milk yield and it increased ($P<0.05$) Zn concentrations in milk. Experimental diets had no significant influence on the content of Ca, Mg, Na and K in milk. The results of this study are encouraging and pave the way for future research.
The study was carried out in the framework of the project under a program BIOSTRATEG funded by the National Centre for Research and Development No. 1/270745/2/NCBR/2015 'Dietary, power, and economic potential of *S. hermaphrodita* cultivation on fallow land.' Acronym SIDA.

Fatty acid profile of cheese from the milk of mountain breeds of sheep in terms of human health

A. Kawęcka, I. Radkowska and J. Sikora
National Research Institute of Animal Production, Krakowska 1, 32-083 Balice, Poland; aldona.kawecka@izoo.krakow.pl

Raising sheep for milk in Poland has regional significance and is practically limited to the mountain regions. The milk is collected between May and September from Polish Mountain Sheep (PMS), Coloured Mountain Sheep (CMS) and Podhale Zackel (PZ), and the products there of enjoy enormous popularity among tourists visiting the area. All the milk products from mountain sheep have been included in the National List of Traditional Products, and three of them (bryndza, oscypek and redykołka) have received the EU protected status (PDO). The present study compared the fatty acid profile of traditional mountain cheeses made from the milk of three sheep breeds. Samples of cheeses (n=12) were analysed on fatty acid composition by gas chromatography. The obtained data were statistically analysed with Statistica using Student's t-test. Among long-chain fatty acids, the most important from the human physiology perspective are PUFA, also known as essential fatty acids (EFA), which are not produced in the human body. These acids exhibit anticancer, antiatherogenic, antiinflammatory and immune-boosting activity. No differences were found between the breeds in the content of vaccenic acid (C18:1 *n-7*), whereas the content of oleic acid (C18:1 *n-9*) in CMS sheep (22.5%) was significantly higher than in the other breeds. CLA content in cheese fat was the highest in CMS sheep (2.3 vs 2.1%). Milk fat contains components detrimental to consumer health: lauric (C12:0) and myristic acids (C14:0) increase the risk of cardiovascular diseases, and palmitic acid (C16:0) increases the total cholesterol content. C12:0 content was lowest in cheese from the milk of CMS sheep (2.6 vs 2.8% in PZ and 2.9 in PMS). A similar tendency was observed for the other two fatty acids. It is concluded that breed had an effect on the composition of fatty acids in sheep milk cheese. From the human health perspective, the most favourable composition was observed for cheese made from the milk of Coloured Sheep. Project 'The uses and the conservation of farm animal genetic resources under sustainable development' co-financed by the National Centre for Research and Development within the framework of the strategic R&D programme nr BIOSTRATEG2/297267/14/NCBR/2016.

Starter culture affects chemical parameters of a traditional Portuguese smoked sausage?

D. Barros, R. Pinto, R. Pinheiro, S. Fonseca and M. Vaz Velho
Instituto Politécnico de Viana do Castelo, Av. Atlântico, Viana do Castelo, Portugal; rpinto@ipvc.pt

Nowadays, consumers' demands for food without additives are increasing revealing concerns about possible adverse health effects associated to the presence of additives in foods. In recent years, new strategies, such as biopreservation, have gained special prominence due to the possibility to naturally control the growth of pathogenic and spoilage microorganisms, with a growing interest in the use of lactic acid bacteria (LAB) and their bacteriocins in traditional cured meat products preservation, DOP and IGP. The aim of this study was to determine the effects of a starter culture addition during the processing of a Portuguese traditional smoked sausage – 'Alheira', on its chemical parameters over storage time at 4 °C. *Lactobacillus plantarum* ST153ch, isolated from a traditional meat cured product, was used as starter culture and the effect of its addition in 'Alheira' processing was studied along 90 days of storage (sampling times at 0, 15, 30, 60, 75 and 90 days). The variables analysed were pH, water activity (aw), moisture content, peroxide and acidity indexes. In previous studies, the microbiological behaviour of the culture was analysed and validated. Samples were analysed in triplicate. ANOVA test was applied followed by Tukey test. Differences were considered significant at the confidence interval of 95% ($P<0.05$). During the 90 days of storage, values of pH, aw, moisture content and acidity index decreased whereas peroxide index increased (in both, sample and control). It was found that 'Alheira' with LAB, compared to the control, presented higher aw (at 75 and 90 days), lower pH (at 30 days), lower acidity index (at 0 and 90 days). Peroxide index was lower at 0 and 90 days, but higher at day 90. No significant differences were found on moisture content over the storage time. Despite of *L. plantarum* ST153ch addition to 'Alheira' changed some of its chemical characteristics it is possible that those changes are acceptable from the consumers' point of view. Sensory analysis should be performed to validate this assumption. Acknowledgments POCI-01-0247-FEDER-017626 – DEM@Biofumados, co-financed by FEDER through COMPETE 2020 – Operational Program for Competitiveness and Internationalization.

The fatty acid composition of rabbit, chicken and pig meat depend upon dietary oil supplementation

T. Pirman, A. Levart, V. Rezar, J. Leskovec and J. Salobir
University of Ljubljana, Biotechnical Faculty, Department of Animal Science, Groblje 3, 1230 Domžale, Slovenia; tatjana.pirman@bf.uni-lj.si

Meat has always been important food in balanced human nutrition because it is a good source of quality proteins, essential amino acids, minerals, vitamins as well as fats. The composition of meat fats could be changed with appropriate dietary strategy in animal nutrition to meet human nutritional recommendation such as optimizing n-3/n-6 PUFA ratio and increase long chain n-3 PUFA, such as EPA and DHA. Since the population in the developing countries does not consume enough n-3 PUFA and the n-6/n-3 PUFA ration in the human nutrition could be even 15:1, consuming meat with better fatty acid composition this could be improved. Some experiments in chickens, rabbit and pig were performed. The aim of our study was to analyse the changes in fatty acid composition of chicken, rabbit or pig meat by changing the fat source in a diet with the emphasis on PUFA content and n-3/n-6 PUFA ratio. Animals using in the experiments were randomly divided into groups and assigned into different dietary treatments and different duration of the experiment, according to the experimental protocols. Experimental diets were fortified with n-3 PUFA compared to the control diet. The origin of n-3 PUFA was linseed oil or linseeds, which was added/substitute in a diet in different levels, depend on the experimental protocol. In the diets and meat, the fatty acid composition was determined by the gas chromatography after the *in situ* transesterification of fats. The data were analysed by the GLM procedure, taking into consideration the linseed supplementation in the diet as the main effect. Enrichment of the diets with the linseed oil or linseeds leads to a production of meat with higher content of n-3 PUFA, better n-6/n-3 PUFA ratio and significantly increased long chain n-3 PUFA EPA at least 2.8 times as compared to non-fortified treatment. The improvement depends on the duration and level of fortification, as well as animal species. Improvement of animal diet with n-3 PUFA could lead to a production of functional foods (meat), but the addition of appropriate antioxidants in the animal diet is also necessary to prevent the fats against rancidity, since PUFA have higher susceptibility to lipid oxidation.

Native breeds and their products in breeders' opinion

P. Radomski, M. Moskała and J. Krupiński
National Research Institute of Animal Production, Krakowska 1, 32-083 Balice, Poland; pawel.radomski@izoo.krakow.pl

A growing demand for products of animal origin and the resulting increasingly rapid expansion of industrial production methods constitute a direct threat to local breeds. Therefore, in order to ensure sustainable development of agricultural landscape, it is essential to conserve the native breeds of farm animals. The genetic resources conservation programme has been developing very rapidly in Poland during recent years. As a result, native breeds included in conservation programmes now stand at almost 8,000 cows (3 breeds), more than 6,500 mares (7 breeds), 62,000 mother sheep (15 breeds) and almost 2,400 mother pigs (3 breeds). The aim of the conducted research was to elicit the opinions of breeders of native breeds on their breeding, knowledge of the derived products market, and the expectations for development. The survey was carried out among 430 breeders of the native breeds of cattle, sheep and pigs from the 7 voivodeships. Over 80% of those surveyed started breeding native breeds after 2004, when Poland joined the EU and subsidies to breeders were provided from the Rural Development Programme as part of agri-environmental payments. Over 97% of the surveyed breeders benefit from the subsidies. Likewise, the choice of breed was dictated mostly by economic reasons (77%), followed by low nutritional and environmental requirements (44%), and, to the smallest extent (12%), by the possibility of generating products of unique/high quality. The breeders are aware of the assets and functional traits of the maintained breed because they most frequently mentioned very good adaptation to local environmental conditions, low nutritional requirements and resistance to disease as the characteristic traits. The breeders are also aware of the quality of the products obtained from native breeds since nearly 87% believe that the breeds can deliver high quality raw materials, ideally suitable for converting into certified traditional or regional products. In addition, 78% think that promoting products obtained from native breeds as certified traditional or regional products among the consumers would allow obtaining higher prices for them.
The research was realised within the project co-financed by the programme BIOSTRATEG, agreement number: BIOSTRATEG2/297267/14/NCBR/2016.

Session 23 Poster 17

Effect of duration of linseed supplementation on nutritional properties of beef

R. Marino, A. Della Malva, M. Caroprese, A. Santillo, A. Sevi and M. Albenzio
university of Foggia, Department of the Sciences of Agriculture, Food and Environment, via Napoli 25, 71122 Foggia, Italy; rosaria.marino@unifg.it

This study was conducted to assess the effect of linseed supplementation and the duration of feeding on nutritional properties of beef. 54 Friesian steers were randomly allocated during finishing period into six experimental treatments following a 2×3 factorial arrangement. The 6 treatments consisted of 2 diets, control (CO) vs linseed diet (LS) containing 10% whole linseed and 3 different time of administration of the diet: 40, 75, or 120 days before slaughter. Fatty acids, and bioactive compounds (creatine, carnosine, creatinine and anserine concentrations) were determined on longissimus thoracis muscle after 8 days of aging. Data were subjected to an analysis of variance, using the GLM procedure of the SAS statistical software The mathematical model included fixed effect due to diet, duration of feeding, interaction of diet × duration and random residual error, all effects were tested for statistical significance (to $P<0.05$) Linseed supplementation increased significantly the percentage of total monounsaturated fatty acid (MUFA), polyunsaturated fatty acids (PUFA), total CLA and n-3 polyunsaturated fatty acids, while reduced saturated fatty acid (SFA) and n-6 polyunsaturated fatty acids. Regarding to the effect of duration of feeding a decline of total PUFA and n-6 PUFA was observed in control diet and in linseed diet, respectively. In addition, meat from steers fed with linseed supplementation showed an increased percentage of n-3, linolenic and EPA fatty acids passing from 40 to 75 days of administration remaining constant thereafter, while, vaccenic acid, CLA 9c,11t, total CLA increased ($P<0.001$) between 40 and 75, but declined at 120 days. Diet significantly affected the profile of bioactive substances, in particular, meat from linseed group showed the greater concentration of creatine, carnosine and anserine than control meat. Duration of feeding significantly affected only creatine concentration, The present result suggested that supplying 10% linseed for 75 days before slaughter is a sufficient feed duration to reach the saturation plateau for the synthesis of the PUFA n-3 metabolites and for enrich meat of bioactive compounds.

Session 24 — Theatre 1

A review of the development of novel dairy cattle traits based on modern techniques in North America

A. De Vries
University of Florida, 2250 Shealy Dr, Gainesville, FL 32608, USA; devries@ufl.edu

Modern techniques such as sensors, image analysis, and genome analysis are leading to new traits that help to improve breeding and management of dairy herds. Developments of these novel traits, and their application, in North America appear to be similar to those in other developed dairy countries. One example is the analysis of bulktank milk fat with mid-infrared analysis tools. Differences in the fatty acid composition lead to insight as to how well the cows are using the feed or interacting with management practices. The ultimate goal of this work is a sensor in the milking system to test milk continuously. In genetics, genomic testing has led to differentiation in milk proteins, with some known to lead to higher cheese yields or healthier milk. Some leading dairy cattle breeders start to emphasize these proteins in their sire selection. Other developments in genetics are the recent availability of novel traits for immunity and disease resistance, such as mastitis, ketosis and metritis. Some private companies have developed proprietary health evaluations that are only available to their customers. Another line of work is a better understanding of data coming from activity and rumination sensors. The novelty here is development of algorithms and decision tools to better interpret and use such data. For example, the time that result in the highest fertility can be determined. Combined with data on various available service sires, insemination values can be calculated to quantify the financial advantage of the optimal mating at the optimal time. Machine learning techniques are also employed to interpret combined data coming from multiple sensors. Automatic feeding systems for dairy calves have resulted in traits that have only recently become available, such as drinking speed and amount of milk consumed. These traits could be used to detect diseases and to predict the future performance as a cow. Work is also underway to measure the environmental impact of individual animals. In summary, many novel dairy cattle traits based on modern techniques are being developed and are becoming available for dairy farmers in North America.

Session 24 — Theatre 2

Endocrine and classical fertility traits of Swedish dairy cows based on in-line milk progesterone

G.M. Tarekegn[1], P. Gullstrand[1], E. Strandberg[1], R. Båge[1,2], E. Rius-Vilarrasa[2], J.M. Christensen[3] and B. Berglund[1]
[1]Swedish University of Agricultural Sciences, Department of Animal Breeding and Genetics, Department of Clinical Sciences, 750 07, 7070, Sweden, [2]Växa Sverige, 750 07, 7070, Sweden, [3]Lattec, Hillerod, 325, Denmark; getinet.mekuriaw.tarekegn@slu.se

The aim was to assess classical, insemination-based fertility traits and endocrine traits based on in-line milk progesterone records collected by DeLaval Herd NavigatorTM (HN). A total of 228,792 observations from 3922 lactations of 1175 Swedish Red (SR) and 1647 Swedish Holstein (SH) cows in 14 herds were used. Endocrine traits were interval from calving to commencement of luteal activity (CLA) and from calving to first heat (CFH), first inter-luteal activity interval (ILI), inter-ovulatory interval (IOI), first luteal phase length (LPL), proportion of samples with luteal activity (PLA) and luteal activity during the first 60 d in milk (LA60). The classical traits included interval from calving to first service (CFS), calving interval (CI) and interval from commencement of luteal activity to first service (CLAFS), all recorded in days. The traits were evaluated using a linear model with explanatory variables: breed, parity, herd, parity × breed and year × season. Overall LSM±SE values of 36.2±0.7, 78.0±1.1, 7.0±0.3, 23.2±0.6 and 10.1±0.6, were found for CLA, CFH, ILI, IOI and LPL, respectively. PLA and LA60 were 55.6 and 89.1%, respectively. There was no difference ($P>0.05$) between SR and SH cows for the endocrine traits. This may indicate that classical explanatory variables have little power in explaining variation in endocrine traits, as supported by low R2-values of the models fitted to all endocrine traits. The overall LSM±SE values of CFS, CI and CLAFS were 48.1±0.8, 386.1±2.9 and 42.6±1.1, respectively. CFH and CFS increased ($P<0.001$) with parity number, however, no influence of parity was observed for CLAFS. SH had significantly ($P<0.001$) longer CFS (82.2±1.3), CI (392.9±3.8) and CLAFS (46.1±1.3) compared to SR (CFS: 75.4±1.4; CI: 379.2±3.7; CLAFS: 39.2±1.4). In conclusion, SH had worse fertility than SR for classical fertility traits but there was no breed difference for endocrine traits.

In-line milk progesterone profiles to estimate genetic parameters for atypical fertility in cows

R. Van Binsbergen, A.C. Bouwman and R.F. Veerkamp
Wageningen University & Research, Animal Breeding and Genomics, P.O. Box 338, 6700 AH Wageningen, the Netherlands; rianne.vanbinsbergen@wur.nl

Fertile cows are cows that return early to cyclicity after calving and have cycles without abnormal hormonal patterns, e.g. caused by cysts. These cows are easy to manage, do not need hormone treatment, and are therefore of interest for farmers. The classical fertility traits are based on calving and insemination records and are influenced by recording errors and farm management. A less biased source to define the reproductive status of cows are progesterone measurements. Until recently it was very labour intensive to collect a sufficient number of progesterone measurements on large number of cows. With the development of in-line systems, milk progesterone levels can be recorded automatically on a daily basis. In this study in-line milk progesterone profiles are used to investigate atypical luteal activity for a large number of cows. Our objective was to estimate genetic parameters for these atypical fertility traits, and to estimate their genetic correlation with milk production traits. Milk production traits, calving records, insemination records, and progesterone records for 15 Dutch dairy herds were available. Classical fertility traits were defined for 5,051 cows and 14,366 lactations in the period from September 1998 until April 2016. Endocrine fertility traits were defined for 1,827 cows and 3,075 lactations in the period from July 2012 until April 2016. Atypical fertility was defined by four binary traits: prolonged commencement of luteal activity (DOVI); prolonged inter-luteal interval (DOVII); delayed luteolysis during first cycle (PCLI); and delayed luteolysis during subsequent cycles before insemination (PCLII). A pattern was considered as atypical if the interval was exceeding the mean plus one standard deviation (calculated from the whole dataset). DOVI was shown in 10% of the lactations; DOVII in 12%; PCLI in 11%; and PCLII in 18%. In total, 40% of the lactations showed one or more of these atypical patterns. Genetic parameters for these atypical fertility traits and their genetic correlation with other fertility traits and milk production traits will be presented and discussed.

Prediction of test-day body weight from dairy cow characteristics and milk spectra

H. Soyeurt[1], F.G. Colinet[1], E. Froidmont[2], I. Dufrasne[1], Z. Wang[3], C. Bertozzi[4], N. Gengler[1] and F. Dehareng[2]
[1]University of Liège, Place du 20 août 7, 4000 Liège, Belgium, [2]Walloon Research Centre, Chaussée de Namur 24, 5030 Gembloux, Belgium, [3]University of Alberta, 1428 College Plaza, Alberta, Canada, [4]Walloon Breeding Association, Rue des Champs Elysées 4, 5590 Ciney, Belgium; hsoyeurt@uliege.be

The knowledge of individual body weight (BW) is a management key in terms of feed efficiency and to assess the environmental footprint of dairy production. From 6 farms, BW were measured on 735 Holstein cows. Daily milk samples were collected on these weighed cows and analysed by mid-infrared spectrometry. The stage and number of lactation were also collated. A spectral cleaning was conducted by calculating GH distances from 17 principal components. Spectra with a GH greater than 3 were discarded. The final dataset contained 720 records. Predicting equations were based on Partial Least Squares regressions. Cross-validation coefficient of determination (R^2cv) and root mean square error (RMSEPcv) of the equation including only spectral data were of 0.19 and 65 kg. Then, days in milk, month of test and lactation stage were added. The obtained R^2cv and RMSEPcv increased (0.43 and 54 kg). The part of the information derived from the spectral data was equal to 6%. By adding the daily milk yield, the BW prediction was slightly improved and showed a R^2cv of 0.45 and a RMSEPcv of 53 kg. The use of Legendre Polynomials to regress the spectral data following the day in milk did not improve the predictability. By deleting samples showing a squared residual higher than its mean + 3 times of its standard deviation, the final equation included 668 samples (93% of the initial set) and had a R^2cv of 0.58 and RMSEPcv of 42 kg. A herd cross-validation was then performed to assess the robustness of the developed equation. RMSEPv ranged from 40 to 58 kg. This preliminary study showed the potentiality to predict an indicator of body weight. As this prediction uses easy to record explicative variables and if a larger validation confirmed the obtained results, this prediction equation could be used to develop large scale study about feed efficiency. Moreover, this method allows to consider the past information if spectral data are available.

Session 24 Theatre 5

Genetic analyses of mid-infrared predicted milk fat globule size and milk lactoferrin

S. Nayeri[1], F.S. Schenkel[1], P. Martin[1,2], A. Fleming[1], J. Jamrozik[1,3], L.F. Brito[1], F. Malchiodi[1], C.F. Baes[1] and F. Miglior[1,3]

[1]University of Guelph, CGIL, N1G2W1 Guelph, Canada, [2]INRA, GABI, 78352 Jouy en Josas, France, [3]Canadian Dairy Network, 660 Speedvale Avenue W, N1K1E5 Guelph, Canada; miglior@cdn.ca

Genetic parameters for mid-infrared (MIR) predicted milk fat globule (MFG) size and lactoferrin were estimated for first parity Canadian Holsteins. A total of 109,029 records from 22,432 cows (105,737 pedigree animals) for lactoferrin and 109,212 records from 22,424 cows (105,070 pedigree animals) for MFG size were available. The linear animal model included fixed effects of herd-test day and days in milk, fixed regressions for age-season of calving effect, random regressions for herd-year of calving, permanent environment and additive genetic effects. Regression curves (on DIM) were modelled using orthogonal Legendre polynomials of order 4 for fixed and order 5 for random effects. Co-variance components were estimated by Bayesian methods via Gibbs sampling using custom-written software and 270,000 samples after a burn-in period of 30,000 samples. Multi-trait analyses using the same model with the MIR-predicted milk components along with test-day milk, fat and protein yields, fat and protein percentage, and somatic cell score were used to estimate genetic correlations among the traits. Average daily heritability estimates over the entire lactation were 0.34 for lactoferrin and 0.50 for MFG size. Milk lactoferrin content was found to have a moderate lactation average genetic correlation with protein % (0.41) and a positive genetic correlation with fat % (0.21). Weak genetic correlations were estimated between lactoferrin and fat and protein yields, and somatic cell score at 0.07, 0.08, and 0.07, respectively. Milk fat globule size had average genetic correlations of -0.23, 0.36, and 0.54 with milk yield, fat yield, and fat %, respectively. Posterior standard deviations for all estimates were low (below 0.001). The moderate genetic correlation with fat percentage suggested that MIR prediction for MFG size may not be perfectly related to fat percentage and thus the prediction equations are not simply predicting fat percentage.

Session 24 Theatre 6

Improving phosphorus efficiency using infrared predicted milk phosphorus content

H. Bovenhuis[1], I. Jibrila[1] and J. Dijkstra[2]

[1]Wageningen University & Research, Animal Breeding and Genetics, Droevendaalsesteeg 1, 6708 PB Wageningen, the Netherlands, [2]Wageningen University & Research, Animal Nutrition Group, De Elst 1, 6708 WD Wageningen, the Netherlands; henk.bovenhuis@wur.nl

A cheap and accurate method for estimating milk P content of individual cows would better allow farmers to feed their cows according to their P requirements. This study aimed at predicting milk P content based on different information sources: routinely recorded milk composition traits, genotypic data and infrared spectra. Data of 1,400 Dutch Holstein-Friesian cows was used. Prediction models were developed using the Partial Least Squares Regression and validated using test set validation. Prediction of milk P content based on protein content has an R^2_v of 41%. Prediction based on genotypes for the DGAT1 K232A polymorphism and the SNP rs29019625 (BTA1, close to SLC37A1) result in R^2_v of 8.7 and 4.7%, respectively. Based on the infrared spectrum the R^2_v for milk P content was 84%. We quantified that phosphorus efficiency can be improved with 17% when feeding cows based on the developed infrared prediction for milk P content.

Session 24 Theatre 7

Genome wide association study of bovine respiratory disease in dairy calves with thoracic ultrasound

A.E. Quick[1], T.L. Ollivett[2], B.W. Kirkpatrick[3] and K.A. Weigel[1]
[1]University of Wisconsin, Department of Dairy Science, 1675 Observatory Drive, Madison, WI 53706, USA, [2]University of Wisconsin, Department of Medical Sciences, 2015 Linden Drive, Madison, WI 53706, USA, [3]University of Wisconsin, Department of Animal Sciences, 1675 Observatory Drive, Madison, WI 53706, USA; kweigel@wisc.edu

Bovine respiratory disease (BRD) is one of the leading causes of morbidity and mortality in dairy calves. The objective of this study was to establish a protocol for objective and efficient assessment of BRD phenotypes in dairy calves to identify markers associated with BRD in a genome wide association study (GWAS) and build a reference population for whole genome selection (WGS). Trained evaluators assessed 1,107 calves on 6 Wisconsin dairy farms at 3 and 6 weeks of age. Clinical scores were assigned by visual appraisal of eyes, nose, ears, attitude, cough, and temperature, and subclinical scores were assigned by measurement of lung lesions using thoracic ultrasonography. Overall BRD scores were computed as the interaction of clinical and subclinical assessments. Single nucleotide polymorphism (SNP) genotypes were available for 1,016 calves, and after quality control and imputation data from 28,696 SNPs and 1,014 individuals remained. A linear mixed model was used for GWAS analysis, with SNP genotype as a fixed effect and polygenic background of the animal as a random effect. Phenotypes at 3 and 6 weeks of age were analysed separately, and BRD scores were considered as binary (healthy or affected) or ordinal (six levels with increasing severity). At 3 weeks of age, 8 and 6 significant SNPs ($P<0.00005$) were detected in the binary and ordinal analyses, respectively, with common SNPs on chromosomes 1, 7,17, and 18. At 6 weeks of age, 3 significant SNPs were detected in each of the binary and ordinal analyses, with common SNPs on chromosomes 8 and 9. Assessment of BRD using visual (clinical) and ultrasound (subclinical) scores allows objective and efficient measurement of BRD for identifying important SNPs by GWAS or implementing WGS. Further analysis is needed to identify putative genes affecting BRD and to assess the reliability of whole genome predictions.

Session 24 Theatre 8

Cardiovascular monitoring towards novel proxies for feed efficiency in the bovine

J. Martell[1], J. Munro[2], P. Physick-Sheard[3] and Y. Montanholi[1]
[1]Harper Adams University, Edgmond, TF10 8NB, Newport, United Kingdom, [2]AgSights, 294 Mill St East, Suite 209, N0B 1S0, Elora ON, Canada, [3]University of Guelph, 50 Stone Rd, N1G 2W1, Guelph ON, Canada; janel.martell@gmail.com

Cattle producers are under pressure to satisfy growing meat demand without compromising animal husbandry. Phenotyping for feed efficiency with complementary investigation of stress inducing procedures could encourage further development of novel proxies to be integrated into breeding programs. The objective was to evaluate cardiovascular function recorded during the breeding soundness evaluation to infer about the efficiency of feed utilization. Feed efficiency was measured via residual feed intake (RFI; kg DM/day) in a total of 107 beef bulls (age; 13.7±0.65 mo) using feed intake, body weight and composition measured throughout 112 days of a performance assessment. During the full duration of the breeding soundness evaluation, heart rate (HR; bpm) was monitored via an electrode-based HR recorder attached around the heart girth. The HR recording was segregated by procedure, including blood collection (BC), measure of scrotal circumference (SC), testis ultrasound (TU), rectal palpation (RP), and electroejaculation (EE). In preliminary analysis on a subset of data (n=50), means of the HR during each procedure were determined, as well as the least square means comparing the 25 most (RFI_{LOW}) and 25 least (RFI_{HIGH}) feed efficient bulls by the proc GLM in SAS. The HR values for BC, SC, TU, RP, EE were; 98.0±17.9, 93.8±15.8, 88.9±16.6, 98.7±22.7, and 223±17.5 bpm, this suggests that stress severity was variable dependent on the handling procedure based on HR monitoring. Comparison of cardiovascular function upon classification of bulls into RFI_{HIGH} and RFI_{LOW} groups (0.89±0.09 vs -0.80±0.09; $P<0.01$), suggested an increased average HR for RFI_{HIGH} bulls compared to RFI_{LOW} bulls during EE (166±6.09 vs150±6.09 bpm; $P<0.01$). Results suggest differences in cardiovascular response and severity to stressors, further analysis may encourage a stronger connection with feed efficiency, as well as other important health and welfare concerns. With further refinement, HR could be an advantageous phenotype to implement in precision farming technologies that could serve to support genomic selection.

Session 24 Theatre 9

Genetic parameters of novel milkability and temperament traits in automatic milking systems

K.B. Wethal[1] and B. Heringstad[2]
[1]Norwegian University of Life Sciences, Department of Animal and Aquacultural Sciences, Faculty of Biosciences, P.O. 5003, 1430 Ås, Norway, [2]Norwegian University of Life Sciences, Department of Animal and Aquacultural Sciences, Faculty of Biosciences, P.O. 5003, 1430 Ås, Norway; karoline.bakke@nmbu.no

The objective and repeated measurements from each visit in automatic milking systems (AMS) gives information about cows milkability and behaviour. In Norway, number of AMS are increasing and in 2016 41% of the total milk production came from farms with AMS. This confirms the importance of incorporating information recorded by milking robots in the genetic evaluation of Norwegian Red dairy cattle. A cow with the ability to milk without interrupting the milking process by kicking off teat cups or hindering the AMS to milk every teat properly is more efficient. In addition, her ability to stay away from the milking unit between milkings and make way for other cows with milking permission is also a desired character. Our aim was to examine how data registered by AMS can be used to define new milkability and temperament traits. Data from 77 commercial AMS herds from year 2015-2017 with information on 4,877 Norwegian Red cows were analysed. We used a mixed linear animal model and the DMU package to estimate genetic parameters for milkings with ≥1 teat cup kick off (KO), incomplete milkings (IM), milkings with one or more teat not found to be milked (TNF) and rejected visits (RV). Two different definitions of these four traits were examined; either as daily records with repeated measurements on each cow, or as the proportion of occurrences with one record on each cow and lactation (pKO, pIM, pTNF, pRV). Heritabilities for daily observations of KO, IM, TNF and RV ranged from 0.01 to 0.06. For pIM, pKO, pTNF and pRV heritability was 0.14, 0.13, 0.12 and 0.02 respectively. Genetic correlations for pIM, pKO, pTNF and pRV varied from 0.16 to 1, with a correlation on one between pTNF and pIM suggesting that these two traits contribute with exactly the same genetic information. Results from our study confirm that genetic variation exists for the new traits that may give additional information in a genetic evaluation. Further, we will estimate genetic correlations with traditionally scored temperament and milkability.

Session 24 Theatre 10

Phenotyping of Hungarian Grey bulls with the optometric VATEM method and launch of the PHENBANK

A. Beck[1], I. Bodó[1], L. Jávorka[1], A. Gáspárdy[1], M. Szemenyei[2] and Á. Maróti-Agóts[1]
[1]University of Veterinary Medicine, Budapest, Department Animal Breeding, Nutrition and Laboratory Animal Science, István u. 2, 1078, Hungary, [2]Budapest University of Technology and Económic, Control Engineering and Information Technology, Budapest, Műegyetem rkp. 3, 1111, Hungary; maroti-agots.akos@univet.hu

The first time the body parameters of Hungarian Grey bulls were measured was back in 1968, with the assistance of the brave herdsmen of the time. Measuring the body parameters of extensively kept beef cattle with traditional methods is very dangerous, usually it is practically impossible. Since 2001 the body parameters of the Hungarian Grey cows have been measured several times, using the optometric VATEM method. However, a population-scale measurement of bulls has not taken place yet. We assessed the bulls in three Hungarian bull breeding centres: in Hortobágy, Apaj and Sarród, in 2016 and in 2017. Since 2015, besides the breeding bulls we have also been measuring the 2-year-old young bull population. For the video footages we have used cameras with HD and 4K resolution. The side camera was placed 15 metres, while the upper camera was placed 5 metres from the axis of the corridor. The measurement process was performed using the VATEM2 software (vatem.hu), while the statistical processing was performed using EXCEL and R. We have measured the following [body] parameters of breeding bulls older than 5 years (n=76) (cm, sd): wither height (138, 6.3), height of back (136, 6.8), rump height (137, 6.4), trunk length (169, 12.9), body length (164, 12.8), chest depth (80, 6.4), shoulder width (52, 5.4), rump1 (53, 6.1), rump3 (24, 2.3), and rump length (43, 5.7). After a statistical analysis and a hierarchical selection we were able to identify bull types. Genetic research and also genome sequencing steers animal research towards the direction of genetics. The methods of phenotyping are unable to keep up with the fast-developing technologies of genetic research. With the help of the IT intense optometric technology, we have created a simple PHENBANK, based on the results of the phenotyping VATEM method. By using the NGS-genome of two Hungarian Grey bulls and publishing their VATEM data, standard and measurement frames, we would like to foster new, in-silico research.

Methane production rate in relation with feed efficiency traits of beef heifers fed roughage diets

G. Renand[1], D. Maupetit[2], D. Dozias[3] and A. Vinet[1]
[1]INRA, UR1313 GABI, Domaine de Vilvert, 78352 Jouy-en-Josas, France, [2]INRA, UE332 Bourges, Domaine de la Sapinière, 18390 Osmoy, France, [3]INRA, UE326 Domaine Experimental du Pin, Domaine de Borculo, 61310 Le Pin-au-Haras, France; gilles.renand@inra.fr

Individual methane emission was measured with GreenFeed systems during 8 to 12 weeks on two groups of 2-year old Charolais heifers while they were tested for feed intake and growth. The 326 heifers were *ad libitum* fed a roughage diet: 252 heifers received herb silage in the first farm and 74 heifers received natural meadow hay complemented with 1 kg concentrates in the second farm. Mean performance were: 534 and 508 kg live weight (LW); 8.75 and 7.89 kg/d dry matter intake (DMI); 932 and 360 g/d average daily gain (ADG) respectively in the two groups respectively, and the same 206 g/d methane production rate (MPR). Feed efficiency (FEff = ADG/DMI) was 109 and 46 g/kg and methane yield (MPY = MPR/DMI) was 24 and 26 g/kg respectively. Methane emission was predominantly correlated with body weight (r≈ +0.7). Correlation coefficients between MPR and DMI or ADG were moderate (r≈ +0.3 to +0.5). There was no correlation between MPR and DMI when both traits were adjusted for metabolic body weight (MBW) and ADG (r= +0.05 and +0.12), i.e. residual feed intake (RFI) was not related with differences in methane emission among heifers of similar body weight and similar daily gain. Residual gain (RGain), that is ADG adjusted for MBW and DMI, was slightly and significantly associated with methane emission among heifers of similar body weight and similar feed intake in both populations (r= +0.23 and +0.22). Although these relationships were faint, this experiment showed that the more efficient heifers could not be classified as low methane emitters.

Empirical analysis of SCC impact on milk production and carry over

J.A. Baro and B. Sañudo
Universidad de Valladolid, ETSIIAA, 34047 Palencia, Spain; baro@agro.uva.es

Milk recording schemes have been in operation under ICAR guidelines for several decades now in most of the European countries. Huge data sets are available for the characterization of many traits recorded individually on a monthly basis. We have analysed the whole set of records from one of the largest milk recording organizations in Spain (Castilla y León, 6.8 million records), exploring the evolution of distributions of yields across the set of recorded factors and, particularly, regarding somatic cell counts (SCC). Logarithmic transforms of SCC provide acceptable symmetry and normality and we adopted the regression of milk yield on logSCC across a set of quantiles (10, 30, 50, 70 and 90% regarding milk yield) as the main analysis tool. Impact of SCC on milk yield was estimated at 7.6 kg/d per decimal logarithmic unit regardless of milk yield level, but this trend ends at different SCC values: 398, 251, 200, 158, and 126 for milk yield quantiles 10, 30, 50, 70 and 90%, respectively. As SCC increases, the plateau in milk yield extends in a similar pattern to decline sharply at lower SCC values for higher yield levels. The effect of SCC on mastitis related milk yield losses differs depending on actual level of milk yield. Our results allow for accurate and repeatable prediction of mastitis related milk yield losses and carry over effects from previous records. Very low cell counts on the previous month do not show a persistent effect on milk yield, but for lower yields there is small persistence. Conversely, very high cell counts in the previous month do show persistent detrimental effects on milk yield, regardless of milk yield level. These results are useful for extension and for expert systems to predict future yields. Traditional SCC is not the best systematic indicator of mastitis available anymore and the foreseeable availability of better choices, such as differential cell counts, must improve the power and scope of this approach.

Session 24 Poster 13

Lameness scoring for genetic improvement of claw health based on data from the Efficient Cow project

A. Köck[1], B. Fuerst-Waltl[2], J. Kofler[3], J. Burgstaller[3], F. Steininger[1] and C. Egger-Danner[1]
[1]ZuchtData EDV-Dienstleistungen GmbH, Dresdner Str. 89, 1200 Vienna, Austria, [2]University of Natural Resources and Life Sciences, Department of Sustainable Agricultural Systems, Gregor-Mendel-Str. 33, 1180 Vienna, Austria, [3]University of Veterinary Medicine Vienna, University Clinic for Ruminants, Veterinärplatz 1, 1210 Vienna, Austria; koeck@zuchtdata.at

The specific objective of this study was to evaluate the use of lameness scoring to genetically improve claw health in Austrian dairy cows based on data from the 'Efficient Cow' project. In the year 2014 a one-year data collection was carried out. Data of approximately 5,400 cows, i.e. 3,100 Fleckvieh (dual purpose Simmental), 1,300 Brown Swiss, 1,000 Holstein kept on 167 farms were recorded. At each time of milk recording, lameness scores were assessed by trained staff of the milk recording organizations. Claw trimming was documented and recorded on these farms as well. Veterinarian diagnoses and culling due to feet and leg problems were available from the routine recording system. Breeding values for lameness were estimated and cows were ranked according to them. A low breeding value for lameness (the bottom 10% of the cows) was associated with a significantly higher frequency of trimmed cows, which indicates that the cows selected by the farmer to be trimmed is not completely random. Additionally, a high breeding value for lameness (the top 10% of the cows) was associated with lower frequencies of claw diseases recorded at trimming, claw and leg diagnoses and culling due to feet and leg problems. Therefore selecting for a better lameness score leads to an improvement of claw health. However, the mean frequency of claw diseases of the top 10% of cows was still quite high (36.8% in Fleckvieh, 18.1% in Brown Swiss and 30.8% in Holstein). In order to reduce the occurrence of claw diseases that are less associated with lameness (e.g. less severe claw diseases), direct information from claw trimmers is necessary. Overall, the results emphasize the great importance of recording claw trimming data for genetic improvement of claw health.

Session 24 Poster 14

Risk factors for dystocia and perinatal mortality in extensively kept Angus cattle

T. Hohnholz[1], N. Volkmann[2], K. Gillandt[2], R. Wassmuth[1] and N. Kemper[2]
[1]University of Applied Sciences Osnabrueck, Faculty of Agricultural Sciences and Landscape Architecture, Am Kruempel 31, 49090 Osnabrueck, Germany, [2]University of Veterinary Medicine Hannover, Foundation, Institute of Animal Hygiene, Animal Welfare and Farm Animal Behaviour, Bischofsholer Damm 15, 30173 Hannover, Germany; t.hohnholz@hs-osnabrueck.de

The objective of the present study was to determine risk factors for dystocia and perinatal mortality in extensively kept Angus cattle. From April 2015 to March 2017 calving ease and calf survival were recorded for 785 births on five Angus cattle farms in Germany. Data such as date of calving, sex and birth weight of the calf, lactation number of the dam, externally measured pelvic parameters of the dams and measurements derived from the calves were documented. The prevalence of dystocia and perinatal mortality was 3.8 and 5.3%, respectively. A logistic regression model was used to predict both calving ease (unassisted or assisted) and calf survival (alive or dead within 48 hours after parturition). The logistic regression model for dystocia included effects of parity of dam (first or later) and calf birth weight (kg). First-parity dams had an 8.4 times higher probability of dystocia than later-parity dams. An 18% increase in odds for dystocia was associated with a one kilogramme increase in calf birth weight. In the perinatal mortality model calving ease (unassisted or assisted), parity of dam (first or later), length of pelvis (cm) and calf birth weight (kg) remained as significant effects. Assisted births tended to result in perinatal mortality 17.8 times more often than unassisted births. The likelihood of perinatal mortality was 3.4 times higher in first- than in later-parity cows. A 24% increase in odds for perinatal mortality was associated with a one centimeter increase in the length of the pelvis. The odds of perinatal mortality decreased by 17% per kilogramme of calf birth weight. In conclusion, the study points out that dystocia and perinatal mortality are closely related traits that are more critical in primiparous than in multiparous cows. Moreover, calf birth weight seems to be a crucial factor for both issues. In addition, the investigation indicates that externally measured length of pelvis has an impact on perinatal mortality.

Session 24 Poster 15

Genetic relationships between dystocia, stillbirth and type traits in Polish Holstein-Friesian cows

A. Otwinowska-Mindur[1], M. Jakiel[2] and A. Zarnecki[2]
[1]University of Agriculture in Krakow, Department of Genetics and Animal Breeding, al. Mickiewicza 24/28, 30-059 Krakow, Poland, [2]National Research Institute of Animal Production, ul. Krakowska 1, 32-083 Balice, Poland; magdalena.jakiel@izoo.krakow.pl

The objective of this study was to estimate direct genetic correlations between calving ease, stillbirth and type traits in Holstein-Friesian primiparous cows. The data were 19,450 Polish Holstein-Friesian primiparous cows with dystocia scores (26.97% unassisted calvings, 66.84% easy pull, 5.97% difficult, 0.23% very difficult; about 7% of the calves were born dead or died within 24 h of birth) and conformation traits (stature, body depth, chest width, rump width, rump angle, rear legs – side view, udder support, dairy form). The multiple-trait REML procedure using an animal model was applied for (co)variance component estimation. For calving ease and stillbirth, the linear model included fixed effects of herd-year-season (HYS), age of calving, and sex of calf. For type traits, the linear model included fixed regression on age at calving, fixed effect of herd-year-season-classifier (HYSC), and lactation stage. There were 3,152 HYS subclasses, 30 age of calving subclasses, 3,077 HYSC subclasses and 19 lactation stages. Genetic correlations between calving ease, stillbirth and type traits were generally very low. For calving ease, the genetic correlation was lowest (0.01) for dairy form, and highest (0.21) for body depth. Calving ease was negatively correlated with stature (-0.02) and udder support (-0.12). For stillbirth, the genetic correlation was lowest (0.03) for rump width, slightly higher (and negative) for chest width (-0.05), stature (-0.07) and rump angle (-0.10), and highest (-0.21) for udder support. The results of this study show that the association of type traits of first-parity cows with their calving performance and calf death are weak or nil.

Session 24 Poster 16

Evaluation of health traits of dairy cows in the Czech Republic

E. Kašná[1], L. Zavadilová[1], P. Fleischer[1,2], S. Šlosárková[2], Z. Krupová[1] and D. Lipovský[3]
[1]Institute of Animal Science, Přátelství 815, 10400 Prague 10, Czech Republic, [2]Veterinary Research Institute, Hudcova 296/70, 62100 Brno, Czech Republic, [3]Czech Moravian Breeding Corporation, Benešovská 123, 25209 Hradištko, Czech Republic; kasna.eva@vuzv.cz

Health problems are the most frequent cause of culling dairy cows in the Czech Republic. Selection indices of Czech dairy cattle already include indicator traits such as somatic cell count or legs and udder conformation. Although, genetic evaluation of direct health traits would enable more precise selection on disease resistance, currently we have no continuous national system for the collection and processing of health data. Data on veterinary prescription-based medical treatment have to be recorded and stored by farmers. The purpose of our study was to determine if these data are usable for genetic evaluation. We gathered the data of 27 common health traits retrospectively from farmers via on-line survey for time period from July, 2016 through to June, 2017. To identify the farms with incomplete data, the owner of each herd with more than 20 lactations was required to report at least 1 record of particular health trait to be included in its evaluation. Reported health events were expressed as lactational incidence rate *LIR* (number of affected lactations / number of lactations at risk) × 100. The study covered 281,914 cows (76% of Czech dairy cows). One or more disorders were reported in 55% of cows. Clinical mastitis was the most common trait (*LIR* 17.9%) followed by metritis (10.9%) and cystic ovary disease (10.9%). Digital dermatitis (5.1%) and interdigital phlegmon (3.5%) were the most frequently reported locomotory apparatus disorders. Primary ketosis (2.2%) was the most frequently reported clinical metabolic disease. Our study showed that some farmers failed to report all health events. In fact, more than 70% of them sent data on reproduction disorders and udder diseases, over 50% reported on orthopaedic disorders and fewer than 30% on metabolic diseases. Despite certain limitations, and after cautious and careful editing, the data from farmers are usable for further processing and genetic evaluation of cattle health. The study was supported by the projects QJ1510217 and MZE-RO0718 of the Ministry of Agriculture of the Czech Republic.

Session 24 Poster 17

Genome-wide association analyses of ovarian cysts and metritis of HF cows – preliminary study

P. Topolski, K. Żukowski, A. Żarnecki and M. Skarwecka
National Research Institute of Animal Production, Department of Genetics and Cattle Breeding, Krakowska 1, 32-083 Balice near Cracow, Poland; piotr.topolski@izoo.krakow.pl

A genome-wide association study (GWAS) approach was applied to find SNPs associated with two cow diseases: ovarian cysts and metritis. Cystic ovaries are recognized as one of the main reasons for reproductive failure in cows, and postparturient metritis has a negative effect on cow fertility. Both diseases reduce the economic performance of dairy herds in many ways. Data for the study were drawn from the PLOWET database used for recording health traits in 4 experimental dairy farms of the National Research Institute of Animal Production. The occurrence of ovarian cysts was recorded in 270 cows and the occurrence of metritis was found in 380 cows. The cows were genotyped using the Illumina EuroG10K chip and the genotype data were stored in the cSNP database system. Next, imputation findhap software was used to generate an HD genotype dataset. As the final dataset we used SNPs with MAF>0.001, with linkage disequilibrium pruning (r2<0.8). The GWAS data were analysed using a logistic regression model assuming an additive SNP effect and models with full dominance (or recessive) for the minor allele. After applying Benjamini-Hochberg multiple testing correction, only the P-values of 28 SNPs were significant. For ovarian cysts, 15 significant SNPs located on 8 autosomes (13 SNPs) and the X chromosome (2 SNPs) were identified. For metritis, 13 significant SNPs located on 10 autosomes (11 SNPs) and the X chromosome (2 SNPs) were found. The corresponding genes within 10 kbp distance of these SNPs were surveyed. This resulted in finding 6 genes that may be associated with ovarian cysts and 5 genes that may be associated with metritis. The pathway analysis showed that some of these genes were classified as 'cytoskeletal regulation by Rho family GTPases'. The remaining genes were classified as small RNA structures involved in regulation of a multitude of genes.

Session 24 Poster 18

Prediction of the energy balance of Holsteins in Japan using milk traits and body weight

A. Nishiura[1], O. Sasaki[1], T. Tanigawa[2] and H. Takeda[1]
[1]Institute of Livestock and Grassland Science, NARO, Tsukuba Ibaraki, 3050901, Japan, [2]Hokkaido Research Organization, Nakashibetsu Hokkaido, 0861135, Japan; akinishi@affrc.go.jp

It is important to improve the energy balance (EB) of dairy cows in lactation to prevent the deterioration of health and fertility. However, number of available data for genetic evaluation of EB is limited because it is difficult to measure EB from energy input and output practically. Our objective was to propose the multiple regression equation to predict EB using the traits measured in dairy herd performance tests. Records of 191 Holstein cows (1st to 9th lactation, days in milk (DIM) 6 to 305) from the Hokkaido Research Organization, which were dried off in 2015-2017, were used. We measured milk yield and dry matter intake daily (61,824 records) and milk components and body weight weekly (7,895 records), and calculated the average values of these traits per 10 DIM. The EB estimate per 10 DIM was calculated by subtracting energy output (milk, basal metabolism, growth) from energy input (digestible energy); the number of EB estimates was 30 per lactation. Multiple regression analysis was used to predict EB. Independent variables were DIM, milk yield (MY), fat % (F), fat yield (FY), protein % (P), protein yield (PY), lactose % (L), lactose yield (LY), fat to protein ratio (FPR), fat to lactose ratio (FLR), protein to lactose ratio (PLR) and body weight (W). These variables except DIM were analysed together with 'd' variables, which were the current minus the previous value. Model reduction was carried out by stepwise regression. The EB estimates at 10 DIM were -29.0, -23.6 and -23.0 MJ/day in the 1st, 2nd and 3rd and over lactation respectively. EB turned positive around 30 DIM and increased until 150 DIM, then fell because of the feed switch, and increased slowly again. Eight variables were selected in the equation to predict EB; DIM, FPR, L, W, dLY, F, dP and dPY. The R-square of this equation was 0.484 and that fitted the data well enough. Our results indicated that EB could be predicted from test day records using this equation.

Session 24 Poster 19

Behavioural patterns of cows depending on age and number of milking sessions

P. Wójcik, A. Szewczyk and I. Radkowska
National Research Institute of Animal Production, Department of Cattle Breeding, Sarego 2, 31-047 Kraków, Poland; piotr.wojcik@izoo.krakow.pl

Behaviour of dairy cows is closely related to their mode of life. Depending on age, housing system and welfare level, they show behavioural patterns in which typical behaviours can be distinguished. The aim of the study was to determine some behavioural patterns of dairy cows depending on their age. In a loose-housing system, cows should spend around 53% of their time resting. Deviations from this time and absence of the resting phase may overload the legs, have a negative effect on ruminal digestion, and disturb blood circulation, thus reducing the milk yield. Daily activity of the first to eighth lactation cows was analysed within lactations and their phases at 50-day intervals. The results showed that the most active are young, first-, and especially second-lactation cows with mean activity of 115.65 steps/h; in the subsequent lactations, the mean activity of the cows decreased to 96 steps/h. The highest within-lactation activity was observed from 150 to 200 days in first and second lactation cows, and after 200 days in older cows. The lowest first and second lactation activity was found up to 50 days, and in older cows from 50 to 150 days of lactation. The highest resting frequency occurred in first lactation cows and decreased with the age of the cows. Regardless of age, all peak-lactation cows (100 days) had the highest indicators of distress and shortest rests. Older cows beyond the fourth lactation were characterized by the longest total resting time and highest resting frequency. In this age category, seventh lactation cows spent by far most time resting. It is concluded from these results that 24-h activity of the cows depends on the number of milking sessions. With three times daily milking, cows were more active in the second and third milking session. Among all the age groups under study, second lactation cows were the most active. When analysing cow activity depending on lactation time, higher activity was observed with twice daily milking of cows beyond 300 days of lactation in the second session.

Session 24 Poster 20

Genetic correlation of conception rate with growth and carcass traits in Japanese Black cattle

K. Inoue[1], M. Hosono[1], H. Oyama[1] and H. Hirooka[2]
[1]National Livestock Breeding Center, Nishigo, 961-8511 Fukushima, Japan, [2]Kyoto University, Sakyo, 606-8502 Kyoto, Japan; k1inoue@nlbc.go.jp

Growth (birth weight (BW) and body weight at 9 month of age (9MW)) and carcass traits (carcass weight (CW) and beef marbling score (BMS)) are of importance for farmers' profitability in Japan. Heavy birth weight occasionally causes calving difficulty, leading to severe calving difficulty and thereby increased calf's and/or cow's death. Calves for feedlot are auctioned at calf markets at 9 months of age on average and body weight at the age is one of significant factors in determining calves' price in the markets. Similarly, CW and BMS for feedlot animals have significant roles in determining carcass price in beef carcass markets. On the other hand, reproductive traits such as conception rate (CR) are important for efficient production of calves. The objective of this research was to estimate the genetic relationships of first service CR with BW, 9MW, CW and BMS in Japanese Black cattle. Genetic parameters were estimated using AI-REML method with a linear bi-variate animal model. The heritability estimates for BW, 9MW, CW and BMS were moderate to high (0.38 to 0.60), whereas the estimates for CR were low (0.02 to 0.05). The negative genetic correlations were found between CR and body growth traits and carcass weight (-0.38, -0.13 and -0.56 for BW, 9MW and CW, respectively), but also between CR and BMS (-0.28). This result shows antagonistic relationships that selecting sires/dams with superior breeding values for CW or BMS would decrease CR of cows.

Session 24 Poster 21

Evaluation of a gait scoring system for dairy cows by using cluster analysis

N. Volkmann and N. Kemper
University of Veterinary Medicine Hannover, Foundation, Institute for Animal Hygiene, Animal Welfare and Animal Behavior, Bischofsholer Damm 15, 30173 Hannover, Germany; nina.volkmann@tiho-hannover.de

An important factor for animal welfare in cattle farming is an early and easy detection of lameness. Locomotion Scoring is a common system to detect lame cows. Presently several different score systems for evaluating lameness exist, which vary in gait and posture traits observed, as well as in type of scale and number of levels. Therefore, the aim of the present study was to validate a previously developed three-point locomotion score for classifying lameness in dairy cows. Cows (n=144) were locomotion scored by a trained observer. Additionally, data on claw lesions detected during hoof trimming was collected. Based on latter data a cluster analysis (PROC CLUSTER) was performed to objectively classify cows in three groups showing a comparable status of claw diseases. Finally, the congruence between scoring system and clustering was tested using Krippendorff's Alpha-Reliability (KALPHA macro). In total, 63 cows (43.7%) were classified as not lame (score 1), 38 (26.4%) as mildly lame (score 2) and 43 (29.9%) as clearly lame (score 3). In comparison, 64 cows (44.4%) showed no diagnoses, 37 (25.7%) one diagnosis, 33 animals (22.9%) two diagnoses and 10 (7.0%) more than two. Digital dermatitis was the most common disease concerning 57 cows (42.2%). Furthermore, 26 cows (19.3%) showed sole haemorrhages and 21 (15.6%) sole ulcers. Most diagnoses affected the hind limbs (76.2%), and in a quarter of cases (25.2%) on both claws of the hind limbs a lesion was detected. The locomotion score for non-lame animals as well as for clearly lame cows and the calculated clusters showed high correspondence (79.4 and 83.7%). The score for cows with uneven gait (locomotion score 2) only had partial agreement (21.1%) to the middle cluster 2. However, generally, Krippendorff's ordinal α was 0.7460, a good degree of reliability. The confidence interval for α_{true} was 0.6755 to 0.8097. The results of this study suggest, that the used three-point locomotion score is suitable to classify cows reflecting their claw diseases correctly.

Session 25 Theatre 1

Perspectives in sustainable aquaculture

S. Čolak, R. Barić, D. Desnica and M. Domijan
Cromaris d.d., Gaženička cesta 4b, 23000 Zadar, Croatia; slavica.colak@cromaris.hr

The aquaculture production in Croatia has grown for over 70% during the last 10 years but still, Croatia is ranked low among EU countries, both for fisheries and for aquaculture production. In Croatia, 72% of fish is farmed in cages in sea and brackish waters and 28% in freshwater ponds. Currently there are three measures important for growth of aquaculture: investments in production in aquaculture, investments in placing products on the market, and innovation. Apparent consumption of fishery and aquaculture products in Croatia amounted to 18.4 kg/per capita in 2015, an increase of 4% compared to the previous year. Main consumed species are small pelagics, mainly sardines, and hake. Breeding Register of the Ministry of Agriculture has 31 entities producing sea bass / sea bream (and small quantities of other species, e.g. meagre) at 59 locations. Over 70% of the production is located in Zadar County area. National Strategic Plan for Aquaculture 2014-2020 (NSPA) predicts continuation of increase of production, although regulative framework in EU and in Croatia for aquaculture development is currently very complex and process for new concessions is complex and long. Development of aquaculture and the improvement of fishery techniques as well as modern fish storage and processing technologies will positively reflect with incensement in consummation of fish and other seafood products. Increasing co-operation within sector between producers, scientific organizations, Universities are necessary for the implementation of new knowledge.

Gilthead seabream predation on mussel farms: a growing conflict

T. Segvic-Bubic, L. Grubisic, I. Talijancic and I. Zuzul
Institute of Oceanography and Fisheries, Laboratory of aquaculture, Setaliste I. Mestrovica 63, 21000 Split, Croatia; tsegvic@izor.hr

The cultivation of mussels (*Mytilus galloprovincialis*) is a growing industry worldwide. However, predation by gilthead seabream *Sparus aurata* has become a significant challenge to mussel growers in Mediterranean causing extensive financial losses. In recent years, a decline in Croatian mussel production has been recorded. Among the many factors that have contributed to the decline of mussel production, i.e. small domestic market, great variability in larval dispersion and settlement, it is suspected that fish predation plays an important role in shellfish sustainability. Recently, wild stock of gilthead seabream has been significantly increased in the eastern Adriatic where effects of global warming, fish escapement through farming and spawning in cages can partially explain such fish expansion. Mussels are a dominant prey item for gilthead seabream, which take advantage of mussel farms that provide a highly abundant and easily accessible food source. Considering that no recent scientific studies have quantified the amount of mussels removed by fish on mussel sites, we conducted one-year long survey aiming to quantify the predation impact. Six most productive mussel farms were seasonally monitored by measuring the percentage of consumed or destroyed ropes of longlines. The results from 680 monitored mussel ropes have revealed that predation is most pronounced during worm seasons (summer, autumn). In average, 75% of mussel ropes were destroyed within the first week of mussel deposition into the sea. Predation varied from 65 to 85% among farming locations. In contrast, during cold seasons (winter, spring) fish predation has contributed to only 15% of mussel seed loss at mussel farms. Currently, attempts to reduce predation are limited to anti-predators bags or predator control nets placed around culture structures. However, results have indicated that additional net protection significantly affects mussel growth rates and has limitations during production and harvest. In conclusion, new protecting methods are needed (underwater repellent against seabream or allowing recreational fishermen to fish in mussel farming area) for supporting farm management stability.

Session 25 Theatre 3

Growth of the Great Mediterranean scallop (*Pecten jacobaeus*) in the Novigrad Sea (Croatia)

B. Baždarić[1], M. Peharda[2], T. Šarić[3] and I. Župan[3]
[1]Agency for Rural Development of Zadar County, Glagoljaška ulica 14, 23000 Zadar, Croatia, [2]Institute of Oceanography and Fisheries, Šetalište Ivana Meštrovića 63, 21000 Split, Croatia, [3]University of Zadar, Trg kneza Višeslava 9, 23000 Zadar, Croatia; branimir.bazdaric@agrra.hr

The scallop *Pecten jacobaeus* (Linnaeus 1758) is commercially valuable bivalve present throughout Mediterranean coastal waters. It lives on sand, mud, and gravel bottoms between 3 and 250 m depth and can grow over 200 mm in diameter. Due to the excellent meat quality and high market price its populations is under high harvesting pressure. Novigrad Sea, a Natura 2000 area, is a registered bivalve production zone, where aquaculture has a high priority. Due to its unique ecological and geographical characteristics (unpolluted sea with high inputs of fresh water from the Zrmanja river), the whole Novigrad sea basin is suitable for intensive bivalve production. The aim of this study was to determine possibilities for commercial production of *P. jacobaeus* in the Novigrad Sea, by monitoring growth parameters and mortality in the farming conditions. The bivalves were marked individually (Hall print tags) and placed at 20 m depth in lanterns in June 2014 and the growth was measured every six months until July 2017.The results showed that *P. jacobaeus* grew to a commercial size after 18 months. In contrast, the literature data indicate that in North Adriatic natural population of *P. jacobaeus* reaches commercial size usually at age of three years. According to our growth data and the absence of mortality, we can state that commercial farming of *P. jacobaeus* in the Novigrad Sea has good potential. However, insufficient amount of available spat in nature and the lack of hatcheries limits the further development of this form of aquaculture.

Growth and mortality of oysters (*Crassostrea gigas*, Thunberg 1793) in Sacca degli Scardovari (Italy)

A. Trocino[1], C. Zomeño[1], F. Gratta[2], M. Birolo[2], A. Pascual[1], F. Bordignon[2], E. Rossetti[3] and G. Xiccato[2]
[1]University of Padova, Department of Comparative Biomedicine and Food Science, Viale dell'Università 16, 35020 Legnaro, Padova, Italy, [2]University of Padova, Department of Agronomy Food Natural Resources Animals and Environment, Viale dell'Università 16, 35020 Legnaro, Padova, Italy, [3]Consorzio Cooperative Pescatori del Polesine O.P. Scarl, Via della Sacca 11, 45018 Rovigo, Italy; cristina.zomenosegado@unipd.it

This study assessed growth and mortality of oysters reared in suspended ropes, and tested the effect of rope emersion time throughout fattening (8 months; October 2016-June 2017). A total of 4,320 triploid oysters were stuck to 36 ropes (320 oysters/rope) and subjected to three emersion systems (12 ropes/system): standard of the farm (F), i.e. variable emersion duration changing according to daily atmospheric conditions; long (L), i.e. 14 emersion hours per day; short (S), i.e. 7 emersion hours per day. Biometric traits were collected at sticking, and after 2, 4, and 8 months. Biometric data were analysed with PROC MIXED (SAS), with emersion system, sampling time and their interactions as fixed effects, and rope as random effect; mortality was analysed with PROC CATMOD. At sticking, oysters exhibited an average weight of 6.04±2.63 g, a length of 39.8±8.38 mm and a width of 23.9±4.39 mm. Oyster length and width were similar from 2 to 4 months (46.1 mm and 33.3 mm on average, respectively), and increased after 8 months (76.0 mm length and 59.6 mm width) ($P<0.001$). After 8 months, oysters subjected to F and L emersion programs were heavier (68.3 g and 66.3 g vs 56.8 g; $P<0.01$), longer (78.1 mm and 77.2 mm vs 71.6 mm; $P<0.01$) and wider (55.6 mm and 55.0 vs 50.8 mm; $P<0.10$) than those subjected to S emersion. Total mortality reached 44.3% in F, 63.3% in L, and 66.8% in S system ($P<0.001$). To conclude, under the tested conditions, oysters fattening appeared feasible and promising, but the short fixed emersion system was the least favourable due to higher mortality and lower growth. Acknowledgements: La Perla del Delta; Veneto Region, Reg. (UE) 508 15/05/2014, DGR 213 28/02/2017.

Novigrad mussels in process of PDO labelling: opportunities and challenges

T. Šarić[1], B. Baždarić[2], N. Perović[1], F. Dadić[1], L. Mulić[1] and I. Župan[1]
[1]University of Zadar, Trg kneza Višeslava 9, 23000 Zadar, Croatia, [2]Agency for Rural Development of Zadar County, Glagoljaška ulica 14, 23000 Zadar, Croatia; tosaric@unizd.hr

Collection of wild mussels (*Mytilus galloprovincialis*, Lamarck 1819) in Novigrad Sea has a long history and it is a well-documented tradition. Nowadays the natural populations of shellfish in Novigrad Sea are significantly reduced due to uncontrolled harvesting and changes in environmental conditions, and mussel farming is becoming increasingly important in this area. Through the implementation of the INOVaDA project, led by the University of Zadar, the quality and production parameters (growth rate, condition index, chemical composition, stable carbon ($\delta^{13}C_{VPDB}$) and nitrogen ($\delta^{15}N_{AIR}$) isotope compositions and microbiology analyses) of mussels from the Novigrad Sea and other commercial production areas from Adriatic Sea were compared. Also, environmental parameters in Novigrad Sea (temperature, salinity, chlorophyll a, stable carbon ($\delta^{13}C_{VPDB}$) and nitrogen ($\delta^{15}N_{AIR}$) isotope compositions) were monitored. Results show how unique environmental factors in Novigrad Sea reflect in higher quality of mussels compared to the other investigated site. Therefore, the application for PDO was logical initiative by the Novigrad mussel cooperative and it was financed by the Croatian ministry of agriculture in 2017. For the PDO purposes more detailed natural/geographical conditions that are transmitted to mussels and particular quality characteristics of Novigrad mussels where stated and justified and will be presented as part of Product Specification in PDO application procedure.

Session 25 Theatre 6

Genetic gain from genomic evaluation in fish breeding programs with separate rearing of families

S. García-Ballesteros, J. Fernández and B. Villanueva
Instituto Nacional de Investigación y Tecnología Agraria y Alimentaria, Departamento Mejora Genética Animal, Ctra de la Coruña km 7.5, 28040 Madrid, Spain; silviagarbg@gmail.com

In order to record the pedigree in aquaculture breeding programs two types of larval rearing are practised. Families can be reared in separate tanks until the fish is large enough to be physically tagged or they can be reared together and genotyped subsequently to reconstruct the pedigree. A particular problem with separate rearing of families is that an environmental effect common to the members of the same family (the so called 'tank effect') is introduced. With standard BLUP, this effect is difficult to disentangle from the genetic effect reducing thus the expected genetic gain. The higher accuracies of estimated genetic effects achieved with genomic evaluation could lead to a better separation of tank and genetic effects. The objective of this study was to compare, through computer simulations, the selection response for an additive trait with a heritability of 0.4 achieved when tank effects are present and genomic evaluations are performed. Three different levels of tank effects were simulated (c2=0.0, 0.1 and 0.3, where c2 is the proportion of phenotypic variance due to tank effects). Both full (GE) and within-family (WFGE) genomic evaluations and BLUP were considered. Also, scenarios ignoring or accounting for the tank effect in the genetic evaluations were compared. In a scenario with low-density genotyping (500 SNPs) that is the most realistic in aquaculture, the evaluation method giving the highest gain was WFGE. This method gave up to 10% more gain than both GE and BLUP for c2=0.0, up to 14 and 31% more gain than GE and BLUP, respectively, for c2=0.1 and up to about 23% more gain than GE and BLUP for c2=0.3. Also, WFGE was the only approach that led to higher gains when the tank effect was included in the model than when this effect was ignored for both c2=0.1 and c2=0.3 (3-15% higher gains when included in the model). In contrast, both GE and BLUP led to higher gains when the tank effect was ignored for c2=0.1.

Session 25 Theatre 7

Transcriptome analysis and classifications of sex using neural network in domesticated zebrafish

S. Hosseini[1], S. Herzog[2], N.T. Ha[1], C. Falker-Gieske[1], B. Brenig[1], J. Tetens[1], H. Simianer[1] and A.R. Sharifi[1]
[1]Georg August University, Department of Animal Sciences, Albrecht-Thaer-Weg 3, 37075 Goettingen, Germany, [2]Max Planck Institute for Dynamics and Self-Organization, Am Fassberg 17, 37077 Goettingen, Germany; shahrbanou.hosseini@agr.uni-goettingen.de

The mechanisms of sex determination (SD) in zebrafish (*Danio rerio*) and its related species are controlled by the interaction between fish genotype and environmental setting. Besides the genetic factors, water temperature is one of the most important physical parameters which leading to alter the sex. Expression of SD genes might be associated with colour pattern (CP) genes in respect to sexual attraction. The main objectives of this study was to investigate the underling molecular mechanism of SD and CP genes and their interaction to gain new insights into the genetic control of sexual dimorphism and colour polymorphism. For the experimental setup eggs of full sib family divided in two treatment groups (28 and 35 °C) were used. Based on this setup 48 samples from gonad and caudal fin tissues were randomly chosen for the transcriptomic analysis. The result showed the effect of high elevated temperature leads to 25% more male in treated group compared to non-treated group. Further to classify the sex two methods were investigated: Deep convolutional neural network (DCNN) for a whole image classification and some handcrafted features based on CP for a support vector machine (SVM) classification. The predicted sex from the DCNN were highly in agreement with the real sex using urogenital papilla inspection (phi-coefficient>0.96). While using SVM to classify the sex by only CP of caudal fin shows a lower association with the real sex (phi=0.71) in temperature treated group. These animals might be 'neomales' regarding masculinization which classify in female group due to lower pigmentation. Transcriptomic analysis of gonad and caudal fin demonstrated highly significantly differentially expressed genes (DEGs) using negative binomial model and exact test in both tissues of comparison groups: (i.e. male control gonad vs female control gonad: DEGs=14,060, male treatment gonad vs female treatment gonad: DEGs=12,932, male control fin vs female control fin: DEGs=419 and male treatment fin vs female treatment fin: DEGs=801).

Session 25 — Theatre 8

Deltamethrin efficacy in controlling parasite *Ceratothoa oestroides* in *Dicentrarchus labrax* farming

S. Čolak[1], R. Barić[1], M. Kolega[1], D. Mejdandžić[1], B. Mustać[2], B. Petani[2] and T. Šarić[2]
[1]Cromaris d.d, Gaženička ul. 4b, 23000 Zadar, Croatia, [2]University of Zadar, Trg kneza Višeslava 9, 23000 Zadar, Croatia; slavica.colak@cromaris.hr

In the cage farming of sea bass (*Dicentrarchus labrax* Line, 1758) parasite isopod *Ceratothoa oestroides* (Risso, 1826) causes significant production losses, which are observed in reduced fish growth and low survival rate. Increased mortality were recorded on young fish up to 20 g. Deltamethrin is a synthetic pyrethrin showing good results in the control of the *C. oestroides*. The aim of this study was to determine the efficacy of deltamethrin (3.5 mg/m^3 of sea water) in the control of the *C. oestroides* in production conditions. The experiment was carried out at Cromaris d.d. within 4 commercial net cages of 16 m diameter and per 200,000 fish individuals of the same origin and initial weight. After 6 months of farming deltamethrin treatments were performed in two cages, while other two were control cages. The average fish weight during antiparasitic bath was 20 g. Two months after the treatments visual control of the entire population was carried out, and total of 100 infested individuals were excluded. Afterwards, biometric analyses were performed on infested fishes and on *C. oestroides* males and females. Sexual maturity and fecundity on *C. oestroides* females was determined. In addition, daily records of fish mortality and sea temperature were recorded. The results showed that the number of *C. oestroides* in treated cages was significantly lower ($P<0.01$) than the number of *C. oestroides* in control cages. Also, *C. oestroides* were smaller in treated cages than in control cages and they were not sexually mature. In control cages the number of sexually mature *C. oestroides* was 22.3 and 21.4%. The obtained results indicated the successful sea bass treatment with deltamethrin on *C. oestroides* removal. However, after 2 months period, new generation of immature *C. oestroides* appeared. These results showed possibility of *C. oestroides* infestation on sea bass greater than 20 g, but the absence of sexually mature females points the importance of this treatment to reduce the dynamics of *C. oestroides* population.

Session 25 — Theatre 9

Efficiency of different genomic coancestry matrices to maximize genetic variability in turbot select

E. Morales-González[1], M. Saura[1], A. Fernández[1], J. Fernández[1], S. Cabaleiro[2], P. Martinez[3] and B. Villanueva[1]
[1]INIA, Mejora Genética Animal, Crta. A Coruña Km. 7.5, 28040 Madrid, Spain, [2]CETGA, Punta de Couso sn, 15695 Aguiño-Ribeira, Spain, [3]Universidade de Santiago de Compostela, Xenética, Campus de Lugo, 27002 Lugo, Spain; elisa.em89@gmail.com

The main objective of aquaculture breeding programs is to increase genetic gain for production traits. However, the control of inbreeding is fundamental to preserve the genetic variability and to prevent inbreeding depression. Indeed, aquaculture species are particularly susceptible to the accumulation of inbreeding as a large number of candidates can be obtained from few selected individuals. The recent development of genomic tools allows a more accurate estimation of the relationships among individuals and therefore a more efficient control of inbreeding. The aims of this study were to estimate genomic inbreeding coefficients and to evaluate the use of different genomic coancestry matrices to maximize genetic variability using the optimal contribution method. We used data from a commercial population of turbot (*Scophthalmus maximus*) comprising 1,437 individuals from 36 families genotyped for 18,097 SNPs. Coancestry matrices evaluated included those based on the (1) proportion of shared alleles (θ_{IBS}); (2) deviation of the expected homozygosity ($\theta_{L\&H}$); and (3) proportion of shared segments (θ_{SEG}). In addition, genomic matrices commonly used in genomic evaluations – VanRaden (θ_{VR}) and Yang (θ_{YANG}) matrices – were also considered. Although the magnitude of the average estimates of inbreeding obtained from θ_{IBS}, $\theta_{L\&H}$ and θ_{SEG} (0.77, –0.01 and 0.08, respectively) differed greatly, the correlations between them were high ($\rho>0.98$). The correlation between each of these coefficients and those obtained from θ_{YANG} and θ_{VR} was respectively about 0.75 and -0.4. Correlations between coancestry coefficients computed from the different matrices were all high (>0.7) and positive. Higher correlations between coancestry coefficients computed from different matrices did not translate in more similar results from the optimisation. Although they led to very similar results in terms of expected heterozygosity, θ_{IBS} and θ_{VR} led to higher numbers of segregating loci and this was achieved by selecting a higher number individuals.

Session 25 Theatre 10

Growth of *Chlamys varia* (Linnaeus, 1758) in experimental cages in the south Adriatic sea

N. Antolovic, V. Kozul, N. Glavic, J. Bolotin and M. Rathman
University of Dubrovnik, Institute for Marine and Coastal Research, Damjana Jude 12, 20000 Dubrovnik, Croatia; nenad.antolovic@unidu.hr

Juveniles of variegated scallop *Chlamys varia* (Lin., 1758) were grown in suspension in Mali Ston Bay to study the growth rate. For the experiment, we used juveniles from last year spawning season. Experiment lasted for 21 months, cages with juveniles of variegated scallop were placed at tree depths: 1, 3 and 5 m. At the end of experiment at 1 m depths the length were 43-59 mm, at 3 m depths 42-53 mm and at 5 m depths, the lengths were 44-55 mm. The average monthly growth of the shell height for the total experiment period was 1.7, 1.6 and 1.7 at 1, 3 and 5 m depth respectively.

Session 25 Poster 11

Cryopreservation of Black sharkminnow, *Labeo chrysophekadion* (Bleeker, 1849) spermatozoa

S.. Ponchunchoovong
Suranaree University of Technology, Institute of Agricultural Technology, 111 University Avenue, Muang, Nakhon Ratchasima, 30000, Thailand; samorn@sut.ac.th

Black sharkminnow is one of the important freshwater cyprinid fish due to its palatability and high market price. They are a widely distributed in the Mekong and Chao Phraya river basins, Thailand to the Malay Peninsula, western Indonesia and Borneo. In Thailand, pickled fish made from black sharkminnow is a famous product of community enterprises in Khongchaim district, Ubon Ratchathani Province. The seasonal change in Thailand causes a lack of adult fish in the Mekong Basin. Black sharkminnow is also difficulty for artificial insemination management due to its low mean semen volume (1.15±0.69 ml per fish), however mean sperm concentration is high ($13.00 \pm 2.62 \times 10^9$ sperm/ml) and wild stocks of black sharkminnow have been reduced due to habitat degradation. Cryopreservation of fish sperm is a valuable technique to assist artificial insemination and conservation of wild stock of black sharkminnow. The objective of present study was to evaluate the effect of extenders and cryoprotectants on motility, viability and fertilization of frozen black sharkminnow spermatozoa. Mature black sharkminnow broodfish (50 males and 15 females) were used. Sperm were diluted 1:1 in two types of extender (modified extender and MC; Modified Cortland solution) containing three types of cryoprotectants (DMSO, MeOH and glycerol) at concentration of 5, 10 and 15%. Our finding found that the combination of MC and 15% DMSO had the best motility (56%) and viability (64%). This was not significantly different from using modified extender and 10% DMSO (P>0.05). The highest fertilization rate (45%) was achieved with the combination of modified extender and 10% DMSO, similar to that of the combination of MC and 15% DMSO (P>0.05). In conclusion, modified extender and 10% DMSO was the most successful for cryopreservation of black sharkminnow spermatozoa.

Gonad development in adult giant grouper (*Epinephelus lanceolatus*)

S. Kubota[1], S. Boonanuntanasarn[1], P. Bunlipatanon[2], N. Saen-In[3], O. Mengyu[2], S. Detsathit[2], R. Yazawa[4] and G. Yoshizaki[4]

[1]Suranaree University of Technology, Institute of Agricultural Technology, 111 University Avenue, Suranaree, Muang, Nakhon Ratchasima, 30000, Thailand, [2]Department of Fisheries, Krabi Coastal Fisheries Research and Development Center, 141 M.6 Saithai, Muang, Krabi, 81000, Thailand, [3]Department of Fisheries, Pang-Nga Coastal Fisheries Research and Development Center, 164 M.9 Thai Muang, Pang-Nga, 82120, Thailand, [4]Tokyo University of Marine Science and Technology, Department of Aquatic Biosciences, 4-5-7, Minato-ku, Konan, Tokyo, 108-8477, Japan; skubota@sut.ac.th

Giant grouper (*Epinephelus lanceolatus*) is the largest fish which has been found in coral reefs. Also, giant grouper has been important as farm grouper which is in high demand as aquaculture species throughout Asia and Europe. So far, farm-raised giant grouper has been developed. However, seed propagation of this species in the captive pond has not been well-controlled because its information of gonad development has been limited for this protogyny species. Therefore, this study aimed to conduct preliminary investigation of gonad development in mature giant grouper. Fish were sampled, and their gonads were performed histological study. The gonad of female fish which has body weight of 7.4-19.6 kg (GSI~0.07-0.20%) contained several stages of oocytes and oogonial cells. In addition, testis was observed in fish at body weight of 11.2-14.4 kg (GSI~0.08-0.12%) which contained spermatogonia, spermatocyte and spermatid. Moreover, intersex gonad was observed in fish at body weight of 14.0-19.0 kg (GSI~0.07-0.12%) which contained both ovarian and testicular germ cells. Therefore, there were an overlap of male and female body size. These findings suggested that male fish could develop directly from juvenile phase as well as by sex change from female. Taken together, the reproductive of giant grouper could be hypothesized to be diandric protogynous hermaphrodite.

Morphological pattern of the gills in various body mass rainbow trout reared by different technology

J. Szarek, E. Strzyżewska-Worotyńska, A. Dzikowski and K. Wąsowicz

University of Warmia and Mazury in Olsztyn, Department of Pathophysiology, Forensic Veterinary Medicine and Administration, Oczapowskiego St. 13, 10-719 Olsztyn, Poland; szarek@uwm.edu.pl

The aim of the present study was to assess the impact of extensive fishery with flow-through system (FTS) and intensive with recirculating aquaculture system (RAS) on the morphological gills' pattern of rainbow trout (in various stage of harvesting). Research was carried out in spring and autumn, on 144 rainbow trout in 6 fish farms, 3 in the FTS technology, the other – RAS. The fish came from the same breeding facilities. For testing were used: 6 trout (n=6), body mass 350-500 g (A-fish), and 6 trout, body mass of 501-850 g (B-fish). Gills sections were fixed for 48 h in 7% buffered formalin and specimens were stained with haematoxylin and eosin. Microscopic lesions were subjected to semi-quantitative assessment by calculating Histopathological Indexes (HAI) presented in Poleksic and Mitrovic-Tutundzic modifications. Microscopic changes were evaluated in accordance with the recommendations of Bernet *et al.* The statistically significant differences between morphological changes of the FTS and RAS fish were calculated using the U Mann-Whitney test with the Bonferroni correction. Low HAI values indicated in most cases the correct gills' structure or slight morphological lesions. Trout from FTS technology had slightly higher dispersion of HAI irrespective of the research season. Trout from the RAS showed statistically significant higher number of moderate and deep morphological lesions than those from the FTS. This assessment concerns especially fish caught in spring and in B-fish. Circulatory disorders occurred statistically more frequently in B-fish, particularly from the RAS than in A-fish. The applied gills' rating allows to determine the impact of technological conditions used for rearing of the rainbow trout on the structure of the organ studied. Gills' status is therefore a biomarker and an evaluation tool of the impact of farming technology on fish organism.

Session 26 Theatre 1

Genomic selection in populations with low connectedness between herds

A. Kasap[1,2], B. Mioc[1], J.H. Hickey[3] and G. Gorjanc[2,3]
[1]University of Zagreb, Croatia, [2]University of Ljubljana, Slovenia, [3]University of Edinburgh, United Kingdom; gregor.gorjanc@roslin.ed.ac.uk

Low connectedness between herds reduces accuracy of estimated breeding values and ranking of animals across herds. This is a common issue in breeding programs that do not use artificial insemination or do not actively exchange sires between herds. Availability of genomic data may present an opportunity to mitigate this issue by providing information on realised genome sharing between animals in different herds. In this study we used simulation to compare the accuracy of pedigree and genomic based estimates of breeding values in populations with varying level of connectedness. The simulation mimicked the structure of a typical sheep breeding program in Croatia. For genomic analyses, we used the single-step method and a training population that comprised either 1000 sires each with 20 phenotyped progeny or 1000 phenotyped progeny. Heritability was 0.3. Accuracy was evaluated for sires and for non-phenotyped progeny. Both pedigree and genomic based models included herd-year as a fixed effect. When the training population comprised sires with phenotyped progeny and each sire had progeny only in one herd both pedigree and genomic estimates had practically null accuracy. This was due to complete confounding between sire breeding values and herd-year effects. When sires had progeny in two herds accuracy increased. The increase was larger for genomic than for pedigree based estimates. Testing sires in more than two herds increased accuracy marginally. When the training population was comprised of phenotyped progeny and each sire had progeny only in one herd the accuracy of genomic estimates was higher than when training population was comprised of sires. When sires had progeny in two herds the accuracy was the same for both training population designs. In conclusion, when training population is comprised of sires low connectedness affects accuracy of pedigree and genomic based estimates in a similar way. In such cases training population for genomic selection should comprise phenotyped individuals rather than sires with phenotyped progeny in disconnected herds.

Session 26 Theatre 2

Genetic connectedness in the U.S. sheep industry

R.M. Lewis[1], L.A. Kuehn[2], H. Yu[1], G. Morota[1] and M.L. Spangler[1]
[1]University of Nebraska-Lincoln, Department of Animal Science, Lincoln, NE 68583, USA, [2]USDA-ARS, Roman L. Hruska U.S. Meat Animal Research Center, Clay Center, NE 68933, USA; rlewis5@unl.edu

Within agricultural systems, groups of animals are often raised under distinct management conditions. If these groups differ genetically, confounding of genetic merit and management unit may bias genetic evaluation. However, if ancestral relationships or connections among different management groups are sufficiently strong, bias in the genetic evaluation is reduced improving the accuracy of selection decisions. In small ruminants, where use of sires across-units can be limited by biological and geographical constraints, serendipitous connections may be inadequate. Such is true within the U.S. sheep industry. Routine monitoring and, where practicable, strengthening of genetic relationships is therefore considered a priority. Connectedness can be evaluated from functions of prediction error (co)variances of estimated breeding values between animals. Using data from the U.S. National Sheep Improvement Program from as early as the 1980s, connectedness was assessed in terminal sire (Suffolk; n»99,000), range (Targhee; n»110,000) and hair (Katahdin; n»50,000) sheep. The extent of connectedness varied among breeds – strongest in Targhee and weakest in Suffolk – reflecting differences in the geographical dispersion of flocks and the prevalence of ram exchange programs. Within breeds, clusters of well-connected flocks were clearly distinguishable likely identifying groups of breeders sharing similar goals. Uptake of genomics is at its infancy in the U.S. sheep industry, although enthusiasm is growing. Using simulation, genomic relatedness has been shown to strengthen estimates of genetic connectedness across-units thereby enhancing the accuracy of genetic evaluation. Given the dispersed structure of the U.S. sheep industry, and challenges in parentage-determination in extensively managed flocks, such information may be particularly valuable. In the meantime, pedigree-based connectedness is proving useful for targeting animals for sampling for genome-wide association studies and for developing resource populations for genomic prediction. Since the relative cost of genotyping in sheep remains high, only such strategic approaches will likely be economic.

Measuring connectedness among herds in French breeding programs for suckling sheep

V. Loywyck
Institut de l'Elevage, BP 42118, 31321 Castanet-Tolosan Cedex, France; valerie.loywyck@idele.fr

Genetic connectedness in BLUP genetic evaluation is becoming increasingly important in animal breeding to ensure genetic comparison of animals raised in different environments. A clustering method, that can handle a large amount of comparisons, was applied for on-farm French suckling sheep genetic evaluation: the aggregation criterion (*Caco*), which reflects the average level of connectedness of each herd was obtained. *Caco* values are low within each breed since knowledge of paternity is low in French suckling sheep breeds, mainly due to the large use of naturally-mated sires: Caco value decreases to zero when paternity rate is less than 70% in a herd. Two criteria were taken into account to define the connected status of a herd within each breed: *Caco* criterion and a percentage of ewes as results of artificial insemination. For each criterion, a common and consensus threshold for the 22 French breeding programs for suckling sheep was applied. These information do not impact the published EBVs but may be useful for technical support of herds and management improvement of the breeding programmes.

Effect of the level of artificial insemination on the genetic gain for a meat sheep breeding program

J. Raoul[1,2] and J.M. Elsen[1]
[1]INRA, GenPhySE, 24 Chemin de Borde Rouge, 31326 Castanet-Tolosan, France, [2]Institut de l'Elevage, BP 42118, 31321 Castanet-Tolosan, France; jerome.raoul@inra.fr

Numerous breeding programs of meat sheep breeds use both artificial insemination (AI) and natural mating sires. With AI, less sires are needed and their EBVs (Estimated Breeding Values) can be computed based on a larger number of progeny records. Thus, AI leads to achieve higher selection differential than natural mating. In addition, the use of AI sires across flocks creates genetic connections among these flocks. In France, AI are mainly achieved with fresh semen and require to synchronize the ovulation of ewes. The control of oestrus in ewes is based on hormones although their use in animal production is more and more disputed. The ban on hormones would lead to a decrease in AI use and thus affect both the selection differential and the genetic connectedness. The aim of this study was to investigate the consequences of a reduced use of AI on the genetic progress (ΔG) for a repeated maternal trait. Using a stochastic model, we simulated a breeding program of 8,000 ewes. In all designs, young males were selected on parent average EBVs. Young AI males were selected prior those used for natural mating. Various levels of the rate of AI, *ai_r* (i.e. the percentage of ewes that were inseminated per year), were considered for three designs: (1) young AI sires were preselected on their parent average EBVs and proven AI sires were then selected on their progeny records (*ai_r*=50,80), (2) young AI sires were selected on their parent average EBVs and used during two reproductive cycles (*ai_r*=5,10,20,50,80), (3) no AI sires were selected (*ai_r*=0). The phenotypes of ewes were simulated based on their true breeding values, the repeatability of the trait, a flock-year random effect and a residual effect. The EBVs were computed by a BLUP animal model using the BLUPF90 software. We computed the ΔG for a period of 15 years. Preliminary results show that a decrease in ΔG occurred from a rate of AI below 20% with a maximal reduction of 30% (*ai_r*=0). Ongoing study will attempt to disentangle the effect of the rate of AI on both the genetic connectedness and the selection differentials.

Genome-based inbreeding in French dairy sheep breeds

S.T. Rodriguez-Ramilo and A. Legarra
INRA, GenPhySE, 24 Chemin de Borde Rouge, 31326 Castanet Tolosan, France; silvia.rodriguez-ramilo@inra.fr

Traditionally, inbreeding has been estimated from pedigree-based information. Nevertheless, in the last years the interest for estimating inbreeding using genomic information has considerably grown. The objective of this study is to evaluate whether inbreeding estimated from ROHs or calculated on a SNP-by-SNP basis provide an accurate measure of inbreeding. The data set available included individuals from French dairy sheep breeds and subpopulations (Basco-Béarnaise, BB; Manech Tête Noire, MTN; Manech Tête Rousse, MTR; Lacaune Confederation, LACCon; and Lacaune Ovitest, LACOvi). Those animals were genotyped with the Illumina OvineSNP50 BeadChip. After filtering, the genomic data included 38,287 autosomal SNPs and 8,700 individuals. Pedigree from those genotyped animals comprised 72,803 animals. Genome-based inbreeding was higher than pedigree-based inbreeding. This can be explained considering that pedigree-based inbreeding assumes that loci are neutral and it does not consider the potential bias resulting from selection. The highest rates of increase in inbreeding per generation were observed when using pedigree (0.0022 – 0.0099; BB, MTN and LACOvi) and ROHs (0.0022 – 0.0046; MTR and LACCon). The lowest rates of increase in inbreeding per generation were observed when calculating inbreeding on a SNP-by-SNP basis (0.0016 – 0.0044), except for LACOvi where the lowest rates were also observed with ROHs (0.0014). Both genome-based measures rank breeds and subpopulations in the same order regarding the effective population size: LACOvi (357), LACCon (227 – 313), MTR (109 – 200), MTN (81 – 179) and BB (59 – 114). Effective population size obtained from pedigree information provided similar estimates. ROHs yield a correct estimate of inbreeding at the genomic level and make it possible to identify specific IBD regions. These regions can be used to improve mating decisions and to minimise the unfavourable effects of inbreeding. Work financed by the ARDI project of Poctefa program (European Union FEDER funds).

Profile of PRNP gene in 3 dairy goat breeds and association with milk production and udder health

S. Vouraki[1], A.I. Gelasakis[2], P. Alexandri[1], E. Boukouvala[2], L. Ekateriniadou[2], G. Banos[1,3] and G. Arsenos[1]
[1]Laboratory of Animal Husbandry, School of Veterinary Medicine, Faculty of Health Sciences, AUTH, University Campus, 54124 Thessaloniki, Greece, [2]Veterinary Research Institute of Thessaloniki, Thermi, 57001 Thessaloniki, Greece, [3]Scotland's Rural College/Roslin Institute, University of Edinburgh, Easter Bush, EH25 9RG, Scotland, United Kingdom; svouraki@vet.auth.gr

The objectives were to determine the genetic profile of scrapie codons 146, 211 and 222 in three dairy goat breeds in Greece and assess its impact on milk production and udder health. A total of 766 dairy goats from seven farms were used. Animals belonged to two native (Eghoria; n=264 and Skopelos; n=287) and one foreign breed Damascus; n=215.Genomic DNA was extracted from individual blood samples. Polymorphisms were detected using Real-Time PCR analysis and four different Taqman assay mixes. Milk production and udder health traits of individual animals were recorded monthly for two consecutive milking periods. The former included daily and lactation milk yield and composition; fat, protein, lactose and solids-not-fat (SNF) content. Udder health traits included udder fibrosis, abscess and asymmetry, milk somatic cell count, and number of colony-forming units. Genotypic, allelic and haplotypic frequencies were calculated and genetic distances between breeds were estimated using fixation index FST with Arlequin v3.1 software. The effects of genotype and allele on the studied traits were assessed with mixed linear models with R package 'lme4'. All resistance-associated alleles (146S, 146D, 211Q and 222K) were detected in the studied goats. Significant differences ($P<0.01$) in genetic profile were observed between the two native breeds and Damascus. Alleles 222K and 146S had the highest frequency (ca.6%) in the native and Damascus breeds, respectively. Resistance-associated alleles and genotypes did not affect animal traits ($P>0.01$) except for a mild adverse effect of codon 146 on milk SNF in native goats. Results suggest that separate breeding programs should apply to native and Damascus goats aiming at enhancing scrapie resistance. Selection for the latter is not expected to compromise animal productivity and udder health. This work was funded by the SOLID project (FP7-266367) and Onassis Foundation.

Application of genomics in breeding schemes for the genetic improvement of sheep and goats

D.J. Brown[1,2], A.A. Swan[1,2], J.H.J. Van Der Werf[3] and R.G. Banks[1,2]
[1]Australian Cooperative Research Centre for Sheep Industry Innovation, University of New England, 2351 Armidale, Australia, [2]Animal Genetics and Breeding Unit, University of New England, 2351 Armidale, Australia, [3]School of Environmental and Rural Science, University of New England, 2351 Armidale, Australia; dbrown2@une.edu.au

The main benefit of genomic selection for sheep and goats is to increase the accuracy of estimates of genetic merit for hard-to-measure traits earlier in life, including carcass, adult wool, milk production and reproduction traits. An extensive genotyped reference population, combined with a significant number of genotyped and phenotyped animals from progeny test and ram-breeding flocks enables genomic predictions with improved accuracies. Key technical challenges for genetic evaluation include weighting of pedigree and genomic information, achieving acceptable run times, and estimation of breeding value accuracy from genomic contributions. Recent analyses enhancements to implement single step analysis to include genomic information have been shown to significantly improve the prediction of progeny performance across most traits. Challenges for the practical application of genomic selection include; cost and design of the reference population, genotyping costs in general, utilisation of data from commercial ram breeding and industry flocks, accommodating smaller breeds and crossbred animals and breeding program design constraints. While genomic information now routinely contributes to increase accuracy of selection in many ram breeding flocks in Australia it is not yet a widespread or mainstream practice. It works best for flocks aiming to reduce generation intervals and focusing on hard-to-measure traits. Experience in Australia demonstrates that for flocks not phenotyping, genomic prediction accuracies are generally low unless they are closely genetically related to the reference population. The benefits from genomic selection are best captured by flocks prepared to evolve their breeding program designs to do so rather than simply superimposing genomics onto their current program. Future developments will evaluate alternate models for incorporating genomic information, ability to estimate more accurate breeding values for animals without any pedigree and phenotypic information, and alternate industry uses for genomic testing.

Predicted accuracies of GEBV from female nucleus or male reference population in Sarda dairy sheep

M.G. Usai, S. Salaris, S. Casu, T. Sechi, S. Miari, P. Carta and A. Carta
AGRIS Sardinia, Genetics and Biotechnologies, Loc. Bonassai, 07100, Italy; acarta@agrisricerca.it

Traditional applications of genomic selection are based on male reference populations which are used to produce GEBV of young males. In many dairy sheep breeds this approach is limited by the small amount of males with a number of daughters sufficient to include them in the reference population. In a previous study we showed that the theoretical accuracy of GEBV for milk yield of young rams can be predicted by a function of a parameter measuring their information on relatives in a female reference population. In the present work we compared the theoretical accuracies of GEBV of 119 young rams based on a female nucleus population (3,731 ewes) with those obtained with a GBLUP model using 414 DYD of rams with daughters in the herd book. The average accuracy was higher for the female reference population (0.49 vs 0.42). Similarly to the previous work, a specific function relating a parameter derived from the numerator relationship matrix, that measures the information on relatives in the male reference population, with the theoretical accuracies of the 119 rams was estimated. The two functions were used to predict GEBV accuracies of 310 young males of the herd book born in 2013 either from the female nucleus or from 4,189 old rams born from 2003 to 2010. The average predicted accuracies were similar (around 0.49). These results show that the female nucleus is crucial to trigger the genomic selection for milk yield in the Sarda breed. In the future, the pileup of genotyped males with daughters in the herd book will increase the accuracy of GEBV by using ssGBLUP methodology which is able to combine information coming from the female nucleus with the herd book male reference population. The precision of both the parameters measuring the information on relatives and the functions linking them to the accuracies of GEBV may be improved allowing the *a priori* selection of young males to genotype and the optimal management of the flow of rams from and toward the female nucleus.

Genomic selection on Latxa dairy sheep

I. Granado-Tajada and E. Ugarte
NEIKER-Tecnalia, Department of Animal production, Agrifood Campus of Arkaute s/n, 01080 Arkaute, Spain; igranado@neiker.eus

Several studies have showed that also in dairy sheep the use of genomic information makes possible the estimation of more accurate breeding values. The implementation of genomic selection on Latxa sheep was studied for the first time in 2014 with other Western Pyrenees breeds. In the specific case of the Latxa breed and contrary to the observed in the rest of the races, the studies did not show a clear advantage of genomic selection. In that moment, the available genomic data were limited and the relationship between animals of training and validation groups was weak. Results were not conclusive and therefore was considered that the study had to be repeated when more complete and stronger genotyping data were available. Since then, the breeding program has continued collecting phenotypic and genotypic data and the current work shown the results obtained in 2018. Using data proceeding of milk recording program (639,588 of Latxa Black faced (LBF) and 392,295 of Latxa Red faced (LRF)), pedigree information (263,241 and 150,167 for LBF and LRF respectively) and genotypes of artificial insemination males (374 and 451 for LBF and LRF respectively) pedigree index estimations (by BLUP) and genomic estimations (by ssGBLUP) were obtained for the 2014 data set. The accuracy was estimated as the correlation between these estimations and genetic or genomic evaluation obtained by progeny test into validation group in 2017. The results were similar to the obtained in the previous study and do not show higher accuracy for genomic selection (0.48±0.09 in LBF and 0.43±0.07 in LRF). It is necessary a more detailed study of data and pedigree structure to understand these results and further studies must be done to analyse the interest of genotyping natural service males and females.

Genome-wide study finds a QTL with pleotropic effects on semen and production traits in Saanen goats

C. Oget[1], V. Clément[2], I. Palhière[1], G. Tosser Klopp[1], S. Fabre[1] and R. Rupp[1]
[1]INRA, 24 Chemin de Borde Rouge, 31326 Castanet-Tolosan Cedex, France, [2]Institut de l'Elevage, Chemin de Borde Rouge, 31326 Castanet-Tolosan Cedex, France; claire.oget@inra.fr

Economic sustainability of dairy ruminant farms mainly relies on production, reproduction, and health of livestock. In this study, we investigated the genomic regions controlling seventeen production traits (eleven type traits, milk yield, four milk composition traits and lifespan of livestock trait), five reproduction traits (semen production and quality), and one trait related to health (Somatic Cell Score, as a proxy for mastitis resistance). Daughter yield deviation values were computed for 597 (672) Alpine and 460 (519) Saanen AI rams from Capgènes testing station for production traits, mastitis (and reproduction). For lifespan, only the oldest rams had yield deviations to allow large progeny and good accuracy, i.e. 341 Alpine and 298 Saanen. All AI rams were genotyped with the Illumina GoatSNP50 Bead Chip. We conducted genome wide association studies using polygenic mixed models and genomic relatedness matrix with the GEMMA software. The two breeds were analysed separately. Results showed a highly significant QTL on CHI 19 (22.8-28.9 Mb) in the Saanen breed for 12 of 23 studied traits. The most significant traits were udder floor position (P=7.31e-24), volume of semen production (P=8.43e-21) and protein yield (P=6.64e-17). We did not find this QTL in the Alpine breed suggesting a breed-specific control. Analyses of sequencing data of 11 Alpine and 9 Saanen rams highlighted a candidate causal mutation in the Sex Hormone Binding Globulin (SHBG) gene. We will determine whether these pleiotropic relationships are further supported by one single mutation or resolved into closely linked loci. Altogether, these results identify a major pleiotropic QTL in the Saanen breed. They will be helpful for goat selection in the future. They indicate that a within-breed genomic selection should be favoured for those traits and that including QTL information might prove useful. The study further pave the way for the addition of reproduction traits in the French goat breeding programs.

Genetic parameters for semen traits in French Alpine and Saanen bucks

V. Clement[1], C. Charton[2] and H. Larroque[2]
[1]Institut de L'Elevage, Chemin de Borde Rouge, 31326 Castanet Tolosan, France, [2]INRA, UMR 1388 GenPhySE, Chemin de Borde Rouge, 31326 Castanet Tolosan, France; virginie.clement@idele.fr

Breeding scheme of French dairy goats is based on creation of genetic progress and its dissemination by frozen artificial insemination. Therefore, management of young males who will become elite bucks is fundamental and determines selection efficiency. However, a lot of animals are bred and are not kept for semen production disorders. The French project 'Maxi'mâle' have the objective to produce indicators to optimize reproductive males management. One aim of the project was to carried out a genetic evaluation on semen production traits in order to improve the number and the quality of doses for bucks retained in the selection scheme. To this end, genetic parameters were estimated using restricted maximum likelihood with an animal model using 254,000 ejaculates of 2,086 bucks of the two main breeds, Alpine and Saanen, collected from 1995 to 2016. Traits considered were: concentration, volume of ejaculates, number of spermatozoa, and two traits recorded after thawing: motility score and percentage of living spermatozoa. Heritability estimates were low to moderate, ranging between 0.05 and 0.10 for traits measured after thawing and between 0.12 and 0.22 for the others traits. These estimates were lower considering ejaculates of young bucks bred in natural condition than for adult bucks bred with an artificial photoperiodic treatment. Genetic correlation between motility score and percentage of living spermatozoa was close to one, suggesting a common genetic determinism. Some traits are positively correlated, like concentration with motility score (around 0.40). Other are genetically opposite, like volume and concentration (-0.65 to -0.39) or volume and motility score (-0.31). Estimated breeding values shown different trends for traits according to breed. Number of spermatozoa shown a genetic gain of 150 million in 20 years in Alpine breed, whereas the motility score shown a genetic gain of 0.13 points in Saanen breed during the same period. The next steps will be to carried out a genomic evaluation for these traits and to investigate the best way of taking into account this information in the selection scheme in order to improve semen production of elite' bucks.

Genetic variation of lactoferrin gene and its association with productive traits in Egyptian goats

O.E. Othman[1], H.R. Darwish[1] and A.M. Nowier[2]
[1]National Research Centre, Cell Biology, 33 El Buhouth St., Dokki 12622, Egypt, [2]Agriculture Research Center, Animal Production Research Institute, Dokki, Giza 12611, Egypt; othmanmah@yahoo.com

Lactoferrin (LF) is a multifunctional protein involved in economically production traits like milk protein composition and skeletal structure in small ruminants including sheep and goat. So, LF gene – with its genetic polymorphisms associated with production traits – is considered a candidate genetic marker used in marker-assisted selection in goats. This study aimed to identify the different alleles and genotypes of this gene in three Egyptian goat breeds using PCR-SSCP and DNA sequencing. Genomic DNA was extracted from 120 animals belonging to Barki, Zaraibi and Damascus goat breeds. Using specific primers, PCR amplified 247-bp fragments from exon 2 of LF goat gene. The PCR products were subjected to single strand conformation polymorphism (SSCP) technique. The results showed the presence of two genotypes GG and AG in the tested animals. The frequencies for both genotypes varied among the three tested breeds with the highest frequencies for GG genotype in all tested goat breeds. The sequence analysis of PCR products representing these two detected genotypes declared the presence of a SNP substitution (G/A) among G and A alleles of this gene. The association between different LF genotypes and milk composition as well as body measurement was estimated. The comparison showed that the animals possess AG genotypes are superior over those with GG genotypes for different parameters of milk protein compositions and skeletal structures. This finding declared that allele A of LF gene is considered the promising marker for productive traits in goat. In conclusion, the Egyptian goat breeds will be needed to enhance their milk protein composition and growth trait parameters through the increasing of allele A frequency in their herds depending on the superior production traits of this allele in goat.

Session 26 — Poster 13

Estimation of genetic parameters in a local sheep breed using random regression animal model

R.S. Pelmus, C. Lazar, E. Ghita, C.M. Rotar and M.A. Gras
National Research-Development Institute for Animal Biology and Nutrition, Laboratory for the management of animal genetic resources, Calea Bucuresti, no. 1, 077015 Balotesti, Ilfov, Romania; pelmus_rodica_stefania@yahoo.com

There is a national interest for the preservation and improvement of local sheep breeds. The local Romanian Teleorman Black Head sheep breed fits best the current economic requirements for milk production. The objective of this study was to determine the genetic parameters represented by the heritability for test-day milk yield and the breeding value for Teleorman Black Head ewes. The genetic parameters were estimated using the Restricted Maximum Likelihood method with a test-day random regression animal model. The data set consists of 631 test-day records from the first lactation of 81 ewes. The pedigree covered 168 animals. The heritability estimates for test-day milk yield ranged from 0.017 on 180 day in milk to 0.148 on 20 day in milk. Over the lactation length, daily breeding values were computed for each animal from the random regression coefficients. The results revealed that the heritability estimates for test-day milk yield were low. The genetic parameters are important in selection program on this local breed sheep for genetic improvement.

Session 26 — Poster 14

Genetic differentiation of the population of Polish Mountain sheep

A. Miksza-Cybulska, A. Kawęcka and A. Gurgul
National Research Institute of Animal Production, Krakowska 1 St., 32-083 Balice, Poland; anna.miksza@izoo.krakow.pl

The mountain sheep found in the Polish Carpathians are derived from their ancestor Valachian (Zackel) sheep, which were brought from the Balkans in the 14th century. As a result of breeding work and directional selection, three breeds of mountain sheep were established: Polish Mountain, Coloured Polish Mountain, and Podhale Zackel. Mountain sheep are versatile utility animals, well adapted to harsh mountain conditions, owing their uniqueness to the role they play in landscape and culture. It is essential to know the biodiversity status to conserve farm animal genetic resources. In order to estimate genetic variation in the three breeds of mountain sheep, a study was performed at the National Research Institute of Animal Production using modern tools of SNP microarrays. They allow simultaneous genotyping of individuals at thousands of polymorphic sites, making it possible to estimate indicators of genetic diversity such as heterozygosity, the ratio of polymorphic loci, and the coefficient of inbreeding. The ovine genome was analysed using the Ovine SNP50 BeadChip array (www.illumina.com), designed for domestic sheep. The study involved three breeds of mountain sheep, 100 animals per group. The mean minor allele frequency (MAF) for the mountain sheep population was 0.294 to 0.299, the mean expected heterozygosity ranged from 0.383 to 0.389, and the mean observed heterozygosity from 0.389 to 0.394. Further analyses are being performed concerning the parameters of inbreeding in a subpopulation (Fst) and runs of homozygosity (ROH), which will allow estimating genetic variation in the mountain sheep populations as well as identifying selection signatures within and between breeds. Identifying allele frequency changes in genes responsible for phenotypic traits that differentiate the breeds will confirm correct classification of the animals into separate breeds as well as showing the genetic background of selected productive traits. Project 'The uses and the conservation of farm animal genetic resources under sustainable development' co-financed by the National Centre for Research and Development within the framework of the strategic R&D programme 'Environment, agriculture and forestry' – BIOSTRATEG, contract number: BIOSTRATEG2/297267/14/NCBR/2016.

A Genomic association study of gastrointestinal parasites in Ghezel sheep breed of East-Azerbaijan

S.A. Rafat[1], P. Ajmone Marsan[2], M. Del Corvo[2], M. Barbato[2], R. Valilou[1], D. Notter[3], R. Pichler[4] and K. Periasamy[4]
[1]university of Tabriz, Animal Science, Bd. 29 Bahman, University of Tabriz, Faculty of Agriculture, 5166616471, Iran, [2]Institute of Zootechnics, Università Cattolica del Sacro Cuore, Piacenza, Italy, [3]Virginia Polytechnic Institute and State University, Virginia Polytechnic Institute and State University, Virginia Polytechnic Institute and State University, USA, [4]Animal Production and Health Laboratory, Joint FAO/IAEA Division of Nuclear Techniques, International Atomic Energy Agency, Vienna, Austria; rafata@tabrizu.ac.ir

Sheep husbandry plays a prominent role for the economy of Iran, with 45 million sheep currently reared in the country. The present study – part of the International Atomic Energy Agency CRP projects on 'Genetic Variation on the Control of Resistance to Infectious Diseases in Small Ruminants for Improving Animal Productivity genetic resistance to nematodes in small ruminants' – aims to investigate the association between a selected panel of SNPs (n=157) in candidate genes and gastrointestinal nematode (GIN) infections. A total of 211 weaned lambs from ten flocks in East Azerbaijan province, Iran, were included in the study. Nematode eggs were counted in lamb's faeces (faecal egg count EPGG) and further classified into four groups: (1) EPGO: Strongyles, (2) EPGN: *Nematodirus* spp., (3) EPGT: *Trichuris* spp., and (4) EPGM: *Marshallagia marshalli*. Lambs were measured for body weight, FAMACHA test, and Packed Cell Volume (PCV). Genomic association analysis was performed using the 'RepeatABEL' R package. The model applied was Yijklm= BW + Si + Tj+ Fk+ SNPl + eijklm. We identified 118 SNPs in 39 genes located on 19 chromosomes significantly associated with either EPGO (72), EPGG (16), EPGN (14), EPGM (9), EPGT (5) and PCV (2). SNPs in CLEC1B and NLRC4 genes were significantly associated with PCV and EPGG. Our results improve the current understanding of gastrointestinal nematodes resistance at the genomic level and will be of aid in defining better breeding plans for lambs within extensive-sheep farming systems.

Phenotypic characterization of the parasite resistance in Morada Nova sheep

A.C.S. Chagas[1], L.G. Lopes[2], M.H. Silva[2], L.A. Giraldelo[2], J.H.B. Toscano[3], S.C.M. Niciura[1], C.H. Okino[1], M.V. Benavides[4] and S.N. Esteves[1]
[1]Embrapa, Southeast Livestock Unit (CPPSE), Rod. Washington Luiz km 234, São Carlos, SP, 13560970, Brazil, [2]UNICEP, R. Miguel Petroni, 5111, São Carlos, SP, 13563-470, Brazil, [3]UNESP, Via Prof. Paulo D. Castellane, Jaboticabal, SP, 14884-900, Brazil, [4]Embrapa, Southern Livestock Unit (CPPSUL), Rod. BR-153, Km 632,9, Bagé, RS, 96401-970, Brazil; carolina.chagas@embrapa.br

Gastrointestinal nematodes (GIN) are a major constraint in small ruminant production, in particular *Haemonchus contortus*, due to parasitic resistance spread worldwide. Morada Nova, a Brazilian hair-sheep breed adapted to tropical conditions, is considered more resistant to GINs. This study aimed to characterize a Morada Nova flock in relation to its resistance against *H. contortus* through phenotypic tools. A total of 151 lambs (divided into 2 groups according to the order of birth) were dewormed (monepantel) after weaning and 15 days later were submitted to the first parasitic challenge by oral artificial infection of 4,000 L_3 of *H. contortus*, isolate Embrapa2010 (day zero; D0). Faecal samples were collected individually for eggs per gram counts (EPG) on days 0, 21, 28, 35 and 42, when the body weight (BW) was also measured. Blood collections for packed cell volume (PCV) were performed each 14 days. After D42, the animals were dewormed again and submitted to the 2nd parasitic challenge, followed by the same collection scheme. The 151 animals were categorized as resistant/R (20%), intermediate/I (60%) and susceptible/S (20%) to *H. contortus* by means of the ranking obtained using the formula = (BW×0.4) – (EPG×0.4) + (PCV×0.2). There was statistical difference ($P<0.05$) for BW, and EPG between the challenge groups (G1 and G2). ANOVA ($P<0.05$) indicated for BW: R(a) = I(a) ≠ S(b) (G1) and I(a) ≠ R(b) ≠ S(c) (G2). For EPG: R(a) ≠ I(b) ≠ S(c) in both groups. The averages for PCV were R(a) = I(b) ≠ S(c). The strongest negative Pearson correlation was between LogEPG and PCV (r=-0.68) and strongest positive one between final weight and weaning weight (r=0.86). The ranking scheme allowed the differentiation between the animals categorized as R, I and S. High EPGs were related to lower PCVs and the period between birth and weaning had an impact on the final live weight.

Estimating genetic parameters for faecal egg count, dag score and weight in Scottish Blackface sheep

A.F.F. Pacheco[1], T.N. McNeilly[2], G. Banos[1,3] and J. Conington[1]
[1]Scotland's Rural College, SRUC, Roslin Institute Building, Easter Bush, Midlothian EH25 9RG, United Kingdom, [2]Moredun Research Institute, Pentlands Science Park, Bush Loan, Midlothian EH26 0PZ, United Kingdom, [3]The Roslin Institute, Roslin Institute Building, University of Edinburgh, Midlothian EH25 9RG, United Kingdom; antonio.pacheco@sruc.ac.uk

The goal of this study was to estimate the genetic parameters of common disease traits associated with natural nematode and coccidian infections in Scottish Blackface sheep. Data on faecal egg counts (FEC) on various species of parasites was collected together with a faecal soiling score (DAG) and live weights (LWT). FEC's were obtained for strongyles (FEC_S), nematodirus (FEC_N) and coccidia (FEC_C) using the McMaster technique. Data from 3,951 lambs sampled twice per year (July and August) over seven years (2011 to 2016) were analysed. FEC's and DAG were log-transformed prior to analysis. Year, sex, lab, birth-rearing rank, age, grazing locations and interactions were used as fixed effects. Average age at sampling was 95 and 129 days with mean live weights of 24.5 and 28.0 kg, respectively. Prevalence of eggs in faeces was lower at 2nd sampling. LWT heritability estimates were moderate at 0.33±0.04 and 0.37±0.05 for the two sampling occasions. FEC heritability estimates for the two sampling occasions were 0.14±0.03 and 0.16±0.04 for FEC_S, 0.17±0.03 and 0.15±0.03 for FEC_N, and 0.09±0.03 and 0.09±0.03 for FEC_C. DAG heritabilities were respectively 0.09±0.03, 0.14±0.03. Genetic correlations between sampling occasions were positive ($P<0.05$) for FEC_S and LWT. At first sampling, FEC_S was genetically (positively) correlated with FEC_N, FEC_C and LWT ($P<0.05$) and FEC_S was phenotypically (positively) correlated with FEC_N and FEC_C ($P<0.05$), but negatively correlated with LWT ($P<0.05$). All other genetic correlations for the 1st sampling were not significant. Correlations among traits at 2nd sampling were inconclusive: only the genetic positive correlation among FEC_N and LWT at 2nd time point was significantly different from zero ($P<0.05$). The significant positive genetic correlations between FEC_S and FEC_N and FEC_C at 1st sampling show that selection of sheep for resistance is feasible and will result in increased resistance against other parasites.

Development of a SNP parentage assignment panel in some North-Eastern Spanish meat sheep breeds

J.H. Calvo[1,2], M. Serrano[3], F. Tortereau[4], P. Sarto[2], M.A. Jiménez[3], J. Folch[2], J.L. Alabart[2], S. Fabre[4] and B. Lahoz[2]
[1]ARAID, Av. de Ranillas 1-D, 50018 Zaragoza, Spain, [2]CITA-IA2, Producción y Sanidad animal, Av. Montañana 930, 50059 Zaragoza, Spain, [3]INIA, Mejora Genética Animal, Ctra. La Coruña km 7.5, 28040 Madrid, Spain, [4]Université de Toulouse, INRA, ENVT, 4GenPhySE, Chemin de Borde Rouge, 31326 Castanet-Tolosan, France; jhcalvo@aragon.es

Accurate pedigree information is an essential tool in genetic breeding programs to ensure the highest rate of genetic gain and allows management of inbreeding. However, the proportion of known sires can be very low in Spanish meat sheep populations, particularly in breeds reared in high mountain areas as in the Pyrenees. Single nucleotide polymorphisms (SNPs) are now the DNA markers of choice for parentage assignment. The objective of this study was to develop a SNP assay to use in some North-Eastern Spanish meat sheep populations for accurate pedigree assignment. Nine sheep breeds were sampled: Rasa aragonesa (n=38), Navarra (n=39), Ansotana (n=41), Xisqueta (n=41), Churra Tensina (n=38), Maellana (39), Roya bilbilitana (n=24), Ojinegra (n=36) and Cartera (n=39). We used SNP genotypes from the Illumina OvineSNP50 BeadChip array. Firstly, we selected the 249 SNPs published from the French panel for parentage assignment due to the high values of the Minor Allele Frequency (MAF) reported in the South-West European breeds. In total, 159 SNPs in Hardy-Weinberg equilibrium, displaying a MAF>0.3 and a call rate>0.97 in all the nine populations, and not associated with Mendelian errors in verified family trios or duos were selected. The average MAF was 0.43, ranging from 0.41 (Churra Tensina) to 0.44 (Xisqueta and Navarra). The probability (PI) that two randomly selected individuals having identical genotypes within breed was very low: it reached its lowest and highest value in the Cartera (8.81×10^{-67}) and the Roya bilbilitana populations (2.29×10^{-64}), respectively. The exclusion probabilities of either one or the two randomly selected parent(s) were close to 1 in all populations. The parentage assignment procedure was tested using KASP technology for genotyping in a final panel of 192 SNPs that included 159 SNPs for parentage assignment and 33 functional SNPs (*PrnP*, *BMP15*, *MTNR1A*…).

Session 26 Poster 19

Genetic parameters of reproductive traits in a prolific w line of barbarine sheep

C. Ziadi[1] and S. Bedhiaf-Romdhani[2]

[1]Institut National Agronomique de Tunisie, Cité Mahrajène, 1082 Tunis, Tunisia, [2]Institut National de la Recherche Agronomique de Tunisie, Laboratoire des Productions Animales et Fourragères, Rue Hédi Karray, 2049 Ariana, Tunisia; ziadichiraz4@gmail.com

This study aims at estimating genetic parameters of ewe reproductive traits in a prolific «W» line of Barbarine sheep. This breed is well adapted to the hard local conditions in low input production systems. The «W» line was created by screening prolific ewes among conventional flocks at the experimental station of the Tunisian National Institute for Agronomic Research (INRAT) and was selected for prolificacy since 1979. A total of 2,726 reproductive records was collected between 1989 and 2016. Traits analysed were litter size at birth (LSB), litter size at weaning (LSW), litter weight at birth (LWB) and litter weight at weaning (LWW). Genetic parameters were estimated with bivariate linear animal model using Gibbs sampling methodology of Bayesian inference. Heritabilities estimates were 0.11, 0.05, 0.04, and 0.05 for LSB, LSW, LWB and LWW, respectively. Genetic correlations estimates between traits were positive and ranged from 0.70 between LSW and LWW to 0.92 between LSB and LSW. Despite the low heritability of reproductive traits, selection on litter size at birth may results in a genetic improvement of other traits since they are highly correlated.

Session 26 Poster 20

Immune cell profiles associated with resistance against *Haemonchus contortus* in Morada Nova sheep

C.H. Okino[1], J.H.B. Toscano[2], L.G. Lopes[3], M.H. Silva[3], L.A. Giraldelo[3], S.C.M. Niciura[1], M.V. Benavides[4], S.N. Esteves[1] and A.C.S. Chagas[1]

[1]Embrapa, Southeast Livestock Unit (CPPSE), Rod. Washington Luiz km 234, São Carlos, SP, 13560970, Brazil, [2]UNESP, Via Prof. Paulo D. Castellane, Jaboticabal, SP, 14884-900, Brazil, [3]UNICEP, R. Miguel Petroni, 5111, São Carlos, SP, 13563-470, Brazil, [4]Embrapa, Southern Livestock Unit (CPPSUL), Rod. BR-153, Km 632,9, Bagé, RS, 96401-97, Brazil; carolina.chagas@embrapa.br

Effective mucosal immune response is essential for the development of resistance to *Haemonchus contortus* in sheep. This study aimed to better elucidate the immune mechanisms involved in the resistance against *H. contortus* in the Brazilian Morada Nova breed. Two groups of 5 sheep, previously characterized as resistant or susceptible to infection by this parasite (extremes from 151 phenotyped animals), were challenged with 4,000 L_3 of *H. contortus* Embrapa2010 isolate, and euthanized at 7 days post-infection. Blood samples were collected for blood count, including differential analysis for granulocytes. Two fragments of abomasal pyloric gland and fundic regions were collected and subjected to histological procedures, and to immunoperoxidase using monoclonal antibodies against CD3, CD79a, MUM1 and lysozyme. The Mann Whitney test was performed to compare all evaluated parameters from resistant and susceptible groups ($P<0.05$). Pathological changes in the fundic abomasal region consisted mainly of moderate lymphoplasmacytic inflammatory infiltrate, including edema, presence of slight secretory and fibrinoid material, while in the pyloric region the main findings included slight mucosa irregularity and heterogeneous inflammatory infiltrate in the mucosa and lamina propria. No significant differences were observed for immunoperoxidase assays, mastocytes for both abomasal regions, and also for eosinophils in the pyloric region, nor for different subtypes of leucocytes in whole blood. Significantly higher levels were observed in the resistant group for eosinophils in the fundic region and for erythrocytes, haemoglobin and total leucocytes. Our partial results indicate an important role of eosinophils in the development of resistance against *H. contortus*, especially those located in the mucosa.

Session 26 Poster 21

SNP based genetic diversity and structure of sheep population in Tunisia

I. Baazaoui[1], E. Ciani[2], S. Mastrangelo[3], M. Ben Sassi[4] and S. Bedhiaf[5]
[1]Faculty of Scienes Bizerte Tunisia, Jarzouna Bizete, 7021, Tunisia, [2]Università degli Studi di Bari Aldo Moro, Università di Bari, Dipartimento di Bioscienze, Biotecnologie e Scienze Farmacologiche Bari, Via G. Amendola, 165/a, 70126 Bari, Italy, [3]University of Palermo, Dipartimento di Scienze Agrarie e Forestali, Viale delle Scienze, 90128 Palermo, Italy, [4]OEP, 30 Rue Alain Savary, Belvédère Tunisie, 1002, Tunisia, [5]INRA Tunisia, Animal and fodder production, Rue Hédi Karray1004, El Menzah, Tunis 1004, Tunisia; imen_baazaoui@hotmail.fr

The study of genetic diversity is essential to update the status of genetic variability of native sheep breeds through dynamic genetic and demographic events. This study present the first genetic investigation of Tunisian sheep population using genome wide SNP markers. To finely reconstruct genetic structure and relationships among Tunisian sheep, 59 samples belonging to 4 local breeds (Barbarine, Queue Fine de l'Ouest, Noire de Thibar and D'man), were genotyped using ovine 50K illumina bead chip. Within breed genetic analysis revealed that Tunisian breeds have a high level of diversity (proportion of polymorphic loci=0.98; He=0.37). The study of genetic relatedness was estimated using pairwise fixation index Fst, showed that the Barbarine and Queue fine de l'Ouest breeds are the most related groups (Fst=0.013) despite their phenotypic differences in tail fatness. The study of genetic structure of breeds using Bayesian clustering via Admixture and Multi-dimensional based on allele sharing distance (SAD) was consistent with Fst result. The findings confirms the genetic closeness and high level of admixture between Barbarine and Queue fine and a clear isolation of Noire de Thibar breed from remaining breeds since this breed was recently created via introgression of European gene flow. Based on above evidence, the closeness between Barbarine and Queue fine is due to extension of anarchic crossbreeding between two populations by small farmers. This long-term practice will present a high risk of genetic erosion of local breeds especially the unique fat tail Barbarine breed, which needs an urgent conservation management of local breeds.

Session 27 Theatre 1

Effects of phosphorus (P) and energy intakes on plasma markers of P deficiency in pregnant heifers

S.T. Anderson[1], M.A. Benvenutti[2], N. Steiger[1], K.L. Goodwin[2], L.J. Kidd[1], M.T. Fletcher[1] and R.M. Dixon[1]
[1]The University of Queensland, Brisbane, 4072 QLD, Australia, [2]Department of Agriculture and Fisheries, Gayndah, 4625 QLD, Australia; stephen.anderson@uq.edu.au

Phosphorus (P) deficiency is common in cattle grazing rangelands of northern Australia, and elsewhere. However late dry season pastures are also low in metabolizable energy. Such concurrent nutritional deficiencies are a major challenge to pregnant heifers, and severe loss of liveweight (LW) and body condition score (BCS) often occurs. This study examined the effects of diet P (adequate versus deficient) on performance of heifers during late pregnancy given sub-maintenance energy intakes. Here we report on plasma biomarkers of P deficiency. Droughtmaster heifers (n=42, initial LW 419±5 kg, BCS 3.9±0.1) housed in individual pens during the last 14-18 weeks of pregnancy were fed restricted amounts of wheat straw and molasses-urea. In a 2×3 factorial design heifers were fed diets of adequate/high (HP) or deficient/low (LP) phosphorus each with high (HE), medium (ME), or low (LE) metabolizable energy. Such E diets were designed to provide nil, moderate or substantial conceptus-free LW loss. Plasma inorganic P (PiP) and total calcium (Ca) were measured by biochemistry, carboxy-terminal telopeptides of type I collagen (CTX-1) and bone alkaline phosphatase (BALP) by immunoassays. During late pregnancy, there was a main effect (P<0.001) of diet P, but not diet E or diet P×E interaction, on PIP, Ca and CTX-1. Heifers fed LP diets had low PiP (0.9 mM), high Ca (2.4 mM), high plasma Ca/P ratio (>2.0) and increased CTX-1 (3.4 ng/ml). Heifers fed HP diets had mean PiP (2.1 mM), Ca (2.2 mM), Ca/P ratio (about 1.0), and CTX-1 (1.5 ng/ml). Plasma BALP concentrations were also increased (P<0.001) under low P diets (LP 37 versus HP 24 ng/ml), and increased (P<0.02) with E intake (LE 25, ME 31 and HE 35 ng/ml). These results support the hypothesis that plasma CTX-1 and total Ca concentrations are indicative of bone resorption responses due to P deficiency, and are not markedly influenced by metabolizable E intake. Plasma BALP as an indicator of defective bone mineralisation is increased by low P diets, but also increased by energy intake. Research supported by Meat and Livestock Australia and Qld DAF.

Impact of menthol supplementation on calcium absorption in ruminants

K.S. Schrapers[1,2], H.-S. Braun[1,2], J. Rosendahl[1,2], G. Sponder[2], K. Mahlkow-Nerge[3], A.K. Patra[2], J.R. Aschenbach[2] and F. Stumpff[2]
[1]PerformaNat GmbH, Hohentwielsteig 6, 14163 Berlin, Germany, [2]Institute of Veterinary Physiology, Freie Universität Berlin, Oertzenweg 19b, 14163 Berlin, Germany, [3]Agrarwirtschaft, FH Kiel University of Applied Sciences, Grüner Kamp 11, 24783 Osterrönfeld, Germany; braun@performanat.de

Subclinical hypocalcaemia is a major metabolic disorder in dairy cows. Although more than 50% of calcium (Ca^{2+}) absorption occurs from the rumen, the underlying molecular mechanisms have not yet been fully elucidated. Recent studies implicate non-selective cation channels of the transient receptor potential family (TRP), several of which are modulated by plant-derived compounds. Therefore, the involvement of TRP channels in ruminal Ca^{2+} absorption and the effect of the TRPV3 agonist menthol were investigated in four trials. Addition of 10 μM menthol increased absorptive Ca^{2+} fluxes across isolated ovine ruminal epithelia by 54% ($P<0.05$; n=4, n=9) in Ussing chamber studies. Following the detection of TRPV3 mRNA in the ruminal epithelium, TRPV3 was evaluated for its potential contribution to ruminal Ca^{2+} transport. HEK cells overexpressing the bovine TRPV3 expressed a conductance for Ca^{2+} and responded to the addition of 1 mM menthol with increased Ca^{2+} influx ($P=0.003$; n=10), confirming the ex vivo results. To investigate the effect of menthol on Ca^{2+} absorption *in vivo*, a feeding study was performed with 72 lactating cows in a replicated 2×2 cross-over design. Cows fed with a menthol-containing blend of essential oils (BEO; 1 g/d) for 20 d showed increased serum Ca^{2+} levels ($P<0.001$). In a second experiment *in vivo*, 24 growing sheep were fed either no BEO (control), low-dose (80 mg/d) BEO or high-dose (160 mg/d) BEO in a randomized block design. After 28 d, absorptive Ca^{2+} fluxes across isolated ruminal epithelia were measured in Ussing chambers. Sheep fed with low-dose BEO showed increased Ca^{2+} transport rates ($P=0.051$; n=8). Our studies provide evidence that menthol may improve ruminal Ca^{2+} absorption and suggest pathways that include the stimulation of TRPV3 channels. Further studies are in progress to evaluate the potential impact of menthol supplementation for hypocalcaemia prevention. Research was supported by ESF, BMWi, A. v. Humboldt Foundation, PerfomaNat, DFG and AfT.

Flavonoid quercetin as a potential regulator of ovarian functions *in vitro*

A. Kolesarova, K. Michalcova, S. Baldovska and M. Halenar
Slovak University of Agriculture in Nitra, Department of Animal Physiology, Tr. A. Hlinku 2, 949 76 Nitra, Slovak Republic; adriana.kolesarova@uniag.sk

The area of nutritional research has an increasing focus on the study of natural substances, which can affect the ovarian functions and reproduction of animals. Quercetin (Q) is a major polyphenol existing in many fruits and vegetables. It can be found in medicinal herbs, plant-derived foods and beverages, fruits as grapes, berries, apples, citrus, vegetables especially onions. Processes of secretory activity (steroid hormones progesterone, 17β-estradiol), proliferation (markers PCNA, cyclin B1) and apoptosis (markers caspase-3, p53) of porcine ovarian granulosa cells after Q treatment (≥98% purity, Sigma-Aldrich) at the doses 0.01, 0.1, 1, 10 and 100 μmol/l were studied. Steroid hormones were assessed by ELISA, markers of proliferation and apoptosis by immunocytochemistry. Progesterone secretion was significantly ($P\leq0.01$) stimulated at 10 μmol/l, but 17β-estradiol was not affected by the Q treatment. In addition, significant ($P\leq0.05$) stimulation in the number of cyclin B1-positive cells was induced by Q at the doses 0.1, 1, 10 and 100 μmol/l. On the other hand, it was not established in the case of PCNA. Presence of caspase-3 in porcine ovarian cells was not significantly ($P\geq0.05$) influenced by Q treatment. In contrast, Q at the dose 10 μmol/l significantly ($P\leq0.05$) inhibited number of porcine ovarian cells containing apoptotic marker p53. The results of our *in vitro* study indicate potential dose-dependent effect of Q on the steroidogenesis, proliferation and apoptosis of porcine ovarian granulosa cells through steroid hormone (progesterone) secretion and the crucial markers of proliferation (cyclin B1) and apoptosis (p53), which could be useful in regulation of female folliculogenesis. This work was supported by APVV-16-0170 and VEGA 1/0039/16.

Session 27 Theatre 4

Mushroom versus mycotoxins in food and feed: mushroom metabolites in control and detoxification

J. Loncar[1,2], A. Parroni[1], P. Gonthier[3], L. Giordano[3], M. Reverberi[1] and S. Zjalic[2]
[1]Sapienza University of Rome, Department of environmental biology, P.le Aldo Moro 5, 00195 Roma, Italy, [2]University of Zadar, Department of ecology, agronomy and aquaculture, trg kneza Višeslava 9, 23000 Zadar, Croatia, [3]University of Torino, Department of agricultural, forest and food sciences, Largo P. Braccini 2, 10095 Grugliasco, Italy; szjalic@unizd.hr

Mycotoxins represent a serious problem in food and feed safety. In animal production this toxins could cause important economic losses both during the breeding, due to impaired growth or major sensibility to the illnesses, and commercialisation of the products, due to their impaired value. In last five decades different strategies for control of mycotoxins in food and feed and decontamination of food and feed stuff have been proposed, but none of them has solved the problem. Most of the applied strategies, both in control and detoxification, are based on use of chemicals and thus could represent environmental hazard. The research on more environmental friendly tools in mycotoxin control is ongoing since decades. The mushroom products showed interesting features in mycotoxin control and detoxification of feed stuff from mycotoxins. Mushroom enzymes can degrade some mycotoxins (aflatoxins i.e.) while mushroom polysaccharides demonstrated the ability to control the production of different mycotoxins simultaneously. In this work an overview of the last results in possible application of mushroom enzymes and polysaccharides in mycotoxin control and detoxification of contaminated feed will be presented.

Session 27 Theatre 5

Influence of Ilex extracts and their fractions on kidney structure and performance

A. Zwyrzykowska-Wodzińska, R. Kupczyński, P. Kuropka, R. Nowaczyk, W. Teodorowicz and A. Szumny
Wrocław University of Environmental and Life Sciences, C. K. Norwida 25, 50-375 Wrocław, Poland; anna.zwyrzykowska@upwr.edu.pl

Infusion Yerba mate is prepared from leaves of *Ilex paraguariensis* which are widely consumed around the world. Traditionally it was used by the Indians of South America before the European colonization. In countries where it is produced (e.g. Argentina, Brazil) mate is not only an important branch of agriculture but has a remarkable status in economy. This species is exported worldwide, including Europe, the United States, and Japan, where it is marketed as a milled plant or extracts used in herbal formulations and functional food products. Among *Ilex* species in terms of phytochemical research *I. paraguariensis* has been the subject of most intensive investigations. Other species were not the subject of the detail study. The aim of this study was to assess the influence of *Ilex* extracts and their fractions on kidney structure. The research material was divided into 2 basic nutritional groups. Group 1 where ordinary fodder was used and Group 2 where animals were fed feed with added cholesterol. As factors modulating the effect of cholesterol on animals, Yerba Mate (YM), *Ilex meserveae* (IM), polyphenols (P), saponins (S) and terpenoids (T) were used. The material after autopsy was fixed in a solution of 4% buffered formalin and then dehydrated in alcohol and embedded in paraffin. Sections 5 µm thick were routinely stained with Hematoxylin and eosin and Alcian blue in own modification. Morphometric studies concerned the size of the glomerulus with the vascular loop and the thickness of the basement membrane of the blood vessels. The study was carried out using a Nikon Eclipse 80i light microscope. Morphometric studies were carried out using the Nis-Elements Ar software. The analysis revealed that there is an influence of the polyphenols and saponins present in YM and IM on the blood-urine barrier, leading to increased ureogenesis where cholesterol significantly reduces this effect.

Session 27 Theatre 6

Animal data: big, or just large?

T. Hamed, J. Schenkels, N. Laundry, B. Szkotnicki and C.F. Baes
University of Guelph, Animal Biosciences, 50 Stone Road East, N1E 2W1, Canada; cbaes@uoguelph.ca

Data in the field of animal science are being generated under a wide variety of conditions (laboratory, research facilities, on farm). Each year, thousands of research trials are carried out on farm animal genetics, metabolism, nutrition, physiology and behaviour. The capture, integration, and valorisation of the information generated represents both a great challenge but also a tremendous opportunity to allow the sector to progress more effectively. Datasets collected in the field of animal science are growing rapidly, and are expected to increase exponentially in size in the near future. Numerous animal phenotypes (immune responses, behavioural observations, nutritional information, bodily fluids such as milk, blood, rumen fluid, etc.) are being recorded. Various cost-effective information-sensing mobile devices are being implemented or will be implemented in the near future, including remote sensing, software logs, cameras, microphones, radio-frequency identification (RFID) readers and wireless sensor networks. Finally, genomic information (SNP-chip genotypes, next-generation sequence data, metagenomic and epigenetic information) are also being collected on a number of animals. The massive volume and the diversity in the format, scope, structure and pattern of this data are creating important challenges. Information is compiled and kept in idiosyncratic formats by different research groups. Finding and valorising existing information is complicated. This leads to overlooking potentially valuable information. Here we present the development of a standardized and curated database, which allows additional analyses, and enables us to address higher-level, more comprehensive theoretical and experimental questions regarding animal science. Our work is currently focused on dairy cattle, but will be expanded to include other livestock species in the near future.

Session 27 Theatre 7

Performance of dual-purpose types, an extensive broiler and a layer type fattened for 67 and 84 days

S. Mueller, R.E. Messikommer, M. Kreuzer and I.D.M. Gangnat
Institute of Agricultural Sciences/ETH Zurich, Universitaetstrasse 2, 8092 Zurich, Switzerland; sabine.mueller@usys.ethz.ch

Layer-type cockerels are culled after hatch. To avoid this practice, dual-purpose type systems could be established, with females used for egg and males for meat production, but a lower performance is to be expected. However, in organic agriculture aiming at a lower performance these types might be competitive. In this study, growth performance of dual-purpose types (Lohmann Dual, LD, and Novogen Dual, ND) was compared with an extensive broiler type (Hubbard S 757, HU) and a layer type (Lohmann Brown, LB), slaughtered either after 67 or 84 days of fattening. The 4×1,350 birds were kept in 20 m^2 compartments of 270 birds each. An organic broiler diet (12.8 MJ metabolizable energy, 230 g crude protein/kg) was fed. Body weight (BW) and feed intake (FI) were determined weekly. Average daily gain (ADG) and feed efficiency (FE, g feed/g gain) were calculated. Analysis of variance considered effects of type, age at slaughter and the interaction. The birds of HU, LD and ND reached a final BW of 1.7 kg after 67 d and 2.4 kg after 84 d of fattening. The LB were about 40% lighter. The ADG was similar for LD and HU for both fattening periods (25 g/d). In ND ADG was in the same range, but was 1.4 g lower for 67 d than for 84 d. The ADG of LB was with 16 g ca. 36% lower compared to the other types. The FI during 67 d was around 68 g/d for LD, ND and HU and increased to 80 g/d during 84 d. In LB, FI and its increase were lower with 56 and 60 g/d during 67 and 84 d, respectively. The FE was most unfavourable for LB (3.6) regardless of age at slaughter. Across 67 d, HU had the best FE (2.6). During 84 d, there was no longer a difference of HU to LD and ND (range: 2.7 to 3.0). The mortality was low (0.9%) and similar for all types and both fattening periods. In conclusion, both dual-purpose types performed at a same level as HU, whereas LB were inferior in all traits. For a comprehensive evaluation for dual-purpose systems, the layer side, slaughter performance and meat quality have also to be taken into account.

Session 27 Theatre 8

Investigation of early feed intake: does suckling rabbit have pellet preferences?

C. Paës[1], L. Fortun-Lamothe[1], T. Gidenne[1], K. Bebin[2], E. Grand[3], J. Duperray[4], C. Gohier[5], G. Rebours[6], P. Aymard[1] and S. Combes[1]

[1]Université de Toulouse, INRA, INPT, ENVT, GenPhySE, Chemin de Borde Rouge, 31326 Castanet Tolosan, France, [2]CCPA, Parc d'activités du Bois de Teillay, 35150 Janzé, France, [3]INZO, Rue de l'Eglise, 02400 Chierry, France, [4]EVIALIS, Talhouët, 56250 Saint-Nolff, France, [5]MiXscience, Rue Courtillons, 35170 Bruz, France, [6]TECHNA, Route de Saint-Étienne-de-Montluc, 44220 Couëron, France; charlotte.paes@inra.fr

Health management is the main challenge facing rabbit breeding. To handle this issue, our objective is to stimulate pups' solid feed intake to accelerate the maturation of digestive microbiota towards a stable state. In order to determine attractive pellet presentation for suckling rabbits, two preference tests were carried out, using a mother-litter separate feeding system. We provided pellets from 3 to 18 d-old in the nest box, and from 15 to 35 d-old in the cage. In the first trial, four pellet diameters were used: (A) 2.0 mm, (B) 3.0 mm, (C) 4.0 mm, (D) 6.0 mm. The pellets were tested in pairs against each other (6 groups of 10 litters). In the second trial, we tested in pairs pellets of same diameter (either 2.5 mm or 4.0 mm) manufactured with different channel lengths (10/12/14 mm for 2.5 mm diameter pellets, 18/20/24 mm for 4 mm pellets – in total 6 groups of 10 litters) to ensure different pellet quality. Feed consumption in the nest was measured daily, and every 4 days after nest removal at 21 d-old. Milk intake was measured twice a week. Birth-weaning mortality, young rabbit growth and milk intake did not differ between treatments (NS). Solid feed intake started from 8 days of age (45 out of 120 litters with a significant consumption). Total feed ingestion in the nest was 2.1±0.6 g DM/rabbit (Trial 1) and 1.5±0.1 g DM/rabbit (Trial 2). In the cage feeders, pellet A showed higher relative consumptions (RC) (61, 67, and 86% when compared with B, C and D), and clear preferences for small pellets over D were found within groups ($P<0.001$). In Trial 2, from 7 to 35 d-old, a lower RC (-24%, $P<0.05$) was found for 4.0 mm diameter pellets processed with the longest die channel (24 mm). Our results indicate that kits solid feed intake can start at 8 days of age and may be modulated according to suitable pellet presentation (e.g. small diameter…).

Session 27 Theatre 9

Carcass characteristics and meat analysis of rabbits feed fungal treated corn stalks

A.A. Abedo, A.A.A. Morad, R.I. El-Kady and A.A. El-Shahat

National Research Centre, Animal Production Department, 33 El Bohouth St., Dokki, Giza, 12622, Egypt; abedoaa@yahoo.com

This work aimed to investigate the effect of feeding fungal treated corn stalks with *Trichoderma ressei* on carcass characteristics and meat analysis of rabbits. Forty-two weaned New Zealand white rabbits, six weeks of age, were equally divided into 7 experimental groups (Each group divided into three replicates of two rabbits in each). The first group was fed on the control diet, the other six groups were fed diets containing corn stalks which replaced clover hay at 33, 66 and 100% biologically treated with *T. ressei* or treated with media only (without *T. ressei*), the experimental lasted for 13 weeks. At the end of the experimental period three rabbits from each group were slaughtered to evaluate carcass characteristics and meat composition. The results showed that values of EBW, CW1, CW2, CW3, DP1, DP2, DP3 and carcass cuts insignificantly differ between treatments and also among different levels of corn stalks. There were no significant differences were recorded ($P>0.05$) either between treated or among levels in respect of chemical analysis of meat, except that of ash%. Statistical analysis between treatments and different level revealed that DM, CP and EE contents of meat were significantly higher with (BTCS) than that of (UTCS). Data of interactions between treatments and levels showed that the DM content was not significantly differed ($P>0.05$) between levels. The level of 66% (BTCS) recorded the highest ($P<0.05$) value of CP content. While level 66% (UTCS) recorded the highest ($P<0.05$) value of EE content, compared to other levels. Statistical analysis revealed that incorporation of biologically treated corn stalks in growing rabbit diets insignificantly increased protein content slightly by 1.63% and decreased fat content by 3.99%, compared with the untreated corn stalks. The differences in ash contents were significantly higher with (BTCS) than that of (UTCS).

Quality of organic eggs as influenced by herbal mixture supplementation

E. Sosnówka-Czajka[1], E. Herbut[1], I. Skomorucha[1] and M. Puchała[2]
[1]National Research Institute of Animal Production, Department of Poultry Breeding, 1, Krakowska Street, 32-083 Balice near Krakow, Poland, [2]National Research Institute of Animal Production, Department of Seep Breeding, 1, Krakowska Street, 32-083 Balice near Krakow, Poland; ewa.sosnowka@izoo.krakow.pl

The aim of the study was to determine the effect of herbal mixture supplemented to the diet on quality of organic eggs from a native breed of hens. The experiment used Rhode Island Red pullets, which were maintained according to organic farming recommendations. At 20 wk of rearing, birds were assigned to two groups. The diet of hens from group II was supplemented with a mixture of three herbs: *Thymus vulgaris* L., *Nigella sativa* L., *Trigonella foenum-graecum* L. At 35 wk of age, 20 eggs were collected from each group to determine egg shell colour, egg weight, albumen height, Haugh units, blood spots, meat spots, yolk colour, yolk weight, and egg shell quality (thickness, weight and density) using EQM equipment (Egg Quality Measurements). The egg shell strength was measured with an Egg Crusher device and the egg shape index was also calculated. The egg yolks were analysed for the fatty acid profile by gas chromatography and for the level of vitamins A and E by liquid chromatography. The results were statistically analysed using one-way analysis of variance, and significant differences were estimated with Duncan's test. The egg quality parameters measured by EQM equipment, egg shell strength, egg shape index and the level of vitamins in egg yolks were at a similar level in both groups. The yolks of eggs collected from group II were lower in linoleic (C18:2) and linolenic acids (C18:3) compared to the control group ($P \leq 0.05$). The same experimental group was also characterized by lower levels of arachidonic (C20:4) acid ($P \leq 0.01$), DHA ($P \leq 0.01$), and PUFA ($P \leq 0.05$), including PUFA-6 ($P \leq 0.05$) and -3 ($P \leq 0.01$), as well as a higher ratio of PUFA6/3 ($P \leq 0.01$) compared to group I. The yolks of eggs from chickens fed the herb supplemented diet had a 2.31% higher MUFA content compared to the egg yolks from the control group. In summary, the herb mixture supplement had an adverse effect on the fatty acid profile of egg yolks.

Effects of *Artemisia herba alba* and olive leaf powder on broiler performance

N. Moula[1], A. Ait Kaki[2], M. Tandiang Diaw[3], P. Leroy[1] and J. Detilleux[1]
[1]Faculty of Veterinary Medicine, University of Liège, Belgium, Département des Productions animales, Quartier Vallée 2, Avenue de Cureghem 6, 4000 Liège, Belgium, [2]Faculty of Science, Mhamed Bougara University of Boumerdes, Boumerdes, 35000, Algeria, [3]University of Thiès, Senegal, Department of Animal Production, Thiès, Senegal, 1000, Senegal; nassim.moula@uliege.be

Like the Mediterranean countries, Algeria has a multitude of plants that can be used in poultry farming. The objective of this study was to evaluate the effects of adding *Artemisia herba alba* and olive leaf (*Olea europaea*) in chickens diets, on their growth performances, blood biochemical parameters and carcass yield. The study was conducted from April to May 2017 in the area of Chemini (Algeria). In a completely randomized design, 60 one-day-old male chicks (Ross-308) were divided into 3 groups, each group containing 2 repetitions of 10 chickens. We fed chickens of the first group a standard commercial diet, the second group the same standard diet with 2% replaced by powder of *A. herba alba* and the third group the same standard diet with 2% replaced by powder of *O. europaea*. Animals were housed inside and fed *ad libitum*. Blood samples were taken from the wing vein on 5 chickens from each group to obtain their glucose, triglycerides, urea, total proteins and cholesterol contents. We removed feed approximately 12 h before slaughter at 42 days old. We recorded daily body weights and feed intake, carcass yield of each chicken. The same design was repeated twice. We used standard analyses of variance to test whether variations between diets were significantly different from null (P-value set at 5%). Results showed that inclusion of *O. europaea* significantly increase mean body weight at 42 days old (2,117.42±26.38 g, 2,230.10±26.38 g and 2,336.66±27.88 g in groups 1, 2 and 3) and decrease glycemia (2.24±0.06, 2.05±0.06 and 1.90±0.06 g/l in groups 1, 2 and 3) and cholesterolemia (1.13±0.05, 1.03±0.05 and 0.95±0.05 g/l in groups 1, 2 and 3). No significant difference were found in the feed conversion ratio (1.79, 1.87 and 1.81 in groups 1, 2 and 3), carcass yield that varied from 67.34 to 68.74%, triglyceride (0.68 to 0.73 g/l), urea (0.03 to 0.04 g/l), and total proteins (26.02 to 27.11 g/l) blood concentrations.

Session 27 Poster 12

Effect of cowpeas and probiotics on broiler chicks' performance and gut microflora populations

G. Ciurescu, I. Sorescu, M. Dumitru, A. Vasilachi and M. Habeanu
National Research & Development Institute for Biology and Animal Nutrition (IBNA), Laboratory of Animal Nutrition & Biotechnology, Calea Bucuresti 1, 077015 Balotesti, Ilfov, Romania; ciurescu@ibna.ro

Price and availability of soybean meal (SBM) on global markets is constantly changing, thereby stimulating interest in maximizing the use of locally produced protein sources is a must. This study aimed to investigate, for a period of 42 d, the effect of different levels (0, 10 and 20%) of raw cowpea seeds (CWP; *Vigna unguiculata* [L] Walp., cv. Ofelia), as replacement of SBM, with and without probiotic addition on growth performance, digestive organ sizes and caecal pH. The impact of the treatments on the intestinal caecum microflora population at d 26 was also evaluated. A total of 720, unsexed 1-d-old Cobb 500 broilers were divided into 6 groups with 4 replicate pens (30 birds/replicate pen). Statistical analysis was carried out using the GLM procedure of SPSS. Data were analysed as a 3×2 factorial arrangement with 3 levels of CWP with and without probiotic (3×10^8 cfu probiotic/kg of diet). The probiotic strain used was the *Lactobacillus plantarum* ATCC 8014 from the IBNA (Balotesti, Romania) bacteria collection. Urease activity (pH change) in CWP was not detected. The results showed that CWP (cv. Ofelia) at 10 or 20% without probiotic in an optimized diet on digestible amino acid contents maintained BW gain (2,764.5-2,805.2 g vs 2,787.7 g in control group; P>0.05). The digestive organ sizes (i.e. gizzard, heart, liver, pancreas, spleen, small intestine, caecum and the small intestine length) and pH of the caecum content were not affected (P>0.05) by treatments. Probiotic addition was beneficially in modulating gut microflora composition. In particular, the caecal *Lactobacillus* ($\log_{10}$ cfu/g of wet digesta) and *Enterococcus* increased (P=0.03; P=0.71, respectively), whereas the coliforms count decreased (P=0.69) compared with those without probiotic. Overall FCR values were similar and tended to increase BW gain (P=0.056). It is concluded that CWP and probiotic addition had a beneficial effect on broiler growth responses and caecal microflora populations.

Session 27 Poster 13

Growth rate and feed conversion ratio in two commercially available slower growing broiler hybrids

A. Wallenbeck, J. Yngveson, S. Gunnarsson, A. Karlsson and K. Arvidsson Segerkvist
Swedish University of Agricultural Sciences, Department of Animal Environment and Health, Box 234, 532 23 Skara, Sweden; anna.wallenbeck@slu.se

Organic broiler production is characterised by a long rearing period (>10 weeks) and diets based on locally available feedstuff. Until recently, conventional fast growing broilers hybrids, unsuitable for long rearing periods, have exclusively been used in organic broiler production in Sweden. At present, two slower growing hybrids are available at the market for Swedish producers. Anecdote reports from producers indicate large variations between batches for these birds. This pilot study compared growth rate (GR) and feed conversion ratio (FCR) between the two slower growing hybrids available on the market for Swedish producers; Rowan Ranger (RR) and Hubbard CYJA57 (H). In total 50 birds fed a commercial standard organic diet (O, 12.0 MJ ME and 196 g protein/kg feed) during a 68 day rearing period were included in the study. The birds were divided into 5 RR groups and 5 H groups with 5 birds per group. Each group was housed in a littered pen (1.0 m × 1.5 m) with bell drinker, hanging tube feeder and perches at two levels. Feed provision and feed residues were recorded and birds were weighed individually once per week. The proportion of female chicks in the groups were on average 56 and 50% in the RR and the H groups, respectively. Preliminary it was found a tendency that RR birds grew faster (49.7±3.93 g/day, mean ± SD) than H birds (45.0±3.03 g/day), P=0.070, but there was no difference in growth variation (SD within group) between hybrids. The current definition by the Swedish organic organization, KRAV, of a slow growing hybrid is a GR below 45 g/day. The RR birds tended to have a more effective FCR (2.59±0.097 kg feed per kg growth) than H birds (2.70±0.079), P=0.098. The results indicate differences in performance between RR and H of importance for commercial organic broiler production and further investigations are needed.

Session 27 Poster 14

Growth performance and carcass characteristics of broiler chicks fed graded levels of carob pod

A.Y. Abdullah and K.Z. Mahmoud
Jordan University of Science and Technology, Animal Production, P.O. Box 3030 Irbid 22110, Jordan; abdullah@just.edu.jo

This study was conducted to evaluate the nutritional value of carob *Ceratonia siliqua* pods (CP) as a corn substitute, and to find the optimal inclusion rates on growth performance, carcass and meat properties of broiler chicks. A total of 600 day-old mixed sex Hubbard chicks were used and randomly distributed into 6 dietary treatments: 0% CP without (T1) or with enzyme (T2), 5% CP without (T3) or with (T4) enzyme and 10% CP without (T5) or with (T6) enzyme. Each dietary treatment was divided into 5 replicates with 20 chicks each. All diets were isonitrogenous and isocaloric. Body weight and growth performance parameters were recorded weekly from 0 to 42 days of age for each pen to determine body weight gain (BWG) and feed conversion ratio (FCR). At the end of the experiment, all broilers were slaughtered to carry out carcass characteristics and meat quality tests. Data were analysed by analysis of variance using SAS general linear models. No significant effects of corn substitution of CP or enzyme supplementation were observed on average feed intake and FCR during all weeks and the overall rearing period (FCR; 1.62, 1.57, 1.66, 1.64, 1.74 and 1.70 for T1 to T6, respectively). Body weight and BWG for week 1, 4 and the entire rearing period decreased (BWG; 2,006, 1,898, 1,800, 1,850, 1,610 and 1,720 g for T1 to T6, respectively) as the CP inclusion level increased ($P<0.001$). No effects of enzyme supplementation were observed. No significant effects of CP inclusion or enzyme supplementation on carcasses cuts, dressing percentages, intestine length and ether extract % were detected. Carcass weight (1,418, 1,432, 1,273, 1,207, 1,100 and 1,270 g for T1 to T6, respectively) and fat pad % (2.1, 1.9, 1.5, 2.0, 1.5 and 1.3 for T1 to T6, respectively) were significantly affected ($P<0.001$) by the substitution levels being higher for T1 and T2 and lower for T5 and T6. There were significant differences ($P<0.0001$) between treatments in meat quality parameters. In general, pH, cooking loss, and lightness were higher while tenderness was lower in T6 compared with T1 and T2. In conclusion, results of the present study show no beneficial effects of carob pod inclusion on growth performance and meat properties of broiler chicks.

Session 27 Poster 15

Effect of varying levels of protein concentrations on the production of growing ostriches

T.S. Brand[1], S.F. Viviers[2], E. Swart[1] and L.C. Hoffman[2]
[1]Animal Sciences, Agriculture: Western Cape, Private Bag x1, 7607 Elsenburg, South Africa, [2]Stellenbosch University, Animal Science, Merriman Road, 7600 Stellenbosch, South Africa; tersb@elsenburg.com

The aim of this study was to investigate the effects of five different levels of protein with corresponding amino acid profiles in the diets of slaughter ostriches (*Struthio camelus* var. *domesticus)* on the feed intake, average daily gain (ADG) and feed conversion ratio (FCR). The trial, with five treatments and three replications per treatment, was structured into 15 camps containing 20 young chicks each. The middle of the five diets was formulated according to specifications predicted by an ostrich nutrition model, and served the purpose of the control for this study. The inclusion levels of diets one through five were thus 11.8, 12.2, 12.6, 13.0 and 13.4% crude protein, with corresponding amino acid profiles, respectively. Weekly feed intake was measured, as well as weekly weighing of the birds was done until they were slaughtered at 13.5 months of age. Results for feed intake had differences ($P=0.002$) between the treatments and a linear function was applied to best fit the data. An incremental increase in diet would result in a subsequent increase in intake by 63 g/bird/day. Diets 3, 4 and 5 had higher intakes than diets 1 and 2. Diet 5 had the highest feed intake of 1,684 g/bird/day, whilst Diet 1 the lowest of 1,432 g/bird/day. Differences ($P=0.0001$) were found between treatments for the ADG, with the trend across the five treatments best fitted by a quadratic function. Diets 3, 4 and 5 did not differ between themselves, with diet 4 (347 g/bird/day) marginally higher than the other two. Diets 1 and 2 had significantly lower ADG's, 259 g/bird/day and 299 g/bird/day respectively, than the other three diets. Therefore, conclusively it can be stated that diets 3, 4 and 5 consistently performed above diets 1 and 2. Increased protein concentrations in the diet above the norms predicted by the ostrich nutrition model did not result in significant increases in performance levels of slaughter ostriches.

Dehulled white lupine seeds as a crude protein source for lactating rabbit does and their litters

L. Uhlířová and Z. Volek
Institute of Animal Science, Department of Nutritional Physiology and Animal Product Quality, Přátelství 815, 10400 Prague, Czech Republic; uhlirova.linda@vuzv.cz

The aim of the present study was to evaluate dehulled white lupine seeds (DWLS, cv. Zulika) as a crude protein source for rabbit diets in comparison with a traditionally used protein sources. Two lactation diets and two weaning diets were formulated, diets were isonitrogenous and isoenergetic within each category. The control lactation diet (L-SBM) contained soybean meal (130 g/kg as-fed basis) and sunflower meal (50 g/kg as-fed basis) as the main crude protein sources, whereas the experimental lactation diet (L-DWLS) was based on DWLS (180 g/kg as-fed basis). Similarly, the control weaning diet (W-SBM) contained soybean meal (70 g/kg as-fed basis) as the main crude protein source, whereas the experimental weaning diet (W-DWLS) was based on DWLS (70 g/kg as-fed basis). All diets were without added fat. A total of 24 Hyplus rabbit does (12 does per treatment; after the 3rd parturition) were fed one of the two lactation diets during the entire lactation period (32 days). Does were housed in modified cages allowing controlled nursing and separate access of does and their litters to feed. The litters were standardized to 8 kits immediately after the birth and offered the weaning diets with the same crude protein source as in lactation diet of their mothers from the d 17 of lactation. Five does per treatment were used to determine the milk composition. Does live weight, feed intake and milk production was not affected by dietary treatment. The milk of does fed the DWLS diet contained less caprylic (P=0.044), capric (P=0.009) and linoleic (P=0.035) acids and more oleic (P<0.001), α-linolenic (P<0.001) and eicosapentaenic (P<0.001) acids resulting in a higher PUFA n-3/arachidonic acid ratio (19.08 vs 11.13 for L-DWLS and L-SBM, respectively; P<0.001). A lower daily solid feed intake (by 3.4 g; P=0.086) leading to higher milk intake/solid feed intake ratio (1.97 vs 1.53 for W-DWLS and W-SBM, respectively; P=0.024) between d 17 and 32 of lactation was observed in litters fed the W-DWLS diet. It can be concluded that dehulled white lupine seeds (cv. Zulika) can replace traditionally used crude protein sources in rabbit diets. Moreover, favourable effect of DWLS on rabbit milk fatty acid composition was detected. The study was funded by project NAAR QJ 1510136.

Effect of feeding fresh forages as basis for TMR on dairy cows' rumen characteristics and nutrients

D. Enriquez-Hidalgo, K. Barrera, S. Peede, D.L. Teixeira and E. Vargas-Bello-Pérez
Departamento de Ciencias Animales, Pontificia Universidad Catolica de Chile, Santiago, Chile; daniel.enriquez@uc.cl

Fresh forages alter rumen characteristics, but also can improve the system sustainability. Unlike alfalfa, berseem clover (BC) does not cause bloat and grows well during the winter especially when it is sown with grasses. The study aimed to evaluate the effect of using fresh alfalfa (ALF), fresh oat/BC (MIX) and alfalfa hay/maize silage (CON) as forage basis for TMR during the winter on dairy cows' rumen characteristics and nutrients degradability. Three non-lactating rumen-cannulated cows were allocated to each TMR for a 14-d period in a 3×3 Latin Square design. Cows received 13 kg DM/d of the TMR (55:45 forage-to-concentrate). In each period: forage samples (3/wk) were analysed for dry matter (DM), crude protein (CP) and neutral detergent fibre (NDF) contents; forage samples were rumen incubated (d9) for 0, 3, 6, 9, 12, 24, 48 and 72 h and nutrients degradability was determined, and rumen fluid was collected (d12 and d14) and analysed for pH, VFA and N-NH_3 contents. Forages differed in DM (CON 55.0, ALF 19.3, MIX 21.0; ±0.49%; P<0.001), CP (CON 10.2, ALF 23.8, MIX 14.9; ±0.32%; P<0.001) and NDF (CON 45.8, ALF 34.8, MIX 39.2; ±0.46%; P<0.001) contents. Ruminal fluid had similar pH (6.41±0.176), total VFA (126±12.1 mmol/l) and butyric acid (10.2±0.58%). The fresh forage diets increased the contents of ruminal N-NH_3 (CON 6.3, ALF 14.5, MIX 7.5; ±0.88 mmol/l), propionic (CON 14.2, ALF 15.2, MIX 15.5; ±0.55%) and valeric (CON 1.2, ALF 1.5, MIX 1.5; ±0.09%) acids (P<0.01), but decreased the acetic acid (CON 71.3, ALF 68.7, MIX 68.5; ±0.72%; P<0.01). Forages had similar rapidly degradable fractions (a) of MS (41.8±0.47%) and FDN (8.5±1.29%) and the slowly degradable fractions (b) (MS: 47.9±2.36%; FDN: 74.2±16.8%), but not for CP (a: CON 53.4, ALF 43.8, MIX 37.0; ±0.85%; b: CON 37.5, ALF 50.4, MIX ±57.2; 1.2%; P<0.001). Forages had different (P<0.05) degradation rates (c) of MS (CON 0.033, ALF 0.116, MIX 0.08; ±0.017%) and CP (CON 0.045, ALF 0.144, MIX 0.110; ±0.016%), but not for NDF (0.064±0.017%). Overall, the use of fresh forages changed rumen characteristics probably due to their degradation characteristics where different. The study was supported by FONDECYT 11160697.

Session 27 Poster 18

Effect of dietary microalgae on rumen fermentation and methanogenesis in an *in vitro* simulation

C.A. Moran[1], G. Jurgens[2], J.D. Keegan[1] and J. Apajalahti[2]
[1]Alltech SARL, Rue Charles Amand, 14500 Vire, France, [2]Alimetrics Ltd., Koskelontie 19 B, Espoo, 02920, Finland; cmoran@alltech.com

Dietary supplementation of cows with microalgae has previously been associated with a decrease in the number of methanogens present in the rumen. The objective of this study was to determine the effect of dietary supplementation with *Aurantiochytrium limacinum* (AURA; CCAP 4087/2) on methanogenesis using an *in vitro* model. Five inclusion levels of AURA (0, 0.6, 1.2, 2.4 or 5.0%) were added to fermentation vessels containing 1 g (DM) of a 50:50 grass silage: commercial feed diet. Methanogen inhibitors, 2-bromoethanesulphonate (BES) and monensin (MON) were included as positive controls and each treatment was replicated 5 times. The vessels were flushed with CO_2 after which 36 ml of anaerobic buffer solution (+38 °C) was introduced under oxygen free CO_2 flow. The simulation was initiated following the addition of 4 ml of freshly strained rumen fluid from a rumen fistulated cow and continued for 14 h at 38 °C. Total gas production was analysed at 3, 6, 9, 12 and 14 h, volatile fatty acids (VFA) and lactic acid concentrations at 9 and 14 hours, and methane concentration at 14 h. Numbers of total eubacteria, methanogens, lactate producers lactate utilizers, rumen protozoa, and pH of the medium were determined after 14 h. Methane and VFA were analysed using gas chromatography. Bacterial numbers were determined using 16S rRNA gene-targeted quantitative RT-PCR. Data were analysed using a Dunnett's post hoc test. AURA had no effect on total gas production; however, the positive controls BES and MON caused significant decreases in gas production (P<0.01) indicating that the *in vitro* model functioned. Methane production was not affected by supplementation with 0.6, 1.2 and 2.4% AURA, while for the 5% group, production increased slightly. The concentrations of total VFA and acetic, propionic, butyric and lactic acids were not affected by supplementation with algae. AURA had no effect on the pH of the medium, the total numbers of microbes, methanogens or numbers of individual microbes that were tested. Overall, using this short term batch culture *in vitro* rumen model, AURA had no effect on rumen fermentation or the numbers of rumen microbes.

Session 27 Poster 19

Replacing corn by crude glycerol in diets of grazing: fibre utilization and ruminal parameters

M.A. Bruni[1], M. Carriquiry[1], J. Galindo[2] and P. Chilibroste[1]
[1]Faculty of Agronomy. UdelaR, EEMAC, Route 3. km 363, 60000, Uruguay, [2]Institute of Animal Science, San Jose de las Lajas, Mayabeque, 32700, Cuba; mbruni@fagro.edu.uy

Eighteen Holstein cows (6 ruminally cannulated) were used in the first 60 days of lactation to evaluate the effect of the replacement of corn grain(CG) by crude glycerol (CGly) on fibre utilization and ruminal parameters (RP). Cows were blocked [by parity (3.17±1.42), body condition score (3.0±0.3), expected calving date], and assigned randomly to two treatments (T). T1= partially mixed diet (PMR) containing sorghum silage (SS) plus concentrate (CG, wheat grain, soybean expeller, minerals, and vitamins) and T2= PMR containing SS plus concentrate where CGly -76.5% glycerol, ALUR Uruguay- replacing the CG. Both groups received PMR in individual feeders and grazed during the afternoon in a temperate pasture with herbage allowance-above the ground level- 60 kg DM/cow/day. Chemical composition of PMR (OM, CP, NDF, and EE in g/kg DM) respectively was for T1 [934, 117, 317, and 5] and [927, 106, 297 and 46] for T2. *In- situ* fibre degradation and RP were measured in two periods. In two consecutive days of each period, ruminal fluid samples were collected for pH and VFA. *In- situ* data was fitted to nonlinear model $D(t)=b(1-e^{-c(t-l)})$, where D= disappearance of FDN, (b) degradable fraction, (c) degradation rate and (l) Lag time. Ruminal parameters were analysed with a mixed model with repeated measures. The statistical analyses were performed with SAS, data is expressed as lsmeans (s.e) and was considered to differ when P<0.05. The parameters of In situ ruminal NDF degradation of pasture 'b' 'c' 'l' were not affected by T, period or interactions and averaged 752 (2.6) g/kg DM, 0.03 (0.003) h^{-1} and 2.88(0.5) h. Ruminal pH was not affected by T, period or interactions and averaged 6.09 (0.18), but as expected it was affected for h. Total rumen volatile fatty acid concentrations (mmol/l) did not differ between treatments 99.55 (20), but proportions of rumen propionate was greater for cows fed glycerol (24.0 vs 28.5 of propionate, T1 vs T2). The data indicates that glycerol is a suitable replacement for corn grain in dairy cows diet at the beginning of lactation.

Magnesium absorption as influenced by rumen passage kinetics in lactating dairy cows

J.-L. Oberson[1,2], S. Probst[1] and P. Schlegel[2]
[1]Bern University of Applied Sciences, School of Agricultural, Forest and Food sciences, Länggasse 85, 3052 Zollikofen, Switzerland, [2]Agroscope, Tioleyre 4, 1725 Posieux, Switzerland; patrick.schlegel@agroscope.admin.ch

The K dependent Mg absorption through the rumen wall may be influenced by other dietary properties, such as forage type or forage to concentrate ratio which are likely associated to rumen passage kinetics. The study aimed to assess the effects of rumen passage kinetics on apparent Mg absorption and retention in lactating dairy cows. The passage kinetics were modified by feeding early or late harvested grass silages containing 341 and 572 g NDF/kg DM, respectively. Six lactating dairy cows, including 4 fitted with ruminal cannulas were randomly assigned to a 3×3 cross over design. The experimental diets consisted of early harvested low NDF (neutral detergent fibre, LOW) and late harvested high NDF (HIGH) grass silage and of concentrate (20% of DM intake). The diets containing late harvested silage were formulated to be either balanced in digestible protein with diet LOW (High+CP) or not. All diets were formulated to contain iso-Ca, -P, -Mg, -K and -Na. Passage kinetics of solid and liquid phase of rumen digesta were evaluated using marker disappearance profile of respectively, ytterbium labelled fibre and Co-EDTA. Cows fed LOW had compared to High+CP and HIGH, an up to 40% lower solid and 26% lower liquid phase volume and a higher liquid passage rate of rumen digesta. Rumen soluble Mg concentration doubled when cows were fed LOW. Faecal Mg excretion was up to 14% higher in cows fed LOW and Mg absorbability was 12% compared to up to 19% in other diets. Urinary Mg excretion in cows fed LOW was half of the ones in the other treatments, but Mg retention was not affected. Protein excess neither affected rumen passage kinetics nor Mg absorption and retention. Mg absorption correlated with rumen liquid volume which correlated with daily NDF intake. Consequently, daily Mg absorption decreased with decreasing NDF intake. To conclude, in addition to the known antagonistic effect of dietary K, present data indicate that Mg absorption was dependent from NDF intake which modified rumen liquid volume, but was independent of dietary protein excess likely associated to low NDF herbages.

Effects of dietary sainfoin on feeding, rumination, and faecal particle composition in dairy cows

A. Kapp[1], M. Kreuzer[2], G. Kaptijn[1] and F. Leiber[1]
[1]Research Institute of Organic Agriculture (FiBL), Ackerstrasse, 5070 Frick, Switzerland, [2]ETH Zurich, Universitätstrasse 2, 8092 Zurich, Switzerland; florian.leiber@fibl.org

An experiment was conducted to test whether dietary sainfoin (*Onobrychis viciifolia*), a plant containing condensed tannins, affects feeding and rumination behaviour of dairy cows fed different roughage-only diets. Out of a herd of sixty lactating Swiss Fleckvieh cows, twenty-nine were chosen for a 6-week experiment. Cows were fed pasture, fresh cut grass and hay. During week 1 (baseline) and weeks 2-4, they had access to pasture for 4 hours during day, and received approximately 6 kg DM/day fresh grass and 4 kg DM/day hay in barn. In week 5 and 6, daily pasture allowance was 2 h; hay offer was doubled. In week 2-6, each cow individually received respectively 1 kg of pelleted feed in the morning and in the evening. Ten cows received ryegrass pellets in week 2, 3 and 5, and sainfoin pellets in week 4 and 6 (sainfoin short term; SST). Nine cows were fed sainfoin (Sainfoin long term; SLT) and another ten ryegrass pellets (Control; C) during the whole term. Rumination and feeding behaviour was measured with RumiWatch® sensor halters during week 1, 4, and 6. Individual faeces samples (4 per cow/week in week 1, 4 and 6) were analysed for particle fractions (wet sieving at 4.0, 2.0, 1.0, and 0.3 mm). Data was evaluated with a general linear model using group and week as fixed factors; baseline data served as covariate. Compared to week 4, all cows showed an increased duration of feeding (15.5%) and a decreased ruminating activity (14.2%) in week 6 ($P<0.001$). Feeding sainfoin pellets (SLT and SST) led to decreased feeding time (2.0-4.6%) and increased rumination time (4.4-6.5%) compared to C ($P<0.05$) in both weeks. Total particle proportion in faecal DM was lower in SST and SLT by 11.7% compared to C ($P<0.05$). Also the proportion of the 1.0-0.3 mm fraction was lower in faeces of SST and SLC ($P<0.05$). Milk yields, fat and protein contents were not affected. Results indicate effects of sainfoin supplements on eating behaviour and digestion even when roughage composition varies. This approach could be further used to develop targeted feeding aimed at improving digestion and feed efficiency in low-input dairy systems.

Reconstituted alfalfa hay in starter feed improves skeletal growth of dairy calves during preweaning

S. Kargar[1], M. Kanani[1], M.G. Ciliberti[2], A. Sevi[2], M. Albenzio[2], R. Marino[2], A. Santillo[2] and M. Caroprese[2]
[1]Shiraz University, Department of Animal Sciences, College of Agriculture, Fars Province, 65, 71441-65186 Shiraz, Iran, [2]University of Foggia, Department of the Sciences of Agriculture, Food and Environment (SAFE), Via Napoli, 25, 71122 Foggia, Italy; maria.ciliberti@unifg.it

Twenty 3-d-old Holstein dairy calves were used to investigate the effects of feeding starter feed containing dry (AH) vs reconstituted (RAH) alfalfa hay at 10% of dietary dry matter on health status during the pre-weaning period (d 1 to 49; n=10 calves per treatment). Dietary treatment of calves were represented by 6 l/d of milk from d 1 to d 43, by 4 l/d of milk from d 44 to d 46, and by 2 l/d of milk from d 47 to d 49 of study and weaned on d 50. Hay was reconstituted with water 24 h before feeding. The offered and refused amounts of starter feed were individually and daily recorded. Calves at birth and every 10 d for the entire experiment were weighed. Average daily gain [ADG; kg of body weight (BW)/d] and gain to feed ratio [kg of BW gain/total dry matter (DM) intake (DMI = milk DM + starter feed DM)] were calculated. In addition, hip width, hip height, body barrel, body length, withers height, and heart girth measurements were recorded every 10 d throughout the experiment. Data were subjected to ANOVA using the MIXED MODEL procedure of SAS (PROC MIXED, SAS 9.4, SAS Inc., Cary, NC) with time (day) as repeated measures for starter feed intake, total DMI, ADG, feed efficiency, and skeletal growth. Data on BW were analysed using the same model without the time effect. Initial BW was considered as a covariate for the weight at weaning and skeletal growth analyses. Feeding AH vs RAH did not affect starter feed intake (0.34 vs 0.42; P=0.39), total DMI (1.03 vs 1.11; P=0.38), ADG (0.56 vs 0.61; P=0.42), feed efficiency (0.54 vs 0.55; P=0.81), and weaning BW (68.4 vs 70.6; P=0.49). Calves fed AH vs RAH gained less hip width (15.8 vs 16.4; P=0.03) and body barrel (93.9 vs 96.7; P=0.05); however, other skeletal measurements were not affected by treatment. As a result, feeding RAH had a minimal effect on growth performance of pre-weaned dairy calves despite an improvement in hip width and body barrel gains.

Efficacy of Virginia fanpetals (*Sida hermaphrodita*) silage in dairy cattle nutrition

C. Purwin[1], M. Fijalkowska[1], Z. Nogalski[2] and Z. Antoszkiewicz[1]
[1]University of Warmia and Mazury in Olsztyn, Department of Animal Nutrition and Feed Science, Oczapowskiego 5, 10-719 Olsztyn, Poland, [2]University of Warmia and Mazury in Olsztyn, Department of Cattle Breeding and Milk Evaluation, Oczapowskiego 5, 10-719 Olsztyn, Poland; purwin@uwm.edu.pl

Virginia fanpetals (*Sida hermaphrodita* Rusby L.) is a perennial plant characterized by high yield potential on infertile soils, and resistance to freezing. The chemical composition of Virginia fanpetals biomass harvested in the budding stage is similar to that of alfalfa. The aim of this study was to determine the efficacy of Virginia fanpetals silage added to diets for dairy cows as a substitute for alfalfa silage. The experiment had a Latin square design (3×3) and was performed on 9 cows. The cows were housed in tie-stalls and were fed individually. Feed intake was monitored daily. Each feeding period lasted 45 days, including a 30-day preliminary period and a 15-day experimental period proper. The control group received maize silage and alfalfa silage in a ratio of 1:1 DM. In experimental groups, alfalfa silage was completely or partially (1/2) replaced with Virginia fanpetals silage (on a DM basis). All diets were supplemented with a commercial concentrate and a mineral premix. During the experimental period proper, milk samples were collected for analyses three times after morning milking. Milk yield from each cow was weighed. Each sample of fresh milk was assayed for proximate chemical composition, i.e. the percentage content of fat, protein, lactose, dry matter, solids non-fat, casein and urea, as well as active acidity, total acidity and density. The inclusion of Virginia fanpetals silage in dairy cattle diets, as a substitute for alfalfa silage, had no significant effect on the yield, chemical composition or physicochemical properties of milk. The results of the study indicate that Virginia fanpetals silage can constitute supplementary feed for dairy cows when combined with maize silage, and can be a viable alternative to alfalfa silage. The study was carried out in the framework of the project under a program BIOSTRATEG funded by the National Centre for Research and Development No. 1/270745/2/NCBR/2015 'Dietary, power, and economic potential of *S. hermaphrodita* cultivation on fallow land.' Acronym SIDA.

Session 27 Poster 24

Dose effects of linseed and rapeseed oils on bovine rumen microbial metabolism in continuous culture

L.P. Broudiscou[1], A. Quinsac[2], P. Carré[3], P. Schmidely[1] and C. Peyronnet[4]
[1]UMR MoSAR, AgroParisTech INRA, 75005 Paris, France, [2]Terres, Inovia, 33600 Pessac, France, [3]OLEAD, Rue Monge, 33600 Pessac, France, [4]Terres, Univia, 75008 Paris, France; laurent.broudiscou@agroparistech.fr

Whereas the digestive consequences of high amounts of lipids in cattle diets have been investigated for many years, little quantitative information is available on how dietary lipids alter concurrently the different rumen functions in relation with their incorporation level. In our trial which was part of the program VACOMET, linseed oil (LO, rich in linolenic acid) and rapeseed oil (RO, rich in oleic acid) were added to dual outflow fermenters inoculated with bovine rumen microbiota. The effects on the degradation of feed constituents, fermentation yields and microbial biomass synthesis were quantified using polynomial regression models. The 5 treatments – control, LO or RO at 2 doses (4% substrate Dry Matter Input (DMI) or 8% DMI) – were randomly assigned to fermenters during 3 experimental periods of 7 days (5 adaptation days and 2 sampling days). The experimental substrate was composed of maize silage 73% DMI, soybean meal 11%, wheat grain 6%, rapeseed meal 4.9% and wheat straw 4.2%. Both oils influenced several fermentation parameters in a curvilinear way. The fermentation rate and the specific productions of acetate and propionate increased to a maximum at the calculated doses of 4% LO and 6% RO. The net productions of isovalerate and ammonia were linearly increased by both oils, LO being twice more active than RO. the efficiency of microbial protein synthesis was decreased in a curvilinear way by both oils and reached a minimum at the dose of 6%, LO being 1.6 times more active than RO. For several metabolic parameters, the type of variation depended on the oil. The specific production of methane decreased linearly with LO and curvilinearly with RO, along with a rise in di-hydrogen production. The true OM degradability began to decrease only with doses of LO higher than 4%. The microbial N daily outflow was linearly decreased only by LO. The comparison of LO with RO suggested that metabolic parameters differed in their response to fatty acids saturation. Moreover, most effects were present at doses below 4% and diminished or even plateaued at higher doses.

Session 27 Poster 25

Bioavailability of α-tocopherol stereoisomers in sheep fed synthetic and natural α-tocopherol

S. Lashkari[1], S.K. Jensen[1] and G. Bernes[2]
[1]Aarhus University, Department of Animal Science, AU Foulum, 8830 Tjele, Denmark, [2]Agricultural Research for Northern Sweden, Animal Husbandry Unit, Department of Agricultural Research, 901 83 Umeå, Sweden; s.lashkari@hotmail.com

Synthetic vitamin E as commercial source of supplementation of α-tocopherol has eight isomeric configurations including four 2R (RSS, RRS, RSR, RRR) and four 2S (SRR, SSR, SRS, SSS). Only the RRR stereoisomer has the highest biological activity. A ratio of 1.36:1 in relative bioavailability of RRR-α-tocopheryl acetate as a natural (Nat-α-T) source to *all-rac*-α-tocopheryl acetate as a synthetic (Syn-α-T) source is generally accepted. However, studies indicate either bioavailability of α-tocopherol stereoisomers or relative bioavailability between them is not constant, but rather dose dependent and differs between organs. However, no information is available about how different ratios between synthetic and natural α-tocopherol affect bioavailability. Thirty lambs were randomly assigned to five different Syn-α-T to Nat-α-T ratio including 100:0, 75:25, 50:50, 25:75 and 0:100%. The sheep were slaughtered after 75 days of feeding and liver, heart, lungs, muscle and abdominal fat was collected for analysis for α-tocopherol stereoisomers. The data were analysed using completely randomized design. Amount of RRR-α-tocopherol generally increased in plasma and all organs, with increasing the proportion of Nat-α-T in the diet ($P \leq 0.05$). The plasma RRR-α-tocopherol amount was the lowest (0.67 μg/ml) and highest (1.66 μg/ml) in diets containing 100:0 and 0:100% of Syn-α-T to Nat-α-T ratio, respectively. The RRR-α-tocopherol content of liver was 2.79, 4.10, 6.25, 8.20 and 7.45 μg/g in diets containing 100:0, 75:25, 50:50, 25:75 and 0:100% of Syn-α-T to Nat-α-T ratio. The bioavailability relative to diet of RRR-α-tocopherol in diets containing 100% Syn-α-T varied from 2.0 in plasma to 3.0 in lung. The bioavailability relative to diet of RRR-α-tocopherol in plasma, all organs and abdominal fat increased with increasing the proportion of Syn-α-T in the diet ($P \leq 0.05$). In conclusion, the results showed that sheep express discrimination against synthetic stereoisomers of α-tocopherol in plasma and other organs.

Analysis of competition stress on the microbial population in the equine hindgut

A. Carroll
Hartpury University Centre, HE Animal & Agriculture, Hartpury College, Hartpury, Gloucester, Gloucestershire, GL19 3BE, United Kingdom; aisling.carroll2@hartpury.ac.uk

The horse is a nonruminant herbivore where fermentation of chyme occurs in the hindgut (cecum and colon). Within this area, there is a dynamic microbial population that serve the host through immune stimulation, energy extraction, detoxification and pathogen exclusion. The impact of stress on this diverse microbial ecosystem has been seen to have an effect on the numbers of microflora that populate this area. This stress could affect health, welfare, and/or performance. Despite this, scientific literature currently provides limited details on the hindgut microflora response to stress. In particular, competition stress which is a regular experience for most performance horses. The aim of this study was to assess if the stress experienced while performing at competition had an effect on the numbers of hindgut microorganisms in faecal samples. Faecal samples were obtained from 8 horses, aged 6-19 (mean=10.38±1.95), over a 5 day period. All horses participated in a showjumping competition on Day 2 of sampling. Samples were collected 3 times daily. 10-fold serial dilutions were conducted on blood agar. Total bacterial counts were enumerated. The cultured bacteria were identified to species level using biochemical tests. 14 bacteria species were identified over the 5 day sampling period. A 37% decrease in total bacterial numbers occurred in a 24 h period post competition. From an average of 87.9 to 55.5 cfu/ml. Bacterial numbers dropped by 54% in total from day 1 to day 5. A significant difference was identified on total bacterial numbers between baseline values of day 1 and day 5 and between day 3 and day 5 ($P<0.05$). No significant difference was found on total bacterial numbers between baseline values of day 1 and day 3 ($P>0.05$). In conclusion, competition stress had a significant impact on bacterial numbers over the 3 day period post-competition. This could suggest that prolonged exposure to competing or competition environments could have a detrimental effect on a horse's health and performance. Further research is required to establish the impact of these microbial variations on the horse.

Virginia fanpetals (*Sida hermaphrodita* Rusby L.) as a source of carotenoids and tocopherols

Z. Antoszkiewicz[1], M. Fijalkowska[1], M. Mazur-Kuśnirek[1], Z. Nogalski[2] and C. Purwin[1]
[1]University of Warmia and Mazury in Olsztyn, Department of Animal Nutrition and Feed Science, Oczapowskiego 5, 10-719 Olsztyn, Poland, [2]University of Warmia and Mazury in Olsztyn, Department of Cattle Breeding and Milk Evaluation, Oczapowskiego 5, 10-719 Olsztyn, Poland; zocha@uwm.edu.pl

The content of β-carotene and tocopherols in animal feed is determined by plant species, harvest date, maturity stage (leaf/stem ratio, the ratio between vegetative and generative shoots), wilting degree and preservation method. The aim of this study was to determine the effect of harvest date (first-cut and second-cut) on the vitamin content of Virginia fanpetals (*Sida hermaphrodita* Rusby L.) herbage and silage. First-cut and second-cut Virginia fanpetals herbage was harvested in the first and second year of cultivation (11 June and 23 September 2015, 11 June and 16 September 2016). Herbage was cut, chopped (10 mm) and ensiled in vacuum-sealed polyethylene bags (600 g per bag, n=3). Samples of fresh herbage and silage (after 90 days of ensiling) were assayed for the content of β-carotene and tocopherols (α, β, γ, δ) (PN-EN ISO 6867: 2002). The significance of differences between means was determined by Duncan's test. In both years of the study, first-cut Virginia fanpetals herbage had lower β-carotene content ($P\leq0.05$) and a higher content of α-tocopherol and total tocopherols than second-cut herbage. Second-cut silage in the second year of cultivation had higher concentrations of β-carotene, α-tocopherol and total tocopherols (by 14 and 23%, 13 and 36%, 10 and 5%, respectively) than first-cut silage in the first year of cultivation. An analysis of the vitamin content of Virginia fanpetals herbage and silage revealed that first-cut herbage should be fed directly to animals, and silage should be made from second-cut herbage. The study was carried out in the framework of the project under a program BIOSTRATEG funded by the National Centre for Research and Development No. 1/270745/2/NCBR/2015 'Dietary, power, and economic potential of *S. hermaphrodita* cultivation on fallow land.' Acronym SIDA.

Session 27 Poster 28

Mycotoxin contamination in maize kernels: electronic nose as a screening tool for the industry

M. Ottoboni[1], M. Tretola[1], L. Pinotti[1], S. Gastaldello[2], V. Furlan[3], C. Maran[2], V. Dell'Orto[1] and F. Cheli[1]
[1]Department of Health, Animal Science and Food Safety, University of Milan, Via Celoria 10, 20133 Milano, Italy, [2]ATPr&d S.r.l., Via Ca' Marzare, 3, 36043 Camisano Vicentino, Italy, [3]CerealDocks S.p.A., Via dell'Agricoltura, 20, 30026 Summaga di Portogruaro, Italy; matteo.ottoboni@unimi.it

The aim of this study was to evaluate the potential use of an electronic nose (e-nose) for rapid mycotoxin detection in maize, with a focus on aflatoxin (AFLA) and fumonisin (FUM) occurrence and co-occurrence. Twenty-five maize samples were analysed by commercial lateral flow immunoassays (LFIAs) and classified as non-contaminated (NC), single-contaminated (SC), and co-contaminated (COC) by AFLA and FUM according to the detection ranges of LFIA kits. The same samples were analysed by a PEN3 e-nose equipped with 10 MOS sensors (Airsense Analytics GmbH). E-nose data were statistically analysed by Discriminant Function Analysis (DFA) (IBM SPSS 22.0 predictive analytics software). Stepwise variable selection was done to select the e-nose sensors for classifying samples by DFA. The overall leave-out-one cross-validated (LOOCV) percentage of samples correctly classified by the quadrivariate DFA model for AFLA was 67%. The overall LOOCV percentage of samples correctly classified by the single-variate DFA model for FUM was 70%. To test the potential of the e-nose in detecting co-contaminated samples, a discriminant function including five e-nose sensors, was used. The overall LOOCV percentage of samples correctly classified for NC, SC, and COC classes was 65%. In the case of NC samples, the percentage of samples correctly classified was 77%, while it drops to 54 and 61%, for SC and COC samples, respectively. Results indicate that e-nose could be a promising rapid/screening method to detect single or co-contaminated maize kernels. However, e-nose is still far from replacing commercial rapid kit assays, which are quite well defined and broadly used.

Session 27 Poster 29

Characterization of molasses composition

A. Palmonari[1], G. Canestrari[1], L. Mammi[1], D. Cavallini[1], L. Fernandes[2], P. Holder[2] and A. Formigoni[1]
[1]DIMEVET, Via Tolara di Sopra, 50, 40064 Ozzano Emilia, Italy, [2]ED&F Man Liquid Products, 3 London Bridge Street, SE1 9SG, London, United Kingdom; alberto.palmonari2@unibo.it

A common strategy to improve diet palatability, homogeneity and sugar content is molasses addition. Molasses are produced worldwide, mainly from sugarcane and beet. However, 'classic' analyses are unable to characterize all the DM components, which could impact rumen and animal performance. Moreover, no data are available on possible differences among molasses. Objective of this study was to apply an integrate approach to properly characterize molasses. 16 cane and 16 beet molasses were sourced worldwide and analysed for dry matter, crude protein, single sugars, starch, minerals, organic acids, DCAD, and anionic compounds. This approach was able to characterize 97.4 and 98.3% DM of cane and beet molasses, respectively. Cane showed a lower dry matter content compared to beet molasses (76.8±1.02 vs 78.3±1.61% a.f.), as well as CP content (4.8±1.7 vs 10.5±1.1% a.f.), with a minimum value of 1% a.f. in cane to a maximum of 12% a.f. in beet molasses. Lactic acid was more concentrated in cane compared to beet (4.69±2.16 vs 3.48±1.37% a.f.), varying from 9.77% max. to 1.23% min. within cane molasses. Aconitic acid was found only in cane molasses, while glycolic acid in beet. The total sum of acids ranged from 2 to 14% a.f. Sulphates, phosphates, and chlorides had a higher concentration in cane molasses, which showed a lower DCAD compared to beet (4.47±4.97 vs 53.94±33.36 meq/100 g a.f.). Within the cane group, it varied from +117.63 to -58.59 meq/100 g a.f., while in beet from +129.20 to +3.24 meq/100 g a.f. The amount of sucrose was higher in beet compared to cane molasses (48.4±1.5 vs 37.5±4.8% a.f.), with high variability within cane (51.00 max to 33.31 min, % a.f.) and beet. Glucose and fructose were detected in cane molasses (4.06±2.07 vs 6.20±2.17% a.f., respectively). In conclusion, obtained results demonstrate the important differences in molasses composition, even within the same group (cane or beet). This highlights the potential problems with using 'typical values', and thus that a more accurate description and characterization of molasses is possible and strictly required, especially if its use in animal feed is to be fully optimized.

Effect of wax-fatty acid complex on adipogenesis of 3T3-L1 cells

S.K. Park[1], U. Issara[1], S.H. Park[1], S.Y. Lee[1], J.H. Lee[2] and J.G. Lee[3]
[1]Sejong University, Food Science & Biotechnology, 209 Neungdong-ro, 05006 Seoul, Korea, South, [2]Sungkyunkwan University, Food Science & Biotechnology, 2066 Seobu-ro, 16419 Suwon, Korea, South, [3]Korea Tourist Supply Center Inc., 636 Cheonho Daero, 04987 Seoul, Korea, South; sungkwonpark@sejong.ac.kr

The aim of this study was to investigate the effect of wax and fatty acid complex on adipogenesis of 3T3-L1 cells. Fatty acids (FA) analysed included Palmitic acid (PA), Stearic acid (SA), Oleic acid (OA), Linoleic acid (LA; ω-6) and Linolenic acid (ALA; ω-3), and natural waxes; Bees wax (BW) and Carnauba wax (CW). After fully confluent, 3T3-L1 cells were treated with combination of FA, BW, and CW for 9 d, and cell toxicity (by MTT assay), triglyceride accumulation (by Oil-red-O staining), and expression of genes related to adipogenesis were analysed by qPCR. Accumulation of lipid droplets was greatest ($P<0.05$) in SA at 50 μM (1.21 nm), but lowest (($P<0.05$) in ALA at 50 μM (0.78 nm) among treatment groups. Treatments of a wide range (0.5~3 ppm) of BW and CW did not affect cell viability. Gene expression of FA synthase tended to be increased in SA by 2.1 folds, but decreased by 0.2~0.5 folds in ALA, BW, and CW when compared with control. Carnitine palmitoyltransferase 1 (CPT1), which plays an important role in mitochondrial activity and fatty acid oxidation, was increased ($P<0.05$) in cells treated with CLA (1.4 fold), BW (1.16 fold) compare to control. Genes including PPARγ and C/ECPα acting in adipocyte maturation and lipid droplet accumulation were increased by 1.5 and 1.2 fold in SA group, but decreased ($P<0.05$) in BW group compare to control. Results of our current study indicate that poly unsaturated fatty acids, such as ALA, and natural waxes, especially BW, could help reduce obesity by activating fat utilization and limiting fat accumulation.

Microbiological assessment of canine drinking water and the impact of bowl construction material

C. Wright and A. Carroll
Hartpury University Centre, Department of Animal and Land Science, Hartpury College, Hartpury, Gloucester, GL19 3BE, United Kingdom; coralie.wright@hartpury.ac.uk

The number of pet dogs (*Canis lupus familiaris*) in the common household is continually rising. The increasingly close contact between humans and cohabitant pets is leading to concerns regarding bacterial transmission of zoonoses. The dog water bowl has been identified as the third most contaminated item within the household, suggesting that it is able to act as a fomite for bacterial transmission, particularly where young or immunocompromised individuals are present. Studies in livestock have identified that water trough construction material influences bacterial count; however no similar research has been conducted for dog water bowls. The objectives of the current study were to identify which dog bowl material – plastic, ceramic or stainless steel – harbours the most bacteria over a 14 day period and whether the species identified varies between bowl materials. The study took place over 6 weeks. A sample of 6 medium sized (10-25 kg) dogs, aged 2-7 (mean=3.8±1.95), was used. All dogs were clinically healthy, housed individually and located within a rural environment. All bowls were purchased brand new and sterilised prior to a two week sampling period. On day 0, day 7 and day 14 swabs were taken from each bowl and 10-fold serial dilutions were conducted on blood agar. The cultured bacteria were subjected to biochemical testing and the most prominent bacteria from day 14 were further identified using PCR. A significant difference was identified for all bowl materials when comparing total cfu/ml between day 0 and day 7 and day 0 and day 14 ($P<0.05$). No significant difference was identified between total cfu/ml and bowl material ($P>0.05$), however descriptive statistics suggest that the plastic bowl material maintains the highest bacterial count after 14 days. Several medically important bacteria were identified from the bowls, including MRSA and *Salmonella*, with the majority of species being identified from the ceramic bowl. This could suggest that harmful bacteria may be able to develop biofilms more successfully on ceramic materials. Further research is required to identify the most suitable or alternative materials for dog water bowls.

Breeding for temperament – possibilities and challenges in dogs and horses

K. Nilsson
Swedish University of Agricultural Sciences, Department of Animal Breeding and Genetics, P.O. Box 7023, 75007 Uppsala, Sweden; katja.nilsson@slu.se

Temperament of dogs and horses has a large influence on the relationship between the animal and the owner. A very scared and/or aggressive animal can cause large problems in every-day life for the owner and could be dangerous to handle. Mentality is also important for the well-being of the animal, and high stress levels could have a negative effect on performance. Today, behaviour problems is one of the most common causes for euthanizing dogs. Studies have shown that riders rank temperament as the most important performance trait. Temperament is heritable in the sense that it is controlled by genes, and there is a large interest for including temperament traits in breeding programs for horses and dogs. A major challenge lies in finding methods and protocols for recording temperament that adequately reflects the animal's genetic potential and has a high enough correlation with the desired trait. In order to get genetic progress breeders will also have to agree on the type of temperament that is preferred. This will be highly influenced by the use of the animal – a competition horse will need to be different from a leisure horse, and a working dog will need different mental qualities from a companion dog. It is probably easier to agree on what is not desired – high fear and aggression. Several protocols have been developed, some focused on measuring working ability or performance in certain situations and others more targeted at recording innate temperament traits. There are several promising examples of temperament assessment protocols showing both high heritabilities for the measured traits and strong correlation with important temperament traits in the animals' everyday life.

A temperament test developed for the Norwegian horse breeds

H.F. Olsen and G. Klemetsdal
Norwegian University of Life Sciences, Dep. of Animal and Aquacultural Sciences, P.O. Box 5003, 1432 Aas, Norway; hanne.fjerdingby@nmbu.no

Imported, specialized breeds outnumber the National horse breeds Døle, Fjord and Nordland/Lyngen in Norway, and to survive in the future, their competitive edge has to be improved. To win market shares, the breeding organizations have the later years started to market these breeds as being mentally robust. Still, the definitions of the temperamental traits in the breeding plans are currently quite vague, and there is a need for a more precise description. Through an earlier study, the same authors discovered five common factors describing the temperament of the National breeds. These factors formed the basis for developing a field test, where the aim was to identify individual performance linked to these factors. In this study, 63 horses were tested at six different farms, through a proposed test procedure consisting of seven test moments composed of different obstacles/tasks. An unfamiliar handler led the horses through the test course while video recording and recording the heart rate of the horse, which together made the basis for a validation of the test. In total, 43 traits were scored through a developed ethogram, and then related to the heart rate registrations and to the five factors from the former study. The scorings were supported by the heart rate results and corresponded to four of the five earlier discovered temperamental factors; anxiousness, conscientiousness, openness, and dominance. The fifth factor agreeableness was not identified through this test. In addition, a factor analysis was performed on the 43 traits in the ethogram, giving eight factors, which also were related to the former five factors and to the heart rate registrations. The expression of traits differed between type of obstacle, separating dynamic and static novel objects, and the consistency between judges was found to be fair to high. The proposed test is assumed to be a good basis for further development of a field test with linear scoring system for temperament in horses, with emphasis on detailed scale description of the temperamental traits that maintains the high consistency between judges. In addition, future field data would also prove the stability of these traits over time and give better knowledge of the genetic correlation between these traits.

Genetic counselling in horse breeding

J. Detilleux
University of Liege, Faculty of Veterinary Medicine, Fundamental and Applied Research for Animal & Health (FARAH), Animal Production Department, Quartier Vallée 2, Avenue de Cureghem 6, 4000 Liege, Belgium; jdetilleux@ulg.ac.be

Genetic disorders are common in horses. Breeding risk evaluation is essential to reduce mutation transmission, design treatment and management strategies and possibly prevent future complications. Risk evaluation must be based upon genetic (e.g. genetic test results, pedigree, breed incidence) and non-genetic (e.g. diet, performance, prevention measures, age) factors, and upon the animal's economic value. This evaluation, and client counselling, are not straightforward, potentially leading to misunderstanding/wrong decisions. Bayesian networks (BN) are especially suited for such situations due to their transparency, ability to handle uncertainty, and event prediction combining sources of information. As an example, I propose a BN to evaluate the risk of a foal suffering from Polysaccharide Storage Myopathy (PSSM1) when mating two phenotypically healthy horses. Rarely fatal, clinical signs of PSSM1 include muscular dysfunctions that prevent normal use of a horse. The PSSM1 mutation is quite common (more than 90% in Belgian horse breeds). The transmission is autosomal dominant with incomplete penetrance. Total recovery is rare and treatment requires management of diet and exercise for life. In the proposed BN, data may include diet, test result and parental disease status. I derived other necessary information from current knowledge of the PSSM1 mode of inheritance and sampled prior probabilities from Beta distributions when necessary. Arbitrary utilities (0 to 100) were given to the mating in line with the offspring's expected genotype. All computations were done on Netica 5.12. Results showed the offspring of a mating of normal parents has a 4.6% probability of showing symptoms similar to PSSM1 with an adequate diet, and 27.3% if inadequate. If one parent is healthy and one unknown, the probability of unaffected offspring is 10.5%, and if both parents are healthy, 62%. Expected utility of the mating increases from 64.5 to 78.7 if one parent is healthy and one unknown, or both parents are healthy. This example shows that the method is useful to objectively discuss issues in genetic counselling. It is fast, easy to implement and allows new information to be included during discussions with decision partners.

The use of genomic data for the analysis of companion animal populations

V. Kukučková, N. Moravčíková and R. Kasarda
Slovak University of Agriculture in Nitra, Department of Animal Genetics and Breeding Biology, Tr. A. Hlinku 2, 94976, Slovak Republic; veron.kukuckova@gmail.com

Genome-wide association studies can be used to reveal disease risk loci in dog breeds. The increasing occurrence of hereditary disorders and diseases in specific breeds has forced breeders and researchers to search for the causes. Run of homozygosity (ROH) analyses were performed on small populations of three breeds of retrievers – Golden, Labrador and Nova Scotia Duck Tolling (NSDTR). The estimation of 20% autozygosity in purebred dogs was confirmed: segments of ROH>1 Mb show an average genome coverage of 25.62% Golden Retrievers, 22.17% in Labrador Retrievers, and 28.% in NSDTRs. Inbreeding estimated by ROH >16 Mb reached 5.84, 5.02 and 10.57%, for each breed, respectively, indicating recent inbreeding. Analysed retriever breeds are sufficiently genetically distinct to present as three separate clusters in admixture analysis. The population of Labrador Retrievers was characterized by relatively low genetic variability, while some level of admixture with the Golden Retriever is also visible. The population of NSDTR created one separated cluster characterizing the sufficient genetic variability based on genetic distances. However all individuals in this population are genetically connected and with extremely high level of recent as well as historical inbreeding. With this study we have started to address the increasing interest of dog breeders in specific analyses, especially the minimization of inbreeding. With increased genotype data, there will be unlimited possibilities to explore the population.

Selection against complex diseases using genomics based breeding schemes in horses

B.J. Ducro[1], A. Schurink[1] and S. Janssens[2]
[1]Wageningen University & Research, Animal Breeding and Genomics, P.O. Box 338, 6700 AH Wageningen, the Netherlands, [2]Catholic University Leuven, Division Animal and Human Health Engineering, P.O. Box 2456, 3001 Leuven, Belgium; bart.ducro@wur.nl

Selection against complex diseases in horses often requires tests on large progeny groups to achieve breeding values with sufficient accuracy. Progeny testing can be challenging in the smaller herdbooks and for diseases that are only recordable later in life or at high costs. Complex diseases are therefore highly suitable for genomic selection programmes. Creation of large reference populations however is often not feasible in horse breeds for reasons similar to performing progeny testing. Meanwhile, knowledge on genomic regions involved in inheritance of complex diseases in horses has increased and considering these regions as quantitative trait loci (QTL) in breeding programs will likely increase efficiency in selection against complex diseases. The aim of this study was therefore to explore how information on QTL best can be included in a standard breeding program to select against a complex disease in horses. Selection response from different breeding scenarios were compared using deterministic simulation. The scenarios differed in heritability of the disease, size of population and selection pressure. Selection against a complex disease was based on combinations of progeny testing and a QTL that explained varying percentage of the additive variance. Simulations indicated that progeny testing schemes could be replaced by QTL selection when the QTL explained 50% of the additive genetic variance. When QTL explained less genetic variance then adding a stage of selection on QTL prior to progeny testing would increase efficiency. QTL enabled additional selection in mares, which would increase selection response by ca. 5%. This study has shown that inclusion of QTL information on a complex disease into a breeding program can improve selection against a complex disease in horses either by an increased selection response or by replacing progeny testing.

Influence of training, age at training onset and management on longevity in Icelandic horses

C. Weiss, K. Brügemann and U. König V. Borstel
University of Giessen, Animal Breeding and Genetics, Leihgesterner Weg 52, 35392 Giessen, Germany; uta.koenig@agrar.uni-giessen.de

The purpose of the present study was to identify factors affecting the success and duration of the competition career (longevity) in Icelandic horses. Competition data of 18,111 horses with 296,723 competition starts registered between 2006 and 2016 in the central registry of the Icelandic horse riders' and breeding association of Germany were used in the present analysis. Furthermore, 9,348 owners, primary caretakers or riders of these horses were contacted with a survey to obtain information on various training, management and husbandry-related factors for these horses. 1,480 complete surveys were available for the present analysis. Survival analysis accounting for right censoring of data revealed that overall median length of competition career was 5 years. However, horses starting in breeding horse performance test classes had a higher probability to remain registered for at least 5 years, compared to horses starting in other classes only (57.4 vs 40.7%, $P<0.05$). The majority of training and management factors as reported by the horses' owners/trainers showed no significant influence on longevity. However, horses that had >5 rather than 1-5 hours/day pasture turnout had a reduced risk of terminating their competition career ($P<0.05$). Horses trained only 1-4 times/week had a higher probability to remain registered for at least 5 years (49.1%), compared to horses trained 5 (35.8%) or 6-7 times/week (39.7%). Age at onset of training did not ($P>0.1$) influence longevity. However, unlike with results from various previous studies on dressage, show-jumping or racing horses, horses that were older (>10 years) at their first start at competition had a longer competition life (7 years) compared to horses that were younger (9-10 years: 6 years, 7-8 years: 5 years, 5-6 years: 4 years, $P<0.05$). These results require further investigations regarding the effect of competition onset on horse health, but bias due to left-censoring of data and possible confounding factors need to be kept in mind when interpreting these results. Overall, results show that adequate turnout and training intensity are more important to longevity than age at first training.

Progressive weaning reduces negative behavioural and biological effects in horses

M.P. Moisan[1], A. Foury[1], M. Vidament[2], F. Levy[2], F. Reigner[3], N. Mach[4] and L. Lansade[2]
[1]INRA, Nutrineuro, 33076 Bordeaux, France, [2]INRA, CNRS, IFCE, PRC, 33380 Nouzilly, France, [3]INRA, Physiologie Animale de L'Orfrasière, 37380 Nouzilly, France, [4]INRA, GABI, 78352 Jouy-en-Josas, France; marie-pierre.moisan@inra.fr

In domestic horses, foals generally are separated from their mothers suddenly and definitively between 4 and 6 months of age. In contrast, under feral conditions foals gradually stop suckling between 11 to 12 months of age but the mother-foal bond remains until sexual maturity. Early weaning has known negative effects on foal behavioural, neuronal and hormonal responses. This study compared sudden and definitive weaning as commonly employed in horse rearing (Sudd group, n=16 mare-foal pairs) to progressive weaning involving slow habituation to separation in the month preceding weaning (Prog group, n=18 pairs). At 6 months of age, mare-foal pairs were moved to large pens with adjacent paddocks. A fencing panel separated Prog foals from their mothers for an increasing amount of time each day. Definitive weaning with complete separation occurred at 8.2±0.4 months old. Foals remained in their boxes while mares were transported to a familiar stable 2 km away. Each method's impact on mares and foals was assessed at set timepoints using behavioural observations and biological measurements of stress markers, incl. salivary cortisol, telomere length, and blood transcriptome profiles and these data were integrated by Principal Component Analysis. Immediately after weaning, Sudd foals were more gregarious, active, reactive to humans, fearful, and had higher salivary cortisol levels. Prog foals were more inquisitive and this trait was associated with longer telomeres and a higher expression of 22 genes. Effects lasted more than 3 months, with noticeable differences in gregariousness, activity, fear, and some gene expression values. Although less marked, Prog mares also had lower stress markers, rested more on weaning day, associated with longer telomeres while most Sudd mares had higher cortisol levels and up-regulation of 11 genes and displayed more alert postures, neighs and general activity on weaning day, indicating higher stress levels. This study shows that compared to sudden weaning, progressive weaning has beneficial effects for both mare and foal.

The use of operant conditioning behaviour analysis in animal training for husbandry and performance

M.L. Cox
CAG GmbH – Center for Animal Genetics, Paul-Ehrlich-Str. 23, 72076 Tuebingen, Germany; melissa.cox@centerforanimalgenetics.com

Operant Conditioning Behaviour Analysis (OCBA) is a science-based method of training animals and analysing animal behaviour. It is the science underlying 'clicker training', although some people use the terms interchangeably. The method is based upon simple principles of learning, and is humane, effective, and applicable to any species. For these reasons, it is becoming the gold standard in all facets of animal training and husbandry. This presentation will give an overview of OCBA and its practical applications in multiple species in the areas of research, industry, and animal performance. The goals are two-fold: (1) to highlight the current use of OCBA in animal training and husbandry, and (2) to encourage further research in these areas to demonstrate the efficacy and welfare benefits of this method.

The centaurs of law and order

V. Deneux
INRA, UMR Innovation, 2 place Viala, Bat 27, 34000 Montpellier, France; vanina.deneux@inra.fr

Until now, animal's natural behaviour and their interactions with humans were mostly studied by Biological Sciences. My research is in Human and Social Studies and aims at analysing the intersubjective work relationships between horses and humans. In this paper, I study work relationships between policemen and their patrol horses in urban environments, to show how each party invests in the joined realization of the work dedicated to enforcing law and order. The present study builds on a main theoretical framework: the Psychodynamics of work, which I use to analyse intersubjectivity of relationship between humans and animals in concrete work situations. The studies based on this theoretical framework already helped highlight the animals' subjective relationships to work and to confirm the existence of an 'animal working'. My study compares two research fields of work of police and horses. The results are expected to highlight the impacts of national or local contingencies, of the historicity of police units and of urban environments on life and the work relationship of policemen with their horses and horses with policemen. The first research field is situated within the French Republican Guard, which included some 460 horses. The Guard performs three missions: law and order, military protocol of the state, and international prestige of France by promoting the French tradition of horse riding. For this study, I focus on their public safety missions. The second research field is situated within the municipal equestrian brigade of the French city of Tours. This is a rather small unit of six horses and six riders, which was created fifteen years ago. First, I will present the living and working conditions of horses in these two structures. Second, based on semi-structured interviews and observations of trainings and patrols, I will show how work is organized between humans and horses in both cases and how this organization impacts on the subjective commitment of horses in the work. Finally, I will attempt to determine whether work situations within same mission of public safety vary according to the national or local contingencies.

The effect of two different housing conditions on horse behaviour

K. Olczakc[1], A.S. Santos[2,3] and S. Silva[3]
[1]University of Agriculture in Krakow, Department of Swine and Small Animal Breeding, Faculty of Animal Breeding and Biology, Institute of Animal Science, Krakow, Poland, [2]CITAB, Animal Science, Universidade de Trás-os-Montes e Alto Douro, Quinta de Prados, 5000-801 Vila Real, Portugal, [3]CECAV, Animal Science, Universidade de Trás-os-Montes e Alto Douro, Quinta de Prados, 5000-801 Vila Real, Portugal; assantos@utad.pt

Horses are naturally free-ranging, social, grazing herbivores. Confinement of stabled horses can result in distress which is associated with the manifestation of abnormal behaviours such as stereotypies, and it has been suggested that increased stable comfort and space may reduce the incidence of stereotypic behaviours. The present study aimed to compare the influence of housing systems on horse behaviour. Two different housing systems with different area and space layout were studied. A total of 8 horses were randomly allocated to one of two different systems. Group A: 4 horses housed individually in boxes of 4×3 m, with straw bedding, and Group B: 4 horses housed in boxes of 4×3 m, with straw bedding and access to an outside area of 15 m^2. The experiment was carried out over a 3-week period. All animals were observed by direct observation and the ethogram was performed. Observations were made during four periods: early morning (8:00-9:00 h); mid-morning (10:00-11:00 h); mid-afternoon (14:00-15:00 h); and late afternoon (16:00-17:00 h). Behaviour data was recorded using the instantaneous scan sampling technique. Each horse was recorded on one scan every 10 minutes for the duration of each observation period. The housing system presented a highly significant influence ($P<0.0001$) on several behaviours. Behaviours typically associated with boredom and stress in confined horses were observed in higher occurrences in horses that where housed without access to the exterior. Results from this study show that housing has an influence on horse behaviour. Horses that had no access to an exterior paddock presented significantly higher number of manifestations of stereotypic behaviours than horses that were allowed to go outside, even in a small paddock of 15 m^2. Access to this small area is enough to decrease the manifestation of stress related behaviours, increasing horse welfare.

Session 28 Poster 11

Foal weaning in Italy: management and factors associated with the development of abnormal behaviours

F. Martuzzi[1], G. Dellapina[2] and F. Righi[2]
[1]University of Parma, Food and Drug, Via del Taglio 10, 40126 Parma, Italy, [2]University of Parma, Department of Veterinary Science, Via del Taglio 10, 40126 Parma, Italy; francesca.martuzzi@unipr.it

The increase of stress at weaning may lead to the development of abnormal behaviours (AB) in the foal. An online survey was carried out with 22 questions concerning husbandry methods and behaviour of both foals and mares. Mares are defined dominant, middle or low ranking based on the number of threats given and received in the herd. Horses with dominant behaviour are commonly considered less willing to learn in some disciplines, whereas dominant stallions in dressage eventing obtain generally higher scores than others. Therefore, factors which could affect dominant attitudes in foals were also analysed. Data from 74 breeders of Central-Northern Italy, describing 148 horses, were analysed to identify husbandry techniques and management risk factors associated with development of AB. Data were compared with surveys in other countries: higher percentages of Italian foals (16.2%) and mares (10.9%) showing AB (vs an average of 5.2% of other studies) and a higher age average at weaning (7 vs 5-6 months) were observed. No effect of breed, feeding, paddock size or housing were found in this study. Taming period was critical for the onset of AB for 4 foals out of 12 presenting AB. Generally, oral-ingestive AB were more frequent (65%) than other AB. Among them, wood chewing was the most frequent (35%). Expression of foal's normal behaviour is negatively affected by dam aggressiveness toward humans ($P<0.05$, chi square test). The following trends were observed ($P<0.1$): development of AB in foals may be related to AB in mares; foal dominant behaviour may be related with mare's dominant attitude in the herd and negatively associated with early handling, which seems favourable for normal behaviour expression. The sample may be biased (respondents may be only breeders interested in improving welfare) however this is an inherent problem with all similar studies. The main difference with North Europe countries is the lower competence of the Italian equine sector. Despite the low number of animal samples, the survey can give some indications on the origin of AB in horses.

Session 28 Poster 12

Positive effects of enriched environment in horses: a behavioural and genomic assessment

A. Foury[1], L. Lansade[2], M. Valenchon[2], C. Neveux[2], S.W. Cole[3], S. Layé[1], B. Cardinaud[4], F. Lévy[2] and M.P. Moisan[1]
[1]NutriNeuro, INRA, Université de Bordeaux, Bordeaux, France, [2]PRC, INRA, CNRS, IFCE, Université de Tours, Nouzilly, France, [3]Division of Hematology-Oncology, David Geffen School of Medicine, University of California, Los Angeles, CA, USA, [4]Bordeaux INP, Bordeaux, France; aline.foury@inra.fr

Enriched environment (EE) is known to promote well-being. Nonetheless, effects of enrichment on personality and gene expression is less known. Therefore, we assess behaviour, personality, learning performance and gene expression in 10-month-old horses maintained in impoverished environment (IE) or EE for 12 weeks. The IE-treated horses had no specific stimulation (individual stalls, standard alimentation) whereas the EE-treated horses were highly stimulated (pasture 16 h/day and large stalls with many sensory stimulations). The EE modified three dimensions of personality: fearfulness, reactivity to humans, and sensory sensitivity. Some of these changes persisted >3 months after treatment. EE-treated horses also show better learning performance in a Go/No-Go task. The whole-blood transcriptomic analysis showed that the genes upregulated in the EE-treated group were involved in cell growth and proliferation, while, in the IE-treated group, activated genes were related to apoptosis. Changes in both behaviour and gene expression may constitute a psychobiological signature of the effects of enrichment and result in improved well-being.

Session 29 Theatre 1

Genetic bases of individual performance variation in pigs

L. Bodin[1], N. Formoso-Rafferty[2] and H. Gilbert[1]
[1]INRA, GenPhySE, 31326 Castanet-Tolosan, France, [2]Universidad Complutense de Madrid, Fac. de Veterinaria, Dept. of Animal Production, Madrid, Spain; loys.bodin@inra.fr

Part of the variability of performances in animal production is due to the genetic variation in the level of a trait between animals; that is the keystone of all genetic improvement programs and has to be maintained to guarantee the long-term sustainability of the selection progress. Another part of the performance variability comes from the individual variation of responses to various environmental conditions, called environmental sensitivity, which can have a highly negative impact on the final profitability of the production. Joint estimations of breeding values (EBV) and genetic parameters of the individual environmental sensitivity of a trait can be obtained along with those of the level of the trait. Moreover, use of these two kinds of EBV in genetic programs partly depends on their correlations as well as their genetic relationships with other traits. In pigs, homogeneity of litter size is desired and in many cases its mean level should not decrease. Similar relationships between objectives exist for the birth weight components (mean and variability); homogeneity is desired to ensure good perinatal viability and subsequent growth, but the mean birth weight should also not decrease. The genetic parameters and relationships between components (mean and variability) of birth weight and litter size in pigs will be presented along with similar parameters estimated in a mice experiment. Furthermore, similar estimations were obtained for production traits, such as residual feed intake and growth, which will be discussed in the perspective of producing more efficient and robust animals.

Session 29 Theatre 2

Description and consequences of variability in sows and piglets

N. Quiniou[1], M. Marcon[1], Y. Salaün[1], J.Y. Dourmad[2], F. Gondret[2], H. Quesnel[2], J. Van Milgen[2] and L. Brossard[2]
[1]IFIP-Institut du Porc, BP 35, 35650 Le Rheu, France, [2]PEGASE, INRA, Agrocampus-Ouest, 35042 Rennes, France; nathalie.quiniou@ifip.asso.fr

Even though animals are from the same genetic line, farmers have to cope with variability both in sows and piglets. In sows, variability is observed in traits such as parity, prolificacy, appetite, body weight (BW) and back fat thickness (BF). For instance, at the beginning of gestation, variability in body condition among sows can be high due to parity and age. In addition, at a given age, variability in litter size, milk potential, and appetite results in different nutrient requirements and consequently in variability of changes in maternal body reserves. Variability in BF can be a problem as several studies have indicated that too high or too low BF values are to be avoided at farrowing as well as at weaning. In both cases, the longevity of the sow is impaired, and farmers are advised to manage the sows toward a target BF depending on the physiological stage, associated with an age-dependant BW, increasing with age up to mature BW. In addition, variation in sow's body condition at farrowing and in prolificacy influences the new born and weaning piglet traits. Compared to less prolific sows, high-prolific sows farrow more piglets, which are both lighter on average and more heterogeneous. Compared to normal birthweight piglets, the survival rate of low birthweight piglets is lower. Providing additional care around birth helps these piglets to survive, but subsequent housing and feeding management have to be adapted to deal with the variability in their growth potential. Nutritional strategies (based on modelling approaches that take into account criteria that influence requirements) are suggested to optimise the expression of the animals' potential, but most often without an intention to reduce inter-individual variability in growth performance. In order to control or reduce variability, other solutions have been evaluated in experimental studies that focus on the level and the dynamic of the feeding plan and the quality of the diet. The challenge is now to validate these solutions in production units, which will be more or less easy depending on the existing housing and feeding systems, and the economic, welfare and environmental context.

Precision feeding of lactating sows: development of a decision support tool to handle variability

R. Gauthier[1], F. Guay[2], L. Brossard[1], C. Largouët[3] and J.Y. Dourmad[1]
[1]Pegase, INRA-Agrocampus Ouest, 35590 Saint Gilles, France, [2]Université Laval, G1V0A6, Québec, Canada, [3]IRISA, Agrocampus Ouest, 35000 Rennes, France; raphael.gauthier@inra.fr

Nutritional requirements of lactating sows mainly depend on milk yield and greatly vary across individuals. Moreover, because the same diet is generally fed to all sows, and feed intake is low and highly variable, nutrient supplies are often insufficient to meet the requirements, especially those of primiparous sows. Conversely, sows with high appetite may be fed nutrient in excess. Acquiring data on sows and their environment at high-throughput allows the development of new precision feeding systems with the perspective of improving technical performance and reducing feeding cost and environmental impact. The objective of this study was thus to design a decision support tool that could be incorporated in automated feeding equipment. The decision support tool was developed on the basis of InraPorc® model. The optimal supply for a given sow is determined each day according to a factorial approach considering all the information available on the sow (i.e. parity, litter size, milk production, body condition…) or predicted from real-time data (i.e. expected feed intake). The approach was tested using data from 817 lactations. Precision feeding (PF) with the mixing of two diets with different nutritional values was then simulated in comparison with conventional feeding (CF) with a single diet. In sows fed in excess PF reduced average digestible lysine excess from 10.9 to 2.7 g/d, whereas in deficient sows the deficiency was reduced from -5.7 to -2.1 g/d. Overall, PF reduced average lysine intake by 6.8%. At the same time, with PF, lysine requirement was met for a higher proportion of sows, especially in younger sows, and a lower proportion of sows, especially older sows, received excessive supplies. PF also reduced average phosphorus intake while limiting the occurrence of excess and deficiency. This study confirms the potential of precision feeding in order to better achieve nutritional requirements of lactating sows and reduce their nutrient intake and excretion. This project has received funding from the European Union's Horizon 2020 research and innovation program, grant agreement no. 633531.

Identification of biological variables associated with robustness of piglets at weaning

A. Buchet[1,2], E. Merlot[1], P. Mormède[3], E. Terenina[3], B. Lieubeau[4], G. Mignot[4], J. Hervé[4], M. Leblanc-Maridor[4,5] and C. Belloc[5]
[1]UMR PEGASE, Agroampus, INRA, 35590 Saint Gilles, France, [2]Cooperl Innovation, BP 60238, 22403 Lamballe, France, [3]Université de Toulouse, INPT ENSAT, INRA, 31326 Castanet-Tolosan, France, [4]IECM, INRA, Oniris, Université Bretagne loire, 44307 Nantes, France, [5]BIOEPAR, INRA, Oniris, Université Bretagne Loire, 44307 Nantes, France; arnaud.buchet@cooperl.com

The robustness of a piglet at weaning can be seen as its ability to express optimal growth without any health problems and regardless of weaning conditions. The identification of the level of the robustness of piglets at weaning could allow implementing targeted cares and treatments. The aim of this study was to identify some biological markers measured around weaning that could be associated with the growth of the piglet after weaning. Blood variables (n=62) describing immunity, stress, oxidative status and metabolism were measured at 26 and 33 days of age on piglets (n=288) from 16 commercial farms selected with contrasting sanitary statuses (deteriorated: SS- or good: SS+). The sanitary status of the farm was significantly associated with 37 of 67 variables measured ($P<0.05$). Thus, piglets reared on SS- farms showed higher activation of the immune system, mobilization of body reserves and oxidative stress after weaning than SS+ piglets. In order to evaluate differences in robustness within farms, the relative ADG from 26 to 47 days of age was calculated (RADG = ADG from 26-47 days of age divided by live weight at 26 days), and piglets were then classified according to the median of their farm in classes of low or high RADG (RADG- or +). This classification was considered as a proxy of robustness. After weaning, RADG+ piglets showed greater immune activation (neutrophil count), lower mobilization of body reserves (non esterified fatty acids and creatinine) and a higher concentration in vitamin A, an antioxidant vitamin, compared to RADG- piglets.

The genetic background of second litter syndrome in pigs

E. Sell-Kubiak[1], E.F. Knol[2] and H.A. Mulder[3]
[1]Poznan University of Life Sciences, Department of Genetics and Animal Breeding, Wolynska 33, 60-637 Poznan, Poland, [2]Topigs Norsvin Research Centre, Schoenaker 6, 6641 SZ Beuningen, the Netherlands, [3]Wageningen University & Research, Animal Breeding and Genomics, P.O. Box 338, 6700 AH Wageningen, the Netherlands; sell@up.poznan.pl

In pig breeding, it is observed that a substantial part of sows have lower performance (i.e. litter size and/or farrowing rate) in parity 2 than in parity 1. This physiological phenomenon is known in literature as second litter syndrome (SLS), which often leads to early culling of the sows. The aim of this analysis was to study the genetic background of SLS. Litter size data from Large White sows kept on commercial farms were provided by Topigs Norsvin. The records of 246,799 litters (total number born) came from 53,794 sows with at least two observations in parities 1-10. The statistical analysis was performed in ASReml 4.1. From all sows,15,394 had larger litters in parity 1 than parity 2 and were kept for at least one more parity. The SLS sows had on average 14.03 piglets in the first litter and 10.86 in the second litter, whereas non-SLS sows had 11.62 piglets in the first litter and 14.45 in the second litter. First, variance components for SLS treated as binominal trait were analysed. The heritability of SLS was 0.09, comparable to litter size. Secondly, the bivariate analysis of SLS with litter size in parity 1 and 2 was performed. The genetic correlation between SLS and litter size in parity 1 was r_g=0.45, whereas between SLS and litter size in parity 2 was r_g=-0.41. The r_g indicates that selection for large litter size in parity 1 increases the incidence of SLS, while selection on large litter size in parity 2 reduces the incidence of SLS. In both cases, r_g are moderate, thus increase (or decrease) in SLS is not rapid, but it is clear that high emphasis on litter size in parity 1 will increase SLS. Nonetheless, with the magnitude of correlations it would be possible to select for increased litter size across parities without simultaneous increase in SLS or even a decline in SLS. Careful breeding program design is needed to optimize genetic gain in litter size, while at same time reducing SLS. This project was financed by Polish National Science Centre grant no. SONATA-2016/23/D/NZ9/00029.

Coping with animal variation using precision feeding techniques

C. Pomar and A. Remus
Agriculture and Agri-Food Canada, 2000, rue Collège, Sherbrooke, Québec, J1M 0C8, Canada; candido.pomar@agr.gc.ca

Variation is essential and inherent to living systems and variation among animals reduces the efficiency with which nutrients are used in conventional group-fed livestock production systems. Nutrients are provided to satisfy the animal needs. When feeding one animal, these needs can be associated with body maintenance, growth and production, and the animal's nutrient requirements can be estimated by adding the amounts of nutrients needed for each of these functions. Individual animals (e.g. pigs) within a given population differ significantly in terms of body weight (BW), BW gain, health, etc. and therefore, differ in terms of the amount of nutrients they each need at a given point in time. Therefore, when feeding a group of animals, the concept of maintenance, growth and production requirements may not be appropriate. When feeding populations of animals, nutrient requirements should be seen as the optimal balance between the proportion of animals that needs to be overfed and underfed. Given that, for most nutrients, underfed animals exhibit reduced performance while overfed animals exhibit near-optimal performance, nutrients have to be provided in sufficient amounts to satisfy the requirements of the most demanding animals in the group in order to obtain optimal herd performance (i.e. growth). Precision feeding is proposed to cope with animal variation as it involves the use of feeding techniques that allow the right amount of feed with the right composition to be provided at the right time to each pig in the herd. Recent research indicates that precision feeding is essential to maximize nutrient efficiency while ensuring the sustainability of the livestock industry by reducing the excretion of nutrients and lowering feeding costs. This new feeding approach represents a paradigm shift in animal feeding, because the optimal dietary nutrient level is no longer considered a static population attribute, but rather a dynamic process that evolves independently for each animal. Precision feeding is a highly promising avenue for improving resource use efficiency in comparison with conventional group phase-feeding systems, and coping with the challenges of managing animal variation.

Accepting or managing pig weight variability at slaughter through optimal marketing?

F. Leen[1], A. Van Den Broeke[1], M. Aluwé[1], L. Lauwers[1,2], J. Van Meensel[1] and S. Millet[1]
[1]Flanders Research Institute for Agriculture, Fisheries and Food, Burg. Van Gansberghelaan 115 b2, 9820, Belgium, [2]Ghent University, Agricultural Economics, Coupure Links 653, 9000 Ghent, Belgium; frederik.leen@ilvo.vlaanderen.be

Variability in birth weight and growth rate between growing finishing pigs results in a distribution of pig weights at marketing. Weight variability at slaughter can provoke economic 'sorting losses', especially when pig pricing schemes have narrow optimal carcass weight ranges. Growth rate variability is a multi-factorial phenomenon that can partly be reduced through management. Completely eliminating growth rate variability is biologically and economically infeasible. Previous research has indicated that with best management practices in both the breeding and the growing-finishing phase, variability in pig weights can be limited to a coefficient of variation (CV) of 8-12% at the first marketing of a batch of pigs. The question then remains: should the remaining variability be managed by optimizing marketing strategies or should it just be accepted?. Therefore, an optimization model for pig marketing was developed assuming normal distributions in pig weights (one for male and one for female pigs) to optimize the weight at and timing of marketing for a total of 10 deciles per sex. The model was applied on animal performance data of a farrow-to-finish farm with a CV of 10.7% for entire males and 9.9% for gilts at 23 weeks of age. In general multi-staged marketing mainly prevented price discounts for overweight pigs, since optimal marketing strategies still include a proportion of pigs (25%) sold underweight to optimize finishing barn turnover rate. The optimal marketing strategy is affected by pig performance (growth and feed efficiency) and market conditions. In our simulations, the optimal strategy consisted of 3 to 4 deliveries with weekly intervals depending on pig prices. The diminishing economic returns from including more delivery stages provide limited leeway for investing in accurate sorting technology and extra labour. Still, redesign of slaughterhouses' pricing schemes towards optimal distributions in primal cuts might require on-farm *in vivo* carcass classification based on camera technology in addition to merely weighing the pigs ready for slaughter.

Variability in growth rates within a farm affects pig carcass composition

P. Aymerich[1,2], D. Solà-Oriol[2], J. Bonet[1], J. Coma[1] and J. Gasa[2]
[1]Vall Companys Group, Pol. Ind. El Segre, Parc. 410, 25191 Lleida, Spain, [2]Universitat Autònoma de Barcelona, Animal Nutrition and Welfare Group, Department of Animal and Food Sciences, Edifici V Travessera dels Turons, 08193 Bellaterra, Spain; paymerich@vallcompanys.es

The aim of the present study was to determine if different growth rates within a pig farm result in differences in carcass composition. During the study, 888,697 growing-finishing pigs of the same breed (Du × (LR×LW)) were followed from the farm to the slaughterhouse. The experimental unit was each truck leaving the farm to the slaughterhouse, which had on average 175 pigs (mix of castrated males and females) and a slaughter weight of 111.8±5.0 kg. The dataset consisted of productive and carcass measurements. On the one hand, the average daily gain (ADG, g/d) during the whole growing-finishing period was measured for each group of pigs leaving the farm in the same truck. On the other hand, 3 carcass parameters were measured at the slaughterhouse. Those measurements were cold carcass weight (CCW, kg) and AutoFom (Carometec Food Technology) measurements such as ham fat thickness (HFT, mm) and carcass leanness (CL, %). Data was analysed with the stats package in R 3.4.1 by performing simple linear regressions and multiple linear regressions to predict HFT and CL, with ADG and CCW as main effects. On average, ADG was 808±95 g/d, CCW 86.2±4.1 kg, HFT 11.4±0.9 mm and CL 59.6±1.1%. The difference on ADG between the fast growing and slow growing pigs within each farm was 250±84 g/d on average. The effect of both regressors by simple regressions on HFT was positive ($P<0.001$). There was an increase of 0.148 mm HFT/kg CCW ($R^2=0.41$) and 0.546 mm for each 100 g/d of ADG ($R^2=0.30$). The multiple regression model with both regressors showed again that increasing both CCW and ADG results in a carcass with higher HFT ($P<0.001$; $R^2=0.53$). As expected, a negative relationship was found between CL and both CCW (-0.098%/kg; $P>0.001$; $R^2=0.14$) and ADG (-0.499% for each 100 g/d of ADG; $P>0.001$; $R^2=0.20$). Additionally, the multiple regression model showed that increasing CCW and ADG decreases CL ($P<0.001$; $R^2=0.25$). In conclusion, within the same farm, pigs that grow faster have a higher carcass fat content than those with a lower ADG.

Linear models versus machine learning to improve monitoring of pneumonia in slaughter pigs

K.H. De Greef and B. Hulsegge
Wageningen Livestock Research, P.O. Box 338, 6700 AH Wageningen, the Netherlands; karel.degreef@wur.nl

Slaughter line inspection generates large amounts of data on lung health, but use is limited. Combined analysis (across farms) is more powerful in detecting patterns, trends and deviations than analysis for farms separately. Machine learning may further provide a step forward for this compared to linear statistics. A data set from 119 farms, comprising 1.694.158 pigs slaughtered in 12.419 batches was used. Linear models (LM: full and stepwise reduced model) and machine learning methods (ML: *RandomForest, ExtremeLearning, GradientBoosting, NeuralNetworks*) were compared in predicting batch incidence of pneumonia. Parameters available comprised info on farm, slaughterhouse, period (week, season, etc.), batch size and incidences from previous batches. Random Forest (RF) proved to be the best ML technique with r^2 of 0.42, r^2 of the full linear model (FLM) was 0.29. Further comparison is based on prediction errors (observed-predicted batch pneumonia%). The Mean Absolute Error of the null-model amounted 7.0, and was reduced to 5.1 and 4.6 by FLM and RF, respectively. Practical application requires a distinction between positive and negative prediction errors. Positive errors (higher incidence than predicted) relate to missed alarms for high incidences, whereas negative prediction errors indicate false alarms. Overall, mean positive prediction errors (MPE) amounted 6.6 and 5.8 for FLM and RF. Mean negative prediction errors (MNE) were 3.3 and 3.1, respectively. To explore differences between farms, the 119 farms were split into four (2×2) quartiles on basis of incidence (50% highest vs 50% lowest) and relative between batch variability (50% highest vs 50% lowest). Positive prediction error was 1.5 higher for farms with high variability compared to farms with low variability (irrespective of pneumonia incidence). Across these categories, RF outnumbered FLM by 10-13%. Results demonstrate that across farm analysis enhances quality of pneumonia monitoring substantially. In this, machine learning is superior to classic linear statistics. Furthermore, distinction into categories of farms on basis of pneumonia incidence and also variability aids in predictive quality; again better with machine learning methods than with linear models.

Genetic parameter estimates for farrowing traits in purebred Landrace and Large White pigs

S. Ogawa[1], A. Konta[2], M. Kimata[3], K. Ishii[4], Y. Uemoto[1] and M. Satoh[1]
[1]Graduate School of Agricultural Science, Tohoku University, 468-1 Aramaki Aza Aoba, Aoba-ku, Sendai, 980-0845, Japan, [2]Faculty of Agriculture, Tohoku University, 468-1 Aramaki Aza Aoba, Aoba-ku, Sendai, 980-0845, Japan, [3]CIMCO Corporation, 2-35-13 Kameido, Koto-ku, Tokyo, 136-0071, Japan, [4]Division of Animal Breeding and Reproduction, Institute of Livestock and Grassland Science, NARO, 2 Ikenodai, Tsukuba, Ibaraki, 305-0901, Japan; shinichiro.ogawa.d5@tohoku.ac.jp

A breeding goal for female pigs is the genetic improvement of the number of piglets weaned per sow. Hence, selection on dam breeds or lines should consider the genetic correlation between the component traits of farrowing and mortality. In the present study, genetic parameters for six reproductive traits related to farrowing events; total number of piglets born (TNB), number of piglets born alive (NBA), number of stillborn piglets (NSB), total litter weight at birth (LWB), mean litter weight at birth (MWB), and gestation length (GL), were estimated for Landrace and Large White pigs in a Japanese breeding company. The number of farrowing records used were 62,534 litters for Landrace and 49,817 litters for Large White. Heritabilities were estimated using a single-trait repeatability model, and genetic correlations were done using a two-trait repeatability model. Heritability estimates were 0.12 for TNB, 0.12 for NBA, 0.08 for NSB, 0.18 for LWB, 0.19 for MWB, and 0.29 for GL in Landrace, and 0.12, 0.10, 0.08, 0.18, 0.16, and 0.34, respectively, in Large White. Genetic correlations between NBA and NSB were unfavourable: 0.20 in Landrace and 0.33 in Large White. Genetic correlations of GL with the other five traits were weak: from −0.18 with NSB to −0.03 with NBA in Landrace, and −0.22 with NSB to −0.07 with NBA in Large White. LWB had a highly favourable genetic correlation with NBA (0.74 in both breeds), indicating the possibility of using LWB for the genetic improvement of NBA.

Session 29 Poster 11

Swine genetic line and sex influence the relationship between carcass weight and fat content

P. Aymerich[1,2], D. Solà-Oriol[2], J. Bonet[1], J. Coma[1] and J. Gasa[2]
[1]Vall Companys Group, Pol. Ind. El Segre, Parc. 410, 25191 Lleida, Spain, [2]Universitat Autònoma de Barcelona, Animal Nutrition and Welfare Group, Department of Animal and Food Sciences, Edifici V Travessera dels Turons, 08193 Bellaterra, Spain; paymerich@vallcompanys.es

The aim of this work was to evaluate the relationship between carcass weight and fat content in different swine genetic lines and sexes. The genetic lines consisted of different crossbreds between a common sow (LR×LW) and different sire lines with different fat deposition rates: a Duroc (DU), a synthetic (SY) and a Pietrain (PI). Evaluated sexes were castrated males and females for DU, castrated males, females and entire males for SY, and finally, females and entire males for PI. To carry out the work, AutoFom (Carometec Food Technology) data of 74,418 pigs from 3 different sire lines was collected from the same slaughterhouse. Carcass measurements included were cold carcass weight (CCW, kg), subcutaneous ham fat thickness (HFT, mm) and carcass leanness (CL, %). Data was analysed with the stats package in R 3.4.1 by performing a multiple regression analysis to predict HFT. Main effects were sire line, sex, CCW, and the interactions between the 3 main factors. In addition, the slopes (variation of mm HFT for each kg of CCV) for genetic lines and sexes were estimated by least square means and compared pairwise. At similar slaughter weights, DU, SY and PI averaged 91.7±4.1, 89.5±4.6 and 89.8±3.8 kg of CCW; and 15.2±1.9, 10.3±2.3, and 9.2±1.8 mm of HFT; and 55.6±2.9, 62.0±2.3 and 63.9±1.6% of CL, respectively. In the multiple regression model, all the factors were significant ($P<0.001$; $R^2=0.68$) and a positive relationship was observed between CCW and HFT. The comparison of slopes showed differences between genetic lines and sexes. Although similar slopes were observed for the genetic lines, DU and SY (0.182 and 0.192 mm HFT/kg CCW; $P=0.095$), those were higher than PI (0.165 mm HFT/kg CCW; $P<0.001$). Regarding sexes, entire males and castrated males showed a similar slope (0.168 and 0.160 mm HFT/kg CCW; $P=0.216$), but their slopes were lower than females (0.214 mm HFT/kg CCW; $P<0.001$). Summarizing, differences in the relationship between carcass weight and fat content are related to genetic lines and sexes.

Session 29 Poster 12

Do small pigs stay small or can they catch up?

C. Vandenbussche, S. Millet and J. Maselyne
ILVO (Flanders research institute for agriculture, fisheries and food), Burg. Van Gansberghelaan 92 bus 1, 9820 Merelbeke, Belgium; jarissa.maselyne@ilvo.vlaanderen.be

The performance of individual pigs can vary a lot, even in the same barn or pen. This leads to management challenges and economic losses due to suboptimal feeding and a trade-off between optimal payment in the slaughterhouse versus the cost for extra delivery moments. To limit variation different strategies can be applied like grouping litters of pigs with the same starting weight or feeding based on weight. In research, a lot of effort is put into precision feeding. In light of this it is interesting to know how pigs grow under *ad libitum* feeding conditions. Do the smallest pigs at the start of the fattening round stay small? And what about the largest pigs? Data of 1,116 pigs from 6 fattening rounds (2012-2018) in the experimental barn of ILVO were used. In 5 rounds housing was in mixed groups (gilts and barrows) of 59, 38 or 35 pigs per pen. In the last round, pens of 15 pigs with separated boars or gilts were used. Pigs of different litters were mixed to have an equal average and variation in the starting weight between pens. Average daily gain between 10 weeks of age (21.9±3.7 kg) and slaughter (106.7±14.7 kg) was 0,66±0.11 kg across all pigs (mean ± SD). Per fattening round, pigs were ranked based on their weight at the start, independent of pen and gender. Of the 15% heaviest pigs at the start, on average 29% were still in the 15% heaviest group at the end of the fattening round, 52% dropped to the 70% average pigs, 8% ended up with the 15% lightest and 11% died or had to be removed. Of the 15% lightest pigs at the start, 37% stayed small, 54% grew to more average weights, while 2% even reached the 15% heaviest weight class and 7% died or had to be removed. Growth within pens and between pigs of the same sex is analysed further. Weight at start of the fattening period is thus an indicator of future weight, but it is by no means a reliable predictor. While most of the 15% lightest pigs remain among the lightest half of the group and the heaviest pigs tend to stay in the heaviest half, some pigs manage to switch to the other side of the spectrum. Hence, management strategies based on weight at start-up have potential, but drastic selection should be avoided to fulfil the potential of all pigs.

Session 29 Poster 13

A pan-European computed tomography procedure for measuring the new EU lean meat content of pigs

G. Daumas and M. Monziols
IFIP-Institut du Porc, BP 35104, 35601 Le Rheu Cedex, France; gerard.daumas@ifip.asso.fr

A new EU regulation on carcass classification applies from July 2018. The new reference to calibrate the pig classification methods is a lean meat percentage based on total dissection (LMPtd). Manual dissection can be replaced by an unbiased computed tomography (CT) procedure. If the national pig population to be sampled has the same characteristics as the population for which a CT procedure has been previously corrected, no additional dissection is required. In most of the national applications for authorisation of classification methods the population characteristics are managed via a stratification on a fat depth, mimicking the LMP variation. The aim of this paper is to propose a pan-European CT procedure to calibrate the pig classification methods without any additional manual dissection. A sample of 29 half-carcasses was CT scanned with 3 mm slice thickness and then fully dissected according to the EU regulation. The CT muscle volume was calculated by thresholding in the Hounsfield range 0-120. It was converted into muscle weight by applying a density of 1.04. The weight was divided by the carcass weight to obtain the lean meat percentage from CT (LMPct), in the same way as done for the LMPtd. LMPtd was regressed on LMPct. Only the slope was significant and was estimated at 0.965 (s.e.=0.002). The RMSE was 0.81. The plot of residuals against fitted values showed no pattern and no heterogeneity of variances. The main source of measurement error is the thresholding of the rind. As the thickness of the rind is very thin (2-3 mm), most of the voxels including rind are mixed voxels, either with air or with fat. Their Hounsfield values are therefore less than 0 and these voxels are classified in non-muscle. Only a few rind voxels have a Hounsfield value in the range [0-120 HU] and are thus misclassified in muscle. This is taken into account by the slope value which is slightly less than 1. The LMPtd range in the sample (53-68) covers the S+E classes (≥55) which represent more than 90% of the EU pigs. It covers too more than 80% of all the national populations, excepted Italy. This robust CT procedure can therefore be applied in 27 Member States, by using a pan-European multiplicative factor of 0.965, without any additional national dissection.

Session 30 Theatre 1

Agricultural sustainability metrics based on land required for production of essential human nutrien

M.R.F. Lee[1,2], T. Takahashi[1,2], G. McAuliffe[1,2] and M. Tichit[3]
[1]University of Bristol, Bristol Veterinary School, Langford, BS40 5DU, United Kingdom, [2]Rothamsted Research, North Wyke, Okehampton, Devon, EX20 2SB, United Kingdom, [3]INRA, Université Paris-Saclay, Paris, France; michael.lee@rothamsted.ac.uk

Agricultural land and food security are coming under great pressure leading to a need to re-assess the role of ruminant livestock in delivering key nutrients, especially in the broader context of global warming potential and land use change. To date, several studies have examined environmental consequences of different food consumption patterns at the diet level; however, few have addressed nutritional variations of a single commodity attributable to on-farm strategies, leaving limited insight into how agricultural production can be improved to better balance environment and human nutrition. Using farm-level data encompassing cattle, sheep, pigs and poultry production systems, we propose a novel approach to incorporate nutritional values of meat products into two metrics: (1) global warming potential (GWP) per nutrient provision; and (2) arable land use (ALU) per nutrient provision. To holistically assess the value of the product with regards to associated with human nutrition, we developed a nutrient index based on approaches proposed by Saarinen *et al.* and McAuliffe *et al.*, where the overall quality of food is represented by a single scalar value to combine information on multiple nutrients, both beneficial and detrimental to human health. The results showed that the relative rankings of the livestock products dramatically altered between the two metrics, with monogastric systems preferred under GWP (poultry < pigs < cattle < sheep) and ruminant systems preferred under ALU (sheep < cattle < pigs < poultry). The trade-off between the two metrics suggests that the globally optimal composition of livestock species is likely to be a mixture of monogastric and ruminant animals and, therefore, for livestock systems to contribute to global food security, multiple aspects of agricultural sustainability should be carefully and simultaneously considered.

Meeting the dual demand for animal products and climate change mitigation by narrowing yield gaps

A. Van Der Linden[1], P.J. Gerber[1,2], G.W.J. Van De Ven[3], M.K. Van Ittersum[3], I.J.M. De Boer[1] and S.J. Oosting[1]
[1]Wageningen University & Research, Animal Production Systems group, De Elst 1, 6708 WD Wageningen, the Netherlands, [2]Food and Agriculture Organization, Animal Production and Health Division, Viale delle Terme di Caracalla, 00153 Rome, Italy, [3]Wageningen University & Research, Plant Production Systems group, Droevendaalsesteeg 1, 6708 PB Wageningen, the Netherlands; aart.vanderlinden@wur.nl

The livestock sector faces the demand for increasing animal production while reducing its greenhouse gas (GHG) emissions and emission intensity (GHG emissions per unit of product). The bio-physical scope to increase animal production is defined as the yield gap. This research aims to assess the effect of narrowing yield gaps on emission intensities for beef production systems in Uruguay, and for dairy production systems in Ethiopia and Bangladesh. Potential cattle production (defined by genotype and climate) and feed-limited production (additionally limited by feed quality and quantity) was simulated with livestock models. Water-limited production of feed crops was obtained from databases, and was combined with potential cattle production (WL-P) and feed-limited production (WL-FL) to calculate beef or milk production per hectare. Actual production was obtained from FAO reports. The yield gap was defined as the difference between WL-P or WL-FL and the actual production. Emission intensities were calculated with IPCC equations (Tier 1). Relative yield gaps were 80-87% of WL-P and 49-62% of WL-FL for different beef production systems in Uruguay at the national level. Emission intensities were lower under WL-P (74-75%) and WL-FL (56-57%) than under actual production. For Ethiopia, yield gaps in dairy systems were 61-81% of WL-P and 39-66% of WL-FL. Emission intensities were lower under WL-P and WL-FL than under actual production. Trends in yield gaps and emission intensities were similar for Ethiopia and Bangladesh. We conclude that results show synergies between narrowing yield gaps and reducing emission intensities. This research can be used to identify and explore feasible intervention options that increase livestock production and reduce emission intensity.

Session 30 Theatre 3

Constraints and opportunities of dairy production to provide ecosystem services in urban India

M. Reichenbach[1], A. Pinto[2], S. König[2], P.K. Malik[3] and E. Schlecht[1]
[1]University of Kassel, Animal Husbandry in the Tropics and Subtropics, Steinstrasse 19, 37213 Witzenhausen, Germany, [2]Justus Liebig University Gießen, Animal Breeding and Genetics, Ludwigstrasse 21B, 35390 Gießen, Germany, [3]National Institute of Animal Nutrition and Physiology, Adugodi, 560030 Bangalore, India; marion.reichenbach@uni-kassel.de

While first human settlements were linked to the beginning of agriculture, in the urbanising modern world, agricultural production systems become increasingly detached from the urban consumption-oriented sphere. As an emerging megacity in India, Bangalore combines rapid urbanisation with a high demand for dairy products, which are supplied in part by urban dairy farms. The aim of this analysis was to identify key features of urban dairy farms, and constraints and opportunities to their provision of ecosystem services. Qualitative and quantitative survey data on farm socio-economics, resources availability, and in- and output markets were obtained from 300 dairy farms in the rural urban interface of Bangalore; data were analysed using ANOVA and Chi Square Frequency Test. Urban farms typically kept 7.2±4.0 cattle, of which 61% were productive dairy animals – milking or dry cows. In the inner city, mainly Holstein Friesian or Jersey (66%) were kept, followed by crossbred cows (33%), while native breeds were an exception. Milk was marketed directly, through a middleman, delivered to a dairy cooperative, or through a combination of marketing channels. In 80% of the cases, cows were let to pasture in the streets or on surrounding green spaces. More than 90% of farmers supplied concentrates to their cows but only 10% produced green or dry forage. All others relied on bought forage or cheap/ free market waste. Manure was mainly sold but also just dumped in public space. Despite limited land availability to maintain cows or grow fodder, the urban environment benefits from the provisioning ecosystem services of green market and household waste (i.e. fodder) conversion into the highly demanded milk; in exchange dairy farmers profit from the variety of direct marketing channels. Positive environmental services on the production side are however offset by a compromised nutrient cycle due to difficult manure management, and risk of water pollution via excreta and nutrients.

Towards field specific phosphate applications norms with machine learning

H. Mollenhorst, M.H.A. De Haan, J. Oenema, A.H. Hoving-Bolink, R.F. Veerkamp and C. Kamphuis
Wageningen University and Research, Droevendaalsesteeg 1, 6708 PB Wageningen, the Netherlands; erwin.mollenhorst@wur.nl

Efficient use of manure is an important link in the nutrient cycle in agricultural systems. On Dutch dairy farms, most manure is applied on grass and cropland, with maize as main crop. With the aim of balancing P input and output at field level, which is the idea behind the currently used, but rather fixed, legal manure application norms, predicting future yields could be a first step to move towards flexible application norms. Machine learning techniques might be useful to predict nutrient harvest more precisely than current norms, because they can be trained based on previous yields and other related variables, without modelling the detailed relationship. This study's objective is to predict future maize yields based on farm data and open source weather data. Data were available from a Dutch dairy research facility, containing 162 records of maize fields between 1996-2014 with information on N and P input and output, irrigation, and P status at field level as well as local weather data. Records covered 24 different fields, with on average 7 times maize in 19 years. Maize yields ranged from 13 to 36 kg P per ha with a mean of 22 kg P. Generalized boosted regression was used to predict maize yields for the years 2010-2014 in kg P per ha per year, for each year based on information from all previous years. By doing so, the model was validated on independent data. Model performance was evaluated by computing the RMSE and correlation coefficient, and compared to currently used legal norms. Results showed an RMSE of 4.6 kg P per ha per year and r was 0.37 for the model, whereas the current, rather fixed norm resulted in an RMSE of 4.9 kg P without a correlation to P yield. The positive correlation shows that the model, although to a minor extent, was able to show a trend in P yields with a somewhat lower RMSE compared to fixed norms. In conclusion, with the limited data available, prediction of P yield, and therewith, defining flexible P application norms for maize, is marginally better than current fixed application norms. This approach, therefore, will be explored further for, e.g. grassland, and for predicting animal manure production in order to replace fixed norms as well.

Past intensification of livestock led to mixed benefits for social and environmental sustainability

J.P. Domingues[1,2], A.H. Gameiro[1,3], B. Gabrielle[1,3] and M. Tichit[2]
[1]USP-FMVZ, 05508-270, Pirassununga, Brazil, [2]INRA/SAD-APT, 75005, Paris, France, [3]UMR EcoSys, 78850, Thiverval-Grignon, France; domi.joaopedro@gmail.com

Livestock production, through its direct and indirect impact on landuse and resource uses, is an important driver of the provision of multiple services. Although a few studies provide an account on the multiple services in different livestock systems, there is still an important knowledge gap on the drivers that contribute to the differentiation of services provisioning across areas. We investigated the hypothesis that the current level of services derives from past intensification of livestock. The objective of this study was to understand the influence of changes in livestock, land use and socio-economic variables on the current provision of social, environmental and cultural services by livestock in France. We took advantage of a long-term country-wide database covering the livestock intensification process and a recent database on services provisioning. We used a set of multivariate methods to analyse simultaneously the changes in livestock intensification from 1938 to 2010, and the current level of services provisioning. Our analysis focused on a set of 60 French departments where livestock play a significant role. Our study revealed that the provision of services was spatially structured and based on three groups of departments characterized by different rates of change in intensification variables. In the first group – transition areas, the low provision of services was mainly associated with mixed crop-livestock areas changing into crop specialised. In the second group – extensive livestock areas, the high levels of environmental and cultural services were mainly associated with higher rates of change on herbivores stocking rate (+95%) and relative increase of grasslands (+13%). In the third group – intensive livestock areas, the high level of social services was mainly associated with high rates of change in monogastrics stocking rates (+1,045%) and milk productivity (+451%). This study provided knowledge to understand how past changes determined the current contribution of livestock areas in providing differentiated bundles of services, which might help steering development of today livestock sector towards more sustainable trajectories.

A systematic review of research on biodiversity in European livestock systems

A. Kok, I.J.M. De Boer and R. Ripoll-Bosch
Wageningen University & Research, Animal Production Systems group, P.O. Box 338, 6700 AH Wageningen, the Netherlands; akke.kok@wur.nl

The decline of biodiversity is a major concern to scientists, society, and policy makers in the European union (EU). The EU Biodiversity Strategy (2010) aims to halt the loss of biodiversity in the EU by 2020. One target to reach this aim is to increase the contribution of agriculture to maintain and enhance biodiversity. According to reports, however, no significant overall progress has been made to meet this target. Moreover, agriculture is both reported to reduce and to enhance biodiversity. To enhance biodiversity in agriculture, we need to understand how biodiversity is measured in the existing studies, and to map relations between agricultural land use and biodiversity. We aimed to (1) review indicators used in science to measure biodiversity in EU livestock systems, and (2) to review described effects of livestock on biodiversity. We conducted a systematic review in Scopus and Web of Science. The search for research articles that assessed impacts of livestock on biodiversity yielded 857 articles after deduplication, which was narrowed down to 163 relevant articles. Species abundance and species diversity were commonly used state indicators of biodiversity across scales. Modelling studies also used aggregated indicators with biodiversity values that were directly linked to land use. Most studies focussed on the impact of grazing ruminants on biodiversity, either for food production or nature conservation purposes. Pigs and poultry were mainly studied in relation to local ammonia emissions. Only few studies considered impacts of land use for feed production on biodiversity. We argue that the traditional pressure-state-response framework to categorize indicators of biodiversity does not provide clear actions to enhance biodiversity in agriculture. Instead, we propose the use of comparators (e.g. grazing intensity levels) in relation with state measures of biodiversity. This review can help to identify commonly used indicators of biodiversity, and provide insight in quantitative relations between agricultural land use and biodiversity.

The full climate effects of grasslands for livestock production

J. Chang[1], P. Ciais[1], T. Gasser[2], D.S. Goll[1], P. Havlik[2] and M. Obersteiner[2]
[1]Laboratoire des Sciences du Climat et de l'Environnement (LSCE-IPSL), Orme des Merisiers, 91191 Gif sur Yvette, France, [2]International Institute for Applied Systems Analysis (IIASA), Schlossplatz 1, 2361 Laxenburg, Austria; jinfengchang@gmail.com

The terrestrial biosphere was assessed to be a net source of biogenic greenhouse gases (GHG; CO_2, CH_4, and N_2O) above pre-industrial emission levels, thus driving global warming. Agriculture is the largest and the second largest contributor to the land biogenic CH_4 and N_2O emissions respectively. Grassland ecosystems support the world's livestock production, and contribute a significant share of global CH_4 and N_2O emissions but also sequestering carbon in some regions. The carbon balance of grassland soils is deeply impacted by the addition of organic and mineral fertilizers, grazing and mowing. In parallel with the increased number of domestic animals, human activities also caused a large reduction in wild grazers, which resulted in a reduction of CH_4 emissions. Although estimates of GHG emissions have been produced for the whole terrestrial biosphere and for cropland cultivation around the year 2000, a specific assessment of the GHG balance of the grassland biome under human management and climate change is still lacking. To address this research gap, we quantified over the period 1901-2010, the net annual carbon balance of grassland biomass, soils and grazing animals, domestic and wild ones, CH_4 emissions from manure and animals and N_2O emissions from manure and mineral nitrogen fertilizers application, using the spatially explicit process-based ecosystem model ORCHIDEE-GM which includes parameterizations of grassland management. The model was run over the whole globe with variable CO_2, climate, nitrogen deposition, land use and management intensity. Pre-industrial GHG fluxes were subtracted from present-day values to quantify anthropogenic carbon sinks and non-CO_2 sources, and to calculate their radiative forcing. Climate change caused by this radiative forcing was computed using the compact Earth System Model OSCAR v2.2. In this presentation, we will discuss uncertainty sources and regional differences of the impacts on climate of grasslands.

Quantifying the land displacement of the french livestock sector

M. Tichit, B. Silva Barboza, T. Bonaudo, J. Domingues Santos and F. Accatino
INRA, Science for Action and Development, Université Paris-Saclay, 75005 Paris, France; muriel.tichit@agroparistech.fr

In Europe, soy meal is an important source of protein in feed. A large share of soy meal is imported from outside Europe leading to land displacement. A key area for innovation is to develop feeding systems based on local feed protein. Objective of this study was to quantify the spatial variation of land displacement due to livestock production and its link other environmental impacts. The method combines different data source (data from agricultural statistics and data from industries and extension services) to calculate five indicators describing: land displacement, nitrogen surplus, nitrogen emission, animal source food, nitrogen conversion efficiency. Indicators were computed for 571 livestock farming regions (LFR). A principal component analysis (PCA) was performed on the five indicators followed by hierarchical ascendant clustering based on the coordinates of 571 LFR in PCA dimensions. Results showed three groups of LFR corresponding to a gradient of land displacement intensity. The first group corresponded to LFR (n=347) where land displacement achieved 6.5 ha /100 livestock units (LU). The second group corresponded to LFR (n=12) showing a doubling of land displacement (12 ha / 100 LU). The third group (n=212) corresponded to LFR with the lowest level of land displacement (5.7 ha / 100 LU). The relationship between land displacement and local nitrogen surplus was strongly positive. The dependency to soy meal import made it possible the development of livestock densities beyond reasonable boundaries for the sustainability of some livestock regions. Beyond 1.5 LU/ha, livestock production was based on transforming imported feedstuff into the LFR rather than locally produced feedstuff. Nearly 40% of LFR may require targeted adjustments of feeding systems and livestock densities in order to enhance local and global sustainability.

Methane emission and performance of steers in pasture and integrated crop-livestock-forest systems

A. Berndt[1], P. Meo Filho[2], A.P. Lemos[1], L.S. Sakamoto[2], A.F. Pedroso[1], J.R. Pezzopane[1] and P.P.A. Oliveira[1]
[1]Embrapa, Rod. Washington Luiz, km 234, P.O. Box 339, 13560-970, Sao Carlos, Brazil, [2]USP–FZEA, Av Duque de Caxias Norte, 13633-900, Pirassununga, Brazil; alexandre.berndt@embrapa.br

Farmers are increasingly using integrated crop-livestock-forest systems due to environmental and economic benefits, including product diversification, better use of the area throughout the year and recovery of degraded areas. These systems can ensure animal welfare by providing shade and superior quality feed, resulting in improved feed efficiency compared to traditional systems. Another benefit relates to environmental sustainability since cattle in integrated systems tend to display lower GHG emissions intensities. The aim of this study was to measure methane emissions from Canchim cattle (5/8 Charolais) in integrated crop-livestock-forest systems. The experiment took place at Embrapa Southeast Livestock, using 30 Canchim steers with an average age of 15±2 months and weight of 245±26 kg, distributed between 5 pasture systems: intensive grazing (IGS), silvopastoral (SPS), integrated crop-livestock (ICL), integrated crop-livestock-forest (ICLF) and extensive grazing (EXT) systems. Paddocks were grazed for 6 days followed by a 28-day recovery period, except for the extensive pasture which was grazed continuously. Enteric methane was measured using the SF_6 technique over five consecutive days, 24 hours per day. Data were analysed using the MIXED procedure of SAS (version 9.3) and averages were compared using Tukey's test with significant differences at $P<0.05$. Differences were observed in the average live weight, with animals in the IGS system weighing more than those in the SPS (286.4 vs 269.2 kg); IGS and ICLF steers gained more weight per day than animals in the ICL system (0.652; 0.647 vs 0.349 g). No differences in daily enteric methane emissions were observed between the systems assessed. However, there were differences in the intensity of methane emissions in relation to daily weight gain, with the ICL animals emitting more methane in comparison to SPS and ICLF animals (415.4 vs 208.4; 205.3 gCH_4/kg DWG). Other studies are required to monitor these systems for a number of years and identify the factors that cause these differences between the integrated systems.

Session 30 Theatre 10

Agroecological levers to improve feed self-sufficiency and to drive farms towards sustainable LFS

V. Thenard[1], E. Morin[2], M.A. Magne[3] and J.P. Choisis[4]
[1]INRA UMR 1248 AGIR, ch borde rouge, 31326 Castanet Tolosan, France, [2]IDELE, Ch Borde Rouge, 31326 Castanet Tolosan, France, [3]ENSFEA UMR 1248 AGIR, Ch Borde Rouge, 31326 Castanet Tolosan, France, [4]INRA UMR 1201 DYNAFOR, Ch Borde Rouge, 31326 Castanet Tolosan, France; vincent.thenard@inra.fr

Livestock farming is an essential activity in many rural areas, mainly in the less-favoured areas. Environmental issues, climate change and economic context weaken the farms' performances and sustainability. Agroecology can provide a framework to ensure the sustainability of the livestock farming systems. Building self-sufficiency of animal feed is a privileged way to star this transition. In the low favoured area it's not so easy to achieve self-sufficiency but farmers experiment different levers for action to improve farm adaptability. In such a context, over the past few years, researchers, farmers, advisors have built together projects to identify and to assess innovative practices for sustainable land use, and the use of diversity resources. The present study is based on around 50-dairy-sheep-farms in two regions in southern France where farmers produce ewe milk for PDO cheeses. We have collected (1) data from farmers‘ interviews about practices, (2) agronomic and economic data. We have identified four patterns describing the management of self-sufficiency. These patterns match with four agroecological levers improving sustainability: (1) using natural resources limiting chemical input; (2) managing feed self-sufficiency producing the whole animals' feed; (3) reducing tillage using pesticides; (4) applying conservation farming increasing plant diversity. Finally to assess practices of farmers we have built a set of indicators concerning three main environmental stakes: soil preservation, no renewable resources use, and cultivated biodiversity. We mobilized these indicators to describe each pattern with its different trade-offs. To improve the self-sufficiency of livestock farming systems, farmers must to redefine the objectives of animal production. This study is an important milestone for dairy sheep production insight to improve the agroecological management of farms. These results should be used to produce tools for agroecological transition.

Session 30 Theatre 11

Linking practices changes and performances to assess agroecological transition in dairy farms

A. Vidal, E. Vall, A. Lurette, M.O. Nozières-Petit and C.H. Moulin
SELMET, Univ Montpellier, CIRAD, INRA, Montpellier SupAgro, Montpellier, France, 2 place Pierre Viala, 34000 Montpellier, France; arielle.vidal@cirad.fr

Agro-ecological transition could be a way to address ecological, economic and social challenges faced by dairy farmers. Performances of livestock system are impacted by changes in farming practices. The purpose is to assess how those changes enable the transition toward agroecological systems. In order to assess agro-ecological transition process in progress, an original and general assessment framework was developed linking an analysis of farms trajectories with set of performance indicators. The framework was tested in milk farms with cows or ewes on two local territories in Burkina Faso and France. Performance indicators were based on the agroecology framework referring to productivity (i.e. litre of milk / hours of work with dairy cows), efficiency (i.e. milk added value / intermediate consumption), level of crop-livestock integration (i.e. carbon recycling or organic fertilizer coverage), and diversity to estimate robustness (i.e. products and users diversities through Shannon index). Indicators were mobilized at two state of the trajectory of the livestock farming system, before and after a period of transformation. This approach help us to determine links between changes in practices and performance of the livestock systems to assess agroecological transition in dairy farms. Here are some results from one of the sixth Burkinabe farm. Between period 1 and period 2, feed storage increased from 0 to 320 kg dry matter / dairy cow. This improvement allowed for an increase in the milk production during dry season from 273 to 865 litres/dairy cow. Economical productivity increased from 32,196 to 56,482 FCFA/dairy cow/year. Production costs increased from 9 to 106 FCFA / litre. However, farmers tended to use more inputs. The continuous presence of dairy cows on family farms improved the agriculture-livestock integration, due to their ability to provide 50% of the crops' organic fertilization needs. This application permit with a diversity of farms several paths towards agro-ecological transition in dairy farms in Burkina Faso.

Diversity in farm activities layout and contribution of agriculture to the territory performances

A. Lurette, L. Lecomte, J. Lasseur and C.H. Moulin
SELMET, INRA, CIRAD, Montpellier SupAgro, Univ Montpellier, 2 place Pierre Viala, 34060 Montpellier, France; charles-henri.moulin@supagro.fr

Although the specialization of agriculture decoupled crop and livestock production systems, the integration of agricultural activities at a farm scale providing a relevant alternative for sustainability of farming systems. However, integration can also be observed between farms at territory scale. This study aims at assessing how the diversity in the layout of farm activities (crop, livestock and mixed farming systems) can play a role in performances building for several dimensions of the sustainability within a small territory. A modelling tool simulates the layout of farm activities and their interactions within a small area of Alpes de Haute Provence in France. The diversity in farms is based on 6 types: 3 sheep farm types(LIV), 2 mixed crop-livestock farm types (MCL) and 1 crop farm type (CRP). From a reference situation (REF) involving CRP and MCL, two scenarios are tested: SPEC, simulates the current evolution toward a specialization, and MIX, simulates an evolution towards the diversification within the landscape. The interactions between farms occur by manure and straw sales. The total territory income remains the highest under REF. SPEC, which leads to the highest number of farms but the lowest number of workers (30), exhibits the highest income per worker (8% higher than REF). MIX exhibits the highest number of workers (34.5) but also the lowest total income (7% less than REF). From the Shannon diversity index calculation, MIX displays the highest diversity index for products providing (1.52) compared to SPEC (0.52) and REF (1.2). Subject to price volatility, MIX is the less sensitive scenario. It also leads to tripled the surfaces of rangelands maintained by flocks compared with SPEC. However, REF reaches the highest diversity index for land use, but with 10% less in rangelands use than MIX. In SPEC, the settlement of sheep farms results in the creation of manure fluxes towards crop farms. These fluxes occur within the same farm in MIX scenario. As simulated, the diversity of farm activities and their interactions influence the evolution of dimensions of the sustainability, including robustness, of a small territory.

Dynamic of litter in palisade grass in integrated crop-livestock and monoculture systems

F.F. Simili[1], J.G. Augusto[1], G.G. Mendonça[2], P.M. Bonacim[1], L.S. Menegatto[1], R. Sechinatto[1] and C.C.P. Paz[1]
[1]Instituto de Zootecnia, APTA/SAA, Sertãozinho, SP, 14174000, Brazil, [2]FMVZ/USP, Pirassununga, SP, 13635900, Brazil; flaviasimili@gmail.com

The objective of this study was to evaluate the Litter deposition (LD) and litter disappearance rate (k) on integrated crop-livestock (ICL) and monoculture systems (single sowing), during the four seasons of the year. It was used a complete randomized block with three replications and five treatments: palisade grass (G), corn and palisade grass sown simultaneously (CG), corn and palisade grass sown simultaneously plus herbicide (CGH), palisade grass sown at topdressing corn (CGT) and palisade grass sown at the line and inter-line of corn (CGL). The pasture was submitted by beef cattle during the Jan until Dec 2017. Existing litter (LE_0) in the pasture was monitored at 28 day intervals by sampling quadrats (0.5×1.0 m) which were randomly positioned in the paddocks (4 quadrats). All litter in each quadrat was collected and dried, cleaned and then weighed. After 14 days, it returned to the same place to collect the deposited material (LD 14). The equation $k = [\ln (LE_0 + LD14) - \ln (LE_0)] / t$ was used to make the estimates of LD and k. The PROC MIXED by SAS was used to analyse the data. There was no difference between treatments ($P>0.05$). The seasons of the year influenced LD and k ($P<0.0001$). The average values of LD were 77.76, 78.98, 101.67 and 41.10 (g/m^2). To the k rate were 0.065, 0.11, 0.045 and 0.052 (g/g/day), to winter, autumn, spring and summer, respectively. High temperatures and rainfall at the summer could provide high forage production without senescent material to be deposited, which explains the low amount of LD. Higher LD values were concentrated in the spring due to the low rainfall and the high amount of senescent material. The highest k rate was obtained in the autumn, with values much higher than in other seasons, due to the fertilization carried out in the previous month. The supply of N increases the activity of the microorganisms and consequently, the values of k, recycling the N faster to the pastures. The deposition of the litter is essential for the protection, structuring and reduction of soil erosion and decomposition of the material represents an increase of nutrients in the ecosystems.

Session 30 Poster 14

Structural and qualitative characteristics in palisade grass in integrated crop-livestock systems

F.F. Simili[1], P.M. Bonacim[1], G.G. Mendonça[2], J.G. Augusto[1], L.S. Menegatto[1], R. Sechinatto[1], K.R.S. Barbosa[1] and C.C.P. Paz[1]

[1]Instituto de Zootecnia, APTA/SAA, Sertãozinho, SP, 14174000, Brazil, [2]FMVZ/USP, Pirassununga, SP, 13635900, Brazil; flaviasimili@gmail.com

The objective of this study was to evaluate the palisade grass establishment in integrated crop-livestock (ICL) and monoculture systems (single sowing) during the four seasons of the year. The characteristics studied were: plant height, leaf area index (LAI) and light interception (LI) and SPAD index. It was used a complete randomized block with three replications and five treatments: palisade grass (G), corn and palisade grass sown simultaneously (CG), corn and palisade grass sown simultaneously plus herbicide (CGH), palisade grass sown at topdressing corn (CGT) and palisade grass sown at the line and inter-line of corn (CGL). The pasture was submitted by beef cattle in continuous stocking management during the Aug 2016 until Dec 2017. The PROC MIXED by SAS was used to analyse the data. The results were significant in all the characteristics studied for treatments and seasons. The plant height was higher for treatment G (52 cm) compared to ICL systems. The results to IAF and LI were higher between the treatments G, GCT and GCL: 5.21 and 0.90, respectively. Higher SPAD index was obtained on ICL treatments (38.7), which can be explained by corn fertilization provides to these systems. The plant height was higher in the summer (61 cm) while in the winter the plant height did not reach more than 44 cm. High temperatures and rainfall in the summer could provide high forage. The LAI and LI was higher in the autumn 5.20 and 0.93, respectively, with values much higher than in other seasons, due to the fertilization carried out in the previous month. The supply of N increases the leaf yield production. Therefore, in relation to the structural characteristics, it is possible to conclude that the monoculture system presented a difference in relation to the integrated systems; which means that will be determinant for decision making between the systems. The pasture establishment on ICL systems is qualitatively better than the monocrop systems. Thus, the ICL systems can represent an alternative to palisade grass establishment, allowing animal and plant production and maximization of soil used.

Session 30 Poster 15

Cultural grazing of sheep as a means of increasing the species diversity of plant communities

I. Radkowska[1], A. Kawęcka[1], M. Szewczyk[2], M. Kulik[3] and A. Radkowski[4]

[1]National Research Institute of Animal Production, Krakowska 1, 32-083 Balice, Poland, [2]Jan Grodek State Vocational Academy in Sanok, Mickiewicza 21, 38-500 Sanok, Poland, [3]University of Life Sciences in Lublin, Akademicka 13, 20-618 Lublin, Poland, [4]University of Agriculture in Krakow, A.Mickiewicza 21, 33-332 Kraków, Poland; iwona.radkowska@izoo.krakow.pl

The aim of the study was to determine the effect of grazing farm animals on species diversity of plants in environmentally valuable areas. The study was performed in the Pieniny National Park (south Poland), where sheep were grazed. The first conservation measures to preserve the existing plant communities were undertaken at this site in 1993, when a soil expert's report was prepared and a phytosociological inventory was carried out. Vegetation was classified by Braun-Blanquet's method and the results were used to mark off valuable plant enclaves. Initially, the south part of the pasture was abundant with a Pieniny thermophilic meadow complex *Anthyllidi-Trifolietum montani*, as well as *Hieracio(vulgati)-Nardetum*. Later on, various meadow communities emerged which were dominated by *Dactylis glomerata* and pastures that were similar to species composition of meadows. There were also eutrophic mires and a tall-herb meadow. Improper practices and variable uses caused the number of species to decrease. The dominant species were *D. glomerata, Trisetum flavescens, Cynosurus cristatus, Poa pratensis, Phleum pratense* and, to a lesser extent, *Festuca pratensis, Festuca rubra, Poa trivialis* and *Agrostis capillaris.* The introduction of cultural grazing led to favourable changes that increased species diversity. The phytosociological inventory taken in 2017 showed the occurrence of 18 to 32 species of plants. The species with the highest coverage in the relevés are: *F. rubra, T. flavescens, P. pratensis, Alopecurus pratensis, Trifolium repens, D. glomerata, F. pratensis, C. cristatus, Ranunculus acris, Alchemilla* sp., *Trifolium pratense.* To sum up: the current, rather stable, economy will restore previous diversity. Grazing of sheep has a positive effect on the number of plant species and limits secondary succession, thus providing active protection of environmentally valuable areas. BIOSTRATEG2/297267/14/NCBR/2016.

Yield of selected types of grass communities in southern Poland

I. Radkowska[1], A. Kawęcka[1], M. Szewczyk[2], M. Kulik[3] and A. Radkowski[4]
[1]National Research Institute of Animal Production, Krakowska 1, 32-083, Poland, [2]Jan Grodek State Vocational Academy in Sanok, Mickiewicza 21, 38-500 Sanok, Poland, [3]University of Life Sciences in Lublin, Akademicka 13, 20-033 Lublin, Poland, [4]University of Agriculture in Krakow, A. Mickiewicza 21, 33-332 Kraków, Poland; iwona.radkowska@izoo.krakow.pl

Because the establishment and existence of grass communities in our climate originate from human activity, their stability and dynamics depend on natural habitat conditions and proper pastoral management practices. This results from the rate of movement of some constituent plant species as well as the susceptibility to different abnormalities under habitat conditions. Grass communities are a group of species with specific biological and ecological properties, which are highly effective in gaining space but also are of major economic and natural significance. For centuries they have also served cultural functions associated with livestock grazing. The aim of the study was to determine the yield of selected types of grass communities located in southern Poland, which are grazed by animals. Botanical composition of the sward was estimated using the Braun-Blanquet method. The yield of the communities was determined with the Różycki method. The highest yield was observed in pastures grazed by cattle: wet community dominated by *Agrostis capillaris* yielded 11.47 t/ha d.m., and community dominated by *Dactylis glomerata* and *Trifolium repens* yielded 10.13 t/ha d.m. Slightly lower yields were found in a horse-grazed pasture dominated by *Arrhenatherum elatius* and *Ranunculus acris* (9.60 t/ha) and in dry community with *Agrostis capillaris* (9.62 t/ha d.m.) grazed by cattle. The lowest yield was noted in a pasture grazed by sheep: *Festuca rubra-Cynosurus* community (6.88 t/ha) and communities of the *Anthyllidi-Trifolietum montani* type (5.72 t/ha d.m.). In terms of the yield, these results allow classifying the studied pastures as intermediate and good. Proper grazing system as well as good yields of the studied communities provide the animals with sufficient fodder during the grazing season. Project 'The uses and the conservation of farm animal genetic resources under sustainable development' BIOSTRATEG2/297267/14/NCBR/2016.

The impact of slaughter weight on the carbon footprint of the total feed intake by pigs

C. De Cuyper, A. Van Den Broeke, V. Van Linden, F. Leen, M. Aluwé, J. Van Meensel and S. Millet
Flanders research institute for agriculture, fisheries and food (ILVO), Scheldeweg 68, 9090 Melle, Belgium; frederik.leen@ilvo.vlaanderen.be

Pork production significantly contributes to the livestock sector's greenhouse gas emissions, with feed (including production, processing and transport) representing the bulk of emissions in pig supply chains. In this study, the feed intake and its impact on climate change was evaluated for pigs with different slaughter weights. In total, 384 animals (Danish type sow × Piétrain sire) were studied and housed with four pigs per pen. All animals were fed *ad libitum* on a three-phase feeding regime and slaughtered at different bodyweights. When only the fattening period was considered, the carbon footprint of the feed intake (CFP_{feed}) was expressed per kg carcass growth. When the production of piglets was also taken into account, the CFP_{feed} was calculated per kg carcass weight. A higher slaughter weight implied a higher feed conversion ratio and thus a higher feed intake per kg bodyweight. Since the CFP of the last phase diet was only slightly lower than that of the first and second phase, slaughter weight had a strong impact ($P<0.001$): the heavier the pig, the higher the CFP_{feed}/kg carcass growth. Per kg increase in slaughter weight, the CFP_{feed} increased by 4.4 g CO_2-eq/kg carcass growth. A higher slaughter weight also implies that fewer piglets have to be produced per year due to a reduction in number of fattening rounds per year. When the feed intake of the sow and piglet was considered additionally, slaughter weight still had a significant effect ($P<0.01$). Per kg increase in bodyweight, an increase of 2.5 g CO_2-eq/kg carcass weight was predicted. However, diet plays an important role as well. Because of a lower protein demand with increasing bodyweight, a hypothetical soybean free diet (assuming identical production results) could be formulated for the third phase. In this case, slaughter weight no longer had an effect on the CFP_{feed}/kg carcass growth or the CFP_{feed}/kg carcass weight. These results illustrate that a higher slaughter weight implies a lower sustainability if no low CFP diet is provided in the finisher phase.

Session 30 Poster 18

Co-grazing of sheep and goats preserves plant diversity and value of Apennine sub-alpine grasslands

R. Primi, B. Ronchi, L. Cancellieri and G. Filibeck
Università degli Studi della Tuscia, Dipartimento di Scienze Agrarie e Forestali (DAFNE), Via S.C. de Lellis, snc, 01100 Viterbo, Italy; ronchi@unitus.it

Grazing induces changes in plant species composition, through a wide variety of indirect and direct mechanisms that act at different levels. The effects of over- or under-grazing on pasture biodiversity, floristic composition and structure are well studied; conflicting and less detailed information is available on the effect of simultaneous grazing by several animal species. We conducted a study to assess patterns of primary productivity, sustainable stocking load and floristic composition in a 1,500-ha rangeland in the subalpine belt in Central Italy (Abruzzo, Lazio e Molise National Park). Within the study area, three grazing management condition can be identified, characterized by: sheep/goat co-grazing; cattle grazing; lack of domestic grazing. Sheep and goats co-grazing is the traditional kind of livestock system that has been practiced throughout centuries in the region, and derived primarily from complementary differences in feeding behaviour. We determined botanical composition, pasture forage value and available forage mass in each of the three case-studies, and sheep/goat co-grazing appeared to be the management regime that provided the highest temporal stability of grassland physiognomy and forage value, preventing both shrub encroachment and unpalatable herb invasions. The results obtained suggest that, at appropriate stocking-rates, co-grazing of sheep and goats can help preserve floristic composition and forage value of Southern European mountain grasslands because of the complementarity of the two herbivores, that allow to obtain an even level of biomass consumption across botanical species – thus reinforcing and optimally managing the well-known beneficial effects of grazing on biodiversity: the maintenance of high levels of light intensity in the community, the prevention of biomass accumulation and competition with other dominant species, both in the long-run and through temporary pulsations of dominance that decrease evenness.

Session 30 Poster 19

The net waterfootprint of cow and sheep milk

A.S. Atzori, M.F. Lunesu, F. Correddu, F. Lai, A. Nudda, A. Cannas and G. Pulina
University of Sassari, Department of Agriculture, V.le Italia 39, 07100 Sassari, Italy; asatzori@uniss.it

Water is classified among blue (man-managed freshwater); green (soil rainwater which evaporated); and grey (polluted). WaterFootPrint (WFP) measures the water used to produce a unit of goods or services. WFP determination refers to the water footprint network based on green and blue water (WFPN) and LCA approaches based on blue water. WFPN estimates of cow milk were 1000 l of freshwater. LCA estimates of WFP of US dairy production ranged from 517 to 0.9 L of blue water/kg of milk. Accounting for green and blue water, we propose the alternative method of the Net Waterfootprint (WFPnet) which estimates green water as differential evapotranspiration (ΔET) between the total ET of a crop or pasture used for animal feeding (local ET values) and the ET of a reference natural cover (e.g. natural grassland/shrubland; 3,200 m^3/ha of ET) from the same area, representing natural substitutes to the land use change. Consumed blue water is from irrigation (local ET values) and for animal drinking and servicing. The WFPnet of sheep and cow milk was simulated for two intensive (INT) and extensive (EXT) farm scenarios (250 vs 150 and 9,000 vs 5,500 lt/year of sheep and cow milk per head; respectively). INT vs EXT showed high feed efficiency, higher consumption of irrigated concentrate and forages. Three scenarios of increasing water use efficiency (WUE; 1.4, 3.0 and 8.0 kg DM/m^3) for corn irrigation were also considered. WFPnet of sheep milk for decreasing values of (WUE) was equal to 218, 421, 796 L of water/kg in INT, and to -135, 10, 280 L of water/kg in EXT, respectively. WFPnet of cow milk was equal to 166, 244, 387 L of water/kg in INT and to 127, 261, 510 L of water/kg in EXT, respectively. WFPnet showed: (1) increasing WFP for increasing use of blue water; (2) decreasing WFP for increasing animal efficiency; (3) lower WFP for EXT scenarios with low DM conversion efficiency but less use of blue water than INT; (4) negative values for scenarios where the irrigated crops had higher WUE than grasslands and pastures had lower WUE than natural covers, allowing leaching and erosion. WFPnet is under calibration in the actions of Forage4Climate and SheepToShip life projects of EU LIFE+15 program on Climate Change Mitigation.

Session 31 Theatre 1

Leveraging on high-throughput phenotyping technologies to optimize livestock genetic improvement

G.J.M. Rosa[1,2], J.R.R. Dorea[2,3], A.F.A. Fernandes[2], V.C. Ferreira[2] and T.L. Passafaro[2]

[1]University of Wisconsin-Madison, Department of Biostatistics & Medical Informatics, 600 Highland Ave, 53792 Madison, WI, USA, [2]University of Wisconsin-Madison, Department of Animal Sciences, 1675 Observatory Dr., 53706 Madison, WI, USA, [3]University of Wisconsin-Madison, Department of Dairy Science, 1675 Observatory Dr, 53706 Madison, WI, USA; grosa@wisc.edu

The advent of fully automated data recording technologies and high-throughput phenotyping (HTP) systems has opened up a myriad of opportunities to advance breeding programs and livestock husbandry. Such technologies allow scoring large numbers of animals for novel phenotypes and indicator traits to boost genetic improvement, as well as real-time monitoring of animal behaviour and development for optimized management decisions. HTP tools include, for example, image analysis and computer vision, sensor technology for motion, sound and chemical composition, and spectroscopy. Applications span from health surveillance, precision nutrition, and control of meat and milk composition and quality. However, the application of HTP requires sophisticated statistical and computational approaches for efficient data management and appropriate data mining, as it involves large datasets with many covariates and complex relationships. In this talk we will discuss some of the challenges and potentials of HTP in livestock. Some examples to be presented include the utilization of automated feeders to record feed intake and to monitor feeding behaviour in broilers, milk-spectra information to predict dairy cattle feed intake, and image analysis and computer vision to monitor growth and body condition in pigs and cattle. HTP and big data will become an essential component of modern livestock operations in the context of precision animal agriculture, boosting animal welfare, environmental footprint, and overall sustainability of animal production.

Session 31 Theatre 2

Unravelling genetic architecture of complex traits using multi-omics approaches

P. Ajmone Marsan[1,2]

[1]Università Cattolica del Sacro Cuore, Proteomic and Nutrigenomic Research Center – PRONUTRIGEN, via Emilia Parmense 84, Piacenza, 25122, Italy, [2]Università Cattolica del Sacro Cuore, Department of Animal Science Food and Nutrition, via Emilia Parmense 84, Piacenza, 25122, Italy; paolo.ajmone@unicatt.it

In livestock, in cosmopolitan dairy cattle in particular, the use of DNA information for genomic selection is well established in all countries having an industrialized agriculture. Conversely, the understanding of biological mechanisms and the identification of genes influencing complex traits is still problematic. Even in humans, where many thousand individuals have whole genome sequences and hundred thousand dense SNP genotypes, Quantitative Trait Loci (QTL) variants identified account for only a limited proportion of complex trait variability. Strategies are being explored to overcome this limitation. These include the investigation of 'intermediate phenotypes' components of complex traits and considered simpler and closer to the biology of animals, and the integration of genomic data with other biological data that connect the genome to the phenotype (transcriptomics, proteomics, metabolomics) and to the environment (epigenomics). Also, gut microbiota is now considered an additional organ playing a role in animal health, feed efficiency and environmental impact, particularly in ruminants, in the production of Green House Gas (GHG). The integration of multi-omic data is still in its infancy, but perspectives are promising. The improvement of omic technologies in terms of cost and reliability is one of the main challenges, as their level of maturity still lags behind that of genomic methods. Others are the annotation of livestock genomes with sites that have important regulatory functions and the development of models and pipelines able to analyse, integrate, visualize and finally facilitate the interpretation of multi-omic data.

Session 31 Theatre 3

Machine learning transcriptome analysis to identify genes associated with feed efficiency in pigs

M. Piles[1], C. Fernandez-Lozano[2], M. Velasco[1], O. González[1], J.P. Sánchez[1], R. Quintanilla[1] and M. Ballester[1]
[1]Institute of Agriculture and Food Research and Technology (IRTA), Animal Breeding and Genetics, Torre Marimon s/n, 08140 Caldes de Montbui, Spain, [2]University of A Coruña, Department of Computation, Faculty of Computer Science, 15071 A Coruña, Spain; miriam.piles@irta.es

The aim was to identify genes associated with feed efficiency using transcriptomic (RNA-Seq) data from pigs phenotypically extreme for residual feed intake (RFI), which was computed considering within-sex regression on mean metabolic body weight, average daily gain, and average backfat gain. The RNA-seq analysis was performed on liver and duodenum tissues from 32 and 33 high and low RFI pigs, respectively. Genes were ranked using permutation accuracy importance score in an unbiased Random Forest algorithm based on conditional inference. Support Vector Machine, Random Forest, Elasticnet (EN) and Nearest Shrunken Centroid algorithms were tested in different subsets of the most informative genes. All the experiments were carried out using a robust experimental design (nested resampling with cross-validation for hyperparameter tuning). AUROC was the performance criterion. ML techniques were tested using as predictors those sets of RNA-seq data adjusted (liver) or not (duodenum) by the batch effect and including the first principal component. The ML algorithm that led to the best performance with all adjusted datasets was EN. In liver, the best classification was obtained using 200 genes. AUROC and accuracy in the test set were 0.85 and 0.78. In duodenum, the best performance was also obtained with EN using 100 genes (AUROC: 0.76; accuracy: 0.69). Both lists of genes were submitted to IPA for functional characterization, identifying canonical pathways and candidate genes previously reported as associated to feed efficiency in several species. It is worth mentioning the NRF2-mediated oxidative stress response and aldosterone signalling in epithelial cells with the DNAJC6, DNAJC1, MAPK8, PRKD3 genes in duodenum, and melatonin degradation II, PPARα/RXRα activation, and GPCR-mediated nutrient sensing in enteroendocrine cells with genes such as SMOX, IL4I1, PRKAR2B, CLOCK and CCK in liver.

Session 31 Theatre 4

Metabolomic phenotypic prediction of growth in pigs

P. Sarup[1], J. Jensen[1], T. Ostersen[2], O.F. Christensen[1] and P. Sørensen[1]
[1]Center for Quantitative Genetics and Genomics, Molecular Biology and Genetics, Blichers Allé 20, 8830 Tjele, Denmark, [2]SEGES, Axeltorv 3, 1609 Copenhagen V, Denmark; pernille.sarup@mbg.au.dk

To provide accurate estimation of breeding values a large number of high quality phenotypic measurements is needed. However, some phenotypes such as feed intake in pigs, are expensive or difficult to measure. Thus, a large part of animals in breeding herds have breeding values based on correlated phenotypes, such as average daily gain and additive genetic relationships with tested animals. An alternative option is to use large scale molecular phenotypes, e.g. proton nuclear magnetic resonance (H^1 NMR metabolomics) for prediction of relevant phenotypes in large cohorts. NMR spectra are affected by both the genotype and the environment, but only the genetic component is useful for prediction of breeding values. To assess the scope for using phenotypes predicted by NMR spectra for breeding purposes, we investigated phenotypic and genetic correlations between total feed intake (TFI), average daily gain (ADG), and back fat (BF) measured directly (y) and predicted by NMR spectra (y_m). We analysed 60 K SNP genotypes, finisher phenotypes and H^1 NMR spectra of blood serum samples taken at end of test from 2,314 3-way crossed pigs (Duroc × (Yorkshire × Landrace)). We predicted y_m using a standard chemometric approach, partial least squares (PLS), or a best linear unbiased prediction method building a metabolite relationship matrix akin to the standard genomic relationship matrix (MBLUP). We used half-sib cross validation sets masking the phenotypes of all offspring from one sire at a time. For all tree phenotypes the MBLUP method had higher average correlation between y and y_m (TFI=0.682, ADG=0.589, BF=0.588) than PLS (TFI=0.496, ADG=0.449, BF=0.484). The genetic correlations between y and y_m predicted by MBLUP estimated by a bivariate model using standard genomic relationship matrix were 0.759, 0.727 and 0.658 for TFI, ADG, and BF respectively. For y_m predicted by PLS the corresponding genetic correlations to directly measured y were 0.729, 0.662, 0.648. We conclude that data from phenotypic prediction by H^1 NMR spectra can be very useful in pig breeding for finisher traits that are difficult to measure.

Session 31 Theatre 5

Integrating blood transcriptome and immunity traits to identify markers of immune capacity in pigs

T. Maroilley[1], F. Blanc[1], G. Lemonnier[1], J. Lecardonnel[1], J.P. Bidanel[1], A. Rau[1], Y. Billon[2], M.J. Mercat[3], J. Estellé[1] and C. Rogel-Gaillard[1]
[1]GABI, INRA, AgroParisTech, Université Paris-Saclay, Domaine de Vilvert, 78350 Jouy-en-Josas, France, [2]GENESI, INRA, Le Magneraud, 17700 Surgères, France, [3]IFIP – Alliance R&D, La Motte au Vicomte BP 35651, 35650 Le Rheu, France; claire.rogel-gaillard@inra.fr

Understanding individual variability of immune capacity in livestock has become a priority to improve sustainability, with the aim to increase disease resistance and resilience in breeding programs. In this study, 550 60-day-old French Large White pigs vaccinated against *Mycoplasma hyopneumoniae* (*M hyo*) were monitored for 55 immunity traits (ITs) measured from blood samples (SUS_FLORA ANR funded project). The ITs included two types of parameters. First, parameters directly measured from blood: complete blood counts, counts of various cell subsets by flow cytometry, serum dosage of anti-*M hyo* IgG and haptoblobin. Second, parameters measured after *in vitro* stimulation of total blood: phagocytosis, production of cytokines (IL-1β, IL-8, IL-10, IL-17, TNFα, IFNγ) after stimulation by LPS or mitogenic agents. All animals were genotyped with 60K Illumina SNP chips. A subset of 243 piglets was chosen for blood transcriptome analysis using Agilent microarrays. We explored covariations between blood expression profiles and IT levels, and could draw lists of the most correlated genes with each IT. Each list represented candidate blood biomarkers potentially predictive of IT variations. As an example, we found 134 genes associated with phagocytosis capacity and we identified a subset of genes that could significantly predict levels of eight ITs related to phagocytosis by a sPLS approach. This gene subset included *CXCR1*, *CCR1* and *TLR2*. Few candidate biomarkers were previously shown to be genetically controlled for their transcription in blood by eGWAS. Thus, our results provide new data to decipher the genetic architecture of IT variations. A next step will be to understand how IT variations could reflect individual robustness while facing pathogens, and how blood biomarkers could be used as predictors of immune capacity.

Session 31 Theatre 6

Hunting for intermediate blood phenotype associated markers in pigs using phenomics and genomics

S. Bovo[1], F. Bertolini[2,3], G. Mazzoni[3], G. Schiavo[1], G. Galimberti[4], M. Gallo[5], S. Dall'Olio[1] and L. Fontanesi[1]
[1]University of Bologna, Department of Agricultural and Food Sciences, Viale Fanin 46, 40127, Bologna, Italy, [2]Iowa State University, Department of Animal Science, 2255 Kildee Hall, 50011, Ames, Iowa, USA, [3]Technical University of Denmark, Department of Bio and Health Informatics, Kemitorvet, Building 208, 2800, Kgs. Lyngby, Denmark, [4]University of Bologna, Department of Statistical Sciences 'Paolo Fortunati', Via delle Belle Arti 4, 40126, Bologna, Italy, [5]Associazione Nazionale Allevatori Suini (ANAS), Via Nizza 53, 00198 Roma, Italy; samuele.bovo2@unibo.it

Basic physiological and biochemical measures or other intermediate animal phenotypes might be related to economic traits. These traits are closer to basic biological functions and for this reason they could be useful to dissect complex production traits. Haematological and blood clinical-chemical parameters reflect the physiological and health status of the animals and are used as biomarkers to describe pathological or sub-pathological conditions. They are considered as indicators of immune functions and components of the adaptive immune system in both humans and livestock. In this study a phenomic approach included high throughput analysis of 30 blood-derived traits in 843 performance tested Italian Large White pigs. These intermediate phenotypes were analysed using Gaussian Graphical Models (GGM) and phenotype networks were constructed using Cytoscape. All pigs were genotyped with the Illumina PorcineSNP60 BeadChip and genome wide association studies were carried out with a single marker approach (GEMMA) and Bayesian analyses (GenSel) with 1 Mbp genomic windows. A total of 50 QTLs were identified some of which were confirmed with the two approaches. Whole genome resequencing data of individual Italian Large White pigs and DNA pools from the same breed and other breeds identified putative causative mutations for some of these QTLs. By dissecting genetic factors affecting intermediate phenotype (i.e. blood parameters) variability, this study opened new opportunities to use proxies of complex production traits in pig breeding.

Session 31 Theatre 7

Discover potential regulatory mechanisms involved in rumen functional changes under high grain diet

L.L. Guan[1], K. Zhao[1,2], Y.H. Chen[1] and G.B. Penner[3]

[1]University of Alberta, Agricultural, Food and Nutritonal Science, T6G2P5, Canada, [2]Shaanxi Normal University, College of Food Engineering and Nutritional Science, 710119, China, P.R., [3]University of Saskatchewan, Department of Animal and Poultry Science, S7N5A8, Canada; lguan@ualberta.ca

Rumen health is essential for achieving the high productivity. It has become a common practice in the dairy and beef industry to use intensive feeding strategies such as high energy and high concentrate diets to maintain and/or enhance the production. However, not all the cattle will develop the sub-acute ruminal acidosis (SARA) or acute ruminal acidosis (RA), the metabolic dysfunctions, when they were fed with high concentrate diet or during the rapid high fermentable dietary transition. To date, the host mechanisms that regulate such individualized responses are largely unknown. In this study, we aimed to identify the host mechanisms through studying: (1) whether rumen microbiome was associated with acidosis-resistance; and (2) whether the difference in expression of genes in the rumen epithelial wall drives the varied adaptation to high grain diet. Two studies were performed at the University of Alberta with 17 cannulated steers (~1 year) in Study 1 and 24 Angus-Hereford cross-bred yearling heifers (~8 months) in Study 2. In Study 1, the microbial profiles were compared between steers with the highest (n=3) and lowest (n=3) acidosis indices using amplicon sequencing and quantitative PCR (qPCR). In Study 2, RNA-seq and the microbial profiles of the rumen contents and tissue collected from cattle under the rapid grain (from 0 to 89% grain over 26 days). Based on these analysis, genes, upstream regulators, SNP as well as microbial markers were identified to be associated with SARA as well as rapid dietary adaptation. These could be potential genetic/microbial markers that may account for the varied individual ruminal pH responses to the dietary transition stress. The application of advanced omics approach has led that both rumen microbes and host gene expression can adapt to the high fermentable dietary transition with varied mechanisms among individuals. This could help industry to develop more selective management to improve the animal productivity through maintaining and/or enhancing the rumen health.

Session 31 Theatre 8

Genome-wide study of structural variants in French dairy and beef breeds

M. Boussaha[1], C. Grohs[1], R. Letaief[1], S. Fritz[1,2], J. Barbieri[3], C. Klopp[4], R. Philippe[5], D. Rocha[1], A. Capitan[1,2] and D. Boichard[1]

[1]GABI, INRA, AgroParisTech, Université Paris-Saclay, Domaine de Vilvert, 78350 Jouy-en-Josas, France, [2]Allice, Maison Nationale des Eleveurs, 75012 Paris, France, [3]GenPhySE, INRA, Université de Toulouse INPT ENSAT, Université de Toulouse INPT ENVT, 52627 Castanet-Tolosan, France, [4]SIGENAE, INRA, SIGENAE, 52627 Castanet- Tolosan, France, [5]GMA, INRA, Université de Limoges, 87060 Limoges Cedex, France; mekki.boussaha@inra.fr

Deep whole genome sequencing coupled with the development of several computational approaches provide opportunities to investigate chromosomal alterations between individual samples across the genome. These structural variants (SVs) affect DNA segments greater than 50 base pairs and correspond to deletions, inversions and tandem duplications and translocations. Several studies revealed involvement of structural variants in phenotypic changes in many species including cattle. In the present study, we performed genome-wide study of SVs using whole-genome sequence data from 360 bulls corresponding to 20 dairy and beef breeds. Bioinformatics detection of potential SVs was performed using Pindel, Delly and BreakDancer. Predicted SVs were filtered using different strategies in order to minimize false positive results. Filtered SVs were subsequently merged in order to define potential SV regions. A panel of SV regions were genotyped using the bovine LD beadchip. Genotyping data were used for validation studies. Furthermore, a genome-wide association study (GWAS) was performed on several thousands of dairy animals in order to assess the impact of validated SVs on routinely measured traits in dairy cattle.

Multi-omics data integration approach for resilience of dairy cattle to heat stress

H. Hammami[1], F.G. Colinet[1], C. Bastin[2], S. Vanderick[1], A. Mineur[1], S. Naderi[1], R.R. Mota[1] and N. Gengler[1]
[1]Gembloux Agro-Bio Tech, University of Liège, TERRA Teaching and Research Centre, Gembloux, 5030, Belgium, [2]Walloon Breding Association, R&D, Ciney, 5590, Belgium; hedi.hammami@uliege.be

Breeding for resilience to heat stress (HS) is a topic where associating multiple omics data has the potential to get a better view of the issues and to allow significant advances to overcome undesirable consequences of future extreme weather scenarios. An example of omics is here epigenomics (e.g. early programming due to heat-stress) allowing new insights to explain biological mechanisms of resilience to HS and G×E interactions. Even if biological mechanisms are complex and still elusive, this study tried to use a holistic approach integrating milk-based biomarkers, climate conditions, and genomics. Data used included 65,907 third-lactation test-day records for production traits (milk, fat and protein yields), specific fatty acids (FA) and metabolites predicted from mid-infrared spectra (C4:0, C18:1cis9, long chain 'LCFA', mono- and unsaturated FA 'MUFA and UFA', acetone and BHB) of 9,327 Holstein cows. Phenotypes were merged with a temperature humidity index (THI) from public weather stations. For each trait, the response to THI was estimated via days in milk (DIM) × THI combination, and for each cow by using a random regression model with a common threshold of THI=62. The slope (heat tolerance)-to-intercept (general) genetic variance ratios increased as THI increased. They were higher during mid-lactation (140-245 DIM) for C18:1 cis9, acetone, BHB and for production traits, whereas higher in early lactation (≤125 DIM) for C4:0, LCFA, MUFA, and UFA. At extreme high THI scale, slope-to-intercept ratios for C18:1 cis9, MUFA, UFA, and LCFA were 3.8, 3.4, 3.1, 2.8 fold higher than milk yield. These findings indicate that tolerance to HS and traditional production trait responses to THI are marginally related, and changes in milk-based biomarkers under high THI better elucidate physiological and metabolic pathways in HS dairy cows. Ongoing genomic wide association studies will better explain genetic markers unravelling the biological background of resilience to HS.

Tackling methane emission through a (pheno, geno and metageno)-omics approach

O. Gonzalez-Recio[1,2], I. Goiri[3], R. Atxaerandio[3], E. Ugarte[3], R. Ruiz[3], R. Alenda[1], J.A. Jiménez-Montero[4] and A. García-Rodríguez[3]
[1]Escuela Técnica Superior de Ingeniería Agronómica, Alimentaria y de Biosistemas. UPM, Producciones Agrarias, Ciudad Universitaria s/n, 28040 Madrid, Spain, [2]INIA, Mejora Genética Animal, Ctra La Coruña km 7.5, 28040 Madrid, Spain, [3]NEIKER-Tecnalia, Producción Animal, Granja Modelo de Arkaute. Apartado 46, 01080 Vitoria-Gasteiz, Spain, [4]CONAFE, Dpto Técnico, Ctra. de Andalucía km 23600, 28340 Valdemoro, Spain; gonzalez.oscar@inia.es

Methane emissions have been tackled in dairy cattle during the last years from a multidisciplinary approach, with more emphasis from nutrition and genetics. Reducing methane emissions is important both from an environmental and economic perspective, as it is a sink of reduced equivalents in the rumen. Sensors are becoming common tools to monitor CO_2 and CH_4 production, yielding large volumes of data. Our project aims a multi-omics and multidisciplinary approach to mitigate methane emissions and improve feed efficiency using sensor, genomic and metagenome information. Average methane ppm was 212 with a coefficient of variation of 0.71. Methane production heritability ranged between 0.18 and 0.30, allocating possibilities to genetically improve this trait. We have estimated that the emphasis on reduced methane production must be between 2 and 8% in a global merit index. We also highlight the importance of the bioinformatics analyses of the microbiota composition based on 16S rRNA gene sequences. The analysed tools (QIIME and MOTHUR) produced comparable results at assigning gene reads to the most abundant genera in rumen samples, important differences were found for less common microorganisms. SILVA seemed a preferred reference dataset for classifying OTUs from bovine rumen microbiota. We found the genetic background of the host animal associated to the relative abundance in 50% of the bacteria and archaea, and in 43% of ciliate analysed. Further, the results showed a significant association between DGAT1, ACSF3, AGPAT3 and STC2 genes with the relative abundance Prevotella genus with a false discovery rate lower than 15%. The analyses of the microbiota can discriminate the most efficient animals. The links between the host genome, methane production and microbiota composition need to be further disentangled.

Genome-wide association and biological pathway analysis of cheese volatilome in dairy cattle

S. Pegolo[1], M. Bergamaschi[1], F. Biasioli[2], F. Gasperi[2], G. Bittante[1] and A. Cecchinato[1]

[1]University of Padova, Department of Agronomy, Food Natural resources, Animals and Environment (DAFNAE), Viale dell'Universita' 16, 35020 Legnaro (PD), Italy, [2]Fondazione Edmund Mach (FEM), Department of Food Quality and Nutrition, Research and Innovation Centre, Via E. Mach 1, 38010 San Michele all'Adige (TN), Italy; sara.pegolo@unipd.it

Cheese quality represents an important driver of consumers' preferences. Volatile organic compounds (VOCs) contribute to the determination of cheese flavour which plays a major role among cheese quality traits. The aim of this study was to exploit an integrated phenomic-genomic approach based on the use of direct-injection mass spectrometry (PTR-ToF-MS), genome-wide association studies (GWAS) and pathway analysis to disentangle the genomic background and the biological functions controlling the characteristics of cheese volatilome. We performed GWAS for 173 spectrometric peaks obtained by using the PTR-ToF-MS analysis of 1,075 model cheeses produced using raw whole-milk samples collected from individual Brown Swiss cows. Animals were genotyped with the Illumina BovineSNP50 Bead Chip v.2. A single marker regression model was applied for GWAS using the GenABEL R package and the GRAMMAR-GC approach. A gene-set enrichment analysis was carried out on GWAS results for 45 identified VOCs, using the Gene Ontology (GO) and Kyoto Encyclopaedia of Genes and Genomes (KEGG) pathway databases, to reveal ontologies/pathways associated with the cheese volatilome. In total, 170 SNPs were significant ($P<5\times10^{-5}$) for 113 traits spanning all *Bos taurus* autosomes (BTAs). The largest windows of consecutive SNPs were located on BTA6 (~81.65-88.07 Mbp) and BTA21 (~40.72-45.33 Mbp) covering the casein cluster and QTL for milk fat yield and percentage, respectively. Gene-set enrichment analyses showed overrepresentation of pathways related to purine and nitrogen metabolism, tight junction and long-term potentiation (Fisher's exact test, false discovery rate <0.05). These results represent a first assessment for a potential link between cow's genes and cheese flavor providing new insight about potential candidate loci and biological functions responsible for the variation of cheese volatilome.

Single nucleotide polymorphisms underlying mastitis predisposition in Polish-Holstein Friesian cows

J. Szyda[1,2] and M. Mielczarek[1,2]

[1]National Research Institute of Animal Production, Krakowska 1, 32-083 Balice, Poland, [2]Wroclaw University of Environemntal and Life Sciences, Department of Genetics, Kozuchowska 7, 51-631 Wroclaw, Poland; joanna.szyda@upwr.edu.pl

Mastitis is an inflammatory disease of the mammary gland, which is one of the most important diseases of the dairy sector, mainly due to high economic importance and increased awareness of animal welfare. The purpose of this study was to characterize links between single nucleotide polymorphisms and the incidence of clinical mastitis. Using information from whole genome DNA sequences of 16 Holstein-Friesian cow half-sib pairs we searched for genome regions differing between a healthy (no incidence of clinical mastitis) half-sib and a mastitis-prone (multiple incidences of clinical mastitis) half-sib. Raw reads were aligned to the UMD3.1 reference genome using BWA-MEM and processed using Picard and SAMtools. After data editing, three cows with average depth of coverage below ten were excluded, what resulted in 13 half-sib pairs available for comparisons. In order to better explore the familial structure of the data, each SNP comparison was performed between two half-sibs without consideration of estimated SNP genotypes, but comparing the number of reads containing a reference and an alternative allele at the SNP position. For this purpose a somatic mutation caller Varscan2, implementing the Fisher's exact test, was applied. As a result 676,168 SNPs were significant across the compared half-sib pairs, but no SNP was common among all 13 of them. 27 SNPs were however common for 6 half-sib pairs. Their genomic annotation revealed that three of them were located within genes, albeit in gene introns. On BTA04 a SNP was located in the intron of the Mitochondrial inner membrane protease subunit 2, on BTA27 a SNP was within the intron of the ankyrin repeat and SOCS box containing 5 gene, and on the BTX a SNP was within an intron of the nuclear RNA export factor 3. The latter appeared as the most promising candidate gene since it is contained in the immune response related KEGG pathways – Influenza A and Herpes simplex infection. The project was supported by the FP7-228394 grant, by the Polish National Science Centre grant 2014/13/B/NZ9/02016, and by The Leading National Research Centre programme for 2014-2018.

Session 31 Poster 13

Analysis of porcine IGF2 gene expression in adipose tissue and its effect on fatty acid composition

L. Criado-Mesas[1,2], *D. Crespo-Piazuelo*[1,2], *A. Castelló*[1,2], *A.I. Fernández*[3], *M. Ballester*[4] *and J.M. Folch*[1,2]
[1]*Universitat Autònoma de Barcelona (UAB), Departament de Ciència Animal i dels Aliments, Facultat de Veterinària, Bellaterra, 08193, Spain,* [2]*Centre de Recerca en Agrigenòmica (CRAG), Plant and Animal Genomics, Consorci CSIC-IRTA-UAB-UB, Campus UAB, Bellaterra, 08193, Spain,* [3]*Instituto Nacional de Investigación y Tecnología Agraria y Alimentaria (INIA), Departamento de Genética Animal, Madrid, 28040, Spain,* [4]*IRTA, Genètica i Millora Animal, Torre Marimon, Caldes de Montbui, 08140, Spain; lourdes.criado@cragenomica.es*

The polymorphism g.3072G>A located in the intron 3 of the Insulin-Like Growth Factor 2 (*IGF2*) gene was described as the causal mutation for a paternally expressed Quantitative Trait Locus for muscle growth and backfat thickness. The objective of this work was to study the association between *IGF2*:g.3072G>A polymorphism and both *IGF2* gene expression and fatty acid composition measured in adipose tissue to comprehend the regulation mechanism of this gene. An expression GWAS with 354 animals from three different experimental backcrosses and 38,639 SNPs revealed that the *IGF2*:g.3072G>A polymorphism was the most significantly associated SNP for *IGF2* mRNA expression in adipose tissue. In addition, *IGF2*:g.3072G>A polymorphism was significantly associated with palmitic, heptadecanoic, hexadecanoic, oleic, linoleic, α-linoleic and arachidonic fatty acids measured in adipose tissue. Finally, the previously described imprinting model in muscle was tested in adipose tissue for the *IGF2*:g.3072G>A genotype, showing than animals carrying AA genotype have the highest *IGF2* gene expression level, whereas homozygous GG and heterozygous animals, evenly which allele is paternally inherited, presented a lower and similar gene expression levels. Hence, the mechanism proposed for *IGF2* gene expression in adipose tissue is a dominant effect of the G allele. As a conclusion, *IGF2*:g.3072G>A polymorphism may play a key role in the regulation mechanism of *IGF2* gene expression and in the determination of fatty acid composition in adipose tissue.

Session 31 Poster 14

Genome-wide association study on ribeye area measured by ultrasonography in Nelore cattle

R. Silva[1], *F. Carvalho*[1], *J.B. Ferraz*[1], *J. Eler*[1], *L. Grigoletto*[1], *R. Lôbo*[2], *L. Bezerra*[2], *M. Berton*[3], *F. Lopes*[3] *and F. Baldi*[3]
[1]*University of São Paulo, Veterinary Medicine, Rua Duque de Caxias Norte, 225, 13635-900, Brazil,* [2]*Nacional Association of Breeders and Researchers (ANCP), Rua João Godoy, 463, 14020-230, Brazil,* [3]*University of São Paulo State, Departament of Animal Science, Via de Acesso Prof. Paulo Donato Castellane s/n, 14884-900, Brazil; jbferraz@usp.br*

This study was carried out to estimate genetic parameters and to identify genomic regions associated with ribeye area (REA) in the Longissimus dorsi muscle ultrasonically measured at the 12th and 13th ribs. Data from 60,325 animals participating in the Nelore Brazil breeding program of National Association of Breeders and Researchers (ANCP) were used. A total of 8,652 animals were genotyped using the Bovine HD Beadchip 777k. The analyses were performed using the BLUPF90 family of programs. The SNP effects were estimated using a single-step procedure (ssGWAS). The model included the fixed effect of the contemporary group and animal age at evaluation (linear covariable); the additive genetic effect and residual effects were considered as random. The GWAS results were reported based on the proportion of additive genetic variance explained by windows of 10 adjacent SNPs. The identification and positioning of the SNPs in the bovine genome were performed using the NCBI, Ensembl and DAVID database. The heritability was moderate for REA (0.32±0.02). The genomic regions located on chromosomes 9, 10, 11, 12, 13, 14, 26 and 28 explained 11.9% of the additive genetic variance. Surveying candidate genes in those regions associated with REA revealed genes such as GPAT2, PLTP and TNNC2. The GPAT2 gene catalysed the biosynthesis of triglycerides and glycerophospholipds. The PLTP gene is related to fat metabolism and intramuscular fat. The TNNC2 gene is expressed in fast skeletal muscle affecting meat quality traits. The identification of these regions would contribute to a better understanding and evaluation of ribeye area muscle in Nelore cattle. Acknowledgments: São Paulo Research Foundation (FAPESP) grant #2017/03221-9.

Genome-wide association study of post-weaning growth in rabbits

M. Ballester, M. Velasco, M. Piles, O. Rafel, O. González and J.P. Sánchez
Institute for Research and Technology in Food and Agriculture (IRTA), Animal Breeding and Genetics Program, Torre Marimon, 08140, Caldes de Montbui, Spain; maria.ballester@irta.cat

The aim of this work was to identify chromosomal regions and candidate genes associated with post-weaning growth in rabbits using a genome-wide association study. A total of 436 weaned rabbits from the Caldes line were distributed in two feeding regimes, 230 were fed *ad libitum* and 206 were restricted to 75% of *ad libitum* intake. Animal weight was recorded from 28 d to 56 d of age. Genotyping was performed using the Axiom rabbit array (Affymetrix) and after standard quality control a subset of 114,604 annotated SNPs were retained for subsequent analysis. The maximum likelihood approach implemented in Qxpak 5.0 was used to test one snp at a time, separately for each trait; ADG on restricted (ADG_r) or full feeding (ADG_f). The model accounted for additive (polygenic) genetic effect, litter and cage effects, as well as the systematic effect of batch (5 levels). The R package *q*-value was used to adjust raw P-values to a False Discovery Rate of 0.05. While no significant associations were found at whole-genome level for any trait; a total of 55 significant trait-associated SNPs (TAS) distributed in *Oryctolagus cuniculus* Chromosomes (OCC) 3, 5 and 21 were declared to be associated with ADG_f at chromosome-wise level. No associations were found for ADG_r at this level. In order to identify positional candidate genes related to ADG_f, four genomic QTL intervals (OCC3:101-114 Mb, OCC5:8-10 Mb, OCC5:18-20 Mb and OCC21:6-9 Mb) were annotated using Biomart. Out of the 55 TAS, 27 (49%) were mapped in protein-coding genes. Remarkably, several genes related to growth such as *ATXN2*, *ACAD10*, *TRAFD1*, *PTPN11*, *NDUFAF6* and *FTO* were mapped in these QTL regions. Polymorphism in the rabbit *FTO* gene has previously been reported to be associated with growth and meat quality traits. Further analysis will be performed to determine the role of these candidate genes for rabbit growth.

Genome-wide association study for birth weight in Nelore cattle using single step GBLUP procedure

F. Carvalho[1], R. Silva[1], L. Grigoletto[1], J. Eler[1], J.B. Ferraz[1], F. Mendonça[2], R. Lôbo[3], R. Medeiros[4] and F. Baldi[5]
[1]University of São Paulo, R. Duque de Caxias Norte, 13635-900, Brazil, [2]Goiás State University, R. da Saudade, 56, 76100-000, Brazil, [3]Association of Breeders and Researchers, R.João Godoy,463, 14020-230, Brazil, [4]University of Georgia, 425 River Road, 30602, USA, [5]São Paulo State University, Acesso Prof. Paulo Donato Castelane, 14884-900, Brazil; jbferraz@usp.br

The aim of the present study was to estimate genetic parameters and identify genomic regions associated with birth weight (BW) to better understand the genetic and physiological mechanisms regulating in this trait in Nelore cattle. A total of 49,475 records for BW and 192,483 animals in the pedigreed provided by the National Association of Breeders and Researchers (ANCP) were used. A total of 3,893 animals were genotyped with the *777k BovineHD BeadChip (Illumina®)*. Analysis were performed using the single step genomic BLUP method. The model included the fixed effects of contemporary group (herd, birth season, sex and management group) and dam age at calving as covariable, and the random direct genetic additive, maternal genetic additive and maternal permanent environment effects. The chromosomic segments, based in 10 consecutive SNP's, that explained 1% or more from total direct or maternal additive genetic variance were selected as candidate regions and were considered for gene prospection analysis using DAVID and GeneCards databases. The direct and maternal heritability estimates for BW were 0.23±0.04 and 0.05±0.04, respectively. For the direct genetic additive effect, associations were detected in BTA1;3;6;7;14;16 and 21 and for maternal genetic additive associations were detected in BTA 3;4;6;10;12;13;14;16;17;21;22 and 26. The most relevant candidate genes found in the reported regions for the direct genetic additive effect were *RYK* (BTA1) and *H2S2* (BTA14), which were associated with growth and muscular growth development, respectively. For the maternal genetic effect, candidate genes like *CNOT4* (BTA4) and *TASP1* (BTA13), associated with β-lactoglobuline production and foetal growth were identified. The results obtained would help to unravel the genetic architecture for the direct genetic additive and maternal genetic additive effect for BW in Nelore cattle.

Session 31 Poster 17

Transcriptomics insights into hormonal regulation of feed efficiency in Nellore cattle

P.A. Alexandre[1,2], L.R. Porto-Neto[1], J.B.S. Ferraz[2], A. Reverter[1] and H. Fukumasu[2]
[1]CSIRO, Agriculture & Food, Brisbane, QLD, Australia, [2]University of São Paulo, FZEA, Pirassununga, SP, Brazil; jbferraz@usp.br

Increase of productivity and reduction of the environmental impact of livestock is some of the selection goals that can be achieved by better understanding the biological regulation of feed efficiency (FE). We aimed to analyse multiple-tissue transcriptomic data from Nellore cattle evaluated for FE in order to uncover the main biological mechanisms associated to this phenotype. Ninety eight animals were evaluated for FE and 9 individuals of each extreme had samples of adrenal gland, hypothalamus, liver, muscle and pituitary collected. RNA was extracted from those samples and sequenced in an Illumina HiSeq2500 equipment (average 13 million reads per sample, 100 pb, pair-ended). Samples were aligned to bovine genome (UMD3.1) using STAR, filtered using Samtools and gene read counts was performed using HTseq. Expression values were estimated by RPKM, keeping only genes presenting an average value of 0.2 FPKM across all samples and tissues. RPKM values were base-2 log-transformed and the mean expression value of each gene, for each group and each tissue was calculated and then the expression in low FE group was subtracted from the expression in high FE group. Next, genes were ranked according to their mean expression for each tissue and divided in five bins. Genes were considered differentially expressed (DE) when the difference between the expression in high and low FE groups were ±3.1 SD from the mean in each bin ($P<0.001$). A total of 471 genes were identified as DE, of which eight were hormones: AMH, PRL, OXT and TSHB were up-regulated in high FE group while PMCH, SST, ADM and FSHB were down-regulated. Up-regulated hormones are mostly related to growth and enhanced metabolism. For instance, TSHB, which stimulates production of T3 and T4 in thyroid of high FE animals, is inhibited by SST, a down-regulated gene in this group. Likewise, FSHB, responsible for spermatozoa production, is less expressed in high FE group and is inhibited by FST, a gene found to be up-regulated in high FE animals. These results indicate the existence of a complex hormonal regulation of FE and further research is needed to fully characterise the role of the endocrine system in cattle FE.

Session 31 Poster 18

Pituitary transcriptomic changes dependent on feed conversion in pigs

K. Piórkowska[1], K. Ropka-Molik[1], M. Tyra[2], K. Żukowski[3] and A. Gurgul[1]
[1]National Research Institute of Animal Production, Department of Animal Molecular Biology, Sarego 2, 31-047 Kraków, Poland, [2]National Research Institute of Animal Production, Department of Pig Breeding, Sarego 2, 31-047 Kraków, Poland, [3]National Research Institute of Animal Production, Department of Cattle Breeding, Sarego 2, 31-047 Kraków, Poland; katarzyna.piorkowska@izoo.krakow.pl

The goal of the present study was to indicate the differentially expressed genes of the whole pituitary between pig groups characterized by significantly different feed conversion. Whole pituitary samples were collected from 16 Puławska pigs (8 pigs per two groups; with high and low feed conversion) maintained at the same environmental and feeding condition. The RNA-seq analysis was performed in 90 single-end cycles on HiScanSQ platform (Illumina) and differentially expressed genes (DEGs) were determined using Deseq2 software. The preliminary study showed that in pigs characterized by better feed conversion 274 genes were differentially expressed. Genes showed decreased expression encode proteins associated with cholesterol metabolism (APOA1, CH25H, APOE) and regulation of hormone levels (POMC, CTSZ, ADM, AGTR1, SOX8, AGT, CPT1, APOA1, GJA1, MYRIP, CYP1B1, NR1D1). Four genes showed over the 10-fold decrease of expression (CRYM, ISM1 and WNK4), where the CRYM protein binds thyroid hormone for regulatory and developmental roles and WNK4 regulates the balance between NaCl reabsorption and K(+) secretion. In turn, genes showed an increase of expression in the pituitary of pigs having better feed conversion encode proteins were related to cellular response to corticotropin-releasing hormone stimulus (NR4A1, NR4A2, NR4A3) and regulation of type B pancreatic cell proliferation (BIRC5, NR4A3, NR4A1). The present study indicates new gene panel related to feed conversion.

Identification of genes related with carcass fatness in pigs based on RNA-seq results

K. Piórkowska[1], K. Ropka-Molik[1], M. Tyra[2], K. Żukowski[3] and A. Gurgul[1]
[1]National Research Institute of Animal Production, Department of Animal Molecular Biology, Sarego 2, 31-047 Kraków, Poland, [2]National Research Institute of Animal Production, Department of Pig Breeding, Sarego 2, 31-047 Kraków, Poland, [3]National Research Institute of Animal Production, Department of Cattle Breeding, Sarego 2, 31-047 Kraków, Poland; katarzyna.piorkowska@izoo.krakow.pl

The transcriptome sequencing (RNA-seq method) of subcutaneous fat tissues was applied to identify gene expression variation between pigs with significant different average backfat thickness. The RNA-seq analysis was performed on fat tissue collected from 16 Pulawska pigs (8 pigs per two groups; with high and low backfat thickness) maintained at the same environmental and feeding condition. The RNA-seq was performed in 90 single-end cycles on HiScanSQ platform (Illumina) and differentially expressed genes (DEGs) were determined using Deseq2 software. In the group of pigs characterized by thicker subcutaneous fat, 166 genes with increased expression were identified, including genes involved in biological processes associated with negative regulation of long-chain fatty acid transport (GO: 0010748) (IRS2, THBS1, AKT2), regulation of metabolic lipids (GO: 0019216) (EEF1A2, IRS2, ACACA, SIK1, ADGRF5, IDH1, FGF2, PDE3B, FASN, AKT2, CYR61, CHD9, ME1, SCD), cellular response to hormonal stimuli (GO: 0032870) (IRS2, ACACA, SIK1, GHR, NR4A1, PDE3B, PTGER2, AKT2, AACS, ADCY6, KAT2B, RHOQ, FOS). Genes involved in the signalling pathway responsible for the activation of gene expression by SREBF (R-SSC-2426168) (ACACA, FASN, CHD9, SCD) and the insulin signalling pathway also were identified (RHOQ, SOCS3, IRS2, AKT2, PDE3B, PYGM, FASN, ACACA). The presented results allowed us to pinpoint the interesting candidate genes related with carcass fat content in pigs.

Integrative systems genetics analysis of bovine endometrium for the receptivity of IVP embryos

G. Mazzoni[1], H.S. Pedersen[2], S. M. Salleh[3], P. Hyttel[3], H. Callesen[2] and H.N. Kadarmideen[1]
[1]Technical University of Denmark, Kemitorvet 208, 2800 Kgs. Lyngby, Denmark, [2]Aarhus University, Blichers Alle 20, 8830 Tjele, Denmark, [3]University of Copenhagen, Grønnegårdsvej 7, 1870 Frederiksberg C, Denmark; gianmaz@bioinformatics.dtu.dk

In vitro embryo production (IVP) and embryo transfer (ET) have increasing impact in cattle production. Traits associated with IVP embryo recipient receptivity are determined by the interplay of multiple genetic and environmental factors. Therefore, we performed a number of systems genetics analyses namely differential-expression analysis (DESeq2), differential weighted gene co-expression network analysis (WGCNA) and weighted interaction SNP hub (WISH) analysis to identify respectively differentially expressed (DE) genes, hubs genes and genome-wide interactions predictive for quality of the embryo recipient cow. Next, we performed an expression quantitative trait loci (eQTL) mapping (Matrix eQTL) that integrates transcriptomic and genomic data to identify variants associated with the expression of the candidate genes. Finally, we compared the eQTLs to previous association studies. RNA-seq data was generated for endometrial biopsies from 23 cows on day 6-8 in the oestrus cycle. On day 6-8 in the following cycle, one IVP blastocyst was transferred to each recipient cow, which were grouped into high (HR) and low receptive (LR) endometrium based on their pregnancy status determined at slaughter (day 26-47 after oestrus). The animals were genotyped with 777k BovineHD BeadChip. A total of 111 DE genes (60 up- and 51 down-regulated in HR cows) were involved in several biological functions for control of histotroph composition. Another 81 differentially co-expressed genes (highly co-regulated in HR cows) were involved in oxidative phosphorylation and cytoskeleton organization. We identified significant local eQTLs and distant eQTL targeting respectively three and six candidate genes. A total of seven eQTLs overlapped with QTL regions associated with reproductive traits such as: daughter pregnancy rate, fertility index, stillbirth, birth index, calf size and calving ease. The eQTLs offer themselves for further analyses for development of genetic markers to identify good quality recipient cows for IVP produced embryos.

Session 31 Poster 21

Validation of multi-series analysis of RNAseq data using chicken *in vitro* model

K. Żukowski[1], J. Dulska[1] and A. Sławinska[2]

[1]National Research Institute of Animal Production, Krakowska 1, 32-083 Balice, Poland, [2]UTP University of Science and Technology, Department of Animal Biochemistry and Biotechnology, Mazowiecka 28, 85-084 Bydgoszcz, Poland; kacper.zukowski@izoo.krakow.pl

Identifying temporal changes in gene expression using RNAseq data allows to characterize the whole spectrum of dynamic changes in the transcriptome. It also helps to understand development of progressive biological mechanisms. In contrast to static, simple pairwise group comparison like DESeq, edger or bayseq, time-series gene expression analysis interrogates biological responses to specific stimulant at each time-point. Following Oh *et al.*, the time-series analysis could be used as (1) single-series to recognize development of treatment pattern; (2) multi-series to recognize biological responses to specific stimuli at each time point and finally as (3) time-course response of cell cycle. The multi-series method was used to address the goal of this study, which was to identify temporal changes of the chicken macrophage-like cell line (HD11) stimulated with immunomodulatory prebiotic (galactooligosaccharides) and probiotic (L. lactis). To collect the data, HD11 cell line was stimulated with the two bioactives and the molecular responses of the cells were analysed at 3, 6 and 9 h post-treatment by RNAseq (Illumina). Temporal changes of differentially expressed genes were determined by regression-based approach implemented in maSigPro R package. The identified genes were assigned to pathways with real-time based functional enrichment GeneSCF tool. The results of multi-series analyses of RNAseq data were compared with step-by-step study completed with conventional bioinformatics tools. The results of this study strongly support previously reported conclusions and allow to better understand the immunomodulatory characteristic of prebiotics and probiotics at the gene and pathway level. Hereby we validated multi-series model for time-course RNAseq data, which can expand the dynamic potential of transcriptomic analyses. Acknowledgements: National Science Centre (NSC) in Cracow, Poland (project no. UMO-2013/11/B/NZ9/00783).

Session 31 Poster 22

Mineral content and immune system association revealed by Nelore muscle gene co-expression network

W.J. Da Silva Diniz[1,2], A.S.M. Cesar[3], L.L. Coutinho[3], H.N. Kadarmideen[1] and L.C.A. Regitano[4]

[1]Technical University of Denmark, Department of Bio and Health Informatics, Kemitorvet, 2800, Lyngby, Denmark, [2]Federal University of São Carlos, Department of Genetic and Evolution, Washington Luis Highway, km 235, 13560-97, São Carlos-SP, Brazil, [3]University of São Paulo – ESALQ, Department of Animal Science, Av. Pádua Dias, 13418-900, Piracicaba- SP, Brazil, [4]Embrapa Pecuária Sudeste, Washington Luis Highway, Km 234, 13560-970, Fazenda Canchim, São Carlos-SP, Brazil; wjarles09@gmail.com

Minerals are essential for several biological processes in the body. Regulatory and functional mechanisms were revealed by analysing single mineral content effects on phenotypes. However, the mineral metabolism is regulated by a complex network in which genetic and environmental factors are involved. It is important to understand the complex relationship among genes and their roles in metabolic pathways and mineral homeostasis in order to identify relevant biomarkers and candidate genes that could be used in selection of elite animals. To fill this knowledge gap, we investigated the gene co-expression patterns, and its relationship with beef mineral content, biological pathways, and hub genes. To this end, *Longissimus dorsi* muscle gene expression profile from 115 Nelore steers was measured by RNA-Seq. After quality control and data normalization, we employed the weighted gene co-expression network (WGCNA) approach to estimate the module-trait relationship and gene significance for 14 minerals. WGCNA revealed 22 modules associated with at least one mineral ($P<0.05$, $|r|>0.2$), which were used to functional enrichment analysis (Module Membership >0.7) using Webgestalt software. Among these modules, the ME_{purple} (441 genes) showed a significant correlation with Ca, K, Mg, P ($r=-0.2$), Na and S ($r=-0.3$). Functional enrichment analysis identified the ME_{purple} as being strongly associated with adaptative immune response, immune system process, and regulation of immune system process. Thus, co-expression network approach and pathway analysis revealed a complex relationship among Ca, K, Mg, P, Na, S and immune genes. These minerals act directly or indirectly in the immune system related pathways identified, and further functional studies are needed to clarify these effects.

Bioinformatic analysis and targeted sequencing – tools for unraveling regulation of immune responses

M. Siwek, A. Dunislawska and A. Slawinska
University of Science and Technology, Animal Biochemistry and Biotechnology, Mazowiecka 28, 85-084 Bydgoszcz, Poland; siwek@utp.edu.pl

There are two types of immune responses: innate and acquired. Both types were measured in unique reference population obtained by crossing Green-legged Partridgelike and White Leghorn. Innate immune response was defined by the level of natural antibodies against two environmental antigens: lipoteichoic acid (LTA) and lipopolysaccharide LPS. Acquired immune response was measured for non-pathogenic antigen: keyhole lympet heamocyanin (KLH). Classical approach based on linkage analysis followed by SNP selection and association study did not unravel entire complexity of this quantitative trait. Therefore, a new tool was applied, which is Next Generation Targeted Sequencing. This tool will allow to define new and rare allelic variants in the chicken genome. To maximize the chances for selection of most probable biological candidate genes associated with innate and acquired immune responses, a thorough in silico analysis must be done. Therefore we aimed to define the most likely candidate genes based on bioinformatics. Panel of genes was selected based on their position in the QTL region linked with antibody response to LPS, LTA and KLH antigens and biological function. Toll-like receptor signalling pathways were analysed to define candidate genes responsible for immune functions related to LPS and LTA. In silico analysis were performed using NCBI, AmiGO, InnateDB and KEGG, Reactome. Pathway analysis was based on Gene Ontology terms, DAVID, overrepresentation analysis (ORA) and WebGestalt online software. CateGOrizer was used for the clustering and visualization of GO terms. As a statistical significant threshold P<0.05 was adopted. All together 119 candidate genes were selected (43 genes for LPS, 39 genes for LTA and 37 genes for KLH). Selected genes take part in immune system processes such as regulation of immune response, leukocyte proliferation or cytokine production. In addition, these are genes actively participate in the proteins and lipids metabolism or in the stress response for external stimulus. These genes will be further sequenced and used for variant discovery. The research was supported by the National Science Center (UMO-2014/13/B/NZ9/02123).

TPI and HSP27 proteins changes of bovine muscle based on ultimate pH during aging

M. Pereira, L. Fonseca, A. Rosa, M. Poleti, E. Mattos, J.B. Ferraz and J. Eler
University of São Paulo, Veterinary Medicine, Rua Duque de Caxias Norte, 225, 13635-900, Brazil; joapeler@usp.br

The ultimate pH (pHu) has been widely discussed in the recent literature because of its potential for predicting other meat quality attributes by exerting a direct influence on proteolysis and biochemical mechanisms that occur in the muscle during *rigor mortis*. In this context, the study of proteins, with metabolic and stress response function, expressed in muscle contributes to meat science by increasing our understanding of the underlying molecular and biochemical processes. Thus, the aim of this work was to evaluate the changes in the expression of proteins triosephosphate isomerase (TPI) and heat shock protein 27 kDa (HSP27) in the *Longissimus thoracis* muscle of F1 South African Simmental × Nellore cattle during aging. Four hundred fifteen cattle were evaluated for pHu and selected three normal pHu (pHu≤5.8) cattle and three intermediate pHu (pHu: 5.80-6.19) in two aging periods (1 and 14 days), totalling 12 samples. Protein extraction was performed and samples were analysed by western blotting assays. The primary antibodies were diluted at 1:5,000 (TPI) and 1:20,000 (HSP27) (Santa Cruz Biotechnology) and the secondary HRP-conjugated anti-mouse antibody (Sigma-Aldrich) diluted at 1:5,000. The images were captured with the ChemiDoc MP Image System (Bio-Rad). The protein band volumes were calculated using the Image Lab v.6.0 software (Bio-Rad) and normalized to glyceraldehyde-3-phosphate dehydrogenase (GAPDH, 31 kDa). Means were analysed by F test (P<0.05) (SAS 9.3). The protein metabolic function, TPI, was less expressed at intermediate pHu at the two aging times. Reduction in the glycolysis rate, which increases the electrostatic repulsion between muscle filaments, contributing to the lateral enlargement of muscle fibres and consequent sarcomere extension. HSP27 protein, stress response function, was less expressed at intermediate pHu in meat aged for 1 day, indicating their protective activity of myofibrillar proteins decreases and meat tenderization increases. These results confirm the relevance of the proteins metabolic and stress response function involved with pHu and beef meat quality. Financial support by FAPESP (2014/12492-8), CNPq (303659/2014-9) and CAPES.

Session 31 Poster 25

Hepatic proteomic analysis of feed efficiency in Nellore cattle

L. Fonseca[1], J. Eler[1], M. Pereira[1], P. Alexandre[1], A. Rosa[1], E. Mattos[1], G. Palmisano[2] and H. Fukumasu[1]
[1]University of São Paulo, Veterinary Medicine, Rua Duque de Caxias Norte, 225, 13635-900, Brazil, [2]University of São Paulo, Parasitology – Biomedical Sciences Institute, Av. Prof. Lineu Prestes, 1374, 05508-900, Brazil; joapeler@usp.br

Feeding is one of the most relevant cost of beef cattle production, accounting for up to 70% in feedlots. Improving nutrient utilization efficiency is essential for the viability of production, given the current scenario of increasing demand for animal protein and highly competitive market. In this context, proteomics appears as an important tool in the search for alternatives that increase the feed efficiency (FE) of beef cattle. Thus, the aim of this work is to characterize the hepatic proteome of beef cattle select for divergent and to associate it with the feed efficiency. Ninety-eight animals were evaluated for FE and individuals of each extreme (six for high and four for low efficiency) had liver samples collected at slaughter. Protein extraction and digestion were performed and samples were analysed in a high-performance liquid chromatograph Easy nanoLC II coupled with tandem mass spectrometer LTQ-Orbitrap Velos ETD (Thermo). The acquired MS/MS data were analysed with the software MaxQuant 1.5.8.3 against the *Bos taurus* UniprotKB database (48,738 entries, 2018) and the search results analysed with Perseus 1.6.0.7. A total of 529 proteins were identified and after quality control 380 proteins were submitted to statistical analysis and gene ontology (GO) enrichment. Twenty-seven proteins were differentially expressed ($P \leq 0.1$, permutation-based FDR) between high and low animals. These proteins were distributed in 13 classes, most of which were oxidoreductases and transferases. GO showed intracellular protein transport, pteridine-containing compound metabolism, coenzyme metabolism, cofactor metabolism and one-carbon metabolism. Molecular function corresponded to oxidoreductase activity acting on the CH-CH group of donors, and terms for cellular component were related to clathrin coat of trans-Golgi network vesicle, cytoplasm, microvillus, and vesicle. These results confirm the relevance of the amino acid and energy metabolism pathways involved with feed efficiency in beef cattle. Financial support by FAPESP (2014/04937-0), CNPq (303659/2014-9) and CAPES.

Session 32 Theatre 1

Pathways to innovation in precision livestock farming

S. Ingrand[1], F. Medale[2] and X. Vignon[3]
[1]INRA, UMR 1273 Territoires, 63122 Saint-Genes-Champanelle, France, [2]INRA, UMR 1419 NUMEA, 64310 Saint-Pée-sur-Nivelle, France, [3]INRA, UMR 1198 BDR, 78352 Jouy-en-Josas cedex, France; stephane.ingrand@inra.fr

Investigations in precision livestock farming (PLF) are a potential for innovations for improving the economic, environmental, welfare and product quality performances of livestock systems. Examples will be shown to exemplify how drawing information collected from monitoring devices and giving them a biological signification allows applications useful for geneticists, nutritionists and physiologists as well as farmers or manufacturers. The researches can be implicated from the hardware for acquisition, to the translation of raw data into biological data, then to the elaboration of complex models for a return to the final users (farmers, breeders, industry,..). Phenotypical data for instance can rapidly be derived from the systematic acquisition of weight or body composition score in a system linked to a milking robot, allowing thus to generate biologic indicators linked to feed efficiency. This information can be used finally by the geneticists, breeders, and by the farmer to manage each individual as well. Researches combining sensors and models for precision feeding show that methodologies for real time acquisition of parameters can be applied to combine knowledge on feed with characterization of animal traits (feed efficiency, digestion, metabolism, behaviour, welfare) in order to modelling mechanisms in variable conditions and generating decision support tools. In the field of monitoring, detection of animal activity via video systems or individual tags/antenna systems allow the analysis of fine behaviours and the detection of variations in level of activities that can be linked to stress situations or disease. The challenge for the future is to extend real-time information feedback to the entire value chain to optimize production from end to end of the chain, with a return to the farm for improved animal welfare and farmer comfort. In France, the interdisciplinary laboratory #DigitAg has been created with this aim, by working around the digital tools and services to be transferred to the agriculture sector. This overview will thus illustrate how PLF may enhance not only the technology (the means), but also the ecological processes in farming systems (the goal).

Computer vision system for heifer height and weight estimation

G. Adin[1], O. Nir[2], Y. Parmet[2], D. Werner[1] and I. Halachmi[3]
[1]Extension Service, The Ministry of Agriculture and Rural Development, Animal Production, Beit Dagan 50250, P.O. Box 28, Israel, [2]Department of Industrial Engineering and Management, Ben-Gurion, Faculty of Engineering Sciences, Ben-Gurion University of the Negev, Beer Sheva, 84105, Israel, [3]The Institute of Agricultural Engineering, Agricultural Research Organization, Precision Livestock Farming (PLF) Lab., The Volcani Center, Beit Dagan 50250, P.O. Box 6, Israel; gaby.adin1@gmail.com

Physical body measurement is an important methodology at the livestock farm as indicator of heifer development. Manual measurements are expensive and introduce high levels of stress to heifers. Computer Vision was used as non-intrusive approach for body measurement to overcome limitations of conventional systems. Several studies attempted to replace conventional with automatic machine vision system however no system was adopted. Monitoring the height and weight of dairy heifers during development automatically, frequently, objectively and reliably is cardinal process. In previous project, the team developed a CV algorithm with three-dimensional camera to identify limping cows. This study used modules of previous algorithm. The idea is based on a 3 dimensional camera and electronic identification of each cart that passes under the camera. In each photograph we measure both the height and the volume of the body. The research included selection of camera hardware suitable for the cowshed development of the algorithm; testing the accuracy of the device compared to a reliable human perspective; and developing a CV and an image processing algorithm. In the model conducted in dairy barn at Volcani Institute, the wither height was R^2 94.9% and the body weight estimation was R^2 94.8%. The validation experiment conducted at the Ein Hahoresh dairy farm R^2 92.2% was obtained for wither height and R^2 97.0% for body weight. In conclusion, a system was developed to measure stable and reliable over time of the dimensions of the body of dairy heifers with a high level of accuracy. In future these principles could be applied to other farm animals: cattle, pigs, horses. Placing the system above water troughs or transit passages will be examined to enable full automation and accessibility of the product to each dairy farm.

Beef Monitor: tracking beef cattle growth and predicting carcass characteristics of live animals

G.A. Miller[1], J.J. Hyslop[1], D.W. Ross[1], D. Barclay[2], A.R. Edwards[3], W. Thomson[4], S. Troy[1] and C.-A. Duthie[1]
[1]SRUC, West Mains Road, Edinburgh, EH9 3JG, United Kingdom, [2]Innovent Technology Ltd, Markethill, Turriff, AB53 4PA, United Kingdom, [3]Ritchie Ltd, Carseview Road, Forfar, DD8 3BT, United Kingdom, [4]Harbro Ltd, Markethill, Turriff, AB53 4PA, United Kingdom; gemma.miller@sruc.ac.uk

The performance of beef cattle is currently evaluated through visual assessment or by weighing through a crush. Video imaging analysis (VIA) is increasingly used in abattoirs to grade carcasses and there is potential for 3D imaging to be used on farm to predict carcass characteristics of live animals. The objectives of this study were to validate the use of a water trough system with automated weighing platform and 3D camera technology to track growth and predict carcass characteristics of live animals using artificial neural networks (ANNs). A variety of breeds (steers and heifers) were placed behind systems on five finishing units for 1-3 months pre slaughter. Images and weights were passively collected each time an animal came to the trough and cattle were tracked through the abattoir at slaughter. An abattoir trial was also conducted where live animals were weighed and 3D images taken in the lairage immediately pre-slaughter. Cold carcass weight (CCW) was provided by the abattoir. Saleable meat yield (SMY) and fat and conformation grades were determined by VIA of carcass images. The relationship between weights measured in a crush and the average of weights measured in the crate on the same day had an R^2 of 0.99. ANN prediction performance was assessed by regression for liveweight (R^2=0.72), CCW (R^2=0.91) and SMY (R^2=0.80). ANNs predicted EUROP fat and conformation grades with 63 and 69% accuracy, respectively. Performance of individual animals can be tracked through accurate, daily weights obtained without the need for manual handling. Carcass characteristics can be predicted for live animals on farm using 3D imaging technology. This system presents an opportunity to reduce a considerable inefficiency in beef production enterprises through marketing of animals at the optimal time.

Session 32 Theatre 4

Use of hoof digital images in estimation of genetic parameters connected with health

R. Kasarda, M. Vlček, O. Kadlečík and N. Moravčíková
Slovak University of Agriculture, Department of Animal Genetics and Breeding Biology, Tr. A. Hlinku 2, 949 76 Nitra, Slovak Republic; radovan.kasarda@uniag.sk

The objective of this study was to estimate the genetic parameters of individual hoof lesions and claw morphometric parameters in Holstein cows. According to this, the heritability and genetic correlations for milk fat to protein ratio were estimated as one of the indicator of the risk of metabolic disorders in dairy cattle. A total of 382 hoof-trimming records from 299 cows, kept in two farms in west part of Slovakia from TOP10, were used in the study. The hoof health data and morphometric parameters were collected immediately after regular functional trimming between 2015 and 2017. Eight claw morphometric parameters (angle, length, heel depth, height, diagonal, width, total and functional areas) were obtained by using digital image analysis. Images were analysed by using NIS Elements 3.0 software. Three types of hoof lesions were included in the analysis; interdigital dermatitis and heel erosion (IDHE), digital dermatitis (DD) and sole ulcer (SU). All of hoof lesions included in the analysis were analysed as binary traits. The frequency of hoof disorders in analysed herds ranged from 83% (IDHE) to 17% (DD). It can be concluded that more than half of analysed cows had at least one type of hoof lesion. To estimate the genetic parameters, the multi-trait animal models and Bayesian approach was used. The estimates of direct heritabilities for claw morphometric parameters ranged from 0.41 (heel height) to 0.62 (claw length). Heritability of total claw area and functional claw area were moderate (0.62). Average estimates of direct heritabilities for IDHE, DD and SU were relatively low 0.01, 0.03 and 0.04, respectively. For the F/P ratio moderate level of direct heritability was found (0.52). The genetic correlations between morphometric parameters were generally moderate to high. Genetic correlation between F/P ratio and claw disorders was very low. Estimated genetic parameters of morphometric traits provide base for the future selection and automation of claw data evaluation by use of machine learning. Observed parameters on F/P ratio could be successfully used in selection for high producing metabolically resistant and healthy cattle.

Session 32 Theatre 5

Diagnosis of ovarian activity: a powerful tool to manage beef cows under extensive grazing condition

G. Quintans[1], J.I. Velazco[1] and C. Lopez Mazz[2]
[1]National Institute for Agricultural Research, Ruta 8 km 281, 33000 Treinta y Tres, Uruguay, [2]School of Agronomy, Garzón 780, 12900, Uruguay; gquintans@inia.org.uy

In South America most of the beef production systems are developed under extensive grazing conditions on native pastures. For this reason, feed intake pattern is irregular and commonly insufficient, and farmers must cope with this variability to maintain moderate to high levels of productivity. During the service period, body condition score (BCS) of sucker cows may not be the best indicator of ovarian cyclicity. In this context, the ovarian activity diagnosis (OAD) was developed to classify each beef cow by ovarian status, to predict the probability of getting pregnant but more importantly, to advice farmers about management decisions to improve reproductive performance. OAD is obtained by ultrasonography performed between the onset and the middle of the service period. Both ovaries are examined and cows are classified in: Pregnant (if the embryo is observed); Cycling (if a corpus luteum (CL) is observed); Superficial Anoestrous (SA, if the maximum follicle diameter is ≥8 mm); Deep Anoestrous (DA, if the maximum follicle diameter is <7 mm). Results of research on commercial farms (>5,000 cows) indicate that early weaning is recommended in DA cows whereas temporary calf removal is in SA cows. For example, recently an experiment was done at INIA using 64 primiparous beef cows that were classified by OAD at the beginning of the service period (68±16 days pp). Cows had (average± SD) 380±29 kg and 4.3±0.33u (scale 1-8) of LW and BCS; 41% of the cows were in DA and 59% in SA. Within DA, cows were assigned to two treatments: control (CON, without any management) and early weaning (EW, calves were removed definitively from cows). Within SA, cows were assigned to two treatments: control (CON, without any management) and temporary weaning (TW, calves were fitted with nose plates for 14 days). One month later, OAD revealed that within DA, 0% CON cows presented a CL while 64% EW did. Within SA, 8% CON cows presented a CL while 61.5% TW did. The OAD is a powerful tool in the management of a beef cow-calf system and farmers are now increasing the usage of this technology with high confidence and success.

Session 32 Theatre 6

Role of fat and muscle mass and mobilization of transition dairy cows on milk yield and reproduction

N. Siachos[1], N. Panousis[1], G. Oikonomou[2] and G.E. Valergakis[1]
[1]Aristotle Univesity of Thessaloniki, Faculty of Veterinary Medicine, School of Health Sciences, Box 393, 54124 Thessaloniki, Greece, [2]University of Liverpool, Institute of Veterinary Science, Faculty of Health and Life Sciences, Leahurst, Neston CH64 7TE, United Kingdom; geval@vet.auth.gr

The objective was to assess the role of body fat and muscle reserves and mobilization during the transition period on milk yield and reproduction of dairy cows. Eighty-five cows from 2 dairy farms (n=32 and 53, respectively) in different parities (1:n=14; 2:n=35; ≥3:n=36) were included in the study. Backfat thickness (BFT) and longissimus dorsi thickness (LDT) were measured by ultrasonography and body condition score (BCS) was assessed at 6 time points relative to calving (day 0): -21d; -8d; 0d; +8d; +21d and +28d. The association among BFT, LDT and BCS was assessed with linear correlation and regression. Milk yield records for the first 100 DIM (MY100), DIM at first artificial insemination (DIM_1stAI) and calving-conception interval (CCI) were available and linear and quadratic regression was used to identify relations with BFT and LDT. On average (SD), cows lost 0.56 BCS units (0.45) (19% of total) and mobilized 5.88 mm (2.88) of BFT (39% of total) from -8d to +28d and 12.48 mm (5.24) of LDT (37% of total) from -21d to +21d. Overall pairwise correlations were: BCS/BFT, r=0.831; BCS/LDT, r=0.695 and BFT/LDT, r=0.570 (P<0.001). A model with both BFT and LDT explained better the BCS variation (R^2=0.768, P<0.001) than either alone. BFT at -8d and BFT mobilization from 0d to +21d was positively related to MY100; a 1 mm increase of BFT at -8d and a 1 mm BFT mobilization from 0d to +21d were associated with 35.1 l (95% CI: 5.5-64.7) and 42.9 l (95% CI: 11.3-74.5) more MY100, respectively (P<0.05). All LDT measurements from -21d to +21d were negatively ralated to CCI and most were also negatively related to DIM_1stAI; a 1 mm increase of LDT at 0d was associated with 0.91 (95% CI: 0.15-1.68) and 2.23 (95% CI: 0.96-3.49) less days in DIM_1stAI and CCI, respectively (P<0.05). Muscle mobilization started at the beginning of the transition period, 2 weeks earlier than fat. Fat mobilization was positively related to milk yield; cows with higher muscle reserves around calving had improved reproductive performance.

Session 32 Theatre 7

Comparable non-invasive techniques for measuring animal-based enteric methane emission on farm

J. Rey[1], A. Garcia-Rodríguez[1], O. González-Recio[2,3], R. Atxaerandio[1], R. Ruiz[1], J.A. Jiménez[4] and I. Goiri[1]
[1]Neiker-Tecnalia, Producción animal, Granja Modelo de Arkaute Apartado 46, 01080 Vitoria-Gasteiz, Spain, [2]Escuela Técnica Superior de Ingeniería Agronómica, Alimentaria y de Biosistemas. Universidad Politéc, Departamento de Producciones Agrarias, Ciudad Universitaria s/n, 28040 Madrid, Spain, [3]Instituto Nacional de Investigación y Tecnología Agraria y Alimentaria, Departamento de Mejora Genética Animal, Ctra La Coruña km 7.5, 28040 Madrid, Spain, [4]Confederación de Asociaciones de Frisona Española (CONAFE), Departamento Técnico, Ctra. de Andalucía, km. 23,600, 28340 Valdemoro (Madrid), Spain; igoiri@neiker.eus

The objective of the present study was to assess the agreement between two breath sampling devices: the hand-held laser methane detector (LMD) and the infrared methane analyser (SNIFFER). On different days, 31 dairy cows in mid-lactation were restrained to measure methane emission simultaneously with the two devices during a 5 min sampling period by placing the laser bean and the infrared methane analyser sampling tube at the cows' nostrils, obtaining a total of 75 paired data. Measurements with LMD were performed at 1 m distance and recorded every 0.5 seconds. LMD takes remote measurements of column density for gas plumes containing methane (ppm-m), while SNIFFER measured methane concentration of the exhaled breath (ppm) every second. For each sampling period and device overall mean methane concentration was calculated. The relationship between methane values for the LMD and methane concentration for SNIFFER was examined using regression with linear and quadratic terms. Data were also analysed using the CORR procedure of SAS, and Pearson and Spearman correlation coefficients were obtained. The mean methane value for the LMD was 82.5 ppm-m (SD=39.9) and the mean methane concentration for the SNIFFER was 828.2 ppm (SD=555.9). A significant but low (P<0.001) linear and quadratic relationship was observed between mean methane concentrations obtained with the two devices (r^2=0.53 and 0.54, respectively). The Pearson and Spearman correlation coefficients were high (r=0.73 and r=0.76, respectively). In conclusion data from both devices could be interchangeable to rank animals for breeding purposes or to evaluate methane reduction strategies.

Prediction equations for manure nitrogen output in beef cattle

A. Angelidis[1], L. Crompton[1], T. Misselbrook[2], T. Yan[3], C. Reynolds[1] and S. Stergiadis[1]
[1]University of Reading, School of Agriculture, Policy, and Development, P.O. Box 237, Earley Gate, RG6 6AR, Reading, United Kingdom, [2]Rothamsted Research, North Wyke, EX20 2SB, Okehampton, United Kingdom, [3]Agri-Food and Biosciences Institute, Large park, BT26 6DR, Hillsborough, United Kingdom; a.angelidis@pgr.reading.ac.uk

Mitigating manure nitrogen (N) output (MNO) from beef cattle benefits farm profitability (by reducing production costs) and the environmental footprint of meat production (by reducing groundwater and atmospheric pollution). Reliable prediction equations for MNO can inform management decisions related to N mitigation. The aim of this study was to develop prediction equations for MNO from growing and finishing beef cattle, using diet chemical composition, body weight (BW) and intakes of dry matter (DM) or N data as predictors. Data were collected from 93 published studies and included 302 treatment means. Prediction equations were developed using residual maximum likelihood analysis in GenStat. Random model accounted for experiment, breed and production stage, and fixed model was developed using the Wald statistic. Accuracy of the new and previously published equations were assessed by internal validation. For this, 2/3 of the data were used to develop equations, which were validated against the remaining 1/3, generating a mean prediction error (MPE) to describe prediction accuracy. Equations using DM or N intakes as primary predictors, in conjunction with diet chemical composition, had better accuracy, than those developed using BW and diet chemical composition. The optimum model (MPE=0.35) was: MNO (g/d) = –92.33 + 16.34 DM intake (kg/d) + 4.346 N (g/kg DM) + 0.224 ADF (g/kg DM) – 0.141 NDF (g/kg DM). When using N intake as sole predictor, the new models had marginally increased prediction accuracy (MPE=0.47) than existing literature models (MPE=0.50). However, new equations using more readily available predictors (BW and diet contents of N and organic matter) had higher prediction accuracy (MPE=0.39).

Relationship between age and body weight at farrowing over 6 parities in Large White × Landrace sow

N. Quiniou
IFIP, BP35, 35650 Le Rheu, France; nathalie.quiniou@ifip.asso.fr

At the beginning of the gestation, parity and back fat (BT) thickness are frequently used by farmers to choose among different feeding plans the most adapted one to feed each sow if individual feeding is possible. Otherwise, BT is used to allocate the sows to one of the available pens and to adapt the feed allowance at the group level. Usually a single target of BT at farrowing is retained at the herd scale, and each sow is expected to make up its BT for the difference between the target and its own initial BT. According to the factorial approach used to assess nutrient requirements, achieving an expected BT gain implies that enough energy is supplied above maintenance. Based on equation published by Dourmad *et al.*, expected energy retention is obtained by the difference between the initial (calculated from measured BT and BW) and the final amount of energy (calculated from expected BT and BW). When no information is collected on BW during the gestation, the expected final BW is also used to assess daily BW and corresponding maintenance requirement and to take into account impacts of housing conditions (temperature, activity level). Then, adequacy of the energy supply depends not only on initial and final BT, but also on initial BW and final BW. Individual BW have been collected over successive parities in the IFIP facilities (n=6,288 from Large White × Landrace sows born since 2000). Data of 90 sows born between 2012 and 2015, studied over at least 6 parities and group-housed from the 28th to the 108th day of gestation, were used to characterize the relationship between age and BW after farrowing based on BW at the first farrowing (BW_P1) and BW gain afterwards, with BW at the 6th parity considered as the mature BW: $BW_i(Age)$, kg = 145.6 + 0.171 × Age_P1_i+ 111.4 × (1 – exp(-1.453/1000 × $(Age\text{-}Age\_P1_i)^{1.084}$)), RMSEP=16 kg; with Age_P1_i: the individual age at the first farrowing used to adjust the mean BW_P1. Based on this equation, the individual expected BW gain (both growth and recovering parts) can be calculated and, combined with BT gain, used to assess the corresponding energy requirement for precision feeding during gestation. Its calibration for other lines or farms will require specific measurements due to interactions with management.

Join dynamics of voluntary feed intake, glycaemia and insulinaemia in *ad libitum* fed pigs

K. Quemeneur[1,2], M. Le Gall[1], Y. Lechevestrier[1], L. Montagne[2] and E. Labussiere[2]
[1]Provimi France, Cargill, 35320 Crevin, France, [2]PEGASE, INRA, Agrocampus-Ouest, 35042 Rennes, France; katia.quemeneur@inra.fr

Short term regulation of voluntary feed intake depends on energy status of the animal. The latter can be assessed by plasma metabolite concentrations but lipogenesis and nutrient oxidation can also modify the respiratory quotient. This study aims to link the within-day dynamics of voluntary feed intake and those of blood metabolites in pigs fed *ad libitum*. Thirty-six pigs (mean BW 35 kg) were fed 6 diets with 2 levels of dietary fibre (13 or 18% NDF for LF and HF by addition of wheat bran, soybean hulls and sugar beet pulp) and 3 levels of aleurone (0, 0.2, or 0.4% for A0, A2 or A4, respectively). After a 2-week adaptation period, they were individually housed during 1 week in respiration chamber to measure feeding behaviour and gas exchanges. On the last day, the dynamics of blood metabolites was followed between 2 voluntary meals. Total feed intake was not modified by HF diet (1,452 g DM) but decreased with aleurone supplementation (-270 g/d for A4 vs A0; P<0.01). Daily heat production (1.37 MJ/kg $BW^{0.60}$/day) and respiratory quotient (1.13) were not modified by diets. The HF diet resulted in less but heavier meals compared with LF diet (8.2 meals per d of 178 g DM vs 9.4 meals of 150 g, P<0.01). Aleurone supplementation decreased daily number of meals (-1.5 for A4 vs A0, P<0.01). Pre-meal plasmatic glucose concentration tended to decrease with HF diet (-7%) and pre-meal insulin concentration tended to increase with aleurone (3.8 vs 6.9 µU/ml for A0 vs A2). During the first hour after the meal, glucose concentration was lower with HF compared with LF (P<0.01). Inter-individual variability of glucose profiles was important and glycaemia peaks ranged from 100 to 150% of the pre-meal level. Insulin and glucose profiles agreed and insulin profiles linked with the size of the first meal (P<0.01). To conclude, dynamics of voluntary feed intake cannot be explained only by profiles of glucose and insulin in *ad libitum* fed pigs.

Effects of feeding systems and maturity type on gonadal development and semen quality of rams

A.M. Du Preez, E.C. Webb and W.A. Van Niekerk
University of Pretoria, Animal and Wildlife Science, Pretoria, 0002, South Africa; amelia.may.dp@gmail.com

This study investigated the effects of feeding systems during the growing phase (five to 12 months old) on gonadal development and semen quality of rams of different maturity types. The effects of rangeland feeding (C), rangeland followed by intensive feeding (TR1) and intensive feeding (TR2) were evaluated in early (Merino; M), medium (Dohne Merino; DM) and later maturity type (SA Mutton Merino; SAM) rams. Growth, size, scrotal measurements, semen quality and post mortem gonadal measurements were recorded. Data was analysed by GLM ANOVA procedures, using IBM SPSS V 23. Differences between treatment means for breeds, feeding systems and interactions were tested by Bonferroni's multiple range test at P<0.05. Final growth and size differed between breeds and feeding systems (P<0.05). Treatment-breed interactions were significant for final mass, ADG, body length and subcutaneous fat (SCF), with highest growth for M-rams fed diet TR2. SAM in TR2 had the heaviest scrotal mass (SM; 905.6±120.9 g), but DM and M had the heaviest SM in TR1 (880.8±142.8 g and 815.4±114.8 g respectively). All breeds deposited most scrotal fat in TR1. M fed TR1 had the least scrotal neck fat (SNF; 1.4±0.08 cm) while DM and SAM had similar SNF (1.7±0.19 and 1.8±0.17 cm respectively), and in TR2 M had most SNF (1.7±0.46 cm). Correlations between percentage normal sperm (PNS) and scrotal fat mass (P<0.05, r=-0.71) was negative in M in TR2 and between PNS and SCF (P<0.05, r=-0.81) in C. In SAM fed TR2, SCF and semen volume (SV; P<0.01, r=-0.88) correlated. In DM fed C, SV and SNF correlated (P<0.05, r=0.75), as well as PNS and SNF (P<0.05, r=-0.79), and SV and PNS (P<0.05, r=-0.71). SV and PNS (P<0.05, r=-0.76) correlated in DM fed TR2, while PNS and final scrotal circumference (P<0.05, r=-0.87) and PNS and scrotal mass (P<0.05, r=-0.69) correlated for DM in TR1. Intensive feeding of young rams improved growth and gonadal development, but adversely affected fertility of early maturing rams due to earlier accumulation of excess scrotal fat, impairing thermoregulation and increasing the percentage abnormal sperm. Efficient feeding programs should make provision for maturity types, fattening rate and semen quality.

Session 32 Theatre 12

Debating common PLF challenges in nutrition, genetics, and in physiology

R. Kasarda[1] and J. Maselyne[2]
[1]Slovak University of Agriculture, Department of Animal Genetics and Breeding Biology, Tr. A. Hlinku 2, 949 76 Nitra, Slovak Republic, [2]ILVO, Belgium; radovan.kasarda@uniag.sk

Debating common PLF challenges in nutrition, genetics, and in physiology The debate will discuss common issues that will be raised during the session. The session speakers and posters' authors will form a panel-of-experts that will answer questions from the audience and identify common research challenges and potential engineering solutions. 15 minutes just before coffee/diner.

Session 32 Poster 13

Effects of new generation elements on performance of beef cattle during background phase

M. Cortese, M. Chinello and G. Marchesini
University of Padova, Dipartimento di Medicina Animale, Produzioni e Salute, Viale dell'università 16, 35020, Italy; martina.cortese.1@phd.unipd.it

The aim of this study was to compare the effect of a mixture of new generation (NTC) trace element (zinc chelate of amino acids hydrate, zinc chloride hydroxide monohydrate, cupric chelate of amino acids hydrate and dicopper chloride trihydroxide) with a control mixture (CON) of inorganic trace elements (zinc oxide and cupric sulphate pentahydrate) on performance and health of young bulls imported from France, during their backgrounding phase (the first 46 days) at the fattening unit. Charolaise bulls (n=107), 300±53 days old and weighing on average 407±23 kg were evenly grouped in 11 pens: the animals of the first 5 pens (n=47) and the remaining 6 pens (n=60) were fed CON and NTC mixtures, respectively. The total mixed ration (TMR), that except for the mix of trace elements, was the same for both groups, was changed after the first half of the trial (Period 1) to cover the increasing nutritional needs. Feed intake (per pen), individual rumination time and activity were measured daily. Bulls were individually weighed at the beginning and at the end of each of the two periods. TMR and leftover samples were collected weekly, submitted to chemical analysis and sieved with a modify Penn State Particle Separator. Weight, average daily gain (ADG), activity and rumination time data were submitted to ANOVA. Mineral integration, pen (within diet) and period were used as fixed factors, while the initial weight was used as a covariate. DMI was not affected (P=0.567) by mineral integration, showing average values of 8.29 and 8.00 kg/d for CON and NTC, respectively, whereas NTC led to a 10% higher ADG than CON (1.72 vs 1.56 kg/d, P=0.05). There were no significant differences in feed conversion rate (5.74 vs 4.79, P=0.188) and rumination time (416 vs 427 min., P=0.185) between CON and NTC, whereas daily activity was higher in NTC (415 vs 424 bit) showing a tendency to significance (P=0,085). The outcomes suggest that NTC favoured cattle performance compared to CON, likely because of its higher trace elements bioavailability.

In vitro effects of 12-oxoeicosatetraenoic acid on trophoblast cells derived from bovine blastocyst

H. Kamada
NARO, Animal Production, Ikenodai-2, Tsukuba, Ibaraki, 305-0901, Japan; kama8@affrc.go.jp

Our previous investigations showed that 12-oxoeicosatetraenoic acid (12-KETE) is a strong candidate of signal for placenta separation at delivery. In this study, effects of 12-KETE on trophoblast cells derived from bovine blastocyst were investigated to presume the function of 12-KETE in the placentome at delivery. Trophoblast cells were formed its cell colony on the monolayer of fibroblast derived from bovine placentome or without feeder cells. The 12-KETE treatment was as follows. Culture medium was changed to DMEM (-FBS) containing 12-KETE, and the morphological changes of colony were observed by phase-contrast microscope. Dye-quenched collagen (DQ-Col) was used to detect matrix metalloproteinase activity. After the incubation of DQ-Col/DMEM for 2 hr, the culture dish was washed by DMEM(-FBS) without DQ-Col and added new DMEM(-FBS) containing 12-KETE. DQ-Col cut by MMP emits green fluorescence. The early apoptosis was detected by double staining of propidium iodide (PI) and YoPro-1 (Apoptotic cell was stained by YoPro-1, but not PI.). We have presented that 40μM of 12-KETE induced an exfoliation of trophoblastic cell sheets from fibroblast monolayer or the bottom of culture dish. Then many apoptotic cells and matrix metalloproteinase (MMP) activation were observed. These results were consistent with *in vivo* observations. In this study, we found out that a lower concentration of 12-KETE (5-10μM) induced a different type of cell exfoliation (Trophoblastic cell colony fell in pieces by 12-KETE addition.). Then the number of apoptotic cells was a few and MMP activity was low. On the other hand, Pefabloc (protease inhibitor) addition with 12-KETE inhibited the falling in pieces of cell colony. These results suggested that protease activity may be a key factor at the process of placenta separation at delivery. This work was supported by JSPS KAKENHI Grant Number 15K07700.

Modelling the feed intake of six commercial South African sheep breeds in a feedlot

T.S. Brand[1], D.A. Van Der Merwe[2] and L.C. Hoffman[2]
[1]Animal Sciences, Agriculture: Western Cape, Private Bag X 1, 7607 Elsenburg, South Africa, [2]University of Stellenbosch, Animal Science, Merriman Road, 7600 Stellenbosch, South Africa; tersb@elsenburg.com

This study investigated the growth and feed intake of six sheep breeds, which are considered popular in large-scale feedlot operations in South Africa; namely the Dohne-merino, Dormer, Dorper, Meatmaster, Merino and South African Mutton Merino (SAMM). Eight lambs (four rams and four ewes) per breed were weaned at ~90 days of age and housed in individual pens until the lambs attained one year of age, when it was assumed that the lambs had attained mature live weight. The lambs were fed a pelleted finisher ration ad *libitum*. The lambs were weighed weekly along with feed intake, to obtain 1,653 records during the study period. These records were used to model the sigmoidal growth of the sheep using the Gompertz function ($W=a.exp(-exp(-b(t-c)))$). Cumulative feed intakes, as well as feed intake as a percentage of body weight, were modelled against the live weight of the sheep. Model parameters were then compared with breed and sex as main effects. No interactions were observed between the main effects for any of the Gompertz function parameters. Significant differences were observed for the asymptotic weight (a) parameter for sex, with rams being higher than ewes (109.4±3.50 vs 89.2±3.50 kg; P<0.001). No differences were observed between sex or breed for the maturation rate constant (b) (0.01±0.004), however the parameter describing the age at maximum growth (c) differed between breeds with Dormer being larger than Dohne, SAMM and Dorper (146 days vs 109, 105 and 98 days, respectively; P<0.001). Cumulative intake (CI) increased linearly with live weight (w) of the sheep (CI=6.41w – 205.74, R^2=0.90,) while percentage intake (PI) was found to decrease linearly as sheep grew. (PI=5.22 -0. 038w, R^2=0.47). The growth of the sheep was accurately modelled using the Gompertz function, while cumulative feed intake could be modelled accurately with live weight, with the slope indicating the feeding efficiency per unit weight gain.

Session 32 Poster 16

Effect of feeding strategy on endometrial gene expression of dairy cows during early luteal phase

A.L. Astessiano[1], F. Peñagaricano[2], J. Laporta[2], A. Meikle[3] and M. Carriquiry[1]
[1]Facultad de Agronomia, UDELAR, Montevideo, 12900, Uruguay, [2]University of Florida, Gainesville, 32611, USA, [3]Facultad de Veterinaria, UDELAR, Montevideo, 11600, Uruguay; lauaste@gmail.com

Primiparous Holstein cows (n=18, 528±40 kg BW and 3.2±0.2 BCS) were used in a randomized block design to evaluate gene expression changes in the endometrium during the early-luteal phase of the oestrous cycle induced by 2 different nutritional treatment during early lactation. Cows were fed either (1) TMR *ad libitum* (17 kg DM/d offered; 70% forage, 30% concentrate; T0) or (2) grazing of *Medicago sativa* (6-h am grazing in 3-d strips; pasture allowance=20 kg DM/d) plus TMR (70% of *ad libitum* TMR; 12 kg DM/d offered; T1) during the first 65 d of lactation. At 45±1 d, cows were synchronized and at d 7 of the oestrous cycle (d 0 = oestrous) endometrial biopsies (n=5/treatment) were collected. Blood samples were obtained at -3, 0, and 7d to determine progesterone (P4) and estradiol 17β (E2) concentrations. Endometrium RNA samples were analysed using RNA sequencing. Plasma P4 and E2 concentrations were not affected by nutritional treatment (1.2±0.3 ng/ml and 8.1±0.8 pg/ml, respectively). However, plasma P4 concentrations and the ratio P4:E2 increased from -3 to 7 d of the oestrous cycle (0.4 to 2.8±0.1 ng/ml and 0.06 to 0.4±0.01, respectively; P<0.01). A total of 102 genes were differentially expressed (DEG; FDR=0.10; fold change >2) between treatments, 20 genes were upregulated, and 82 genes were downregulated in T0 when compared to T1 cows. Functional enrichment analysis, using MeSH terms from category Chemicals and Drugs database (using Fisher's exact test; P<0.01), revealed terms associated with calcium phosphate, muscle protein, protein kinase, insulin and progesterone. Particularly, out of 71 genes in the Progesterone MeSH term, only 5 were DEG between treatments. Two genes were upregulated (OXT and CSN3) and 3 genes were downregulated (NOS2, SPADH1, MMP9) in T0 cows. Although P4 concentrations were similar, the differences in gene expression observed during early luteal phase point towards a different endometrial environment suggesting that these genes are involved in the remodelling of uterine tissue and may be required to modulate the physiological status of the uterus.

Session 32 Poster 17

Effect of parity on metabolic-endocrine response to a grazing event in lactating dairy cows

A.I. Trujillo, J.P. Soutto, A. Casal, M. Carriquiry and P. Chilibroste
Facultad de Agronomia, UDELAR, Produccion Animal y Pasturas, Montevideo, 12900, Uruguay; anatruji@fagro.edu.uy

The aim of this study was to characterize the metabolic-endocrine status of multiparous (M) and primiparous (P) Holstein cows during the first am grazing event (GE). Eighteen cows (M=9 and P=9; 540 and 501±32 kg BW; 28.5 and 23.3±3 kg milk yield, M and P, respectively, and similar DIM=73±7 and BCS=2.75±0.25), were assigned in a randomized block design. Cows were allowed, for 20 days (15 of adaptation and 5 of measurements), to graze an oat pasture (DM=14%, NEL=1.45 Mcal/kg DM; herbage allowance= 30 kg/DM/d) in daily strips (from 8 to 16 h) and complemented with a mixed ration after pm milking. Herbage intake at the first GE (HI, kg DM) was estimated by weighing animals before and after GE with correction for insensible loss and the duration of the first GE was measured. Blood samples were collected before (S0) and immediately after finished the first GE (S1) to determine serum concentrations of NEFA, BHB, and insulin. Data were analysed with a mixed model that include parity as fixed effect and block as random effect. Pearson's correlation was used to study associations between selected variables. The HI was greater (P=0.047) for M than P (5.03 vs 4.02±0.3 kg DM). The BHB and NEFA concentration at S0 were greater (P<0.04) for M than P cows (0.55 vs 0.37±0.04 and 0.36 vs 0.29±0.02 mmol/l, respectively). However, BHB concentration increase 84±20% from S0 to S1, maintaining differences between parity at S1 (P=0.009), while NEFA concentration decrease 50±6% and did not differ between parity at S1 (P=0.37). Insulin did not differ (P>0.46) between parity at S0 (10.96±0.85 uU/ml), but increased 18.6±12% from S0 to S1 without differences between parity (P=0.72). In addition, for P cows, concentration of BHB and insulin at S0 were positively correlated to HI (r=0.81, P=0.008; r=0.80, P=0.009; respectively), while variation of insulin from S0 to S1 was negatively correlated to HI (r=-0.65, P=0.06). Moreover, NEFA variation from S0 to S1 was positively correlated (r=0.74, P=0.03) with duration of GE, for M cows. Metabolic-endocrine status would be signals involved in feed intake regulation of M and P dairy cows grazing pastures and would be probably associated to changes in energy balance.

Development of maintenance energy requirement for sheep using calorimeter chamber data

Y.G. Zhao, A. Aubry and T. Yan
Agri-Food and Biosciences Institute, Agriculture Branch, Hillsborough, County Down, BT26 6DR, United Kingdom; tianhai.yan@afbini.gov.uk

The energy feeding system for sheep currently adopted in the UK is largely determined by the Agriculture and Food Research Council. However, there is evidence indicating this system underestimated metabolizable energy (ME) requirements for maintenance (ME_m) for the current sheep breeds. Therefore, the objective of the present study was to evaluate if the ME_m values recommended by AFRC were still valid for the current sheep flock in the UK. The data (n=88) used in the present study were collated from 3 sheep experiments with energy metabolism data measured using indirect open-circuit respiration calorimeter chambers. The animals were at age of 4 to 16 months including growing lamb (female, entire male and castrated male) and dry ewe and had a mean live weight of 45.7 kg (s.d.=9.68, from 29.5 to 71.5 kg). Sheep were selected from 4 breeds (Belclare, Lleyn, Texel and Highlander) and offered either fresh grass only diets or fresh grass and concentrate diets in restricted feeding or *ad libitum* feeding. Fresh grass was cut daily in the morning from perennial ryegrass swards which were managed to simulate grazing conditions. All animal were housed in individual pens for 19 days, then transferred to calorimeter chambers and housed there for 5 days with feed intake, faeces and urine output and gaseous exchange measured during the final 4 days. The data obtained were analysed using the linear regression technique with effects of experiments, breeds and sex removed. A linear relationship of energy balance (EB, $MJ/kg^{0.75}$) against ME intake (MEI, $MJ/kg^{0.75}$) was developed: $EB = 0.668_{(0.0636)}\ MEI - 0.327_{(0.0511)}$ (R^2=0.71). This equation indicates a ME_m of 0.489 $MJ/kg^{0.75}$. This value is 48% higher than that (0.33 $MJ/kg^{0.75}$) recommended by AFRC, indicating that using the energy feeding system of AFRC to ration the current sheep flock would underestimate their maintenance feed intake, and thus impact the production efficiency.

Association of BCS with backfat and longissimus dorsi muscle thickness in transition Holstein cows

N. Siachos[1], N. Panousis[1], G. Oikonomou[2] and G.E. Valergakis[1]
[1]Aristotle University of Thessaloniki, Faculty of Veterinary Medicine, School of Health Sciences, Box 393, 54124 Thessaloniki, Greece, [2]University of Liverpool, Institute of Veterinary Science, Faculty of Health and Life Sciences, Leahurst, Neston CH64 7TE, United Kingdom; geval@vet.auth.gr

The objective was to assess the relationship of body condition score (BCS) with ultrasound measurements of backfat (BFT) and longissimus dorsi muscle thickness (LDT) in transition Holstein cows under field conditions. Eighty-five Holstein cows from 2 dairy farms (n=32 and 53, respectively) in different parities (1: n=14; 2: n=35; ≥3: n=36) were assessed at 6 time points: 21 and 8 days prepartum; at parturition; and 8, 21 and 28 days postpartum. Measurements of BCS (1-5 scale), BFT and LDT (5−7.5-MHz linear transducer) were simultaneously recorded for each cow (in total 488 records). Pairwise linear correlation and regression analyses for BCS, BFT and LDT were done. A multivariate linear regression model was employed to assess the effect of BFT and LDT on BCS. Predictor importance was assessed with the automatic linear modelling function. Mean (±SD) BCS (3.17±0.37) and LDT (31.66±7.58 mm) peaked at 21 days prepartum, while BFT peaked at 8 days prepartum (13.77±4.28 mm). Mean (±SD) BCS (2.67±0.32) and BFT (9.52±2.50 mm) declined until 28 days postpartum, while LDT (21.80±5.55 mm) declined until 21 days postpartum. Differences in BCS, BFT and LDT between prepartum peaks and postpartum lows were significant ($P<0.05$). Overall pairwise correlations were: BCS/BFT, r=0.831; BCS/LDT, r=0.695 and BFT/LDT, r=0.570 ($P<0.001$). BFT and LDT were related better quadratically to BCS (R^2=0.717 and 0.483, respectively, $P<0.001$) than linearly. A model with BFT and LDT combined explained better the variation in BCS (R^2=0.768, $P<0.001$) than either alone. BFT had a higher predictor importance (0.79) compared to LDT (0.21). In primiparous cows BCS was better correlated with LDT than with BFT (r=0.789 and 0.698, respectively) compared to multiparous ones (r=0.702 and 0.848, respectively). Both BFT and LDT significantly affected BCS estimates in transition Holstein cows. Primiparous cows have lower fat and higher muscle reserves prepartum than multiparous at the same BCS, thus requiring a different management approach.

Transcriptomic impact of rumen epithelium induced by butyrate infusion in dairy cattle in dry period

M.J. Ranilla[1], R.L. Baldwin[2], R.W. Li[2], Y. Jia[3] and C.J. Li[2]
[1]Universidad de León, IGM ULE-CSIC1, Campus de Vegazana, s/n, 24071 León, Spain, [2]AGIL, BARC, USDA, Beltsville, MD 207052350, USA, [3]Johns Hopkins University, The Lieber Institute for Brain Development, Baltimore, MD 21205, USA; mjrang@unileon.es

Using next-generation sequencing and bioinformatics we sought to develop a better understanding of regulation in rumen epithelial transcriptome of cattle in the dry period induced by butyrate infusion at the level of the whole transcriptome. Butyrate, a byproduct of ruminal fermentation of feedstuffs is also a central element in control of ruminal epithelial differentiation. Following baseline control sampling at 0 h (CON), a 2.5 M solution of butyrate was continuously infused into the rumen for 168 h at a rate of 5.0 l/day. The infusion was then stopped and cows were maintained on the basal ration for an additional 168 h. Rumen epithelial samples were serially collected via grab biopsy through rumen fistulae at 0, 24, 72, 168 h (D0, D1, D3, D7) and 168 h post infusion (D 14). Compared to CON (pre-infusion at 0 h), a total of 3,513 genes were identified to be impacted in the rumen epithelium by butyrate infusion at least once at the different sampling time points at a stringent cutoff of FDR<0.01. The maximal effect of butyrate was observed at D7. Among these impacted genes, 117 genes were responsive consistently from D1 to D14, and another 42 genes only through D7. Temporal effects induced by butyrate infusion indicate that the transcriptomic alterations are very dynamic. Gene ontology enrichment analysis revealed that in the early stage of rumen butyrate infusion (on D1 and D3), the transcriptomic effects in the rumen epithelium were involved with mitotic cell cycle process, cell cycle process, and regulation of cell cycle. Bioinformatic analysis of cellular functions, canonical pathways and upstream regulator of impacted genes underlie the potential mechanisms of butyrate-induced gene expression regulation in rumen epithelium. The nutrigenomics approach may eventually lead to more precise management of utilization of feed resources in a more effective health and nutritional practices by better apprehending the whole animal response to nutrition, physiological state, and their interactions.

Relationship between TMR particle size distribution and digestibility of hay-based diets

M. Simoni[1], F. Righi[1], A. Foskolos[2], P. Formaggioni[1] and A. Quarantelli[1]
[1]University of Parma, Department of Veterinary Science, Via del Taglio 10, 40126 Parma, Italy, [2]Aberystwyth University, Institute of Biological, Environmental and Rural Sciences (IBERS), Gogerddan, SY23 3EE, Gogerddan, Aberystwyth, United Kingdom; marica.simoni@studenti.unipr.it

The aim of present work was to assess the relationship between hay-based Total Mixed Ration (TMR) particle size distribution and diet apparent digestibility (D). The trial was performed in 5 farms located in the area of Parmigiano Reggiano cheese production -North of Italy- where TMR was feed once a day. Three sampling days/farm were conducted at 15 days intervals at 0, 12, 24 hours after TMR distribution (t0, t12 and t24), at the beginning, middle and at the end of the manger. Five faecal samples were collected at t12 from fresh healthy lactating cows (60 to 90 days in milk). Physical-chemical and D evaluations were performed on TMR samples. Particle size was determined using the Penn State Particle Separator (PSPS). Apparent dry matter D (DMD) and neutral detergent fibre D (NDFD) were estimated using 240 hours undigested NDF (uNDF) as a marker, considering t12 faecal uNDF as representative of the daily excretion and using the weighted average dietary uNDF from samples collected at the 3 intervals assuming a 60% TMR intake in the in the first 12 h after distribution. The relationship between particle size distribution at t0 in the different sieves and estimated NDFD and DMD were studied through a curve fitting procedure. Significant daily variations in TMR chemical analysis and sieves %, probably related to sorting, were found in 4 farms. Residues on the middle sieve appeared strongly related to D with R^2=0.6197 and 0.5274 respectively for DMD and NDFD while residues on the upper sieve showed a weaker relationship (R^2=0.2567 and 0.3361 for DMD and NDFD). Both NDFD and DMD were estimated to be maximised when upper sieve residue was at 12% of the sample as fed, middle sieve residue was around 25%, lower residue was at 37% and bottom residue was higher than 40% of the sample. It appears that the careful preparation of the hay-based TMR diet – taking into account the suggested values of particle size distribution- can be a tool to improve its apparent digestibility.

Reasons for variations in residual feeding intake in lactating Nellore cows

J.N.S.G. Cyrillo[1], L.L. Souza[1], M.F. Zorzetto[1], N.D.C. Silva[1], J.A. Negrão[2] and M.E. Mercadante[1]
[1]Instituto de Zootecnia, Sertãozinho, SP, 14174-000, Brazil, [2]Universidade de São Paulo, FZEA, Pirassununga, SP, 13635-900, Brazil; cyrillo@iz.sp.gov.br

In order to assess growth and milk production traits related to the residual feeding intake variation (RFI) in lactating cows, 27 Nellore first calving cows (1,164±25 d of age and 506±40 kg of BW) were evaluated in a performance test over 177 d (GrowSafe Systems Ltd.). Cows were assigned into two groups according to the order of parturition, with one group consisting of 17 females that calved in November (G1) and another group with 10 females that calved in December 2016 (G2). The RFI was calculated for two stages of lactation as follows: the first stage from 21 to 100 d postpartum (RFI 1) and the second stage from 100 to 188 d postpartum (RFI 2). The following traits were evaluated: dry matter intake (DMI), average daily gain (ADG), metabolic body weight (BW0.75), ultrasonic fat thickness (FT) obtained in five anatomic sites, and energy-corrected milk production in 24 h (ECMP). RFI was obtained as the difference between DMI and predicted DMI obtained by multiple regression models including the following covariates: parturition group, ADG, BW0.75, FT and ECMP. The model used for the CAR1 calculation captured 29% of the variation in dry matter intake (P=0.1758), being 1% of the variation was explained by the parturition group, 3% by ADG, 18% by BW0.75, 1% by FT and 6% by ECMP. The model used for the CAR2 calculation was able to identify 40% of the variation in dry matter intake (P=0.1758), (P=0.0462), being 0.03% of the variation explained by the parturition group, 21% by ADG, 17% by BW0.75, 0,09% by FT and 1% by ECMP. Spearman correlations between DMI, BW0.75 and ADG at G1 and G2 were 0.60, 0.68, and 0.85 respectively, while -0.28 was verified between RFI1 and RFI2. We can conclude that at the beginning and in the intermediate period of lactation, ADG and BW0.75 were the main factors responsible for the variation in DMI of lactating Nellore cows, while FT and EMCP did not influence this variation.

Approaches to improve dual-purpose breeds' meat performance

D. Kohnke[1], H. Hamann[2] and P. Herold[1,2]
[1]Universität Hohenheim, Institut für Tropische Agrarwissenschaften, Garbenstr. 17, 70599 Stuttgart, Germany, [2]Landesamt für Geoinformation und Landentwicklung Baden-Württemberg, Referat 35, Zuchtwertschätzteam, Stuttgarter Str. 161, 70806 Kornwestheim, Germany; pera.herold@lgl.bwl.de

Modern dual-purpose breeds make an important contribution to the economic efficiency in farms. Different studies have shown the benefits of these breeds in comparison to dairy-type breeds regarding climate and resource efficiency. For a good performance of dual-purpose breeds, it is of importance to improve recording of traits of meat performance continuously. So far, only data of slaughtered young bulls have been included in the breeding value estimation. In this study, new data sources such as marketing data of living calves as well as data from slaughtered fattening heifers will be explored to what extent they can improve the quality for breeding value estimation for meat performance. Additional traits are daily gain and auction price (for living calves) as well as trading class and net gain (for fattening heifers). The gender specific traits for daily gain, trading class and net gain correlate strongly for the Fleckvieh breed (German dual purpose Simmental), which is why, unlike the auction price, a specific trait for each gender is not useful. The heritability for the daily gain is 30%; for auction price of males and females, it is very low: 4 and 3% respectively. For net gain (h^2=21%) and trading class (h^2=18%), the heritabilities are lower compared to the routine breeding value estimation for young bulls (h^2=29 and 21%). The genetic correlation between daily gain and net gain or rather trading class is 0.49 and -0.05 respectively. The same investigations will be carried out for Braunvieh cattle. In a next step, the influence of these new traits on the existing breeding values will be examined and the gain of information and reliability will be evaluated.

Contrasting metabolic indicators of energy and stress status in slaughter lambs and beef cattle

D.W. Pethick[1], S.M. Stewart[1], K.M.W. Loudon[1], G.E. Gardner[1], P. McGilchrist[2], J.M. Thompson[2], F.R. Dunshea[3], R. Polkinghorne[4], I.J. Lean[5] and G. Tarr[6]
[1]Murdoch University, Murdoch, 6150 WA, Australia, [2]University of New England, Armidale, 2350 NSW, Australia, [3]The University of Melbourne, Parkville, 3010 Vic, Australia, [4]Birkenwood, Murrurundi, 2338 NSW, Australia, [5]Scibus, Camden, 2570 NSW, Australia, [6]Sydney University, Darlington, 2006 NSW, Australia; d.pethick@murdoch.edu.au

Under Australian pre-slaughter conditions, cattle and sheep undergo similar management which leads to increases in blood glucose, lactate, non-esterified fatty acids (NEFA) and D-3-hydroxy butyrate (BHOB) as a result of fasting, physical exertion and adrenergic stress. It was hypothesised that lamb and beef cattle would demonstrate a similar metabolic response at slaughter. The lambs of mixed breed and sex (n=2,877), approximately 9 months old, hot carcase weight (HCW) 23±2.9 kg (mean, SD), were from the Meat and Livestock Australia Genetic Resource flock extensively raised and slaughtered at 2 different farms/abattoirs after 24-36 hours of feed deprivation. Pastures fed beef cattle (HCW=290±40 kg) were from 3 studies (1) *Bos indicus* cross steers (n=343) from 4 Central Queensland farms and slaughtered at one abattoir after 24-48 hours from dispatch (2) *Bos taurus* steers and heifers (n=240) sourced from King Island and slaughtered approximately 24 hours from dispatch (3) *B. taurus* steers and heifers (n=244) sourced from mainland Tasmania and slaughtered at the same abattoir as (2) approximately 48-72 hours from dispatch. Blood was collected at exsanguination immediately after stunning. Fat mobilisation was substantially higher for lambs with plasma NEFA (1.2±0.5 mM) and BHOB (0.4±0.1 mM) levels 2.8 and 1.7 times higher respectively while plasma glucose (6.8±1.1 mM) and lactate (12.0±3.1 mM) responses were 1.5 and 3.4 time higher respectfully in the beef cattle. It is concluded that the liver and muscle of beef cattle are under stronger adrenergic influence at the point of slaughter while lambs are mobilising more adipose tissue. The mechanisms underpinning these differences and any effects on meat quality are under further investigation.

Meat, past and future: will our children eat animal protein substitutes?

H. Huang[1], J. Serviere[1] and J.F. Hocquette[2]
[1]International Meat Secretariat (IMS), 5 rue Lespagnol, 75020 Paris, France, [2]INRA, UMRH, Theix, 63122 Saint-Genès Champanelle, France; jacquesserviere1@gmail.com

The relationship between man and domestic animals for food is rapidly evolving. What are the key outcomes, and what could the future bring? In the first part, the improvement of the main characteristics of meat 'quality' are summarized by a brief historical overview, using as specific examples data obtained from meat associations in least three different beef producing regions (France, USA, Uruguay): genetic improvement of breeds, carcass quality and organoleptic characteristics. The second part is on the challenges facing the industry. There are increased demands for accountability in production efficiency, animal care, and impact on the environment. Much has been written on this topic -we will summarize the main points. Challenges are context specific. There are important differences in production related to geography (physical endowments such as land, feed, water) and climate which influences choice of production system and species. Regarding consumption, developed countries have a very different pattern compared to developing countries. There is broad overlap between efficiency, animal care, and environment. They can be mutually re-enforcing (sustainable). However, there are also conflicts from societal demands (e.g. animal welfare, organic) from an increasingly urbanized public. More generally, there are concerns about climate change and increased stress on environmental resources. The third part focuses on future possibilities. How can conventionally produced meat respond to growing population and resource constraints? Will consumers continue to accept animals are raised and killed for meat? Is it more sustainable to produce vegetable protein compared to animal protein? Can meat be artificially produced at affordable prices and will consumers accept this alternative? The analysis will draw from on a range of research, including at INRA, data gathered from IMS membership, the FAO hosted Global Agenda for Sustainable Livestock, and the FAO-OIE-WHO One Health approach. This analysis provides a global overview, including well known institutions (FAO, OIE) and the only international private sector red meat association, the IMS.

Session 33 Theatre 4

Modelling beef meat quality traits during ageing by early post-mortem pH decay descriptors

C. Xavier[1], U. Gonzales-Barron[2], A. Muller[1] and V.A.P. Cadavez[2]
[1]ICBAS Institute of Biomedical Sciences Abel Salazar, University of Porto, R. Jorge de Viterbo Ferreira 228, 4050-313 Porto, Portugal, [2]CIMO Mountain Research Center, School of Agriculture, Polytechnic Institute of Braganza, Animal Science, Campus de Santa Apolónia, 5300-253 Braganza, Portugal; vcadavez@ipb.pt

Previous work has demonstrated that beef carcasses can be promptly and accurately classified into optimal quality and cold-shortened in accordance to the concept of pH/temperature 'ideal window' by using carcass characteristics and early *post-mortem* pH/temperature decay descriptors. The objective of this study was to assess the combined effects of the aforementioned variables on the two main eating quality attributes of meat – namely, tenderness (measured as shear force) and juiciness (measured as cooking loss) – during chill ageing. The pH and temperature in *longissimus thoracis* muscle of 51 beef carcasses were recorded during 24 h *post-mortem*, and decay descriptors were then obtained by fitting exponential models. Measures of Warner-Bratzler shear force and cooking loss were obtained from cooked meat after 3, 8 and 13 days of cold ageing. The fitted mixed-effect models revealed that both meat tenderisation and cooking loss increased with ageing ($P<0.01$) although their rates slowed down in time ($P<0.05$). Beef carcasses with a higher pH (obtained at different endpoints: 1.5, 3.0, 4.5 or 6.0 h *post-mortem*) produced aged meat with increased tenderness ($P=0.013$) and increased water retention during cooking ($P=0.016$) than those of lower pH. Nonetheless, the slower the pH decay rate, as happens in a cold-shortened carcass, the lower the potential for tenderisation ($P=0.038$) and water retention ($P=0.050$) during ageing. Whereas sex affected shear force, with females producing meat of higher tenderness, aged meat of increased water retention was produced by heavier beef carcasses ($P<0.001$). The good fitting quality of the shear force ($R^2=0.847$) and cooking loss ($R^2=0.882$) models and their similarity among the different endpoints post-mortem indicated that both eating quality attributes can be approached by recording the pH decline of a beef carcass during the first 3.0 hours after slaughter.

Session 33 Theatre 5

Improving meat yield and quality in cattle – stand back and let the geneticists take over!

D.P. Berry[1], M. Judge[1], S. Conroy[2], T. Pabiou[2] and A.R. Cromie[2]
[1]Teagasc, Moorepark, Co Cork, Ireland, [2]Irish Cattle Breeding Federation, Bandon, Co Cork, Ireland; donagh.berry@teagasc.ie

The contribution of genetic improvement to advances in animal performances is well recognised. A well-designed study in broilers clearly illustrated how genetic improvement contributed up to 90% of the gains in efficiency over a 40 year period with advances nutrition contributing the remainder. Geneticists are (rightfully) blamed with causing the erosion of reproductive performance in dairy cows at the end of the 21st century; geneticists are now (rightfully) acknowledged for reversing the trend in dairy cow reproductive performance with the expectations (substantiated by on-going experimental studies) in countries like Ireland, that the reproductive performance of the modern dairy cow will revert back to the 'good old days' of the 1980's despite phenomenal increases in milk output per cow during that period. All this was achieved because there was an (international) willingness, underpinned by market signals, to correct past mistakes. The benefit of breeding is that it is cumulative and permanent meaning that the gains achieved are compounded with each generation. To be considered for inclusion in a breeding program a trait must fulfil three criteria: (1) importance, (2) exhibit genetic variation, and (3) be measurable or correlated with a measurable traits (ideally at a low cost). Meat yield and quality are certainly important and large exploitable genetic variability in known to exist. The remaining hurdle is the willingness to improve these traits which is largely dictated by the resources required to feed a sustainable, efficient, and effective breeding program. This is largely dictated by access to data from which to generate accurate genetic evaluations. Several alternative strategies exist, one of which includes international collaborative efforts as underway for feed intake and efficiency (also a difficult to measure trait). Although the capital and running cost is actually large, on a per kg of meat produced nationally, the cost is minuscule. Breeding has proven itself to be successful in (rapidly) improving performance; it's time now to apply this skillset to rapidly improving meat yield and quality.

Session 33 — Theatre 6

Crossbreeding with beef bulls in Swedish dairy herds – analysis of calving and carcass traits

S. Eriksson[1], P. Gullstrand[1], W.F. Fikse[2], E. Jonsson[3], J.-Å. Eriksson[2], H. Stålhammar[4], A. Wallenbeck[1,3] and A. Hessle[3]

[1]Swedish University of Agricultural Sciences, Dept. Animal Breeding and Genetics, Box 7023, 75007, Sweden, [2]Växa Sverige, Box 288, 75105, Sweden, [3]Swedish University of Agricultural Sciences, Dept. Animal Environment and Health, Box 234, 53223, Sweden, [4]VikingGenetics, Örnsro, Skara, 53294, Sweden; susanne.eriksson@slu.se

Crossbreeding with beef bulls has potential to increase beef production from dairy herds, but effects on calving ease must be considered. We compared crossbreds (dairy dams × beef sires) and purebred dairy cattle regarding field-recorded calving (n>1.3 million) and carcass traits (N>1.1 million). The dairy breeds Swedish Red and Swedish Holstein, and the beef breeds Angus, Hereford, Limousin, Simmental and Charolais were included. Differences in least square means were estimated between calf breeds for calving difficulty and stillbirth in separate analyses for first and later calvings, with and without sire EBV nested within breed in the models. For carcass gain, carcass weight, EUROP conformation and fatness classifications, breed differences were estimated separately for heifers, steers and young bulls. The lowest frequency of calving difficulty was found for dairy × Angus, followed by purebred dairy and dairy × Hereford calves. Crossbreeding with the late maturing breeds Charolais and Simmental gave more calving difficulties, especially in primiparous cows. Frequency of stillbirth was lower for dairy × beef crosses than for purebred dairy calves. Differences between offspring of bulls with different EBVs were most pronounced in Charolais. Crosses sired by late maturing beef sires had higher carcass gain, carcass weight and conformation score compared with purebred dairy cattle or crosses with early maturing beef breeds. Carcass gain and weight were highest in crosses with Charolais and Simmental sires, but Limousin crosses were superior in conformation. Advantages of dairy × beef crosses for these traits were largest in young bulls. Crosses with early maturing beef breeds gave highest fat scores. In conclusion, the use of beef sires in crossbreeding with dairy dams can improve beef production, and the choice of both breed combination and individual beef bull is of considerable importance for optimizing the outcome.

Session 33 — Theatre 7

Multi-trait analysis of meat quality traits in Limousin, Charolais and Blonde d'Aquitaine breeds

A. Michenet[1,2], D. Rocha[1], G. Renand[1], R. Saintilan[1,2], Y. Ramayo-Caldas[1] and R. Philippe[3]

[1]GABI, INRA, AgroParisTech, Université Paris-Saclay, Domaine de Vilvert, 78350 Jouy-en-Josas, France, [2]ALLICE, 149 rue de Bercy, 75012 Paris, France, [3]GAMAA, INRA, Université de Limoges, 87000 Limoges, France; alexis.michenet@inra.fr

Meat quality traits in beef cattle are complex traits which are difficult and expensive to measure. They are not yet selected in French beef cattle selection programs. However, the interest for a genomic evaluation is growing. A better understanding of the genetic determinism, including relation between traits, and between breeds, is of major importance in order to set an efficient model for genomic selection. In this study, Genome-wide association studies (GWAS) were performed for six traits (tenderness, shear force, juiciness, flavour, colour, intra-muscular fat). An association weight matrix (AWM) was performed to analyse GWAS results. This systems biology approach made it possible to identify candidate genes affecting one or more traits. The dataset was made of 1,215 Limousin, 1,059 Charolais and 943 Blonde d'Aquitaine animals genotyped and phenotyped. Genotypes from the Bovine SNP50 BeadChip® were imputed on Bovine HD BeadChip®. GWAS were then performed with GCTA software. The AWM was done considering successively one of the six phenotype as the key phenotype. The P-value threshold of 0.001 was used in AWM. A correlation threshold of 0.30 was retained to include other traits in the AWM. For the three breeds, shear force, juiciness and flavor were correlated with tenderness. A total of 1,586 genes were identified for the 6 key phenotypes in the 3 breeds. Previously known genes (*CAPN1, CAPN5, CAST, GHR, MSTN*) affecting the traits were detected and many extra genes were discovered. Most genes were breed-specific which reduce the gain of a multi-breed evaluation. There were 29 genes detected for at least one trait and in two breeds. The false positive discovery rate was lower for these genes. Further studies using sequence data of key ancestors could lead to the identification of causal mutations, which could be used as fixed effect for genomic selection.

Session 33 Theatre 8

A new model for the genetic evaluation of carcass traits in Switzerland

S. Kunz[1], S. Strasser[2], U. Schnyder[1], U. Schuler[1], M. Berweger[1], F. Seefried[1] and P. Von Rohr[1]
[1]Qualitas AG, Chamerstrasse 56, 6300 Zug, Switzerland, [2]Mutterkuh Schweiz, Stapferstrasse 2, 5201 Brugg, Switzerland; sophie.kunz@qualitasag.ch

Since 2004, genetic evaluations for carcass conformation and net gain until slaughtering for calves (c) and adults (a) have been published for the six major beef cattle breeds in Switzerland (Angus, Braunvieh, Charolais, Limousine, Simmental and Aubrac). So far, carcass fatness was included in the model as a fixed effect. In recent years, Mutterkuh Schweiz – The Swiss Beef Cattle Association – observed a decline in average carcass fatness. This is the reason why they wanted to provide new breeding values for this trait as a selection tool in beef cattle. The objective of this project was to develop a new model and add carcass fatness grade for c and a as newly evaluated traits. Fattening phenotypes (carcass conformation grade CC, carcass fatness grade CF and carcass weight CW for calves a and adults a, respectively) and pedigree data have been processed for estimating genetic parameters applying a 6-trait (CCc, CCa, CFc, CFa, CWc, CWa) multivariate animal model. Animal's sex and slaughterhouse were modeled as fixed effects. CC and CF have the grader as an extra fixed effect. Herd×year and animal were treated as random effects. In order to correct for age at slaughter, linear and quadratic regressions on age were fitted as covariates. In this multi-breed evaluation, genetic groups are defined by breed and birth year accounted for unknown parents. Estimated heritabilities were moderate to high: 0.22 (CWc), 0.30 (CWa), 0.31 (CFc), 0.36 (CFa), 0.50 (CCc) and 0.56 (CCa). Genetic correlations within trait-classes between calf and adult traits were high (0.84 for CC, 0.73 for CF and 0.63 for CW), but markedly different from 1. Genetic correlations between CFc and all other traits were close to zero. Low correlations are obtained between CFa and CCc (-0.17), CFa and CCa (0.15), CFa and CWc (-0.20), CFa and CWa (0.26), CCc and CWa (0.29). Moderate correlations are obtained between CCc and CWc (0.52), CCa and CWc (0.36), CCa and CWa (0.38). Hence, selection on CF should not sacrifice too much genetic progress in CC and CW.

Session 33 Theatre 9

Accelerating the development of carcase measurement technologies in Australia

G.E. Gardner, C.G. Jose, H.B. Calnan, S.M. Stewart, D.W. Pethick, P. McGilchrist, D.J. Brown, W.S. Pitchford, C. Ruberg and J. Marimuthu
Advanced Livestock Measurement Technologies Project, Murdoch University, Murdoch, WA 6150, Australia; g.gardner@murdoch.edu.au

Within the Australian Lamb and Beef industries, the pricing of livestock at slaughter poorly reflects the amount of saleable meat in carcases. Likewise, within a cut price has little association with eating quality for lamb. This lack of transparency in livestock trading is largely due to the lack of carcase measurement to properly quantify these traits. To address this, the Advanced Livestock Measurement Technologies project was initiated to accelerate the development of technologies that can measure both carcase composition and eating quality at chain-speed within Australian abattoirs. Where possible, synergies with automation are sought, facilitating the path to adoption by simply modifying existing hardware, rather than imposing new hardware and processes requiring additional labour. The flow of data resulting from these new technologies will be captured within genetic databases and producer feedback systems, and tools provided to enhance decision making in response to this feedback. In many cases technologies are being developed that measure traits by different means than is traded upon within industry, or that measure new traits that previously didn't exist. For example, photonic devices are being developed that determine chemical fat percentage in muscle, yet in the beef industry only visual marbling is graded and this trait is not measured in lamb and pork industries. Similarly, new on-line DEXA systems can measure whole carcase lean meat yield in lamb and beef, yet only GR tissue depth or P8 fat depth are currently traded on. On this basis work is required to not only accelerate the development of these measurement technologies, but also to develop the industry systems required to underpin their on-going calibration and auditing. These industry systems will be structured around 'gold-standard' measurements of carcase composition determined by computed tomography and chemical fat composition determined by Soxhlet fat extraction. Close interaction will be required between producers, processors, auditors and regulating authorities for industry to keep pace with this technological transformation.

Session 33 Theatre 10

Are there any global indicators of early and late stress response in beef cattle?

J.O. Rosa[1], M.J. Carabaño[2], J. De La Fuente[3], C. Meneses[2], C. Gonzalez[2], C. Perez[3], D. Munari[1] and C. Diaz[2]
[1]FCAV/UNESP, Jaboticabal, Sao Paulo, Brazil, [2]INIA, Dpto. Mejora Genética Animal, Ctra de la Coruña km 7,5, 28040 Madrid, Spain, [3]UCM, Facultad de Veterinaria, 28040 Madrid, Spain; cdiaz@inia.es

In livestock production, animals are exposed to management conditions that induce stress responses. Understanding the complexity of stress responses may help to select animals less reactive to external stimuli and therefore, with less negative effects of stress on their welfare, easiness of handling, performance and beef quality. The aim of the present study was to characterize the response to stress in two different periods, fattening (F) and slaughter (S) and to evaluate if the early response in F could be used to predict response at S using a set of biomarkers of stress in beef cattle. Blood samples of 80 *Avileña-Negra Iberica* male calves were collected in the feedlot (between 4 and 7 days before finishing) and at the slaughterhouse during exsanguination. Afterwards, 7 biomarkers (albumin, cortisol, creatine kinase, glucose, lactate, lactate dehydrogenase and globulin) were determined in blood plasma. Statistical analyses were performed using principal components and discriminant analyses. During F, lactate dehydrogenase has the most important contribution while at S, lactate was the main indicator of stress, which suggests that at slaughter the main attribute is a muscle response, with possible negative impact on meat quality. Albumin and lactate were the most discriminant biomarkers during both periods. Using all biomarkers, 91 and 86% of the animals were correctly assigned to feedlot and to slaughter conditions, respectively. Removing albumin, the discrimination capacity was reduced to 54% for the F group and to 47% for the S group. Removing lactate, 100% of the individuals were correctly attributed to the F group while the S group showed 24% of miss-matching. Results indicate that this set of markers discriminate the two different stress conditions. Therefore, biomarkers linked to stress during fattening may not be a good predictor of stress at slaughter.

Session 33 Theatre 11

Bioeconomic modelling of specialist beef finishing systems in Scotland

C. Kamilaris[1], R.J. Dewhurst[1], B. Vosough Ahmadi[1] and P. Alexander[2]
[1]Scotland's Rural College (SRUC), Kings Buildings, West Mains Road, EH9 3JG, Edinburgh, United Kingdom, [2]University of Edinburgh, School of Geosciences, Drummond Street, EH8 9XP, Edinburgh, United Kingdom; harry.kamilaris@sruc.ac.uk

Scottish beef finishing enterprises are potentially impacted by the political developments and continuous regulatory revisions, such as CAP reforms and Brexit. Bioeconomic farm-level modelling can support decision making by identifying optimized practices, and providing information for farmers, policy-makers and other relevant stakeholders. One example is the Grange Scottish Beef Systems Model (GSBSM), which was designed to simulate Scottish beef finishing enterprises using data from a Scotland's Rural College (SRUC) study of beef and numerous agricultural input and output prices. Here we use the model to determine the cost-effectiveness for different management practices and slaughter ages (i.e. short: 14-17 months, medium: 18-23 months, long: 24-35 months). Systems were modelled on the basis of a typical 40 hectare farm. Analysis revealed negative net margins for all the scenarios included. On an average basis, reported in £000's per farm, steer systems lost 3.75 for the short, 11.79 for the medium and 60.26 for the long duration systems, while heifers reported as losses of 28.48, 21.88 and 58.74, respectively. The highest financial returns were recorded for the 14 month steer system and the 18 month heifer system; while, in contrast, the least profitable systems were the 24 month for both steers and heifers. Beef fattening farmer returns were simulated with genetic selection of stock for feed efficiency under different systems. Outcomes showed higher net margins for using higher genetic merit stock, under all systems examined, allowing the 14, 15 and 20 month steer systems, as well as the 18 and 19 month systems for heifers to become profitable. These results imply that there are opportunities for profitable and sustainable beef production in Scotland, for both cereal and forage based systems, particularly for steer finishing when aiming for a younger age profile at slaughtering. Furthermore, optimizing genetic selection for feed efficiency in cattle could be used to improve profitability for beef finishing systems.

Session 33 Poster 12

Sustainable production of high quality beef and pork in Slovenia – focus on SLO-ACE project

M. Čandek-Potokar[1], V. Meglič[1], B. Lebret[2] and J.-F. Hocquette[3]
[1]Agricultural Institute of Slovenia (KIS), Hacquetova ul. 17, 1000 Ljubljana, Slovenia, [2]INRA-Agrocampus, UMR 1348 Pegase, St.Gilles, 35590 St. Gilles, France, [3]INRA VetAgroSup, UMR 1213 Herbivores, Theix, 63122 St. Genes Champanelle, France; meta.candek-potokar@kis.si

Sustainable food production is one of key priorities in the national strategy of Slovenia, as it is a very small country with limited resources and agricultural land availability. Over half of its territory is covered by forests, 3/4 of agricultural land is categorized as less favorable areas with 2/3 of permanent grassland. The country is mainly hilly and despite its small size has diverse climates (Mediterranean, alpine to continental). This specific geo-agro-climatic context provides favourable conditions for cattle breeding, which is the most important agricultural segment. While dairy sector productivity has progressed in the past decades, the beef sector remains unexploited and its innovation potential needs to be increased to improve its sustainability. The sector is characterized by predominance of double-purpose breeds with emphasis on milk production, leading to unexploited meat potential with low growth intensity (<900 g/day) and carcass conformation (over half in class R) and low fatness (60% in class 2). The key challenges are to better exploit the potentials of traditional breeds and permanent grassland, to include meat quality in the breeding programs, and to better consider consumer and societal expectations regarding meat quality, animal welfare, and environment. Pork is the primary meat consumed in Slovenia (40 kg/capita). However, self-sufficiency has dropped from 80 to 35% in the last 20 years, leading to high importations for consumption and domestic transformation. Current challenge is to increase domestic pig production and transformation of added-value pork products. Recently approved H2020 coordination and support action project SLO-ACE will pursue these challenges and endeavour to improve the structural organization, human resources and knowledge transfer capacity needed for high impact research devoted to sustainable high quality beef and pork production in Slovenia. On this area of research, SLO-ACE will be supported and mentored by the French institute INRA. SLO-ACE is financed through H2020 program (grant agreement no.763655).

Session 33 Poster 13

Factors affecting growth of suckling Angus calves

R. Wassmuth[1], T. Hohnholz[1], K. Gillandt[2] and N. Kemper[2]
[1]University of Applied Sciences Osnabrueck, Am Kruempel 31, D 49090 Osnabrueck, Germany, [2]University of Veterinary Medicine Hannover, Bischofsholer Damm 15, 30173 Hannover, Germany; r.wassmuth@hs-osnabrueck.de

The objective of this study was to evaluate suckling performance of Angus calves in different regions of Germany in relation to traits of their mothers. Data of 600 single born calves in four herds were collected in northeast lowlands and in low mountain ranges of Germany. The mean weights at birth and at day 200 were 36.2 kg (±4.1 kg) and 232.0 kg (±33.5 kg), respectively. Average daily gain from birth to the 200th day of life was 1,139.4 g (±164.0 g). Udder volume and teat circumference were derived from 316 German and Aberdeen Angus cows. Both traits were classified in 3 categories according to volume and circumference, respectively. In statistical model 1, the fixed factors herd (4), calving season (2) and parity (3) had a significant influence on birth weight. Calves of the lowland herd and the warmer calving season showed the lower birth weight. First parity cows gave birth to the lightest calves (35.5 vs 37.5 and 37.9 kg). Statistical models 2 and 3 included the fixed effects of herd, calving season, parity, birth weight in 3 classes and udder volume (model 2) or teat circumference (model 3). A heavier birth weight led to an increased daily gain ($P<0.05$) and a higher weight at day 200 by trend ($P>0.05$). Udder volume had a significant influence on daily gain and medium-sized udders led to the highest daily gain (1,108.2 vs 1,062.3 and 1,079.1 g). A similar result was found for the weight on day 200 but the influence of the udder volume was not significant. Higher teat circumferences led to higher daily gain and weaning weight but results were not significant. According to the present results, breeding for a medium-sized udder could influence growth of calves positively without any obvious disadvantage as opposed to breeding for higher birth weights or teat circumferences. Increased birth weights could adversely affect calving ease, and greater teat circumferences could make suckling/sucking more difficult. Region, calving season and parity of cows influenced growth of calves. It could be concluded that medium-sized udders led to an increased growth rate which could be implemented in breeding programmes.

Session 33 Poster 14

Comparison of Hereford and Hereford × Charolais bulls in conventional beef production in Croatia

A. Ivanković, M. Pećina, M. Konjačić and N. Kelava Ugarković
University of Zagreb Faculty of Agriculture, Department of Animal Science and Technology, Svetošimunska cesta 25, 10000 Zagreb, Croatia; aivankovic@agr.hr

Beef production is a very important part of animal production in Croatia. It is constantly faced with the problem of small number and poor quality of calves for beef production. One of the solutions is to produce more crossbreed calves and choose appropriate genotype considering production technology. The aim of this research was to determine production characteristics of Hereford bulls and crossbreed Hereford × Charolais bulls in conventional fattening conditions. Twenty-five Hereford and twenty-five crossbreed Hereford × Charolais bulls were used in the study. The bulls were fattened in the same housing and feeding conditions, fed *ad libitum* with diet based on maize silage and concentrates. The bulls were slaughtered at the age of 15±0.2 months according to the standard procedure. Carcasses were chilled under commercial conditions at 4 °C/24 h. EUROP conformation score of carcass and subcutaneous fat thickness were determined at 24 h *post-mortem*. Subcutaneous fat thickness was measured with a calliper over the 12th rib of the *m. longissimus dorsi*. Minolta Chroma CR-410 colorimeter was used in order to evaluate meat colour. Statistical analysis was performed using the General Linear Model. Average daily gains during fattening period were greater in crossbreed Hereford × Charolais bulls than in Hereford bulls (977 vs 1,031.5 g; P<0.05). Hereford bulls reached lower final live weight (452 vs 472 kg; P<0.05) and hot carcass weight (270 vs 281 kg; P<0.05) than crossbreed Hereford × Charolais. Dressing percentages of Hereford bulls and crossbreed Hereford × Charolais (59.7 vs 59.5%) were not significantly different. Subcutaneous fat thickness was greater (+4.1 mm; P<0.001) in Herefords than crossbreed bulls. No significant influence of breed on meat colour was observed. The results confirm good performance of Hereford bulls and Hereford × Charolais crossbreed bulls for beef production.

Session 33 Poster 15

Assessment of the equation for predicting enteric methane production from feedlot beef cattle

T.T. Berchielli, A.K. Almeida, G. Fiorentini, J.D. Messana and Y.T. Granja-Salcedo
Unesp, Animal Science, Via de Acesso Prof. Paulo Donato Castellane, 14884-900, Jaboticabal, SP, Brazil; ttberchi@fcav.unesp.br

The objective of the present study was to evaluate the Intergovernmental Panel on Climate Change equation for predicting enteric methane (CH_4) production from feedlot beef cattle. We evaluated the IPCC ($CH_{4},_{MJ/d}$=DMI, $_{kg/d}$ × GE,$_{MJ/kg}$ × 0.065, $_{\%\ of\ GE}$) equation for estimating methane production from feedlot beef cattle using individual data, from four studies, of dry matter intake (DMI) and gross energy (GE) content of the diet of 137 Nellore steers (mean body weight 513±47.5 kg and 637±79.9 days of age). These animals were fed *ad libitum* over 20% of corn silage (DMI of 8.39±1.84 kg/d; 17.7±1.19 MJ GE/kg, and 158±13.7 g of crude protein/kg in DM basis) and gained 1.25±0.391 kg/d. Studies included in this evaluation used the Sulphur hexafluoride (SF_6) tracer technique to measure CH_4 production. The MIXED procedure of SAS was used to evaluate the equation by regressing residual (observed minus predicted) values on the predicted values centred on their mean values. The intercept of the linear regression equation was used to estimate mean biases, whereas linear bias (i.e. systematic bias) was assessed using the slope of the linear regression equations. The study was considered as RANDOM effect, also the fixed effect of fat addition on the diet was tested (Fischer's protected LSMEANS). Observed CH_4 production was 7.82±2.29 MJ/d. The model evaluation revealed that the IPPC equation overestimated the enteric CH_4 production in 2.19±0.497 MJ/d (P=0.012). Moreover, residual enteric CH_4 production was greater (P<0.01) in animals supplemented with fat (means bias=-3.20±0.551 MJ/d) than those fed diets with no fat addition (means bias=-1.17±0.521 MJ/d). Thus, fat addition considerations should be included in the next review of IPCC guidelines. A systematic bias was also detected in the present evaluation (P<0.01; slope =-0.532±0.0804, unitless). In this sense, it is expected an increase in the mean bias of 0.532 units, per unit of increase in the predicted values. This means, that the overestimation is aggravated as predicted value increases (increased DMI and/or GE content in the diet).

Assessment of equation for predicting enteric methane production from grazing beef cattle

T.T. Berchielli, A.K. Almeida, G. Fiorentini, J.D. Messana and Y.T. Granja-Salcedo
Unesp, Department of Animal Science, Via de Acesso Prof. Paulo Donato Castellane, 14884-900, Jaboticabal, SP, Brazil; ttberchi@fcav.unesp.br

The objective this study was to evaluate the Intergovernmental Panel on Climate Change equation for predicting enteric methane (CH_4) production from grazing beef cattle. We evaluated the IPCC (CH_4,MJ/d=DMI$_{,kg/d}$ × GE$_{,MJ/kg}$ × 0.065$_{,\% GE}$) equation for estimating methane production from grazing beef cattle using individual data from four studies of dry matter intake (DMI) and gross energy (GE) content of the diet of 97 Nellore steers, (mean body weight (BW) of 496±67.5 kg, and 19.3±5.04 months of age). Animals were kept in paddocks of *Brachiaria brizantha* pasture, supplemented from 0.5 to 3% BW (DMI of 10.3±1.86 kg/d; 18.3±1.23 MJ/kg, and 164±10.4 g of crude protein/kg, DM basis) and gained 762±176 g/d. All studies included in this evaluation used the Sulphur hexafluoride (SF_6) tracer technique to measure CH_4 production. The MIXED procedure of SAS was used to evaluate the abovementioned equation by regressing residual (observed minus predicted) values on the predicted values centred on their mean. The intercept of the linear regression equation was used to estimate mean biases, whereas linear bias (i.e. systematic bias) was assessed using the slope of the regression equations. The study was considered as RANDOM effect, also the effect of fixed effect of fat addition on the diet was tested. Observed CH_4 production was 7.68±2.91 MJ/d. The present model evaluation revealed that the IPPC equation overestimated the enteric CH_4 production in 3.79±1.18 MJ/d (P=0.03), irrespective of fat addition in the supplement (P=0.14). Moreover, the observed enteric CH_4 represented 4.57±2.91% GE, this value is 30% lower than that assumed in the IPCC equation. The value of 6.5% of GE losses in enteric CH_4 should be revisited for future review of IPCC guidelines. A systematic bias was also detected in the present evaluation (P<0.01; slope =-0.939±0.127, unitless). In this sense, it is expected an increase in the mean bias of 0.939 unit per unit of increase in the predicted value. This means, that the overestimation is aggravated as predicted value increases.

Selection index for prediction of dressing percentage in Nguni cattle using RTU measurements

J. De Vos, H. Theron and E. Van Marle-Köster
University of Pretoria, Animal and Wildlife Sciences, Lynnwood road, Hatfield, Pretoria, 0002, South Africa; janidevos94@gmail.com

The Nguni is an indigenous small framed breed adapted to sub-tropical conditions and mostly farmed under extensive production systems. A trail was conducted to evaluate growth and carcass traits of Nguni cattle in feedlots as an alternative production system. Real time ultrasound (RTU) measurements are routinely used for prediction of carcass traits on live animals. It is a cost effective and non-invasive measurement compared to methods based on post slaughter evaluations. The aim of this study was to compile a selection index for predicting Dressing Percentage (DP) at a 120 days slaughter (A2 grade) based on RTU measurements. Live weight, slaughter weight and warm carcass weight of 200 animals were recorded. RTU measurements were performed to estimate back fat (BF) and eye muscle area (EMA) at 72 and 91 days on test. Animals were slaughtered when an A2 (1-3 mm fat) carcass grade was reached. The slaughter age and weight at A2 grade were 429 (±44.3) days and 347 (±33.7) kg respectively and the average DP was 56,1% (±0.12). EMA measured at 72 days was 49.2 (±0.5) cm^2 and BF was 3.2 (±0.1) mm at 72 days on test. At 91 days on test, EMA was 52.4 (±5.7) cm^2 and BF 3.5 (±0.9) mm. An index was compiled using ANOVA and regression for modelling DP at an average slaughter date of 120 days. At 72 days on test, DP = (51.4 + 0.14 BF + 0.15 EMA – 0.01 Slaughter weight) and at 91 days on test, DP = (52.3 + 0.17 BF + 0.12 EMA – 0.008 Slaughter weight). The regression coefficient (R^2) for the selection index was 0.27 and 0.17 at 72 days and at 91 days respectively. Thus, RTU measurements at 72 days on test improved the prediction compared to the RTU measured at 91 days. Slaughter weight had no significant effect with only a small contribution to the overall prediction of dressing percentage. The results suggest that the earlier RTU measurements are better predictors of dressing percentage for Nguni cattle in a feedlot. Other studies have shown that real-time ultrasound measurements are good predictors of meat yield. It has also been reported that accuracy of prediction tend to decrease with RTU measurements recorded closer to slaughter, similar to this study.

Feed additives as rumen fermentation modulators during transition of beef calves

S. Yuste, Z. Amanzougarene, M. Fondevila and A. De Vega
Universidad de Zaragoza-IA2, M.Servet 177, 50013 Zaragoza, Spain; avega@unizar.es

Tannins and medium-chain fatty acids (MCFA) have been suggested as rumen fermentation modulators slowing down bacterial activity and, thus, preventing ruminal acidosis. The aim of this work was to assess the effect of inclusion of these additives on animal performance, and rumen fermentation, during transition of beef calves to high-concentrate diets. Eighteen 5-month old (212±27.0 SD kg) ruminally fistulated Limousine crossbred male calves that had been fed only on milk and grass were randomly assigned to three diets over a 30-day period. Treatments were a highly fermentable diet with (kg/kg, as fed) 0.59 barley and 0.15 maize as control (C), C plus 2% of a commercial 65:35 chestnut and quebracho extract of tannins (T), and C plus 0.6% of a commercial mixture of MCFA (K). Dry matter intake (DMI, daily) and live weight (LW, weekly) were recorded. Rumen fluid was sampled on days 0, 1, 7, 14, 21 and 30 before the morning feeding (0 h), and after 3, 6 and 9 h. Data were analysed with PROC MIXED of SAS, and a repeated measures model was used for analysis of rumen fermentation data. No differences were detected on average daily gain (P=0.96), feed to gain ratio (P=0.87) or LW at the end of the trial (P=0.83). Similarly, total concentrate DMI was not affected by diet, although it was numerically lower in group T than in groups C and K (126, 149, 138 kg DM, respectively; P=0.56). Diets promoted a similar rumen fermentation pattern with no differences in rumen pH (P=0.59), ammonia (P=0.35), lactic acid (P=0.28), and total volatile fatty acids (P=0.55) concentrations, or in acetate (P=0.49), propionate (P=0.22) and butyrate (P=0.49) proportions. At the level of inclusion of tannins and MCFA reported in this study no effect was found on rumen fermentation and animal performance. It seems that control diet did not promote a drop in rumen pH, which averaged 6.32, and this could be the reason for the lack of effect of the studied additives.

Session 33 Poster 19

Insulin responsiveness increases with maturity of beef cattle

G.E. Gardner[1,2], J.M. Thompson[1] and H.B. Calnan[2]
[1]Australian CRC for Beef Genetic Technologies, Armidale, NSW 2351, Australia, [2]Murdoch University, South St, Murdoch 6150, Australia; honor.calnan@murdoch.edu.au

Adequate muscle glycogen at slaughter is crucial to producing premium quality meat, particularly to reduce the incidence of dark cutting in beef. Insulin is the hormone primarily responsible for the uptake and storage of glucose by muscles, and increased sensitivity to insulin will increase muscle glycogen storage. Selection for muscling was shown to increase insulin responsiveness in steers, though whether this effect is consistent at different stages of animal maturity is unclear. This experiment challenged 16 cattle (5 Wagyu, 5 Angus and 6 Piedmontese) with two doses of insulin (0.5 and 2IU/kg live weight) at earlier (15 month) and later (36 month) stages of maturity. The cattle were individually stalled and acclimatised for 14 days before they were challenged with IV insulin. Blood was sequentially sampled from a jugular catheter at -30, -15, -10, -5, 0, 2.5, 5, 10, 15, 20, 30 and 120 minutes relative to the insulin injections and analysed for glucose concentrations. On sampling days cattle were fed in small portions hourly to ensure steady-state blood metabolites. Blood glucose responses over time were modelled with a customised exponential function that provided the area under the curve (AUC) between 0 and 30 minutes. Response to the insulin challenges differed between breeds (P<0.05), and contrary to expectations, less muscled Angus demonstrated a greater response to insulin than more muscled Piedmontese cattle. At 15 months, Angus cattle responded most to the 2IU/kg insulin challenge with a glucose concentration AUC of 32.4±3.6 mM/30 min, compared to the Piedmontese response of 22.0±3.3 mM/30 min (P<0.05). Generally, the response to the insulin challenges increased with animal maturity, particularly in Piedmontese cattle, whose response to 2µg/kg insulin increased from 22.0±3.3 mM/30 min at 15 months to 33.7±3.3 mM/30 min at 36 months of age (P<0.05). This increased insulin responsiveness was unexpected, given that fatness increases with maturity and has been linked to insulin insensitivity. These results instead suggest that insulin responsiveness will not negatively influence meat quality in beef cattle aged up to 36 months.

Session 33 Poster 20

Fatty acid profile in beef from animals in long term feeding with dried distiller grain in the diet

J. Fonseca Lage[1], A. Hoffmann[2], A. Cristina Ferrari[2], L. Delevatti[2] and R. Andrade Reis[2]
[1]Trouw Nutrition, José Rocha Bonfim, 214, 13080650, Campinas, SP, Brazil, [2]UNESP, Animal Science, Paulo Donato Castelane, s/n, 14884900, Jaboticabal, SP, Brazil; josilage@gmail.com

In the Brazilian beef production, dried distiller grains with solubles (DDG's) represent an alternative to replace protein sources in supplements for grazing cattle. The growth trial included 2 seasons, wet and dry. First growing in pasture (Brachiaria brizantha, Marandu grass) for 5 months, and finishing in the dry season (100 d). Two different feeding approaches were applied in the dry season: Express Feedlot® (EF), with a high level of concentrate in pasture (1.5% of BW), or conventional feedlot (CF) (mixed ration with 70% concentrate). In the wet season, the pastures were divided into 9 paddocks totalling 9.9 ha. Sixty-nine Nellore bulls (12 mo; 249.03±36.84 kg), were divided in 3 treatments: conventional supplement (CON) with cottonseed meal as a protein source; DDG's replacing 50% of protein source (DDG50) and DDG's replacing 100% of protein source (DDG100). In the dry season, the animals maintained the same protein source treatment (CON, DDG50 and DDG100) applied either on EF or CF. EF bulls (n=33; 392±33.18 kg BW) were finished on Marandu grass in 9 paddocks (3 by treatment). CF bulls (n=36; 391±35.2 kg BW) were confined in 9 collective pens (3 by treatment). After the finishing phase, animals were slaughtered and the carcasses were refrigerated (0 °C; 48 h). Samples were collected from the longissimus to determine the fatty acid (FA) profile (extraction and methylation) and performed by gas chromatography. Animals in EF with DDG100 had greater amount of linoleic acid (C18:2, cis-9 cis-12; P=0.045) compared to all other treatments. Meat from animals in EF had greater concentrations of CLA (C18:2, cis-9 trans-11; P<0.01), saturated FA (P=0.015) and polyunsaturated FA (PUFA; P=0.015). However, monounsaturated FA (P<0.01), unsaturated FA (P=0.015) were lower in the meat from animals in EF. Meat from animals on DDG100 had greater concentrations of PUFA (P=0.017) compared to animals fed CON or DDG50. The use of Express Feedlot to finishing animals and DDG replacing the total amount of protein in supplements positively influences the presence in meat of desirable fatty acids to human health.

Session 33 Poster 21

Genes related to meat colour of Nellore cattle

M.E. Carvalho[1], F.R. Baldi[2], M.H.A. Santana[1], A.F. Rosa[1], R.V. Ventura[3], R.S. Bueno[4], M.N. Bonin[5], F.M. Rezende[6], J.P. Eler[1] and J.B.S. Ferraz[1]
[1]University of São Paulo, Duque de Caxias Norte, 13635900, Brazil, [2]São Paulo State University, Via Paulo Donato Castelane, 14884-900, Brazil, [3]University of São Paulo, Duque de Caxias Norte, 13635900, Brazil, [4]University of Sao Paulo, Av. Duque de Caxias Norte, 13900635, Brazil, [5]Federal University of Mato Grosso do Sul, Av. Senador Felinto Muller, 79070900, Brazil, [6]Federal University of Uberlandia, Av. Pará, 38400902, Brazil; jbferraz@usp.br

Colour is an important physical attribute of meat, due to the consumer's attraction for a bright red meat. Usually, the consumer associates meat colour with freshness and wholesomeness. The meat colour can be measured by the Cielab system, and its parameters are: L*, a* and b*. The aim of this study was to identify, by ssGWAS, genomic regions that potentially have association with meat colour in Nellore cattle. Phenotypes were obtained from *Longissimus thoracis* muscle between the 12th and 13th ribs on the right half-carcass and aged for 7 days. Data from 909 Nellore bulls were analysed. Those animals were genotyped with Illumina beadchip BovineHD® (777K). Analyses were performed using a pedigree composed by 6,276 animals and with contemporary groups (farm and slaughter batch) as fixed effects and age at slaughter as a covariate. Single step analyses were realized by Blupf90 program considering windows of 10 markers (SNP) to estimate their effects. This procedure enables the identification of regions associated with meat colour along the chromosomes. After quality control (MAF <0.05%, call rate <90%), 463,995 SNPs in autosomal chromosomes were used in the association analyses. For the parameters L*, a*and b*, respectively, we found 16, 19 and 17 candidate regions of genome, that explained more than 1% of the additive variance. These regions were explored and some genes were identified in these regions such as, for the L* parameter: *AMOTL2, DARS, PHACTR2, C8orf34, U6, ZNF536, FBXO3, XAF1, TRIM39-RPP21;* for a*: *IQCJ, ARHGAP15, BICD1, CALR3, EPS15L1, SH3GL2, CHURC1-FNTB, CMC1, CCK;* and b*: *ARPC2, CEP128, GAS7, RPTOR, EFTUD1, USP31, SORCS1*. With ssGWAS method using high density panel, it was thus possible to identify regions related to meat colour in Nellore cattle. Later on, these genes and their pathways will be investigated.

Preliminary results of beef production of Spanish Berrenda cattle breeds at test station

A. Gonzalez[1], P. Valera[2], C. Bartolome[3], M.E. Muñoz-Mejias[1], A. Domingo[4], F.J. Cuevas[3] and E. Rodero[1]
[1]University of Cordoba, Animal Production, Campus de Rabanales, Ctra. madriz-Cadiz, km 398, 14071 Cordoba, Spain, [2]ANABE, C/ Registros, 48, 28470, Cercedilla Madrid, Spain, [3]IMIDRA, CENSYRA, Ctra. de Colmenar Viejo a Guadalix de la Sierra, km 1, 28770 Colmenar Viejo (Madrid), Spain, [4]CENSYRA, Junta de Extremadura, Camino Santa Engracia S/N, 06007 Badajoz, Spain; celia.bartolome.criado@madrid.org

Two hundred and seven calves of two endangered Spanish breeds, Berrenda en Colorado (BC) and Berrenda en Negro (BN), were used to analyse their meat production. The average age of calves at the beginning of the study was 325.8 days. They are Spanish cattle breeds grown up in extensive breeding at Dehesa ecosystem, where they get low supplementary feeding. The breeding programme included selective actions to improve their meat production traits. This control has been carried out in two test stations, located in Colmenar Viejo (CV) and Badajoz (B). The mean live weight gain was 163.0 and 160.3 kg for BC and BN respectively at CV and 148.6 and 134.9 for BC and BN at B. Differences in live weight gain between BC and BN were significant ($P<0.05$) only at B. The average daily gain (ADG) for both breeds was higher for animals tested in B than in CV (1.2 and 1.1 kg, respectively). The total feed consumption was lower for BC in both stations (797.9 in CV and 820.1 kg in B) than for BN (1,181.9 and 1,293.2 kg, respectively). There were differences ($P<0.05$) between breeds for ADG in B and for total feed consumption in CV. The highest feed conversion ratio was observed for BN in CV (8.5) and the lowest for BC in B (5.5). This great difference added to the high coefficient of variation could be due to the high variability of animal origins. Differences were significant ($P<0.05$) only for B. All meat production traits in both breeds and test stations had significant influence ($P<0.05$) based on the different test series. When ADG of both Berrenda and Retinta breeds, a Spanish native breed grown up in similar conditions, were compared, BC tested in B presented higher ADG than Retinta (1.2 kg).

Physicochemical and sensory evaluation of meat of the bulls fed diet with rapeseed or linseed meal

E. Sosin-Bzducha[1] and M. Puchała[2]
[1]National Research Institute of Animal Production, Deparment of Nutrition Physiology, ul. Krakowska 1, 32-083 Balice, Poland, [2]National Research Institute of Animal Production, Department of Sheep and Goat Breeding, ul. Krakowska 1, 32-083 Balice, Poland; michal.puchala@izoo.krakow.pl

The aim of the study was physicochemical and sensory evaluation of meat from young bulls fed diet with rapeseed or linseed meal. The experiment used meat from 30 bulls of the Polish Red-and-White (ZR) (n=14) and Simmental (SM) (n=16) breeds. Bulls were fed diet with rapeseed or linseed meal, formulated according to IZ-INRA recommendations. At the age of 18 months, bulls were slaughtered and samples of the *longissimus lumborum* muscle (MLL) were analysed for physicochemical (pH, heating loss, colour – lightness, redness and yellowness, shear force) and organoleptic properties (colour, taste and aroma intensity, taste and aroma desirability, structure, juiciness, delicacy and tenderness). The sensory evaluation following heat treatment (roasting) was performed on 5-point scale (1 point the worst; 5-points-the best). There were no statistical differences in meat pH. The meat colour did not differ between breeds although the redness was greater for the Polish Red-and-White breed and for linseed fed group. The heating loss was greater for meat of ZR bulls (33.6 vs 32.6%), and for bulls fed diet with rapeseed meal (34.6 vs 31.8%). The shear force was lower for meat from the Polish Red-and-White breed (120.1 N) and linseed meal group (116.9 N) than for Simmental breed (124.6 N) and meat of bulls fed with rapeseed meal (128.0 N). Sensory assessment showed that meat from ZR breed received higher scores due to greater taste (4.42 vs 3.9) and aroma intensity (4.32 vs 4.12) and better tenderness (3.96 vs 3.85). The sensory panel evaluated higher the colour of meat from ZR breed than from SM breed (4.1 vs 3.9). The aroma intensity was greater for rapeseed meal group (4.24 vs 4.1) but less desirable comparing to linseed group (4.0 vs 4.2).

Mitigation of methane emissions in beef cattle using bioeconomic models

J. Lopez-Paredes[1], *R. Alenda*[1] *and O. González-Recio*[1,2]

[1]*Universidad Politécnica de Madrid, Producción Agraria, Senda del Rey s/n, 28040, Spain,* [2]*Instituto Nacional de Investigación y Tecnología Agraria y Alimentaria, Ctra. de La Coruña, km 7,5, 28040, Spain; gonzalez.oscar@inia.es*

This work shows an approach of bio economic models to include CH_4 mitigation in beef cattle breeding goal. Three different scenarios were developed: (1) Current replacement index for Blonde d´Aquitaine: 'Current situation'; (2) Carbon tax; (3) Establishment of a methane quota per farm. Data on 4,573 cows and 7,498 calves from the National Association of Blonde d'Aquitaine were used to estimate CH_4 kg per year per unit of product (slaughtered calf). Economic weights were estimated under the three scenarios using bio economic model, which led to three selection indices accounting for three groups of traits, functional, production and (for scenarios 2 and 3) methane traits. For Scenario 1, functional traits and production traits accounted for 48 and 52% of the economic importance. For scenario 2, methane traits supposed 4.6% of economic importance, whereas functional traits and production economic importance had a relative weight of 49.2 and 46.2%, respectively. Relative importance of methane traits was lower in scenario 3 (1.8%); in contrast to the enhanced weight for production traits (52.4%). The importance on functional traits decreased to 45.8% in a quota situation. The expected economic response in scenario 1 was 69.1€/year (97% of this was from production traits and 3% from functional traits). This expected genetic gain would decrease to +65.3€/year if a penalty on carbon tax was applied, placing more emphasis on functional traits (10%). A quota scenario resulted in a similar response per animal as Scenario 1 (69.94€/year), with production traits accounting for the 96% of this total economic response, however a reduction in the number of animals per farm woul be expected to accomplish with the quota, and therefore lower benefits. Selection for cows with lower carbon footprint involves changes in the animal, while ensuring profitability in future generations. This study shows different strategies to tackle emissions through breeding. Any strategy to tackle methane emission in beef cattle needs to be carefully considered because it will have an impact on the type of cows in future generations.

Pre-weaning growth of calves of Brahman cross cows sired with Wagyu and Belgian Blue bulls

P. Panjono[1], *A. Agus*[1], *T. Hartatik*[1], *S. Bintara*[1], *I. Ismaya*[1], *B.P. Widyobroto*[1], *I.G.S. Budisatria*[1], *D.A. Priyadi*[1], *P. Leroy*[2] *and N. Antoine-Moussiaux*[2]

[1]*Universitas Gadjah Mada, Faculty of Animal Science, Jln. Fauna No. 3, Kampus UGM Bulaksumur, Yogyakarta, 55281, Indonesia,* [2]*University of Liège, Faculty of Veterinary Medicine, Liège, 4000, Belgium; panjono@ugm.ac.id*

This study was conducted to observe the pre-weaning growth performance of calves of Brahman Cross (BX) cows sired with Wagyu and Belgian Blue (BBB) bulls. Nine heads of BX cows and three heads of Sumba Ongole cows were artificially inseminated with BBB bulls, thirteen heads of BX cows were artificially inseminated with Wagyu bulls, and twenty five heads of BX cows were naturally mated with BX bulls. The calves were raised together with their dams until they reached their weaning age, i.e. until six months of age. The data collected were analysed by one way analysis of variance. The BBB-crossed, Wagyu-crossed and BX calves showed, respectively, a birth body length of 63.50±3.90, 59.31±9.61 and 58.93±6.43 cm, a withers height of 69.38±3.77, 69.38±3.77 and 73.87±4.73 cm, a heart girth of 72.08±4.38, 71.54±8.66 and 74.67±6.80 cm, and a body weight of 32.17±5.17, 29.15±10.95 and 28.20±6.52 kg. There was no significant difference in birth body size and weight among groups, except the withers height. The withers height of BBB-crossed calves was lower ($P<0.05$) than that of BX calves. The BBB-crossed, Wagyu-crossed and BX calves showed, respectively, an average body length of 0.15±0.03, 0.17±0.03 and 0.11±0.03 cm/day, a withers height of 0.18±0.02, 0.14±0.02 and 0.10±0.02 cm/day, heart girth of 0.32±0.05, 0.24±0.03 and 0.16±0.03 cm/day, and a body weight daily gain of 0.74±0.13, 0.62±0.06 and 0.39±0.11 kg/day. The average body length daily gain of BBB-crossed and Wagyu-crossed calves were higher ($P<0.01$) than that of BX calves. The average withers height, heart girth and body weight daily gain of BBB-crossed calves were higher ($P<0.05$) than those of Wagyu-crossed calves and those of Wagyu-crossed calves were higher ($P<0.001$) than those of BX calves. It is concluded that crossing BX cows with BBB and Wagyu bulls improves body size and weight gain of calves and crossing with BBB bull give higher improvement than that with Wagyu bulls.

Session 33 Poster 26

Katerinis: in extinction! Meat's physical and chemical parameters of this greek cattle

D.K. Karatosidi and D.D. Dimos

Panhellenic Association of 7 indigenous livestock breeders, Perraivvou 20, 42100 Trikala, Greece; despinakaratosidi@yahoo.com

The Katerinis is an autochthonous cattle breed of Steppe type, resulting from the *Bos primigenius*. Katerinis breed is registered in the list of FAO 'World List of Domestic Animal Diversity, WWL-DAD: 3' along with 114 other breeds of cattle and 33 goats. Nowadays in Greece, there are only 217 animals remaining of this breed raised in the region of Thessaly (central Greece). The individuals of this local breed are characterized by a small size and low milk and meat productions whereas, in the past, they were mainly used for their labour capacity. However, in particularly difficult inland environments, they are still used as productive animals, as they are able to use food resources that would not otherwise be used, and they can fit very well in extensive management conditions. The rescue of endangered indigenous genotypes, including Katerinis, highly depends on an appropriate economic collocation of the industries of production. This may be achieved only through the chemical and gastronomical evaluation and characterization of products, which have to be associated with their territory and cultural traditions of local people. For these reasons, it seemed useful to us to investigate some qualitative aspects of Katerinis' meat. We determined the physical and chemical composition of meat with the aim to establish a future rescue plan based on potential high quality and desirable nutritional characteristic of Katerinis breed. For this research, 6 bulls were reared outdoors and slaughtered at 18 months of age. Samples of *Longissimus dorsi* muscle were sampled and their chemical and fatty acid compositions were determined. All analyses were realised according to ASPA methods. The results of this research showed that the meat of Katerinis breed is genuine, and seems to correspond to the nutritional needs of the modern consumer. Therefore, it is necessary to continue the research in order to characterize the chemico- nutritional and gastronomical identity of this meat, in relationship with the rearing system, the type of feeding and age of animals in order to create a quality mark that could enhance, definitively, the production and preservation of the Katerinis breed.

Session 34 Theatre 1

Sustainability of animal agriculture is affected by crop management

J.L. Vicini[1] and G. McNunn[2]

[1]Monsanto Company, Regulatory, 800 N. Lindbergh Blvd., Chesterfield, MO 63167, USA, [2]EFC Systems, 2321 N Loop Drive, Ames, IA 50010, USA; john.l.vicini@monsanto.com

Animal agriculture is challenged with societal issues such as animal welfare, antibiotic use, food vs feed and sustainability. These issues can be emotional and accurate information is required to engage on these topics. Consumers are increasingly interested in sustainability of meat, milk and eggs because agriculture, both crop and animal ag, is considered a major contributor to greenhouse gasses (GHG); however, agriculture can be a solution for storing GHG. Cropping systems are major decisions for dairy farms and most farms in the US have relied in a large part on genetically modified (GM) crops and no-till (NT) planting, for 20 yr. In the US, more than 90% of corn, soybean and cotton are GM and livestock are major consumers of these crops. Since feed production is the largest component of animal agriculture it is important to understand how this technology impacts farm sustainability. We hypothesized that GHG would be reduced by planting GM crops and adoption of NT and tested this hypothesis by developing a model-based analysis across 12 states in the U.S. corn-belt. The analysis generated potential erosion and GHG emissions estimates associated with the implementation of several practices based on public SSURGO data, NASS crop productivity data, and spatially explicit weather data. Rill and inter-rill erosion resulting from rainfall and surface runoff were modelled and sediment loss and deposition due to wind was simulated. In addition to erosion, soil organic carbon (SOC) change and nitrous oxide flux (N_2O) were modelled using the Denitrification-Decomposition model (DNDC; U NH, USA). Based on the analysis, average GHG emissions associated with corn production across the US corn belt could be reduced by an estimated 8.7 kg CO_2e/bu and 14.3 kg CO_2e/bu by switching from conventional hybrid production to a reduced till and NT GMO system, respectively. A corresponding average reductions in soil loss of 10.9 tn/ac and 13.6 tn/ac due to water and wind erosion were also estimated based on a change to a reduced till or NT GMO system. Based on the model estimates, feeding GM crops in a no-till system makes production of meat, milk and eggs more sustainable with respect to GHG emissions and erosion.

Session 34 Theatre 2

Future cow barn in relation to manure quality

P.J. Galama[1] and W. Van Dijk[2]
[1]Wageningen University and Research, Animal Sciences Group, De Elst 1, 6708 WD Wageningen, the Netherlands, [2]Wageningen University and Research, Plant Sciences Group, Edelhertweg 1, 8219 PH Lelystad, the Netherlands; paul.galama@wur.nl

The design of a cow barn does not only have impact on animal welfare but also on manure quality, which is important in relation to use as fertiliser or soil improver on grassland and arable land and to decrease the ammonia emission. The most common housing system in the Netherlands is a Freestall with cubicles and a slatted floor with slurry storage. New developments in housing systems are based on different floor types in Freestalls to reduce ammonia emission and Freewalk housing systems to increase space per cow for more animal welfare and natural behaviour and reduce ammonia emission. The development of Freewalk housing with a bedding of wood chips or straw and an artificial floor which separates urine from faces will be shown. The faeces are picked up by a manure robot. The effect of these housing systems on manure quality (also solid and liquid fractions) will be shown based on case studies on the research station Dairy Campus and commercial farms. In addition, the economic value of ten different manure products has been evaluated from the perspective of a dairy and arable farmer. The revenues concern fertiliser savings (nitrogen, phosphorus, potassium) and organic matter supply for improving the soil fertility and crop production. The costs involve cost of different housing system, (mechanical) separation, sampling & analysing, transport and application on the field. Organic manures from a bedding with straw are interesting as fertiliser and soil improver but must be weighed against the high cost of straw, although the housing system is cheaper. Primary separation of manure with a different floor type in a cubicle stable or with an artificial floor in a Freewalk barn can make a dairy farm more sustainable and may offer arable farmers more opportunities to select the right fertiliser or soil improver. The results are based on research in the projects Bedded Pack Barns in Netherlands since 2012 and interaction Arable and Livestock production since 2017. Also different housing systems in the European project Freewalk will be shown, which started in 2017.

Session 34 Theatre 3

Cattle welfare across free walk and cubicle housing systems in six European countries

I. Blanco-Penedo[1], A. Kuipers[2], M. Klopčič[3] and U. Emanuelson[1]
[1]SLU, Institutionen för kliniska vetenskaper Box 7054, 750 07 Uppsala, Sweden, [2]WUR, De Elst 1, 6708 WD Wageningen, the Netherlands, [3]UL, Groblje 3, 1230 Domžale, Slovenia; isabel.blanco.penedo@slu.se

Innovative Free Walk Systems (FWS) are promising housing from point of view of animal welfare, environment and economics. But it has not yet been scientifically evaluated in a cross country framework. The aim of this work was to evaluate the effects of FWS compared to cubicle housing (C) on cow hygiene, integument alterations and lameness, performing the comparison across a wide range of European production conditions (more extensive than done before): The Netherlands (NL), Germany (DE), Italy (IT), Slovenia (SL), Austria (AT) and Sweden (SE); https://subsites.wur.nl/en/freewalk.htm. The selected 44 dairy and suckler farms (half FWS with innovative practices and half C) were visited during Winter 2017-18 to conduct an adaptation of the Welfare Quality® protocol on 2013 cows by a trained observer. Statistical analysis on welfare was performed using STATA software. Herd prevalence results indicated that cows on C presented a significantly ($P<0.05$) higher frequency of integument alterations throughout the body such as hairless patches (on tarsus, hindquarters, carpus); and lesions (on tarsus, hindquarters) than cows on FWS. In contrast, more lesions on carpus and swelling on tarsus were observed in FWS. Housing types did not generally differ on cow dirtiness, neither claw conditions nor lameness. By country, the dirtiness on lower legs, hindquarters and teats was lower ($P<0.05$) in NL, and on the udder ($P<0.05$) in AT. Cows in IT and SL presented the highest prevalence of integument neck alterations. The overall prevalence of lame cows was 19.6%, with large differences between and within the countries. The median (range) herd prevalence of lameness was 6.9% (4.1-11.4) (NL), 20.9% (0-25) (DE), 23.3% (7.3-29.5%) (IT), 22.2% (0-34.8) (SL), 34.4% (22.4-44.4) (AT) and 14.8% (11.5-18.2) in SE. NL and SE had the highest prevalence of not lame cows and IT on severe lameness and claw overgrowth. It is planned to complete the analysis comparing seasons, classifying mix systems, examining farm resources and management factors to conclude about welfare across housing systems and countries.

Session 34 Theatre 4

Climate and bedding in different barn systems

L. Leso[1], M. Klopčič[2], P. Galama[3], A. Kuipers[3] and M. Barbari[1]
[1]University of Firenze, Department of Agricultural, Food and Forestry Systems, Via San Bonaventura, 1, 50145 Firenze, Italy, [2]UL-Biotechnical Faculty, Dept. of Animal Science, Groblje 3, 1230 Domžale, Slovenia, [3]WUR, Wageningen Livestock Research, De Elst 1, 6708 WD Wageningen, the Netherlands; lorenzo.leso@unifi.it

In FreeWalk project (www.freewalk.eu) a variety of barn systems is studied. Lying space in a cubicle housing system is around 3 m^2 per cow, while lying space in a bedded pack barn or on an artificial floor is 12-15 m^2 per cow. Housing without cubicles or stalls is defined as Freewalk housing. 'Waste' products are used as bedding material, which are expected to increase the C-content of manure and consequently the soil structure. To function as bedding, and later on as fertilizer, aerating and cultivation of these materials mixed with the animal manure takes place until the compost bedding is matured. Bedding materials examined (physical characteristics and cost) are straw, manure, sawdust, wood shavings and green garden waste. Artificial floors are built up of different layers. Climate (temperature, humidity, air velocity and light) in 44 of those barns spread through Europe is being measured at intervals as well as temperature, dry matter and bacteria composition of bedding during 2018/2019. A few barns are monitored continuously. Also, the framework of a water balance model for Freewalk housing is in development. Dry bedding is a key factor, affecting hygiene level of the bedding and animal. Optimal moisture level for a cultivated pack ranges between 40-65%. Drying rate is kg $H_2O/m^2 \times day$. Insight in factors determining the moisture content of bedding is explored. Faeces and urine represent the main input. The spatial distribution of excreta in the lying and feeding alley allows estimating the quantity of water reaching the pack. Observations will be carried out on research centres with cameras and one case farm in Italy, and trials are carried out in a Climatic chamber, using samples with different moisture content, levels of air temperature, velocity and humidity, at different pack depths. The model will be validated on basis of the field data obtained from the composted beddings. Preliminary results will be presented.

Session 34 Theatre 5

Improving the dairy farm efficiency with the milk carbon footprint assessment

C. Brocas[1], S. Danilo[1], S. Moreau[1], A. Lejard[2] and J.B. Dolle[1]
[1]Institut de l'élevage, Maison Nationale des Eleveurs, 149 Rue de Bercy, 75595 Paris cedex 12, France, [2]France Conseil Elevage, 42 Rue de Châteaudun, 75009 Paris, France; catherine.brocas@idele.fr

Milk carbon footprint represents a challenge and an opportunity for the dairy sector to highlight its current and future accomplishments. Although environmental drivers are not well received by farmers, evidences are available to illustrate that lower GHG emissions are associated with reduced operational costs. The French Livestock Institute (Institut de l'Elevage), in association with key players in the French dairy sector i.e. dairy advisory enterprises and French dairy board (CNIEL), has launched the LIFE CARBON DAIRY project with the main objective to promote an approach to reduce the milk carbon footprint at farm level by 20% over 10 years. In order to reach the goal, project's partners developed a Life Cycle Assessment tool named CAP'2ER® aiming at measuring the milk carbon footprint in dairy farms in France. Answering the LCA approach, the milk carbon footprint assessed in CAP'2ER® is covering the greenhouse gases (GHG) emissions to determine the Carbon Footprint of milk (CF). Applied on 3,348 farms representing various milk production systems in France, the project provides a good overview of the average national CF. On these farms, the average CF is 1.04 kg CO_2e per litre Fat and Protein Corrected Milk. Variations in CF are explained by discrepancies in farm management. Practices with the largest impact on CF average are milk yield, age at first calving, quantity of concentrate, N-fertilizer used and fuel consumed. The project show that it exist a difference of 30 €/1000 l between the lowest 10% milk carbon footprint and the highest 10%. This reinforces the fact that improving production efficiency and reducing the carbon footprint of milk production are highly complementary. It's why, the milk carbon footprint assessment is a good means to provide farmers with information about GHG emissions from dairy system and the link with farming practices.

Sound hooves: detection of lameness in dairy cows by acoustic analysis

N. Volkmann[1], U. Richter[2], R. Hölscher[3] and N. Kemper[1]
[1]University of Veterinary Medicine Hannover, Foundation, Institute for Animal Hygiene, Animal Welfare and Animal Behaviour, Bischofsholer Damm 15, 30173 Hannover, Germany, [2]University of Kassel, Section of Agricultural and Biosystems Engineering, Nordbahnhofstr. 1a, 37213 Witzenhausen, Germany, [3]Hölscher + Leuschner GmbH & Company KG, Siemensstraße 15, 48488 Emsbüren, Germany; nina.volkmann@tiho-hannover.de

An important factor for animal welfare in cattle farming is the detection of lameness, which is caused by diseases of the claws and limbs. Therefore, the aim of this project is to develop a system which is capable of an automated and early diagnosis of lameness in cattle by analysing the walking sounds. As a first step, data was generated from lactating cows (n=77) in a freestall barn. Cows walked consecutively along a runway with slatted floor. In one part of this alley (length: 3.1 m; wide: 0.8 m) eight piezoelectric sensors recorded the walking sounds of the cows. Simultaneous the locomotion of the animals was scored with a three-point system by a trained observer and the run was recorded by video. Subsequently the cows were examined by a professional hoof trimmer to detect claw lesions. The average time for passing the measuring section was 3.72 seconds (SD=0.8) and the mean standard deviation in volume was 0.019 decibels (SD=0.007). The mean time walking across the test track was significantly higher ($P<0.0001$) in cows with a locomotion score >1 (3.94s) compared to animals with score 1 (3.21s) and thus, showing a smooth and quick gait pattern. The standard deviation of volume (SDV) in the recorded sound signal was considered as a factor for the force of cow's footsteps. Therefore, a higher value of SDV describes an increased difference between sound signal and no sound signal. The type of claw diseases (1=none, 2=not infectious, 3=infectious, 4=both) showed a significant effect on SDV ($P=0.0105$). Cows with non-infectious diseases such as sole ulcers showed significantly lower SDV ($P=0.0407$) than cows with no disease (0.015 vs 0.021 dB). This result confirmed the assumption, that in particular cows with non-infectious diseases have a greater sensitivity to pain and demonstrate a less forceful gait pattern. These first results clearly show the potential of walking sound analysis for individual lameness detection.

Assisting dairy farmers with the use of genomic testing and reproductive technologies in the USA

A. De Vries[1], J.C. Dalton[2] and D.A. Moore[3]
[1]University of Florida, 2250 Shealy Dr., Gainesville, FL 32608, USA, [2]University of Idaho, 1904 E. Chicago St., Caldwell, ID 83605, USA, [3]Washington State University, Bustad, Pullman, WA 99163, USA; devries@ufl.edu

Improved reproductive performance of dairy cattle compared to two decades ago and the use of sexed semen are now leading to an abundant supply of replacement heifers in the USA. This large supply of heifers has dairy farmers wondering if they should raise fewer heifers. For many US farmers this is a paradigm shift. One of the objectives of our USDA funded project is to assist dairy farmers with decisions on how many and which heifers to raise. Farmers can adopt many potential strategies. Strategies include some combinations of genomic testing, use of sexed semen, in-vitro produced embryos and embryo transfer, beef semen, and culling. Our project has resulted in a tool that calculates profitable strategies for individual farms. Constraints may be imposed, such as raising enough dairy heifers to replace culled cows, or limiting the use of sexed semen. The tool was developed in collaboration with stakeholders from the allied dairy community, who helped direct the user-interface, and inputs and outputs. Using general sensitivity analysis, we found, for example, that in herds with good reproduction, the most profitable strategy was a combination of genomic testing, beef semen, sexed semen, and conventional semen. More than the minimum number of required dairy heifer calves were produced to allow for some genetic selection among dairy calves. The tool has also been used with individual farmers to determine the value of genomic testing, which groups of dairy cattle should be inseminated with sexed semen, and the maximum price for in-vitro produced embryos. To help with the actual implementation of a selected strategy, we developed a second tool, which calculates insemination values for beef, sexed, and conventional semen for individual cattle. These economic values help prioritize heifers and cows for the three insemination options. Cull values are also calculated. These tools have been introduced to dairy farmers at workshops in multiple locations throughout the USA. Further discussions with the allied dairy community are underway to accelerate uptake and impact of both tools.

On the international transferability of genomic breeding

L. Chavinskaia
INRA, LISIS, LISIS, Université Paris-Est Marne-la-Vallée, 5 boulevard Descartes, 77454 Marne La Vallée, France; lidia.chavinskaia@inra.fr

Currently, the accelerated development of genome-wide breeding technologies questions their world-wide propagation. Livestock production is a historical human activity present almost everywhere in the world and is one of the most technology driven fields in agriculture. The predictive technologies represent the basis of modern animal breeding scientifically grounded and objectified. As 'modern' and 'science' usually and implicitly mean in this context a prerogative of western countries, the question is: how these technologies do circulate across countries including non-hegemonic developing countries? The scientific knowledge is traditionally considered, especially in the economic theory, as a public good as its use by one does not reduce availability to others and as no one can be effectively excluded from its use. But is it a reason to think that anybody can use the genomic breeding methodology published in 2001 by Meuwissen, Hayes and Goddard to implement the technology of genomic breeding evaluation? Michel Callon, a French sociologist of science and technology argues against this classical economists' point of view that the scientific knowledge cannot be considered as a public good as it is completely useless unless it is integrated into the complete network of elements (other linked knowledge, human skills, instruments, infrastructure, etc.) which make it transformable into technology, and possibly useable. According to Callon, science can be defined as a public good for a different reason which consists in considering it as a source of technological diversity. Regarding this claim and using the theoretical frame of the Actor-Network Theory (ANT), we provide here a sociological vision of breeding evaluation technology as a socio-bio-technical network. Its dynamics is presented as a permanent reconfiguration of its elements. We demonstrate how science pushes the technology to circulate across countries creating a useful diversity of evaluation scales. This diversity is objectified by the knowledge about the intrinsic biological nature of the internationally marketed genetic resources expressed in the 'genotype by environment' interaction.

Economic importance of traits of Angus breed in organic system

Z. Krupová, E. Krupa, J. Přibyl, E. Kašná and M. Wolfová
Institute of Animal Science, Přátelství 815, 104 00 Prague, Czech Republic; krupova.zuzana@vuzv.cz

Aberdeen Angus breed is one of the most numerous beef breed in the Czech Republic. In many herds it is running under the organic production system. Breeding values are estimated for growth traits, calving performance and exterior but comprehensive economic selection index of the breed has not been established until now. Economic weights (EWs) of traits currently used as selection criteria and of further traits (totally 16 traits) were calculated to be applied for selection decisions under the actual production and economic conditions. In accordance with the estimated breeding values, also the EWs for direct and maternal components of the appropriate traits were calculated. EWs of traits under the intensive and extensive rearing of heifers (age at first calving of 2 and 3 years, respectively) was investigated. In both alternatives, the most important traits were weight gains of calves at 120, 210 and 365 days of age (the three traits made 67% of the total economic importance of all evaluated traits), productive lifetime of cows (11%) and survival rate of calves till weaning (9%). Residual feed intake of breeding heifers and of cows (new traits) counted 3.6% of the total economic importance of all included traits. Low EWs calculated for calving performance (1%) and carcass conformation (0.2%) correspond to predisposition of the breed for easy calving and to selling of animals at a live weight base, respectively. In the extensive system, the EWs of mature weight was doubled (from -0.59 to -1.07 € per kg live weight per cow and per year) due to higher feeding costs for maintenance. Nevertheless, relative economic importance of the trait remained low (3 and 4%) in both systems. EWs for the direct and maternal components of traits actually used as selection criteria were as follows: 158 and 116 € per class of calving performance, 2.12 and 1.55 € per kg of birth weight of calves and 4.36 and 3.19 € per kg of weight gain at 365 age of calf, all expressed per cow and year. EW of the productive lifetime of cows (50 € per year) suggests that this trait should be a further selection criterion for the breed in the future. The study was supported by project MZE-RO0718 of the Czech Republic.

Does *Salmonella* sp. survive in cow slurry during storage?

A.S. Soares[1], C. Miranda[1], H. Trindade[1] and A.C. Coelho[2]
[1]Universidade de Trás os Montes e Alto Douro / CITAB, Quinta de Prados, 5000-801 Vila Real, Portugal, [2]Universidade de Trás os Montes e Alto Douro / CECAV, Quinta de Prados, 5000-801 Vila Real, Portugal; anasoares@utad.pt

Salmonella sp. is a genus of pathogenic microorganisms belonging to the *Enterobateriaceae* family, with a wide distribution in nature. With the gastrointestinal tract of man and animals as its main reservoir, the occurrence of contamination of the environment and water is based on the release of *Salmonella* microorganism by the faeces of the infected animals. Birds and cattle are considered the main disseminators, and the pathogen can be isolated from animal effluents. Salmonellosis is one of the main forms of infection associated with production animals, with a significant economic impact on the commercialization of these animals and their by-products. The aim of the study was to evaluate the survival capacity of *Salmonella* sp. in the liquid fraction of dairy cow slurry stored ate 20 °C and 4 °C during 90 days. Three different treatments were applied to the liquid fraction obtained after mechanical separation of the slurry: addition of biochar, addition of sulfuric acid (pH≈5.5) and the combination of the addition of biochar and sulfuric acid. Each treatment was repeated in triplicate and the samples for detection of *Salmonella* sp. (by microbiological methods) were collected on the third day after the start of the test and repeated every fortnight until the end of the ninety days of storage. In this study, there was a trend toward a decrease in bacterial survival over the course of the assay but the bacteria tended to survive longer in the treatments stored at 4 °C. Being salmonellosis, a disease considered the most widespread zoonosis in the world, control in slurry during storage represent a major challenge for Public Health.

Genetic and phenotypic correlations between workability and production traits in HF cattle

B. Szymik[1], P. Topolski[1], W. Jagusiak[2] and K. Żukowski[1]
[1]National Research Institute of Animal Production, Department of Genetics and Cattle Breeding, Krakowska 1, 32-083 Balice near Cracow, Poland, [2]Agricultural University of Cracow, Department of Genetics and Animal Breeding, Al. Mickiewicza 24/28, 30-059 Kraków, Poland; bartosz.szymik@izoo.krakow.pl

The aim of this study was estimated genetic and phenotypic correlations between workability (WT) and milk production traits with and without genomic information in Polish Holstein-Friesian dairy cattle: milking speed (MS) and temperament (MT) and five milk production traits: milk, fat and protein yield (kg), fat and protein percentage (%). Records of 13,280 cows (milking from 2007 to 2014) were collected in the database system SYMLEK belonging to Polish Federation of Cattle Breeders and Dairy Farmers. The cows were scored for WT at the second test-day of the first lactation. Genetic and phenotypic correlation coefficients between WT and milk production traits were estimated using Gibbs sampling method which has been implemented in the BLUPF90 package. Multi trait linear model of observation was applied to estimate (co)variance components. The model included the fixed effects of HYS and lactation stage, fixed regressions on percent of Holstein-Friesian genes and age of calving and random genetic effect. Genetic and phenotypic correlations between WT and milk production traits were estimated first only with pedigree data information. Next, data has been extended with genomic information about 2,228 sires of cows. For MT and MS we estimated heritabilities coefficients: 0.12 and 0.25, respectively. Genetic correlation coefficients between MS and milk production traits ranged from -0.13 for protein percentage to 0.40 for fat yield. Genetic correlation coefficients between MT and milk production traits varied from -0.41 for protein yield to 0.15 for fat percentage. Phenotypic correlation coefficients between MS and milk production traits ranged from -0.01 for fat to 0.04 for fat percentage, and between MT and milk production traits were slightly lower (from -0.02 for protein percentage to 0.01 for fat yield.

Session 34 Poster 12

From data integration to practical application: pathogen-specific udder herd health management tool

M. Suntinger[1], W. Obritzhauser[2], B. Fuerst-Waltl[3], C. Firth[2], M. Mayerhofer[1] and C. Egger-Danner[1]
[1]ZuchtData EDV-Dienstleistungen GmbH, Dresdnerstraße 89/19, 1200 Vienna, Austria, [2]Institute of Veterinary Public Health, University of Veterinary Medicine, Veterinärplatz 1, 1210 Vienna, Austria, [3]Division of Livestock Sciences, University of Natural Resources and Life Sciences, Gregor-Mendel-Str. 33, 1180 Vienna, Austria; suntinger@zuchtdata.at

Due to advancing technology on dairy farms, data integration is becoming increasingly important with regard to professional herd management. The aim of this study was to develop pathogen-specific evaluations to optimize the web-based udder health program for herd management to promote strategies to improve udder health in dairy cattle. Investigations were preceded by the integration of results of culture milk samples from laboratories into the Central Austrian Cattle Database. This included harmonisation of analysing methods and nomenclature of milk culture results within Austria, while taking international state-of-the-art research in the field into account. Research and development was based on an observational study conducted in cooperation with 250 farms, 17 veterinarians, 6 milk laboratories and research institutions. Pathogen-specific udder health reports on individual cows, current and serial herd infection reports, and parameters allowing benchmarking within and across herds were developed, visualised in clearly arranged figures. Evaluations provide vital information on farm-specific pattern(s) of pathogens. In addition, the combination of bacteriological data and routinely-recorded animal production and health data provide details on period(s) of risk of infection as well as the cow group(s) at risk. Failures in management and sources of infection can be identified more easily and eliminated at an earlier stage. Assessing the infection status of the udder, by means of milk culture results, can assist in decision-making processes regarding more precise control and prevention measures to improve udder health. The more information available, the more targeted a treatment can be: this tool could, therefore, play a crucial role in the prudent use of antimicrobials on dairy farms. Results will be implemented in routine use in herd management within the Central Cattle Database in Austria and Germany (RDV) shortly.

Session 34 Poster 13

Population structure of Holstein cattle in Brazil

V.B. Pedrosa[1], M.F. Sieklicki[1], H.A. Mulim[1], A.A. Valloto[2] and L.F.B. Pinto[3]
[1]Ponta Grossa State University, Department of Animal Science, Av Gen Carlos Cavalcanti 4748, 84030-900 Ponta Grossa PR, Brazil, [2]Paraná Holstein Breeders Association – APCBRH, Rua Professor Francisco Dranka 608, 81200-404 Curitiba PR, Brazil, [3]Federal University of Bahia, Av. Adhemar de Barros s/n, 40170-115 Salvador BA, Brazil; vbpedrosa@uepg.br

The objective of this study was to evaluate the population structure of Holstein herds in Brazil. The data used in this research were provided by the Paraná Holstein Breeders Association (APCBRH), with a pedigree file of 206,796 Holstein animals born between 1970 and 2014. Results were obtained for pedigree integrity, effective number of founders (fe) and ancestor (fa), generation interval. The integrity level of the pedigree was estimated by the mean of the sum of (1/2)n, where n is the number of generations that separate the individual from each known ancestor. The estimates of the effective number of founders (fe) and number of ancestors (fa) were calculated using the formulas: fe=1/$\sum$k=1^{qk2}) and fa=1/($\sum$j=1^{qj2}), where qk is the origin probability of the genetic contribution of an ancestor k and qj is the marginal combination of an ancestor j. The generation interval was obtained based on the average age of the parents at the birth of the progeny. Population parameters were calculated using the software ENDOG v.4.5. The pedigree showed animals with known relationship up to the sixth generation, presenting a degree of completeness of 78.02% for the first generation. The effective number of founders (fe) was 418 animals and for the effective number of ancestors (fa) was 400 animals, with a corresponding fe/fa ratio of 1.04. The mean generation interval was 6.3 years. The ratio between founders and ancestors demonstrated no reduction of genetic variability in the population and no bottlenecks with loss of the origin genes. The generation interval was slightly higher than that presented in countries with high milk production, demonstrating that there is a need to reduce this indicator to accelerate the genetic progress. Funded by Fundação Araucária.

Dormant alfalfa (*Medicago sativa* L.) as an alternative for milk production in Puno, Peru

R. Rivera, J. Vargas and C. Gomez
Universidad Nacional Agraria La Molina, Animal Science, Av. La Molina s/n, Lima12, Peru; rrivera4@ncsu.edu

Peruvian highlands have 73% of cattle population of the nation and 52% of cattle in Puno is for milk production. Puno is the fifth milk producer region in Peru and dairy has been developing at a fast rate during the last decade. An alternative implemented within the last years by national, regional and NGO's initiatives includes the introduction of dormant alfalfa, a cultivar adapted to high altitude (4,000 m.a.s.l.) and extreme cold weather conditions. Alfalfa with a dormancy type of 3.8 is able to survive under harsh climate conditions remaining dormant during the months of May-October and then start growing for the rainy season from November to April. This study was conducted with a semi-structured survey of 24 dairy producers from 7 districts in the provinces of Puno, Huancane, Lampa, and Melgar. The aim of the study was to identify the milk cost of production and the main components of dairy farm cost in a system using alfalfa as the main forage. SPSS software was used and generated three groups by size and milk cost of production EUR 0.22, 0.19 and 0.19/l of milk for the small, medium and large producer respectively. The average milk production per cow was ~8.0, 9.9 and 10.2 for the small, medium and large-scale farm respectively. The main cost components in milk production are labour and feeding with 44 and 39% respectively. The cost of milk production in Puno where alfalfa is used for grazing was less in comparison with the intensive system developed in the Peruvian coast of Lima or Arequipa.

Practice of milking cows three times daily on Danish dairy farms

J.O. Lehmann, L. Mogensen and T. Kristensen
Aarhus University – Foulum, Department of Agroecology, Blichers Alle 20, 8830 Tjele, Denmark; jespero.lehmann@agro.au.dk

Milking high-yielding dairy cows 3 times daily in traditional milking systems is widely practiced in many countries, and the objective with this study was to gather information on how this is practiced in Denmark. Around 90% of Danish dairy farmers are members of the Danish milk recording federation, and we sent an electronic questionnaire to the 157 of their members, who practice milking 3 times daily. This corresponds to 8% of their members after correcting for 761 members with milking robots. Ninety dairy farmers answered the questionnaire (57%). The 78 conventional dairy farmers averaged 349 annual cows and 11,700 kg ECM per annual cow whereas the 12 organic (13%) averaged 372 annual cows and 9,250 kg ECM per annual cow. In contrast, the average member herd in the milk recording federation has 188 annual cows. Forty-nine herds started practicing milking 3 times daily before 2015, and hence before the milk quota was abolished. Across all respondents, they used an average of 3.11 hours to milk all cows once, which equals 106 cows per hour. The minimum milking interval was 7.5 and the maximum was 8.5 hours. One milking is on average handled by 1.5 persons, and 4.5 different persons handle milkings during a week. Twenty farms milk cows in a rotary milking parlour, 67 in a standard milking parlour and 3 in a stanchion barn. Fifty-four herds split lactating cows in 3 or more groups whereas 18 had only 1 group. All lactating cows received the same TMR in 54 herds whereas 9 herds received 3 or more different TMRs. In fact, 36 of the 54 herds with only 1 TMR had nowhere else to automatically provide concentrate. Farmers were asked to indicate which of a number of given positive and negative effects they had experienced after switching to milking 3 times daily. Of positive effects, 93% experienced a higher yield, 58% a better udder health and 61% a more coherent working day for employees. In contrast, 50% indicated no negative effects of the switch, albeit 19% had more difficulties with finding enough employees, and 8% saw some cows loose more weight after calving. Furthermore, 29% of farmers perceived a higher number of different milkers in relation to consistent milking routines to be a negative effect.

Session 34 Poster 16

Ten years evolution of dairy cattle herds: fertility, production and management

P. Feyjoo[1], J.L. Pesántez[2,3], A. Heras[1], V. Sanz[3], R. Patrón[4], N. Pérez[5], J. Mesías[1,2], M. Vázquez-Gómez[1], C. García-Contreras[3], O. Fargas[6] and S. Astiz[3]

[1]UCM, Animal Production, 28040, Madrid, Spain, [2]Cuenca University, 0101220, Cuenca, Ecuador, [3]INIA, 28040 Madrid, Spain, [4]TRIALVET, 28721 Madrid, Spain, [5]European University of Madrid, 28670 Madrid, Spain, [6]VAPL S.L., 08551 Barcelona, Spain; martavazgomez@gmail.com

This study described the productive data evolution from 71 dairy herds over ten years (2008 to 2017). During this period, two global management strategies changed: timed Artificial Insemination (AI) programs since 2011 onwards and composting bedding systems since 2014. The results showed an increase of total cows/ herd (293±299 in 2008 vs 498±558 cattle in 2017), and maximal herd size (980 in 2008 vs 2,505 cattle in 2017). The percentage of heifers (based on the total of animals) was 69.0% without a tendency of change over the years. Regarding production, average daily milk yield per lactating cow increased with time (30.3±2.5 in 2008 vs 33.5±2.2 l/cow in 2017). While calving to first AI interval and estrous detection rate remained stable (91 days and 51.2%, respectively), average days in milk decreased with time (204±14d in 2008 vs 195±16d in 2017). Similarly, the total of AIs per pregnancy decreased from 4.4±1.2 in 2008 to 3.6±1.1 AI/P in 2017. Accordingly to these data, conception and pregnancy rates increased (CR=28.7±5.0% and PR=15.0±3.5% in 2008 vs CR=32.0±5.6% and PR=16.4±4.3 in 2017). Age at first AI decreased from 16.1±1.4 m of age in 2008 to 14.8±51.5 m in 2017, with a reduced age at first calving from 26±1.7 m to 25.1±1.6 m of age in 2017. Neonatal mortality was 8.5%/year during the whole period, with a minimum of 7.2% in 2013 and a maximum of 13.4% in 2008. The annual percentage of culled cows remained stable (29.1±6.8 in 2008 vs 29.8±7.2% in 2017). Dry period length decreased from 64±10 in 2008 to 62±9d in 2017, without an apparent decreasing tendency over the period. In conclusion, the increase in the size of farms is confirmed during this 'after milk quota period'. The historically impaired fertility with an increasing individual production could not be observed. Moreover, cow's fertility and heifers reproductive efficiency (age at first calving) improved during this period.

Session 35 Theatre 1

Challenges for livestock breeding in developing countries

B. Besbes, G. Leroy, P. Boettcher and R. Baumung

Food and Agriculture Organization of the United Nations (FAO), Viale delle Terme di Caracalla, 00153 Rome, Italy; roswitha.baumung@fao.org

According to the Second Report on the State of the World's Animal Genetic Resources for Food and Agriculture, breeding goals have been defined for 53% of the world's locally adapted national breed populations and 35% are under genetic evaluation. These percentages are significantly lower in developing countries and may be overestimated, considering countries that did not respond. Such information suggests that establishing and sustaining structured livestock breeding programmes remain challenging in many countries, particularly for low-input systems. The State of the World Report also identified gaps in terms of institutional capacities and supporting breeding policies. To circumvent those gaps and to improve the performance of local livestock, importation and introgression of exotic breeds are frequently considered as the solution by decision makers. However, the success of crossbreeding programmes depends on different prerequisites such as continual access to adequate breeding stock, adaptability of the stock to the local production environment, the management and level of inputs required for the improved livestock to express their genetic potential or integration within a reliable market chain. Furthermore, a lack of a long-term breeding strategy will result in indiscriminate crossbreeding threatening subsequently the diversity of locally adapted breeds, without obtaining desired gains in productivity and profitability. For any kind of breeding programme, priorities need to include capacity-building at all levels from livestock-keepers to policy-makers, as well as strengthening the organizational structures.

Session 35 Theatre 2

How animal breeding can contribute to sustainable pig production

P.W. Knap
Genus-PIC, Ratsteich 31, 24837 Schleswig, Germany; pieter.knap@genusplc.com

Sustainable pig production safeguards the Quadruple-Bottom-Line interests of People Planet Profit Pigs: (1) food safety, access-and-benefit-sharing; (2) biodiversity, environmental efficiency; (3) food security, farmer profitability; (4) animal welfare. Animal breeding can contribute to improvement of all these, we focus here on 3 of them. Environmental efficiency is about efficient use of scarce resources (particularly feed) and mitigation of pollution; improvement of net feed efficiency and reduction of production losses are important breeding objectives and good genetic progress is being made; N excretion per slaughter pig was reduced by ~0.6%/year since 1970. Farmer profitability was a central objective since Hazel's (1943) selection index and gross value per slaughter pig increased by ~9%/year since 1975. Also, possible G×E issues with using temperate-bred genotypes in hot conditions can be overcome by proper data recording in the response environment and processing these for the selection environment. Animal welfare is about (1) invasive treatments (tail docking, castration) where the genetic background of tailbiting is now being unravelled in terms of actor- and victim-propensity, and selection for low boar taint compound levels is underway; (2) behavioural deprivation leading to frustrated coping behaviour, where the genetic architecture of the HPA stress axis is now being researched; (3) robustness & adaptability, where selection against mortality is improving commercial survival rates by ~0.7%/year since 2000, genomic technology focused on specific disease resistance is under fast development, and the genetic background of aggressive dominance behaviour is now being unravelled. An important part of all above issues seems to be directly influenced by the gut microbiome; we may expect substantial contributions to improvement from that angle. It will certainly be one of the hot R&D topics in pig breeding of the coming decade. Significant trade-offs among the main sustainability elements exist, most notably among environmental efficiency, farmer profitability, and animal welfare; it follows that much of the focus of pig breeding goal design will be determined by political issues rather than technical ones. The same holds for a trade-off between animal welfare and animal integrity.

Session 35 Theatre 3

Sustainable intensification of global cattle production using high-input genetics

S. Van Der Beek
CRV, P.O. Box 454, 6800 AL Arnhem, the Netherlands; sijne.van.der.beek@crv4all.com

FAO has defined five guiding principles for sustainable food and agriculture. The first principle is: 'Improving efficiency in the use of resources'. Improved efficiency is required since agricultural production has to grow significantly in the next 30 years while at the same time environmental load has to decrease. Growth especially has to occur in low-income countries where resources are limited and the environment for animal production is challenging. In the last decades, breeding programmes of high-income countries have contributed to development in low-income countries via the (subsidized) export of semen. The impact of these exports is often limited, and the effect of the use genetics from high input breeding programmes varies. Brazil is an example of a country with an enormous growth of the animal production sector in the last decades. Cow numbers have increased, production per animal has increased, and technology has successfully been implemented. Differences between production systems and production environments are however huge, and in many cases the yield gap is large. CRV has been a main supplier of genetics throughout this period of growth in Brazil. The lessons learnt could be applied more widely: (1) first, farmer needs have to be identified; (2) knowledge transfer is extremely important and precedes the use of improved genetics; (3) uptake of reproduction techniques is crucial; (4) production systems vary widely; each system requires other genetics; (5) in most production systems crossbreeding between high input genetics and locally adapted genetics is optimal; (6) which high input genetics should be used depends on the intensity of the system; (7) to minimize the yield gap, genetics has to fit the production system and constant knowledge transfer is required. In conclusion, high input genetics can contribute to sustainable intensification. It requires the right mix of high input and locally adapted genetics, and a strong emphasis on knowledge transfer.

Session 35 Theatre 4

Leveraging new technologies and novel breeding strategies to sustainably grow global beef production

M.A. Cleveland and F.B. Mokry
Genus ABS, De Forest, WI 53925, USA; matthew.cleveland@genusplc.com

There has been significant progress in the genetic improvement of beef production traits in the last 50 years which, combined with improved management practices, has led to large increases in the amount of beef produced, even as cattle numbers declined. Increasing population growth and a growing middle class is driving demand for high quality beef, particularly in the developing world. To meet this growing demand, the global beef industry will need to produce more beef using the same, or fewer, resources. To remain a competitive and affordable source of protein, beef production must be economically sustainable for farmers, i.e. input costs are minimized and beef outputs are maximized. Sustainability is achieved in large part through genetic improvement of traits directly related to farmer profit in his or her particular environment, which allows for investment in new technologies that drive further improvement. This improvement must be ultimately underpinned by increases in production efficiency (e.g. feed efficiency or robustness) to remain sustainable and thus environmental sustainability is a natural result of economic sustainability. The use of new technologies, e.g. genomics, sexed semen, gene editing, will be critical in accelerating genetic gain. While rapid genetic progress has the potential to benefit all stakeholders, strategies to disseminate improved genetics to commercial beef producers are needed to ensure the supply chain can capitalize on the improvement delivered. Many commercial beef programs, however, are not optimally aligned with the needs of the supply chain for the efficient production of beef. Alternative breeding strategies, including those adopted in other species such as pigs and poultry, will allow production systems to fully benefit from improved genetics and existing resources. The key to a sustainable increase in beef production is the efficient dissemination of improved genetics coupled with approaches to ensure farmers realize maximum genetic potential and genetic value that is captured across the supply chain.

Session 35 Theatre 5

Poultry breeding for sustainable increase in production

V.E. Olori
Aviagen Group, 11 Lochend Road, Newbridge, EH28 8SZ, United Kingdom; veolori@aviagen.com

Demand and production of poultry meat and eggs is set to increase in many regions of the world with marginal environments. Investment in research and development by breeding companies is presently undertaken under the premise that, increase production has to be done efficiently with robust animals suitable for various environments and production systems, to maximise output while minimising impact on the environment and ensuring the welfare of the animals. Sustainable increase in poultry production requires highly efficient and resilient birds, appropriately raised by skilled producers. While selective breeding is focussed on evolving these birds of the future, breeding companies know that managing poultry to minimise environmental impact of production and optimizing the environment for biological efficiency, health and wellbeing, requires adequate knowledge and skill in poultry husbandry. Poultry managers and carers need to adapt management practices to the changing genetic makeup of poultry populations. Hence, in addition to genetic improvement and sustainable supply of poultry stock, breeding companies are increasingly involved in training and education schemes. These are aimed at equipping the current as well as next generation of stockmen and women with the knowledge and practical skills required to raise poultry birds efficiently. This capacity building (train the trainer principle), along with investment in production technologies is pertinent for the future and necessarily in many of the regions were production is set to increase. Application of technologies by highly skilled professionals is a panacea for sustainable increase in supply of poultry products with high standard of animal welfare and minimal impact on the environment. It will allow producers to capture the full genetic potential of superior genotypes from balanced breeding. This is a key factor in meeting the future need for poultry products especially in regions where demand and production is set to increase in the next three decades.

Session 35 Theatre 6

Breeding for sustainable laying hens

J.A.M. Van Arendonk, J. Visscher, K. Peeters and P. Van As
Hendrix Genetics, P.O. Box 114, 5830 AC Boxmeer, the Netherlands; johan.van.arendonk@hendrix-genetics.com

Animal genetics has played a key role in improving sustainability of poultry production and will continue to do so in the future. In this presentation we describe how Hendrix Genetics, through our sustainability and breeding program, contributes to improving sustainability for all partners in the animal protein value chain. We have established a sustainability program comprising of three building blocks: animals, people and planet. Animal welfare, biosecurity and genetic resources are the key priorities within the building block animals. Ensuring animals are treated with care and respect, and are kept under the highest standards of welfare, is essential. We ensure that taking good care of animals is embedded in our company culture. As global suppliers of breeding stock, we have a responsibility for ensuring biosecurity and animal health. In addition, we also have an obligation to protect our genetic resources. People make our business, that's why we strive to enhance the quality of life for consumers, customers and colleagues. Consumers in different countries across the world vary, but purchasing with sustainability in mind has increased overall during the last years. Reducing the use of antibiotics is a key part of the building block planet. In our breeding program for laying hens we select for improving productivity and reducing the environmental impact of laying hens through improving input efficiency and increasing the length of the laying period. Further, we see increased attention for other elements of sustainability, in particular for animal health and welfare. In response, we are continuously exploring opportunities for improving our breeding program on those aspects by adapting the breeding goal and adding novel selection traits. In laying hens, we have introduced selection for social effects for survival of hens with intact beaks. In addition, we have added feather cover as a selection trait. The development of an automatic nest to record egg production in a group house setting, helps us to select laying hens that will perform well in cage-free systems. In our breeding program, selection for improved health and welfare is combined with selection for improved resource efficiency and product quality. Breeding for sustainable laying hens is all a matter of finding the right balance.

Session 35 Theatre 7

Plenary discussion

J.A.M. Van Arendonk
Hendrix Genetics, P.O. Box 114, 5830 AC Boxmeer, the Netherlands; johan.van.arendonk@hendrix-genetics.com

In 2050 the world population will be over 9 billion people. Especially Africa and India will be confronted with huge population growth. To feed all these people, the production of food must grow substantially. This will also be the case for animal food, despite a possible decline of animal food consumption in Europe. International animal breeding has contributed substantially to the increase of animal food production in the past decades. The question is how animal breeding could contribute to sustainable global food production in the coming 30 years. The speakers of the session will participate in a plenary discussion concerning this topic. The main questions involve what do breeding companies do, and what can be done to increase the food production in a sustainable way, especially in developing countries (e.g. Africa, India, Latin America), and how can research contribute? Is there *e.g.* any opportunity for using New Breeding Techniques?

Session 36 Theatre 1

Research and development efforts on optimizing key parameters in industrial insect production

L.-H. Heckmann
Danish Technological Institute, Bioengineering and Environmental Technology, Kongsvang Alle 29, 8000, Denmark; lhlh@dti.dk

Since 2012, the Danish Technological Institute (DTI), an international private and self-owned research and technology organization, has worked with insects for feed and food. DTI covers the full insect value chain from production to products; and has over the last six years led and/or participated in more than 15 completed or ongoing research and development (R&D) projects involving *Musca domestica*, *Hermetia illucens*, *Alphitobius diaperinus* and *Tenebrio molitor*. In this presentation, we will highlight a range of examples from our R&D work on optimizing key parameters in industrial insect production focusing on reproduction, feed development and temperature with special emphasis on *T. molitor* and *H. illucens*. This will include published and unpublished results from national and international publicly-funded projects such as: (1) SUSMEAL (https://www.dti.dk/projects/project-sustainable-mealworm-production-for-feed-and-food/38305?cms.query=susmeal) (2) WICE (https://www.dti.dk/projects/project-can-insects-convert-organic-household-waste-into-valuable-mink-feed/38308?cms.query=wice) (3) inVALUABLE (https://www.dti.dk/specialists/invaluable/38118?cms.query=invaluable). Running from January 2017 to December 2019, inVALUABLE is, to-date, one of the largest R&D projects in Europe on insects as feed and food. The project is led by DTI and involves 10 partners and has a total budget of 3.7M EUR. The vision of inVALUABLE is to create a sustainable resource-efficient industry for animal protein production based on mealworms. The partners span the entire value chain and include entrepreneurs, experts in biology (entomology and nutrition), biotech, automation, processing and food tech and -safety. This interaction of competences is key to lifting insect production to an industrial level. Ongoing activities of the inVALUABLE project with reference to mealworm production will get special attention during the presentation.

Session 36 Theatre 2

Automated insect rearing

T. Spranghers[1], S. Schillewaert[1], F. Wouters[1], C. Coudron[2], D. Deruytter[2], J. Claeys[2] and T. Anthone[3]
[1]VIVES Hogeschool, Agro- and Biotechnology campus Roeselare, Wilgenstraat 32, 8800 Roeselare, Belgium, [2]Inagro, Ieperseweg 87, 8800 Rumbeke-Beitem, Belgium, [3]VIVES Hogeschool, Industrial science and technology campus Kortrijk, Doorniksesteenweg 145, 8500 Kortrijk, Belgium; filip.wouters@vives.be

Due to the recent change in law, there is an increasing demand of insects and insect derived products for feed and food purposes. To fulfil this demand and lowering production costs an upscaling of insect rearing is required by focusing on automation of the labour intensive tasks. A modular concept of an automated insect farm was the start to focus on the automation and optimisation of those different tasks. Within this concept, an automated feeding line for mealworm was developed. This feeding line makes it possible to implement a strict feeding regime and monitoring of the feeding. The line makes optimal use of space and there is minimal movement of the crates with the insects. In addition, several sieving techniques were compared for both mealworm (*Tenebrio molitor*) and the black soldier fly (*Hermetia illucens*). An overview of those techniques is given scoring the separation result and the cost price. Research results are part of projects financed by VLAIO Tetra and Interreg Vlaanderen-Nederland.

Early prediction of final harvest for the mealworm

D. Deruytter, C.L. Coudron, J. Claeys and S. Teerlinck
Inagro, Applied insect breeding research center, Ieperseweg 87, 8800, Belgium; david.deruytter@inagro.be

In an industrialized mealworm farm it is important to know the number of individuals in each container as early as possible in a fast and reliable way. Combined with a growth chart, this information can lead to a more efficient feeding regime and hence a faster growth and more revenue. Therefore different possibilities were assessed to predict the number of offspring (Tenebrio molitor) based on the beetle density and or egg density. Three different techniques were assessed to predict the final harvest. In the first two techniques the beetles were allowed to deposit their eggs on the bottom of a black 600×400 mm container. A full factorial design was assessed with eleven beetle densities (between 2.2 and 250 mg beetles/cm^2) and five deposit times (between 1 and 14 days). In the first method a linear regression model was constructed based on the density, deposit time and interaction to predict the number of offspring. In the second method image processing was used to calculate the area covered by eggs and this was used as proxy of number of offspring. Finally in the last method attempts were made to separate the eggs from the substrate. Hence, perfect dosing of the containers with eggs would be possible. The initial results indicate a clear correlation between the beetle density, deposit time and the final number of mealworms. Furthermore, it was possible to construct a formula based on a single new parameter: beetle-days, similar to the concept of degree-days. The main advantage is that one can *a priori* determine the final harvest. Using ImageJ it was possible to determine the area covered by the eggs and construct a model to correlate this number to the final number of mealworms. This may be useful in automated insect production systems to *a posteriori* determine the number of mealworms or to check the first method (e.g. changes in egg deposit due to an illness). Finally, it was possible to gather individual eggs when beetles deposited through a mesh in flour. However, this results in an additional handling step of the delicate eggs which may result in additional mortality. In conclusion, it is possible to determine the final harvest at a very early stage via the 3 presented techniques potentially resulting in more optimal feeding and growth.

Insect biomass quality and safety: basic concepts, recent issues, and future challenges

L. Pinotti[1], M. Ottoboni[1], M. Tretola[1], I.V. Boccazzi[2], S. Epis[2] and M. Eeckhout[3]
[1]Department of Health, Animal Science and Food Safety, University of Milan, 20133 Milan, Italy, [2]Department of Biosciences, University of Milan, 20133 Milan, Italy, [3]Department of Applied Biosciences, Ghent University, 9000 Gent, Belgium; luciano.pinotti@unimi.it

The use of alternative feed ingredients in farm animal's diet can be an interesting choice from several standpoints, even though their nutritional value, quality and safety should be always kept in mind. Among different alternative feed ingredients, insect based ingredients can be considered a hot topic. Insects are looked as an interesting alternative protein source for feed and are expected to be increasingly used in Europe as replacers for animal-derived proteins especially in aquaculture, even though their potential for other farm species and as fat sources, cannot be excluded. Although the insect growing substrate is determinant in defining the insect's composition, both insect species and life stage (larvae vs pre-pupae vs adult form) represent further source of variation. These latter aspects could impact not only on feed formulation, but also on feed production (pre-processing steps, such us drying and defatting) and feed technology. From a circular economy point of view, insects are commonly reared on biomass like leftover of vegetable and fruit food (organic waste streams). This implies that micro-livestock are used to upgrade low value organic waste streams into high-value biomass. But, several concern are moved against organic waste, mainly due to safety issues. In fact, although vegetable waste streams already constitute the rearing material for insect growing, to date, the possible safety hazards associated with the use of these ingredients are hardly known. Accordingly, the aim of the present review is to address recent advances in term of insect material feed technology and selected safety concerns.

Session 36 Theatre 5

Influence of feeding substrate and strain on protein, fat and carotenoid content in silkworm

C. Chieco[1], L. Morrone[1], G. Bertazza[1], S. Cappellozza[2], A. Saviane[2], F. Gai[3], N. Di Virgilio[1] and F. Rossi[1]
[1]Institute of Biometeorology, CNR, via P.Gobetti 101, Bologna, 40129, Italy, [2]Honey bee and silkworm research unit, CREA, Via Eulero, 6a Padova, 35143, Italy, [3]Institute of Sciences of Food Production, CNR, L.go P. Braccini 2, Grugliasco, 10095, Italy; c.chieco@ibimet.cnr.it

The overexploitation of fishmeal and soy for feedstuff industry have poses the attention on unconventional protein sources more environmental friendly, as insects. We evaluate the feasibility to optimize and control the quality of Silkworm pupae for feedstuff industry through the choice of appropriate strains and their feeding with different substrates (fresh mulberry leaves and artificial diet). Feeding substrate composition strongly influence fat and protein content of silkworm pupae. The two tested strains (White Polyhybrid and Golden Yellow Nistari), had higher fat and lower protein contents when fed with fresh mulberry leaves compared to the artificial diet. Analysis showed that also the n3/n6 ratio was affected almost exclusively by the feed substrates factor: silkworm pupae grown on fresh leaves had 3 times greater n-3 content than n-6 respect to those feed with artificial diet. On the contrary, pigment content in pupae is specifically determined by the strain; Golden Yellow Nistari strain showed always higher carotenoids with respect to the White Polyhybrid strain regardless of the rearing substrate. Our results establish that silkworm pupae composition well reflects the quality of the food ingested by the juvenile instars, thus offering the possibility to obtain ad hoc final products destined to different employments in the feedstuff industry.

Session 36 Theatre 6

Role of temperature and waste type on the development and survivorship of *Hermetia illucens*

M. Shumo
Center for Developmet Research, University of Bonn, Ecology and Natural Resources Managment, Genscherallee 3, 53113 Bonn, Germany; mshummo@hotmail.com

Black soldier fly (BSF) *Hermetia illucens* feed on various organic materials, which make them suitable for waste management. Nevertheless, the conditions for their optimum production are challenging to achieve. This study investigated temperature effects on development and survivorship of BSF using brewer spent grain and cow dung as rearing substrates. The results showed higher and significantly different growth rate in experimental substrate (22±0.98-13±0.67 days) versus the control substrate (54±2.56- 52±2.78 days) at 25-30 °C. The interaction was also significant (ANOVA: $F (4, 355) = 8.65$, $P<0.0001$), indicating that the temperature effect on development duration was different for the two organic wastes. Similarly, the BSF development to 5th Instar was significantly different with a range of 0.69±0.06-0.54±0.05 mg vs 0.41±0.03-0.43±0.02 mg at the same temperature. The BSF length up to the 5th instar in both experimental and control substrates increased with temperature, with the longest instar (2.15±0.06 mm) in brewer spent grain at 25 °C versus 1.90±0.05 mm in the control substrate. The highest width (0.52±0.01 mm) was observed in the control substrate verses 0.49±0.02 mm in the experimental substrate at 35 °C. The highest prepupae length (2.26±0.03 mm) was observed in control substrate at 15 °C and continued to reduce with temperature compared to that of experimental substrate (22.22±0.01 mm), which generally remained constant. The general observations showed that experimental substrate provided relevant nutrients and better burrowing ground for the BSFL than the control substrate, which accounts for relatively higher growth and egg production. Furthermore, there was positive association between larval survival, biomass produced, and development time with the type of rearing substrate. The study recommends further investigation on the feasibility of using BSF in waste management in terms of economic, environmental as well as social factors prior to implementation in order to optimize the benefits.

A shining star in Turkey as a new profitable animal production opportunity: the Red Wiggler

O. Yılmaz[1] and N. Cicek Atikmen[2]

[1]Cankiri Karatekin University, Vocational High School of Cerkes, Department of Plant and Livestock Production, Cerkes, 18600 Cankiri, Turkey, [2]Cankiri Karatekin University, Faculty of Forestry, 18000 Cankiri, Turkey; zileliorhan@gmail.com

Vermicomposting can be called as a kind of organic fertilizer of nature. The Red Wiggler (*Eiseniafetida*) yields vermicast by using some organic materials such as animal manure, vegetables, kitchen waste, agricultural crop residues, or organic by-products from industries. Vermicomposting is based on earthworms and microorganisms to help stabilize active organic materials. It also convert active organic materials to a valuable soil amendment and source of plant nutrients. If vermicompost is added to soil, it enhances the nutrients available to plants. Vermicompost has also been shown to increase plant growth and depress plant disease and insect pest attacks. Vermicomposting products have many applications, including vegetable production, home gardening, fruit gardening, landscaping, vine growing, and in agriculture. This survey study is conducted in July, August and September 2016. During survey about 19 vermicomposting enterprises were visited and a questionnaire was asked to business owners. According to data vermicomposting enterprises were not organized. Business owners complained about some difficulties such as lack of information in public, universities and Ministry of Food, Agriculture, and Livestock.

How to design your career for the livestock industry?

C. Lambertz[1,2]

[1]Free University of Bolzano, Faculty of Science and Technology, Universitätsplatz 5, 39110 Bolzano, Italy, [2]Young EAAP, Via G.Tomassetti 3 A/1, 00161 Rome, Italy; christian.lambertz@unibz.it

This session is especially dedicated to young scientists during their early stages of their scientific career independent whether at master, PhD or PostDoc level. Important decisions with impacts on our future career have to be taken during these stages. For young scientists a variety of options arise with jobs opportunities becoming more and more diverse. So do job requirements change constantly and may even change more dramatically in the future in view of developments such as digital farming. Consequently, young scientists have to adjust their skills and expertise to get fit for the future. 'How to best prepare for a career in the livestock industry?' will be the main focus of this session. Experts in the field of recruiting new staff in companies from different fields of livestock production will give us an overview of what to consider when developing a career for the industry. In the second part of the session, the pan-European initiative 'EURAXESS – Researchers in Motion' will be presented. It is a network of more than five hundred Service Centres located in 40 European countries. The network delivers information and support services to professional researchers. Backed by the European Union and its Member States, it supports researcher mobility and career development, while enhancing scientific collaboration between Europe and the world. Furthermore, Marie Skłodowska-Curie actions, which provide grants for researchers of all stages during their careers, will be presented.

Foot pad infections in broiler breeders – significance and prevalence

I. Thøfner, L.L. Poulsen, R.H. Olsen, H. Christensen, M. Bisgaard and J.P. Christensen
University of Copenhagen, Department of Veterinary & Animal sciences, Stigbøjlen 4, 1870 Frederiksberg, Denmark; jpch@sund.ku.dk

In intensive poultry production systems good health and good management is crucial to obtain high levels of animal welfare and high production yields throughout the production cycle. In broiler breeders, decreased foot pad integrity mainly caused by poor litter conditions may subsequently result in an increase in mortality due to septicaemia, endocarditis and arthritis as the major manifestations, over time. Although the pathogenesis is not fully elucidated, the aetiology of these infections is often Gram positive (G+) cocci, such as *Staphylococcus* spp., *Enterococcus* spp. and *Streptococcus* spp. It is hypothesized that foot pad lesions serve as port of entry for systemic or localised bacterial infections. In a recent Prohealth study we investigated the causes of the so-called normal mortality in four broiler breeder flocks from 20 to 60 weeks of age. Furthermore the individual foot pad health was recorded for each bird. Foot lesions were defined as presence of hyperkeratosis, ulcerations/necrosis or pododermatitis. Mortality caused by G+ cocci was low in young birds but increased significantly during production resulting in approximately 13% of the total mortality. During the observation period, we observed a strong correlation between foot pad health and age of dead birds with foot pad lesions increasing significantly from below 40% in young birds (20-29 weeks) to almost 80% in birds more than 50 weeks old. To further investigate these findings an experimental infection model using foot pads lesions as port of entry in old broiler breeders was established. Inoculation resulted in systemic lesions (sepsis, endocarditis and arthritis), corresponding to natural cases under field conditions, as well as injection site abscesses. This work is part of the EU-FP7 funded PROHEALTH project (grant no. 613574).

The effect of enrichment on broiler leg health – a systematic review

I.J. Pedersen and B. Forkman
University of Copenhagen, Department of Veterinary and Animal Sciences, Groennegaardsvej 8, 1870, Denmark; bjf@sund.ku.dk

Leg problems is one of the major welfare problems in modern broiler production. Environmental enrichment is one of the strategies that a broiler producer may use to improve leg health. A number of papers have been published that have evaluated possible types of environmental enrichment. The current paper is a systematic review on the effect of enrichment on leg health in broiler chicken. We selected six different types of environmental enrichment for inclusion in the review, these were; light program, intensity of light, stocking density, perches, straw bales and separation of resources. We did a systematic literature search for each of the enrichment types. This yielded a total of 65 studies, 59 randomized trials and 6 cross-sectional studies, that were considered suitable for inclusion in the review. A quality assessment of the methodological quality of all 59 randomized trials was performed, specifically concerning blinding, randomization and reliability of measures. The most common leg health measures from the studies were gait score, foot pad dermatitis, hock dermatitis and tibial dyschondroplasia. The most common types of enrichment investigated were light programs (n=24) and stocking density (n=24). Provision of perches and increased intensity of light were only effective in improving leg health to a limited extent and both mainly affected foot pad- and hock dermatitis. In contrast there was evidence that both a lowered stocking density and an intermittent light schedule can improve leg health. Few studies were available on the effect of provision of straw bales and separation of resources; however the existing literature showed that both types of enrichment could be effective in improving leg health. In conclusion there is good evidence that appropriate environmental enrichment can decrease the prevalence of leg problems in broilers. This research was funded by the EU FP7 Prohealth project (no. 613574).

Evaluation of two laying hen strains kept in different cage systems for liver health

E.E. Onbaşılar[1], E. Erdem[2], N. Ünal[1], A.S. Tunç[3], A. Kocakaya[1] and B. Yaranoğlu[4]
[1]Vet. Med., Dept of Anim Sci, Ankara University, 06110, Turkey, [2]Vet. Med., Dept of Anim Sci, Kırıkkale University, 71450, Turkey, [3]Vet. Med., Dept of Pathology, Ankara University, 06110, Turkey, [4]Vet. Med., Dept of Anim Sci, Balıkesir University, 10145, Turkey; akocakaya@ankara.edu.tr

Alternative cage systems for laying hens have been spread around the world due to raising welfare awareness. Fatty liver disorder (FLD) is a major problems in commercial flocks and it is affected by the cage system. This study was conducted to evaluate the differences in blood parameters and FLD of two laying hen strains kept in conventional (CC) and enriched (EC) cages, respectively. A total of 532 Lohmann Brown Classic (LB) and 532 Lohmann LSL Classic (LW) hens were kept in CC and EC from 16 to 73 weeks of age. The CC and EC were in same facility and designed to keep 20 and 18 hens/cage, respectively. At 16 weeks of age, 252 and 280 birds of each strain were weighted and assigned to EC and CC. Each group consisted of 14 cages. Hens were weighted at 73 weeks of age and 14 hens per group with an average weight were selected. Levels of alanine amino transferase (ALT), aspartate amino transferase (AST), cholesterol and triglycerides were determined from blood samples collected at slaughter. Relative liver weights were calculated after they were slaughtered and macroscopic evaluation of liver was performed. Two-Way ANOVA was used to determine the effect of cage types and strain and their interactions on blood parameters. Differences in fatty liver scores (FLS) were evaluated by Chi-square test. The ALT, AST and cholesterol levels of hens kept in CC were greater than in EC hens. Serum ALT levels in LB hens were lower than LW hens, indicating strain differences with respect to ALT levels. Serum AST, cholesterol and triglycerides levels did not differ between strains. Interactions were found between cage types and strain for serum ALT level. Relative liver weights were not affected by cage type and strain. In this study, most hens showed high FLS at higher age but cage type and strain had no effect on FLS. The present study revealed that enrichment was not sufficient to reduce FLS. This study was supported by Ankara University (BAP-10A3338005).

A genetic biomarker panel for diagnosis of coccidiosis and necrotic enteritis in broilers

T. Giles[1], T. Van Limbergen[2], P. Sakkas[3], D. Maes[2], I. Kyriazakis[3], P. Barrow[1] and N. Foster[1]
[1]University of Nottingham, School of Veterinary Medicine and Science, Sutton Bonington Campus, Leicestershire NG7 2NR, United Kingdom, [2]Ghent University, Department of Reproduction, Obstetrics and Herd Health, Salisburylaan 133, 9820 Merelbeke, Belgium, [3]Newcastle University, Department of Agriculture, School of Natural and Environmental Sciences, King's Road, Newcastle upon Tyne NE1 7RU, United Kingdom; n.foster@nottingham.ac.uk

Coccidiosis and Necrotic Enteritis (NE) remain important diseases throughout the EU broiler industry, impacting on health and economic loss. Diagnosis of Coccidiosis and NE often relies upon the clinical presence of disease and post-mortem investigations. Diagnostic platforms which are specific and sensitive for Coccidiosis and NE may, therefore, be very useful tools to monitor flock health and detect early/pre-clinical infection. During the Prohealth study, we used whole genomic profiling to detect differential expression of >1000 genes (<0.05) in the terminal ileum of Ross 308 chickens naturally or experimentally infected with Eimeria spp. Further Genespring analysis and multiple testing correction allowed us to accurately determine a panel of 6 genes which were up-regulated in all intestinal samples from infected chickens following PCR. These genes encode for proteins involved in cell proliferation, cell integrity and gene expression. Further analysis indicated that the panel may be used to differentiate sub-clinical (no clinical disease) and clinical infection (clinical symptoms and lesions) in different genetic lines (Ross 308 and Ranger Classic). When the specificity of the biomarker panel was tested in intestinal tissues isolated from 20 day old Ross 308 with NE (experimentally infected with C. perfringens), the 6 genes were still differentially expressed but in the opposite direction (down-regulated). Thus, the gene panel may also differentiate between broilers infected with Eimeria and those infected with C. perfringens. This work is part of the EU-FP7 funded PROHEALTH project (grant no. 613574).

Novel maternal traits affecting piglet survival

S.M. Matheson[1], G.A. Walling[2], R.J. Thompson[1], I. Kyriazakis[1] and S.A. Edwards[1]
[1]Agriculture, School of Natural & Environmental Sciences, Newcastle University, Newcastle upon Tyne, NE1 7RU, United Kingdom, [2]JSR Genetics Limited, Southburn, Driffield, East Yorkshire, YO25 9ED, United Kingdom; stephanie.matheson@ncl.ac.uk

Selection for hyperprolific sows has resulted in larger litter sizes and a greater variation in birth weight of piglets, with increased number of lightweight piglets that have an increased potential for death by crushing. The aim of this study was to investigate two novel maternal traits affecting piglet survival: the proportion of IUGR piglets born in a litter, and sow lying characteristics (as determined by rump-mounted accelerometers). Piglet births (21,159 piglets; 1,574 litters) were recorded for 52 weeks in a population of Landrace sows (862) crossed with either White Duroc or Large White sires. Sow data included parity, litter size, flooring of the farrowing crate, leg conformation, sensor-derived lying characteristics and week of farrowing. Piglet data gathered 18-24 hours after birth were: weight (as a proxy for birth weight), sex, head shape, and date and reason for any death. IUGR was assessed visually from head morphology. For each litter, the proportion of IUGR piglets in a litter (pIUGR) and the proportion of piglets surviving to processing (pSURV) were calculated. Genetic correlations between pIUGR and pSURV were moderately-high and negative, suggesting that selection for reduced proportion of IUGR will give a corresponding increase in the proportion of piglets surviving to processing (SURVPROP-IUGRPROP, g= -0.88±0.273). The heritability (0.23±0.06) and repeatability (0.30±0.036) for pIUGR piglets in a litter was moderate, indicating this may be a good trait to include in breeding programmes. Sows with a high maximum acceleration in the stand-to-lie transition crushed more piglets, although the trait did not appear to be heritable, while other lying behaviour traits, such as the mean duration of the downward transition had a slightly higher heritability (0.07±0.213). This research was funded by the EU FP7 Prohealth project (no. 613574).

Genomic analysis of immune traits of two maternal pig lines

K. Roth[1], C. Dauben[1], E. Heuß[1], M.J. Pröll[1], H. Henne[2], A.K. Appel[2], K. Schellander[1], E. Tholen[1] and C. Große-Brinkhaus[1]
[1]Institute of Animal Science/University Bonn, Animal Breeding and Husbandry group, Endenicher Allee 15, 53115 Bonn, Germany, [2]BHZP GmbH, An der Wassermühle 8, 21368 Dahlenburg-Ellringen, Germany; katharina.roth@itw.uni-bonn.de

During its entire lifetime, each animal is exposed to a wide range of interactions within its environment, where the immune system is supposed to protect it against pathogenic invasions. In the last decade the interest in resilience, robustness and tolerance investigations for future livestock breeding has been substantially increased. Therefore, this study focused on the genetic improvement of survivability, health and immune status of piglets and growing pigs. Within the 'pigFit' project complete blood count (CBC) was performed with a haematology analyser and serum haptoglobin was measured by classic photometric method. In addition, cytokine levels (IL-1β, IL-6, IL-4, IL-8, IL-12, IL-10, TNFα and IFNγ) were quantified using 'Porcine Cytokine/Chemokine Multiplex Magnetic Bead 8-plex Panel'. A uni- and bivariate approach were applied in ASReml 4.0 to calculate the heritabilities (h^2), as well as the phenotypic (r_p) and genetic (r_g) correlations for 22 examined complex immune traits in Landrace (LR) (n=606) and Large White (LW) (n=502) piglets. For LR, 14 traits like cytokines and erythrocytes showed a moderate $h^2 \geq 0.4$. LW piglets showed only in three traits a $h^2 \geq 0.4$ related to erythrocyte characteristics as HBE and MCV. Moreover, 27 significant $r_g \geq 0.4$ were detected in both breeds indicating possible differences in immune mechanisms between the investigated lines. A univariate genome-wide association study (GWAS) using the R Package 'GenABEL' detected several significant Single Nucleotide Polymorphisms (SNPs) associated with the examined immune parameters. A TNF receptor associated factor 5 (TRAF5) located on chromosome 9 for inflammatory cytokines IL-1β and IL-6 and anti-inflammatory IL-10 provides an indication for pleiotropic genetic structure, which will be further analysed with multivariate statistical approaches. First results help to understand the complexity within the pigs' immunity to improve the robustness of piglets by breeding.

Sow management interventions to improve piglet survival

V.A. Moustsen[1], D. De Meyer[2], L. Vrielinck[2], T. Van Limbergen[3] and S.A. Edwards[4]
[1]SEGES Pig Research Centre, Agro Food Park 15, DK 8620 Aarhus N, Denmark, [2]Vedanko bvba, Keukelstraat 66A, 8750, Wingene, Belgium, [3]Faculty of Veterinary Medicine, Ghent University, Salisburylaan 133, B-9820 Merelbeke, Belgium, [4]University of Newcastle, Kings Road, Newcastle upon Tyne NE1 7RU, United Kingdom; vam@seges.dk

Management interventions which result in calmer sows might be beneficial for piglet survival. A first experiment tested the effect of two interventions on behaviour of sows in two Danish commercial herds with farrowing pens. In a split-plot design, background music (M) was the main plot and sow handling (scratching, S) was a subplot, with 111/110 sows in each group. +S-sows were scratched by farm staff once daily for 15 s. Classical music was played 06.00-18.00 from 5d before to 5d after expected farrowing. Scratching resulted in a significant decrease in avoidance behaviour (0-2 scale) in a forced approach test (+S=0.63+0.03, – S=0.74+0.03, P=0.02) whereas music had no significant effect. However, farm staff stated that sows in all treatment groups were less reactive and easier to handle than sows in the non-treatment group. The combined treatments were therefore tested in a commercial trial in Belgium. Treatment sows experienced background music (commercial radio station) 06.00-18.00 from entry to the farrowing unit and for the whole lactation period. Staff also performed daily 15 s of backscratching per sow from entry until farrowing. Treated groups were interspersed with control groups to give 3 groups of treated sows (n=140 sows) and 7 contemporary control groups (n=314 sows). Piglet mortality was significantly reduced by the intervention (treated 9.83%, control 11.91%, P<0.05). Batch weighing of weaners from 3 groups per treatment also suggested higher weaning weight of piglets from treated sows (6.1 kg, n=1,296) vs piglets from control sows (5.35 kg, n=1,296). Positive handling of the sows in the farrowing rooms proved to be beneficial for the sows and their piglets but also for staff, who reported that the sows were calmer and also preferred working with music in the farrowing rooms to offer a win-win situation. This research was funded by the EU FP7 Prohealth project (no. 613574).

Enriching the sow environment and diet during gestation reduced piglet neonatal mortality

H. Quesnel[1], E. Merlot[1], B. Peuteman[1], A. Prunier[1], D. Gardan-Salmon[2] and M.C. Meunier-Salaün[1]
[1]INRA, PEGASE, 35042 Rennes, France, [2]CCPA group, ZA du Bois de Teillay, 35150 Janzé, France; helene.quesnel@inra.fr

Sow environment during gestation can generate maternal stress which could influence piglet health and survival after birth. The study aimed to investigate a strategy of environmental and nutritional enrichment to reduce maternal stress and its consequences on piglet mortality. Gestating sows were group-housed in a conventional system on a slatted floor (C, n=26), in the same conventional system with environmental and nutritional enrichment (CE, n=30) or in larger pens enriched with straw bedding (E, n=27). The enrichment of the CE group consisted of pieces of oak attached to a chain (three per pen) and straw pellets provided in the trough at a rate of 200 g/d from 3-30 days of gestation (DG) and 400 g/d from 31-104 DG. On DG 105, sows were transferred into farrowing pens and housed in identical individual stalls on a slatted floor. Cortisol concentration was measured in sow saliva during gestation, sow behavioural and investigative activities were recorded on DG 101 and piglet mortality was recorded. Cortisol concentration was greater (P<0.05) in C and CE than in E sows on DG 14 while it was intermediate in CE sows compared with C and E sows on DG 105 (before transfer to farrowing pens), and no longer differed among the 3 treatments on DG 107 (after the transfer). On DG 101, CE sows exhibited a lower proportion of stereotypies compared to C sows (22 vs 34%, P<0.05) but a greater proportion compared to E sows (7%, P<0.05). On this same day, CE sows had more investigative sequences than C sows (7.3±7.0 vs 1.7±1.8, P<0.01) but less than E sows (20.3±13.8; P<0.01). Rate of early mortality (i.e. piglets dead at birth + piglets that died within 12 h of birth) was lower in groups CE and E (6.6 and 6.3%, respectively) than in group C (11.1%, P<0.05), but overall mortality (stillbirth + preweaning death) did not differ significantly among the 3 groups (23.2, 19.1 and 19.3% in groups C, CE, and E, respectively, P=0.35). Enriching the sow environment and diet during gestation therefore improved sow welfare and reduced piglet mortality at and soon after birth. Research was funded by the EU FP7 Prohealth project (no. 613574).

Assessment of natural vs mechanical farm ventilation using daily registered data in fattening pigs

I. Chantziaras[1], M. Klinkenberg[1], T. Van Limbergen[1], J. Dewulf[1], D. De Meyer[2], L. Vrielinck[2], C. Pineiro[3], M. Jimenez[3] and D. Maes[1]
[1]Ghent University, Faculty of Veterinary Medicine, Salisburylaan 133, 9820, Belgium, [2]Vedanko bvba, Wingene, 8750, Belgium, [3]PigCHAMP Pro Europa, Segovia, 40003, Spain; ilias.chantziaras@ugent.be

Climatic conditions such as temperature, stable gases and humidity may influence performance, welfare and health of the pig. The present study aimed to assess the effect of ventilation type and air quality parameters on respiratory disease and welfare in fattening pigs. Two fattening pig units were used, unit A (1,256 pigs) with mechanical ventilation, unit B (1,264 pigs) with natural ventilation. Animal genetics, nutrition, stocking density and health management were the same for both units. Within each unit, 3 batches of fattening pigs (08/2015 to 12/2016) were monitored throughout the fattening period. Parameters that were daily measured included: CO_2, NH_3, relative humidity, water consumption, temperature and prevalence of respiratory disease. The welfare status of the animals was scored twice per production round with a simplified version of the Welfare quality® protocol. In unit B (natural ventilation), there were higher levels of NH_3 (16 vs 13 ppm, $P<0.01$), CO_2 (1,929 vs 1,483 ppm, $P<0.01$) water consumption (2,490 vs 2,151 l/day, $P<0.01$), temperature (27.1 vs 23.4 °C, $P<0.01$) but no differences in humidity levels (61.1 vs 61.5%, $P=0.26$). Overall, more pigs in unit B showed respiratory disease (1.96 vs 1.41%, $P<0.01$). Concerning the welfare assessment, higher scores -indicating welfare problems- were noted on the animals from unit B (average score 21 vs 12.5). When analysed univariably, the prevalence of respiratory disease was strongly positively associated with the levels of NH_3 ($P<0.01$), CO_2 ($P<0.01$) and temperature ($P<0.01$), and numerically positively linked with water consumption ($P=0.1$) or humidity ($P=0.6$). In a multivariable mixed model, higher levels of CO_2 ($P<0.01$) and the presence of natural ventilation ($P<0.01$) both increased the point prevalence of respiratory disease. In conclusion, our findings suggest that the mechanical ventilation assured better environmental conditions for the fattening pigs and improved the health and welfare status of the animals. A research funded by the EU FP7 Prohealth project (no. 613574).

Animal based indicators as tools to study production disease in pigs

L. Montagne[1], A. Boudon[1], M.C. Meunier-Salaün[1], M. Karhapää[2], H. Siljander-Rasi[2] and N. Le Floc'h[1]
[1]INRA, UMR PEGASE, 35000 Rennes, France, [2]Luke, Koetilantie 5, 0790 Helsinki, Finland; lucile.montagne@agrocampus-ouest.fr

The incidence and severity of production diseases (PD) are influenced by environmental conditions and animal predisposition. The control of PD requires the identification of risk and predisposing factors at several levels from epidemiological to infra animal studies. The identification of pigs affected by diseases, and at risk to be, is one of the objectives of the PROHEALTH project. Firstly, we present the results of a literature review aiming at identifying animal-based traits characterizing the physiological and health status of pigs. Those traits are potential indicators of diseases. This review aimed also at quantifying the influence of factors relating to animal or husbandry practices on these traits. In a second time, pigs were experimentally challenged to enhance the risk of leg, digestive and respiratory diseases and some of these indicators were used to evaluate animal health status. The literature analysis showed that indicators relating to performance characteristics or immunological response are potential tools to prognosis diseases. Accordingly, experimental studies showed that immune indicators such as acute phase proteins are associated with depressed growth performance and feed efficiency in challenged pigs. The influence of genetic background and sex was not evaluated by the meta-analysis, this information often lacking in the literature. Yet, experimental trials clearly revealed that different lines of pigs (French Large White or Finish Landrace) had different susceptibility of lameness and osteochondrosis, respiratory and inflammatory disorders, and digestive disorders at weaning. Moreover, osteochondrosis lesions were more severe in boars than in gilts. Altogether our results showed that some of these indicators could help to identify production diseases or to test practice limiting the disease. This work has received funding from the EU-FP7 PROHEALTH project (grant agreement no. 613574).

The public, animal production diseases and policy; what lessons can be learnt from consumer research

B. Clark[1], L. Panzone[1], G. Stewart[1], I. Kyriazakis[1], J. Niemi[2], T. Latvala[2], R. Tranter[3], P. Jones[3] and L. Frewer[1]
[1]Newcastle University, School of Natural and Environmental Sciences, Newcastle uponTyne, NE1 7RU, United Kingdom, [2]Natural Resources Institute Finland, Kampusranta 9, 60320, Seinäjoki, Finland, [3]School of Agriculture, Policy and Development, University of Reading, Whiteknights, P.O. Box 237, RG6 6AR, United Kingdom; beth.clark@newcastle.ac.uk

The use of intensive animal production systems is increasing globally. Problematically, animals in these systems are more prone to production diseases. To ensure the acceptability of animal production systems in the future, and to maintain public trust in stakeholders across the food chain, public attitudes towards these systems, including interventions utilised to both prevent and treat production diseases, need to be understood, and addressed in policy. Two systematic reviews of 80 and 54 studies explored public attitudes and willingness-to-pay for farm animal welfare (FAW), with a survey (n=2,330) of 5 European countries used to explore attitudes to production diseases. Findings show that the public have concerns about intensive production systems, potentially in relation to concerns about FAW standards, linked to the requirement for humane treatment and naturalness within animal husbandry. The acceptability of interventions to prevent animal diseases in production systems were linked to these concerns, with consumers indicating greater references for more proactive interventions utilising housing and hygiene strategies. The use of medicine based interventions raised concerns in relation to food safety, human and health and antimicrobial resistance. Industry and policymakers need to consider these concerns in future policy recommendations, potentially applying a combination of market and policy based solutions to align consumer preferences for FAW with standards applied within intensive production systems. Legislation is required to ensure the interventions to deliver safe food, as well as promote policies focused on the reduction of anti-microbial resistance. The research highlights the need for effective communication and knowledge exchange with all stakeholders involved in the supply chain, as well as transparency in decision-making linked to policies focused on FAW in intensive animal production systems.

Relationship between farrowing and piglet traits, and mortality in a free-farrowing system

T. Han, J. Yun, M. Nystén, S. Hasan, S. Börkman, A. Valros, C. Oliviero and O. Peltoniemi
University of Helsinki, Department of Production Animal Medicine, P.O. Box 57, Koetilantie 2, 00014, Finland; taehee.han@helsinki.fi

Free farrowing systems have been related with increased piglet mortality in case of large litter size. On the other hand, public concern about animal welfare is guiding pig industry towards production with free farrowing pens. Aim of the study was to identify the relationship between piglet mortality during the first 24 h after start of farrowing in a free farrowing system and farrowing and piglet traits. The births of 259 piglets [Duroc × (Danish Yorkshire × Danish Landrace)] from 15 sows were video recorded. The sows were kept in free farrowing pens (crate size: 220×80×180 cm, pen size: 250×240 cm). At birth, body weight of each piglet was recorded by sampler. The birth order, birth interval, vitality score and latency from birth to first udder touch (latency to udder) of each piglet were determined from the recording. The average number of live-born piglets per litter was 17.3±2.7 and total live-born mortality was 17.3%. Forty five piglets died, of which 39 were crushed and 6 died due to other causes. There were no significant differences between piglets that died or survived in their birth interval, cumulative farrowing duration or vitality score. However, died piglets were tended to have earlier relative birth order than survived piglets (42.9±4.9 vs 51.4±2.1%; P=0.073). In addition, piglets that died had lower birthweight (1,088.5±60.1 vs 1,315.2±23.7 g; P<0.001) and longer latency to udder (52.3±10.7 vs 34.4±2.9 min; P=0.033) compared to piglets that survived. Our study suggests that low birthweight and long latency to udder, which might affect colostrum intake and increase risk for hypothermia and therefore may lead to crushing, were the major causes of piglet mortality in free farrowing system with large litter size.

Session 38 Poster 13

Effects of supplementing organic acids to the diet of prepartum sows on neonatal piglet mortality

J. Yun, S. Hasan, S. Saha, C. Oliviero and O. Peltoniemi
University of Helsinki, Production Animal Medicine, Koetilantie 2, 00790, Finland; jinhyeon.yun@helsinki.fi

Despite benefits of larger space for achieving prepartum natural behaviour in loose-housed sows, an increase in piglet mortality mainly due to crushing has been of great concern. The primary goal of the present study was to examine the effect of supplemental organic acids to the late gestation diet on neonatal piglet mortality. An additional goal was to study the effect of the loose-housed farrowing system on neonatal piglet mortality. A total of 60 sows were moved to farrowing units 7 days before the expected parturition date. The sows and their offspring were allocated to a factorial design with two factors, diet [CON (normal sow diet) vs ORG (normal sow diet supplementing tall oil fatty and resin acids)] and housing [CRATE (crate size: 225×65×65, pen size: 325×250) vs FREE (crate size: 225×159×191, pen size: 325×250)]. The live-born piglet mortality rate was remarkably higher in FREE than in CRATE (11.8±1.9 vs 3.3±1.8%, P<0.01). This was due to the higher rate of crushed piglets seen in FREE compared to CRATE (11.5±1.8 vs 2.4±1.8%, P<0.001). In FREE, the sows with ORG diet had a lower rate of crushed piglets (5.6±2.5 vs 17.5±2.6%, P<0.01), and thus the lower live-born piglet mortality rate (6.0±2.5 vs 17.5±2.7%, P<0.05) than the sows with CON diet, whereas among sows in CRATE, the live-born piglet mortality rate was not affected by different diets (1.3±2.6% for CON vs 5.3±2.6% for ORG). Consequently, these data indicate that supplemental tall oil fatty and resin acids to the diet of the prepartum sow could reduce mortality rates of neonatal piglets in loosed-housed system within 24 h postpartum.

Session 38 Poster 14

Microbiota after weaning in pigs from two lines divergently selected on residual feed intake

L. Montagne[1], N. Le Floc'h[1] and I. Rychlik[2]
[1]PEGASE, INRA, Agrocampus-Ouest, Rennes, 35042, France, [2]Veterinary Research Institute, Hudcova 70, 621 00 Brno, Czech Republic; lucile.montagne@agrocampus-ouest.fr

Microbiota is a key element in feed digestive efficiency and in the development of diseases such as post weaning diarrhoea. This study aims to characterize the microbiota composition and activity through analysis of fermentation products in the faeces of post-weaned pigs from two lines of divergently selected for residual feed intake (RFI, LRFI and HRFI for low and high), a measure of feed efficiency. Faeces were collected on 72 pigs on week 1 and 3 after weaning at 28 d of age. One week after the weaning, *Acidaminococcus*, *Anaerovorax*, *Blautia*, *Collinsella*, *Erysipelotrichaceae incertae sedis*, *Megasphaera*, *Mitsuokella*, *Butyrivibrio*, *Faecalibacterium*, *Olsenella* and *Prevotella* were significantly more abundant in HRFI line whereas *Bacteroides* and *Clostridium* XlVa were more abundant in the LRFI line. Within *Bacteroidetes*, the replacement of *Bacteroides* with *Prevotella* has been previously reported in pig after weaning. Faecal microbiota of the two lines was very similar at week 3. Total VFA and ammonia concentration increased similarly in the two lines between week 1 and 3. Despite difference in microbiota composition, there was only slight difference between lines in individual VFA on week 1. The proportions of acetate and branched chain fatty acids decreased whereas that of propionate and butyrate increased with time. These changes were more marked in the LRFI (-4.5 and + 4.2% for acetate and propionate) than HRFI lines (– 1.4 and +1.7%). To conclude, the microbiota profile and activity was similar between lines 3 weeks after the weaning but the microbiota of the HRFI line seems to adapt faster to solid and complex diet. This work has received funding from the EU-FP7 PROHEALTH project (grant agreement no. 613574).

Session 38 Poster 15

Genomic analyses of hereditary defects in porcine dam line populations

C. Große-Brinkhaus[1,2], I. Brinke[2], H. Henne[3] and E. Tholen[2]
[1]Association for Bioeconomy Research (FBF e.V.), Adenauerallee 174, 53113 Bonn, Germany, [2]University of Bonn, Institute of Animal Science, Endenicher Allee 15, 53115 Bonn, Germany, [3]BHZP GmbH, An der Wassermühle 8, 21368 Dahlenburg-Ellringen, Germany; cgro@itw.uni-bonn.de

Beside economic losses, the occurrence of hereditary defects in pig populations has a great impact on animal welfare concerns in pig production. In the last decades, breeding against defects is mainly focused on atresia ani, hernias (scrotal, umbilical, and inguinal), cryptorchidism, shivering, hermaphrodites and splay legs. However, the inheritance of these traits is still unclear. This knowledge gap might be explained by a low frequency as well as problems in phenotyping. Often, these traits are recorded immediately after birth, whereas some phenotypes like umbilical and inguinal hernia appear late in the pigs live which makes recording error prone. Against this background, the aim of this study was to evaluate the data of two maternal lines of a commercial breeding company, to estimate variance components and to investigate the genomic background. Data were collected in two maternal pure bred lines and their crosses on 60 farms. Depending on the type of trait, data of 180,000 litters recorded from 2005 to 2017 were available. In a first step, consequences of different strategies to remove unreliable data were evaluated. For the variance component estimation, all traits were investigated as litter traits, considering the full pedigree within the model. Heritabilities (h^2) ranged between 0.009 and 0.17. In a second approach the hereditary defects were treated as a binary individual trait of the piglet itself. Analysing these data by logit threshold models revealed higher h^2 in a range of 0.25 to 0.44. Whether these high estimates reflect the true genetic background or result from an overestimation, as shown in human genetic studies, will be discussed. Estimated breeding values from the different models will be used as depended variables to identify relevant SNP-markers. For more than 1,500 pigs per line genotypic information are available. Beside these classical genome-wide association studies, the available variants will be checked for missing homozygosity. The identified genetic markers can be implemented in breeding programs and will allow to support animal welfare.

Session 38 Poster 16

Effect of an isotonic protein drink with milk replacer on growth rate of suckling and weaned piglets

A. Firth[1], C. Van Der Peet-Schwering[2], R. Verheijen[2], L. Van Luiten[3] and A.V. Riemensperger[1]
[1]Tonisity International Limited, 16 Fitzwilliam Place, Dublin 2, Ireland, [2]Wageningen Livestock Research and Swine Innovation Centre (VIC), Vlaamseweg 17, 6029 PK Sterksel, the Netherlands, [3]HAS University of Applied Sciences, Spoorstraat 62, 5911 KJ Venlo, the Netherlands; angela@tonisity.com

Aim of the study was to test the effect of (1) an isotonic protein drink (IPD; Tonisity Px, Tonisity International), (2) an IPD and a milk replacer (MR) from day (d) 2 to 8 after birth and from d3 before to d3 after weaning or (3) a MR from d2 to 8 after birth on the growth rate of suckling and weaned piglets compared to piglets that are only fed dry feed and no IPD or MR. Litters (n=80) used in the study were assigned to the following treatments: a control group (CG) receiving no IPD or MR, an IPD group receiving 500 ml IPD per litter per d, an IPD+MR group and a MR group. Pigs were weaned at d26 and followed for 5 weeks after weaning. Parameters recorded were: number of live and stillborn piglets, individual piglet weights (birth, d9, d1 before weaning, week (w) 1, 2 and 5 post weaning), mortality, daily intake of IPD and MR at different growth stages as well as dry feed intake, diarrhoea scores and veterinary treatments. Weight of sows and backfat was measured when entering the farrowing room and at weaning. Data was statistically analysed by means of F-tests using generalized linear models. IPD pigs weighed 7.76 kg at weaning, which was significantly higher than other treatment groups (IPD+MR 7.22 kg, MR 7.20 kg and CG 7.13 kg, $P<0.05$).) Daily gain from birth to weaning (g) tended ($P=0.09$) to be highest in the IPD group (237 g) compared to the control (219 g) and the MR group (223 g). Compared to the IPD+MR group daily gain in the IPD group was only numerically higher (237 vs 229 g). The same was seen for daily gain from d 9 to weaning ($P=0.08$; control group: 237 g, IPD group: 258 g, IPD+MR group: 245 g; MR group: 238 g). Litters given the IPD from d2 to 8 of age and d3 before to d3 after weaning had a more than 0.5 kg higher weaning weight compared to the control, IPD+MR and MR group. Despite a very low energetic value of the IPD, performance was best in this group, indicating its enterotrophic properties, which lead to improved nutrient absorption and resulted in increased weaning weight.

Session 39 Theatre 1

The role of Interstallion in supporting international sport horse breeding by improved transparency

K.F. Stock[1], Å. Viklund[2], I. Cervantes[3], A. Ricard[4], K. Christiansen[5], O. Vangen[6] and S. Janssens[7]
[1]IT Solutions for Animal Production (vit), Heinrich-Schroeder-Weg 1, 27283 Verden (Aller), Germany, [2]SLU, Dept. of Anim. Breed. & Genet., Box 7023, 75007 Uppsala, Sweden, [3]UCM, Avda. Puerta de Hierro, 28040 Madrid, Spain, [4]INRA, UMR 1313 GABI, Domaine de Vilvert, 78352 Jouy-en-Josas, France, [5]Danish Warmblood, Vilhelmsborg Allé 1, 8320 Maarslet, Denmark, [6]NMBU, Dept. of Anim. and Aquacult. Sciences, P.O. Box 5003, 1432 Ås, Norway, [7]KU Leuven, Livestock Genetics, Kasteelpark Arenberg 30, 3001 Heverlee, Belgium; friederike.katharina.stock@vit.de

Interstallion (IntS) is the permanent working group of the EAAP Horse Commission focused on breeding aspects. Aims are to improve understanding and use of information in sport horse breeding and to support decision making and selection on an international level. To achieve transparency across countries, practices of studbooks regarding data recording and evaluations in the breeding programs are compiled and published. IntS has actively supported research initiatives on across country genetic evaluations and comparisons between nationally available genetic proofs for traditional and new traits. Warmblood studbooks from around the world have been surveyed in 2002 and again in 2015 to provide an update overview of status and development of sport horse breeding. Continued engagement of IntS in promoting and harmonizing the use of linear traits in sport horse breeding has started in 2011. Base work included information collection for scientific review and online publication of comprehensive material which enables overviewing and comparing available applications. To further support knowledge transfer and allow regular exchange between science and practice and across studbooks, a series of international workshops has been established. Activities related to linear profiling and refined data recording are particularly important in connection with introducing new genomic applications in sport horse breeding and accordingly highly appreciated by the studbooks. The coordinative and bridging role of IntS has already benefitted collaborative research which is expected to grow further in importance with increasing interest in using new traits and new modern technologies to strengthen the breeding programs for sport horses worldwide.

Session 39 Theatre 2

Experiences from the Nordic Interstallion project

Å. Viklund[1], S. Furre[2], J. Philipsson[1] and O. Vangen[2]
[1]Swedish University of Agricultural Sciences, Animal Breeding and Genetics, P.O. Box 7023, 75007 Uppsala, Sweden, [2]Norwegian University of Life Science, Department of Animal and Aqua Sciences, P.O. Box 5033, 1432, Ås, Norway; asa.viklund@slu.se

Sport horse breeding is international and when Interstallion was formed in 1998 the aim was to study methods to compare genetic evaluations across countries. Two PhD projects concluded that a joint international genetic evaluation would be feasible, but the practical implementation was not realized due to distrust and competition among studbooks. Therefore, in the following Nordic Interstallion project we studied the potential and willingness of a joint genetic evaluation in the Nordic countries. These countries have similar tests for young horses and records of competition performance, and stallions are largely of the same origin. In Sweden and Denmark, national genetic evaluations are conducted annually, but in the small populations of Norway and Finland no EBVs are estimated. Thus, a joint Nordic genetic evaluation would have to be based on raw data. Combining data was challenging and time consuming as use of unique identification numbers was not consequent among studbooks. Results showed that all studbooks would benefit from a joint genetic evaluation. Breeding values could be estimated earlier with higher accuracy for many more stallions than in national evaluations. In this project both researchers and representatives of the studbooks worked towards a common goal, showed engagement, trust and transparency, which are prerequisites for success. However, the project would not have been possible without granting from a research fund. Although the scientific proofs of the benefits are unquestionable, a joint routine evaluation is not yet in place. That decision, including resource allocation, now lies in the hands of the studbooks. When sport horse breeding is entering the genomic era, collaboration between studbooks and researchers from different countries will be the key for success, and then traditionally estimated BVs across countries are a prerequisite.

Collaboration of studbooks advancing development of genomic selection for sport horses

M. Wobbe[1,2], K.F. Stock[1,2], S. Neigenfind[3], N. Krattenmacher[3], W. Schulze-Schleppinghoff[4], M. Von Depka Prondzinski[5], E. Kalm[3], R. Reents[2], C. Kühn[6], J. Tetens[7] and G. Thaller[3]
[1]University of Veterinary Medicine Hannover (Foundation), Inst. for Anim. Breed. & Genet., Buenteweg 17p, 30559 Hanover, Germany, [2]IT Solutions for Animal Production (vit), Heinrich-Schroeder-Weg 1, 27283 Verden, Germany, [3]Christian-Albrechts-University of Kiel, Inst. of Anim. Breed. & Husbandry, Olshausenstr. 40, 24098 Kiel, Germany, [4]Oldenburger Pferdezuchtverband e.V., Grafenhorststr. 5, 49377 Vechta, Germany, [5]Werlhof-Institut MVZ, Schillerstr. 23, 30159 Hanover, Germany, [6]Leibniz Institute for Farm Animal Biology (FBN), Inst. of Genome Biology, Wilhelm-Stahl-Allee 2, 18196 Dummerstorf, Germany, [7]University of Goettingen, DNTW Functional Breeding, Burkhardtweg 2, 37077 Goettingen, Germany; mirell.wobbe@vit.de

High-quality phenotypes and a strong reference population (gRP) are crucial for exploiting the potential of genomic selection. In 2017, horse breeding associations and experienced research partners have joined forces and started an initiative for promoting equine genomic R&D in Germany. Harmonized linear profiling with a comprehensive scheme is delivering the phenotypic data basis. Candidates for gRP have so far been recruited from DNA sampled mares of five studbooks and were required to have linear profiles including performance in gaits, preferably also in jumping. Phenotypes were qualitatively ranked according to proximity to the breeding goal (highest value of assessments under rider). To further optimize gRP composition, pedigree relationships among the horses were considered for selecting horses for SNP genotyping with moderate density (70k+). From the first genotyping cohort, 755 horses and 55,170 SNPs remained after quality testing and were used for preliminary genomic analyses with PLINK software. Discipline and breeding policy related stratification patterns were confirmed, underlining the importance of careful interpretation of first results of association analyses. These revealed significances for several traits including specific aspects of trot and canter like front limb mechanics and stride length. Increase of sample size and the target of 5,000 horses in the gRP will allow refined analyses with the aim of reliable genomic predictions for important traits in sport horses.

The French surveillance network of equine mortality causes: a new way of monitoring major diseases

J. Tapprest[1], M. Linster[1,2], N. Foucher[1], J.P. Amat[1,3], N. Cordonnier[2] and P. Hendrikx[3]
[1]French agency for food, environmental and occupational health & safety (Anses), Laboratory for equine diseases, 14430 Goustranville, France, [2]National veterinary school of Alfort, Pathological anatomy unit, 94704 Maisons-Alfort, France, [3]Anses, Coordination and support unit for surveillance, 69364 Lyon, France; jackie.tapprest@anses.fr

Autopsy is recognized as one of the activities that can effectively contribute to the surveillance of major diseases in animal populations. The French surveillance network of equine mortality causes (Resumeq) was created in 2015 for the qualitative surveillance of equine mortality through the centralization of autopsy data in a national database and their overall epidemiological analysis. Its objectives are the qualification of the causes of equine mortality, the monitoring of their evolution over time and space and the early detection of emerging diseases. It is an event-based surveillance system that involves a diversity of actors and structures. An institutional organization has been defined that includes a steering body, a scientific and technical support committee and a coordination unit. Different specific tools have been developed such as standardized autopsy protocols, thesaurus for the anatomo-pathological terms and the causes of equine death and an interactive web application for the visualization of data analysis results by the network contributors. The four French veterinary schools, in addition to 15 veterinary laboratories and 9 veterinary clinics already contribute to the production and centralization of standardized data. To date, around 1000 cases of equine autopsy, mainly located in the West part of France, have been collated and the geographical coverage is gradually improving. Data analysis allows the ranking of the main causes of death and the identification of particularly threatening causes of death at a local, regional or national level. These first results demonstrate the feasibility and interest of this surveillance at a national level. Moreover, in the future, this surveillance could take a European dimension if several countries decide to jointly value their autopsy data. *Acknowledgments to all participants in the Resumeq network, to the Regional Council of Normandy and to the Fonds Eperon.*

A proposal for the Spanish horse breeds as a subpopulation within a European genomic metaanalysis

I. Cervantes[1], J.P. Gutiérrez[1], M.D. Gómez[2], A. Molina[2] and M. Valera[3]
[1]Departamento de Producción Animal, Universidad Complutense de Madrid, 28040 Madrid, Spain, [2]Departamento de Genética, Universidad de Córdoba, 14071 Córdoba, Spain, [3]Departamento de Ciencias Agroforestales, Universidad de Sevilla, 41013 Sevilla, Spain; icervantes@vet.ucm.es

Nowadays, molecular genetics and bioinformatics advances have increased and are available to perform new methodologies to optimize breeding programs and maximize the genetic progress in livestock breeds. Sport horse breeding programs are characterized by long generation intervals and low genetic gains. Another particularity in horse breeding is the high proportion of male foals castrated at a very young age, resulting in low selection intensities in subsequent steps of selection. Genomic selection helps to increase the accuracy in artificial selection; it allows selecting individuals in an early stage and in traits difficult to be measured. Moreover, the genome wide association analysis is a complement for breeding programs and the high density panels are recently available for this specie. This project is aimed at applying these techniques into Horse Breeding Programs in Spain. During this project two objectives will be developed: the application of genomic selection to predict genomic breeding values comparing with the traditional ones and the development of genome wide association analyses in performance and diseases traits. Performance datasets from five main breeds: Pura Raza Español (PRE), Arab (A), Anglo-Arab (AA), Spanish Trotter (STH) and Spanish Sport Horse (SSH) will be used. These breeds include as selection objective the dressage, show jumping, eventing, trotting, and endurance performance. The linear profiling information will be included. In addition, records for the most important diseases in Spanish horse breeds will be collected. A total of 5,750 individuals will be sampled and genotyped (1,500 PRE, 1,500 SSH, 1000 A, 1000 STH and 750 AA). This, jointly with the pedigree and the phenotypes will allow designing the most efficient strategy to include genomic information in the current breeding program steps. This proposal will produce enough knowledge for the use of this methodology in Spanish Horse breeding programs and will prepare them to the connection to other European Horse Breeding programs.

The AGRIHORSENET COST action attempt

R. Mantovani
Dept. of Agronomy Food Natural resources Animals and Environment, Viale dell'Universita', 16, 35020 Legnaro (PD), Italy; roberto.mantovani@unipd.it

The AGRIHORSENET has been an attempt of building a COST action around a network of researchers involved in horse studies. The idea of a COST came during the summer of 2013, thinking to a possible information sharing on horse breeds diffused across Europe. Particularly, the focus of the AGRIHORSENET was to build a general inventory of European native horse breeds, aiming at preserving traditions, the cultural heritages and the economic role of horses in the European faming system. A great help to AGRIHORSENET came from a previous meeting held in Geneva at the end of 2010 organized by the Horse Heritage Foundation (HHF) and managed by Bertand Langlois (France), who recruited a group from 10 EU countries connected to heavy draught horse breeds. Because of the lack of subsequent development in the preservation project from HHF, the opportunity of replacing the HHF with COST was considered and proposed during the 2013 EAAP meeting in Nantes. The network of participants recruited included 14 researchers from 11 EU countries, and the project was enlarged to all native European horse breeds (i.e. not only heavy draught breeds). A challenge proposal (10,000 words + summary and keywords) to COST call was submitted twice reaching a better score in the first than in the second attempt, although with no success in both cases. Main reasons of failure can be attributed to the actual lack of connections between working groups and a perceived low impact of deliverables on agriculture and food sectors. Perspective for a successful project could come in my view only from an effective collaboration and team work between researchers from different countries, considering the COST call now requires a full proposal of 15 pages.

Organization of the saddle horse industry: a prospective research comparing European countries

G. Bigot
Irstea, UMR Territoires (Université Clermont Auvergne, AgroParisTech, INRA, Irstea, VetAgro Sup), centre de Clermont-Ferrand, 9 avenue Blaise Pascal, CS 20085, 63178 Aubiere, France; genevieve.bigot@irstea.fr

Numerous references highlight a recent growth of the horse industry in most developed countries, explained by a keen interest in leisure and sport with horses. However, studies mainly refer to disparate sources and focus on difficulties to assess precisely this sector. In fact, the saddle horse industry groups together various activities from horse breeding to leisure riding or sporting events as well as different actors: riding schools, livery yards, trainers, breeders, professional riders, private owners… The status and the national organization vary broadly among countries explaining the lack of homogenous databases, and difficulties to analyse the real development of this industry. Therefore, our aim is to overcome these methodological difficulties in building a project that would analyse the organization of the saddle horse industry in some European countries. With a concerted approach between researchers, institutions and professionals, we could first realize an inventory of existing national data and references about saddle horses productions, activities, and firms directly involved in this industry. According to the specificities of researchers that could be involved in the project, this work could also precise impacts of this industry on economic and environmental development of various territories. First exchanges with European colleagues in recent years underlined their interest in such a subject. To this aim, a starting point could be the writing of an article comparing some different countries about the saddle horse industry. The second step could concern the constitution of a working group to select axes that could be of interest for research at the European level in view to develop a collaborative study. Such a project could associate researchers from different countries and different approaches. It would give rise to a European overview of the horse industry that would be useful for professional and institutional stakeholders in view to have a better knowledge of this sector at the European level.

How can we measure equid welfare during long-distance transportation to slaughter?

A. Chapman
World Horse Welfare, Campaigns, Anne Colvin House, Snetterton, Norfolk, NR16 2EP, United Kingdom; alanachapman@worldhorsewelfare.org

Horses, ponies and donkeys are transported across the European Union (EU) for slaughter. Some of these journey will be short, less than 8 hours; however many cover long distances with some journeys lasting days. We have limited knowledge and understanding of the impact that these long-journeys have on the overall welfare of any equid, with particular reference to behavioural measures, and as a result it is difficult to identify the main factors that may affect their welfare during long-distance transportation. In order for their welfare to be managed it must first be measured. Desktop research has been used to identify peer reviewed 'welfare assessment' methodologies with relevant parameters selected, alongside field based trials to test the practicality of these different approaches in this specific context. Numerous welfare assessment methodologies and protocols exist that effectively allow for equids to be assessed in various environments and situations; however there is not one that allows for the welfare of the equid being transported long-distance to be fully assessed. Abnormal behaviours seen in equids being transported long-distances do not appear in literature although they have been routinely observed during field visits. Currently available protocols can be adapted by removing parameters that will not be utilised further and adding additional parameters that are not currently included. Robust welfare assessment protocols that can be easily carried-out in-situ need to be developed to allow the individual horse as well as group assessments to be undertaken and findings shared and utilised by key stakeholders across the EU. In order to fully move our understanding of the impact that long-distance transportation has on horses, ponies and donkeys going to slaughter, a multi-disciplinary collaborative approach must be used to further develop effective welfare protocols that relate to equids being transported long distances.

Session 40 Theatre 1

You'll miss the best things if you keep your eyes shut: the art and science of animal behaviour

J.N. Marchant-Forde
USDA-ARS, LBRU, West Lafayette IN 47907, USA; jeremy.marchant-forde@ars.usda.gov

Every stockperson knows the importance of animal behaviour as a husbandry tool. However, it is also a scientific discipline. One positive characteristic of animal behaviour is that it is accessible. Any animal scientist working with whole animals instantly becomes an anecdotal observer of behaviour. However, its accessibility can also be problematic in that animal scientists with other specialties think it is simple to measure, and incorporate it into their studies without the necessary knowledge to measure it appropriately. Animal behaviour is no different from any other aspect of biology, in that the more you learn about it, the more complex it becomes. We can define behaviour as the actions and reactions of whole organisms, and its interactions with its environment and with other animals of the same or other species. How we design our study will largely depend on the question we are asking, but that question is the essential starting point. It may be broad or it may be quite specific, but it will drive the behavioural component of our study, and establish a series of methodological questions that will ultimately help us fix our approach. For example, will our study be purely observational or will we be applying experimental treatments or asking more complex behavioural questions? What will be our level of analysis – are we interested in generalized behaviour or more detailed components of specific behaviours? Where will we carry out the study – in a large, complex environment or a small, simple pen? How many animals can we observe and do we need to identify individuals? When will we observe – all day or just a snapshot? How will we observe – directly or record and analyse later? What are the behaviours of interest – are they long-duration behavioural states or short-duration behavioural events? What will be our sampling rule – will we use *ad libitum*, focal, scan or behaviour sampling? What will be our recording rule – will we use continuous recording, instantaneous sampling or one-zero sampling? These aspects all need consideration if the behaviour component of a study is to produce valid and worthwhile data. All these questions will be covered. Done correctly, behaviour can add significant value and information to studies in other specialties of animal science.

Session 40 Theatre 2

Social network analysis allows a novel interpretation of livestock social interactions

S.P. Turner[1], S. Foister[1], R. Roehe[1], L. Boyle[2] and A. Doeschl-Wilson[3]
[1]SRUC, Easter Bush, EH25 9RG, United Kingdom, [2]Teagasc, Moorepark, Fermoy, P61 C997, Ireland, [3]Edinburgh University, Easter Bush, EH25 9RG, United Kingdom; simon.turner@sruc.ac.uk

Animal social interactions are conventionally recorded as dyadic traits between two individuals when in reality those individuals often live in a larger social group. Dyadic interactions are unlikely to occur independently of other interactions in the group; information that is lost from most behavioural analyses. Social network analysis (SNA) quantifies an individual's position in its social group using both direct dyadic interactions and indirect ones involving intermediary animals. These direct and indirect interactions also combine to quantify the group-level network structure. In animals, SNA has rarely been used to predict the outcomes of individual network position or group-level network structure and its use to address livestock production issues is only beginning. Using aggressive interactions between pigs as an example, we illustrate how SNA can use individual network position and overall network structure to predict the level of injury following regrouping. At 8 weeks of age, 1,170 pigs were mixed into 78 single sex groups of 15 pigs. The frequency and duration of all aggressive interactions were extracted from videos for 24 h post-mixing. Skin lesions were counted 24 h and 3 weeks (SL-3wk) post-mixing as measures of the outcome of acute and chronic aggression. The effect of individual network position and overall network structure on skin lesions was analysed using REML mixed models and stepwise regression. Individual network position correlated poorly with conventional interaction measures (e.g. duration of fighting). At the group level, pens with a divided network structure (high betweenness centralisation) at regrouping had significantly higher SL-3wk ($F_{1,57}$=4.38, P<0.05). Those pens with a large clique of pigs that fought directly at mixing had significantly lower SL-3wk ($F_{1,58}$=6.25, P<0.01) irrespective of the behaviour of other pigs. The position of pigs within their network was lowly but significantly heritable and affected their own level of injury. The results demonstrate that SNA can complement conventional methods to describe social interactions and has power to predict the outcome of those interactions.

Session 40 Theatre 3

Towards an automated assessment of pig behaviours on farm

E.L. Labyt[1], G.L. Lagarrigues[2], O.S. Sakri[1], C.G. Godin[1], C.T. Tallet[3] and A.P. Prunier[3]
[1]CEA LETI, Department Systems and Solutions Integration, 17 rue des Martyrs – MINATEC Campus, 38054 Grenoble, France, [2]Probayes, 180 avenue de l'Europe, 38330 Montbonnot, France, [3]INRA, UMR 1348 PEGASE, 16 Le Clos, 35590 Saint Gilles, France; armelle.prunier@inra.fr

Tail biting and aggression in finishing pigs are injurious behaviours affecting health and welfare of pigs as well as productivity of the farms. In the PIGWATCH European project (ERANET Anihwa), INRA and CEA are working on development of an automated technique, based on the use of sensors and machine learning algorithms, to detect injurious behaviours or abnormal patterns of activity. A wireless ear tag was developed, including a triaxial accelerometer and an Android application for data recording, processing and alert sending to the farmer on his smartphone when injurious behaviours are detected. Pigs were housed in groups of 8 on solid floor covered daily with fresh straw. Twelve pigs, i.e. 4 per group, were fitted with these ear tags. Their activity was recorded with the sensors during a period of two months. Their behaviour was analysed using video records on selected days. They were subjected to straw deprivation followed by food restriction in order to stimulate injurious behaviours or changes in the behavioural pattern of activity. In a first step, 24 hours of video records were analysed and synchronized with signals from sensors for each pig. Relevant mathematical features were extracted from signals to predict various pig's behaviours and notably, discriminate injurious behaviours from normal activity. These features were used in machine learning algorithms to build a model, able to automatically predict pig's behaviours and detect injurious ones. Regarding 'marked' fights (>3 aggressive acts within 10 s), the model has a sensitivity of 42% and a specificity of 62%. This model has been implemented in an Android App and will be assessed in farms in Germany, notably in terms of true and false alerts. We will get the feedback from farmers on the usefulness and how to improve the system ergonomic. As a third step, the whole database collected at INRA is currently processed with this model to predict other pig's behaviours (e.g. resting, feeding) and assess individual and nycthemeral variations.

Session 40 Theatre 4

Considerations in the conversion of automatically generated behavioural data to useful information

B. O'Brien[1], J. Werner[2] and L. Shalloo[1]
[1]Teagasc, Animal & Grassland Research and Innovation Centre, Moorepark, Fermoy, Co. Cork, Ireland, [2]University of Hohenheim, Institute for Agricultural Engineering, 70599, Stuttgart, Germany; bernadette.obrien@teagasc.ie

The application of the 'Smart Farming' concept in the livestock areas of agricultural production requires a closing of the knowledge gap in methodology between data collection and generation of meaningful information and insights in real time, to fulfil its potential to improve or maintain animal welfare and performance. In order to create a management tool that is as effective as a decision made by a person conducting visual monitoring, the sensor technology must be able to identify a change, compare to a previously set standard value, and decide on the correct action if sufficiently different from that standard value. Currently, at the most rudimentary level, real time data recording and real-time management of different types of data (from animals, their feed consumption, atmosphere in the shed) can be analysed in real-time to flag critical values which are important for animal well-being decision-making. The major issue is to learn from this multiplicity of data and convert it to information that may be used, e.g. to create and develop appropriate livestock management tools for the farmer. Firstly, the data may be generated in very large volumes at very frequent time points. The compilation of data into manageable and meaningful sets is necessary. Secondly, data may be retrieved from different sources, e.g. sensors, weather, geo-location. A case in point may be described by cow grazing behaviour data recorded through an accelerometer and pedometer together with cow localization (by triangulation). Finally, data may be generated in different formats, such as images or high resolution spatial maps of e.g. temperature. Relevant questions include what is the optimum frequency of data recording for different subjects/activities, degree of summary and possible integration of the data, relation/comparison to the developed standard, and feed into a decision support tool. New considerations and technologies are required for management and analysis of such data.

Session 40 Theatre 5

Behavioural monitoring of livestock on pasture based on motion sensor data

J. Maxa, G. Wendl and S. Thurner
Bavarian State Research Centre for Agriculture, Institute of Agricultural Engineering and Animal Husbandry, Voettingerstr. 36, 85354 Freising, Germany; jan.maxa@lfl.bayern.de

Systems monitoring animal behaviour and health status are nowadays successfully used in indoor housing systems as well as on paddocks primarily by cattle. On the contrary monitoring systems are not yet commonly implemented for livestock especially on wide range pastures. Nevertheless such techniques can help to reduce daily workload and to improve pasture management and welfare of grazing animals. Therefore, the main aim of this study was to (1) develop and test a tracking system for grazing cattle and sheep on alpine pastures (2) evaluate the usage of such a system by the farmer and (3) develop classification algorithms for detection of the specific cattle and sheep behaviours with possible future implementation into a pasture monitoring system. Hence, a prototype of a GPS-GSM tracking system was developed and tested over the pasture seasons 2015-2017. The workload, divided into 34 activities, was recorded on a total of 13 alpine farms located in Bavaria, Germany. Data for the development of classification algorithms were obtained from three independent trials where in total 16 heifers and 5 ewes were fitted with a tracking collar prototype, collecting GPS positions and motion sensors data (3-axis accelerometer and magnetometer) at 1 and 3 to 10 Hz, respectively. Up to seven behavioural activities were recorded based on continuous sampling observations of random animals in trial 1 and video-based observations in trials 2 and 3. Mixture models were applied in order to obtain thresholds among behaviours of classification algorithms. These algorithms resulted in moderate to very high sensitivity (56-99%) and specificity (81-98%) for the most important behaviours regarding animal and pasture management: grazing, ruminating and lying. Furthermore, workload analysis on alpine farms showed a potential through the usage of livestock monitoring systems by reducing the time needed to control and search animals on wide-range pastures.

Session 40 Theatre 6

Patterns of affiliative and agonistic interactions in dairy cattle

B. Foris[1], J. Langbein[2] and N. Melzer[1]
[1]Leibniz Institute for Farm Animal Biology, Institute of Genetics and Biometry, Wilhelm-Stahl-Allee 2, 18196, Germany, [2]Leibniz Institute for Farm Animal Biology, Institute of Behavioural Physiology, Wilhelm-Stahl-Allee 2, 18196, Germany; foris@fbn-dummerstorf.de

Investigations of socio-positive and socio-negative interactions in animal groups are relevant for animal welfare. Social tension from group composition may negatively affect individuals even in a well-designed and healthy environment. New technologies allow the automated acquisition of data suitable to assess social behaviour. However, for the evaluation of such data, it is necessary to first study in detail the social interactions that are potentially captured. In dairy cattle there is no guideline regarding the relevant sampling approach to assess affiliative and agonistic behaviour in free-stall barns. Hence, we performed continuous video analysis of social behaviour over five successive days in a group of 11 lactating Holstein cows to describe the spatiotemporal occurrence of affiliative (grooming) and agonistic (displacement) dyadic interactions. The frequency and location of social interactions were determined. To explore the stability of social behaviour over days the correlations of daily aggregated grooming or displacement actor-receiver matrices were tested using the Mantel test. We performed the same analysis considering matrices accumulated over two days. On average 48.2±3.9 grooming and 136.2±49.6 displacements were observed daily. Each day most displacements happened during daytime and at the feeder. In contrast, grooming was almost evenly distributed over the day and mostly observed at the lying stalls. The correlations between daily matrices were positive, however not all significant, for grooming as well as for displacement. For accumulated matrices without overlapping days we found only significant correlations of moderate strength for both interaction types. Our results indicate consistent spatiotemporal distribution of social interactions within the group. However this pattern may depend on stocking density and barn. In our setup, two days interactions provided a reliable picture of the social structure. These results could be used as a basis for the development of robust automated methods for the assessment of social behaviour in dairy cattle groups.

Standing of dairy cows in early lactation before and after milking in AMS

A. Herlin[1], E. Sivertsson[2] and L. Borgenvall[3]
[1]Swedish University of Agricultural Sciences, Biosystems and Technology, P.O. Box 103, 23053 Alnarp, Sweden, [2]Länsstyrelsens Kronoberg, na, 351 86 Växsjö, Sweden, [3]Växa Sverige, Klustervägen 11, 590 76 Vreta Kloster, Sweden; anders.herlin@slu.se

The period around calving is the most critical for a dairy cow as she is going through changes both environmentally, socially, hormonally and metabolically. The transition from the dry period, via calving to early lactation is very important and cows are sensitive and vulnerable to disease. In conventional milking systems, milking parlours, long waiting times prior to milking have been addressed as unnecessary and that it inflicts on the available time budget of cows. It can have consequences on eating and resting times of the animals, which can affect both production and welfare. In automatic milking systems (AMS), cows are not affected by management decisions such as how many cows are collected to milking at the same time. But they will be affected by feeding management and if the cow has free or controlled access to the feed and the milking unit. Therefore, it would be of interest to monitor and understand the factors that influences cows' activity level, mainly standing, before and after milking in AMS cubicle housing systems and further understand how it influences production, health and welfare. Cow activity was studied in four dairy farms with AMS. Two farms had controlled cow traffic and two had free cow traffic. A total of 19 cows, distributed over the farms including primi- and multiparous cows were studied regarding standing and lying during the first 21 days of lactation by use of IceTag units (IceRobotics Ltd, Edinburgh, Scotland, UK). Overall daily standing was on average 60%. Standing prior to milking was on average 83±13 min and standing post milking was on average 84±24 min. This suggests that cows will stand on average 2 h and 47 min per milking and in AMS systems where cows generally are milked more than twice a day suggests daily standing to be more than 5 hours per day linked to milking. However, some of that time, cows will spend on feeding. A few cows became ill during the period and they had very much different standing and lying than healthy cows. Cases with long time standing (>4 h) prior to milking occurred most often in one of the controlled cow traffic herds.

Group structure in sows – a social network approach

M.K. Will[1,2], K. Büttner[2], T. Kaufholz[1,3], C. Müller-Graf[1], T. Selhorst[1] and J. Krieter[2]
[1]Federal Institute of Risk Assessment, Max-Dohrn-Str. 8-10, 10589 Berlin, Germany, [2]Institute of Animal Breeding and Husbandry, Christian-Albrechts-University, Olshausenstr. 40, 24098 Kiel, Germany, [3]Institute for Theoretical Biology, Humboldt-University, Invalidenstr. 43, 10115 Berlin, Germany; kbuettner@tierzucht.uni-kiel.de

Social interactions between sows influence welfare and productivity. Therefore, this study aims at exploring the impact of weekly rehousing events on the network structure of group-housed sows with special regard to centrality parameters. Data were recorded in a gestation unit with 200 sows equipped with ear tags recording the sow's position each second. Based on this position data, affiliation networks for different location areas in the gestation unit were built, i.e. sows which were at the same time in the same location were connected with an edge in the network representation. Degree, closeness and betweenness were calculated to analyse the group structure and the position of each sow in the network. The influence of different sows' characteristics (e.g. gestation class, weight class, newly rehoused, genetics) on the centrality parameters was also evaluated. Over a four week period, networks of different time window sizes (24, 12, 6, 3, 2 and 1 h) were created and analysed separately. For all sows, a daily rhythm could be derived from the calculated centrality parameters regarding the networks of 6 h or shorter. This rhythm was hardly influenced by the rehousing events. For the 6 h snapshot, the mean centrality varied in the following range: degree: 89-105; closeness: 0.66-0.84; betweenness: 0.0011-0.0029. The subgroup of newly rehoused sows had a wider fluctuation for mean degree (68-150), mean closeness (0.61-0.88) and mean betweenness (0.0009-0.0048), i.e. they had to find their place in the consisting hierarchy. Furthermore, gestation and weight class affected the centrality parameters significantly. Here, the time-of-day influences the results. Heavier and older sows had higher centrality parameters when the feeding time started indicating their high position in the rank and feeding order. To sum up, social network analysis gives new insights in the contact structure of group-housed sows which can be used for management decisions enhancing both, animal welfare and productivity.

Session 40 Poster 9

Characteristics of behaviours and growth of weaned piglets in different social experiences

H.S. Hwang[1], J.K. Lee[1], T.K. Eom[1], J.K. Hong[2], T. Choi[2] and S.J. Rhim[1]
[1]Chung-Ang University, School of Bioresource and Bioscience, Ansung, 17546, Korea, South, [2]National Institute of Animal Science, Swine Science Division, Cheonan, 31000, Korea, South; sjrhim@cau.ac.kr

This study was conducted to clarify behaviour and body weight of weaned piglets between different social experiences during suckling period. Control (no social experience, n=4) and treatment (social experience, n =4) groups of weaned piglets (4 ind./pen) were housed and observed with the aid of video technology for 8 consecutive hours in early (days 1 and 2) and late (days 22 and 23) periods after mixing. Inactive (Mann-Whitney U test Z=-3.09, P<0.01), locomotion (Z=-3.30, P<0.01) and exploration (Z=-2.59, P<0.05) were significantly different between control and treatment groups in early mixed period. Feeding differed between groups in late mixed period (Z=-3.39, P<0.01). In social behaviours, agonistic (Z=-3.19, P<0.01) and interaction with adjacent pen (Z=-2.49, P<0.05) were significantly higher in treatment group than in control group during early mixed period. However, social behaviours dramatically decreased in late mixed period (Z=-4.15, P<0.01). There was negative correlation between average daily gain of body weight and agonistic behaviour (Spearman's rank correlation analysis, r=-0.33, P<0.01). In social experienced weaned piglet groups, there were less agonistic behaviour (Z=-3.56, P<0.01) and more daily gain of body weight (Z=-2.14, P=0.03) in this study. Social experience during suckling period would be useful for stable group management in swine industry.

Session 41 Theatre 1

Resource-use efficiency of animal-source food: the level of analysis matters

I.J.M. De Boer
Animal Production Systems group, Wageningen University, P.O. Box 338, 6700 AH Wageningen, the Netherlands; imke.deboer@wur.nl

It is widely recognized that the food system is a major user of natural resources, and that the contribution of livestock is significant. Many studies, therefore, aim at improving the efficiency of the use of natural resources during the production and consumption of food, especially animal-source food. Resource-use efficiency is commonly defined as a ratio of the use of natural resources in a production activity over the output of that production activity, e.g. kg of phosphorus used per unit of wheat crop produced, or ha of land mobilized per unit of milk consumed. My aim is to show that the level at which you define resource-use efficiency, or in other words your system boundary (i.e. animal, farm, herd, chain, region or food system), affects your conclusions, and hence your pathways for sustainable development of food systems. Most studies in literature define resource-use efficiency at animal, herd, farm, or chain level, and conclude that resource-use efficiency can be improved by, among others, improving feed digestibility and efficiency, increasing reproductive rates and animal yields, and reducing diseases. Only a few studies address resource-use efficiency of the entire food system. These studies show that we should use animals for what they are good at, namely converting biomass that we cannot or do not want to eat into valuable products, such as nutrient-dense food (meat, milk, and eggs) and manure. By converting these biomass streams, livestock recycle nutrients back into the food system that otherwise would have been lost in food production. These studies suggest that we should no longer focus on improving life-time productivity, but on improving the efficiency with which animals recycle biomass unsuited for human consumption back into the food system. Acknowledging this recycling role of animals in the food system also offers most potential to produce in an environmentally-friendly and animal-friendly way. It however also implies that developed countries have to significantly reduce their consumption of animal-source food.

Indicators of efficiency and environmental footprint for Eastern Alps mountain dairy systems

M. Berton[1], S. Bovolenta[2], M. Corazzin[2], L. Gallo[1], S. Pinterits[3], M. Ramanzin[1], W. Ressi[3], C. Spigarelli[2], A. Zuliani[2] and E. Sturaro[1]

[1]University of Padova, DAFNAE, Viale dell'Università 16, 35020 Legnaro, Italy, [2]University of Udine, DI4A, Via Sondrio 2, 33100 Udine, Italy, [3]Umweltbüro, Bahnhofstraße 39, 9020 Klagenfurt, Austria; marco.berton.1@unipd.it

In this study, we present the preliminary results of a project (IR VA Italia-Österreich 'TOPValue') aiming at identifying the added value of livestock 'mountain products' in terms of multifunctionality. We analyse synergies and trade-offs between production efficiency (gross energy conversion ratio, ECR) and environmental footprint (Life Cycle Assessment) indicators in mountain dairy farms. Data originated from 38 farms (45±26 Livestock Unit, 20.6±4.9 kg Fat Protein Corrected Milk – FPCM/cow/day), associated to 6 cooperative dairies in the eastern Alps. System boundaries included impacts due to herd and manure management, on-farm feedstuffs production, purchased feedstuffs and materials. Impact categories assessed were Climate Change (CC), Eutrophication (EP), Cumulative Energy Demand and Land Occupation. Two functional units were used: 1 kg of FPCM and 1 m^2 of farming land. Milk vs meat biophysical allocation (IDF method) was used. Mean impact values were 1.2±0.3 kg CO_2-eq, 6.6±2.0 g PO_4-eq, 3.1±1.5 MJ, 1.8±0.7 m^2/y per 1 kg FPCM, and 0.8±0.2 kg CO_2-eq, 4.1±1.3 g PO_4-eq, 1.8±0.7 MJ per 1 m^2. Mean ECR was 7.6±1.4 MJ feed/MJ milk, with 89% of gross energy deriving from non-human-edible feedstuffs, nearly totally produced on-farm. All impact categories per 1 kg FCPM were positively correlated, whereas only CC and EP per 1 m^2 were correlated. Positive correlations were found between ECR and impact categories per 1 kg FPCM, whereas no significant correlations were found between ECR and impact categories per 1 m^2. The inclusion of the dairy factory in the system boundaries permitted to obtain the CC values per 1 kg of cheese, with values ranging from 9.7 to 14.4 kg CO_2-eq (from fresh to ripened ones). The results evidenced an efficient use of local resources; the differences due to the use of different functional units suggest to consider different indicators for the development of strategies aiming at improving the sustainability of mountain livestock chains.

Are labour productivity, specialisation and efficiency of livestock production systems, compatible?

P. Veysset

Inra, UMRH, 63122 St Genès Champanelle, France; patrick.veysset@inra.fr

As a major source of growth, productivity is a key variable in economics. In agriculture, the productivity gains are often assessed at the field (crop yield), animal (milk yield) or worker (labour productivity) scale. But, total factor productivity (TFP) is a variable of interest because its time changes is an indicator of technical and managerial efficiency. TFP measures the growth from quantity changes (all output and inputs together); while partial factor productivity measures the volume of output (agricultural products) per unit of considered input e.g. land, animals, or workers. The technical efficiency of a production system can be assessed by the volume of output over the volume of intermediate consumption and equipment used. Based on a database of 164 suckler-cattle farms in France from 1980 to 2015, we observed that, over the 36-year period, the TFP increased at a low rate of +0.17%/year (the cumulative volume of outputs increased little more than the cumulative volume of inputs used). This small increase in the TFP was linked to the constant increase in labour productivity (+2.03%/year), while the productivity of the other factors (intermediate consumptions, land, and equipment) decreased. Every year, these farmers use more intermediate consumption and more equipment per unit of agricultural products. Despite heavy investments in equipment's and despite the huge increase in labour productivity the volume of output per ha of agricultural area decreased and farmers' incomes remained stable. In an another study, technical efficiency of 71 organic livestock farms (cattle, sheep and goats for meat or milk production) was calculated for 2015. Multivariate analysis have shown that efficient farms are those which are specialized with an important part of grass in the system, self-sufficient in forages; less efficient systems are structure with a diversified cropping activity, with a high labour productivity level. It seems that increasing the size of a livestock farm, at constant workforce, while increasing the crops activities, burdens the technical efficiency of the system. The concept of economies of scale does not work, what about the concept of economies of scope?

Session 41 Theatre 4

Relationship between feed efficiency and physiological stress parameters in Duroc × Iberian pigs

W.M. Rauw[1], E. De Mercado[2], L.A. Garcia Cortés[1], L. Gomez Raya[1], L. Silió[1], M.C. Rodriguez[1] and E. Gómez Izquierdo[2]

[1]INIA, Dept Mejora Genética Animal, Ctra de La Coruna km 7, 28040 Madrid, Spain, [2]ITACyL, Centro de Pruebas de Porcino, Ctra. Riaza-Toro, s/n, 40353 Hontalbilla, Spain; rauw.wendy@inia.es

Selection for improved feed efficiency may reduce the ability of animals to respond to environmental challenges. Alternatively, selection may result in animals with a reduced stress response. Fifty-three crossbred Duroc × Iberian barrows were used in this study. During the fattening period, feed consumed and body weight were recorded 22 times every 3 to 8 days over a period of 124 days. Body weight gain (BWG) was calculated and residual feed intake (RFI) was estimated for three consecutive periods of 56 (P1), 35 (P2), and 33 (P3) days. Blood samples were collected on day one of the experiment (P1), on day 78 of the experiment (P2) and at slaughter (day 125; P3); plasma cortisol, lactate and glucose levels were measured. In P2, pigs with lower RFI (high efficient) had lower levels of glucose ($r=0.32$, $P<0.05$) and tended to have lower levels of cortisol ($r=0.29$, $P=0.06$). In P3, Pigs with lower RFI had lower levels of lactate at slaughter ($r=0.37$, $P<0.01$). Overall, pigs with higher levels of lactate also had higher levels of glucose. This was significant in P2 ($r=0.53$, $P<0.001$) and a tendency in P1 ($r=0.23$, $P=0.09$). At slaughter, pigs with higher levels of cortisol had higher levels of lactate ($r=0.48$, $P<0.001$) but lower levels of glucose ($r=-0.35$, $P<0.05$). The latter may reflect a faster exhaustion of glycogenic stores at the time of slaughter resulting from stress. The results indicate that increased levels of physiological stress parameters negatively affected feed efficiency in Iberian pigs. Conversely, selection for improved feed efficiency may result in a reduced stress response. This work was funded by the Ministerio de Economía y Competitividad of the Spanish Government (AGL2016-75942-R, IBERFIRE), in support of SusAn ERA-Net 'SusPig'.

Session 41 Theatre 5

Integrating feed efficiency into uk beef cattle breeding objectives, and the impact on ghg emissions

T.W.D. Kirk[1], T.J. Byrne[1], P.R. Amer[2] and E. Wall[3]

[1]AbacusBio International Limited, Roslin Innovation Centre, Easter Bush, Midlothian, EH25 9RG, United Kingdom, [2]AbacusBio Limited, P.O. Box 5585, Dunedin, 9058, New Zealand, [3]SRUC, Roslin Institute Building, Easter Bush, Midlothian, EH25 9RG, United Kingdom; tkirk@abacusbio.co.uk

Breeding objectives can play an important role in determining the optimal size and direction of genetic changes in traits. They enable both breeders and commercial farmers to direct breeding emphasis towards specific market outcomes or address key production aspects of their farming system. Previous formalised breeding objectives were last developed for maternal and terminal UK beef cattle in 2005 and 1998, respectively. In the time since, there has been a significant increase in the desire, from both industry and market, to address economic and environmental aspects of ruminant livestock production. Genetic improvement offers a way to reduce the greenhouse gas (GHG) emissions of beef production while still improving performance of the animal in other key areas, such as welfare and profitability. Animals with relatively lower emissions per unit of productive output have better emissions intensity. In part, this is expressed through improvement in a feed efficiency trait directly. Development of a new breeding objectives for the UK beef industry presented the opportunity to include new traits which account for genetic differences in feed efficiency, as well as an update to, and expansion of, other on-farm profit drivers. This paper discusses the development of new breeding objectives for the UK beef industry, including feed efficiency, and reports the impact of selection for maternal and terminal index improvement on GHG emissions at an industry level. Consideration has been given to integrating total feed intake and residual feed intake into the breeding objective. The way in which the trait influences industry understanding and therefore adoption rates, and thus the potential impact of any reduction in emissions due to genetic improvement.

Session 41 Theatre 6

Linking environmental models and economic tools for trade-off analysis – a German case study

M. Zehetmeier[1], A. Reindl[1], V. Karger[1], M. Strobl[1], U.K. Müller[2] and G. Dorfner[1]
[1]Bavarian State Research Center for Agriculture, Menzinger Straße 54, 80638 München, Germany, [2]California State University, 2555 E San Ramon Ave, 93740 Fresno, USA; monika.zehetmeier@lfl.bayern.de

While researchers integrate economic and environmental models, the modeling tools provided to stakeholders most often address only economic or environmental issues. Therefore, these tools cannot predict trade-offs and synergies between economic problems and environmental issues. We linked two economic tools to a GHG emission model and applied these new integrated models to two cases, dairy cattle and winter wheat in Bavaria. The objective of this study is twofold: (1) building two integrated models to build trade-off and synergy curves and (2) explore these curves for the two integrated models. We linked one benchmarking and one predictive economic tool (i.e. an accounting and a partly mechanistic simulation tool), both developed at the Bavarian State Research Center for Agriculture, each with a GHG emission model. To test the developed integrated benchmark model, we used data from 100 dairy farms for the year 2013. To test the developed integrated predictive model, we simulated a typical dairy cow production system in Bavaria with dual purpose dairy cows. We studied trade-offs and synergies between profitability and GHG emissions of dairy cow production systems by varying milk yield. Similar tests for winter wheat production by varying nitrogen input are ongoing. The predictive model finds a strong synergistic effect between profitability and GHG emission reductions: profitability increases and GHG emission decreases with higher milk yield depending on the choice of method to account for co-products in GHG modelling. In contrast, the benchmark model found that milk yield has only a relatively small impact on both profitability (6%) and GHG emissions (16%). Integrating economic tools with environmental models allows us to expand the usefulness of existing economic tools to stakeholders by identifying trade-offs and synergies. However, our study showed that different economic models can yield contradictory results. Future studies must validate integrated models and explore the robustness, sensitivity, and scope (usefulness in different case studies) of predicted trade-offs and synergies.

Session 41 Theatre 7

Exploring options to recycle and prevent phosphorus waste in a food system

H.R.J. Van Kernebeek[1,2], S.J. Oosting[2], M.K. Van Ittersum[3], R. Ripoll-Bosch[2] and I.J.M. De Boer[2]
[1]Livestock Research, Wageningen University & Research, De Elst 1, 6708 WD Wageningen, the Netherlands, [2]Animal Production Systems, Wageningen University & Research, De Elst 1, 6708 WD Wageningen, the Netherlands, [3]Plant Production Systems, Wageningen University & Research, Droevendaalsesteeg 1, 6708 PB Wageningen, the Netherlands; heleen.vankernebeek@wur.nl

Recycling and preventing phosphorus (P) waste in the food system are essential for food security. The aim of our modelling exercise was to assess the potential of recycling and preventing P waste in a food system, in order to reduce the dependency on phosphate rock. We modelled a hypothetical food system designed to produce sufficient food for a fixed population with a minimum input requirement of phosphate rock. This model included representative animal and crop production systems, and was parameterised using data from the Netherlands. We assumed no import or export of feed and food. We first assessed the minimum requirement of phosphate rock in a baseline situation in which humans derive 60% of their dietary protein from animals (PA), and 42% of crop waste is recycled. Results showed that about 60% of the P waste in this food system resulted from wasting P in human excreta. We then evaluated the requirement of phosphate rock for alternative situations to assess the (combined) effect of (1) recycling waste of animal products, (2) recycling waste of crop products, (3) preventing waste of animal and crop products, and (4) recycling human excreta and industrial processing water. Recycling of human excreta showed most potential to reduce P waste, followed by prevention and finally recycling of agricultural waste. Fully recycling P waste could reduce the input of phosphate rock by 90%. Finally, for each situation, we studied the impact of consumption of PA in the human diet from 0 to 80%. The optimal amount of animal protein in the diet depended on whether P waste from animal products was fully recycled or prevented: if it was, then a small amount of animal protein in the human diet resulted in the most sustainable use of P; but if it wasn't, then the most sustainable use of P would result from a vegan diet. The principles in our model also hold for food systems with different farming practices and climatic and soil conditions.

Session 41 Theatre 8

Relationship between resource use, efficiency and sustainability of sheep-crop farming systems

T. Rodríguez-Ortega[1,2], A. Bernués[1,2] and A. Olaizola[1,3]

[1]Instituto Agroalimentario de Aragón – IA2 (CITA-Universidad de Zaragoza), Av. Montañana 930, 50059 Zaragoza, Spain, [2]Centro de Investigación y Tecnología Agroalimentaria de Aragón (CITA), Av. Montañana 930, 50059 Zaragoza, Spain, [3]Universidad de Zaragoza, Departamento de Ciencias Agrarias y del Medio Natural, C/ Miguel Servet 177, 50013 Zaragoza, Spain; abernues@aragon.es

Farming systems determine the resource embeddedness in agricultural products. Quantification of efficiency and sustainability is challenging in Mediterranean sheep-crop farming systems due to their different specialization, crop-animal integration and intensification. Emergy analysis computes the different qualities of energies involved in a production process and expresses them in equivalent solar energy. It provides emergy-based indicators such as intensity (emergy used per hectare and year), efficiency (emergy invested per energy obtained), and sustainability (ratio of self-sufficiency to environmental loading). Our objective was to evaluate these indicators on three representative farms of Mediterranean sheep-crop farming systems of Aragón (Spain) at two different levels: the product and the farm level (including lamb meat, permanent crops, and rainfed and irrigated arable crops). Our results show that the pasture-based sheep farm had the lowest intensity and efficiency and the highest sustainability, as opposite to the partially-integrated farm, while the fully-integrated farm obtained intermediate scores. Sheep products were less intensive and more sustainable than crops due to their capacity to use renewable natural resources (i.e. pastures). Since emergy increases with the hierarchical level of the food chain (e.g. from crops to meat), efficiency should be compared between products of similar trophic level. Lamb meat production was 1.9 and 1.3 times more intensive and efficient, respectively, in the partially-integrated farm than in the pasture-based sheep farm, but it was 5.1 times less sustainable. Increasing the boundaries of emergy analysis from the farm to the regional level could show strategies of resource sharing among farms for increased regional sustainability.

Session 41 Theatre 9

Producing lambs while limiting concentrates in various pedoclimatic contexts: which performances?

M. Benoit[1], R. Sabatier[2], J. Lasseur[3], P. Creighton[4] and B. Dumont[1]

[1]INRA, UMRH, 63122 Saint Genès-Champanelle, France, [2]INRA, Unité Ecodéveloppement, 84914 Avignon, France, [3]INRA, UMR Selmet, 34060 Montpellier, France, [4]Teagasc, Athenry, Co.galway, Ireland; marc-p.benoit@inra.fr

This study aims at evaluating the multi-performances of meat-sheep farming systems optimized for the adequation between production objectives and feed input levels. It is based on five farm types in contrasted environments, in Ireland (« Irl » system) and France: « Graz » (grazer system), « 3×2 » (accelerated reproduction) and « OF » (organic farming) in Massif Central uplands, « DT » (Double transhumance) in Mediterranean rangelands. A model was used to simulate farm operations based on consistent parameters (economic situation, equipment) and to calculate indicators of economic and environmental performances, and feed/food competition. Farmers' strategies led to a wide range of ewe productivity (0.82-1.66 lambs/ewe/yr) and concentrate consumption (0-148 kg/ewe/yr). On one hand, pasture-based French farming systems ('Graz' and '*DT*') get the best economic and environmental performances, but lamb production is highly seasonal and does not meet the sheep industry expectations. That is also the case for the Irish system that is moreover penalized by a rather high nitrogen fertilization (in relation with the high price of the land) that is deteriorating environment performances. On the other hand, the *3×2* and *OF* systems better meet the sheep industry requirements with lambs sold all along the year; however, they require more concentrates, which decreases economic and environmental (GHG emissions and non-renewable energy) performances and leads to higher feed/food competition. These results reveal different priorities between farmers' goals and sheep industry expectations. They also illustrate the great diversity of potential adaptations of meat-sheep farms to a wide range of pedoclimatic conditions, with specific strategies leading to contrasted but sustainable systems.

Improving the farm-scale livestock production efficiency through field-scale soil measurements

T. Takahashi[1,2], P. Harris[2], A.S. Cooke[2], M.J. Rivero[2] and M.R.F. Lee[1,2]
[1]University of Bristol, Langford House, Langford, Somerset, BS40 5DU, United Kingdom, [2]Rothamsted Research, North Wyke, Okehampton, Devon, EX20 2SB, United Kingdom; taro.takahashi@rothamsted.ac.uk

The preceding literature examining the relationship between 'soil health' and its potential benefits to agricultural productivity can be broadly classified into two groups: agronomic studies based on field-scale randomised controlled trials and socioeconomic studies based on farm-level data collection. Recent research on pasture-based livestock production systems has demonstrated, however, that grassland production suffers from a considerable level of spatial heterogeneity, in terms of both yield and forage quality, even within a relatively small geographical area. This finding suggests that neither of the aforementioned two approaches is able to capture the complete mechanism surrounding the soil-pasture-animal-income nexus in a socially relevant manner, as the vast majority of commercial enterprises around the world adopt rotational grazing systems. Using high-resolution primary data from the North Wyke Farm Platform, an intensively instrumented farm-scale research facility located in southwest UK, this study proposes a novel analytical framework to evaluate short-term and long-term benefits of soil health at each individual field that constitutes an entire livestock system. The life-cycle liveweight gain for each cattle and sheep was attributed and allocated to individual fields according to their contribution through (a) grazing time, (b) silage harvest, and (c) nutrient retention for following seasons, producing a unique field-level dataset that encompasses soil, pasture, animal and emissions-related variables. The results of regression analyses showed that an increase in the soil organic carbon (SOC) stock led to a greater liveweight gain, while intra-field heterogeneity in SOC was detrimental to animal production (both $P<0.05$). However, these relationships could not be detected when the spatial resolution of soil data was artificially and randomly reduced prior to regression, suggesting the importance of capturing within-field variability in production efficiency to improve farm-scale economic and environmental efficiencies of commercial livestock producers.

Economic efficiency of suckler cow herds in the Czech Republic

J. Syrůček[1], L. Bartoň[1], J. Kvapilík[1], M. Vacek[2] and L. Stádník[2]
[1]Institute of Animal Science, Cattle breeding, Přátelství 815, 104 00 Praha Uhříněves, Czech Republic, [2]Czech University of Life Sciences Prague, Department of Animal Husbandry, Kamýcká 129, 165 00 Praha 6 Suchdol, Czech Republic; syrucek.jan@vuzv.cz

Cattle production is one of the most important animal industries in the Czech Republic (CR). Suckler cows are the only cattle category with increasing numbers in a long-term. In 2017, there were totally 216,095 suckler cows in the CR. The aim of every breeder is to maximise profit given by the difference between income and expenses per herd and year, and suckler cow operations are no exception. The economic aspect of suckler cow farming is important because reaching profitability is a prerequisite for maintaining and further expansion of herds. Input data were collected to using a questionnaire containing 95 questions from 20, 20, 22, and 19 farms of different size in 2013, 2014, 2015 and 2016, respectively. It was calculated that in the period from 2013 to 2016 the average cost per cow and year was 30,313 CZK and the profitability ranged between +2.1 and +8.7%. The profit was reached only when the subsidies accounting for 47% of total revenues were provided. The data obtained proved the economy of scale. Average break-even points were determined for the number of weaned calves, selling price of calves, and subsidies (77 calves, 56 CZK/kg live weight an 12,169 CZK/cow/year respectively). Break-even points are minimum requirements to achieve zero profitability. However, to reach e.g. 10% profitability, it would be necessary to wean more than 92 calves per 100 cows and year. Based on the results of the sensitivity analysis, the selling price of calves, the number of weaned calves, the length of calving interval and the amount of support payments were identified as the factors with the highest impact on overall herd profitability, and can be considered as key for further improvement of economic efficiency.

A model-based approach to characterize individual autozygosity at global and local genomic scale

T. Druet[1] and M. Gautier[2]
[1]GIGA-R, University of Liège, Unit of Animal Genomics, Avenue de l'Hopital, 11, 4000 Liège, Belgium, [2]INRA, UMR CBGP, 755 avenue du Campus Agropolis, 4988 Montferrier-sur-Lez cedex, France; tom.druet@uliege.be

When two alleles from the same individual descend from a single allele in an ancestor, they form an homozygous-by-descent (HBD) or autozygous segment. Increased autozygosity is frequently associated with negative effects on fitness. Therefore, characterization of individual autozygosity is of high importance in many livestock populations. The HBD segments are unfortunately not observed and must be identified with marker or sequence data. This is often performed with simple rule based methods (e.g. runs of homozygosity) that are efficient in most cases although HBD segments result from a complex process. Indeed, frequency and size of HBD segments are random variables influenced by the number of generations to the common ancestor and by the local recombination rate. In addition, the genome of an individual contains HBD segments inherited from multiple ancestors tracing back to different generations in the past. Ideally, the model should take into account the marker marker allele frequencies, account for genotype calling errors and also handle sequencing data, including low-fold or genotype-by-sequencing experiments. We herein describe such a model-based approach estimating global and local HBD probabilities. After studying some properties of the model, we illustrate its application on real data sets. First, we assess the realized inbreeding levels in the restored European Bison using a relatively limited number of markers. Next, we apply our model to identify HBD segments in cattle using whole-genome sequencing data.

A combined physical-genetic map for dairy cattle

A. Hampel, F. Teuscher and D. Wittenburg
Leibniz Institute for Farm Animal Biology, Institute of Genetics and Biometry, Wilhelm-Stahl-Allee 2, 18196 Dummerstorf, Germany; wittenburg@fbn-dummerstorf.de

Molecular markers can be used to assess the genetic variation in a livestock population. Not only are the physical but also the genetic position of markers important for characterising a livestock population and for planning mating decisions in livestock breeding. Though the physical position is available from the current genome assembly, the genetic position can be inferred from the probability that the molecular variants at marker pairs are transmitted jointly to the next generation, i.e. the recombination rate. For several species (e.g. human), a combined physical-genetic map connecting both scales exists but such a map is not available in high quality for dairy cattle. The population structure, which may be caused by family stratification, has an influence on the estimates of genetic parameters. As half-sib families are typical in livestock, and particularly for dairy cattle, we studied and further extended an expectation-maximization algorithm considering the marker genotypes of half sibs for estimating the recombination rate between pairs of markers. Then, a linear model approach respecting the physical distance between SNPs was developed and applied to all estimates of recombination rate. The concrete location of each marker and the density of estimates along a chromosome were considered; this allowed the identification of local phenomena like hot- and coldspots. We analysed the genotypes at 39,780 single nucleotide polymorphisms of 265 Holstein-Friesian cows, leading to 12,759,713 estimates of intrachromosomal recombination rate. A physical mapping function was achieved for each chromosome, and it was validated using the known physical and genetic position of 761 microsatellites which were retrieved from NCBI (National Center for Biotechnology Information). The pattern of curves strongly varied among chromosomes, and conspicuous regions for a deeper post-analysis appeared.

Posterior probabilities of genome-wide associations in backcross families

M. Reichelt, M. Mayer, F. Teuscher and N. Reinsch
Leibniz Institute for Farm Animal Biology, Institute of Genetics and Biometry, Wilhelm-Stahl-Allee 2, 18196 Dummerstorf, Germany; reichelt@fbn-dummerstorf.de

In a Bayesian setting, posterior probabilities of association between markers and QTL (Quantitative Trait Loci) can be inferred by comparing the posterior distribution of genetic variance for a sequence of markers with their expectation. Thereby this expectation is derived from the data under the assumption of an equal contribution of all markers to the total genetic variance. In certain family types, such as backcross families, there exists a high level of linkage disequilibrium, which can be described by a covariance matrix, reflecting prior knowledge on the underlying genetic map and parental linkage phase. By capitalising on this covariance matrix it is possible to also draw inferences on genetic covariability between chromosome segments. For this purpose we have developed a generalisation of the aforementioned test procedure and investigated its application to a variety of simulated backcross data. The analysis showed that posterior probabilities of covariances to be large provide additional information on the existence of further QTL on a chromosome when a major QTL is present. By analysing the covariances in addition to the variances the identification of QTL positions may be improved in families with high linkage disequilibrium.

A sparse-group lasso variant for whole-genome regression models in half sibs

J. Klosa[1], N.R. Simon[2] and D. Wittenburg[1]
[1]Leibniz Institute for Farm Animal Biology, Institute of Genetics and Biometry, Wilhelm-Stahl-Allee 2, 18196 Dummerstorf, Germany, [2]University of Washington, Department of Biostatistics, Health Sciences Building 650, Box 357232, 98195 Seattle, WA, USA; klosa@fbn-dummerstorf.de

For the genomic evaluation of domestic animals, whole-genome regression methods are applied which use extensive information about genomic markers (e.g. single nucleotide polymorphisms; SNPs). As the number of model parameters increases with a still growing number of SNPs, multicollinearity between covariates can affect the precision of marker-effect estimates. Furthermore, it is desired to select those markers that are relevant to trait expression. Hence, selection-and-shrinkage approaches are a promising option to generate sparse solutions with higher precision. The objective of this study is to develop a statistical method following the sparse-group lasso, which builds a model that automatically selects a sparse set of predictor variables. Because a particular source of dependence between SNPs is due to linkage and linkage disequilibrium (LD), an extension is developed that considers the spatial genome structure and allows grouping according to those measures. The sparse-group lasso variant selects a solution that is sparse among and within these groups. In addition, when appropriate for a given phenotype, this method can promote the inclusion of SNPs from contiguous genomic regions. Thus, genomic regions that affect a trait can be more effectively identified. As the span of linkage and LD strongly depends on the population structure, this study focusses on a population consisting of half-sib families which is typical, for instance, in dairy cattle. The method is evaluated using simulated data resembling a dairy cattle population. In simulated scenarios, the new method is more effective than a standard lasso regression, which ignores relevant structural information.

Session 42 Theatre 5

Theoretical basis to extend single-step genomic prediction of dominance in a pig population

R.R. Mota[1], S. Vanderick[1], F.G. Colinet[1], G.R. Wiggans[2], H. Hammami[1] and N. Gengler[1]
[1]Gembloux Agro-Bio Tech, University of Liège, TERRA Teaching and Research Centre, Passage des Deportes, 2, 5030 Gembloux, Belgium, [2]Council on Dairy Cattle Breeding – CDCB, One Town Center, 4201 Northview Drive, Suite 302, 20716, Bowie, USA; rrmota@uliege.be

Single-step methods predict implicitly unknown gene content information of non-genotyped from known gene content for genotyped animals. This theory is well derived in an additive setting. There are reasons not to ignore the dominance context when working with partially genotyped populations. This study addressed several outstanding issues in this context. First, it presented the theoretical basis for dominance single-step genomic best linear unbiased prediction theory. A specific and important issue in all dominance setting is the handling of inbreeding. A total of five dominance single-step inverse matrices were tested and described as C^1 to C^5 by considering different parameterization (e.g. different ways to account for inbreeding) for pedigree-based and genomic relationships matrices. We simulated genotypes for real crossbred pig population (n=11,943 animals). The SNP effects were assumed to be equal to calculate true dominance values. We added random noise and used them as phenotypes. Accuracy was defined as correlation between true and predicted dominance breeding values. We applied five replicates and estimated accuracies between three situations: all (S_1); non-genotyped (S_2) and inbred non-genotyped animals (S_3). Potential bias of predicted dominance values was assessed by regressing the true dominance values on predicted values. Accuracies of each tested matrix (C^1 to C^5) were 0.75, 0.33 and 0.35 in average, for S_1, S_2 and S_3, respectively. The matrix C^5 better performed and breeding values from C^1 and C^2 were more biased than those obtained by using C^3, C^4 and C^5. We showed a useful approach to predict dominance gene contents for non-genotyped from genotyped animals. Better matrix compatibility can be obtained by re-scaling the pedigree-based and the genomic relationship matrices to obtain standardized diagonal elements equal to 1 minus the inbreeding coefficient, i.e. the C^5 matrix.

Session 42 Theatre 6

Genomic mating allocation model with dominance to maximize overall genetic merit in Landrace pigs

D. Gonzalez-Dieguez[1], C. Carillier-Jacquin[1], A. Bouquet[2,3], L. Tusell[1] and Z.G. Vitezica[1]
[1]INRA, GenPhySE, Université de Toulouse, 31326 Castanet Tolosan, France, [2]France Génétique Porc, BP35104, Le Rheu, 35651, France, [3]IFIP Institut du Porc, BP35104, Le Rheu, 35651, France; david-omar.gonzalez-dieguez@inra.fr

Mating allocation strategies that account for dominance can be of interest for maximizing the overall genetic merit of future offspring. In a genomic context, accounting for dominance effects in genetic evaluations is easier than in a classical pedigree-based context. The objective of the present study was to evaluate, in terms of genetic gain, the efficiency of a genomic mating allocation model accounting for dominance in a Landrace pig population. Genetic variance components were estimated for three traits (age at 100 kg, backfat depth and average piglets weight per litter) using an additive and dominance GBLUP model with inbreeding. The estimated additive and dominance genetic variances were used to obtain additive and dominant SNP effects using a BLUP-SNP model. Then, additive breeding values (BV) and total genetic values (TGV, those including dominance) were predicted for the offspring of all possible matings between 40 boars and around 1,500 sows (the number of available sows depended on the trait). Following a traditional breeding scheme, the best matings resulting from 40 boars and 600 sows, were selected based either on BV or TGV using linear programming. The expected genetic gain was calculated as the difference between the mean BV (or the mean TGV) of selected matings and the mean BV (or the mean TGV) of all possible matings. Results show that, for the analysed traits, mating allocation is a feasible and a potential strategy to improve the productive performance of the offspring (i.e. to improve their TGV) without compromising the additive genetic gain in this Landrace pig population.

Construction of a genetic relationship matrix using linked markers

L. Gomez-Raya, W.M. Rauw, L. Silio, M.C. Rodriguez and L.A. Garcia Cortes
INIA, Dept Mejora Genética Animal, Ctra de La Coruña km 7, 28040 Madrid, Spain; gomez.luis@inia.es

Genomic BLUP (GBLUP) is a linear mixed model incorporating a marker-based genomic relationship matrix (G-matrix) instead of a relationship matrix based on a known pedigree. The G-matrix is constructed assuming that (1) the contribution of each SNP to the trait is zero (centring), (2) each SNP contributes equally to the variance of the trait (scaling), and (3) Hardy-Weinberg conditions are held. Therefore, construction of the G-matrix ignores that markers are linked reducing part of the genomic information contributed by relatives. The goal of this study was to construct a G-matrix using linked markers with the following steps: (1) estimation of linkage disequilibrium between all pairs of consecutive markers, (2) computing all haplotype frequencies at each of two consecutive loci, and (3) develop formulae for centring and scaling the G-matrix without the requirement of Hardy-Weinberg equilibrium. The assumptions of the model were (1) gene action is fully additive, (2) there is no recombination between linked markers and (3) each marker contributes equally to the additive mean and variance of the trait. The centring was performed by subtracting to the genotype effect (1, 0.5, 0, -0.5, -1 depending on the alleles of the genotype of each individual) the mean of the trait (f11-f22, where f11 and f22 are the frequencies of the haplotypes 11 or 22 at the two loci). The scaling was performed by summing over loci the contribution of each loci of $0.5[(f11+f22)-(f11-f22)^2]$ to the total additive variance. A total of 435 Iberian sows were genotyped with the Illumina 60K array. After editing, there were 26,360 Single Nucleotide Polymorphism (SNP) for constructing of the G-matrix. Linkage disequilibrium was estimated using the EM-algorithm for each pair of consecutive SNPs. Only pairs of loci with at least haplotype frequencies of 0.01 for f11 and f22 were used reducing the total number of SNPs to 13,021. The average of the diagonal elements of the newly constructed G-matrix using linked markers was 0.988. The average of the off-diagonal elements was -0.002. The eigenvalues ranged from 0 to 16.6. The extension of this method to three or more loci as well as to another general departures of Hardy-Weinberg equilibrium is discussed.

Efficient computational strategies for multivariate single-step SNPBLUP

J. Vandenplas[1], H. Eding[2] and M.P.L. Calus[1]
[1]Wageningen University and Research, Animal Breeding and Genomics, P.O. Box 338, 6700 AH Wageningen, the Netherlands, [2]CRV BV, P.O. Box 454, 6843 NW Arnhem, the Netherlands; jeremie.vandenplas@wur.nl

Computational requirements for solving a multivariate single-step SNPBLUP (ssSNPBLUP) that fits three types of additive genetic effects, that is SNP markers and residual polygenic effects for genotyped animals, and polygenic effects for non-genotyped animals, were investigated. When the system of equations of ssSNPBLUP is solved iteratively with a preconditioned conjugate gradient method, the main computational cost is the computation of the product of the coefficient matrix with a vector at each iteration. For this computation, only submatrices are needed. If they can be stored in core memory, efficient and optimized libraries and parallel processing can be exploited. After rearrangement of ssSNPBLUP equations, only two multiplications of a dense matrix, that is the SNP matrix, by a vector, in addition to multiplications of sparse matrices, are required. Implicit imputation of genotypes of non-genotyped animals can be avoided by implicitly imputing direct genomic values instead. Memory and time requirements for implicit imputation were further reduced by deriving an equivalent formulation that involves only, e.g. a few hundred thousand, ancestors and non-genotyped progeny of genotyped animals, instead of, e.g. several millions non-genotyped animals. Computational demands for the proposed strategies are therefore almost linearly proportional to the number of genotyped animals. These computational strategies were implemented in a shared-memory Fortran program, and tested on a four-trait dataset with >6 million animals in the pedigree and >90,000 genotyped animals. Using 5 threads, wall clock time per iteration for ssSNPBLUP was about five seconds, and more than 55% of the time was used for the multiplication of the SNP matrix with a vector. Similar, or slightly smaller, wall clock times per iteration were observed for single-step GBLUP with a directly inverted genomic relationship matrix, or with an approximated sparse inverse of the genomic relationship matrix. In conclusion, the implemented computational strategies allow to perform multivariate ssSNPBLUP with many thousands of genotyped animals within manageable amounts of memory and time.

Session 42 Theatre 9

Ordination methods to build microbiota similarity matrices for complex traits prediction

A. Saborío-Montero[1] and O. González-Recio[2,3]

[1]UPV, Animal Science and Technology, Camino de Vera s/n, 46022 Valencia, Spain, [2]ETSIAAB. UPM, Produccion Agraria, Ciudad Universitaria s/n, 28040 Madrid, Spain, [3]INIA, Mejora Genetica Animal, Crta. de la Coruña, km 7,5, 28040 Madrid, Spain; alesabor@gmail.com

There is evidence supporting a microbiome effect over phenotypic expression of complex traits, however few studies has included this effect in statistical models to estimate the proportion of phenotypic variance attributable to microbial systems. The aim of this study was to compare performance of ordination methods for build microbiota similarity matrices, to predict microbiome contribution in complex traits. A population of 1000 individuals was simulated; a matrix comprising genomic relationships between 1000 genotyped individuals (G) was used, and a matrix of microbial relative abundance, having 83 operational taxonomic units (OTU) and 1000 individuals was simulated from real microbiome data. The ordination methods of multidimensional scaling (MDS), principal correspondence analysis (PCoA), detrended correspondence analysis (DCA), and non-metric multidimensional scaling (NMDS) were independently used according to Phyloseq package in R to build the microbiota similarity matrices (K), from the simulated microbial relative abundance matrix. A genomic and microbiomic BLUP was performed and solved using the semi-parametric method of Bayesian reproducing kernel Hilbert spaces regressions (RKHS) as described in BGLR package of R. The statistical model, was fitted including a simulated complex trait as response variable, and the G matrix and the output K matrices from ordination methods as RKHS factors. Simulated heritabilities and microbiabilities of 0.10, 0.25 and 0.50 were tested. Some of this matrices achieved acceptable precision in the estimation of simulated variances, heritabilities and microbiabilities. The inclusion of microbiome information in statistical models to predict complex traits might be an alternative to enhance performance of genomic evaluations in animal breeding programs.

Session 42 Theatre 10

Integrating gene expression data into genomic prediction

Z.C. Li[1], N. Gao[2], W.R.M. Johannes[3] and H. Simianer[1]

[1]Goettingen university, animal science, Albrecht Thaer Weg 3, 37075 Goettingen, Germany, [2]South China Agricultural University, Animal Science, South China Agricultural University, 510000 Guangzhou, China, P.R., [3]KWS SAAT SE, Grimsehlstraße 31, 37574 Einbeck, Germany; lizhengcao2008@sina.com

Gene expression profiles potentially hold valuable information for the genetic analysis and breeding value and phenotype prediction. In this study, we tested the utility of transcriptome data in phenotype prediction with 185 inbred lines of *Drosophila melanogaster* for 9 traits in two sexes. We incorporated the transcriptome data into genomic prediction via reproducing kernel Hilbert space regression (RKHSR), combining both a sequence-based (2.8 mio. SNPs) and a transcriptome-based (18,140 gene expressions) covariance matrix in the prediction equations (GRBLUP). For most traits GRBLUP and sequence-based GBLUP provided similar predictive ability, while GRBLUP could capture more phenotypic variance explained by transcriptome data. There was only one trait (olfactory perceptions to Ethyl Butyrate in females) in which the predictive ability of GRBLUP (0.23) was significantly higher than the predictive ability of GBLUP (0.21). We conclude that accounting for transcriptome data has the potential to improve genomic predictions if a more targeted collection of transcriptome data on a larger scale can be implemented.

Session 42 — Theatre 11

Recursive binomial model to analyse piglet survival in a diallel cross between Iberian pigs

L. Varona[1], J.L. Noguera[2], J. Casellas[3], J.P. Rosas[4] and N. Ibañez-Escriche[5]

[1]Universidad de Zaragoza, Departamento de Anatomía, Embriología y Genética, 50013 Zaragoza, Spain, [2]IRTA, Genetica i Millora Animal, 25198 Lleida, Spain, [3]Universitat Autónoma de Barcelona, Departament de Ciència Animal i dels Aliments, 08193 Bellaterra, Spain, [4]INGA FOOD S.A., Programa de Mejora Genética CASTUA, 06200 Almendralejo, Spain, [5]Universitat Politècnica de Valencia, Departamento de Ciencia Animal, 46071 Valencia, Spain; lvarona@unizar.es

Piglet mortality is an important factor in pig production. Nevertheless, its statistical analysis presents serious difficulties because of its categorical nature. In recent studies, it has been suggested that the most appropriate model is a binomial model that modelled the survival of each specific piglet with a logit approach. In this study, we applied a population specific recursive binomial model to obtain estimates of the Dickerson crossbreeding parameters in a diallel cross between three strains of Iberian pigs (Entrepelado; EE, Torbiscal; TT and Retinto; RR). A total of 18,193 records from 3,800 sows distributed as follows: EE (2,843 records, 707 sows), ER (2,336, 527), ET (942, 177), RE (806, 196), RR (4,472, 874), RT (2,450, 488), TE (193, 36), TR (1,993, 359) and TT (2,158, 452). The average litter size (total number born) and stillbirth were 8.292±2.284 and 0.285±0.787, respectively. In addition, we used a pedigree of 4,609 individuals to analyse these data with a Bayesian approach through a Gibbs Sampler. The recursive relationship between litter size and pig mortality was clearly non-linear for all crosses. The results were examined at 5, 8, 11 and 14 piglets as reference points and they indicate that the direct line effect of the EE population is higher that RR and TT with posterior probabilities over 0.97 at all reference points. Nevertheless, we were not able to detect any relevant differences between maternal effects and only the heterosis effect between TT and RR was clearly positive with posterior probabilities over 0.97 at all reference points. The main advantage of the proposed model is that allows to calculate crossbreeding parameters (line, maternal and heterosis) conditioned to each specific litter size taking into account the population specific non-linear relationships with piglet mortality and litter size.

Session 42 — Theatre 12

Structured antedependence model for longitudinal analysis of social effects on ADG in rabbits

I. David[1], J.P. Sanchez[2] and M. Piles[2]

[1]GenPhySE, Université de Toulouse, INRA, ENVT, Chemin de borde rouge, 31326 Castanet tolosan, France, [2]IRTA, Torre Marimon s/n, 08140 Caldes de Montbui, Barcelona, Spain; ingrid.david@inra.fr

Even if there are evidences that the intensity of social interactions between partners vary with time, very few genetic studies have investigated how social genetic effects (SGEs) vary over time. To overcome this issue, the objectives of the present study were to analyse longitudinal records of average daily gain (ADG) in rabbits and to evaluate, by simulation, the response to selection for such longitudinal trait. Five weekly ADG records from 3,096 rabbits under feed restriction after weaning and raised in pen of 8 were used for the analysis. A linear animal mixed model including SGEs with week specific random effects that follow structured antedependence (SAD) functions was fitted to the data using ASReml and the Fortran program that we have developed (freely available on zenodo). The social heritability was higher in week 1 (0.44) than in weeks 2 to 5 (ranging from 0.16 to 0.23). The correlation between the SGEs of different weeks was moderate to high for weeks 2 to 5 (0.62 to 0.91) and weaker between the first week and the other weeks (0.33 to 0.47). The direct-social genetic correlations were negative at any time. Based on the same data design, the same variance components but considering 3 different sets of direct-social genetic antagonism (strong, moderate, weak); we simulated 7 generations of selection using a SAD model including SGEs or not to estimate breeding values. Results obtained showed that the increase in ADG with selection decreased with the direct-social genetic antagonism and was improved (~by 30%) when SGEs were taken into account. In conclusion, results confirmed that SGEs vary over time and do not correspond to the same trait after mixing than later in life, probably as a consequence of social hierarchy establishment observed at that time. Accounting for SGEs in the selection criterion maximizes the genetic progress.

Modelling of grouped recorded data for different population structures in pig breeding

M. Shirali[1], B. Nielsen[2], T. Ostersen[2], O.F. Christensen[1], P. Madsen[1], G. Su[1] and J. Jensen[1]
[1]Center for Quantitative Genetics and Genomics, Department of Molecular Biology and Genetics, Aarhus University, Blichers alle 20, 8830, Denmark, [2]SEGES, Pig Research Centre, 1609 Copenhagen, Denmark; mahmoud.shirali@mbg.au.dk

The aim of this study was to investigate the effect of population structure on estimation of genetic parameters and accuracy of predicting breeding values in models for group recorded data. Data were simulated by mimicking the structure of two real pig breeding datasets. In the first case, a central testing station was assumed with average daily feed intake records of 23,824 individual animals representing a low between and within pen relationship due to pigs originating from different breeder farms. In the second case, a breeder' herds was assumed with average daily body weight gain of 100,285 individual animals representing a large relationships within pen due to on farm testing. The trait simulated had heritability of individual records of 0.33, and variances for genetic of 300, pen effect of 70, and litter effect of 40. In simulation, pedigree and data structure of real datasets were used, while breeding value for each individual animal followed Mendelian rules. The grouped records were obtained by summing the individual records of all animals in each pen in the datasets. The results showed that modelling of the grouped data results in similar estimates of genetic parameters as modelling of individual data, however, standard errors were substantially larger. A reduced model ignoring litter and pen effects suggested that the variance from these effects are captured by the residual effect in the model. The case with animal in pens having low relationships resulted in larger standard error of genetic parameter estimates compared to the case with animals in pens closely related. The average accuracy of estimated breeding value decreased from 0.67 in individual recorded data to 0.27 when using grouped records in less related pen data, while with closely related animals in each pen, average accuracy of estimated breeding value decreased from 0.70 in individual recorded data to 0.48 in grouped records. The study showed the feasibility of using grouped records in pig breeding programs but preferably with higher relatedness among animals in each group.

A modified penalised regression approach to precision-related questions in genomic evaluations

M. Doschoris and D. Wittenburg
Leibniz Institute for Farm Animal Biology (FBN), Institute of Genetics and Biometry, Wilhelm-Stahl-Allee 2, 18196 Dummerstorf, Germany; doschoris@fbn-dummerstorf.de

Modern genome-wide association studies as well as methods for genomic selection frequently involve linear models where the number of genetic markers vastly exceeds the number of observations. In this high-dimensional setting, exploring the precision of SNP-effect estimates leading to a reliable identification of causative variants is complicated. What is more, an exact analysis of the precision of estimates remains an open question. Current studies tackle high-dimensionality employing ridge regression and related shrinkage methods, incorporating a covariance matrix between marker genotypes into the penalty term. Notwithstanding, the postulated pattern for the structure of the covariance matrix has followed so far only simplified assumptions. We propose a modified version of ridge regression with penalty terms including more complex structures consisting of functions of the theoretical covariance matrix and/or the corresponding precision matrix. We analyse, theoretically as well as numerically, the methods in order to determine marker-specific shrinkage factors for the penalty terms under consideration. The method was evaluated utilizing realistically simulated datasets of half-sib families consisting of 1000 individuals each with a total of 3,320 markers equally distributed among two chromosomes. Investigating their properties with reference to the accuracy of estimated effects will provide reliable evidence for the identification of causative variants. This opens the possibility to detect or to confirm causative variants.

A novel three-step approach to ascertain individual levels of genetic diversity within populations

M. Neuditschko[1,2], T. Druml[1], G. Grilz-Seger[3], M. Horna[4], A. Ricard[5,6], M. Mesarič[7], M. Cotman[8], H. Pausch[9] and G. Brem[1]
[1]University of Veterinary Sciences Vienna, Institute of Animal Breeding and Genetics, Veterinärplatz 1, 1210 Vienna, Austria, [2]Agroscope, Swiss National Stud Farm, Les Longs Prés, 1580 Avenches, Switzerland, [3]Pöckau 41, 9601 Arnoldstein, Austria, [4]Slovak University of Agriculture in Nitra, Department of Animal Husbandry, Tr. A. Hlinku 2, 949 76 Nitra, Slovak Republic, [5]Institut Français du Cheval et de l'Equitation, Recherche et Innovation, La Jumenterie, 61310 Exmes, France, [6]Institut National de la Recherche Agronomique, UMR 1313 Génétique Animale et Biologie Intégrative, Allée de Vilvert, 78352 Jouy-en-Josas, France, [7]University of Ljubljana, Clinic for Reproduction and Large Animals, Cesta v Mestni log 47, 1000 Ljubljana, Slovenia, [8]University of Ljubljana, Institute of Preclinical Sciences, Cesta v Mestni log 47, 1000 Ljubljana, Slovenia, [9]ETH Zürich, Animal Genomics, Eschikon 27, 8315 Lindau, Switzerland; markus.neuditschko@agroscope.admin.ch

Within the scope of current genetic diversity analyses, population structure and homozygosity measures are independently analysed and interpreted. To enhance the analytical power of genetic diversity analyses we combined the visualization of population networks with admixture and runs of homozygosity (ROH). We applied our novel three-step approach to illuminate the fine-scale population structure of the Haflinger horse. Therefore, we collected high-density genotypes of 531 horses originating from seven different breeds which were involved in the formation of the Haflinger population, namely 32 Italian Haflingers, 78 Austrian Haflingers, 190 Noriker, 23 Bosnian Mountain Horse, 20 Gidran, 33 Shagya Arabians and 155 Purebred Arabians. Implementing the total genomic length of ROH segments and individual levels of admixture into the high-resolution population network, it became feasible to clearly identify outbred (admixture) and inbred (ROH segments) horses within the respective breeds and to visualize fine-scale population structures of the Haflinger breed. We demonstrated that the combination genealogical aspects (population networks) with gene pool characterization (admixture) and diversity measures (ROH) is a highly efficient method to ascertain the complex aspects involved in the development of a composite breed.

Predicting inbreeding depression load for litter size in Iberian pigs

J. Casellas[1], N. Ibáñez-Escriche[2], L. Varona[3], J.P. Rosas[4] and J.L. Noguera[5]
[1]Universitat Autònoma de Barcelona, Departament de Ciència Animal i dels Aliments, Campus UAB, 08193, Spain, [2]Universitat Politècnica de València, Departament de Ciència Animal, Camí de Vera s/n, 46071 València, Spain, [3]Universidad de Zaragoza, Departamento de Anatomía, Embriología y Genética Animal, C. Miguel Servet 177, 50013 Zaragoza, Spain, [4]INGA FOOD SA, Programa de Mejora Genética Castúa, Avda. A Rúa 2, 06200 Almendralejo, Spain, [5]Instituti de Recerca i Tecnologia Agroalimentàries, Genètica i Millora Animal, C. Alcalde Rovira Roure 191, 25198 Lleida, Spain; joaquim.casellas@uab.cat

Inbreeding depression can be defined as the reduced biological fitness in the offspring of related individuals. The objective of this research was to characterize the inbreeding depression load for litter size in two Iberian pig lines. The mixed model approach has been implemented on *Entrepelado* (EE; n=2,407; 7.94±0.04 piglets) and *Retinto* (RR; n=3,501; 8.40±0.04 piglets) data sets. The operational model accounted for parturition (1st, 2nd, 3rd, 4th, 5th and 6th and upper), line of the insemination boar (EE, RR or *Torbiscal*), and three random effects (herd-year-season, infinitesimal additive genetic background, and inbreeding depression load). Pedigree data included 805 (EE) and 909 individuals (RR), whereas only 430 (1.6 to 25.0%) and 354 (0.2 to 25.0%) of them were inbred, respectively. After Mendelian decomposition, the EE pedigree provided 1,192 partial inbreeding coefficients (0.2 to 12.5%), and the RR pedigree generated 810 partial inbreeding coefficients (0.1 to 0.25%). Mixed model equations were solved by Bayesian inference and the statistical relevance of the inbreeding depression variance component (σ_d^2) was tested by a Bayes factor (BF) approach. The modal estimate of σ_d^2 was higher in EE (0.81 vs 0.38 piglets^2), although both lines provided a BF larger than 10 units. It is important to note that 68.7% (EE) and 74.8% (RR) individual-specific inbreeding depression load estimates were negative. The genetic background for inbreeding depression distributed heterogeneously across Iberian pigs and inherited generation by generation. Its impact can be anticipated and current analysis provide reliable information to be properly integrated into current breeding programs.

Session 42 Poster 17

Principal component analysis of factors effecting reproductive performance of prolific sheep

E. Emsen[1], M. Kutluca Korkmaz[2] and B.B. Odevci[3]

[1]Ataturk University, Animal Science, Ataturk University, Department of Animal Science, Erzurum, Erzurum, Turkey, [2]Ataturk University, Ataturk University, Ispir Hamza Polat Vocational School, Ispir, Erzurum, Ispir Hamza Polat Vocational School, Ispir, Erzurum, 25900, Turkey, [3]Kadir Has University, Management Information Systems, Kadir Has University, Istanbul, 34083, Turkey; ebruemsen@gmail.com

Romanov breed has been selected in crossbreeding with Turkish native breeds as a source of increasing reproductive efficiency by crossbreeding is due to the other advantages, such as high adaptability, suitability of being managed successfully in different systems and higher survivability rates of its crossbred lambs, which are lacking in other prolific breeds. Reproductive efficiency in a prime lamb producing ewe flock is a key driver of profitability. In this study, principal components analysis was applied to a set of reproductive traits of prolific sheep in order to reduce the number of traits and understand their relationship for breeding purposes. Multivariate techniques, other than path coefficients and multiple regression, have been used only to a very limited extent in the field of animal science. Principal component analysis is a multivariate technique for reducing p correlated measurement variables to a smaller set of statistically independent linear combinations of the original measurements. This technique attempts to find linear compounds of the original variables which can account for the dependency structure existing among the original measurements. A total 100 Romanov crossbreed ewes (2-4 age) was used to record pregnancy rate, birth type, sex, litter size at birth and weaning age and total productivity. Correlation coefficient and Principal Component Analysis (PCA) were used to determine relationships in reproductive performance. The highest correlation coefficient was computed between litter size at birth and total productivity (r=0,72) and this value was continued up to litter size at birth with r=0,51. The power of explaining of variance of these factors was found 67.49%. The indicators for explaining the total variance were birth type, total productivity, birth weight and weaning age. In the result of PCA 3 components out of 9 was explained.

Session 42 Poster 18

Weighting SNP panels for genomic prediction by simulated annealing

M. Martín De Hijas-Villalba[1], L. Varona[2], J.L. Noguera[3], N. Ibáñez-Escriche[4] and J. Casellas[1]

[1]Universitat Autònoma de Barcelona, Departament de Ciència Animal i dels Aliments, Edifici V Travessera dels Turons, 08193 Bellaterra, Spain, [2]Universidad de Zaragoza, Departamento de Anatomía Embriología y Genética Animal, C/ Miguel Servet 177, 50013 Zaragoza, Spain, [3]Institut de Recerca i Tecnologia Agroalimentàries, Genètica i Millora Animal, Av. Rovira Roure 191, 25198 Lleida, Spain, [4]Universitat Politècnica de València, Departament de Ciència Animal, Camí de Vera s/n, 46071 València, Spain; melani.mhv@gmail.com

Genomic evaluation strategies in livestock populations are mainly based on genomic BLUP (gBLUP), which compiles the genetic information from thousands of single nucleotide polymorphisms (SNPs) at equal level into the genomic relationship matrix G. The exclusion of some SNPs has been previously suggested to improve accuracy, although this simplifies the contribution of each genetic marker to 'all or nothing'. This study proposed a new approach where SNPs were weighted by a value ranging between 0 and 1 (model W) and compared against standard gBLUP models (model G) by analysing simulated data. Simulated data consisted on diploid individuals with two autosomal chromosomes (100 cM each) with more than 1,500 polymorphic SNPs and 300 polymorphic QTL scattered randomly among the genome (h^2=0.1, 0.25 and 0.4), after 1,001 non-overlapping generations of random mating and accounting for recombination, mutation and linkage disequilibrium. QTL effects were sampled and rescaled from a standard normal distribution. Optimum SNP panels were characterized by simulated annealing and mixed model equations were solved by the iterative Gauss-Seidel procedure. The performance of model W was evaluated under two different statistical parameters, the correlation between simulated and predicted breeding values (i.e. accuracy) and mean squared error (MSE). Model W have reported better fit to data than model G for both correlation and MSE. Using correlation as reference statistic, accuracy improved >15% regardless of the heritability. With MSE as reference, model fit increased >10% under low heritability (h^2=0.1) and ~15% for remaining genetic backgrounds. As expected, the implementation of weights for SNP when constructing G provided higher accuracies than former approaches where all SNPs had the same influence.

Session 42 Poster 19

Modeling heterozygosis by principal component in tropical beef composite

J.B.S. Ferraz[1], L. Grigoletto[1], E.C. Mattos[1], J.P. Eler[1] and F. Baldi[2]
[1]College of Animal Science and Food Enginner, University of Sao Paulo, 225 Duque de Caxias Norte Avenue, 13635-900 Pirassununga, SP, Brazil, [2]Sao Paulo State University, School of Agrarian and Veterinary Science, Pro. Paulo Donato Castellane road, 14884-900 Jaboticabal SP, Brazil; jbferraz@usp.br

Montana Tropical beef composite presents maximum genetic diversity since it was developed through the combination of four different biological groups, grouping in N (zebu – *Bos indicus*), A (*Bos taurus* adapted to tropical conditions), B (*B. taurus* of British origin) and C (*B. taurus* of European Continental origin) to explore as much heterozygosity and breed complementarity as possible. Therefore, the most important benefit from using composite breed is the retained heterosis that provides great genetic and phenotypic gains to improve the industry productivity. However, the models used to estimate components of variance and breeding values can cause some confounding between the additive and non-additive components. The study aim was to integrate the genomic information through the G matrix and use the PCA to reduce the number of variables in the model. Data from post weight gain from weaning to yearling (WG) of 1,035 animals with genomic information from the GGP LD indicus 30K Array were used. The models considered were: (1) Heterozygosis model (HM) which include contemporary group as fixed effect, outcrossing percentages for direct and maternal effects for all biological groups combination effects (covariates), direct additive genetic and residual as a random effect; (2) Principal components model (PCAM) including the same fixed and random effects, but the biological type combinations, including the first and second PCA components as covariates. To select which model better fit the components and data used, the Akaike information criteria (AIC) was applied. For components of variance, both models present similar values for additive, also to heritability coefficients 0.12±0.06 and 0.11±0.05 for HM and PCAM, respectively. However, look the AIC values, where the smaller is better, PCAM obtained criteria from 9,138.4 against 9,169.8 to HM. Thus, the results obtained in this study reveals the benefit of using genomic information and the improvement of modelling with PCA may advance genetic and genomic selection for Tropical beef composite.

Session 42 Poster 20

Genetic correlation between gastrointestinal parasites resistance traits in sheep using the ssGBLUP

F. Carvalho[1], R. Silva[1], L. Grigoletto[1], J. Eler[1], J.B. Ferraz[1], G. Banchero[2], P. Oliveira[1], M. Berton[3] and F. Baldi[3]
[1]University of São Paulo, Veterinary Medicine, Rua Duque de Caxias Norte, 225, 13635-900, Brazil, [2]Instituto Nacional de Investigación Agropecuaria, INIA La Estanzuela, Andes 1365, piso 12 CP, 11100, Uruguay, [3]São Paulo State University, Via de Acesso Prof. Paulo Donato Castellane s/n, 14884900, Brazil; jbferraz@usp.br

The objective of this study was to estimate (co)variance components between gastrointestinal parasites resistance indicator traits in Santa Inês breed. The phenotypic records were collected from 700 naturally infected animals of Santa Inês breed belonging to four flocks located in the Minas Gerais and São Paulo southeast states and Sergipe northeast state of Brazil. The relationship matrix was composed of 32,292 animals. Genetic correlations(r_g) were estimated between famacha (FAM) method anaemia indicator and the egg counts per gram of faeces (EPGlog) with haematocrit (HCT), white blood cell count (WBC), red blood cell count (RBC), haemoglobin (HGB) and platelets (PLT), respectively. A total of 576 animals were genotyped with the Ovine SNP12k *BeadChip* (*Illumina, Inc.*), that contains 12,785 biallelic SNP markers. The (co)variance components were estimated using a two-trait model by single step genomic BLUP procedure. The model included the fixed effects of contemporary groups (farm and management group), month of sample collection and sex. The heritability estimates for FAM, EPGlog, HCT, HGB WBC PLT and RBC were 0.27, 0.11, 0.23, 0.14, 0.69, 0.01 and 0.90, respectively. It is expected a higher genetic gain for FAM when compared to EPGlog. The r_g estimated between FAM with EPGlog, HCT, HGB, WBC, PLT and RBC were -0.03, -0.88, -0.94, -0.95, 0.03, -0.67, and between EPGlog with HCT, HGB, WBC, PLT and RBC were -0.18, -0.16, -0.63, 0,03, -0.48, respectively. These results showed that the FAM method could be a better indicator of parasite-infected animals and is important for the selection of animals with higher resistance to gastrointestinal parasites. The results obtained were important to identify an indicator method of parasite infestation and could be an adequate selection criterion to identify more resistant animals to gastrointestinal parasites.

Genetic parameters for original and transformed interval reproduction traits of sows

E. Žáková, E. Krupa, J. Wolf and Z. Krupová
Institute of Animal Science, Přátelství 815, 10400 Prague, Czech Republic; krupa.emil@vuzv.cz

Heritabilities and genetic correlations were estimated for four reproduction interval traits of sows (farrowing interval (FI), interval from weaning to the first mating (IW1M), interval from weaning to next farrowing (I)WF, and interval from the first mating to next farrowing (I1MF)). The transformation procedure, where only part of data needed to be transformed (observations greater than the median) to overcome the skewness and to increase the heritability of the interval traits was applied. The analyses were based on performance test data of the Czech Large White (CLW) breed provided by the Czech Pig Breeders Association. Variance and covariance components were estimated using the restricted maximum likelihood and were optimised using a quasi-Newton algorithm with analytical gradients implemented in the VCE 6.0 program. Proportion of explained variability captured by used statistical models was significantly higher for all transformed traits. Data transformation in all interval reproduction traits changed the distributions of data to more closely approximate the normal distribution. Spearman's correlation coefficients between original and transformed traits were high (0.82-1.00). Estimated heritabilities for original traits were low (0.02-0.04). The transformation of traits increased the proportion of the additive genetic effect compared to the corresponding original traits. The increase ranged from +0.01 (IW1M and IWF) to +0.11 (I1MF). The transformation procedure has positive impact on estimation process. There were obtained lower standard errors for genetic parameters of transformed traits. A significantly shorter computational time (a smaller number of iterations and less computational memory) was needed to estimate genetic parameters for transformed traits. Selection for transformed reproduction interval traits is therefore expected to yield favourable correlated selection responses in all other transformed interval traits. Finally, optimising the farrowing interval traits will be beneficial for effective utilisation of inputs and sustainability of production.

Principal components regression for genetics analysis in Montana Composite cattle

B. Abreu Silva[1], J. Eler[1], F. Bussiman[1], E. Mattos[1], M.L. Santana Júnior[2] and J.B. Ferraz[1]
[1]University of São Paulo, Veterinary Medicine, Rua Duque de Caxias Norte, 225, 13635-900, Brazil, [2]Federal University of Mato Grosso, GMAT, Rodovia Rondonópolis-Guiratinga, KM 06 (MT-270), 78735-901, Brazil; joapeler@usp.br

In genetic breeding programs an obstacle to obtaining accurate estimates of genetic parameters and breeding value (EBV) is multicollinearity (MCL), defined as the presence of strong linear relationships between explanatory variables leading to erroneous inferences. The aim of this study was to evaluate the impact of correction for MCL with the use of principal component methodology (PCA). The trait used was post-weaning weight gain (PWG – 215d). The data contains about 116,558 phenotype and 256,180 pedigree information of composite beef cattle from the Montana Composite program. The model included the fixed effects of contemporary group, mother's age class at calving, age at measurement and as covariables: age at measurement, effects of racial composition and heterozygosis (direct and maternal). The covariables referring to the effects of racial composition and heterozygosis (direct and maternal) were grouped and the principal component analysis was applied in order to correct the problem of MCL. Five components was obtained and they explained 97% of variance. Thus those components were replaced in the model as covariables for the genetic analysis in a Bayesian approach (GIBBS1F90). A single chain of 550,000 rounds with a bur-in period of 50,000 and a thinning interval of 100 samples was used. The heritability estimates were 0.13 and 0.11 (PCA and without correction respectively). The Spearman's correlations among the models were 0.94; the rank correlation was 0.05. The result presented here support the idea of corrections; once the values of heritabilities do not differ between models, but the EBV do differ. Considering the low result of the rank correlation found between the models, the correction for MCL is of utmost importance. For PCA we suggest additional studies on MCL in the specific case of this methodology and a future cross-validation.

Sow maternal characteristics of two genetic selection lines in an outdoor farrowing system

L.J. Pedersen[1], S.-L.A. Schild[1], M.K. Bonde[2] and T. Serup[3]
[1]Aarhus University, Animal Science, Blichers Allé 20, 8830, Denmark, [2]Center of Development for Outdoor Livestock Production, Marsvej 43, 8960 Randers, Denmark, [3]SEGES, Organic farming, Agro Food Park 15, 8200 Aarhus, Denmark; lene.juulpedersen@anis.au.dk

In Danish organic pig herds with outdoor farrowing, total mortality was 29% in 2015. Both large litter size and low birth weight were identified as important risk factors for early death. Litter size and birth weight are strongly influenced by choice of genetic and thus we investigated maternal characteristics of the currently used LY sows from DanBred (n=22) and of Topigs Norsvin (TN70) (n=25). All animals were purchased from one multiplier herd per line at an age of 20-24 weeks. Gilts were housed, fed and managed under similar conditions at AU's research farm in DK. They were inseminated with DanBred Duroc semen balanced between lines. Animals farrowed in four batches each representing 11-13 animals of each line. Batch 1 and 2 comprised first parity and batch 3 and 4 comprised second parity. The number of functional teats determined the nurse litter size. Surplus piglets and piglets with a birth weight <700 g were euthanized. Piglets were weaned at 49 d of age. Birth and weaning weight of each piglet were sampled and analysed in a MIXED model with sow as experimental unit. Litter size and number of functional teats were analysed in a GLM model, while litter mortality was analysed in a GLIMIXED model for binomial-distributed data. Piglets of TN70 vs DanBred had significantly higher birth weight (1st parity: 1,403 vs 1,168 g; 2nd parity: 1,652 vs 1,359 g) ($P<0.0001$) and higher weaning weight ($P<0.0004$). TN70 vs DanBred sows had significantly more functional teats (15.4 vs 14.1. teats) ($P<0.0001$), gave birth to fewer piglets per litter (1st parity: 14.7 vs 16.5 piglets; 2nd parity: 16.8 vs 20.1 piglets) ($P=0.003$), had fewer dead ($P=0.003$), mainly due to more euthanized piglets in DanBred, while number of weaned was equal. In conclusion, TN70 had lower litter size, more functional teats and piglets were heavier at birth and at weaning. Based on these results from a limited number of animals in first and second parity, TN70 appears more suited for outdoor farrowing with low management input.

Cognitive bias and group preference when housing fattening pigs in a small vs very large group

U. König Von Borstel[1], A. Kiani[1] and H. Schwanhold[2]
[1]institute of animal husbandry and biology, Department of Animal Breeding and Genetics, Leihgesterner Weg 52, 35392 Gießen (Giessen), Germany, [2]institute of livestock production systems, Department of animal sciences, Albrecht-Thaer-Weg 3, 37075 Göttingen, Germany; alikianee@gmail.com

Cognitive bias and group preference when housing fattening pigs in a small vs very large group Early career competition It has been suggested that pigs are unable to individually recognize all animals when living in groups >ca. 50 individuals, raising the question, if this potentially unnatural social situation negatively affects animal welfare. Therefore, the aim of the present study was to investigate the emotional state of fattening pigs when housed in a small vs very large group. Fourteen pigs were trained on a go/no-go task to discriminate two visual stimuli, a positive cue that predicted a food reward, if the pig approached a bucket, and a negative cue that predicted no reward. After six weeks of training, 91 pigs originally housed together in one large group were split into a small (11 pigs) and a large (80 pigs) group with equal densities (0.8 m^2/pig) and 7 trained pigs per group. Cognitive bias and preference tests were conducted with the trained pigs at -1, 1, 3,7, 10 and 13 days in relation to splitting the original group. For all control comparisons, according to mixed model analysis results were as expected, i.e. pigs of both groups approached the positive cue equally fast and more rapidly than the negative cue, and before splitting the group there were no significant differences ($P>0.1$) in latency to approach the ambiguous cues between pigs that were later housed in the small vs large group. However, on day 3, 7 ($P<0.05$) and 10 ($P<0.1$) after splitting, pigs of the small group showed greater latencies to approach the ambiguous cue, compared to pigs of the large group, indicating a more pessimistic bias in pigs housed in the small group. The preference test revealed no clear preference ($P>0.05$) for the small or large group for either pigs from the small or large group, but pigs were generally very reluctant to return to the group housing, indicating that no active choice was made. Results suggest that pigs' welfare in small groups is negatively affected, likely due to the reduced total space availability.

Session 43 Theatre 3

Do farrowing and rearing systems affect the agonistic behaviour of pigs at regrouping?

A. Lange[1], C. Lambertz[2], S. Ammer[1], M. Gauly[2] and I. Traulsen[1]

[1]Georg-August-University, Livestock Systems, Albrecht-Thaer-Weg 3, 37075 Göttingen, Germany, [2]Free University of Bozen-Bolzano, Faculty of Science & Technology, Piazza Università 5, 39100 Bolzano, Italy; aszulc@gwdg.de

Besides separation from the sow, the relocation to a new environment and mixing with unknown conspecifics are factors altering welfare for piglets around weaning and during fattening. On two experimental farms (farm1, farm2) the agonistic behaviour of growing and finishing pigs born in three farrowing systems (farrowing crates (FC), free farrowing pens (FF) or group housing of lactating sows (GH)) was studied (farm1=8 batches, farm2=7 batches). Conventional two-phase rearing (regrouping at the start of the rearing and finishing period, CONV, farm1+2) was compared to two-phase rearing with the rearing phase in the farrowing pen (FP, farm1) or to single-phase rearing (wean-to-finish (WTF), farm2). In half of the batches tails were undocked. Males were castrated on farm1 and raised intact on farm2. To assess the agonistic behaviour, skin lesion (SL) scores on the front body (scale 0 to 2) for a total of 3,650 piglets in the first week after weaning and 1,664 pigs at beginning of the fattening period were analysed. On both farms, CONV reared GH piglets (farm1=24.4%, farm2=17.3%) showed less moderate and severe SL, compared to FF (farm1=47.8%, farm2=52.6%) and FC piglets (farm1=51.3%, farm2=50.7%) after weaning. Single-phase reared piglets showed a comparable amount of moderate and severe SL after weaning (GH_{WTF}=23.6%, FF_{WTF}=61.4%, FC_{WTF}=54.6%). When piglets were reared in the farrowing pen, the farrowing systems did not seem to impact moderate and severe SL (GH_{FP}=25.2%, FF_{FP}=20.7%, FC_{FP}=16.8%). At the beginning of the finishing period, WTF pigs showed less moderate and severe SL than both two-phase rearing systems (WTF=15.9%, CONVfarm1=60.7%, CONVfarm2=59.4%, FP=62.2%). While the farrowing system did not affect SL in WTF and CONV fattening, GH finishing pigs reared in the farrowing pen showed less moderate and severe SL than FF and FC finishing pigs (GH_{FP}=33%, FF_{FP}=80%, FC_{FP}=73.7%). No sex differences were found. To conclude, GH positively influenced the incidence of SL after 1st regrouping, especially for FP pigs. WTF rearing lowered the incidence of SL drastically in the finishing period.

Session 43 Theatre 4

Tail position as pig welfare indicator in commercially raised pigs with intact tails

T. Wallgren, A. Larsen and S. Gunnarsson

Swedish University of Agricultural Sciences, Animal Environment and Health, Box 234, 53223 Skara, Sweden; torun.wallgren@slu.se

Rearing pigs with intact tails, without risking large tail biting outbreaks, is one of the biggest animal welfare challenges within todays' pig production. During this experiment, tail lesion scoring of 460 undocked finishing pigs was carried out weekly with respect to tail shortening, damage and freshness of lesions. The experiment was conducted during one batch of finishers from approx. 30-115 kg LW in a commercial Swedish herd. Along with lesion scoring, tail position (hanging or curled) was scored at feeding. The pigs were fed with wet feed in pens with long through (11 pigs/pen). The aim was to investigate if tail position correlate to presence of tail lesions and hence could be used as an easily detected sign of tail biting for farmers. The hypothesis was that a hanging tail is an indicator of tail lesions. Preliminary analyses show that 124 of the pigs had hanging tails; 50.4% at one occasion, 23.2% at two occasions, 12% at three occasions and 14.4% between 4 and 8 occasions (approx. 70% of the production period has passed). In 33.2% of the cases, hanging tail did not correspond to o tail lesions at the current occasion. In 97.6% of the cases pigs with hanging tails had tail lesions later on. Some degree of lesions was found in 41.1% of pigs with curled tail. As a preliminary analysis, total number of hanging tails and total lesion score was calculated and divided into 4 percentiles each, ranging from low to high, on individual and pen level. On individual level 36.3% of the pigs belonged to the same percentile in in both groups and 32.3% belonged to adjacent percentile. In 8.9% of the cases pigs belonging to the top percentile in one trait belonged to the bottom percentile in the other trait, suggesting that tail posture is an intermediate indicator of tail lesions on individual level. On pen level, 40.5% of the pens belonged to the same percentile in both groups and 45.2% to the adjacent group. This suggests that hanging tails might indicate tail biting in the group, also in pigs with not yet affected/hanging tails. Hence hanging tails might be a better indicator on group level than in individual level. In these preliminary results, type or severity of lesion has not been taken into consideration.

Session 43 — Theatre 5

Rearing undocked pigs on fully-slatted floors using multiple enrichment and variety: a pilot

J.-Y. Chou[1,2,3], C.M.V. Drique[3,4], D.A. Sandercock[2], R.B. D'Eath[2] and K. O'Driscoll[3]
[1]University of Edinburgh, Royal (Dick) School of Veterinary Studies, Easter Bush, EH25 9RG, United Kingdom, [2]SRUC, Animal and Veterinary Science Research Group, Easter Bush, EH25 9RG, United Kingdom, [3]Teagasc, Pig Development Department, Moorepark, Fermoy, Co. Cork, P61 P302, Ireland, [4]Agrocampus Ouest, L'ingénieur Agronome, Rennes, 35042, France; jenyun.chou@ed.ac.uk

In fully-slatted systems tail-biting is a difficult issue to manage when pigs' tails are not docked, because loose enrichment material can obstruct slurry systems. This pilot study seeks to determine whether: (1) intact-tailed pigs can be reared with a manageable level of tail biting by using multiple enrichment sources compatible with fully-slatted systems; and (2) whether variation of enrichment provision has an effect. Ninety-six undocked pigs were given the same multiple enrichment items from one week after birth until weaning. At weaning, four different combinations of 8 enrichment items were selected based on the following traits: rootability, durability, edibility, form of presentation (attached, suspended or moveable), texture, and location in the pen. These were randomly assigned to 8 pens (n=12 pigs/pen), totalling two pens per combination. Four pens had the same combination (SAME) from assignment to finish and four pens switched combinations every two weeks (SWITCH). Individual lesion scores and behaviour observations of enrichment use and harmful behaviours were recorded every two weeks. MIXED procedure in SAS 9.4 was used to analyse the data. The average tail damage score during the course of the experiment was low (0.93±0.02; on a 0-3 scale; 0 = no damage, 1 = bite marks), but SAME pigs had worse tail lesions inspected post-mortem than SWITCH pigs (1.00±0.07 vs 0.80±0.07, $P<0.05$). Only one severe tail biting event involving a single victim was recorded in a SAME pen. The overall interaction with the enrichment did not show a gradual decline during the production cycle. Pigs interacted with a rack of loose material most frequently ($P<0.001$). When given grass, SAME pigs consumed more materials in the rack than SWITCH pigs ($P<0.001$). The study showed promising results for keeping undocked pigs on a fully-slatted floor using slat-compatible enrichment with a manageable level of tail biting.

Session 43 — Theatre 6

PigWatch: early automated detection of tail biting and aggression

H.A.M. Spoolder[1], M. Zebunke[2], E. Labyt[3], C. Godin[3], C. Tallet[4], A. Prunier[4], S. Dippel[5], B. Früh[6], M. Dall Aaslyng[7], H. Daugaard Larsen[7] and H.M. Vermeer[1]
[1]Wageningen Livestock Research, P.O. Box 338, 6700 AH Wageningen, the Netherlands, [2]Leibniz Institute for Farm Animal Biology, Wilhelm-Stahl-Allee 2, 18196 Dummerstorf, Germany, [3]CEA LETI, Minatec Campus, 38054 Grenoble, France, [4]PEGASE, Agrocampus Ouest, INRA, 35590 Saint-Gilles, France, [5]Friedrich-Loeffler-Institut, Dörnbergstr. 25/27, 29223 Celle, Germany, [6]Research Institute of Organic Agriculture, Ackerstrasse 113, 5070 Frick, Switzerland, [7]Danish Meat Reasearch Institute, Gregersensvej 9, 2630 Taastrup, Denmark; hans.spoolder@wur.nl

Despite decades of intensive research, aggression and tail biting in pigs still persist on many farms. Remedial measures are dependent on early diagnosis of these injurious behaviours. The European PigWatch Project (https://pigwatch.net, Anihwa ERA-Net) aims to sensitize stock persons to early behavioural signs of pigs. It also develops 3 automated techniques to record injurious behaviours. To start we developed and tested an on-farm observation protocol for animal handlers, requiring them to observe their pigs differently. Farmers in the 5 participating countries of the project confirmed that the position of the tail (hanging or curled) is a good early indicator. For some the use of the protocol changed the way they look at their animals. Pigwatch also develops technological solutions. A first study focused on increased behavioural activity as a sign of ongoing aggressive acts. A sensor will send an alarm to an app when activity is conspicuous. The first prototype detects 42% of fights, with 62% true positives. Its sensitivity and specificity is still being improved. In a second study, a multispectrum camera was developed to detect blood (Hb) through the use of LEDs emitting 6 different wavelengths, of which the reflection is caught on separate digital photos. The principle was tested with success on fresh and old blood and compared with red ink. It is now applied on group housed pigs. Finally, in a third study, tail length and lesions are automatically detected by camera in-line at the abattoir. The system was used on pigs from 225 herds (250+ carcasses / batch). It can generate a report per herd on the number of pigs, the prevalence of tail lesions and the fraction of short and very short tails.

Boosted regression trees to predict pneumonia, growth and meat percentage of slaughter pigs

H. Mollenhorst, K.H. De Greef, B.J. Ducro, B. Hulsegge, A.H. Hoving-Bolink, R.F. Veerkamp and C. Kamphuis
Wageningen University and Research, Droevendaalsesteeg 1, 6708 PB Wageningen, the Netherlands; erwin.mollenhorst@wur.nl

Abnormalities, like pneumonia or deviating growth, reduce production efficiency due to multiple deliveries per round and pigs delivered with a low meat yield. Adapting management based on early prediction of abnormalities (at the onset of the fattening period) can lead to higher revenues or decreased costs of production. Existing datasets from, e.g. the slaughter plant or the production cycle, can be used for prediction. These data, however, are often incomplete, especially at individual level. Machine learning techniques are able to deal with incomplete data and might be more suitable for predictions on large volumes of data. Aim of this study was to predict deviant slaughter pigs based on routine data available at the onset of the fattening period. Data from a Dutch pig research facility, containing 65,208 records of individual pigs born between 2004-2016, of which 33,628 had information on pneumonia status were used. Information was available on offspring, litter, transfer dates between and locations during production stages, and individual live-weights at several production stages. Slaughter data were used to define three binary traits indicating pneumonia (average occurrence 8%), 10% lowest growth rate and 10% lowest meat percentage for each individual pig. Generalized boosted regression was used to predict these traits at the onset of the fattening period. Area under the ROC curve (AUC) and sensitivity at a fixed specificity of 90% were used as performance criteria. Validation on an independent dataset showed that AUCs were poor to fairly good for prediction of low meat percentage (0.53), pneumonia (0.61) and low growth rate (0.77). Sensitivity at 90% specificity was 11, 15 and 44%, respectively. This means that no reasonable prediction could be made for low meat percentage and pneumonia. On the other hand, for predicting pigs with low growth rate a three times increase in positive predicted value could be achieved, which means that problem pigs can be identified a bit better, using routine data. Therewith, this study can be considered a first step towards an early warning system for abnormalities.

Automatic estimation of slaughter pig live weight using convolutional neural networks

D.B. Jensen[1], K.N. Dominiak[2] and L.J. Pedersen[3]
[1]Wageningen University and Research, Hollandseweg 1, 6706 KN Wageningen, the Netherlands, [2]University of Copenhagen, Groennegaardsvej 2, 1870 Frederiksberg C, Denmark, [3]Aarhus University, Blichers Allé 20, 8830 Tjele, Denmark; dan.jensen@wur.nl

An accurate estimate of the live weight of slaughter pigs is useful to the farmer in several different ways. First of all, knowing the weight of the pigs in a pen will let the farmer know the optimal time to send his pigs to the slaughter house. Second, being able to accurately monitor the growth curves of the pigs can be used to detect issues with the feeding formulation or emerging diseases, which can manifest in a reduced growth rate. Finally, an accurate estimate of the weights of the pigs can be used for more accurate dosing of medicine, which can potentially lead to a lower use of antibiotics and other medicines. Some big pig producers have staff employed for the sole task of performing manual weightings. This practice is however very laborious and time consuming, making it unfeasible for most small producers. We demonstrate that a convolutional neural network (CNN) can be trained on a selection of pigs to automatically estimate the individual live weight of other pigs. Our CNN takes an image of an individual pig, taken with a camera fixed above the drinking nipple of the pen, as its input. The output is a numerical estimate of the weight of the pig in the image. The input images are scaled down from 768×576 pixels to 58×58 pixels. The CNN is trained and tested on images of pigs from one pen during one grower/finisher period, with the pigs weighing between 25 and 85 kg. The weight measurements and the images were made once per week. The pen contained a total of 18 pigs, but for each of the weekly observations, the number of individually identifiable pigs varied from 2 and 18. Images of a given pig on a given day is not used in both the training and the test set, thus ensuring that the CNN was tested on unseen images. The overall mean absolute error of our CNN is 4.4 kg. Further research will attempt to improve the performance further by including additional image data in the training process and by optimizing the rescaling of the image resolution.

Session 43 Theatre 9

Hacking CASA to predict boar semen fertility

C. Kamphuis[1], B. Visser[2], P. Duenk[1], G. Singh[1], A. Nisch[1], R.M. De Mol[1], R.F. Veerkamp[1] and M.L.W.J. Broekhuijse[3]
[1]Wageningen University and Research, Droevendaalsesteeg 1, 6708 PB Wageningen, the Netherlands, [2]Hendrix Genetics Research, Technology & Services B.V., Spoorstraat 69, 5831 CK Boxmeer, the Netherlands, [3]Topigs Norsvin Research Center, Schoenaker 6, 6641 SZ Beuningen, the Netherlands; roel.veerkamp@wur.nl

Computer Assisted Semen Analysis (CASA) is an objective method to assess semen quality. CASA systems generate >400 parameters per measurement, but since their relationship with fertility traits has not been studied extensively, artificial insemination (AI) laboratories only use a limited number of parameters to decide how to process an ejaculate. A 24 h hackathon was organised to study whether CASA could be exploited better to predict boar semen fertility. Four teams were challenged to model proprietary datasets of CASA and ejaculate data, and open access weather data to predict three fertility traits (total number piglets born, total stillborn, gestation length). Developed models were evaluated by computing correlation coefficients between predicted and actual values of the three traits using an independent validation set. One team was able to predict all three fertility traits by combining biological knowledge with Gradient Boosting Machine, a machine learning algorithm building regression trees on all parameters in a fully distributed way. Their models used 59 parameters, including 53 CASA parameters beforehand marked as potentially relevant by the company, four boar parameters (age, temperature at ejaculation, interval between ejaculates, and total morphological abnormalities), and two weather parameters (mean outside temperature, and daily outside temperature range). Twenty-five parameters were consistently important in predicting fertility. Surprisingly, parameters related to temperature ranked high in importance. Moreover, 19 CASA parameters were included in this shortlist, including the motility parameters currently used by AI laboratories. Correlation coefficients were low (0.11, -0.09, and 0.03 for total number born, stillborn and gestation length, respectively) but in line with expectations. Hacking CASA in just 24 h revealed new insights in using more CASA data in daily AI routine practice, and in temperature playing an important role in predicting semen fertility.

Session 43 Theatre 10

Improving prediction accuracy of meat tenderness in Nelore cattle using artificial neural networks

F.B. Lopes[1,2], C.U. Magnabosco[1], T.L. Passafaro[3], L.F.M. Mota[2,3], M.G. Narciso[4], G.J.M. Rosa[3] and F. Baldi[2]
[1]Embrapa Cerrados, BR-020, 18, Sobradinho, Brasilia, DF, 70770-901, Brazil, [2]São Paulo State University – Júlio de Mesquita Filho (UNESP), Departament of Animal Science, Prof. Paulo Donato Castelane, Jaboticabal, SP, 14884-900, Brazil, [3]University of Wisconsin-Madison, Department of Animal Sciences, 436 Animal Sci. Building 1675 Observatory Dr. Madison, WI, 53706, USA, [4]Embrapa Rice and Beans, GO-462, km 12, Santo Antônio de Goiás, GO, 75375-000, Brazil; camult@gmail.com

Traditional genetic selection for meat tenderness in beef cattle is constrained by the cost and difficulty of its measurement, and genomic selection (GS) has arisen then as an alternative to improve it. Artificial Neural Networks (ANN) are gaining prominence in GS as it can fit complex relationships, potentially increasing predictive accuracy. A special case of ANN is the Deep Neural Networks (DNN), which have multiple hidden layers. DNN have been successfully used in the field of image analysis and speech recognition, but few studies have applied it on GS. Thus, this study was carried out to compare the predictive ability of ANN and DNN with Bayesian Ridge Regression, Bayesian Lasso, Bayes A, Bayes B, and Bayes Cπ in estimating genomic breeding values for meat tenderness on 575 Nelore cattle. It was used 219,863 SNP located on autosomes, with minor allele frequency >5%, deviation from HWE ($P>10^{-6}$), and with linkage disequilibrium <0.8. Meat tenderness was moderately heritable (0.17), indicating that it could be improved by direct selection. Prediction accuracies were similar for all Bayesian models, ranging from 0.20 (Bayes A) to 0.22 (Bayes B), and higher for ANN (0.33). This might be explained by the fact that ANN can capture non-linear relationships such as non-additive genetic effects, while the Bayesian models used fit only linear functions. Although DNN are recognized to delivery more accurate predictions, in our study ANN with one hidden layer and 105 neurons was enough to increase prediction accuracy. Hence, these results indicate that an ANN with relatively simple architecture was more efficient in predicting genomic breeding values for meat tenderness in Nelore cattle. Acknowledgments: São Paulo Research Foundation grant #2017/03221-9.

Session 44 Theatre 1

Compound feed costs for dairy cattle affected by local sourcing

G. Van Duinkerken and G.J. Remmelink
Wageningen Livestock Research, De Elst 1, 6708 WD Wageningen, the Netherlands; gert.vanduinkerken@wur.nl

European dairy production tends to increase its focus on local sourcing of feed materials. Exclusion of protein sources imported from outside Europe is mainly targeted on soybean meal from Latin America. However, there also is an overall tendency to increase European self-sufficiency for feed materials and reduction of import of feed materials. A simulation study was performed to calculate the impact of exclusion of soybean products, or exclusion of all non-European feed materials, on the prices of compound feed. Monthly prices of feed materials in 2017 were averaged and these averages were used as reference prices in the simulations. Three common compound feeds for dairy cattle ('standard low protein', 'medium protein rich' and 'protein rich') were formulated using a least cost linear programming tool. Three scenarios were compared: (1) business as usual with full availability of imported protein rich feed materials such as Latin American soybean meal; (2) exclusion of all non-European soybean products; (3) compound feed with exclusion of all non-European feed materials. For the low protein compound feed, scenario 2 did not result in higher compound feed costs, compared to scenario 1. For medium protein rich and protein rich compound feeds, scenario 2 resulted in higher compound feed costs of €0.67 to €5.71 (excl. VAT) per 100 kg, respectively. In scenario 3, compound feed costs increased compared to scenario 1, with €1.17, €2.54 and €7.38 (excl. VAT) per 100 kg, for low, medium and high protein compound feeds, respectively. If local sourcing would be implemented at large scale in Europe, drastic changes in availability of feed materials would occur, also severely affecting feed material prices, or resulting in shortage of compound feed ingredients. Increased European protein self-sufficiency, further advances in application of low-protein livestock diets, and innovations in novel feed proteins, are necessary steps for further implementation of local sourcing in Europe. Sustainability assessments for local sourcing concepts are desired, because local sourcing does not necessarily improve key performance indicators for sustainability, such as the carbon footprint.

Session 44 Theatre 2

Factors associated with feed efficiency traits in Italian Jersey cows

F. Monti[1], G. Visentin[2], M. Marusi[2], R. Finocchiaro[2], J.B.C.H.M. Van Kaam[2], G. Civati[2] and R. Davoli[1]
[1]University of Bologna, Department of Agricultural and Food Sciences (DISTAL), viale Fanin 46, 40127 Bologna (BO), Italy, [2]Associazione Nazionale Allevatori Frisona Italiana (ANAFI), via Bergamo 292, 26100 Cremona (CR), Italy; giuliovisentin@anafi.it

Feed efficiency in lactating cows is one of the major aspects in modern dairy farm management. Purpose of the present study was to identify sources of variation of body live weight (LW) and feed efficiency traits in a population of 8,516 primiparous Jersey cows. The variables tested included LW, fat corrected milk (FCM), predicted dry matter intake (pDMI), energy corrected milk (ECM), and predicted feed efficiency (pFE_ECM), defined as ECM/pDMI. Linear scoring and age at evaluation were used to estimate LW. Factors associated with such traits were investigated using linear mixed models which included the fixed effects of origin of the paternal grandsire (USA, Denmark, Canada), stage of lactation, year of birth, and age at first calving. Random terms were cow, contemporary group (herd-year-season at linear scoring), and the residual. Mean of LW and pFE_ECM were 435 kg and 1.36, respectively. Feed efficiency was weakly negatively correlated (-0.15) with LW, and pDMI was positive correlated with ECM (0.88). Values of LW ranged from 444.01 to 449.88 kg in cows born between 2006 and 2014. Furthermore, cows with paternal grandsire born in Canada were the heaviest, but they had the lowest values for FCM, pDMI, ECM and pFE_ECM. Feed efficiency, FCM, ECM, and pDMI increased in the first three months of lactation, and then decreased. Kilos of pDMI and FCM increased concurrently with age at first calving and systematically decreased as year of birth proceeds. Energy corrected milk of cows born between 2011 and 2014 was lower than ECM of animals born between 2008 and 2010, consequently resulting in lower pFE_ECM. Late calving Jerseys were less feed efficient compared to early calving contemporaries. Systematic environmental factors associated with feed efficiency will be used to adjust phenotypes for routine genetic evaluation.

Session 44 Theatre 3

Selection for claw health and feed efficiency in the Czech Holstein

Z. Krupová, E. Krupa, M. Wolfová, J. Přibyl and L. Zavadilová
Institute of Animal Science, Přátelství 815, 10400 Prague, Czech Republic; krupova.zuzana@vuzv.cz

Breeding of the Czech Holstein cattle (CH) is currently based on estimated breeding values for the complex of production, functional and exterior traits. Breeding objective for CH includes 10 traits; 17 traits are considered as selection criteria. Claw disease incidence (CLD) and residual feed intake (RFI) were added to the breeding objective and selection criteria to improve health status and feed efficiency and finally enhance the long-term sustainability of the breed. Economic weights (EWs) and selection responses for the new traits and for the current breeding objective traits were calculated. Selection response was estimated assuming standardised selection intensity. EWs of the new traits expressed per cow and year and calculated under the actual production and economic conditions were: -100.1 € per case of CLD and -79.37, -37.16, and -6.33 € per kg of DMI per day for RFI of cows, breeding heifers, and fattened animals, respectively. Milk yield, milk components, conception rate and productive lifetime remained the most important breeding objective traits. Participation of the selection index traits was (A) settled on actual trait ratio, but making space for the new traits that were given 5, 5, 1, and 1% from the sum of the relative trait weights in the index, and (B) optimised to reach the maximal selection response in breeding objective and the highest index reliability. In the alternative A, genetic selection response calculated per generation interval was -0.029 cases of CLD and -0.015, -0.005 and -0.004 kg of DMI per day of cows, heifers and fattened animals, respectively. The appropriate overall economic selection response in the new traits was 4.39 €. In alternative B, a slightly higher response in RFI (e.g. -0.020 kg of DMI per day of cows) was obtained and the index reliability increased from 46 to 61%. Selection response in the current breeding objective traits remained positive and almost the same in both alternatives. Selection response expected in the new evaluated traits was favourable and, therefore, these traits should be considered as the new breeding objective traits for the local dairy cattle population. The study was supported by project MZE-RO0718 of the Czech Republic.

Session 44 Theatre 4

Relationships between cow performance with milk urea yield and efficiency of protein utilization

M. Correa-Luna, N. Lopez-Villalobos, D.J. Donaghy and P.D. Kemp
Massey University, School of Agriculture and Environment, 4442 Palmerston North, New Zealand; m.l.correaluna@massey.ac.nz

Phenotypic correlations of cow productive performance with lactation average of milk urea (MU) and total lactation MU yield (MUY) were estimated in two herds during an entire season under grazing. Productive performance traits were total lactation yields of milk (MY), fat (FY), protein (PY) and lactose (LY), lactation averages of live weight (LW) and body condition score (BCS) and change of LW and BCS from calving to 100 days in milk. Between July 2016 and May 2017, 210 cows were milked twice-daily with high supplementary feed inclusion and 258 cows were milked once-daily with low supplementary feed inclusion. In early, mid and late lactation, milk samples were collected to measure MU. At every herd-test date, daily milk urea yield was calculated as the quantity of MU in a given daily milk yield, and efficiency of crude protein utilization (ECPU) was calculated as daily protein yield (PY) divided by daily crude protein intake (CPI); this last variable derived from daily intake estimations of metabolizable energy requirements. Partial phenotypic correlations among traits were computed using a model that included the fixed effects of herd, lactation number, and the covariables deviation from median calving date, proportion of Friesian and Friesian × Jersey heterosis. Productive traits were strongly correlated ($P<0.001$) with MUY (0.62 to 0.79) and MU concentration (0.27 to 0.39). The ECPU was moderately and positively correlated with MY, PY, LY and MUY, but not with FY. Live weight and BCS were negatively correlated ($P<0.001$) with ECPU (-0.20 and -0.38). Changes in BCS were negatively correlated ($P<0.001$) with productive traits (-0.20 to -0.25), MUY (-0.26), MU (-0.21) and ECPU (-0.28). The correlation coefficient between MU and ECPU was 0.15 ($P<0.001$), indicating that cows with low MU concentration are not the most efficient at to utilising crude protein. It is possible that mobilization of body reserves in early lactation is contributing not only as a source of energy, but also as a source of protein, to an increasing demand for nutrients towards peak milk yield.

Differences in productivity of grazing dairy cows with low versus high genetic merit for milk urea

N. Lopez-Villalobos[1], M. Correa-Luna[1], J.L. Burke[1], N.W. Sneddon[1,2], M.M. Schutz[3], D.J. Donaghy[1] and P.D. Kemp[1]
[1]Massey University, School of Agriculture and Environment, 4442 Palmerston North, New Zealand, [2]Fonterra, On-Farm Research and Development, 4442 Palmerston North, New Zealand, [3]Purdue University, Department of Animal Sciences, 47907-2054 West Lafayette, IN, USA; n.lopez-villalobos@massey.ac.nz

Genetic selection for low milk urea nitrogen has been proposed as a tool to reduce nitrogen excretion; but potential effects on production and quality of milk have not been evaluated under grazing conditions. The present study provides estimates of genetic parameters for milk urea (MU) concentration and evaluates productivity differences between cows with low versus high genetic merit for milk urea in two dairy herds. Estimates of heritability and breeding values for MU were obtained using 1,284 herd-test records of 66 Holstein-Friesian (F), 55 Jersey (J) and 137 crossbred (FxJ) cows milked once daily in Herd 1, and 51 F, 3 J and 156 FxJ cows milked twice daily in Herd 2. Cows calved in spring 2016 and were milked until May 2017. A repeatability animal model was used to estimate variance components for MU. The model included the fixed effects of herd, lactation number, herd-test-date; the covariables deviation from median herd calving date, proportion of F and FxJ heterosis, and the random effects of animal and permanent environment of the cow. The estimate of heritability for MU concentration was 0.24 and breeding values (BV) ranged from -4.0 to 6.3 mg/dl. Cows were ranked into three classes (low, medium and high) of genetic merit for MU concentration within each farm and lactation number. Compared with low MU BV cows, high MU BV cows produced an additional 351 kg milk, 7.9 kg protein, and 14.4 kg lactose per lactation ($P<0.05$), but total lactation fat yield, live weight at calving and live weight losses after calving were not significantly different. The high MU BV cows had a greater efficiency of crude protein utilization (kg of milk protein/kg crude protein intake). These results indicate that within the range of MU BV in these herds, cows genetically disposed to having lower levels of MU were less productive in terms of milk and protein yield. Therefore, great caution must be used when selecting against MU as a tool to reduce nutrient excretion.

Feeding of rumen-protected fat limited by Dutch industry to avoid negative effects on cheese quality

A. Elgersma
independent scientist, P.O. Box 323, 6700 AH Wageningen, the Netherlands; anjo.elgersma@hotmail.com

After the abolishment of the milk quota system on 1 April 2015, limits on the production of kilograms fat disappeared and dairy farmers in The Netherlands increased the use of bypass fats in cow rations. Dietary fat that resists lipolysis and bio-hydrogenation by rumen microorganisms, but gets digested in the lower digestive tract, is known as bypass, inert, or rumen protected fat. Feeding bypass fat to high producing dairy cows enhances the energy density of the ration. When milk prices increased, feed producers indicated that feeding (rumen-protected) fat should be reconsidered. Farmers were encouraged by advertisements, magazine articles, and by their feed advisors to purchase rumen-protected fat. The use of rumen-protected fat increased exponentially. Nutrition experts were interviewed, but no mention was made of effects on milk fat composition and possible risks for dairy product quality in terms of nutritional value or technological properties of milk. A farmer magazine article of 2 September 2017 estimated that 30% of farmers used bypass fats. This led to changes in the milk fatty acid profile, i.e. higher levels of palmitic acid (C16:0), as well as a higher solid fat content. As a result, processing of milk into fat-containing products, e.g. cheese, became negatively affected. Upmarket cheese producers noticed that cheeses were harder and became more prone to mechanical damage. This is not surprising as saturated fats have a high melting point, and should have been considered by feed manufacturers and advisors before promoting rumen-protected fats for dairy cow diets. Per the 1st of November 2017, the Dutch dairy (NZO) and animal feed industries (Nevedi) took a joint decision to stop the use of fractionated fats in dairy cow rations, at least for the winter period 2017-2018. In conclusion, changing farming practices in 2016 and 2017 were driven by profit and one-sided information to boost milk and milk fat production, ignoring effects on product chain and consumers.

Session 44 Theatre 7

Effects a of reduced day milking interval on dairy cows performances and labour costs

V. Brocard[1], A. Caille[1] and E. Cloet[2]

[1]Institut de l'Elevage, Monvoisin BP85225, 35652 Le Rheu Cdx, France, [2]Chambre d'Agriculture de Bretagne, 2 allée St Guénolé, CS 26032, 29322 Quimper Cdx, France; valerie.brocard@idele.fr

Dairy farmers are reluctant to change their milking interval (MI), which is in average over 10 hours between morning milking and evening milking in Western France. However, a shorter interval could help them to have more flexibility in their work and improve their living conditions. It could also make the job of dairy worker more attractive if the duration of their working day remained below 8 to 9 hours. The aim of the experiment led in Trévarez experimental farm and Knowledge Transfer Center (Brittany, France) was to assess the effects of a reduced day MI on dairy cows performances. The tested interval was 6 h 30 min compared with a classical control MI of 10 hours. The study lasted four months, during two winters, with 2 groups of 30 Holstein cows indoors. Over a complete day period, cows of the 6h30 group produced 4.5% milk less than the control group during the first month, then the cows adapted to the new MI and the difference disappeared. The effects on fat content and SCC were not significant but the protein content was significantly reduced by 1 g per kg with the shorter MI. Health and feeding behaviour were not impacted. At the morning milking, the 6h30 group produced significantly more milk but at evening milking it produced significantly less milk with less protein content and higher fat content. According to the literature, after 16 hours of udder filling, physiological changes impact synthesis of milk components for the following two milkings. Hence, to improve the socio economic resilience of his farm, the dairy producer can choose between reducing the economic loss by keeping the same MI, or improving his living conditions. If he hires an employee, a reduced MI could prevent him from paying overtime and it could compensate the economic loss due to the reduced protein content. Nevertheless performances should not be affected if the night interval remains below 16 hours, and the day MI longer over 8 hours. It can bring some flexibility compared with the current average MI in farms.

Session 44 Theatre 8

No dry period: consequences for dairy cows, cash flows, and climate

A. Kok, A.T.M. Van Knegsel, C.E. Van Middelaar, H. Hogeveen, B. Kemp and I.J.M. De Boer

Wageningen University & Research, P.O. Box 338, 6700 AH Wageningen, the Netherlands; akke.kok@wur.nl

Farmers traditionally stop milking a cow 6-8 weeks before next calving (dry period), to maximise milk production in the next lactation. The resulting high milk production in early lactation, however, results in a negative energy balance and is associated with reduced health and fertility. Omitting the dry period improves the energy balance in early lactation at the cost of milk production. This project aimed to evaluate and integrate sustainability impacts of omitting the dry period, with a focus on cow welfare, cash flows at farm level, and greenhouse gas (GHG) emissions per unit milk. Welfare was addressed by monitoring lying and feeding behaviour of 81 cows with no dry period or a 30-day dry period in weeks -4 and 4 relative to calving. Using mixed models, statistical analyses showed that cows with no dry period had a 1 hour per day shorter lying time in week -4 than cows with a dry period, whereas they had a 1 hour longer lying time and a greater feed intake in week 4. The absolute daily lying time (12.6 h) and relatively constant feeding rate suggest that welfare of cows with no DP was not impaired by milking in late gestation. A dynamic stochastic simulation model at herd level was developed to estimate the economic impact for the farmer, and GHG emissions per unit fat-and-protein-corrected milk (FPCM; cradle-to-farm-gate analysis). The impact of dry period length on milk yield, days open, and fertility culling was derived from empirical data of Dutch dairy farms. Omitting the dry period reduced overall milk yield at herd level by 3.5%, compared with a dry period of 8 weeks. Accounting for feed costs, youngstock rearing, sale of surplus calves and culled cows, omitting the dry period cost 16 euros per cow per year. These costs might be offset by a reduction in veterinary costs, due to the improved metabolic status in early lactation. Omitting the dry period increased GHG emissions per unit FPCM by 0.5%. A 5% reduction in culling rate could offset this negative impact. In conclusion, omitting the dry period can improve cow welfare with a small negative impact on cash flows and GHG emissions, which may be offset by improved cow health.

Session 44 Theatre 9

Aggregating dairy cattle welfare measures into a multicriterion welfare index

F. Tuyttens[1], B. Ampe[1], S. Andreassen[2], A. De Boyer Des Roches[3], F. Van Eerdenburg[4], M.J. Haskell[5], M. Kirchner[2], L. Mounier[3], M. Radeski[6], C. Winckler[7], J. Bijttebier[1], L. Lauwers[1,8], W. Verbeke[8] and S. De Graaf[1,8]

[1]ILVO, Scheldeweg 68, 9090 Melle, Belgium, [2]Univ. Copenhagen, Fac. Health & Medical Sciences, Groennegaardsvej 8, 1870 Frederiksberg, Denmark, [3]Univ. Lyon, VetAgro Sup, UMR1213 Herbivores, 69280 Marcy-L'Étoile, France, [4]Utrecht Univ., Herd Animal Health, 3508 TD Utrecht, 3508 TD Utrecht, the Netherlands, [5]SRUC, Animal & Veterinary Sciences, West Mains Road, EH9 3JG Edinburgh, United Kingdom, [6]Ss. Cyril & Methodius Univ. Skopje, Fac. Veterinary Medicine, Lazar Pop-Trajkov 5-7, 1000 Skopje, Macedonia, [7]Univ. Natural Resources & Life Sciences, Sustainable Agricultural Systems, Gregor-Mendel Straße 33, 1180 Vienna, Austria, [8]UGent, Fac. Bioscience Engineering, Coupure links 653, 9000 Ghent, Belgium; frank.tuyttens@ilvo.vlaanderen.be

Although crucial for evaluating the (social) resilience of cattle husbandry, aggregating data from welfare measures into a balanced 'welfare index' (WI) remains a challenge. We developed a user-friendly WI for dairy cattle with a minimal number of key animal-based measures, and which is simple, transparent and discriminative, and corresponds with expert opinion. The WI is calculated as the sum of the severity score (i.e. how severely a welfare problem affects cow welfare) multiplied with the herd prevalence for each measure. The selection of measures (lameness, leanness, mortality, hairless patches, lesions/swellings, SCC) and their severity scores were based on expert surveys (14-17 trained users of the Welfare Quality® cattle protocol). The prevalence of the welfare measures was assessed in 491 European herds. Experts allocated a welfare score (from 0-100) to 12 fictitious herds for which the prevalence of each welfare measure was benchmarked against all 491 herds. Quadratic models indicated a high correspondence between these subjective scores and the WI (R^2=0.91). The WI allows both a numerical (0-100) as a qualitative ('not classified' to 'excellent') evaluation of welfare. Although it is sensitive to those welfare issues that most adversely affect cattle welfare (as identified by EFSA), the WI should be accompanied with a disclaimer which lists adverse effects that cannot be detected adequately by the current selection of measures.

Session 44 Theatre 10

Does welfare improvement jeopardize emission mitigation: model calculations for Austrian dairy farms

A.C. Herzog[1], S. Hörtenhuber[2], C. Winckler[1] and W. Zollitsch[1]

[1]University of Natural Resources and Life Sciences Vienna, Department of Sustainable Agricultural Systems, Gregor-Mendel-Straße 33, 1180 Vienna, Austria, [2]Research Institute of Organic Agriculture, Seidengasse 33-35/13, 1070 Vienna, Austria; a.herzog@students.boku.ac.at

The dairy farming sector is a significant contributor to global anthropogenic greenhouse gas emissions (GHGE). Along with numerous measures to mitigate the sector's environmental impact, good health and welfare are suggested to keep emission levels low. However, potential effects of welfare improvement measures on the environmental impact of dairy farms have hardly been investigated so far. The aim of this study was therefore to model and quantify the potential effects of welfare improvement measures on the GHGE level of dairy farms in Austria, using each farm as its own reference base. Models for three dairy production systems were built, representing typical Austrian farms in different production areas (alpine, uplands, lowlands). Production intensities (6,000-8,000 kg ECM), feeding regimes as well as housing and manure management systems (tie-stalls vs free-stalls; solid vs liquid) were defined accordingly. For each model farm, total GHGE were calculated (in CO_2-eq/cow.year and CO_2-eq/kg ECM) on the farm level. Using life-cycle assessment the global warming potential (GWP_{100}) of each farm will be determined. In a second step, for each farm the effect of selected welfare improvement measures such as access to pasture or increases in space allowance and cleaning frequency will be modelled. Emission levels will be recalculated, based on the results of a detailed literature review, i.e. regarding the expected effects of the measures on parameters of emission formation. The GWP_{100} will then be recalculated and compared to the reference values. The results will provide a first insight into the quantitative effects of welfare improvement measures on the environmental impact of dairy farms. We expect the results to show, that in the case of pasture access and cleaning frequency positive effects of good welfare can outweigh potential costs of welfare improvement measures in terms of emissions. Regarding changes in spatial allowances, we expect an increase in the GWP_{100} due to emissions from construction remodelling.

Incorporating methane emissions into the breeding goals in dairy cattle under different scenarios

L. Ouatahar[1], J. Lopez-Paredes[2], N. Charfeddine[3] and O. González-Recio[2,4]
[1]Universidad Politécnica de Valencia, Camino de Vera s/n, 46022 Valencia, Spain, [2]Universidad Politécnica de Madrid, Ciudad Universitaria s/n, 28040 Madrid, Spain, [3]Spanish Holstein Association (CONAFE), Ctra. de Andalucía km 23600 Valdemoro, 28340 Madrid, Spain, [4]INIA, Crta. de la Coruña km 7.5, 28040 Madrid, Spain; latifa.ouatahar@gmail.com

One of the main challenges of dairy farming in Europe is to obtain economically a profitable and sustainable production with lower greenhouse gas emissions (GHG). Livestock is responsible for 14% of GHG emissions. Methane (CH_4) from enteric fermentation represents the main source of these emissions, which also means a loss of energy for the cow and a waste of money for the producer. However, at present, GHG emissions are not included into the breeding goals in any livestock specie. This study aims to include enteric CH_4 emissions in the breeding goal of Spanish dairy cattle and to evaluate the expected genetic response of traits in the selection objective under different scenarios in the Spain: (1) current situation as benchmark (without applying an economic value on CH_4 emissions), (2) carbon tax on CH_4 emissions, (3) quota of CH_4 emissions and (4) CH_4 as an energy loss. A bio-economic model was developed to estimate economic values of milk production traits and CH_4 emissions. The estimated economic values were 0.01, 1.94, and 4.48 (€/kg) for milk yield, fat and protein in scenario 1. For CH_4, the economic values were -1.21, -6.44 and -0.45 (€ per kg CH_4 per lactation) for scenarios 2, 3 and 4, respectively. The expected genetic response for each trait included in the genetic selection index under the four scenarios, showed that placing a negative economic value on CH_4, leads to a reduction of CH_4 emissions (kg/year) in scenarios 2, 3 and 4 by -0.83, -4.26, and -0.33, respectively, while in scenario 1 (reference) the response was 0.06. On the other hand, the benefit of using the total index (€/year) decreases when an economic value is attributed to CH_4 by -0.4, -8.7 and -0.1% for scenarios 2, 3 and 4 respectively, compared to scenario 1. This study showed that there is a potential in mitigating CH_4 emissions by genetic selection through inclusion of CH_4 in selection objectives of dairy cattle, at expenses of lower genetic gain on overall profit.

Technical and economic impacts of concentrates according to stage and row of lactation of dairy cows

V. Brocard[1], E. Tranvoiz[2] and B. Portier[2]
[1]Institut de l'Elevage, Monvoisin BP85225, 35652 Le Rheu Cdx, France, [2]Chambre d'Agriculture de Bretagne, 2 allée St Guénolé, CS 26032, 29322 Quimper Cdx, France; valerie.brocard@idele.fr

In a high volatility context both for milk and crop prices in western Europe, dairy farmers need to be more flexible in terms of herd management. In the same time, they need to secure their incomes. One of the possible lever to reach this objective is to change feeding practices and especially to decrease concentrate levels, as this part of the diet is the most expensive one. Thus two experimental trials were led on the effects of concentrate supplementation during 3 years at the experimental station of Trévarez, Brittany, France. They aimed at assessing the technical and economic interest of bringing extra production concentrates (4 kg during 3 months) either in early or mid lactation of Holstein cows. One of the 3 years-experiments was led on a grazing system, the other on a maize silage based system. Dairy performances, health status, reproduction performances and animal condition score were assessed. The results show that the concentrate efficiency remains close to 0.5 kg of extra milk produced milk by extra kg of concentrate with little effect on milk solids. This efficiency does not statistically differ according to stage or row of lactation, or type of forage base. Moreover, the concentrate supplementation does not affect the health, reproduction performances and body condition scores. The economic analysis shows that the extra amount of production concentrate is not improving the farmers' incomes in the actual context of low milk prices. From the French farmer's point of view, only the milk produced from home grown forages (balanced with the right amount of protein cakes in case of maize diets) has a positive margin over feeding cost and should be produced.

Session 44 — Poster 13

Economic assessment of mountain dairy farms considering concentrate supplementation and breed

S. Kühl, L. Flach, C. Lambertz and M. Gauly
Free University of Bolzano, Universitätsplatz 1, 39100 Bolzano, Italy; sarah.kuehl@unibz.it

Small-scale dairy farms contribute valuably to the conservation of traditional landscapes in the Alps. Nevertheless, achieving economic success is one of their main challenges. Especially extensive farms are often not able to make a living from their farm work what increases the risk of farm abandonment. The aim of this study is to analyse the financial viability of different farm types (intensive vs extensive feeding and Tyrolean Grey [TG] vs Brown Swiss [BS] cattle) of mountain dairy farms in South Tyrol (Northern Italy). The four groups 'TG extensive' [TG_E] (n=10), 'BS extensive' [BS_E] (n=5), 'TG intensive' [TG_I] (n=12) and 'BS intensive' [BS_I] (n=17) have been surveyed with respect to their profitability, farm structure, husbandry and feeding system. The results show that the profit without subsidies and labor costs per cow and year is similar between the two extensive groups (233.3 vs 290.1€; $P \geq 0.05$) and the two intensive groups (820.3 vs 821.8€; $P \geq 0.05$). However, the standard deviation is very high why a cluster segmentation (Ward-Clustering) based on the farms profit without subsidies, costs for concentrates per cow and grazing days was conducted to get a better description of profitability. Four clusters were identified whereas farms in cluster 1 show a negative income per cow and year (-805.1€, n=5), cluster 2 and 3 farms have a low to medium income (103.2€, n=15; 972.0€, n=18) and cluster 4 farms have a high income (2,096.6€, n=6). In the first and the latter cluster, mostly BS_I farms are included showing that intensive feeding can lead to very high and very low income. Extensive farms are mainly in cluster 2 and 3. Cluster 4 further contains farms with the highest percentage of loose housing systems (66.7%) but no farm in this cluster keep their cows on pasture. Summing up, the results show that there is a high variation within the mountain dairy farms and that a positive income not only depend on the breed or feeding system. However, intensive (mainly BS_I) farms have the chance to achieve higher profits whereas extensive farms more often are depending on subsidies to be viable in the current situation.

Session 44 — Poster 14

Breeding values for live weight of dairy cattle

C. Fuerst[1], B. Fuerst-Waltl[2], C. Pfeiffer[2] and C. Egger-Danner[1]
[1]ZuchtData EDV-Dienstleistungen GmbH, Dresdner Strasse 89/19, 1200 Vienna, Austria, [2]University of Natural Resources and Life Sciences Vienna (BOKU), Gregor-Mendel-Strasse 33, 1180 Vienna, Austria; fuerst@zuchtdata.at

This study is part of a larger project whose overall goal is to evaluate the possibilities for genetic improvement of efficiency in Austrian dairy cattle. With regard to efficiency, live weight is a trait that might be considered in breeding programs. In the year 2014 a one-year of extensive data collection was carried out. In addition to routinely recorded data (e.g. milk yield), data of novel traits such as body condition score, live weight, body measurements as well as individual feeding information were recorded. The specific objective of this study was to analyse whether the recorded data combined with results from linear scoring are sufficient to develop a genetic evaluation for live weight. In a first step, genetic parameters were estimated for live weight and body measurements and the breeds Fleckvieh (FV; 3,329 cows; 20,905 records) and Brown Swiss (BS; 1,428 cows, 9,071 records). The fixed effects region-year-season, lactation-age of calving, lactation-days since calving, herd-year and scorer-year as well as the random permanent environmental and genetic effect of the animal were included. The heritabilities for live weight were 0.42 (FV) and 0.43 (BS). Highest genetic correlations to live weight were found for chest and belly girth ($r>0.80$). In a multivariate genetic evaluation test-run, routine linear conformation traits since the year 2000 were also considered; resulting data sets comprised records from 170,091 FV and 62,560 BS cows. Genetic trends for live weight were more or less stable. For both breeds, the breeding value (BV) correlation to the total merit index (TMI) was about 0. Hence no marked change should be expected for live weight when selecting for TMI. BV correlations to the dairy index were also close to 0 while those to the beef index were notably positive. In tendency, fitness related traits were negatively correlated to live weight, in particular longevity, direct calving ease and vitality of young stock. The development of a genetic evaluation is feasible. All possible data sources should however be utilized, e.g. results from linear scoring but also weights from AMS systems or auction sales.

Session 44 Poster 15

Dairy cow behaviour and stall cleanliness when using flexible vs conventional free stall dividers

E. Grotelüschen[1], K. Geburt[1] and U. König V. Borstel[2]
[1]University of Göttingen, Albrecht-Thaer-Weg 7, 37075 Göttingen, Germany, [2]University of Giessen, Leihgesterner Weg 52, 35392 Giessen, Germany; uta.koenig@agrar.uni-giessen.de

Conventional steel free stall dividers are a common cause of technopathies in dairy cattle and a potential solution to alleviate these problems might be flexible free stall dividers. The aim of the present study was to compare cow's behaviour and stall cleanliness when given the choice between stalls with flexible dividers and steel dividers. For this purpose, video recordings were taken from a total of 8 head-to-head free-stalls chosen as focus stalls within in a compartment of a free stall barn for 46 dairy cows (stall: animal ratio= 1:1). Four of the focus stalls were equipped after a control period of 4 days with flexible stall dividers (FLEX) while the remaining stalls remained equipped with conventional (STEEL) free stall dividers. Generalized (binary and Poisson distributed data) linear mixed model (normally distributed data) analysis accounting for repeated observations per stall revealed that, while there were no significant differences ($P>0.1$) between stalls before installation of FLEX dividers (control period), after an acclimatization period of 6 weeks, cows had longer visits and lying bouts (47.1±2.2 vs 40.5±1.9 min/ h visiting) in FLEX rather than STEEL stalls ($P<0.05$). There were no differences in frequencies of different head postures ($P>0.1$), but cows were found more often lying in postures with three as well as four feet stretched out in FLEX compared to STEEL (both $P<0.05$). Furthermore, cows were faster to get up in FLEX compared to STEEL stalls (4.8±0.08 vs 5.3±0.09 sec; $P<0.05$), suggesting that they were less reluctant to make contact with the FLEX dividers. Surprisingly, cows lay more straight in FLEX rather than STEEL (mean angle of lumbar spine relative to dividers ± SD: 10.6±8.9° vs 15.5±10.8°, t-test: $P<0.05$), but there were no differences in scores for cleanliness of stalls ($P>0.1$). However, the probability of attempts to displace a cow in a stall by another cow were higher with FLEX rather than STEEL ($P<0.05$). Overall, results suggest that flexible free-stall dividers might be a good alternative to conventional steel dividers, in particular with regard to cow comfort when lying and getting up.

Session 44 Poster 16

Environmental impacts of sulphuric acid and added biochar on cattle slurry – a life cycle assessment

C. Miranda[1], A.S. Soares[1], A.C. Coelho[2], H. Trindade[1] and C.A. Teixeira[1]
[1]CITAB, Universidade de Trás-os-Montes e Alto Douro, Quinta de Prados, 5000-801 Vila Real, Portugal, [2]CECAV, Universidade de Trás-os-Montes e Alto Douro, Quinta de Prados, 5000-801 Vila Real, Portugal; carlisabelmi@hotmail.com

The growth of livestock production worldwide has contributed to increase the environmental problems from animal slurry and its management. The gaseous emissions to the environment may occur during slurry storage and field application. Although slurry treatment technologies are being developed to reduce the overall environmental impacts of slurry, it is important to analyse these impacts in other life cycle stages or impact categories. The aim of this study was to investigate the environmental effects of biochar and sulphuric acid addition and the combination of them into cattle slurry when compared to the conventional slurry management practices (untreated), during storage. Three relevant and commonly environmental impact categories were quantified using the Life cycle assessment (LCA): acidification potential (AP), eutrophication potential (EP) and global warming potential (GWP). This analysis was carried according to ISO 14040/44 using GaBi software and CML 2001 impact category indicators. Comparisons were based on a functional unit of 1000 kg of cattle-slurry. The preliminary results of environmental analysis demonstrated that the treated scenarios have the lowest impact in all categories compared to the reference scenario in which slurry was not treated. A major impact of storage was observed on GWP. However, the acidification without the addition of biochar showed a reduction of impact in all categories on the comparison with the reference, biochar and biochar and acid combination scenarios. In addition, the acidification of the dairy cattle slurry showed a reduction of CH_4, CO_2 and NH_3 cumulative emissions during storage compared to other treatments and control. This process allowed reducing 81% of the EP, 89% of the AP and 90% of the GWP. In conclusion, the LCA can be an important tool of decision to identify the environmental consequences around slurry management, during its processing and storage.

Session 44 Poster 17

Is pre-milking teat disinfection worthwhile if you have low herd SCC?

D.E. Gleeson and B. O'Brien

Teagasc, Animal & Grassland Research and Innovation Centre, Moorepark, Fermoy, cork, Ireland; david.gleeson@teagasc.ie

Pre-milking teat disinfection has been shown to reduce bacterial numbers on teat skin and to be most effective against infections caused by environmental pathogens such *Streptococcus uberis*. The most prevalent mastitis-causing pathogen identified in quarter milk samples in Irish dairy herds is *Staphylococcus aureus*, accounting for 60% of all mastitis cases. Pre-milking teat disinfection is being adopted by farmers in Ireland (14%), particularly where there are increased herd infection rates and raised somatic cell count (SCC). Pre-milking teat disinfection may have little benefit when *S. aureus* is the predominant pathogen and when herd SCC is $<200\times10^3$ cells/ml. The objective of this study was to investigate if teat disinfectant applied pre-milking, would have any benefit when the herd SCC was $<200\times10^3$ cells/ml. Statistical analysis of data was performed using SAS software. Teat disinfectant was applied to the left front and right hind teats of all cows in two research herds (A=97 cows: B=166 cows) and the right front and left hind teats received no disinfectant prior to milking over a complete lactation. Quarter milk samples were obtained on 5 occasions for microbiological analysis and all clinical cases were recorded during the lactation. Teat skin from a sample of 20 cows was swabbed for *Staphylococcus* and *Streptococcus* bacteria on 3 occasions during lactation. Pre-milking teat disinfection had no significant impact on quarter SCC and new infection rates ($P>0.05$). The mean SCC was 168×10^3 cells/ml and 169×10^3 cells/ml for disinfected teats and non-disinfected teats, respectively, and there was no differences in SCC observed between farms (A=161×10^3 cells/ml: B=177×10^3 cells/ml). The number of clinical cases was lower for disinfected teats (n=18) compared to the non-disinfected teats (n=26). However, the number of sub-clinical incidences was higher for disinfected teats (n=18) compared to non-disinfected teats (n=14). Bacterial counts were lower on disinfected teats compared to non-disinfected teats prior to cluster application ($P<0.001$). In conclusion, routine application of pre-milking teat disinfectant in pasture-grazed herds is unlikely to be of benefit where herd SCC is below 200×10^3 cells/ml.

Session 44 Poster 18

Influencing factors on hair cortisol concentrations in pigs and cattle

S. Heimbürge, E. Kanitz, A. Tuchscherer and W. Otten

Leibniz Institute for Farm Animal Biology, Wilhelm-Stahl-Allee 2, 18196 Dummerstorf, Germany; heimbuerge@fbn-dummerstorf.de

There is growing interest in the use of hair cortisol concentration (HCC) as a potential indicator for long-term stress or increased hypothalamic-pituitary-adrenal (HPA) axis activity in farm animals. Before evaluating HCC as an indicator of chronic stress, however, it is necessary to elucidate animal-related sources of variation in each species of interest. Therefore, the aim of this study was to investigate age, sex, reproductive cycle, hair colour and body region as potential sources of variation in HCC in pigs and cattle. For this aim, hair samples were collected from German Landrace pigs, Holstein cattle and crossbreeds at different ages from birth to adulthood and both sexes. In addition, the influence of reproductive cycle and different body regions on HCC was investigated in pigs, and the effect of hair colour was examined in Holstein cattle. Hair samples were taken by electric clippers from the neck, caudodorsal or pelvic region and tail tip. Samples were washed twice with isopropanol, grounded with a ball mill and cortisol was extracted with methanol. HCC was then analysed with a commercially available enzyme immunoassay. Data were analysed by multiple comparisons of the least square means by Tukey Kramer tests using MIXED procedures in SAS/STAT software. Results show higher HCC in calves compared to young cattle, heifers and cows of the Holstein race ($P<0.001$). Likewise, 2-week-old piglets had higher HCC than pigs aged 10 or 27 weeks or sows ($P<0.001$). Sex had no effect on HCC in pigs or cattle. In addition, no difference in HCC was found between black or white hair from Holstein cattle. Furthermore, there were significant differences in HCC between body regions with higher HCC in tail hair compared with hair from the neck or pelvic region of sows ($P<0.001$). The reproductive cycle seems to be an additional source of variation with increased HCC in samples representing late gestation, birth and lactation. In conclusion, our findings indicate a substantial variation of HCC depending on age, reproductive cycle and body region in pigs and cattle, which have to be considered for hair sampling procedures and interpretation of HCC as an indicator for long-term stress.

An on-farm algorithm to guide selective dry cow therapy

A.K. Vasquez[1], C. Foditsch[1], M. Wieland[1], R.A. Lynch[2], P.D. Virkler[1], S. Eicker[3] and D.V. Nydam[1]
[1]Cornell University, Dept. of Population Medicine and Diagnostic Sciences, 144 East Ave., Ithaca, NY 14853, USA, [2]Department of Animal Science, Cornell University, Ithaca, NY, USA, [3]Valley Ag. Software, Tulare, CA, USA; amy.kristin.vasquez@gmail.com

A selective-dry-cow therapy algorithm was evaluated for microbiological cure risk, new infection risk, culling and occurrence of clinical mastitis before 30 DIM, and 1st-test milk yield and linear score (LS) in a randomized on-farm clinical trial including 611 dairy cows. An algorithm using DC305 and test-day data was used to identify cows as 'low risk' (cows that likely will not benefit from dry cow antibiotics) or 'high risk' (cows that will benefit). Low risk cows were those that had all of: ≤200k SCC at last test, an average SCC ≤200k on the last 3 tests, no signs of mastitis at dry-off, and have not had more than 1 clinical mastitis event in the current lactation. Low risk cows were randomly assigned to receive either intramammary antibiotics and external teat sealant (DCT) or teat sealant only (TS). Quarter milk samples were obtained from cows at dry-off and 1-7 DIM to determine cure and new infection at the quarter level. Samples from high risk cows were used to determine positive and negative predictive values (PPV, NPV) of the algorithm. Mastitis events, milk production, LS, and culling data were retrieved from DC305. Data analysis was performed in SAS 9.4: binary outcomes were analysed with logistic regression models while continuous outcomes were compared with linear regression models. PPV and NPV were each 70%. Of cultures eligible for cure analysis (n=171), 97.7% of DCT cured, while 93.2% of TS did (least squares means). Odds ratio of cure TS:DCT=0.32; 95%CI: 0.11-0.96). Positive cultures for coagulase negative *Staphylococcus* (CNS) at dry-off accounted for 95% of the non-cures (n=19/20). Risk ratio for new infection was 1.29 for TS:DCT (95%CI: 0.90-1.85). CNS accounted for 58% of new infections (n=77/134). There were no statistical effects of treatment group on culling (DCT n=18; TS n=15), clinical mastitis (DCT n=9; TS n=5), milk (kg) (DCT=40.5; TS=41.2), or LS (DCT=2.5; TS=2.7). The impact of CNS to increased new infection risk and decreased bacteriological cure needs to be further investigated. These results suggest that the employed algorithm decreased dry cow antibiotic use by approximately 60% without adversely impacting production outcomes.

Testing trace elements in bulk milk as a tool in providing the optimal amount to dairy cattle

J. Muskens and G.H. Counotte
GD Animal Health, Arnsbergstraat 7, 7418 EZ Deventer, the Netherlands; j.muskens@gddiergezondheid.nl

Since 2013, bulk milk of dairy cattle can be tested in the Netherlands for the contents of different trace elements. In the first year, selenium (Se) and iodine (I) were tested in bulk milk samples of about 2600 dairy herds. In 2014 copper (Cu) and zinc (Zn) were added to the testing protocol. Since 2018, bulk milk is also tested for the mineral phosphorus (P). The number of herds, of which bulk milk samples were tested, increased to more than 3,000, which is about 20% of all Dutch dairy farmers. In this voluntary program, farmers can choose for testing their bulk milk two, four or six times per year. The bulk milk is collected automatically, the farmer does not need to send samples to the laboratory. In all these schedules the bulk milk is tested at least once during winter and summer time. This is important, because farmers can have different feeding regimes during summer (pasture or not) and winter, possibly with different mineral contents of the total ration. The results of testing Se, I and Zn are described as: shortage, low-normal, normal, high-normal and possible excess. For Cu the results are described as shortage, low-normal and no indication for shortage. The test result of P is expressed as g/l, and this amount is compared with the average of all tested samples and the amount of P that is the basis for calculating the P content of the ration of dairy cattle. These results are sent to the farmer and his veterinarian. If the results are not-normal, they can be used to improve the mineral status of the herd (feeding more, less and/or other mineral(s)). And 2-6 months later, by the next bulk milk testing, automatically the farmer gets a good indication if changes in the mineral supply have led to improvement in the bulk milk test results. The results of testing Cu, Zn, Se, I and P in bulk milk samples will be discussed, like percentages of herds with divergent test results for the five trace elements and differences of test results during winter and summer time, and possible reasons for these differences.

Session 45 Theatre 3

Effects of temperament on growth, physiology, and puberty attainment in *Bos indicus* beef heifers

R.F. Cooke[1] and J.L.M. Vasconcelos[2]
[1]Texas A&M University, Department of Animal Science, 2471 TAMU, College Station, TX 77845, USA, [2]São Paulo State University (UNESP), School of Veterinary Medicine and Animal Science, Fazenda Experimental Lageado, Botucatu, SP 18610, Brazil; reinaldocooke@tamu.edu

This experiment evaluated the effects of temperament on growth, plasma cortisol concentrations, and puberty attainment in *Bos indicus* beef heifers. A total of 170 Nelore heifers, weaned 4 months prior to the beginning of this experiment (day 0 to 91), were managed in 2 groups of 82 and 88 heifers each (initial weight=238±2 kg, initial age=369±1 days across groups). Heifer temperament was evaluated via exit velocity on day 0. Individual exit score was calculated within each group by dividing exit velocity into quintiles and assigning heifers with a score from 1 to 5 (1 = slowest; 5 = fastest heifer). Heifers were classified according to exit score as adequate (ADQ, n=96; exit score ≤3) or excitable temperament (EXC, n=74; exit score >3). Heifer weight, body condition score (BCS), and blood samples were obtained on days 0, 31, 60, and 91. Heifer exit velocity and score were recorded again on days 31, 60, and 91. Ovarian transrectal ultrasonography was performed on days 0 and 10, 31 and 41, 60 and 70, 81 and 91 for puberty evaluation. Heifer was declared pubertal at the first 10-day period in which a corpus luteum was detected. Exit velocity and exit score obtained on day 0 were positively correlated ($r \geq 0.64$, $P<0.01$) with evaluations on days 31, 60, and 91. During the experiment, ADQ heifers had greater ($P<0.01$) mean BCS and weight gain, and less ($P<0.01$) mean plasma cortisol concentration compared with EXC heifers. Temperament × day interactions were detected ($P<0.01$) for exit velocity and exit score, which were always greater ($P<0.01$) in EXC vs ADQ heifers. A temperament × day interaction was also detected ($P=0.03$) for puberty attainment, which was delayed in EXC vs ADQ heifers. At the end of the experiment, a greater ($P<0.01$) proportion of ADQ heifers were pubertal compared with EXC heifers. In summary, *B. indicus* heifers classified as EXC had reduced growth, increased plasma cortisol concentrations, and hindered puberty attainment compared to ADQ heifers.

Session 45 Theatre 4

Does season alter dairy cows' preference for pasture and their behavioural pattern?

K.S. Stanzel[1], J. Ingwersen[1], C. Lambertz[2], I. Traulsen[1], D. Albers[3] and M. Gauly[2]
[1]University of Göttingen, Department of Animal Sciences, Livestock Systems, Albrecht-Thaer-Weg 3, 37075 Göttingen, Germany, [2]Free University of Bozen, piazza Università 5, 39100 Bolzano, Italy, [3]Chamber of Agriculture of Lower Saxony, Mars-la-Tour-Straße 6, 26121 Oldenburg, Germany; katharina.stanzel@agr.uni-goettingen.de

Because access to pasture for dairy cows is perceived positively by consumers, free-choice studies on the preference for pasture and seasonal influences on changes of their behaviour have been conducted, but only focused on single factors. Therefore, the aim of the study was to estimate the impact of various parameters on cows' preference, their locomotive (LB) and ingestive behaviour (IB) during the pasture season. In 2016, the behaviour of 12 high-yielding Holstein Friesian cows on a farm in Germany was assessed during 3 periods of 18 days each; Mai/June (P1), July/August (P2) and September (P3). Cows could choose freely to stay either outdoors (OD) or indoors (ID) for 13 h per day (free choice time, FCT). Transits were registered automatically and the time spent ID and OD was calculated. Climatic parameters (ID+OD: temperature, relative humidity; OD: solar radiation, wind speed) were measured to calculate the temperature humidity index (THI) and heat load index (HLI). The individual IB and LB was recorded continually by fitting cows with RumiWatch® noseband sensors and pedometers. Data were checked for normal distribution and, if necessary, transformed and statistically analysed with a mixed model. LB during FCT was spend lying (51.4%), standing (43.4%) and walking (5.2%). IB was differentiated into eating (36.3%), ruminating (36.0%) and other head movements (27.7%). During FCT, cows spend on average 78.9% of the time OD. The duration of FCT spend OD increased (<0.05) over the pasture season from 71% (P1) to 85.4% (P2) and 87.9% (P3). Daytime had an impact on the individual activities, e.g. rumination time was higher at night (<0.001). The mean HLI during FCT had an influence on rumination time (<0.0001). The results depict that cows could convey their natural behaviour, i.e. they preferred the pasture over the whole season and spend most of the FCT resting, eating and ruminating. Yet, seasonal, diurnal parameters and their interactions had an effect on the cows' behaviour.

Developing sustainable permanent grassland systems and policies across Europe: SUPER-*G*

J.P. Newell Price[1] and J.R. Williams[2]
[1]ADAS Gleadthorpe, Meden Vale, Mansfield, Nottinghamshire NG20 9PD, United Kingdom, [2]ADAS Boxworth, Cambridge, Cambridgeshire CB23 4NN, United Kingdom; paul.newell-price@adas.co.uk

The efficient management of permanent grasslands (PG) is key to the productivity and profitability of many livestock grazing systems. The persistence of these systems in terms of their economic, environmental and social sustainability is also dependent on the policies that support and underpin not only the maintenance of livestock farming skills, but also the management practices that deliver public goods and services to society. The overall objective of the SUPER-*G* project is to identify and define sustainable PG systems and policies that will be effective in optimising productivity, whilst supporting biodiversity and delivering a number of other ecosystem services (ES), such as climate regulation, water quality, mediation of water flows and erosion control; and help retain permanent grasslands and the rural communities that depend on them. The SUPER-*G* project aims to achieve: (1) better understanding of the importance and functioning of PG; (2) benchmarking of PG performance across Europe (within Mediterranean, Atlantic, Continental, Alpine, Pannonian and Boreal biogeographic regions); (3) development of integrated approaches for profitable and sustainable PG management; and (4) development of tools and policy mechanisms to support the maintenance and sustainable management of PG. This paper will outline the multi-actor and transdisciplinary approach used in the project and provide examples of farm decision support tools, including software and smart phone applications (apps) that are currently available in some parts of Europe; and on which the project will build to provide farm business efficiency support to livestock grazing farmers.

Access to grazing for high yielding dairy cows: a novel experience in Northern Italy

L. Leso and M. Barbari
University of Firenze, Depart. of Agricultural, Food and Forestry Systems (GESAAF), Via San Bonaventura 13, 50145 Firenze, Italy; lorenzo.leso@unifi.it

In recent years, lack of pasture access revealed to be a major area of concerns for consumers regarding dairy farming. Despite that, the practice of grazing dairy cattle has been almost completely abandoned in Italy. The objective of the present study was to evaluate the effects of grazing for a limited part of the day on milk production, milk composition and cattle welfare. The study was conducted during spring 2017 in a commercial dairy farm located in the Po Valley (45.17 N, 10.75 E). A total of 133 lactating Holstein cows were assigned to either continuous housing in a free stall barn (CONF) or housing in a free-stall barn and grazing a ryegrass-based pasture 6 h/day after the morning milking (PAST). Cows in both treatments were fed the same mixed ration *ad libitum* inside the barn. Intake of mixed ration and pasture were measured weekly at group level. Individual milk production and composition were measured monthly. Cows were scored for body condition (BCS; 1=thin, 5=fat) monthly and for lameness (1=not lame, 5=severely lame) at the beginning and at the end of the experiment. Data were analysed with linear mixed models for repeated measures. Cows in the PAST group consumed 3.41 kg DM of herbage but had lower total DMI compared with cows in CONF (22.34 vs 24.99 kg DM, $P<0.01$). Milk production of primiparous cows did not differ between treatments (32.11 vs 31.61 kg/day, $P=0.77$) while multiparous cows in PAST produced significantly lower milk than in CONF (33.85 vs 37.65 kg/day, $P<0.01$). The PAST cows tended to have lower BCS compared with CONF counterparts (3.28 vs 3.40, $P=0.07$). Part-time grazing did not affect SCC significantly (199 vs 275 cells × 1000/ml) but reduced milk fat percentage (4.03 vs 4.29%, $P<0.05$). Overall, milk protein content did not differ between treatments (3.35 vs 3.42%, $P=0.13$) but a significant reduction was observed in PAST during the course of the study. At the end of the experimental period, cows in the PAST group had better gait than those in CONF (1.40 vs 1.62, $P<0.05$). Results indicate that proving access to grazing has the potential to improve cattle welfare but may pose some challenges in maintaining adequate milk quality and body condition.

Reliability of breeding values for dry matter intake by adding data from additional research farms

G.C.B. Schopen[1], Y. De Haas[1], J.J. Vosman[2], G. De Jong[2] and R.F. Veerkamp[1]
[1]Wageningen University & Research, Animal Breeding and Genomics, Droevendaalsesteeg 1, 6708 PB Wageningen, the Netherlands, [2]CRV, Animal Evaluation Unit, Wassenaarweg 20, 6843 NW Arnhem, the Netherlands; ghyslaine.schopen@wur.nl

Dairy feeding cost are above 50% of all costs. Breeding for efficient cows, i.e. cows that efficiently use their dry matter intake data to produce milk, is very important to reduce the costs. The Netherlands developed a genetic evaluation for dry matter intake (DMI) in dairy cows to obtain breeding values for DMI to be used for selecting the most efficient cows. This study shows the reliability of genomic breeding values (gEBV) for DMI after combining data from experimental farms and Dutch feeding companies. About 129,000 weekly DMI records with ~7,000 lactations of ~4,300 Holstein-Friesian cows born between 1979 and 2016 were available of which ~1,500 cows had been genotyped. DMI data consisted of historic data from experimental farms of Wageningen University & Research (WUR), Trouw Nutrition and Schothorst Feed Research. Using a test-day model adjusted for lactation stage, gEBV for DMI were estimated using the inverse of the combined pedigree and genomic relationship matrix (H-1). For bulls without daughter information for DMI, reliabilities of gEBV for DMI averaged 33% and ranged from 13 to 54%. For bulls with daughter information for DMI, reliabilities of gEBV for DMI averaged 50% and ranged from 24 to 90%. Combining gEBV with predictor traits, like milk production traits and live weight, the reliability of gEBV for DMI for bulls without daughter information for DMI averaged 66% and ranged from 59 to 73%. For bulls with daughter information for DMI, the reliability of gEBV for DMI combined with predictor traits averaged 72% and ranged from 64 to 91%. At the moment, ~5,000 additional weekly DMI records with ~700 lactations of ~500 cows with genotypes from a WUR experimental farm and Dairy Campus are added to the genetic evaluation for DMI. These numbers increase even further when more DMI data from Trouw Nutrition, Schothorst Feed Research, Flemish feeding companies and a commercial farm equipped with the roughage intake control (RIC) system becomes available in spring 2018. Higher reliabilities of breeding values for DMI are expected.

The economic value of saved feed in dairy breeding goals

R.S. Stephansen[1], J. Lassen[2], J.F. Ettema[3] and M. Kargo[1,4]
[1]SEGES, Agro Food Park 15, 8200 Aarhus N, Denmark, [2]VikingGenetics, Ebeltoftvej 16, 8960 Randers SØ, Denmark, [3]SimHerd A/S, Asmildklostervej 11, 8800 Viborg, Denmark, [4]Aarhus University, Blichers Allé 20, 8830 Tjele, Denmark; rass@seges.dk

Implementation of feed efficiency (FE) into dairy breeding schemes is of great importance and value. Focus the past decades has been on creating phenotypic registrations and on genetic evaluation of such a trait; however, little research has been carried out on deriving the economic value (EV) for the trait(s). Saved feed is composed of two components: (1) Feed saved due to lower maintenance/body weight and (2) Feed saved through improved efficiency. The objective of this study was to derive the EV of FE at different feed efficiency levels and within two different lactation periods. The simulation tool SimHerd (www.simherd.com) was used to simulate feed efficiency. We defined FE as Residual Feed Intake (RFI), calculated as the difference between actual and predicted feed intake, to avoid double counting in the breeding goal. Actual feed intake was derived from simulated energy requirements (milk production, maintenance, etc.) divided by a variable feed efficiency. Predicted feed intake was derived by using RFI regression coefficients from Li *et al.* Predicted feed intake was calculated as one or two traits within lactation, to simulate the effect of considering mobilization in relation to RFI. The SimHerd simulations had four scenarios: (1) one RFI trait – no variance for RFI, (2) one RFI trait – variance introduced, (3) two RFI traits – no variance and (4) two RFI traits – variance introduced. Variance was introduced on RFI in scenario 2 and 4 to meet reported phenotypic levels by Li *et al.* Within each scenario, daily feed intake was differed one or two kg of dry matter (DM) from a default scenario, by changing feed efficiency in SimHerd. The economic value of RFI was derived by comparing total farm profit, calculated from SimHerd output, within scenarios. The economic value of RFI varied from 0.16 to 0.18 €/kg DM across all scenario simulations. There was not an effect of feed efficiency level, introduced variance and partial RFI ($P > 0.05$) within scenario. These results suggest the economic value of feed efficiency corresponds to the applied feed price (0.17 €/kg DM).

Selection for feed efficiency in Holstein cows based on data from the Efficient Cow project

A. Köck[1], M. Ledinek[2], L. Gruber[3], F. Steininger[1], B. Fuerst-Waltl[2] and C. Egger-Danner[1]
[1]ZuchtData EDV-Dienstleistungen GmbH, Dresdner Str. 89, 1200 Vienna, Austria, [2]University of Natural Resources and Life Sciences, Department of Sustainable Agricultural Systems, Gregor-Mendel-Str. 33, 1180 Vienna, Austria, [3]Agricultural Research and Education Centre, Raumberg 38, 8952 Irdning-Donnersbachtal, Austria; koeck@zuchtdata.at

This study was part of the project 'Efficient Cow' whose overall objective was to evaluate the possibilities for genetic improvement of efficiency in Austrian dairy cattle. In the year 2014 a one-year data collection was carried out. Data of approximately 5,400 cows, i.e. 3,100 Fleckvieh (dual purpose Simmental), 1,300 Brown Swiss, 1,000 Holstein kept on 167 farms were recorded. In addition to routinely recorded data (e.g. milk yield), data on novel phenotypes like body weight, body condition score, lameness, claw health, subclinical ketosis and data about feed quality and feed intake was collected. The specific objective of this study was to estimate genetic parameters for feed efficiency traits and to investigate their relationships with fertility and health in Holstein cows. The following feed efficiency traits were considered: ratio of milk output to metabolic body weight (ECM/BW0.75), ratio of milk output to dry matter intake (ECM/DMI) and ratio of milk output to total energy intake (ECM/INEL). Heritabilities of feed efficiency traits were moderate and ranged from 0.11 for ECM/INEL to 0.14 for ECM/BW0.75. More efficient cows were found to have a higher milk yield, lower body weight, slightly higher dry matter intake and lower body condition score. Cows with a higher efficiency had a higher fat-protein-ratio, a longer calving interval and a higher frequency of fertility disorders. Higher efficiency was, however, associated with a lower somatic cell count, less lameness and a lower culling rate. Overall, cows with a medium efficiency combine both, a high milk yield with good fertility and health.

Feed from the sea: a move towards sustainable ruminant livestock production using brown seaweed

M. Campbell[1], A. Foskolos[2] and K. Theodoridou[1]
[1]Institute for Global Food Security, Queen's University Belfast, 13 Stranmillis Road, BT9 5AF Northern Ireland, United Kingdom, [2]Institute of Biological, Environmental and Rural Sciences, Aberystwyth University, Penglais, Ceredigion, SY23 3DA Aberystwyth, United Kingdom; mcampbell105@qub.ac.uk

Seaweed could provide an alternative, locally available, animal feedstock that is currently underexploited in the UK. According to FAO, livestock are accountable for 14.5% of global anthropogenic greenhouse gas emissions; methane accounts for approximately 44% of this. It was shown previously that adding seaweed to the ruminant diet could reduce methane emissions associated with rumen fermentation. Brown seaweeds contain valuable bioactive compounds known as phlorotannins (PT) which are linked to the methane abating behaviour of seaweed in the rumen. The current study aims to investigate the nutritional value of brown seaweeds and their effect on methane production. Two brown seaweed species, *Fucus vesiculosus* (FVS) and *Saccharina latissimi* (SAC), were collected in spring 2017 from Bangor, Co. Down, N. Ireland. Nitrogen (N) content was analysed by Dumas method and crude protein (CP) was estimated using Nx4.17. The Folin Ciocalteu method was used for PT content determination. Gas production was assessed using the *in vitro* gas production technique at 3, 6, 12, 24, 48, 72, 96 and 120 h for the determination of cumulative gas production; Lucerne, a non-PT containing plant, was used as a control. Statistical analyses were performed using JMP®, Version 13.2.1. FVS and SAC had a CP content of 8.1 and 11.0% DM, respectively. The *in vitro* gas production results showed significant differences in total methane production between FVS and SAC ($P<0.0001$); compared to Lucerne, FVS and SAC reduced methane by 95 and 60%, respectively. The PT analysis revealed that FVS had a significantly higher PT content (3.17%DM) compared to SAC (0.368%DM). Although brown seaweeds are lower in CP compared to conventional feeds (40-50% CP in soybean), the environmental benefits should be considered. SAC is a species with high potential as a sustainable feed in ruminant nutrition. Further work is required to establish optimum inclusion rates in the diet.

Validation of methane measurements in dairy cows obtained from two non-invasive infrared analysers

M. Sypniewski[1], M. Pszczoła[1], T. Strabel[1], A. Cieślak[2] and M. Schumacher-Strabel[2]
[1]Poznan University of Life Sciences, Department of Genetics and Animal Breeding, Wolynska 33, 60-637 Poznan, Poland, [2]Poznan University of Life Sciences, Department of Animal Nutrition and Feed Management, Wolynska 33, 60-637 Poznan, Poland; mbee@up.poznan.pl

In our study we conducted comparison of two non-invasive techniques used to measure methane concentrations in dairy cattle kept in commercial conditions: Fourier transform infrared spectroscopy (FTIR) analyser and nondispersive infrared (NDIR) analyser which is normally used in respiration chamber – a golden standard method for CH_4 measurements. The analysers were connected to the same feed bin placed in an automatic milking system. Both analysers sampled simultaneously air exhaled by the cow during milking. Measurements were carried out for 5 days. Measurements were averaged per milking giving 467 observations of methane (CH_4) and carbon-dioxide (CO_2) concentrations of 44 Polish Holstein – Friesian cows. The average daily CH_4 concentration for FTIR was 548 ppm and 541 ppm for NDIR. The average daily CO_2 for FTIR was 6,646 ppm and 6,430 ppm for NDIR. The differences between measurements were statistically insignificant. Pearson's correlation between all observations obtained from the two analysers was 0.86 for CH_4 and 0.84 for CO_2. The repeatability of FTIR (0.53 for CH_4 and 0.57 for CO_2) was higher than of NDIR (0.57 for CH_4 and 0.47 for CO_2). The coefficient of individual agreement for all observations equalled 0.98 for CH_4 and 0.89 for CO_2, concordance correlation coefficient was 0.48 for both gases, and the correlation between random cow solutions obtained with the two analysers was 0.98 for CH_4 and 0.97 for CO_2. This study has shown that FTIR and NDIR analysers for measuring gas CH_4 and CO_2 provide similar results when used under commercial farm conditions. High level of agreement with sensor used in golden standard confirms the ability of FTIR method to deliver reliable measurements of CH_4 and CO_2 concentrations over the milking period in order to generate a larger data set which could be further used for genetic evaluation.

Artificial neural networks for prediction of BW of male cattle by using real-time body measurements

Y. Bozkurt[1], T. Aydogan[2], C.G. Tuzun[1] and C. Dogan[1]
[1]Suleyman Demirel University, Faculty of Agriculture, Department of Animal Science, Isparta, 32260, Turkey, [2]Suleyman Demirel University, Faculty of Technology, Department of Software Engineering, Isparta, 32260, Turkey; yalcinbozkurt@sdu.edu.tr

It was aimed to predict BW of Holstein and Brown Swiss cattle grown in a 12-month Intensive beef production system by Artificial Neural Networks (ANN). For this purpose, 40 animals were used in total and composed of 20 animals of the Brown Swiss breed and 20 animals of the Holstein breed with the age of about 4-5 months at the beginning of the experiment. Associations between bodyweight and some real-time body measurements such as heart girth (HG), wither height (WH), body length (BL), body depth (BD), hip width (HW) and hip height (HH) were examined for prediction ability, using the data with 1,068 observations for each traits.. There were no significant differences ($P > 0.05$) in the predicted values between breeds. Therefore, their data were combined to develop models. It was found that the model included all traits as input gave the highest R2 value of 93.4. The results also showed that BWs can be predicted from body length and wither height as single inputs by ANN (R2=81 and 72% respectively). However, unexpectedly chest girth gave the lowest R2 value of 60.7%. It can be concluded that ANN models showed that in management situations where BW cannot be measured it can be predicted accurately by measuring BL alone and even WH and different ANN models. More training of the models may be needed for prediction purposes to be used in different feeding and environmental conditions and for different breeds.

Performance and carcass traits of steers fed increasing levels of sorghum dry distiller's grains

A. Simeone, V. Beretta, J. Franco, M. Novac, V. Panissa and V. Rodriguez
University of the Republic, Ruta 3 km 363, Paysandu, 60000, Uruguay; asimeone@adinet.com.uy

The present study evaluated the effect of increasing levels of inclusion of sorghum dry distiller's grains plus soluble (DDGS) in the diet on the performance and carcass traits of lot-fed finishing steers. Thirty-six Hereford steers (331.0±33.5 kg) were randomly allocated to four total mixed rations differing in the level of DDGS (30.7% CP, 55,3% NDF, 2.1% ADIN, 9.9% fat): 0, 15, 30 or 45% of ration DM in replacement of sorghum grain (SG) and sunflower meal (SFM) in the control diet (20% *Setaria italica* hay, 0% DDGS, 65% SG, 10% SFM, 5% premix). Chemical composition of 0% DDGS ration was 12.3% CP, 29.1% NDF, 15.6% ADF and 2.84% fat. As DDGS level in the ration was increased CP, NDF, ADF and fat, gradually increase up to 20.1, 43.9 20.9 and 6.0%, respectively in the 45% DDGS ration. Animals were gradually introduced to experimental diets (2 weeks) and then fed in individual pens at 2.8 kg DM/ 100 kg liveweight, during 12 weeks. Feed was delivered in 3 meals. Animal were weighed every 14 days, and dry matter intake (DMI) was determined daily. Feed to gain ratio (FGR) was calculated based on mean values for the experimental period. Steers were all slaughtered on same date and carcass traits were measured. The experiment was analysed according to a randomized plot design with repeated measures for LW and DMI. Linear and quadratic effects associated to DDGS level were tested. Increasing DDGS level in the ration did not affect liveweight gain (0% 1.5, 15% 1.44, 30% 1.54, 45% 1.62 kg/d, SE 0.06; P>0.05), DMI (0% 10.5, 15% 10.6, 45% 10.6, 30% 11.4 kg/d, SE 0.33; P>0.05) or FGR (0% 6.8, 15% 7.6, 30% 7.0, 45% 8.4, SE 0.48; P>0.10). No differences due DDGS level in the ration were observed in carcass weight (240 kg SE 5.4), carcass yield (54.2% SE 0.6) or ribeye area (49.1 cm^2 SE 2.5), while a negative linear tendency was observed for subcutaneous backfat (SBFmm=9.64-0.054DDGS; P<0.10). Results suggest that, when feed offer is controlled, including DDGS from sorghum in up to 45% of DM in high grain finishing rations would not affect FCR and carcass weight, although differences in carcass fat deposition might be observed.

An analysis of factors affecting success of artificial insemination in Wagyu cattle

T. Oikawa
University of the Ryukyus, Faculty of Agriculture, 1 Senbaru, Nishihara-cho, Okinawa, 903-0213, Japan; tkroikawa@gmail.com

With decrease of Wagyu (Japanese Black) farms, the number of Wagyu cows deceased consistently in Japan. Therefore efficient production of calves is a key to maintain Wagyu production. However, reproductive performance of cows was reported to be declining. The objective of this study is to investigate factors which determine success of artificial insemination of heifers/cows. Records in study were taken from artificial insemination (AI) notes recorded by seven AI technicians. Area and period of the survey were northern part of Okinawa and twelve years from 2005 to 2016, respectively. Those records with the following criteria were deleted; inconsistent birth records with calf registry book, birth records by embryo transfer, records of twin birth, records of farms less than ten cows, records with more than ten parity. After editing, the total number of records was 20,928. The records were AI records in 945 cows. Generalized linear mixed model was applied for the analysis using GLIMMIX procedure of SAS. A linear model included random animal effect (heifer/cow effect), fixed effects including farm (122), year of insemination (12), season of insemination (4), order of insemination (max=17), parity (10), AI technician (7), position effect of ovulation (right and left) and covariates (linear and quadratic regression on age). Success of AI was set for a response in a binary observation. Link function was logit. Success rate of AI was 0.57. Estimated variance of animal effect was low (0.013±0.009; estimate±se). Fixed effect of AI season and order of AI service were statistically insignificant. Odds of AI year had small fluctuation and no directional change. Odds of farm effect varied from 0.2 to 1.0 mostly. However a few farms showed Odd more than 1.0. Effect of AI technician showed moderate variation. The position effect of ovulation indicated that left ovary had better conception rate than right ovary. Reason of this result seems to be either sampling error or positional effect of internal organ. Odds of parity had upward trend with a progress of parity, suggesting that the trend is observed due to selection of a cow. In contrast, odds of AI success declined significantly at older ages.

Session 45 — Poster 15

Enteric methane emissions from heifers fed grass-clover silage or pulp silage made from grass-clover

A.L.F. Hellwing, V.K. Damborg, P. Lund, S.K. Jensen and M.R. Weisbjerg
Aarhus University, AU Foulum, Blichers Allé 20, 8830 Tjele, Denmark; annelouise.hellwing@anis.au.dk

Screw pressing of grasses and legumes for production of protein for mono-gastric animals has regained interest during the last years. The remainder from the extraction (pulp) has lower concentrations of water soluble protein and sugar and higher concentration of neutral detergent fibre (NDF) compared with the original forage. The aim with this experiment was to compare feed intake, enteric methane emissions and rumen fermentation for heifers fed grass-clover silage (GCS) or pulp silage (PS). Eight Holstein heifers with an average age of 17.2±1.1 (mean ± standard deviation) months were divided into two blocks according to days in pregnancy. The heifers were 121±14 and 80±10 days in pregnancy in block 1 and 2, respectively. The experiment was designed as a cross-over experiment. Within block, two heifers were randomly allocated to start on GCS and the other two on the PS. The grass-clover used for extraction and thereby production of pulp silage and for grass-clover silage was harvested at the same field, grass-clover silage 6 days later than the grass-clover used for the pulp. Both silages were ensiled in bales without addition of silage additives. The heifers were fed silage as the sole feed *ad libitum*. Each period in the experiment lasted 14 days with the first 11 days for adaptation, and the remaining three days for measuring methane emissions using respiration chambers. A stomach tube was used for rumen fluid sampling the last day of each period just before feeding. The data was analysed in PROC MIXED with silage and period as fixed effects, and animal and block as random effects. The NDF concentrations were 450 and 394 g/kg dry matter (DM) in the PS and GCS, respectively. The DM intake was 8.5 kg for both the GCS and PS. The methane emission per day (P=0.01) and per kg DM intake (P=0.02) were higher for the PS (261 and 30.9 l, respectively) than for the GCS (249 and 29.2 l, respectively). The rumen pH (P=0.002) and the proportion of volatile fatty acids (VFA) from acetate (P=0.01) were higher for GCS than PS whereas the proportion of VFA from propionate (P=0.01) and butyrate (P=0.02) were higher for PS than GCS. In conclusion, feeding of PS resulted in higher enteric methane emissions.

Session 45 — Poster 16

Does residual sward height impact on animal performance and pasture characteristics ?

P. Chilibroste[1], P. Soca[1], M. Oborsky[1], G. Menegazzi[1], P.Y. Giles[1], D.A. Mattiauda[1], C. Genro[2] and F. Lattanzi[3]
[1]Facultad de Agronomía, Universidad de la República, Ruta 3 km 363, Paysandú, 60000, Uruguay, [2]Empresa Brasileña de Investigación Agropecuaria, Rodovia BR-153, km 632,9, 242, Bagé, Brazil, [3]Instituto Nacional de Investigación Agropecuaria, Ruta 50, km 11, Colonia, 11100, Uruguay; pchili@fagro.edu.uy

An experiment was carried out to evaluate the effect of three grazing intensities applied from March the 1st to July the 31st of 2017 on sward mass (SM), sward height (SH), green cover (GV) and milk production and composition of autumn calving dairy cows, during spring (evaluation period). The grazing treatments were applied based on residual sward height: Lax (L, 15 cm), Medium (M, 12 cm) and Control (C, 9 cm). During August and September treatments were grazed down to 4 cm to prevent tall fescue early flowering. The statistical design was RCBD with 4 spatial blocks of 3 parcels each (total=12 parcels) grazed by 3 milking cows per parcel. Each treatment was grazed when three new extended leaves were developed. Instantaneous stocking rate was the same across treatments (15 cows/ha). Sward characteristics were determined in 12 point per parcel (square 0.1 m^2) at the beginning (BG) and end (AG) of the grazing cycle. Thirty six's milking cows with 618±48 kg BW, 2.6±0.8 BCS, 2.6±0.8 parity number and 224±7 DIM were used. Response variables were analysed with Proc Mixed of SAS and means values were declared different when Tukey test <0.05. Sward mass was not different between treatments (2,302±129 kg DM ha-1) but was higher BG than AG (2,779 vs 1,826 kg DM/ha). Sward height was different between treatments, determinations (BG, AG) and their interaction: 18.0a, 20.0a, 16.9a (BG) and 9.3c, 12.1bc and 13.1b (AG) cm for C, M and L, respectively. Green cover was not different between treatments (72.2±2.35%) but higher BG (82.7%) than AG (61.7). Milk production was different between treatments (12.1b, 17.4a and 17.7a, for C, M and L, respectively) without differences in milk components (3.3±0.12 and 3.9±0.3 for protein and fat %, respectively). High residual SH (>12 cm) supported higher levels of milk and solid production probably linked to changes in sward structure.

Session 45 Poster 17

Field pea can partially replace soybean in the fattening diets of ruminants

M. Blanco, I. Casasús and M. Joy
CITA – IA2, Unidad de Producción y Sanidad Animal, Avda. Montañana 930, 50059 Zaragoza, Spain; icasasus@aragon.es

There is an interest to replace the use of soybean by local legumes in Southern Europe to increase the protein self-sufficiency. The aim of the study was to analyse the effect of the inclusion of field pea *(Pisum sativum)* in the concentrate on the performance and carcass weight of: (1) light lambs and (2) young bulls. In both trials, the concentrates were iso-energetic and iso-proteic and pea replaced gradually soybean. Concentrates and straw were fed on *ad libitum* basis. The fattening concentrate offered to lambs had 0, 10, 20 and 30% pea (11.8 MJ ME/kg, 175 g/kg crude protein). Weaned male lambs (n=54; 13.4 kg LW; 31 d of age) received concentrates until 23 kg LW, when they were slaughtered. The concentrate fed to young bulls had 0, 15, 30 and 45% pea (11.6 MJ ME/kg and 130 g/kg crude protein). Weaned male calves (n=31; 239 kg LW; 150 d of age) received concentrates until 508 kg LW. Hot carcass weight was registered just after slaughter. In the light lambs, the inclusion of pea did not affect the weight gains (average 245 g/d), the total concentrate intake (24.3 kg DM), the feed conversion ratio (2.44 g/g) or the duration (42 days) of the fattening period. However, the inclusion of 10% of pea increased hot carcass weight when compared to the inclusion of 20% pea (10.54, 10.93, 10.45 and 10.63 kg for 0, 10, 20 and 30% pea, respectively) and the dressing percentage compared with 0 and 20% pea (47.0, 45.5, 45.4%, respectively; $P<0.05$). The inclusion of pea in the concentrate of young bulls did not affect weight gains (1.46 kg/d), the feed conversion ratio (4.82 kg/kg) or the duration of the fattening period (183 d) ($P>0.05$). Nevertheless, the inclusion of pea had a cubic effect on the concentrate intake (7.34, 7.07, 7.63, 6.75 kg FM/d for 0, 15, 30 and 45% pea, respectively; $P<0.05$). The inclusion of pea did not affect carcass weight (293 kg) or dressing percentage (57.7%) of young bulls ($P>0.05$). Consequently, soybean can be replaced by pea in the fattening concentrates of both light lambs and young bulls, however, the effect on carcass and meat quality should be evaluated. The percentage of inclusion of pea should be decided depending on the prices of each feedstuff as there were no relevant effects on the performance during the fattening period.

Session 45 Poster 18

Effects of rations with high amount of crude fibre on rumen fermentation in suckler cows

H. Scholz[1], P. Kühne[1], A. West[2], R. Staufenbiel[2] and G. Heckenberger[3]
[1]Anhalt University of Appled Scienes, Faculty LOEL, Strenzfelder Allee 28, 06406 Bernburg, Germany, [2]Free University Berlin, Königsweg 65, 14163 Berlin, Germany, [3]State Institute for Agriculture and Horticulture Saxony-Anhalt, Lindenstraße 18, 39606 Iden, Germany; heiko.scholz@hs-anhalt.de

Body Condition of suckler cows at the time of calving has an important effect of calving ease. At the end of the grazing period (often after early weaning), however, an increase of BCS can often be observed under German conditions. In the last 8 weeks before calving the body condition should be reduced or at least not increased. Rations with a higher amount of crude fibre can be used (rations with straw or late mowed grass silage). 8 suckler cows (Charolais) were feeding a total mixed ration (TMR) in the last 8 weeks before calving and grass silage after calving. By the addition of straw (30% [TMR1] vs 60% [TMR2] of dry matter) was varied the amount of crude fibre in the TMR (Grass silage, straw, mineral) before calving. After calving was grass silage [GS] feeding *ad libitum*. Last measurement took place on the pasture [PS]. Rumen fluid, plasma, body weight and back fat thickness were collected. Rumen fluid pH was assessed immediately after collection using an electronic pH meter. Volatile fatty acids (VFA), sedimentation, methylene-blue and amount of infusorians were measured. From 4 key figures an 'index of rumen fermentation' [IRF] in the rumen was formed. Statistical analysis took place with ANOVA with fixed effects of treatment (TMR1, TMR2, GS and PS) and number of lactations (3-7 lactations) using SPSS Version 25.0 for windows. Rumen fluid pH had significant differences between variants (TMR 1 by 6.6; TMR 2 by 6.9; GS by 6.6 and PS by 6.9), but was not affected by other effects. The IRF showed a disturbed fermentation in the rumen by feeding the TMR 1+2 with high amount of crude fibre (Score: >10.0 points) and a very good situation for fermentation during grazing the pasture (Score: 6.9 points). Furthermore significant differences could be found for VFA, methylene blue and the amount of infusorians. The long-term use of crude fibre-rich rations in the period before calving may cause deviations from undisturbed fermentation in the rumen and adversely affect the utilization of the feed in the rumen.

Session 45 Poster 19

Feed intake of suckler cows in the period of calving with different amount of crude fibre

P. Kühne[1], H. Scholz[1] and G. Heckenberger[2]
[1]Anhalt University of Applied Sciences, Faculty LOEL, Strenzfelder Allee 28, 06406 Bernburg, Germany, [2]State Institute for Agriculture and Horticulture Saxony-Anhalt, Lindenstraße 18, 39606 Iden, Germany; petra.kuehne@hs-anhalt.de

Around calving time feed intake of beef cows is important for milk production and metabolic situation. 35 beef cows were used in these investigations to measure the daily feed intake from 2 weeks ante partum to 4 weeks postpartum. There were 3 variants tested: [1] Total Mix Ration (TMR) from drying dairy cows, [2] grass silage *ad libitum* and [3] grass silage mixed with straw. The beef cows weighted [1]: 872±72 kg, [2]: 868±72 kg and [3]: 935±95 kg live weight and showed an average back fat thickness of 22 mm. A feed intake of TMR could be measured in the time before calving of 13.8±3.6 kg DM (1.6 kg DM/100 kg) and postpartum of 18.7±3.7 kg DM per day (2.2 kg DM/100 kg). The consumption of grass silage was 9.4±3.2 kg DM (1.1 kg DM/100 kg) before calving and 14.3±4.9 kg DM per day (1.7 kg DM/100 kg) after calving. Feed intake before calving was 12.3±1.8 kg DM per day (1.3±0.2 kg DM/100 kg) and 16.1±2.1 postpartum (1.4±0.2 kg DM/100 kg) by grass silage mixed with straw. The intake of crude fibre was different between TMR (4,146 g/d), grass silage (2,911 g/d) and grass silage mixed with straw (4,677 g/d). The effect of a lower intake of crude fibre in the ration of grass silage could measure in the NSBA (Net-Acid-Base-Excretion). With 211±22 mmol/l there was a significant higher NSBA in the group of TMR [1] than the cows in the ration of grass silage [2] with an average of 150±83 mmol/l or grass silage mixed with straw 122±66 mmol/l [3]. After calving suckler cows fed TMR quickly reach their maximum intake of dry matter and crude fibre, which is much higher than the standards published by the DLG. In the grass silage feeding and feeding the grass silage mixed with straw a delayed increase in feed intake was observed and reached only in the 4th week postpartum the recommendations of DLG.

Session 45 Poster 20

Nitrogen-use-efficiency of beef cows fed total mixed ration, grass silage or grazed on pasture

P. Kühne[1], H. Scholz[1] and G. Heckenberger[2]
[1]Anhalt University of Applied Sciences, Faculty LOEL, Strenzfelder Allee 28, 06406 Bernburg, Germany, [2]State Institute for Agriculture and Horticulture Saxony-Anhalt, Lindenstraße 18, 39606 Iden, Germany; petra.kuehne@hs-anhalt.de

Nitrogen-use-efficiency (i.e. N-efficiency) is defined as percentage of feed nitrogen (N) that is converted into milk and meat protein. Optimizing N-efficiency has the potential to improve the productivity of livestock systems while reducing their environmental impact. During calving period (2 weeks antepartum and 4 weeks postpartum), 14 beef cows were monitored for daily feed intake. Group 1 received a Total Mix Ration (TMR) containing 52% grass silage, 22% alfalfa silage, 21% corn silage, 4% straw and 1% mineral supplement (all dry matter [DM] basis) Crude protein (CP) and crude fibre content of the TMR was 14 and 27%, respectively. Group 2 received *ad libitum* grass silage cut to two different chop lengths (15% CP; 27% crude fibre). Mean TMR intake was 17.7 kg DM/d. Intake of grass silage was 14.8 kg DM/d (>10 cm chop length) and 16.9 kg DM/d (≤2 cm chop length), respectively. The percentage share of faeces was 25% of dry matter intake and with an amount of N by 26.6 g/kg DM can be calculated N-loss of 27% of N-intake. Mean milk yield of 16.0 kg/d was determined by Weight-Suckle-Weight method. Average milk protein concentration was 3.37%, while milk fat concentration was 3.56%. Milk urea concentration was 237 mg/litre. In this study, N-efficiency from feed to milk was 22%.During the grazing period of suckling cows (Mai to September) increased N excretion has been observed. Based on estimated N requirements of 85 g/d for maintenance and 87 g/d for milk production (19% N-efficiency) there we found N losses in urine and faeces by an average of 358 g/d (urine: 78%; faeces: 22%). It could be shown that there is the N-efficiency of suckler cows at a moderate level compared to dairy cows. The investigation in N-efficiency during the grazing season show differences between an extensive (<1.0 GV/ha) and intensive (>1.0 GV/ha) use of grassland.

Eating behaviour and milk production of high efficient vs low efficient high yielding lactating cows

Y.B. Ben Meir[1,2], H. Levit[1,3], I. Halachmi[3], J. Miron[2] and S.J. Mabjeesh[1]
[1]The Hebrew University of Jerusalem, Animal science, Faculty of Agriculture, Food and Environment Robert H. Smith, 7610001 Rechovot, Israel, [2]Agricultural Research Organization, Animal science, Derech Hamacabim 68, 7505101 Reashon Lezion, Israel, [3]Institute of Agricultural Engineering, Precision Livestock Farming (PLF), Derech Hamacabim 68, 7528809 Reashon Lezion, Israel; yehoshavbm@gmail.com

This study aimed to identify individual characteristics differing between high efficient (HE, upper 20%, n=31) and low efficient (LE, lower 20%, n=31) lactating cows. The entire herd (155 milking cows) was fed individually a low-roughage diet. Daily dry matter intake (DMI), rate of eating, meal size and daily rumination time were significantly higher in LE compared to HE cows. On the other hand, HE cows exhibited higher digestibility of DM, crude protein and neutral detergent fibre, and longer rumination per kg DMI than the LE cows. Daily eating time was similar in the HE and LE groups and higher than that of the remaining 93 mid-efficient (ME) cows. Number of meals /d, average meal duration, daily lying time, and pedometer activity were similar in the HE, LE and ME groups. The HE cows produced 1.75% more milk but similar energy corrected milk (ECM) yield compared with the LE cows. Milk fat, protein and lactose content were slightly lower in the HE cows than in the LE group. Body weight (BW) and BW gain were similar in the three efficiency groups. Diurnal distribution of DM intake showed five distinct major waves of eating between 05:00 to 23:00 and an additional flat eating wave during night. Higher peaks (greater meal size) were found in the LE cows compared with the HE group. Daily DMI was positively correlated with ECM production (r=0.61), BW (r=0.4), eating rate (r=0.57), meal size (r=0.54) and negatively correlated to rumination/kg DMI (r=-0.56). Energy balance calculation showed that the lower efficiency of the LE cows was attributed to their excess heat production and energy losses.

Investigating genetic and functional relationships among growth traits in beef cattle

F.B. Lopes[1,2,3], C.U. Magnabosco[1], T.L. Passafaro[2], L.F.M. Mota[2,3], G.J.M. Rosa[2], J.L. Ferreira[4] and F. Baldi[3]
[1]Brazilian Agricultural Research Corporation, BR-020, 18, Sobradinho, Brasilia, DF, 70770-901, Brazil, [2]University of Wisconsin-Madison, Departament of Animal Sciences, 436 Animal Sci. Building 1675 Observatory Dr. Madison, WI, 53706, USA, [3]São Paulo State University, Júlio de Mesquita Filho (UNESP), Department of Animal Sciences, Prof. Paulo Donato Castelane, Jaboticabal, SP, 14884-900, Brazil, [4]Federal University of Tocantins, Avenida Paraguai Bairro Cimba, Araguaína, TO, 77824838, Brazil; camult@gmail.com

In cattle production, knowledge of causal relationships among growth traits can contribute to more efficient genetic selection and management decisions. A Structural equation model (SEM) approach was used to analyse gestation length (GL), birth weight (BW) and weight at 120 (W120), 210 (W210), 365 (W365) and 450 (W450) days of age in Nellore cattle. The dataset comprised information on 18,000 animals raised on pasture in the Northern Brazil. A directed acyclic graph (DAG) describing causal relationships between the traits was inferred using a Bayesian implementation of the Inductive Causation (IC) algorithm. A SEM was subsequently fitted conditionally on the inferred DAG. Estimates of direct and maternal heritability for GL, BW, W120, W210, W365 and W450 were 0.61, 0.47, 0.29, 0.25, 0.37 and 0.46, and 0.15, 0.16, 0.40, 0.37, 0.21 and 0.17, respectively. Estimates of genetic correlations among these traits ranged from 0.17 (BW and W365) to 0.96 (W365 and W450). The IC algorithm retrieved five directed relationships: GL-W365, GL-W450, W120-W365, W210-W365 and W365-W450, and three other links were directed based on biological or temporal knowledge: GL-BW, GL-W210 and W120-W210. The inferred DAG indicates that GL and W120 are key variables upstream the causal network affecting W450. The network indicates also that W210 and W365 are intermediate variables through which GL and W120 affect W450. The estimates of heritability and genetic correlations suggest that genetic improvement can be obtained through direct indirect selection on intermediate phenotypes in the network. The causal network provides also information on potential targets for management interventions to improve the market weight in beef cattle. Acknowledgments: São Paulo Research Foundation grant #2017/03221-9.

Session 45 Poster 23

Whole oats as an alternative fibre source for growing cattle fed high grain diets

V. Beretta, A. Simeone, M. Cedres and N. Zabalveytia
University of the Republic, Ruta 3 km 363, Paysandu, 60000, Uruguay; beretta@fagro.edu.uy

Whole oats (WO), because of the high relative weight of hulls in the kernel, is lower in energy and higher in fibre content (32%) compared to other cereal grains. This makes WO a possible fibre source for high grain diets fed to cattle. An experiment was conducted to evaluate the effect of graded levels of substitution of Lucerne hay (LH) for WO in a high grain diet, on physical effective fibre (peNDF) supply, feed utilisation and performance of lot-fed growing beef cattle. Twenty four Hereford early-weaned females calves (77±13 kg, 64±10 days old) were randomly assigned to one of four total mixed rations differing in WO:LH ratio (on a DM basis): 0:30%, 10:20%, 20:10%, 30:0%, plus 70% of a commercial concentrate. Rations were balanced for 17% CP across treatments, while NDF content decreased from 36.1 to 27.1% with WO increase in the ration. Calves were gradually introduced to the experimental diets and then fed *ad libitum* in individual pens during 10 weeks. Liveweight (LW) was recorded every 14 days without fasting, dry matter intake (DMI) was determined daily, and mean feed to gain ratio (FGR) for the experimental period was calculated. Content of peNDF was measured on ration and orts samples, and apparent DM digestibility (DMD) was estimated on week seven through total faeces collection. The experiment was analysed according to a randomized plot design with repeated measures. Linear and quadratic effects associated to WO level in the ration were tested. LW gain showed a quadratic response (LWG, kg/d=$1.12+0.018x-0.0005x^2$, $P<0.05$) with maximum gains of 1.30 kg/d for 18.5% of WO and 22.2% peNDF in the diet. No differences due to WO:LH ratio were observed in DMI (4.50±0.31 kg/d; 3.53±0.12% LW) or peNDF intake (1.04±0.31 kg/d). However increasing WO level in the ration linearly increased ($P<0.01$) DMD of ingested feed (DMD %= $64.0+0.44x$, $R^2=0.82$) and digestible DM intake (DDMI, kg/d=$2.27+0.017x$, $R^2=0.0.84$). On the overall, as WO in the diet increased, less feed was necessary per unit of gain (FGR= $3.85-0.0139x$; $P<0.05$). Results suggest that when LH represents up to 30% of a total mixed ration fed to growing beef calves, it is viable to substitute this fibre source for WO.

Session 46 Theatre 1

Effect of censoring on parameter estimation and predictive ability using an indirect genetic model

M. Heidaritabar
Aarhus University, Blichers Allé 20, Postboks 50, 8830 Tjele, Denmark; marzieh.heidaritabar@mbg.au.dk

Social interactions, for which the phenotype of an individual is influenced by the direct genetic effect (DGEs) of an individual, as well as the indirect genetic effects (IGEs) of its group mates, are important in group-housed animals, such as pigs. The observations for traits may be incomplete (censoring) due to e.g. mortality from diseases and involuntary culling. Censoring may be a problem for indirect genetic models (IGM), because removal of individuals change the group size and affect the phenotypes of remaining group mates. We investigated the effect of censoring on (co) variance estimates and predictive ability of an IGM. We simulated a population including 6 discrete generations of random selection, where in each generation 50 sires and 200 dams were selected randomly as parents for next generation. Model validation was conducted in the sixth generation where individuals were randomly allocated to groups of size 8 to train the model in the fifth generation. Moderate direct and indirect heritability (0.3) was simulated. Genetic correlations between DGE and IGE (rDI) were assumed to be 0, 0.5, or -0.5. Four datasets were simulated: (1) complete data, (2) 50% groups in the training data had missing values, (3) 50% groups in the validation data had missing values, (4) 50% groups in both training and validation data had missing values. Largest phenotypic value in each group was set to missing. Predictive ability was the correlation between true and predicted phenotypes for both individual and group phenotypes. A classical animal model and an IGM was applied. Results showed larger total heritability compared with classical heritability, when rDI was 0 and 0.5. However, the negative rDI reduced the total heritability compared with classical heritability (0.12 vs 0.28). Moreover, results showed that compared to the full dataset, censored data only slightly affected the predictive ability. Ignoring censoring may not be an issue for an IGM.

Session 46 Theatre 2

Realized autozygosity and genetic differentiation of Landrace × Large White crossbreds

L. Gomez- Raya[1], W.M. Rauw[1], J. Dunkelberger[2] and J.C.M. Dekkers[2]
[1]INIA, Dept Mejora Genética Animal, Ctra de La Coruña km 7, 28040, Madrid, Spain, [2]Iowa State University, Dept of Animal Science, Kildee Hall, 50011, USA; gomez.luis@inia.es

Crossbreeding is a common breeding practice to increase performance by heterosis in farm animals. The use of molecular techniques can provide new insights into conserved and diverged chromosomal regions across the genome of genetically distant breeds of swine. Autozygosity, estimated with the probability of autozygosity F_L, and genetic differentiation, estimated by F_{ST}, were evaluated in 1,173 Landrace × Large White crossbreds. Short-term changes can be viewed as the maintenance of conserved regions of autozygosity (not eroded by recent recombination events). Long-term changes are caused by the dispersive effect of genetic drift since the split-up of the two breeds and were evaluated by estimating the F_{ST} statistic. In some cases, both approaches could be confounded if a region of autozygosity persisted since the genetic split-up of the breeds. Allele frequencies in the parental populations were estimated with a newly developed maximum likelihood approach, which also makes use of crossbred genotype information. Probabilities of autozygosity were higher in the central parts of chromosomes. There was much variation in autozygosity between chromosomes – the correlation of F_L with distance to the telomere ranged from 0.02 to 0.70 across chromosomes. Long-term genomic changes, as estimated by F_{ST}, were evenly distributed across the swine genome. The average F_{ST} was 0.038 (SD=0.059). There were only 159 SNPs with F_{ST}>0.30 and they were distributed evenly along the chromosomes. The correlation between estimates of F_L and estimates of F_{ST} across the genome was -0.095 (S.E.=0.006). Analysis of gene content in the chromosomal regions with the 2,000 SNPs with the highest likelihood ratio test for F_L and high F_{ST} with the software PANTHER showed overrepresentation of genes with a regulatory function. Genes with biological functions associated with production, such as tissue development, anatomical structure, and animal organ development, were also overrepresented in regions with high F_{ST}. In conclusion, short- and long-term changes in the genetic make-up of isolated breeds can be revealed by genetic analysis of crossbreds.

Session 46 Theatre 3

Behaviour of method LR (linear regression) to measure bias and accuracy

F.L. Macedo and A. Legarra
INRA, GenPhySE, 31326 Castanet Tolosan, France; fernando.macedo@inra.fr

Cross validation is the most extended method to estimate the prediction ability in animal selection schemes; however it present problems on the quality of results. The Linear Regression (LR) method compares EBVs obtained with old ('partial') and old+new ('whole') data to infer biases and accuracies. In this work, we present preliminary results on the behaviour of LR method using simulated data. Based on heritabilities of 0.5 and 0.1, 20 populations were simulated with the QMSim software. In the simulation individuals were selected by BLUP evaluation and mating system were performed to reduce the average inbreeding and kinship of the population. Only fathers born in generation 5 with at least 5 daughters in generation 6 were used. BLUP pedigree evaluations were performed using a partial data set (without daughter's information) and a whole data set (with daughter's information). Statistics were obtained between the estimated breeding values of the partial (EBVp) and whole data sets (EBVw) and between the EBVp and true breeding values (TBV) obtained from the simulation. Five statistics were considered: (bias) difference between average EBVp and EBVw, with an expected value of 0, (slope) regression of EBVw on EBVp with an expected value of 1, (accuracies) correlation between EBVp and EBVw (ρ_{pw}) (proportional to increase in accuracy), covariance between EBVp and EBVw (proportional to accuracy on partial) and regression of EBVp in EBVw (proportional to increase in reliabilities). All metrics were also calculated by substituting EBVw for TBV to ascertain if metrics using EBVw predict results for TBV. The statistics comparing EBVp-EBVw and EBVp-TBV are almost identical for both values of h^2. The most important differences were observed in ρ_{pw} (in h^2 of 0.1: EBVp-EBVw=0.36 and EBVp-TBV=0.27; h^2 of 0.5: EBVp-EBVw=0.32 and EBVp-TBV=0.27) and it was always overestimated on EBVp-EBVw. In short, biases and accuracies were correctly estimated using statistics from 'whole' and 'partial' genetic evaluations. The similarity of the results obtained using EBVw or TBV suggest that statistics proposed in method LR could be useful for measure of bias and accuracies in breeding schemes. Further work will include genomic information.

Influence of age on variance components for bodyweight in commercial male and female broiler chicken

T.T. Chu[1,2], P. Madsen[2], L. Wang[2], J. Henshall[3], R. Hawken[3] and J. Jensen[2]
[1]Wageningen University, WIAS, Research Animal Breeding and Genomics, P.O. Box 338, 6709 PG Wageningen, the Netherlands, [2]Aarhus University, Center for Quantitative Genetics and Genomics, Department of Molecular Biology and Genetics, Blichers Allé 20, 8830 Tjele, Denmark, [3]Cobb-Vantress Inc., Siloam Springs, AR 72761-1030, USA; chu.thinh@mbg.au.dk

The main objective of this study was to estimate variance components for body weight (BW) at different ages in male and female commercial broiler chicken. Weekly BW until 6 weeks of age were obtained from pure-line broilers tested in a normal commercial production environment. In total, the dataset comprised 18,161 broilers. A multivariate reduced rank model using pedigree-based BLUP was developed to estimate parameters of BW traits at different weeks of age. Further reductions of the model using random regression models were not successful as it led to significantly lower fit. It was found that there was no sex-by-genotype interaction for BW traits at different weeks of age, but heterogeneous residual variances of BW for males and females from 2-6 weeks of age were found. The residual variances for male BW were significantly higher than for female BW. The direct additive genetic, maternal permanent environmental and residual variances increased sharply as age of the broilers increased. However, with increasing weeks of age, the ratio of the maternal permanent environmental variance to the total phenotypic variance reduced gradually from 0.11 at 1 week of age to below 0.05 at 5 weeks of age and disappeared completely at 6 weeks of age. Heritability of BW traits at different weeks of age was 0.22-0.30 for males and 0.25-0.36 for females. The direct additive genetic effects on two consecutive weekly BWs were highly correlated with genetic correlations ranging from 0.85-0.99. However, the genetic correlation between early and late BW were only 0.34-0.55. In conclusion, there were heterogeneous residual variances for male and female BW in all weeks of age from 2-6 weeks; the maternal permanent environmental effect on BW of the broilers reduced with increasing age; and genetic correlations between BWs at early and later ages are low.

Multi-trait mixed modelling of milk infrared spectral data for better accuracy of prediction

T.K. Belay[1], B.S. Dagnachew[2] and T. Ådnøy[1]
[1]Norwegian University of Life Sciences (NMBU), Department of Animal and Aquacultural Sciences (IHA), Arboretveien 6, 1432 Ås, Norway, [2]Norwegian Institute of Food, Fisheries and Aquaculture Research (Nofima), Department of Aquaculture breeding and genetics, Osloveien 1, 1430 Ås, Norway; tesfaye.kebede.belay@nmbu.no

Fourier-transform mid-infrared (FT-MIR) spectral information for prediction of breeding values and phenotype could be exploited by using two approaches. One approach (Indirect Prediction – IP: the conventional method) transforms spectra to a single-trait (like fat content or other derived traits – 'fat') and then applies genetic analysis on the predicted phenotype together with pedigree information and estimated variance components. The second approach (Direct Prediction – DP) uses a multi-trait mixed model on (dimension reduced) spectral variables to obtain multi-trait breeding value predictions that later are combined into prediction of breeding values for the intended trait ('fat'). Similar approaches may be used for predictions of phenotype ('fat') based on the spectra. The two prediction approaches were evaluated using both real and simulated data for their ability to predict phenotype and breeding values from milk FT-MIR spectra. Links between the spectra and traits of interest ('fat') were developed using partial least square (PLS) regression. An animal model was fitted to predict breeding values and phenotype. Accuracy of prediction was estimated as the correlation between measured or simulated and predicted values, or based on coefficient matrix to find prediction error variance. Accuracies of breeding value predictions were higher in the DP than in the IP approach (3-5% improvement in real data and 4.1-56.4% in simulated data). The reverse was true for accuracy of phenotype prediction. Performance of the DP approach was conditional on genetic and residual correlation structures, number of observations, and accuracy of calibration model, type of PLS regression coefficients and dimension reduction techniques used. In conclusion, the DP approach is the method of choice for breeding value prediction directly from heritable parts of spectra, while the classical PLS regression based prediction equation or the IP approach is so far preferable for phenotype prediction.

Chromosomal partitioning of correlations among dairy and beef traits in dual purpose Fleckvieh breed

M. Špehar[1], C. Edel[2], R. Emmerling[2], K.U. Götz[2], I. Curik[3] and G. Gorjanc[4,5]
[1]Croatian Agricultural Agency, Zagreb, Croatia, [2]Bavarian State Research Centre for Agriculture, Grub, Germany, [3]University of Zagreb, Zagreb, Croatia, [4]University of Edinburgh, Edinburgh, United Kingdom, [5]University of Ljubljana, Domžale, Slovenia; mspehar@hpa.hr

Understanding genetic correlations among traits is important for multi-trait breeding. In this study we partition correlations among dairy and beef traits by chromosomes in dual purpose Fleckvieh breed. The data comprised 4,105 progeny-tested bulls genotyped with Illumina BovineSNP50K. Dairy traits were milk yield (MY) and fat yield (FY), while beef traits were net daily gain (NG) and carcass grading (CG). To partition correlations we performed a two-step analysis. First, we estimated allele substitution effects of SNP markers with a multivariate marker model (ridge regression). For this step we used a Monte Carlo Markov Chain method and saved samples from posterior distribution of allele substitution effects. Second, we used these samples and marker genotype data to obtain samples from posterior distribution of breeding values, which were in turn summarized to obtain samples from posterior distribution of genetic covariances and correlations among traits. We performed the second step for the whole genome as well as for different chromosomes, which enabled us to partition overall correlations among traits by chromosomes. Allele substitution effects were positively correlated among all traits (rMY:FY=0.30, rMY:NG=0.17, rMY:CG=0.15, rFY:NG=0.17, rFY:CG=0.15, and rNG:CG=0.28). Overall correlations within dairy and beef traits were high and positive (rMY:FY=0.76, rNG:CG=0.46), while they were low between these two groups of traits (rMY:NG=0.16, rMY:CG=-0.06, rFY:NG=0.11, and rFY:CG=-0.10). Chromosome specific correlations ranged between -0.03 and 0.02 for all pairs of traits. The chromosomal partitioning of the overall trait correlations indicated a positive contribution from within chromosomes and a substantial negative contribution from between chromosomes. For example, the negative correlation between MY and CG of -0.06 had the contribution of 0.41 from within chromosomes and the contribution of -0.47 from between chromosomes. This methodology will be used for further fine partitioning in the future.

Impact of the mitogenome inheritance on the milk production traits in Holstein cows

V. Brajkovic[1], M. Ferenčaković[1], M. Špehar[2], D. Novosel[1], V. Cubric-Curik[1], I. Međugorac[3], E. Kunz[3], S. Krebs[4], J. Sölkner[5] and I. Curik[1]
[1]University of Zagreb Faculty of Agriculture, Svetosimunska 25, 10000 Zagreb, Croatia, [2]Croatian Agricultural Agency, Ilica 101, 10000 Zagreb, Croatia, [3]LMU München, Veterinärstraße 13, 80539 München, Germany, [4]LMU München, Gene Center, Feodor-Lynen-Straße 25, 81377 Munich, Germany, [5]University of Natural Resources and Life Sciences Vienna, Gregor Mendel Strasse 33, 1180 Vienna, Austria; vbrajkovic@agr.hr

Mitochondrial genome is a part of the oxidative phosphorylation metabolic pathway and is responsible for the production of energy in an organism. Still, the impact of mitogenome inheritance on the production traits in livestock is rarely studied. We sequenced (NGS) complete mitogenomes of cows representing 109 maternal lineages. This enabled us to assign mitogenome sequence information to 3,040 cows with 7,576 milk production records (milk yield, fat yield; and protein yield). Thus, we were able to apply quantitative genetic model and estimate the proportion of total variance explained by mitohondrial inheritance (m^2). We estimated m^2 with three different models: (1) cytoplasmic model with maternal lineages (m^2_{CYTO}), (2) haplotypic model with mitogenome haplotypes (m^2_{MITO}) and (3) amino-acid model with unique amino-acid combinations (m^2_{AMIN}). Effects of animal, parity, calving season, region, year, heard and age at first calving were also considered in each model. Estimated proportions of phenotypic variances explained by m^2_{CYTO} and m^2_{MITO} were almost identical ranging from 0.04 to 0.05 for all three milk traits. In amino-acid model, the explained proportion of total variance was higher for protein yield (m^2_{AMIN}=0.07), equal for milk yield (m^2_{AMIN}=0.05), and lower for fat yield (m^2_{AMIN}=0.03). Obtained results show that considerable proportion of the phenotypic variance in milk traits is explained by mitogenome variation. While our further research is targeted towards identification of causal mutation, the utilisation of mitogenome inheritance in practical animal breeding remains challenging.

Genetic mechanisms associated with host defence to infection in the cow mammary gland

V.H. Asselstine[1], F. Miglior[1,2], P.A.S. Fonseca[1], A. Islas-Trejo[3], J.F. Medrano[3] and A. Cánovas[1]
[1]University of Guelph, Animal Biosciences, 50 Stone Rd. East, Guelph, ON, N1G 2W1, Canada, [2]Canadian Dairy Network, 660 Speedvale Avenue West, Suite 102, Guelph, ON, N1K 1E5, Canada, [3]University of California- Davis, Department of Animal Science, 1 Shields Ave, Davis, CA 95616, USA; vasselst@uoguelph.ca

Mastitis is currently one of the most challenging and profit limiting problems in lactating dairy cows. Cows have certain defence mechanisms in place, such as the mucus plug in the mammary gland that forms after milking. The faster this plug forms, the less susceptible the cow is to having bacteria enter her mammary gland, thus, preventing some instances of mastitis. The study of milk transcriptome from healthy and mastitic samples using RNA-Seq technology can provide measurements of transcript levels associated with the immune response to the infection. This will aid in understanding the development of the disease and the associated host defence. Transcriptome analysis using RNA-Seq was performed in Holstein milk somatic cells from 6 cows; two samples were taken from each cow from two separate quarters, one classified as healthy (n=6), one as mastitic (n=6). In total, 449 genes were differentially expressed between the healthy and mastitic quarters. In general, the genes with the highest expression (Reads Per Kilo base per Million mapped reads; RPKM≥500) in the healthy category were associated with milk components (i.e. *CSN2* and *BLG*), and in the mastitic category they were associated with immunity (i.e. *B2M* and *CD74*). Functional analysis identified 55 significant metabolic pathways (FDR<0.05) associated with the immune system and diseases, such as interleukin 17 (*IL-17*) and mastitis. Next, a list of candidate genes for SNP discovery was established, which included *NFKBIA, CD74, FCER1G, B2M, SDS* and *GLYCAM1* genes. In conclusion, the identification of genes and biomarkers associated with host defence to infection will aid in improving the sustainability of agricultural practices, by facilitating the selection of cows with improved immune system and resistance to mastitis.

Glycemic index in finishing diet of Iberian pigs and its effect on muscle transcriptome

A. López-García[1], L. Calvo[2], Y. Núñez[1], R. Benítez[1], J. Ballesteros[1], C. López-Bote[3], J. Segura[3], J. Viguera[4] and C. Óvilo[1]
[1]INIA, Ctra La Coruña km 7.5, 28040 Madrid, Spain, [2]INCARLOPSA, Ctra N-400 km 95.4, 16400 Tarancón, Spain, [3]UCM, Avda Puerta de Hierro, s/n, 28040, Madrid, Spain, [4]Imasde Agroalimentaria, C/ Nápoles 3, 28224 Pozuelo de Alarcón, Spain; adrian.lopez@inia.es

Swine farming mainly focuses on fresh meat production, and genetic selection for commercial breeds usually favors efficiency and neglects intramuscular fat (IMF) deposition. In Spain, a particular case occurs with Iberian pigs, as they have high tendency to fatness and IMF content. But, as Iberian pig is frequently crossbred with Duroc, final dry-cured products become heterogeneous for IMF, as in other commercial breeds. New knowledge in transcriptomics has opened another way to modulate IMF content, as gene expression may be modified via diet composition. Thus, it is worth to know how the pig transcriptome changes with diet modifications along its different life stages. In this experiment we studied the influence of glycaemic index (GI) of the finishing diet on Iberian crossbred pigs' muscle transcriptome. 200 animals were randomly distributed in 2 fattening dietary groups, high (H) and low (L) GI, with values of 16.0 and 13.0 respectively. Both diets were isoenergetic. Carcass performance and meat quality and composition measures were taken, as well as IMF measures for all high-quality cuts. Samples from Biceps femoris of 24 animals, 6 males and 6 females from each treatment, were collected for transcriptomic analysis. After RNA sequencing, the standard Tuxedo Protocol was employed for the data analysis. ANOVA showed little differences for phenotypic traits: ADG was significantly higher in HGI group for both sexes. In males, backfat thickness was higher in HGI group, also affecting noble cuts performance. 55 differentially expressed (DE) genes between diets were found, some of them related with carbohydrate or lipid metabolism (CHI3L1, PFKFB3, B3GALT1, FADS2) and with skeletal muscle development (ANKRD1, CSRP3). Most of these genes were DE only in females, which points to a strong interaction between diet and sex regarding to transcriptome. Phenotypic data also show differences between sexes. qPCR validation and additional analysis of related biological functions and metabolic pathways is needed to deepen in our findings.

Sequence-based association study of resistance to paratuberculosis in Holstein and Normande cattle

M.P. Sanchez[1], R. Guatteo[2], A. Davergne[3], C. Grohs[1], S. Taussat[1,4], P. Blanquefort[5], A. Delafosse[6], A. Joly[7], C. Fourichon[2] and D. Boichard[1]
[1]GABI, INRA, AgroParisTech, Université Paris Saclay, CRJ, 78350 Jouy-en-Josas, France, [2]BIOEPAR, Oniris, INRA, Atlanpole-Chantrerie, CS 40706, Nantes, France, [3]GDS Haute Normandie, Agropole, 76235 Bois-Guillaume, France, [4]Allice, 149 rue de Bercy, 75595 Paris, France, [5]GDS Pays de la Loire, Quantinière, 49800 Trélazé, France, [6]GDS Orne, Chemin de Maures, 61000 Alençon, France, [7]GDS Bretagne, Av E Degas, 56000 Vannes, France; didier.boichard@inra.fr

A case-control genome-wide association study was applied at the whole genome sequence level to identify QTL of resistance to paratuberculosis in French Holstein (HOL) and Normande (NOR) cattle. Infected cases included either confirmed clinical cases or non-clinical shedder cows confirmed with both positive blood ELISA test and faecal PCR. Control cows were at least 60 months old and born in the same herd and during the same month as infected cows. They were required to have two repeated negative blood ELISA tests, confirmed by a third negative ELISA test and a negative faecal PCR test obtained in one unique laboratory. A total of 1,644 HOL (806 cases / 838 controls) and 649 NOR (416 cases / 233 controls) cows with confirmed infectious status were genotyped with the Illumina BovineSNP50 BeadChip. Genotypes were imputed to the high density (HD) level with FImpute and then up to whole genome sequence with Minimac using 1,466 sequenced bulls of the '1000 bull genomes' Run 6. GWAS was carried out within breed with GCTA, accounting for the population structure through a HD-based genomic relationship matrix. The most significant QTL was detected in HOL on chromosome (BTA) 13 at 63.5 Mb. Two other regions were genome-wide significant on BTA 12 at 70.7 Mb in HOL and BTA 23 in the major histocompatibility complex region in both breeds. Results were markedly improved by the analysis at the sequence level with more significant and results and accurate locations. Candidate genes were identified in each detected region and a gene network was proposed.

Unraveling genomic regions associated with environmental variance of litter size in rabbits

C. Casto-Rebollo[1], M.J. Argente[2], M.L. García[2], A. Blasco[1], R. Pena[3], L. Fontanesi[4] and N. Ibáñez-Escriche[1]
[1]Institute for Animal Science and Technology, Universitat Politècnica de València, 46022 València, Spain, [2]Departamento de Tecnología Agroalimentaria, Universidad Miguel Hernández de Elche, 03202 Orihuela, Spain, [3]Departament de Ciència Animal, Universitat de Lleida, 25003 Lleida, Catalonia, Spain, [4]Department of Agricultural and Food Sciences, Division of Animal Sciences, University of Bologna, 40127 Bologna, Italy; cricasre@posgrado.upv.es

Genetic determination of environmental variance (Ve) is becoming of increasing interest in animal breeding. Ve can be decreased by selection, leading to animals with greater capacity to cope with changes of the environmental conditions. A ten-generation divergent selection experiment for Ve of litter size in rabbits was performed at Miguel Hernández University of Elche. Does of the low Ve line tolerated external stressors more effectively than those of the high Ve line. In order to identify genomic regions associated to Ve of litter size, 288 does of the high (149) and low (139) lines at generations 11th and 12th were genotyped using the Affymetrix Axiom OrCunSNP (94,282 SNPs after quality control). Ve was calculated as the within-doe variance of litter size, after litter size pre-correction by year-season and parity-lactation effects. Single marker regression analyses were performed using three different approaches: (a) with line as fixed effect, (b) corrected by genotype relatedness matrix (GRM) and (c) corrected by GRM excluding the chromosome of the SNP tested. In all cases, a region (1.3 Mb) with 54 relevant SNPs ($P<6.26e-04$) was identified in chromosome 3. In this region, three genes highlighted as candidate genes (*DOCK2*, *LCP2* and *SLIT3*). Functions of these genes are involved in immunological processes such as macropinocytosis, thymic T cell selection, mast cell activation and response to cortisol. Further studies are underway to validate and understand the biological function of these candidate genes, before they can be applied to rabbit breeding schemes.

Session 46 Theatre 12

Genetic analysis for production and health traits in a commercial rabbit line

H. Garreau[1], M. Maupin[2], J. Hurtaud[2] and M. Gunia[1]
[1]GenPhySE, INRA, INPT, ENVT, Université de Toulouse, Castanet Tolosan, 31326, France, [2]HYPHARM SAS, La Corbière, 49450 Sèvremoine, France; herve.garreau@inra.fr

Genetic parameters and trends were estimated for the selection criteria body weight at 70 days (BW70), carcass yield (CY) and resistance to infectious diseases (ID) in the paternal commercial rabbit line AGP59 of the Hypharm breeding company. The ID criteria is a binary trait based on presence (1) or absence (0) of clinical signs of diseases. This study included data recorded on 39,726 selection candidates and 5,372 slaughtered sibs between 2008 and 2016. All animals were weighed at 70 days. Clinical signs of disease were systematically recorded for all weighed animals and also for those which died between weaning and the end of the growing period. The sib-testing population is created at each weaning by collecting young rabbits in primiparous does litters. Their live body weight and carcass weight was recorded at 71 days of age and carcass yield was then estimated. Genetic parameters and genetic trends were estimated using REML and BLUP methodology. Heritability estimates were 0.28±0.02, 0.44±0.05 and 0.03±0.01 for BW70, CY and TI, respectively. Genetic correlations between the three traits were not significantly different from zero. Phenotypic correlations were also low, except for the negative value (i.e. favourable) between BW70 and ID (-0.42±0.05). The annual genetic gain, estimated in genetic standard deviation units (traits units) was 0.49 (92 g), 0.32 (0.38 points of carcass yield) and -0.12 (-0.004 points of diseased animal frequency) for BW70, CY and TI, respectively. The results demonstrate that selection for both production and health traits is possible.

Session 47 Theatre 1

Insect production: current EU regulations and sector's 1-3 year roadmap for widening opportunities

T. Arsiwalla
IPIFF / Protix, Industriestraat 3, 5107 NC Dongen, the Netherlands; tarique.arsiwalla@protix.eu

1. ANIMAL FEED: Current situation: (1) Feed for the insects: only plant origin sources (pre consumer) and milk or egg products (strictly no meat, fish, manure or post consumer organics) (2) Use of insect proteins (PAP): only in pet food or aquaculture; use of insect lipids: allowed for all farmed animals as well as pet food IPIFF roadmap for next 1-3 years: (3) Feed for the insects: allow the use of pre-consumer 'former foodstuffs, including meat and fish'; this comprises for example beyond-expiry date products from supermarkets (4) Use of insect proteins (PAP): allow the use of insect PAPs in feeds for chickens and pigs. 2. HUMAN CONSUMPTION: IPIFF takes an active role in informing and guiding its members in relation to building and filing Novel Food applications, which are required to go or stay on the EU market.

Session 47 Theatre 2

Safety aspects when rearing insects for feed or food consumption

H.J. Van Der Fels-Klerx[1], R. Andriessen[2], J. Van Schelt[3], R. Van Dam[1], T.C. De Rijk[1] and L. Camenzuli[1]
[1]RIKILT Wageningen UR, Toxicology, Novel Foods and Agrichains, Akkermaalsbos 2, 6708 WB Wageningen, the Netherlands, [2]Proti-farm R&D BV, Harderwijkerweg 141B, 3852 AB Ermelo, the Netherlands, [3]Koppert BV, Veilingweg 14, 2651 BE Berkel en Rodenrijs, the Netherlands; ine.vanderfels@wur.nl

Nowadays, several species of insects, such as larvae of *Alphitobius diaperinus* (lesser mealworms, LMW) and *Hermetia illucens* (black soldier flies, BSF) are considered a valuable source of novel proteins for feed and food production in Europe. Before insects can be introduced on large scale in the European market, possible safety issues need to be addressed, such to ensure product safety. This study aimed to investigate the potential accumulation of four different mycotoxins in BSF and LMW larvae, using substrate that was artificially contaminated with four different mycotoxins (aflatoxin B1, deoxynivalenol (DON), ochratoxin A or zearalenone) and the mixture. Substrate was spiked at three different concentrations, and two controls were included, resulting into 17 treatments. Spiked concentrations were 1-25 times the current EC limits for mycotoxin present in feed ingredients. BSF and LMW were reared on the substrates, and then starved for two days to clean their gut. Samples were collected from the harvested larvae, and the residual material both before and after starvation of the insects. All samples were analysed for the presence of the mycotoxins and for some metabolites, using LC-MS/MS. No differences were observed in survival rate and growth rate of BSF and LMW in the 16 out of the 17 different treatments. Results of the chemical analyses showed that the two insect types did not accumulate the four mycotoxins and most of the metabolites investigated. From the mass balances, it was shown that the four mycotoxins were metabolized by the insects to varying extent, with not only difference between the toxins but also between BSF and LMW. Metabolites investigated did account for the mass balance only to a small extent except for zearalenole metabolites in BSF. It was concluded that BSF and LMW excreted or metabolized the four mycotoxins that were present in their substrate, and that these four mycotoxins do not seem to be a concern in insect considering current EC limit for feed production.

Session 47 Theatre 3

Legal challenges and opportunities for insects as novel food and their use in animal feed

N. Carbonnelle
twoBirds, Avenue Louise 235 box 1, 1050 Brussels, Belgium; nicolas.carbonnelle@twobirds.com

The classification of insects as novel food has been clarified through the adoption of Regulation (EU) no. 2015/2283, which replaces Regulation 258/97. Insects have also a strong potential, and have started being used, in animal feed, albeit subject to rather strict conditions. The novel food status of edible insects implies that they are subject to safety assessment and pre-market approval before being placed on the market. The new Regulation provides two different procedures that can be used to place edible insects in the European market. In both cases, the applicant is required to submit a set of information to the Commission that may involve the EFSA in the food safety assessment. This presentation analyses the procedures provided by the new novel food regulation. In that perspective, the guidelines issued by EFSA in September 2016 will be critically analysed, highlighting the aspects that appear to be more relevant when preparing an application for authorising an insect species, and the legal framework as a whole, including the novel food regulation and its implementing rules, will be presented. As regards insects used in feed, our research shows that the use of PAPs in animal feed is still strongly restricted at European level because of the risk of BSE, although some changes have been introduced for aquaculture animals. The results also show that strong barriers remain to the use of organic waste (catering and household waste) to feed insects, making their use as animal feed less competitive in terms of cost. It follows that, in order to use them within a circular economy, further amendments to rules currently in place are needed.

Session 47 Theatre 4

Effect of post-harvest starvation and rinsing on microbial numbers in mealworm larvae

E. Bragason and A.N. Jensen
Technical University of Denmark, National Food Institute, Kemitorvet, bygning 202, 2800 Kgs. Lyngby, Denmark; anyj@food.dtu.dk

Mealworms (*Tenebrio molitor*) reared for feed and food are commonly starved 1-2 days after harvest before being killed to empty their gut and presumably lower the microbial load for sanitary reasons. This study aimed to assess the bacterial numbers in mealworms before and after starvation 24 h or 48 h and also the effect of rinsing the mealworms with water. At start of the experiment, mealworm larvae close to pupation were separated from their flour based substrate. Sub-samples of 8 g mealworm were collected 0 h, 24 h and 48 h after separation and either homogenized directly in a sterile mortar or rinsed twice beforehand in 100 ml water for 1 min at 200 rpm (each treatment sampled in duplicate). For each sample, ten-fold dilutions series were prepared from 5 g of homogenized mealworm in 45 ml saline (0.9% NaCl, 0.1% peptone). Each dilution was plated on Plate Count Agar (PCA) (0.1 ml) and Enterobacteriaceae Count Plate (ECP) (Petrifilm, 3M Denmark) (1 ml). PCA and ECP were incubated overnight at 37 °C before counting of colonies. The aerobic count was 8.3±0.1, 7.9±0.3 and 7.9±0.1 log cfu/g in average (±SD) for 0, 24 and 48 h of starvation, respectively, with no effect of rinsing in water. Also the effect of starvation did not seem to affect the resulting microbial load markedly. The *Enterobacteriaceae* numbers were 6.9±0.4, 5.9±0.3 and 7.0±0.4 log cfu/g in average (±SD) for 0, 24 and 48 h of starvation. The explanation for the observed reduction of *Enterobacteriaceae* of approx. 1 log cfu/g after 24 h but not after 48 h starvation is uncertain. The *Enterobacteriaceae* family of Gram-negative bacteria includes both harmless bacteria as well as human pathogens like *Salmonella* and *Escherichia coli*. Although the taxonomic level was not resolved further in this study, this finding emphasizes that means to reduce the bacterial need to be established as post-harvest starvation alone has little effect on the microbial load in mealworms. More knowledge about how the microbial composition and load of the feed substrate influences the microbial content in the mealworm larvae may help to ensure the microbial quality of mealworm products.

Session 47 Theatre 5

Consumers' acceptance of insects as food, feed and dog food: a comparative study

T. Von Jeinsen and H. Heise
University of Goettingen, DARE, Platz der Goettinger Sieben 5, 37073 Goettingen, Germany; hheise@gwdg.de

In recent years, insect farming on industrial scale has grown strongly. Insects have great potential to convert organic matter into high value sources of protein and fat efficiently. They need far less feed than conventional sources of meat such as poultry, pigs and cattle and generate only a fraction of greenhouse gases, those animal species do. For this reason, insects appear to be a true future alternative to traditional meat products. Furthermore insects could replace other protein sources such as soy beans in livestock feed. Due to the favourable composition of nutritional substances and their hypoallergenic effects, insects are also considered to be an attractive component of pet food. While there is already a considerable number of studies which found that the idea of eating insects has not yet become very popular among European consumers, there is little known about consumers' acceptance of insects as feed for livestock or as pet food. To close this research gap, 500 dog-owning German consumers were questioned via an online survey about their acceptance of insects as food, feed and dog food. The determinants of acceptance of different food, feed and dog food products was analysed using a food innovation adoption model by Ronteltap *et al.* First results show that consumers are most willing to accept insects indirectly, as they show the highest acceptance to try meat or eggs from animals that have been fed with insects. They also expressed cautious optimism towards trying insect based dry fodder or treats for their dogs. However, these consumers were rather undecided whether they would be willing to accept insect based canned fodder as dog food and insect based food products for human consumption remain rather unacceptable. A comparison of different food products showed that consumers were least willing to try whole mealworms followed by insect based flour, chocolate containing whole mealworms and burgers containing insect flour. Consumers were rather undecided whether they should try cookies and noodles containing insect flour. Our results have important managerial implications and can help to establish a market segment for insect based products which is accepted by society.

Insects in animal feed: beyond the protein concept

L. Gasco
University of Turin, Department of Agricultural, Forest, and Food Sciences, largo P. Braccini 2, 10095 Grugliasco, Italy; laura.gasco@unito.it

Animal products are of fundamental importance in human nutrition mainly because of the high biological value of proteins they provide. In the last years, the demand for animal proteins has continued to increase worldwide and estimates forecast that by 2050, as the demand for food will grow by 60%, the production of animal proteins will grow by around 1.7% / year leading to an increase in animal feed production. Sustainable feed production essentially requires sustainable feed ingredients, and proteins are one on the major issues. Insect products are one of the most promising sources to generate proteins sustainably and with low footprint. Research highlighted how insect-based proteins could efficiently be included in animal feeds in partial substitution of conventional proteins commodities in fish, poultry and pigs. Recently the focus has moved to animal health as insects seems to be a promising source of various bioactive substances (antimicrobial peptides, chitin, lauric acid) with pharmacological functions, able to modulate the animal microbial communities or to stimulate the immune response. Therefore, beyond the protein contribution in animal feeds, insects could be viewed as source of valuable compounds with positive effects on growth, health and resistance against pathogens.

Feeding on larvae of black solider flies doesn't substantially modify chicken caecal microbiota

N. Moula, J.-F. Cabaraux, E. Dawans, B. Taminiau and J. Detilleux
University of Liege, Sart Tilman, 4000, Belgium; jdetilleux@ulg.ac.be

Although insects have been proposed to feed poultry, few studies have assessed their impact on microbiota even though it has an important role in chicken performance. Here, we aimed at characterizing the change in caecal microbiota of chicken fed larvae of black soldier fly (BSF) raised on horse manure. We collected larvae at the pre-pupal stage and stored them directly at – 20 °C. In parallel, we raised 40 one-day-old male Ross chicks and individually fed them a commercial feed with either 0% or 8% of de-frosted BSF larvae. We collected bacterial DNA from samples of larvae, their growing substrate, and from the caecal content of the chickens for microbiota characterization with the software Mothur. We used analyses of variance to determine whether percent abundances of caecal operational taxonomic units (OTUs) were different in chicken receiving one or the other diet, after adjusting for the effect of replication. Around 20% of the genus *Dysgonomonas* were retrieved in de-frosted BSF larvae and in the growing substrate after passage. This may be one of the mechanisms used by the larvae to transform manure because these bacteria have a fermentative metabolism producing acids and no gas. A total of 37 families and 5,275 species were detected in chicken caecal microbiota. *Firmicutes* were the most abundant phylum (90.83%) followed by the Bacteroidetes (6.93%). The majority of sequences within the Firmicutes phylum belonged to the families *Ruminococcaceae* (33.01%) and *Lachnospiraceae* (46.17%). No significant differences were found between experimental and control groups in the mean relative abundances of bacterial OTUs. However, relative abundances of both *Bacillaceae* and of *Rhodobacteraceae* were low (<0.5%) and significantly lower in the caeca of birds receing BSF than control diets. Within *Bacillaceae*, 50% were *B. thuringiensis*, a bacteria known to produce metabolites highly toxic to insect which suggests larvae contain antimicrobials that affect growth of *B. thuringiensis*. These preliminary results are helpful to better understand the effects of insect feed on poultry microbiota.

Nutritive value of black soldier fly (*Hermetia illucens*) larvae reared with onion residues

O. Moreira[1], B. Nardozi[1], R. Nunes[2] and D. Murta[2,3,4]
[1]INIAV Santarém, Quinta da Fonte Boa, 2005-048, Portugal, [2]Ingredient Odyssey LDA, Quinta das Cegonhas, Apartado 577, 2001-907 Santarém, Portugal, [3]CIISA-Faculdade de Medicina Veterinária – ULisboa, Avenida da Universidade Técnica, 1300-477 Lisboa, Portugal, [4]FMV – Universidade Lusófona de Humanidades e Tecnologia, Campo Grande, 376, 1749-024 Lisboa, Portugal; olga.moreira@iniav.pt

This research is part of the Project ENTOVALOR (POCI-01-0247-FEDER-017675), based on Circular Economy, aiming to develop the knowledge and processes that will allow organic residues to be reintroduced in the value chain as a nutrient source, using black soldier fly (BSF) as the linkage element. The specific objectives are related with the reintroduction of the nutrients in the value chain, the contribution to the establishment of quality and biosafety standards for secondary raw materials originated by waste recovery and also the production of new products and services originated by waste valorisation and insect meal production. Among the expected results, the creation of novel protein sources, such as insect meal and insect protein extract are foreseen, developing an alternative nutritional source for animal feed. BSF larvae were reared in two different substrates: (1) a commercial meal, composed by wheat bran (50%), alfalfa (30%) and maize bran (20%) and (2) a mixture of the commercial meal and onion residues (1:1.5) to achieve an equal relative humidity of 60%. The preliminary results to be presented will concern larvae nutritional value for monogastric: chemical composition (protein, fat, fibre, energy, minerals and fatty acid composition) and *in vitro* digestibility. The effects of the onion inclusion and of the larvae development stage on the nutritive value of larvae will be evaluated. The results will contribute to the selection of larvae rearing substrates based in agricultural byproducts and to larvae efficient use as an alternative protein rich ingredient for the monogastric compound feed industry.

***In vitro* rumen fermentation of black soldier fly larvae reared on different sources of food waste**

K. Di Giacomo[1], P.A. Giraldo[1], C. Leach[2] and B.J. Leury[1]
[1]The University of Melbourne, Faculty of Veterinary and Agricultural Sciences, Parkville, Vic. 3010, Australia, [2]Hatch Biosystems, Abbotsford, Vic. 3067, Australia; kristyd@unimelb.edu.au

Organic waste e.g. vegetable waste, can be redirected away from landfill and bioconverted into a valuable insect (black soldier fly, BSF) larvae derived livestock protein source. However, there is limited research on feeding BSF to ruminants. This experiment investigated *in vitro* rumen fermentation of BSF larvae reared from three waste sources. The BSF larvae were fed either seaweed (S) waste from local beaches, potato waste (PW) sourced from a local potato product manufacturer or commercial horse pellets (C). Each treatment was incubated at either 100, 50 or 30%, with the remainder of the fermentation substrate provided as Lucerne hay, and 100% Lucerne hay included as control (L). Samples of each treatment (1 g) were added to serum flasks (n=5 per treatment) containing buffered rumen fluid obtained from rumen-cannulated lactating dairy cows. The flasks were purged with carbon dioxide and maintained in a water bath at 39 °C. Gas production was monitored every 5 min for 48 h using the ANKOM wireless *in vitro* gas production system. Gas production was modelled using a Gompertz equation to determine the rate (slope, β) and maximum volume of gas production (max). Statistical analysis was conducted in GenStat using a two-way ANOVA with the main effects of treatment and inclusion rate and the interactions between them. There was an interaction between treatment and inclusion rate such that β was lowest in C100% and greatest in PW30% (L=0.15; C=0.15, 0.17 and 0.16; PW=0.16, 0.18 and 0.27 and S=0.23, 0.18 and 0.17 ml/g for 100, 50 and 30% respectively, SED 0.039; P=0.025). Similarly, the max was lowest for 100% BSF treatments and was greatest for PW (L=59; C=23, 64 and 64; PW=66, 71 and 69; S=36, 57 and 60 ml for 100, 50 and 30% respectively, SED 4.3, $P<0.001$). These data demonstrate that *in vitro* fermentation rate (β) and max are modified by the larvae rearing substrate and by the larvae inclusion level. This suggests that BSF larvae will have a different nutritional value for ruminants depending on the larvae rearing substrate.

Effects of black soldier fly meal in poultry and fish diets on performance and product quality

F. Leiber, T. Stadtlander, J. Wohlfahrt, C. Sandrock and V. Maurer
Research Institute of Organic Agriculture (FiBL), Ackerstrasse, 5070 Frick, Switzerland; florian.leiber@fibl.org

Insect-based protein meals are in discussion as sustainable components in livestock and fish diets, with a potential to recycle wasted food materials and to replace partly feeds from arable land. Black soldier fly (BSF; *Hermetia Illucens*) is of particular interest in this context. Since insect production for animal feed appears as an additional trophic level in the food chain, it should either base on substrates, which are not directly edible for livestock and fish, or significantly improve feed efficiency or product quality to justify its application. Feed efficiency and product quality after replacing conventional protein sources by BSF meal were therefore tested in recent experiments with layers, broilers and trout. In a layer experiment, respectively four groups of 10 hens were fed either a control diet, a diet containing 12 g/100 g or a diet with 24 g/100 g defatted Hermetia meal for four weeks. Neither laying performance nor feed efficiency (g/g egg weight) nor egg composition differed by diet. In a fattening experiment with 15 broilers per group the partial replacement of soybean meal by mixtures of either alfalfa or peas with Hermetia meal (7.8 g/100 g Hermetia) did neither affect weight gains nor carcass weights compared to a control diet. Compared on a group level, neither feed intake nor feed efficiency differed with the diets. Also weights of meat cuts, shear force and meat colour were not affected by diet. Only cooking loss increased in meat from broilers provided with the Hermetia-pea mixture (P>0.001). In a further feeding trial, young rainbow trout (body weight 67 to 125 g) were fed for 7 weeks with either a usual control diet or a feed, where 46% of the fishmeal was replaced by Hermetia meal. Initial and final body mass were equal with both diets and no differences were found for growth rate, weight gain. Chemical fish composition was not affected. A degustation panel did not reveal differences in taste, odour or texture of the trout filets. If also in future no advantageous effects of feeding Hermetia meal on performance, feed efficiency or product quality would be found, production of BSF should prove sustainability in itself to justify its use as animal feed component.

Evaluation of reusable hiding units for rearing house crickets (*Acheta domesticus*)

M. Vaga, E. Gustafsson and A. Jansson
The Swedish University of Agricultural Sciences, Department of Anatomy, Physiology and Biochemistry, Ulls väg 26, 75007 Uppsala, Sweden; merko.vaga@slu.se

In order to improve hygiene and to reduce the material waste in production of house crickets (*Acheta domesticus*), two housing methods were evaluated that could substitute non-cleanable cardboard materials often used in cricket rearing. Hiding units were prepared out of tubes of approx. 6 cm long black PEM pipe (∅ 25 mm) or transparent silicone water pipe (∅ 25 mm), horizontally stacked to 5 layers (floors) high with silicone. Both units were placed inside a 20 l plastic box containing 30 adult crickets bought from a local pet store. Feed and water were placed close to both units. Twice a day on days 3-6 and days 29-31 of the observation period the number of crickets inside and on the hiding units was counted. None of the crickets died during the observation period. Overall the crickets preferred to use the black hiding unit more than the transparent one (P<0.01). In fact only 2 times it was observed crickets using the transparent unit. The floor had significant effect (P<0.001) as floor 1 and 3 were used 41 and 25% of the time, respectively, whereas the 5th floor was used only 3% of the time. There were no significant differences between climbing on the hiding unit and being inside the unit, except for the 1st floor where crickets preferred to spend 74% of the time inside. This however can also be due to lower available outer surface of the lowest floor. It can be concluded that house crickets prefer dark hiding places to either shade them from excess light or due to natural hiding behaviour from predators. Reusable and cleanable black water piping or similar dark materials can provide long term solutions for providing extra floor space in cricket rearing facilities for safe food production.

Session 47 Poster 12

Optimizing growth during first weeks of Tenebrio molitor rearing

C.L. Coudron, D. Deruytter, J. Claeys and S. Teerlinck
Inagro, Ieperseweg, 8800, Belgium; carl.coudron@inagro.be

For scientific experiments with mealworm the following conditions are considered optimal: a uniform temperature of 26 °C with a relative humidity (RH) of 70% and daily addition of a moisture source like carrot. However, little is known about specific growth conditions for mealworm hatchlings and their needs during the first weeks. Nevertheless, in other production animals like broilers optimal growth is achieved when the temperature steadily decreases from 32 to 20 °C before slaughter. Personal contact with several large scale breeders indicates that they also alter their climate depending on the growth stage of mealworm. Moreover there seem to be major differences between breeders on when to start the supply of a moisture source. All these factors indicate that the growth rate can be accelerated significantly by determining size specific optimal conditions (like temperature, humidity and moisture source supply). To assess the influence of different parameters a comparison was made with a standard breeding method at 26 °C, 70% RH and supply of moisture starting 28 days after ending oviposition. In contrast to most lab scale tests, the experimental design is at a semi-industrial scale where rearing trays are commonly 600×400 mm to account for potential heat production and density effects. Starting at a higher temperature of 29 °C young mealworm development can be accelerated, while temperature needs to decrease to 26 °C for older mealworm. Literature indicates that higher humidity (RH>70%) is beneficial for growth in all stages of mealworm development as they can absorb atmospheric water. Initial results indicate that by nursing them at 80% or even 90% RH during their first weeks, growth accelerates, dehydration of eggs can be prevented and need for an external moisture source might be postponed. Young mealworm eat a minimal amount of the moisture source and the remainder may be a source for fungal infections. However starting the supply of moisture source soon after hatching, has a positive influence on the growth of mealworm. By determining the optimal humidity and the minimal amount of moisture source necessary where everything is consumed and growth rate is enhanced, lifecycle can be shortened. Thus making mealworm rearing more profitable for industrial breeding.

Session 47 Poster 13

Morphology and localization of carbonic anhydrase in the digestive tract of the house cricket

E. Thorsson, A. Jansson and L. Holm
Swedish University of Agricultural Sciences, Anatomy, Physiology and Biochemistry, Box 7011, 750 07 Uppsala, Sweden; lena.holm@slu.se

Crickets are omnivores and can efficiently convert plant material into valuable protein. That, and a higher percentage of edible weight compared to cattle and chicken, makes them highly interesting as a protein source to both humans and animals. Knowledge about crickets is limited and to gain insight into their digestive functions the aim of this study was to describe the morphology of the digestive tract and the localization of the enzyme carbonic anhydrase (CA). Conventionally bred *Acheta domesticus* (20) at the age of 40-45 days were divided into two groups, five males and five females in each. One group was starved for 36 h and controls were fed chicken feed, both groups had free access to water. After euthanization (CO_2 followed by decapitation) the digestive tract was dissected, anatomical data collected and tissues were prepared for histological evaluation and histochemical localization of CA following Ridderstråle´s resin method. The digestive tract consists of foregut, midgut and hindgut and shows a morphology similar to other species of the *Gryllidae* family. The surface epithelium is simple and varies from flattened or low cuboidal to high columnar cells with immature cells in clusters called nidi. The results suggest presence of goblet cells and possibly gastric glands in the midgut. Striated muscle was detected along the entire length of the digestive tract varying in thickness from a few fibres to a double layer, with several fibres in each layer. Membrane-bound staining for CA was detected in the muscle cells along the entire digestive tract. In the midgut basolateral membrane-bound CA was shown in the columnar epithelial cells but not in goblet cells or cells of nidi. There were no differences in CA staining, gut or body size between sexes or treatments. The gut contents of the foregut is acidic and we suggest that alkaline secretions from the midgut epithelium, mediated by CA, neutralize the chyme as it enters the midgut.

Insects for feed: descriptive statistics on proximate and nutritional composition from a review

D. Meo Zilio and M. Guarino Amato
Council for Agricultural Research and Economics, Salaria 31, 00015 Monterotondo (RM), Italy; david.meozilio@crea.gov.it

Insects for feed are object of growing debate. Their suitability has still to be fully accepted. Insects composition is variable and could be modified via manipulation of rearing substrates and growing conditions. This work aims to report the results of a number of papers to describe the variability and highlight differences in composition and nutritional quality of some insects, reared under varied conditions. Data on composition, fatty acids, mineral content and amino acids (AA) were collected from 60 original papers out of 150. Four species, included in EU regulation on animal protein (893/2017), *Acheta domestica* (AD), *Tenebrio molitor* (TM), *Hermetia illucens* (HI), *Musca domestica* (MD) were considered. Data not consistent or not sufficiently described were excluded. Following descriptive statistics were calculated: mean, median, coefficient of variation (CV), range, standard error (SE). Crude protein content ranges between 62% (AD) and 43.2% (HI) with a coefficient of variation between 11 and 20%. AD and MD are similar for fat (16.5 and 17.5%), HI ranks intermediate (24%) and MD is the fattest (30%). MD shows the highest CV (47%). Minerals and ashes are in general related to the diet. The highest dispersion (CV>100%) pertains Ca, Fe, Cu, Zn, Mn. HI ranks first for saturated fatty acids content, but n6/n3 is most favourable. For the other species n6/n3 is unbalanced towards the n6 group. Fatty acids are supposed to be very diet dependent. CV is low for saturated/unsaturated ratio, but much higher for n6/n3, ranging from 33% for MD (only 3 observations were acquired) to 135% for HI. Among the most limiting AA, all species show good content of lysine, a content of methionine higher than the soybean meal, low content of tryptophan. The most variable AA are: tryptophan (CV up to 173% in MD) and cysteine with a CV ranging between 67% for AD e HI and 112% (TM). The highest CV for methionine is in HI (78%). The high dispersion within species supports the possibility to change the nutritional value of insects according to diet and growth conditions. Further in depth studies are needed to confirm the hypothesis and to set up protocols in order to modify the nutritional value as a function of target species.

Black soldier fly larvae reared on contaminated substrate by *L. monocytogenes* and *Salmonella*

F. Defilippo, A. Grisendi, V. Listorti, M. Dottori and P. Bonilauri
IZSLER, RE, Pitagora 2, 42124, Italy; paolo.bonilauri@izsler.it

The safety of protein from insects and subsequently the safety of meat and fish from animals fed on such a diet requires further assessment. Black soldier fly (*Hermetia illucens* L.) larvae (BSFL) have been noted to reduce the microbial load of substrates, decreasing concentrations of bacteria in compost and faecal material. The purpose of our research was to study the behaviour of BSF larvae reared on diets artificially contaminated with *Salmonella* typhimurium and *Listeria monocytogenes*. Two sets of experiments were conducted in order to investigate: (1) the reduction of pathogens in contaminated substrate without larvae (control); (2) the presence of bacteria in larvae, prepupae, pupae and their substrate. BSFL were reared under controlled conditions (RH 70%, photoperiod 14:10 h (L:D) and temperature 25 °C) on two substrates (Gainesville diet and a homemade artificial diet). The larval substrate was contaminated with *S. typhimurium* and *L. monocytogenes* (about 1×10^8 cfu/g). For each trial substrate and larvae were tested for pathogen enumeration, pH, and aw. In controls the pH remained stable until the end of experiments (6.59-7.95) while aw remained stable until 7 days (0.977) but decreases reaching 0.650 after this period. In substrates with larvae pH and aw remained stable through the duration of the study. The contamination of *L. monocytogenes* remained constant in control experiments while a slow decline of *Salmonella* (-0.0077 log.conc/h D=130 h) was observed. When larvae were present, *Salmonella* showed a fast decline in the first 8 days (-0.02 log.conc/h D=45 h) but after this initial reduction the pathogens remained constant in the substrates, while *L. monocytogenes* showed a very slow decline rate (0.0042 log.conc/h D=239 h). When larvae enter in pupal stage their contamination with *Salmonella* and *L. monocytogenes* was significantly lesser then the substrate (-2 log). This study confirms that larvae show the same pathogens concentration of their growing substrate and their feeding activity could reduce pathogens contamination in the growing substrate, but the rate of their action seem to be largely insufficient to reach the food and feed security objective.

Session 47 Poster 16

Nutritional evaluation of meal from earthworm reared on fruit and vegetable waste

A. Tava[1], E. Biazzi[2] and D. Tedesco[3]

[1]Centro di Ricerca Zootecnia e Acquacultura CREA-ZA/Research Centre for Animal Production and Aquacul, viale Piacenza 29, 26900 Lodi, Italy, [2]Centro di Ricerca Zootecnia e Acquacultura CREA-ZA/Research Centre for Animal Production and Aquacul, viale Piacenza 29, 26900 Lodi, Italy, [3]Dipartimento Scienze e Politiche Ambientali Università Milano, via Celoria 10, 20133 Milano, Italy; aldo.tava@crea.gov.it

One main challenge in global food production is how to provide enough protein for human consumption and some alternative way has been recently considered. Insects have received a growing deal of attention as a promising way to cope with some of the major food and nutrition challenges facing the world. Among the terrestrial invertebrates, utilization of earthworms may represent another answer ecologically, economically and socially acceptable as alternative protein source. Nutritional properties of earthworms has been known since a long time; they are also used in traditional Chinese medicine for their anti-inflammatory, antioxidant and depurative properties due the presence of several biologically active compounds. In this investigation, earthworm (*Eisenia fetida*) were grown and reared on fruit and vegetable waste in compliance with EU regulations referred to the safety of food of animal origin. The freeze-dried and powdered earthworms were analysed according to standard methods for evaluation of different parameters. The samples contains large amount of protein (63.8-72.4% DM), essential amino acids, total fat (9.2-10.3% DM) and fatty acid, including mono- and polyunsaturated, very low content of carbohydrates, minerals (including Ca, Mg and Fe), sterols (6-7 mg/g DM), total polyphenols (3.6-3.8 mg/g DM) and their antioxidant activity. Based on these results, earthworms can represent an excellent raw material source of nutrients for animals and human consumption, especially for feed/food integrators. This work was supported by 'Fondazione Cariplo' project n. 2015-0501, 'Bioconversion of fruit and vegetable waste to earthworm meal as novel food source'.

Session 48 Theatre 1

Associations between the hepatic lipidome and feed efficiency in Holstein dairy cows

A.B.P. Fontoura, J.E. Rico, A.N. Davis and J.W. McFadden

Cornell University, Animal Science, 507 Tower Rd, 14850 NY, USA; abf63@cornell.edu

Factors underlying differences in metabolic efficiency, nutrient partitioning and lipid metabolism during the transition period are poorly understood. Hepatic lipid metabolism may be an important contributor of the biological variation in feed efficiency. Thus, our objectives were to evaluate performance, steatosis and the hepatic lipidome in dairy cows diverging in feed efficiency. Twenty-three multiparous gestating Holstein cows (BW=646±86 kg) were evaluated for feed efficiency using residual feed intake (RFI) during the transition period from late gestation to lactation. At the end of experiment, cows were categorized as efficient and inefficient. Body-weight was recorded weekly throughout the experiment. Liver biopsies were performed pre- (d=-12) and post-partum (d=10). Milk composition was analysed during the first 28 d post-partum. Liver samples were analysed for lipid content and submitted to TOFMS. Log transformed, auto-scaled metabolites were analysed using MetaboAnalyst. Least square means comparison of RFI groups regarding performance, liver lipid content and lipids were calculated using the GLM procedure of SAS. Partial correlations between traits were measured through the MANOVA/PRINTE statement within the GLM procedure of SAS. Between efficient and inefficient cows, no differences among milk composition (P=0.67) or liver lipid content pre- (P=0.29) and post-partum (P=0.30) were observed. However, clear distinctions in the hepatic lipidomic profile were observed. Efficient animals displayed a pattern of elevated monoalkyl-diacylglycerol (MADAG, P=0.03) in the pre-partum period and diminished values of MADAG and phosphatidic acid (PA) in the post-partum (P=0.07; P=0.02, respectively). In addition, total phosphatidylcholine (PC, r=-0.40, P=0.07) and phosphatidylethanolamine (PE, r=-0.43, P=0.05) in pre-partum were negatively correlated with RFI. Overall, our results suggest fat export through the liver might be associated with feed efficiency and in this context, timing of fat mobilization might be of great importance. The differences in the hepatic lipidomic profile of cows with divergent efficiencies highlight new opportunities in the development of efficiency biomarkers related to lipid metabolism.

Session 48 Theatre 2

Variation in the solubilization of nitrogenous compounds in wheat straw by different white-rot fungi

N. Nayan[1], J.W. Cone[1], A.S.M. Sonnenberg[2] and W.H. Hendriks[1]
[1]Wageningen University & Research, Animal Nutrition Group, Building nr. 122, De Elst 1, 6708 WD Wageningen, the Netherlands, [2]Wageningen University & Research, Plant Breeding, Droevendaalsesteeg 1, 6708 PB Wageningen, the Netherlands; nazri.nayan@wur.nl

White-rot fungi are known besides their unique ability to depolymerize cell wall components, to assimilate nitrogenous compounds from substrates. This modification may change the solubility of protein for fermentation in the rumen. To investigate this, the nitrogen in fungal treated wheat straw (3 fungal species, 2 strains each) was fractioned according to the Net Carbohydrate and Protein System and assessed for *in vitro* protein fermentation using a modified gas production technique ($IVGP_N$). Fractionation of CP showed an increase in non-protein nitrogen, buffer soluble protein and partly buffer insoluble protein. A decrease was seen in cell wall bound protein fractions. The $IVGP_N$ of straw treated with *Ceriporiopsis subvermispora* strains were not different to the control, but increased by 30.2 to 47.1% in *Pleurotus eryngii* and *Lentinula edodes* strains. The $IVGP_N$ was positively correlated to easily (buffer) soluble protein and negatively to cell wall bound protein. All fungi also increased the arginine (~56%) and lysine (~15%) contents. This study shows the importance of assessing the protein solubilization by different fungal strains, which can uncover unique mechanisms in the cell wall depolymerization.

Session 48 Theatre 3

The effect of silage Wet Brewers' Grains using new technologies filling the bag

R. Loucka, P. Homolka, F. Jancik, P. Kubelkova, Y. Tyrolova and A. Vyborna
Institute of Animal Science, Nutrition and Feeding of Farm Animals, Pratelstvi 815, CS 104 00 Prague Uhrineves, Czech Republic; loucka.radko@vuzv.cz

Wet Brewers' Grains (WBG) is a quality and nutritious feed for farm animals. Because fresh it spoils very quickly (not more than 2 weeks) ensilaging is recommended. The use of additives or absorbents is problematic as the temperature is reduced as well as the contamination of the microorganism. A new way is to preserve WBG in the bag in which is tipped straight from the truck. The aim of research was to determine how long WBG, preserved without any additives, can be stored in a bag with a capacity of 25 tons. The WBG, still warm (60 °C), was tipped out of a truck directly into the bag, affixed to the back of a truck. The bag was immediately anaerobically closed. The WBG was stored more than 3 months in the bag. The temperature was continuously monitored by Thermochron sensors every 1 hour. Samples for fermentation, chemical and microbial analyses were taken from the bag every 2 weeks, at a depth of 5, 20 and 35 cm below the tarp. After opening the bag, aerobic stability of WBG silage samples was determined. If the WBG still warm tipped out of a truck directly into the bag and anaerobically closed, it can keep good quality even 3 months, although they have not been preserved using additives. After opening the bag the aerobic stability declined rapidly, and the number of yeasts and molds significantly increased. After 7 days WBG silage was evaluated very poorly. It can be concluded the technology of WBG stored in lorry-bag is recommended, even if no additives are used. If the air does not get inside the bag, the silage will last for two months. There is recommended to feed the silage after opening bag as soon as possible. Research was supported by project MZE QJ1510391.

Session 48 — Theatre 4

Free and rumen-protected essential oils incubated *in vitro*: stability and fermentation parameters

N. Amin[1], F. Tagliapietra[2], N. Guzzo[1] and L. Bailoni[1]

[1]University of Padua, Department of Comparative Biomedicine and Food Science (BCA), Viale dell'Università, 16, 35020 Legnaro (PD), Italy, [2]University of Padua, Department of Agronomy Animals Food Natural Resources and Environment (DAFNAE), Viale dell'Università, 16, 35020 Legnaro (PD), Italy; nida.amin@phd.unipd.it

The aim was to test the susceptibility of 3 essential oils (EOs: Olistat-Cyn, Olistat-G, and Olistat-P), produced by SILA Srl (Noale, VE, Italy), in 2 different forms (free: fEOs and rumen-protected: rpEOs) to *in vitro* ruminal degradation using the Ankom Daisy^{II} technique. Three vessels were prepared for each incubation with buffer (1,596 ml), filtered rumen fluid (400 ml) and a standard diet for dry cows (15.96 g). The EOs (0.2 g of fEOs or 1 g of rpEOs) were placed in Ankom F57 filter bags (25 µm pore size). In each incubation, bags containing fEOs and rpEOs were added in the same vessel. Experimental design was as follows: 3 incubationsx6 incubation timesx6 EOsx4 replicates, plus 108 blanks (filters without EOs), for a total of 540 filters. Dry matter disappearance (DMD) of EOs was determined at 0, 2, 6, 12, 24, and 48 h. The fermentation fluid of each vessel was analysed for pH, volatile fatty acids (VFAs), and protozoa No. at 0 and 48 h. As expected, fEOs were highly degraded at 48 h (64.0, 80.5, and 84.6% DMD for Olistat-Cyn, Olistat-G, and Olistat-P, resp.) in comparison to rpEOs. (13.4, 13.7, and 12.4% DMD, resp.). Due to this stability of all rpEOs, we can hypothesize that the observed changes of the rumen fermentation parameters were mainly due to the fEOs, being incubated in the same vessel with rpEOs. In comparison to the fermentation profile at 0 h, Olistat-G after 48 h caused a significant decrease in the pH (6.93 vs 6.36; $P<0.001$) and the total protozoa No. (4.73 vs 4.40 $\log_{10}$/ml; $P<0.05$) while a significant increase in the total VFAs (17 vs 33 mmol/l; $P<0.05$). However, the other EOs (Olistat-Cyn and Olistat-P), caused no significant ($P>0.05$) differences between the rumen fermentation parameters at 0 and 48 h. In conclusion, the protection of EOs from ruminal degradation by microencapsulation was found to be very effective to ensure the rumen by-pass. Among EOs, Olistat-G was capable of changing rumen fermentation, with potential effects on methane reduction.

Session 48 — Theatre 5

Development of a microwave sensor application for online detection of corn silage dry matter content

V. Perricone[1], A. Agazzi[1], A. Costa[1], M. Lazzari[1], A. Calcante[2], M. Baiocchi[3], E. Sesan[4], G. Savoini[1] and F.M. Tangorra[1]

[1]Università degli Studi di Milano, Dipartimento di Scienze Veterinarie per la Salute, la Produzione Animale e la Sicurezza Alimentare, Via Celoria 10, 20133 Milano, Italy, [2]Università degli Studi di Milano, Dipartimento di Scienze Agrarie e Ambientali, Produzione, Territorio, Agroenergia, Via Celoria 2, 20133 Milano, Italy, [3]PTM s.r.l., Via per Isorella 22/A, 25010, Visano (BS), Italy, [4]Sgariboldi s.r.l., Via Pietro Nenni 15, 26845 Codogno (LO), Italy; vera.perricone@unimi.it

The moisture content of silages can be subjected to variations in time, leading to a potential decrease in dry matter (DM) content in the total mixed ration (TMR) and consequent impaired performance and health status in cows. With the purpose to maintain and improve the overall efficiency and the health of dairy cows, this study aimed to develop a new application in agricultural system for a commercial microwave sensor (MS) to online detect corn silage (CS) DM content. This would lead to the daily adjustment of the total amount of CS provided in the TMR based on its actual DM content, lowering performance fluctuations of cows. A calibration straight line for MS over CS samples was developed, following the sensor manufacturer instruction, from 75 dried samples experimentally rehydrated to reach 50-70% of moisture and the relative MS *Quotient* (emitted/residual energy) over four readings per sample was recorded. A Linear regression analysis showed a good correlation between *Quotient* and moisture content (r^2=0.79). MS calibration was then tested against 22 CS fresh samples, subsequently analysed for DM (Reg. CE 152/2009). The linear regression analysis evidenced a good relationship between MS DM and lab DM results (r^2=0.73), even if an offset of 9.60% with a slight difference in the slope of the straight lines (0.03 rad) was observed. A MS could be a promising tool to daily online monitor the DM content of CS, although further implementation in the number of samples for the calibration and the development of a correction coefficient for offset and slope of the straight line are needed to improve the accuracy. Funding was provided by the project P.L.U.S.- Precision Livestock Unifeed System (ID: 145923 CUP: E77H16001570009).

Multiomics analysis of feed efficiency in danish breeding boars

V.A.O. Carmelo and H.N. Kadarmideen
Danish Technical University, DTU Bioinformatics, Department of Bio and Health Informatics, Bygning 208, Kemitorvet, 2800 Kongens Lyngby, Denmark; vaocar@bioinformatics.dtu.dk

Feed efficiency (FE) is the most economically important trait in pig farming, and consequently is the main breeding goal in commercial pig farming. Since the advent of genomic breeding methods, the improvement and accuracy of breeding has greatly improved. In contrast, the gain in trait related functional biological knowledge has not increased in the same way. Gain in functional biological knowledge of FE in not only scientifically interesting, but can also lead to improved selection through identification of biomarkers or through the use of weighted genomic breeding models. Here we have performed a multi-omics study on Danish performance tested young boars with a wide array of FE related phenotypes and accurate FE measurements available, enabling us to identify novel biological knowledge about FE in production pigs. FE is not a high heritability trait, making it is difficult to identify the biological background behind it. In total, 113 (61 Duroc Purebreed and 52 Landace Purebreed) performance tested young boars were included in the study. Blood samples were collected at 30 and 100 kg and tissue samples from liver and muscle were collected at slaughter. Blood Metabolomics, liver and muscle transcriptomic and genomic profiles were generated. We applied linear models to individual blood metabolites to identify potential FE associated metabolites. In Duroc, 62 out of 682 metabolites at 100 kg were significantly associated with FE ($P<0.05$). In landrace, 75 out of 686 metabolites were significantly associated at 100 kg ($P<0.05$). Modelling the changes in metabolite concentration from 30 to 100 kg, 75 and 59 out of 734 metabolites were significantly related to FE from Duroc and landrace boars respectively($P<0.05$). Applying uniform distribution based bootstrapping, we are able to show that all of the above associations between FE and blood metabolites are extremely unlikely to be random ($P<10^{-4}$). By integrating genomic, transcriptomic and metabolic data and using network-based methodologies, we also present strategies for identifying pathways and genes associated with FE in breeding boars.

Improved cattle growth by methionine-balanced diets does not result from lower protein degradation

G. Cantalapiedra-Hijar[1], L. Bahloul[2], C. Chantelauze[1], V. Largeau[1], N. Khodorova[3], H. Fouillet[3] and I. Ortigues-Marty[1]
[1]UMR 1213 Herbivores INRA-VetAgro Sup, INRA, 63122, France, [2]ADISSEO France S.A.S., Adisseo, 92160, France, [3]UMR PNCA, AgroParisTech, INRA, Université Paris-Saclay, 75005, France; isabelle.ortigues@inra.fr

Methionine (Met) is a limiting amino acid, and balancing diets for Met enhances animal performances. We hypothesized that this improvement is due to a reduction in protein degradation. The impact of rumen protected Met on whole-body protein degradation was tested in 36 young bulls fed iso-energy diets formulated at 2 levels of metabolizable protein (MP) [120% (High) vs 100% (Normal) recommendations] and 2 levels of Met [control (1.9%MP) vs balanced (2.4%MP)] using Smartamine M. Animal's proteins were initially enriched in ^{15}N by including ^{15}N labelled urea in the diet during 35 days. Urinary spot samples were collected from each animal over 5 months (n=13) after the last administration of ^{15}N-labelled urea and analysed for their ^{15}N enrichment. The rate of whole-body protein degradation was assessed through the rate of urinary ^{15}N depletion after having stopped the ^{15}N labelling, the latter reflecting mainly the rate at which the free plasma amino acids, one precursor of urinary urea, are renewed in the whole-body by degradation of initially ^{15}N-enriched proteins. Fitting adequately the urinary ^{15}N depletion kinetics required a bi-exponential mixed-effect model, considering animals as random and the other experimental factors as fixed. The need of a bi-, rather than a mono-, exponential model indicated that urinary N originates from two different nitrogen pools of distinct turnover rates, rapid vs slow, and mainly reflecting the protein degradation rates in splanchnic and peripheral tissues, respectively. High protein diets increased the average daily gain (ADG, +16%, $P<0.001$) as well as both the fast (88.0%/d vs 69.5%/d, $P<0.001$) and slow (10.2%/d vs 7.99%/d, $P=0.006$) protein degradation rates. In contrast, Met balanced diets tended to increase the ADG (+9.2%, $P=0.09$) but had no effect on the protein degradation rates ($P>0.72$). To conclude, this suggests that growth improvement by Met balanced diets does not involve a protein degradation reduction, but more likely a protein synthesis stimulation.

Session 48 Theatre 8

Size and density influence of concentrates to increase by-pass protein fraction in dairy cows' diet

F. Dufreneix[1,2], P. Faverdin[2] and J.-L. Peyraud[2]
[1]Agrial, 14 rue des Roquemonts, 14000 Caen, France, [2]PEGASE, Agrocampus Ouest, INRA, 35590 Saint Gilles, France; florence.dufreneix@inra.fr

Increasing by-pass protein is one of the challenges of ruminant nutrition to both cover the protein needs of animals and reduce the nitrogen excretion in the environment. Industrial process such as heat treatment, formaldehyde tanning and vegetable extract are focused on the protection of proteins against ruminal fermentations but show some limits (low intestinal digestibility, toxicity, low or variable efficiency). The reduction of the mean retention time (MRT) by varying size and density of particle is another way to increase by-pass protein fraction of concentrates. Plastic particles of four sizes (0.5, 1, 2 and 3 mm) and four densities (0.9, 1.1, 1.3 and 1.5) were used to study the combined effect of size and density on MRT of particles without microbial fermentation interactions. The kinetics of faecal excretion of particles (17 sampling over 106 hours) were studied in a Latin square experiment including four lactating cows. Data were adjusted to a double exponential model. Density has a quadratic response with the fastest escape for densities 1.1 and 1.3 in the total digestive tract (TMRT=29.5 and 31.2 hours respectively) as well as in the compartments isolated by the model compared to densities 0.9 and 1.5 (TMRT=64.0 and 51.2 hours respectively). Size has a linear effect on the total digestive tract transit time (+12.9 hours for the 3 mm size compared to the 1 mm) and in the time of first appearance of particles but no effects were observed in the others compartments. A combined effect of size and density is observed: particle size has no effect on TMRT when density is between 1.1 and 1.3 but outside this range, an increase of particle size induces an increase in the TMRT. In conclusion a density in a range 1.1 to 1.3 will be optimal for the by-pass of particles of concentrate. These particles will be submitted to ruminal fermentations, and their optimal size must be small enough to pass the reticulo-omasal orifice but large enough to delay the start of fermentation on their surface and thus the loss of density by fermentation gases production. Size 3 to 4 mm would be a good compromise between these two constraints to allow the shortest escape from the rumen.

Session 48 Theatre 9

A first approach to predict protein efficiency of dairy cows through milk FT-MIR spectra

C. Grelet[1], A. Vanlierde[1], M. Hostens[2], L. Foldager[3], M. Salavati[4], E. Froidmont[1], M.T. Sorensen[3], K.L. Ingvartsen[3], M. Crowe[5], C. Ferris[6], C. Marchitelli[7], F. Becker[8], G.plus.e. Consortium[9] and F. Dehareng[1]
[1]CRA-W, Gembloux, Belgium, [2]Ghent University, Merelbeke, Belgium, [3]Aarhus University, Tjele, Denmark, [4]RVC, London, United Kingdom, [5]UCD, Dublin, Ireland, [6]AFBI, Belfast, United Kingdom, [7]CREA, Roma, Italy, [8]FBN, Dummerstorf, Germany, [9]http://www.gpluse.eu/index.php/project/partners, UCD, Dublin, Ireland; c.grelet@cra.wallonie.be

Protein efficiency has become a key factor in dairy production for both environmental and economic reasons. Cost effective and large-scale phenotyping methods are required to improve this trait through genetic selection or feeding and management of cows. The aim of this study is to evaluate the possibility of using MIR spectra of milk to predict protein efficiency of dairy cows. Data were collected from 133 cows, from calving until 50 days in milk, in 3 research herds distributed in Denmark, Ireland and UK. For two herds, diets were designed to challenge cows and induce production diseases. Amounts of protein ingested (kg/day) and fat and protein corrected milk (FPCM, in kg/day) were measured daily. Protein efficiency to produce milk was 'quantified' by using the ratio 'FPCM/protein ingested'. MIR milk spectra were recorded twice weekly and were standardized into a common format to avoid bias between apparatus or periods. Regression models between protein efficiency and MIR milk spectra have been developed on 1,145 observations using PLS or SVM methods and a cross-validation was realized using 10 subsets. The model was better in terms of R^2 of cross-validation and error when using SVM method compared to PLS method. Inclusion of milk yield and lactation number as predictors, in combination with the spectra, also improved the calibration. The best model was obtained by using spectra, milk yield and lactation number as predictors, and SVM modelling with R^2cv of 0.75. These preliminary results show that there is a possibility to have information on protein efficiency to produce milk through milk MIR spectra. This could allow large-scale predictions for both genetic studies and farm management.

Intestinal stem-cell organoids as experimental models to investigate feed efficiency

E.D. Ellen[1], N. Taverne[2], A.J. Taverne-Thiele[2], O. Madsen[1], H. Woelders[1], R. Bergsma[3], E.F. Knol[3], S. Kar[4], Y. De Haas[1], M.A.M. Groenen[1] and J.M. Wells[2]
[1]Wageningen University & Research, ABG, P.O. Box 338, 6700 AH Wageningen, the Netherlands, [2]Wageningen University & Research, HMI, P.O. Box 338, 6700 AH Wageningen, the Netherlands, [3]Topigs Norsvin, Schoenaker 6, 6641 SZ Beuningen, the Netherlands, [4]Wageningen University & Research, P.O. Box 338, 6700 AH Wageningen, the Netherlands; esther.ellen@wur.nl

Improving feed efficiency is an important trait in livestock production. So far, feed conversion ratio (FCR) has been used to improve feed efficiency. However, the underlying biological mechanisms of feed efficiency are still not clear and very complex. Adult stem-cell derived organoids have already proven to be powerful experimental models of mammalian biology, and offer possibilities to advance our molecular understanding of feed efficiency. A proof of principle experiment was performed on organoids isolated from intestine samples of 12 pigs with divergent phenotypes for feed efficiency; 6 efficient pigs (FCR=2.19±0.03) and 6 less efficient pigs (FCR=2.61±0.04). Tissue was collected from the ileum and the colon organoids for transcriptomics by RNAseq and histology. A new method for generating an enhanced, near physiological 2D culture system from porcine intestinal organoids was used to perform a transport study on confluent small intestinal monolayers. The results of histology and glucose and amino acid transport will be presented. First preliminary results indicate that there are distinct differences in gene expression between the efficient and less efficient group. However, further analysis are needed. First findings show that organoids can be used to accurately measure nutrient transport function. As such they offer new possibilities to understand the potential role of intestinal functions in feed efficiency and provide new phenotyping tools for future genomic selection analysis.

Effects of microencapsulated complex of organic acids and essential oils on pig

H.J. Oh[1], C.H. Lee[1], J.S. An[1], W.K. Kwak[1], Y. Won[1], J.H. Lee[1], S.Y. Oh[1], S.D. Liu[1], S.Y. Cho[2], D.H. Son[2] and J.H. Cho[1]
[1]Chungbuk National University, Animal Sciences, Cheongju, Chungbuk, Republic of Korea, 286-44, Korea, South, [2]Eugene Bio, Animal Healty & Nutrition, Sunwon, Gyeonggi-do, Republic of Korea, 1-610, 306, Korea, South; dhgkswls17@naver.com

A total of 90 pigs (6.47±0.27 kg; 21 d of age) were used in this 22-wk feeding trial to evaluate the effect of microencapsulated complex of organic acids and essential oils (MOE) on growth performance, nutrient digestibility, blood profile, faecal microflora and lean meat percentage in weaning to finishing pigs. Pigs were randomly distributed into one of three treatment groups (6 replicate pens/treatment, 5 pigs/pen). Dietary treatments were: CON, basal diet; MOE 1, basal + 0.1% MOE, MOE2, basal diet + 0.2% MOE. In growth performance, The ADG, ADFI and G/F ratio were a significantly difference in treatment (P<0.05). In nutrient digestibility, ATTD of DM was linearly increased (P<0.05) in MOE2 compared with other treatment. In blood profiles, RBC and IgG showed significantly difference in pigs fed with MOE (P<0.05). in faecal microflora, *Lactobacillus* concentration increased (P<0.05) in MOE2 treatments compared with other treatments. In meat quality, the meat colour (a*) and drip loss decreased linearly (P<0.05) with increasing level of MOE. In conclusion, MOE supplementation could improve growth performance, nutrient digestibility, blood profile, faecal microflora and carcass trait in weaning to finishing pigs.

Session 48 Theatre 12

Group recordings accounted for drop out animals

B. Nielsen[1], H. Gao[2], T. Ostersen[1], G. Su[2], J. Jensen[2], P. Madsen[2] and M. Shirali[2]
[1]SEGES, Pig Research Centre, Breeding and Genetic, Axeltorv 3, 1609 Copenhagen V, Denmark, [2]Aarhus University, Center for Quantitative Genetics & Genomics, Blichers Allé 20, 8830 Tjele, Denmark; bni@seges.dk

Selection for lower unit of feed intake per unit of body weight gain is one of the most important traits in pig breeding programs. Feed intake is often recorded during a well-defined period of performance test. During the performance test pigs are kept in pens and electronic feeders allow for feed intake recorded of each pig in the pen. The use of electronic feeders makes the measurement of feed intake very costly. To reduce the phenotyping cost of feed intake, feed intake recorded at pen level can be an alternative approach. Statistical models have been developed to obtain the genetic parameters and breeding values (BV) for each individual animal utilizing group recorded data. Feed data of 24,434 boars on test station shows a correlation of 0.36 between BV if either individual or group recorded feed intake were used. Though, the models are ignoring the effect of animals dropped the test period due to e.g. sickness or death. Here, we are extending the group models to longitudinal analysis of group recorded traits, and at the same time taking into account animals dropped out of the performances test. The longitudinal model allows prediction of the individual BV of feed intake for all pigs in the pen. Pigs that were dropped out before end of performers test were recorded for weight and time. The longitudinal model covers the variation between repeated measurements within animals and variation between animals, in addition to the dynamics of feed intake during the test period in pigs. The results shows that longitudinal analysis of group recorded traits are feasible. This allows genetic parameter estimation of the trait of interest, and prediction of breeding values of individual animal. The developed method can be implemented into breeding programs if group records of animals are available.

Session 48 Theatre 13

Chemical and nutritional value of organic feedstuffs: a need to address in monogastric feeding

A. Roinsard[1], V. Heuze[2], H. Juin[3], D. Renaudeau[4], D. Gaudré[5] and G. Tran[2]
[1]ITAB, 9, rue André Brouard, 49100 Angers, France, [2]AFZ, 16 rue Claude Bernard, 75005 PARIS, France, [3]INRA, Domaine du Magneraud, 17700 Surgeres, France, [4]INRA, UMR PEGASE, 35590 Saint-Gilles, France, [5]IFIP, La Motte au Vicomte, 35650 Le Rheu, France; valerie.heuze@zootechnie.fr

As it is mandatory since January 2019 to feed organic pigs and poultry with 100% organic feeds in the EU, a better knowledge of the nutritional value of organic raw materials is, more than ever, a key for farmers to develop efficient nutrition systems for organic livestock farming. Studies are being conducted in the French CASDAR programme SECALIBIO to compare organic raw materials to conventional ones, to establish the composition and nutritional value of organic feeds, and to evaluate their variability. A database of 6,577 samples of 94 organic raw materials, most of them commonly used in monogastric feeding (soybean meals, maize, faba bean, etc.), was constituted using data from the French Feed Database of the AFZ, from previous research programmes and from feed companies participating in SECALIBIO. *In vivo* digestibility trials on pigs and poultry were conducted to evaluate the specificities of organic raw materials and provide new data (for forage peas for example). The variability of organic raw materials was generally comparable to that of conventional ones. However, the fat content of organic sunflower (6.2 to 25.2% MS) and soybean (5.37 to 22.6% MS) meals varied a lot due to processing technologies, and it was suggested to delineate categories with consistent crude protein and fibre content. In connection with it, for poultry, the digestibility of protein could vary a lot for sunflower (76.5-83.0%) and soybean (74.4-87.32%) meals. Organic soybean meals was reported to have a lower lysine/crude protein ratio than the conventional one (5.75 vs 6.14%): further investigations will determine the impact of this difference on feed formulation. The first practical result of this study will be the creation of specific tables of composition and nutritional value of organic feeds for pigs and poultry which represents a key for farmers to develop an efficient nutrition system in organic farming.

Effects of supplementing humates on ruminal parameters of Nellore steers

R.R.S. Corte, A.G. Lobo, T. Briner Neto, G.S. Abiante, J.S. Silva Neto, S.O. Pietriz, L.F.M. Azevedo, L.S. Martello and P.H.M. Rodriguez
Sao Paulo University, Animal Science, Av Duque de Caxias Norte, 225, 13635900, Brazil; rscorte@usp.br

The banning of antibiotics use as animal growth promoters in the European Union in 2006 has increased demand for alternative feed additives that can be used to improve animal production. Humate substances, or humic acids, are geological deposits in the earth's surfacecan, used as a organic source of feed additive to cattle. Eight rumen-cannulated Nellore steers were distributed to four diets, each of which differed in humate level, in a replicated 4×4 Latin square experimental design to access the ruminal parameters. Treatments were crescent levels of humic acid (0, 0.7, 1.4 and 2.1%) on dry matter basis of a 50:50 roughage concentrate ratio diet. Each experimental period consisted of 19 days; the first 14 days for diet adaptation and the last 5 days for data collection. On experimental day 19, ruminal pH was analysed by a continuous measurement probe. In order to quantify short-chain fatty acids (SCFAs) production, rumen contents were sampled prior to feed and 3, 6, 9 and 12 h after morning feeding. The pH remained in the ideal range (6.3) for ruminal functioning, with no difference between the evaluated treatments. The SCAFs molar proportion were affected by dietary treatments: isobutiric (0: 1.5, 0.7: 1.4, 1.4: 1.3 and 2.1: 1.2), isovaleric (0: 2.5, 0.7: 2.3, 1.4: 2.3 and 2.1: 2.2) and valeric (0: 1.6, 0.7: 1.4, 1.4: 1.4 and 2.1: 1.3) SCFAs decreased as level of humate increased (linear effect, P<0.0001, P=0.04, P=0.001, respectively). The acetic/propionic relation (0: 3.4, 0.7: 3.5, 1.4: 3.7 and 2.1: 3.66) increased as humate level increased (linear effect, P=0.0015). For acetic, propionic and butiric SCAFs no differences were observed. The humates use as additive for beef altered SCFAs molar concentration, providing adequate condition for rumen functioning and avoiding metabolic disorders as conventional additives.

Effect of digestible calcium level on the digestible phosphorus requirement of fattening pigs

A. Samson[1], E. Schetelat[2], C. Launay[1] and E. Janvier[1]
[1]NEOVIA, R&D, Rue de l'Eglise, 02402 Chateau Thierry, France, [2]INZO°, 1 rue de la Marébaudière, 35766 Montgermont, France; ejanvier@neovia-group.com

Phosphorus (P) is a key nutrient in swine diets as it is implicated in many metabolic pathways and it is concerned by environmental issues. Finding new strategies to optimize P supply in swine husbandry is then of particular interest. Phytases improve P bioavailability but there are probably other considerations that have to be taken into account. Recent data showed that calcium (Ca) fed in excess may reduce growth performance in fattening pigs. In 1996, Cromwell suggested that the mode of action explaining the negative impact of Ca seems to be related to a disturbance of P metabolism before. This study was thus conducted to determine whether the digestible phosphorus (digP) requirement of fattening pigs is affected by the digestible calcium level (digCa). One-hundred and twelve pigs (females and barrows), 70 days of age, fed restricted and according to a two-phase strategy (grower diet for 35 days and then a finisher diet) were used in this study. Pigs were housed individually and allocated to one of eight treatments, compared in a 2×4 factorial design: two digCa levels (0.30 and 0.40%) and four digP levels (grower/finisher diet: 0.15%/0.12%, 0.20%/0.17%, 0.25%/0.22% and 0.30%/0.27%). The experimental diets did not contain exogenous phytase. In the growing phase, no effect of the digCa level on growth performance was observed. In the finishing phase, independently of the digP level, increasing digCa induced a significant reduction in growth and feed efficiency (P<0.001). Regardless the digCa level, the digP requirement to maximize growth was approximately 0.23% in both growing and finishing phases. Bone mineralization was not significantly affected by digCa level, while the quantity of P deposited was significantly affected by the digP level. P deposition in the phalanx peaked at 0.25% of digP. These data confirm that excess dietary Ca may be detrimental for the performance of fattening pigs. Nevertheless, the mode of action explaining this negative impact remains unclear. Indeed, our results suggest that P metabolism (absorption and utilization) is probably not the only physiological aspect to be affected by digCa level.

Effect of enzyme complex on nutrient digestibility and growth performance of growing-finishing pigs

S.K. Park[1], N. Recharla[1], J.H. Ryu[1], M.H. Song[2], D.W. Kim[3], Y.H. Kim[3] and J.C. Park[3]
[1]Sejong University, Food Science & Biotechnology, 209 Neungdong-Ro, 05006 Seoul, Korea, South, [2]Chungnam National University, 99 Daehak-ro, 34134 Daejeon, Korea, South, [3]National Institute of Animal Science, RDA, Cheonan, 31000 Sunghwan, Korea, South; sungkwonpark@sejong.ac.kr

This study was conducted to investigate the effect of enzyme complex (EC) on nutrient digestibility, growth performance of growing-finishing pigs. Enzyme complex was composed of xylanase, a-amylase, b-glucanase, and protease. A total of 6 feed ingredients including wheat meal, soy meal, fish meal, oriental herbal extract, Italian ryegrass, and peanut hull with or without enzyme complex were examined. Supplementation of EC numerically increased the ileal and total tract digestibility of wheat, soybean meal, and oriental herbal extracts *in vitro*. A total of 36 weaned pigs were fed corn-soybean meal based diets without (control) or with EC for 6 wk, digestibility, growth performance, and faecal metagenomics were analysed. Pigs in EC group showed greater ($P<0.05$) growth performance (ADG 484 g/d) compare to those in control group (ADG 452 g/d), but ileal and total tract digestibility of DM were not different between control and treatment groups. Metagenomic analysis of faeces showed that bacterial sequences of both control and treatment groups were comprised predominantly of *Firmicultes* and *Bacterioidetes*. Abundance of *Treponema* and *Barnesiella* was higher ($P<0.05$) in EC than control. Genera including Meniscus, *Fibrobacter*, *Butyricicoccus*, *Clostridium*, and *Succinicvibrio*, however, were lower ($P<0.05$) in EC when compared with control group. Results from our current study demonstrate that diets supplement with muti-enzyme might improve growth performance of pigs via regulating microbial society in growing-finishing pigs.

***In vitro* assessment of the use of formate salts to enhance rumen digestibility of roughage**

A. Foskolos[1], Z.C. Morgan[1], M.D. Holt[2] and J.M. Moorby[1]
[1]Aberystwyth University, IBERS, Campus Gogerddan, SY23 3EE, United Kingdom, [2]Symbiont Nutrition, Chandler, Arizona, 85226, USA; anf20@aber.ac.uk

Formate salts (FS) have been suggested as digestion enhancers for ruminants. Our objective was to investigate the effects of a FS product on rumen fermentation. Two different samples of five feeds were used: (1) cottonseed hulls (CH), (2) cotton gin trash (GT), (3) wheat straw (STR), (4) nutritionally improved wheat straw (NIS), and (5) perennial ryegrass silage (GRS). Formate salts were added at 0, 600, 1000, and 2000 ppm. Samples were incubated in duplicate with buffer and rumen fluid from three cannulated cows, and treated independently. Gas volume was recorded at 3, 6, 12, 24, 48, 72, 96, and 120 hours (h) of incubation, and analysed in a gas analyser. At 120 h, the *in vitro* dry matter digestibility (IVDMD) was measured. For CH, total gas, CO_2 and CH_4 volumes were higher at early stages of fermentation (12 h) with FS addition (10 vs 27 ml of total gas / g DM, for 0 vs 2,000 ppm, respectively). Then, CO_2 equilibrated resulting at no difference at 120 h (on average 107 ml CO_2 / g DM), while total and CH_4 volume were higher at the two highest doses. In the case of GT, total gas, CO_2 and CH_4 volumes were higher for FS treatments from 3 h to 120 h (84 vs 124 ml of total gas / g DM, for 0 vs 2,000 ppm, respectively). For STR, CO_2 volume were unaffected, but CH_4 volume increased with increasing addition of FS from 6 to 120 h that also affected total gas volume at 120 h (160 vs 194 ml of total gas / g DM, for 0 vs 2,000 ppm, respectively). For NIS, total gas, CO_2 and CH_4 volume increased as FS addition increased at 3 h. However, CO_2 volume equilibrated at 72 h, resulting in higher total gas volume at 120 h only for the highest dose (184 vs 215 ml of total gas / g DM, for 0 vs 2,000 ppm, respectively). Moreover, CH_4 volume was higher at 6 h with the two highest doses of FS when added to GRS. This resulted in higher total gas volumes from 24 to 120 h. Finally, IVDMD was not affected by FS addition. In conclusion, the addition of FS resulted in apparent shifts in fermentation dependent on the substrate feed. Acknowledgements: Support was provided through the Sêr Cymru NRN-LCEE project Cleaner Cows.

Utilization of the continuous measurement technology to control the feeding of dairy cows

R. Loucka, P. Homolka, F. Jancik, P. Kubelkova, Y. Tyrolova and A. Vyborna
Institute of Animal Science, Animal feeding and nutrition, Pratelstvi 815, CS 104 00 Prague Uhrineves, Czech Republic; loucka.radko@vuzv.cz

Using of new technologies based on automatic continuous measurement of different variables (pH, temperature, chewing activity) can be taken in real time to improve the performance and welfare of animals. The aim of the experiment was to investigate the effect of external intervention (mechanical stimulation of rumen or changes of TMR) on the work of the rumen in 22 high-performance Holstein dairy cows, and their performance. Feed consumption was measured continuously using automatic feeding boxes. Two mixed diets (TMR) were tested. The main differences between tested diets were in the structure and in the composition. The experimental diet contained the by-products of the brewing industry, the control diet was without them. The pH and rumen temperatures were measured with eCOW boluses every 15 minutes. Each trial period lasted 3 weeks. At the beginning of the second period, 3 artificial brushes were orally inserted to rumen. At the end of each period, milk samples were taken from dairy cows to determine the percentage of fat, protein and urea content. The external intervention (RF) did not show any significant effect on the monitored indicators. The individual differences between the dairy cows or during the experiment were more pronounced, both in feed intake, milk yield, chewing activity, and in changes in pH and temperature in the rumen. After the application of the RF, the P-group increased the time of feeding by 15 minutes/day, and C-group by 9 minutes. After the application of the RF, P-group increased the chewing frequency by 4.4%, and C-group by 6.4%. In the first period milk contained 3.6% of fat and 3.3% of protein, in the second period in P-group 3.75% of fat and in C-group 3.68% of protein. Research was supported by project TG01010082.

Reducing sugar inclusion and extrusion conditions influence rumen degradability of faba bean blends

P. Chapoutot[1], O. Dhumez[1], A. Geramain[2], E. Certenais[2] and G. Chesneau[2]
[1]UMR MoSAR, INRA, AgroParisTech, Université Paris-Saclay, 16 Rue Claude Bernard, 75005 Paris, France, [2]Valorex, La Messayais, 35210 Combourtillé, France; patrick.chapoutot@agroparistech.fr

Reduction in nitrogen (N) ruminal degradability of proteaginous or oilseeds by different treatments of extrusion has already been described, but studies combining effects of reducing sugars and extrusion conditions on proteaginous/oilseed mixtures are rare. Moreover, precise technological parameters were seldom reported in extrusion trials. 16 treatments were studied to test the influence of 2 types of sugars, with different dextrose equivalent (red+, 30% vs red++, 95%), included at 2 doses (5%, D5 vs 10%, D10) in faba bean/rapeseed mixtures (90/10%) during extrusion. Several technical conditions were tested for maturation (duration of 1/4 vs 1 h; 5% of added water, W5, vs 15%, W15) and extrusion (weak constraints, WK, vs medium, MD; product temperature of 110 °C, T110 vs 140 °C, T140). *In sacco* N ruminal degradation was measured on 3 mm grounded blends, and used to calculate N effective degradability (NED, %). Analysis of variance allowed to adjust NED for main factors of the incubation design (period, animal and feed effects) and to subdivide feed effect into main studied factors (sugar and dose, maturation duration, added water, mechanical and thermic effects) and their interactions. NED slightly (P<0.001) decreased with red+ sugar (84.9 vs 87.2% for red++) and when sugar is included at 5% (84.2 vs 88.0 for D10). No significant influence of temperature was observed, but NED was significantly (P<0.001) lower for ¼ h duration (85.2 vs 87% after 1 h), W5 (83.2 vs 89.0% for W15) and MD (84.6 vs 87.6% for WK). Nevertheless, all double interactions were significant (P<0.05, at least), except sugar × dose (P<0.10) and water × temperature (P<0.20). Moreover, few particular combinations of different factors lead to NED lower than 80% (78% for Red+_D5%_1/4h_W5%_WK_T110; 76% for Red+_D5%_1h_W5%_MD_T110 and Red++_D5%_1/4h_W5%_MD_T140). These NED variations were not precisely related with *in vitro* N degradability but could be explained by a quantitative synthetic technological index resuming the different extrusion conditions.

Variability of nitrogen degradability of High Protein sunflower meal

P. Chapoutot[1], B. Rouille[2], C. Peyronnet[3], E. Tormo[3] and A. Quinsac[4]
[1]UMR MoSAR, INRA, AgroParisTech, Université Paris-Saclay, 16 Rue Claude Bernard, 75005 Paris, France, [2]Institut de l'Elevage, Monvoisin B.P. 85225, 35652 Le Rheu, France, [3]Terres Univia, 11 rue de Monceau, CS 60003, 75008 Paris, France, [4]Terres Inovia, Rue Monge, Parc Industriel, 33600 Pessac, France; patrick.chapoutot@agroparistech.fr

Around 900,000 tons of high protein (HiPro) sunflower meals (SM) are imported per year in France since 2012-2013, essentially from Black See area. These HiPro SM are obtained after a dehulling process and a mechanically- and solvent-extraction. The aim of this study was to quantify the level and the variability of nitrogen (N) degradation of these new feeds compared to well-known low protein (LoPro) SM. For that, 65 samples of LoPro and HiPro SM coming from various countries were collected in 2016 by French feed companies in order to get the largest dispersion of chemical content (n=65; crude protein, CP=39.0±4.2% DM; crude fibre, CF=21.0±4.2% DM; N *in vitro* degradability, DE1=54.7±4.8%). *In vitro* and *in sacco* measures were conducted on 15 samples chosen among the 65 group to be representative of the whole variability and the different origins (12 HiPro SM: France 4, Ukraine 5, Argentina 2, Hungary 1; 2 LoPro SM: France; 1 totally dehulled SM: Bulgaria). *In vitro* DE1 was measured and in sacco standardized method was applied on the 1 mm-grounded samples to calculate their N effective degradability (NED, %) according to a 6%/h particle outflow rate. Analysis of variance allowed to adjust NED values for the main factors of the incubation design (period, animal and feed effects). Average CP, CF and DE1 values of the 15 SM group were 39.0±4.7, 20.7±4.9% DM and 53.1±6.3%, respectively. The NED varied from 82.1 to 92.2% (86.0±2.6). No differences ($P<0.05$) were observed between origins and types of SM, even if the experimental design was not achieved for that purpose. The observed relationship between DE1 and NED was: NED = 79.8 (±4.4) + 0.29 (±0.08) × DE1 (n=15; R^2=0.48; ETR=1.9) When compared to the predictive model actually used in France and applied to SM, the slope of this regression was not statistically different from 0.36 but the intercept was about 7 points higher.

Chestnut tannin inclusion and extrusion conditions influence rumen degradability of faba bean blends

O. Dhumez[1], P. Chapoutot[1], A. Geramain[2], E. Certenais[2] and G. Chesneau[2]
[1]UMR MoSAR, INRA, AgroParisTech, Université Paris-Saclay, 16 Rue Claude Bernard, 75005 Paris, France, [2]Valorex, La Messayais, 35210 Combourtillé, France; ophelie.dhumez@agroparistech.fr

The reduction in nitrogen (N) ruminal degradability of proteaginous or oilseeds by different treatments of extrusion has already been described, but studies combining the effects of tannins and extrusion conditions on proteaginous/oilseed blends are rare. Moreover, precise technological parameters were seldom reported in extrusion trials. 16 treatments were studied to test the inclusion or not (T00) of chestnut tannins (0.2%, T02) during the extrusion of faba bean/rapeseed blends (90/10%) combined with different processes varying according to maturation/extrusion conditions: 2 levels of maturation duration (1/4 vs 1 h), of additional water (5%, W5 vs 15%, W15) and of extrusion mechanical constraints (weak, WK vs strong, ST). *In sacco* N ruminal degradation was measured on blends grounded on a 3 mm mesh, and used to calculate their N effective degradability (NED, %). Analysis of variance allowed to adjust NED values for the main factors of the incubation design (period, animal and feed effects) and to subdivide feed effect into the main studied factors (tannin inclusion after maturation, maturation duration, added water and mechanical effects) and their interactions. When studying the main effects, NED slightly but significantly ($P<0.001$) decreased for T02 (82.5 vs 85.3% for T00), after ¼ h of maturation (83.6 vs 84.2% for 1 h), with W5% (82.1 vs 85.7% for W15%) and, surprisingly, with weaker extrusion conditions (83.0 vs 84.7% for ST). Nevertheless, all the double interactions were significant ($P<0.001$), except tannin × mechanical one. Moreover, for most of the treatments NED≥80%, and only few particular combinations of the different factors lead to NED = 78% (T02-1/4h-T60-W5%-ST, 77.8 for T02-1h-T45-W15%-WK and T00-1/4h-T60-W5%-WK). These NED variations were not precisely predicted by *in vitro* N degradability. Moreover, they could not be explained by a quantitative synthetic technological index resuming the different extrusion conditions.

Estimation of utilizable crude protein in tropical ruminant feedstuffs by three different methods

K. Salazar-Cubillas and U. Dickhöfer
Animal Nutrition and Rangeland Management in the Tropics and Subtropics, Institute of Agricultural Sciences in the Tropics, Fruwirthstraße 31, 70599 Stuttgart, Germany; khaterine.salazar-cubillas@uni-hohenheim.de

Advanced techniques exist to estimate the utilizable crude protein (uCP) supply from ruminant feedstuffs as the sum of rumen undegradable feed crude protein (CP) and the microbial CP reaching the duodenum. However, these methods and related algorithms have been developed for feedstuffs commonly used in temperate husbandry systems. Hence, the aim was to compare three methods for estimating uCP supply from tropical ruminant feeds. Samples of 6 concentrates, 7 forage grasses, and 13 forage legumes were incubated in duplicate in the rumen of three cows for 2, 4, 8, 16, 24, and 48 h to determine ruminal CP degradability and organic matter digestibility. The uCP was then estimated according to Lebzien and Voigt (i.e. reference method). Additionally, non-protein nitrogen, soluble protein, neutral detergent insoluble nitrogen, and acid detergent insoluble nitrogen were analysed in duplicate and used to estimate uCP according to Zhao and Cao (i.e. chemical method). Finally, samples were incubated *in vitro* in triplicate during two runs and uCP estimated from mean ammonium concentrations in incubation medium according to Steingaß *et al.* (i.e. *in vitro* method). Mean uCP estimates of all samples (n=26) from different methods were compared by linear regression analysis. The uCP values ranged from 79.9 to 253.8 g/kg dry matter. There were linear relationships between uCP values from the reference method and those estimated based on chemical (slope: 93.2, standard error (SE) 9.9; intercept 0.41, SE 0.05; R2=0.73) or *in vitro* procedures (slope: 92, SE 14.1; intercept: 0.55, SE 0.09; R2=0.57). Results suggest that uCP values estimated from concentrations of chemical protein fractions in tropical ruminant feedstuffs are more accurate than those determined *in vitro*.

Effect of feeding red osier dogwood on feed digestibility and acute phase response in beef heifers

W.Z. Yang[1], L.Y. Wei[1], W.M.S. Gomaa[1], B.N. Ametaj[2] and T.W. Alexander[1]
[1]Lethbridge Research and Development Centre, 5403, 1 Ave S, Lethbridge, T1J 4B1, Canada, [2]University of Alberta, 4-10G Agriculture/Forestry Centre, Edmonton, T6G 2P5, Canada; wenzhu.yang@agr.gc.ca

Red osier dogwood (ROD) is an abundant native shrub plant across Canada and it is rich in bioactive compounds with total phenolic concentration varying between 4 to 22%. Objective of this study was to determine the effect of increasing substitution of ROD for barley silage in high-grain diet on feed intake, nutrient digestibility and acute phase response in beef heifers. Five rumen cannulated Angus heifers (live weight=705±66.8 kg) were used in a 5×5 Latin square design with each period including 14 d of adaptation and 7 d of data collection. Treatments were control diet containing 15% barley silage and 85% barley concentrate (dry matter basis), diet that substituted with 3, 7 or 15% of ROD for equal proportion of silage, respectively, or diet supplemented with antibiotics (330 mg monensin + 110 mg tylosin/day per head). Digestibility was determined using Cr_2O_3 as external marker by sampling faeces from the rectum. Blood samples were collected during last day of each period for determining acute phase protein concentrations. Data were analysed using the MIXED procedure of SAS with model included treatment as fixed effect and the random effects of heifer and period. The effect of increasing ROD inclusion was examined through linear and quadratic orthogonal contrasts of SAS. Dry matter intake quadratically (P<0.02) increased with increasing the replacement of silage by ROD from 0, 3, 7 to 10% (11.5, 12.1, 12.4 and 11.8 kg/d). Digestibility of dry matter linearly (P<0.04) increased with increasing ROD from 0, 3, 7 to 10% (72.1, 74.2, 76.1 and 74.9%). Feeding ROD versus monensin/tylosin increased intake (P<0.01) and digestibility of dry matter. Plasma concentration of haptoglobin linearly (P<0.01) increased (0.92, 1.03, 1.21 and 1.78 mg/ml) and concentration of serum amyloid A tended (P=0.06) to increase (39.6, 39.8, 43.7 and 50.5 μg/ml) with increasing ROD from 0, 3, 7 to 10%. These results indicate greater feed value of ROD than barley silage. The ROD could also be used to improve immune status in growing beef heifers fed high-grain diet.

Session 48 Poster 24

Predicting maize starch degradation rates and ruminal starch disappearance with NIR spectroscopy

C.W. Cruywagen and J.H.C. Van Zyl
Stellenbosch University, Animal Sciences, Private Bag X1, Matieland, 7602 Stellenbosch, South Africa; cwc@sun.ac.za

The ratio of vitreous to floury endosperm in grain determines vitreousness. Low vitreous maize, also referred to as soft maize, contains more floury and loosely packed starch granules in the endosperm compared to high vitreous maize where the endosperm contains more densely packed starch granules embedded within a complex protein matrix. Maize vitreousness is negatively correlated with rate and extent of ruminal starch fermentation. The objective of the study was firstly to determine if near-infrared spectroscopy (NIR) can be used to predict vitreousness of milled maize accurately. A second objective was to determine if NIR hardness index values of maize could be used to predict the rate (k_d) and extent of *in vitro* starch disappearance. Six maize samples that varied significantly in terms of vitreousness were selected from a set of 90 samples with known vitreousness. The selected samples were incubated in buffered rumen liquor for 0, 3, 6, 12, 24 and 48 h, whereafter ruminal starch disappearance parameters were determined by a non-linear model. The first derivatives were then used in a secondary model to determine predicted ruminal starch disappearance from the rumen (PRD). All the kinetic coefficients were subjected to a main effects ANOVA with the aid of Statistica. Relationships between NIR and k_d and PRD were determined with regression analysis, whereafter a one-way ANOVA was used to determine significance between parameters. Significance was declared at $P \leq 0.05$. Hardness indexes calculated from NIR analyses at a single absorbance of 2230 nm showed inverse linear and quadratic relationships for both k_d and PRD. Linear coefficients were $r^2=0.819$ for k_d and 0.946 for PRD. Quadratic responses showed adjusted r^2 to be 0.917 for k_d and 0.993 for PRD. The *in vitro* rate of starch degradation decreased significantly (from 45 to 11%/h) as maize vitreousness increased from very soft to very hard. Similarly, PRD decreased (from 79 to 61%) as vitreousness increased. It was concluded that NIR spectroscopy can be used to accurately predict the rate and extent of ruminal starch disappearance. The technology can be used by the feed industry to increase accuracy in dynamic feed formulation models.

Session 48 Poster 25

Mineral metabolic profile of lactating Nellore cows evaluated for residual feed intake

L.L. Souza[1], J.N.S.G. Cyrillo[1], M.F. Zorzetto[1], N.C.D. Silva[2], J.A. Negrão[3] and M.E.Z. Mercadante[1]
[1]Instituto de Zootecnia, Sertãozinho, SP, 14174-000, Brazil, [2]Universidade José do Rosário Vellano, Alfenas, MG, 37130-000, Brazil, [3]Universidade de São Paulo, FZEA, FZEA, Pirassununga, SP, 13635-900, Brazil; cyrillo@iz.sp.gov.br

The objective of this study was to evaluate the relationship between residual feed intake (RFI) and the metabolic indicators of the mineral status of lactating Nellore cows. Twenty-seven Nellore cows (1164±25 days of age and 506±40 kg of initial body weight) were submitted to the feed efficiency test (GrowSafe® Systems) started 21 days postpartum, during 79 days. The roughage:concentrate ratio of the diet was 90:10. The cows were weighed every 21 days to obtain the average metabolic body weight (BW0.75) and the average daily gain (ADG). Ultrasonic fat thickness was evaluated at 20±5 and 83±5 days postpartum in five anatomic sites and an average of fat thickness (FT) was obtained. Cows were milked by machine milking at 63±5 and 85±5 days postpartum to estimate milk production in 24 h, which was subsequently corrected for energy (MMc). Blood samples were collected at 15±5, 41±5 and 62±5 days postpartum, and the serum calcium, phosphorus and magnesium dosages were evaluated for mineral metabolic profile. RFI was obtained as the residue of the dry matter intake (DMI) regression equation on ADG, BW0.75, FT and MMc (R2=0.29). Cows were classified as negative RFI (RFI<0, more efficient) and positive RFI (RFI>0, less efficient). Significant effects of ADG, BW0.75, FT and MMc on DMI were not observed. Calcium and magnesium concentrations were similar in the two classes of RFI (P>0.05), however, there was a difference in the phosphorus concentration between the RFI classes (P=0.003). More efficient cows had a higher blood phosphorus concentration (7.82±0.3 mg/dl) than less efficient cows (7.27±0.3 mg/dl). In conclusion, Nellore lactating cows of different RFI have distinct mineral metabolic profile. Phosphorus blood concentration can be used as an indicator of feed efficiency in lactating Nellore cows.

Using self-feeders with high concentrate diets for finishing beef cattle

V. Beretta, A. Simeone, J. Franco, A. Casaretto, S. Mondelli and G. Valdez
University of the Republic, Ruta 3, km 363. Paysandú, 60000, Uruguay; beretta@fagro.edu.uy

In Uruguay, lot-fed finished cattle receive better price from the industry. However, operative aspects related to roughage manipulation and ration delivery may impair implementing feedlot feeding systems in low scale farms. An experiment was conducted to evaluate the effect of removing long fibre from the ration and replacing daily hand-feeding in conventional troughs (HF) for self-feeders (SF), on feedlot performance and carcass traits of Hereford steers. Two total mixed rations: 20% *Setaria italica* hay/ 80% concentrate (HR), or 8% wood chips/ 92% concentrate (WR), were formulated for similar ME (2.85 Mcal/kg), CP (11.8%) and NDF (HR: 31.8%; WR: 28.7%) content. Thirty six steers (362±48.5 kg) blocked by liveweight (LW) were randomly allocated to 9 pens and 3 treatments: (1) HF of HR ration; (2) HF of WR ration; (3) SF of WR ration. Animals were gradually introduced to diets and then fed *ad libitum* during 10 weeks. LW was recorded every 14 days, dry matter intake (DMI) was determined daily on even weeks and mean feed to gain ratio (FGR) was calculated. Steers were all slaughtered on same date and carcass traits were measured. The experiment was analysed according to a randomized block design, with repeated measures for LW and DMI. Fibre source and delivery system effects were tested through orthogonal contrasts. Replacing HR for WR did not affect physically effective fibre supply ($peNDF_{>1.18mm}$: 15.3 vs 13.1%; $P>0.05$) nor LW gain (1.77 vs 1.81 kg/d; $P>0.05$), but it reduced DMI ($P<0.01$) improving FGR (7.6 vs 6.0; $P<0.01$). Replacing HF for SF did not affect LW gain (1.81 vs 1.71 kg/d; $P>0.05$), but it increased DMI ($P<0.05$) and FGR value (6.0 vs 6.8; $P<0.05$). Higher FGR could be related to observed larger variation in DMI between weeks and between days within a week for SF ($P<0.10$) compared to a more stable intake pattern in HF. No differences due to fibre source or feed delivery system were observed in carcass weight (265.6±15.6 kg), backfat (11.3±2.1 mm) or ribeye area (54.4±4.9 cm^2). Results suggest that, as long as required peNDF intake is assure, it is feasible removing long fibre from finishing rations as well as its supply in SF. Given that FGR increases with SF, decision should be mediated by a cost/ price analysis.

Effect of *Pleurotus* species pre-treatment of lignocellulose on a reduction in lignin of rice stubble

P. Paengkoum[1], T. Vorlaphim[2], S. Paengkoum[3] and C. Yaungklang[4]
[1]Suranaree University of Technology, School of Animal Production Technology, Institute of Agricultural Technology, Suranaree University of Technology, Muang, Nakhon Ratchasima, 30000, Thailand, [2]Suranaree University of Technology, School of Animal Production Technology, Institute of Agricultural Technology, Suranaree University of Technology, Muang, Nakhon Ratchasima, 30000, Thailand, [3]Nakhon Ratchasima Rajabhat University, Program in Agriculture, Nakhon Ratchasima Rajabhat University, Muang, Nakhon Ratchasima, 30000, Thailand, [4]Rajamangala University of Technology Isan, Applied Biology in Animal Production, Faculty of Sciences and Liberal Arts. Rajamangala University of Technology Isan, Muang, Nakhon Ratch, 30000, Thailand; pramote@sut.ac.th

The aim of this study was to examine the bioconversion of rice stubble fermentation with *Pleurotus ostreatus* (POT), *Pleurotus sajor-caju* (PSC) and *Pleurotus eous* (PE). The rice stubbles was inoculated with the fungi and incubated in the dark cupboard in the laboratory at 30 °C and 75% relative humidity (RH). The chemical composition and *in vitro* degradability of untreated rice stubble and treated rice stubble with *Pleurotus* species were analysed at day 20, 25, 30, 35 and 40th inoculation. Results shown that all of fermentation by *Pleurotus* fungi treatments were apparently increased ($P<0.001$) in crud protein (CP) content when compared with the control. Whereas significant decreased in neutral detergent fibre (NDF), acid detergent fibre (ADF), acid detergent lignin (ADL), hemicellulose, and cellulose contents of rice stubbles by fungal fermentation. *In vitro* gas production was significantly increased at day 25th fermentation in all fungal treatments from 24-96 h incubation. The estimated organic matter digestibility (OMD) of *Pleurotus* species fermented at 25 days was improved from 52.02 to 62.12, 63.75, and 65.27% (control, PSC, PE and POT) respectively. For the estimated of ME was similarly trend with organic matter digestibility 7.44, 8.95, 9.19 and 9.43 MJ/kg DM (control, PSC, PE and POT). It was implied that the period time was effected to fungi fermentation.

Session 49 Theatre 1

Outlook on local pig breeds as drivers of high quality pig production-ambitions in project TREASURE

M. Čandek-Potokar[1], L. Fontanesi[2], J.M. Gil[3], B. Lebret[4], R. Nieto[5], M.A. Oliver[6], C. Ovilo[7] and C. Pugliese[8]
[1]KIS – Agric. Inst. of Slovenia, Hacquetova 17, 1000 Ljubljana, Slovenia, [2]Univ. di Bologna, DISTAL, Viale Fanin 46, 40127 Bologna, Italy, [3]CREDA-UPC-IRTA, Terrades 8, 08860 Castelldefels, Spain, [4]UMR PEGASE, INRA Agrocampus-Ouest, 16 Le Clos, 35590 St.Gilles, France, [5]EEZ-CSIC, Camino de jueves s/n, 18100 Granada, Spain, [6]IRTA, Finca Camps i Armet, Monells, Spain, [7]INIA, Ctra. Coruña Km 7.5, 28040 Madrid, Spain, [8]Univ. di Firenze, Via delle Cascine 5, 50121 Firenze, Italy; meta.candek-potokar@kis.si

Project's idea is a new paradigm of pig production systems and development of sustainable pork chains based on European local pig breeds which are held in less intensive systems, use locally available feeding resources and are adapted to local agro-climatic conditions. Local pig breeds provide products with typical, generally high sensory quality and regional identity searched by consumers. Despite revived interest, these breeds are mainly untapped and often remain endangered. The few successful examples in Europe demonstrate that the best conservation strategy is to ensure breed is self-sustaining resulting from good valorisation of pork products. As their productivity is low, local pork chains can become sustainable only when their genetic potential is benefited, production systems optimised and their products viable on the market. To enhance sustainability of pork chains based on local breeds, it is essential to gain scientific proofs of their genetic singularity, productive potential and product qualities, to develop genetic tools for authentication and breeding, to optimize pig nutrition and management (enhance welfare, use of local feeding resources), to evaluate their environmental impact, to assess consumers' attitudes, acceptability and purchase intentions, and to develop adapted marketing strategies. In addition, project promotes knowledge exchange and building of functional networks among regions esp. by means of creating an 'umbrella' trade mark as exploitation booster. These challenges are addressed by the project and will be presented. Funded by European Union's H2020 RIA program (grant agreement no. 634476).

Session 49 Theatre 2

Environmental impacts of pig production systems relying on European local breeds

A.N.T.R. Monteiro[1], A. Wilfart[2], V.J. Utzeri[3], N.B. Lukač[4], U. Tomažin[4], L. Nanni-Costa[3], M. Čandek-Potokar[4], L. Fontanesi[3], J. Faure[1] and F. Garcia-Launay[1]
[1]PEGASE, INRA, Agrocampus Ouest, 35590 Saint-Gilles, France, [2]SAS, INRA, Agrocampus Ouest, 35000 Rennes, France, [3]UNIBO, Viale G.Fanin 44, 40127 Bologna, Italy, [4]KIS, Hacquetova ulica 17, 1000 Ljubljana, Slovenia; florence.garcia-launay@inra.fr

Traditional pig productions systems, relying mainly on local pig breeds and outdoor rearing, have been poorly investigated so far in terms of environmental impacts. The few existing studies did not account for possible sequestration of carbon and emissions consecutive to grazing. Twenty-five farms of Gascon breed in France (FR), 8 with Mora Romagnola breed in Italy (IT), and 15 of Krškopolje breed in Slovenia (SI) were evaluated while accounting for the emissions from pasture intake and the potential for carbon sequestration. Pig production system in SI presented the lowest impacts per kg of live weight, due to better feed conversion ratio caused by indoor production and due to lower impacts of feeds – most diets were based on grains, vegetables, tubers and roots produced on farm. Among the systems, acidification potential (AP) was 13% higher in IT than the average for FR and SI, due to higher dietary crude protein content (+9% than the average), while the eutrophication potential (EP) was 27% higher in FR system than the average, as a result of higher phosphorus content of feeds (+28% than the average). When the potential of carbon sequestration was taken into account, the GWP impact was reduced 4% on average. Conversely, when accounting for the emissions from pasture intake the GWP was increased by 2%, mainly when a high digestible grass was considered. The use of high digestible grass provided lower AP and EP impacts than low digestible grass. The large variability between farms in terms of environmental impacts suggests that the margins for improvement of local breeds' production rely on improvement of feed composition and supply, and origin of feed ingredients. There is a great need for better estimation of digestibility of grasses and of carbon sequestration, in order to reduce the uncertainties associated with the environmental impacts evaluated of outdoor pigs' systems.
Funded by European Union's H2020 RIA program (grant agreement no. 634476).

Session 49 Theatre 3

Digestibility and nitrogen balance in Cinta Senese growing pigs fed different protein levels

C. Aquilani, F. Sirtori, R. Bozzi, O. Franci, A. Acciaioli and C. Pugliese
University of Florence, DISPAA – Department of Agrifood Production and Environmental Sciences, Via delle Cascine 5, 50144 Firenze, Italy; chiara.aquilani@unifi.it

The aim of this study (of TREASURE project) was to evaluate the protein digestibility and the N retention of four diets, containing 12, 14, 16 and 18% of crude protein (CP) in Cinta Senese growing pigs. Bentonite as internal indicator was added to each diet (2% as fed). Eight castrated males, weighing 55 kg of lw on average, were cyclically fed with the four diets according a Latin-square design. The animals were regularly weighed before each experimental cycle to adjust the daily amount of feed, according to their metabolic weight (90 g DM/kg MW). Every cycle consisted in 10 days of diet adaptation in box, two days of adaptation in metabolic cage and three days of trial, when faeces and urines were collected. The whole trial lasted a total of 8 weeks. Sampling took place at fixed hours, once a day for urines and twice a day for faeces. On feed and faeces, the following analysis were carried out: moisture, protein, ether extract, ash, NDF, ADF, ADL and acid insoluble ash (AIA). On urines, nitrogen content was determined. Total tract apparent digestibility (TTAD), balance and efficiency of nitrogen utilization and energy partition of the experimental diets were calculated. Data were analysed by SAS using the GLM model and considering sampling day and animal as fixed effects and diet and metabolic weight as continuous effects. TTAD of dry matter, organic matter and protein showed a parabolic trend with maximum at 14.5-15.0% of CP. Concerning the N-related parameters, results showed that the N intake, N adsorbed and total N excreted increased with the dietary CP content. Contrariwise, the biological value (BV) of the diet and the N retained/intake ratio linearly decreased as the CP increased. Indeed, the lowest BV was observed for the 18% CP diet (45.28%), while the highest was showed by 12% CP diet (59.82%). Energy digestibility followed a parabolic trend, while metabolizable energy decreased linearly from 12%CP to 18%CP diets. In conclusion the protein requirement for Cinta Senese growing pig (from 50 to 80 kg lw) can be fulfilled by the 12% CP diet. Funded by European Union's H2020 RIA program (grant agreement no. 634476).

Session 49 Theatre 4

Effects of immunocastration on performance and nitrogen utilization of Iberian pigs

P. Palma-Granados, N. Herrera, L. Lara, M. Lachica, I. Fernández-Fígares, I. Seiquer, A. Haro and R. Nieto
Estación Experimental del Zaidin, CSIC, Cno. del Jueves sn, 18100, Armilla, Granada, Spain; rosa.nieto@eez.csic.es

Immunocastration – vaccination against gonadotropin releasing hormone (GnRH) has been shown to prevent sexual development and boar taint in pigs being a feasible alternative to surgical castration (SC). Studies in conventional pigs point out that immunocastrated (IC) pigs show better performance than SC pigs. Apart from the benefits on animal welfare, this fact could be of interest for Iberian pigs and other fatty pig types, characterized by long productive cycles and low capacity for lean tissue growth. There is a question on whether protein requirements might be increased in IC compared to SC Iberian pigs. The purpose of this study (in project TREASURE) was to examine the effects of immunocastration on Iberian pig growth and nitrogen (N) retention under 3 dietary protein concentrations. Fifty-four pure Iberian pigs were used (3 sexes: IC males, IC females, SC males; 3 isoenergetic diets: 150, 130 and 110 g CP/kg DM, 6 pigs/treatment combination). Vaccination against GnRH was at 4.3 (40 kg) and 6 months of age (70-80 kg). Pigs were individually housed consuming the experimental diets from 40 to 105 kg-BW with a slight restriction. Digestibility and N balance assays were performed at 50 and 90 kg BW, respectively. The IC males showed higher overall growth rate (+13%; $P<0.001$) and feed efficiency (+9.6%; $P<0.001$) than SC males and IC females, and slightly higher daily feed intake (+3%, $P<0.001$). No significant effect of dietary protein level was found on growth or feed efficiency. Before the second vaccination, IC pigs showed higher N retention (13 to 39%; $P<0.001$) and efficiency of N utilization (22 to 47%; $P<0.001$) than SC males and IC females. Pigs fed the lower protein content diet retained less N ($P<0.05$). However, no significant differences among sex groups in N retention parameters were detected after the second vaccination. Higher performance of IC male pigs seems to be related to a greater capacity for N retention before receiving the second dose of the vaccine. Funded by European Union's H2020 RIA program (Grant agreement no. 634476).

Development of long-term, pre-finishing immunocastration protocols for male Iberian pigs: 2. carcass

F.I. Hernández-García, M. Izquierdo, M.A. Pérez, A.I. Del Rosario, N. Garrido and J. García-Gudiño
Center of Scientific & Technological Research of Extremadura (CICYTEX), Animal Production, Autovía A-5, Km 372, 06187-Guadajira (Badajoz), Spain; francisco.hernandez@juntaex.es

Male immunocastration (IC) should be adapted to the long life cycle of Iberian (IB) pigs. We previously developed long-term, 3-dose, pre-finishing IC protocols for male Iberian pigs whose efficacy seemed to be influenced by nutritional and stress level. Study 1 aimed to improve the efficacy of protocols by a short-term increase in feeding intake. IC males (ICM; n=47) were fed commercial concentrate in an extensive system, immunized against GnRH at 11, 12 and 14 months (m) of age and slaughtered at 16 m. The treated subgroup (23 ICM) was submitted to a 15-day *ad libitum* feeding period starting at the 3rd vaccination, and the remaining ICM were the Control subgroup. The treated ICM group reached 100% efficacy as all its animals had <150-g testes (which was the threshold for blood testosterone presence in our earlier studies). In contrast, 4/24 Control ICM had >150-g testes. Study 2 aimed to adjust the protocol to suit the chronology of the acorn-feeding free-range period (*montanera*; MT). We tested whether improving body condition homogeneity at the start of MT would homogenize testicular atrophy. Control IB pigs (C; n=18) were immunized at 10.5, 12 and 13.5 m. Treated pigs (T; n=17 IB males) were immunized at 10.5, 11.5 and 13 m, with a 15-day *ad libitum* (AL) feeding period starting at the 3rd dose. *Montanera* started at 13.5 m, and slaughter took place at 16 m. The AdLib group (IB × Duroc; n=15) were fed AL with concentrate during growth and finishing phases in an extensive system, immunized at 8, 9 and 10 m, and slaughtered at 13 m. The Adlib and T treatments showed a 100% efficacy. Backfat androstenone and skatole levels were basal. Ham yield and backfat thickness were similar for the 3 groups. Unlike in Study 1 and our previous studies, intramuscular fat content, backfat thickness and meat texture did not seem different from those usually reported for surgically castrates. We conclude that nutritional level can be used to improve the IC efficacy. Longer finishing after the last vaccination may increase fat deposition. Funded by European Union's H2020 RIA program (grant agreement no. 634476).

Gut microbiota analyses for sustainable European local porcine breeds: a TREASURE pilot study

J. Estellé, M. Čandek-Potokar, M. Škrlep, Č. Radović, R. Savić, D. Karolyi, K. Salajpal, M.J. Mercat, G. Lemonnier, O. Bouchez, J.M. García-Casco, P. Palma-Granados, R. Nieto, A.I. Fernández, B. Lebret and C. Óvilo
Treasure consortium, Hacquetova 17, Ljubljana, Slovenia; jordi.estelle@inra.fr

The study of gut microbiota and its effects on hosts has emerged as an essential component of host homeostasis and global efficiency. Besides host's influence on gut microbiota, major quantitative and qualitative changes may occur in the composition of the intestinal microbiota due to the influence of diet and other environmental factors. In accordance with the TREASURE project global aim of enhancing sustainability of production systems for local pig breeds, the objective of our task was to conduct a pilot characterisation of intestinal microbiota in order to test its usefulness to characterize several local European pig populations and their production systems. This approach has been applied to populations belonging to the following European traditional breeds: Gascon (France), Iberian (Spain), Krskopolje (Slovenia), Mangalitsa (Serbia), Moravka (Serbia) and Turopolje (Croatia). For each breed, faecal samples have been collected along different experiments performed in the TREASURE project targeting the comprehension of a particular traditional production system (e.g. open-air farming), management practice, or the comparison of breeds. In all experiments, the metagenomics technique employed is the re-sequencing of the bacterial 16S in an Illumina MiSeq system. Overall, the results have shown that the gut microbiota analysis is a promising approach for the characterisation of these local breeds, by allowing a deeper understanding of their production systems and potentially allowing the development of new certification approaches. Preliminary results will be summarized in this communication. Funded by European Union's H2020 RIA program (Grant agreement no. 634476).

Modelling study with InraPorc® to evaluate nutritional requirements of growing pigs in local breeds

L. Brossard[1], R. Nieto[2], J.P. Araujo[3], C. Pugliese[4], Č. Radović[5] and M. Čandek-Potokar[6]
[1]PEGASE, INRA, Agrocampus-Ouest, 35042 Rennes, France, [2]EEZ, CSIC, Camino de Jueves s/n, 18100 Granada, Spain, [3]IPVC, Praça General Barbosa, 4900 Viana de Castelo, Portugal, [4]UNIFI, Via delle Cascine 5, 50144 Firenze, Italy, [5]IAH, Autoput za Zagreb 16, 11080 Belgrade-Zemun, Serbia, [6]KIS, Hacquetovaulica 17, 1000 Ljubljana, Slovenia; ludovic.brossard@inra.fr

Models as InraPorc® have been developed to simulate pig growth and to determine nutrient requirements. They are largely applied to conventional breeds but so far not to local breeds. Our study aimed to use InraPorc® to determine nutrient requirements of growing pigs from local breeds in H2020 EU project TREASURE. Data on feed composition, allowance and intake, and body weight (BW) were extracted from literature reports or experiments conducted within the project. They were used to calibrate parameters defining a growth and intake profile in InraPorc®. We obtained 15 profiles from 9 breeds (Alentejano, Basque, Bísaro, Calabrese, Cinta Senese, Iberico, Krškopolje, Mangalitsa Swallow Bellied and Moravka). Breeds had 1 to 3 profiles depending on experimental conditions or data sources. Conditions of the study affected calibration results. The mean protein deposition (PD) was low for all breeds from 39.9 to 91.0 g PD/day vs over 110 g/d in conventional breeds. For 40-100 kg BW range, the age of the pigs at 40 kg BW was between 110 and 206 days, denoting different feeding management in addition to genetic differences. Average daily gain (ADG) and feed intake curves showed similar shape. Protein deposition rate was the highest in breeds with the highest ADG. Lysine requirements were largely covered in all studies and breeds, the highest requirements being observed with the highest ADG. In all breeds a low part of total body energy retention was dedicated to protein, conversely to lipids. Despite some methodological limitations, this study provides a first insight on nutrient requirements for some local breeds. Funded by European Union's Horizon 2020 RIA program (grant agreement no. 634476).

An alternative to restricted feeding in Iberian pigs using an agro-industrial by-product of olive oi

J.M. García-Casco[1], M. Muñoz[1], M.A. Fernández-Barroso[1], A. López-García[1], C. Caraballo[1], J.M. Martínez-Torres[2] and E. González-Sánchez[2]
[1]INIA, Centro Cerdo Ibérico – Dpto. Mejora Genética Animal, Ctra EX101, km 4.7, 06300 Zafra, Spain, [2]Universidad de Extremadura, Escuela de Ingenierías Agrarias, Avda. Adolfo Suárez, s/n, 06007 Badajoz, Spain; garcia.juan@inia.es

Traditional Iberian pig production is characterized for having a fattening period with a feeding based on acorn and pastures. During the previous growing period, the feeding is restricted to avoid undesirable weight gains. However, this procedure could cause feeding stress. The use of *ad libitum* diets based on olive by-products during the growing period may be an alternative to avoid this stress. Two diets based on olive by-products, one incorporating dry olive pulp in the feed (DD) and the other one incorporating olive cake in wet form (WD) were compared with a control standard diet group (CD). CD and DD diets were supplied once a day and WD diet was supplied *ad libitum* and supplemented with a specific feed given once a day. Comparisons were performed using ANOVA for: growth, backfat fatty acid profile, carcass composition, percentage of intramuscular fat (%IMF) and other quality meat traits (thaw, cook and centrifuge force losses, shear force, marbling, Minolta colours and myoglobin concentration). No significant differences between the treatment groups were observed for most of the traits. Although animals fed with DD and WD diets grew slower than those fed with CD during the growing period, no differences in the total average daily gain were observed. DD animals showed a higher carcass yield and less %IMF in loin. Olive-cake diets caused higher levels of unsaturation than CD one after the growing period. Lower centrifuge force losses were observed in WD than in CD. DD samples were paler and less red-coloured and has lower myoglobin content than CD and WD ones. Results (of this study within project TREASURE) suggest that WD diet could be a suitable feeding for growing period in traditional Iberian pig production since negative effects on growth, carcass and quality traits were not observed in the current study. Funded by European Union's H2020 RIA program (Grant agreement no. 634476).

Effects of pasture keeping and acorn feeding on growth, carcass- and meat quality of SH pigs

M. Petig[1], C. Zimmer[1], M. Čandek-Potokar[2], S. König[3] and H. Brandt[3]
[1]BESH, Haller Str. 20, 74549 Wolpertshausen, Germany, [2]KIS, Hacquetova ulica 17, 1000 Ljubljana, Slovenia, [3]Institute of Animal Breeding & Genetics, Ludwigstraße 21B, 35390 Gießen, Germany; matthias.petig@besh.de

The Schwäbisch-Hällisch pig (SH) is the oldest autochthonous pig breed in Germany. It has been rescued through a regional marketing program. The traditional feedstuff acorn was revitalized through a premium meat program where outdoor reared pigs get an acorn supplementation. In this study (made in TREASURE project), the effects of pasture keeping and acorn supplementation on growth, carcass composition and meat quality traits of purebred SH pigs were investigated. In 2015 and 2016 in total 305 pigs were introduced to the trial when entering the fattening barn with an average weight of 34+-6 kg. The final live weight was 138+-14 kg with an average slaughter weight of 107+-12 kg. Up to a live weight of 91 kg on average, all pigs where fattened under equal conditions. They were kept in a barn with outdoor access, a total place allowance of 1.7 m^2/animal and a cereals soya bean mixture as feed. After division in three trial groups one group went to pasture where every pig had a place allowance of 400 m^2. The pasture was equipped with huts and a water-/ feeding station. The outdoor group (OA), represented by 57 pigs, was fed with a cereals soya bean mixture which was supplemented with dried acorns (20%). The indoor group with acorn feeding (IA) consisted of 58 pigs and got the same feed as the OA group. As a control group (IC) 190 pigs were fattened indoor without acorns. All animals were fed *ad libitum*. All feeds had common energy- and protein levels (appr. 13 MJ-P, appr. 17% CP). Regarding growth performance the OA pigs showed significant lower daily gain than the other groups in the last fattening period (704 vs 789 g (IA) and 785 g (IC)). The OA group showed with 2.9% a significant higher intramuscular fat content (IMF) than the IC group (2.3%) while the IA animals reached 2.7% IMF. The OA group showed significant lower drip loss values than the others (OA=09.%, IA=1.7%, IC=1.7%). In conclusion pasture keeping reduces growth performance but improves meat quality. Acorn supplementation also has a positive effect on meat quality traits. Funded by European Union's H2020 RIA program (grant agreement no. 634476).

Effects of rice husk supplementation during pre-finishing on production traits of Iberian pigs

M. Izquierdo[1], F.I. Hernández-García[1], J. García-Gudiño[2], J. Matías[1], N. Garrido[1], I. Cuadrado[1] and M.A. Pérez[1]
[1]Center of Scientific & Technological Research of Extremadura (CICYTEX), Autovía A5, Km 372, 06187-Guadajira (Badajoz), Spain, [2]IRTA, Camps i Armet, 17121 Monels, Gerona, Spain; mercedes.izquierdo@juntaex.es

Iberian pigs must be feed-restricted during the pre-finishing period *(premontanera)* due to their adipogenic nature. Supplementation with rice husk (a fibrous, abundant and inexpensive by-product) before finishing may increase satiety, decrease stress and competition, increase animal welfare and might reduce weight gain variability and improve carcass uniformity. This study evaluated the effect of rice husk supplementation on weight gain, carcass composition and hematological traits. Castrate male Iberian pigs were assigned to 3 treatment groups (n=15/group), which, during *premontanera*, from 10 to 14 months of age, were fed concentrate-based diets differing in fibre content, namely Control (C; 5.0%), Medium Fibber (MF; 8.5%) and High Fibber (HF; 12.0%) groups. The MF and HF diets included rice husk (integrated into the concentrate) as a supplemental source of fibre. Rations were isocaloric. Five pigs from each treatment were slaughtered at the end of *premontanera*, whereas the remaining animals were submitted to free-range acorn-feeding (*montanera*) and slaughtered at 16 months of age. Body weight (BW) and *in vivo* ultrasonographic body composition were monitored. A blood sample was collected at the end of *premontanera* to determine the neutrophil/lymphocyte ratio as a chronic stress index. None of the groups had diarrheic problems, regardless the fibre level. Average daily gain from the 12th to the 14th month of age was greater for the HF group, which also exhibited a more homogeneous growth rate among animals. Moreover, the HF group had the thickest carcass backfat and smallest loin area among the animals slaughtered before *montanera*. Regarding backfat fatty acid composition *premontanera*, HF pigs had the lowest C17:0 and C18:3 and the greatest C20:1 percentages, and a trend for greatest MUFA and smallest PUFA values. In conclusion, supplementation with high levels of fibre (12%) from rice husk did not have any negative effect on health, growth or carcass composition.

Session 49 Theatre 11

Determination of fatty acid groups in intramuscular fat of various local pig breeds by FT-NIRS

R. Bozzi[1], S. Parrini[1], A. Crovetti[1], C. Pugliese[1], A. Bonelli[1], S. Gasparini[1], D. Karolyi[2], J.M. Martins[2], J.M. Garcia-Casco[2], N. Panella-Riera[2], R. Nieto[2], M. Petig[2], M. Izquierdo[2], V. Razmaite[2], I. Djurkin Kušec[2], J.P. Araujo[2], M. Čandek-Potokar[2] and B. Lebret[2]

[1]Animal Science, DISPAA – Universiy of Firenze, Via delle Cascine, 5, 50144 Firenze, Italy, [2]TREASURE consortium, Hacquetova ul. 17, 1000 Ljubljana, Slovenia; riccardo.bozzi@unifi.it

The objective of the present study is to evaluate the potential use of FT-NIRS for predicting intramuscular fat (IMF) and fatty acid groups (MUFA; PUFA; PUFA n-3, PUFA n-6; SFA) on pig grounded muscles. The research considered 160 fresh samples of Longissimus collected from 12 European local pig breeds (TREASURE project). For every sample, lipids were extracted from IMF and fatty acid profile was determinate by gas chromatography. Two aliquots of each sample were scanned using FT-NIRS Antaris II model. Mathematical pre-treatments (multiplicative scatter correction, 1st and 2nd derivate) were applied and outliers' spectra were identified and removed when necessary. Partial least square regression was used on the average spectrum and the models validated using an external data set. Results are evaluated in terms of coefficient of regression and root mean square errors in calibration (R^2-RMSE) and validation (Rp^2-RMSEP). As expected, the best results were obtained for IMF with R^2 higher than 0.99 and RMSE lower than 0.2. Unsaturated fatty acids, probably due to the absorption of the cis double bond in a specific region of near infrared spectra, obtain acceptable R^2 (0.89 for MUFA and 0.75 for PUFA n-3 and PUFA n-6). SFA achieved a R^2 of 0.81 that is lower than values reported in other studies probably because of the large variability of genotypes used. The validation models achieved both lower coefficients of determination and higher RMSEP than the calibration models; however, R^2 differences between calibration and validation were smaller than 5%, except for SFA. Hence, the FT-NIRS seems promising to estimate the principal parameters of fatty acid groups on muscle samples from different European autochthonous pig breeds. Inclusion of other samples can improve the accuracy and the robustness of the models, especially considering the high variability of the samples. Funded by European Union's H2020 program (grant agreement no. 634476).

Session 49 Theatre 12

Consumers' study on traditional pork products from local breeds: expectations and hedonic evaluation

B. Lebret[1], Z. Kallas[2], H. Lenoir[3], M.H. Perruchot[1], M. Vitale[4] and M.A. Oliver[4]

[1]PEGASE, INRA, Agrocampus-Ouest, 35042 Rennes, France, [2]CREDA-UPC-IRTA, C Esteve Terradas 8, 08860 Castelldefels, Spain, [3]IFIP-Institut du Porc, La Motte au Vicomte, 35651 Le Rheu, France, [4]IRTA, Finca Camps i Armet, 17121 Monells, Spain; benedicte.lebret@inra.fr

Assessing consumers' acceptability of traditional pork products (TPP) from local pig breeds is essential to ensure sustainability of regional pork chains. As part of TREASURE project, expectations and hedonic perception of Protected Designation of Origin (PDO) Noir de Bigorre (NB) dry-cured ham produced from pure Gascon breed were assessed. The study was conducted in Toulouse (France) as expanding market for TPP with 124 consumers, regular purchasers of TPP (quota sampling method). Three products were used, described as follows: 'NB-PDO dry-cured ham 24 months ripening – local pig breed in extensive system' (NB24), 'NB-PDO dry-cured ham 36 months ripening – local pig breed in extensive system' (NB36) as innovative TPP to enhance sensory quality, and 'Iberian ham – 50% Iberian pig' (IB) as competing product. Following the expectation disconfirmation theory, the sensory test included three phases: blind (tasting without information), expected (product description only: no tasting) and actual liking (tasting with information), assessed using a 9-point scale from 1: extremely dislike to 9: extremely like. Data were submitted to ANOVA (mixed model). Blind hedonic test showed no differences ($P>0.05$) between products that all displayed high liking score (6.7 to 6.8). Product description strongly influenced expected liking ($P<0.001$) with higher score for NB36 than NB24 (7.8 vs 7.2, $P<0.05$) and lowest score for IB (5.2). Hedonic test with information showed higher actual liking for NB36 and NB24, that were similar, than for IB (7.4 and 7.2 vs 5.9, $P<0.001$). Both NB hams displayed higher actual than blind liking ($P<0.05$), denoting positive effect of information on acceptability. Actual liking was similar to expected for NB24, indicating fulfilment of hedonic expectations. By contrast, actual liking was lower than blind for IB ($P<0.05$). Results will be completed with consumer preferences and willingness to pay for TPP and innovations in TPP. Funded by European Union's H2020 RIA program (grant agreement no. 634476).

Economic valuation of social demand for key features of the Noir de Bigorre pork production chain

L. Brossard[1], F. Garcia-Launay[1], B. Lebret[1], J. Faure[1], E. Varela[2] and J.M. Gil[2]
[1]PEGASE, INRA, Agrocampus-Ouest, 35042 Rennes, France, [2]CREDA-UPC-IRTA, C Esteve Terrades 8, 08860 Castelldefels, Spain; ludovic.brossard@inra.fr

Extensive farming systems produce for the society goods and services definable as public goods (biodiversity…), or having public good features (landscape attractiveness…). The provision of these public goods may not be guaranteed due to the lack of recognition of their values in markets and policies that ultimately can put in risk the future of these systems. Within TREASURE project, a choice experiment (CE) survey was applied to assess the social demand for relevant attributes of the Noir de Bigorre (NB) French regional pork chain producing Gascon local pig breed in extensive system with public good character. CE is an economic valuation method estimating the social demand for a given attribute or for combinations of them in management scenarios. Five relevant attributes of the NB chain and their current and potential levels in alternative management options were identified from focus groups: probability of existence of the breed in next 25 years, farm size, feedstuff origin, geographical availability of the products and type of selling places. A monetary attribute was included to assess the social demand for the previous attributes. A valuation questionnaire was administered to 418 individuals (365 through web-based survey, 53 face-to-face), half of them located in the South West of France, i.e. the production area of the NB chain. Results of the CE showed that the respondents had a distinctively urban profile, with no agricultural family background. Almost 40% of the respondents did not know or consume products from the NB chain. On average guaranteeing the survival of the breed achieved the highest willingness to pay (112.37 €/household/year). Respondents were willing to pay 42.35€ to maintain equal number of small and medium farms and 21.86€ to have feedstuff produced in the proximity of the farms. Geographical availability and selling places of products contributed to a lesser extent to shape their preferences. Funded by European Union's H2020 RIA program (grant agreement no. 634476).

Majorcan Black Pig: a sustainable production system for high quality meat products

J. Tibau[1], N. Torrentó[1], J. Jaume[2], J. González[3], M. Čandek-Potokar[4], N. Batorek-Lukač[4] and R. Quintanilla[1]
[1]IRTA, Animal Breeding and Genetics, Monells, 17121, Spain, [2]SEMILLA-CAIB, C/ d'Eusebi Estada, 145, 07009 Palma, Spain, [3]IRTA, Product Quality, Monells, 17121, Spain, [4]Agricultural Institute of Slovenia, Hacquetova ulica, 17, 1000 Ljubljana, Slovenia; joan.tibau@irta.cat

Majorcan Black Pig (MBP) is a native breed from Mallorca, in the Balearic Islands, characterized by its high rusticity and adaptation to Mediterranean climatic conditions, with ability to exploit the scarce natural resources. This pig population had a great importance in the economy and cultural heritage until mid-20th century. The introduction of leaner pig breeds, the impact of different diseases and rural migration led to a progressive decline of this breed. Recognized as an endangered autochthonous pig breed in Spain, MBP has a conservation and improvement program supported by Balearic Government technical services. The current census is close to 80 farms with more than 1,300 breeding pigs. MBP farms are managed in extensive conditions and the feeding regime is based on pasture, cereals, legumes, figs, almonds, acorns and Mediterranean shrubs, with eventual supplementation based on barley and green peas. Performances are largely dependent on available natural resources; mean productivity is 16 piglets/sow-year and a post-weaning growth around 500 g/day until the slaughtering target weight (140-160 kg). Compared with commercial breeds, MBP has a largely higher subcutaneous fat depot, with carcasses showing back fat depths reaching 7 cm and flare fat weights usually up to 6 kg. Important differences are also observed in meat quality, with the loin of MBP presenting higher intramuscular fat content (~8%), darker colour, slightly lower shear force, higher levels of MUFA (~50%) and lower levels of PUFA (<10%). The main products are the 'Sobrassada de Porc Negre Mallorquí', a specialty fat-rich cured sausage granted with a PGI certification, and the roasted 'porcella' (3 month purebred piglets). The proportion of animals devoted either to produce 'porcella' or to be fattened until a heavy slaughter weight depends upon the available natural resources. This practice represents a really sustainable production system. Funded by European Union's H2020 RIA program (grant agreement no. 634476).

Session 49 Poster 15

Innovative patties from Majorcan Black Pig meat: results of a consumer study in Barcelona

J. Dilmé[1], E. Rivera-Toapanta[1], M. Vitale[1], Z. Kallas[2], M. Gil[1] and M.A. Oliver[1]
[1]IRTA, Monells, 17121, Spain, [2]CREDA, Castelldefels, 08860, Spain; mariaangels.oliver@irta.cat

The Majorcan Black Pig (MBP) is an untapped breed from Mallorca Island, in the Mediterranean Sea. The aim of the study was to innovate new, healthier products to expand demand and ensure the survival of MBP farms by identifying new market niches. To achieve this, bioactive compounds such as β-glucans and polyphenols were added to the patties, by including mushrooms (*Boletus edulis*) or blueberries (*Vaccinium corymbosum*), respectively in their composition. A study with 120 consumers was carried out in Barcelona. The consumers had to test five types of MBP patties, three innovative treatments and two controls: MBP patties (A), MBP with porcini mushrooms (B), MBP with blueberries (C); Pork and beef (D) and beef (E). The experimental design consisted of three tests: (1) Blind test: consumers tasted the five types of patties (overall liking on a scale from 1 (dislike extremely) to 9 (like extremely)). (2) Expected test: Consumers indicated general acceptance to the description of the five treatments, with the information about the origin of the meat, production system and potential beneficial effects of the added healthy ingredients. (3) Informed test: Consumers repeated the sensory test, but with the same information as the previous test in each tasting. There were no significant differences according to gender in the blind test. MBP patty treatment had a significantly higher sensory acceptance than the rest (in both the Blind and Informed test). In addition, the average score obtained by the MBP patty was significantly higher in the Blind and Informed tests – which include a sensory evaluation – implying that it was the preferred one by the consumers. The type of patties with vegetal ingredients were scored significantly lower than the other types of patties, and there was no significant effect of the information that had been given to consumers, indicating that consumers did not like so much the sensory characteristics of these patties (texture and flavor). There is a need to provide clear and understandable information to the consumers about the differential characteristics of the products, ensuring sensory quality. Funded by European Union's H2020 RIA program (grant agreement no.634476).

Session 49 Poster 16

Some results on performances of Krškopolje pigs in project TREASURE

N. Batorek-Lukač, U. Tomažin, M. Škrlep and M. Čandek-Potokar
KIS-Agric. Inst. Slovenia, Hacquetova 17, 1000 Ljubljana, Slovenia; nina.batorek@kis.si

Krškopolje pig (KKP), a Slovenian autochthonous breed, has been little studied and more knowledge is needed about their growth potential and performances in different production systems. Present work aims to overview the preliminary results of H2020 project TREASURE, evaluating performance in different rearing systems: conventional (CON) organic (ORG) and traditional (TRD). Growth was evaluated in lactation (study 1, CON vs ORG, n=156), growing and fattening period (study 2, CON vs ORG n=36; and study 3, CON vs TRD n=12); main carcass traits were evaluated after slaughter at approximately 230 days of age (study 2 and 3). Results show that weight at weaning and growth rate in lactation were ≈20% lower in ORG than CON farms, whereas no major differences in growth rate were observed between pigs in two systems in growing/fattening period when similar nutritional value of the diet and feeding was provided. However, in fattening period pigs in ORG system had 13% higher daily gain than in CON system, which could be ascribed to high consumption of alfalfa hay and perhaps less feed dissipation. Yet no differences in body weight and carcass traits were noted between the two groups at the end of the trial. Similar daily gain was observed in KKP pigs raised in CON and TRD system during growing period (until app. 45 kg), whereas in the period between 45-90 kg pigs fed TRD meal (*ad libitum*), which is based on mix of cereals and root crops, achieved 49% lower daily gain ($P<0.01$) compared to those fed complete feed mixture (CON). Consequently CON pigs were heavier at slaughter (+32 kg, $P<0.001$), had increased backfat (+7 mm, $P<0.05$) and muscle (+18 mm, $P<0.01$) thickness, but were not different in meat % ($P>0.05$). Preliminary results presented provide some information about performance of KKP, however further studies are needed to know better its growth potential and nutritional needs. Funded by European Union's H2020 RIA program (Grant agreement no. 634476).

Effect of acorn feeding on quality and aromatic profile of dry sausages produced from Turopolje pigs

D. Karolyi[1], N. Marušić Radovčić[2], H. Medić[2], Z. Luković[1], U. Tomažin[3], M. Škrlep[3] and M. Čandek Potokar[3]
[1]University of Zagreb, Faculty of Agriculture, Svetošimunska 25, 10000 Zagreb, Croatia, [2]University of Zagreb, Faculty of Food Technology and Biotechnology, Pierottijeva 6, 10000 Zagreb, Croatia, [3]Agricultural institute of Slovenia, Hacquetova ulica 17, 1000 Ljubljana, Slovenia; dkarolyi@agr.hr

Turopolje pig (TP) is an endangered Croatian autochthonous breed typically reared in an outdoor production system linked to local oak forests and marsh meadows in Turopolje region in Central Croatia. To self-sustain TP breed the utilization of meat products with added value has recently been proposed in TREASURE project. So far little information is available on the quality attributes of TP meat products, including the effect of locally available feeding resources. Hence present study aimed to examine the quality and aromatic profile of dry-fermented sausages of TP that were reared in similar conditions but fed either conventional (CF) or acorn supplemented (AF) finishing diet. The quality of end product, pH and aw values, moisture, fat and protein content, fatty acid (FA) profile, oxidative stability, texture profile analysis (TPA), sensory evaluation and aromatic profile were determined. Data were analysed by TTEST or NPAR1WAY (for sensory data) procedure at an alpha level of 0.05. Compared to CF, AF sausages tended ($P<0.1$) to have higher moisture content and TPA chewiness, but generally had similar physicochemical, rheological and sensorial traits and FA composition, except for higher share of C14:0. Lipid oxidation, measured as 2-thiobarbituric acid-reactive substances, was more pronounced in AF than CF sausages. Aromagram showed terpenes as the most abundant volatiles (around 52% of the total area) in both types of sausages, followed by aldehydes (11-14%), aromatic hydrocarbons (6-10%), phenols (6-7%), alcohols (5-6%), acids (4-6%), sulphur compounds (2-5%), ketones (around 3%), esters (less than 2%) and aliphatic hydrocarbons (less than 1%). Compared to CF, AF sausages had less aromatic and aliphatic hydrocarbons and sulphur compounds, and more esters. These results suggest that acorn feeding may affect some properties of TP dry-fermented sausages but more research is needed. Funded by European Union's H2020 RIA program (grant agreement no. 634476).

Consumers' acceptance of health-related innovations in dry-cured ham from Turopolje pig breed

D. Karolyi[1], M. Cerjak[1], M.A. Oliver[2], J. Dilme[2], M. Vitale[2], Z. Kallas[3], J.M. Gil[3] and M. Čandek Potokar[4]
[1]University of Zagreb, Faculty of Agriculture, Svetošimunska ulica 25, 10000 Zagreb, Croatia, [2]IRTA, Finca Camps i Armet s/n, 17121 Monells, Spain, [3]CREDA UPC IRTA, C/Esteve Terrades 8, 08860 Castelldefels, Spain, [4]Agricultural institute of Slovenia, Hacquetova ulica 17, 1000 Ljubljana, Slovenia; dkarolyi@agr.hr

Turopolje pig (TP) is a local Croatian breed which nearly extinct in the second half of the 20th century. Currently, despite the state support, the TP is still endangered and to self-sustain the breed a new marketing strategy, based on the meat products with an extra added value, is needed. As consumers nowadays increasingly demand for more convenient and healthier types of products, in present work (within TREASURE project) we investigated consumers' acceptance of health-related innovations associated with the reduction of salting or smoking of TP dry-cured ham. A consumer (n=120) sensory test was carried out in Zagreb city area with the three types of TP hams (typically salted and smoked, less salted or less smoked) and two types of standard hams (conventional and premium) from modern pig breeds. Effect of information on innovation and/or breed on ham preferences was tested using three-step procedures as blind, expected and actual (informed) test on liking scale from 1 (dislike extremely) to 9 (like extremely). Data were analysed by GLM procedures at 0.05 α-level. In the blind test, in the absence of information, no significant differences between ham liking scores were found. In the expectancy test, when only information is given, all types of TP ham were more preferred than conventional ham, but only typical TP ham was preferred over the premium ham. Finally, when tasting is repeated with the information, all TP hams were scored higher than premium ham, while innovative TP hams were scored similar as conventional ham. This results suggest the preference of TP hams over the standard hams and a good acceptance of health-related innovations in TP ham by Croatian consumers.
Funded by European Union's H2020 RIA program (grant agreement no. 634476).

Session 49 Poster 19

Majorcan Black Pig carcass, meat and fat quality parameters assessed by a standardised toolbox

J. Gonzalez[1], M. Gil[1], M.A. Oliver[1], J. Jaume[2], B. Lebret[3], I. Díaz[1] and J. Tibau[1]
[1]IRTA, Monells, 17121, Spain, [2]SEMILLA-CAIB, C/ d'Eusebi Estada, 145, 07009 Palma, Spain, [3]PEGASE INRA, Agrocampus-Ouest, 35042 Rennes, France; joan.tibau@irta.cat

TREASURE project aims to develop sustainable pork chains based on European local pig breeds. The Porc Negre Mallorquí (Majorcan Black Pig, MBP) is an untapped pig breed from Mallorca, in the Balearic Islands, included in the project due its high rusticity, adaptation to the environment and meat quality differentiation. Particularities of MBP are mostly unknown out from Mallorca, with only one representative meat product, the 'sobrassada de Mallorca de Porc Negre', a fat-rich cured sausage owning a PGI trademark. Carcass, meat and fat quality of MBP was evaluated, according to a standardised toolbox of parameters. The toolbox aimed for a harmonised knowledge of untapped breeds, made from the contribution of researcher's expertise and the literature. The mean values and standard deviation for carcass traits (n=29) were: 156.8±11.8 kg for live weight, 125.3±12.9 kg for carcass weight, 75.1±3.2% for carcass dressing, 88.7±3.7 cm for carcass length and 55.2±10.1 mm for ZP fat. The most remarkable results for the meat quality parameters toolbox (n=58) were an ultimate pH of 5.58±0.12 and electrical conductivity of 6.99±3.44 mS, a lightness value (L*) of 44.10±3.59, a drip loss of 1.49±1.46%, a shear force of 3.46±1.07 kg, and an intramuscular fat (IMF) content of 6.11±3.09%. The maximum value of IMF was 19%, a hint of the wide variability within this untapped breed, representing a disadvantage for the product standardisation in case of fresh meat, but not for 'sobrassada', mainly made of grounded ham, shoulder and loin. The results for the major fatty acid (FA) groups in back fat (n=48) samples were 41.00±1.42% of saturated FA, 51.27±1.64% monounsaturated FA and 6.88±0.71% polyunsaturated FA. 'Sobrassada' includes 40-50% of back fat, thus the FA composition is critical for product quality. The FA composition, with a high proportion of MUFA and low of PUFA contributes to an optimal technological quality of MBP back fat. Funded by European Union's H2020 RIA program (grant agreement no. 634476).

Session 49 Poster 20

Adipose tissue transcriptome in Iberian and Duroc pigs fed different energy sources

R. Benítez, B. Isabel, Y. Núñez, E. De Mercado, E. Gómez Izquierdo, J. García-Casco, C. López-Bote and C. Óvilo
INIA, UCM and ITACYL, Madrid, 28040, Spain; rmbenitez@inia.es

Diet and breed are main factors influencing animal body and tissue composition. In this study, we evaluated the effects of a diet supplemented with 6% high oleic sunflower oil (HO) or carbohydrates (CH) as energy source on subcutaneous ham fat composition and gene expression in growing Iberian and Duroc pigs. A comparative study of the ham subcutaneous fat transcriptome between animals fed with both diets was carried out in the two breeds. The study comprised 30 Iberian and 19 Duroc males. These animals were kept under identical management conditions and the two isocaloric and isoproteic diets were provided *ad libitum*, starting with 19.9 kg of average LW. All animals were slaughtered after 47 days of treatment, with 51.2 kg of average LW. Twelve animals of each breed were randomly selected for RNAseq analysis (six of each diet). In both breeds, the diet induced changes in the fatty acid composition of subcutaneous fat samples. The HO diet had higher monounsaturated fatty acids and oleic acid, and lower SFA than the CH diet. We detected 218 differentially expressed (DE) genes conditional on diet in Iberian pigs and 68 DE genes in Duroc pigs, 29 of these were common. Out of these, 27 genes were upregulated in HO diet in Duroc but upregulated in CH diet in Iberian (i.e. *PDLIM3, ANK1* and *PYGM*), one gene showed the opposite regulation (*SERPINE1*) and only one DE gene was upregulated in HO diet in both breeds (*CYP1A1*). These results indicate a strong interaction breed×diet on transcriptome. We performed a functional analysis (metabolic pathways and GO enrichment) of the 29 common genes, which showed the enrichment of functions related to carbohydrate metabolic process (digestion and absorption), starch and sucrose metabolism, insulin signalling pathway, glycogen catabolic process and purine nucleotide metabolic process. The bioinformatic analysis also allowed the prediction of potential regulators (such as PPARGC1A, MEF2C, DMD, MYOCD, or MYOD1) for the expression differences observed. The results indicate the direct deposition of nutrients and a profound and different effect of the diet on adipose tissue gene expression between breeds, affecting relevant biological pathways.

Session 49 Poster 21

Effects of breed, tissue and gender on cholesterol contents of pork from local Lithuanian pig breeds

V. Razmaite, S. Bliznikas and V. Jatkauskiene
Institute of Animal Science of Lithuanian University of Health Sciences, Department of Animal Breeding and Genetics, R. Žebenkos 12, 82317 Baisogala, Lithuania; violeta.razmaite@lsmuni.lt

The objective of this study (within TRESURE project) was to examine the effects of breed, tissue location and gender on the cholesterol content in different tissues of local Lithuanian pig breeds. The study was carried out on 24 Lithuanian White and 18 Lithuanian indigenous wattle pigs, 28 females and 14 castrated males. All pigs received a complete concentrate feed. Pork composition and properties were detected according to the EU reference methods. The cholesterol content was determined by HPLC analysis using the system Shimadzu 10Avp. The data were subject to the analysis of variance in general linear (GLM) procedure in SPSS 17. The model included the fixed effect of breed, gender, tissue location and feeding level. LSD significance test was used to determine the significance of differences of means between the groups. The comparisons of local conserved Lithuanian breeds showed negligible differences in the longissimus muscle fat content, however semimembranosus muscle of Lithuanian White pigs had higher fat content than the same muscle of Lithuanian indigenous wattle pigs ($P<0.05$). Although longissimus muscle from both Lithuanian pig breeds has lower contents of cholesterol ($P<0.01$) compared with the conventional hybrids, the differences between Lithuanian breeds were not large. Only the backfat of Lithuanian White pigs had higher content of cholesterol than the backfat from Lithuanian indigenous wattle pigs ($P<0.05$). The lowest contents of cholesterol were found in the backfat ($P<0.001$) and the highest contents in the semimembranosus muscle from both breeds, however, the differences between the muscles were not significant. The gender of pigs showed its effect on the semimembranosus cholesterol content ($P<0.05$): females had higher content of cholesterol than castrated males. Cholesterol contents in both muscles showed negative correlations with pig age ($P<0.001$). In the semimembranosus muscle cholesterol content showed negative correlation with pig weight ($P<0.05$). Negative correlations were also found between backfat thickness and cholesterol content in backfat. Funded by European Union's H2020 RIA program (grant agreement no. 634476).

Session 49 Poster 22

Breed effects on adipose tissue and muscle transcriptome in growing Iberian and Duroc pigs

R. Benítez, B. Isabel, Y. Núñez, E. De Mercado, E. Gómez-Izquierdo, J. García-Casco, C. López-Bote and C. Óvilo
INIA, UCM and ITACYL, Madrid, 28040, Spain; rmbenitez@inia.es

Iberian pig production is based on both purebred Iberian and crossbred Duroc × Iberian pigs. Iberian and Duroc breeds show important phenotypic differences in growth, fattening, tissue composition and meat quality. The study of breed effects on gene expression, could explain phenotypic and metabolic differences between breeds. The objective of this study was to evaluate breed effects on phenotype and ham subcutaneous adipose tissue and Biceps femoris muscle transcriptome in growing Iberian and Duroc pigs with RNAseq technology. The study comprised 30 Iberian and 19 Duroc males. These animals were kept under identical management conditions and were slaughtered with 51.2 kg of average LW. Twelve animals of each breed were randomly selected for RNAseq analysis. The Iberian pigs showed greater feed intake, backfat thickness, and saturated fatty acids content, whereas the Duroc pigs had greater ham weight and polyunsaturated fatty acids content. We detected 349 differentially expressed (DE) genes in ham subcutaneous fat (such as *LEP, PCK1, IGBP3* or *PLIN2*) and 347 DE genes in Biceps femoris (such as *FASN, RBP4, INSIG1, PCK2* or *ME1)* (FDR 0.01) conditional on breed. Out of these, thirty-five genes were found DE in both tissues, including candidate genes such as *LYZ, RBP7, FOS* or *FMOD*. We performed a functional analysis (metabolic pathways and GO enrichment) which showed enrichment of functions related to cellular movement and development in both tissues. Also, we found enrichment of functions related to carbohydrate metabolism in adipose tissue and functions related to lipid metabolism in muscle. Relevant biological pathways as LXR/RXR activation were affected in both tissues. The bioinformatic analysis also allowed the prediction of potential regulators (such as ATF4, ERBB2, INS1 or TNF) for the expression differences observed. The results indicated a strong effect of the breed on gene expression in both tissues affecting relevant molecular functions related to the phenotypic differences observed.

Session 49 Poster 23

Major differences in gut microbiota composition of Iberian pigs in montanera vs commercial systems

J.M. García-Casco[1], M. Muñoz[1], G. Lemonnier[2], J.M. Babilliot[2], O. Bouchez[3], A.I. Fernández[4], F.R. Massacci[2], M.A. Fernández-Barroso[1], A. López-García[1], C. Caraballo[1], C. Óvilo[4] and J. Estellé[2]
[1]INIA, Centro Cerdo Ibérico, Ctra EX 101, km 4,7, 06300 Zafra, Spain, [2]INRA-Université Paris-Saclay, AgroParisTech-UMR1313 GABI, Domaine de Vilvert, 78352 Jouy-en-Josas, France, [3]INRA-Genotoul, US 1426 GeT-PlaGe, Chemin de Borderouge – Auzeville, 31320 Castanet-Tolosan, France, [4]INIA, Dpto. Mejora Genética Animal, Ctra La Coruña, km 7,5, 28040 Madrid, Spain; garcia.juan@inia.es

The traditional raising system for Iberian pigs includes an open-air fattening period (*montanera*) where animals have a diet dominated by acorns. This system allows obtaining dry-cured products of unmatched quality. The most usual method for the verification of *montanera* products relies on assessing the particular backfat fatty acid composition after slaughter. Given that the advances on nutrition science makes feasible to emulate *montanera* meat products in an industrial farming environment, there is a growing need for the identification of complementary certification approaches. The objective of this study of TREASURE project was to evaluate the potential of gut microbiota composition analyses from faecal samples collected at slaughter as a new discrimination method of the *montanera* production system. Gut microbiota composition of 131 Iberian pigs (92 *montanera* and 131 commercial feeds) was determined by re-sequencing the bacterial 16S gene in an Illumina MiSeq device. Bioinformatic analyses were performed using Qiime's subsampled open-reference OTU calling method. NMDS and PERMANOVA analyses performed using the Vegan R package showed significant effects of diet and sampling batch on gut microbiota composition. The microbiota of *montanera*-raised pigs showed a tendency towards a reduced microbial diversity. Differential abundance analyses performed with the metagenomeSeq R package confirmed these differences by identifying 997 out of 1,703 OTUs whose abundance was significantly different between both systems. Our results suggest that the gut microbiota composition of Iberian pigs sampled at the slaughter could be used as a supplementary certification tool of traditional *montanera* Iberian pigs. Funded by European Union's H2020 RIA program (grant agreement 634476).

Session 49 Poster 24

Gut microbiota composition in Iberian pigs fed with olive oil by-products during the growing period

M. Muñoz[1,2], J.M. García-Casco[1,2], G. Lemonnier[3], D. Jardet[3], O. Bouchez[4], M.A. Fernández-Barroso[1,2], F.R. Massacci[3], A.I. Fernández[1], A. López-García[1,2], C. Caraballo[1,2], E. González-Sánchez[5], C. Óvilo[1] and J. Estellé[3]
[1]INIA, Mejora Genética Animal, Crta. de la Coruña, km 7,5, 28040 Madrid, Spain, [2]Centro de I+D en Cerdo Ibérico INIA-Zafra, Ctra Ex-101, km 4.7, 06300 Zafra, Spain, [3]INRA, UMR1313 GABI, AgroParisTech, Université Paris-Saclay, Jouy-en-Josas, France, [4]INRA, US 1426 GeT-PlaGe, Genotoul, Castanet-Tolosan, France, [5]Universidad de Extremadura, Escuela de Ingenierías Agrarias, Avda. Adolfo Suárez, s/n, 06007 Badajoz, Spain; mariamm@inia.es

The traditional Iberian pig production system is characterized by a final open-air fattening period (montanera), in which the animals are fed acorns and grass, preceded by a growing period where the feeding is restricted to avoid undesirable weight gain. New growing diets based on olive agro-industrial by-products could be an alternative to avoid this restriction. The objective of the current study (of TREASURE project) was to analyse the effect of two growing alternative diets on the gut microbiota composition from faecal samples collected before and after montanera. Three diets, one incorporating dry olive pulp in the feed (DD), one incorporating olive cake in wet form (WD) and a control diet (CD) were supplied to 45 animals (15 per diet) during the growing period (45 to 95 kg of body weight). The gut microbiota composition of each individual was evaluated at two time points: before transition to montanera (95 kg) and at slaughter (160 kg). Microbiota analyses were performed by re-sequencing the bacterial 16S gene (V3-V4) in an Illumina MiSeq. Bioinformatics analyses were performed by using Qiime's open-reference subsampled OTU calling approach. The effect of diets on microbiota composition and diversity was evaluated using the Vegan package in R. Bray-Curtis distances, NMDS and PERMANOVA analyses showed significant effects on microbiota composition. WD caused an increase in microbiota diversity. In the second sampling point, differences in composition and diversity could be also observed after acorn supplementation. Funded by European Union's H2020 RIA program (Grant agreement no. 634476).

Lithuanian consumer preferences towards products of different fatness

R. Sveistiene[1] and S. Marasinskiene[2]
[1]Lithuanian University of Health Sciences, Animal Science Institute, R. Zebenkos 12, 82317, Lithuania, [2]Lithuanian Endangered Farm Animal Breeders' Association, R.Zebenkos 12, 82317, Lithuania; ruta.sveistiene@lsmuni.lt

The objective of the study (made in TREASURE project) was to analyse how Lithuanian consumers perceive pork products from fatty pigs and identify their attitudes, and the possibility to include special branded products from Lithuanian local pigs to the marketing scheme. Since ancient times Lithuanians have been cultivating animal husbandry and eating fresh, salt-cured and smoked meat. The interest in traditional and healthy food has increased in recent years encouraged artisans and restaurants to revive old food preparing traditions. The breeding of animals from rare breeds (genetic pool) and income from their production is not competitive compared with industrial breeds because of high amount of fat. One of the first steps to promote breeding of local pig breeds and their traditional products was their demonstration and knowledge dissemination during different agricultural exhibitions and shows. Lithuanian Endangered Farm Animal Breeders' Association contributed to producing and testing of two products: dried sausages and a kind of lard sausages 'Lašiniuotis'. To describe the preferences of consumers, the test of frequencies and chi-square tests were applied using the SPSS Statistics program. Respondents of different genders, age and employment presented consumption frequencies of different meat, including pork. Obviously, Lithuanians justified the reputation of pork eaters (X^2=78.7, df=2 P<0.001). Although only 25.2% of the respondents were involved in animal production, more than 80% of them answered that they possess information on pig growing and this shows that many people did not lose touch with the countryside. Public testing of the presented products showed that 227 (48.4%) consumers of different age and occupation preferred and voted for dried sausages and 242 (51.6%) respondents preferred and voted for 'Lašiniuotis'. It can be concluded that more fatty products could find consumers among Lithuanian population. Funded by European Union's H2020 RIA program (grant agreement no. 634476).

Gut microbiota composition in Krškopolje pigs under conventional and organic production systems

J. Estellé[1], F.R. Massacci[1], D. Esquerré[2], D. Jardet[1], G. Lemonnier[1], C. Óvilo[3], M. Skrlep[4], K. Poklukar[4] and M. Čandek-Potokar[4]
[1]GABI, INRA, AgroParisTech, Université Paris-Saclay, Domaine de Vilvert, Jouy-en-Josas, France, [2]INRA, US 1426 GeT-PlaGe, Genotoul, Castanet-Tolosan, Toulouse, France, [3]INIA, Dpto. Mejora Genética Animal, Ctra. de la Coruña, Madrid, Spain, [4]Agricultural Institute of Slovenia, Hacquetova 17, Ljubljana, Slovenia; klavdija.poklukar@kis.si

The Krškopolje pig is the only indigenous Slovenian pig breed being still farmed. The breed is characterised by its rusticity and adaptation to the traditional farming system and has a black coat colour with a white belt across shoulders and forelegs. Despite the limited number of animals, the breed is on the way of becoming more sustainable due to the growing interest for local traditional breeds, in particular for organic farming. The objective of our study (made within project TREASURE) was to characterise the gut microbiota composition in 36 Krškopolje pigs raised by using three different diets: conventional feed, conventional feed supplemented with alfalfa pellets and organic feed supplemented with alfalfa hay. The gut microbiota composition of all pigs at two time points (155 and 228 days-of-age) was determined by re-sequencing the bacterial 16S gene in an Illumina MiSeq. Bioinformatics analyses were performed by using Qiime's open-reference OTU calling subsampled method following author's recommendations. NMDS and PERMANOVA analyses performed by the Vegan R package showed significant effects of dietary regime, showing that the organic feed produces a differentiation of the gut microbiota composition when compared with the other two diets and the two time points analysed. This pilot study illustrates the power of metagenomic analyses for the identification of effects of diets in the farming conditions of local European pig breeds. Funded by European Union's H2020 RIA program (Grant agreement no. 634476).

Gut microbiota composition ofTuropolje pigs in outdoor production and acorn supplementation

J. Estellé[1], F.R. Massacci[1], D. Esquerré[2], D. Jardet[1], G. Lemonnier[1], C. Óvilo[3], M. Čandek-Potokar[4], K. Salajpal[5] and D. Karolyi[5]
[1]INRA, UMR GABI, AgroParisTech, Université Paris-Saclay, Domaine de Vilvert, Jouy-en-Josas, France, [2]INRA, US 1426 GeT-PlaGe, Genotoul, Castanet-Tolosan, Toulouse, France, [3]INIA, Dpto. Mejora Genética Animal, Ctra. de la Coruña, Madrid, Spain, [4]Agricultural Institute of Slovenia, Hacquetova 17, Ljubljana, Slovenia, [5]University of Zagreb, Faculty of agriculture, Zagreb, Croatia; ksalajpal@agr.hr

The Turopolje pig is a local porcine breed originating from Croatia characterised by its rusticity and adaptation to the traditional outdoor farming system linked to local oak forests. While this once-widespread breed was almost completely replaced by industrial breeds in the second half of the 20th century, there is growing interest on promoting its production in the framework of sustainable porcine production. The objective of our study (made in project TREASURE) was to characterise the gut microbiota composition in 24 Turopolje pigs raised outdoor by using a conventional feed vs a conventional feed supplemented with acorns. The gut microbiota composition of all pigs was determined by re-sequencing the bacterial 16S gene in an Illumina MiSeq device. Bioinformatics analyses were performed by using Qiime's open-reference OTU calling subsampled method following author's recommendations. NMDS and PERMANOVA analyses performed by using the Vegan R package showed significant effects of diet, showing that the acorn supplementation has a relevant impact on gut microbiota, notably by reducing richness and diversity. This result was confirmed in the differential abundance analysis which showed that 152 out of 1,466 OTUs were significantly different between the two groups. Interestingly, the predicted microbiota metabolic functions that were differentially abundant involved fatty acid and amino-acid metabolisms. Funded by European Union's H2020 RIA program (Grant agreement no. 634476).

Changes in carcass and meat traits during the montanera finishing period of Iberian pigs

M. Izquierdo[1], F.I. Hernández-García[1], N. Garrido[1], J. García-Gudiño[2], J. Matías[1], I. Cuadrado[1] and M.I. Pérez[1]
[1]Center of Scientific & Technological Research of Extremadura (CICYTEX), Autovía A5, Km 372, 06187 Guadajira (Badajoz), Spain, [2]IRTA, Finca Camps i Armet, Monells, Gerona, Spain; mercedes.izquierdo@juntaex.es

Iberian pigs are usually slaughtered at an average body weight (BW) of 160 kg. The free-range, acorn-feeding montanera starts at an average weight of 100 kg, lasts for 2.5-3 months, and pigs gain around 60 kg during this period. During this extensive finishing, the weight gain is mostly based on subcutaneous and intramuscular fat accretion, but other changes in carcass and meat composition may occur as well. This study (made in TREASURE project) evaluated the changes in carcass and meat composition occurred in Iberian pigs between the end of premontanera (pre-finishing period) and the end of montanera, as assessed in animals slaughtered at 110 and 160 kg BW, respectively. Castrate male Iberian pigs (n=45) were randomly assigned to 2 treatments (2 slaughter times) and raised on concentrate up to 110 kg BW. At this point, 15 pigs were slaughtered, and the remaining 30 pigs were finished in montanera with acorns and grass and were slaughtered at an average BW of 158 kg. Carcass and several meat traits were evaluated in both cases. In general, carcass fat (subcutaneous and intramuscular) increased significantly during the finishing period, with the exception of loin weight and area and the outer backfat layer, that remained unchanged after the finishing period. Hip height and ham length did not change, but carcass length increased significantly after finishing. In relation to meat colour, only the b index (blue to yellow) was significantly greater in finished pigs. Regarding backfat fatty acid profiles, PUFA and SFA decreased and MUFA increased significantly after the montanera finishing period, but C18:3 remained unchanged. In conclusion, pre-finishing Iberian pigs (at 110 kg BW) had similar muscle accretion, hip height, outer backfat layer thickness and C18:3 proportion than the finished ones. The BW gain during montanera was essentially based on fat deposition, and finishing increased the proportion of MUFA and decreased that of PUFA and SFA. Funded by European Union's H2020 RIA program (grant agreement no. 634476).

Session 49 Poster 29

The effect of birth-weight on growth performance and meat quality in Iberian pigs

M. Vázquez-Gómez[1], C. García-Contreras[2], S. Astiz[2], A. Olivares[1], E. Fernández-Moya[3], A. Palomo[3], A. Gónzalez-Bulnes[2], C. Óvilo[2] and B. Isabel[1]
[1]UCM, Animal Production, 28040, Madrid, Spain, [2]INIA, 28040 Madrid, Spain, [3]Ibéricos de Arauzo 2004 SL, 37408 Salamanca, Spain; martavazgomez@gmail.com

The variability of birth-weight (BIW) in lean commercial swine breeds is known to cause heterogeneity in growth patterns and meat quality affecting profitability. This effect may be more severe in local breeds because of its reproductive characteristics. There is scarce information for traditional breeds, so we evaluated BIW effects on postnatal development during the growing-fattening phase (from 72 d-old to slaughter) and on carcass and meat quality at slaughter in Iberian crossbred pigs. Males and females (120 pigs each) classified by BIW into very low, low and average BIW (VLBIW, LBIW and ABIW) were distributed by sex and BIW. At 110 d-old, VLBIW and LBIW females showed a lower feed conversion rate (FCR, $P<0.05$) than the other groups. However, the VLBIW females had the highest FCR values from 111 d-old to the slaughter ($P<0.05$). At slaughter, VLBIW pigs showed less average daily gain weight (ADGW) and weight ($P<0.0001$) than heavier groups due to the lowest values of females. Age to slaughter was negatively correlated with BIW and decreased by approximately 28 days per kg of BIW increased ($P<0.0001$, for both). Regarding carcass quality analysis, VLBIW pigs showed shorter carcass than heavier groups ($P<0.05$). Meat quality showed that LBIW and ABIW males had the highest and lowest values of intramuscular fat values, respectively. The analysis of fatty acid (FA) profile of loin showed that sex modulated the neutral fraction (triglycerides) while BIW modulated the polar fraction (phospholipids). The polar fraction of LVBIW showed lower values of unsaturation index and polyunsaturated FA (PUFA; $P<0.05$, for both), but higher concentrations of monounsaturated and saturated FA (MUFA and SFA; $P<0.05$, for both). The neutral fraction of males showed higher desaturation indexes (MUFA/SFA and C18:1/C18:0 ratios) and MUFA content than females, but lower SFA content. Our results support the adverse effects of low BIW, modulated by sex-related effects, on postnatal growth traits and carcass and meat quality of Iberian pigs.

Session 49 Poster 30

Integrated evaluation of the sustainability of the Noir de Bigorre pork chain

A.N.T.R. Monteiro[1], J. Faure[1], M.-C. Meunier-Salaün[1], A. Wilfart[2] and F. Garcia-Launay[1]
[1]PEGASE, INRA, Agrocampus Ouest, 35590 Saint-Gilles, France, [2]SAS, INRA, Agrocampus Ouest, 35000 Rennes, France; florence.garcia-launay@inra.fr

The social expectation for more sustainable livestock farming systems is growing. However, multidisciplinary approaches to assess the sustainability of pig farming systems are not sufficiently addressed. Twenty-five farms of local pig breed's production in France (Noir de Bigorre chain – Gascon breed) were evaluated altogether in terms of environmental impacts (EN), economic sustainability (EC) and animal welfare (AW), using multiple factor analysis (MFA) and hierarchical clustering (HC). The first dimension of MFA (22.7% the total variance) was mainly associated to the EN (r=0.56) and AW (r=0.52), while the second (20.8% of the total variance) was linked to EC (r=0.41). The HC resulted in the identification of four groups: Group 1 comprised farrow-to-finish farms, with a farmer aged between 20-30 years-old, characterized by high EN, high dietary crude protein (CP) of fattening feeds and feed conversion ratio (FCR), low AW, and usually low EC. It could be described as 'sustainability unfavourable', with an overall inefficient management of the farm, when the high FCR could be a result of the high feed waste. Group 2, composed by 75% of feeder-to-finish farms, was characterized by high transferability, high land occupation (LO), and low AW. It could be described as 'AW unfavourable and Transferability favourable'; maybe due to the high LO, animals were raised with lower level of care. Group 3 comprised farrow-to-finish farms managed by a man, and characterized by high EC, high number of sows and low EN per ha. It could be described as 'EC and EN favourable', described by high farm size and good management practices. Group 4, characterized by high AW and low EN, could be described as 'AW and EN favourable', with more attention to AW and lower surface available to pig production. The use of an integrated evaluation highlights different profiles of farmers which are associated with various results in the different themes considered. Based on complementary themes, it provides a broader representation of the sustainability of pig farming systems than the use of one theme. Funded by European Union's H2020 RIA program (grant agreement no. 634476).

Session 49 Poster 31

Assessing performance and management of European local pig breeds in project TREASURE

R. Nieto, M. Čandek-Potokar, C. Pugliese, J.P. Araujo, R. Charneca, J. García-Casco, E. González-Sánchez, F.I. Hernández-García, M. Izquierdo, D. Karolyi, B. Lebret, V. Margeta, M.J. Mercat, M. Petig, C. Radovic and R. Savic
TREASURE Consortium, Hacquetova ul. 17, 1000 Ljubljana, Slovenia; rosa.nieto@eez.csic.es

Unlike in modern breeds, the knowledge on performance and management of local pig breeds is very limited. These are key aspects to be developed for their successful exploitation that requires strategies adapted to their specific productive and metabolic characteristics, quite different from those of modern pig breeds, to ensure their preservation and future expansion, particularly for those more endangered. These are some of the challenges addressed by the project TREASURE. For this purpose, a series of experiments involving 11 European local breeds -differing in the level of development- have been carried out covering various and complementary aspects like nutritional requirements in different productive phases (Cinta Senese, Iberian), feeding practices involving locally available resources (Krskopolje, Schwäbisch-Hällisches, Turopolje, Mangalitsa, Bísaro, Iberian, Gascon), and innovative management and housing practices to enhance product quality or improve animal welfare (Cinta Senese, Alentejano, Bisaro, Iberian, Mangalitsa, Moravka, Krskopolje, Schwäbisch-Hällisches, Black Slavonian). All these activities have been designed to obtain essential information to develop future productive strategies for each of the involved breeds, taking into account to obtain local high-quality products, and seeking for optimum pig performance and high animal welfare conditions in the production systems under consideration. The information provided constitutes a unique and valuable set of data for the management of these breeds and the further development of local pork chains. In some cases it is the first available data for the breeds concerned. A general overview of the developed activities will be presented highlighting some of the achievements obtained. Funded by European Union's H2020 RIA program (Grant agreement no. 634476).

Session 50 Theatre 1

State of equine data available today at European level

C. Cordilhac
French Horse Institut – Institut Français du Cheval et de L'équitation, Service des relations internationale. International Department, Lieu dit La Veillere, 28240 Saint Victor de Buthon, France; claire.cordilhac@ifce.fr

Knowledge of equine industries in Europe is deficient in the field of socio-economic data. The specificity of the horse is one of the main causes. Indeed, compared to other animal species, the equines are very specific. Their different values, the reasons for their breeding, their different life cycles, adapted to the heterogeneity of relationships that men have with them, makes them a species apart from the others. For this reason, they cannot be counted the same as cattle, sheep or poultry. Within the European institutions, two organizations manage data on the equine sector. Eurostat, and within the Health DG, the TRACES service. In both cases, the data are incomplete. For Eurostat, the data come from each Member States and from the FAO, which have only pure agricultural data (slaughter horses). In both cases, data are very incomplete and depend on the size of the sector in the country concerned, the collection or not of the equine data, the scope and method chosen. The TRACES service records the health movements of certain horses in Europe. Again, the data are very partial because not all horses are registered. The actors of the equine industry need urgently precise and comparable figures. Efforts must therefore be made at European level on the raw knowledge itself, on the definition of the scope and on the method of collection and classification. This work is likely to be even more difficult at European level as Eurostat plans to no longer integrate equine data into its agricultural statistics after 2020.

Session 50 Theatre 2

Importance of having updated equine data and how research can help data collection

C. Vial[1] and R. Evans[2]

[1]Ifce (French Institute for Horse and Riding), INRA, UMR 1110 MOISA, 2 place Pierre Viala, 34060 Montpellier, France, [2]Norwegian University College for Agriculture and Rural Development, Høgskulen for landbruk og bygdeutvikling, 4340 Bryne, Norway; celine.vial@inra.fr

In the European Union, the horse industry has undergone significant changes since the middle of the 20th century. It has changed from a primarily agricultural and industrial sector activity to one firmly rooted in sports, leisure and consumption. These evolutions generate the development of new kinds of activities and the growth of the whole horse industry. Consequently, new questions arise about the role of equines in economic dynamism, culture, social links, rural development… However, little is known about the horse industry. Few studies have been conducted and in most countries, statistical data on the horse sector is lacking or incomplete. This lack of equine data at an international level but also at national levels is a problem often encountered by researchers, institutions, professionals and policy makers, moreover in the current context of the preparation of the next CAP EU Agriculture Policy (2020-2024) such data is vital. That's why it is important that the different actors in the horse industry join their efforts to promote better production of statistics on equine activities. This could help researchers to develop international studies on this sector, institutions to highlight the growing importance of this sector and improve its recognition as a real agricultural actor, professionals to be more legitimately included in the agricultural world, and policy makers to create more effective policies. Through this, researchers can participate in the improvement of data collection in different ways: promoting it in their country, providing local data to international institutions but also creating new data collection processes and methodologies that could be used uniformly in every country. This presentation addresses these issues in order to give rise to further discussions amongst attendees.

Session 50 Theatre 3

Equine data collected by the European Horse Network

F. Grass

European Horse Network, Bruxels, Belgium; florence.gras@parimutuel-europe.org

The European Horse Network (EHN) needs to show the importance of the equine industry through relevant and credible figures. It is necessary for our advocacy work and to be influent, to demonstrate the maturity of the industry, its impact on the economy and its benefits for society. In 2010, EHN, recently established network, made the exercise to collect relevant existing data from national representatives and recent surveys. As a reference and comparison, the US economic study made in 2005 by the American Horse Council as well as the Economic impact of the European Horseracing industry (EPMA survey 2009) were used. In October 2010, EHN published the Key Figures of the European Horse industry (still available on EHN web site: http://www.europeanhorsenetwork.eu/horse-industry) Since then, EHN has made many attempts to update the horse industry figures (combine it with EEF annual collection of data, support World Horse Welfare report overview in 2015 'Removing the blinkers', create EHN Data group to adopt a common methodology with the help of the French IFCE observatory…). EHN needs more help, support and involvements from members as well as direct contacts inside national ministries/agencies before obtaining accurate and significant updated figures. Moreover, EHN needs advice and expertise from researchers in order to agree on a common methodology to collect and aggregate comparable data. Can EAAP's experts bring forward their knowledge and expertise?

Use of two French equine databases for a better knowledge of mortality and other demographic issues

J. Tapprest[1], E. Morignat[2], X. Dornier[3], M. Borey[1], P. Hendrikx[4], B. Ferry[3], D. Calavas[2] and C. Sala[2]
[1]French agency for food, environmental and occupational health & safety (Anses), Laboratory for equine diseases, 14430 Goustranville, France, [2]Anses, Epidemiology Unit, 69364 Lyon, France, [3]French Horse and Riding Institute, IFCE, 75013 Paris, France, [4]Anses, Coordination and support to surveillance Unit, 69364 Lyon, France; jackie.tapprest@anses.fr

Imperfect knowledge of the equine population at regional, national or international levels is a limiting factor for accurate socio-economic, animal health and welfare studies. In France, two complementary databases centralize national equine population data: the French equine census database SIRE, managed by the French Horse and Riding Institute, collects individual equine data while information on dead equines are centralized in the Fallen Stock Data Interchange database (FSDI) managed by the French Ministry of Agriculture. The objective of our study was to evaluate whether the combined use of the FSDI and SIRE databases can provide relevant estimates on equine mortality and a better knowledge of equine demography. Annual mortality ratios were calculated for the French equine population and survival analyses performed. Spatial and temporal variations in equine mortality were described. The mean annual mortality ratio for the French equine population was estimated at 3.17%. Survival rates differed according to breeds with the highest median age at death for ponies. The spatial distribution of deaths varied according to breeds and age categories. The weekly description of mortality highlighted marked seasonality of deaths. Thus, the joint analyses of the two databases are very promising in terms of knowledge of equine mortality and its monitoring in time and space. This study also underscored the possibility of reciprocal corrections of the databases resulting in an improvement of the demographic data quality of each database. However, a better traceability between the two bases is desirable in order to enable their comprehensive interoperability and synergistic use. Acknowledgements to the French Ministry of Agriculture, the French Horse and Riding Institute, the French fallen stock companies, the Regional Council of Normandy and the Fonds Eperon.

The agricultural census: a database to analyse the horse industry actors: example of France

E. Perret, E. Scozzari and G. Bigot
Irstea, UMR Territoires (Université Clermont Auvergne, AgroParisTech, INRA, Irstea, VetAgro Sup), 9 avenue Blaise Pascal – CS 20085, 63178 Aubiere, France; genevieve.bigot@irstea.fr

The agricultural census is carried out in each European country under the request of the European community. However, each country manages this survey according to its means and national context. In France, all firms with horses for breeding and/or for use, can access to the agricultural status. Thus, the question is how to use this national database to analyse actors of the equine industry and their evolution in agriculture, and also outside this sector. In France, the national equine population was estimated at 1 million and less than a half was kept in farms. But the analysis of the agricultural census highlighted similitude between features of farmers keeping equines and more generally, horse keepers in France. Many firms kept a very small herd: more than 50% of farms keeping equines had from 1 to 3 equines and their main activity was not necessarily related to the horse industry. Largest keepers with more than 12 equines gathered the majority of the herd present in the agricultural sector. However, current inquiries of the French agricultural census were insufficient to distinguish the major production sectors of the French horse industry as racing, leisure and sports as well as the meat production. Also, we developed a method to complete national data from the agricultural census with information collected in the French system SIRE (which recorded specifically each horse in France, with breed and location). First results pointed out that the draught horse livestock was mainly kept on farms. This trend was no so obvious for the saddle horse breeding. Whatever the breed, equines were particularly present in grassland farms. Agricultural features as for example the total meadow area, the stocking rate (livestock unit/ha) and their combination could characterize the main activity of horse specialized farms as breeding, training or livery boarding. So if similar methods could be applied in each European country, the agricultural census could become a European database to analyse the actors of the horse industry in Europe.

Outcomes of a web-survey to collect stakeholders' opinion on welfare requirements for horses

F. Dai[1], M. Tranquillo[2], E. Dalla Costa[1], S. Barbieri[1], E. Canali[1] and M. Minero[1]
[1]Università degli Studi di Milano, via Celoria 10, 20133 Milano, Italy, [2]IZSLER, Sezione Diagnostica di Bergamo, Via Bianchi 9, 25124 Brescia, Italy; francesca.dai@unimi.it

The Animal Welfare Indicators (AWIN) project aimed at developing animal-based welfare assessment protocols for different species, including horses. To ensure a good acceptance of protocols, stakeholders were invited to participate in a multi-language web-survey. Participants answered 14 open questions about appropriate requirements to guarantee high levels of horse welfare on farm. Text mining was used to analyse answers. Participants properly completed 122 surveys. Most of them were women (85%), veterinarians (35%) and horse owners (34%), coming from Italy (30%) and United Kingdom (15%). To describe welfare requirements, the words 'water' and 'feed' was the most frequently used (40 and 35 times, respectively): participants considered the welfare principle Good Feeding as the most relevant. As for the principle Good Housing, the shelter appears of primary importance (31); the presence of pasture is mentioned only 18 times, reflecting the habit of keeping horses in single boxes all over the year. It has been demonstrated that the possibility of free grazing prevents abnormal behaviours and enhances welfare, but owners frequently do not perceive it as feasible or safe for horses. The principle Good Health was linked to 'care' (24) and 'health' (20). To describe the principle Appropriate Behaviour, respondents used the words 'training' (23) and 'company', mentioned only 16 times. However, research demonstrated that social isolation is one of the main predisposing factors for behavioural problems development. The results showed that horse stakeholders consider welfare primary linked with appropriate feeding. At the same time, allowing the possibility to interact with conspecifics and spend time at pasture is paramount to guarantee horse welfare. Stakeholders' involvement is fundamental for any action intended to improve animal welfare; this work portrays the stakeholders' perception, highlighting the need of proper dissemination of scientific knowledge. *Acknowledgements* The Animal Welfare Indicators (AWIN) Project has been co-financed by the European Commission, within the VII Framework Program (FP7-KBBE-2010-4, Grant n. 266213).

Is horse meat tender enough to grill?

B. Luštrek[1], S. Žgur[1], A. Kaić[2] and K. Potočnik[1]
[1]University of Ljubljana, Biotechnical Faculty, Department of Animal Science, Groblje 3, 1230 Domžale, Slovenia, [2]University of Zagreb, Faculty of Agriculture, Department of Animal Science and Technology, Svetošimunska cesta 25, 10 000 Zagreb, Croatia; klemen.potocnik@bf.uni-lj.si

Due to lack of recent data horse meat is believed to be less tender type of meat, however, meat tenderness represents one of the most important factors of eating quality, which influences consumer's purchase decision. The belief stems from the past, when horses used for human consumption were older, weaken or injured working animals, and their meat was tougher than the meat of bred slaughter horses today. To assess horse meat tenderness and establish its palatability regarding the consumer expectations, a study was made, comparing the results to the beef tenderness standards as no standards yet exist for the horse meat. Meat samples from 60 slaughter horses of Slovenian and Croatian origin were obtained from a specialised commercial butchery 'Hot horse'. *M. longissimus thoracis* (LT) and *M. semitendinosus* (ST) samples were taken from each carcass aged for 14 days, and Warner-Bratzler shear force measurement (WB) was performed on each sample. The horses belonged to different breed groups (warm-blooded, draft and draft crossbred) and their age varied from 7 to 35 months. Model in statistical analysis included effects of age, divided in 4 classes (7-10 months, 11-18 months, 19-24 months and >24 months), type of muscle, sex and breed group. The analysis of variance showed that age, sex and breed group did not significantly affect muscle tenderness ($P>0.05$), while type of muscle did ($P<0.0001$). Total mean WB was 3.7 kg and least square means for LT and ST were 2.7 kg and 4.6 kg, respectively. Muscles were categorised according to the measured WB in 3 standard tenderness groups – tender ≤3.9 kg; intermediate 3.9 kg $<x\leq$4.6 kg; and tough >4.6 kg. In the literature beef LT was categorised as 'intermediate' and ST as 'tough'. Horse LT and ST in this study were categorised as 'tender' and marginally 'intermediate', respectively. It can be concluded that investigated parts of horse meat are tender enough to satisfy consumer's palate. To confirm the results more similar research should be made, leading to formation of horse meat quality standards.

Horse industry in Lithuania

R. Šveistienė and A. Račkauskaitė
Lithuanian University of Health Scienes, Animal Science Institute, R. Zebenkos st. 12, Baisogala, Radvilskis distr. Lithuania, 82317, Lithuania; alma.rackauskaite@lsmuni.lt

Unfortunately, the horse-related information is currently either lacking or distributed across different sectors. This research is based on a literature review, analyses of secondary data and interpretative analysis. The horse population of Lithuania is decreasing steadily from 29 thousands in 2010 to 16 thousands in 2017. 47% of all horses in 2017 were involved in studbooks and 12% of them were owned by Lithuanian National Stud. The human population in Lithuania is swiftly decreasing from 3.1 million in 2010 to 2.8 million in 2017. The number of horses per 1000 persons was 5.7 in 2017. This trend is decreasing when comparing the data of 2010, which are 9.2. Number of rider licenses of the National Equestrian Federation was 420 in 2017 and it was stable in last few year. On the other hand, the number of riding schools has greatly increased. In 2010 there was only 42 riding schools in Lithuania, while in 2017 there was already 105 of them. Also the number of licensed trotter trainers and harness race drivers has increased from 24 in 2010 to 57 in 2017. Lithuanians do not have a tradition to consume horses meet as a food. The slaughtered horses are exported. Totally 2250 slaughtered horses exported for in 2010 and 637 horses exported in 2017. In Lithuania there is a niche for horse meat and milk products.

Study on endurance riding activities in Romania

L.T. Cziszter, F.L. Bochis, S.E. Erina and I.M. Balan
Banat's University of Agricultural Sciences and Veterinary Medicine 'King Michael I of Romania', from Timisoara, Calea Aradului 119, 300645, Romania; cziszterl@animalsci-tm.ro

The aim of the present study was to describe the endurance riding activities carried out in season 2017 in Romania. Endurance riding is the second popular discipline in Romania, after obstacles, by the number of registered horses (120) and riders (97). According to the National Federation rules, there are six trails with imposed speed: 10 km (poneys), 20 km (amateurs), 30 km (beginners), 40 km (novices), 60 km (novices), and 80 km (novices). Trails are divided into several loops, in practice 2 or 3, according to the entire distance, so that a loop has between 15 and 22 km. For year 2017, we take into study five events. Eighty-eight couples were registered for competing in the endurance events, but only 74 couples took the start. In the 30 km trail, 41 couples took the start, but only 37 couples finished. Average speed for the first loop was 11.47 km/h and horses need 752 sec. to recover. The second loop was done at 11.68 km/h speed with 893 sec. recovery time. For the 40 km event, a number of 14 couples took the start, but only nine couples finished the trail. The average speed was 11.93 km/h, with 11.1 km/h in the first loop and 12.69 km/h in the second loop. The time needed for the horses to recover was, on average, 545 sec. after the first loop and 643 sec. after the second loop. In the 60 km race, 13 couples took the start and 12 couples finished the second loop. The average speed was 13.41 km/h, with 13.03 km/h for the first loop and 13.81 km/k for the second loop. The average time for horses to recover was 454 sec. after the first loop and 740 sec. after the second loop. For the 80 km event, five couples finished the ride out of six couples at start. The average speed was 14.08 km/h, with 14.01 , 14.07 and 13.97 km/h for the first, second and third loop, respectively. The average times to recover were 360, 540 and 560 sec. for the first, second and third loop, respectively. In all endurance rides the speed, as well as the time horses needed to recover, increased from first to the second loop, except for the 80 km race, where the speed was similar for all three loops, while time to recover was higher in second and third loops compared to the first loop.

Characterisation of reproductive parameters in different horse breeds

J. Poyato-Bonilla[1], M.D. Gómez[1], M.J. Sánchez-Guerrero[1], E. Bartolomé[1], I. Cervantes[2], S. Demyda-Peyrás[3] and M. Valera[1]

[1]ETSIA, Universidad de Sevilla, Departamento de Ciencias Agroforestales, Ctra. Utrera, km 1, 41013 Sevilla, Spain, [2]Universidad Complutense de Madrid, Departamento de Producción Animal, Avda. Puerta de Hierro s/n, Madrid, Spain, [3]IGEVET–Instituto de Genética Veterinaria Ing. Fernando N. Dulout, UNLP-CONICET La Plata, Facultad de Ciencias Veterinarias UNLP, La Plata, Argentina; juliapb92@gmail.com

Reproductive traits are crucial for the survival of species in natural populations. In domestic species, they have major influence on economical effectiveness of breeding. Reproductive efficiency in horses is usually described as low due to several limits in reproductive techniques and fertility problems lead to profitability difficulties as breeding animals are expensive. Thus, the objective of this study was the characterisation of the main reproductive parameters in 8 Spanish horse breeds, 5 of them used for sport performance: Pura Raza Española (PRE), Pura Raza Árabe (PRÁ), Spanish Sport Horse (SSH), Anglo-Arab Horse (AAH) and Spanish Trotter Horse (STH) and 3 endangered populations: Purebred Menorca Horses (PRMe), Jaca Navarra (JN) and Burguete (BUR). Pedigree data used ranged from 2,315 (JN) to 307,831 (PRE) horses. A total of 10 reproductive parameters were estimated for each breed. Age at first foaling, which varied from 1,937.63 (PRE) to 3,446.99 (STH) days in dams and 1,789.93 (BUR) to 3,744.76 (STH) days in stallions. Age at last foaling, ranging from 3,770.40 (PRE) to 4,179.00 (PRMe) days in dams and 3,395.32 (BUR) to 4,926.55 (STH) days in stallions. Average interval between foaling was estimated exclusively in dams and ranged from 431.83 (PRE) to 1,086.22 (AAH) days. Interval between first and second foaling, calculated only in dams, showed results from 598.85 (PRE) to 1,097.76 (AAH) days. Average reproductive life ranged from 1,255.78 (BUR) to 2,657.72 (PRÁ) days in dams and from 1,564.34 (BUR) to 2,225.55 (STH) days in stallions. Generational intervals ranged from 8.01 (via stallion-son, BUR) to 13.26 years (via stallion-daughter, STH). Sexual rate of the descendants was approximately 50%, except in the case of JN and BUR, where the percentage of daughters was higher. Differences observed between breeds could be explained more by zootechnical than by physiological reasons.

Genetic parameters and breeding values for linear type traits in the Czech sport horse population

A. Novotná, A. Svitáková, Z. Veselá and L. Vostrý

Institut of Animal Science, Pratelstvi 815, 105 00 Praque, Czech Republic; novotna.alexandra@vuzv.cz

Linear profiling of horses has been used in the Czech Republic for 20 years but selection of individuals into breeding based on predicted breeding values are missing. The aim of this study was analyse the quality of input database, estimate for the first time the genetic parameters and predicted breeding values of a linear profiling of the sport horse population. The linear profiling database was obtained for the period 1997-2015 and comprised 12,455 horses. Each horse had only one record for linear profiling and measurement. In total 22 linear type traits and 3 measured traits (height at withers, heart girth, cannon bone circumference) were evaluated. The database comprised a relatively high number of incomplete data, e.g. places of evaluation (33%) or classifiers (14.9%) therefore only 8,194 records were used for estimate of the genetic parameters. The heritabilities and genetic correlations were estimated using a BLUP AM with AIREML. The statistical model included fixed effects of gender, age, place classifier×year and random effect of individual. The heritability estimates obtained for linear type traits and movements show a lower value (0.05-0.31) while heritabilities for measured traits were medium to high (0.43-0.67). Genetic correlations were estimated by a two-trait analysis and ranged from 0.0-0.93. Breeding values were standardized on a scale with the mean 100 and standard deviation of 20 points and the base population included all animals with a linear profiling. Breeding values of sires with a reliability greater than 50% were published. This study provides essential information for the development of a routine genetic evaluation of linear description that can provide objective information to breeders which will lead to an improvement in conformation traits according to the defined breeding objectives. Estimates of heritabilities and breeding values indicate that selection of individuals is feasible even if the current database quality seems to be inadequate. In the future an important part of the improvement of database quality is systematic supply of new information. Supported by Project No. QJI510141, QJ1510139 and MZE-RO0718 of Ministry of Agriculture of the Czech Republic.

Session 50 Poster 12

The influence of milking on the behaviour of Sokolski mare and foals

G.M. Polak

National Research Institute of Animal Production, Department of Horse Breeding, Wspolna Str. 30, 00-930 Warsaw, Poland; grazyna.polak@izoo.krakow.pl

The breed of Sokolski coldblooded horses was created in Poland in the beginning of XX century. They were used for agriculture purposes. Currently efforts are being made to use them for different purposes like for milk production. The aim of this study was to assess the capacities of milk production, the impact of milking on behaviour, welfare state and health of Sokolski mares and foals. In this study used five Sokolski mares where used. Mares were milked mechanically 1 time a day for 5 days/ week using a milking machine. Foals were separated from mothers for about 3-4 hours before milking and kept in the neighbouring box, having eye contact with the mares. During the first 50 days of milking, we obtained 258 l of milk, which means that the average amount was 1.3 l/mares/day: max 3.0; min 0.8. The horses were kept in the stables boxes and on the paddock (at least 8 hour/day) during the whole research period, without a access to the pasture. They received from 1.5 to 2 kg/mare/day of concentrate and hay and straw *ad libitum*. The preliminary research carried out showed that period of milking does not negatively affect the behaviour, condition and welfare of mares and foals. Separation of foals from mares during milking did not affect their health or behaviour. Observations indicated standard increase of weight of the foals and very fast adaptation of cold-blooded mares for milking process. However, it was observing a rapid drop down of milk production influenced by unexpected events, increased noise and a presence of groups of foreign persons in the stable.

Session 51 Theatre 1

Comparison of two methods based on meta-analyses for updating post-weaning lamb growth requirements

V. Berthelot, B. Fança and D. Sauvant

Modélisation Systémique Appliquée aux Ruminants, AgroParisTech, INRA, Université Paris-Saclay, 75005 Paris, France; valerie.berthelot@agroparistech.fr

Two independent methods were compared to estimate the energy and protein growth requirements of post-weaning lambs, both based on meta-analyses. Two databases ('LambGrowth' and 'LambComp') were developed from studies published in peer-reviewed journals or in institutional reports (IDELE). The 'LambGrowth' database (npub=88, nexp=255), used for method 1, compiles data from studies assessing the effect of various diets on lamb performance with precise data on diet composition and feed intake. The 'LambComp' database (npub=21, nexp=129), used for method 2, includes data on lamb body composition (lipids and proteins) at different body weights (BW). The method 1 was used to update lamb requirements in the INRA system. Inter-publications regressions of UFV (=1.76 Mcal of net energy) and PDI intakes (Y) on average daily gain (ADG, X) were adjusted. The intercept is the estimation of maintenance requirement and the slope the estimation of the growth requirement. With method 1, the UFV and PDI requirements for 100 g of ADG were 0.19 UFV and 25.3 g PDI respectively for a mean BW of 30 kg in the database. For method 2, after modelling growth curves, growth rate of whole BW and empty BW, daily lipid and protein gains were estimated from the 'LambComp' database and the net energy and protein requirements calculated. In line with the evolution of the composition of the BW gain (increase in lipid and decrease in protein) as the lamb gets older, the energy requirement per 100 g of ADG increases from 0.13 to 0.18 UFV and the protein requirement decreases from 23.5 to 21.0 g PDI for lambs ranging from 20 to 40 kg BW. Method 2 provided lower requirements for gain than method 1 (respectively -14% and -12% for UFV and PDI requirements for a 30 kg Lamb) but with differences getting lower for energy and higher for PDI when the lamb gets heavier. Method 1 allowed estimation of maintenance requirements for energy (0.019 UFV/d/kg BW) and PDI requirement for non-productive function (1.5 g/d/kg BW) but gives a fixed estimation of growth requirements contrary to method 2. Neither methods allowed us to separate requirements according to gender.

Distribution of sheep litter size: a worldwide survey

L. Bodin[1], J. Raoul[2], K.L. Bunter[3], A.A. Swan[3], S. Janssens[4], L. Brito[5], E.I. Bjarnason[6], O. Keane[7], S. Salaris[8], S. McIntyre[9], J.F. Jakobsen[10], J.L. Alabart[11], J. Folch[11], B. Lahoz[11], E. Fantova[12], L.F. de la Fuente[13], A. Molina[14], M.D. Perez Guzman[15], L. Mintegi[16], S. Maatoug[17], J. Conington[18], G. Ciappesoni[19], D. Gimeno[20] and R.M. Lewis[21]
[1]INRA GenPhySe, UMR1388, 31326 Castanet-Tolosan, France, [2]IDELE, 31321 Castanet-Tolosan, France, [3]Animal Genetics and Breeding Unit (AGBU), Univ. New England, Armidale NSW 2351, Australia, [4]Univ. Leuven, Department of Biosystems, Leuven, Belgium, [5]Univ. Guelph, Canada, [6]Icelandic Agricultural Advisory Centre, Iceland, [7]Teagasg, Animal & Bioscience Department, Grange, Dunsany, Co. Meath, Ireland, [8]DIRPA-AGRIS Sardegna, Research Unit Genetics and Biotechnology, Olmedo, Italy, [9]Beef + Lamb New Zealand Genetics, Dunedin, New Zealand, [10]Norwegian Association of Sheep and Goat Breeders, Ås, Norway, [11]CITA, Inst. Agroalim. de Aragón, Unidad de Producción y Sanidad Animal, Zaragoza, Spain, [12]Carnes Oviaragón S.C.L., 50014 Zaragoza, Spain, [13]Univ. León, Facultad de Veterinaria, Dpto de Producción Animal, León, Spain, [14]Univ. Cordoba, Dpto de Genetica, Cordoba, Spain, [15]CERSYRA, Valdepeñas, Ciudad Real, Spain, [16]ARDIEKIN, Vitoria-Gasteiz, Spain, [17]INAT, Laboratoire des Ressources Animales et Alimentaires, Tunis, Tunisia, [18]Scotland's Rural Coll, Edinburgh, Midlothian, Scotland, United Kingdom, [19]INIA Las Brujas, Canelones, Uruguay, [20]Uruguayan Wool Secretariat, Department of Animal Genetic Improvement, Montevideo, Uruguay, [21]Univ. Nebraska, Lincoln, USA; loys.bodin@inra.fr

Litter size (LS) in sheep (total lambs born including still born) is a categorical trait which ranges from 1 to 3 and exceptionally up to 7. A survey has been conducted on a large number of breeds from all over the world to gather the LS distribution of adult ewes per breed from 1995 to 2017 removing years with too few data. A simple analysis showed that there is a general law linking the different LS classes of a given breed to its mean prolificacy. However few breeds deviate from the standard norm; this variability will be discussed.

Seasonally anestrous ewe ovarian response to CIDR and eCG with estradiol-17β or estradiol cypionate

L.R.Y. Trucolo, R. Khomayezi, B. Makanjuola, M. Payne, K. McQueen, B.M. Thibault, M.S. Mammoliti and D.M.W. Barrett
Dalhousie University, 58 Sipu Awti, B2N 5E3, Truro, NS, Canada; david.barrett@dal.ca

Reproductive performance in seasonally anestrous ewes is poor even after using current controlled breeding protocols. Anestrous ewes treated with a medroxyprogesterone acetate sponge, for 12 or 14 d, and estradiol-17β (E_2) have synchronized follicular wave emergence. The objective of this study was to compare the ovarian response of seasonally anestrous ewes treated with a CIDR and eCG with either an E_2 or estradiol cypionate (ECP) injection 6 d after CIDR insertion. Ewes received CIDRs (Day -12) and an eCG injection (500 IU; Day 0) at CIDR removal and an injection of E_2 (350µg; n=5) or ECP (350µg (ECP350; n=4) or 70 µg (ECP70; n=5)) 6 d before CIDR removal (Day -6). Ovarian ultrasonography was done daily from Day -6 to 0 and twice daily starting 24 h after CIDR removal until confirmed ovulation. A ram was introduced every 6 h from 24 to 84 h after CIDR removal to observe estrus. Data was analysed by ANOVA (Two Way RM; One Way) and then by Tukey Test. There was a treatment, day, and interaction effect on daily maximum follicle diameter (MFD; $P<0.005$). Daily MFD was larger in E_2 (3.7±0.3 mm) and ECP70 (3.1±0.3 mm) than ECP350 (1.8±0.3 mm) ewes ($P<0.05$). Daily MFD decreased from Day -5 to -3 and increased from Day 0 to 2 ($P<0.05$). Daily MFD was larger in E_2 than ECP350 ewes from Day -1 to 3 and larger in ECP70 than ECP350 ewes from Day 1 to 3 ($P<0.05$). Follicles were ≤2 mm in ECP350 ewes from Day -3 on. Follicle wave emergence was earlier in E_2 (3.2±0.7 d) than ECP70 (6.0±0.7 d) ewes after E_2 or ECP treatment ($P<0.05$); wave synchrony was similar ($P>0.05$). Estrus onset was earlier for ECP350 (30.0±0.0 h) than ECP70 (55.5±9.0 h) ewes after eCG treatment ($P<0.05$); E_2 ewes were intermediate (36.0±3.3 h; $P>0.05$). Estrus was longer than 60 h for two ECP350 ewes and one ECP70 ewe. Estrus ended at the same time for E_2 and ECP70 ewes ($P>0.05$). Ovulations occurred in four E_2, no ECP350, and three ECP70 ewes. Ovulation tended to be sooner in E_2 (2.5±0.0 d) than ECP70 (3.3±0.8 d) ewes after eCG treatment ($P=0.073$). The ECP doses in this study do not appear to be suitable replacements for E_2 in a CIDR-eCG estrus synchronization protocol for seasonally anestrous ewes.

Phenotypes and genetic parameters for Swedish sheep breeds in conventional and organic production

L. Rydhmer[1], E. Jonas[1], E. Carlén[2] and K. Jäderkvist Fegraeus[1]
[1]Swedish University of Agricultural Sciences, Dept of Animal Breeding and Genetics, Box 7023, 75007 Uppsala, Sweden, [2]Växa Sverige, Box 288, 751 05 Uppsala, Sweden; lotta.rydhmer@slu.se

In the genetic evaluation of Swedish sheep, traits measured in conventional and organic production systems are treated as the same trait. However, both genetic parameters and breeding goals may differ between the production systems. Our aim is to develop breeding values including several traits, e.g. growth rate and litter size, for different production systems and breeds. We used data from the field-recording scheme, and analysed differences between organic and conventional production for purebred Swedish Pelt (n=259,641, 18% organic) and Fine wool (n=73,491, 9% organic) sheep. In general, the organic herds were larger than the conventional herds (121 vs 77 animals). For Pelt sheep, average birth weight was slightly higher in organic than conventional herds (3.95 vs 3.80 kg), but birth weight did not differ between production systems for Fine wool sheep (3.37 kg). Early growth rate, recorded between birth and 60 days of age (around 20 kg), was lower in organic herds (P<0.01) for both breeds. For Pelt sheep, growth rate from birth to performance test (around 33 kg) was slightly higher in organic than conventional herds (236 vs 231 g/day, P<0.001). For Fine wool sheep, growth rate from birth to performance test was lower in organic herds (238 vs 250 g/d, P<0.001). There was no difference between the production systems in litter size (1.8 lambs born for Pelt and 2.2 lambs born for Fine wool). Longevity recorded as number of litters per ewe was higher in organic than conventional herds for both Pelt and Fine wool sheep (P<0.001). Genetic parameters for selection traits will be presented for different breeds in conventional and organic production systems.

Major genes effect estimation and characterisation of a local Romanian sheep breed

M.A. Gras, M.C. Rotar, C. Lazar, R.S. Pelmus, E. Ghita and H. Grosu
National Research & Development Institute for Biology and Animal Nutrition, Genetic Resources Management, Calea Bucuresti nr. 1, Balotesti, Ilfov, 077015, Romania; gras_mihai@yahoo.com

Milk production improvement was always a challenging task on sheep breeds. Issues like: absence of animal identification, distance between farms, pedigree and recording system, multiple and unknown mating, low heritability of sex linked traits, make impossible traditional breeding programs appliance. However, a marker assisted selection (MAS) program should be feasible in local conditions. The genomic evaluation methods can improve accuracy of breeding value estimation, especially at ovine breeds, and can accelerate selection response. Besides that, the estimation methods that use just phenotypes from genotyped animals (genomic-BLUP and BLUP-SNP) have a limited accuracy because reference populations are usually small. From this reason, is necessary the expansion of GBLUP methodology to single-step GBLUP (ssGBLUP) and validation of that on the larger population. Our previous studies underline importance of major genes like α-casein (CSN), β-lactoglobulin (LGB) and prolactin (PRL) in milk production traits and the increase of accuracy for genomic evaluation when those genes are used. Animal breeding and genetics group from our institute studied previously effect of major genes over production traits, but ssGBLUP methodology was studied with simulated datasets. Also, on previous papers, we studied introduction of molecular marker information in different models used for breeding value estimation. Due to specific particularities and breeding technology, it is an increased probability of major differences between real and simulated datasets. Starting from these data, this paper aims to characterise genetically a local sheep breed (Teleorman Black Head) from Romania, and estimation of genetic effect for major genes related with milk and meat production, health and reproduction.

Session 51 Theatre 6

Population structure of five Swedish sheep breeds

C.M. Rochus[1], E. Jonas[2] and A.M. Johansson[2]

[1]Uppsala University, Department of Medical Biochemistry and Microbiology, Box 582, 751 23 Uppsala, Sweden, [2]Swedish University of Agricultural Sciences, Department of Animal Breeding and Genetics, Box 7023, 75007 Uppsala, Sweden; anna.johansson@slu.se

Swedish sheep breeds are part of the group of North European Swedish sheep breeds. In this study, five of the Swedish breeds were analysed with high density SNP chip for the first time. Some of the Swedish sheep breeds have previous been studied with microsatellite markers and endogenous retroviruses. One of the five studied breeds is a commercial breed selected for pelt and meat production (Gotland sheep). The other four studied breeds (Gute sheep, Dalapäls sheep, Klövsjö sheep and Fjällnäs sheep) are local heritage breeds. They come from different geographic areas in Sweden and differ in population size. Our objectives were to better understand the population structure of native Swedish sheep breeds and to compare them with breeds from other countries. We studied population structure using principal components analysis, a mixture model to estimate individual ancestry coefficients and a population tree model. We also estimated historic effective population size. The results showed that native Swedish sheep breeds are related to other north European short-tailed sheep (sheep from Iceland, Scotland, Norway, Finland and Russia) but were genetically unique. All the five Swedish breeds were very distinct and had long branch lengths in the population tree indicating high amounts of genetic drift. Among the studied Swedish breeds Gotland sheep and Gute sheep was most closely related, which was not surprising since they both originate from the island Gotland. The breed Fjällnäs sheep was of particular interest as this breed was recently discovered in Nortern Sweden and has not been included in any publications before. The samples from this breed were taken only a few years after it was discovered in northern Sweden. Our results showed that these animals grouped together based on their genotypes and were not close to other breeds. However, we could detect at least two groups within this breed. The analyses of historical effective population size shows that all five breeds have a declining effective population size and that the four heritage breeds have consistently smaller effective population sizes than the commercial breed Gotland sheep.

Session 51 Theatre 7

Effects of birth and rearing type on lamb performance to slaughter on a rotational grazing system

T.W.J. Keady[1], N. McNamara[1] and J.P. Hanrahan[2]

[1]Teagasc, Athenry, Co Galway, Ireland, [2]University College Dublin, School of Veterinary Medicine, Dublin, Ireland; tim.keady@teagasc.ie

The objective of this paper is to present the effects of birth and rearing type on the performance, over a 12-year period, of lambs reared on a grass-based system from birth to slaughter. A total of 7,964 lambs (989 singles, 4,462 twins, 2,151 triplets, 312 quads, 50 quintuplets) were produced in a rotational-grazing system between 2006 and 2017. Their dams (overall litter size 1.98) were on various studies, predominantly examining the effects of nutrition during mid and late gestation. Ewes were joined with rams for lambing in March; they were shorn and housed in mid-December, and offered grass silage. Level of concentrate was offered in late gestation varied according to litter size. Ewes went to pasture within a few days of lambing. Ewes rearing singles and twins, and their lambs received no concentrate at pasture. Ewes rearing triplets were offered 0.5 kg concentrate daily for 5 weeks post lambing and their lambs had access to concentrate (maximum of 300 g daily) until weaning (WN) at 14 weeks old. All lambs were grazed as one flock post WN, without concentrate supplementation. Sward height, measured in each paddock pre- and post-grazing. Mean pre- and post-grazing sward heights for April, May and June were 5.4, 7.4 and 7.9 cm; and 3.5, 4.6 and 5.1 cm, respectively. Lamb birth weight for lambs born as singles, twins and triplets and quadruplets were 5.6, 4.5, 3.7 and 3.2 kg ($P<0.001$); corresponding values for total mortality to weaning were and 9.8, 8.8, 19.3 and 35.7% ($P<0.01$), respectively. For lambs born and reared as singles, twins or triplets lamb daily LW gain birth to WN and WN to slaughter (g/d), WN weight (kg), age at slaughter (d) and carcass weight were 330, 271 and 279 (3.5, $P<0.001$); 183, 178 and 163 (10.3, $P<0.001$); 38.0, 31.2 and 31.0 (0.35, $P<0.001$); 154, 196 and 196 (4.8, $P<0.001$); and 20.0, 19.8, and 19.9 (0.20; $P<0.05$), respectively. Mean LW at WN was within 0.7 kg (2.5%) of the overall average for 10 of the 12 years. All lambs were slaughtered prior to the end of the grazing season (early December). Mean carcass weight was 19.9 kg. It is concluded that high levels of performance are consistently achievable from grass-based systems of prime lamb production.

Genetic parameters estimation for reproductive traits of goat breeds in Croatia

M. Špehar[1], D. Mulc[1], D. Jurković[1], T. Sinković[1], A. Kasap[2] and Z. Barać[1]
[1]Croatian Agricultural Agency, Ilica 101, 10000 Zagreb, Croatia, [2]University of Zagreb, Faculty of Agriculture, Department of Animal Science and Technology, Svetošimunska 25, 10000 Zagreb, Croatia; mspehar@hpa.hr

The objective of this study was to estimate genetic parameters for reproductive traits – litter size (LS) and birth weight (BW) in Alpine, Saanen, and German Improved Fawn goat breeds in Croatia. Pedigree information for 141,384 animals and 123,450 phenotypic records for each trait collected on 12,410 does from year 2000 to 2017 were included in the analysis using a single-trait repeatability animal model. Fixed class effects in the model were breed, parity, season of kidding as year-month interaction (and litter size for BW), while age at lambing was treated as covariate and fitted using quadratic regression nested within parity. Direct additive genetic effect, flock-year of lambing, and permanent environment effect within parity were included in the model as random effects. Variance components were estimated using Residual Maximum Likelihood method as implemented in the VCE-6 program. Estimates of heritability, flock-year of lambing, and permanent environment effect for LS were 0.075±0.003, 0.111±0.004, 0.677±0.003, respectively. For BW, estimates of heritability, flock-year of lambing, and permanent environment effect were 0.098±0.004, 0.406±0.007, 0.184±0.004, respectively. Regardless of being similarly heritable traits, large discrepancy in estimates of permanent environment and flock-year of lambing effect among the traits indicate a significantly higher phenotypic plasticity of BW in comparison to LS. The obtained results should contribute to better understanding of phenotypic variability of BW and LS in dairy goat populations. Fitting these sources of phenotypic variability in BLUP models for prediction of BVs should provide high accuracy and unbiased ranking of the animals, especially in joint across flock evaluation systems with considerably different environmental conditions among flocks.

Reducing the oxygen content in modified atmosphere packages to meet consumer demands

M.T. Corlett[1,2], L. Pannier[1,2], K.R. Kelman[1,2], R.H. Jacob[2,3], D.W. Pethick[1,2] and G.E. Gardner[1,2]
[1]Veterinary and Life Sciences, Murdoch University, Murdoch, WA 6150, Australia, [2]Australian Cooperative Research Centre for Sheep Industry Innovation, Armidale, NSW 2351, Australia, [3]Department of Primary Industries and Regional Development, South Perth, WA 6151, Australia; g.gardner@murdoch.edu.au

Colour is an important attribute influencing the purchasing decisions made by consumers. High oxygen modified atmosphere packaging (MAP) is used by the retail industry to display meat as it enhances the colour of meat by increasing the proportion of Oxymyoglobin (OMb). However, recent evidence has shown that beef and lamb cuts stored under high oxygen MAP have reduced tenderness. Therefore work is required to establish a reduced concentration of oxygen MAP that minimises the impact on tenderness yet still meets the colour requirements for consumer acceptability. This study investigates a range of oxygen concentrations in MAP to identify the minimum levels required to maintain suitable colour. Lamb loins (n=50) from the Meat & Livestock Australia genetic resource flock in Katanning, Western Australia, were aged at 2 °C in vacuum for either 5 or 20 days and then repackaged in MAP containing 0, 20, 40, 60 or 80% oxygen (MAP0, MAP20, MAP40, MAP60 or MAP80), 20% carbon dioxide and the balance met by nitrogen gas. Samples were left under simulated retail display for 6 days after which OMb% was determined using a Hunterlab® instrument. At 6 days of retail display the OMb% for samples aged 5 days were 36.1±1.0, 49.4±1.0, 57.7±1.0, 61.5±1.0, 62.4±1.0 for MAP0, MAP20, MAP40, MAP60 and MAP80. For samples aged 20 days the trend was similar, with values of 27.2±1.0, 52.8±1.0, 62.5±1.0, 63.5±1.1, 64.2±1.0, although differences between oxygen containing MAP and MAP0 were greater. The best results for OMb% were achieved by the MAP80 mixture, yet importantly the MAP40 mixture produced a similar level of OMb% in both 5 day and 20 day aged product. This represents an opportunity to reduce oxygen levels within MAP to 40% to counter the negative effects of high oxygen MAP on meat toughness. Future studies will explore the impact of MAP40 on tenderness.

Which factors influence lifetime effectivity and length of productive life of dairy goats?

M.-R. Wolber[1], H. Hamann[2] and P. Herold[1,2]
[1]Universität Hohenheim, Institut für Tropische Agrarwissenschaften, Garbenstr. 17, 70599 Stuttgart, Germany, [2]Landesamt für Geoinformation und Landentwicklung Baden-Württemberg, Referat 35, Zuchtwertschätzteam, Stuttgarter Str. 161, 70806 Kornwestheim, Germany; mrwolber@uni-hohenheim.de

Dairy goat husbandry is becoming more and more important in Germany. Increasingly, dairy goats are milked for more than a year and are not mated seasonally. Yet, the extent to which prolonged lactations influence lifetime effectivity and length of productive life of the individual animal has not been studied. The present study investigates the effects of milk yield and milk ingredients within the first lactation or total milk yield and total milk ingredients on lifetime effectivity or length of productive life. In this study, milk recording data were analysed from 9,190 dairy goats (German Fawn and German White goats) from the birth years 1988-2006. The traits included milk yield (kg), milk ingredients (kg and %), persistence and extended milking at the end of lactation. Significant effects are identified by an analysis of variance. Fixed effects are breed, birth year and age class at first kidding. Random effects are sire and herd effect. Fixed effects within first lactation are birth type, milk yield, fat and protein content or fat-protein quotient, urea content in the first 120 days in milk, persistence (2:1) and status of extended milking. Fixed effects within life are birth type, persistence (2:1), fat and protein content or fat-protein quotient and urea content and the proportion of days in milk are extended milking in the life of a goat. First results show that high persistence (>100) within the first 240 lactation days of the first lactation has a significant positive effect on the milk yield (2.53 kg/day-3.08 kg/day) per day in the life of an animal and the longevity (605.54-859.62 days). Extended milking in first lactation has a significant negative effect on milk yield per milking day (2.72 kg/day vs 2.82 kg/day) in a goat's life and a significant positive effect on length of productive life (607.83 vs 709.24 days) of an animal.

Session 51 Theatre 11

Transcriptome variation in response to gastrointestinal nematode infection in goats

H.M. Aboshady[1,2,3], M. Bederina[3], N. Mandonnet[3], R. Arquet[4], A. Johansson[2], E. Jonas[2] and J.C. Bambou[3]
[1]AgroParisTech, 16 rue Claude Bernard, 75005 Paris, France, [2]SLU, Ulls väg 26, 75007 Uppsala, Sweden, [3]INRA-URZ, Petit Bourg, 97170 Guadeloupe, France, [4]INRA-PTEA, Petit Bourg, 97160 Guadeloupe, France; hadeer.moursy@slu.se

Gastrointestinal nematode (GIN) infection is one of the most economically important constraints in small ruminant production. Due to the rise of anthelmintics resistance throughout the world, alternative control strategies have to be developed. The development of breeding programs for resistance to GIN is one promising strategy. However, there is a need to improve knowledge on the physiological mechanisms underlying genetic resistance to GIN infection to identify pertinent biomarkers. In this study we compare the immune response of resistant (res.) and susceptible (sus.) Creole goats. A total of 24 kids, 12 sus. and 12 res. to GIN on the basis of the estimated breeding value, were infected twice with 10,000 L3 *Haemonchus contortus*. Physiological and parasitological parameters were monitored during the infection. Seven weeks after the second infection extreme kids (n=6 res. and 6 sus.) chosen on the basis of the faecal egg counts (FEC) and 3 uninfected control animals were slaughtered. Sus. kids had significantly higher FEC compared with res. kids during the second infection but no differences in worm burden, male worm count, female worm count or establishment rate was observed. A higher number of differentially expressed genes (DEG) were identified for the comparison of uninfected vs infected animals in both abomasal mucosa (n=792 DEG) and lymph nodes (n=1,726 DEG) compared with sus. vs res. groups (n=342 and 450 DEG, respectively). 'Cell cycle' and 'cell death and survival' were the main identified networks in mucosal tissue when comparing uninfected vs infected kids. Antigen processing and presentation of peptide antigen via major histocompatibility complex class I were in the top biological functions for the DEG identified in lymph nodes for the comparison of sus. vs res. kids. The TGFβ1 gene was one of the top 5 upstream regulators DEG in the comparison of mucosa tissue of infected vs uninfected and sus. vs res. kids. Our results are one of the few investigating differences of the expression profile induced by GIN infection in goats.

Genome-wide structure and dynamics of Nubian goats from Northeast Africa

N. Khayatzadeh[1,2], G.T. Mekuriaw[3], A.R. Elbeltagy[4], A. Aboul-Naga[4], A. Haile[1], B. Rischkowsky[1] and J.M. Mwacharo[1]
[1]Small Ruminant Genomics, ICARDA, Addis Ababa, Ethiopia, [2]University of Natural Resources and Life Sciences, Vienna, 1180, Austria, [3]Department of Animal Production and Technology, Bahir Dar, University, Ethiopia, [4]Animal Production Research Institute, Dokki, Giza, Egypt; j.mwacharo@cgiar.org

The Nubian goat breed-group includes the Algerian Mzabite, Egyptian Zaraibi and Nubian goats found in Ethiopia, Sudan and Eritrea, and is one of the ancestral founders of the Anglo-Nubian. Whether the breed-group comprises one or separate genepools remains unknown as is their genetic contribution to the Anglo-Nubian dairy traits. We used Caprine 52K SNP Beadchip generated data to investigate the genome structure of Nubian goats from Egypt, Sudan and Ethiopia. Population structure analysis revealed two genetic groups (A and B) in Egypt. These were distinct from the Ethiopian and Sudanese Nubians, which occurred, in close proximity suggesting close genetic relationship. TreeMix analysis however, revealed one weak migration event involving Egyptian group-A with Sudanese Nubian suggesting possible historical connections. From hapFLK analysis, eight selection sweeps were revealed on chromosomes 6, 8, 9, 11, 13, 18, 23 and 25, respectively. The one on chromosome 6 was present in both Egyptian groups, while the ones on chromosomes 9 and 18, and 11 and 25 were specific to group-A and -B, respectively. The sweeps on chromosomes 8, 13 and 23 were common to Sudanese and Ethiopian Nubians. Functional annotation of the candidate regions in the Egyptian groups revealed several genes implicated in dairy traits (*ABCG2*), fat metabolism (*PPARGC1A, ACMS3*) and reproduction (*BMPR1B; LCORL*) as well as *ARL6IP1* and *TNRC6A* in group-A with roles in thermo-tolerance during lactation and climatic stress tolerance, respectively. The ones in Sudanese and Ethiopian Nubian were associated with adaptation and immune traits (*RAP1*, *BPIFA/B* gene family, *HS3ST2*, *CDR2*). Our findings suggest that the three genepools present in Northeast African Nubians, as a repository of diverse genes, are an integral component of agro-biodiversity that may be required at present, and in future, to respond to different production and climate change scenarios.

Genome-wide association study for production traits in Zandi sheep

H. Mohammadi[1], S.A. Rafat[1], H. Moradi Shahrebabak[2], J. Shoja[1] and M.H. Moradi[3]
[1]University of Tabriz, Faculty of Agriculture, Animal Science, Bd. 29 Bahman, 5166616471, Iran, [2]University of Tehran, Department of Animal Science, University College of Agriculture and Natural Resources, Karaj, Iran, [3]University of Arak, Department of Animal Science, University College of Agriculture and Natural Resources, Arak, Iran; rafata@tabrizu.ac.ir

Growth and wool production traits are significant economic traits in sheep. The objective of this study was to identify the genomic regions affecting growth and wool production traits in Iranian Zandi fat-tailed and carpet-wool sheep. The phenotypic data included birth weight, weaning weight, 6-, 9-, and 12-month weight, pre-weaning ADG, post-weaning ADG, fibre diameter, fibre diameter coefficient of variation, prickle factor, staple length, kemp and, outer coat fibre. The hair and blood samples were collected from 100 Iranian Zandi Sheep based on estimated breeding values (EBVs) of body weight using two-tailed strategy. All individuals were genotyped using the medium-density Illumina Ovine SNP50 BeadChip.The total number of 100 animals belonged to 36 half-sib families with unrelated dams as possible (1-9 progenies per sire).After quality control, 47,411 SNPs were analysed by TASSEL program in a mixed linear model (MLM). Twenty three regions, in which four were associated with more than one trait, located in 12 chromosomes were associated with growth and wool traits ($P<5\times10^{-6}$). Gene annotation was implemented with the latest sheep genome (Ovis_aries_v4.0) and the results showed that these genomic regions overlapped with some genes such as KCNIP4, PPARGC1A, ASAP1, ANK2, WWOX, SYNE1, FBXO5, AKAP6, FABP3, ANGPTL4, ATP6V1B2, PARK2, and KRTAP11-1, that were in associate with postnatal growth, regulate metabolic pathways, adipogenesis, gluconeogenesis, skeletal muscle differentiation, bone growth, and structural proteins. Gene ontology term enrichment analysis revealed that genes involved in positive regulation of muscle structure, muscle tissue development, and fatty acid metabolic were over-represented in the identified candidate genes.

Session 51 Poster 14

The effect of neutral detergent fibre source on lambs growth performance and meat nutritional value

J. Santos-Silva[1], A. Francisco[1], T. Dentinho[1], J. Almeida[1], A. Portugal[1], S. Alves[2], E. Jerónimo[3] and R. Bessa[2]
[1]INIAV, Fonte Boa, 2005-048 Vale de Santarém, Portugal, [2]Faculdade de Medicina Veterinária, U. Lisboa, Av. Universidade Técnica, 1300-477 Lisboa, Portugal, [3]CEBAL, R. Pedro Soares, 7800-295 Beja, Portugal; jose.santossilva@iniav.pt

The objective of this study was to evaluate the effect of changing the source of neutral detergent fibre (NDF) in complete ground diets on lambs growth, carcass quality and meat nutritional value. Twenty lambs were individually housed and assigned to 3 diets with low starch (5.6% DM), high oil (7.5% DM) and the same NDF content (42.9%DM) but differing in NDF composition. The main NDF source was dehydrated pelleted alfalfa (*Medicago sativa*) with levels of 200, 400 and 600 g/kg DM, that were balanced with the proportions of soyhulls (*Glicine* max), and dehydrated citrus and beat pulps. The experiment lasted 6 weeks. Intake was controlled daily and lambs weight weekly. After slaughter, carcass quality parameters were determined and *Longissimus dorsis* (LD) muscle was sampled for chemical and physical analysis. Intake increased ($P<0.05$) with alfalfa proportion in the diet but growth rate was unaffected (290 g/day). Carcass traits and LD and fat colour, (L*,a*, b*), were not changed by the diet. Meat shear force measured 7 days after slaughter presented low values (2.75 kg/cm^2) and diets did not affect the LD colour stability and lipid oxidation during storage. The composition of NDF affected rumen biohydrogenation (BH) pattern and the proportion of some bioactive fatty acids in meat. Increasing the forage proportion in diets increased the main healthy BH intermediates vaccenic (*t*11-18:1; $P=0.003$) and rumenic acids (*c*9,*t*11-18:2; $P=0.021$) and decreased the proportion of *t*10-18:1 ($P=0.058$) and *t*10,c12-18:2 ($P=0.007$). Moreover, increasing alfalfa proportion in diets reduced the frequency and intensity of the *t10* shift, as evaluated by *t*10-/*t*11-18:1 ratio in the tissues. Concluding, the NDF source is a determinant factor on the results of the feeding strategies designed to improve the nutritional value of lamb meat.

Session 51 Poster 15

Effects of parity and litter size on milk yield of commercial dairy goats in Australia

F. Zamuner[1], K. Digiacomo[1], A.W.N. Cameron[2] and B.J. Leury[1]
[1]The University of Melbourne, Faculty of Veterinary and Agricultural Sciences, Parkville, Victoria 3010, Australia, [2]Meredith Dairy Pty Ltd, 106 Cameron Rd, Meredith, Victoria 3333, Australia; fzamuner@student.unimelb.edu.au

Predictions of how parity number and litter size (number of foetuses) influence cumulative milk yield in commercial dairy goats can assist farmers in making management decisions and contribute to increased efficiency of animal selection. This study aimed to investigate the effects of litter size and parity on cumulative milk yield during the first 270 days in milk (CMY) in Australian commercial dairy goats. This experiment was conducted at Meredith Dairy commercial farm (Meredith, Australia, 37°50'S; 144°04'E), during four consecutive kidding seasons, from June 2016 to March 2017. In total, 1000 (~250 per kidding season) Saanen and Saanen-cross goats (1-7 years, LW; 66±17.0 kg, and BCS; 2.5±0.3) were enrolled in the study. Does were housed in a single barn and *ad libitum* fed a total mixed ration (9.7 MJ/kg ME and 160 g/kg CP DM). Litter size was determined 60-80 days post-breeding using transabdominal ultrasound. Does were machine milked twice daily in a 2×36 parlour with electronic milk meters (De Laval, SE). Does were culled in line with commercial practice and only the does that achieved 270 DIM (n=682) were analysed. Data obtained was analysed using the Linear Mixed Models procedure of GenStat (version 18.1). Litter size, parity and kidding season were included as fixed effects and animal was included as a random effect. The effects of parity (primiparous, PP; n=257 vs multiparous, MP; n=425) and litter size (singletons, SG; n=252 vs twins, TW; n=430) on CMY were studied. The herd average for CMY was 723.4±6.66 L. The CMY was markedly increased (+27%) by parity (MP vs PP, 838±8.4 vs 609±10.1 L; $P<0.001$), and to a lesser extent (+8%) by litter size (TW vs SG, 752±8.7 vs 695±10.4 L; $P<0.001$). Further, twin-pregnancy increased CMY of both PP (+11%) and MP goats (+5%), (PP-TW vs PP-SG, 644±12.2 vs 575±12.5 L; $P<0.001$) and (MP-TW vs MP-SG, 858±10.0 vs 814±15.4 L; $P=0.016$). Results demonstrate that although both parity and litter size affected the CMY of dairy goats, the response was more pronounced for the effects of parity.

Impacts of early life nutrition on fat tissue morphology and gene expression in adult sheep

S. Ahmad[1], L.K. Lyngman[1], R. Dhakal[1], M. Mansouryar[1], M. Moradi[1], J.S. Agerholm[2], P. Khanal[3] and M.O. Nielsen[1]
[1]University of Copenhagen, Grønnegårdsvej 3, 1870 Frederiksberg C, Denmark, [2]University of Copenhagen, Dyrlægevej 68, 1870 Frederiksberg C, Denmark, [3]Nord University, Kongens Gate 42, 7713 Steinkjer, Norway; sharmila.ahmad@sund.ku.dk

We assessed the long-term impacts of pre- and early postnatal malnutrition on fat tissue in sheep. The fat tissue samples were obtained from 2½ years old males (M) and females (F) sheep, born to twin-pregnant dams fed with NORM (100% of energy (E) and protein (CP) requirements), HIGH (150% E and 10% CP) or LOW (50% of NORM) diets during the last trimester. Postnatally the twins were fed moderate (CONV) or high-fat-high-carbohydrate (HCHF) diet until 6 months of age. Thereafter, they were fed with a moderate diet for 2 years and then slaughtered. Adipocyte (AP) morphology and mRNA expression for a range of genes were assessed in subcutaneous, perirenal, mesenteric and epicardial fat and linear mixed model was used for data analyses. Even after 2 years of dietary correction, there was a clear shift towards larger AP and altered gene expression patterns for adipogenic and lipid metabolism genes in HCHF compared to epicardial fat from CONV sheep. HCHF sheep appeared to be tolerant towards impacts of early postnatal obesogenic diet, since HIGH-HCHF sheep attained similar expression for many genes to those of NORM-CONV. M had overall higher expression levels for a wide range of genes compared to F. Long-term implications of prenatal programming were observed more frequently and pronounced in M than F. In subcutaneous fat, LOW sheep of both sexes had distinct changes in AP shape and size compared to NORM and HIGH; LOW and HIGH sheep had reduced expression levels for a range of genes compared to NORM sheep. Perirenal fat was the most sensitive among all tissues to early life malnutrition, and prenatal implications were observed most in LOW sheep irrespective of the postnatal diet. In conclusion, even 2 years of dietary correction could not reverse impacts of early obesity on adipocyte morphology. Long-term differential implications of prenatal malnutrition on fat tissues development may depend on timing of the nutritional insults. Changes in AP morphology could not readily be associated to expression of genes involved in a wide range of functional traits.

Effect of early shearing on the productive performance of offspring in their first 18 months of age

C. López Mazz[1], F. Baldi[2], G. Quintans[3], M. Regueiro[1] and G. Banchero[4]
[1]Fac. de Agronomia, Garzón 780, 12900, Uruguay, [2]Facultad de Ciencias Agrarias, SP, 11884100, Brazil, [3]INIA, Treinta y Tres, 33000, Uruguay, [4]INIA, Colonia, 70006, Uruguay; tatolopezmazz@gmail.com

Nutritional and environmental stress at critical moments of foetal development can alter the physiology and metabolism of the organs and tissues of the offspring and influence its growth after weaning. The effect of shearing at 50 days of gestation on body condition (BCS) and body weight (BW) of male offspring in the first 18 months of life were evaluated. Seventy-nine male lambs (Polwarth) born in spring, whose mothers were assigned to two treatments factors: shearing time (prepartum (PS) and postpartum (U)) and litter size (single (S) and twin (T)) resulting in four treatments: SPS (n=23) single lambs born to PS ewes; SU (n=21) single lambs born to U ewes; TPS (n=18) twin lambs born to PS ewes and TU (n=17) twin lambs born to U ewes were used. From weaning (104±0.6 d) male lambs were handled on improved pastures, and BCS (scale of 1-5) and BW were recorded every 14 days. Data were analysed using a repeated-measure analysis with PROC MIXED procedure of SAS (SAS 9.3). Means were compared by Tukey-Kramer test ($P<0.05$). The weight of the lambs at birth ($P=0.0002$, 4.8±0.3 vs 4.4±0.2 kg) and at weaning ($P=0.01$, 21.5±0.9 vs 19.2±0.9 kg) was greater in PS lambs than U. BCS was not different (P=0.10, 1.9±0.05 vs1.9±0.06 units PS and U) among treatments. BW of PS was higher than U male lambs (P=0.02, 34.6±0.7 vs 32.2±0.7 kg) but was not accompanied by a higher (P=0.42) daily gain of BW (0.73±0.04 vs 0.78±0.04 kg). BW of S and T male lambs was not different (P=0.10, 34.3±0.8 vs 32.5±0.8 kg, PS and U). The maximum difference (P=0.007) in BW among treatments was reached at 285 days old (33.9±0.9 vs 30.3±0.9 kg, PS and U) but the difference was lost after the age of 12 months. U male lambs could not compensate after weaning the lower BW they had at birth and at weaning. The high BW reached by PS lambs up to 12 months of age is of great importance for sheep meat producers, since it is the period where the products have the greatest commercial value.

Session 51 Poster 18

Effect of condensed molasses solubles on performance, digestibility and rumen parameters in sheep

K. O'Reilly, W.A. Van Niekerk, R.J. Coertze and L.J. Erasmus
University of Pretoria, Dept Animal and Wildlife Sciences, Lynnwood Road, 0001 Pretoria, South Africa; lourens.erasmus@up.ac.za

This study investigated the effect of different inclusion levels of condensed molasses solubles (CMS) when replacing sugarcane molasses in high concentrate diets on feed intake, growth performance, digestibility and rumen parameters of sheep. The treatments were 0% CMS (control), 4% CMS, 8% CMS and 12% CMS inclusion on an as is basis. Diets were formulated on an iso-energetic and iso-nitrogenous basis. Experiment 1 was a 4×4 Latin Square design with four rumen cannulated Merino wethers. Data was analysed using the Proc GLM model (SAS) and significance between means was declared at <0.05 using the Fischers test. Feed intake did not differ (P >0.05) between treatments. Organic matter digestibility was lower (<0.05) for the 0% CMS and 4% CMS treatments compared to the 12% CMS treatment. Total volatile fatty acid production was lower (<0.05) for the 8% CMS treatment compared to the 12% CMS treatment. Apparent nitrogen retention and other rumen fermentation parameters (rumen pH, rumen ammonia N) did not differ (P >0.05) between treatments. Experiment 2 was a randomised complete block design with 200 Mutton Merino lambs (27.6±4.8 kg) with 10 animals in each of 5 pens per treatment. Data was analysed as a randomised complete block design using the Proc GLM model (SAS) for the average effects over time. Dry matter intake did not differ (<0.05} but the 0% CMS treatment had a lower (<0.05) average daily gain and final live body weight compared to the 4% CMS treatment. The 0% CMS treatment had a higher (<0.05) feed conversion ratio compared to the 4% CMS, 8% CMS and 12% CMS treatments. The 0% CMS treatment resulted in a lower dressing % and carcass fat code compared to the other treatments (P<0.05). The copper concentration in the liver samples of the 0% CMS treatment was lower (<0.05) than the 8% CMS treatment; however, all treatments had liver copper concentrations within the normal range. Results suggest that condensed molasses solubles can be included up to 12% on an as is basis to replace sugarcane molasses in high concentrate diets without any adverse effects. Further research needs to be conducted on higher inclusion levels of CMS in sheep feedlot diets..

Session 51 Poster 19

Comparison of lactation models in pasture-based dairy ewes in Bosnia and Herzegovina

V. Batinic[1], D. Salamon[2], S. Ivankovic[1], N. Antunac[2] and A. Dzidic[2]
[1]University of Mostar, Faculty of Agriculture and Food Technology, Biskupa Čule bb, 88000 Mostar, Bosnia and Herzegowina, [2]University of Zagreb, Faculty of Agriculture, Svetošimunska 25, 10000 Zagreb, Croatia; adzidic@agr.hr

Ewes milk from Kupres, Privor and Stolac dairy sheep breeds is mainly used for the production of fine cheese varieties. To the best of our knowledge there is no information about the milk production and composition of these pasture-based dairy ewes. Therefore, the aim of this article is to determine the best model to describe lactation curves in different pasture-based dairy ewes in Bosnia and Herzegovina. All pasture-based dairy ewes are kept with their lambs for at least 30 days of lactation. Milk production and milk composition samples (milk fat, protein, lactose and dry matter percentage) were collected during early, mid and late lactation in Kupres (n=267), Privor (n=204) and Stolac (n=226) pramenka dairy ewes. For the statistical analysis, a repeated measures model was used with ewe as a random effect and breed, lactation number and day of lactation, defined as fixed effects. Four lactation models (Wilmink, Cubic, Ali-Shaeffer and Guo-Swalve) were compared. Models were selected based on the lowest error variance component. The Guo-Swalve model was found to be the best for all measured variables. Kupres pramenka dairy ewe was the highest producing dairy ewe breed with 151 kg of milk during 187 days of lactation (0.81 kg/d), compared to Privorska with 99 kg of milk during 163 days of lactation (0.61 kg/d) and Stolac with 112 kg of milk during 190 days of milk (0.59 kg/d). Fat, protein and dry matter percentage increased throughout lactation, while lactose showed small decrease in all mentioned breeds. The prediction of the milk yield and milk composition from the Guo-Swalve model could be used by the national breeding program for the Kupres, Privor and Stolac breed. Higher producing Kupres pramenka dairy ewe showed the shape of the standard lactation curve compared to Privor and Stolac dairy ewe breed which showed the atypical constantly decreasing shape of the lactation curve which is common in low producing dairy ewes. The effects of the day of lactation had great influence on the total milk yield, milk composition and on the shape of the lactation curve.

Population genetic indices in Hungarian Cikta sheep

E. Kovács[1], K. Tempfli[1], A. Shannon[2], P. Zenke[2], Á. Maróti-Agóts[2], L. Sáfár[3], Á. Bali Papp[1] and A. Gáspárdy[2]
[1]Széchenyi István University, Department of Animal Sciences, Vár tér 2, 9200 Mosonmagyaróvár, Hungary, [2]University of Veterinary Medicine, Department of Animal Breeding Nutrition and Laboratory Animal Science, István st. 2, 1078 Budapest, Hungary, [3]Hungarian Sheep and Goat Breeders' Association, Lőportár st. 16, 1134 Budapest, Hungary; tempfli.karoly@sze.hu

Samples of 72 animals from three flocks were analysed using nine microsatellites in order to evaluate the connection between existing populations of Cikta sheep, an indigenous breed in Hungary. Cikta was a well-known and popular breed until 1960-70. Nearly 30,000 animals were recorded in 1947 whereas the breed came close to extinction in 1975 when only 112 animals remained and a genetic preservation programme was launched. Today's population is considered stable with approximately 1000 recorded animals. Fluorescent-labelled oligonucleotides and fragment length analysis was applied for microsatellite genotyping. Overall, observed and effective allele numbers were 5.63±1.71 and 3.76±1.10, respectively. Mean Fis (-0.18±0.12) and Fit (-0.13±0.11) values indicated heterozygous excess. Considerably low mean Fst (0.04±0.03) and discriminant analysis revealed lasting effects of the 1970s bottleneck as three analysed populations showed slight genetic differentiation. Three of the nine microsatellites significantly (P<0.05) deviated from Hardy-Weinberg equilibrium, namely *BM8125*, *CSSM47*, and *MAF214*. Results can be applied in mating plans to maintain existing diversity in the breed. This publication was supported by the EFOP-3.6.1-16-2016-00017 project.

Effect of soybean meal treated with *Cistus ladanifer* tannins on lambs growth and meat quality

T. Dentinho[1], K. Paulos[1], A. Francisco[1], E. Jeronimo[2], J. Almeida[1], R. Bessa[3] and J. Santos-Silva[1]
[1]INIAV, Fonte Boa, 2005-048 Vale de Santarém, Portugal, [2]CEBAL, Rua Pedro Soares, 7800-295 Beja, Portugal, [3]Faculdade de Medicina Veterinária, ULisboa, Av. Universidade Técnica, 1300-477 Lisboa, Portugal; teresa.dentinho@hotmail.com

Condensed tannins (CT) may improve the digestive utilization of feed in ruminants, mainly due to a decrease of protein rumen degradability and a subsequent increase of amino acid flow to the intestine. A productive trial was conducted during 6 weeks to evaluate the effect of treating soybean meal (SBM) with an extract of *Cistus ladanifer* CT on lamb´s growth, carcass and meat quality. Twenty four Merino Branco lambs were individually housed and assigned to three dietary treatments based on hay and concentrate in a proportion of 15/85 (W/W) and fed at 4% live weight. Concentrates were formulated to contain: 16% of crude protein (CP) with untreated SBM (Control); 12% of CP with untreated SBM (Restricted protein (RP)); 12% of CP with SBM treated with 15 g/kg on DM of *C. ladanifer* CT (RPCT). Intake was controlled daily and lambs were weighed weekly. Blood samples were collected to determine total protein, glucose and urea N (BUN). After slaughter carcass dressing percentage and high priced joints proportion were determined and the shoulders were dissected. *Longissimus* muscle was sampled for chemical and physical analysis. Lambs fed with RPCT diet had average daily weight gain (ADG) similar (P>0.05) to lambs fed with Control and higher (P<0.05) than lambs fed with RP diet (212 vs 168 g). Animals fed with RP and RPCT diets showed similar BUN level, but lower (P<0.05) than animals fed with Control diet (14.6 vs 22 mg/dl). Carcass traits and meat quality parameters were not affected by treatments (P>0.05). Concluding, *C. ladanifer* CT can be used as feed additive to reduce the protein content of diets, and hence the feed costs, without compromising lamb performance and meat quality.

Session 51 Poster 22

The role of BCS evolution across milking period on milk production traits in low-input dairy goats

N. Siachos[1], G.E. Valergakis[1], R. Giannakou[1], C. Squires[1], A.I. Gelasakis[2] and G. Arsenos[1]
[1]Aristotle University of Thessaloniki, Faculty of Veterinary Medicine, Box 393, 54124, Greece, [2]ELGO-Demeter, Veterinary Research Institute, Thermi, 57001, Greece; arsenosg@vet.auth.gr

The objective was to assess the role of BCS and its evolution patterns across milking period on milk traits in low-input dairy goats. A dataset of 4,890 records from 644 dairy goats (298 for 2 successive lactations) from 7 low-input farms and 3 different breeds (Skopelos, Eghoria Greek and Damascus) was used. It included monthly measurements of BCS, milk yields and the gross chemical composition of milk during milking period. Total milk yield (MY), fat yield (FY) and protein yield (PY) were calculated according to ICAR recommendations. A two-step cluster analysis to establish BCS patterns was used. Resulting clusters were grouped in 4 distinct major clusters (C1 to C4) based on BCS at weaning and its changes thereafter, as follows: C1: n=117, at a permanently low BCS (ca. 2.0); C2: n=351, increasing from a low BCS (ca. 2.0); C3: n=198, at a permanently medium BCS (ca. 2.5) and C4: n=276, increasing from a medium BCS (ca. 2.5). General linear models were used to assess the relationship of BCS clustering with the studied milk traits, adjusting for breed effects. Comparisons among clusters were made using Kruskal-Wallis test, with SPSS v.21. Mean (±SD) MY, FY and PY were 226.2 kg (±136.23), 10.1 kg (±5.27), and 7.6 kg (±4.24). BCS clustering was significantly associated ($P<0.05$) with ln-transformed MY ($R^2=0.108$), FY ($R^2=0.094$), and PY ($R^2=0.137$). Goats in C4, followed by C3, had significantly higher median MY, FY and PY compared to C2 and C1 ($P<0.05$). In C4, MY was increased by 23.3, 32.1 and 42.3%, FY was increased by 16.2, 32.2 and 42.9% and PY was increased by 19.7, 44.2 and 54.6%, compared to C3, C2 and C1, respectively. Similarly, in C3, MY was increased by 7.1 and 15.4%, FY was increased by 13.7 and 23.0% and PY was increased by 20.5 and 29.2%, compared to C2 and C1, respectively. Goats with BCS of 2.5 at weaning and in positive or even neutral energy balance thereafter, significantly outperformed those with BCS of 2.0 at weaning regarding, total MY, FY and PY. Hence, management should focus on limiting BCS loss pre-weaning. The work was funded by the SOLID project (FP7-266367).

Session 51 Poster 23

Effect of feeding dried distillers grain with solubles on performance of growing Awassi lambs

S.M.D. Hatamleh and B.S. Obeidat
Jordan University of Science and Technology, Animal Science, Faculty of Agriculture, 22110 Irbid, Jordan; bobeidat@just.edu.jo

A 91-day fattening study (7 days used as adaptation period and 84 days for data collection) was conducted to evaluate the effect of partial replacement of soybean meal and barley grain by dried distiller's grain with solubles (DDGS) on performance and carcass and meat characteristics of Awassi lambs. Twenty seven newly weaned Awassi ram lambs (2 to 3 months of age) were housed individually (1.5×0.75 m) and assigned randomly to one of three isonitrogenous dietary treatments. Treatments were: (1) fed a control diet (CON; n=9), (2) fed 12.5% DDGS (DDGS12.5; n=9) or (3) fed 25.0% DDGS (DDGS25; n=9) of dietary dry matter (DM) to replace part of soybean meal and barley grains. Diets were fed *ad libitum* and formulated to contain 16.0% CP (DM basis). On day 56 of the study period, 6 lambs from each group were chosen at random and were housed individually in metabolism crates (1.05×0.80 m) to evaluate nutrient digestibilities. At the end of the feeding period, fifteen lambs were slaughtered to evaluate carcass characteristics and meat quality. Intakes of DM, crude protein (CP), neutral detergent fibre (NDF) were similar among treatment diets. However, EE intake was greater ($P<0.05$) in DDGS25 compared to the DDGS12.5 and CON groups. Average daily gain was similar among treatment diets. Nutrient digestibilities did not differ among treatment diets except for the EE which was lower ($P<0.05$) in the CON diet compared to DDGS250 diet while DDGS125 diet was intermediate. Carcass characteristics were not different among diets. With the exception of shear force and redness, which were greater ($P<0.05$) in DDGS250 compared to DDGS125 and CON diets, meat quality parameters [i.e. pH, water holding capacity, colour coordinates (whiteness and yellowness)] were unaffected by treatment. Results of the current study demonstrate that dried distillers grains with solubles can partially replace soybean meal and barley when fed at 12.5 or 25% without impacting the growth performance and with minimal effects on carcass characteristics and meat quality.

Session 51 Poster 24

Effect of an alternative rearing method on milk production and lamb weight gain

S.A. Termatzidou, N. Siachos, P. Kazana, I. Rose and G. Arsenos
Aristotle University of Thessaloniki, Laboratory of Animal Husbandry, Panepistimioupoli, 54124 Thessaloniki, Greece; arsenosg@vet.auth.gr

Removing lambs from their mothers at early stages of lactation for small periods during the day may increase milk yields, because ewes have time to recover. Therefore, we tested if an alternative rearing methods affects milk production of ewes and weight gain of lambs. Forty primiparous Lacaune ewes were used. For the first 2 weeks after lambing, they were milked once a day and were kept with their lambs all day. Thereafter, ewes were allocated into 2 groups (control C and test T; n=20 in each), balanced for lambing date and prolificacy. Housing and feeding management were the same for both groups and the trial lasted 2 weeks. In group C, lambs were constantly kept with their mothers. In group T, lambs were separated for 12 h during the night and reunited again after the morning milking. Lambs had *ad libitum* access to feed and water. Ewes in both groups were milked once a day and milk yield (MY) was recorded using volumetric milk meters. All lambs were weighed at the start of the trial and weekly to calculate average daily gain (ADG). Mean daily MY and total MY for ewes and ADGs for lambs were log-transformed to achieve normality. Comparisons between C and T groups were done with t-test for equal variances or Mann-Whitney test for unequal variances. T ewes had higher daily MY(±SD) (0.87±0.35 vs 0.54±0.35 l, P=0.001) and produced 61% more milk (15.62±6.38 vs 9.71±7.40 l, P=0.001) than C ewes during the trial period. Moreover, mean(±SD) weight at the start of the trial was 7.36 kg(±1.4) and 8.02 kg(±1.4) for C and T lambs, respectively. C group lambs had higher ADG compared to T (295.7±102.5 vs 214.6±89.0 g, P<0.05) during the first week. ADG during the second week and in total were not significantly different between lambs of C and T groups, although C lambs grew faster (by 25.4%). than T lambs. Therefore, alternative lamb rearing produces a trade off between ewe milk production and lamb growth as well as higher feed and labour costs. However, when the main source of farmer's income is milk, alternative lamb rearing may be a preferable option to optimize saleable milk production instead of producing light lamb carcasses.

Session 51 Poster 25

Regulation of anti-mullerian hormone (AMH) by oocyte specific growth factors in ovine granulosa cell

V. Richani, D. Kalogiannis and S. Chadio
Agricultural University of Athens, Animal Science and Aquaculture, Iera odos 75, 11855 Athens, Greece; shad@aua.gr

Oocyte secreted factors, belonging to the transforming growth factors-β (TGF-β) superfamily regulate ovarian function via paracrine and/or autocrine action, but regulation of AMH production is yet poorly understood. The aim of the present study was to investigate the role of oocyte derived factors GDF9 and BMP15 on AMH and estradiol production from ovine granulosa cells, *in vitro*. Granulosa cells were harvested from small (1-3 mm diameter) and large (>3 mm diameter) follicles from ovine ovaries and cultured in serum free conditions for 48 hours with or without GDF9 and BMP15 in the presence or absence of FSH. Concentration of 17β-estradiol and AMH in culture medium were determined by RIA and ELISA methods, respectively. Statistical analysis was performed by one way Anova, followed by LSD test, using Statgraphics program. The results showed that addition of GDF9 significantly (P<0.05) decreased the production of AMH from small follicles, whereas BMP15 had no effect. In large follicles, GDF9 also caused a decrease in AMH concentration, without reaching significance. Estradiol production from granulosa cells from both small and large follicles was also found to be significantly (P<0.05) decreased by GDF9 alone or in combination with FSH, compared to control. In conclusion, these findings show for the first time that the oocyte-secreted factor GDF9 reduces the production of AMH, thus suggesting an attenuation of its inhibitory action on the progression of small follicles in ewes.

Prepartum grazing with oat pasture and its effect on mother-lamb behaviour at parturition

M. Regueiro[1], F. Baldi[2], C. López Mazz[1] and G. Banchero[3]
[1]Faculty of Agronomy, UDELAR, Garzón 780, 12900, Uruguay, [2]UNESP, SP, 14884-900, Brazil, [3]INIA, Ruta 50, Km 11, 70006, Uruguay; marielregueiro@gmail.com

Nutrition during gestation has an important role at time of parturition for both mother and offspring. The effect of different nutritional management during the last month of pregnancy on body condition (BCS), body weight (BW) and mother-lamb behaviour at parturition was evaluated. One hundred and forty multiparous Corriedale ewes gestating a single lamb were offered native pasture (7% crude protein (CP)) during the first four months of gestation and were assigned to two differential nutritional management during the last month of pregnancy: (1) ewes grazing oat pasture (14% CP) (GO, n=71); (2) ewes fed native pasture (GNP, n=69), regarding BW and BCS at the beginning of the treatment (49.1±0.7 vs 49.7±0.7 kg and 3.74±0.05 vs 3.77±0.05 GO and GNP respectively). At parturition BW and BCS were registered, as well as duration of labour, delivery assistance and Maternal Behaviour Score (MBS, range 1-5). In lambs, birth weight, Apgar test for newborns (score 0-10), and time elapsed to stand and suck was recorded. Data were analysed using PROC MIXED and GENMOD of SAS, (mean ± SEM; $P \leq 0.05$). At parturition GO ewes had higher BW (56.7±0.8 vs 52.7±0.7 kg $P<0.0001$) and BCS (3.7±0.04 vs 3.4±0.06 $P=0.0007$) than GNP. Duration of labour was longer (32.5±4.7 vs 24.03 min, $p<0.001$) and percentage of delivery assistance was significant greater (22.5 vs 4.3%) in GO group. Lambs from GO mothers were heavier (4.77±0.07 vs 4.28±0.09 kg, $P=0.0037$), and took longer to stand (36.4±4.2 vs 28.5±3.9 min, $p<0.001$) and suck (56.4±5.2 vs 51.5±4.9 min, $P=0.0011$) than GNP lambs. There was no effect on MBS or Apgar test. The increment in BW and BCS of GO animals resulted in heavier lambs, that led to longer time of labor and less vigor compared with lambs of GNP ewes. Higher percentage of birth assistance in GO ewes suggests that the benefit of a greater weight at birth can be exploited only if control of parturition is performed. However, grazing oat pasture should be an alternative to be used with twin bearing ewes since lambs are usually lighter and have less vigor than single ones.

Infra-red thermography as a monitoring tool for detection of sub clinical mastitis in dairy ewes

M. Odintsov-Vaintrub[1], R. Di Benedetto[1], M. Chincarini[1], G. Giacinti[2], M. Giammarco[1], I. Fusaro[1], A. Merla[3] and G. Vignola[1]
[1]University of Teramo, Faculty of Veterinary Medicine, Località Piano D'Accio, 64100 Teramo, Italy, [2]Istituto Zooprofilattico del Lazio e della Toscana M. Aleandri, CReLDOC, 00178 Roma, Italy, [3]University of Chieti-Pescara, Department of Neuroscience, Imaging and clinical science, Via dei Vestini, 33, Chieti, 66100, Italy; modintsovvaintrub@unite.it

Infra-red thermography (IRT) is a no contact measurement technique of superficial temperatures (ST). It is used as a diagnostic or screening tool. In this study, sterile milk samples were taken from 236 clinically healthy milking ewes, one sample for each half-udder. They were analysed both for SCC as well as for bacteriological positivity. Sub-clinical mastitis was defined as: >500,000 SCC, and bacteriological test (+). Based on these results, animals with doubtful or incomplete data were ruled out, and the rest were divided into 3 groups: *A: both half-udders (-),102 ewes. B: one half-udder (-) one half-udder (+), 50 ewes. C: Both half-udders (+), 30 ewes*. IRT evaluation was done before sampling using a FLIR IR- vet, 420 esc. camera. The evaluation included 4 measurement areas; right and left half-udders with lower and upper area for each. Irrespective of the individuals, positive half-udders showed a significant difference of ST in the lower area compared to negative ones ($P<0.02$). However, it was not possible to define a temperature clearly indicative of a suspected subclinical mastitis. Nevertheless, comparing the delta temperature expressed in absolute value between the half udders in the same ewe showed a highly significant difference ($P<0.0001$) between the 3 groups. The ewes were then divided into 2 groups by placing a delta temperature *cut-off* (0.5 °C) in the lower mammary area. The results were as follows; 110 ewes resulted as 'mastitis unsuspected', 87 were correctly identified (based on the gold standard of bacteriological positivity) and 23 were false negatives. A further 72 ewes resulted as subclinical mastitis suspects (delta temp>0.5 °C), with 15 false positives and 57 correctly identified cases. Hence, adopting delta temperature as a criterion for animals suspected as having subclinical mastitis in at least one half udder showed a sensitivity equal to 71% (57/80), a specificity of 85.3% (87/102) and an accuracy of 79% ((57 + 87) / 182), while setting a cut-off at 0.5 °C.

Session 51 Poster 28

Analysis of voluntary intake of Virginia fanpetals (*Sida hermaphrodita* Rusby L.) silage

M. Fijalkowska[1], Z. Nogalski[2], Z. Antoszkiewicz[1] and C. Purwin[1]
[1]University of Warmia and Mazury in Olsztyn, Department of Animal Nutrition and Feed Science, Oczapowskiego 5, 10-719 Olsztyn, Poland, [2]University of Warmia and Mazury in Olsztyn, Department of Cattle Breeding and Milk Evaluation, Oczapowskiego 5, 10-719 Olsztyn, Poland; maja.fijalkowska@uwm.edu.pl

Virginia fanpetals (*Sida hermaphrodita* Rusby L.) can be a valuable protein source in ruminant diets due to its desirable agronomic characteristics and a high content of digestible protein. The aim of this study was to compare the palatability of Virginia fanpetals silage vs alfalfa silage and grass silage based on a preference test in sheep. Virginia fanpetals silage, alfalfa silage (first-cut herbage, budding stage) and grass silage (first-cut herbage, heading stage) were analysed. Voluntary feed intake was evaluated in the preference test. The feeding trial was performed on 8 adult rams, placed in individual pens. In the experimental period proper, the rams were divided into two equal groups (of 4 animals each). The voluntary intake of *Sida* silage was compared with alfalfa silage in group 1, and with grass silage in group 2. During the test, the animals had free access to all types of silage, offered at 5 kg in identical containers. Leftovers were collected and weighed 2, 4, 6, 8, 12 and 24 hours after the first feeding. The results of the preference test revealed that total daily intake and the rate of voluntary intake were higher ($P<0.001$) in Virginia fanpetals silage (1,427 g DM/d) than in wilted alfalfa silage (710 g DM/d) and in Virginia fanpetals silage (1,468 g DM/d) than in grass silage (601 g DM/d). An analysis of the feeding trial, and the chemical composition and quality of silage indicates that the higher voluntary intake of *Sida* silage resulted from its lower content of structural carbohydrates (ADF 328 g/kg DM vs 384 and 360 g/kg DM in alfalfa and grass silage, respectively). Positive effect of higher intake of *Sida* silage could also result of its palatable flavor, which is characteristic for this plant. The study was carried out in the framework of the project under a program BIOSTRATEG funded by the National Centre for Research and Development No. 1/270745/2/NCBR/2015 'Dietary, power, and economic potential of *S. hermaphrodita* cultivation on fallow land.' Acronym SIDA.

Session 51 Poster 29

Effect of different dietary protein sources on growth performance of Egyptian growing lambs

H. Metwally[1] and S. El-Mashed[2]
[1]Faculty of Agriculture, Animal Production, Ain Shams University – Shubra, Cairo, Egypt, [2]Animal Production Research Institute, Nutrition and Husbandry Department, Ministry of Agriculture, 11241, Egypt; sh.elmashed@hotmail.com

Fifty growing Barky male lambs were used to study the effect of different dietary protein sources on growth rates, rumen parameters and feed efficiency. Animals were divided into five experimental groups, ten lambs each. Groups were fed five different iso-nitrogenous, iso-energetic diets containing one of each protein sources. Protein sources were cotton seed meal (C.S.M.), bean beans (BB), Alfalfa hay (AH), sunflower seed meal (SSM) or linseed meal (LSM). Protein sources were evaluated for its solubility in three different solvents, for its in-vitro nitrogen disappearance and for electrophoresis analysis. Results ranked protein sources according to their solubility in distilled water, Mc.Dougal's buffer and rumen fluids. Significant differences were detected in in-vitro nitrogen disappearance between experimental protein sources. Results of electrophoresis analysis of protein sources showed that (SSM) contained more bands on the column (16) than other sources while Alfalfa hay included highest low molecular weight protein (112) and SSM had the highest high molecular weight Protein (135). Linseed meal contained the widest range between its contents of protein with a low molecular weight and that of high molecular weight (107). Total gain was significantly ($P>0.05$), the lowest (15.9 kg) for diet 2(BB) and the highest for diet 4(SSM) (17.5 kg). Feed efficiency was the best for diet 4 (containing SSM) (7.5 kg DM/kg group) while it was 8.392 kg DM/ kg growth in diet 5 (LSM).Rumen fluids analyses showed no significant differences in pH values. Protein sources had no significant effect on DM digestion in all groups. It was concluded that differences caused by protein sources in growth parameters of male growing lamps can be neglected and the effective factor should be the price of each source.

Session 52 Theatre 1

The contribution of livestock farming to the provision of ecosystem services

R. Ripoll Bosch[1] and E. Sturaro[2]

[1]Wageningen University, De Elst 1, 6708 WD Wageningen, the Netherlands, [2]University of Padova, DAFNAE, Viale dell'università 16, 35020 Legnaro, Italy; raimon.ripollbosch@wur.nl

Livestock production is at the spotlight in the debate of sustainable development. Livestock production is usually associated to environmental impacts, such as greenhouse gas emissions, land use or degradation of ecosystems. However, a growing body of literature acknowledges that certain livestock farming systems are multifunctional. Aside from food, they also provide a range of other functions and services that are valued and demanded by society, such as landscape maintenance, biodiversity preservation or forest fire prevention. The different nature of these services makes their inventory and accounting difficult. Moreover, the continued supply of services, which do not have a market value, is currently threatened. The ecosystem service (ES) framework is drawing attention as a method to evidence the importance of agro-ecosystems in delivering services and public goods to society. Nevertheless, its use and application is challenging. On the one hand, there are methodological challenges to overcome such as identification of all ES delivered, harmonisation in methodologies for accounting and measuring, understanding relationships between different ES or relate the delivery of ES in agroecosystems to farming intensity. On the other hand, there is the challenge to design and implement proper measures and policies to ensure the continued supply of ES that the society demands. The aim of this session is to set the scene on the contribution of livestock farming to the provision of multiple ecosystem services. The session will bring in a round table researchers on the field of ES and livestock production. The debate will revolve around the current research on ES and future prospects, the strengths and limitations of the ES framework and the methodologies to assess ES, how to move from a theoretical framework to the application in practice and the pathways for decision makers on how to implement measures to ensure the provision of ES by livestock farming.

Session 53 Theatre 1

Exploring farmer attitudes and preferences to inform the development and implementation of breeding

D. Martin-Collado[1], C. Diaz[2], A. Maki-Tanila[3], M. Wurzinger[4] and T. Byrne[5]

[1]CITA, Avda. Montañana 930, 50059 Zaragoza, Spain, [2]INIA, Ctra. Coruña, km. 7,5, 28040, Madrid, Spain, [3]Helsinki University, Yliopistonkatu 4, 00014 Helsinki, Finland, [4]BOKU, Borkowskigasse 4, 1190 Vienna, Austria, [5]AbacusBio Int.Ltd., Roslin Innov. Centre, Edinburgh, United Kingdom; dmartin@cita-aragon.es

Farmers' attitudes towards genetic improvement tools, and the consideration to farmers' preferences for improvements in animal traits in its design, are key drivers of uptake and of farmer participation in breeding programmes. In recent years, farmers' attitudes and views about genetic improvement, their trait preferences, and the link between these and implementation and application of genetic improvement programmes have gained increasing importance. The determination of farmers' trait preferences is not trivial and therefore many different method and approaches have been used to analyse them. On the other hand, farmers´ attitudes towards breeding tools have only been specifically studied in few occasions. We present and critically discuss the different approaches and methods that have been used to date to explore farmer's attitudes towards breeding tools and to analyse their traits preferences. We present two works as a framework for the discussion. The first work focuses on the analysis of Australian dairy farmers' preferences for improvements in cow traits to inform the 2014 review of the national breeding objective for the Australian dairy industry. The results of this work have been published elsewhere. The second work focused the development of a standard method to assess farmers' attitudes towards genetic improvement tools, which aim to tackle the lack of a reference measure of attitudes in this regard. The tool will allow to benchmark attitudes over time, and across different groups of farmers. It consists of a fixed set of attitudinal statements that comprises all the attitudinal positions existing in the animal breeding discourse. Farmers score the statements based on their agreement with them and their answers determine their attitude toward breeding tools. The set of statements is the core of a survey which also includes other additional data about the farm and the farmer. The survey has been tested with 4 sheep breeds and 10 beef breeds farmers in Australia, New Zealand and Spain and 625 surveys have been collected. We present the tool, analyse the results of the survey comparing farmer´s attitudes across breeds, livestock species, countries, and discuss how farming system and farmer profile influence farmers´ attitudes towards genetic improvement tools.

The value of government funded genetic improvement programs and why they make sense

T.J. Byrne[1], *P.R. Amer*[2], *C.D. Quinton*[2] *and F.S. Hely*[2]

[1]*AbacusBio International Limited, Roslin Innovation Centre, Easter Bush, EH25 9RG Midlothian, United Kingdom,* [2]*AbacusBio Limited, P.O. Box 5585, 9058 Dunedin, New Zealand; tbyrne@abacusbio.co.uk*

Many industrialised dairy, beef, and sheep information systems for genetic improvement programs are supported, in part, by government funding. This contradicts the more privatised models that exist in pig, poultry, and aquaculture breeding. There are several reasons why this is the case. Firstly, there is widespread market failure in sheep and beef and, to a lesser extent, dairy gene stock markets. That is, commercial users of improved genetic material often do not have the scale, or are not willing, to pay enough to support commercially functioning genetic improvement programs. This justifies governments stepping in. More recently there is an increased value proposition in governments funding information systems for genetic improvement because of increased awareness and need to deliver industry and/or social goods from genetic improvement. Examples of social good outcomes include reducing greenhouse gas and nitrate emissions from agriculture or providing the means to improve animal welfare outcomes such as polled in cattle. In this context, government-supported information systems present a viable option as an audit point for industry good traits (particularly when coupled with genomics); they are low administration for the farmer and can reduce capital costs associated with implementing auditing systems, for example, in greenhouse gas emissions on an individual farm basis. It is also important to recognise the potential for government funded information systems to enable a much quicker response to adverse situations and/or the means to help capitalise on opportunities. Examples include changes in the market (e.g. a geographical shift in demand or greater competition from other countries in export markets), changes in the production environment (e.g. due to new disease challenges, a more variable climate, farming within environmental regulatory limits), and changes in technologies (the applications of genomics with data from commercial flocks and herds). Government funded information systems fill a market gap and can provide industries with tools and options to increase economic and environmental efficiency and deliver social benefit.

Session 53 Theatre 3

Breeding objectives for honeybees in a honey production vs a commercial pollination operation

G.E.L. Petersen[1,2], *P.R. Amer*[1], *P.F. Fennessy*[1] *and P.K. Dearden*[2,3]

[1]*AbacusBio Ltd, P.O. Box 5585, Dunedin, New Zealand,* [2]*University of Otago, Department of Biochemistry, 710 Cumberland Street, Dunedin 9016, New Zealand,* [3]*Genomics Aotearoa, 710 Cumberland Street, Dunedin 9016, New Zealand; gpetersen@abacusbio.co.nz*

Western honeybees, *Apis mellifera*, are an important agricultural species, but have received considerably less attention in animal breeding than more traditional livestock species. In most cases, selection is performed *ad hoc* and based mainly on beekeeper intuition and experience. Structured breeding programs with the necessary understanding of the economic framework to guarantee long-term success are rare. Due to a stark difference in honey- and pollination fee-driven industries, profitability of commercial beekeeping operations is likely to be increased by different parameters. Here we present a breeding objective for honey production as well as a preliminary breeding objective for a pollination system. New breeding objectives were developed for a national honeybee genetic improvement project in New Zealand. In both cases, the basis for the objective was a profit function describing the influence of several traits on revenue and costs within the specified system. The honey production traits included honey production, temperament, brood viability, colony survival, queen survival and winter survival. The preliminary pollination traits included spring population build-up, temperament, brood viability, colony survival, winter survival and honey production. Linear breeding objective equations were derived using standard quantitative genetics methodology.

Session 53 Theatre 4

Genetic gain for production and maternal traits in a two-stage optimum contribution selection scheme

B. Ask, H.M. Nielsen, M.M. Castillero and M. Henryon
SEGES, Breeding and Genetics, Axeltorv 3, 1609 Copenhagen V, Denmark; bas@seges.dk

We showed that a two-stage selection scheme with optimum contribution selection (OCS) at both stages realizes more genetic gain for both production and maternal traits compared to a scheme with truncation selection in the initial selection stage and OCS in the final selection stage. This is important because preselection by truncation may assign more selection pressure to production than maternal traits, as maternal traits differ from production traits. Maternal traits are often recorded late in life after selection, they have low heritabilities, and they can be unfavourably correlated with production traits. To maximize genetic gain in two-stage selection schemes with both production and maternal traits, it may, therefore, be necessary to use OCS in both the preselection and final selection stage. We tested our premises by simulating two selection schemes with each two selection stages for sires. In the preselection stage, sires were selected by either truncation or OCS. The preselected proportion was 50, 25, 10, 5, 3, or 1%. In the final selection stage, sires were selected by OCS. Dams were truncation selected. Simulations were run for 25 generations. Each generation, there were 700 matings with 3 male and 3 female offspring per mating. All candidates were phenotyped for the production trait (h^2=0.20) prior to preselection, and selected dams were phenotyped for the maternal trait at the birth of the next generation of offspring. The heritability of the maternal trait was 0.20, 0.10, or 0.05, and the genetic correlation between the production and maternal trait was -0.5, 0, or 0.5. Preliminary results show that preselection by OCS results in higher aggregate genetic gains than preselection by truncation. This was achieved by increased genetic gain for the production trait, but without improvements in the genetic gain of the maternal trait. The heritability of the maternal trait did not influence this result. Consequently, preselection in a two-stage selection scheme should be done by OCS to improve aggregate genetic gains without loss of genetic gain in maternal traits.

Session 53 Theatre 5

Effects of changes in the breeding goal on genetic improvement for maternal traits in Landrace pigs

E.M. Eriksen[1], D. Olsen[2], I. Ranberg[2] and O. Vangen[1]
[1]Norwegian University of Life Sciences, Department of Animal and Aquacultural Sciences, P.O. Box 5003, 1432, Ås, Norway, [2]Topigs Norsvin Breeding Company, Storhamargata 44, 2317 Hamar, Norway; odd.vangen@nmbu.no

Genetic changes in seven traits related to maternal ability in Norwegian Landrace has been calculated based on breeding values from calculated by DMU, including genetic correlations between the traits. In 1992, the first maternal trait, litter size, was introduced, and BLUP breeding values became the basis for selection. Previously, in the breed's history, selection was phenotype-based for growth, efficiency and leanness traits. In 2003, 3-week weight (litter wt. adjusted for litter size) and in 2007 survival from birth to 3 weeks were introduced, and from 2010 the two last maternal traits shoulder sore (SS) and body condition score (BCS) at weaning were included in the breeding goal. The economic weights for litter size has increased from 18 to 28% in the breed through the period, survival and litter weight from 16 to 21% since being introduced, while SS and BCS has increased from 7 to 21% since introduction. Additionally, the variation in 3-week weights has been recorded from 2010 but not given breeding values. Genetic trends were calculated based on about 395,000 records for the first traits, down to about 150,000 records for the last introduced traits. The pedigree file includes 936,000 animals of which 40,000 were genotyped. The genetic trend for litter size have been positive the whole period, while the correlated responses in the other maternal traits were negative until introducing 3-week weights. Then the weights increased and the decline in survival stopped. However, survival increased only after recording of that trait started in 2007. When SS and BCS were included in the breeding goal, the genetic trends in these traits changed from negative to positive (from undesirable to desirable). The correlated responses in variation of litter weights is with today's breeding goal positive, meaning lower variation. Detailed results will be given on all traits, however, the results show that it is possible to obtain sustainable improvements in all maternal traits when reliable records are available and the traits are given relevant economic weighting in the selection program for a maternal line.

Development of an ecologic chicken breeding programme in Germany

B. Zumbach, M. Neyrinck and I. Günther
Ökologische Tierzucht gemeinnützige GmbH, Auf dem Kreuz 58, 86152 Augsburg, Germany; birgit.zu@gmx.com

Chicken breeding for both layers and broilers is in the hands of a few private breeding companies supplying hybrid chicken internationally. Intensive specialized breeding with the goal to increase market share has led to welfare problems, e.g. most of the broiler lines are not able to mate naturally any more, and male layer chicks are culled at hatch because of lacking economic value. An ecological chicken breeding programme has been set up by the company Ecologic Animal Breeding GmbH in Germany, which was founded in 2015 by the associations Bioland and Demeter. It is based on the following premises: (1) considering breeding a transparent, accessible and free cultural asset of common welfare; (2) ecologic feeding as a basis for all breeding animals without the addition of synthetic amino acids; (3) group husbandry according to ecological conditions (i.e. barns with roofed outside climate areas and natural light; (4) no individual cages for breeding animals, no artificial insemination; (5) in-ovo sexing is not considered a solution for the culling male layer chicks instead, male layer chicks are to be raised; (6) no preventive use of antibiotics; (7) no manipulation of beaks, combs and wings, (8) breeding maintaining intact animals (i.e. ability to fly, walk, mate naturally, no behavioural anomalies). Three lines with about 800 hens each are being kept as parent populations at the nucleus farm Bodden: White Rock (WR) and New Hampshire (NH) originating from lines bred for more than 25 years under ecological conditions at the University of Halle, and Bresse-Gauloise (BR) from the Demeter farm with breeding activities Rengoldshausen. Individual egg performance testing is based on trap nest control, where hens trigger the trap, i.e. closure of the door, when they enter the nest. The goal of this investigation is to characterize the performance of these three populations as a starting point for different breeding strategies towards a dual-purpose chicken and/or chicken with emphasis on egg performance, as well as to point out the challenges encountered. Results regarding egg performance and body weight will be presented.

Breeding programs to improve survival time in laying hens

E.D. Ellen and P. Bijma
Wageningen University & Research, ABG, P.O. Box 338, 6700 AH Wageningen, the Netherlands; esther.ellen@wur.nl

Mortality due to feather pecking is a worldwide problem in the laying hen industry, resulting in reduced survival time. Reduced survival time is a result of social interactions between group members, which has important consequences for breeding strategies. To improve survival time, it is important to use a selection method that takes into account both the genotype of the individual itself (direct genetic effect) and the genotype of its group mates (indirect genetic effect). So far, however, selection responses (ΔG) in practice were small, because (1) survival time is only known at end of laying and many hens are still alive (2) for selection candidates, their own performance records on survival time under field conditions are not available. Furthermore, following the phenomenon when selecting on a threshold trait, selection for improved survival could result in a reduction in heritable variance in survival time, which can have large consequences for ΔG. The aim of this study was to investigate different breeding approaches to improve survival time in laying hens and to investigate the effect of level of mortality on ΔG in survival time. A deterministic simulation based on selection index theory, using SelAction software, was used. Traditional recurrent testing design and genomic selection (GS) were compared. An index based on survival time and egg yield (which contains traits related to egg production) was devised, genetic correlation (r_g) was 0 or -0.3. To generate response curve of survival time and egg yield, weights varied from selection for survival time only to selection for egg yield only. Model predictions show that GS can yield a rapid improvement in survival time. Reducing ΔG in egg yield with 20% yielded an improvement in survival time of ~46 days/yr (r_g=0) or ~16 days/yr (r_g=-0.3). However, level of mortality had a large impact on total heritable variance in survival time. Improving survival yielded a substantial reduction in ΔG. In conclusion, our results show that it is in principle feasible to improve survival time by genetic selection. It depends, however, on the weight breeding companies give to survival time, relative to other traits, and on mortality rate in the population, whether such an improvement will be realized in practice.

Session 53 Theatre 8

Conservation of a native dairy cattle breed through crossbreeding with commercial dairy cattle breed

J.B. Clasen[1], S. Østergaard[2], E. Strandberg[1], W.F. Fikse[3], M. Kargo[4] and L. Rydhmer[1]
[1]Swedish University of Agricultural Sciences, Dept. of Animal Breeding and Genetics, Box 7023, 750 07 Uppsala, Sweden, [2]Aarhus University, Dept. of Animal Science, Blichers Allé 20, 8830 Tjele, Denmark, [3]Växa Sverige, Box 288, 751 05 Uppsala, Sweden, [4]Aarhus University, Dept. of Molecular Biology and Genetics, Blichers Allé 20, 8830 Tjele, Denmark; julie.clasen@slu.se

Over the last centuries, many native cattle breeds all over the world have become extinct or are close to extinction due to increased popularity of a few breeds. This has led to a remarkable loss of genetic diversity, which ought to be conserved in order to overcome potential bottlenecks that comes with lack of genetic variation. One way of conserving native breeds is by utilising them in crossbreeding schemes, which has shown favourable effects on especially functional traits and herd economy in dairy cattle, because of heterosis. Systematic crossbreeding may create a demand for purebred animals and thus an opportunity to avoid a shrinking population size for the native breed. However, native breeds are often not able to compete economically with modern dairy breeds, such as e.g. Holstein, thus the creation of such a crossbreeding strategy must be made attractive to farmers. Terminal crossbreeding of Swedish Polled Cattle with Swedish Red or Swedish Holstein may be a suitable strategy to conserve the Swedish Polled Cattle, whilst maintaining profitable herds that can survive under Swedish conditions. In terminal crossbreeding, the crossbred animals are only kept as production animals while only purebred animals are used for breeding. Simulation of management and economic consequences of this strategy will be performed with the SimHerd model, with data on the animals' performance, costs and incomes from Växa and other Swedish databases as input. Different scenarios will be compared. In the simulated herd, some of the cows are purebred Swedish Polled and the others are crossbred cows with Swedish Holstein or Swedish Red sires. Also, a herd will be simulated with only purebred Swedish Polled for comparison. Both conventional and organic farming systems will be simulated. In this study, Swedish Polled Cattle is used, but the results may be applicable to other native European dairy breeds.

Session 53 Theatre 9

Inclusion of candidate mutations in genomic evaluation for French dairy cattle breeds

P. Croiseau[1], C. Hozé[1,2], S. Fritz[1,2], M.P. Sanchez[1] and T. Tribout[1]
[1]GABI, INRA, AgroParisTech, Université Paris Saclay, Domaine de Vilvert, 78350 Jouy en Josas, France, [2]Allice, 149 rue de Bercy, 75012 Paris, France; pascal.croiseau@inra.fr

The 1000 Bull Genomes project offers an excellent opportunity to move from QTL detection to causal mutation identification. In this aim, all French genotyped animals were imputed to the sequence level and we performed GWAS for all the available traits in the three main dairy cattle breeds (Holstein, Normande and Montbéliarde). Based on the results of these analyses, main candidate mutations were selected and designed on the custom part of the 6th version of the EuroG10K SNP chip (10KV6). The aim of this study was to assess whether the use of the genotypes for these candidate mutations could improve the accuracy of genomic predictions, compared to the use of the Illumina Bovine SNP50BeadChip® (50K). For that purpose, the pool of genotyped animals was split into training and validation populations. The three following alternative scenarios were compared to the reference situation (use of the 50K chip): (1) Virtual 50K chips (V50K) were designed by replacing all or part of the 50K by candidate mutations identified from GWAS on imputed sequences from the run4 population of the 1000 Bull Genome project. Different strategies to define these V50K were proposed (scenario 1). (2) The 50K was augmented with the candidate mutations of the custom part of the 10KV6 chip. The genotypes of the bulls for these latter SNP being imputed using as reference population either the run4 population of the 1000 Bull Genomes project (scenario 2) or (3) the animals currently genotyped for the 10KV6 chip (scenario 3). At best, the various strategies tested in scenario 1 led to very limited gains in accuracy compared to the 50K. However, scenarios 2 and 3 resulted in substantial improvement in prediction accuracy for some traits. In addition, we applied scenario 3 on three regional dairy cattle breeds (Abondance, Tarentaise and Vosgienne) and we observed similar results.

Session 53 — Poster 10

Aceh cattle: between tradition and the need of sophisticated breeding strategy

T.S.M. Widi[1], N. Widyas[2] and I.G.S. Budisatria[1]

[1]Faculty of Animal Science, University of Gadjah Mada, Animal Production, Jl. Fauna No. 3 Kampus Bulaksumur, 55281 Yogyakarta, Indonesia, [2]Faculty of Agriculture, University of Sebelas Maret, Animal Husbandry, Jalan Ir. Sutami No.36A, Jebres, 57126 Surakarta, Indonesia; widi.tsm@ugm.ac.id

Aceh cattle is another Indonesian native breed with unknown origin. They carry a mixture of *Bos indicus*, *Bos javanicus* and a little bit of *Bos taurus* in their genetic constitution. Aceh cattle play important role for farmer livelihoods. This article discusses the phenotypic variation within Aceh cattle breeding center population as well as between the breeding centre and the common Aceh cattle owned by farmers. In total 995 records were obtained from Aceh Cattle Breeding Centre (BC) and common cattle (CC, owned by farmers) from both sexes and various ages (retrieved around year 2011-2013). The records contained information on observed body weight (BW), chest girth (CG), body length (BL) and wither height (WH). We divided age cattle into 3 classes: AG1 (<1 yr old), AG2 (1-2 yrs) and AG3 (>2 yrs). A linear model incorporating the systematic effect of age and groups (CC or BC) was built to analyse the data (for each sex separately). In AG1, CC were heavier (99.8±37.4 kg) compared to BC (92.3±25.5 kg) with $P<0.05$. The average BW for AG1 were 146.1±28.7 and 124.1±27.4 kg; AG2 were 169.4±27.0 and 108.7±26.4 kg for CC and BC, respectively. Also for female cattle, CC were heavier compared to BC (103.4±49.3 and 82.1±25.5; 187.0±48.3 and 116.5±40.2; 182.5±3.5 and 100.8±26.7) for AG1, AG2 and AG3, respectively. Aceh cattle has relatively small body with remarkable variations among individuals. The environment and management factors were suspected to play role in this. Moreover, the *'meugang'* (beef-feast event) requires large amount of male cattle to be slaughtered 3 times/yr. Traditionwise, people would only consume Aceh cattle. Without a proper breeding strategy, the event will eventually exhaust the superior male genetic resources. The breeding centre was offered as a solution to bring a better genetic resources management. However, it seems that the role of it has yet to reach the optimum; thus, breeding system's evaluation and improvement are needed.

Session 54 — Theatre 1

Combining data from multiple sensors to estimate daily time budget of dairy cows

M. Pastell and L. Frondelius

Natural Resources Institute Finland (Luke), Production Systems, Koetilantie 5, 00790, Finland; matti.pastell@luke.fi

Several sensors are currently available for measuring cow behaviour. The data is often aggregated to daily averages and those summary values are used for further analysis. The aim of this pilot study was to combine the data from three sensor system on more detailed time resolution to calculate the time budget of dairy cows in a freestall barn. The behaviour of 34 dairy cows housed in a freestall barn was monitored during 7 days using Ubisense UWB indoor positioning system, IceQube pedometers and Insentec RIC system for measuring feeding time. IceQube sensors were set up to record the timestamps of lying bout transitions, and the Insentec system recorded start and end time of each feeding bout. The positioning data was median filtered and interpolated to account for missing data. The data from all sensors were interpolated to 1 second time resolution and time aligned. The data was used to calculate daily time budget for each cow including derived measures: time that cows spend standing in lying stalls and lying time in the manure alleys. Cows spent (mean ± SD) 13.6±2.4 h lying, 3.8±0.84 h feeding and 2.0±0.46 h standing in stalls. In total these behaviours accounted for 19.5±1.5 h of daily cow behaviour. The rest of the time was either spent walking or standing, but additional processing of positioning data is needed to differentiate between these two behaviours. This study developed the methodology and software to combine data from multiple sensors. The results show that combining pedometer and positioning data gives additional information on stall use. Interpolated high accuracy behavioural data can also be used to calculate behavioural synchrony of the cows and may reveal new behavioural patterns. In the future we will utilize this data to link this information for monitoring health problems.

Session 54 Theatre 2

Modelling lactation curve with explicit representation of perturbations

P. Gomes[1,2], A. Ben Abdelkrim[2], L. Puillet[2], V. Berthelot[2] and O. Martin[2]
[1]NEOVIA, Talhouët, 56250 Saint-Nolff, France, [2]UMR Modélisation Systémique Appliquée aux Ruminants, INRA, AgroParisTech, Université Paris-Saclay, 16 Rue Claude Bernard, 75005 Paris, France; pierre.gomes@agroparistech.fr

Animal husbandry aims at managing the environment of animals to enable them to express their production potential. However, ruminants are often confronted with perturbations that affect their performance. Evaluating the effect of these perturbations on animal performance could provide metrics to quantify how animals cope with their environment and therefore better manage them. For dairy herds, milk production is an accessible data that has been extensively used to predict lactation curve. However, a typical milk yield time series rarely follow the ideal lactation curve. It often presents several stalls, with variable frequency, intensity and shape. In this study a lactation curve model with explicit representation of perturbations was developed. Fitting the model on milk time series generates two curves in only one run. The first represents a potential theoretical curve without perturbation and the second corresponds to the perturbed curve (explicit individual perturbations modelled by way of exponential functions). Perturbation characteristics (starting dates, intensity and shape) were estimated on complete lactation data from 10 individual goats which have undergone digestive perturbations. Results showed that 7±2.2 perturbations were detected by goat. Residual standard deviations were 52±6% lower than a classical milk production model. Loss of production due to perturbations was 16.7±12.7%. By incorporating explicit representation of perturbations, the model allowed the characterization of both potential milk production and deviations induced by perturbations, and thereby comparison of animals. These indicators are likely to be useful to move from raw data to decision solutions in dairy production.

Session 54 Theatre 3

A 2-step dynamic linear model for milk yield forecasting and mastitis detection

D.B. Jensen, M. Van Der Voort and H. Hogeveen
Wageningen University and Research, Hollandseweg 1, 6706 KN Wageningen, the Netherlands; dan.jensen@wur.nl

Milk yields which deviate from their expected patterns can be indicative of dairy cow diseases, including mastitis. In general, the milk yield models currently described in the scientific literature implicitly assume constant time intervals between milking sessions. When milking robots are used, however, milking intervals are not fixed. We expect that a model, which considers how milk yield is affected by the intervals between milkings as well as the milk production curves, is likely to provide more useful health information than the more naive models currently described in the literature. Here we present a dynamic linear model (DLM) designed to forecast milk yields of individual cows as they are milked in milking robots. The DLM forecasts the yield for a milking in two steps: first a Wood's lactation function is used to model expected daily milk yield. Secondly, the expected fraction of the total daily milk yield is estimated for a given milking by a second degree polynomial function of the time interval since the last milking. We wanted to investigate whether this DLM could be used as part of an online mastitis detection system, and whether or not farm-specific implementations would be needed for this purpose. To this end, farm-specific implementations were created for 5 different dairy farms which had at least weekly on-line measurements of somatic cell count (SCC) for the individual cows. A mixed effects model was then used to describe the standardized forecast errors of the DLM given the lactation stage, SCC level, and whether or not the proper farm-specific version of the DLM was used. Only the SCC level and interactions between SCC level and lactation stage significantly affected the value of the standardized forecast errors. We therefore conclude that forecasts of milk yield, made using our DLM, can be a useful part of a mastitis detection system. Furthermore, we conclude that it is not necessary to create a farm-specific version of the DLM for this purpose. Further research should look into comparing the mastitis prediction utility of this model with those of other models currently described in the scientific literature.

Session 54 — Theatre 4

Monitoring growth of identified and unidentified pigs using data from an automatic weighing system

A.H. Stygar[1,2] and A.R. Kristensen[2]
[1]Natural Resources Institute Finland (Luke), Economics and society unit, Latokartanonkaari 9, 00790 Helsinki, Finland, [2]University of Copenhagen, Department of Veterinary and Animal Sciences, Grønnegårdsvej 2, 1870 Frederiksberg C, Denmark; anna.stygar@luke.fi

In pig farms equipped with frequent body weight (BW) monitoring, such as walking through scales, more than 60,000 BW observations can be collected per batch. So far, farmers were using collected BW information mostly for pig marketing purpose (through a BW sorting system). However, these observations can also be used to alert the farmer about a sudden decline in pigs' growth (at individual or batch level), and to provide current and historical growth statistics for batches. The main obstacle in providing alarms was the fact that most farms are not identifying pigs. An additional difficulty was selection bias caused by the fact that farmers continuously sell the heaviest pigs from a batch as soon as they reach a defined BW threshold. In this study we have developed a tool which can be used to provide growth alarms at individual level (identified pigs) and batch level (identified or unidentified pigs). Moreover, the developed tool can provide current statistics on growth and perform retrospective analyses. The tool was built as a dynamic linear model (DLM). The growth of pigs was described by parameters at batch level representing a quadratic growth function with daily harmonic fluctuations in BW and an autoregressive effect at pig level. In the version for unidentified pigs, we have removed parameters describing individuals. However, we have accounted for the variance in the pig effect by adding it to the random residual. After the first delivery, BW of the remaining pigs was described by a truncated normal distribution. The forecast errors obtained from the DLM were standardized and monitored with a tabular cusum. To test the tool we have used data from 3 batches, 1,058 pigs and 146,926 individual weighings. We have demonstrated that frequent observations of unidentified pigs can provide meaningful alarms on growth deterioration at batch level. Frequent BW information can be useful in informing farmer about unexpected events influencing growth e.g. outbreaks of diseases. The historical information on growth might be valuable in making decisions regarding management.

Session 54 — Theatre 5

Precision feeding with a decision support tool dealing with daily and individual pigs' body weight

N. Quiniou[1], M. Marcon[1] and L. Brossard[2]
[1]IFIP-Institut du Porc, BP35, 35650 Le Rheu, France, [2]PEGASE, INRA, Agrocampus-Ouest, 35042 Rennes, France; nathalie.quiniou@ifip.asso.fr

Nutritionists, feed companies and equipment manufacturers look for solutions that help farmers to improve sustainability of pig production. Based on experimental results obtained in silico or *in vivo*, a better adequacy between amino acid supplies and requirements increases feed efficiency and farmer's income and reduces the environmental impact of growing pigs, highlighting the interest for precision feeding. Data are collected to characterize daily animal traits (e.g. body weight, BW) and their variation from one day to another (e.g. growth rate, ΔBW). They are used to determine the requirement for maintenance and growth on the next day, respectively. Therefore, adequacy between requirements and supplies depends on these predicted BW and ΔBW. The double exponential smoothing (Holt-Winters) method with a smoothing parameter α=0.6 ($HW_{0.6}$), presents a low sensitivity to the number of latest values used to forecast BW. It seems to allow for a secured prediction of BW soon after the beginning of the growing phase (at least after 4 days). A group of pigs was used in restricted feeding conditions to compare results obtained either with a 2-phase feeding strategy, considered as the control treatment, or a precision feeding strategy based on BW forecasting with the $HW_{0.6}$ method. Pigs allocated to both treatments were group-housed in the same pen, equipped with the decision support system built in the Feed-a-Gene project to manage the data, to determine in real-time the corresponding nutritional requirements, and to adapt the feed characteristics provided to each pig through the blend of two diets (9.75 MJ net energy/kg, 0.5 or 1.0 g of digestible lysine per MJ). Available results from 24 pigs per treatment indicate that overall average growth performance were not influenced by the feeding strategy ($P>0.58$ for both average daily gain and feed conversion ratio) but digestible lysine intake was reduced by 6% (1,774 vs 1,879 g, $P<0.01$) and N output by 7% ($P<0.01$) with precision feeding. Results will be completed by a second group using the same treatments. This study is part of the Feed-a-Gene project and received funding from the European Union's H2020 program under grant agreement no. 633531.

From raw data to optimal sow replacement decisions – an integrated solution

J. Hindsborg and A.R. Kristensen
University of Copenhagen, Dept. of Veterinary and Animal Sciences, Grønnegårdsvej 2, 1870 Frederiksberg C, Denmark; jehi@sund.ku.dk

Several sow replacement models (SRM) have been proposed in literature, aiming to offer a decision support tool for pig farmers. The application of SRMs in practice however, has been a challenging procedure as its implementation requires significant knowledge about the calibration process to the individual herds. Despite the comprehensive amount of livestock data that is often available, it has not been combined with a replacement model in order to yield an integrated and data-driven solution, until now. We present an application which connects directly to the farmer's local management database, from where data is extracted and applied in three different dynamic models. Using historical data, the models yield production estimates on litter size, farrowing and mortality rates and are updated in line with new data being recorded, which subsequently allows for the automatic calibration of the SRM. Each sow is evaluated in the SRM with its individual history, and is assigned an economic value and a potential which expresses whether the sows' litter size history surpass or fall behind the herd average for its corresponding parity. In the application, all sows are organized into two tables, where the first is a list of all active sows which can be utilized to optimize the current culling strategy, by replacing sows that generate negative value (i.e. decision support). The second is a list of all culled sows with an associated economic assessment of the historical losses due to non-optimal culling. This assessment is divided into three categories: culled too early, dead sows (lost slaughter value) and culled gilts. Both tables are supported by sorting functions in terms of dates, value, parity, etc. Additionally, the three dynamic models also serve as a dynamic production monitoring tool. The results of these three models are locally stored in a database (specific to each herd), which allows the user to evaluate historical progress or decline in various parameters.

Machine learning algorithms for lamb survival and weaning weights

B.B. Odevci[1], E. Emsen[2] and M. Kutluca Korkmaz[3]
[1]Kadir Has University, Management Information Systems, Istanbul, 34083, Turkey, [2]Ataturk University, Department of Animal Science, Erzurum, 25240, Turkey, [3]Ataturk University, Ispir Hamza Polat Vocational School, Ispir, Erzurum, 25900, Turkey; bahadir.odevci@gmail.com

Lamb survival is a complex trait influenced by many different factors associated with management, climate, behaviour of the ewe and lamb, and other environmental effects. Lamb survival and weaning weights of lambs are important issues in high altitudes and cold climates. In this study, machine learning algorithms were applied into the behavioural (dam-lamb) and productive traits effecting lamb survival and weaning weights of crossbreed lambs produced in high altitude and cold climate region of Turkey. The data set included 69 Romanov F1 lambs from 55 fat tailed indigenous ewes (Awassi=23 and Morkaraman=37; 2-5 ages) managed in semi intensive system at the Ataturk University Experiment Station, Erzurum, Turkey. Lambs were born in February-March (winter) under shed lambing conditions. Sources of data on dam and lambs behavioural factors were used as stated by Emsen *et al.* and Dwyer. Lambs were weaned at 60 days of age. Factors included were dam body weight at lambing, age of dam, litter size at birth, lamb birth weight, lamb sex, lambing assistance, maternal and lamb behaviours, weaning weights of lambs. In lamb survival and weaning weights, we used classification algorithms which were implemented using The Waikato Environment for Knowledge Analysis (WEKA). The most successful classification algorithms applied for lamb survival and weaning weights was MultilayerPerceptron with 100 and 41% accuracy rates, respectively. Trees. Variables were categorized for lamb survival, lamb behaviour, mothering ability. Within the classification Trees, Ramdomtree for lambs survival and J48 for weaning weights of lambs clearly outperformed all other methods with 96 and 41% accuracy rates, respectively. Our results showed that machine learning algorithms we used have better predictive power in classifying lamb survival than those for weaning weights of lambs.

Session 54 Theatre 8

Towards an integrated system from data to a solution or decision

M. Pastell[1] and R. Kasarda[2]
[1]luke.fi, Finland, [2]uniag.sk, Slovak Republic; matti.pastell@luke.fi

Towards an integrated system from data to a solution or decision; a debate There are quite a few sources of data: PLF sensors generates huge amount of data on animal individual level, farm management software, data along the chain, and big data from other sources. The debate will discuss common issues that will be raised during the session. The session speakers and posters' authors will form a panel-of-experts that will answer questions from the audience and identify common research challenges and potential engineering solutions. 15 minutes just before coffee/diner.

Session 55 Theatre 1

Effects of heat stress in dairy cattle on the performance of their offspring

A. De Vries[1], S. Tao[2], J. Laporta[1], F.C. Ferreira[1] and G.E. Dahl[1]
[1]University of Florida, 2250 Shealy Drive, Gainesville, FL 32608, USA, [2]University of Georgia, 2360 Rainwater Road, Tifton, GA 31793, USA; devries@ufl.edu

Exposure to elevated temperature and humidity during late gestation in dairy cattle can program reduced yields in the subsequent lactation largely through compromised mammary growth in the dry period. In a review of our own work and the literature, we found that cows exposed to heat stress during the dry period produced up to 5 kg/d less milk in the subsequent lactation than cows that received evaporative cooling. There is also emerging evidence that adult capacity for milk synthesis can be programmed in the calf that the dam is carrying by events during the calf's foetal life occurring at least two years before she starts milk production. Specifically, calves born to dams that are heat stressed for the final 6 weeks of gestation produced 19% less milk in the first lactation compared to calves from dams provided with evaporative cooling. The increased milk production in animals born to dams under evaporative cooling occurred without a greater decline in body weight that accompanies negative energy balance during early lactation. Thus, the increase in milk production suggests an increase in the efficiency of conversion of feed to milk. These data indicate that a brief period of heat stress late in development reduces the physiological efficiency of the cow to result in a substantial decline in productivity. There is also some evidence that the granddaughters of cows under heat stress in late gestation produce less milk, and further that heat stress during the time of conception could affect mature production of the resulting calf. The economic losses associated with not cooling dry cows have long been underestimated in the USA, given that farmers in hot climates often focused on cooling their lactating cows but not their dry cows. Investment in cooling of dry cows is profitable everywhere in the USA except Alaska when only the effects on milk production in the subsequent lactation are included. The importance of cooling dry cows in hot climates is even greater when the effects on their offspring are included.

Session 55 Theatre 2

Beef cattle thermoregulation in response to naturally occurring heat stress on pasture

R.M. Mateescu, H. Hamblen, A.M. Zolini, P.J. Hansen and P.A. Oltenacu
University of Florida, Animal Sciences, 2250 Shealy Drive, Gainesville, FL 32611, USA; raluca@ufl.edu

Heat stress is a cause of major economic losses to cattle producers, especially in tropical and subtropical environments. The objectives of this study were to assess the phenotypic variability in core body temperature and sweating rate and to evaluate the effect of coat type, temperament, and weight on core body temperature and sweating rate in Brangus heifers. During August and September of 2016, 725 Brangus heifers on pasture were evaluated in four separate groups (n=200, 189, 197, and 139). Environmental measurements of dry bulb temperature (Tdb) and relative humidity (RH) were measured every 15 min during the entire time of data collection and the temperature-humidity index (THI) was used to quantify heat stress conditions. Coat score, sweating rate, chute score, exit score, and live weight were recorded as the animals passed through the chute. Vaginal temperature was recorded every 5 minutes for 5 consecutive days. There was significant variation in vaginal temperature between heifers in the same environmental conditions (σ^2u=0.049), suggesting opportunities for selective improvements. A repeatability of 0.47 and 0.44 was estimated for sweating rate and vaginal temperature, respectively, suggesting that one measurement would be able to adequately describe the sweating capacity or ability to control the body temperature of an individual. Vaginal temperature increased as THI increased, with approximately one hour lag time in the animal's response. Vaginal temperature was lower (P=0.015) and sweating rate was lower (-5.49±2.12 g/(m^2·h), P<0.01) for heifers that demonstrated a calmer behaviour in the chute. Animals with shorter, smoother hair coats had significantly lower vaginal temperatures when compared to animals with longer hair coats (P<0.01). Heifers weighing more maintained a significantly lower vaginal temperature (<0.0001). Our results showed that hair coat, temperament, and weight influenced vaginal temperature regulation.

Session 55 Theatre 3

Physiological and production responses of Tunisian Holsteins under heat stress conditions

H. Amamou[1], M. Mahouachi[2], Y. Beckers[1] and H. Hammami[1]
[1]Gembloux Agro-Bio Tech, University of Liège, AgroBioChem, Gembloux, 5030, Belgium, [2]High School of Agriculture of Kef, University of Jendouba, Le Kef, 7119, Tunisia; hamamou@doct.uliege.be

The objective of this study was to measure the physiological and production responses of dairy cattle under high thermal stress conditions of the Mediterranean climate. Two experiments were conducted in four farms located in sub-humid, semi-arid, and arid regions of Tunisia respectively. The first experiment was conducted for 45 days from July to August under heat stress conditions (HS, 22 to 35 °C) and the second was run for similar days (from November to December) under thermal neutral conditions (TN, 5 to 17 °C). During each experiment, 20 cows in each farm were used. Respiration rate (RR), skin shoulder (SS) and skin rump (SR) temperatures were measured at 09:00, 11:00, and 14:00 h. Rectal temperature (RT), milk yield and milk samples were recorded only at 14:00 h. Meteorological data (temperature and relative humidity) were measured inside the barns along both experiments periods at every 10'mn. Under HS conditions, RR (61 breaths/min), SS (37.6 °C), and SR (37.7 °C) were 2.3, 1.3, and 1.3 fold higher compared with TN conditions. When compared to TN conditions, milk yield, fat and protein percentages were lower under HS conditions (6.7 kg, 2.4%, 2.7% vs 8.4 kg, 3.6%, 3.2%), whereas cows exhibited higher somatic cell scores under the HS period (5.6 vs 4.6). An increasing in HS from THI 71 to THI 87 (maximal THI) simultaneously increased RR by 2 breaths/min, SS by 0.5 °C, SR by 0.5 °C, and RT by 0.2 °C, and decreased milk yield by 0.1 kg per increase in one THI unit.THI was highly correlated with all physiological traits used in this study. In conclusion, HS sensitivity was identified for all cows through alterations of physiological parameters, reduced milk yield and altered milk composition of Holstein cows. Studying fluctuations of these parameters might help to develop adaptation strategies facing the climate change challenges in the Mediterranean region.

Session 55 Theatre 4

Relationship between foot and claw health traits and conformation traits in Czech Holstein cattle

L. Zavadilová, E. Kašná, P. Fleischer, M. Štípková and L. Vostrý

Institute of Animal Science, Genetics and Breeding of Farm Animals, Přátelství 815, 104 00 Prague 10, Czech Republic; zavadilova.ludmila@vuzv.cz

The heritabilities for different types of foot and claw disorders and their association with selected conformation traits were estimated in the Czech Holstein cattle by applying logistic models. The disorders from 23,850 lactations of 9,422 Holstein cows were recorded on 7 farms in the Czech Republic from 1999 to 2016. Three groups of foot and claw disorders were defined: skin diseases (INF), including digital and interdigital/superficial dermatitis, and interdigital phlegmon; claw disorders (CD) including ulcers, white line disease, horn fissures, and double sole; and overall foot and claw disorders (OFCD) comprising all the recorded disorders. Phenotypic data on linear scoring were available for conformation traits of cows: body condition score (BCS), stature, rear legs rear view, rear leg set (side view), foot angle, locomotion, and feet & legs score. Dichotomous response variables were the presence or absence of foot and claw disorders in lactation. Lactational incidence rates (number of affected lactations / number of lactations at risk × 100) for claw disorders (CD), skin diseases (INF), overall foot and claw disorders (OFCD) were 19.2, 25.3, and 44.5%, respectively. A linear logistic model with a binary variable was used for subsequent statistical analysis. Effects of herd of calving, year of calving, season of calving and age at calving were highly significant for all disorder groups. Estimates of heritability were 0.194 for CD, 0.168 for INF, and 0.136 for OFCD. The groups of foot and claw disorders were strongly associated with body condition score (BCS), rear legs rear view, rear leg set (side view), locomotion, and feet & legs score, weak associations were found with the foot angle and stature. The highest risk for CD was associated with poor BCS (odds ratio, OR=1.76), sickled legs (OR=1.66), 65-70% feet & legs score (OR=1.58), and severe abduction and/or short stride (OR=1.46). INF was mostly associated with poor BCS (OR=1.50), 65-70% feet & legs score (OR=1.48), severe toe out (OR=1.32), and severe abduction and/or short stride (OR=1.35). The work was supported by the project QJ1510144 and MZE-RO0718 of the Ministry of Agriculture of the Czech Republic.

Session 55 Theatre 5

Survival analysis for health traits in German Holstein dairy cows

T. Shabalina, T. Yin and S. König

Justus-Liebig-University of Giessen, Institute of Animal Breeding and Genetics, Ludwigstrasse 21B, 35390 Giessen, Germany; taisiia.shabalina@agrar.uni-giessen.de

In routine genetic evaluations for longevity, type traits are widely used as early predictors for cow survival. As an alternative, health traits from images, sensor technology or diagnosis keys according to ICAR guidelines, might be valuable predictors in survival analyses. Such health traits have been recorded in large-scale German test herds since ~10 yrs. In this study, we used survival analysis with a Weibull proportional hazard model to evaluate the effects of 13 different disease traits on functional longevity (FL). Culling data and diseases were from 129,386 German Holstein cows, kept in 57 test herds. Calving years of cows ranged from 2004 to 2017. The no. of censored cows was 39,171 (30.27% of the total dataset). FL was defined as the period from the first calving until culling date. Average productive life of cows with lifetime data was 977.10 d (±626.77 d). Within lactation, we created three different stages: From calving to DIM 59, form DIM 60 to DIM 299, and from DIM 300 to the next calving date. The Weibull model included time-independent effects of the herd and age at first calving, and time-dependent effects including calving year, calving season, milk yield from the first test-day, and the cow health status within lactation stage. The relative risk for involuntary culling for cows with clinical mastitis was moderate for diseases in parity 1, but was an obvious culling reason for infected cows at second stages in parities 2 and 3. Same trends were observed for subclinical mastitis. Interestingly, cows with the claw disease 'dermatitis digitalis' during the first stage in parity 1 had lower culling risks compared to healthy cows. Impact of metabolic diseases on culling risks was not significant. By trend, ketosis and acidosis from the first stage were associated with shorter FL. Female fertility disorders from parities 2 and 3 had negative impact on FL. The heritability of FL was 0.33. Genetic correlations between FL and all diseases were favourable in the range from -0.41 (clinical mastitis from second stage in parity 1) to -0.002 (subclinical mastitis in first stage in parity 1).

Disposal reasons as potential indicator traits for direct health traits in German Holstein cows

J. Heise[1], K.F. Stock[1], S. Rensing[1] and H. Simianer[2]
[1]IT Solutions for Animal Production (vit), Heinrich-Schroeder-Weg 1, 27283 Verden, Germany, [2]University of Goettingen, Animal Breeding and Genetics, Albrecht-Thaer-Weg 3, 37075 Goettingen, Germany; johannes.heise@vit.de

Disposal reasons of dairy cattle have been routinely recorded via the German milk recording system for decades and are available almost population-wide for Holsteins. Studies on patterns of distributions of disposal reasons across different lactations have indicated that disposal reasons might be meaningful indicators for some health traits like udder diseases, metabolic diseases, and claw and leg disorders. Using disposal reasons in genetic evaluations of direct health traits is therefore expected to increase the accuracy of estimated breeding values (EBV). In this study, the following disposal reasons were considered: 'poor performance', 'infertility', 'udder diseases', 'poor milkability', 'claw and leg disorders', 'metabolic diseases', 'other diseases', and 'other reasons'. The reasons 'sold for dairy purposes' and 'age' were not considered. Survival of the first lactation was defined, dependent on the respective disposal reason (0, if the cow was culled for this reason; 1, if the cow was culled for another reason or survived the first lactation). Holstein cows (black-and-white and red-and-white) were included, if they were born after 1994. A total of 11,263,102 cow records were available for the analyses. Based on 2,893,212 cow records from Eastern-German herds, variance components were estimated for each trait separately, using a sire model with a fixed herd × year effect. Genetic evaluations were also performed univariately, using the same animal model with fixed effects of herd × year × season and region (individual federal states of Germany, Luxembourg, Austria) for each trait. Estimated heritabilities ranged from 0.58% ('poor milkability') to 1.88% ('udder diseases'). EBV showed generally plausible patterns of correlations to routinely evaluated traits, e.g. 'udder diseases' had a high correlation to the RZS (somatic cell score, 0.54), and 'infertility' to the RZR (reproduction index, 0.54). Multivariate analyses with direct health traits will provide correlation estimates that help verifying the obvious potential of using disposal reasons as indicators for health in German Holstein cows.

Combining clinical mastitis with somatic cell indicators for udder health selection in Spanish cows

M.A. Peréz-Cabal[1] and N. Charfeddine[2]
[1]Complutense University of Madrid, Department of Animal Production, Madrid, 28040, Spain, [2]CONAFE, Spanish Holstein Association, Valdemoro, 28340, Spain; mapcabal@vet.ucm.es

A recording system at a large scale for clinical mastitis (CM) events is not generally available and therefore, mastitis resistance has been improved in most countries by indirect selection using somatic cell count (SCC). In Spain, CM have been recorded since 2012 in 393 herds in the region of Galicia. Test-day SCC data were used to define traits that could be used in a multi-trait evaluation combined with CM. The first one is an average of somatic cell score (ASCS), a log-transformation of SCC. The other traits were defined as binary depending on if any test-day SCC exceeded or not a threshold (peak) during the time period: 100,000 cells/ml (SCC100), 200,000 cells/ml (SCC200), and 400,000 cells/ml (SCC400). All traits were defined for first and later lactations, and by periods of DIM. For first lactation, three periods were defined (P1: from 15 days before calving to 60 DIM; P2: from 61 to 150 DIM; P3: from 151 to 305 DIM). For later lactations, only two periods were defined (P1: from 15 days before calving to 150 DIM, and P2: from 151 to 305 DIM). The heritability of CM was 2.8 and 2.3% and heritability of ASCS was 11.2 and 14.3%, for first and later lactations, respectively. The heritabilities of peaks in SCC showed a trend such that the higher the threshold the higher the heritability. Genetic correlations between CM in first and later lactations was 0.99, indicating that in our data both were the same trait. Correlations between CM in different periods in first lactation ranged from 0.56 to 0.9, while for later lactations was 0.98. Genetic correlations between CM and ASCS were 0.85 and 0.94 for first and later lactations, respectively. In first lactation, the correlations between CM and the peaks of SCC were higher than 0.90 and ranged between 0.78 and 0.87 for later lactations, such that the higher the threshold the higher the correlation. The present study has shown that the definition of different traits for CM in first and later lactations is not necessary but for different periods within the same lactation would be worthy. Both ASCS and SCC200 are useful traits which should be included with CM in multi-trait evaluation for udder health improvement.

Session 55 Theatre 8

Investigations of the variation of horn phenotypes and the genetic architecture of scurs in cattle

L.J. Gehrke[1], D. Seichter[2], I. Ruß[2], I. Medugorac[3], C. Scheper[4], S. König[4], J. Tetens[5] and G. Thaller[1]
[1]Christian-Albrechts-University Kiel, Olshausenstraße 40, 24098 Kiel, Germany, [2]Tierzuchtforschung e.V. München, Senator-Gerauer-Str.23 a, 85586 Grub, Germany, [3]LMU Munich, Veterinaerstr. 13, 80539 Munich, Germany, [4]University of Kassel, Ludwigstraße 21B, 35390 Gießen, Germany, [5]Georg-August-University, Burckhardtweg 2, 37077 Göttingen, Germany; lgehrke@tierzucht.uni-kiel.de

Due to serious animal welfare concerns regarding dehorning of calves, the interest in breeding genetically hornless (i.e. polled) cattle has increased considerably. Polledness is supposed to be inherited in an autosomal dominant manner, but occasionally individuals with an unexpected phenotype (scurs) emerge. Scurs are described as incomplete developed horns, which are not fused to the frontal bone. Intriguingly, scurs occur in genetically polled animals only. The *polled* locus was identified a few years ago. Nevertheless, a conclusive explanation for the occurrence of scurs has not yet been found and the mode of inheritance of horns is still under debate. The most accepted model assumes a second locus, the *scurs* locus, which is interacting with the *polled* locus. Interestingly, two additional phenotypes seem to be affected by the *polled* locus, atypical eyelashes and poll shape. To investigate the variation of horn phenotypes, we surveyed 854 polled Holstein Friesian cattle. The horn phenotypes of polled cattle appeared to be more complex than expected. The animals were classified into four categories: 'smoothly polled', 'small frontal bumps', 'frontal bumps' and 'scurs'. Furthermore, the poll shape and the eyelashes were surveyed and were categorized in 'flat', 'slightly peaked', 'peaked', 'extremely peaked' and in 'ordinary', 'atypical' (e.g. double rows of eyelashes, bushy eyelashes), respectively. For all animals a direct gene test for polledness was performed. Subsequently, to study the genetic architecture of scurs, a data set of 240 animals were HD-Chip genotyped. The HD-genotyped animals were all female and heterozygous polled. Of those animals, 31 had scurs, 132 individuals had frontal bumps and 77 were smoothly polled. A case control approach will be used for genome-wide association studies.

Session 55 Theatre 9

Relationships between subclinical ketosis, BCS, fat-protein-ratio and other diseases in Fleckvieh

B. Fuerst-Waltl[1], A. Köck[2], F. Steininger[2], C. Fuerst[2] and C. Egger-Danner[2]
[1]University of Natural Resources and Life Sciences Vienna (BOKU), Gregor-Mendel-Strasse 33, 1180 Vienna, Austria, [2]ZuchtData EDV-Dienstleistungen GmbH, Dresdner Strasse 89/19, 1200 Vienna, Austria; birgit.fuerst-waltl@boku.ac.at

This study is part of a larger project whose overall goal is to evaluate the possibilities for genetic improvement of efficiency in Austrian dairy cattle. In the year 2014 a one-year of extensive data collection was carried out. In addition to routinely recorded data (e.g. milk yield, fertility, disease data, etc.), data of novel traits such as subclinical ketosis (detected by using the milk Keto-Test from ELANCO), body condition score, body weight, lameness, claw disorders, body measurements, mid-infrared-spectra as well as individual feeding information and feed quality were recorded. The specific objective of this study was to analyse phenotypic and genetic associations between subclinical ketosis, body condition score, fat-protein-ratio and other diseases in Austrian Fleckvieh cows. Phenotypic relationships revealed that cows with a positive milk Keto-Test result had a lower body condition score during lactation, a higher fat-protein-ratio and a higher risk of other diseases. Heritability of subclinical ketosis was 0.05. Clinical ketosis and milk fever had a heritability of 0.007 and 0.020, respectively. For body condition score and fat-protein-ratio higher heritabilities of 0.17 and 0.14, respectively, were found. Genetic correlation estimates between traits were consistent with phenotypic associations. Metabolism with subclinical and clinical symptoms is complex and different parameters are used to describe this complex. The results showed that different information sources and traits can be used to improve the metabolic disease resistance of dairy cows.

Session 55 Poster 10

Thermal comfort indexes and physiological variables of Caracu cattle under thermal stress

S.B.G.P.N.P. Lima[1], A.P. Freitas[1,2], N.T. Bazon[1], R.P. Savegnago[1], R.H. Branco[1] and C.C.P. Paz[1,2]
[1]Institute of Animal Science, Beef Cattle Research Center, Rodovia Carlos Tonanni, km 94 Sertãozinho – SP, 14160-900, Brazil, [2]University of Sao Paulo, Medical School, Av. Bandeirantes, 3900 Monte Alegre, Ribeirão Preto, SP, 14049-900, Brazil; sergio.bgpnpl@hotmail.com

The objective of this study was to evaluate the effect of heat stress in Caracu cattle for animals under the shade and exposed to the sun using thermal comfort indices. The study was conducted at Instituto de Zootecnia, Sertãozinho, SP, Brazil. A total of 40 animals were used to measure the heat stress when the black globe temperature was higher than 40 °C. The environmental variables were dry bulb temperature, black globe temperature, and relative humidity. The environmental variables were transformed to Temperature-Humidity Index (THI), Black Globe-Humity Index (BGHI), and Heat Tolerance index (HTI). The variables measured on the animals were rectal temperature and back surface temperature. Four evaluations were carried out: two in October and two in December 2017. The records were collected from 08 to 10 h for all the 40 animals, and from 10:30 to 15 h in which a group of 20 animals were exposed to the sun for two hours (stress period), and the other group of 20 animals were submitted to the shade for the same time. The animals that were exposed to the sun in October were submitted to the shade in December and vice-versa. Data were analysed by GLM procedure of SAS. The THI values ranged from 87 to 93 under the sun and from 80 to 84 under the shade. The BGHI values ranged from 87 to 89 under the sun and from 81 to 84 under the shade, indicating that the animals were in heat stress in both spots, but in worse conditions of stress when were exposed to the sun. On the other hand, the HTI values indicated that the animals under the shade presented greater heat comfort (9.84) compared to animals under the sun (9.59) (<0.01). The back surface temperature was 41.2 °C under the sun and 38.0 °C under the shade ($P<0.01$), indicating that the sun exposure, due to the incidence of solar radiation, increased the body surface temperature on 3.2 °C, compared to animals under the shade. These results indicated that is important for Caracu animals be placed in the shade when they are under tropical climate conditions.

Session 55 Poster 11

Correlation between heat tolerance and residual feed intake in Caracu cattle under thermal stress

S.B.G.P.N.P. Lima[1], A.P. Freitas[2], R.P. Savegnago[1], N.T. Bazon[1], J.N.S.G. Cyrillo[1], J.A. Negrão[3] and C.C.P. Paz[1,2]
[1]Instituto de Zootecnia, Beef Cattle Research Center, Rodovia Carlos Tonani, Km 94, Zona Industrial, Sertãozinho, SP, Brazil;, 14160-900, Brazil, [2]Ribeirão Preto Medical School, University of Sao Paulo, Av. Bandeirantes, 3900 Monte Alegre, Ribeirão Preto, SP, 14049-900, Brazil, [3]Faculty of Animal Science and Food Engineering, University of São Paulo, Av. Duque de Caxias Norte, 225, Pirassununga, SP, 13635-900, Brazil; sergio.bgpnpl@hotmail.com

The objective of this study was to evaluate if residual feed intake (RFI) could be related to heat tolerance of Caracu animals exposed to the sun and under the shade. A total of 40 Caracu animals at Instituto de Zootecnia, Sertãozinho, SP, Brazil, were evaluated for residual feed intake (RFI), in which negative RFI means more efficient animals with respect to feed conversion and vice-versa. After recording RFI, the animals were exposed to heat stress on days that the black globe temperature reached 40 °C. Four evaluations were carried out: two in October and two in December 2017. The records were collected from 08 to 10 h for all the 40 animals, and from 10:30 to 15 h, in which a group of 20 animals were exposed to the sun for two hours (stress period), and the other group of 20 animals were submitted to the shade for the same time. The variables measured on the animals with respect of heat tolerance was rectal temperature, transformed into Heat Tolerance Index (HTI). The animals that were exposed to the sun in October were submitted to the shade in December and vice-versa. The Tukey test from GLM of SAS was used to compare HTI values between positive RFI and negative RFI animals, exposed to the sun and under the shade ($P<0.05$). The mean (μ) and the standard deviation (s) of HTI values in the sun were, 9.58 and 0.09, respectively, for negative RFI animals. For positive RFI animals under the sun, the μ and s were 9.28 and 0.10. For negative RFI animals under the shade, the μ and s were 9.95 and 0.8. For positive RFI animals under the shade, the μ and s were 9.79 and 0.09. The difference in HTI for the RFI categories in both situations (under the sun and shade) was not significant ($P>0.05$). Thus, animals with high RFI and low RFI are heat tolerant in a similar way.

Session 55 Poster 12

Combined selection for milk and weigth in cattle in the tropical environment

R.R. Rizzi[1], M.O. Oropeza[2], F.C. Cerutti[2] and J.C.A. Alvarez[2]
[1]Department of Veterinary Medicine, Via Celoria, 10, 20133 Milan, Italy, [2]Asociación de Criadores de Raza Carora, Av. Francisco de Miranda, 3050 Carora, Venezuela; rita.rizzi@unimi.it

In dairy cattle, production is positively related to the size of the animal; in fact, the selection for some type traits, such as stature, body depth and chest width improves milk production. However, large cows have larger nutritional requirements than small cows. In areas where food availability is low, it might be useful to improve milk production, taking into account the live weight. In particular, in breeds reared in a tropical environment, more emphasis should be placed on the production of milk per unit of live weight (or metabolic weight) in order to improve biological efficiency. The purpose of this note is to estimate genetic parameters for milk production as a function of live weight and selection response of the live weight. For this analysis 16,759 weights and 12,718 lactations of 3,050 Carora cows were considered. For each lactation milk per 500 kg of live weight were calculated. An Animal Model Multiple trait was used to estimate the components variance for weight, milk in a standard lactation and milk/weight ratio were estimated. The (co) variances and genetic parameters were used to calculate the genetic gain for weight. The heritability of weight, milk in a standard lactation and milk/weight ratio were 0.44, 0.20 and 0.18, respectively. Weight was positively correlated both to milk (rG=0.33) and to milk/weight (rG=0.22). Given a selection intensity equal to 1, in Carora cattle a decrease in weight in one generation was observed if direct selection is made on mil/weight. In order to obtain an animal with good milk yield but of reduced size, it is possible to select for cows that have higher productions at the same weight. The selection for this trait must however be monitored taking into account also other morphological and reproductive aspects.

Session 55 Poster 13

Heat tolerance or extensive ability to acclimate

A. Geraldo[1,2], F. Silva[1], C. Pinheiro[1,2], L. Cachucho[1], C. Matos[1], E. Lamy[1,2], F. Capela E Silva[1,2], P. Infante[1] and A. Pereira[1,2]
[1]University of Évora/ECT, Apartado 94, 7002-554, Portugal, [2]ICAAM, Apartado 94, 7002-554, Portugal; ageraldo@uevora.pt

Heat thermal stress is a major concern environmental stress for dairy cattle, it limits animal growth, metabolism, and productivity. Taken this, the joint selection for productivity and adaptability should be considered in the actual dairy farms programs. This study aimed to evaluate the seasonal acclimatization process of cows with different milk yield potential. From a dairy farm located in Alentejo, Portugal, 13 Holstein-Friesian cows were chosen, 7 with high milk yield potential (HMP), ≥9,000 kg of milk at 305 days of lactation, and 6 with low milk yield potential (LMP), <9,000 kg. The trail was separated in 3 periods: (P1) Summer: acclimated cows in heat stress; (P2) Summer: acclimated cows in thermoneutrality; (P3) Winter: acclimated cows in thermoneutrality. Respiratory frequency (RF), rectal temperature (RT) milk composition and plasma triiodothyronine levels (T3) were collected. No differences were found in RF and RT between HMP and LMP cows in any of the periods. RF and RT values were significantly higher in P1 (64.13±12.78 mov./min. and 38.82±0.68 °C) than in P3 (36.13±7.67 mov./min. and 38.06±0.52 °C). Although, in P1, some HMP cows had RT values that indicated heat stress. We found no differences between groups in the lactose, protein, fat, β-Hydroxybutyric acid and somatic cell count. Urea was significantly higher in P1 in the HMP (293.62±35.97 mg/kg) than in LMP (253.69±33.81 mg/kg). T3, in both groups, gradually increased from P1 to P3 (P1-142.00±13.77; P2 – 157.36±10.72; P3 – 170.69±17.78 ng/dl). During summer, HMP had T3 values significantly lower than the LMP cows (P1: HMP-133.33±8.14, LMP-152.40±11.97; P2: HMP-146.50±7.64; LMP-170.40±12.29 ng/dl). Despite these results had revealed that HMP and LMP cows did not show significant differences in RF and RT variables, ongoing the acclimatization process, the HMP presented a lower metabolic activity as well a change in the nitrogen metabolic pathways.

Session 56 Theatre 1

Multidisciplinary approaches to livestock production

J. Van Milgen[1], M.H. Pinard-Van Der Laan[2], E. Schwartz[3], Ç. Kaya[4] and V. Heuzé[5]
[1]INRA-Agrocampus Ouest, UMR Pegase, Le Clos, 35042 Rennes, France, [2]INRA, UMR GABI, Domaine de Vilvert, 78352 Jouy-enJosas, France, [3]INRA, UR VIM, Domaine de Vilvert, 78352 Jouy-enJosas, France, [4]EFFAB, Dreijenlaan 2, 6703 HA Wageningen, the Netherlands, [5]Association Française de Zootechnie, 16, rue Claude Bernard, 75231 Paris, France; jaap.vanmilgen@inra.fr

Livestock production is an essential component of a sustainable food supply. The dimensions of sustainability combined with the complexity of biology call for multidisciplinary approaches to assess the functioning of (components of) livestock production systems. Horizon 2020 projects such as Saphir and Feed-a-Gene are funded by the European Commission and have a multidisciplinary approach towards livestock production (e.g. nutrition and genetics, genetics and health, novel management strategies and socio-economic aspects). The objective of this session is to address and discuss the challenges and opportunities in multidisciplinary research in livestock production. As there is no 'one-size-fits-all' solution, how can different actors and stakeholders make collectively best use of (disciplinary) knowledge and levers to make livestock production more sustainable?

Session 56 Theatre 2

Twists and turns of interdisciplinary work in resarch projects: which conditions and achievements ?

M. Cerf
INRA, UMR LISIS, BP1 Bâtiment EGER, 78 850 Thiverval Grignon, France; marianne.cerf@inra.fr

The emphasis on interdisciplinary research is far from new: promoting it was at the core of the DGRST French research plan in the 70's and of the Hawkesbury University experience in Australia. It resulted in the emergence of the INRA SAD division in the 80's. Today, many funding agencies ask for interdisciplinary work teams. Well recognized journals such as Science give room to interdisciplinary results. Doctoral schools value interdisciplinary training even if this remains controversial regarding recruitment standards. More epistemological reflexion is available to address the foundations of interdisciplinary scientific work. Anyway, practising interdisciplinary work and managing interdisciplinary projects remain a big challenge. There is no clear-cut recommendations or on-the shelf tools which can be transferred to newcomers. Many researchers who joined interdisciplinary projects find it difficult, disappointing. They acknowledge that the project is more a collection of disciplines working on separate tasks with loose coordination in terms of renewed understanding of the issue addressed in the project. Why is it so? I will rest on my experience at the interface between agronomic and social sciences to address the dynamic of interdisciplinary work in practice. I will put emphasis on various dynamic patterns and stress the need for intermediary objects, shared learning and reflexivity along the interdisciplinary journey. I will illustrate that fruitful interdisciplinary work takes place at the border of the disciplines but often questions core assumptions underlying its main stream. Therefore, taking part to interdisciplinary work put the participants in a risky position in their own community as the legitimacy of their work becomes discussed. To escape this uncomfortable position, researchers often trigger a debate on the balance between excellence and relevance of the scientific work. I will then conclude by focusing on the management of interdisciplinary projects. I will argue that people involved in the management of interdisciplinary work act as brokers. As for any brokering work this implies to have the ability and the curiosity to navigate between different points of view on a reality and to support the cross fertilization between these points of view.

Session 56 Theatre 3

Detection and characterization of the feed intake response of growing pigs to perturbations

H. Nguyen Ba[1], M. Taghipoor[2] and J. Van Milgen[1]

[1]INRA, UMR Pegase, Le Clos, 35590 Saint-Gilles, France, [2]INRA, UMR MoSAR, 16 rue Claude Bernard, 75231 Paris, France; hieu.nguyen-ba@inra.fr

Improving robustness for farm animals is seen as a new breeding target. However, robustness is a complex trait and not measurable directly. Robustness can be characterized by examining the animal's response to environmental perturbations. Although the origin of environmental perturbations may not be known, the effect of a perturbation on the animal can be observed, for example through changes in voluntary feed intake. We developed a generic model and data analysis procedure to detect these perturbations, and subsequently characterize the feed intake response of growing pigs in terms of resistance and resilience as elements of robustness when faced with perturbations. We hypothesize that there is an ideal trajectory curve of cumulative feed intake, which is the amount of feed that a pig desires to eat when it is not facing any perturbation. Deviations from this ideal trajectory curve are considered as a period of perturbation, which can be characterized by its duration and magnitude. It is also hypothesized that, following a perturbation, animals strive to regain the ideal trajectory curve. A model based on differential equations was developed to characterize the animal's response to perturbations. In the model, a single perturbation can be characterized by two parameters which describe the resistance and resilience potential of the animal to the perturbing factor. One parameter describes the immediate reduction in daily feed intake at the start of the perturbation (i.e. a 'resistance' trait) while another describes the capacity of the animal to adapt to the perturbation through compensatory feed intake to rejoin the ideal trajectory curve (i.e. a 'resilience' trait). The model has been employed successfully to identify the ideal trajectory curve of cumulative feed intake in growing pigs and to quantify the animal's response to a perturbation by using feed intake as the response criterion. Further developments include the analysis of individual feed intake curves of group-housed pigs that can be exposed to the same environmental perturbing factors to quantify and to compare different pigs. This study is part of the Feed-a-Gene project and was funded by the European Union under grant agreement no. 633531.

Session 56 Theatre 4

Layers response to a suboptimal diet through phenotype and transcriptome changes in four tissues

F. Jehl[1], M. Brenet[1,2], A. Rau[1], C. Désert[1,2], M. Boutin[1,2], S. Leroux[1], D. Esquerré[1], C. Klopp[1], D. Gourichon[1], A. Collin[1], F. Pitel[1], T. Zerjal[1] and S. Lagarrigue[1,2]

[1]INRA, 147 Rue de l'Université, 75007 Paris, France, [2]Agrocampus Ouest, 65 Rue de Saint-Brieuc, 35000 Rennes, France; frederic.jehl@inra.fr

Poultry meat and eggs are major sources of nutrients in the human diet. The long production career of laying hens expose them to biotic or abiotic stressors, lowering their production. Understanding the mechanisms of adaptation to stress is crucial for selecting robust animals and meeting the needs of a growing human population. In this study, financed by the French ChickStress and the European Feed-a-Gene (grant agreement no. 633531) programs, we compared the effects of a 15%-energy-reduced diet (feed stress, FS) vs a commercial diet (control, CT) on phenotypic traits and adipose, blood, hypothalamus and liver transcriptomes in two feed-efficiency-diverging lines. Phenotypic traits showed differences between lines or diets, but no line × diet interaction. In the FS group, feed intake (FI) increased and hens had lower body- and abdominal adipose weight, compared to CT group. We found no differences in egg production or quality. At the transcriptomic level, 16,461 genes were expressed in one or more tissues, 41% of which were shared among tissues. We found differentially expressed genes between lines or diet in all tissues, and almost no line × diet interactions. Focusing on diet, adipose and liver transcriptomes were unaffected. In blood, pathways linked to amino acids, monosaccharides, and steroid metabolism were affected, while in the hypothalamus, changes were observed in fatty acid metabolism and endocannabinoid signalling. Given the similarities in egg production, the FS animals seem to have adapted to the stress by increasing FI and by mobilizing adipose reserves. Increase in FI did not appear to affect liver metabolism, and the mobilization of adipose reserves was apparently not driven at the transcriptomic level. In blood, the pathways linked to metabolic processes suggest a metabolic role for this tissue in chicken, whose erythrocytes are nucleated and contain mitochondria. FI increase might be linked to the hypothalamic pathway of endocannabinoid signalling, which are lipid-based neurotransmitters, notably involved in the regulation of appetite.

What potential of genome-wide integrative approaches to predict vaccine responses?

F. Blanc[1], T. Maroilley[1], M.H. Pinard-Van Der Laan[1], G. Lemonnier[1], J.J. Leplat[1], E. Bouguyon[2], Y. Billon[3], J.P. Bidanel[1], B. Bed'hom[1], J. Estellé[1], S. Kim[4], L. Vervelde[4], D. Blake[5] and C. Rogel-Gaillard[1]

[1]GABI, INRA, AgroParisTech, 78350 Jouy-en-Josas, France, [2]VIM, INRA, 78350 Jouy-en-Josas, France, [3]GenESI, INRA, 17700 Surgères, France, [4]The Roslin Institute and Royal (Dick) School of Veterinary Studies, University of Edinburgh, Midlothian, United Kingdom, [5]Department of Pathobiology and Population Sciences, Royal Veterinary College, University of London, Hatfield, United Kingdom; claire.rogel-gaillard@inra.fr

The impact of host genetic variations in shaping innate and adaptive immune responses is an emerging lever to consider in new vaccination strategies. Merging genetic and genomic data to identify prospective biomarkers that could predict individual's immune capacity and response to vaccines is a challenging question addressed by the H2020-funded SAPHIR project, both in chickens and in pigs. Large White pigs (48 families) were vaccinated against *Mycoplasma hyopneumoniae* (*M. hyo*, 182 piglets) or Influenza A Virus (IAV, 98 piglets) at weaning (around 28 days of age) with a booster vaccination three weeks later. The humoral vaccine response was measured by following the dynamics of seric *M. hyo*- or IAV-specific IgG every week during five weeks post-vaccination, and before slaughtering at 21 weeks of age. For chickens, vaccine responses were measured on vaccinated commercial broilers (Cobb 500) and on a subset of animals challenged with *Eimeria maxima* (from 96 to 36 chickens). Animal responses were evaluated by the measure of serum levels of IL-10 with an in-house developed ELISA system, body weight gain, lesion scores, and parasite load. For each species design, blood was sampled before vaccination on the vaccine day for high-density SNP genotyping and RNAseq analysis. We have identified significant associations between gene expression in blood before vaccination and vaccine responses in pigs or body weight as a measure related to the vaccine follow-up in chickens. Thus, we provide a proof of concept that blood could be used as a relevant source of biomarkers predictive of vaccine responses. We will further discuss the potential of integrating multi-level genomic and phenotypic data to better understand individual vaccine responses and identify levers of action.

Immune responses after administration of innovative *Myocoplasma hyopneumoniae* bacterins in pigs

D. Maes[1], A. Matthijs[1], G. Auray[2], C. Barnier-Quer[3], F. Boyen[1], I. Arsenakis[1], A. Michiels[1], F. Haesebrouck[1] and A. Summerfield[2]

[1]Ghent Univ., Salisburylaan 133, 9820 Merelbeke, Belgium, [2]Institute of Virology and Immunology, Sensemattstr. 293, 3147 Mittelhäusern, Switzerland, [3]Univ. Lausanne, Chemin des Boveresses 155, 1066 Epalinges, Switzerland; dominiek.maes@ugent.be

Vaccination against *Myocoplasma hyopneumoniae* is widely used, but available vaccines provide only partial protection. This study aimed to screen innovative bacterins based on a highly virulent *M. hyopneumoniae* field strain for their ability to induce potent immune responses. Nine groups, each consisting of 6 *M. hyopneumoniae*-free piglets, were primo- (D0; 39 days of age) and booster (D14) vaccinated with 7 different experimental bacterins, a commercial bacterin as a positive control or PBS as a negative control. The experimental bacterin was formulated either with dmLT (group A), DDA:TDB liposomes (B), DPPC:DC-Chol liposomes+C-di.AMP (C), DPPC:DC-Chol liposomes+CpG ODN, resiquimod and Pam3Cys-SK4 (TLR ligands; D), PLGA:CTAB microparticles+TLR ligands (E), O/W emulsion+TLR ligands (F) and DOPC:Chol liposomes+TLR4 agonist and QS-21 (G). The specific immune response was assessed by the levels of specific antibodies in serum and in bronchoalveolar lavage fluid (BALf), and by T-cell specific responses measuring TNF, IFN-γ and IL-17 in CD4 T cells. On D28, 6/6 pigs from groups B, C, D, F, G and the commercial vaccine group, and 2/6 pigs from group E were seropositive. Group B, C and the commercial vaccine group had significantly higher OD-values for IgG in serum than group A and the negative control group, and the OD-value from group E was significantly lower compared to group C and the commercial vaccine group ($P \leq 0.05$). Serum IgA ELISA results did not differ over time or among groups. In group F, 1/6 pigs tested positive for *M. hyopneumoniae* specific IgA in BALf on D28. At D14, there was an upregulation of both TNF and IFN-γ double positive as well as IL-17$^+$ CD4 T cells in the commercial vaccine and F groups. At D28, a strong TNF and IFN-γ response was observed in CD4 T cells from groups B and F, while a significant IL-17 response was seen in cells from the group E compared to the negative control group. Formulation B, E and F seem to be promising *M. hyopneumoniae* vaccine candidates.

Effect of heat stress on faecal microbiota composition in swine: preliminary results

M. Le Sciellour[1], I. Hochu[2], O. Zemb[2], J. Riquet[2], H. Gilbert[2], M. Giorgi[3], Y. Billon[4], J.-L. Gourdine[5] and D. Renaudeau[1]
[1]PEGASE, INRA, Agrocampus-Ouest, 35042 RENNES, France, [2]GenPhySE, Université de Toulouse, INRA, INPT, INP-ENVT, 31320 Castanet Tolosan, France, [3]PTEA, INRA, 97170 Petit-Bourg, France, [4]GenESI, INRA, 17700 Surgères, France, [5]URZ, INRA, 97170 Petit-Bourg, France; mathilde.lesciellour@inra.fr

Gut microbiota plays a central role in health and nutrient digestion and would help the host for better coping with environmental perturbations. In tropical conditions or in temperate countries during Summer, elevated ambient temperatures can cause economic losses to the pig industry. During heat stress (HS), the reduction in voluntary feed intake is the main adaptation response for reducing heat production. This lower feed intake has subsequent negative effects on pig performance. The main purpose of this study was to investigate the relationships between HS and gut microbiota composition. A better understanding of the microbiota response to HS could allow the selection for animals well adapted to HS. Genetically related pigs were raised under temperate or tropical farm conditions with mean thermal humidity indexes respectively 23 and 25.5 from 11 to 23 weeks of age. In temperate conditions, pigs were submitted to a 3-week HS challenge at 30 °C. Faecal samples were collected in all pigs at 23 weeks of age in both environments (n=1,200 samples) and at 26 weeks of age in the temperate environment (n=600). Therefore, it was possible to compare microbiota from pigs raised in a temperate environment, a tropical climate, and exposed to HS. Microbiota extracted from pigs under temperate and tropical climate had different compositions whereas pigs exposed to heat challenge or raised in tropical conditions tended to share a common microbiota. HS challenge drastically modified gut microbiota and the groups before and after the challenge could be predicted in a multilevel sparse partial least square discriminant analysis with 30 OTUs and a mean classification error rate of 14%. Our experiment suggests that microbiota can be used as biomarkers of HS exposition. This study is part of the Feed-a-Gene Project funded by the European Union's H2020 Program (grant 633531), and of the PigHeat project funded by the French National Agency of Research (ANR-12-ADAP-0015).

The socio-economic evaluation of vaccines in livestock systems

C. Bellet and J. Rushton
Institute of Infection and Global Health / University of Liverpool, IC2 Building, 146 Brownlow Hill, L3 5RF, Liverpool, United Kingdom; camille.bellet@liverpool.ac.uk

Socio-economic evaluations provide a foundation to define resource allocation priorities and disease management strategies, which in turn help to tailor interventions to the specific context of a livestock system. This presentation examines how socio-economic evaluations of vaccines are performed in animal health research in relation to three European cases of respiratory and digestive diseases in cattle, pig and poultry. It explores their limitations and the potential of transdisciplinary approaches to support more robust evaluations of livestock vaccination interventions. Vaccines not only help to protect against acute diseases but also against chronic diseases that may not hit the headlines, but which are important for livestock businesses and society. Vaccines also benefit unvaccinated animals, their owners and consumers by decreasing disease transmission, which in turn reduces the risk of epizootics and the pressure generated by these epizootics on health care providers and market economy. In addition, vaccines impact on drug use reducing selection pressure for drug resistance, and improving drug efficacy in humans. However, for vaccines to generate these benefits, the associated vaccination strategies need to protect critical proportions of animals on a given time scale. These strategies depend on vaccine quality, availability, accessibility and acceptability to farmers, their animal health advisors, and increasingly consumers. The growing complexity of vaccines, vaccine delivery and their relationship to long and diverse food systems demand interdisciplinary and integrative approaches that address human, animal and ecosystem dimensions. Yet current socio-economic evaluations of livestock vaccines remain mostly limited to one of the many dimensions of a disease problem such as the reduction of disease incidence. The presentation concludes that there is a need to unpack the dynamics of farm practices and the food system to understand how these shape animal health management and vaccine intervention. This requires a transdisciplinary approach with a depth of different forms of expertise and data sources in order to achieve informed evaluations on the trade-offs of vaccine intervention.

Session 56 Theatre 9

Panel discussion 1: opportunities and difficulties in multi-disciplinary and multi-actor research

J. Van Milgen[1] and M.H. Pinard-Van Der Laan[2]
[1]INRA-Agrocampus Ouest, UMR Pegase, Le Clos, 35042 Rennes, France, [2]INRA, UMR GABI, Domaine de Vilvert, 78530 Jouy-en-Josas, France; jaap.vanmilgen@inra.fr

Panel discussion with stakeholders and interaction with participants on the opportunities and difficulties of multidisciplinary and multiactor research.

Session 57 Theatre 1

Development of and Imputation with a SNP map derived from the latest reference genome sequence

X. Yu[1], J.C. McEwan[2], J.H. Jakobsen[3] and T.H.E. Meuwissen[1]
[1]Norwegian University of Life Sciences, IHA, Arboretveien 6, 1433 Ås, Norway, [2]AgResearch Limited, Invermay Agricultural Centre, 19053, Mosgiel, New Zealand, [3]Norwegian Association of Sheep and Goat Breeders, P.O. Box 104, 1431 Ås, Norway; xijiang.yu@nmbu.no

SNP chips of different densities are often used together on farm animals for economic reasons. The missing genotypes on the low-density (LD) chips can be imputed with genotype results of high-density (HD) chips and a linkage map of HD loci. Genome references and SNP chips are under continuous development. This may introduce several problems. For example, different chips may be based on different versions of the reference sequence. The SNP names can come from different naming systems. There are also typically many duplicated loci, because of SNP types and importance. Any one of these problems can increase imputation errors. One solution for these problems is to derive a SNP map from the most up-to-date reference sequence. The probes that come with the manifest of a chip-design can be used to position the SNPs on this map. Each of these probes is 50 base pairs (bp) in length, and can uniquely define a mutation position. Because the ends of the probes may also be the SNP, one bp was trimmed from both ends of a probe. All possible 48-bp sequences of a chromosome were then indexed by their positions and sorted in alphabet order. This can speed up the search procedure thousands of times, for sequential searches were converted to binary ones. This method was tested on data from Norwegian White Sheep genotyped with an 8k LD chip and a 600k HD chip. Random animals who were genotyped with the HD chip were masked at the missing loci on the LD chip. Only autosome loci were considered. Using the provided 600k linkage map, the correct allelic imputation rate was only 85%. Using the map derived from the sheep genome reference version 3.1.91, the imputation rate increased to 92%. When the number of animals with known HD genotypes increased from 120 to 617, the imputation rate increased to 94%, indicating the accuracy improvement was mainly from the new map. In conclusion, using a SNP map derived from the latest reference can greatly increase imputation rate. The described binary search algorithm makes such map construction feasible with limited computation costs.

Light-treated rams and bucks abolish reproductive seasonality in sheep and goats

P. Chemineau[1], J.A. Abecia[2], M. Keller[1] and J.A. Delgadillo[3]
[1]INRA, CNRS, PRC, 37380 Nouzilly, France, [2]U Zaragoza, IUCA, Zar, Ar, Spain, [3]U A Narro, CIRCA, UL, Torreon, Mexico; philippe.chemineau@inra.fr

Seasonality of breeding is trait of temperate and subtropical sheep and goats, which has major technical and economical consequences. In males and females, photoperiod imposes male sexual rest and cessation of ovulatory and oestrus activities in spring and summer in breeds of these latitudes. Recently, we showed the high efficiency of light-treated rams and bucks to abolish the seasonality of breeding in females. While maintained in natural light, rams and bucks received extra-light (7 h daily to reach 16 h) during 2 consecutive months in winter. One and half month after the end of extra-light, males showed intense libido, high plasma testosterone concentrations, and are ready to be joined with females for mating. The permanent presence of these treated males completely suppresses anoestrous and anovulatory season of ewes and goats, which ovulate continuously all the year round. In ovariectomized ewes and goats, the light-treated males prevent the decrease of LH plasma concentrations, suggesting that these treated males act centrally to neutralize the seasonal inhibitory effect of estradiol on LH pulsatility. This was the first time, for any seasonal species of mammals, that an external factor was able to neutralize the inhibitory effect of photoperiod. When used in farms to induce out-of-season ovulations and oestrus by the 'male-effect', light-treated bucks and rams were much more efficient than control males, which displayed weak sexual behaviour. Fertility of females joined with light-treated males was much higher than that of females maintained with control males. An economic approach of using these bucks in Mexico has shown that, due to reduction in kid mortality, increase in lactation length and kids price at sell, the annual economic benefit can reach 3,000€ per flock. It is concluded that these simple and cheap light treatment of rams and bucks to induce high out-of-season sexual activity is a very efficient tool to abolish reproductive seasonality in ewes and goats.

Evaluation of accelerometers as an effective tool to measure sheep behaviour in a pastoral context

P.G. Grisot[1], A. Philibert[1], F. Demarquet[2] and A. Aupiais[1]
[1]Institut de l'Elevage, 149 rue de Bercy, 75595 Paris, France, [2]Ferme Expérimentale de Carmejane, Carmejane, 04510 Le Chaffaut St Jurson, France; pierre-guillaume.grisot@idele.fr

The present work was performed within a project which aims to specify the technic and functional characteristics of a decision-support tool for small ruminant pastoral systems. Our objective was to assess whether accelerometers could be effective sensors to measure sheep behaviour in pastoral systems. To achieve this, we designed an algorithm to predict animal behaviour from accelerometer data. Within a batch of 25 préalpes du sud ewes, 9 ewes were equipped with collars containing accelerometers; their individual behaviour at pasture (on a rangeland comprising grassland and bushland patches) was observed 6 hours a day during 3 consecutive days. The behaviour was recorded as: 'Lying-Ruminating', 'Lying-Still', 'Lying-Sleeping', 'Standing-Grazing', 'Standing-Eating Bush', 'Standing-Eating High', 'Standing-Running', 'Standing-Walking'. Observations were compared to accelerometer data with a capture frequency of 100 or 25 Hz. Accelerometer data was divided in segments of 5 seconds, and segments tagged with only one behaviour were held. For each segment, 18 statistical indicators, like mean and standard error, were computed on the 3 axis and analysed with the Random Forest algorithm to predict sheep behaviour. Both with the 100 Hz and the 25 Hz frequencies, the prediction rates were good (92.5 and 92.4% of successful predictions, respectively). Predictions were good for 'Standing-Grazing' and lying (without discrimination of the 3 types of lying) behaviours (98.0% and over 90.7%, respectively), while 'Standing-Ruminating' has a poor prediction rate because of a confusion with 'Lying-Ruminating'. 'Standing-Running' and 'Walking' are predicted with 73 and 62% of good prediction respectively. Due to rare occurrence 'Standing-Eating Brush' was never predicted. We consider these first results to be very encouraging. To reduce the volume of data to manage further developments will be carried on using a 25 Hz frequency. The algorithm of prediction will be improved by additional sequences of observations from other animals in the flock or with animals of different breeds from two other flocks. Project CLOChèTE, supported by CASDAR funds.

Session 57 Theatre 4

Low cost portable microwave system for non-destructive measurement of carcass fat depth

J. Marimuthu and G.E. Gardner
Murdoch University, School of Veterinary and Life Sciences, Murdoch, WA 6150, Australia; jayaseelan.marimuthu@murdoch.edu.au

Non-destructive methods of measuring carcase fatness are essential for production animal industries as they enable optimised carcase boning, producer feedback, and value-based trading. However, the suitability of the technology that takes these measurements depends upon numerous factors including accuracy, reliability, cost, portability, speed, ease of use, safety, and for *in vivo* measurements the need for fixation or sedation. Working within these constraints a low-cost portable Microwave System (MiS) has been developed at Murdoch University for use in lamb abattoirs. Using custom-made probes, this system has been designed to measure back fat depths in lamb carcases. This study details the calibration and design of this prototype MiS system, testing the hypothesis that it will provide a reliable estimate of fat depth in lamb carcases. One hundred and ten mixed sex lamb carcasses with hot carcase weights ranging between 17 and 39 kg were scanned at the *C* site (5 cm from the midline at the 12th or 13th rib) using the prototype MiS system at 1, 2 and 4 hours post-mortem in an abattoir chiller. Measurements were recorded at 531 frequencies ranging from 100 MHz to 5.4 GHz at 10 MHz intervals at a power of -10 dBm. Carcase *C* site fat depth was measured manually using Vernier callipers, and varied between 0.6-8.5 mm (average of 3.56 mm). The calibrated and processed frequency domain MiS signals across 531 frequencies (100 MHz-5.4 GHz) were used to estimate *C* site fat depth using Partial Least Squares (PLS) regression coupled with leave-one-out cross-validation. At 1 h post-mortem the prototype MiS system demonstrated a prediction R^2 of 0.62 with root-mean-square-error of the prediction (RMSEP) of 1.08 mm. This precision improved at the latter readings with prediction R2 and RMSEP at 2 h post-mortem of 0.68 and 0.99 mm, and 0.65 and 1.04 mm at 4 h post-mortem. The predicted values at the 3 measurement times were highly correlated with correlation coefficients ranging between 0.85-0.91, and differences between predictions across the measurement times at most differing by ±1.5 mm. These results demonstrate the capacity of this prototype MiS system to estimate fat depth at the *C* site in lamb carcasses.

Session 57 Theatre 5

Benefits for sheep farmers of monitoring grass growth, quality and utilisation

A.E. Aubry
Agri-Food and Biosciences Institute, Agriculture, Large Park, BT26 6DR, United Kingdom; aurelie.aubry@afbini.gov.uk

Grazed grass is often poorly utilised on sheep farms. Increasing the uptake of grass measurements by producers is key to better inform grass management and monitor progress towards more ambitious targets. One of the aims of this study, funded by DAERA and AgriSearch, is to illustrate the benefits achieved by sheep producers when monitoring their grazing fields in terms of grass growth and quality throughout the grazing season. During the 2017 grazing season, six sheep producers throughout Northern Ireland measured grass heights from their grazing fields every week using rising plate meters and took grass samples once every 2 weeks for NIRS analyses to estimate nutritional quality. Participating farmers also used tools such as AgriNet to enter grass and animal data and inform grazing management. The results of this on-farm work improves our understanding of spatial and temporal variations in key parameters such as DM and ME contents. It also establishes clear linkages between grass heights and corresponding DM covers on a range of sheep farms, and illustrates how the data can be used to inform grazing rotations throughout the season, by better quantifying and predicting grass surplus and shortages.

Investigating factors affecting lifetime performance in ewes on a network of commercial farms

E. Genever, H. King and N. Wright
Agriculture and Horticulture Development Board, Stoneleigh Park, Kenilworth, Warwickshire, CV8 2TL, United Kingdom; liz.genever@ahdb.org.uk

The Challenge Sheep project is tracking the performance of around 5,000 replacements mated in autumn 2017 in 13 commercial flocks in England. A further 5,000 replacements will be recruited into the project in autumn 2018. The project runs until 2024. It was established as a previous project found that a disproportionate number of light lambs (less than 17 kg) at eight weeks of age came from young ewes. The aims of the project are to investigate the factors, particularly during the rearing phase, affecting lifetime performance through detailed analysis of commercially collected data and to limit the impact of these factors to reduce replacement costs. Around 50% of the tracked replacements are being mated as ewe lambs, while the remainder mated as shearlings (around 18 months of age). A range of replacement policies are covered in the project flocks and include home-bred and bought-in purebred and crossbred replacements. Data are being collected at five points over the year – mating, scanning, lambing, eight weeks of lactation and weaning (90 days). Body condition score (BCS) and weights are being collected on the replacements regularly through the project. They will be compared against established targets – 60 and 80% of mature weight for ewe lambs and shearlings at mating respectively – to understand what is the consequence on lifetime performance of not hitting early targets. Lambs are being linked to their dams via an electronic identification tag near birth, and BCS and weight of dams will be analysed to understand its impact on lamb performance to weaning. Early results are demonstrating the impact of weight gain between mating and scanning on scanning percentage, with heavier replacements having more lambs at scanning. Data on lamb losses and reasons for the replacements leaving the flock are being collected, alongside detailed health plans. The results are being regularly communicated to the Challenge Sheep farms to ensure decisions can be made based on data. Discussion groups of other local sheep farmers have been established around the Challenge Sheep farms to ensure results are communicated out widely. As more data are collected, more detailed statistical analysis will be conducted.

Use of electronic identification (EID) associated technologies in marginal sheep farming systems

C. Morgan-Davies[1], J.M. Gautier[2], B. Vosough-Ahmadi[3], P. Creighton[4], A. Barnes[3], R. Corner-Thomas[5], S. Schmoelzl[6] and D. McCracken[1]
[1]SRUC, Hill & Mountain Research Centre, SRUC Kirkton, Crianlarich, FK20 8RU, United Kingdom, [2]Institut de l'Elevage, BP 42118, 31321 Castanet Tolosan Cedex, France, [3]SRUC, Land Economy, Environment and Society Research, Kings Buildings, West Mains Road, Edinburgh, EH9 3JG, United Kingdom, [4]Teagasc, Animal & Grassland Research and Innovation Centre, Athenry, Co. Galway, Ireland, [5]Massey University, Private Bag 11-222, Palmerston North, 4442, New Zealand, [6]CSIRO, Chiswick, Armidale, NSW 2350, Australia; claire.morgan-davies@sruc.ac.uk

The use of precision livestock farming is well recognised in the cattle, dairy, pig and poultry sectors. However, in the sheep farming sector, it is often limited to using compulsory electronic identification (EID) only for identification purpose. Research in Scotland has highlighted how EID associated technologies, such as EID readers and auto-sorters, could be used to address other important needs within the marginal sheep sector (e.g. health, welfare, efficient management). Research in other parts of Europe and in Australia/New Zealand also shows potential benefits as well as barriers to the use of the EID associated technologies in sheep systems, especially in marginal or extensive areas. Strengthening links at international and European levels between institutes and research farms would be invaluable. It could encourage exchanges, inform the wider sheep industry on the benefits of these technologies, and overcome uptake barriers. This paper presents results from workshops between different overseas groups (UK, Ireland, Norway, France, Australia and New Zealand), led in 2017-18 by researchers in Scotland, which highlighted potential opportunities and challenges of EID associated technologies for marginal sheep systems. Examples are labour savings, performance management, improving health of indoor and outdoor animals by detection & prevention of diseases, locating animals & reducing predation for outdoor animals. Main issues that require more focused research and discussions are: uptake by the end users (i.e. farmers and advisors), education, knowledge transfer, and the development of the next generation of EID and sensors-based technologies.

Session 57 Theatre 8

The value of information from commercial livestock in multi-tier sheep breeding schemes

B.F.S. Santos[1,2], J.H.J. Van Der Werf[2], T.J. Byrne[1], J.P. Gibson[2] and P.R. Amer[1]
[1]AbacusBio Limited, 442 Moray Place, Dunedin 9010, New Zealand, [2]University of New England, ERS, Natural Resources Bldg (W055), UNE, Armidale NSW 2351, Australia; bsantos@abacusbio.co.nz

In breeding schemes forming a part of a multi-tier integrated production system, genetic progress is disseminated from the nucleus to the commercial tier through the transfer of males. There might be opportunity to use commercial information from lower tiers to increase the rate of progress, reduce the genetic lag between tiers and improve prediction accuracy in individuals from the nucleus. Through deterministic simulation, this study estimated the economic benefits that can be generated by implementing performance recording in conjunction with either DNA pedigree assignment or genomic selection. The overall economic benefits of improved performance in the commercial tier offset the costs of recording the multiplier tier. The net cumulative benefit per commercial ewe over a period of 40 years ranged from $117 to $249 of additional genetic progress depending on the level of G×E and breeding objective of the production system. There is evidence of significant sire re-ranking and differences in spread of EBVs and heritability estimates on economically important traits across environments. The opportunity to expand the benefits of commercial data were reliant on the use of these data to effectively improve selection accuracy of nucleus individuals. This has potential to benefit larger numbers of animals across tiers, as it allows accurate selection of candidates that perform well across environments. The extent of genomic relationships between nucleus and commercial individuals was critical to predict the improved accuracy contributed by the commercial information, as was the number of commercial individuals tested. In practice, the benefits of commercial information are much larger if data is used to increase accuracy of prediction in breeding individuals. Commercial performance records and genotypes may strengthen the links between breeders and farmers leading to more relevant genetic progress through the industry. Novel business agreements will be required for implementation in situations where commercial tiers are separate business entities to the nucleus and/or multiplier tiers.

Session 57 Poster 9

Productive and selenium status in lambs affected by selenium biofortified corn-preliminary results

Z. Antunovic, Z. Klir, Z. Loncaric and J. Novoselec
Faculty of Agricutlure in Osijek, University of J. J. Strossmayer in Osijek, Department of Animal Husbandry, V. Preloga 1, 31000 Osijek, Croatia; zantunovic@pfos.hr

The widespread selenium deficiency in soil, thus in the feedstuffs from such soil, needs to be prevented by various methods. In recent years biofortification of selenium has been carried out by soil fertilization and foliar application of crops. The aim of this study was to present the preliminary results of the research carried out with the addition of selenium biofortified corn (BC) in diets of fattening lambs. The research was conducted with 20 Merinolandschaf lambs, average age of 70 days, during 30 days of fattening. Lambs were divided in two groups. Feed mixture of the control group (CG) contained 23% corn (0.014 mg Se/kg DM), while in experimental group (EG) was 23% BC (0.278 mg Se/kg DM). Feed mixture, hay and water were offered to lambs *ad libitum*. Body weight and body measurements (withers height, trunk length, circumference, width and depth of chest, circumference of shin bone, length and circumference of leg) as well as daily weight gain and indices of body development were determined. After lambs´ slaughter, samples of muscle (m. semimembranosus), liver, kidney, lungs, spleen, peritoneum and heart tissues were taken. Determination of selenium concentrations in feed and tissues were obtained by inductively coupled plasma. No significant differences were obtained in production traits between CG and EG (body weight: 29.80 vs 30.98 kg; daily weight gain: 232 vs 258 g and feed conversion: 3.54 vs 3.47 kg/kg), as well as most of body measures and indices of body development, although better values were obtained in EG. However, higher concentrations of selenium in lungs and liver (CG: 0.260 and 0.745 mg/kg, respectively; EG: 0.326 and 0.916 mg/kg, respectively; $P<0.05$) as well as higher ($P>0.05$) concentrations in kidney, liver, heart, spleen and peritoneum were determined in EG compared to CG. The results of the study indicate the possibility of using BC in lambs´ diets. It is necessary to include biofortified corn containing different concentrations of selenium in diets, longer duration of experiment and determination of metabolic status. This will contribute to more comprehensive conclusions when using BC in fattening lambs.

Session 58 Theatre 1

Faecal biomarkers for intestinal health in nutritional studies

T.A. Niewold
KU Leuven, Kasteelpark Arenberg 30, 3001 Heverlee, Belgium; theo.niewold@kuleuven.be

Gut health is central to animal health, which is particularly true in production animals. The type of nutrition and its components are important factors influencing gut health. Hence there is a definite need for biomarkers for intestinal health *in vivo* for the evaluation of the effects of nutrition and additives. Previously, intestinal health was determined by taking mucosal biopsies which is invasive and unpractical. Preferably, biomarkers should be non-invasive, and faecal sampling is such a method. There are ample examples in humans, up to fifteen biomarkers have been proposed, and assays and reagents are available, whereas this is hardly the case for production animals such as pigs, chicken, and calves. Most promising are biomarkers of small intestinal inflammation because the latter is reciprocal to health and growth. It has become clear that certain biomarkers may not be present in all species (especially chicken), or if present, are immunologically different. The latter means that specific antibodies to be developed. Alternatively, non-immunological assays could be used such as the enzymatic assay for the inflammatory cell product myeloperoxidase (MPO) in mammals, but not in chicken. In chicken, neopterin in excreta was suggested because of the availability of antibodies; however, no relationship with inflammation was evident as opposed to what was found in mammals. A very promising biomarker for small intestinal inflammation is Pancreatitis Associated Protein (PAP), also known as regenerating islet-derived 3 alpha (Reg3α) in mammals. In faeces of rats levels reflect the severity of small intestinal damage. Available reagents were not cross-reactive with pig PAP because pig PAP is Reg3γ instead of Reg3α. We have developed our own antibodies against pig PAP, and first results in pig faeces show promising results. It thus appears to be a feasible non-invasive biomarker for determining the effect of nutrition on intestinal health. The availability of such biomarkers will greatly contribute to the development of improved feed formulation, and can be used for diagnosis and prevention of intestinal problems which are associated with poor intestinal health and retarded growth.

Session 58 Theatre 2

Development of a new ELISA test for pancreatitis associated protein detection in pig

E. Mariani[1,2], G. Savoini[2] and T.A. Niewold[1]
[1]KU Leuven, Faculty of Bioscience Engineering, Department of Biosystems, Kasteelpark Arenberg 30, B-3001 Heverlee, Belgium, [2]University of Milan, Department of Health, Animal Science and Food Safety, Via Celoria 10, 20133 Milan, Italy; elena.mariani1@unimi.it

New biomarkers, non or minimally invasive, are needed in order to test the effect on animal health and intestinal diseases of feed additives used to improve pig performance. Two interesting biomarkers for intestinal health in pigs are Myeloperoxidase (MPO) and Pancreatitis associated protein (PAP). The first one is an enzyme used to quantify the number of inflammatory cells present in tissue and in faeces and the second one is a protein produced by Paneth cells upon inflammatory stimuli and has anti-inflammatory and bactericidal activity. The main aim of this study was to develop and validate a new sandwich ELISA test for the detection and quantification of PAP in faecal samples of pig. Our study consisted of two main phases: development and validation of the test. For the development of the ELISA, we previously used polyclonal antibodies from rabbit serum immunized using a pure peptide containing the N-terminus of pig PAP. For the validation of the test we used extracts from faecal samples derived from animals with known high or low growth performance. Furthermore, the temperature stability of PAP in faeces was tested with and without protease inhibitor cocktail. After optimization of the ELISA preliminary results show a good relationship between health and PAP faecal concentrations, in a reciprocal way, similar to what has been found for faecal MPO. Furthermore, PAP seems to be exceptionally stable in faeces. The present results suggest that PAP is a very promising candidate for a non-invasive (faecal) biomarker for intestinal health and growth.

Session 58 Theatre 3

Faecal Pig DNA: a potential non-invasive marker of gut cell loss

K.R. Slinger[1], K. May[1], A. Chang[1], M.R. Bedford[2], J.M. Brameld[1] and T. Parr[1]
[1]University of Nottingham, School of Biosciences, LE12 5RD, United Kingdom, [2]AB Vista, Marlborough, SN8 4AN, United Kingdom; sbxkrs@nottingham.ac.uk

Detection of pig DNA in faecal samples could be a non-invasive marker of gut cell loss. The objectives were to optimise pig DNA detection from faeces and investigate any effects of a feed containing an enzyme additive. Sixteen newly weaned female Camb12 (Landrace × Duroc × Large White) piglets (8.8±1.38 kg) were allocated to control (n=8) and xylanase treatment groups (n=8), the latter received the same feed but including 26,000-17,800 BXU/kg Econase XT 25. Faeces were collected on days 15, 29 and 42; DNA was extracted using a phenol-chloroform method and assessed by NanoDrop™ 2000 spectrophotometer. Primers were designed for both the pig actin (conserved across α, β & γ isoforms) and cytochrome b (CYTB) genes. Bacterial DNA was detected using published 16S gene primers. Following PCR to an end-point (40 cycles), semi-quantitative assessments of DNA amplicons were made by non-denaturing gel electrophoresis and quantified using Quantity One® software. Quantitative PCR (qPCR) was carried out on a LightCycler® 480 (Roche) instrument using SYBR Green methodology. Data was analysed by 2-way ANOVA (treatment × time) using Genstat 16th Edition, significance at P<0.05. Semi-quantitative PCR indicated a trend for a decreased quantity of pig actin DNA in the faeces of the xylanase group (P=0.084); but qPCR using the same primers was unsuccessful. Using qPCR pig CYTB was detected. There was significantly less pig CYTB amplicon (P=0.039) in the faeces of the xylanase treated group, with a trend (P=0.087) for an effect of time, but no interaction. Using qPCR there was no effect of treatment or time on the quantity of the bacterial 16S DNA. The pig mitochondrial CYTB gene appears to be present at a higher concentration in pig faeces than actin genomic DNA, therefore is a better gene target for quantitative analysis of faecal pig DNA. Xylanase inclusion lead to reduced quantities of pig DNA in faeces, indicating that it may reduce gut cell losses, as this DNA presumably originates from pig gut cells. These results indicate that the assessment of pig DNA in faeces may potentially be a non-invasive marker of gut cell loss.

Session 58 Theatre 4

Impact of weaning age on gut microbiota composition in piglets

F.R. Massacci[1,2,3], M. Berri[4], M. Olivier[4], J. Savoie[5], G. Lemonnier[3], D. Jardet[3], M.N. Rossignol[3], F. Blanc[3], M. Revilla[3,6], M.J. Mercat[7], J. Doré[8,9], P. Lepage[9], C. Rogel-Gaillard[3] and J. Estellé[3]
[1]Research and Development Department, IZSUM 'Togo Rosati', Via Salvemini, Perugia, Italy, [2]DISTAL, Bologna University, Via Fanin, Bologna, Italy, [3]GABI, INRA, AgroParisTech, Université Paris-Saclay, Domaine de Vilvert, Jouy-en-Josas, France, [4]ISP, INRA, Université Tours, Val de Loire, Nouzilly, France, [5]UE PAO, INRA, INRA Tours, Nouzilly, France, [6]MoSAR, INRA, AgroParisTech, Université Paris-Saclay, rue C. Bernard, Paris, France, [7]IFIP-Institut du porc and Alliance R&D, La Motte au Vicomte, Le Rheu, France, [8]MetaGenoPolis, INRA, Université Paris-Saclay, Domaine de Vilvert, Jouy-en-Josas, France, [9]MICALIS, INRA, AgroParisTech, Université Paris-Saclay, Domaine de Vilvert, Jouy-en-Josas, France; francesca.massacci@inra.it

Weaning is a crucial period of pigs, accompanied by nutritional, environmental and social stresses. Studies comparing different ages at weaning have shown that increasing weaning age improves wean-to-finish growth performances and reduces mortality. However, the impact of weaning age on the early-life establishment of the gut microbiota remains under-investigated in pigs. Our objective was to compare the gut microbiota composition of piglets weaned at different ages. 48 piglets were divided in 4 groups of 12 animals weaned at either 14, 21, 28 or 42 days-of-age. Faecal samples were collected at 3 different time points: day of weaning, 7 days after weaning and at 60 days of age. Faecal DNA bacterial composition was assessed by sequencing the V3-V4 regions of the 16S rRNA gene. Bioinformatic and biostatistical analysis showed that each weaned group had significant differences between the sample points through weaning transition, confirming that the gut microbiota changes before and after weaning. In addition, microbiota diversity increased according to weaning age, with piglets weaned at 42 days-of-age having a highest alpha diversity and richness. Interestingly, piglets weaned at 42-days maintained a more stable diversity until day 60. We show that late weaning leads to a higher diversity of potentially beneficial microbes prior to the crucial challenge of weaning and might thus provide a competitive advantage to piglets.

Coccidiostatic effects of tannin rich diets in rabbit production

H. Legendre[1,2,3], K. Saratsi[2], N. Voutzourakis[2], A. Saratsis[2], A. Stefanakis[2], P. Gombault[4], H. Hoste[1], T. Gidenne[3] and S. Sotiraki[2]
[1]Université de Toulouse, INRA/ENVT, UMR 1225 IHAP, 23 Chemin des Capelles, 31076 Toulouse, France, [2]Hellenic Agricultural Organization Demeter, Laboratory of Parasitology, Veterinary Research Institute, Campus Thermi P.O. Box 60272, 57001 Thermi, Greece, [3]Université de Toulouse, INRA, INPT, ENVT, GenPhySE, ENVT, 31326 Castanet-Tolosan, France, [4]MULTIFOLIA, 1bis grande rue, 10380 Viâpre Le Petit, France; smaro_sotiraki@yahoo.gr

The potential anticoccidial effect of tannin containing resources such as sainfoin and carob in rabbits feed given to does at pre-weaning and to growing rabbits were tested. The trial started at parturition (D0), 24 does and their litters were assigned into three groups. They were fed either with a control (Group CO), a carob (containing 10% carob pods meal) (Group RO) or a sainfoin diet (containing 34% dehydrated sainfoin pellets) (Group SA). All diets were made isonitrogenous and isoenergetic and also balanced for crude fibre, but differed by their tannins content. Weaning occurred at D37, and growing rabbits remained in the same cage until D51, and then were transferred to fattening cages until the end of the trial. Weight gain of young rabbits among the three groups did not differ. Economical FCR post-weaning was reduced for rabbits of group SA compared to CO and RO groups. Faecal oocyst count in group SA was 60% lower than in CO and RO groups. Areas under the curve (AUCs) for oocysts counts calculated when rabbits remained in maternity cages after weaning did not differ according to diet. AUCs calculated after rabbit transfer in fattening cage was 62% lower in group SA than in CO and RO groups. The main species identified (from D59 to D83) was E. magna (53% of oocysts). AUCs for E. magna did not differ between diets. In conclusion, incorporation of sainfoin in a balanced rabbit diet can lower oocyst excretion of coccidia by 60%, and ameliorated the economical FCR.

Milk metabolites are non-invasive biomarkers for nutritional and metabolic disorders of Dairy Herds?

C.M.S.C. Pinheiro[1], I. Domingues[1], P. Vaz[2], R. Moreira[2] and P. Infante[3]
[1]University of Évora/ICAAM/ECT, Animal Science Department, Apartado 94, 7002-554 Évora, Portugal, [2]EABL, Quinta da Medela, 3810-455 Aveiro, Portugal, [3]University of Évora/CIMA/ECT, Mathematic Department, Apartado 94, 7002-554 Évora, Portugal; ccp@uevora.pt

The interpretation of milk metabolites from milk recording can be indicative of nutritional and metabolic disorders. The nutrient imbalances as the relationship between carbohydrates fermentability and protein degradability in the rumen can be diagnosed by milk urea nitrogen (MUN), protein and relation of fat/protein (F/P) in milk. The metabolic imbalances, as the negative energy balance, hyperketonemia, ketosis and acidosis can be diagnosed by β-hydroxybutyrate (BHB), fat and the relation of F/P in milk. Thus, milk metabolites can be indicators of health and welfare of the cow. This study analysed 110,461 individual milk samples of 9,523 lactating dairy cows collected monthly from January 2015 to March 2017 from 27 herds of South of Portugal, with an official milk recording. The mean of lactating cows per herd was 353±270 (mean ± SD) and milk production per cow was 35.08±9.80 kg/day. During the first 30 days of lactation 7.7% of milk recording had BHB concentration over 0.2 mmol/l, indicating that these cows had high possibility of being with clinical ketosis. 44.8% of milk recording had the relation of F/P over 1.4 and 49.3% had milk fat over 4.5% indicating that about 45% of the cows were probability mobilizing body fat. 86.7% of milk recording had MUN concentration between 101 and 299 mg/kg indicating that the relation between carbohydrates and protein of the diet was appropriate. On the other hand, 11.9% of milk recording had the relation of F/P above 1.4 and 21.6% had milk protein above 3% indicating that some animals are ingesting a small proportion of protein in the diet comparing with the quantity of carbohydrates. In conclusion, these non-invasive biomarkers can reflect nutritional and metabolic disorders, but the interrelation between them must be taken into account. The thresholds of this milk metabolites to indicate health disorders are not consensual among the authors.

CH_4 estimated from milk MIR spectra: model on data from 7 countries and 2 measurement techniques

A. Vanlierde[1], F. Dehareng[1], N. Gengler[2], E. Froidmont[1], M. Kreuzer[3], F. Grandl[4], B. Kuhla[5], P. Lund[6], D.W. Olijhoek[6], M. Eugene[7], C. Martin[7], M. Bell[8], S. McParland[9] and H. Soyeurt[2]

[1]CRA-W, 5030, Gembloux, Belgium, [2]ULg-GxABT, 5030, Gembloux, Belgium, [3]ETH, 8092, Zürich, Switzerland, [4]Qualitas AG, 6300, Zug, Switzerland, [5]FBN, 18196, Dummerstorf, Germany, [6]AU Foulum, Tjele, 8830, Denmark, [7]INRA, 63122, Saint-Genès-Champanelle, France, [8]AFBI, BT26 6DR, Hillsborough, United Kingdom, [9]Teagasc, Moorepark, Cork, Ireland; a.vanlierde@cra.wallonie.be

Availability of a robust proxy to estimate individual daily methane (CH_4) emissions from dairy cows would be valuable especially for large scale studies, for instance with genetic purpose. Milk mid infrared (MIR) spectroscopy presents potential to meet this aim as spectra can be obtained routinely at reasonable cost through milk recording process. Development of a prediction equation requires as much variability as possible in the reference data set to improve the accuracy and ensure the robustness of the model. So, two datasets including CH_4 measurements and corresponding milk MIR spectra have been merged: the first contained 532 data from 156 cows of Ireland and Belgium with CH_4 measurements obtained with SF_6 tracer technique; the second reached 584 data from 147 cows of Switzerland, United Kingdom, France, Denmark and Germany with CH_4 measurements obtained with respiration chambers. In addition to the Partial Least Squares (PLS) equation using the raw CH_4 values, a second equation was performed with a reduction of 8% to CH_4 values from chambers to evaluate the need to correct the potential method bias in accordance with literature. A 5-group cross-validation was performed to test the robustness of the models. R^2 and the standard error of cross-validation were 0.63 and 62 g/day from raw data and 0.65 and 59 g/day when CH_4 respiration chamber values were adjusted. This slight improvement due to the adjustment of chamber measurement does not permit to conclude that this correction is needed. The study of residuals showed a non-significant effect due to the CH_4 measurement technique. In conclusion this new equation combining CH_4 from 2 different techniques covered more variability (cows, diets and country specific information) and shows potential as a proxy especially for genetic evaluation.

RNA based amplicon sequencing: an emerging approach to study diet related shifts in rumen microbiota

N. Amin, B. Cardazzo, L. Carraro and L. Bailoni

University of Padua, Department of Comparative Biomedicine and Food Science (BCA), Viale dell'Università, 16, 35020 Legnaro (PD), Italy; nida.amin@phd.unipd.it

This study is aiming to analyse the diet dependent shifts in the metabolically active rumen (bacterial, archaeal and protozoal) communities of Italian Simmental cows, based on 16S and 18S rRNA amplicon sequencing. Seven lactating cows were used as rumen fluid donors. At the beginning of the experiment, all cows were pregnant (from 148 to 203 days in gestation). The experiment was divided into 2 collection periods depending on different dietary treatments and physiological stages: late lactation (from 248 to 332 DIM) and dry period (from 2-52 days before calving). Cows were fed a specific total mixed ration (TMR) during each collection period, formulated to cover the nutritional requirements. The rumen fluid was collected by esophageal probe from the cows receiving the specific TMR at least 10 days prior to collection. RNA was extracted in duplicates from each cow rumen fluid sample and processed for cDNA synthesis that was used as template in PCR amplification of V3-V4 region of bacterial and archaeal 16S rRNA genes and V9 region of eukaryotic 18S rRNA gene, followed by sequencing using 300 bp paired-end Illumina Miseq platform. Rumen fluid samples were also analysed microscopically for recording the total No. and motility of protozoa. A diet dependent shift was observed in the most commonly detected rumen bacterial phyla (*Bacteroidetes, Firmicutes,* and *Proteobacteria*) and archaeal families (*Methanomassiliicoccaceae, Methanobacteriaceae*, and *Methanosarcinaceae*). Rumen protozoal communities, total protozoa No. and motility was variable among different cows and dietary treatments. This emerging approach allowed to study the diet dependent modifications of the total metabolically active rumen microbial communities of cows belonging to Italian Simmental breed. In future, this innovative procedure can also be applied to samples obtained using non-invasive techniques, i.e buccal fluid, chewed bolus (regurgitated digesta), or faeces to explore rumen microbial communities with possible advantages of large-scale periodic sampling, and avoiding the need of surgically modified animals.

Session 58 Poster 9

Effects of cinnamaldehyde on microbial protein synthesis and bacterial diversity in Rusitec

J. García-Rodríguez[1], M.D. Carro[2], C. Valdés[1], S. López[1] and M.J. Ranilla[1]
[1]Universidad de León, IGM CSIC-ULE, Departamento de Producción Animal, Campus de Vegazana s/n, León, 24071 León, Spain, [2]Universidad Politécnica de Madrid, Dpto. Producción Agraria, ETSIAAB, 28040 Madrid, Spain; mjrang@unileon.es

Improving microbial protein synthesis in the rumen is a very important aim in ruminant feeding, not only in productive terms but also for the environmental impact derived from a low protein utilization efficiency. The objective of this study was to evaluate the effect of cinnamaldehyde (CIN) on microbial protein synthesis (MPS) and bacterial diversity in Rusitec fermenters fed a dairy sheep diet (50:50 alfalfa hay:concentrate). Four Rusitec fermenters were used in a cross-over design in two 14-day incubation periods, and in each run 2 fermenters received one of the experimental diets: unsupplemented (CON) and supplemented daily with 180 mg CIN/l ruminal liquid. On day 14, both liquid and solid digesta were sampled to determine MPS using 15N, and additional samples were taken for DNA extraction and automated ribosomal intergenic spacer analysis (ARISA). The addition of CIN increased ($P<0.05$) MPS in the solid digesta (127 and 160 mg N/d for CON and CIN, respectively) and tended to increase ($P=0.06$) MPS in the liquid digesta. Consequently, both total MPS and MPS efficiency were higher ($P<0.05$) in CIN-treated vessels compared with CON (237 vs 289 mg N/d, and 23.0 vs 28.5 g microbial N/kg OM apparent fermented for CON and CIN, respectively). Supplementation with CIN increased ($P<0.05$) both the number of peaks detected by ARISA (31.3 vs 36.0 for CON and CIN) and the Shannon index (3.42 vs 3.58 for CON and CIN) in the solid phase, whereas no effects were observed in the liquid phase. These results suggest that cinnamaldehyde can be an interesting additive in ruminant feeding, increasing the amount and efficiency of microbial protein synthesis and the bacterial diversity in the rumen digesta. Funding from the Spanish Ministry of Economy and Competitiveness (Project AGL2008-04707-C02-02) is gratefully acknowledged.

Session 58 Poster 10

Nutrients' digestibility in sheep rations containing legume seeds

K. Zagorakis[1], V. Dotas[2], M. Karatzia[2], M. Koidou[1], E. Sossidou[3] and M. Yiakoulaki[4]
[1]Aristotle University of Thessaloniki, Department of Animal Nutrition, Thessaloniki, 54124, Greece, [2]Hellenic Agricultural Organization-DEMETER, Research Institute of Animal Science, Giannitsa, 58100, Greece, [3]Hellenic Agricultural Organization-DEMETER, Veterinary Research Institute, Thessaloniki, 57001, Greece, [4]Aristotle University of Thessaloniki, Department of Range Management, Thessaloniki, 54124, Greece; karatzia@rias.gr

Determination of *in vivo* apparent total tract digestibility of nutrients is a time demanding, expensive and not completely in line with the EU regulations method. Finding alternatives and animal-friendly methods is very important as long as it is accompanied by a relatively high repeatability. The aim of this study was to compare the total tract digestibility of nutrients to that estimated using lignin as an internal marker. Use of lignin is an animal-friendly and low-cost method since trials with animals are not required. For this purpose, an *in vivo* digestibility trial was carried out using castrated rams of the Chios breed in a 4×4 Latin Square experimental design. The apparent nutrient digestibility coefficients of the four rations containing as a main protein source either soybean meal or ground seeds of lupine or faba bean or chick pea were determined over four experimental periods. For each nutrient, 16 digestibility values with the *in vivo* experiment were determined and 16 values with lignin were estimated, respectively. Data were analysed using SPSS-23 statistical package. Linear regression equations ($y = a + bx$) concerning on digestibility of organic matter, crude protein, ether extract, nitrogen free extracts, NDF, ADF, hemicelluloses and cellulose were created and correlation coefficients were determined. The results showed high reliability and significant correlation between experimentally determined and estimated values ($R^2=0.9413 - 0.9866$ & $P\leq0.001$, for the digestibility of all nutrients). It was concluded that the use of lignin as internal marker is a reliable and rapid method for nutritive value estimation of sheep rations containing legume seeds as main protein source.

Exploiting genomic data of autochthonous pig breeds: conservation genetics comes of age

L. Fontanesi[1], G. Schiavo[1], S. Bovo[1], A. Ribani[1], C. Geraci[1], S. Tinarelli[1,2], M. Muñoz[3,4], A. Fernandez[3], J.M. García-Casco[3,4], R. Bozzi[5], P. Dovc[6], M. Gallo[2], B. Servin[7], J. Riquette[7], M. Čandek-Potokar[8], C. Ovilo[3] and the Treasure Consortium[8]
[1]University of Bologna, Dept. of Agricultural and Food Sciences, Viale Fanin 46, 40127 Bologna, Italy, [2]ANAS, Via Nizza 53, 00198 Roma, Italy, [3]INIA, Dept. Mejora Genética Animal, Crta de la Coruña km. 7,5, 28040 Madrid, Spain, [4]Centro de I+D en Cerdo Ibérico INIA-Zafra, Ctra. Santos 1, 06300 Zafra, Spain, [5]University of Florence, Dept. of Agrifood Production and Environmental Sciences, Via delle Cascine 5, 50144 Firenze, Italy, [6]Biotechnical Faculty, Dept. of Animal Science, Groblje 3, 1230 Domžale, Slovenia, [7]INRA, GenPhySE Lab, 24 chemin de Borde-Rouge Auzeville Tolosane, 31326 Castanet Tolosan, France, [8]Kmetijski Inštitut Slovenije, Hacquetova 17, 1000 Ljubljana, Slovenia; luca.fontanesi@unibo.it

Animal genetic resources are important reservoirs of genetic diversity derived by distinct selection pressures or as result of adaptation to production conditions. The TREASURE project has investigated genetic variability in 20 European autochthonous pig breeds with the aim to describe their singularity, evaluate their adaptation, develop new methodologies for their management and identify DNA markers for breed allocation and meat authentication. Genomic data have been obtained by genotyping candidate gene markers and high density single nucleotide polymorphism arrays in ~48 animals from each breed and by whole genome resequencing. Description of genetic diversity has been obtained using several parameters. Runs of homozygosity and genomic inbreeding measures have been correlated with pedigree inbreeding coefficients. A few breed specific markers have been identified and applied. Genome wide association studies have identified genomic regions affecting unique phenotypes. This project represents one of the few examples of exploitation of genomic information that not only benefits the investigated animal genetic resources but also can provide useful information that could impact commercial populations. Funded by European Union's H2020 RIA program (grant agreement no. 634476).

Structural differences among pig genomes illustrate genetic uniqueness of breeds

M. Zorc[1], J. Ogorevc[1], M. Škrlep[2], R. Bozzi[3], M. Petig[4], L. Fontanesi[5], C. Ovilo[6], Č. Radović[7], G. Kušec[8], M. Čandek-Potokar[2] and P. Dovč[1]
[1]University of Ljubljana, Biotechnical Faculty, Department of Animal Science, Groblje 3, 1230 Domzale, Slovenia, [2]Agricultural institute of Slovenia, Hacquetova 17, 1000 Ljubljana, Slovenia, [3]DISPAA, Animla Sciences Section, Via delle Cascine 5, 50144 Firenze, Italy, [4]BESH, Raiffeisenstraße 18, 74523 Schwäbisch Hall, Germany, [5]University of Bologna, Department of Agricultural and Food Sciences, Laboratory of Livestock Genomics, Viale Fanin 46, Bologna, Italy, [6]INIA, 6Department of Animal Breeding, Ctra. De la Coruña, km 7.5, 28040 Madrid, Spain, [7]Institute for Animal Husbandry, Auto put 16, 11080 Belgrade-Zemun, Serbia, [8] J.J. Strossmayer University, Faculty of Agriculture in Osijek, Vladimira Preloga 1, 31000 Osijek, Croatia; peter.dovc@bf.uni-lj.si

The availability of high-throughput whole-genome sequencing (WGS) data illustrating differences among different pig breed genomes opened a new area of genomic research focused on variation caused by single nucleotide polymorphisms (SNP), small scale variation and structural variants which may all contribute to phenotypic variation among pig breeds. In our study (performed within TREASURE project) we re-analysed WGS-based data sets from more than 20 breeds, including commercial and local breeds as well as some wild boar genomes, deposited in publicly available databases. This bioinformatics tool enables discovery of new SNPs, estimation of allele frequencies (genotyping by sequencing) at candidate loci and identification of structural variation in a wide range of pig breeds. The analysis underlined the relevance of structural differences at KIT and MC1R locus involved in colour pattern formation, as well as LEPR locus associated with fatness, fatty acid metabolism and intramuscular fat composition. This approach allows discovery of important genomic differences between commercial breeds and local breeds which are analysed in the frame of the TREASURE project. Extensive mining of publicly available genomic data can together with the newly generated genomic information from local breeds, significantly contribute to the detailed characterisation of animal genetic resources present in local pig breeds. Funded by European Union's H2020 RIA program (grant agreement No. 634476).

Genetic structure of autochthonous and commercial pig breeds using a high-density SNP chip

M. Muñoz[1], J.M. García-Casco[1], A.I. Fernández[1], F. García[1], C. Geraci[2], L. Fontanesi[2], M. Čandek-Potokar[3] and C. Óvilo[1]

[1]INIA, Departamento Mejora Genética Animal, Crta. de la Coruña, km 7,5, 28040, Madrid, Spain, [2]University of Bologna, Department of Agricultural and Food Sciences, Viale Fanin 46, 40127, Bologna, Italy, [3]Kmetijski Inštitut Slovenije, Hacquetova 17, 1000 Ljubljana, Slovenia; mariamm@inia.es

One of the main objectives in TREASURE project is the genetic characterisation of European autochthonous pig breeds through genetic and genomic tools. The aim of the current pilot study was to evaluate the usefulness of the GeneSeek Genomic Profiler porcine SNP chip to describe genetic diversity of five Mediterranean autochthonous and three commercial pig breeds. 44 animals from Iberian (12), Krskopolje (4), Casertana (5), Cinta Senese (5), Apulo Calabrese (5), Duroc (5), Large White (4) and Landrace (4) breeds were genotyped. All DNA samples were successfully genotyped (call rates ≥0.98). A total of 59,193 SNPs were used in the genetic analyses after QC filtering. Observed (Ho) and expected (He) heterozygosities, FIS statistic and genetic distances (DS and FST) were computed. The overall FST value was 0.15. Ho and Hs values per breed ranged from 0.25 to 0.36 (Ho) and from 0.26 to 0.40 (Hs). Calabrese and Duroc were the breeds with the highest (0.16) and lowest (-0.01) FIS values, respectively. DS and FST genetic distances were very similar. Duroc and Large-White were the closest breeds, since DS and FST were 0.09 and 0.19, respectively and Landrace and Duroc the furthest ones, being DS and FST values equal to 0.22 and 0.34. Distribution of minor allele frequencies (MAF) showed that Iberian is the breed with the highest number of monomorphic SNPs (25.3%) and Landrace and Krskopolje have more alleles at intermediate frequencies. These results provide insights on the genetic diversity and relationships among the investigated breeds. This study will be enlarged to characterize the structure of European autochthonous pig populations using the same SNP chip. Funded by European Union's H2020 RIA program (grant agreement no. 634476).

Impact of merging different commercial lines on the genetic diversity of the Dutch Landrace pigs

B. Hulsegge[1,2], M.P.L. Calus[2], A.H. Hoving[1,2], M.S. Lopes[3], H.J.W.C. Megens[2] and J.K. Oldenbroek[1]

[1]Centre for Genetic Resources, the Netherlands, Wageningen University & Research, P.O. Box 338, 6700 AH, Wageningen, the Netherlands, [2]Wageningen University & Research, Animal Breeding and Genomics, P.O. Box 338, 6700 AH, Wageningen, the Netherlands, [3]Topigs Norsvin Research Center, P.O. Box 43, 6640 AA Beuningen, the Netherlands; ina.hulsegge@wur.nl

The pig breeding industry has experienced a number of mergers in the past decades. In these mergers, various traditional breeds and commercial lines were merged or discontinued, which is expected to narrow down the genetic diversity of commercial lines. The objective of the current study was to assess the level of genetic diversity that is still present in the commercial Topigs Norsvin Dutch Landrace in comparison to other Landrace lines conserved in the Dutch gene bank collection. Samples from six Dutch Landrace lines (Dutch lines from Fomeva, Dumeco and Stamboek, and Dutch Norwegian/Finnish lines from Cofok, Dumeco and Stamboek; (n=11 to 49)) for which gene bank collections were available, and the commercial Topigs Norsvin Dutch Landrace line (n=34) were genotyped with the 80K porcine SNP chip. Principal components analysis clearly divided the Landrace lines into two main clusters represented by Norwegian/Finnish Landrace lines, introduced in the Netherlands between 1970 and 1980, and Dutch Landrace lines. The current commercial Topigs Norsvin Dutch Landrace was positioned between the Norwegian/Finnish and the Dutch lines. Structure analysis revealed that each of the six gene bank lines has a unique genetic identity. The Topigs Norsvin Dutch Landrace showed a high admixture level and relationships with the six conserved lines. The 7 different lines all contributed to the genetic diversity of the Landrace breed. As expected, the current Topigs Norsvin Dutch Landrace contains only little unique genetic diversity not present in the other lines. Total genetic diversity, measured by Eding's core set method, was 0.99, while the diversity of the Topigs Norsvin Dutch Landrace was 0.89. These results indicates that merging commercial Landrace lines did reduced the genetic diversity of the Landrace population in the Netherlands. It is, therefore, advisable to conserve selection lines in a gene bank before merging them.

Demographic structure and genetic diversity of the Bísaro pig: evolution and current status

G. Paixão[1], N. Carolino[2] and R. Payan-Carreira[1]
[1]Universidade de Trás-os-Montes e Alto Douro, CECAV, Quinta dos Prados, 5000-801 Vila Real, Portugal, [2]National Institute of Agrarian and Veterinarian Research, Strategic Unit for Biotechnology and Genetic Resources, Quinta da Fonte Boa, 2005-048 Vale de Santarém, Portugal; gus.paixao@utad.pt

Sustainable use of genetic resources, in adaptive breeding and conservational programs, depend on good management of the genetic diversity. Bísaro pig is a local endangered swine Portuguese breed, descending from the Celtic line, well known by its high valued smoked-cured products. This study intends to access the evolution and current demographic structure and genetic status of the Bísaro population. The complete pedigree information contained in the Bísaro herd book, including records from 206,507 animals born from January 1994 to June 2017, was used to evaluate the population structure and investigate the current breed's genetic variability. Since the breed's foundation, the number of registered animals, producers and farrowing records have steadily increased. The mean progeny size for sires was 113.30±211.42 and 2.92% of all breeding boars originated 25.94% of all registered births indicating a marked unbalanced use of certain sires. The calculated generation interval was 1.92±1.12. The mean calculated equivalent generations was 4.45, and 97.8% of all the animals had known parents, indicating a good degree of pedigree completion and depth. The average inbreeding coefficient and average relatedness was 10.27%, higher than most of those calculated for other local and commercial pig breeds worldwide. A low f_e/f_a ratio was obtained (1.07), showing a well-balanced founder/ancestor contribution, and subsequent genetic transfer from generation to generation. Yet, the f_e/f ratio was particularly low (0.02) compared with the other native Portuguese pig breeds. The estimated loss of genetic diversity due to unequal founder contribution was 0.43%, and represented a greater relative proportion when compared to random genetic drift. The Genetic Conservation Index steadily grew over the years with a mean value of 6.63±5.09. The effective population size for the entire population calculated from the inbreeding increase of two successive generations and from log regression method ranged from 15.42 to 68.54, respectively.

The effects of selection on the genomic architecture of the Italian Large White and Duroc heavy pigs

F. Bertolini[1,2,3], G. Schiavo[2], S. Tinarelli[2,4], M. Gallo[4], M.F. Rothschild[1] and L. Fontanesi[2]
[1]Iowa State University, Department of Animal Science, 2255 Kildee Hall, 50011 Ames, Iowa, USA, [2]University of Bologna, Department of Agricultural and Food Sciences, Viale G. Fanin 46, 40127 Bologna, Italy, [3]Technical University of Denmark, Department of Bio and Health Informatics, Kemitorvet, Building 208, 2800 Kgs. Lyngby, Denmark, [4]Associazione Nazionale Allevatori Suini (ANAS), Via Nizza 53, 00198 Roma, Italy; fbert@iastate.edu

Artificial selection may cause an increasing of homozygosity in circumscribed regions of the genome and lead to allele drift. This is particularly important for selection nuclei of pure breeds. In this work we analysed the genome of the Italian Large White (ILW) and Italian Duroc (ID) breeds, that are purposely selected for the heavy pig breeding industry, to identify regions of high homozygosity. A total of 1,953 Italian Large White and 460 Italian Duroc pigs were genotyped with the Illumina PorcineSNP60 BeadChip. Filtering was performed to remove poor quality single nucleotide polymorphisms (SNPs) having call rate<0.9 that were then imputed whereas no other filter was applied. The Runs of Homozygosity (ROH) analyses were performed with the sliding window-based option of the Plink 1.9 software, allowing a minimum of 30 SNP to define a ROH. Then, for each SNP, the percentage of the number of animals that have a ROH containing the SNP was calculated and the top 0.994 percentile of the SNPs empirical distribution was considered. The analyses of the ILW detected four regions, three on porcine chromosome (SSC) 1 and one region on SSC14, with size from ~1 to 5 Mb and with maximum peak of shared regions of 70% on SSC1. The analysis of the ID pigs showed several SNPs shared by more than the 50% of the animals, particularly two regions on SSC9 and SSC15, with maximum peak of 95% on SSC15. The detected regions contained genes involved in functions such as DNA repair and biosynthesis of cellular amino acid. The possible effects of these regions and the genes included in relation to production traits still need to be analysed. Partially funded by European Union's H2020 RIA program (grant agreement no. 634476).

Session 59 Theatre 7

Incidence of RYR1 genotype and its effect on meat quality in Slovenian Krškopolje pigs

M. Škrlep[1], U. Tomažin[1], N. Batorek Lukač[1], M. Prevolnik Povše[2], J. Ogorevc[3], P. Dovč[3] and M. Čandek-Potokar[1]
[1]KIS -Agric. Inst. Slovenia, Hacquetova 17, 1000 Ljubljana, Slovenia, [2]Univ. of Maribor, FKBV, Pivola 10, 2311 Hoče, Slovenia, [3]Univ. of Ljubljana, BF, Groblje 3, 1230 Domžale, Slovenia; martin.skrlep@kis.si

In this study (of project TREASURE*), two objectives were pursued; to investigate the incidence and effect of *RYR1* genotype on carcass and meat quality traits in the only Slovenian local pig breed Krškopolje (KK). Thirty-six castrates originating from 12 litters were reared on the same farm and slaughtered in a commercial abattoir at the average age of 228±6 days and weight of 121±14 kg. Carcass properties and longissimus dorsi (LD) quality traits and chemical composition were determined. Pigs were genotyped for c. C1843 (p. Arg615Cys) at RYR1 locus by PCR-RFLP method (mutant allele further denoted as 'n' and wild as 'N'). Data were analysed using GLM procedure of SAS/STAT module, with fixed effect of RYR1 genotype in the model. Genotype frequencies were 2.7, 41.6 and 55.5% for n/n, N/n and N/N respectively. Compared to N/N animals, pigs with mutant recessive allele (N/n) exhibited lower growth rate and leaner carcasses (i.e. higher lean meat %, muscle thickness and loin eye area, thinner backfat and smaller fat area over LD, $P<0.05$). A marked effect of RYR1 on meat quality was also observed. In N/n pigs, the rate of pH fall was faster ($P<0.05$). Meat of N/n pigs exhibited also higher shear force resistance ($P<0.05$) in addition to lower water holding capacity (higher drip, thawing and cooking loss, $P<0.05$). The present results demonstrate relatively high incidence of RYR1 mutation in KK pigs calling for more breeding effort for its elimination and thus improvement of meat quality and processing aptitude of the breed. Funded by European Union's H2020 RIA program (grant agreement no. 634476).

Session 59 Theatre 8

Poster summary

S. Millet[1] and M. Čandek-Potokar[2]
[1]Flanders research institute for agriculture, fisheries and food (ILVO), Scheldeweg 68, 9090 Melle, Belgium, [2]Agricultural Institute of Slovenia (KIS), 7, Hacquetova ul. 1, 1000 Ljubljana, Slovenia; sam.millet@ilvo.vlaanderen.be

In this 30 minute overview, the posters describing different aspects of local pig breeds will be summarised. For each poster, the key message will be given in one or two slides.

Session 59 Poster 9

Small variation in diet energy content affects muscle gene expression in Iberian pigs

Y. Núñez[1], L. Calvo[2], R. Benítez[1], J. Ballesteros[1], J. Segura[3], J. Viguera[4], C. López-Bote[3] and C. Ovilo[1]
[1]INIA, A6, Madrid, Spain, [2]Incarlopsa, Tarancón, Madrid, Spain, [3]UCM, A6, Madrid, Spain, [4]IMASDE, Pozuelo, Madrid, Spain; nunez.yolanda@inia.es

Modulation of dietary energy content may be a tool to influence pork fat quantity and quality. The metabolic response to different diets depends on individual factors as genetic background, age or sex and treatment factors as intensity or duration. The aim of this study was to analyse the effects of small variations in the energy content of growing and finishing diets on ham muscle transcriptome in Iberian commercial crossbred pigs. One hundred and fifty animals of both sexes (82 females and 68 males) were employed. These were subjected to three dietary protocols: a control diet with low energy in growing and finishing periods (LL, n=44), a low energy diet in growing and high energy diet in finishing (LH, n=54) and a high energy diet in both growing and finishing periods (HH, n=52). High and low energy diets were isoproteic and differ in energy and fat content (5% higher digestible energy in H diet). Scarce effects of the diets were observed at the phenotypic level with a trend for higher and quicker fattening in the animals fed high energy in growing period, especially in males. Animals were sacrificed with a mean live weight of 158 kg. Biceps femoris muscle samples from 30 animals (5 males and 5 females from each treatment) were collected for transcriptome analysis with RNAseq technology. Differential expression analyses showed 8 genes differentially expressed (DE) between HH and LL and 23 DE genes between LH and LL, when both sexes were jointly analysed. The differential expression results obtained in the separate analysis of males and females suggested a strong interaction diet×gender. For example, in males relevant genes involved in lipid metabolism, such as *SCD* and *PLIN1*, were DE between HH and LL, and other ones involved in glucose metabolism were DE between LH and LL, such as *PPP1R3B* and *PDK4*; which were not DE in females. These interaction effects may be related with the phenotypic results and should be further explored. Results show interesting diet effects, conditional on sex, on the expression of relevant genes after slight modifications of diet composition in different periods.

Session 59 Poster 10

An online phenotype database: first step towards breeding programs in local pig breeds

M.J. Mercat[1], E. Zahlan[1], M. Petig[2], H. Lenoir[1], P. Cheval[1], M. Čandek-Potokar[3], M. Škrlep[3], A. Kastelic[4], B. Lukić[5], J. Nunes[6], P. Pires[6] and B. Lebret[7]
[1]IFIP-Institut du Porc, La Motte au Vicomte, 35651 Le Rheu, France, [2]BESH, Haller Str. 20, 74549 Wolpertshausen, Germany, [3]KIS, Hacquetova ul. 17, 1000 Ljubljana, Slovenia, [4]KGZS-Zavod NM, Šmihelska cesta 14, 8000 Novo mesto, Slovenia, [5]PFOS, TRG SV TROJSTVA 3, 31000 Osijek, Croatia, [6]IPVC, Praça General Barbosa, 4900 Viana do Castelo, Portugal, [7]PEGASE, INRA, Agrocampus-Ouest, 35042 Rennes, France; benedicte.lebret@inra.fr

In order to further allow implementation of breeding programs in local pig breeds, with selection objectives defined for each local breed, we aimed at developing a standardised recording of carcass and meat quality traits. These data have to be connected with herdbooks to estimate genetic parameters of the traits (heritabilities and genetic correlations) which are necessary to define breeding objectives. Today the situation is very different from one local breed to another. No or very few phenotypes are recorded in some of them, while breeding programs already exist for a few breeds. To promote phenotyping, a dedicated database and a website were developed in the frame of the TREASURE project. First, the required variables have been collected for six local breeds: Basque (FR), Bísaro (PT), Crna slavonska (HR), Gascon (FR), Krškopoljski (SI) and Schwäbisch-Hällisches (DE). In total 74 variables have been identified dealing with animal herdbook information (10), rearing and growth (22), carcass (22) and meat quality (20) attributes. The database is compatible with the various identifiers used in the different countries: animal IDs, breed, farm… codifications. Major attention has been paid to the description of measurement methods of traits. Thus, each carcass and meat quality phenotype is associated to a method description representing 35 additional variables. The website can be easily translated into several languages. The website and database are currently on test until the end of the TREASURE project. All the breeds studied in TREASURE are free to use these tools. The database can be duplicated so that each partner can host its own data. Funded by European Union H2020 RIA program (grant agreement no. 634476).

Session 59 Poster 11

Variance components for semen production traits in Swiss pig breeds

A. Burren[1], M. Zumbrunnen[1], H. Jörg[1] and A. Hofer[2]
[1]Bern University of Applied Sciences, School of Agricultural, Forest and Food Sciences HAFL, Länggasse 85, 3052 Zollikofen, Switzerland, [2]SUISAG, Allmend 8, 6204 Sempach, Switzerland; marion.zumbrunnen@bfh.ch

The objective of this study was to estimate variance components (VC) for the semen production traits ejaculate volume, sperm concentration and number of semen portions in the Swiss pig breeds Swiss Large White (SLW), Swiss Landrace (SLR), PREMO® (PR) and Duroc (DU). For this purpose, the semen production traits from 821 boars (SLW=243; SLR=49; PR=437; DU=92) with 48,175 records (SLW=6,804; SLR=1,880; PR=34,356; DU=5,135) were used. The data were collected in the years 2009-2016. Genetic variances and covariances were estimated by REML using the software ASReml 3.0. The model for genetic parameter estimation across all breeds included the fixed effects birthyear of boar, age of boar × month of collection, year of collection, collection interval, breed and boar handler as well as a random additive genetic component and a permanent environmental effect. The same model without a fixed breed effect was used to estimate VC separately for each of the breeds SLW, PR and DU. Estimated heritabilities across all breeds were 0.11, 0.15 and 0.18 for ejaculate volume, sperm concentration and number of semen portions, respectively. Similar heritabilities were estimated for ejaculate volume (0.10; 0.14; 0.32), sperm concentration (0.19; 0.14; 0.06) and number of semen portions (0.07; 0.23; 0.12) in SLW, PR and DU breed, respectively. The phenotypic and genetic correlations across all breeds between ejaculate volume and sperm concentration were negative (-0.52; -0.29). The other correlations across all breeds were positive. The phenotypic and genetic correlations were 0.21 and 0.32 between number of semen portions and ejaculate volume, respectively. Between number of semen portions and sperm concentration the phenotypic and genetic correlations were 0.61 and 0.77, respectively. The results seem plausible and show, that genetic improvement trough selection would be possible in semen production traits.

Session 60 Theatre 1

Ethical questions in equine research/training

R. Evans[1] and A.S. Santos[2]
[1]Hogskulen for Landbruk og Bygdeutvikling, 22 Arne Garborgsveg, 4340 Bryne, Norway, [2]Universidade de Trás-os-Montes e Alto Douro, CECAV, Animal Science, Qta de Prados, 5000 Vila Real, Portugal; rhys@hlb.no

This session focuses on the broad range of ethical challenges faced by animal researchers, from veterinary science, to the social causes of animal welfare issues, from the perspective of the Horse Commission. It will be a Challenge session where invited speakers will raise topics and both the speakers and, in some cases, the audience will be invited to discuss the issues raised, bringing their own unique experiences to the dialogue. The purpose is not to set ethical guidelines but rather to expand our thinking about the ethical implications of what we do.

Session 61 Theatre 1

Host genotype or performance: what makes host animal less tolerant to nematode infections?

G. Das, M. Stehr and C.C. Metges
Leibniz Institute for Farm Animal Biology, (FBN-I7.0), Wilhelm-Stahl-Allee 2, 18196, Dummerstorf, Germany; gdas@fbn-dummerstorf.de

Considerable differences exist between various breeds of host animals in terms of resistance and tolerance to parasites. Tolerance to infections is assessed through host animal performance, which is strongly influenced by host genetic. To demonstrate whether host genotype influences the ability of an animal to tolerate parasite infections, performance and genetic effects need to be separated. Although both male and female birds of broiler lines are capable of producing meat efficiently, laying activity is sex-dependent. Chicken lines that have been specifically designed to produce either meat (e.g. Ross-308; R) or egg (e.g. Lohmann Brown+; LB+) or both (e.g. Lohmann Dual; LD) provide an interesting model that might help separate genotype differences from those of performance dependent effects in challenged host. Since genetic effects are expected to exist in both sexes (except for sex-related traits), differences between male and female birds of layer genotypes might partly be attributed to their performance direction and the associated metabolic costs. In two separate experiments (E), we compared nematode-infected or -uninfected male birds for growth (E1) and female birds for laying performance (E2). In E1 male birds of R were compared with those of LB+ and LD. In E2 female birds of LB+ and LD were used. In male birds, overall *A. galli* burden was higher in R than in LB+ ($P<0.05$), whereas higher numbers of *H. gallinarum* were found in LB+ and LD than in R ($P<0.05$). Overall *A. galli* and *H. gallinarum* burdens of LD did not differ from those of R and LB+ ($P>0.05$). In female birds, worm burden with either nematode did not differ between LB+ and LD ($P>0.05$). Performance data (e.g. feed intake, body weight, laying performance) indicated that male birds of R and female birds of LB+ were less tolerant to nematode infections as compared to their male or female counterparts in their respective experiments. In both cases, growth (E1) or laying performance (E2) was impaired due to a nematode-infection induced lower feed intake. Our data collectively suggest that tolerance to nematode infections may not directly depend on host genotype, but its performance level.

Session 61 Theatre 2

Walking on the tiptoe: Is footpad dermatitis in turkeys a problem of the metatarsal foot pad only?

J. Stracke, B. Spindler and N. Kemper
University of Veterinary Medicine Hannover, Foundation, Institute for Animal Hygiene, Animal Welfare and Animal Behavior, Bischofsholer Damm 15, 30173 Hannover, Germany; jenny.stracke@tiho-hannover.de

Foot pad dermatitis (FPD) is an accepted indicator for animal welfare, giving evidence of the animal's health on the one hand and providing information of the management of animal husbandry on the other hand. Various scoring systems exist to evaluate the severity of FPD, with most of them based on describing lesions of the metatarsal foot pad (FP). However, there is only little information on the incidence of lesions on the digital pads (DP). This study therefore aimed to describe the occurrence of lesions on the DP in turkey feet, as well as their interaction with lesions on the FP. At the slaughterline, a total of 250 feet were scored by an automated camera system, providing a 5-scale scoring level with rising severity (0-4), based on the size of the lesion on the FP. Digital pictures of these 250 feet were used for further examination. Each DP was scored using a 5-scale score to evaluate the size of the affected necrotic area (0-4, with rising severity) and a 3-scale score was used providing information on the grade of swelling (0-2). Furthermore, the amount of affected digitals was evaluated. Descriptive statistics is provided for data analysis. Feet which were expected to be intact according their scoring of the FP (n=50), were observed to show lesions on up to every digital, providing scoring levels for each state of severity in both, necrotic areas and level of swellings. Other than expected no linear relationship between FP lesion and DP lesion became obvious (i.e. rising severity in FP lesion score results did not result in rising severity of DP lesion). Each of the parameters were found to show a decrease in severity from FP lesion score 0 to score 1, whereas severity increased from score 1 to score 2. The severity in DP lesions did not increase with rising severity in FP lesions (2, 3 and 4), but stagnated. As a conclusion lesions of the digitals do not develop equivalent to the lesions of the metatarsal foot-pads. Therefore, it is supposed that they might be worth further studies, as one might assume lesions on the digitals to be a relevant indicator of animal welfare, too.

Early warning system for enteropathies in intensive broiler farming

V. Ferrante[1], E. Tullo[1], S. Lolli[1], F. Borgonovo[1], M. Guarino[1] and G. Grilli[2]
[1]Università degli Studi di Milano, Environmental Science and Policy, via Celoria, 2, 20133 Milano, Italy, [2]Università degli Studi di Milano, Veterinary Medicine, via Celoria, 10, 20133 Milano, Italy; valentina.ferrante@unimi.it

In intensive commercial poultry farms, stocking density and environmental conditions can be considered risk factors causing health and welfare problems, mainly represented by enteric diseases. Today, on farms, the enteric pathologies are not preventable and the diagnosis is possible only when the pathology is full-blown. For these reasons, the use of real time monitoring represents a simple way to measure variables that can give an early warning for the farmer providing clear and suitable alerts to help them in their routine. The prompt reaction to any change in health, welfare and productive status is the key for the reduction in drugs usage and for the improvement of animal wellbeing. This study aimed to the application of a PLF diagnostic tool, sensible to the variation of volatile organic compounds, to promptly recognise coccidiosis in intensive farming, supporting veterinarians and enabling specific treatments in case of disease. The experimental trial was carried out in the facilities of the Università degli Studi di Milano and lasted for 45 days. One hundred and twenty Ross 308 one-day-old chicks were placed at day 0 and split in two separate boxes (A and B) with standardised ventilation and rearing conditions. Both boxes measured 2×3 m, the floor was covered with wood shavings and the stocking density was 33 kg/m^2. The oocysts count in the faeces was used as gold standard for the diagnosis of coccidiosis. According to the gold standard, the score plot showed that the air sample could be clustered into two different groups: healthy and sick animals. Furthermore, the discrimination power showed a value close to the unit (0.70-0.80) at the level of 250 oocysts/g of faeces, indicating the good distinction of air samples. The results showed the capability of volatile compounds to reveal coccidiosis in a very early stage and several days before the presence of diarrhoea. These preliminary results are promising for the automated and early detection of coccidiosis in poultry farming contributing to reduce the impact of veterinary drugs. The results showed in this study have been patented (PCT/IB2017/053397).

Breeding for high natural antibody levels reduces impact of *E. coli* (APEC) challenge in chickens

T.V.L. Berghof[1,2], M.G.R. Matthijs[3], J.A.J. Arts[2], H. Bovenhuis[1], R.M. Dwars[3], J.J. Van Der Poel[1], M.H.P.W. Visker[1] and H.K. Parmentier[2]
[1]Wageningen University & Research, Animal Breeding & Genomics, P.O. Box 338, 6700 AH Wageningen, the Netherlands, [2]Wageningen University & Research, Adaptation Physiology, De Elst 1, 6708 WD Wageningen, the Netherlands, [3]Utrecht University, Faculty of Veterinary Medicine, Department of Farm Animal Health, Yalelaan 7, 3584 CL Utrecht, the Netherlands; tom.berghof@wur.nl

Natural antibodies (NAb) are antibodies recognizing antigens without previous exposure to these. Keyhole limpet hemocyanin (KLH)-binding NAb levels in chickens are heritable, and higher KLH-binding NAb levels were associated with higher survival during laying period. Selective breeding for higher NAb levels might therefore increase survival. The objective of this study was to investigate difference in disease resistance between chickens selectively bred for high or low levels of total KLH-binding NAb levels. Two infection experiments were performed to test for differences in disease resistance against avian pathogenic *E. coli* (APEC). At 8 days of age, chickens (n=100 chickens/treatment/line) were intratracheally inoculated with: 0.2 ml phosphate buffered saline (PBS) mock inoculate (both experiments), 0.2 ml PBS containing $10^{8.20}$ colony-forming units (cfu)/ml APEC (experiment 1), $10^{6.64}$ cfu/ml (experiment 2), or $10^{7.55}$ cfu/ml APEC (experiment 2). Mortality was recorded for 7 days after inoculation. In experiment 1, the High line had a significant lower mortality compared to the Low line. In experiment 2, similar trends were found, but not significant, likely due to the lower APEC doses used. For all APEC doses, 50-60% reduced mortality was observed in the High line compared to the Low line. In addition, morbidity (i.e. colibacillosis lesion scores) was determined on various organs of the surviving chickens at 15 days of age. The lowest APEC dose resulted in significant lower lesion scores in the High line compared to Low line. We conclude that selective breeding for high KLH-binding NAb levels increases APEC resistance in early life. This gives opportunity for reducing the impact of diseases by selective breeding for increased disease resistance.

Session 61 — Theatre 5

Chilean citizen opinion regarding laying hen production systems

D. Lemos Teixeira[1], M.J. Hötzel[2], M. Azevedo[1,2], L. Rigotti[1], L. Salazar-Hofmann[1], R. Larraín[1] and D. Enríquez-Hidalgo[1]

[1]Pontificia Universidad Católica de Chile, Departamento de Ciencias Animales, Santiago, Chile, [2]Universidade Federal de Santa Catarina, Laboratório de Etologia Aplicada, Florianópolis, 88034-000, Brazil; dadaylt@hotmail.com

The aim of this study was to assess the attitudes of Chilean citizens for layers' production scenarios differing in housing and feather pecking levels. Citizens (n=404) were recruited in public places and were informed that laying hens from all type of egg production systems may be affected by feather pecking, and that most of laying hens in Chile are kept in barren, 550 cm^2 cages/hen. Participants were randomly assigned to one out of six different short hypothetic scenarios that described, in different combinations, one of three types of egg production systems [CON: conventional cages (550 cm^2/hen); FUR: furnished cages (750 cm^2/hen); FRE: free-range (2 m^2/hen)] and one out of two levels of feather pecking [(50 to 80% of the groups (High) or very low (Low)]. Participants were then asked about their level of agreement using Likert scale (1 to 5): (Q1) 'How much do you approve this type of production system?'; (Q2) 'How likely these hens are suffering under these conditions?']. The effects of the scenarios on responses (dependent variable) were analysed with Kruskal Wallis and Mann Whitney. Scenarios affected ($P<0.05$) the responses of Q1 (CON-High: 1.5a, CON-Low: 2.2b, FUR-High: 2.4b, FUR-Low: 2.9c, FRE-High: 3.4c, FRE-Low: 3.8d) and Q2 (CON-High: 1.4a, Con-Low: 1.9b, Fur-High: 1.8b, Fur-Low: 2.4c, FR-High: 2.9d, FR-Low: 3.4e). As expected, participants had more positive attitudes towards scenarios with lower feather pecking and with hens in a free-range system than in cages (furnished and conventional). However, feather pecking influenced participant attitudes towards the scenarios, especially when hens were housed in cages, as CON-Low had similar approval than the FUR-High and FUR-Low than FRE-High. Therefore, these results suggest that citizens prefer free-range systems even when informed of the risk of other animal welfare challenges.

Session 61 — Theatre 6

***In vitro* prebiotic effect of fermentable feed ingredients on the intestinal health of weaned piglets**

J. Uerlings[1], M. Schroyen[1], J. Bindelle[1], E. Willems[2], G. Bruggeman[2] and N. Everaert[1]

[1]Gembloux Agro-Bio Tech, ULiège, Precision livestock and Nutrition Unit, Passage des déportés, 2, 5030 Gembloux, Belgium, [2]Royal Agrifirm Group, Landgoedlaan 20, 7325 AW Apeldoorn, the Netherlands; julie.uerlings@ulg.ac.be

Dietary strategies such as the inclusion of prebiotics have been suggested to modulate intestinal microbiota, promoting the proliferation and metabolic activity of beneficial bacteria. In piglets, this strategy could result in a reduction of post-weaning associated disorders and thus, the use of antibiotics. To date, mainly purified fractions have been tested for their prebiotic effects at weaning while trials on potential health promoting effects of whole grain cereals or roots and corresponding by-products remain rare. In this study, inulin- and pectin-based ingredients, with different levels of purity (cereal/root, by-products and purified fractions) have been screened *in vitro* for their fermentation kinetics as well as for their SCFA and microbiota profiles. Briefly, after an hydrolysis with porcine pepsin and pancreatin, undigested residues were submitted to a batch fermentation using preweaning piglet' faeces as inoculum. In addition to gas measurements, fermentation juices were collected for SCFA production kinetics and microbiota composition. Inulin, oligofructose and chicory root exhibited an extensive and rapid fermentation in comparison to chicory pulp although butyrate levels of root and pulp did not reach the ones of the purified fractions on a dry matter basis. Nevertheless, chicory pulp showed significantly higher levels of *Lactobacillus* ssp., *Bifidobacterium* ssp., *Clostridium* cluster IV and butyryl-CoA:acetate-CoA transferase gene abundance in comparison to chicory root. Sugar beet pulp as well as orange and lime by-products displayed extensive fermentation patterns equivalent to the ones of purified pectin as well as the greatest butyrate production. Moreover, several orange and lime by-products also displayed significantly higher levels of *Bifidobacterium* ssp. and a higher *Firmicutes* to *Bacteroidetes* ratio in comparison to purified pectin. Therefore, chicory root and pulp as well as orange and lime by-products appear to be promising ingredients for piglet' diets to modulate intestinal fermentation for health purposes.

Is it possible to reduce antimicrobial consumption by improving biosecurity and welfare at pig farms

A.H. Stygar[1], I. Chantziaras[2], I. Toppari[3], D. Maes[2] and J. Niemi[1]
[1]Natural Resources Institute Finland (Luke), Bioeconomy and Environment Unit, Latokartanonkaari 9, 00790 Helsinki, Finland, [2]Ghent University, Faculty of Veterinary Medicine, Unit of Porcine Health Management, Salisburylaan 133, 9820 Merelbeke, Belgium, [3]Animal Health ETT, PL 221, 60100 Seinäjoki, Finland; anna.stygar@luke.fi

This study aimed to describe how the number of antimicrobial treatments (expressed as number of treatments per pig) was associated with biosecurity, welfare and meat inspection results of a farm. The data was collected between 2011 and 2013 from 406 Finnish farms, and recorded in the pig health and welfare classification system (SIKAVA). Each time a veterinarian was prescribing an antimicrobial, the number of animals treated, the dose, and the main reason for treatment were registered in SIKAVA. In addition, at least 4 times per year pig farms had to be visited by the herd veterinarian who assessed biosecurity and animal welfare parameters (air quality, condition of facilities, cleanliness, enrichment, stocking density, symptoms of diseases). Finally, data from slaughterhouse inspections was collected e.g. condemnations due to lameness, abscesses and pleurisy. For analysis, these datasets were aggregated at the farm-level to quarter of a year. In total, only around 4% of all pigs received antimicrobial treatments. The main reasons for antimicrobial treatments were musculoskeletal diseases (42%), tail biting (33%) and respiratory disorders (12%). A standard zero-inflated negative binomial model was used to quantify the number of antimicrobial treatments per pig. Statistically significant parameters (and their interactions) depended on the reason for treatment. Here only results for treatment due to musculoskeletal diseases will be presented. The count of antimicrobial treatment per pig increased with the size of a farm, poor condition of water equipment, poor enrichment, increased condemnations due to lameness or pleurisy (lung condemnation) and a combination of poorer condition of pens, and stocking density. This study suggests that by improving biosecurity and welfare at pig farms antimicrobial consumption can be reduced. This research is based on the EU FP7 funded PROHEALTH project (no. 613574).

The effect of disease resistance gene polymorphisms on production traits in Italian Large White pigs

C. Geraci[1], A. Reza Varzandi[1], G. Schiavo[1], M. Gallo[2], S. Dall'Olio[1] and L. Fontanesi[1]
[1]University of Bologna, Department of Agricultural and Food Sciences, Viale Fanin 46, 40127 Bologna, Italy, [2]Associazione Nazionale Allevatori Suini(ANAS), Via Nizza 53, 00198 Roma, Italy; claudia.geraci3@unibo.it

Pathogens derived diseases have a large negative impact on pig production chains. A few genetic markers associated with disease resistance have been recently identified and used in pig breeding plans in several populations as part of disease control programs. Neonatal Diarrhoea (ND), Post-weaning Diarrhoea (PWD) caused by enterotoxigenic *E. coli* and Porcine Reproductive and Respiratory Syndrome caused by PRRS virus are diseases with high priority for the pig industry. Since interactions or antagonism between growth and disease resistance traits could exist, we investigated the association between markers already reported (fucosyltransferase 1, *FUT1* associated with resistance to PWD; mucin 4, *MUC4* associated with resistance to ND; and guanylate binding protein 5, *GBP5*, associated with resistance to PRRS) with production traits (average daily gain, ADG; back fat thickness, BFT; lean meat cuts, LC; visible intermuscular fat, VIF; ham weight loss at the first week of salting, HWLFS) in 778 performance tested Italian Large White pigs. We also evaluated allele frequencies of these polymorphisms in 134 pigs from five local breeds (Apulo-Calabrese, Casertana, Cinta Senese, Mora Romagnola and Nero Siciliano). The frequency of the resistance-associated alleles for the 3 polymorphisms was higher in local breeds, supporting a potential adaptation or natural selection to disease resistance in these populations. All investigated markers were significantly associated with ADG and *FUT1* was also associated with BFT for Italian Large White pigs. Pigs with the resistance genotype of *FUT1* had higher ADG while pigs with the resistance genotypes at *MUC4* and *GBP5* showed lower ADG. A retrospective evaluation of allele frequencies of these polymorphisms did not show any significant changes in the Italian Large White population over the last 20 years. Our results indicate that the implementation of selection programs should integrate information of markers considering their potential pleiotropic or antagonistic effects on different traits.

Session 61 Theatre 9

Preliminary study of parasitic nematodes in farmed snails (Helix aspersa maxima) in Greece

K. Apostolou[1], M. Hatziioannou[1] and S. Sotiraki[2]

[1]University of Thessaly, Department of Ichthyology and Aquatic Environment, School of Agricultural Sciences, Futokou Street, 38446 Volos, Greece, [2]Veterinary Research Institute, Hellenic Agricultural Organization Demeter, Thermi, 57001 Thessaloniki, Greece; apostolou@uth.gr

Parasitic nematodes that have terrestrial gastropods as main hosts are found in eight different families. They are to be found in all organs of the general body cavity (digestive tract, reproductive organs), the pallial cavity (kidney, pulmonary sac), and in the pedal musculature (foot tissue). The aim of this study is to register parasite species in farmed snails in Greece. Parasites are a risk to a snail farm, as they may spread to the farmed snail population, especially when it is dense or hygiene rules are not met. A total of 1,100 gastropod was collected from 22 different farms of 4 different types (mixed system, open field, net-covered greenhouse and elevated sections) in two different periods (summer and autumn 2017). The farms are located in Central, Western and North Greece and bred snails (*Helix aspersa maxima*) for human consumption. Samples of faeces were analysed using a McMaster method for the number of nematode eggs. The presence of nematode larvae or adult was identified through morphological examination and dissection of 440 snails of commercial size (average weight 14.02±2.48 g). The results showed an average number of eggs per gram of faeces of 4,205.18. We found all farms positive, with a higher value 20,880 EPG (Eggs per Gram, Mixed system) and lower 140 EPG (Open Field). In this first survey in Greece three nematodes species were isolated. The prevalence of *Phasmarhabditis hermaphrodita* ranged from 10 to 40% (intestine), 20% for *Muellerius capillaris* (foot tissues, 3rd stage larvae) and 15% for *Alloionema appendiculatum* (foot tissues). *P. hermaphrodita* can increase the mortality rate of gastropods. *M. capillaris* is the most common lungworm of sheep and has been associated with severe diffuse interstitial pneumonia. *A. appendiculatum* is frequently found in farmed snails, where it can cause substantial mortality of young snails.

Session 61 Theatre 10

Sublethal concentrations of xanthine oxidase for preconditioning of pre-freezing rooster semen

M. Yousefi[1], M. Sharafi[1] and A. Shahverdi[2]

[1]Tarbiat Modares University, Faculty of Agriculture, Department of Poultry Science, P.O. Box 14115-336, Iran, [2]Royan Institute, Department of Embryology at Reproductive Biomedicine Research Center, ACECR, Tehran, P.O. Box 16635-148, Iran; m.sharafi@modares.ac.ir

Unique physiological properties of rooster sperm make it susceptible to different damages during cryopreservation. Using sublethal oxidative stress before cryopreservation of sperm is a new approach used in recent years to improve the viability of sperm after thawing. The aim of this project was to evaluate the effects of very low concentrations of Xanthine Oxidase (XO) as sublethal stress before cryopreservation on the motility, viability, mitochondria activity and DNA stability of rooster sperm after thawing. Sperm samples were collected from 10 roosters, once a week, and different concentrations of XO (0, 0.01, 0.1, 1 and 10 µM) were used to consider the effects of low level of oxidative stress induced by XO on the thawed sperm performance. A significantly higher ($P<0.05$) percentage of total motility, progressive motility and mitochondria activity were obtained in sperm treated with XO-1 compared to X-0, XO-0.01, XO-0.1 and XO-10. DNA stability of rooster semen was not affected by the treatment of semen before freezing with XO ($P>0.05$). Maybe in the future, pre-treatment of rooster semen before cryopreservation with the approach of sub-lethal stress would be a strategy to improve the cryosurvival.

Pecking and piling: behaviour of conventional layer hybrids and dual-purpose breeds in the nest

M.F. Giersberg, B. Spindler and N. Kemper
University of Veterinary Medicine Hannover, Foundation, Institute for Animal Hygiene, Animal Welfare and Farm Animal Behaviour, Bischofsholer Damm 15, 30173 Hannover, Germany; nicole.kemper@tiho-hannover.de

Oviposition in an appropriate nest is a behavioural priority in laying hens, and layer strains are known to differ in their patterns of nest use. The aim of the present study was to assess and compare the nest use and the behaviour in the nest in conventional layer hybrids (Lohmann Brown plus, LB+) and dual-purpose breeds (Lohmann Dual, LD). A total of 1,800 untrimmed hens per genetic strain was housed in four compartments of an aviary system with conventional colony nest boxes on the top tier. Video-based data were recorded in two of 12 nests per compartment on two days per week for a total of 18 weeks during the laying period. The number of hens per nest was detected via time sampling method for 12 h after the lights were switched on. Both the continuous observations of the times focal hens spent in the nest, and the instantaneous time samplings of the hens' behaviour were performed for 6 h after lights-on. On average, 4.2±2.5 LB+ and 6.7±4.4 LD hens were observed per nest box during the first 6 h lights-on. In both hybrids the number of hens/nest decreased after 6 h lights-on (0.7±1.0 LB+; 2.8±2.7 LD). Focal LB+ hens spent more time (13:30±09:33, mm:ss) in the nest than LD hens (11:46±09:11). Hybrid effects ($P<0.05$) were found for all observed behaviours, except for scratching and comfort behaviour. The most frequent behaviour in the LB+ hens was sitting (61.3% of the animals), whereas most of the LD showed piling in the nest (37.1%). In contrast, piling occurred only in 1.6% of the LB+ hens. Pecking behaviour was observed in a total of 5.8% of LB+ and 1.7% of LD. LB+ hens showed more ground-, object-, gentle feather- and severe feather pecking than LD. In contrast, LD performed more aggressive pecking and no severe feather pecking at all. The results indicate that not only the patterns of nest use but also the behaviour in the nest differed between conventional layers and dual-purpose breeds.

Session 61 Poster 12

Predicting MPS resistance breeding value with peripheral blood immunity breeding values

K. Suzuki[1], T. Sato[1], T. Okamura[2], C. Shibata[3], Y. Suda[4], K. Katoh[1], S. Roh[1] and H. Uenishi[5]
[1]Tohoku University, Graduate School of Agricultural Science, Aoba-ku, Sendai, Miyagi, 980-8572, Japan, [2]National Insitute of Livestock Science, Tsukuba, Ibaragi, 305-0901, Japan, [3]Miyagi Livestock Experimental Station, Osaki, Miyagi, 989-6445, Japan, [4]Miyagi University, Hatatate, Sendai, Miyagi, 982-0215, Japan, [5]National Institute of Agrobiological Sciences, Tsukuba, Ibaragi, 305-0901, Japan; keiichi.suzuki.c6@tohoku.ac.jp

Mycoplasma pneumonia of swine (MPS) is responsible for great financial losses within the swine industry. The purpose of the present study was to examine which immunological factors should be adopted as selection traits when aiming to improve MPS morbidity. This experiment used 1,349 Landrace pigs from a line selected for five generations. Selection criteria were daily body weight gain from 30 kg to 105 kg, backfat thickness at 105 kg body weight measured using ultrasound technology, and the area of MPS-associated morbid change in 629 slaughtered sibs. When animals reached 70 kg body weight, the first immunization with approximately 108/ml of sheep red blood cells (SRBC) took place. The second immunization was given 1 week later. Blood samples were collected at 7 weeks of age, when individuals had reached 105 kg of body weight. Phagocyte activity, white blood cell number (WBC) and the ratio of granular leukocytes to lymph cells were all evaluated. Complement alternative pathway activity, specific antibody productions to SRBC, IgG titer of Mycoplasma hyopneumoniae, and, cytokine concentrations (IL-10, IL-13, IL-17, TNF-α, and IFN-γ) were measured by ELISA. Heritability and breeding value of MPS morbidity and immunity traits were estimated using VCE6. PLS regression analysis was conducted using SAS. Heritability estimates of the immunological traits were low (0.03~0.23) without 0.31 of WBC at 7weeks. As a result of the PLS analysis using the breeding values of MPS morbidity and 17 peripheral blood immunity traits, MPS breeding value was predicted by seven immunological traits, at a contribution rate of 85%.

Session 61 Poster 13

Chestnut extract seems to modify intestinal bacterial population in the jejunum of piglets

M. Girard, N. Pradervand, P. Silacci, S. Dubois and G. Bee
Agroscope, Tioleyre 4, 1725 Posieux, Switzerland; marion.girard@agroscope.admin.ch

Owing to the increased occurrence of antimicrobial resistances, alternatives to antibiotics to control post-weaning diarrhoea (PWD), especially in case of prophylactic use, is crucial. This study aimed to investigate whether a standard diet (SD) supplemented with 2% chestnut extract (CE; Silvafeed Nutri P/ENC, Silvateam, Italy) could affect the severity of PWD, organic matter digestibility and intestinal bacterial population of piglets infected or not with enterotoxigenic *E. coli* (ETEC) F4. At weaning, 48 piglets were allocated in a 2×2 factorial design balanced for weaning body weight and litter. Three pigs were housed per pen where they had *ad libitum* access to either the SD or CE diet. In each of the 2 dietary groups, 4 d after weaning, 12 piglets were orally infected (INF) with 5 ml of an ETEC (F4ac, LT+, STb+) suspension (10^8 cfu/ml), whereas the remaining 12 pigs received 5 ml PBS (NINF). Faecal score and faecal dry matter were monitored daily. Piglets were sacrificed 3, 6 or 7 d post-infection and jejunum and colon contents were collected to determine bacterial population and organic matter digestibility. The ETEC infection resulted in greater faecal score and lower faecal dry matter ($P<0.05$ for each) but had no ($P>0.10$) effects on feed intake per pen, average daily weight gain, organic matter digestibility nor bacterial population in the jejunum. Feed intake per pen and organic matter digestibility were similar ($P>0.10$) between SD and CE groups. However, piglets fed the CE diet tended ($P=0.08$) to be 43 g heavier than those fed the SD diet. Compared with the SD diet, supplementation with CE reduced faecal score and increased faecal dry matter ($P<0.05$ for each). Preliminary results on bacteria quantification tend to indicate a lower ($P\leq0.09$) relative quantity of *Clostridium* cluster IV and *Lactobacillus* in the jejunum of CE compared with SD piglets. The results of this study suggest that regardless of ETEC infection, chestnut extract reduces faecal score and tend to modify bacterial population in the jejunum without altering organic matter digestibility.

Session 61 Poster 14

Impact of different group sizes on performance, plumage cleanliness and leg disorders in broilers

A. Kiani and U. König Von Borstel
University of Justus Liebig, Department of Animal Breeding and Genetics, Leihgesterner Weg 52, 35392 Gießen (Giessen), Germany; alikianee@gmail.com

The aim of the present study was to assess welfare and performance of broiler chickens housed in different group sizes. For each of three repetitions, within a large group of 6,400 mixed-sex day-old broiler chicks, birds selected for the present study were housed in four types of rectangular enclosures which provided 10 m^2 [small], 30 m^2 [medium], 100 m^2 [large] and 500 m^2 [very large] floor space. Per pen there were 100, 300, 1000 and 5,000 birds, respectively and therefore constant density for all groups (10 birds/m^2). Fifty birds per group were randomly selected and tagged as focus animals. Performance parameters, as well as mortality and health and welfare parameters [gait score, hock burn, tibial dyschondroplasia (TD), footpad dermatitis, plumage cleanliness], were measured at the end of starter (0-10d), grower (11-24d) and finisher (11-38d) period. Welfare and health parameters were assessed using scoring systems on a scale from 0 (indicating non-affected birds) to 2, 3 or 5 (indicating severely affected birds). Broilers in small and medium group sizes had higher body weight and average daily weight gain in the three rearing phases compared to large and very large groups ($P<0.05$). Results for gait scores, hock burn and plumage cleanliness indicated better welfare of birds in small (except hock burn, $P>0.05$) and medium-sized groups, compared to very large groups ($P<0.05$), and TD was more severe in very large compared to large groups. However, foot pad dermatitis was less severe in very large (0.27+ 0.04) compared to medium sized groups (0.44+ 0.05; $P<0.05$), and there were no significant differences for mortality. Results of this study show that some performance and welfare parameters in broilers are negatively affected by the large group size and suggest that general assumption of detrimental effects of large group sizes might need to be reassessed, especially for new commercial broilers, but further research is needed with regard to commercially relevant group sizes.

Reduction of ammonia and nitrogen oxide emissions from pig production using air biofilter

W. Krawczyk, J. Walczak and P. Wójcik
National Research Institute of Animal Production, Sarego 2, 31-047, Poland; wojciech.krawczyk@izoo.krakow.pl

The negative environmental impact of pig production consists in deposition of biogenic elements in soil and water, leading to their overfertilization, eutrophication and contamination, as well as emission of gases affecting the health and welfare of animals and humans. The aim of the study was to determine the reduction potential of NH_3 and NO_x emissions by filtration of air from indoor pig housing. Determinations were made of the groups of NH_3 and NO_x compounds in unfiltered air from pig housing and from using air biofilters with different mixtures forming the biofilter bed. The experiment used mixtures of 3 biofilter beds and air forced into these beds from climatic-respiration chambers. A total of 120 fattening pigs [(Polish Landrace ′ Polish Large White) ′ Duroc] were kept on litter in groups of 10 pigs per chamber in 3 replications. Biofiltration was performed using three mixtures: (1) 50% peat, 25% chopped straw, 25% sawdust; (2) 50% chopped straw, 25% peat, 25% sawdust; (3) 50% sawdust, 25% peat, 25% chopped straw, which were placed in the biofilter, to which mechanically exhausted air from the building was supplied. The data were analysed using Statgraph software. Biofilter use was effective in reducing NH_3 and NO_x emissions from pig production. The experimental biofilter beds were more efficient in reducing ammonia than nitric oxide emissions. The most effective ammonia reducers were mixture 3 (50% sawdust, 25% peat, 25% straw) and mixture 1 (50% peat, 25% sawdust, 25% straw), which reduced the emissions by 88 and 85%, respectively. The usefulness of biofilter mixture 3 was also confirmed statistically for reduction of NO_x. It reduced nitric oxides by 61% compared to 4% for mixture 2.

Prevalence of fibrinous pericarditis at slaughter inspection on heavy pigs (170 kg)

M. Bottacini[1], A. Scollo[2], B. Contiero[1] and F. Gottardo[1]
[1]University of Padova, Department of Animal Medicine, Production and Health, Viale dell'Università 16, 35020 Legnaro (PD), Italy, [2]Swivet Research, Via Martiri della Bettola 67/8, 42123 Reggio Emilia, Italy; flaviana.gottardo@unipd.it

Fibrinous pericarditis has been detected with different prevalences at post-mortem inspection. Although it's not a specific food safety problem, it affects animal health leading to cardiac dysfunction, with possible negative implications for the production. Many pathogens or circulatory alterations can directly or indirectly cause fibrinous pericarditis. The objective of this study was to assess the prevalence of this lesion at slaughter on pigs reared to 170 kg, verifying the association with pleurisy and pneumonia lesions. In one year 658 pig batches from 236 farms located in the North of Italy were visually scored during slaughter procedures for presence of fibrinous pericarditis. Pleurisy and pneumonia lesions were scored using the grids respectively developed by Dottori *et al.* and Madec and Derrien. A total of 58,099 plucks was individually recorded on the dataset. An annual mean of 5.99% of plucks with fibrinous pericarditis was recorded, with a maximum per batch of 26.32%. No seasonal variability was found, nor a correlation with pneumonia lesions. A positive correlation at batch level ($r=0.59$; $P<0.001$) was found between presence of pericarditis and the most severe stage of pleurisy. This was demonstrated also at pluck level, since half of the plucks with pericarditis (49.01%) had severe pleurisy. Farm of origin accounted for 38.44% of variation between batches. These results don't suggest a higher risk of detecting fibrinous pericarditis at slaughter for heavy pigs if compared to previous studies on lighter productions but highlight the connection with severe stages of pleurisy. The lack of correlation with pneŭmonia lesions questions previous results reporting *M. Hyopneumoniae* as the most frequent isolated bacterium from pericardium. Despite its old chronic appearance, further microbiological studies could add important information for a better pathogenesis comprehension. The rejection of affected hearts from human consumption, and the presumable performance reduction on the live animal make fibrinous pericarditis a not negligible income loss for pig industry.

Effect of sample size and length of observation period on agreement of pig organ lesion prevalences

M. Gertz and J. Krieter
Christian-Albrechts-University, Institute of Animal Breeding and Husbandry, Olshausenstr. 40, 24098 Kiel, Germany; jkrieter@tierzucht.uni-kiel.de

The observations of organ lesion scoring of a particular observation period need to be averaged into organ lesion prevalences in order to achieve the application to an on-farm, animal-health monitoring system. However, only little is known about how many observations one should choose to obtain a valid representation of the on-farm health with regard to the variations in organ lesion scoring. The variability of lesion prevalences can be reduced by increasing the number of observations used for averaging, but as the number of carcasses per farmer is limited, one may also need to increase the length of the observation period. This results in a trade-off between validity and timeliness of prevalences, which this study focuses on. For this purpose, scorings of different organ lesions (pleurisy (*pl*), pneumonia (*pn*), milkspots (*ms*), pericarditis (*pc*)) of approximately 11.6 million carcasses of four German abattoirs from 2014-2017 were analysed. For prevalence calculation, the sampling approach used varied the number of observations per farmer (2-100) from successive periods of various lengths (1, 3, 6 and 12 month). The agreement of consecutive prevalences per farmer and abattoir was evaluated using the Spearman correlation (SP) and intra-class correlation (ICC). Further, the signal-to-noise ratio (SNR) and confidence interval (CI) were calculated as a measure of relative prevalence variability. Results indicate distinct differences between the respective lesions analysed, but also substantial differences between abattoirs. The highest average agreement was observed for *ms*(ICC: 0.52, SP: 0.39) and *pl*(ICC: 0.41, SP: 0.42), whereas *pn*(ICC: 0.19, SP: 0.20) and *pc*(ICC: 0.15, SP: 0.14) were considerably lower. ICC and SP as well as SNR and CI tend to form a plateau if using more than 20-50 observations per prevalence, thus indicating roughly constant agreement. With respect to the observation period, agreement was highest for period lengths of 3 to 6 months. In conclusion, results indicate that sample size and abattoir have more influence on agreement than period length.

Effects of zeolite in aflatoxin B1 contaminated diet on aflatoxin residues in duck's tissues

I. Sumantri[1], H. Herliani[1] and N. Nuryono[2]
[1]University of Lambung Mangkurat, Animal Science, Ahmad Yani Street, KM 36, Banjarbaru, 70714, Indonesia, [2]Universitas Gadjah Mada, Chemistry, Kaliurang Street, Bulaksumur Campus, Yogyakarta, 55281, Indonesia; isumantri@ulm.ac.id

Aflatoxins are toxic metabolites produced mainly by *Aspergillus flavus* and *A. parasiticus*. Among them, aflatoxin B1 (AFB1) is the most toxic and carcinogenic compound both for human and animal. Duck is one of the most sensitive animals to aflatoxin exposure that may related with their liver biotransformation capacity. Therefore, consumption of AFB1 contaminated diet will not only lead to decrease on duck performance but potentially present aflatoxin residues in duck's tissues and egg. This research was conducted to study the effects of zeolite as an AFB1 adsorbent in reducing aflatoxin residues in the liver, meat, and egg of Indonesian local laying duck, *Anas platyrinchos Borneo*. Sixty-four of 7 months laying ducks were randomly allotted to 2 levels of AFB1 (low: 30 ppb; and high: 70 ppb) and 2 levels of zeolite inclusion (0 and 2%). *In vivo* trial was conducted for 28 days and at the end of treatment, the birds were sacrificed. Meat, liver, and egg samples were collected for AFB1 and aflatoxin M1 (AFM1) determination. AFB1 and AFM1 were analysed using ELISA tests. Data were analysed by analysis of variance using the general linear model of SPSS software. Results showed levels of AFB1 significantly ($P<0.05$) increase AFB1 concentration in liver and AFM1 in the egg. Zeolite inclusions tended to reduce aflatoxins residues in the liver and egg, but these were not significantly different ($P>0.05$). In conclusion, zeolite inclusion in the diet could reduce aflatoxin residues in the liver and egg of laying duck.

Associations between management and feather damage in laying hens in enriched cages in Canada

C. Decina[1], O. Berke[1], N. Van Staaveren[2], C.F. Baes[2], T. Widowski[2] and A. Harlander-Matauschek[2]
[1]University of Guelph, Population Medicine, 50 Stone Rd. E, N1G 2W1, Guelph, Canada, [2]University of Guelph, Animal Biosciences, 50 Stone Rd. E, N1G 2W1, Guelph, Canada; aharland@uoguelph.ca

Canada is transitioning from conventional cages to alternative housing systems for laying hens, such as enriched cages. However, feather damage (FD) due to feather pecking, when a bird pecks/pulls at the feathers of another, remains a welfare concern in these systems. This study assessed bird, housing and management associations with FD in laying hens in alternative housing systems in Canada. Information about housing and management practices was collected through a questionnaire administered to 122 laying farms nation-wide (response rate of 52.5%), providing information on 26 flocks in enriched cages and 39 flocks in non-cage systems. In conjunction, farmers visually scored a sample of 50 birds using a feather cover scoring system (0: fully feathered, 1: broken/missing feathers with naked area smaller than a 2-dollar coin, 2: poorly feathered with a naked area larger than a 2-dollar coin) to estimate the prevalence of FD on the back/rump area (percentage of birds with score >0). Preliminary analysis using linear regression modelling in R was used to find associations between predictors related to bird characteristics, nutritional and general management and the prevalence of FD in enriched cage flocks. FD prevalence was estimated at ~53% at 40 weeks of age within the model, and age was positively associated with FD. Higher prevalence of FD was associated with brown feathered birds compared to white feathered birds and in flocks where more frequent feeder runs were used. In contrast, more frequent use of the manure belt was associated with substantially lower FD. Finally, lower FD was also observed when a scratch area was provided or when more than 3 workers performed daily inspections. These predictors reflect different subject areas related to stress or fear in birds which have previously been associated with FD. Further longitudinal investigation into management practices in enriched cages could better elucidate the associations found in this study and help farmers improve laying hen welfare during the transition to alternative systems.

Livestock effluents: farm scale effluent management towards agronomic and energetic valorization

O. Moreira[1], R. Fragoso[2], H. Trindade[3], R. Nunes[4], D. Murta[4,5,6] and E. Duarte[2]
[1]INIAV Santarém, Quinta da Fonte Boa, 2005-048, Portugal, [2]LEAF Instituto Superior de Agronomia, U Lisboa, Tapada da Ajuda, 1349-017 Lisboa, Portugal, [3]Universidade de Trás os Montes e Alto Douro, Quinta de Prados, 5000-801 Vila Real, Portugal, [4]INGREDIENT ODYSSEY, Quinta das Cegonhas, 2001-907 Santarém, Portugal, [5]FMV – Universidade Lusófona de Humanidades e Tecnologia, Campo Grande 376, 1749-024 Lisboa, Portugal, [6]CIISA-FMV, ULisboa, Av Universidade Técnica, 1300-477 Lisboa, Portugal; olga.moreira@iniav.pt

Livestock production is concentrated in certain geographical regions, some of which without enough area for agronomic valorisation of effluents, creating an imbalance between 'manure production and manure use'. Therefore, in order to be competitive and comply with legal requirements, the sector must search for alternative management strategies for its effluents which should promote a circular economy. The objective of this Operational Group, supported by PDR 2020 (Portugal), is to valorise livestock effluents as a resource, focusing on the production and integrated management of the different flows generated at farm scale, optimizing their use as secondary raw materials, recovering energy and nutrients, improving farm nutrient balances and promoting sustainable management. The main activities to achieve these objectives are related with the four main routes: manure processing; feed and fertilizer replacement to develop a sustainable animal husbandry, bioenergy production in a livestock farm and biodegradation by black soldier fly larvae. The main goal of the project is the application of an established roadmap for effluents management and testing the weak and strong points to promote common advances in the nexus biowaste/energy/bio-fertilizer, closing the nutrients' cycle towards a sustainable bioenergy economy, creating a balance between manure production and manure use from 2018 onwards. The beneficiaries will be the animal producers and farmers and policy makers on manure use and production regarding legislation adaptation to new EU rules. This Operational Group (GOEfluentes) includes eleven partners from Research and Teaching, Agri Associations and Agri Enterprises.

Session 62 Theatre 2

Planning tool for calculating carbon footprint of milk and meat

L. Mogensen[1], T. Dorca-Preda[1], M.T. Knudsen[1], N.I. Nielsen[2] and T. Kristensen[1]
[1]Aarhus University, Department of Agroecology, Foulum, 8830 Tjele, Denmark, [2]SEGES, Agro Food Park 15, 8200 Aarhus, Denmark; lisbeth.mogensen@agro.au.dk

If farmers want to reduce the climate impact of their cattle production, they need to know the effect of a given action. Greenhouse gas (GHG) emissions related to feed production and enteric fermentation are the major hotspots in cattle production. However, the complex interaction between herd, feed production, manure management and the various emissions related to that, makes it difficult to predict the overall climate impact of a given action. The existing climate tools are typically suitable for calculating carbon footprint based on historical data e.g. last year's production. In the present project a planning tool for calculating carbon footprint was developed. The aim is that this climate model is implemented in NorFor which is a ration formulation tool used on commercial dairy and beef farms where an optimization of nutritional, economic and climate parameters can take place in an optimization at the same time. In this newly developed tool, carbon footprint is calculated as the sum of the major GHG contributions; feed production, enteric methane emissions, emissions related to manure management and other smaller contributions to GHG emissions like energy use, etc. The most important input data to the tool is the planned feeding per animal per day (kg DM) as well as which technologies that are used for manure management. By using this tool, it was shown that changes in feeding compared with a standard feed ration for milking cow could reduce carbon footprint per kg milk by up to 10% from reduced contribution from feed production and enteric methane. The effect was obtained by choosing feeds with a low carbon footprint. However, it should be kept in mind that some of these feeds are by-products, which will not be available in unlimited amounts. Overall, the planning tool shows a potential for practical implementation of carbon reduction measures at the farm level.

Session 62 Theatre 3

Whole-farm systems modelling of greenhouse gas emissions from suckler cow beef production systems

S. Samsonstuen, B.A. Åby and L. Aass
Norwegian University of Life Sciences, Department of Animal and Aquacultural Sciences, Box 5003, 1432 Ås, Norway; stine.samsonstuen@nmbu.no

Human population growth and climate change are putting pressure on the agricultural production systems, which must secure food production while minimizing greenhouse gas (GHG) emissions. In 2015, agriculture accounted for 10% of the total GHG emissions in Europe. Previous studies have found substantial differences in GHG emissions between continents and between farms within a country dependent on natural resources and management. The whole-farm model HolosNorBeef was developed to estimate net GHG emissions from beef production systems in Norway. HolosNorBeef is an empirical model based on the HolosNor model and the methodology of the Intergovernmental Panel on Climate Change with modifications to Norwegian conditions. The model considers both direct emissions of CH_4, N_2O and CO_2 from on-farm livestock production including soil C changes, and indirect N_2O and CO_2 emissions associated with leaching and inputs used on the farm. The emission intensities from average beef cattle farms in Norway was estimated by considering typical herds of British and Continental breeds located in two different regions, giving four scenarios. In general, location A had lower emissions than location B due to higher carbon sequestration. The total emissions were lower for the British breeds in both locations, however expressed per kg beef the Continental breeds had lower emissions. Emission intensities per kg beef was within the same range as the other Nordic countries.

Session 62 Theatre 4

Comparison of mitigation strategies for GHG emissions from Japanese Black cattle production systems

K. Oishi and H. Hirooka

Kyoto University, Graduate School of Agriculture, Kitashirakawa-Oiwake-Cho, Sakyo-ku, 606 8502 Kyoto, Japan; kazato@kais.kyoto-u.ac.jp

Economic and environmental assessment of multiple mitigation strategies for greenhouse gas (GHG) emissions was carried out for commercial integrated cow-calf-fattening Japanese Black production systems using a life cycle assessment based bio-economic simulation model. The aim of the strategies was to increase or maintain profit as well as to reduce GHG emissions of life-cycle production of a cow. Seven strategies were investigated: three management strategies (reduced fattening period of feedlot animals (FP), increased maximum culling parity of cows (PA) and the number of AI services per parity (AI)), two feeding strategies (reduced CP content of feeds (CP) and supplementation of calcium soaps of linseed oil fatty acid in feeds (LF)) and two breeding strategies (genetic improvement of carcass weight (CW) and conception rate (CR)). Cost and profit were used as economic outputs and GHG emission (CO_2 eq.) was used as an environmental output. Furthermore, integrated indicators such as economic outputs per GHG emissions were also used for evaluation of the strategies. When cost was taken as an economic output, FP and CP could reduce total GHG emissions without increasing costs whereas CW and CR could reduce cost per unit of GHG emissions. In contrast, when profit was used as an economic output, most of the strategies could improve profit per unit of GHG emissions but could not reduce total GHG emissions. The result showed that there were trade-offs between profit and environmental load for these strategies. Among the strategies assessed in the present study, FP and CP were win-win strategies that can increase total profit and reduce total GHG emissions from production.

Session 62 Poster 5

Calculation models for estimation of labour requirement for loose barn dried hay production

J. Mačuhová, B. Haidn and S. Thurner

Bavarian State Research Center for Agriculture, Institute for Agricultural Engineering and Animal Husbandry, Vöttinger Str. 36, 85354 Freising, Germany; juliana.macuhova@lfl.bayern.de

Fodder production and feeding belong to most labour time consuming tasks on dairy farms. To produce a hay of high nutrition quality, new technologies for barn dried hay production were introduced to farmers in the last years. However, less information is available about labour requirement for barn dried hay production by recently used techniques on dairy farms. The aim of this study was to create calculation models for estimation of labour requirement for selected tasks during loose barn dried hay production. Therefore, labour studies were performed (i.e. labour input for individual work elements and appropriate influencing variables were recorded) on dairy farms to determine standard times, i.e. labour requirement for single work elements. To prove if the standard times were really true mean values, they had to reach an accuracy (ε) of 10%. If not, then examination of the required number of observation (n') was performed for required accuracy at a probability of 95% and next measurements will be scheduled. Most standard times already reached required accuracy. Recently, first calculation models for estimation of labour requirement by loose barn dried hay production were created for two subtasks: harvesting of hay with forage wagon and filling of drying boxes with hay crane. Many influence variables can be modified in calculation models (e.g. capacity of forage wagon and grabber, dimension of drying boxes, distances, etc.) and their impact on labour time requirement can be examined. So e.g. by 40 t of harvested hay and the capacity of a crane of 208 kg per grabber, a total 161.36 min would be needed to store the hay in the drying box. Using 2 forage wagons, 294.04, 212.36 and 160.51 min would be needed for harvesting, transport and unloading by the storage capacity of the forage wagons of 1.5, 2.2 and 2.9 t, respectively. It means that the hay crane operator would have no idle time only in the last calculation example and the idle time of forage wagon drivers would be also kept to a minimum. The calculation models can help the farmer to choose the right capacity of applied techniques for his farm condition.

Session 62 Poster 6

Development of daily mean temperature estimation method for certain locations in Republic of Korea

K.I. Sung[1], B.W. Kim[1], K.D. Kim[2] and S.G. Kang[3]
[1]College of Animal Life Sciences, Kangwon National University, Animal Resource Sciences, Lab no. 303, College of Animal Life Sciences, 24341 Chuncheon, Korea, South, [2]Gangwondo Agricultural Research and Extension Services, Pyeongchang, 25300, Korea, South, [3]Rural Development Administration, Jeonju, 54875, Korea, South; kisung@kangwon.ac.kr

The objective of this study was to develop the daily mean temperature estimation method for a certain location based on the climatic data collected from the nearby Automated Synoptic Observing System (ASOS) and Automated Weather System (AWS) to improve the accuracy of the climate data in forage crop yield prediction models. To perform this study, the descriptive statistics of the daily and annual mean temperature data collected from the ASOS (n=75) and AWS (n=278) were analysed, and histograms were generated. In addition, the correlation and multiple regression analysis were applied to the location information (longitude, latitude and altitude) and the temperature data from weather observation stations. The longitude, latitude, and altitude all had a significant effect on temperature. Thereafter, it was considered to be possible to estimate the daily mean temperature based on location information. Based on the location information, daily mean temperature estimation method was developed through multiple regression equations of the nationwide meteorological sites and the triangulation method which uses the coordinate and distance. To evaluate the applicability of the equation developed to estimate daily mean temperature based on location information, the comparative analysis between the estimated daily mean temperature using this method and the observed daily mean temperature of the AWS (18 stations, 5 years) from which the location information and daily mean temperature data could be obtained was performed. The coefficient of determination (R^2) ranged from 0.89 to 0.99, which indicate that the estimated equation fitted well to the observed temperature data. In addition, it could be considered that the daily mean temperature estimation method developed in this study had a statistically high reliability. Therefore, it could be concluded that this method could be used to estimate the daily mean temperature data of certain locations where there is no meteorological station.

Session 62 Poster 7

Efficiency of organic rabbit production in a pasture system

J. Walczak, L. Gacek and M. Pompa-Roborzyński
National Research Institute of Animal Production, Sarego 2, 31-047 Kraków, Poland; jacek.walczak@izoo.krakow.pl

Rabbits are often kept in small and medium-sized farms because they breed easily and are easy to raise. This is one of the species for which no organic farming requirements have been specified at EU level. However, these regulations require that each species should have access to the natural environment in which it was formed. For rabbits, this is pasture. The study was performed with 77 female, 30 male and 2,000 young rabbits of the Termond White (TR) breed and the same numbers and groups of young rabbits of the Popielno White (PR) breed (native breed). Animals were kept in two systems: indoors and in own-made mobile cages on pasture. Rabbits were fed a ration containing 17% CP and 2,400 kcal ME, with constant access to water and also to sward in the pasture system. An average of 1.9 kg green material was obtained per m^2 of pasture area. Nine full production cycles were performed with natural weaning at 35 and 120 days of rearing. There were more rabbits born litter in TR (8.19) compared to PO breed (6.62). At 35 days age, significantly greater litter size was observed for TR compared to PO breed (7.57 vs 6.14), as well as higher litter weight (673.68 g vs 637.53 g). In the pasture system, statistically significant differences were found between the number of rabbits per litter (TR 8.19 vs PO 6.27). At 35 days of age, these differences were not significant (4.59 vs 4.07). No differences were observed between the mean body weights of the rabbits at 120 days (TR 2.46 vs PO 2.47 kg), but there was a difference in feed conversion (TR 4.77 vs PO 5.09 kg/kg). During that period, rabbits raised indoors had a weight gain of 3.06 kg (TR) and 2.96 kg (PO), with a feed conversion of 4.19 kg/kg (TR) and 4.38 kg/kg (PO). These values differed significantly in relation to pasture rearing. Mortality in the TR breed was significantly higher in the pasture system (13 vs 27%) and there were no differences in the PO breed (27 vs 29%). Mortality was mainly due to intestinal disorders. Behavioural observations indicate that until 120 days of rearing, rabbits do not show any burrowing behaviour. Around 3 weeks before kindling, does begin to exhibit such behaviour. The costs of pasture rearing were found to be 97% higher.

New kid on the block: a genomic breeding value to select for heat tolerant dairy cattle

T.T.T. Nguyen[1], *J.B. Garner*[2] *and J.E. Pryce*[1,3]

[1]*Agriculture Victoria Research, Department of Economic Development, Jobs, Transport and Resources, 5 Ring Road, Bundoora, Victoria 3083, Australia,* [2]*Ellinbank Research Centre, Department of Economic Development, Jobs, Transport and Resources, 1301 Hazeldean Road, Ellinbank, Victoria 3821, Australia,* [3]*La Trobe University, School of Applied Systems Biology, 5 Ring Road, Bundoora, Victoria 3083, Australia; thuy.nguyen@ecodev.vic.gov.au*

Heat stress has become a challenge facing the dairy industry. There is a need to identify and breed heat tolerant dairy cattle. We have achieved this through the development of genomic breeding values (GEBV) for heat tolerance (HT). We first merged climate data with milk production records that were collected between 2003 and 2016. The rate of decline of milk, fat and protein yield (namely cow slope) for 424,540 Holstein and 84,702 Jersey cows was calculated when temperature and humidity exceed a defined comfort level. Slope of a sire is the daughter average. A reference population consisting of 11,853 cows and 2,236 sires for Holsteins and 4,268 cows and 506 sires for Jerseys (both having estimated slope and high density genotypes), were used to derive a genomic prediction equation. HT GEBV can then be determined for other animals with genotypes. To validate the HT GEBV, we predicted HT GEBV for 390 Holstein heifers, then selected 24 extreme predicted heat tolerant and 24 extreme predicted heat susceptible heifers for a 4-day heat challenge. The predicted heat tolerant group showed significantly less decline in milk production, lower rectal and intra-vaginal temperatures than the predicted heat susceptible group. This indicates that the HT GEBV will enable selection for cattle with better tolerance to heat stress. We expressed HT GEBV by applying the economic weight of milk, fat and protein to its components (decline in milk, fat and protein). Within each breed, the HT GEBV was then standardized to have a mean of 100 and standard deviation (SD) of 5. The mean reliabilities of HT GEBV among validation sires were 38% in both breeds. HT GEBV was found to be unfavourably correlated to production and favourable to fertility. The HT GEBV was released in Dec 2017 for the first time.

Do dual-purpose cattle react differently than dairy cattle along a continuous environment scale?

B. Bapst[1], *M. Bohlouli*[2], *S. König*[2] *and K. Brügemann*[2]

[1]*Qualitas AG, Genetic Evaluation, Chamerstrasse 56, 6300 Zug, Switzerland,* [2]*Justus-Liebig-University Gießen, Institute of Animal Breeding and Genetics, Ludwigstraße 21B, 35390 Gießen, Germany; beat.bapst@qualitasag.ch*

In the second half of the penultimate century, Brown Swiss cattle (BS) were exported from Switzerland to North America. There the breeding objective was changed from a triple-purpose (milk, meat and work) to milk only and today BS is one of the most important dairy cattle breeds in the world. Fifty years ago Switzerland and other European countries started to cross their dual-purpose Brown Swiss (BS) with North American BS to improve yield traits. In Switzerland a small subpopulation was kept as a dual-purpose breed, without any admixture with the North American BS. This subpopulation is called Original Braunvieh (OB) and has grown rapidly in the last 10 years; nowadays the population harbours about 10,000 live cows registered in the herdbook. This setting makes the BS population ideal for the investigation of different genotype × environment interactions (G×E) in dual-purpose versus dairy cattle breeds. We developed a reaction norm model, where we included meteorological data in a random regression test day model (RRTDM), which was derived from the official Swiss genetic evaluation system. As an environmental descriptor, daily temperature-humidity indexes (THI) from 60 official federal weather stations were assigned to each BS and OB herd in Switzerland. After data editing, 150,545 and 5,384,987 test day (TD) records remained from 8,062 OB and from 272,649 BS cows, respectively. In order to get an unbiased comparison BS TD were only chosen from weather stations, which were also assigned to OB herds. In a first step, daily protein yield was analysed. For fitting the additive genetic and the permanent environment effects, the RRTDM was extended with the covariates days in milk (DIM) and THI, both with Legendre polynomials of order 3. Preliminary results show that the additive genetic variance of OB along the DMI and along the THI axis is much more constant than of BS. Further analysis will investigate different developments of the model and the effect of the inclusion of THI as a covariate in RRTDM on the breeding values and the ranking of sires and dams.

Session 63 Theatre 3

Early-programming of dairy cattle, a potential explanation to the adaptation to climate change

H. Hammami[1], F.G. Colinet[1], C. Bastin[2], S. Vanderick[1], A. Mineur[1], S. Naderi[1], R.R. Mota[1] and N. Gengler[1]
[1]Gembloux Agro-Bio Tech, University of Liège, TERRA Teaching and Research Centre, Gembloux, 5030, Belgium, [2]Walloon Breding Association, R&D, Ciney, 5590, Belgium; hedi.hammami@uliege.be

Breeding for robustness and considering genotype by environment interaction (G×E) is linked to adaptation. Recently, it has been established that gene expression can be affected by the environment during the embryo development. The concept of early programming has been demonstrated in many settings. This study aimed to assess the impact of thermal stress when dairy cows been conceived on their lifetime performances. Studied traits were milk yield and some novel milk-based biomarkers, fertility (days open), health (somatic cell score and ketosis), and heat tolerance. Data used compromised 905,391 test day of 58,297 cows in parity 1 to 3 for production traits, health and ketosis status, 104,635 records of 48,125 cows for days open, and 399,449 test days recorded (linked with temperature humidity index values, THI) of 28,203 cows for heat tolerance trait. Date of conception was estimated using the next calving date of the cow and subtracting 280 d from the calving interval. Cows being conceived in summer (June-August) were considered as influenced by heat stress (environment 1) and those conceived in winter (December-February) as neutral-thermal conditions (environment 2). G×E was analysed by a multi-trait model for days open in which each of the 3 lactations measured in heat stress and thermo-neutral conditions were considered as separate traits. For the rest of the traits, it was analysed using reaction norm models, in which the trait is considered a function of an environmental descriptor (i.e. THI, days in milk) in the two discrete environments. First results showed that genetic correlations across both early-life defined environments and lactations were substantially lower than unity, implying that effects of genes for cows conceived under neutral-thermal conditions may be different of the effects for the same genes for cows conceived under heat stressed conditions.

Session 63 Theatre 4

Characterization of thermal stress in Avileña-Negra Ibérica beef cattle breed at feedlot

M. Usala[1], M. Carabaño[2], M. Ramon[3], C. Meneses[2], N. Macciotta[1] and C. Diaz[2]
[1]University of Sassari, Department of Agricultural Science, Viale Italia 39, 07100 Sassari, Italy, [2]INIA, Department of Animal Breeding, Ctra. La Coruña Km 7.5, 28040 Madrid, Spain, [3]CERSYRA-IRIAF, Av. Del Vino 10, 13300, Valdepeñas, Ciudad Real, Spain; musala@uniss.it

Thermal stress is a research area of growing interest because of its effects on animal performance and animal production practices. The growing interest on thermal stress in feedlot cattle is the result of severe economic losses associated with extreme weather events. Despite this susceptibility, thermal stress in beef cattle has received considerably less attention than in dairy cattle. Therefore, studies on the thermal stress effect on beef cattle might improve well-being and productivity of these animals. This study aimed to analyse the impact of adverse weather conditions on gain of weight in male calves of Avileña-Negra Ibérica local beef breed during fattening. A total of 23,645 weight records belonging to 5,876 animals collected between 2005 and 2017 were used. Meteorological data were collected for the same period from the station of Gotarrendura selected as the most representative of this study due to proximity to the feedlot. The lag effect of each weather descriptor (daily average (Tavg), Maximum (Tmax) and minimum (Tmin) Temperature, Temperature- humidity index (THIavg) and Wind chill index (WCI)) on gain weight was investigated for the recording date and up to 30 days before, using a lag non-linear model for time series implemented in the ´dlnm´ package of R. Thresholds defining the thermoneutral zone and the slopes of performance decay above and below this comfort zone were also estimated using the 'segmented' package of R. Results showed that the impact of heat stress was greater than that of cold stress in this breed. The comfort region ranged between -1 and 25 °C for daily average temperature (Tavg). The thermal load in the day of control (T0) proved to have more impact on the gain of weight at feedlot than thermal loads measured in previous days. Because of the large correlation between Tavg along the 30 d period, T0 could be interpreted as the accumulation of temperatures occurring during the whole period. A similar pattern was not observed for Tmax and Tmin, which presented a more erratic behaviour.

Environmental sensitivity for dairy traits in Slovenian Holstein population

B. Logar[1], M. Kovač[2] and Š. Malovrh[2]
[1]Agricultural Institute of Slovenia, Hacquetova ulica 17, 1000 Ljubljana, Slovenia, [2]University of Ljubljana, Biotechnical Faculty, Zootechnical Department, Groblje 3, 1230 Domžale, Slovenia; betka.logar@kis.si

The ability of organisms to adapt to environmental changes is defined as environmental sensitivity. Also in geographically small regions, cattle can be are reared in heterogeneous environments. The study examines the environmental sensitivity of dairy traits in the Slovenian Holstein population. Milk, protein and fat yields in 305-day lactation were studied. The first and first three lactations data were used. Herd-year averages were treated as environmental variables. Data analyses were done by a reaction norm sire model, including all the observations and implementing a standardized environmental variable (EV). Estimates of heritabilities varied from 0.04 at -2 standard deviation (s.d.) of EV and 0.40 at +2 s.d. The permanent environment and residual variances were heterogeneous, and estimates were increasing by improving the value of environmental variable. Estimates of variance of the reaction norm slopes were greater than zero and indicated the presence of genetic variability of the environmental sensitivity, the highest being up to 43×10^{-4}. Genetic correlation between the level and the slope estimated by reaction norm model was high, in most cases above 0.8, suggesting a scaling effect. There is evidence of scaling and re-ranking of sires, which is more common in the low environment (<-1 s.d.). The results suggest presence of genetic variability of environmental sensitivity of dairy traits in the studied population in which scaling and also re-ranking of animals are anticipated.

Adjusting for macro-environmental sensitivity in growth rate of Danish Landrace and Duroc pigs

M. Dam Madsen[1], P. Madsen[1], B. Nielsen[2], T.N. Kristensen[3,4], J. Jensen[1] and M. Shirali[1]
[1]Aarhus University, Department of Molecular Biology and Genetics, Center for Quantitative Genetics and Genomics, Blichers Alle 20, 8830 Tjele, Denmark, [2]SEGES, Pig Research Centre, Axeltorv 3, 1609 Copenhagen, Denmark, [3]Aalborg University, Department of Chemistry and Bioscience, Fredrik Bajers Vej 7H, 9220 Aalborg, Denmark, [4]Aarhus University, Department of Bioscience – Genetic, Ecology and Evolution, Ny Munkegade 116, 8000 Aarhus, Denmark; mette.madsen@mbg.au.dk

The aim of this study were to: (1) model the macro-environmental sensitivity (macro-ES) in average daily body weight gain (ADG) in Danish Landrace and Duroc pigs; (2) examine breed differences in macro-ES; and (3) investigate the heterogeneity of ADG among sexes. Data for the analysis were records of ADG from 75,021 Danish Duroc (32,297 males and 42,724 females) and 166,652 Danish Landrace (70,521 males and 96,131 females), routinely performance tested by DanAvl. Data were analysed in a Bayesian setting with Gibbs sampling for males and females separately, using a univariate animal reaction norm model with unknown covariate. Results are presented as means of the posterior distribution for each parameter. Significant genetic variance in macro-ES for ADG were found in both sexes of both breeds. We observed significantly higher additive genetic variance of macro-ES in Duroc males and females compared to Landrace males and females (27 and 118% higher additive genetic variance, respectively). The genetic correlation between the additive genetic variance of the general breeding value and the additive genetic variance of macro-ES was -0.13 in Duroc males and -0.21 in Landrace males. In females, the genetic correlation was 0.09 in Duroc and 0.32 in Landrace. The residual variance differed among herds in both sexes of both breeds, showing the importance of considering heterogeneous residual variance in breeding programs. Males had larger variance of the general breeding value than females in both breeds. Males had larger residual variance in all 15 herds that rear Duroc of both sexes and in 22 of the 23 herds rearing Landrace of both sexes, indicating a genetic heterogeneity among sexes. In conclusion, our results show that by modelling macro-ES it is possible to adjust for variation of animals across environments in breeding programs.

Session 63 Theatre 7

Does selection for feed efficiency in pigs improve robustness to heat stress?

W.M. Rauw[1], J. Mayorga Lozano[2], J.C.M. Dekkers[2], J.F. Patience[2], N. Gabler[2], S. Lonergan[2] and L.H. Baumgard[2]
[1]INIA, Dept Mejora Genética Animal, Ctra de La Coruña km 7, 28040, Madrid, Spain, [2]Iowa State University, Dept of Animal Science, Kildee Hall, 50011, Ames, IA, USA; rauw.wendy@inia.es

Heat stress (HS) accounts for over $900 million loss annually in the U.S. swine industry resulting from slower growth rates, reduced feed efficiency and poor reproduction. Pigs with increased feed efficiency have lower basal metabolic rates and reduced metabolic heat production. Therefore, genetic selection for improved feed efficiency may reduce susceptibility to HS. Study objectives were to investigate the effects of selection for residual feed intake (RFI) on growth and feed intake in response to repeated exposures to HS. A total of 35 and 31 pigs from a divergent selection line for low (LRFI) and high RFI (HRFI), respectively, where subjected three separate times to a 4-d HS load, which was preceded by a 9-d thermal neutral (TN) adaptation period and alternated by 7-d TN conditions: 1-TN adaptation, 2-HS, 3-TN, 4-HS, 5-TN, 6-HS, and 7-TN. Average daily body weight gain (BWG) and daily feed intake (FI) were calculated for each period; feed efficiency (FE) was estimated as BWG/FI. Overall, between TN and HS, the change in BWG tended to be larger in LRFI (- 306 g/d) than in HRFI pigs (+ 12 g/d; P=0.09), but the change in FI was similar (a decrease of 542 and 509 g/d, respectively). Overall, FE was not significantly affected by HS in either line. Selection for improved FE did not reduce susceptibility to HS. In contrast, results show that LRFI pigs (high efficient) were slightly, albeit non-significantly, more affected by HS (less robust) than HRFI pigs. Although water intake was not recorded in the present study, other studies suggest that water consumption may be greater in HRFI than in LRFI animals, which may result in a higher capacity for heat dissipation. This work was funded by the Iowa Pork Producers Association (14-243) and USDA NIFA-AFRI award number 2011-68004-30336, in support of SusAn ERA-Net 'SusPig'.

Session 63 Theatre 8

Effects of specific breeding lines for organic dairy production in Denmark

M. Slagboom[1], H.A. Mulder[2], A.C. Sørensen[1], L. Hjortø[1], J.R. Thomasen[1,3] and M. Kargo[1,4]
[1]Aarhus University, Department of Molecular Biology and Genetics, Blichers Alle 20, 8830 Tjele, Denmark, [2]Wageningen University & Research Animal Breeding and Genomics, P.O. Box 338, 6700 AH Wageningen, the Netherlands, [3]VikingGenetics, Ebeltoftvej 16, 8960 Randers SØ, Denmark, [4]SEGES Cattle, Agro Food Park 15, 8200 Aarhus N, Denmark; margotslagboom@mbg.au.dk

The aim of this study was to compare genetic gain in organic dairy production when running an environment-specific breeding program compared to one breeding program for both organic and conventional dairy production. Three different scenarios were simulated: running one breeding program for both environments (scenario 1), running two environment-specific breeding programs with selection of bulls across environments (scenario 2a), and running two environment-specific breeding programs with selection of bulls within each environment (scenario 2w). Genetic correlations between the trait in the two environments varied between 0 and 1. The reliability of predicting GEBVs was 0.65 in both environments. Each scenario was simulated for 30 years using stochastic simulation. In each environment, 1000 bull calves were genotyped. Each year 50 young bulls were selected within each breeding program to keep the rate of inbreeding approximately the same in the scenarios. Breeding animals were selected based on a total merit index. Preliminary results show that genetic gain for the organic breeding goal was higher for scenario 1 when the genetic correlation between the trait in the two environments was higher than 0.65, otherwise scenarios 2a and 2w yielded more genetic gain for the organic breeding goal. This low break-even genetic correlation is similar to previous studies, but it is expected that the break-even genetic correlation will increase when increasing the number of genotyped selection candidates. Below a correlation of about 0.80, scenarios 2a and 2w were similar in terms of genetic gain for the organic breeding goal. In scenario 2a, bulls could be selected across environments, but below a correlation of 0.8 they were not because bulls from the other environment were inferior. The effect of breeding goal differences, size of breeding programs, and a lower GEBV reliability on genetic gain will be further studied.

Whole-genome sequence quest targeting parent-of-origin effects in pigs

I. Blaj[1], C. Falker-Gieske[2], J. Tetens[2], R. Wellmann[3], S. Preuss[3], J. Bennewitz[3] and G. Thaller[1]
[1]Kiel University, Institute of Animal Breeding and Husbandry, Hermann-Rodewald-Strasse 6, 24118 Kiel, Germany, [2]Göttingen University, Department of Animal Science, Functional Breeding Group, Burckhardtweg 2, 37077 Göttingen, Germany, [3]University of Hohenheim, Institute of Animal Science, Garbenstraße 17, 70599 Stuttgart, Germany; iblaj@tierzucht.uni-kiel.de

An epigenetic process refers to any process that alters gene activity without bringing any change to the nucleic acid sequence. Among the many types of epigenetic processes, genomic imprinting is characterized by a locus at which the two alleles are not equivalent from a functional point of view. This phenomenon is regarded to be the primary epigenetic source for parent-of-origin effects (POEs) as imprinted loci display a gene expression dependent on the maternal or paternal allelic inheritance. Preliminary investigations on 2.554 genotyped (Illumina PorcineSNP60 BeadChip) individuals from two F2 porcine crosses originating from various founder breeds were carried out upon a set of fatness and growth related phenotypic traits. The maximum contribution of POEs to the total phenotypic variance was 18% for fat thickness over the loin muscle trait in one of the crosses. Parent-of-origin GWAS models were performed by testing the paternal and maternal alleles separately. For average daily gain, backfat thickness and meat to fat ratio, strong association signals were found via the paternal model on SSC2 in the region of *IGF2*, one of the best-known imprinted genes affecting muscle mass and fat deposition. The founders of the crosses were sequenced. An imputation step led to the availability of whole-genome sequence (WGS) data for the initially genotyped individuals. An extrapolation of the preliminary analysis is envisioned on the quest to target POEs for various production traits using WGS data while advancing the understanding of imprinting and epigenetics.

Comparison of birth weight variability divergent selected mice lines with founder population

N. Formoso-Rafferty, J.P. Gutiérrez and I. Cervantes
Universidad Complutense de Madrid, Avda. Puerta de Hierro s/n, 28040, Spain; n.formosorafferty@ucm.es

Selecting to modify environmental variability of some traits has been shown to be possible, and its reduction has been reported to be related to robustness with benefits for animal production and welfare. Specifically, a successful divergent selection experiment for birth weight environmental variability has been carried out during 17 generations in mice, having been found a significant marked difference between low (LV) and high (HV) variability lines. However, there were signs suggesting that HV line could have been stopped its response as a consequence of a possible natural limitations. The objective of this work was to compare the birth weight variability of the divergent selected lines with that of a non-selected population (PB) with the same origin than the animals starting the experiment, to evaluate the differential response of each of the lines. According to the experimental design 43 females of each line were mated to 43 males within line, one to one having up to two parturitions. Simultaneously mating was carried out in the non-selected population as designed for them, 34 males to 62 females, one or two females each male, having up to two parturitions one of the females mated to each male. Variability parameters for PB always ranked between lines, but clearly closer to HV line. For example, mean birth weight standard deviation was 0.15 for PB, being 0.18 and 0.10 respectively for HV and LV lines. Birth weight differences were more remarkable being 1.70 for PB and 1.77 and 1.43 for the respective HV and LV lines. However litter size was the lowest for PB (8.4) versus 8.7 and 10.1 for HV and LV lines. Some of these results could be partially conditioned by a slightly older age of PB animals. It could be concluded that both lines had a genetic response, but this response was clearly higher for de LV line confirming the natural limits in HV line such as the uterine capacity for too big fetus, protecting the population towards a higher variability in birth weight which would generate less viable individuals.

Session 63 Theatre 11

Body weight variation as indicator for resilience in chickens

T.V.L. Berghof[1], H. Bovenhuis[1], J.A.J. Arts[2], F. Karangali[1], J.J. Van Der Poel[1], H.K. Parmentier[2] and H.A. Mulder[1]
[1]Wageningen University & Research, Animal Breeding & Genomics, P.O. Box 338, 6700 AH Wageningen, the Netherlands, [2]Wageningen University & Research, Adaptation Physiology, De Elst 1, 6708 WD Wageningen, the Netherlands; tom.berghof@wur.nl

Resilient animals have a minimal loss of performance (i.e. low variation) despite environmental perturbations. Environmental perturbations can have different origins, and resilience is therefore composed of several forms of resilience, e.g. heat resilience or disease resilience. The objective of this research was to investigate genetic parameters of resilience based on body weight (BW) variation and its relation with natural antibody (NAb) levels. NAb are antibodies recognizing antigens without previous exposure to these and might be an indication of general disease resistance. BW data of 8,008 individuals selectively bred for high and low NAb levels was collected on a four-weekly interval until around 40 weeks of age. In addition, keyhole limpet hemocyanin (KLH)-binding total, IgM and IgG NAb were measured. Standardized residuals (res_{st}) of individual's BW compared to lines' average BW were used as an estimate for resilience by calculating the log-transformed variance of res_{st} (lnvar), skewness of res_{st} and the autocorrelation between res_{st}. Heritabilities were 0.07 for lnvar, 0.04 for skewness and 0.02 for the autocorrelation, showing some heritable variation in these resilience indicators, although the heritability of the autocorrelation was not significantly deviating from zero. The genetic correlation between lnvar and the autocorrelation was high (0.75), while skewness was not correlated with lnvar and the autocorrelation. Genetic correlations between the three resilience indicators and NAb were all low and not significantly different from zero. The genetic correlations between NAb and the resilience indicators based on BW variation in early life may indicate that both types of resilience indicators capture different aspects of resilience or that there was a lack of disease occurrence to show common genetic variation. Nevertheless, the relatively high heritabilities of resilience indicators based on BW offer opportunity for further study, such as quantifying the predictive value of these resilience indicators for the response to environmental perturbations.

Session 63 Theatre 12

Capturing indirect genetic effects on phenotypic variability

J. Marjanovic[1,2], H.A. Mulder[2], L. Rönnegård[1,3], D.J. De Koning[1] and P. Bijma[2]
[1]Swedish University of Agricultural Sciences, Department of Animal Breeding and Genetics, Ulls väg 2, 75007 Uppsala, Sweden, [2]Wageningen University & Research, Animal Breeding and Genomics, P.O. Box 338, 6700 AH Wageningen, the Netherlands, [3]Dalarna University, School of Technology and Business Studies, Högskolegatan 2, 79188 Falun, Sweden; jovana.marjanovic@wur.nl

Phenotypic variability of a genotype is relevant in domestic populations. Such variability has been studied as a quantitative trait in itself, and is often referred to as inherited variability (IV) or heritable variation in environmental variance. So far, studies on IV have only considered direct genetic effects on IV. Observations from aquaculture and plants, where competition often leads to more variability among individuals, suggest that an additional level of genetic variation in variability may be generated through competition. Social interactions, such as competition, are often a source of Indirect Genetic Effects (IGE). An IGE is a heritable effect of an individual on the trait value of another individual. To exploit genetic variation in variability originating from IGE, we need statistical models to capture these effects. Here we investigate the potential of existing statistical models for IV and trait values, to capture the direct and indirect genetic effects on variability. We conducted a simulation study in which competition between social partners (i.e. IGEs) leads to IV of trait values. Data were analysed with ASReml 4.1. using the following models (1) a direct sire model for IV, (2) an indirect sire model for IV, (3) a direct sire-dam model for the trait, (4) and an indirect sire-dam model for the trait. The potential of models to capture IGE on variability was tested by comparing EBV from each model with simulated direct BV for trait level, and direct and indirect BV for competition. Our results show that a direct model of IV almost entirely captures the direct genetic effect of competition on variability. Similarly, an indirect model of IV captures indirect genetic effects of competition. Models for trait levels, however, captured only a small part of the genetic effects of competition on variability. Capturing genetic effects of competition on variability is possible with direct and indirect models of IV, but may require a two-step process.

Preliminary results for the length of productive life of Holstein cows in Croatia

M. Dražić, M. Špehar, Z. Ivkić and Z. Barać
Croatian Agricultural Agency, Ilica 101, 10000 Zagreb, Croatia; mdrazic@hpa.hr

The length of productive life (LPL) is a trait of considerable importance in dairy husbandry which could be seen as a composite of production, health, and reproduction within specific environmental and farming conditions. Herd structure differs among regions, with small and medium sized herds in the central and large herds in the eastern part of Croatia, having consequently different management. The objective of this study was to investigate the effect of herd size on the LPL of the registered Holstein cows in Croatia. Data included 200,634 records for 87,620 cows calved from 2002 to 2017. Analyses were performed using Proc Lifetest in SAS statistical package. Data were divided in 6 strata by herd size (1 = 1-5 cows, 2 = 6-10, 3 = 11-30, 4 = 31-100, 5 = 101-500, and 6 = >500 cows). This division was justified by the average milk production per group, where lowest average production had group 1 (LS mean 5,632.37) and the production increased linearly to group 6 (LS mean 8,290.18). The following null hypothesis was tested in Proc Lifetest: there are no differences in the overall mean LPL of cows reared in the herds of different size. Differences between curves for LPL were tested using Kaplan-Meier estimator and the log-rank test. Significance was set at $P<0.05$. Herd size influenced LPL and average survival was from 42.35 months (group 6) to 50.88 months of age (group 1). These preliminary results demonstrate the effect of farming conditions on LPL.

Genetic variation due to heat stress for pregnancy of Zebu beef heifers during the breeding season

M.L. Santana Junior[1], G.A. Oliveira Junior[2], A. Bignardi[1], R. Pereira[1], J.B. Ferraz[3], J. Eler[3] and E. Mattos[3]
[1]Federal University of Mato Grosso, GMAT, Rodovia Rondonópolis-Guiratinga, KM 06 (MT-270), 78735-901, Brazil, [2]CRV Lagoa, Rodovia Carlos Tonanni, km 88, 14174-000, Brazil, [3]University of São Paulo, Veterinary Medicine, Rua Duque de Caxias Norte, 225, 13635-900, Brazil; joapeler@usp.br

The objective of the present study was to identify genetic variation due to heat stress tolerance for heifer pregnancy in Nellore cattle during the breeding season using a reaction norm model. The Nellore heifers were born between 2003 and 2013 on six farms located in three Brazilian States. Heifer pregnancy (HP) was analysed as a binary trait, with a value of 1 (success) assigned to females that were diagnosed pregnant and a value of 0 (failure) assigned to those that were not pregnant after breeding season. A threshold model animal model was carried out for HP to estimate variance components and genetic parameters as a function of temperature-humidity index (THI) values. As evidence of the adaptation of Nellore cattle to heat stress, on the phenotypic scale, HP of Nellore cattle was only slightly affected by the increase in THI values (-0.85%/THI). The posterior means of the heritability for the additive genetic effect ranged from 0.17 at THI 82.5 to 0.01 at THI 88. The mean genetic correlations for HP between different THI values were much below unity. The genetic correlation between extreme THI values (82.5 and 88) was -0.47. Wide genetic variation in the response to heat stress and heterogeneity in the variance components and genetic parameters were observed across the THI scale. The breeding values of animals for HP were sensitive to changes in THI, thus demonstrating an important effect of genotype by environment interaction due to heat stress.
Financial support by CNPq (442538/2014-6) and CNPq (303659/2014-9).

Session 63 — Poster 15

Genetic studies on heat stress in dairy cattle in Kenya

J.M. Mbuthia, M. Mayer and N. Reinsch
Leibniz Institute for Farm Animal Biology (FBN), Genetics and Biometry, Wilhelm-Stahl-Allee 2, 18196 Dummerstorf, Germany; mbuthia@fbn-dummerstorf.de

It is apparent that the performance, well-being, and health of an animal are influenced by biometeorological factors. Climate change at global level is considered the major threat to the viability of livestock production systems. Of particular importance is the effect of heat stress and its influence on livestock production. In dairy cattle, environmental conditioning, feeding modifications, and genetic selection for heat tolerance remain the main strategies for alleviating the effects of heat stress. Investigation of the genetic sensitivity to absolute and changing heat stress and genetic evaluation for heat tolerance in cattle is of great importance for improved food security. This study is designed to investigate individual cow's genetic and environmental components as a function of grid interpolated climatological data in Kenya. Random Regression Models were applied because they are more flexible and accurate. They permit inferences regarding the genetic aspects of the lactation curve. From the models, individual heat stress influences are identified. By modelling the cow performance as a function of temperature and humidity data, highly productive individuals with low sensitivity to heat stress potentially could be identified and selected. Data on milk performance and pedigree records for different breeds were obtained from the Kenya Livestock Breeders Organization and Livestock Recording Centre. For the Friesian breed, a total of 38,216 first lactation test-day milk yield records distributed across 189 herds were used. The study herds' locations were characterized into agro-ecological zones, which provide a standardized characterization of climate, soil and terrain relevant to agricultural production. The herds were distributed across high potential, medium potential and semi-arid agro-ecological zones and these exhibited significantly different lactation curves. This data, merged with meteorological data from nearest weather stations is applied in reaction norm models to estimate the genetic components for heat tolerance. Identification and selection of well adapted cattle will not only improve production under heat stress conditions but will also require fewer inputs and environmental interventions.

Session 63 — Poster 16

Reaction norm model for estimation of genetic parameters for yearling weight of Nelore cattle

C.C.P. Paz[1,2], A.P. Freitas[1], M.E.Z. Mercadante[2], J.N.S.G. Cyrillo[2] and M.L. Freitas Santana Jr[3]
[1]Universidade de São Paulo, Faculdade de Medicina de Ribeirão Preto, Av. Bandeirantes, 3900, Ribeirão Preto, SP, 14049-900, Brazil, [2]Instituto de Zootecnia, Centro APTA Bovino de Corte, Rodovia Carlos Tonanni, km 94 Sertãozinho, SP, 14160-900, Brazil, [3]Universidade Federal de Mato Grosso, Instituto de Ciências Agrárias e Tecnológicas, Avenida dos Estudantes, 5055, Rondonópolis, MT, 78736-900, Brazil; aniellypf@hotmail.com

Data from a selection experiment for growth carried out with Nelore cattle were used to evaluate the importance of genotype by environment interaction (G×E) in the estimation of genetic parameters for yearling weight (YW) Nelore. The experiment was started in 1976 and in 1980; three lines of Nelore cattle were established: control (NeC) selected for mean YW; selection (NeS) and traditional (NeT), both selected for higher YW. The NeT was an open line that eventually received bulls from other herds. A total of 8,208 measurements of weight of male and female, born between the years of 1981 and 2015 were used. The herds belonged to the Instituto de Zootecnia of Sertãozinho, SP. Two models were used: animal model (AM) and reaction norm model (RNM) using as environmental descriptor the contemporary group solutions. For the NeC, the heritability for YW was estimated to be 0.56 (0.08) in the AM and ranged from 0.60 (0.08) to 0.77 (0.05) by RNM. For the NeS herd, the heritability for YW was 0.50 (0.04) by AM and ranged from 0.47 (0.06) to 0.70 (0.04) by RNM. For the NeT, the heritability was 0.49 (0.04) by AM and ranged from 0.44 (0.06) to 0.69 (0.04) by RNM. Higher estimates of heritability were observed in favourable environments in the three herds studied. The genetic correlations between the intercept and the slope were 0.11, 0.52 and 0.55 for NeC, NeS and NeT, respectively. This indicates that animals of the herds NeS and NeT with higher genetic level of production respond positively to environmental improvements. The low genetic correlation between the intercept and the slope obtained for the NeC herd may imply re-ranking of animals in different environments. Therefore, G×E should be considered in genetic evaluations.

Session 63 Poster 17

Genetic trends of the yearling weight of Nellore cattle using reaction norm model

A.P. Freitas[1], M.L. Santana Jr[2], M.E.Z. Mercadante[3], J.N.S.G. Cyrillo[3], R.P. Savegnago[3] and C.C.P. Paz[1,3]
[1]Faculdade de Medicina de Ribeirão Preto, Universidade de São Paulo, Departamento de Genética, Av. Bandeirantes, 3900, Ribeirão Preto, SP, 14040-030, Brazil, [2]Instituto de Ciências Agrárias e Tecnológicas, Universidade Federal de Mato Grosso, Avenida dos Estudantes, 5055, Rondonópolis, MT, 78736-900, Brazil, [3]Instituto de Zootecnia, Centro APTA de Bovinos de Corte, Rodovia Carlos Tonani, Km 94, Sertãozinho, SP, 14160-900, Brazil; aniellypf@hotmail.com

A genetic selection experiment for growth with Nelore cattle was carried out to evaluate the genetic trends for environmental sensitivity and general level of production for yearling weight (YW). The experiment started in 1976 and in 1980; three lines of Nelore cattle were established: control (NeC) selected for mean YW, selection (NeS) and traditional (NeT), both selected for higher YW. The NeT was an open line that eventually received bulls from other herds. A total of 8,208 measurements of body weight of male and female, born between the years of 1981 and 2015 were used. The herds belonged to Instituto de Zootecnia of Sertãozinho, SP, Brazil. The reaction norm model was adopted using as environmental descriptor the contemporary group solutions. Genetic trends were estimated for general level of production (intercept of RN) and for environmental sensitivity (slope of RN). The genetic trends obtained for intercept were -0.023; 3.427 and 3.042 kg/year for NeC, NeS and NeT herds, respectively, being that the regression of the NeC herd was not significant. This indicates that populations of NeS and NeT herds tend to have higher production levels, since both are selected for greater YW. The genetic trends for environmental sensitivity were 0.096; 0.536 and 0.478 kg/year for these herds NeC, NeS and NeT respectively, indicating that the three populations are moving towards greater environmental sensitivity, particularly the NeS and NeT herds. The results indicated that genetic variation exists for environmental sensitivity of the three herds and over time animals tend to respond better to improvements in environmental conditions. However, this result must be treated with caution, since greater environmental sensitivity seems to limit the response to genetic selection under sub-optimal conditions.

Session 63 Poster 18

Foetal mortality and fertility in divergently selected lines for birth weight homogeneity in mice

N. Formoso-Rafferty, M. Arias-Álvarez, J.P. Gutiérrez and I. Cervantes
Universidad Complutense de Madrid, Avda. Puerta de Hierro s/n, 28040, Spain; n.formosorafferty@ucm.es

The selection for less sensitivity with respect to environment, as indicated by a low variation around the optimum trait value, has been reported to have benefits in the productivity and animal welfare. A divergent selection experiment for birth weight environmental variability has been successfully conducted during 17 generations in mice. Animals from the low variability line (LV) have been shown to be more robust in the sense of having higher litter size and survival at weaning in a common breeding environment, than those from the high variability line (HV). The objective of this work was to analyse the differences between those divergently selected lines in terms of embryo mortality and the female capacity of becoming pregnant. A total of 43 females per line in the last two generations of the divergent experiment were studied. Ultrasound scans were performed at day 14 of gestation in order to know the number of total foetus and their distribution in the uterine horns and the embryo mortality from day 14 until birth. Mortality was addressed as the difference between litter size at birth and foetal number at 14 days of gestation. The number of pregnant females in the first three days after mating was used to measure fertility. A linear model was fitted to analyse foetal mortality, litter size and number of embryos at 14 days of gestation, and a categorical model was used to study fertility, including line, generation and its interaction as fixed effects. Despite the fact that there were no significant differences in the number of foetuses, litter size was significant higher in LV (9.82 vs 8.36 pups, $P<0.001$). Foetal mortality was significantly lower in LV than in HV, (1.39 vs 2.87 foetus, $P<0.001$). LV females were more fertile than HV females (53.49 vs 23.26%). According to these results, the line selected for low environmental variance would be preferable in terms of reproductive outcome, robustness and welfare.

Session 63 Poster 19

Genome-wide association analysis for days to first calving under different environmental conditions

L.F.M. Mota[1,2], F.B. Lopes[2], A.C.H. Rios[2], G.A. Ferandes Junior[2], R. Espigolan[2], A.F.B. Magalhaes[2], R. Carvalheiro[2,3], G.J.M. Rosa[1] and L.G. Albuquerque[2,3]

[1]University of Wisconsin, 1675 Observatory Doctor, Madison, Wi, 53706, USA, [2]School of Agricultural and Veterinary Sciences, Sao Paulo State University (UNESP), Via de Acesso Prof. Paulo Donato Castelane, Jaboticabal, SP, 14884-900, Brazil, [3]National Council for Science and Technological Development (CNPq), Brasilia, DF, 71605001, Brazil; flaviommota.zoo@gmail.com

Days to first calving (DFC) plays an important role on economic efficiency in beef cattle production. Days to First Calving are affected by environmental factors, which can lead to physiological changes in response to heifers' adaptation to a wide range of environments. This study aimed to identify genomic regions with persistent effect on DFC in Nelore cattle for different environmental conditions (EC). The dataset consisted of 102,076 DFC records on heifers, from which 1,911 were genotyped with the Bovine HD SNP Chip. Genotypic information of 725 sires was also used in the analyses. Animals were raised in farms located in the Midwest, Southeast and Northeast of Brazil. A reaction norm model was used to estimate the animals' response to environmental changes. The SNP effects on DFC were estimated in three EC levels (Low, Intermediate, and High) using a linear transformation of the genomic breeding values. A meta-analysis approach was used to identify persistent genomic regions affecting DFC across environments. SNP effects were considered significant when -log10(P-value) >6.0, and genes located within ±100 kb window were identified using Ensembl. In summary, combining results from the three EC, a total of 145 SNP markers were significant. From the gene network analyses, key genes PLAG1, IGF1 and IGF2 were found to directly affect reproductive pathways, related to female pregnancy, age to puberty and gamete generation. This finding suggests that common variation in DFC trait on different EC is essential to controlling reproductive precocity. The identification of such genomic regions with persistent effects on DFC might aid the genetic improvement of Nelore cattle for sexual precocity in heifers. Acknowledgments: São Paulo Research Foundation (FAPESP) grants #2009/16118-5, #2015/25356-8 and #2017/02291-3 and CNPq # 559631/2009-0.

Session 63 Poster 20

Effect of Piétrain sire and dam breed on the performance of finishing pigs: an intervention study

C. De Cuyper[1], S. Tanghe[1], S. Janssens[2], A. Van Den Broeke[1], J. Van Meensel[1], M. Aluwé[1], B. Ampe[1], N. Buys[2] and S. Millet[1]

[1]Flanders research institute for agriculture, fisheries and food (ILVO), Scheldeweg 68, 9090 Melle, Belgium, [2]Livestock Genetics, KU Leuven, Kasteelpark Arenberg 30, 3001 Heverlee, Belgium; sam.millet@ilvo.vlaanderen.be

Genetic evaluation of Piétrain sires in Flanders occurs under standardized conditions, on test stations with fixed dam breeds, standardized diets and uniform management practices. As environmental conditions vary on commercial farms and differ from the test stations, this study aimed at understanding to what extent the sire, dam breed, and the interaction between both affects the translation of breeding values to practice. Dams of two commercial breeds were inseminated with semen from one of five different sires selected for contrasting breeding values (daily gain, feed conversion ratio and carcass quality). Reproduction parameters were scored, and for each sire by dam breed combination six pen replicates (with three gilts and three barrows per pen) were evaluated for growth performance from nine weeks of age (20 kg) to slaughter (110 kg), and for carcass and meat quality. Both sire and dam breed affected performance, carcass and meat quality traits, but no significant sire × dam breed interactions could be observed. Though, a tendency for interaction on average daily feed intake between 20 and 110 kg (P=0.089) was present, with differences between sires being more pronounced for one dam breed compared to the other. In general, offspring of all tested sires behaved similarly in both dam breeds, indicating that estimated breeding values for Piétrain crosses determined in one dam breed are representative for other dam breeds as well.

Ewes' functional longevity: analysis and modelling in French suckling sheep breeds

E. Talouarn[1], V. Loywyck[2] and F. Tortereau[1]
[1]INRA, UMR 1388 GenPhySE, 24 Chemin de Borde Rouge, 31326 Castanet-Tolosan Cedex, France, [2]Institut de l'Elevage, Chemin de Borde Rouge BP 42118, 31321 Castanet-Tolosan Cedex, France; estelle.talouarn@inra.fr

In the light of an increasing demand for robustness, functional longevity appears to be an interesting criterion to study in addition to the current main selection criteria for French suckling sheep (prolificacy and mothering qualities). This study aims at characterizing genetic and non-genetic effects on functional longevity for French suckling ewes. Functional longevity is approached through length of productive life, using linear models as well as survival analysis methods. Analysis are undertaken on two representative French breeds (one specialized meat breed and a hardy meat breed), using data from the national performance recording system. Linear models are applied on the 62,410 Ile-de-France and 93,679 Blanches du Massif Central ewes born between 1990 and 2004 (uncensored careers). Survival analysis focuses on ewes born after 1990 from a known ram. Main significantly influencing non-genetic factors ($P<0.05$) were flock (mean size, acceleration, induction, renewal rates) and first lambing (age, number of lambs) characteristics. Heritability estimations were low with estimated values around 0.05 regardless of the breed or of the statistical approach. Although, selection could be intended on this trait in regard to its variability in the studied populations and the possibility of defining an early predictor.

Vocalization: a key indicator of health and welfare in future PLF systems for poultry

M.A. Mitchell[1] and J.E. Martin[2]
[1]Scotland's Rural College (SRUC), Kings Buildings, West Mains Road, Edinburgh, EH9 3JG, United Kingdom, [2]University of Edinburgh, Royal (Dick) School in Veterinary Studies, Easter Bush Campus, Midlothian, EH25 9RG, United Kingdom; malcolm.mitchell@sruc.ac.uk

Predicted expansion of intensive animal production requires the development of PLF solutions supporting animal and environmental monitoring and control. Early warning of problems associated with animal health and welfare are key. Acoustic recording and analysis of vocalization may provide the basis for monitoring and early warning systems. Vocalizations in poultry may reflect the birds' affective state and may change during exposure to stress or be associated with altered health or welfare status. The present study characterized the changes in broiler vocalizations associated with imposition of a typical stress through disturbances of the flock. Video and sound recordings were made from two commercial broiler flocks. Data were analysed from 7 identified disturbances. Behaviour was recorded (instantaneous scan sampling at 30 second intervals for 10 minutes) before and after each disturbance. Vocalization analysis was performed on 2s epochs of whole flock vocalisation noise analysed at 60 second intervals 10 minutes before and after each disturbance. Spectrographs for each epoch were produced using Audacity™ software and maximum, minimum and peak frequencies were obtained. Paired t-tests or Wilcoxon Matched Pairs tests were used to compare behaviour and vocalization frequencies before and after disturbance. After disturbance, the number of birds standing, walking and shuffling increased, while sleeping, stretching and dustbathing decreased. There was a significant increase in range and peak frequency of vocalisations in disturbed birds. The prevalence of some behaviours correlated with flock level vocalisation frequencies. It is suggested that acoustic recording and analysis of broiler vocalizations may be developed as a novel, non-invasive welfare measure for poultry for incorporation into whole house monitoring and control systems.

Session 64 Theatre 2

Turkey gait score measured with sensors

B. Visser[1], J. Mohr[2], W. Ouweltjes[3], E. Mullaart[4], P.K. Mathur[5], R.C. Borg[6], E.M. Van Grevenhof[7] and A.C. Bouwman[7]

[1]Hendrix Genetics, Research & Technology Centre, Spoorstraat 69, 5831 CK Boxmeer, the Netherlands, [2]Hendrix Genetics North America, Riverbend Drive Suite C 650, N2K 3S2 Kitchener, Canada, [3]Wageningen University & Research, Animal Health and Welfare, P.O. Box 338, 6700 AH Wageningen, the Netherlands, [4]CRV BV, P.O. Box 454, 6800 AL Arnhem, the Netherlands, [5]Topigs Norsvin, Research Centre, P.O. Box 43, 6640 AA Beuningen, the Netherlands, [6]Cobb Europe, Koorstraat 2, 5831 GH Boxmeer, the Netherlands, [7]Wageningen University & Research, Animal Breeding and Genomics, P.O. Box 338, 6700 AH Wageningen, the Netherlands; aniek.bouwman@wur.nl

Locomotion is an important welfare and health issue in livestock production. Breed4Food aims to improve locomotion through breeding and herd management using precision phenotyping. In most species, locomotion is scored by human observers. However, sensor technology is emerging and has the potential to replace human observers. In turkey breeding there is strong selection on gait scores of breeding candidates. The animals are scored by a human observer while walking through a corridor after weighing. The gait scores are reliable, repeatable, and valuable for breeding. However, the score is a temporal observation, that requires a trained observer, and handling of the animals. The aim of this study is to use sensor technology to predict gait score and assess the accuracy of prediction. Different sensors were applied during the gait scoring of the turkeys by the human observer. Approximately 400 turkeys of a male line of 20 weeks were assessed on six underlying traits, which form a combined gait score. With the output of each sensor, we will try to predict the six underlying scores and the overall gait score using machine learning techniques. The accuracy is assessed as the correlation between the predicted gait score traits and the scores from the human observer. Positive results will lead to the replacement of the human observer by sensors to get an objective scoring, which could also lead to more frequent scoring or even continuous monitoring of the locomotion of turkeys. The knowledge obtained in this Breed4Food project will also enhance potential application of sensor technology to measure locomotion in other livestock species, such as chickens, cattle and pigs.

Session 64 Theatre 3

A new PLF sensor aiming for reducing broilers' body temperature fluctuations

S. Druyan[1] and N. Barchilon[2]

[1]Agricultural Research Organization (A.R.O.) – the Volcani Centre, Animal science Institution, 68 HaMaccabim Road, 15159 Rishon LeZion 7505101, Israel, [2]Agricultural Research Organization (A.R.O.) – the Volcani Cent, PLF Lab. Institute of Agricultural Engineering, 68 HaMaccabim Road, 15159 Rishon LeZion 7505101, Israel; shelly.druyan@mail.huji.ac.il

The continuous genetic selection for performance traits resulted in a considerable enhancement of daily feed consumption, leading to alterations in growth mechanisms and development. These developments were not accompanied by the necessary increases in the size of the cardiovascular and respiratory systems, nor sufficient enhancements of their functional efficiency. This has resulted in a relatively low capability for maintaining adequate dynamic steady-state mechanisms in the body that should balance energy expenditure under extreme environmental conditions. Thus, modern broilers have an elevated metabolic rate and consequently elevated internal heat production that leads to insufficient maintenance of dynamic steady-state of thermoregulation processes, resulting in enhancement of body temperature fluctuations. In this study, a new thermal-camera based sensor was designed, built and validated. This presentation will explore techniques aiming at stabilizing broilers' body temperature fluctuations based on the application of the new sensor.

Debating PLF challenges in poultry, exotic animals, horses, reindeer, rabbits

M. Mitchell[1] and S. Druyan[2]
[1]SRUC, Scotland, United Kingdom, [2]ARO The Volcani Centre, Israel; malcolm.mitchell@sruc.ac.uk

Debating PLF challenges in poultry, aquaculture, less conventional animals, exotic animals, reindeers, horses, rabbits, etc. The debate will discuss common issues that will be raised during the session. The session speakers and posters' authors will form a panel-of-experts that will answer questions from the audience and identify common research challenges and potential engineering solutions. 15 minutes just before coffee/diner.

Thermoregulation of piglets from two genetic lines divergent for residual feed intake

O. Schmitt[1,2,3], Y. Billon[4], H. Gilbert[5], A. Bonnet[5] and L. Liaubet[5]
[1]SRUC, Animal Behaviour and Welfare Team, Animal and Veterinary Sciences Research Group, West Mains Road, Edinburgh, United Kingdom, [2]The University of Edinburgh, Department of Animal Production, Easter Bush Veterinary Centre, Royal (Dick) School of Veterinary Studies, Roslin, Midlothian, United Kingdom, [3]Teagasc, Pig Development Department, Teagasc Animal and Grassland Research and Innovation Centre, Moorepark, Co. Cork, Fermoy, Ireland, [4]INRA, UE1372, GenESI, Le Magneraud, 17700 Surgeres, France, [5]Université de Toulouse, INRA, ENVT, GenPhySE, 24 chemin de Borderouge, 31326 Castanet-Tolosan, France; laurence.liaubet@inra.fr

Hypothermia is a factor of piglet neonatal mortality. This study used Infra-Red Thermography (IRT) to assess thermoregulation abilities of piglets from two lines divergent for residual feed intake (RFI). At birth, body weight and rectal temperature were recorded from piglets of the 11th generation of the low RFI (LRFI, more efficient; n=34) and the high RFI (HRFI, less efficient; n=28). IRT images were taken at 8, 15, 30 and 60 min post-partum. Temperatures of the ear base and tip, and minimum, maximum and average temperatures of the back (*i.e.* shoulders to rumps) were extracted with Thermacam Researcher Pro 2.0, and analysed with linear mixed models (SAS 9.4). All temperatures increased overtime. The rectal temperature of piglets at birth was correlated with the initial temperature of the ear base and the maximum back temperature (0.36 and 0.35, respectively, $P<0.05$). Overall, LRFI piglets had higher minimum (28.0±0.16 vs 26.8±0.16 °C; $P<0.001$) and average (35.5±0.20 vs 34.5±0.13 °C; $P<0.001$) back temperatures than HRFI piglets. The interaction between line and time affected the temperature of the ear tip and the minimum back temperature ($P<0.05$), but pair-wise comparisons were not significant. However, ear tip temperature decreased in HRFI piglets between 8 and 15 min post-partum while it increased in LRFI piglets (-1.1±0.42 vs 0.5±0.45 °C, $P<0.05$). In conclusion, IRT allowed non-invasive assessment of piglets' thermoregulation abilities. Piglets selected for low RFI seemed to have better thermoregulation abilities at birth.

Effects of dietary threonine levels on performance and offspring traits in yellow broiler breeders

S.Q. Jiang, L. Li, Z.Y. Gou, X.J. Lin, Y.B. Wang, Q.L. Fan and Z.Y. Jiang
Key Laboratory of Animal Nutrition and Feed Science in South China, Ministry of Agriculture, Institute of Animal Science,Guangdong Academy of Agricultural Sciences, Dafeng Street 1, Wushan, Tianhe District, Guangzhou, 510640, China, P.R.; 1014534359@qq.com

The effects of dietary threonine (Thr) levels on performance, offspring traits, embryo amino acid transportation and protein deposition in broiler breeder hens was investigated in this study. A total of 720 Lingnan yellow-feathered broiler breeder hens were randomly divided into 1 of 6 dietary treatments with 6 replicates per treatment (20 birds per replicate). The breeder hens were fed either the basal diet (Thr 0.38%) or the basal diet supplemented with 0.00, 0.12, 0.24, 0.36, 0.48 and 0.60% Trp from 29w to 38w. The results showed that graded levels of Thr from 0 to 0.60% in the diet produced quadratic (R^2=0.2176, P=0.0495) positive responses in laying rate. Dietary supplemental 0.12% Thr increased laying rate by 4.67% (P<0.05). Hatchability was higher in breeders fed 0.12 and 0.24% Thr than those of control birds (P<0.05). Dietary supplemental Thr had significant effects on expressions of *mucin 2* (*MUC2*) in duodenum, colon and uterus, and *ZO-1* in duodenum of breeder hens (P<0.05). In chick embryo at embryonic age 18, there were significant upregulations of dietary Thr levels on the transcripts of liver and breast muscle *poultry target of rapamycin*, thigh *threonine dehydrogenase*,duodenum and ileum *aminopepridase* (P<0.05). Chick liveability and serum uric acid nitrogen concentration were increased, and liver glutamic-pyruvic transaminase activity was decreased by dietary Thr supplementation (P<0.05). It concluded that there were positive effects of supplemental dietary Thr on laying production of breeder hens and offspring performance, and this was associated with the regulations of gene expressions related to amino acid transportation and protein desposition. The optimal dietary Thr supplemental level was 0.298% or 0.388 g/d for Chinese yellow-feathered broiler breeder hens aged from 29w to 38w.

Do high-growth-rate rabbits prefer diets richer on amino acids than those recommended?

P.J. Marín-García, M.C. López, L. Ródenas, E.M. Martínez-Paredes, E. Blas and J.J. Pascual
Intituto de Ciencia y Tecnología Animal, Ciencia Animal, Camino de vera s/n, 46022 Valencia, Spain; pabmarg2@upv.es

Current recommendations of limiting amino acids for growing rabbits lead to a reduced amino acids retention (less to that expected) in high growth rate animals. In a previous trial, it was estimated that an increase up to 15% of these amino acids could solve this deficit in high growth rate periods. The objective of this trial was to study the ability of growing rabbits, with high growth rate, to choose between diets differing in their amino acids content to fit their needs. A total of 58 weaned rabbits from R line (selected for growth rate during the fattening period), individually housed, were used from 28 to 63 days of age. Two diets were *ad libitum* offered to each animal in a choice-feeding trial: diet M with the current recommendations for lysine, sulphur amino acids and threonine (7.3, 5.2 and 6.2 g/kg, respectively) and diet H with up to 15% more of them (8.5, 6.0 and 7.1 g/kg, respectively). Growing rabbits of R line preferred the diet H (+18.9% feed intake respect to M diet; P=0.06), especially those rabbits with a high growth rate (more than 53.5 g/d of daily gain) that showed a high preference for diet H (+65%; P<0.05). These results would indicate that the growing rabbit seems to have the ability to choose between diets differing in their amino acid content, and that animals with high growth rate would need more of some of these limiting amino acids.

Effect of the level of lysine, sulphur AA and threonine in diets for rabbits with high growth rate

P.J. Marín-García, M.C. López, L. Ródenas, E.M. Martínez-Paredes, J.J. Pascual and E. Blas
Instituto de ciencia y Tecnología animal, Universitat Politècnica de València, Ciencia Animal, Camino de Vera s/n, 46022, Spain; pabmarg2@upv.es

The reduction of the protein levels in diets for growing rabbits, with some signs of the possible presence of some limiting amino acid (AA), makes necessary the review of the level of inclusion of lysine (Lys), sulphur AA (sAA) and threonine (Thr) in diets for rabbits with high growth rate. In a previous experiment, it was checked that the AA´s combination that showed the lowest plasmatic urea nitrogen's concentration contained 7.3, 6.0 y 5.3 g/kg of Lys, AAs y Thr, respectively (12.96±0.45 mg/dl). The aim of this trial was to compare the growth performance obtained with this AA's combination (Diet MHL) or following current recommendations (Diet MMM), using 126 animals from R line from the Universitat Politècnica de València. Animals fed with Diet MAB showed better average daily gain and feed conversion ratio than those fed with Diet MMM ($P<0.05$). These results would indicate that, for animals with high growth rate, the dietary level of sAA should be increased (to 6 g/kg) and the dietary level of Thr should be decreased (to 5.3 g/kg).

Management of reproduction of red deer under conditions of intensive farming

R. Kasarda[1] and J. Pokorádi[2]
[1]Slovak University of Agriculture, Department of Animal Genetics and Breeding Biology, Tr. A. Hlinku 2, 949 76 Nitra, Slovak Republic, [2]XCELL Slovakia Breeding Services, Inc., Ventúrka 1, 811 01 Bratislava, Slovak Republic; radovan.kasarda@uniag.sk

Analysis of reproduction of Red Deer (*Cervus Elaphus*) was realised based on data obtained from in total 60 females mated with 3 sires at AI station for red deer, fallow deer and mouflon in EU – XCELL Slovakia Breeding Services Inc. Age of all evaluated animals was 2-5 years, females after min. first birth and sires with ejaculate quality protocol. All animals were individually traced (ear tag, Id chip, DNA profile) and bred under similar farming conditions including veterinary service, welfare and nutrition. Efficiency of artificial reproduction using was evaluated according to effect of heat synchronization, accuracy of estimation of time of ovulation and effect of place of deposition of AI dose. Male reproduction was evaluated as quantitative and qualitative parameters of ejaculate during mating season (August-October). After synchronization of females with PMSG (intramuscular) several internal and external signs of ovulation were visible. Reaction on ovaries and cervix was evaluated by palpation (rectum) and by use of speculum intravaginal 54 hours after application. After application of 200 IU PMGS, in 100% of females presence of CL was observed, cervical patency in 93% and vulva repletion in 93% of females. Ovulation occurred 54-56 h after synchronisation with PMGS. Deposition of AI dose into the uterine body resulted in significantly higher pregnancy rate. Optimization of management of assisted reproduction in Red Deer under intensive farming is particular goal in deer breeding of wide importance. Main goal in deer breeding is increase and alignment of daily gain and live weight of offspring as basic quantitative parameters with direct influence on profitability of production. Use of proven sires of known origin and breeding value has positive impact on genetic gain and allows prevention of mating of relatives and preservation of good gene-pool for the future generations.

Session 64 — Theatre 10

Muscle marker and growth factors genes expression in four commercial broiler strains

K.I.Z. Jawasreh and S.I. Al-Athamneh
Jordan University of Science and Technology, Animal Production, Irbid 22110, Jordan; kijawasreh@just.edu.jo

Under the same environmental conditions, growth performance between and within commercial Broiler strains usually differed. A standard feeding trial (28 d) was designed for understanding the observed variations between broiler strains in growth performance and their muscle marker and growth factor genes expression profile. The trial included 800, One-day-old chicks; 200 from each of Ross 308, Indian River, Cobb and Hubbard broiler strains, with 10 replicate pens per strain and 20 birds per pen. At the end of the experiment, 40 birds (10 from each strain) were weighed and slaughtered. The Pectoral muscle samples were taken for the purpose of evaluating the mRNA expression levels of muscle marker genes (MyoD, myogenin) and muscle growth factors (IGF-1 and myostatin) using a real-time thermal cycler (CFX96 Touch TM Real time PCR, BIO-RAD, California, USA). Data were analysed as a completely randomized design using General Linear Model procedure of SAS. Through the rearing period there were significant differences in body weight between the studied strains but at the end of the experiment (28 days), Indian River gained the highest ($P<0.0001$) body weight (1,573.619 g) Compared to Hubbard and Cobb (1,499.032 and 1,459.060 g, respectively) while Ross strain was the lowest (1,355.164 g). Indian River strain had the highest ($P<0.05$) fold expression of IGF-1 compared to Ross, Cobb and Hubbard expression. No significant difference between Ross and Cobb in MYO-D gene expression, while their expression was significantly higher than that in Indian River and Hubbard. Indian River and Ross strains had the highest MYO-G fold gene expression among all strains. Ross strain significantly expressed the highest Myostatin gene compared to the other three strains. As a conclusion the IGF-1, Myo- and Myostatin gene expression tend to associate with the higher growth rate and may be responsible for the higher gain observed in Indian River.

Session 65 — Theatre 1

New breeding goals and role of genomics on adaptation and resilience traits in French dairy sheep

J.M. Astruc[1], D. Buisson[1], H. Larroque[2] and G. Lagriffoul[1]
[1]Institut de l'Elevage, 31321 Castanet-Tolosan cédex, France, [2]GenPhySE, Université de Toulouse, INRA, INPT, ENVT, 31326 Castanet-Tolosan cédex, France; jean-michel.astruc@idele.fr

For several decades, the breeding organizations ruling the French dairy sheep breeds have developed efficient breeding programs aiming at improving the local breeds in their production area and system. This has led to an annual genetic gain ranging between 0.15 and 0.22 standard deviation unit, depending on the breed. The breeding programs have steadily included new selection criteria in addition to milk yield: milk quality, resistance to mastitis, udder morphology, scrapie resistance. Following the development and spread of genomic selection (GS) in dairy cattle in the early 2010, different research × development programs have assessed the feasibility of GS in dairy sheep: reference populations in each breed, genomic evaluation to improve GEBVs' accuracy, conception, modelling and optimization of dairy-sheep-customized genomic breeding schemes. These achievements have resulted in the switch towards GS in 2015 in the Lacaune breed and in 2017 in the Pyrenean and Corsican breeds. This new era of GS has opened the way to great expectations. The increasing efficiency of the programs, even though not as dramatic as promised in cattle (20% additional genetic gain observed in the Lacaune breed 3 years after the starting point of GS) will make easier the inclusion of new traits in the upcoming years. In a context of feed resources autonomy and adaptation to climate changes, news traits will be related to: health resistance and resilience traits such as resistance to internal parasites, adaptation and rusticity traits such as longevity and lactation persistency, functional morphology such as feet and legs traits, feed efficiency, reproduction traits such as semen production, predicted milk fatty acid and protein composition allowed by MIR spectrometry, The breeding goals of the near future will aim at being more balanced and at improving both sustainability and cost-effectiveness of a dairy sheep production anchored in its production system and producing high quality PDO cheese. Finally, both genomic tools and balanced breeding goals will help to maintain the existing biodiversity by better managing the genetic variability of the breeds.

Perspectives of small ruminants in the Mediterranean part of Croatia

Z. Barać and M. Špehar
Croatian Agricultural Agency, Ilica 101, 10000, Croatia; zbarac@hpa.hr

The sheep rearing has a long tradition in Croatia. The small ruminants' population and their economic importance have been significantly changed during the past 60 years. Industry and tourism development have led to significant decline of small ruminants population. The total sheep and goat population is around 650 and 78 thousand animals currently, reared on 23 thousand farms. Around 60% of sheep and 45% of goat population is reared in Mediterranean part of Croatia. Sheep production of this area is based on autochthonous breeds, extensive pasture, and crop production systems. These breeds are used for meat (primarily lamb) and milk production. Milk is mainly used for cheese production either traditionally in small family dairies or industrially in large dairies however in a smaller extent. Wool production has not any economic importance, since only a small proportion is processed by the domestic textile industry and unfortunately most of the wool ends as a waste. The most population of goats consists of indigenous breeds, mainly kept in Mediterranean region. Goat farming has a large importance since they are adapted to weaker conditions for breeding in comparison to other livestock species. The most part of Croatian goat population is reared for meat. The increase of small ruminants' population should be one of the goals in Croatian Mediterranean area not only for economic but also environmental reasons (fire prevention). Production covers only 50% demands on small ruminants' meat in Croatia. Also, demands on local sheep and goat types of cheeses are steadily increasing. For future development of sheep and goat sector, the most important measures are: use of funds available to farmers within the Program of rural development (EFRD), to increase production on existing farms as well as to build new farms. Furthermore, younger and educated people have to be included in this production. Protection of local products by quality designations (marks of quality) also will help in their recognition on the market.

Session 65 Theatre 3

Spanish dairy sheep breeds face new challenges on Mediterranean area

E. Ugarte[1], R.J. Ruiz[1], F. Arrese[2], R. Gallego[3] and F. Freire[4]
[1] NEIKER-Tecnalia, Basque Institute of Agricultural Research and Development, Animal production, Agrifood Campus of Arkaute s/n, 01080 Arkaute, Spain, [2]CONFELAC, Latxa and Carranzana dairy sheep breeders´ associations confederation, Agrifood Campus of Arkaute s/n, 01080 Arkaute, Spain, [3]AGRAMA, National Association of Selected Sheep Breeders of Manchega Breed, Avda. Gregorio Arcos, 19, 02005 Albacete, Spain, [4]ASSAFE, Assaf Spanish dairy sheep breeders´ association, Granja Florencia, 49800, Toro, Spain; eugarte@neiker.eus

The paper presents a general perspective of the actions that the Spanish associations of dairy sheep breeds are carrying on to face new challenges related with climate change, sustainability, consumer requirements, environmental aspects, etc. and the new research needs that these challenges pose to maintain the balance between production, adaptation and sustainability. Despite the diversity of dairy sheep production systems in Spain (Assaf, Manchega, Churra and Latxa), work is being done in similar topics, although with different degrees of intensity. The most important areas on which they are working are: resistance to heat stress, appearance of new diseases, sustainability assessment, use of new raw materials for food, reduction of methane emissions, incorporation of genomic selection, introduction of new traits within breeding schemes, incorporation of TICs, etc. Several of these issues are being implemented jointly within the European ISAGE project.

Session 65 Theatre 4

French conservation programs of sheep and goats in a Mediterranean context: a SWOT analysis

C. Danchin-Burge
Institut de l'Elevage, Génétique et Phénotypes, 149 rue de Bercy, 75012 Paris, France; coralie.danchin@idele.fr

There are a half dozen breeds of sheep and goats that are classified as « rare » in the French Mediterranean region. Their present demographic situations are alternating between booming numbers (Rouge du Roussillon sheep) to more threatening situation (Brigasque sheep). In a context of climate changes, we realized a SWOT analysis to see what the main factors are that will contribute to the future of these breeds. Interestingly, climate changes can actually be seen as a positive factor. All these breeds were neglected for their poor production in an intensive environment. On the other hand, their hardiness (mostly their walking abilities on a calcareous soil, valorisation of coarse fodder) is now becoming an opportunity for producing in this context. These abilities are also valuable for territorial management, with fruitful collaborations with local organizations for fire control. The main threat for these breed is mostly due to land issues: finding land to build a farm; land use in a context of high predation (wolf). One of shared weakness of these breeds is the lack of data, including basic ones such as pedigrees. Also there are almost no scientific data that can acknowledge their hardiness. It is hoped that H2020 project such as SMARTER would be an opportunity to tackle this issue.

Session 65 Theatre 5

Sources of flexibility in replacement and culling practices in dairy-sheep farms of Corsica

L. Perucho[1], C.H. Moulin[2], J.C. Paoli[1], C. Ligda[3] and A. Lauvie[2]
[1]INRA, Quartier Grossetti, 20250 Corte, France, [2]UMR SELMET, INRA-CIRAD, Montpellier SupAgro, Univ Montpellier, 2 place Pierre Viala, 34060 Montpellier, France, [3]HAO, VRI, P.O. Box 60272, 57001 Thessaloniki, Greece; anne.lauvie@inra.fr

The animal resource is increasingly studied for its contribution to the flexibility of farming systems under uncertain environments. Sources of flexibility conferred by the breeding stock include (1) the animal flows in and out the herd, (2) the diversity of species, breeds and animal populations within breeds and (3) the animal adaptive capacities. Nevertheless, those various sources of flexibility may not all coexist on farm. The aim of the study is to understand the ways Mediterranean dairy sheep farmers combine those sources of flexibility at the farm level. A sample of 30 farms was built to consider the diversity of the dairy-sheep systems in the region of Corsica. Action on the breeding animals flows (source 1) was assessed through the practices of female replacement and culling. The use of a diversity of breeding males from the regional Corsican breed sheep populations referred to the source 2. Action on the adaptive capacities of the animals (source 3) was assessed through specific trade-off performed by farmers between production traits and functional traits, as: prolificacy, breed standard, temperament, milking ease, sensitivity to mastitis. Bertin's graphical data analysis was used to group farmers with a similar use of the different sources of flexibility. Flexibility conferred by adjustments of breeding animal flows ranged from low to high across the sampled farms. For example, management of breeding males flow ranged from the systematic replacement of males through exclusive internal replacement on early lambings (n=2) to the use of both internal males (from early and late lambings) and external males coming routinely from several providers. Consideration of functional traits differed according to the adjustments of breeding animal flows in and out the flock. For example, farmers using both female and external male flows as a source of flexibility to take advantage of genetic opportunities were more flexible on the sensitivity to mastitis than other farmers.

Genotype by environment interaction for length of 1st inter lambing interval in sheep

A. Kasap, V. Držaić and B. Mioč
University of Zagreb, Faculty of Agriculture, Department of Animal Science and Technology, Svetošimunska 25, 10000 Zagreb, Croatia; akasap@agr.hr

Seasonal reproductive activity of ewes in temperate climates is a heritage of natural selection which favoured sheep giving birth at the most appropriate time of year in term of food availability. Short photoperiods stimulate pineal gland to synthesize more melatonin which triggers a series of hormonal reactions included in follicle maturation and ovulation. Melatonin exerts its function through specific MT1 G-protein receptor coded by MTNR1A gene (in sheep on 26th chromosome). Two polymorphic (bi-allelic) loci (606 and 612) on the 2nd exon of the gene have been associated with the seasonality. Presence of allele/s C-606 and G-612 have been mainly associated with whole-year breeding and shorter inter lambing period, but little has been known about genotype by environment interactions (GEI) on this issue. The aim of the study was to answer whether the same genotypes exhibit different phenotypes in different environments. The study was conducted on 277 natural-serviced ewes that were genotyped (MTNR1A – 606) and phenotyped in two different environments in Croatia (littoral=ENV1, n=157, continental=ENV2, n=120). The trait of interest was length of 1st inter lambing interval. In addition to the effect of interest (GEI), type of birth, season of lambing, and age at 1st lambing (covariate) were fitted in the ANCOVA statistical model to account for the available sources of phenotypic variability. Frequencies of genotypes CC, CT and TT in the ENV1 were 0.15, 0.55 and 0.30, and in the ENV2 0.26, 0.49, and 0.25, respectively. Estimates (LS means) of inter lambing intervals for CC, CT and TT in the ENV1 were 331.9, 355.4, and 371.3 days, and in the ENV2 363.5, 370.7, and 365.7 days, respectively. Results pertaining to ENV1 go in line with the majority of previous findings on this issue while those pertaining to ENV2 do not. The latter could be due to more intensive rearing system in ENV2 as compared to ENV1 which lead to weaker opportunity of the ewes to capture signals from the nature. Albeit determined re-ranking of genotypes across environments implicated existence of GEI, magnitudes of differences were too small, and intra class variances too high to detect statistical significance ($F_{2, 265}$=0.95, P=0.39).

Adaptation to Mediterranean climate challenge in ruminants: thermal tolerance phenotypes and genetics

M.J. Carabaño[1], M. Ramón[2], M. Usala[1], M. Serrano[1], J.M. Serradilla[3], C. Meneses[1], M.A. Jiménez[1,4], M.D. Pérez-Guzmán[2], A. Molina[3], R. Gallego[5], F. Freire[4] and C. Díaz[1]
[1]INIA, Mejora Genética Animal, Ctra de A Coruña km 7.5, 28040 Madrid, Spain, [2]IRIAF, CERSYRA, Avenida del Vino 10, 13300 Valdepeñas, Spain, [3]University of Cordoba, Producción Animal, Campus de Rabanales, Ctra. Madrid-Cádiz km 3, 14071 Córdoba, Spain, [4]ASSAF.E, Granja Florencia, 49800 Toro, Spain, [5]AGRAMA, Avda. Gregorio Arcos, 19, 02005 Albacete, Spain; mjc@inia.es

Extreme weather events pose a challenge for animal production in the Mediterranean area. In particular, IPCC predictions point at the Mediterranean basin as a hot spot of the climate change effects, which will create harsher conditions and more areas affected by thermal stress effects. Local breeds are expected to be adapted to environmental conditions but heat tolerance has been rarely quantified and selection for productivity may interact with their resilience to the hot summer oror extreme and variable thermal conditions. Moreover, highly selected cosmopolitan breeds, which are expected to be more susceptible to environmental challenges, are also frequently used in Mediterranean farms. One tool to deal with adaptation to thermal stress is genetic selection, but it poses challenges relative to what phenotypes can be used and how selection for thermal tolerance will interact with the current selection objectives. In this work, results of the impact of thermal load on the traits that are selection goals (mainly, milk yield and composition and growth in feedlots) in local Spanish breeds, studied in the context of the iSAGE EU project and other national projects, are presented. Alternative phenotypes to measure tolerance and their relationship with productivity are shown. Antagonistic relationship between some heat tolerance measures and productivity seem stronger in highly selected cosmopolitan breeds than in local populations. Finally, results relative to the genetic background of thermal tolerance are presented. Heritability estimates of alternative measures of heat tolerance range from low to moderate. Genome-wide association studies in some of these populations point at a variety of genomic regions that influence heat tolerance, showing a polygenic control of this trait and coincidence with genes associated with productive traits.

Session 65 Theatre 8

Prospects of local sheep and goat breeds for sustainable farming in a changing mediterranean region

P.Y. Aad[1] and M. Kharrat-Sarkis[2]
[1]Notre Dame University-Louaize, Sciences, S 224, FNAS, NDU-Louaize, Zouk Mosbeh, Lebanon, [2]Universite Saint Joseph, Ecole Supérieure d'Ingénieurs d'Agronomie Méditerranéenne, Taanail, BP 159 Zahlé, Liban, Lebanon; paad@ndu.edu.lb

Small ruminant production systems of transhumance in the Mediterranean region, especially Lebanon, Syria and Jordan, depend greatly on natural pastures, and are highly threatened in a changing climate. Anticipated increase in solar radiation and ambient temperature as well as decreased amount of rain, water quality and pasture availability constitute a major threat to the endurance and sustainability of small ruminant farmers. Local breeds such as Awassi sheep and Baladi goats are highly desirable in the region due to their assumed resilience, however both face current management practices that compromise performance. Studies on both breeds have resulted in very diverse results, possibility due to high genetic diversity. Studies on the resilience of both breeds are very promising and showcase how simple managerial strategies can lead to major benefits to both farmers and flock. In fact, the local Baladi goat provided short flushing during breeding via access to agriculture cropland show marked improvement in productivity, ovulation and gestation. Timing of shearing and travel, housing outdoor, salt supplementation and even fat-tail docking in Awassi sheep have improved reproductive performance. The high resilience of these animals is clearly depicted by low panting scores in Awassi ewes exposed to high solar radiation and drinking salt water, with minimal effect on productivity and fertility. Nevertheless, the reproductive potential of both Baladi and Awassi populations is very variable, and selection is required. Altogether, these strategies can remedy the socio-economic fragility facing the small ruminant sector, particularly in Lebanon, by providing a cheap crop by-product complementary diet at specific timing, ensure body condition maintenance, decrease stress and ensure adequate weight at birth of the litter for best survival and subsequent growth and health. Extension and proper transfer of science needs to become a major area of focus to educate farmers on best managerial practices and the upcoming threats to animal welfare and rural sustainability.

Session 65 Theatre 9

Dynamics and prospects for crop-livestock systems in South and North of Mediterranean

V. Alary[1], J. Lasseur[2], A. Aboulnaga[3], T. Srairi[4] and C.H. Moulin[5]
[1]CIRAD, TA C-112/A Campus International de Baillarguet, 4398 Montpellier cedex 5, France, [2]INRA, 2 place Pierre Viala, 34060 Montpellier cedex 2, France, [3]APRI, Dokki, Giza, Cairo, Egypt, [4]IAV Hassan II, Madinat Al Irfane, B.P. 6202 Rabat, Morocco, [5]Montpellier SupAgro, 2 place Viala, 34060 Montpellier cedex 2, France; charles-henri.moulin@supagro.fr

Mediterranean farming systems evolve in response to multiple changes and have to adapt to the current and future pressures, included demographic growth, urbanization, acute competition for land and water, climate change and agricultural inputs' prices volatility. We aim at exploring future pathways for integrated crop-livestock systems in South and North Mediterranean countries in order to identify for policy makers and developers potential sustainable increase in efficiency and adaptability regarding resources' utilization,. Research activities were performed in three countries: Egypt, France and Morocco, through case studies in a gradient of various socio-ecological contexts, from favourable contexts (plains and irrigation schemes) to harsher contexts (mountains, rain-fed areas). We mobilized various datasets: farm surveys and monitoring, interviews, databases and previous studies. We identified two main trends and five archetypical systems: a centrifugal trend of specialization, toward cash crops or dairy farming in favourable areas, and rangeland systems for meat production in harsher environment, and a centripetal trend of diversification, maintaining mixed crop-livestock systems in irrigated areas and agro-pastoral livestock-crop systems in intermediate rain-fed areas. Main issues of these research pinpointed necessity in: (1) overwhelming antagonism between social vulnerability and ecological efficiency of crop-livestock farming systems through dedicated rural development policies; (2) limiting micro-regional specialization processes through the maintaining of system diversity; (3) taking advantage of spatial mobility abilities of livestock to reinforce crop-livestock integration at regional level, limiting efficiency loss and reinforcing sustainability for vulnerable livestock farmers.

Session 65 Poster 10

Characterization of dairy sheep and goats production systems in France: first step for a G×E study

H. Larroque[1], G. Lagriffoul[2], A. Combasteix[2], J.M. Astruc[2], D. Hazard[1], A. Rolland[1] and I. Palhière[1]
[1]INRA, UMR 1388 GenPhySE, Chemin de Borde Rouge, 31326 Castanet-Tolosan, France, [2]Institut de l'Elevage, Chemin de Borde Rouge, 31326 Castanet-Tolosan, France; jean-michel.astruc@idele.fr

In a fluctuating economic, environmental and societal context, farms of dairy small ruminants attempt to increase their autonomy in feed resources. Due to the diversity of soil and climate conditions in France, breeders develop diverse farming and feeding strategies. In order to investigate presence of G×E interactions, the first step consisted in describing dairy sheep and goats production systems to identify contrasted farms typologies. An exploration was carried out using a large set of variables present in the databases of milk-recording, genetic evaluation, technical support (describing farming and feeding systems, when available), and also geographical and meteorological data (METEOFRANCE, French official service of meteorology and climatology). The most discriminating variables were selected to conduct a multiple correspondence analysis. A classification of herds was then performed within geographic areas for sheep (440, 463 and 98 herds in Roquefort area, Western Pyrenean, and Corsica island, respectively), and within all the country for goats (1,136 herds, whose 514 in the breeding nucleus). For dairy sheep in the Roquefort area, 4 clusters of flocks in Lacaune breed were identified according to their geographical location, precocity of grass growth (in relation to altitude), and amounts of concentrate and forages distributed. In the Western Pyrenean area, 4 clusters in Basco-Béarnaise breed and 7 clusters in Blond-faced Manech breed have been highlighted on the basis of their location-altitude and of herd management criteria (rate of first lactations at 2 years, flock size). In Corsica, the study has highlighted 3 groups of flocks according to their altitude and level of production (linked to the artificial insemination rate). For goats, 4 clusters of herds were discriminated according to area of production (West / East gradient), breeding goals (fat and protein contents or milk yield), system of sales (cheese maker or deliverer to dairy industry), herd size and reproduction organization (out of season or not). This study was carried out with the financial support of the European project iSAGE.

Session 65 Poster 11

Betaine mitigates hyperthermia in heat-stressed, hypohydrated fattening Awassi lambs

H.J. Al-Tamimi[1], B.S. Obeidat[1], K.Z. Mahmoud[1] and M. Daradka[2]
[1]Jordan University of Science and Technology, Animal Production, Irbid, Amman Street, 22110, Jordan, [2]Jordan University of Science and Technology, Veterinary Medicine, Irbid, Amman Street, 22110, Jordan; hosamt@gmail.com

Fattening lambs in desert areas often experience hyperthermia. Betaine, an osmolyte, was shown to improve water metabolism in animals. Thirty six 60-days old weaned Awassi ram lambs were allotted to 4 treatment groups. Following a fattening period of 56 days, all animals were implanted miniature thermologgers at different body locations to measure temperatures of the intraperitoneal cavity (T_{ip}), subcutaneous (T_{sq}), intra-scrotal cavity (T_{sc}) and reticulo-rumen (T_{rum}). All loggers were pre-synchronized with an outer meteorological station [measuring light intensity, solar radiation (SR), ambient temperature (T_a) and percent relative humidity (RH%)]. Half of the lambs ingested daily dietary betaine (BET; top spread) at 8 g/head throughout the fattening and trial period (total of 77 days), while the remaining half acted as controls (CON; BET-free). All animals were initially kept in shaded (SHD) individual pens for a period of 7 days. Then, 9 lambs from each treatment group were moved to unshaded (SUN) individual pens, and exposed to summer solstice direct SR. Shifting from SHD to SUN housing resulted in pronounced rises of T_{sq} and T_{ip} (but not T_{rum}) in both treatments, with BET being responsible for lower degree of displayed hyperthermia than CON. The upsurges in T_{ip} and T_{sq} were 0.29 and 0.30 ±0.03 °C, for CON over BET counterparts (when switched to SUN; but not maintained in SHD), suggesting an improved osmoregulation potential by BET, and hence an enhanced resistance to hyperthermia. Regardless of BET and/or SUN treatments, all lambs similarly maintained a pooled mean daily gradient of T_{sc} below T_{ip} by 3.71-4.93 ±0.02 °C. Interestingly, T_{rum} did not vary among treatment groups. Consistently, based upon mean daily values, T_{rum} exceeded Tc by 0.33 and 0.50 ±0.05 °C, in SHD and SUN exposure, respectively. This suggests the employment of cardiovascular mechanisms, distinct from metabolic pathways within the gastrointestinal tract, in thermoregulation by young Awassi lambs. Betaine inclusion in lamb diets is a promising tool in mitigating hyperthermia in young lambs.

Session 66 Theatre 1

Methods for prediction of breeding values for young animals in single-step evaluations

E.C.G. Pimentel, C. Edel, R. Emmerling and K.-U. Götz
Bavarian State Research Center for Agriculture, Institute of Animal Breeding, Prof.-Dürrwaechter-Platz 1, 85586 Poing-Grub, Germany; eduardo.pimentel@lfl.bayern.de

Single-step genomic evaluations have the advantage of simultaneously combining all pedigree, phenotypic and genotypic information available. However, systems with a large number of genotyped animals show some computational challenges. In many genomic breeding programs genomic predictions of young animals should become available for selection decisions in the shortest time possible, which requires either a very effective estimation or an approximation with negligible loss in accuracy. We investigated different procedures for predicting breeding values of young genotyped animals without setting up the full single-step system augmented for the additional genotypes. The situation envisioned assumed three full evaluations per year with monthly interim prediction of young animals. Methods were based on the transmission of the information from single-step breeding values of genotyped animals that took part in the previous full run to the young animals, either through genomic relationships or through a marker based model. The different procedures were tested on real data from the April 2017 run of the German-Austrian official genomic evaluation for Fleckvieh. The dataset included 62,559 genotyped animals and was used to run single-step evaluations for 24 conformation traits. A further dataset comprising 1,806 young animals was used for prediction and validation. The reference values for validation were the predicted breeding values of the young animals from a full single-step run containing the genotypes of all 64,365 animals. Correlations between the approximated predictions and the ones from the full single-step run containing also genotypes from young animals were on average 0.991 for the best method (from 0.981 to 0.994 across traits). In conclusion, prediction of single-step breeding values for young animals can be approximated using systems of size at most equal to the number of genotyped animals from the previous full run.

Session 66 Theatre 2

Variance estimates from the algorithm for proven and young animals are similar to current methods

M.N. Aldridge[1], J. Vandenplas[1], R. Bergsma[2] and M.P.L. Calus[1]
[1]Wageningen University & Research, Animal Breeding and Genomics, Droevendaalsesteeg 1, 6708 PB Wageningen, the Netherlands, [2]Topigs Norsvin Research Center B.V., Schoenaker 6, 6641 SZ Beuningen, the Netherlands; michael.aldridge@wur.nl

As the amount of genotyped data generated continues to increase, traditional methods of prediction are becoming unsuitable due to computer limitations. The Algorithm for Proven and Young animals (APY), is a method that can potentially solve the issue. The objective of this study was to determine if variance estimates made with single step APY are similar to those made with just the pedigree, and single step using the full inverted G matrix. It was hypothesised that estimates would not be significantly different, provided that the G matrix has the same base as the A matrix. To test the hypothesis, average litter birth weight was analysed. Noncore animals included 9,695 genotyped and phenotyped individuals, with 34,442 trait records. A further 23,079 genotyped but not phenotyped animals were treated as core animals. Using pedigree of noncore and core animals, the estimated additive genetic variance was 13,620 and 13,570 (ASReml and REMLF90 respectively), the residual variances were 14,176 and 14,180, and the permanent environmental variance was 3,478 and 3,511 respectively. Variances were estimated using a subset of genotypes, from noncore animals for the G-1 matrix, and a subset of 6,000 core and 4,000 noncore animals for the APY G-1 matrix. The additive genetic (14,280 and 13,661), residual (14,270 and 14,213) and permanent environmental (4,220 and 4,519) variances, were similar for the subset G-1 and APY G-1. In many current implementations of genomic prediction using an inverse G or APY G matrix, variance components estimated with A are used. In some cases it may be more appropriate using the G and APY G matrices to compute the variances, which would then be used in the equivalent GBLUP model. The alternative would be to either, modify the G and G APY such that the variances estimated with A still apply, or modify the variances estimated from A to be used in the GBLUP models.

Accuracy of genomic prediction using a novel Bayesian GC approach

M.H. Kjetså[1], J. Ødegaard[2] and T.H.E. Meuwissen[1]
[1]Norwegian University of Life Sciences (NMBU), Faculty of Biosciences, Arboretveien 6, 1432 Ås, Norway, [2]AquaGen, Breeding and Genetics, Box 1240, 7462 Trondheim, Norway; maria.kjetsa@nmbu.no

Increasing the accuracy of genomic prediction is important, especially in traits with expensive recording and low heritability. This study is looking at the accuracy of predicting GEBVs for host resistance to lice in Atlantic salmon (*Salmo salar*). It compares a novel Bayesian C approach, 'Bayes GC' with the traditional GBLUP. Bayes GC fits a polygenic effect and a Bayes C term simultaneously. The polygenic effect handles the background relationships and the Bayes C fits SNPs with larger effects by using a prior probability, π, for a SNP having an effect or not in the model. The data used is 1,385 phenotyped and genotyped Atlantic salmon. The phenotyped animals were genotyped with a 200K SNP Chip. They were imputed to 750K SNPs from a reference population of 69 animals from their parent generation. The animals were phased in the imputation process. The phenotype is host resistance to sea lice from salmon reared in sea cages, and had an estimated heritability of 0.14±0.03. By imputing from 200 to 750K there was no increase in the accuracy of prediction using GBLUP. The Bayes GC approach gave a slight increase in accuracy (~0.5 percentage points) compared to GBLUP, however, this increase was not statistically significant.

Single-step genomic REML with algorithm for proven and young and reduced pedigree

D. Lourenco[1], V.S. Junqueira[2], Y. Masuda[1] and I. Misztal[1]
[1]University of Georgia, Animal and Dairy Science, 425 River Rd, 30602 Athens, USA, [2]Federal University of Viçosa, Animal Science, P. H. Rolfs Av, 36570-900 Viçosa, Brazil; danilino@uga.edu

Efficient computing techniques allow the estimation of variance components for virtually any traditional dataset. When genomic information is available, variance components can be estimated using genomic REML (GREML). If only a portion of the animals have genotypes, single-step GREML (ssGREML) is the method of choice. The genomic relationship matrix (G) used in both cases is dense, limiting computations depending on the number of genotyped animals. The algorithm for proven and young (APY), which takes into account the dimensionality of genomic information, can be used to create a sparse inverse of G with close to linear memory and computing requirements. In ssGREML, the inverse of the realized relationship matrix (H-1) also includes the inverse of the pedigree relationship matrix, which can be dense with long pedigree, but sparser with short. The purpose of this study was to investigate whether costs of ssGREML can be reduced using APY with truncated pedigree. The effects of pedigree truncation on EBV were also investigated. Simulations included 22,012 animals from 10 generations, with selection based on EBV. Phenotypes (h^2=0.3) were available for all animals in generations 1-9. A total of 4,000 animals in generations 8 and 9, and 2,000 validation animals in generation 10 were genotyped for 50k SNP. Variance components were estimated with complete pedigree and phenotypes, and removing one generation of data a time, starting from the oldest. Average information REML and ssGREML with APY using 2,700 core animals were compared. Changes in accuracy were also investigated. Using fewer generations of data reduced the ability of pedigree-based model in estimating heritability, however, the decrease in heritability was less when ssGREML was used. Computing time was not greatly reduced because the number of core animals was high compared to the number of genotypes. Prediction accuracy and (G)EBV accuracy for validation animals remained constant with all levels of data truncation. The use of ssGREML with APY and truncated pedigree has minimal impact on variance components and may help to reduce computing resources in larger genotyped populations.

Using the variable effect predictor with WGS data finds one causal variant for early death in calves

G.E. Pollott, M. Salavati and D.C. Wathes
RVC, PPS, Royal College Street, NW1 0TU London, United Kingdom; gpollott@rvc.ac.uk

Whole genome sequence (WGS) data have the potential to assist in finding the site of a new deleterious variant causing a novel Mendelian disease. The challenge is how to find the site of the new variant from the 3 billion base pairs in a mammalian genome. It has been suggested that the search only needs to take place in that proportion of sites showing variation, using say, a Variant Call Format file. Recent studies have found this to be 10 to 20 million sites, depending on the number of animals used. Further reduction in the search area can be made by looking for long runs of homozygosity (ROH) in recent recessive conditions. A review of published reports suggests this may reduce the search area down to anything from 0.6 to 20 Mbp. Upon alignment of the WGS reads to a reference genome, variant calling could be performed on the mapped reads using tools such as the Genome Analysis Toolkit. The variant call set information could be further annotated by the Variant Effect Prediction tool of the ENSEMBL database in order to indicate the potential severe to moderate effects within and around coding regions (e.g. 5 kbp up or downstream from the transcript start site). We report data from 9 animals (4 cases and 5 parental controls) from Irish Moiled cattle with a condition that causes mortality within a few days of birth. From this WGS dataset we located 9,650 variant sites on the whole genome with a 'high-impact' SIFT score. Using these data, in conjunction with SNP ROH results from 58 controls and 17 cases, we were able to locate a 25 MB region on BTA4 likely to contain the site of the novel variant. Inspecting the 24 high-impact sites in this region and looking for heterozygous controls and homozygous cases for a dangerous variant led us to suggest only one site as the possible origin of this condition. This was a single-base change splice acceptor variant in the glucokinase gene (GCK) and is likely to have drastic protein folding changes (PHYRE2 prediction). This enzyme plays a key role in glucose uptake and regulation of insulin secretion. Human mutations are associated with early-onset diabetes. Future work will be undertaken in the breed to confirm these findings and implement a suitable breeding programme for controlling the condition.

Mapping for selection signatures associated with aggressive behaviour in cattle

P. Eusebi, O. Cortés, S. Dunner and J. Cañón
Universidad Complutense de Madrid, Animal production department, Avenida Puerta del Hierro s/n, 28040 Madrid, Spain, Spain; paulinaeusebi@gmail.com

The domestication of the bovine cattle and its subsequent artificial selection processes left signatures of selection that allows for the identification of genomic regions that the evolution of the different bovine breeds has caused. Several studies describe the genomic variability among breeds specialized mainly on meat or milk production however, there are limited studies orientated towards bovine behavioural features. The current study is focused on mapping genomic signatures of selection, which may provide insights of differentiation between neutral and selected polymorphisms with effects on the Lidia bovine breed traditionally selected for agonistic behaviour, compared with two tamed Spanish breeds, the Asturianas de los Valles and Morenas Gallegas, with tame behaviour. Two different approaches, the Bayescan and SelEstim, both using the same methodology based on allele frequencies differences among populations, were applied using genotypic 50K SNP Beadchip data. Two genomic regions in common were detected with both Bayescan and SelEstim, this latter approach having an advantage of distinguishing between selected and nearly neutral polymorphisms and estimating the intensity of selection. In this way, the frequency of the allele within the Lidia breed populations in these regions was opposite from the frequency of the allele selected in the two Spanish tamed breeds. In these genomic regions several putative genes associated to metabolic pathways, such as the serotonergic and dopaminergic signalling pathways, associated to different behavioural patterns, were detected. The difficulty to detect selective sweeps with statistical significance in polygenic traits, in which many loci shift their frequencies moderately, could explain that only two genomic regions were shared with both methodologies. Other reasons may be the limitations of the 50K chip and the sample size of the analysis. The present study allowed identifying genomic regions with opposite direction of selection in the Lidia breed compared to the tamed breeds, corroborating that targeted selection with different goals have left detectable imprints on the genome.

Session 66 Theatre 7

High resolution genomic analysis of four local Vietnamese chicken breeds

E. Moyse[1], V.D. Nguyen[2], A. Dor[1], D.T. Vu[2] and F.P. Farnir[1]
[1]University of Liege, FARAH – Animal Production, Quartier Vallée, 2 Avenue de Cureghem, 6, 4000 Liège, Belgium, [2]VNUA, CIRRD, Ngo Xuan Quang, Hanoi, Viet Nam; evelyne.moyse@uliege.be

Local chicken breeds in Vietnam make up approximatively 85% of the national poultry population. Although these breeds are largely present, their productivity is low and their use is nowadays threatened by the massive importation of foreign productive breeds. In that changing context, conservation programs targeting several emblematic breeds have been set up. The goal of these programs is to characterize the endangered breeds and to keep a pool of characteristic animals in order to preserve this genetic heritage. Our contribution in these programs is a fine characterization of 4 local breeds at the genomic level. 95 animals from 4 breeds (32 Dong Tao (DT), 27 Ho (HO), 18 Mhong (MN) and 18 Mia (MA)) have been sampled in various farms from the Red River delta region in Vietnam. DNA has been extracted and genotypes have been obtained using a 500K Affymetrix SNP chip. Population diversity measures (Fst, PCA) have been calculated, and genomic inbreeding measures at the global and local genomic scales have been obtained. Although geographically close, DT and HO breeds are clearly different at the genomic scale. Both breeds also differ from MN and MA, these last two breeds being more similar to each other. Some individuals provided some evidence of breed intercrossing, while others could be considered as purebred. Genomic inbreeding measures revealed high levels of homozygosity in all 4 breeds, with inbreeding coefficients between 10 and 20%. Analyses using inbreeding age classes also revealed that the buildup of the current inbreeding is due to recent to very ancient (several thousand generations) contributions. The observation of differences at the genomic level supports the presence of distinct breeds, although the occurrence of crossbred animals in an assumed purebred sample triggers the need of genomic tools to unambiguously identify the breed of these animals. High levels of inbreeding also deserve attention for the creation of future nucleus to be used for breeds conservation.

Session 66 Theatre 8

Understanding unmapped reads using *Bos taurus* whole genome DNA sequence

J. Szyda[1,2] and M. Mielczarek[1,2]
[1]National Research Institute of Animal Production, Cattle Breeding, Krakowska 1, 32-083 Balice, Poland, [2]Wroclaw University of Environemntal and Life Sciences, Department of Genetics, Biostatistic group, Kozuchowska 7, 51-631 Wroclaw, Poland; joanna.szyda@upwr.edu.pl

Alignment to the reference genome is one of the most important aspects of studies based on whole genome DNA sequence. It is an intermediate step of all studies using Next Generation Sequence data. Further steps down typical bioinformatic pipelines focus on short reads, which were successfully (i.e. uniquely and with high confidence) aligned to the reference genome, while unmapped reads are disregarded. Nevertheless, such unmapped reads may carry valuable biological information, which can help in finding new genomic regions previously excluded from the reference genome, or to identify specific characteristics of individual genomes. Therefore, in this study we aimed to explain possible reasons of non-aligned reads. The study material consisted of whole genome DNA sequences of 48 Brown Swiss bulls, sequenced in paired-end read mode on the Illumina HiSeq 2000, with a genome averaged coverages varying between 8-28. Short reads were aligned to the UMD.3.1 reference genome. Then, we selected unmapped reads of good overall quality expressed by the read average quality threshold of 20. Such edited reads were compared between bulls in order to identify sequences of high similarity between all individuals. Furthermore, BLAST was used to find location of those reads, based on the NCBI nucleotide data base.

The impact of copy number variants on gene expression

M. Mielczarek[1,2], M. Fraszczak[2] and J. Szyda[1,2]
[1]National Research Institute of Animal Production, Krakowska 1, 32-083 Balice, Poland, [2]Wroclaw University of Environmental and Life Sciences, Biostatistics group, Department of Genetics, Kozuchowska 7, 51-631, Poland; magda.mielczarek@upwr.edu.pl

Copy number variations (CNVs) defined as gains (duplications) or losses (deletions) of longer DNA fragments are the major source of genetic diversity in mammals. Their length ranges from 50 bp to 1,000,000 bp, therefore they have a potential to cover many functional elements of the genome, such as genes or regulatory sequences, which can markedly affect phenotypes. In this study we investigated the impact of copy number variations on gene expression. The analysed material comprises RNA-seq from livers of six pigs generated by deep transcriptome sequencing using HiSeq 2000 platform. The bioinformatic pipeline consisted of: (1) raw data editing based on reads quality (2) alignment of reads to the *Sscrofa*11.1 reference genome, and (3) quantification of gene expression. Furthermore, regression models were used to assess the impact of CNV on the level of gene expression. As a result, we observed a significant correlation between the occurrence of CNVs overlapping with coding regions of genes and gene expression level, what indicates that CNVs are important modulators of gene expression and thus of resulting phenotypes.

Foetal and maternal placenta cells respond differently to a deleterious foetal mutation

K. Rutkowska
Institute of Genetics and Animal Breeding of the Polish Academy of Sciences, Postępu 36A, Jastrzębiec, 05-552 Magdalenka, Poland; k.rutkowska@ighz.pl

Foetal growth is a complex process that depends on balanced communication between foetus, placenta and mother. Placental dysfunction is the main cause of the intrauterine growth restriction (IUGR). We previously identified a 3' deletion in the non-coding *MIMT1* gene that, when inherited from the sire, causes IUGR and late abortion in Ayrshire cattle with variable levels of severity. The aim of this study was to analyse the expression of *DDX1*, *RB1CC1*, *MKI67*, *EMP2*, and *AHNAK* genes expressed in maternal and foetal placenta from 12 cows carrying *MIMT1*$^{Del/WT}$ foetuses and 12 cows carrying wild-type foetuses terminated on day 94±12 of gestation. All foetuses were gestated in wild-type cows and all were fathered by the same *MIMT1*$^{Del/WT}$ sire. The relative gene expression difference between IUGR and wild-type in both tissues was calculated for each animal (Ct). All cDNA samples were assayed in triplicate and relative expression levels normalized to endogenous *GAPDH* expression. Two-way ANOVA was used to estimate the effect of interaction of the *MIMT1* genotype with placenta tissues. The study revealed increased *DDX1*, *RB1CC1*, *MKI67* and *AHNAK* gene expression in the maternal placenta of the *MIMT1*$^{Del/WT}$ foetuses. No expression differences in foetal placenta for the studied genes were found. Our study showed that maternal and foetal placental cells responded differently to the foetal IUGR genotype. Caution is therefore warranted when analysing gene expression changes in non-dissected samples of placenta. The study was funded by National Science Centre, Poland, research project no. 2016/23/N/NZ9/00232.

Genomic regions influencing intramuscular fat in divergently selected rabbit populations

B.S. Sosa Madrid[1], P. Hernández[1], P. Navarro[2], C. Haley[2], L. Fontanesi[3], R.N. Pena[4], M.A. Santacreu[1], A. Blasco[1] and N. Ibañez-Escriche[1]
[1]Institute for Animal Science and Technology, Universitat Politècnica de València, P.O. Box 2201, Valencia 46071, Spain, [2]MRC Human Genetics Unit, MRC Institute of Genetics and Molecular Medicine, University of Edinburgh, EH4 2XU, United Kingdom, [3]Department of Agricultural and Food Sciences, Division of Animal Sciences, University of Bologna, 40127 Bologna, Italy, [4]Departament de Ciència Animal, Universitat de Lleida – Agrotecnio Center, 25198 Lleida, Catalonia, Spain; bosomad@posgrado.upv.es

Divergent selection for intramuscular fat (IMF) in *Longissimus thoracis et lumborum* muscle in rabbits was carried out at the Universitat Politècnica de València, achieving a high selection response for IMF of 2.7 standard deviations in the 9th generation. The aim of this study was to identify genomic regions associated with IMF in these lines and to assess the effect of IMF divergent selection on the rabbit genome structure. Samples from 477 rabbits of the high (240) and low (237) IMF lines were genotyped using the Affymetrix Axiom OrcunSNP Array (199,692 SNPs). Data analysis used three different approaches: (1) single marker regressions accounting for relatedness, (2) as (1) but including a random litter size effect, (3) a multi-marker method (Bayes B) assessing the effect of each SNP and the percentage of genetic variance explained by non-overlapping one megabase (Mb) windows. All methods identified genomic regions associated with IMF on chromosomes 8 and 13 and two of the methods identified regions on chromosomes 1, 12 and 14. A region of 1.95 Mb on chromosome 8 accounted for 6.8% of the genetic variance for IMF. Furthermore, the largest allele frequency difference ($\geq$0.89) between the rabbit IMF lines was on chromosome 8 (55.45-55.76 Mb), encompassing 22 consecutive SNPs. Functional annotations of some genes within the associated genomic regions (e.g. *FAM135A, PLBD1, NPC1, STARD10 and APOLD1*) are related to carboxylic ester hydrolase activity and lipid transport, binding and localization. Further research on these regions is needed to validate these candidate genes in order to advance our understanding of the biological mechanisms underlying IMF in rabbits and potentially other species.

Identification of causal mutations underlying feet and leg disorders in cattle

T. Suchocki[1,2], C.H. Egger-Danner[3], H. Schwarzenbacher[3], M. Mielczarek[1,2] and J. Szyda[1,2]
[1]National Research Institute of Animal Production, Krakowska 1, Krakowska 1, 32-083 Balice, Poland, [2]Wroclaw University University of Environmental and Life Sciences, Department of Genetics, Kozuchowska 7, 51-631 Wroclaw, Poland, [3]ZuchtData EDV-Dienstleistungen GmbH, Dresdner Straße 89/19, 1200 Vienna, Austria; tomasz.suchocki@gmail.com

Feet and leg disorders are traits of increasing importance in dairy cattle because of their impact on animal welfare. The main aim of this study was identification of particular genes responsible for the disorder of feet and leg health. We used a data set consisting of 2,223 cows from Fleckvieh and Braunvieh breeds genotyped using GeneSeek®Genomic ProfilerTM HD panel. After editing, 74,762 SNPs with the average MAF of 0.31 and the average call rate equal to 99.48% were used for statistical modelling of the total number of hoof diseases scored for a cow until day in milk 100. Additionally, 78 whole genome sequences were used to identify the polymorphisms in coding sequences (i.e. exons of genes) located in the proximity of SNPs from the panel. We discovered two significant SNPs at the 5% significance level located on chromosomes 7 and 14. Marker rs109798552 located on chromosome 7 had P-value after FDR correction equal to 0.034, while rs110532594 located on BTA14 had corrected P-value equal to 0.022. Additionally, we found the 'aggregate genotypes' built based on sequence and panel data which significantly influence the trait.

Session 66 Theatre 13

Choice of statistical model highly affects QTL detection in a commercial broiler line

H. Romé[1], W. Mebratie[1], R. Hawken[2], J. Henshall[2] and J. Jensen[1]
[1]Aarhus Universitet, Blichers Allee 20, 8830 Tjele, Denmark, [2]Cobb-Vantress, P.O. Box 1030, 72761-1030 Siloam Springs, Arkansas, USA; helene.rome@mbg.au.dk

For decades QTL have been detected in poultry. At this times, the chicken QTL database includes 1,532 QTL for body weight. With the emerging use of genomic feature models, external accurate knowledge on the genetics architecture of agronomical traits can be exploited. The aim of this study was to investigate the QTL affecting body weight in broilers by analyzing a commercial line that has been genotyped and phenotyped under commercial conditions. In the first part of the experiment, birds were weighed for a part at 6 weeks old and for a second part at 5 weeks old. The group contained male and female birds. Genotypes were obtained with a 50k Illumina array and after quality control (SNP call rate, individual call rate, MAF, HWE) analysis was performed on 46,645 SNP and 22,618 birds. A single marker regression fitting the genetic additive effect and the maternal effect to corrected for population structure has been performed using DMU. In the analysis, the data with sex and period separately (4 GWAS), all the data together (1 GWAS) or with sex or period separately (2 GWAS) were considered. Different effects were included to the model and tested. The threshold was set considering the Bonferroni correction (0.05/46645). Using the model giving less bias and best accuracy of prediction for genomic selection, no QTL were found. Moreover, with this model most significant P-value were underestimated suggesting that our model was too conservative. The simplest model, using all data together, detected 24 significant SNPs, but the P-values were highly overestimated (λ>2). We think that this overestimation could be due to a change in allele frequencies due to pressure of selection segregating the population in two groups. We are trying to identify the factor affecting our models. Our study shows that depending on the model used, results of GWAS could be substantially different. With a non-conservative model the risk of detecting false QTL increase, but with a conservative model no significant QTL are detected. It is necessary to use careful modelling to find a good balance between a too permissive model and over-conservative model for QTL detection.

Session 66 Theatre 14

Multilocus random-SNP-effect genome-wide association studies for health traits in horses

O. Distl[1], J. Metzger[1] and B. Ohnesorge[2]
[1]University of Veterinary Medicine Hannover, Institute for Animal Breeding and Genetics, Buenteweg 17p, 30559 Hannover, Germany, [2]University of Veterinary Medicine Hannover, Clinic for Horses, Buenteweg 9, 30559 Hannover, Germany; ottmar.distl@tiho-hannover.de

Mixed linear model (MLM) approach is widely used for genome-wide association studies. These models control for population stratification (K) and polygenic background (Q). The genotype of each SNP is fitted as a fixed effect, one at a time. Using MLM enables to test a very large number of SNPs in a run. However, power of MLM is low for detecting small effects of SNPs, and significance thresholds have to be corrected for multiple testing. In order to overcome these limitations for traits that are controlled by multiple loci or several major genes plus many small effect loci, multilocus models have been proposed. These models do not require a correction for multiple testing but are only feasible for data with a smaller number of variables due to the multicollinearity issue. Alternatively, a single-locus approach is combined with a multilocus method. In the first stage, SNPs are tested under a random effect linear mixed model and in the second stage, a multilocus random effect mixed mixed model is employed. This approach has been implemented in mrMLM by Wang *et al.* A further development was the integration of least angle regression with empirical Bayes for multilocus models (pLARmEB) proposed by Zhang *et al.* Herein, the LARS algorithm (least angle regression) is employed for selection of the most likely associated SNPs. Both methods are more powerful and accurate in detection and estimation of SNP effects with lower false positive rates. We compared MLM, mrMLM and pLARmEB for genome-wide data using equine SNP50 and whole genome sequencing data for health traits in horses. The equine SNP50 data we used included 250 horses with ERU (equine recurrent uveitis) and controls. Whole genome sequencing data for the 250 horses were imputed using 80 horse genome sequences. According to the comparison of results from the different runs, multilocus models seemed to have more power for detection of quantitative trait nucleotides (QTN). Genotyping of associated QTN in large samples should be done to validate these results in real data.

Session 66 — Poster 15

Covariance estimation for yield and behavioural traits in Italian honey bee by linear-threshold model

S. Andonov[1], C. Costa[2], A. Uzunov[3], P. Bergomi[2], D. Lourenco[4] and I. Misztal[4]
[1]Swedish University of Agricultural Senescence, Department of Animal Breeding and Genetics, Box 7023, 75007 Uppsala, Sweden, [2]Consiglio per la ricerca in agricoltura e l'analisi dell'economia agrarian, Via di Saliceto 80, 40128 Bologna, Italy, [3]Faculty of Agricultural Sciences and Food, P.O. Box 297, 1000 Skopje, Macedonia, [4]University of Georgia, Athens, Department of Animal and Dairy Science, 425 River Road, Athens, GA 30602, USA; sreten_andonov@yahoo.com

In honey bees, colonies are made up of many individuals that belong to different castes and generations and are interdependent on functioning of the super-organism. The challenge in honey bee breeding program is the estimation of non-genetic and genetic effects. In Italian breeding program, mating the queens are with unknown drones. Using 4,003 records for honey yield (kilograms) and defensive and swarming behaviours (scored 1-5), we estimated heritabilities and genetic correlation between traits with two-trait linear-threshold or threshold models. The pedigree had only dam side of 4,160 queens (186 base population, 1,625 dam and 365 grand-dam queens). Model accounted for fixed effects of performance test year (13 levels) and tester apiary (72 levels); random effects of interaction performance test year-tester apiary and additive genetic effect of queen. For (co)variance components the THRGIBBS1F90 software was used with 500,000 Gibbs sampling iterations, discarding the 10% initial iterations. Convergence checks and posterior means were obtained by POSTGIBBSF90. Heritability estimates were 0.25±0.04 for honey yield, 0.30±0.07 for defensive behaviour, and 0.31±0.07 for swarming behaviour. A weak genetic correlation (0.15) was found between honey yield and defensive behaviour, whereas the genetic correlation between honey yield and swarming behaviour was moderate and positive (0.42). Genetic correlation between defensive and swarming behaviours was 0.51. Modelling linear and categorical phenotypes in a two-trait model is a feasible option for estimating breeding values in honey bee breeding programs. This model is especially important given the increasing interest in behaviour traits. Those traits are usually less recorded than honey production, but can benefit from the considerable genetic correlations.

Session 66 — Poster 16

Association between SNPs in LEP and LEPR genes with production traits in Ukrainian Large White pigs

Y. Oliinychenko[1], N. Saranceva[1], A. Saenko[1], V. Vovk[1], V. Balatsky[1] and O. Doran[2]
[1]Institute for Pig Breeding and Agro-Industrial Production, National Academy of Agricultural Sciences, Shvedska Mogila 1, 36013, Poltava, Ukraine, [2]University of the West of England, Faculty of Health and Applied Sciences, Frenchay Campus, Coldharbour Lane, BS16 1QY, Bristol, United Kingdom; olena.doran@uwe.ac.uk

Leptin is a hormone which plays critical role in controlling appetite, fat deposition, and the rate of metabolic processes in general. It is encoded by leptin gene (*LEP*) and its signal transmission is controlled by leptin receptors encoded by leptin receptor gene (*LEPR*). Associations between polymorphisms in *LEP* (g. 2845 A>T, g. 3996 T>C), *LEPR* (c.232A>T, c. 2856C>T) and pig production traits have been reported for a number of commonly used European breeds. Ukrainian traditional pig breeds have superior meat eating quality, but they are not currently used in European breeding programmes. One of the reasons is lack of information on genes controlling meat quality traits in Ukrainian breeds. The aim of this study was to investigate the level of polymorphy of chosen *LEP* and *LEPR* single nucleotide polymorphisms (SNPs) in Ukrainian Large White pigs and to determine associations between the polymorphisms and some of the production traits (backfat thickness at 6-7th rib, 10th rib, and at sacrum, and daily weight gain). The study was conducted on 108 Ukrainian Large White purebred pigs. *LEP* and *LEPR* SNPs were genotyped using Restriction Fragment Length Polymorphism Analysis of PCR-Amplified Fragments technique. The study demonstrated that *LEP* 2845, *LEPR* 2856 and *LEPR* 232 are polymorphic in Ukrainian Large White pigs. The SNP *LEP* 3996 was found to be monomorphic and CC homozygous. No significant associations were found between *LEP* 2845, *LEPR* 2856 SNPs and the production traits. The pigs with TT allele variant in SNP *LEPR* 232 had a significantly lower backfat thickness at the 10th rib, 6th-7th rib, and their daily weight gain was lower when compared to animals with CT and CC alleles variants. The results suggest that the SNP *LEPR* 232 can be considered as a potential selection marker for backfat thickness and daily weight gain in Ukrainian Large White pig breed.

Session 66 Poster 17

Detection of selection signatures in selected pig breeds using extended haplotype homozygosity test

A. Gurgul[1], K. Ropka-Molik[1], E. Semik-Gurgul[1], K. Pawlina-Tyszko[1], T. Szmatoła[1], M. Szyndler-Nędza[1], M. Bugno-Poniewierska[1], T. Blicharski[2], K. Szulc[3] and E. Skrzypczak[3]

[1]National Research Institute of Animal Production, UL. Sarego 2, 31-047, Poland, [2]Institute of Genetics and Animal Breeding, Postępu 36A, Jastrzębiec, 05-552 Magdalenka, Poland, [3]Poznań University of Life Sciences, Wojska Polskiego 28, 60-637 Poznań, Poland; magdalena.szyndler@izoo.krakow.pl

Detection of selection signatures provides a direct insight into the mechanism of selection and allows further discovery of candidate genes related to animals' phenotypic variation. In this study, by using PorcineSNP60 Illumina assay genotypes, we detected selection signatures in 530 pigs belonging to four breeds: Polish Landrace (PL), Puławska (PUL), Złotnicka White (ZW) and Złotnicka Spotted (ZS), the last three of which represent native populations. A relative extended haplotype homozygosity (REHH) test implemented in Sweep v.1.1 software was used to detect within-breed ongoing selection. The analysis showed clear non-random distribution of selection signals across the breed genomes, with visible overrepresentation of long and common haplotypes on, e.g. SSC2 and 14 in PL or SSC4 and 18 in PUL. As many as 171 genes potentially being under selection were found in PL, 116 in PUL, 84 in ZW, and 35 in ZS breed. The genes were associated with *inter alia* skeletal system development and morphogenesis (HOXB1, -2, -5, -6, -7, -9) in PL pigs and muscle fibre development (FLNC, SMO) in PUL pigs. In ZW breed, the genes were involved in, e.g. metal ion homeostasis and regulation (TTC7A, SLC30A10), negative regulation of innate immune response (DUSP10) and negative regulation of T cell apoptotic process (SLC46A2). The study provides several potential candidate genes selected in individual breeds and provides a strong basis for further research aimed at identification of sources of genetic variation of traits characteristic for the studied pig breeds.

Session 66 Poster 18

Analysis of copy number variation regions in bovine genome

M. Frąszczak[1], M. Mielczarek[1,2], E. Nicolazzi[3], J.L. Williams[4] and J. Szyda[1,2]

[1]Department of Genetics, Wrocław University of Environmental and Life Sciences, 7 Kożuchowska Street, 51-631 Wrocław, Poland, [2]National Research Institute of Animal Production, 1 Krakowska Street, 32-083 Balice, Poland, [3]Council on Dairy Cattle Breeding, 4201 Northview Dr, Bowie, MD 20716, USA, [4]Davies Research Centre, University of Adelaide, Roseworthy, SA 5371, Australia; magda.mielczarek@upwr.edu.pl

Genomic structural variations represent an important source of genetic diversity. Copy number variations (CNVs) are gains and losses of genomic sequence, with reported lengths greater than 50 base pairs (bp), between two individuals of a species. In this study, we focussed on CNVs identified in cattle based on two different algorithms: a read depth (RD) and a split read (SR). The material consists of whole genome DNA sequences determined for 132 bulls representing five breeds (Brown Swiss, Fleckvieh, Guernsey, Simmental, and Norwegian Red). Alignment to the UMD3.1 reference genome was carried out using BWA-MEM after that CNVnator and Pindel software were used for CNV detection. The output of CNVnator was used as a baseline data set (validated by CNVs detected by Pindel and overlapping in at least 70%) After validation, the set of CNVs was reduced to 9.53% deletions and 19.41% duplications. Furthermore, to determine the degree of similarity between animals, for each pair of bulls the Jaccard similarity coefficient was calculated. Based on the resulting distance matrix multidimensional scaling was performed. Additionally Nei genetic distance between breeds was estimated. For analysis of the results nonparametric tests were used. The functional annotation of the CNV regions characteristic for a given breed were carried out. Many more deletions than duplications regions were identified, but duplicated genomic regions were longer than deleted regions. The Nei distances (0.084 for deletions and 0.081-0.099 for duplication regions) suggest the highest similarity between Simmental, Fleckvieh and Brown Swiss bulls. The bulls from the Norwegian Red breed were the most distantly related to other breeds. Considering animals within each breed, the lowest inter-individual variation was noticed for Brown Swiss bulls and the highest for Norwegian Red and Fleckvieh bulls.

Session 66 Poster 19

Exploiting extreme phenotypes to investigate haplotype structure and detect signatures of selection

E. Tarsani, A. Kominakis, G. Theodorou and I. Palamidi
Agricultural University of Athens, Animal Science, Iera Odos 75, 11855 Athens, Greece; etarsanh@gmail.com

In the present study we used 700 male broilers with records on body weight at 35 days (BW35) by taking the 5% lower (n=350, average BW35=1,876.5 g, L) and upper (n=350, average BW35=2,417.4 g, H) tails of a broiler population (n=3,500). We performed haplotype blocks (HB) analysis in the two tail populations and detected signatures of selection using the Wright's fixation index F_{ST} and pooled (both tails) data. Number of HB was lower in the H than in the L tail (34,005 vs 38,442) while the average HB length was higher in the H than the L tail (9,878.5 vs 7,597.3 bp). The genome coverage by HB was higher in the H than L tail (0.37 vs 0.32). A total number of 53 signatures of selection dispersed in 18 autosomes were suggested by markers exhibiting F_{ST} values higher than 0.20. A number of 192 QTLs related to growth traits (e.g. carcass weight) were found to lie within genomic regions 50 kb downstream and upstream the selection signatures, with 96 QTLs having been reported to affect body weight. The search for putative candidate genes within the specified regions revealed 67 positional candidate genes, some of them related to growth traits (e.g. PSMB4, MYOM2, ATP8B2, CAMK2D). Three genes (GNAO1, MYOM2 and TPM3) were found to participate in muscle contraction via functional enrichment analysis.

Session 66 Poster 20

IGF1 and IGFBP2 polymorphisms in Hungarian Yellow and broiler chickens

K. Szalai, K. Tempfli, E. Lencsés-Varga and Á. Bali Papp
Széchenyi István University, Department of Animal Sciences, Vár tér 2, 9200 Mosonmagyaróvár, Hungary; tempfli.karoly@sze.hu

Indigenous Hungarian Yellow hens (n=436) and Ross broilers (n=103) were genotyped for the A570C single nucleotide polymorphism (SNP) in insulin-like growth factor (*IGF1*) and for the G645T non-synonymous SNP in insulin-like growth factor binding protein (*IGFBP2*) genes. IGF1 is a member of the polypeptide hormone family with structural similarity to insulin. Concentrations of IGF1 and various polymorphisms of the gene have previously been associated with several growth traits in chickens, whereas IGFBPs can modulate the bioactivity and growth-promoting effects of circulating IGF1. PCR-RFLP method with HinfI and BseGI restriction enzymes was applied for genotyping. Significantly (P<0.05) different *IGF1* and *IGFBP2* allele frequencies were detected between the dual-purpose Hungarian Yellow and the broiler populations, as *IGF1* C and *IGFBP2* G alleles were fixed in Hungarian Yellow, while in broilers their prevalence was 0.10 and 0.92, respectively. Furthermore, *IGBP2* genotype was associated (P<0.05) with live weight at 38 days, carcass weight and breast muscle weight in the broiler population. Despite its low polymorphism information content (0.12), *IGFBP2* polymorphism can be valuable for selection programmes in broilers. This work was supported by the EFOP-3.6.3-VEKOP-16-2017-00008 project co-financed by the European Union and the European Social Fund.

Session 66 Poster 21

Genetic parameters for growth in Pietrain boars with implementation of social genetic effects

T. Flisar, M. Kovač and Š. Malovrh

University of Ljubljana, Biotechnical Faculty, Animal Science Department, Groblje 3, 1230 Domžale, Slovenia; tina.flisar@bf.uni-lj.si

The aim of the study was to estimate the contribution of social genetic effect to the heritable variation in the growth rate in different intervals in Pietrain boars in Slovenia. The focus was on the data structure and the differences in estimates of variance components with and without included social genetic and environmental effects. The growth during the performance test of 983 boars from 540 litters was recorded in the nucleus herd of the Pietrain breed. Boars finishing test in the period from April 2006 to March 2018 were included in the analysis. Each group was formed with from four to nine boars. The majority of groups consisted of boars originating from three to four litters. Weighing was performed every two weeks from around 30 to 100 kg of body weight. A large variability of boars' age and weight was observed. Daily gain on several intervals up to 100 kg of body weight was analysed. Groups differed also in the surface area per animal. On average, twice as minimal required surface was provided per boar. Data sets were prepared with the SAS software, while variance components were estimated using Wombat. Results revealed different contributions of components to the phenotypic variance of the growth rate on different intervals. The estimates of the social genetic variance were non-significant. However, it was suggested to include the social genetic effect in the genetic evaluation due to identification of total genetic variance, which might exceed phenotypic variance. Correlations between direct and social genetic effects were mostly negative, indicating the presence of competitive interactions. It was confirmed, that there is a need to include social genetic effect in the genetic evaluation, despite the small population size.

Session 66 Poster 22

Genetic and phenotypic correlation between survival rate and litter weight at birth in American mink

K. Karimi[1], M. Sargolzaei[2,3], Z. Wang[4], G.S. Plastow[4] and Y. Miar[1]

[1]Dalhousie University, Department of Animal Science and Aquaculture, Truro, NS, B2N5E3, Canada, [2]Semex Alliance, 5653 Highway 6 North, Guelph, ON, N1H6J2, Canada, [3]University of Guelph, Department of Animal Biosciences, Guelph, ON, N1G2W1, Canada, [4]University of Alberta, Livestock Gentec, Department of Agricultural, Food and Nutritional Science, Edmonton, AB, T6G2P5, Canada; karim.karimi@dal.ca

Improvement of reproductive performance has been defined as one of the main objectives in mink breeding. Productivity of mink farming can be influenced by survival rates from birth to pelting. Data of 3,046 litters collected from the Canadian Centre for Fur Animal Research (CCFAR) at Dalhousie Agriculture Campus during the period of 2002-2016 were used to estimate the phenotypic and genetic correlation between survival rate at birth (SB) and average litter weight at birth (AWB) in American mink. A bivariate repeatability model was used to estimate the genetic parameters of the studied traits using ASREML 4.0. Linear covariate of sex ratio and fixed effects of year (2002-2016), age of dam (1 to 4 years), colour (11 colour-types) and number of live kits at birth (1 to 14) were fitted in the model. The estimated heritabilities (±SE) were 0.13±0.03 for SB and 0.28±0.05 for AWB, respectively. Furthermore, repeatabilities (±SE) were estimated to be 0.23±0.04 for SB and 0.31±0.06 for AWB, respectively. A significant ($P<0.05$) genetic correlation was observed between SB and AWB (0.66±0.10). Additionally, a moderate phenotypic correlation (0.34±0.02) was estimated between these traits. These results indicated that selection on higher average litter weight at birth can improve the survival rate of kits at birth. It was recommended to include the average litter weights in selection indices to improve maternal abilities of farmed mink. The genetic parameters estimated in the present study are useful to understand the biology of these traits and to improve them through multi-traits selection programs.

Frequency of genotypes associated with milk synthesis in sows of maternal breeds

M. Szyndler-Nędza, K. Ropka-Molik and A. Mucha
Instytut Zootechniki – Państwowy Instytut Badawczy, Ul. Sarego 2, 31-047, Poland; magdalena.szyndler@izoo.krakow.pl

The aim of the study was to identify the polymorphisms of genes associated with milk synthesis that may influence colostrum and milk lactose levels in Polish Large White (PLW) and Polish Landrace (PL) sows. Using blood samples from 58 PLW and 65 PL sows, polymorphisms were sought in the following genes: progesterone receptor (PGR) gene, insulin receptor (INSR) gene, 11β-hydroxysteroid dehydrogenase type 2 (HSD11β2) gene, and beta 1,4-galactosyltransferase-1 (B4GALT1) gene. All genes were screened to identify polymorphisms using Sanger sequencing. Within INSR locus a SNP was detected in 3'UTR region (rs341480883) and PCR-RFLP method was applied to its detection (XmnI endonuclease). Also, in B4GALT1 gene, 3'UTR variant (rs324864064) was detected which was identified by PCR-RFLP method (BsmFI). Within PGR gene, missense variant (rs343511161) was detected using AhdI restriction enzyme. In HSD11B2 gene, two mutations were genotyped using Sanger sequencing: synonymous mutation (rs326322123) and missense variant – p.Pro130Ser (rs343355675). As regards the PGR gene in the PLW breed, 87.93% animals had the homozygous AA genotype, 12.07% had the heterozygous TA genotype, and no animals presented the TT genotype. All three polymorphic forms of this gene were identified in the PL breed (AA –75.38%, TA – 23.08%, TT – 1.54%). Three versions of the INSR gene occurred in both breeds: CC (5.17% PLW, 52.31% PL), TC (81.03% PLW, 44.62% PL) and TT (13.79% PLW, 3.08% PL). A silent mutation in the HSD11β2 gene was represented by three genotypes in the PLW breed: CC (10.71%), TC (30.36%), TT (58.93%), and by only two genotypes in the PL breed: TC (17.19%) and TT (82.81%). A missense mutation in the HSD11β2 gene in both breeds was represented by only two genetic forms: CC (92.98% PLW, 87.5% PL) and TC (7.02% PLW, 12.5% PL). As regards the polymorphism in the B4GALT1 gene, three genetic forms were identified in both breeds: AA (29.31% PLW, 14.06% PL), AG (50.00% PLW, 45.31% PL), and GG (20.69% PLW, 40.63% PL). Hardy-Weinberg disequilibrium was only found for the INSR gene in the PLW breed.

Effect of *WNT10A c.303C>T* gene polymorphism on reproductive traits in pigs

A. Mucha, K. Piórkowska, K. Ropka-Molik and M. Szyndler-Nędza
National Research Institute of Animal Production, Sarego 2, 31-047 Krakow, Poland; aurelia.mucha@izoo.krakow.pl

Research to find genetic markers for reproductive traits in pigs has been conducted for many years. WNT10A ligand is a member of the WNT family of proteins, which play a vital role in human and animal growth and development. The objective of the study was to analyse the effect of *WNT10A c.303C>T* gene polymorphism on reproductive traits in 452 Polish Large White and Polish Landrace sows. Reproductive traits investigated were: number of piglets born alive (NBA), number of piglets on day 21 (N21) in parity 1 and average from parities 2-4; the age of gilt at first farrowing (AFF); average interval between successive litters (IBTSL). The polymorphism was genotyped by PCR-RFLP methods, the restriction enzyme used in this analysis was *DraI*. For the studied group of sows, the following genotype frequency was found: CC – 0.67, CT – 0.27, TT – 0.06. The smallest number of piglets born and reared (NBA and N21) in parity 1 was observed in the sows of TT genotype and the values of these traits differed highly significantly ($P \leq 0.01$) in relation to the sows of CC (by 0.73 and 0.82 piglet) and CT genotypes (by 0.77 and 0.72 piglet, respectively). Also the smallest size was characteristic of parities 2-4 from the sows of TT genotype in terms of the number of piglets at birth (NBA) and at 21 days of age (N21), but in this case no statistically significant differences in relation to the other sows were found. No significant differences also occurred when comparing the sows of different genotypes with regard to AFF. The shortest IBTSL was characteristic of the sows of TT genotype and it differed significantly ($P \leq 0.05$) in relation to the interval between successive litters in the sows of CT genotype. The obtained results show that *WNT10A c.303C>T* gene polymorphism has an effect on litter size (especially parity 1) and the interval between litters.

Alternative derivation method of MME from BLUP and restricted BLUP of breeding values not using ML

M. Satoh, S. Ogawa and Y. Uemoto
Tohoku University, Aramaki Aza Aoba, Aoba-ku, Sendai, 980-8572, Japan; masahiro.satoh.d5@tohoku.ac.jp

Two mixed model equations (MME) for best linear unbiased prediction (BLUP) of breeding values and for restricted BLUP of breeding values were derived by maximum likelihood from the joint normal probability distribution of the observations and breeding values. As a result, MME are actually more general than maximum likelihood because it can be proved that fixed effect solutions of MME are identical to solutions of generalized least-squares equations and random effect solutions of two MME are identical to BLUP and restricted BLUP of breeding values, respectively, and then it does not depend on normality. The objective of this study is to derive directly each MME from BLUP and restricted BLUP equations algebraically without assuming the joint normal density function of the data and breeding values. In order to achieve this objective, we use Matrix inversion lemma (Sherman-Morrison-Woodbury identity), generalized least-squares equations, and best linear unbiased predictors or restricted best linear unbiased predictors of breeding values. Then we show deriving directly each MME from BLUP and restricted BLUP equations for breeding values without assuming the joint normal distribution of the data and random effects. However, if we cannot assume the multivariate normal density distribution of the estimated aggregate breeding value and each breeding value for selected traits, the response to selection by restricted BLUP may deviate from the expected values.

Meta-analysis of gene expression in chicken stimulated with prebiotic or synbiotic *in ovo*

A. Dunislawska, A. Slawinska, M. Bednarczyk and M. Siwek
UTP University of Science and Technology, Department of Animal Biochemistry and Biotechnology, Mazowiecka 28, 85-084 Bydgoszcz, Poland; aleksandra.dunislawska@utp.edu.pl

Microflora of the gastrointestinal tract in animals, including birds, affects the host organism by regulating the immune response, metabolism, digestive processes and nutrient absorption. Our team has developed a technology for reprogramming the microbiome composition at the early stages of embryo development by prebiotic and synbiotic delivery using *in ovo* technology. Prebiotics are non-digestible nutrients that have a positive influence on selective stimulation, growth and activity of bacteria present in the host's gastrointestinal tract, while the synbiotic refers to the simultaneous administration of pre- and probiotic in a single dose and their synergistic effect. The aim of study was to demonstrate common effects caused by pre-/synbiotic stimulations in chicken transcriptome. Analyses were done on a unique dataset based on three expression microarray experiments. These projects analysed the effects of prebiotic alone (galactooligosaccharides) or synbiotics (strains of *Lactococccus* or *Lactobacillus* bacteria with prebiotic) administered *in ovo*. In each of the project, whole genome microarray analyses based on RNA from immune tissues were performed. Bioinformatics analysis were carried out using the NetworkAnalyst tool and R packages to integrate three data sets and perform meta-analysis. These packages enable data visualization, combining multiple datasets based on P value and effect size. Selected genes were analysed for signalling and common pathways based on gene ontology using R packages and WebGestalt tools. Obtained results showed the significant effect of bioactive substances delivered *in ovo* on changes in gene expression. These changes are associated with improved physiological and production traits of broiler chickens including meat quality, metabolism and immune functions. Meta-analysis of gene expression dataset pointed out genes, pathways, and regulator elements in which activation occurs after administration of the bioactive substance *in ovo*.

Session 66 Poster 27

Comet assay for analysis of DNA damage in sperm of Wistar rats treated with ilex extracts

A. Zwyrzykowska-Wodzińska[1], A. Tichá[2], R. Hyšpler[2], Z. Zadák[2], R. Štětina[2], A. Szumny[1] and P. Kuropka[1]
[1]Wrocław University of Environmental and Life Sciences, C. K. Norwida 25, 50-375 Wrocław, Poland, [2]Charles University, Faculty of Medicine and University Hospital, Sokolská 581, 500 05 Hradec Králové, Czech Republic; anna.zwyrzykowska@upwr.edu.pl

Yerba Mate is an extract of leaves (*Ilex paraguariensis* A. St. Hilaire), enjoying steadily increasing consumption around the world. One of the distinctive characteristics of yerba mate is high concentration of antioxidants, mainly caffeoyl derivatives and flavonoids, which are very important for health care. The consumption of mate extracts may be an effective way to protect against the DNA damage that has been shown to cause mutations through miscoding, therefore responsible for initiating carcinogenesis. To the best of our knowledge only *I. paraguariensis* has been the subject of intensive phytochemical and biological studies, so we concentrated on available other *Ilex* varieties: *I. aquifolium, Ilex aquifolium* 'Argentea Mariginata', and *Ilex meserveae.* The comet assay is a simple and effective method for evaluating DNA damage of cells with and without division capability. This assay detects both single strand and double strand DNA breaks, which causes the relaxation of DNA loops that migrate out of the cell under the influence of an electric field. The aim of study was to assess the application of comet assay for measuring the DNA damage in rat sperm cells treated with *Ilex* extracts. Male adult Wistar rats were treated for 56 days with *Ilex* extracts in DER 50 g/l. Biological material in the form of spermatozoa was collected for analysis from the epididymis. During this experiment, sperm cells were embedded in a thin layer of agarose on a microscope slides and lysed with detergent under high salt conditions. Using alkaline pH conditions resulted in unwinding of double-stranded DNA. The electrophoresis resulted in the migration of relaxed DNA loops towards the anode which forming a comet tail, which were observed under fluorescence microscope. The Lucia software was used for comet scoring, and the following comet measures were recorded tail length, tail DNA. Obtained results suggest that the comet assay is useful technique for evaluating the DNA damage in sperm.

Session 66 Poster 28

Cattle mitogenome sequence analyses revels the presence of selection

D. Novosel[1], V. Cubric-Curik[1], V. Brajkovic[1], S. Krebs[2], E. Kunz[3], C. Vernesi[4], I. Medugorac[3], O. Rota-Stabelli[4] and I. Curik[1]
[1]Faculty of Agriculture University of Zagreb, Department of Animal Science, Svetosimunska 25, 10000 Zagreb, Croatia, [2]LMU Munich, Laboratory for Functional Genome Analysis, Gene Centre Munich, Veterinärstraße 13, 80539 München, Germany, [3]LMU Munich, Population Genomics Group, Department of Veterinary Sciences, Veterinärstraße 13, 80539 München, Germany, [4]Fondazione Edmund Mach, Molecular Ecology, Research and Innovation Centre, Via E. Mach, 1, 38010 S. Michele all'Adige (TN), Italy; dinko.novosel@gmail.com

Mitochondria have DNA maternal inheritance in vertebrates and has 37 genes, 13 encodes subunits of oxidative phosphorylation (OXPHOS) metabolic pathway, 22 for mitochondrial tRNA, 2 for rRNA and highly variable non-functional D-loop region. While mitochondrial genome controls one of the most important processes in cell, production of energy, it is often modelled assuming constant mutation rate and equilibrium neutral model of molecular evolution. We analysed 799 complete mitogenomes, belonging to >100 breeds. Mitogenome was split in 18 partitions: CYTB, COX1-3, ATP6, ATP8, ND1-6, ND4L, 12S rRNA, 16S rRNA, 22 tRNAs merged in one partition, start and end D-loop region. In the calculation of molecular clock BEAST 2.4.8. software package was used. The estimation of selection pressure, calculated by HyPhy software within Datamonkey platform, was analysed by Fixed Effects Likelihood, Detect Individual Sites Subject to Episode Diversifying Selection and Fast, Unconstrained Bayesian AppRoximation for Inferring Selection. The highest molecular clock rate was observed in non-functional D-loop region, specifically in end part while rRNAs and tRNAs seems to be quite conserved. In the OXPHOS subunit coding region the highest molecular clock (3.84×10^{-7} and 3.63×10^{-7} mutations/per base/per year) and percentage of pervasive negative/purifying selection (0.0561 and 0.0525 sites/base) was observed in ND5 and CYTB, respectively. Evidence of pervasive positive/diversifying selection was observed in CYTB (1 site), ND5 (2 sites) and COX3 (1 site). Our results will help better phylogenetic modelling and identification of mitogenome detrimental mutations.

Session 66 Poster 29

The evaluation of PECR, FN1, PNKD and PLCD4 mutation effects on pig traits

K. Piórkowska[1], A. Stuczyńska[2], M. Tyra[1] and K. Żukowski[1]

[1]Natinal Research Institute of Animal Production, Krakowska 1, 32-083, Balice, Poland, [2]University of Huddersfield, School of Applied Sciences, Queensgate, HD1 3DH, Huddersfield, United Kingdom; katarzyna.piorkowska@izoo.krakow.pl

In the present study, the effect of PLCD4, PECR, FN1 and PNKD mutations on pig production traits was analysed. The genes were selected based on targeted enrichment DNA sequencing (TEDNA-seq) of fragment pig chromosome 15. This chromosome region is QTL-rich (quantitative trait loci rich) related to fat content, meat quality and growth traits in pigs. The association study was performed by using GLM model in SAS software and included over 500 pigs representing 5 populations maintained in Poland. The results obtained showed that upstream PECR mutation influenced hardness of semimembranosus muscle ($P<0.001$), variation (C/T) of PLCD4 gene affected feed conversion, intramuscular fat and water exudation, and missense and frameshift mutations of PNKD were associated with texture parameters measured after cooking. Summarizing, the investigated gene variants delivered valuable information that could be used during developing SNP microarray for genomic estimated breeding value procedure in pigs. Moreover, the TEDNA-seq method could be used to preselect the molecular markers associated with pig traits.

Session 66 Poster 30

Transcriptomic changes in broiler chicken hypothalamus depending on growth rate

K. Piórkowska[1], K. Żukowski[1], K. Połtowicz[1], J. Nowak[1], K. Ropka-Molik[1], D. Wojtysiak[2], N. Derebecka[3] and J. Wesoły[3]

[1]National Research Institute of Animal, Sarego 2, 31-047, Kraków, Poland, [2]Agricultural University of Cracow, Institute of Veterinary Sciences, Adama Mickiewicza 21, 33-332, Kraków, Poland, [3]Adam Mickiewicz University, Faculty of Biology, Wieniawskiego 1, 61-712 Poznań, Poland; katarzyna.piorkowska@izoo.krakow.pl

Poultry meat production has been mainly oriented towards obtaining large chickens in a very short time. Therefore, the novel genetic types of broilers are characterized by an increasingly rapid rate of growth and increasing weight of muscles. However, meat-type chicken populations are highly divergent in growth rate (GR), final body weight (BW) and carcass composition. The differences in slaughter weights can reach even 1.5 kg in one population, notwithstanding that the uniformity of the flock is very important. In the present study, it was attempted to estimate the hypothalamic response at the transcriptome level for divergence in chicken growth rate. The measurements considered Ross308 cockerels in two age groups: 3-week-olds and 6-week-olds. In total, in 3-week-old and 6-week-old cockerels 176 and 319 genes were regulated, respectively. In hypothalamus of lower GR cockerels, biological processes and pathways associated with feeding behaviour and thyroid hormone release were regulated because differential expression was observed in BSX, POMC, ALDH1A1, NMU, PMCH, GHRH, MIM1, TBX3 and CGA genes.

Session 66 Poster 31

Donkey individual identification and parentage verification using a specifically designed 22plex PCR

J.A. Bouzada[1], V. Landi[2], J.M. Lozano[3], M.R. Maya[3], A. Trigo[3], M.A. Martínez[2], M.M. Gómez[2], T. Mayoral[1], E. Anadón[1] and L.B. Pitarch[1]

[1]Lab. Central de Veterinaria, Identificación Genética, Ctra M106, Km, 1,4, 28110 Algete, Spain, [2]Animal Breeding Consulting, S.L., Astrónoma Cecilia Payne, 14014 Córdoba, Spain, [3]TRAGSATEC, Julián Camarillo 6 B, 28037 Madrid, Spain; jbouzada@mapama.es

Still today, donkeys and mules are essential for transportation in poor, arid, and rough regions of the world, however, in developed countries, this pack animal is no longer required. As a consequence, not only individual breeds are endangered, but also the whole species is heading for extinction. One of the first stages in the conservation program of endangered breeds is the evaluation of their genetic variability. In this study we propose a multiplex PCR including a set of 22 microsatellite markers (AHT4, AHT5, ASB23, COR112, COR214, HMS2, HMS3, HMS5, HMS6, HMS7, HMS18, HTG6, HTG7, HTG10, HTG15, LEX27, LEX33, TKY297, THY312, TKY321, UCDEQ(CA)425 and VHL20), specifically designed for donkey, to be co-amplified simultaneously in a multiplex PCR. Changes in primers were required to achieve a final configuration allowing all markers to be analysed together. The proposed microsatellites were all well-amplified in donkey samples and all of the microsatellites used in this study were amplified and were polymorphic in blood samples of domestic donkey breeds. DNA Donkey Comparison Test 2016-2017 organized by International Society for Animal Genetics (ISAG) included these markers and propose a standardized nomenclature. The proposed microsatellites were all well-amplified in donkey samples and all of the microsatellites used in this study were amplified and were polymorphic in blood samples of domestic donkey breeds. DNA from whole blood was extracted using BioSprint 96 Blood Kit (Qiagen) and markers were simultaneously amplified using the QIAGEN Multiplex PCR kit. PCR products were analysed in a single run on an ABI PRISM® 3130xl Genetic Analyzer. The mean allele number for the microsatellites included in the proposed panel is 5.50. The cumulative probability of exclusion reached 99.9997725%. This donkey-22plex PCR panel can be used for genetic characterization, parentage verification and individual identification.

Session 66 Poster 32

Expression of miRNAs in placenta is associated with intrauterine growth restriction in cattle

K. Rutkowska[1], K. Flisikowski[2], M. Łukaszewicz[1] and J. Oprządek[1]

[1]Institute of Genetics and Animal Breeding of the Polish Academy of Sciences, Postępu 36A, Jastrzębiec, 05-552 Magdalenka, Poland, [2]Technische Universität München, Lehrstuhl für Biotechnologie der Nutztiere, Liesel-Beckmann Str, 85354 Freising, Germany; k.rutkowska@ighz.pl

Placenta dysfunction is the primary cause of intrauterine growth restriction (IUGR) in mammals. Our previous study showed that paternally inherited deletion of 3' end of the non-coding MER1 repeat containing imprinted transcript 1 (MIMT1) gene causes IUGR and late abortions in Ayrshire cattle. The aim of the study was to analyse the association of miRNA expression in placenta with the IUGR in cattle. The maternal and foetal placenta samples were collected from 10 cows gestating $MIMT1^{Del/Wt}$ foetuses and 13 cows gestating wild-type foetuses terminated on the 94±12th day of gestation. All foetuses were gestated in wild-type cows and all were fathered by the same $MIMT1^{Del/WT}$ sire. We performed small RNA sequencing and found 46 miRNA differentially expressed in foetal and 37 in maternal placenta of IUGR calves. For qPCR validation seven miRNAs, the miR-29a, miR-92b, miR-125b, miR125b-1-3p, miR-7641, miR-4321 and miR-2895 were selected. The analysed miRNAs play a role in immune response, angiogenesis or oxidative stress, regulation of cell growth, embryonic stem cells differentiation, placental development and regulation of endometrial receptivity. The relative miRNA expression differences were assayed in triplicate and normalized to endogenous miR-99b control. The statistical association analysis between miRNA expression and *MIMT1* genotype was performed using two-way ANOVA accounting for genotype of foetus and placenta part effects. The qPCR study revealed differential expression of all studied miRNAs in maternal placenta. Particularly miR-4321 and miR-2895 showed 3.8 and 2.2- fold lower expression in $MIMT1^{Del/WT}$ than in $MIMT1^{WT/WT}$ maternal placenta. Interestingly, these miRNAs showed opposite expression levels in foetal placenta of $MIMT1^{Del/WT}$. Our study suggest that both parts of the placenta respond differentially to deleterious foetal mutation.

Session 66 Poster 33

Preliminary results evaluating horn traits of economic importance in Hippotragus niger niger

G.C. Josling[1], A.A. Lepori[1], F.W.C. Neser[1], P.C. Lubout[2] and J.B. Van Wyk[1]
[1]University of the Free State, Animal, Wildlife and Grassland Sciences, P.O. Box 339, 9300 Bloemfontein, South Africa, [2]Wildlife Stud Services, Postnet Suite 489, Private Bag X025, 0040 Lynwood Ridge, South Africa; neserfw@ufs.ac.za

The wildlife industry in South Africa has shown immense growth since the 1990s brought about by the privatization and commercialization of the game segment. In recent years, extremely high prices were paid for trophy quality breeding animals, especially sable antelope and buffalo. Much of the economic value of these animals can be attributed to horn size, which is an important trait for trophy hunters. The main objective of the study was to estimate genetic parameters for the economically important horn traits currently measured in industry. To date, no quantitative genetic analysis has been done for any traits in sable antelope-simply because no data was available. The total number of records included in the evaluation for each trait were: 1,713 for horn length (SHL), 1,503 for -circumference (SHC), 1,486 for -tip to tip (SHTT), 1,505 for -tip length (SHT) and 1,447 for -rings (SHR). Because it is so difficult to create contemporary groups in game, males and females were considered individually within six month age clusters. A Markov Chain Monte Carlo (MCMC) multi-trait analysis within R was used to estimate (co)variance parameters for the different horn traits. The results indicate a significant (P<0.05) sex effect for all the studied traits under investigation and also suggest that it is not economically viable to measure horn length of males and females after 54 months of age. The horns of females are on average 40% shorter than those of bulls at maturity. Continuous horn growth throughout the lifetime of sable is suggested by the formation of ring posts, but is often masked by horn attrition and inadequate measuring techniques. The low inbreeding coefficient of 0.0043 suggest adequate genetic diversity in the studied population. The heritability estimates of the horn traits varied from 0.085 to 0.52, while the genetic correlations ranged from 0.1 to 0.6, with the highest correlation found between horn length and tip to tip. More data are required to support the results of the current study.

Session 66 Poster 34

Fixed and random regression models for South African Holstein under two production systems

M. Van Niekerk[1,2], F.W.C. Neser[2] and V. Ducrocq[1,2]
[1]AgroParisTech, Université Paris-Saclay, GABI, INRA, 78350 Jouy-en-Josas, France, [2]University of the Free State, Department of Animal, Grassland and Wildlife Sciences, P.O. Box 339, Bloemfontein 9300, South Africa; vanniekerkm2@ufs.ac.za

In fixed regression models (FRM), the shape of the lactation curve is determined by fixed (non-genetic) effects only, while the additive genetic effect is assumed to be constant within lactation. Random regression models (RRM) allow this shape to also vary with the additive genetic and permanent environment effects of the cow. Test-day records from South African Holstein (SAHST) herds were analysed using restricted maximum likelihood under different multitrait (milk production for each of the first three lactations) FRM, including the current FRM officially used for SAHST genetic evaluations. Two separate datasets were considered including herds using either a pasture (PAST) or a total mixed ration (TMR) production system. Results of the various FRM were compared to the current SAHST genetic evaluation model which clusters most of the fixed effects together and considers them as constant over the lactation. An alternative FRM cumulating different lactation curves for different fixed effects was retained based on the results of the analyses of the PAST dataset. This model was also compared to the current SAHST model using the TMR dataset. The alternative FRM was then extended to a RRM combining for each lactation an average production effect and a persistency effect and this model was compared with the current SAHST model for both production systems. Both the RRM for PAST and TMR had a better goodness of fit as well as a lower MSE than the current SAHST model. Overall, RRM had higher heritability estimates for both production systems, especially at the beginning and at the end of each lactation. Estimates of between lactation genetic correlations were also higher for RRM. Results from RRM analyses showed that the genetic correlations between average production and persistency within parity were stronger (and negative) for TMR than for PAST systems. The extra source of information from RRM enables a genetic prediction of persistency and are expected to increase the accuracy of genetic predictions.

The effect of SNPs in estrogen receptor gene on milk production traits in Slovak Spotted cattle

A. Trakovická, N. Moravčíková and R. Kasarda
Slovak University of Agriculture in Nitra, Department of Animal Genetics and Breeding Biology, Tr. A. Hlinku 2, 94976 Nitra, Slovak Republic; anna.trakovicka@uniag.sk

The aim of this study was to identify the population structure based on SNPs in gene encoding bovine estrogen receptor (*ERα*) and to evaluate their impact on milk production traits in Slovak Spotted cattle. In this study two polymorphic sites in the 5' region of the bovine *ERα* gene were analysed, A/G transition upstream to exon C (GenBank no. AY340597) and A/G transition in putative promoter of exon B (GenBank no. AY332655). The genomic DNA was extracted from in total of 150 blood samples of Slovak Spotted cattle by using standard protocol for salting-out method. All animals were genotyped using the PCR-RFLP method and restriction endonucleases *Bgl*I and *Sna*BI. After RFLP analyses the following allele frequencies were found: *ERα/BglI* G 0.890 and A 0.110; *ERα/SnaBI* G 0.067 and A 0.933. The average effective allele number (ENA=1.193) reflecting the decrease of possible variability realization of loci. However, no departure from Hardy-Weinberg equilibrium was found, which indicates only low impact of factors such as migration or inbreeding in the population. Even if the observed heterozygosity reached the level 0.163±0.036, the Wright's F_{IS} index indicated sufficient proportion of heterozygous individuals within population (F_{IS}=-0.029). The impact of SNPs on milk production traits (milk production, protein and fat yield) was tested by using GLM procedure adopted in SAS 9.4. Apart from the genotype effect, the linear model included fixed effects of sire, breeder, age at first calving, calving interval, days open, and number of lactations (on average R^2=0.72). Our study showed that *ERα/SnaBI* polymorphism significantly affected each of analysed milk production traits ($P<0.05$), whereas *ERα/BglI* showed no significant impact. In the future, a genome-wide association study will be carried out to confirm the effect of SNPs in the estrogen receptor gene on milk production in Slovak Spotted cattle.

Evaluation of milk quality of Sokolski cold-blooded and warmblood mares: preliminary results

G.M. Polak
National Research Institute of Animal Production, Department of Horses Production, Wspolna str. 3, 00-930 Warsaw, Poland; grazyna.polak@izoo.krakow.pl

Studies conducted in many countries show that mare's milk has a high biological value and could be used for human consumption. An additional advantage of mare's milk is its composition, more similar to human milk than cow's milk. An analysis was conducted in two herds of horses, located in the National Research Institute of Animal Production experimental centre in the north-west region of Poland. The aim of the research was to analyse the quality of Sokolski mares' milk and to compare it with the milk of warmblood mares. Milk was collected from 5 Sokolski mares and 13 warmblood mares, milking took place once a day. The results show big differences in milk composition, depending on the period of lactation and the individual in the same herd, while only small differences between the two herds were observed. Significant differences in quantity of milk were found between the mares ranging from 0.98 to 1.93 kg/day. Average content of nutrients was (g/100 g): fat 0.2-2.1; protein 3.21-0.67; lactose 5.79-7.42; total solids 8.16-11.09. The unsaturated FA (%) were represented mainly by the palmitic acids (22.68), oleic (22.90) and linoleic acid (14.07). The main components of whey proteins are (% total) β-lactoalbumin (20.7), α-lactoalbumin (15.07), IMG (3.10); lactoferrin (2.13), and Lisozyme (12.11).

Solving single-step GBLUP MME by extended KKT equations and block anti-triangular factorization

M. Taskinen, E.A. Mäntysaari and I. Strandén
Natural Resources Institute Finland (Luke), Myllytie 1, 31600 Jokioinen, Finland; matti.taskinen@luke.fi

In the standard animal evaluations, BLUPs of breeding values are solved from Henderson's mixed model equations (MME) that involve inverses of relationship matrix G and residual variance matrix R. The same fixed and random effects can be alternatively solved from normal equations that contain the variance matrix of the observations' variance-covariance matrix V. Furthermore, the normal equations can be arranged to a Karush-Kuhn-Tucker (KKT) matrix equation. The MME KKT system has the fixed effects model matrix X and its transpose on the off-diagonal and a zero block and V on the diagonal. The basic MME KKT system avoids inverses of the variance structures G and R. The KKT matrix equation is naturally indefinite with both positive and negative eigenvalues, and is not ideal for iterative solution methods and, therefore, is not widely used. The KKT systems can, however, be solved using Block Anti-Triangular (BAT) factorization techniques. We present an extended MME KKT in which major sparse matrices of MME are positioned on the off-diagonals (with X) and all other, possibly full matrix parts of MME are on the diagonals (instead of V). Single-step genomic BLUP (ssGBLUP) can be expressed as an extended KKT so that the very sparse 'animal model' parts and the full genomic information parts can be handled with different approaches. For numerical solutions, BAT factorizations were used to solve BLUPs from the MME KKT systems. Sparsity preserving QR decomposition was used for the sparse parts and the full matrix parts were solved using iterative solving methods. In the latter, the inversion of possible ill-behaving genomic relationship structure was avoided.

Relationships between a TLR4 allele associated with IBK and production traits in dairy cattle

M.P. Mullen[1], M.C. McClure[2] and J.F. Kearney[3]
[1]Bioscience Research Institute, Athlone Institute of Technology, Athlone, Co. Westmeath, Ireland, [2]ABS Global Inc., Wisconsin, WI 53532, USA, [3]Irish Cattle Breeding Federation, Bandon, Co. Cork, Ireland; mmullen@ait.ie

Infectious Bovine Keratoconjunctivitis (IBK), also known as pinkeye, is an infectious bacterial disease attributed to *Moraxella bovis*, impacting on both animal welfare and resulting in significant economic losses in cattle production. An allele (8:g.108833985A>G) in the Toll Like Receptor 4 (TLR4) gene has previously been associated with susceptibility to IBK infection. The objective of the current study was to examine if any evidence could be found for relationships between this mutation and routinely recorded production traits (milk, fertility, carcass and health traits (n=14)) in Holstein Friesian dairy cattle in Ireland. Genotypes and phenotypes (expressed as predicted transmitting abilities (PTAs)) on 14,939 dairy cows were obtained from the Irish Cattle Breeding Federation (ICBF). Following parental contribution removal and deregression only animals with an adjusted reliability >20% were included in the analysis which included n=14,773, 2,467, 12,688, 2,157, 3,237, 4,414, 893, 504, 5,111 and 6,155 cows for milk traits (n=5), calving interval, gestation length, maternal calving difficulty, calving difficulty, carcass weight, carcass conformation, carcass fat, cull cow weight, and somatic cell score, respectively. The association between SNP and deregressed PTA was analysed in ASREML using weighted mixed animal models with the SNP in dosage format and percent Holstein included as a fixed effect. The G allele (minor allele frequency of 0.42) was associated with a reduction in maternal calving difficulty ($P<0.05$) with tentative associations for reduced milk protein composition ($P<0.1$) in addition to increased cull cow weight ($P<0.1$). There was no association ($P>0.05$) with any of the other milk, fertility or carcass traits analysed in this study. The selection of animals less susceptible to infectious diseases, including IBK, is a desirable goal and these results provide little evidence to support maintenance of this IBK susceptibility allele on farm or in the national herd.

Precision gene editing of porcine MSTN by CRISPR/Cas9 and Cre/LoxP

Y. Bi
Hubei Academy of Agricultural Sciences, Institute of Animal Science and Veterinary Medicine, No.4, Nanhu Avenue, Hongshan District, Wuhan, 430064, China, P.R.; sukerbyz@126.com

Clean farm animal transgenics is ever increasingly desirable as the introduction of unwanted sequences such as selectable marker genes (SMGs) incurs public concerns on food safety, biodiversity and agro ecosystem. CRISPR/Cas9-mediated homologous recombination (HR) was exploited to knock out (KO) one allele of MSTN in pig primary cells with an efficiency of 7.7%. Cre recombinase was then used to excise the SMG with an efficiency of 82.7%. The SMG-free non-EGFP cells were isolated by flow cytometery and immediately used as donor nuclei for nuclear transfer. A total of 685 reconstructed embryos were transferred into three surrogates with one delivering two male live piglets. Molecular testing verified the mono-allelic MSTN KO and SMG deletion in these cloned pigs. Western blots showed approximately 50% decrease in MSTN and concurrent increased expression of myogenic genes like MyoD1, MyoG and Myf5 in muscle. Ultrasonic detection showed the increased longissimus muscle size and decreased backfat thickness. Histological examination revealed the enhanced myofibre quantity but myofiber size remained unaltered, indicating that the overgrown muscle mass resulted from hyperplasia other than hypertrophy. We have successfully generated non-mosaic and marker-free pigs with a predictable and precise KO of MSTN combining CRISPR/Cas9 and Cre/LoxP technologies. Our work takes advantage of predictable and clean gene targeting method to prevent the occurrence of inhomogenity that results from NHEJ-induced mutations. Furthermore, deletion of the SMG minimizes potential biosafety and environmental risks and helps encourage wider acceptance.

Session 66 Poster 40

Panel of SNPs in genomic regions of candidate genes controlling milk production traits in cattle

N. Moravčíková, O. Kadlečík, A. Trakovická and R. Kasarda
Slovak University of Agriculture in Nitra, Department of Animal Genetics and Breeding Biology, Tr. A. Hlinku 2, 94976 Nitra, Slovak Republic; ondrej.kadlecik@uniag.sk

The aim of this study was to identify and localize genetic markers in bovine genome controlling milk production traits as well as to analyse the allele and genotype frequencies of those SNPs in Slovak Pinzgau cattle. The genotyping data have been obtained from in total of 152 animals (37 AI sires, 35 dams of sires, and 80 dams of dams), representing the genepool of current Pinzgau population in Slovakia. All animals were genotyped using Illumina BovineSNP50v2 BeadChip. Initially, the database of 149 genetic markers with allelic variants participating in the metabolic processes related to milk production were obtained according to previous GWAS studies. After joining this database with Illumina BovineSNP50 map file, the final database included totally 24 genetic markers localized across 13 autosomes and 18 genes. The highest proportion of common SNPs was found on BTA6 in regions of genes encoding milk proteins (*CSN3*, *CSN1S1*, *ABCG2*), on BTA2 in genes responsible for fertility and early embryonic development (*STAT1*), and on BTA4 in leptin gene (*LEP*) that modulates feed intake, energy expenditure and fertility. Allelic variants of identified genetic markers can be used to predict mainly the quality of milk and production of higher added value products. In term of practice use, the detailed view on animals genetic background focused on those genetic markers can be used to select parents for specific mating purposes, diversification of breeding goal and increased genetic gain based on genomic selection, placing greater prior emphasis on known QTLs associated with milk production traits as well as fertility.

Session 66 — Poster 41

Genetic polymorphism of CSN3 at four ecotypes of Serrana goat

F. Santos-Silva[1], I. Mineiro[2], C. Sousa[1], F. Pereira[3], I. Carolino[1], N. Carolino[1] and J. Santos-Silva[1]
[1]INIAV, Fonte-Boa, 2005-048 Vale de Santarém, Portugal, [2]ESAC, Bencanta, 3045-601 Coimbra, Portugal, [3]ANCRAS, Zona Industrial, R. D, 33, 5370-32 Mirandela, Portugal; fatima.santossilva@iniav.pt

κ-casein protein encoded by CSN3 gene reveals high polymorphism in goats that has been associated to protein level and technological properties of milk. The work was carried out in cooperation of National Breeder Association of Serrana goat and ALTBIOTECH project with the aim to investigate genetic diversity at CSN3, exon 4, in four ecotypes of Portuguese Serrana goat. Blood samples of, 38, 34, 54 and 54 goats were collected from Jarmelista (JAR), Ribatejano (RIB), Serra (SER) and Transmontano (TRA) ecotypes, respectively. DNA was extracted with a commercial kit, and a 645 bp fragment was amplified by PCR, followed by a primer extension analysis (SNaPshot™Kit, Applied Biosystems), with six primers that allow simultaneous identification of single point mutations (SNPs) at 104, 166, 242, 328, 440 and 448 sites, originating seven possible variants classified from A to G. After capillary electrophoresis, SNPs were identified with Genescan and Genotyper software. Chi-square was used to test Hardy Weinberg Equilibrium (HWE). Results reveal polymorphism in four positions (166, 328, 440 and 448). Only three alleles, A, B, and C were identified, in five genotypes AA, BB, CC, AB, BC. Allele B was the most frequent (0.62) followed by A (0.38) and C (0.01), identified in only two animals of SER, at homozygous and heterozygous condition. Most common genotypes were AB (0.47) followed by BB (0.38). AB was the most frequent (0.53-0.56) at RIB and TRA populations, while BB was the most frequent (0.44-0.47) in SER and JAR. AA frequency was near to 0.14 for all and BC and CC appear only at SER with very low frequencies (0.02). Minimum value of Observed Heterozygosity (Ho) or Expected (He), were 0.39 and 0.49, respectively, at JAR and maximum value were 0.56 and 0.49, respectively, at TRA. He was not different from Ho except for population SER that showed significant departures from HWE mostly because of a deficit in heterozygosity. If future results confirm the expected benefits of certain alleles namely B, to improve milk abilities, this information could be a very important contribution to support breeding plans of the Serrana goat.

Session 66 — Poster 42

Genomic evaluation on the French scale of Holstein bulls from a large Indian NGO

V. Ducrocq[1], D. Cruz Hidalgo[1], M. Boussaha[1], P.D. Deshpande[2], A.B. Pande[2] and M. Swaminathan[2]
[1]GABI,INRA, AgroParisTech, Université Paris-Saclay, 78350 Jouy-en-Josas, France, [2] BAIF, Central Research Station, Uruli Kanchan, Maharashtra, India; vincent.ducrocq@inra.fr

BAIF Development Research Foundation (http://www.baif.org.in) is the largest Indian NGO in agriculture. BAIF runs a very large bull stud, which produced in 2016 12.5 million doses from *Bos taurus* bulls (Holstein (HF) and Jersey), *Bos indicus* and crossbred bulls, as well as buffaloes. BAIF also manages a bull dam nucleus herd founded in the late 1970s from Canadian HF, Danish Friesian and Jersey heifers. In 2015, a total of 288 animals from BAIF stud and nucleus farm were genotyped with the Illumina Bovine SNP50 Beadchip® including 59 HF bulls and 23 HF cows. The cluster option of the R package 'adegenet' distributed the HF bulls into 4 groups according to their SNP similarities without using any prior knowledge. Looking at the pedigree of these bulls, the four groups could be interpreted according to their origin as sons of: 'old' Danish bulls (G1), 'old' Canadian bulls (G2), 'more recent' Canadian and Indian bulls (G3), 'recent' (born in the 1990s), and US and French bulls (G4). Based on these genotypes, HF bulls and cows were genomically evaluated on the French scale for all traits evaluated in France, as if they were born in France (recognizing that this scale is not the proper one for selection in India). Group averages showed large differences between groups for production traits, with a strong increase in milk yield and a strong decrease in fat % from G1 to G4, and limited changes for protein%. Quite surprisingly, a large improvement in Functional Longevity was observed from G1 to G4, related to a strong improvement of overall udder type and feet and legs. Udder health, fertility and calving traits GEBVs were globally positive on the French scale in contrast with production and type traits GEBVs, which were substantially lower than the French genetic base (cows born in 2008-2010). Therefore, current imports of HF bulls semen are likely to strongly improve udder and feet and legs traits, but will also lead to a huge increase in genetic level for production compared with the former generation of BAIF bulls. This may be challenging in some areas with limited resources.

Signatures of 'merinization' via principal component analysis

S. Megdiche[1], J.A. Lenstra[2] and E. Ciani[3]
[1]Institut Supérieur Agronomique Chott-Mariem, 4042 Chott-Mariem, Sousse, Tunisia, [2]Utrecht University, Faculty of Veterinary Medicine, Yalelaan 1, 3584 CL Utrecht, the Netherlands, [3]University of Bari ALdo Moro, DBBB, Via Amendola 165/A, 70126 Bari, Italy; elena.ciani@uniba.it

Merino and most Merino-derived sheep breeds worldwide share a peculiar phenotype (long and soft white bright fleece, well-formed staple, consistent and pronounced crimp throughout the fleece, high fibre density in fleece; fine fibre diameter; presence of skin wrinkles and wool-covered face and legs, and presence of spiralled horns) that is worthwhile of investigation. We previously tested for the presence of 'merinization' signatures in various Merino populations using the LAMP package. In this study, we adopted the approach, implemented in the pcadapt package. Two dataset were used, one including 3 Italian Merino-derived breeds (Gentile di Puglia, Sopravissana and Merinizzata) and 3 not-Merino populations from Italy and Spain (Appenninica, Bergamasca and Churra, respectively), and one including Australian Merino and the 3 previously mentioned not-Merino breeds. A number of 125 and 111 candidate SNPs were detected in the two datasets, respectively, when applying a False Discovery Rate (FDR)<0.2. After comparison, 9 loci, distributed over 6 chromosomes (3, 6, 8, 10, 13, 19) resulted to be shared between the two tested scenarios. Interestingly, out of them, 7 loci, distributed over 4 loci (6, 8, 10, 13) were also highlighted as possible selection candidates in a dataset including 7 Merino breeds (Spanish Merino, Australian Merino, Merinolandschaf, Rambouillet, Sopravissana, Gentile di Puglia and Merinizzata) and 6 not-Merino breeds (Bergamasca, Appenninica, Rasa Aragonesa, Ojalada, Churra and Castellana) analysed through genome-wide smoothed F_{ST}. Furthermore, at OAR10, a signal was detected in this study (29,413,536 to 29,776,019 bp) that nicely fits with previous results from LAMP analysis in Chinese Merino (28,628,167-35,035,511) and Australian Merino (28,857,230-30,295,323). Overall, the region encompasses several genes of which the best candidate is the evolutionary conserved Furry gene (*FRY*), whose disruption in *Drosophila* has been shown to provoke abnormally branched bristles and strong multiple-hair cell phenotype. Further studies are ongoing to confirm the role of *FRY* in sheep.

GWAS on whole genome sequences for cheese-making traits and milk composition in Montbéliarde cows

M.-P. Sanchez[1], V. Wolf[2], M. El Jabri[3], S. Fritz[1,4], M. Boussaha[1], S. Taussat[1,4], C. Laithier[3], A. Delacroix-Buchet[1], M. Brochard[5] and D. Boichard[1]
[1]GABI, INRA, AgroParisTech, Université Paris Saclay, Domaine de Vilvert, 78350 Jouy-en-Josas, France, [2]Conseil Elevage 25-90, Roulans, 25640 Roulans, France, [3]Institut de l'Elevage, 149 rue de Bercy, 75012 Paris, France, [4]Allice, 149 rue de Bercy, 75012 Paris, France, [5]Umotest, Ceyzériat, 01250 Ceyzériat, France; marie-pierre.sanchez@inra.fr

A genome wide association study (GWAS) was performed at the sequence level for cheese-making properties (CMP) and milk composition in Montbéliarde (MO) cows. We then used an association weight matrix to identify interacting genes co-associated with CMP and milk composition. Nine CMP traits (three cheese yields, five coagulation traits and milk pH) and proteins, fatty acids, minerals, citrate and lactose contents were predicted from mid-infrared (MIR) spectra. Data from cows with at least three test-day records during the first lactation (1,506,037 test-day records from 194,934 cows) were adjusted for non-genetic effects and averaged per cow. 50K genotypes, available for a subset of 19,862 cows from routine genomic selection analyses, were imputed in two steps: at the HD level using HD genotypes of 522 MO bulls and then, at the sequence level using 27 millions of sequence variants selected from the run6 of the 1000 bull genomes project (1,466 animals including 54 MO bulls). Effect of each sequence variant was tested using a mixed model including also a mean and a random polygenic effect. Numerous QTL were identified for both CMP and milk composition traits. Most of them had pleiotropic effects. We identified a set of genes co-associated with CMP and milk composition with the well-known caseins, PAEP and DGAT1 genes as well as tens of other genes including SLC37A1, ALPL, MGST1, BRI3BP, SCD, AGPAT6, FASN, ANKH, PICALM genes. In each of this genes, GWAS and post-GWAS analyses, combined with functional annotations, led to the identification of a limited number of candidate causative variants.

Session 66 Poster 45

Diversity of innate immunity genes in the Czech Simmental cattle

K. Novák[1], J. Kyselová[1], V. Czerneková[1], M. Hofmannová[1], T. Valčíková[2], K. Samake[3] and M. Bjelka[4]
[1]Institute of Animal Science, Přátelství 815, 104 00 Prague, Czech Republic, [2]Czech University of Life Sciences, Kamýcká 129, 165 06 Prague, Czech Republic, [3]Charles University, Albertov 6, 128 43 Prague, Czech Republic, [4]Breeding Cooperative CHD Impuls, Bohdalec 122, 592 55 Bobrová, Czech Republic; novak.karel@vuzv.cz

The programme of screening of the production population of the Czech Red Pied cattle, a local Simmental breed of combined type, for the diversity in innate immunity genes was started in 2016. In parallel with the standard microarray genotyping, the focus at the candidate genes for infection resistance is expected to provide additional information about the full diversity and its phenotypic effects. The panel of 18 genes for Toll-like receptors and members of the Toll-signalling pathway was resequenced with three different platforms (PacBio, MiSeq and HiSeq) in a population of 150 bulls. These animals represented the development of the breed from 2000 to 2017. Inclusion of the subpopulation of the breed that is conserved in the genetic resources programme since 2010 and reflects the genetic structure around 2000 facilitates the detection of the selection trends. Applying the principle of hybrid sequencing, the full spectrum of the present variants was revealed. The diagnostic SNPs for the haplotypes present (tagSNPs) were used for subsequent genotyping of individual animals where the primer extension was the first choice. In parallel, a system for the online collection of health data from eleven farms and 1,500 cows was established. The disease key is based on the recommendations of ICAR from 2013. In the conserved population, the numbers of polymorphisms and haplotypes (in brackets) in the antibacterial TLRs were 6 (4) in TLR1, 16 (4) in TLR2, 8 (10) in TLR4, 26 (6) in TLR5, and 4 (6) in TLR6. The allelic richness in the production population was slightly reduced, comprising 3 SNPs for TLR1, 13 for TLR2, 7 for TLR4, 5 for TLR5, but 26 for TLR6. The inverted relation between the two populations in TLR6 is surprising and is being investigated. Lower functional importance of this locus is the working hypothesis. The antiviral TLRs (TLR3, -7, -8, -9 and -10) harboured another 40 polymorphisms that are also evaluated for their health effect.

Session 66 Poster 46

Genetic analysis of a temperament test for working dog breeds in Sweden

E. Strandberg[1], H. Frögéli[2] and S. Malm[2]
[1]Swedish University of Agricultural Sciences, Department of Animal Breeding and Genetics, P O Box 7023, 75007 Uppsala, Sweden, [2]Swedish Kennel Club, Box 771, 191 27 Sollentuna, Sweden; erling.strandberg@slu.se

Dogs of working dog breeds in Sweden are required to participate in the Dog Mentality Assessment (DMA) if they are to be used in breeding. The aim of this study was to define behavioural traits that can be used for genetic evaluation of working dogs in Sweden, and estimate genetic parameters for these traits. Data on 33 behavioural scores from the 10 subtests were used to estimate heritabilities for 12 breeds, with a total of 61,434 records. In order to summarize these 33 variables into fewer traits, a factor analysis was applied to the phenotypes. This resulted in factors that could be named Sociability (SOC), Playfulness (PLAY), Distance Play (DPLAY), Chase-proneness (CH), Curiosity/Fearlessness (C/F), and Aggressiveness (AGGR), with the number of factors identified depending on the breed (e.g. AGGR and DPLAY were identified as factors in only 7 and 6 breeds, respectively). Often Distance Play scores were associated with SOC. To simplify a routine genetic evaluation 6 traits were defined for all breeds: SOC, PLAY, CH, C/F, AGGR and Gunshot Avoidance (GUN), however, the exact definition of new trait phenotypes was breed-specific and guided by the within-breed factor analysis. Previously, the first 4 traits have been combined and called Boldness. Heritability estimates of all 33 scores over all breeds averaged 0.13, ranging from 0.03 to 0.23. Heritabilities averaged around 0.25 for SOC, PLAY, and C/F and around 0.13 for the other 3 traits. These heritabilities were higher than the average of the heritabilities for the individual behavioural scores that were components of these new traits. Overall, SOC, PLAY, CH, and C/F were positively genetically correlated with each other (ca 0.4-0.6), weakly correlated with AGGR and negatively correlated to GUN (ca -0.3 to -0.5). In conclusion, the results show that it should be possible to achieve genetic improvement in Boldness traits and a decreased AGGR and GUN.

Session 66 Poster 47

Experiences in estimation of metafounders ancestral relationships of dairy sheep

A. Legarra[1] and J.M. Astruc[2]
[1]INRA, GenPhySE, 31326 Castanet Tolosan, France, [2]IDELE, 31326 Castanet Tolosan, France; andres.legarra@inra.fr

Metafounders are representations of inbred and related base populations. Dairy sheep breeding schemes have steady genetic progress but also unrecorded parentships of 10 to 80% of the individuals depending on the breed. Thus, use of genetic groups is essential to obtain unbiased breeding values and correct genetic trends. However, fitting genetic groups as fixed effects is not tenable in view of genomic information and metafounders are a theoretically appealing solution. This involves estimation of relationships within and across metafounders (Gamma: Γ) using genotyped of individuals that may be several generations removed from them. Here, we present experiences in estimating Gamma in Manech Tete Rousse by a pseudo-EM maximum likelihood method. The number of genotyped individuals is 2,111, and a pedigree with their ancestors included 15,270 individuals and 11 metafounders (Gamma is a 11×11 matrix) defined every 3-4 years from 1975 to 2009. These metafounders only model missing pedigrees, there is no introduction of foreign animals in this breed. The pseudo-EM algorithm always maximized the likelihood and different initial points ended up in the same estimates of Gamma. Results show a self-relationship of 0.47 to 0.54, increasing with time, and high relationships (correlations ranging from 0.89 to 0.98) across metafounders. Closer metafounders are more related, as expected. Results are coherent with a priori knowledge of the dynamics of the breed.

Session 67 Theatre 1

Innovations for sustainable animal nutrition

L.A. Den Hartog[1,2] and F. Brinke[1]
[1]Trouw Nutrition, Research and Development, P.O. Box 299, 3800 AG, Amersfoort, the Netherlands, [2]Wageningen University, Animal Nutrition Group, P.O. Box 338, 6700 AH, Wageningen, the Netherlands; leo.den.hartog@trouwnutrition.com

The overall increase (2010-2050) in animal and aqua protein production expectation is 60% ranging from 38% for pork to 104% for poultry. But there is also a significant number of challenges facing the animal and allied industries with respect to sustainable global production of meat, fish, dairy and eggs where market demands and consumer needs will put more constraints on our production systems and methods. For an optimal utilization of earth's surface for producing food, 35-40% of the recommended daily protein consumption of adults should come from animal protein. In addition, the on average worldwide productivity of farm animals is 30-40% below their genetic potential because of suboptimal conditions and health status. These challenges are dynamic and diverse and solutions and opportunities will require development of appropriate technology and using and advancing our knowledge base. Advances in animal nutrition will contribute to meet these challenges. Environmental and nutritional influences during early life have a profound and long-lasting effect on performance and health. The rapid development of antimicrobial resistance urges the need for effective strategies to reduce antibiotic use in animal production. A drastic reduction of antibiotic use can be achieved by moving to a new farming model based on an integrated and multi-stakeholder collaboration that integrates feed, farm and health management. Targeted feed additive strategies can be applied to control microbial quality of feed and water and support gut health. Precision nutrition methods and tools, such as dynamic feed evaluation and animal models, can be implemented to economically optimize the feed program and reduce emissions into the environment. Sustainable feed supply meeting market demands is feasible and will require a multidisciplinary approach of all stakeholders in the value chain.

Session 67 Theatre 2

Expectations of breeding industry from research projects

J.A.M. Van Arendonk
Hendrix Genetics, P.O. Box 114, 5830 AC Boxmeer, the Netherlands; johan.van.arendonk@hendrix-genetics.com

Animal genetics has already played a key role in improving the sustainability of animal production and needs to continue to do so in the future. Within Hendrix Genetics, we have established a sustainability program comprising of three building blocks: animals, people and planet. Animal welfare, biosecurity and genetic resources are the key priorities within the building block animals. As global suppliers of breeding stock, we have a responsibility for ensuring biosecurity and animal health. In addition, we ensure that we protect the diversity of our genetic resources. People make our business, that's why we strive to enhance the quality of life for consumers, customers and colleagues. Minimizing the environmental impact of livestock through improving input efficiency and helping to reduce the use of antibiotics are key parts of the building block planet. By responding to the needs of our partners in the protein value chain, we are continuously exploring new opportunities for improving sustainability. In addition, the company is investing in minimizing its own ecological footprint to preserve and improve the environment that its activities impact. Responding to the needs of our partners implies that we are continuously exploring innovations to measure and improve health, welfare and productivity of animals. These innovations need to be based on not only a solid understanding of the underlying biology but also an overall view on the issue at stake. Developing a solid understanding is an important but not the only driver to be involved in a research project. Equally important drivers for participation are creating awareness in the scientific community for the issues involved in implementing innovations and training a new generation of researchers. Solving sustainability issues often requires collaboration in multidisciplinary teams. Industry participation in research projects will speed-up innovations and contribute to the training of new talents that are focussed on creating innovations. Collaboration is crucial for realizing the improvements in sustainability.

Session 67 Theatre 3

The importance of research for developing standards to control animal diseases: the role of the OIE

S. Messori and E. Erlacher-Vindel
World Organisation for Animal Health (OIE), Science and New Technologies Department, 12 rue de Prony, 75017 Paris, France; s.messori@oie.int

The World Organisation for Animal Heath (OIE), is the intergovernmental organisation responsible for improving animal health and welfare worldwide. The OIE was established in 1924, and it has today 181 Member Countries. Recognised as reference by the World Trade Organization, the OIE develops standards designed to prevent and control animal diseases including zoonoses, to ensure sanitary safety of world trade. The Sixth Strategic Plan of the OIE for the period 2016-2020 outlines the need for continued development of standards and guidelines based on science for the management, control and/or eradication of disease, including zoonoses, taking into account economic, social and environmental factors. Scientific research is key to find effective ways to prevent, control, and eradicate pathogens posing threats to both animal and human health. National and international research collaborations are fundamental for ensuring the timely delivery of needed disease control tools, greatly contributing to protect animal health. The OIE is actively supporting research collaboration and coordination through its participation in the STAR-IDAZ International Research Consortium on Animal Health (IRC), a global forum of public and private R&D programme owners/managers. The aim of the STAR-IDAZ IRC is to improve the control tools for a list of priority diseases/issues through the delivery of candidate vaccines, diagnostics, therapeutics, procedures, and key scientific information/tools to support risk analysis and disease control. Through its involvement in the STAR-IDAZ IRC, the OIE supports the implementation of research projects on topics of high interest for advancing animal health at global level.

Session 67 Theatre 4

Effects of protein restriction on rumen metagenome and metabolic profile in Holstein calves

S. Costa, J. Balcells, G. De La Fuente, J. Mora, J. Álvarez-Rodríguez and D. Villalba
University of Lleida, Animal Science, Rovira Roure 178, 25198, Spain; scosta@ca.udl.cat

The aim of this work was to study the impact of a crude protein (CP) restriction on productive performance, metagenomics and metabolic profile in growing (120 to 270 days of age) Holstein calves intensively reared. Forty calves were assigned to two dietary treatments: CP in the concentrate was formulated either based on the levels used commercially (CTR: 12% CP on an as-fed basis) or reducing them (LP: 10% CP on an as-fed basis). Concentrate was supplemented with barley straw and both were supplied *ad libitum*. Live weight (LW) and concentrate intake were registered automatically. Ten animals per treatment (220 kg of LW and 155 days of age) were sampled to determine nitrogen balance, rumen metagenome, and urinary/plasma metabolic profiles. Nitrogen balance was estimated by difference between intake and excretion (urinary plus faecal excretions). Rumen bacterial and archaeal community composition was analysed by taxonomic profiling of 16S ribosomal RNA variable regions. Urine and plasma samples were analysed by liquid chromatography coupled to mass spectrometry. The results showed that, at the beginning of the growing phase, CP restriction reduced average daily gain in LP animals (1.65±0.04 vs 1.77±0.04 kg/d, for LP and CTR respectively, P=0.044). Nitrogen excretion (65.9±5.3 vs 81.7±5.3 g/d, for LP and CTR respectively, P=0.049) and retention (30.2±4.2 vs 48.7±4.2 g/d, for LP and CTR respectively, P=0.006) were lower in LP animals than in CTR animals. Only 69% of detected operational taxonomic units were common between both treatments and protein restriction raised richness levels in rumen microbiota (129.9±7.8 vs 101.4±7.8 OTUs/animal, for LP and CTR respectively, P=0.019). Dietary CP restriction led to an increase in *Actinobacteria* phylum, mainly integrated by *Bifidobacterium* genera which is capable of fermenting starch. CP restriction induced the appearance of new discriminant metabolites, being this effect clearer in urine than plasma samples.

Session 67 Theatre 5

Genetic and environmental influence on colostrum quality and absorption in Swedish dairy cattle

J.M. Cordero-Solorzano[1,2], J.J. Wensman[3], M. Tråvén[3], A. Larsson[3], T. De Haan[3] and D.J. De Koning[2]
[1]Animal Health Service of Costa Rica (SENASA), Box 3-3006, 40104 Heredia, Costa Rica, [2]Swedish University of Agricultural Sciences, Department of Animal Breeding and Genetics, Box 7023, 750 07 Uppsala, Sweden, [3]Swedish University of Agricultural Sciences, Department of Clinical Sciences, Box 7054, 750 07 Uppsala, Sweden; juan.cordero@slu.se

Colostrum with sufficient IgG content is essential for the newborn calf, as it requires this passive immunity to survive until weaning. Previous studies have shown a high variation in the amount of colostrum antibodies in dairy cows, with a large proportion having low antibody levels. Failure of passive transfer (FPT) occurs when a calf does not absorb enough antibodies (<10 g/l of IgG in serum) from the colostrum. Some calves absorb antibodies very effectively while others do not. This difference in uptake cannot be explained solely by the time, amount and quality of the colostrum given. The purpose of this study is to identify genetic and environmental factors that can explain this difference in the effectiveness of antibody uptake in calves and variation in colostrum quality in cows. Three experimental farms were included in the study. Colostrum samples from 1,311 cows calving from January 2015 to April 2017 were collected and analysed by Brix refractometer to estimate antibody concentration. For two of the farms, serum from 785 calves was collected at 2 to 7 days after birth and analysed by total IgG ELISA. Brix values ranged from 6.5 to 38.9%, and calves serum IgG from 1.1 to 91.6 g/l. Preliminary results using a linear mixed model show an effect of breed on colostrum quality with Holstein cows displaying significantly higher values than Swedish Red ($P=5\times10^{-12}$). This was also true for samples with shorter calving to colostrum sampling times ($P=2\times10^{-16}$). Genetic parameters will be estimated for colostrum Brix values, serum total IgG and Apparent Efficiency of Absorption (AEA). Early estimates of Herd-Year-Season explain 2.5 to 16% of the variation observed.

Session 67 Theatre 6

A systems genetics approach reveals potential regulators of feed efficiency in pigs

Y. Ramayo-Caldas[1,2], M. Ballester[2], J.P. Sánchez[2], R. González-Prendes[1,3], M. Amills[1,3] and R. Quintanilla[2]
[1]Universitat Autonoma de Barcelona, Facultat de Veterinaria, UAB, 08193 Bellaterra, Spain, [2]IRTA, Animal Breeding and Genetics Program, Torre Marimon, 08140 Caldes de Montbui, Spain, [3]Centre of Research in Agricultural Genomics, Universitat Autònoma de Barcelona, 08193 Bellaterra, Spain; yuliaxis.ramayo@irta.cat

Feed efficiency (FE) has a major impact on the economic sustainability of pig production. In this study, we used the Association Weight Matrix (AWM) approach to identify interacting genes and pathways associated with FE in pigs. Eleven traits related to FE, growth and fat deposition in 352 Duroc pigs were considered, including body weight gain (ADG), feed intake (ADFI), residual feed intake (RFI), and food conversion ratio (FCR). Initially, a genome-wide association analysis was performed on the 11 traits using ~31K SNPs. The additive effects of the significant SNPs located within or ≤10 kb from the nearest annotated gene (Sscrofa11.1 assembly) were used to build the AWM matrix relating genes effects to phenotypes. Pairwise correlations across AWM rows were used to predict gene-gene interactions. The resulting co-association gene network was formed by 704 genes (47 were Transcription Factors (TF)) connected by 35,819 edges. To identify putative regulators, we explored combinations of TF that spanned most of the network topology with minimum redundancy. Among them, it is worth noting that LHX4, POU2AF1 and TCF7L2 were co-associated with 397 genes. SNPs in these genes explained 51, 48, 46 and 35% of the phenotypic variance for RFI, ADFI, FCR and ADG, respectively. The functional annotation of these 397 genes showed that immune response is among the most overrepresented biological processes. In addition, social behaviour, circadian entrainment, and regulation of cell growth processes were also identified as overrepresented. According to the literature, TCF7L2 has been associated with eating behaviour, FCR, and RFI in pigs. LHX4 is involved in the control of differentiation and development of the pituitary gland, whereas POU2AF1 is essential for the response of B-cells to antigens. We hypothesize these TF mediate a highly inter connected regulatory cascade that seems pivotal for FE in pigs.

Session 67 Theatre 7

The PEGaSus project: phosphorus efficiency in *Gallus gallus* and *Sus scrofa*

C. Gerlinger[1], M. Oster[1], H. Reyer[1], E. Magowan[2], E. Ball[2], D. Fornara[2], H.D. Poulsen[3], K.U. Sørensen[3], A. Rosemarin[4], K. Andersson[4], D. Ddiba[4], P. Sckokai[5], L. Arata[5], P. Wolf[6] and K. Wimmers[1]
[1]Leibniz Institute for Farm Animal Biology (FBN), Wilhelm-Stahl-Allee 2, 18196 Dummerstorf, Germany, [2]Agri-Food and Biosciences Institute (AFBI), 18a Newforge Lane, BT9 5PX, Belfast, United Kingdom, [3]Aarhus University (AU), Blichers Allé 20, 8830 Tjele, Denmark, [4]Stockholm Environment Institute (SEI), Linnégatan 87D, 10451 Stockholm, Sweden, [5]Universita Cattolica del Sacro Cuore Piacenza (UCSC), Via Emilia Parmense 84, 29122 Piacenza, Italy, [6]University of Rostock, Justus-von-Liebig-Weg 6b, 18059 Rostock, Germany; oster@fbn-dummerstorf.de

Phosphorus (P) is an irreplaceable component of life and used in all agricultural production systems. It is a finite but recyclable resource which is not efficiently used and reused in agricultural production leading to serious concerns for soil and aquatic ecosystems. The strategic aim of the conducted research is to provide and assess solutions to secure sufficient supplies of high quality animal products from resource-efficient and economically competitive agro-systems that are valued by society and preserve soil and water ecosystems. Therefore, the fate of P in fodder, animals, microbiota, slurry, soil and water is traced. This implies an improved understanding of the dynamic influxes and effluxes of P which might be of critical importance to improve endogenous mechanisms of P utilization and to produce P-resilient phenotypes. In particular, the genomic, epigenetic, and transcriptomic variation is addressed following an intrauterine phosphorus conditioning throughout pregnancy ('Metabolic Programming') and post-weaning dietary mineral challenges in pigs. Current transcriptomic analyses indicate that variable dietary P regimens prompt immunomodulatory implications in intestinal and renal tissue sites while maintaining systemic mineral homeostasis. Therefore, deviations from current dietary recommendations (limited and excess P intake) must be carefully considered, as the endogenous mechanisms that respond to variable P diets may impact important adaptive immune responses. The study contributes to deliver novel approaches of P management to balance economic and environmental sustainability of the European animal production.

Why farmers adopt environmental practices: the case of French dairy farms

T.T.S. Siqueira[1,2], D. Galliano[1] and G. Nguyen[1]
[1]Institut National de la Recherche Agronomique, 75 Voie du TOEC, 31400, France, [2]Université de Toulouse, Ecole d'Ingénieurs de PURPAN, 75 Voie du TOEC, 31400, France; tiago.siqueira@purpan.fr

The adoption of environmental practices is an important issue for livestock production. This study explores the main drivers of the adoption of nine environmental practices in the case of French dairy farms. We first tested the role of internal farm factors related to its structure and governance, followed by the role of external factors related with spatial, regulatory and market features. We used data of 47,562 dairy farms from the 2010 French Agricultural Census to study the correlation between internal and external factors and each one of the nine environmental practices. The results show that among governance and managerial characteristics, attitudes towards uncertainties play a more important role than individual characteristics of the farmer (e.g. age, gender, or training). The results also suggest that farm size has a negative effect on the adoption of permanent grassland, leguminous for forage, no irrigation, crop rotation and no fertilisers and no pesticides practices. However, farm size has a positive effect on the adoption of no-till, manure treatment and agroecological structures as hedges, lines of trees, woods, and fallow lands. In terms of external factors, the paper shows the central role of the spatial environment of the farm and, more specifically, the environmental behaviour of neighbouring farms as a major driver of farm adoption behaviour. The statistical analysis also highlights the strong correlation of positioning on alternative markets, short circuits, organic products, or quality markets on the adoption of these practices. Finally, as the literature suggests, we show that environmental regulations also drive the adoption of farming environmental practices.

Session 67 Theatre 9

Panel discussion 2: user needs and applications

M.H. Pinard-Van Der Laan[1] and J. Van Milgen[2]
[1]INRA, UMR GABI, Domaine de Vilvert, 78530 Jouy-en-Josas, France, [2]INRA-Agrocampus Ouest, UMR Pegase, Le Clos, 35042 Rennes, France; marie-helene.pinard-van-der-laan@inra.fr

Panel discussion and interaction with participants on the needs of stakeholders with respect to the complexity (and thus the multidisciplinarity) of livestock production and animal-derived products and the way research can respond to these needs.

Session 67 Poster 10

Polyphenols and IUGR pregnancies: maternal hydroxytyrosol supplementation and foetal development

C. Garcia-Contreras[1], M. Vazquez-Gomez[2], L. Torres-Rovira[1], J.L. Pesantez[1,3], P. Gonzalez-Añover[2], S. Astiz[1], B. Isabel[2], C. Ovilo[1] and A. Gonzalez-Bulnes[1]

[1]INIA, Crta. de la Coruña, km 7,5, 28040 Madrid, Spain, [2]Faculty of Veterinary, UCM, Av. Puerta de Hierro, s/n, 28040 Madrid, Spain, [3]School of Veterinary Medicine and Zootechnics, University of Cuenca, Avda. Doce de Octubre, 010220 Cuenca, Ecuador; garcia.consolacion@inia.es

Hydroxytyrosol (HT) is a polyphenol present in virgin olive oil with antioxidant, metabolism-regulatory, anti-inflammatory and immuno-modulatory properties. Previous studies have shown positive effects of maternal HT supplementation during pregnancy on final birth weight in Iberian sows. However, there is are few data on the physiopathogenesis of this effect. The present study analysed foetal phenotype after HT supplementation in pregnant sows with a nutritional restriction of 50%. From day onwards of pregnancy, 6 Iberian sows were treated with 2.1 mg/kg of feed day of HT (group HT) whilst 7 other sows were used as control (group C). At day 100 of pregnancy, a total of 100 foetuses were collected. foetal body weight and weights of all viscera were determined. For statistical purposes, litter size was categorized afterwards into two groups (1-7 vs >8 piglets/litter). The final distribution was 55 foetuses in group C and 45 in group HT. The results showed a similar mean litter size (7.8±0.7 and 7.5±1.2 foetuses for groups C and HT, respectively), with no significant differences in the incidence of IUGR between groups (6 IUGR foetuses in both groups). However, foetuses of group C were heavier than foetuses of group H both in small (752.74±36.65 vs 701.25±9.01 g respectively; $P<0.005$) and in large litters (717.56±20.50 vs 652.16±20.89 g respectively; $P<0.01$). These results are opposite of previous data in Iberian and hyperprolific sows and may indicate either inconsistent effects among replicate studies or different effects depending on the stage of pregnancy. Foetuses of large litters in the group HT showed higher weight ratios of the brain ($P<0.05$), kidneys, pancreas, and spleen relative to body weight ($P<0.0001$ for all). These traits suggest an adaptive response to maternal undernutrition of foetuses in the group HT despite a lower body weight.

Session 67 Poster 11

The effect of fertilization on qualitative and *in vitro* fermentation parameters of C3 grasses

C.J.L. Du Toit[1], W.A. Van Niekerk[1], H.H. Meissner[1], L.J. Erasmus[1] and L. Morey[2]

[1]University of Pretoria, Animal and Wildlife Sciences, Private Bag x28, 0028, Hatfield, South Africa, [2]Agricultural Research Council, Biometry, 1134 Park Street, 0087, Hatfield, South Africa; linde.dutoit@up.ac.za

The aim of the study was to evaluate the effect of level of nitrogen (N) fertilization on certain qualitative parameters and *in vitro* total gas and methane (CH_4) production of temperate (C3) grass species commonly used in South Africa. Treatments included three C3 grass species (*Dactylis glomerata, Festuca arundinaceae, Lolium perenne*) with tree levels of N fertilizer (0, 50 and 100 kg N/ha). The experiment was conducted in a greenhouse in 10 l pots using a standardised soil mixture. After seed germination all treatments were thinned to three uniform seedlings per pot. All pots received a single dressing of N fertilizer (LAN, 28%N) as per the experimental treatments. The pots were rotated weekly in the greenhouse to minimize environmental effects and watered to 90% field capacity. Samples for analysis were harvested by hand after an 8 week regrowth period. The data were subjected to an analysis of variance (ANOVA) with two factors and 3 block replications using the GLM procedure of SAS. Student's t-LSD (Least significant difference) was calculated at a 5% significance level to compare means of significant source effects. Increasing the rate of N fertilization increased the crude protein (CP) concentration but had no effect on the fibre fractions within species. *L. perenne* had the highest ($P<0.05$) CP at the 100 kg N/ha treatment and the highest *in vitro* organic matter digestibility (IVOMD) across all N treatments compared to *D. glomerata* and *F. arudinaceae*. The rate of N fertilization had no effect on the *in vitro* total gas and methane (CH_4) production within the species. *L. perenne* and *D. glomerata* had the highest and lowest 48 hour CH_4 production respectively. *D. glomerata* emerged as the specie with the lowest methanogenic potential (CH_4: Total gas) after 48 hours incubation.The data suggests that the stage of physiological development of forages might have a greater influence on the fibre fractions and methanogenic potential of forages compared to the effect of N fertilizer application.

Session 67 Poster 12

Semen and sperm quality parameters of potchefstroom koekoek roosters fed dietary Moringa oleifera

N.A. Sebola
North West University, Animal science, Cnr. Dr Albert Luthuli & University Drive Internal Box 575, Private Bag 2046, Mmabatho 2735, South Africa; 22954457@nwu.ac.za

The study was designed to evaluate semen and sperm quality of Potchefstroom Koekoek roosters fed Moring oleifera leaf meal (MOLM) aged 40 weeks. At four weeks of age, the chickens were randomly allotted to two dietary treatments diets consisting of 0 (control) and 70 g/kg MOLM. A completely randomised design experiment, with 5 birds, replicated 4 times was used for this experiment. Semen was collected six times a week by the dorso-abdominal massage method. Semen was used to evaluate ejaculate volume, sperm concentration, live in total sperm, live normal sperm, sperm quality factor and abnormal sperm using a computer-aided sperm analysis system. The 2 way interaction (diet × day) ($P>0.05$) significantly affected semen volume. Semen of roosters fed MOLM had higher progressive motility than those fed the control control diet (57.73 vs 39.38, respectively). Semen of rooster fed the control diet had higher non-progressive motility and static of 48.59 and 12.57, respectively. Supplementation with MOLM resulted in higher ($P>0.05$) sperm velocity/rapid (89.06%) than the control diet. Supplementation also resulted in higher VCL (106.64 μm/s), VAP (66.19 μm/s) and VSL (45.54 μm/s). In conclusion, MOLM has proven to be a good supplement that can be used to improve fertility of animals.

Session 67 Poster 13

Growth performance and slaughter traits of Baladi and Shami-Baladi kids raised in Summer in Jordan

M.D. Obeidat and B.S. Obeidat
Jordan University of Science and Technology, 221110 Irbid, Jordan; mdobeidat@just.edu.jo

In developing countries such as Jordan, sheep and goats play a major role in household economic through meat and milk production. Due to some health considerations and consumer preferences, the demand for goat meat has increased during the past several years. Lee *et al.* reported that goat meat had lower levels of hypercholesteremic fatty acids and higher levels of unsaturated fatty acids which make it healthier compared to lamb meat. However, meat production of local goat breeds is not adequate to meet the high demand. As a result, some exotic breeds with noticeable meat and milk production have been imported to the country and raised as pure breeds or crossed with local goats. During the last decade, the most famous exotic breed that has been widely used in Jordan is Shami goat. A total of thirty newly weaned kids (15 Baladi (BB) kids and 15 Shami-Baladi (SB) kids) were evaluated for growth performance and slaughter traits. The trial lasted 77 days (7 days for adaptation and 70 days for data collection). Feed intake was measured on daily basis. Body weight of kids was measured at the beginning of the study and biweekly thereafter. At the end of the trial kids were slaughtered to examine carcass traits. Data were analysed using the MIXED procedures of SAS. Initial weight, final weight, and ADG were not affected ($P>0.05$) by the kids' genotype. In addition, Genotype of the kid showed no significant effect on feed to gain ratio. Hot and cold carcass weight, dressing percentage were also not affected by the kid's genotype ($P>0.05$). Offal formed about 13% of the carcass, with no difference between both genotypes. Kid genotype had no significant effect on shoulder, rack and legs percentages. However, SB kids had higher loin cut percentage compered to BB kids ($P<0.05$). Results of this study indicate that crossing Shami and Baladi goat breeds did not have a significant effect on either growth performance or slaughter traits. This could be due to the high temperatures during summer season. Future studies on different seasons may merit further investigation.

Session 67 Poster 14

Genotype × feed interactions for Piétrain sires

S. Palmans[1,2,3], S. Janssens[3], J. Van Meensel[2], N. Buys[3] and S. Millet[2]
[1]Agricultural Research and Education Center, Kaulillerweg 3, 3950 Bocholt, Belgium, [2]Flanders research institute for agricultural, fisheries and food, Scheldeweg 68, 9090 Melle, Belgium, [3]Livestock Genetics, Biosystems, KU Leuven, Kasteelpark Arenberg 30, 3001 Heverlee, Belgium; steven.janssens@kuleuven.be

Piétrain terminal sire estimated breeding values in Belgium are based on performance of progeny raised in test stations. Feeds used in test stations have high energy and amino acid content, which may differ from conventional feeds used on-farm. The question arises whether the genetic evaluation performed on these concentrated feeds is transferable to commercial practice. Therefore we designed an experiment with progeny of 5 terminal sires on two different feeding regimes: a high feeding level (H) with higher energy level (+2.5 to +6.6%) and digestible amino acid:energy ratio (+15.8 to +21%) in comparison with a low feeding level (L). For every sire×feed combination 6 pens of 6 pigs (3 gilts and 3 boars) were raised till slaughter (23.1±4.1 to 112.2±11.5 kg BW). There was a tendency towards a sire×feed interaction for average daily feed intake (P=0.058) and gain:feed-ratio (P=0.086). H-feed resulted in higher BW gain, lower feed intake and better G:F-ratio (P<0.001). No interaction effect on lean meat percentage was observed. Differences between progeny of different sires were more distinct on the H-feed compared to the L-feed. In conclusion, we observed significant differences between sires and between feeds for daily gain, gain:feed-ratio and carcass quality. At most, weak genotype×feed interactions were detected (only for ADG and gain:feed ratio), which are attributed to increased differences between the progeny-groups on the H- diet, but with little re-ranking of sires.

Session 67 Poster 15

Effects of zeolite CPL in-feed supplementation on blood indicators of energy metabolism in cows

R. Turk[1], D. Đirìčić[2], S. Vince[1], Z. Flegar-Meštrić[3], S. Perkov[3], B. Beer-Ljubić[1], D. Gračner[1] and M. Samardžija[1]
[1]Faculty of Veterinary Medicine, University of Zagreb, Heinzelova 55, Zagreb, 10000, Croatia, [2]Veterinary Practice Đurđevac, Malinov trg 7, 48350 Đurđevac, Croatia, [3]Merkur University Zagreb, Zajčeva 23, Zagreb, 10000, Croatia; rturk@vef.hr

Natural zeolites, commonly supplied in a form of clinoptilolite (CPL), are porous aluminosilicates, which crystalline structure enables cation exchange and adsorption of organic compounds, such as mycotoxins. The use of zeolite as a feed additive has been establish in animal nutrition to improve their performance, health and production. The objective of this study was to evaluate serum indicators of energy metabolism of cows supplemented with zeolite CPL. The study was conducted on twenty Holstein-Frisian dairy cows assigned into two groups, control (n=10) and CPL group (n=10). The CPL group received 100 g zeolite CPL (Vibrosorb, Viridsfarm, Podpićan, Croatia) in the ratio on daily basis started from the third month of pregnancy. Blood samples were taken on days -180, -90, -60, -30, -10, 0, 5, 12, 19, 26, 33, 40 and 60 relative to parturition. Serum glucose, lipid parameters (triglycerides, cholesterol and HDL-C), nonesterified fatty acids (NEFA) and β-hydroxybutirate (BHB) concentrations were assayed spectrometrically by commercial kits (Randox, UK). Glucose concentration was significantly higher (P<0.005) in CPL-fed cows on day 12 of lactation (3.6 mmol/l) compared to the control group (2.9 mmol/l). Triglyceride level was significantly higher (P<0.05) in the control cows on day 0 (0.16 mmol/l) than in CPL group (0.12 mmol/l), while cholesterol and HDL-C were not affected. Serum NEFA concentration was significantly lower (P<0.05) in CPL group on days 0, 5, 12 and 19 (0.8, 0.9, 0.8 and 0.6 mmol/l, respectively) than in the control cows (0.5, 0.4, 0.3 and 0.2 mmol/l, respectively). In addition, BHB concentration was significantly lower (P<0.05) in CPL group on day 19 of lactation (0.7 mmol/l) compared to the control group (1.9 mmol/l). Results demonstrated that cows supplemented with zeolite CPL had higher glucose level in early lactation and lower degree of lipid mobilisation following parturition indicating that zeolite CPL in-feed supplementation has beneficial effect on energy metabolism in dairy cows during the transition period.

Session 67 Poster 16

Form of organization and environmental externalities: the case of Brazilian dairy farms

T.T.S. Siqueira[1,2], D. Galliano[1] and G. Nguyen[1]
[1]Institut National de la Recherche Agronomique, Chemin de Borderouge, Castanet-Tolosan, 31320, France, [2]Université de Toulouse, Ecole d'Ingénieurs de PURPAN, 75 Voie du TOEC, 31400, France; tiago.siqueira@purpan.fr

A diversity of organizational forms of agriculture exists. From family to industrial they are all asked to reduce their environmental externalities. The aim of the paper is to study the links between forms of organization of farms and environmental externalities. On the one hand, it explores the adaptation mechanisms and learning processes related with the adoption of environmental practices. On the other hand, it explores the role of regulation, spatial and market aspects in setting up incentive and coordination mechanisms for the adoption of environmental practices. We collected data from six Brazilian dairy farms through semi-directive interviews and farm visiting in 2016. Each interview was translated and the noteworthy facts reported by verbatim. The result shows that there is a strong overlap of the domestic and productive organization that influences the adoption of environmental practices on family farms. They show that family dynamics, preferences, and intergenerational changes are linked to the adoption process. The results also suggest that spatial and market aspects can have a strong influence on the adoption. In fact, they can allow incentive and coordination mechanisms related with learning process taking place in the adoption of environmental practices. Thus, universities, cooperatives, local networks, and high value markets also seem drive these processes. Finally, to be more successful, instead of punitive measures, agro-environmental policies should (1) considering the needs of different forms of organization, (2) promote the creation of coordination between actors from farms to consumers through local and market measures.

Session 67 Poster 17

***In vitro* screening of the anthelmintic effects of by-products from the chestnut industry**

J. Dahal[1], S. Ketavong[1], E. Pardo[1], E. Barbier[2], M. Gay[3], H. Jean[3], V. Niderkorn[4] and H. Hoste[1]
[1]INRA, UMR 1225 IHAP INRA/ENVT, 23 chemin des capelles, 31076 Toulouse, France, [2]Société MG2Mix, France, Route de Rennes, 35220 Chateaubourg, France, [3]Société Inovchâtaigne, St Médard, 24400 Mussidan, France, [4]INRA, UMR Herbivores, Theix, 63122 Saint-Genès-Champanelle, France; h.hoste@envt.fr

Gastro-intestinal nematodes remain a major health threat for the outdoor breeding of ruminants. For more than 50 years, synthetic anthelmintics have been the cornerstone to control these parasites. However, resistance to commercial anthelmintics is now worldwide. There is thus a need for alternative solutions to synthetic anthelmintics. The use of tannin containing forages as nutraceuticals has been widely explored. The possible use of other tannin containing resources is worth to be explored. This study aimed at screening and comparing the potential anthelmintic effects of a range of by-products from the chestnut industry based on an *in vitro* assay (LEIA) applied on the nematode *Haemonchus contortus*. A total of 14 samples were obtained depending on several factors, in particular peels vs other by-products (e.g. leaves, burr), the mode of production (organic or conventional farming), and/or geographic origin. The total tannin content ranged from 1.70 to 8.12% equivalent of tannic acid. The total phenol content ranged from 1.84 to 8.96% equivalent of tannic acid. Inhibitory concentrations 50% (IC50) were calculated from the LEIA test. The effective concentration (EC50) averaged 89.6 µg/ml, ranging from 30.6 to 150 µg/ml. Statistical analyses showed that peels have a significantly higher anthelmintic activity (lower IC50) compared to the other by-products tested. A similar trend was found for by-products issued from organic and conventional farming. Spearman correlations between total phenol, total tannin, and IC50 were negative but not significant. However, the correlation was highly significant between total tannin and total phenol ($r= 0.989$, $P<0.01$). These preliminary *in vitro* results suggest that by products of the chestnut industry, especially peels, have a high potential as an alternative to synthetic anthelmintics to control gastro-intestinal nematodes in small ruminants. These *in vitro* results have to be confirmed by *in vivo* studies.

Session 68 Theatre 1

Genomic predictability of single-step GBLUP for production traits in US Holstein

Y. Masuda[1], I. Misztal[1], P.M. Vanraden[2] and T.J. Lawlor[3]
[1]University of Georgia, 425 River Road, Athens, GA 30602, USA, [2]US Department of Agriculture, Beltsville, MD, 20705, USA, [3]Holstein Association Inc., Brattleboro, VT 05301, USA; yutaka@uga.edu

The objective of this study was to validate genomic predictability of single-step genomic BLUP for 305-day protein yield for US Holsteins. The full data set included phenotypes collected from 1989 through 2015 and pedigrees limited to 3 generations back from phenotyped or genotyped animals. The predictor data set was created by cutting off the phenotypes, pedigree animals, and genotypes in the last 4 years from the full data set. The genomic relationship matrix was created with the Algorithm of Proven and Young (APY) with 18,359 core animals. Genomic PTA (GPTA2011) were calculated for predicted bulls that were young bulls in 2011 but had at least 50 such daughters in 2015. We calculated the daughter yield deviations with the full data (DYD2015) for the predicted bulls (n=3,797). We also used the official GPTA published in 2011 with a multi-step method as a comparison, although the official methods have changed since then. Coefficient of determination (R^2) was calculated from a linear regression of DYD2015 on GPTA2011. We investigated the effect of different unknown parent groups (UPGs) to compensate for incomplete pedigree. When applying QP-transformation to A^{-1} (UPGA), the R^2 was 0.52 compared to 0.51 from the official GPTA. When including UPGs in H^{-1} (UPGH), the R^2 was less than 0.4. When treating UPGs as a random effect with UPGH, the R^2 increased but it is still smaller than UPGA. Without any UPGs, R^2 and the inflation were similar to the official GPTA. For type traits in US Holstein, UPGH has higher than the other options and solved convergence problems. Problems with UPGH could be due to a wrong assumption (no missing parents on genomic relationships) or specific pattern of missing parents in US Holstein for production traits. Several solutions are investigated, including refined UPGH formulas, random genetic groups, indirect evaluation of non-contributing animals, corrections for uncertain pedigree, and a use of metafounders instead of UPGH. Use of the latest data in validation is also crucial. An accurate modelling is required for the current pedigree structure of the U.S. dairy population in single-step GBLUP.

Session 68 Theatre 2

Identification genomic regions associated with characters correlated with fertilizing capacity bulls

G.Molina[1], M.J. Carabaño[2], S. Karoui[2] and C. Diaz[2]
[1]Universidad Politécnica de Valencia, Ciencia Animal, Camí de Vera, s/n, 46022 Valencia, Spain, [2]INIA, Mejora Genética Animal, Ctra. La Coruña km 7.5, 28040 Madrid, Spain; mvzgabrielmolina@gmail.com

The sustainability of dairy farms is at risk due to the decline in the reproductive efficiency of cows with high rates milk production throughout the world. The cow fertility has been included in the genetic improvement programs as a selection goal. However, male fertility has been seen to a lesser extent. Therefore, the objective of this study was to identify single nucleotide polymorphisms (SNPs) significantly associated to the phenotypic evaluation of fertility of the bull and semen quality parameters. Phenotypes included 715 animals with evaluation of artificial insemination (EAI), 199 bulls with data of rate of sperm DNA fragmentation at 0 and 6 hours (SDF_0/6) and 431 bulls with records of individual motility (IM), mass motility (MM), motility post freezing (TPM), volume (VOL), concentration (CONC) and number of spermatozoa (NSPZ). Genotypes of 754 bulls of the Illumina BovineSNP50 Bead Chip have been used in the study. SNPs were edited on their minor allele frequency (>0.0001) and call rate (>0.95). An analysis of association based on a linear mixed model that considers the additive effect of each SNP to be tested for association and a polygenic effect was used to estimate the effect of markers for each of the examined traits. The Benjamini and Hochberg method to control FDR at 10% was used to determine the SNPs with relevant signals for each trait. The SNPs associated with SDF_0 and SDF_6 are in or near genes with biological functions involved in reproduction, such as Spermatid perinuclear RNA-binding protein (SPRB) in BTA 2, ILF2 in BTA 3 (238 kb from SNP) that participates in the gamete generation and WAPL in BTA 28 (432 kb from SNP), involved in meiosis process. The association between SNPs with the bull's ability of fertilization would allow, through genomic selection, an increase of reproductive efficiency in dairy herds.

Session 68 Theatre 3

Deriving dimensionality of genomic information from limited SNP information

I. Pocrnic, D.A.L. Lourenco and I. Misztal
University of Georgia, Department of Animal and Dairy Science, 425 River Road, Athens, GA 30602, USA; ipocrnic@uga.edu

The dimensionality of the genomic, single nucleotide polymorphism (SNP), information can be defined as the number of independent chromosome segments (Me), and therefore as a function of effective population size (Ne) and genome length (L) in Morgan. Knowing dimensionality of genomic information is useful for predicting the theoretical accuracy of genomic selection (GS), finding optimal size of the SNP chips, in computational algorithms as e.g. Algorithm for Proven and Young (APY), and as degrees of freedom in genome-wide association studies (GWAS). When both the number of SNP and genotyped individuals are large, dimensionality can be approximately calculated as the number of non-negligible singular values of gene content, or the number of non-negligible eigenvalues of genomic relationship matrix (GRM) that explain 98% of the variation. The purpose of this study was to determine whether dimensionality can be determined from the data with few SNP and/or genotyped individuals. Populations with varying numbers of SNP markers (6k to 300k), genotyped individuals (1k to 20k), Ne (60 and 120) and L (10, 20, 40 and 50 M) were simulated, and eigenvalue profiles were constructed as the numbers of largest eigenvalues explaining specific percentage of the variation of the GRM. For the 300k scenario, interpolated number of eigenvalues explaining 10, 50, 80 and 98% of variation were 4.1, 66, 310, 2,387 for NeL, 7.7, 121, 561, 3,622 for 2NeL, 4.2, 87, 498, 3,931 for Ne2L, 8.0, 162, 888, 5,687 for 2Ne2L, 4.3, 120, 850, 6,165 for Ne4L, and 8.3, 236, 1,648, 9,163 for 2Ne5L, where Ne=60 and L=10. In the larger datasets and past 80% of the explained variation, the number of eigenvalues was slightly less than a linear function of Ne and L. At 10% of the explained variation, eigenvalue profiles change little with size of data and were proportional to Ne, but independent of L. Based on the simulations in this study, Ne could be approximated as 14 times the number of largest eigenvalues explaining 10% of the variation. For the larger percentages of explained variation, eigenvalue profiles were proportional to combination of Ne and L, and therefore could be used to approximate Me.

Session 68 Theatre 4

Genetic parameters of colostrum qualitative traits in Holstein dairy cows in Greece

A. Soufleri[1], G. Banos[1,2], N. Panousis[1], D. Fletouris[1], G. Arsenos[1] and G.E. Valergakis[1]
[1]Faculty of Veterinary Medicine, School of Health Sciences, Aristotle University of Thessaloniki, Box 393, Thessaloniki, 54124, Greece, [2]Scotland's Rural College and Roslin Institute University of Edinburgh, Edinburgh, EH25 9RG, Midlothian, United Kingdom; geval@vet.auth.gr

The objective was to estimate genetic parameters of colostrum fat (F), protein (PR), lactose (L), total solids (TS) and energy (EN) content. Colostrum F, PR and L content from 1,074 Holstein cows of 10 herds was determined by Milkoscan; EN content was calculated using NRC equations. Colostrum TS were measured with a digital Brix refractometer. Parity number (P), season (S) and age at calving (A), colostrum quantity (Q), time interval between calving and colostrum collection (T), dry period duration (D), BCS and milk yield in previous lactation (M) were recorded. Each trait (TS, F, PR, L and EN) was analysed with a univariate mixed model including the fixed effects of herd, P, S, A, Q, T, D, BCS and M, and the random animal additive genetic effect. All pedigree available was included in the analysis bringing the total animal number to 5,662. Estimates of (co)variance components were used to calculate heritability for each trait. Correlations between TS, F, PR, L, EN and Q were estimated with bivariate analysis using the same model. The ASREML software was used for all statistical analyses. Mean % (±SD) colostrum TS, F, PR and L content was 25.8±4.7, 6.4±3.3, 17.8±4.0 and 2.2±0.7%, respectively; mean EN was 1.35±0.3 Mcal/l. Heritability estimates for TS, F, PR, L and EN were 0.27, 0.21, 0.19, 0.15 and 0.22, respectively ($P<0.05$). Genetic correlations were not different from zero ($P>0.05$). Several significant ($P<0.05$) phenotypic correlations were derived: TS were negatively correlated with Q (r=-0.10) and positively with F (r=0.21), PR (r=0.92) and EN (r=0.70); F was positively correlated with EN (r=0.82); PR was positively correlated with EN (r=0.58) and negatively with Q (r=-0.13); L was negatively correlated with most traits. In conclusion, colostrum content traits are heritable and can be amended with genetic selection. Colostrum TS provides an indirect assessment of immunoglobulin concentration, whose heritability was estimated for the first time, offering interesting perspectives regarding calf health.

Searching for protein biomarkers related to pre-slaughter stress using liquid isoelectric focusing

C. Fuente-García[1,2], N. Aldai[1], E. Sentandreu[2], M. Oliván[3], F. Díaz[3] and M.A. Sentandreu[2]
[1]University of the Basque Country (UPV-EHU), Lactiker Research Group, Department of Pharmacy and Food Sciences, Paseo de la Universidad 7, 01006 Vitoria-Gasteiz, Spain, [2]Institute of Agrochemistry and Food Technology (IATA-CSIC), Carrer Catedràtic Agustín Escardino Benlloch 7, 46980 Paterna (Valencia), Spain, [3]Servicio Regional de Investigación y Desarrollo Agroalimentario (SERIDA), Apdo 13, 33300 Villaviciosa (Asturias), Spain; claudia.fuente@ehu.eus

Proteome changes derived from animals that have suffered pre-slaughter stress are a fact although the discovery of associated biomarkers is still a challenge. In this study, Proteomic analysis was carried out on 16 loin samples of beef from Asturiana de los Valles breed and crossbreds animals previously classified as normal and DFD meat at 24 h post-mortem using pH measurements. Sarcoplasmic sub-proteome of *Longissimus thoracis* muscle was fractionated by the use of liquid isoelectric focusing (OFFGEL) in the pH range of 3 to 10, followed by sodium dodecyl sulphate polyacrylamide gel electrophoresis (SDS-PAGE) of each retrieved fraction. The obtained protein separation profile showed high reproducibility along the different samples. Five different bands showed significant statistical differences ($P<0.01$) in both sample groups, which allowed to compare and distinguish normal and DFD meat. Some proteins present in these bands, which were identified by liquid chromatography coupled to tandem mass spectrometry, were phosphoglucomutase-1 and alpha-crystallin B. The significance of this study relies on the optimization of the OFFGEL technology. This method separates proteins along different fractions according to their isoelectric point; then the obtained fractions can be further separated by SDS-PAGE. This achievement stands out as an alternative to the use of two dimensional gel electrophoresis, enabling a higher resolution in protein separation and shorter analysis times.

Differential protein expression in Nellore cattle divergent for meat tenderness

L.F.S. Fonseca[1], A.F.B. Magalhães[1], L.A.L. Chardulo[2,3], R. Carvalheiro[1,2] and L.G. Albuquerque[1,2]
[1]São Paulo State University, Faculty of Agricultural and Veterinary Sciences, Via de Acesso Prof. Paulo Donato Castellane S/N, 14884-900 Jaboticabal, São Paulo, Brazil, [2]National Council for Scientific and Technological Development, St. de Habitações Individuais Sul Blocos A, B, C, D, 71605-001 Brasília, Distrito Federal, Brazil, [3]São Paulo State University, Faculty of Veterinary Medicine and Animal Science, Rua Doutor Walter Mauricio Correa S/N, 18618-681 Botucatu, São Paulo, Brazil; lgalb@fcav.unesp.br

Brazil has a herd of more than 150 million of Nellore cattle (about 72%), which despite being adapted to Brazilian management conditions, present lower meat tenderness when compared to *Bos taurus*. The knowledge of the genetic action mechanisms and protein expressed in muscle tissue will help to understand the genetic differences between animals with contrasting phenotypes for tenderness. The objective of this work was to find differentially expressed proteins related to meat tenderness. To achieve this goal, 20 extreme samples for meat tenderness (20 with high and 20 with low meat tenderness) were selected. The animals used in this study were non castrated male, with an average age of 731±81 days, belonging to the same contemporary group, and the meat tenderness value was obtained by shear force analyses in Longissimus dorsi muscle not aged. To visualize all the proteins expressed in the samples, we used mass spectrometry LC-MS/MS. We found 150 proteins, among them, 33 were differently expressed (qvalue<0.05). Six proteins were upregulated and 27 downregulated, with respect to the tender meat group. The proteins ACTN1 (Alpha-actinin-1 (Alpha-actinin cytoskeletal isoform)) and MYLK2 (Myosin light chain kinase 2) were expressed more in tender meat. These proteins compose myofibril, an organelle whose function is muscle contraction and have functions related to calcium ion transport. The protein TBA4A (Tubulin alpha-1 chain) was more expressed in the tough meat group and acts as the main composite of the microtubules which, with actin, compose the cytoskeleton. The knowledge acquired through this study, bring new understandings that could help generate tools for developing strategies aiming to improve meat tenderness. São Paulo Research Foundation (FAPESP) grant #2016/23937-6.

Session 68 Theatre 7

Introduction of a software program developed as an automation system for slaughter houses

Y. Bozkurt[1], T. Aydogan[2], C.G. Tuzun[1] and C. Dogan[1]
[1]Suleyman Demirel University, Faculty of Agriculture, Department of Animal Science, Isparta, 32260, Turkey, [2]Suleyman Demirel University, Faculty of Technology, Department of Software Engineering, Isparta, 32260, Turkey; yalcinbozkurt@sdu.edu.tr

In this study, a web-based computer software program based on Artificial Neural Networks models for estimating carcass characteristics such as carcass weight, yield, as well as pre-slaughter bodyweight of beef cattle was developed. This program was integrated with a camera system and can be installed on `Personal Computers` and provide an automation system based on 'internet protocol' system to be used in slaughter houses. The statistical regression and artificial neural network models developed in previous studies were used for the development of the software. For the validation of the models, data regardless of the breeds were collected from the beef cattle brought to a private slaughterhouse. Of the total 3,742 animals evaluated, 2,670 heads are Holstein, 112 heads are Brown Swiss, 516 heads Simmental and 444 heads from other breeds. The prediction ability of the models were evaluated resulting in high prediction levels with ±3% kg error margin for individual carcasses. Then it was decided to develop software as an automation system which includes beef producers information and other managerial information. Therefore, the developed automation system will be able to determine easily the carcass quality and meat yield characteristics and will be beneficial to beef producers and abattoir management. Moreover, it will speed up the slaughtering operations and increase the slaughtering capacity of the enterprises because the information collected by the automation system can be quickly tracked and recorded in a computer environment and is easily reachable wherever there is internet access.

Session 69 Theatre 1

Effect of fermented whole-crop cereals with or without supplementing inoculant in finishing pigs

C.H. Lee[1], M.H. Song[2], W. Yun[1], J.H. Lee[1], W.G. Kwak[1], H.J. Oh[1], S.Y. Oh[1], S.D. Liu[1], H.B. Kim[3] and J.H. Cho[1]
[1]Chungbuk National University, Department of Animal Science, 344, S21-5, 1, Chungdae-ro, Seowon-gu, Cheongju-si, Chungcheongbuk-do, Republic of Korea, 28644, Korea, South, [2]Chungnam National University, Department of Animal Science and Biotechnology, Chungnam National University, Daejeon, Republic of Korea, 34134, Korea, South, [3]Dankook University, Department of Animal Resources and Science, Dankook University, Cheonan, Republic of Korea, 31116, Korea, South; lch8315@naver.com

Two experiments were conducted to evaluate the effect of fermented whole wheat, fermented whole barley with or without supplementing inoculant (probiotics) on growth performance, nutrient digestibility, blood constituents, and faecal microbiota in finishing pigs. In Exp. 1, a total of 20 finishing pigs ((Landrace × Yorkshier) × Duroc, average body weight of 82.3±2.6 kg) were allotted to 4 dietary treatments to check the palatability of the dietary feed in one room. The 20 pigs were conducted a four-choice feeding comparisons during 4 consecutive 28-d experimental periods (1wk per period). The diet treatments were included a basal diet; FW = basal diets + 1% fermented wheat without inoculum, FWI = basal diets + 1% fermented wheat with inoculum, FB = basal diets + 1% fermented barley without inoculum, FBI = basal diets + 1% fermented barley with inoculum. Throughout the experimental period, pigs fed FWI and FBI diets had significantly higher feed palatability compared with FW, FB diets. In Exp. 2, a total of 20 finishing pigs (average body weight of 75.8±0.4 kg) were allotted to 4 dietary treatments (1 pigs/pen, 5 pens/treatment) to evaluate growth performance, nutrient digestibility, blood constituents, and faecal microbiota. Dietary treatments were same as Exp. 1. In nutrient digestibility, pigs fed FBI had higher dry matter digestibility, whereas crude protein digestibility was greater in pigs fed FW compared with other treatments. The number of lactobacilli in faeces was significantly higher in FWI and FBI treatments inoculated with feed microorganisms. In conclusion, our results indicated that dietary supplementation with fermented wheat and barley with supplementing inoculant had a beneficial effect on palatability and DM digestibility, and effect of increasing the number of lactobacilli in faeces.

Session 69 Theatre 2

Effects of restricted feeding with fermented whole-crop barley and wheat in finishing pigs

J.S. An[1], H.B. Kim[2], W. Yun[1], J.H. Lee[1], W.G. Kwak[1], H.J. Oh[1], S.D. Liu[1], C.H. Lee[1], T.H. Song[3], T.I. Park[3], M.H. Song[4] and J.H. Cho[1]

[1]Chungbuk National University, Department of Animal Science, 344, S21-5, 1, Chungdae-ro, Seowon-gu, Cheongju-si, Chungcheongbuk-do, Republic of Korea, 28644, Korea, South, [2]Dankook University, Department of Animal Resources and Science, Dankook University, Cheonan, Republic of Korea, 31116, Korea, South, [3]National Institute of Crop Science, Barley breeder, Crop Breeding Division, 181 Hyeoksin-ro, Iseo-myeon, Wanjugun, Jeollabuk-Do 55365, 55365, Korea, South, [4]Chungnam National University, Department of Animal Science and Biotechnology, Chungnam National University, Daejeon, Republic of Korea, 34134, Korea, South; ajs6@daum.net

A total of 80 pigs ((Landrace × Yorkshire) × Duroc) with average body weight of 72.9±2.6 kg were used in the present study to investigate the effects of fermented whole crop wheat and barley with or without supplementing inoculums throughout restricted feeding in finishing pigs. Pigs were fed *ad libitum* throughout the experiment as (1) the control (CON), and other four groups were restricted to 10% in CON diet and fed *ad libitum* fermented whole crop cereals: (2) fermented whole crop barley with inoculums; (3) fermented whole crop barley without inoculums; (4) fermented whole crop wheat with inoculums; (5) fermented whole crop wheat without inoculums. During the entire experiment, the average daily feed intake (ADFI) decreased in fermented barley and fermented wheat compared to CON, while no difference was observed in average daily gain (ADG), feed efficiency (G: F) between the control and fermented whole crop barley, wheat diets group. In conclusion, restricted feeding with fermented whole crop barley and wheat regardless of supplementing inoculums showed no significant difference in growth performance compared to CON. This suggests that there is a possibility with fermented whole crop barley and wheat to replace part of the conventional diets.

Session 69 Theatre 3

Responses of Boer goats to saline drinking water

R.A. Runa, L. Brinkmann, A. Riek and M. Gerken

University of Göttingen, Department of Animal Sciences, Albrecht-Thaer-Weg 3, 37075 Göttingen, Germany; rukhsana.amin-runa@agr.uni-goettingen.de

Due to global climatic changes, salinization of ground water and soil is an increasing worldwide phenomenon, thus creating new threats for farm animal production. Our study investigates the capacity of goats to differentiate saline water in a free choice system. In this study, 12 non-pregnant Boer goats aged between 1 to 8 years with an average body weight of 46.4±8.3 kg were kept in individual pens under controlled stable conditions for 4 weeks. Animals had access to cut hay and water *ad libitum*. In the control phase (1 week), only fresh water was supplied in five identical buckets for each pen. During the subsequent treatment phases (3 weeks), fresh water and four different concentrations (0.75, 1.0, 1.25, and 1.5% NaCl) of saline water were offered simultaneously in a free choice system. The positions of the salted water were changed daily at random. Individual water, feed and mineral supplement intake were recorded daily, while body weight and body condition score were measured weekly. Data were analysed statistically by using the MIXED procedure of the software package SAS (version 9.3). Total water intake, dry matter intake and total sodium intake were significantly ($P<0.001$) higher during the treatment phase. The total sodium intake of the goats ranged between 0.37 and 0.55 g /kg $BW^{0.75}$/day, being 8 to 11 fold higher than the daily requirements of sodium for body maintenance. Young goats avoided saline water intake more strictly than older animals. All goats showed a significant preference for fresh water (0% salt) over saline water. At the first offering of the simultaneous choice situation (week 2), animals did not differentiate between salt concentrations of 0.75 and 1.0%. However, with proceeding treatments (week 3 and 4), animals distinguished saline water concentrations more distinctly and preferred the 0.75% salt concentration, while concentrations of 1 to 1.5% were avoided. The results suggest that goats are able to differentiate between saline water concentrations and to balance their sodium intake through water by self-selection in a free choice system.

Effect of temperature in the context of climate change on nutrient requirements of lactating sows

J.Y. Dourmad[1], J.L. Gourdine[2] and D. Renaudeau[1]
[1]PEGASE, INRA Agrocampus Ouest, 16 Le Clos, 35590 Saint-Gilles, France, [2]URZ, INRA, Centre de Recherche Antilles-Guyane, 97170 Petit Bourg, France; david.renaudeau@inra.fr

Because of their intense metabolism, lactating sows are highly sensitive to high ambient temperature, which induces a reduction in their voluntary feed intake and milk production. This also results in a decrease in piglet weight gain and an increase in the mobilization of body reserves, which may impair reproduction after weaning. In the context of climate change which increases average temperature and the frequency of periods of heat stress, the aim of this work was to quantify the relationships between ambient temperature and sow and piglet performance, and to use these relationships in a prediction model of nutrient utilization by sows. A database with 46 publications and 254 observations was built in order to adjust prediction equations for different criteria such as feed intake, respiratory frequency, body temperature, litter growth, and milk production. These equations were then incorporated into a simulation model including (1) a bioclimatic module predicting the effect of outdoor temperature on the indoor temperature perceived by the sow and (2) a nutrition module predicting the effect of temperature on feed intake, milk production, energy and amino-acid utilization, and body reserves. This enabled a decision support tool to be developed for the prediction of performance and nutritional requirement of lactating sows in different climatic conditions. The model was used to simulate the effect of climate change on sow's and piglets' performance using CLIMATOR climate database for West and South France. The results show highly significant effects of season and region on animal performance and nutritional requirements. The simulations performed for years 2045-2050 indicate that these effects tend to become more marked with climate change. For instance digestible lysine requirement per kg feed was on average 10% higher during summer than during winter, and the requirement simulated for period 2045-2050 was about 2% higher than for 2000-2005. In practice the model developed should enable feed composition to be better adapted to the season and to the geographical location of farms and to expected changes of climate for the future.

Registered and expected efficacy of the feeding hogs at large enterprises in Croatia 1990-2016

M. Sviben
Freelance consultant, Siget 22B, 10020 Zagreb, Croatia; marijan.sviben@zg.t-com.hr

It was published in veterinary handbook how many grams of the feed had to be given a day and how many kilograms of the live weight could be achieved at the end of the 1st till the 31st week (7-217 days) of the pigs' life. Thanking to The Croatian Agricultural Agency, where the reports for the years 1990-2016 had been saved, it was possible to research the congruence of the indicators of efficacy of the feeding hogs at large enterprises in Croatia with the values expected from the data in above mentioned manual and in the papers published later on (2001, 2005). In the year 1990 494,279 hogs were fed 123 days gaining from 27.84 kg 606.8 g/day with the feed conversion rate of 3.77 kg/kg. Expected means for the pigs fed from 86 till 206 (120) days were 619.9 g/day and 3.80 kg/kg. During the year 2005 246,463 hogs were fed 127.97 days from 25.86 kg gaining 612.5 g/day. Expected mean for the feeding from 81 till 209 days was 616.0 g/day. At the same time mean live weights of contemporarily bred pigs in favourable circumstances were expected according to the equation $Y_C = 49.5292 + 5.1363X_C + 0.1030X_C^2 - 0.0021X_C^3$, where $X_C = (X-105)/7$. Daily gains registered at large feeding enterprises in Croatia for years 2011-2015 were 748.1, 772.8, 827.0, 816.3 and 841.0 g/day doing 94.74, 96.94, 103.62, 102.99 and 105.83% of expected values. In the year 2016 420,971 hogs were fed 120.14 days gaining from 25.98 kg 915.4 g/day. Expected efficacy indicator was the daily gain of 912.5 g/day feeding hogs from 61 till 164 days of age according to the programme of feeding pigs for expected yield, published in 2001, when calculations started to be made using the equation $Y_C = 37.1560 + 5.2680X_C + 0.1546X_C^2 - 0.0034X_C^3$ for the live weight achieved on the day of age, taking $X_C = (X-77)/7$. At the different levels of production during the quarter of century registered magnitudes of the efficacy indicators of feeding hogs were congruent to expected ones.

Session 69 Theatre 6

Effects of feed form and delivery on growth, feed efficiency and carcass quality of finisher pigs

F.M. O'Meara[1,2], G.E. Gardiner[1], J.V. O'Doherty[3] and P.G. Lawlor[2]

[1]Waterford Institute of Technology, Cork Road, X91K0EK, Ireland, [2]Teagasc, Pig Development Department, Moorepark, Co. Cork, P61C996, Ireland, [3]University College Dublin, Belfield, D04V1W8, Ireland; fiona.omeara@teagasc.ie

It has long been reported that pelleting feed for pigs improves growth and feed conversion efficiency (FCE). Reasons for this include improvements in digestibility due to the gelatinisation of the starch component in the diet, increased nutrient density per unit volume and reduced feed wastage during feeding. There is limited information in the literature comparing liquid, dry and wet/dry feed delivery systems, in controlled conditions, on the growth and FCE of pigs and where information is available it is often conflicting. The aim of this study was to compare the effect of feed form (meal and pellet) and delivery (liquid, dry and wet/dry) on the growth, FCE and carcass quality of finisher pigs. This experiment was unique since all treatments were compared in the same facility simultaneously. The experiment was conducted in two batches using a total of 432 pigs. In each replicate 216 pigs (33.8±0.55 kg) were housed in same sex (entire male or female) pens of 6 pigs/pen. The experiment was a 2×3 factorial arrangement with 2 factors for diet form (meal and pellet) and 3 factors for delivery (liquid, dry and wet/dry). The treatments were:1. Meal from dry feeder,2. Meal from wet/dry feeder,3. Meal from liquid system,4. Pellet from dry feeder,5. Pellet from wet/dry feeder,6. Pellet from liquid system. In total there were 12 pens per treatment. The experiment lasted 63 days during which growth and feed intake were recorded every 21 days. Data were analysed by the MIXED procedure of SAS 9.4. Pigs fed the pelleted diet in dry form had the best FCE. Pigs fed the liquid diets had higher average daily gain ($P<0.001$) and average daily feed intake ($P<0.001$) but poorer FCE ($P<0.001$) than those fed dry and wet-dry. It is likely that feed wastage is an issue with liquid feeding. Pigs fed liquid feed were heavier before slaughter than those where feed delivery was dry and wet/dry ($P<0.001$). In conclusion pelleting improved FCE. Liquid feeding increased feed intake and growth to slaughter while dry feeding resulted in a superior FCE compared with all other feed delivery methods.

Session 69 Theatre 7

Implications of climate change on small ruminant systems in Europe

G. Pardo[1], A. Del Prado[1], S. Mullender[2], K. Zaralis[2], M. Dellar[3], D. Yañez-Ruiz[4] and M.J. Carabaño[5]

[1]Basque Centre for Climate Change (BC3), Edificio Sede, Campus EHU, Barrio Sarriena, s/n, 48940 Leioa, Bizkaia, Spain, [2]The Organic Research Centre (ORC), Elm Farm, Hamstead Marshall, Newbury RG20 0HR, United Kingdom, [3]Scotland's Rural College (SRUC), Easter Bush, EH25 9RG Midlothian, United Kingdom, [4]Estación Experimental del Zaidín, CSIC, Profesor Albareda 1, 18008 Granada, Spain, [5]INIA, Ctra. La Coruña Km 7.5, 28040 Madrid, Spain; guillermo.pardo@bc3research.org

Climate projections for Europe indicate a general warming trend and more variable precipitation patterns, with an increase in frequency and length of drought periods. Expected changes in climate will impact on livestock systems, both on the animals directly and the production system more widely, unless adaptation strategies are implemented across the whole food supply chain. In this context, small ruminant production systems are subject to specific challenges regarding their future. On one hand, they could be particularly vulnerable to environmental changes, as a large share of the production is held in marginal lands and/or semi-arid conditions. Yet on the other hand, small ruminants have features that provide competitive advantages against other livestock species in the face of a changing climate. Aiming to understand the implications that climate change could have for small ruminant production in Europe, we conducted a literature review of the best to-date information available on climate change interactions on sheep and goat systems. The review first identifies the main expected impacts at the animal level (direct impacts on productivity, fertility, health and welfare) and at the forage level (changes in quantity and quality). General adaptation strategies to cope with climate impact on pasture production (e.g. increasing multi-species mixtures, reduced tillage, grazing management) and animal heat stress (e.g. nutritional management, heat resistant breeds) are discussed. Finally, for the different bio-climatic areas in Europe, the review identifies specific adaption measures tested at field and animal level in order to analyse their potential to alleviate some of the identified climate threats in every particular context. Authors acknowledge funding support from the iSAGE H2020 project (679302-EU Research & Innovation programme).

Session 69 Theatre 8

Greenhouse gas mitigation from feeding nitrate in an Australian beef farm, a partial life cycle anal

P.A. Alvarez Hess[1], P.J. Moate[2], J.L. Jacobs[2], K.A. Beauchemin[3] and R.J. Eckard[1]
[1]The University of Melbourne, Building 840 Parkville, 3010 Victoria, Australia, [2]Department of Economic Development Jobs Transport and Resources, Ellinbank, 1301 Hazeldean Road, Ellinbank, 3821 Victoria, Australia, [3]Agriculture and Agri-Food Canada, 5403 1st Ave. S., Lethbridge, T1J4B1 Alberta, Canada; pabloah@student.unimelb.edu.au

The Australian carbon offset method recommends feeding nitrate to cattle at a maximum rate of 7 g NO3/kg DMI. This partial farm-gate life cycle assessment (LCA) analysed this offset method by quantifying the effect of replacing urea with nitrate, on a N equivalent basis, on whole farm greenhouse gas (GHG) emissions and revenue of an Australian beef farm. The beef farm modelled was located in Queensland, Australia and considered a cow-calf and grass-fed beef stock operation over a 10 year period. The LCA began with the farm retaining 1,206 female and 21 male calves. Offspring were kept on the farm and were sold for meat at an age of 18 months at a live weight of 370 kg. The Australian National Greenhouse Gas Inventory method was used for estimating GHG emissions. Nitrate was fed to all growing stock and adult cows and was assumed to reduce daily enteric methane emissions by 3.8%. Two scenarios were analysed to estimate the cost of feeding nitrate when used to replace urea: (1) same unit cost of both sources and (2) actual costs of two sources. Prices per kg of N for commercially supplied lick blocks were \$6.8 for a urea block (13% N, 30% urea) and \$23.6 for a calcium nitrate block (6% N, 26% nitrate). A carbon price of \$11.82 per tonne CO_2e was used. Net farm GHG emissions were reduced by 2% by replacing urea with nitrate. Scenario 1 generated a yearly extra revenue of \$0.96/animal finished (AF). Scenario 2 replaced 47% of the urea with nitrate and generated yearly losses of \$155/AF. The price of the calcium nitrate lick block at which profit was neutralized was \$4.3/kg N (\$0.26/kg block). Due to a low calcium nitrate content and increased unit cost, combined with low abatement and low carbon price, feeding the nitrate block generated considerable losses when actual prices of both N sources were used. Use of nitrate as an offset method was not profitable and modifications to the offset method would be required to incentivise its use in the Australian beef industry.

Session 69 Theatre 9

Chemical composition and *in vivo* digestibility of seaweed as a protein source for ruminant nutrition

Ş. Özkan Gülzari[1,2], V. Lind[2], I.M. Aasen[3] and H. Steinshamn[2]
[1]Wageningen Livestock Research, De Elst 1, 6708 WD Wageningen, the Netherlands, [2]Norwegian Institute of Bioeconomy Research, Grassland and Livestock, Post box 115, 1431 Ås, Norway, [3]SINTEF Industry, P.O. Box 4760 Torgarden, 7465 Trondheim, Norway; seyda.ozkan@nibio.no

Norway lacks domestically produced protein-rich feed sources, which leads to the import of soya to support the ruminant production. Marine macroalgae species, including *Saccharina latissima* and *Porphyra* spp., common to the coast of Norway, appear to be potential agents for feeding animals as they contain favourable amounts of protein and amino acid. Furthermore, macroalgae may provide a functional feed ingredient with potential to contribute to the efficiency and carbon neutral process of animal production systems. For this purpose, an *in vivo* digestion trial was conducted to evaluate the effects of inclusion of macroalgae on nutrient digestibility and protein value. Four castrated rams were fed four different diets consisting of a control diet without protein supplement, a diet with extracted soybean meal (SBM), and two diets with macroalgae, a protein rich fraction of *S. latissima* and a commercial *Porphyra* meal (CoDo International Limited), respectively. The protein supplemented diets were planned to be isonitrogenous. The trial was run for four periods, each with eight day adaptation and seven day collection of urine and faeces. There were no differences between the rations in digestibility of the fibre fractions and organic matter. The digestibility of protein was similar in diets containing SBM (65.6%) and *Porphyra spp.* (64.2%), and higher than the diet containing *S. latissima* (57.6%), which was higher than that of control (54.5%). The protein value of the diets will further be evaluated by assessing the amino acid composition in the blood plasma.

Effects of GOS delivered *in ovo* on performance and microbiota in chickens under heat stress

A. Slawinska[1,2], M. Zampiga[3], F. Sirri[3], A. De Cesare[3], G. Manfreda[3], S. Tavaniello[1] and G. Maiorano[1]
[1]University of Molise, via de Sanctis 1, 86100 Campobasso, Italy, [2]UTP University of Science and Technology, Mazowiecka 28, 85-084 Bydgoszcz, Poland, [3]University of Bologna, via del Florio 2, 40064 Ozzano dell'Emilia, Italy; slawinska@utp.edu.pl

Poultry production has been focused on highly productive chicken breeds raised in controlled environments. In meat-type chickens, the desired performance efficiency comes from rapid growth and conversion of nutrients into the body mass. Trade-off of such intensified production is a narrow margin for the animals to deal with sup-optimal conditions. Heat provides particularly difficult challenge in meat-type chickens, due to their underdeveloped circulatory system and lack of efficient thermoregulatory mechanisms. The primary effect of heat is dysbiosis of intestinal microflora and leaky guts leading to influx of pathogens and their toxins from the gut luminal content into the body. Results are: (1) systemic inflammation and immune response and (2) inhibited feed intake to reduce metabolic heat. Our approach to mitigate harmful effects of heat stress in poultry is to stimulate their microbiome in perinatal period with *in ovo* delivered prebiotic (GOS, Galactooligosaccharides). To test our hypothesis, two trials were conducted, using fast-growing and slow-growing meat-type chickens. In both trials, *in ovo* stimulation of the embryonic microflora was performed by injection of GOS (Clasado Biosciences Ltd., Jersey, UK) into the air chamber on day 12 of eggs incubation. The animals were reared until reaching commercial slaughter weight. Heat was applied at the end of rearing (30 °C, 10-14 days). Performance (growth, feed conversion ratio, mortality) and meat quality traits were evaluated. Caecal microbiota composition was determined using 16s rDNA sequencing (Illumina). In this paper we show patterns of chicken growth depending on the genotype and environment. We also demonstrate beneficial impact of GOS on mitigating losses attributed to heat stress. As such, we present a strategy to adjust poultry production to challenging environmental conditions resulting from climate change. Acknowledgements: OVOBIOTIC grant (RBSI14WZCL, MUIR, Rome, Italy).

Feed efficiency components at fattening in rabbit lines selected for different objectives

M. Pascual[1], M. Piles[1], J.J. Pascual[2], L. Rodenas[2], M. Velasco[1], W. Herrera[1], O. Rafel[1], D. Savietto[3] and J.P. Sanchez[1]
[1]Instituto de Investigación y Tecnología Agroalimentarias (IRTA), Programa de Mejora y Genética Animal, Caldes de Montbui, 08140, Spain, [2]Universitat Politècnica de València (UPV), Deparment of Animal Science, Valencia, 46022, Spain, [3]Université de Toulouse, INRA, INPT, ENVT, GenPhySE, Castanet, Tolosan, France; mariam_pascual@irta.es

Feed efficiency (FE) for maintenance and for growth in two rabbit lines with different productive aptitude was studied. A total of 29 rabbits from Prat line (P; selected for litter size at weaning) and 29 rabbits from Caldes line (C; selected for growth rate at fattening), equally distributed between males (M) and females (F) were used. Kits weaned at 30 days of age were kept from 53 to 60 days of age fed *ad libitum* in individual cages. From 60 to 64 days of age (AdLib), animals were also fed *ad libitum*. Maintenance requirements were calculated for each line as 430 KJ kg-0.75 day-1 multiplied by its mean metabolic weight. At day 64, animals from both lines were distributed in groups R90, R80 or R70, feeding at 90, 80 or 70% maintenance requirements, respectively, from 64 to 67 days of age (Re). For each line and sex individual daily gain and individual daily feed intake were related fitting a linear spline regression with a single knot set at (DFIm,0), being DFIm the daily feed intake when animals do not modify live weight. The knot was set at the point yielding the lowest residual variance after a grid search. DFIm was higher in C (110.3 and 109.9 g/day in M and F, respectively) than in P line (82.8 and 77.5 g/day in M and F). FE for growth (slopes at AdLib) was higher in C (0.51 and 0.63 g/g in M and F) than in P line (0.45 and 0.47 g/g in M and F). During Re FE to cover maintenance requirements was higher in F than M, especially in P line (1.46 and 2.77 g/g in M and F for P line vs 1.40 and 1.63 g/g in M and F for C line). Results indicate that growing rabbits of line C have higher maintenance requirements than those from line P, however the formers are more efficient for growth. Moreover, F from line P are more efficient to cover their maintenance requirements than M from the same line or both M and F from line C. These differences could be consequence of the different selection processes conducted in each line.

Nutritional valorization of ginger lily (*Hedychium gardnerianum*) by treatment with urea

C.S.A.R. Maduro Dias, C.F.M. Vouzela, J.S. Madruga and A.E.S. Borba
University of the Azores, Institute of Agricultural and Environmental Research and Technology (IITAA), Rua Capitão João d'Ávila, 9700-042 Angra do Heroísmo, Açores, Portugal; cristianarodrigues@gmail.com

The ginger lily (*Hedychium gardnerianum,* Sheppard ex Ker-Gawl) is an invading herb of the Zingiberaceae family, which has been used in the Azores (Portugal) as unconventional forage in cattle feed in seasons with shortage of grass. Due to its dispersion and quantity, it is a plant that traditionally appears in wood-pastures, which play an important part in the Azores' traditional animal production systems, characterized by the usage of pasture all round of the year. The ginger lily has a low nutritive value, reason why this work intends to study its nutritive enhancement through treatments with urea. After drying, the ginger lily was sprinkled with a 5% dry matter concentration urea solution and placed in a leak-proof container for 4 weeks. The determination of *in vitro* digestibility and chemical composition was performed, in triplicate, over treated and untreated (control) ginger lily samples. The data was analysed according to the t test for independent samples, with differences being considered meaningful whenever P<0.05. The obtained results indicate that the treatment with urea leads to a significant increase (P<0.05) of the crude protein levels (from 6.29 to 14.64 DM%), a decrease of the ADL fraction, from 10.15 to 8.52 DM% and a significant increase (P<0.05) of dry matter digestibility from 38.6 to 43.92% and of organic matter digestibility from 29.41 to 34.78%. We can thus conclude that the treatment with urea, in a 5% DM concentration, may be an effective way of increasing the ginger lily's nutritive value.

Valorisation of Macaronesia invasive vegetal species

C.S.A.R. Maduro Dias, C.F.M. Vouzela, H.J.D. Rosa, J.S. Madruga and A.E.S. Borba
Univeristy of the Azores, Institute of Agricultural and Environmental Research and Technology (IITAA), ECOFIBRAS (MAC/4.6D/040, Rua Capitão João d'Ávila, 9700-042 Angra do Heroísmo, Açores, Portugal; alfredo.es.borba@uac.pt

The environmentally-sustainable use of Macaronesia islands invasive plants may involve the exploration of their natural fibres and residues in animal feed or, if they cannot be used to this end, in composting, e.g. because they contain toxic substances or have low digestibility value. The studied invasive plants were the *Arundo donax* (giant cane), *Pennisetum setaceum* (crimson foutaingrass), *Agave americana* (sentry plant), and *Ricinus communis* (castor-oil-plant), all propagated without control in the three archipelagos (Canary, Azores & Madeira) and which figure in IUCN's TOP100 most dangerous invasive plants. During the first phase of this study, these plants were characterized chemically and biologically. It was observed that *A. donax* and *P. setaceum* show elevated crude protein (CP) values, 12.84 and 16.33 DM%, respectively, and extremely high NDF values, of 75.31 and 80.88 DM%, respectively, which leads to a DM digestibility of 54.99%. The *A. americana*, although with low NDF value (22.78 and 25.95 DM% for Terceira Island and Santa Maria, respectively), presents a very low CP value (4.24 and 5.63% for Terceira and Santa Maria, respectively), lower than the 7% which is usually considered the minimum required value for the normal functioning of microorganisms in rumen. However, its dry matter digestibility is high (86.14 and 78.88 DM% for Terceira and Santa Maria, respectively). The *R. communis* is, of all the studied samples, the one that shows the best values, 24.01 DM% for CP and 25.75 DM% for NDF, however, due to the presence of toxic substances in this plant, it cannot be used in animal feed. As a strategy to up the value of these plants, we propose that a treatment with urea or NaOH is applied to *A. donax* and *P. setaceum*, and enrichment with nitrogen to Agave, for use in animal feed. Due to its toxic properties, *R. communis* must only be used in composting.

Session 69 Poster 14

Effect of broccoli by-product and artichoke plant on blood metabolites and urea of goat's milk

P. Monllor Guerra, G. Romero Moraleda, A. Roca Gumbau, R. Muelas Domingo and J.R. Díaz Sánchez
Universidad Miguel Hernández de Elche, Tecnología Agroalimentaria, Carretera de Beniel km 3.2, 03312 Orihuela (Alicante), Spain; jr.diaz@umh.es

The aim of this experiment was to study the effect of including two ensiled feedstuffs (broccoli by-product and artichoke plant) in three different levels (25%-40%-60%, on dry matter basis) in Murciano-Granadina goats diets on health status during short term. A group of 63 goats were selected and distributed into 7 groups (6 treatments and 1 control) with similar characteristics of milk production, somatic cells count, body weight, state of lactation and parity. All the diets were iso-energetic and iso-proteic. During the pre-experimental period all the animals were fed with a control diet. Then, each group of goats was fed with one of the seven diets and, after two adaptation weeks, three samplings were carried out weekly. In each sampling, individual udder milk samples were obtained in order to analyse milk urea content by near-infrared spectroscopy (MilkoScanTM FT2, Foss). Blood samples were obtained to analyse serum glucose, BHB (β-hidroxybutirate) and urea by enzymatic spectrophotometry. Variables were analysed by a linear mixed model, considering type of diet, sampling and their interaction as fixed effects. The animal was considered as a random effect. The inclusion of broccoli by-product and artichoke plant silages did not affect serum BHB and glucose levels. However, a decrease in serum urea levels of groups fed with broccoli by-product was observed. Finally, any relevant difference was observed in milk urea content between groups. Therefore, the use of these ensiled horticultural by-products in dairy goats feeding is not harmful to their health and, with a suitable formulation of the diet, an appropiate energy/protein balance can be achieved.

Session 69 Poster 15

Grape pomace as a possible bioactive compounds source for animal nutrition: a preliminary result

B. Gálik, R. Kolláthová, D. Bíro, M. Juráček, M. Šimko, M. Rolinec, O. Hanušovský, Ľ. Balušíková, P. Vašeková and S. Barantal
Slovak University of Agriculture, Department of Animal Nutrition, Tr. A. Hlinku 2, 94976 Nitra, Slovak Republic; branislav.galik@uniag.sk

Grape by-products are sources of many potential bioactive compounds, which have a potential to be positive in feed ration for nutrients utilisation increasing point of view. The aim of the study was to determine the nutritive quality and digestibility of grape pomace (*Vitis vinifera* L.), variety *Pinot Gris* as a possible feeding source for animal nutrition. Experiment was realized in cooperation with University Experimental Farm in Kolinany. Chemical analysis and *in vitro* digestibility were determined in the Laboratory of Quality and Nutritive Value of Feeds (Faculty of Agrobiology and Food Resources, Slovak University of Agriculture in Nitra). Standard laboratory procedures and principles were used. *In vitro* organic matter digestibility was analysed via PEPCEL methodology. Grape pomace samples were taken in average dry matter content 39%. For samples, average crude protein content at the level 102.54 g/kg of dry matter was typical. The average content of crude fibre was 164.33 g/kg of dry matter (ADV: 268.38 g/kg of dry matter, NDV: 305.48 g/kg of dry matter). In samples of grape pomace content of nitrogen free extract at the level 691.39 g/kg of dry matter was found in average. In samples, lignin content 236.9 g/kg of dry matter in average was detected. *In vitro* organic matter digestibility of analysed sampled ranged from 38.08 to 39.84%. However, additional research is needed. Feeding trial with different dosage of grape pomace in feed rations will be analysed with the effect on biochemical indicators and nutrients digestibility. This work was supported by the Slovak Research and Development Agency under the contract No. APVV-16-0170.

Session 69 Poster 16

The addition of non-fibre carbohydrates with different rumen degradation rates in high forage diets

A. Foskolos[1], M. Simoni[2], S. Pierotti[2], A. Quarantelli[2], F. Righi[2] and J.M. Moorby[1]
[1]Aberystwyth University, IBERS, Campus Gogerddan, SY23 3EE, United Kingdom, [2]Parma University, Veterinary Science, Parma, 43126, Italy; anf20@aber.ac.uk

It is important to synchronize the availability of energy and protein in the rumen to increase nitrogen (N) use efficiency in ruminant animals. This is especially so in high-forage diets where the main source of energy is slowly degradable neutral detergent fibre and a considerable amount of N is located in the quickly degradable pool. Two studies were conducted to investigate the effect of the addition of non-fibre carbohydrates (NFC) with different degradation rates on rumen fermentation. Treatments were: (1) 100% (on a DM basis) grass silage (GRS), or the substitution of 20% DM of it with (2) corn (CORN), (3) physically processed corn (OZ), and (4) sugar (SUG; sucrose). In study 1, the gas production technique was used, recording gas volumes and composition (CO_2 and CH_4) at 4, 8, 12, 21, 36, 48, 72, and 96 hours of incubation. At 96 hours, the *in vitro* dry matter (DM) digestibility (IVDMD) was calculated. In study 2, the rumen simulation technique (RUSITEC, with 16 vessels) was used. Rumen fluid from 5 and 4 cannulated cows was used in studies 1 and 2, respectively, and treated independently to prepare inoculants for gas production flasks (2 replicates per cow and treatment) and RUSITEC vessels (4 per treatment, one from each animal). In study 1, the addition of NFC did not affect cumulative gas production or its composition, but it increased IVDMD compared with GRS. Similarly, in study 2 the addition of NFC did not affect total gas production or composition, with the exception of CORN that increased both CO_2 and CH_4 concentrations. Concentrations of total volatile fatty acids, acetate, propionate and butyrate of the 24-hour outflow were not different among treatments. However, outflow of NH_3-N was significantly lower for all NFCs with the lowest being that of SUG followed by OZ and CORN. Moreover, 48-hour DM digestibility averaged 42% and was higher for all NFCs compared with GRS. In conclusion, the addition of NFC in high forage diets may improve N utilization without impairing rumen fermentation. Acknowledgement: support was provided through the Sêr Cymru NRN-LCEE project Cleaner Cows.

Session 69 Poster 17

High-fibre diet for Iberian-Duroc crossbred pigs: effects on some meat quality traits

M.A. Fernández-Barroso[1], J.M. García-Casco[1], M. Muñoz[1], A. López-García[1], C. Caraballo[1] and E. González-Sánchez[2]
[1]Centro de I+D en Cerdo Ibérico, INIA, Dpto. Mejora Genética Animal, Crta EX-101, 06300 Zafra (Badajoz), Spain, [2]Escuela de Ingenierías Agrarias, Universidad de Extremadura, Instituto Universitario de Investigación de Recursos Agrícolas (INURA), Avenida Adolfo Suárez, 06007 Badajoz, Spain; fernandez.miguel@inia.es

Last years, intensive Iberian pig production based on crossbred Duroc × Iberian is significantly increasing, with a corresponding higher demand on raw material in order to feed animals. Environmental consequences are also related with this demand; therefore, sustainable strategies are required to minimize the negative effects. Local agro-industrial by-products could be an alternative for sustainable Iberian pig production but it is necessary to analyse their possible effects on growth, carcass composition and meat quality. In this work, we deepen in our previous results, analysing the effect of a high fibre diet on Iberian × Duroc crossbred pigs towards other three meat quality traits: water-holding capacity measured by centrifuge force losses (CFL, %), myoglobin concentration (Mb, mg/g) and collagen content (Col, %). A high-fibre olive cake-based diet group (AF) supplied during growing period (up to 95 kg) was compared with a control standard diet group (C). Fattening diet (up to 160 kg) was the same for both groups. The individuals were slaughtered at two weights (95 and 160 kg) in two trials developed in different farms in two consecutive years. First, Pearson correlations coefficients between traits measured on previous experiments (shear force, thaw and cooking losses and Minolta colour) and CFL or Mb were obtained. An ANOVA for each trial was also performed with a linear model using growing diet and slaughter weight as fixed effects. In the first trial, no significant differences at 90 or 160 kg were observed for any trait between diets. In the second trial, higher values of Mb and Col were observed for AF diet (1.26 vs 1.32 and 0.91 vs 1.12, respectively). The differential effect of the year on these traits could be possibly due to sampling differences or environmental variations. The whole results support our previous findings about the non-influence of a high-fibre diet based on olive by-product during the growing period in the overall meat quality.

Session 69 Poster 18

Effect of urea addition on fermentation quality and dry matter losses of grape pomace silage

M. Juráček, D. Bíro, M. Šimko, B. Gálik, M. Rolinec, O. Hanušovský, P. Vašeková, R. Kolláthová, Ľ. Balušíková and S. Barantal
Slovak University of Agriculture in Nitra, Department of Animal Nutrition, Tr. A. Hlinku 2, 94976 NItra, Slovak Republic; miroslav.juracek@uniag.sk

The aim of the work was to determine the influence of added urea on fermentation quality and the losses of dry matter of grape pomace silage. Grape pomace (*Vitis vinifera* L.) *Pinot Gris* variety in mini silo bags in laboratory conditions was ensilaged. Grape pomace with dry matter content 45% in 2 variants was ensilaged: without silage additives (control) and with addition of urea (2 kg/t). Mini silo bags for 5 weeks in climatized laboratory (t 22±1 °C) were hermetically sealed and stored. The samples of silage from each variant (n=3) were taken for laboratory analysis after opening the mini silo bags. Dry matter content gravimetrically by drying of sample (103±2 °C) was determined. Content of fermentation acids on analyser EA 100 (Villa Labeco, SR) using the ionic electrophoresis method was detected. Active acidity by electrometric method was determined. Statistical parameters using SPSS Statistics 20.0 (IBM) (ANOVA-Tukey test) were evaluated. In the laboratory mini silos dry matter losses during fermentation were significantly lower ($P<0.05$) for the urea silage (11.5%) and higher for the control silage (19.1%). Application of urea at 0.2% of original matter increased the lactic acid content. The content of acetic acid in silage with urea in comparison with silage without additive was lower. There was no significant difference between variants in average content of lactic and acetic content. Better ratio of lactic acid to acetic acid in silage with urea was determined (6.2:1 vs 4.9:1). Undesirable butyric acid was not found in any samples. Grape pomace silage with urea had significantly ($P<0.05$) higher value of pH. The addition of 2% urea at the time of ensiling has proven beneficial to reduce dry matter losses during the fermentation process of the grape pomace silage. The content of lactic and acetic acid by the application of urea were not influenced. The addition of urea significantly increased value of pH in grape pomace silage. This work was supported by the Slovak Research and Development Agency under the contract No. APVV-16-0170.

Session 69 Poster 19

Study on sorghum diet influence on fatty acids concentration in the meat of the fattening steers

I. Voicu, A. Cismileanu and D. Voicu
National Research Development Institute for Animal Biology and Nutrition (IBNA), 1 Calea Bucuresti, Balotesti, Ilfov, 077015, Romania; voicu.ilie@yahoo.com

Sorghum grains can be a useful feed for ruminants as Romania, like other countries in S-E Europe, is confronted with drought more often nowadays. We aimed to study the effect of feed sorghum grains on fatty acids (FA) composition in beef (muscle and the liver tissue). The study was conducted for 12 weeks on 21 Romanian Black Spotted fattening steers (256 kg average body weight) assigned uniformly to 3 groups of 7 steers each. The control group (C) was fed a compound feed without sorghum grains (only corn, barley, wheat and sunflower meal), group E1 had 15% sorghum grains in the compound feed by replacing corn, while group E2 received 25% sorghum grains in the mixture. All the groups received the same alfalfa haylage as forage. The diets has similar energetic and proteic characteristics (7.04 meat feed units, mFU/day and 576 g intestinally digestible protein, IDP/day). Sorghum grains had not a different FA proportion (% FAME) vs corn grains, only minor differences like palmitoleic acid (0.59 vs 0.30), oleic acid (34.92 vs 29.40) and linoleic acid (46.19 vs 49.90). But the rations of the groups C, E1 and E2 had no significant difference for SFA (21.41, 19.35 and 21.86, respectively) or for UFA (78.59, 80.65 and 78.14, respectively). For FA in the muscle (*Longissimus dorsi*) we noticed increased values for stearic acid in E2 (17.07 vs 14.96 in C group), but not statistically significant, so the SFA were higher in E2 (51.95) vs C (46.10), but also not statistically significant. UFA were lower only in E2 (46.64) vs C (53.00) due to the important contributions of oleic and alfa-linoleic acids, but with no statistical significance. The FA from E1 muscle were similar with those in C muscle (SFA 44.43 and UFA 54.19). For liver tissue there were no differences for FA profile between groups: the SFA were 43.68 in C samples, 43.75 in E1 and 41.69 in E2; the UFA were 55.68 for C, 55.88 for E1 and 57.78 for E2. The conclusion is that FA profile of the beef (muscle and liver) is not influenced by replacing corn with sorghum up to 25% of the compound diet of the animals.

The effect of forage to grain ratio on *in vitro* methane production from wheat vs corn

P.A. Alvarez Hess[1], P.J. Moate[2], S.R.O. Williams[2], J.L. Jacobs[2], M.C. Hannah[2] and R.J. Eckard[1]
[1]The University of Melbourne, 161 Barry Street, Parkville, 3010 Victoria, Australia, [2]Agriculture Victoria, Department of Economic Development, Jobs, Transport and Resources, 1301 Hazeldean Road, Ellinbank, 3821 Victoria, Australia; pabloah@student.unimelb.edu.au

In ruminant diets, the inclusion of greater proportions of grain have been reported to reduce methane, but not all grains are equally effective and there are indications that the proportion of grain in the diet may have a non-linear effect on methane production. The aim of the study reported here was to study the *in vitro* methane mitigating effect of different proportions of wheat and corn in the substrate. It was hypothesized that a greater proportion of wheat or corn in the incubation substrate fermented *in vitro* will reduce methane production (MPR) in the fermentation gas. This *in vitro* study included ten treatments, five rates of corn (0, 250, 500, 750 and 1000 mg/g DM) and five rates of wheat (0, 250, 500, 750 and 1000 mg/g DM). The *in vitro* incubation was conducted using the Ankom GP system over 24 hours. Alfalfa hay was used as base forage and was substituted with increasing rates of grain. Treatments that contained wheat were incubated in ruminal fluid from donor cows fed wheat and treatments that contained corn were incubated in ruminal fluid from donor cows fed corn. Ruminal fluid was collected from six cows, three fed wheat and three fed corn. The experiment was repeated in two *in vitro* runs. Each run involved 4 replicates of each grain at each rate. Data were analysed by Anova with a 2×2, factorial treatment structure of grain type by proportion of grain, with polynomial contrasts for that proportion of grain embedded in the ANOVA, and blocking structure of run (1-2) and fermentation bottle (1-4) within run. Greater proportions of wheat had a negative linear effect ($P<0.001$) on MPR, but there was no effect ($P=0.15$) of greater proportions of corn on MPR. Methane production was less ($P<0.001$) in treatments that used wheat as substrate than in treatments that used corn as substrate. It is concluded that a greater proportion of wheat, but not corn, in the substrate, reduces *in vitro* MPR.

Grazing systems production with forestry on Brazilian beef cattle productivity

R.R.S. Corte[1], S.L. Silva[1], P. Méo Filho[1], A.F. Pedroso[2], S.N. Esteves[2], A. Berndt[2], P.H.M. Rodriguez[1] and P.P.A. Oliveira[2]
[1]Universidade de Sao Paulo, Duque de Caxias Norte, 225, 13635-900, Brazil, [2]Embrapa Pecuária Sudeste, Rodovia Washington Luiz, km 234 s/no, Fazenda Canchim, 13560-970, Brazil; rosanaruegger@gmail.com

Sustainable options of intensification and integration of beef cattle production systems increase the productivity, diversifies financial income and reduce the pressure on forests in Brazil. To assess the productivity of different grazing system production scenario, a total of 30 Canchim steers (284.8±6.00 kg of LBW; 15 months old) were allotted to five grazing systems with two replicates each (blocks) during one year: (1) Extensive (EXT) – continuous grazing system; (2) Integrated Crop-livestock System (CL) – rotational grazing system with crop rotation in each paddock in four year cycles (three years with pasture and one year with corn); (3) Integrated Crop-livestock-forestry System (CLF) – the same as CL with eucalyptus trees (15×2 m spacing); (4) Intensive (INT) – dryland rotational grazing system; (5) Integrated Silvopastoral System (SP) – rotational grazing with eucalyptus trees (15×2 m spacing). With exception of the extensive system, all systems were limed, fertilized and manage in a rotational grazing system. Data were analysed as completely randomized block design using PROC MIXED. Animals were slaughtered with a minimum of 450 kg LW. The INT and IPF systems presented greater stocking rate ($P=0.001$) (INT=2.3, SP=2.4 AU/ha) than those grazed the EXT and CL areas (EXT=1.2, CL=1.5 AU/ha). The CLF area presented similar stocking rate (1.9 AU/ha) to other systems. The INT system presented greater productivity in kg/ha/year on live BW ($P=0.045$) (INT =515.9, IPF=446.7, CL=439.4, CLF=393.5, EXT=245.4), carcass ($P=0.038$) (INT =280.7, CL=236.9, IPF=230.7 CLF=203.4, EXT=129.8) and carcass edible portion ($P=0.045$) (INT =199.3, IPF=159.3, CL=170.0, CLF=142.1, EXT=92.9) than EXT area, with no difference for other systems (CL, CLF and SP).The forestry inclusion in integrated systems as SP and CLF provided intermediate productivity in kg/ha/year as conventional system (INT and EXT). It is possible to reach high levels of productivity as well as environmental, social and economic sustainability with Agroecosystems.

Session 69 Poster 22

Use of Brussels sprouts in sheep diets: effects on *in vitro* gas production

T. De Evan[1], C.N. Marcos[1], A. Cevallos[1], C. Jiménez[1], M.J. Ranilla[2] and M.D. Carro[1]
[1]Universidad Politécnica de Madrid, Dpto. Producción Agraria, ETSIAAB, 28040 Madrid, Spain, [2]Departamento de Producción Animal, Universidad de León, IGM ULE-CSIC, Campus de Vegazana, s/n, 24071 León, Spain; mjrang@unileon.es

The use of agroindustrial by-products in animal feeding could contribute to increase the profitability of livestock farms and to reduce environmental pollution, but there is frequently a lack of information on their nutritional value. The objective of this study was to investigate the effects of including increasing amounts of Brussels sprouts (BS; *Brassica oleracea* var. gemmifera) in the concentrate of a dairy diet on *in vitro* gas production kinetics. Brussels sprouts samples were obtained at local markets in Spain, dried at 45 °C and ground to 1 mm. The control diet contained 40:60 alfalfa hay:concentrate, and the second was composed of corn, barley, wheat, soybean meal, wheat bran, calcium soap, calcium phosphate and vitamin-mineral premix in proportions of 32, 30, 15, 14, 7.0, 1.0, 0.50 and 0.50 g per 100 g of concentrate (fresh matter basis). Three experimental diets were formulated to include dried BS at levels of 8 (8BS), 16 (16BS) and 24% (24BS) of the concentrate by partially replacing wheat, corn and soybean meal. All diets were isonitrogenous (16.1% crude protein) and isofibrous (31.5% neutral detergent fibre). Four *in vitro* incubations were carried out with rumen fluid from sheep as inoculum and gas production was measured at different times between 2 and 120 h of incubation. Increasing levels of BS in the concentrate linearly increased the asymptotic gas production (P=0.005; 282, 292, 298 and 296 ml/g DM for control, 8BS, 16BS and 24BS diets, respectively) and the average gas production rate (P=0.015; 7.3, 7.8, 8.2 and 8.0 ml/h, respectively), which was defined as the average gas production rate between the start of the incubation and the time when half of the asymptotic gas volume was produced. In contrast, there were no differences among diets either in the fractional rate of gas production (P=0.320) or the initial delay in the onset of gas production (lag time; P=0.387). These results indicate that BS could be included in the concentrate of dairy sheep diets up to 24% without any adverse effects on *in vitro* rumen fermentation. Funded by AGL2016-75322-C2-1-R and C2-2-R.

Session 69 Poster 23

Effects of level and source of dietary protein on intake, milk yield and nitrogen use in dairy cows

D. Kand[1], N. Selje-Aßmann[1], R. Von Schmettow[2], J. Castro-Montoya[1] and U. Dickhöfer[1]
[1]Animal Nutrition and Rangeland Management in the Tropics and Subtropics, University of Hohenheim, 70593 Stuttgart, Germany, [2]Agricultural Experimental Station, University of Hohenheim, 70599 Stuttgart, Germany; kand.deepashree@uni-hohenheim.de

Lowering rumen-degradable protein intake may effectively increase nitrogen (N) use efficiency (i.e. g milk N/g N intake) and thereby reduce N emissions in dairy cattle; however, kinetics of feed N supply to rumen microbes and efficiency of microbial protein synthesis vary with diets. Hence, aim was to study the effects of rumen nitrogen balance (RNB), dietary protein source and their interaction on feed intake, milk yield and N use efficiency in dairy cows. Four diets were tested in 24 lactating Holstein-Friesian cows during 4 periods (12 d adaptation, 8 d sampling) in a complete Latin square design. Diets were fed *ad libitum* and comprised of grass silage, maize silage, grass hay and barley straw as forages and four different concentrate mixtures at a constant forage:concentrate ratio of 55:45 (on dry matter (DM) basis). The concentrate mixtures mainly contained barley grain, molasses, feed sugar and either faba beans (FB) or SoyPass (SP) as two main protein sources (>35% of total dietary crude protein). Composition of concentrate mixtures were adjusted to create diets with RNB of 0 g N/kg DM (i.e. FB0, SP0) or -3.2 g N/kg DM (i.e. FB-, SP-). Daily DM intake and milk yield were measured and samples analysed for their chemical composition. Data were analysed by mixed-model analysis with RNB, protein source, their interaction and period as fixed effects and animal as random factor. Mean DM intake (P<0.01) and fat-corrected milk yield (P<0.01) were lower for FB- than FB0, but similar for SP diets (P≥0.44). Concentrations of milk fat (P<0.01) and protein (P=0.08) were higher in SP0 than SP-. Milk urea concentrations were lowest (P<0.01) and N use efficiencies highest (P<0.01) in FB- and SP-, with greater differences between RNB levels for FB diets. In conclusion, reducing dietary RNB can improve N use efficiency and thus lower the risk of N emissions from dairy farming. Possible negative effects of low RNB on feed intake and cow performance appear to be more pronounced in diets with rapidly degradable protein sources.

Comparative efficiency of corn silage and sorghum silage given to fattening steers

A. Sava and I. Voicu
National Research Development Institute for Animal Biology and Nutrition (IBNA), 1 Calea Bucuresti, Balotesti, Ilfov, 077015, Romania; voicu.ilie@yahoo.com

The water shortage due to global warming made the draught-adapted forages, such as sorghum, to gain land. The purpose of our study was to compare the feeding efficiency of corn silage (R1) and sorghum silage (R2) on two groups of 8 Romanian Spotted steers each, fattened for 115 days, from 275 to 430 kg. The diet formulations included corn silage or sorghum silage and a concentrate feed (barley and sunflower meal) which provided 7.83 kg DM, 7.78 milkFU and 640 g IDP. The average daily weight gain was 1,307±133 g for R1 and 1,242±74.5 g for R2, with feed conversion ratio of 6.00±0.51 kg DM, 6.08±0.49 milkFU and 475±39.15 g IDP for R1, and 6.28±0.09 kg DM 6.23±0.09 milkFU and 511±7.75 g IDP for R2, with no statistically significant differences between the groups. The nitrogen balance, performed in digestibility stands, showed average values of ingested nitrogen of 170.72 g in R1 and 186.7 g in R2, with no statistically significant differences between the groups. However, the proportion of digested nitrogen was 66±3.37% in R1 and 68±1.25% in R2, significantly different (P=0.004). Also, significant differences (P=0.001) were noticed in the proportion of nitrogen eliminated through urine, with 49.00±3.33% in R1 and 53.67±9.2% in R2, but no differences were noticed between the proportion of retained nitrogen: 17.00±0.13% in R1 and 14.4±0.02% in R2 (P=0.072). This shows that the sorghum silage can replace completely the corn silage for this category of cattle.

Response of weaning piglets to dietary addition of sorghum associated with peas: linseed mixed (3:1)

M. Habeanu, G. Ciurescu, C. Tabuc, N.A. Lefter, A. Gheorghe, M. Dumitru, T. Mihalcea and E. Ghita
INCDBNA Balotesti, Calea Bucuresti 1, Ilfov, 077015, Romania; mihaela.habeanu@ibna.ro

The aim of this work was to investigate whether the addition of 20% sorghum var. Albanus in the diet fed to weaning piglets would alter faeces microbiota, enteritis and plasma biochemical profile. Moreover, we also assessed the influence of the dietary sorghum associated with a mix of peas var. Tudor and linseed var. Lirina (3:1) on faeces bacterial composition and enteritis frequency. Thirty piglets weaned at 28d were randomly assigned for 21d to 3 groups: control 1 (C) based on classical diet (corn and soybean meal), experimental 1 (ES) diet based on partial substitution of corn by sorghum, and experimental 2 (PLS) group fed with ES diet with additional inclusion of peas:linseed (8%). The faeces samples were collected on 1d, 7d, 14d and 21d to assess faeces microbiota changes (*E. coli, Lactobacillus, Enterobacteriaceae*). We used a scoring system of 1 to 3 in order to determine the severity of diarrhoea. The blood samples were collected aseptically at 21d to determine the biochemical profile using Analyser BS-130. The bacteria faeces content, with values express as lg of cfu/g of sample, was similar between groups (P>0.05). Irrespective the diet, the faeces sampling day influenced significantly the bacteria composition: while *E. coli* values were close on 1std compared to 7d-21d, *Lactobacillus* was 1.25 times higher on 14d vs 7d and 1.65 times higher on 7d vs 21d; *Enterobacteriaceae* were 1.02 times higher on 21d vs 7d. Enteritis frequency was significantly higher in the 1st week in group C compared to ES and PLS groups (5.35% C group, 4.08% ES group and 2.04% PLS group). The faecal scores tended to be influenced by diet (2.25 C diet, 2.0 ES diet and 2.21 PLS diet, P=0.058). The plasma lipid, protein, minerals and enzyme indicators, were within the normal limits (P>0.05). In conclusion, the results obtained by the dietary substitution of 14% corn by sorghum suggested a positive influence in terms of health status. Furthermore, using a mix of two feedstuff rich in protein (peas) and oil (linseed) in the sorghum diet, did not alter significantly the health indicators. The sampling day had a significant influence on faeces microbiota.

Session 69 Poster 26

Effect of concentrate feeding level on methane emissions and performance of grazing Jersey cows

J.D.V. Van Wyngaard[1], R. Meeske[2] and L.J. Erasmus[1]

[1]University of Pretoria, Dept. Animal and Wildlife Sciences, Lynnwood Road, 0001 Pretoria, South Africa, [2]Outeniqua Research Farm, Dept. Agriculture, Western Cape, Airport road, 6530 George, South Africa; lourens.erasmus@up.ac.za

Concentrate supplementation has been documented as an effective and practical enteric methane (CH_4) mitigation strategy. However, limited studies evaluated the effect of concentrate feeding level on enteric CH_4 emissions from grazing dairy cows, and none of these studies included a pasture-only diet or reported on rumen fermentation measures. Sixty multiparous Jersey cows (six rumen-cannulated) were subjected to a randomised block design, with the cannulated cows subjected to a replicated 3×3 Latin square design, to evaluate the effect of concentrate supplementation (0, 4, and 8 kg/cow per day; as fed) on enteric CH_4 emissions, milk production, dry matter intake (DMI), and rumen fermentation of dairy cows grazing perennial ryegrass pasture during spring. Experimental cows grazed as one group and were allowed 14 d dietary adaptation. Enteric CH_4 emissions from 30 intact cows were measured with the sulphur hexafluoride tracer gas technique over a single 9-d measurement period with parallel DMI measures. Pasture DMI was calculated from total faecal output and pasture digestibility using TiO_2 and indigestible neutral detergent fibre, respectively. Milk yield and composition, cow condition, and pasture measures were also recorded. Total DMI (13.4 to 18.0 kg/d), milk yield (12.9 to 19.2 kg/d), energy corrected milk (14.6 to 20.7 kg/d), milk lactose content (46.2 to 48.1 g/kg) and gross energy intake (239 to 316 MJ/d) increased, while milk fat content (50.0 to 44.2 g/kg) decreased with increasing concentrate feeding level ($P<0.05$). Minor treatment effects on ruminal pH and dry matter disappearance were observed. Methane production (258 to 302 g/d) and CH_4 yield (20.6 to 16.9 g/kg of DMI) were unaffected, while pasture DMI (13.4 to 10.8 kg/d) and CH_4 intensity (20.4 to 15.9 g/kg of milk yield) tended to decrease with increasing concentrate feeding level ($P<0.10$). Results indicate that concentrate supplementation on high quality pasture-only diets have the potential to effectively reduce CH_4 emissions per unit of milk yield.

Session 69 Poster 27

Composition and digestibility of whole plant triticale as affected by maturity stage and genotype

A. De Zutter[1], L. Douidah[2], G. Haesaert[1], S. De Campeneere[2] and J. De Boever[2]

[1]Ghent University, Department Plants and Crops, Faculty of Bioscience Engineering, Coupure Links 653, 9000 Ghent, Belgium, [2]ILVO (Flanders Research Institute for Agriculture, Fisheries and Food), Animal Sciences Unit, Scheldeweg 68, 9090 Melle, Belgium; johan.deboever@ilvo.vlaanderen.be

Maize is grown in monoculture on many dairy farms in NW Europe. Triticale for use as whole plant (WPT) may contribute to the diversification of crop rotation, resulting in a lower disease and weed pressure and higher yields. Moreover with climate changing to warmer and dryer summers, WPT offers an advantage compared with maize as it can be harvested before the summer heat. A six-year study was started to investigate the potential of WPT as alternative roughage crop for maize silage. An important objective is to study the factors affecting the digestibility of organic matter (OMd) and cell wall digestibility (NDFd). In a first field trial, 36 genotypes were harvested at 5 maturity stages and samples from these 180 WPT objects were scanned by NIRS. Based on spectral variation, 80 samples were selected to determine dry matter (DM), crude protein (CP), crude fat, crude ash, sugars, starch, NDF, ADF and ADL as well as cellulase OMd and rumen fluid NDFd. The DM content of the WPT for the 5 stages averaged 268, 285, 467, 508 and 574 g/kg, respectively. During maturing, sugar content decreased, whereas starch content increased with a clear shift from the second to the third stage. With advancing maturity there was a relative small decrease in NDF, CP and ash content. The NDFd seemed hardly affected by maturity stage with on average 53.9 and 55.4% for the first and fifth stage. On the other hand, mean OMd increased linearly with later harvesting and amounted to 63.8, 65.0, 67.9, 68.1 and 69.1%, respectively. In addition to the effect of maturity stage, there appeared a large variation among genotypes. For the last stage, OMd ranged between 67.3 and 73.2%. This variation opens perspectives to select for WPT with high nutritive value. The reference data from this first trial were used to develop NIRS calibrations in order to allow rapid feed evaluation throughout the study.

Session 69 Poster 28

The effects of fungal treatment on tannins and organic matter digestibility of invasive bush species

M.L. De La Puerta Fernandez and G.W. Griffith

Aberystwyth University, Institute of Biological, Environmental and Rural Sciences, Penglais Campus, SY23 3DA, United Kingdom; mad79@aber.ac.uk

Namibia is affected by bush encroachment on a massive scale. More than 65% of the country (~619,000 km^2) is affected by bush encroachment. This phenomenon consists of an invasion/thickening of undesired woody species with an associated suppression of the palatable grasses. It decreases biodiversity and the carrying capacity of the ecosystem. Bush biomass is composed of trees and shrubs that have leaves and twigs with high protein content that could serve as important components of ruminant rations. Although studies support the use of invasive shrub species as protein supplements, these species also contain variable amounts of different anti-nutritive factors which can cause very unpredictable effects on animals ranging from positive to negative.) A better understanding and management of these antinutritive factors is required to enable the use of this vast animal feed resources. It is against this background that this study aims to evaluate the effectiveness of solid-substrate fermentation of *Acacia mellifera* wood and twigs with a phylogenetically diverse range of fungi, and to examine the effects of these on lignin, crude protein (CP), total tannins (TT) and condensed tannins (CT) degradations and organic matter digestibility (OMD). Ash sawdust and straw were used as control and the effect on acetone extraction on fermentation patterns was also examined Preliminary data suggest that *A. mellifera* twigs and wood have >5% dry matter (DM) content of CT and TT which can have detrimental effects in animal production. Different fungi have a different radial growth rate (RGR) on different substrates and show a clear preference for certain substrates. *A. mellifera* can be a suitable substrate for fungal growth if the appropriate fungi is selected. Extraction of polyphenols positive effects fungal growth.

Session 70 Theatre 1

Production environment realized genetic merit of litter size, but the impact depends on herds

B.G. Poulsen, M.A. Henryon and B. Nielsen

SEGES, Pig Research Centre, Breeding and Genetic, Axeltorv 3, 1609 Copenhagen V, Denmark; bgp@seges.dk

In this study we tested that genetic merit for litter size in purebred pig flows out to production herds and define impact as the realized improvement in production due to purebred selection. We investigate genetic impact of nucleus selection for litter size on production sows and observed litter size at farrow of 84,246 first-parity production sows: 8,074 sows were Danish Landrace × Danish Yorkshire F1 crosses, 75,038 were rotational crosses between the two breeds, 787 were purebred Landrace, and 347 were purebred Yorkshire. Sows were housed in production herds and born in 2015-2016. The quality of the production data was low. To obtain reasonable results of such data a method for filtering herds with valid data was developed. After filtering only 173 herds of 450 herds were left for the analysis. The relation between genetic merit of litter size at day 5 in the two purebred dam lines and the litter size in production sows was estimated using a linear mixed model. The mixed model regress phenotypic litter sizes of production sows to their breeding value for litter size at day 5. The model included dam age, dam breed, sire breed, and farrowing month as fixed effects; and variation between herds as a random effect. As expected the results showed that highest litter size was estimated by F1 crossbred sows having 16.4 piglets per litter, followed by the rotational crossbred sows ranged from 15.4 to 15.9 piglets per litter. The estimated regression coefficient shows that selection for litter size in the purebred breeds significantly affected litter size in crossbred sows. However, we also estimated a significant G×E interaction across production herds. In some production herds, the impact of purebred selection was close to zero, while in other production herds the impacts were twice the expected impact. Our study shows that selection for larger litter size in purebred pig breeds increased litter size in production sows, but the impact significantly dependent on the hosting production herd.

Genetic parameters for litter quality traits of Austrian Large White and Landrace sows

C. Pfeiffer[1], B. Fuerst-Waltl[1], P. Knapp[2], A. Willam[1], C. Leeb[1] and C. Winckler[1]
[1]University of Natural Resources and Life Sciences Vienna (BOKU), Division of Livestock Sciences, Gregor-Mendel Str. 33, 1180 Vienna, Austria, [2]Schweinezucht und Besamung Oberoesterreich, Waldstr. 4, 4641 Steinhaus/Wels, Austria; christina.pfeiffer@boku.ac.at

Since decades breeding goals have exclusively focused on prolificacy. However, large litters can affect farm economics, require more labour due to an intensive birth management and control and additionally the piglets' and sows' welfare is decreased. Increased litter size may decrease individual birth weights of piglets and thus result in impaired vitality and higher mortality rates. Therefore Austrian pig breeders recently revised the breeding goals and are eager to implement a piglet vitality index into routine genetic evaluation, which could consist of litter homogeneity and piglet vitality. The aim of this study was to estimate the first genetic parameters for total born piglets (TBP), individual birth weight 24 h post partum (IBW), litter weight (LW), average birth weight (ABW), standard deviation of birth weight (SDB), number of piglets below 1 kg of birth weight (NB1), the percentage of piglets below 1 kg per litter (NB1P). Altogether, 24 Austrian breeders currently weigh each live born piglet within 24 h post partum. In total, data from 21,980 piglets of 1,479 sows (Large White, Landrace, F1) were used to estimate genetic parameters. A bivariate animal model was set up using VCE6.0. The traits were analysed fitting the fixed effects breed, litter number, year-season and farm as well as the random permanent environmental and genetic effect of the sow and the random effects of litter and service sire. Heritabilities were all significant ranging from 0.08 for NB1, 0.09 for NB1P, 0.15 for IBW, 0.17 for SDB, 0.18 for TBP, 0.18 for LW and 0.33 for ABW. Genetic correlations between TBP and LW, ABW, SDB, NB1 and NB1P were 0.40, -0.25, 0.23, 0.03 and 0.26, respectively. Genetic variance for litter quality traits in the Austrian maternal lines population is given. Results confirm the negative relationship between TBP and litter quality traits. Based on the genetic parameters, an implementation of litter quality traits into routine genetic evaluation would be possible and therefore piglet survival can be also increased through breeding.

Validation of litter quality assessment by pig breeders aiming to develop a piglet vitality index

C. Pfeiffer[1], C. Winckler[1], C. Leeb[1], B. Fuerst-Waltl[1], P. Knapp[2] and A. Willam[1]
[1]University of Natural Resources and Life Sciences (BOKU), Department of Sustainable Agricultural Systems, Gregor-Mendel Str. 33, 1180 Vienna, Austria, [2]Schweinezucht und Besamung Oberoesterreich, Waldstr. 4, 4641 Steinhaus bei Wels, Austria; christina.pfeiffer@boku.ac.at

Since decades breeding goals exclusively focused on prolificacy. However, large litters can have implications on piglets' and sows' welfare. Due to increased litter size individual birth weights of piglets may decrease and vitality is impaired. This can cause higher mortality rates. Therefore Austrian pig breeders have revised the breeding goals thus including a piglet vitality index into routine genetic evaluation. This index comprises litter homogeneity and piglet vitality with the former being in the scope of this paper. Altogether, 19 Austrian breeders are currently assessing litter homogeneity using 4 defined categories, which have been derived from crucial individual birth weights (<1 kg and >1.8 kg) of piglets. For at least one year, breeders have to assess the litter categories at approx. 24 h post partum and then weigh each live born piglet. Before data collection started, breeders were trained twice, jointly and individually on their farm. After 3 months of litter assessment, breeders got the first feedback including Cohen's kappa values for the agreement between subjective assessment and the actual birth weights. Based on 1.587 litters, Cohen's kappa ranged from 0.13 to 0.58. When analysing the training period (2 to 3 farrowing batches, depending on herd size) and subsequent farrowings separately, Cohen's kappa ranged from 0.01 to 0.53 during training and subsequently increased to 0.21 to 0.61, respectively. Regarding the agreement achieved in the training period, farmers were either trained again or received advice in the feedback sheet and on the phone, which already led to an increase of agreement. The aim is to reach Cohen's kappa values of 0.65 for reliable implementation of the assessment of litter homogeneity into routine genetic evaluation and to replace the expensive weighing of piglets. It is concluded that breeders have to be trained continuously to meet requirements for reliable data acquisition.

Nitric oxide precursor: sport nutrition leveraged to increase piglet livability

M. Van Den Bosch[1], E.F.J.M. Van Gelderen[1], D. Melchior[1], H. Van Den Brand[2] and A.A.M. Van Wesel[1]
[1]Cargill, Animal Nutrition, Veilingweg 23, 5334 LD Velddriel, the Netherlands, [2]Wageningen University & Research, Adaptation Physiology Group, P.O. Box 338, 6700 AH Wageningen, the Netherlands; moniek_van_den_bosch@cargill.com

Piglet livability can be defined as the percentage of potential viable piglets that are raised till weaning. A recent trend in sport nutrition is the use of nitric oxide boosting technologies to enhance athletic performance or to increase endurance. We investigated whether maternal dietary nitrate supplementation, leading to nitric oxide (NO) formation, would shorten the duration of farrowing and increase piglet liveability. NO formation is hypothesized to lead to a higher blood flow and therefore oxygen and nutrient flow to the foetuses (due to the vasodilative effect). This may reduce the risk of asphyxiation and thus increase vitality of piglets at birth and/or decrease stillbirth and pre-weaning mortality. A total of 350 sows were allocated to six treatments including a control lactation diet without dietary nitrate and diets containing either 0.03, 0.06, 0.09, 0.12 and 0.15% of dietary nitrate (calcium nitrate double salt). Dietary nitrate supplementation in late gestation and early lactation linearly increased piglet birth weights (P=0.04) as well as body weight at 72 h of age (P=0.00). A tendency for a quadratic effect (P=0.10) of dosage of maternal nitrate supplementation was found on pre-weaning mortality of piglets with the lowest level of mortality at a moderate dosage of nitrate supplementation. From 190 sows, additional information on piglet vitality, placental quality, umbilical cord blood gasses and farrowing were collected. The probability of a higher vitality score of piglets linearly increased with ascending dosage of maternal dietary nitrate (P=0.03). A tendency for a higher pO2 with increasing dosage of nitrate was found (P=0.10). Placenta width increased with increasing dosage of maternal dietary nitrate (P=0.02), but no effect on placenta length and redness was found (P>0.10). Farrowing duration and birth interval were not affected by maternal dietary nitrate supplementation (P>0.10). In summary, NO boosting technology is a promising solution to enhance piglet livability.

Management of large litters using milk supplementation – preliminary results

C.K. Thorsen[1], M.L.V. Larsen[1], V.A. Moustsen[2], P.K. Theil[1] and L.J. Pedersen[1]
[1]Aarhus University, Dept. of Animal Science, Blichers Allé 20, 8830 Tjele, Denmark, [2]SEGES, Agro Food Park 15, 8200 Aarhus N, Denmark; cecilie.thorsen@anis.au.dk

The use of high prolific sows demand a strategy to foster surplus piglets. Using nurse sows may limit survival and growth and compromise welfare. An alternative strategy is to provide milk supplementation in milk cups. The aim of this study was to investigate growth and mortality rate of piglets in large litters with and without milk supplementation. The study was a 2×2×2 factorial design with milk supplementation (+/-), litter size day 1 (14 or 17 piglets) and sow housing (crate/loose) as factors. The study comprised of 98 litters over three batches. Piglets were weighed each week until weaning on d 28 post partum (pp). Behavioural observations of use of milk cups were made for 12 h on d 7 pp and used to divide piglets into categories according to milk use: LOW (≤2 milk cup use/12 h), MEDIUM (3-5 use/12 h), HIGH (≥6use/12 h)). The effect of treatments and their interactions on average daily weight gain to d 7 (ADG7) and 28 (ADG28), and the effect of milk use and birth weight on individual piglets' ADG7 and ADG28 were analysed in mixed models. Preliminary results showed a higher ADG7 and ADG28 in litters with 14 piglets compared to 17 (P<0.001). ADG7 and ADG28 increased with increasing birth weight (P<0.001) and was higher for piglets with MEDIUM/HIGH milk cup use than for LOW (P<0.001). The odds of a piglet being a LOW milk cup user compared to MEDIUM/HIGH was tested in a logistic regression model and increased with increasing birth weight (P<0.01). Further, the effect of treatments and their interactions on litter mortality and of birth weight on individual piglets' risk of dying were tested in logistic regression models. The mortality rate was higher with loose housing (P<0.01). Milk supplementation reduced mortality rate in litters with 17 piglets to similar level as litters with 14 piglets without milk (P<0.01). The risk of dying increased with decreasing birth weight (P<0.01). This preliminary data suggest that milk supplement can lower litter mortality in large litters. It also improves weight gain of those piglets that learn to use the milk cup (mainly piglet with larger birth weight).

Session 70 Theatre 6

The influence of individual birth weight on fattening performance in pigs

K. Nilsson, P. Wahlgren and N. Lundeheim
Swedish University of Agricultural Sciences, Department of Animal Breeding and Genetics, P.O. Box 7023, 75007 Uppsala, Sweden; katja.nilsson@slu.se

Litter size is the major reproduction trait of economic importance in piglet production. However, larger litters are generally associated with higher piglet mortality and a high within-litter variation in birth weight. This might adversely affect later growth performance. The aim of this study was estimate the relationship between birth weight and slaughter pig performance. Records on birth weight from 32,531 purebred Swedish Yorkshire piglets born in 2011-2012 were available from 13 nucleus herds. Of these piglets, 8,827 also had records from field performance testing, including age and backfat thickness at approx.100 kg. Piglets born from first parity sows weighed on average 120 grams less (P<0.001), compared with piglets born from second parity sows, when corrected to the same litter size. Liveborn piglets were on average 210 g (P<0.001) heavier compared with stillborn piglets. Litter size influenced litter mean birth weight with a decrease of 33 grams per each total born piglet in the litter (P<0.001). Piglets that were heavier at birth reached 100 kg live weight faster and with thinner backfat. Pigs born from first parity sows took longer, on average 4.4 more days, to reach 100 kg live weight compared with pigs born from second parity sows (P<0.001), when corrected to the same litter size. Males were 2.2 days younger at 100 kg live weight compared to gilts (P<0.001). However, backfat thickness in pigs born by first parity sows were thinner by 0.2 mm compared to pigs born from second parity sows (P<0.001). Males had thinner backfat thickness than gilts by 0.8 mm (P<0.001). The results indicate that to maximize growth performance, it might be better to breed for smaller litters with even, high birth weight. However, for a farrow-to-finish herd, larger litter size may still be of economic interest.

Session 70 Theatre 7

Discussion on high litter size challenges

E.F. Knol
Topigs Norsvin Research Center b.v., Research, P.O. 43, 6640AA Beuningen, the Netherlands; egbert.knol@topigsnorsvin.com

Six abstracts relating to the subject large litters, low birthweights, uniformity and piglet quality. This slot will be used for general discussion of the six presentations.

Economic importance of maternal and direct pig traits calculated applying gene-flow method

E. Krupa, Z. Krupová, E. Žáková and M. Wolfová
Institute of animal Science, Přátelství 815, 10400 Prague, Czech Republic; krupa.emil@vuzv.cz

The overall aim of breeding is to produce more efficient animals. The relative economic weights of maternal and direct trait components expressed per genetic standard deviation were calculated by EWPIG 2.0.0 software considering the proportions of genes from the individual breeds or lines and the timing of expected genetic changes in the individual genotypes during defined investment period for Large White (LW), Landrace (L) and Pietrain (PN) in three-way crossing system. Influence of discount rate per one reproduction cycle (0.00, 0.02 and 0.05) during 8 years investment period was investigated. Artificial insemination was assumed in the all links; however boars were used for oestrus stimulation. The stationary state of sows herd was calculated by using of the Markov Chains and was based on the sows conception rate, sow mortality rate and sow culling rate for health problems or low litter size. The economically most important trait in both dam breeds was survival rate of young animal after the nursery phase (16-18%). Following the most important traits were feed conversion in finishing and number of piglets born alive with nearly the same relative economic importance (12 -14%). Further most important traits that reached about 10% of the total economic importance of all traits were sow conception rate, piglet survival until weaning and in nursery phase. As expected, the importance of maternal traits in sire breed PN was negligible (1%). The most important direct traits were survival rate of young animal after the nursery phase (25%), feed conversion in finishing (19%), dressing percentage (17%), and survival rate of piglets in nursery (14%). Lean meat content reached about half of the importance of named traits (about 7%) which was similar to the importance of feed conversion of piglets in nursery phase, but less than the importance of lifetime daily gain of finished animals (9%). The relative importance of direct and maternal traits did not changed substantially when discounting the trait expressions with discount rates of 2 to 5% during an 8 year investment period in comparison with no discounting and with taking into account only one generation of progeny in all links of the crossing system. This study was supported by the project MZE-RO01718.

Optimum penalty on average relationships in preselection in a two-stage optimum contribution scheme

H.M. Nielsen, B. Ask and M. Henryon
SEGES, Danish Pig Research Centre, Breeding & Genetics, Axeltorv 3, Copenhagen V, Denmark; hmni@seges.dk

We tested that two-stage optimum contribution selection (OCS) realises just as much genetic gain as single-stage OCS when we restrict rates of inbreeding in the initial stage of two-stage OCS for different preselected proportion of sires. We simulated breeding schemes resembling pig breeding with single-stage and two-stage OCS. In single-stage OCS, sires were selected by OCS, whereas dams were truncation selected. The breeding schemes had a total of 100 matings in each generation. Each dam had six offspring, producing 300 male and 300 female selection candidates (approx.). Selection was for a single trait with heritability 0.25. All candidates were phenotyped for the trait before selection. In single-stage OCS, rate of inbreeding was 0.8% using a penalty on average relationships of -60. Two-stage OCS was carried out as single-stage OCS with the one exception that 25, 15. 10, 5, 3, 2, or 1% of the sires were pre-selected in an initial OCS stage. For each preselected proportion of sires, rate of inbreeding was restricted using 13 different penalties on average relationships (-20, …,-1000) in the initial stage of the two-stage OCS. In the final selection stage, the same penalty as for one-stage OCS was used. We found an optimal penalty on average relationships, which showed that restricting inbreeding too much in the initial selection stage reduced genetic gain. For low preselected proportions (<5%), a higher penalty was needed to attain a low rate of inbreeding compared to higher preselected proportions. Below 5% preselected sires, a limit was reached and it was impossible to realise the same genetic gain and rate of inbreeding as in single-stage OCS. We concluded that two-stage OCS realised as much genetic gain and the same rate of inbreeding as single-stage OCS when we preselected at least 5% of the sires. Thus, two-stage OCS works if selection is not too strict and if inbreeding is not restricted too much in the initial selection stage.

Impact of immunocastration on growth performances and carcass quality of heavy gilts

L. Pérez-Ciria[1], F.J. Miana-Mena[1], R. Gutiérrez[2], L.E. Cobos[3], M.C. López-Mendoza[4] and M.A. Latorre[1]
[1]Univ. Zaragoza, C/ Miguel Servet 177, 50013 Zaragoza, Spain, [2]Coop. Ganadera Caspe, Ctra. de Maella, 50700 Caspe, Zaragoza, Spain, [3]Cartesa, Pol. La Paz, 44195 Teruel, Spain, [4]Universidad CEU-Cardenal Herrera, Edif. Seminario, 46113 Moncada, Valencia, Spain; leticiapcgm@gmail.com

A trial was carried out with 64 Duroc × (Landrace × Large White) gilts of 57.9±4.94 kg of body weight (BW) intended for Teruel dry-cured ham production. For that, minimum levels of fat thickness, measured at *Gluteus medius* muscle (m. GM; >16 mm), are required to improve the ripening process and ham quality. In the last decades, a great problem related to the lack of fat has been detected, which is especially important in gilts, and the immunocastration could be a solution. The objective of this study was to assess the impact of immunization against GnRH on growth performances and carcass quality. There were two experimental treatments: intact gilts (IG) and immunocastrated gilts (IMG). Immunization was carried out with two injections of Vacsincel® (Zoetis): the first one at 102 days of age (approx. 58 kg BW) and the second one at 122 days of age (approx. 76 kg BW). The replicate was the pen with 8 pigs (n=4) for performance traits and the animal (n=16, chosen at random) for carcass traits. The same feeding planning was provided for both groups through the trial. Pigs were slaughtered at 134.4±5.64 kg BW. Data were analysed as a completely randomized block design using the GLM procedure of SAS. From the application of the 1st to the 2nd injection, gilts grew similarly independently of the treatment. The application of the 2nd dose neither had influence in growth performances during the next 28 days. However, from 102 kg BW to slaughter, IMG grew faster (*P*=0.049) and ate more feed (*P*=0.016) than IG with no effect on feed conversion ratio. Regarding to carcass characteristics, only carcass fatness was influenced; the IMG showed thicker fat depth at m. GM than IG (*P*=0.044). It is concluded that immunocastration in gilts increased average daily growth and feed intake from 102 to 134 kg BW (slaughter), with no penalization of feed efficiency. Also, this immunization against GnRH increased the carcass fatness which is desirable in pigs intended for dry-cured ham production.

Session 70 Theatre 11

Association between serology for four respiratory pathogens and pig carcass traits

R. Fitzgerald[1], J.A. Calderón Díaz[2], M. Rodrigues De Costa[2,3], E.G. Manzanilla[2], J.P. Moriarty[4], H. McGlynn[1] and H. O'Shea[1]
[1]Cork Institute of Technology, Bishopstown, Cork, Ireland, [2]Teagasc Grassland Research & Innovation Centre, PDD, Moorepark, Fermoy, Co. Cork, Ireland, [3]Universitat Autònoma de Barcelona, Dept de Ciència Animal I dels Aliements, Facultat de Veterinaria, Barcelona, Spain, [4]Dept of Agriculture, Food and Marine Laboratories, CVRL, Backweston, Celbridge, Co. Kildare, Ireland; rose.fitzgerald@mycit.ie

The aim of this study was to investigate the associations between serology for four respiratory pathogens and pig carcass traits. Blood samples were collected from 128 pigs originating from 5 farrow-to-finish Irish commercial pig farms, 2 to 3 days prior to slaughter. Gender (67 females and 61 males) and slaughter age (average 178.7±4.45 days) were available for all pigs. Cold carcass weight (CCW) and lean meat % (LM%) were obtained from the abattoir. Serology was completed using IDEXX ELISA kits on all samples for *A. pleuropneumonia* (APP), *M. hyopneumoniae* (Mhyo), Porcine Reproduction and Respiratory Syndrome Virus (PRRSv) and Swine Influenza Virus (SIV) to quantify sample to positive (S/P) anti-body ratios. Data were analysed using mixed models in SAS v9.3. For each pathogen, separate models were built including S/P values and slaughter age as linear covariates and gender as fixed effects. Pig within each farm was included as random effect. Results for linear covariates are presented as the regression coefficient ± standard error. Lower (*P*<0.05) CCW was observed in pigs with higher S/P values for APP (-9.6±2.68) and PRRSv (-6.5±1.62). Similarly, there was a tendency (*P*=0.06) for lower CCW in pigs with higher Mhyo S/P values. CCW was not associated with SIV (*P*>0.05). There was no observed association between slaughter age and CCW (*P*>0.05). LM% was not associated with any of the four studied diseases (*P*>0.05). However, younger pigs at slaughter returned higher LM% (*P*<0.05). Gender was not associated with CCW or LM% (*P*>0.05). The results indicate that exposure to three of the respiratory pathogens analysed contributed negatively towards the carcass traits investigated. The recognition of poor CCW in slaughter age animals indicates that disease management strategies require review and intervention.

Herd characteristics and serological indicators associated with the performances of swine herds

C. Fablet[1], N. Rose[1], B. Grasland[1], N. Robert[2], E. Lewandowski[2] and M. Gosselin[3]
[1]ANSES, B.P. 53, 22440 Ploufragan, France, [2]Boehringer Ingelheim France, Santé Animale, Les Jardins de la Teillais, 3 allée de la grande Egalonne, 35740 Pace, France, [3]Univet Santé Elevage, rue Monge, 22600 Loudéac, France; christelle.fablet@anses.fr

This work aimed at identifying infectious and non-infectious factors associated with the growing-finishing performances of 41 French swine herds. Data related to management, biosecurity, husbandry and the main technical performances (average daily weight gain (ADG), feed conversion ratio (FCR) and mortality (MORT) from 8 to 115 kg and carcass slaughter weight (CSW)) were collected by questionnaires. Blood was sampled from 20 pigs from two batches (10 to 12 weeks old and at least 22 weeks old). Infections by Lawsonia intracellularis, Mycoplasma hyopneumoniae, porcine reproductive and respiratory syndrome virus (PRRSV), porcine circovirus type 2 (PCV2) and swine influenza viruses were detected by specific ELISAs. Two groups of herds were identified using a clustering analysis: a cluster of 24 herds with the highest technical performance values (mean ADG=781.1±26.3 g/day; mean FCR=2.5±0.1 kg/kg; mean MORT=4.1±0.9%; and mean CSW=121.2±5.2 kg) and a cluster of 17 herds with the lowest performance values (mean ADG=715.8±26.5 g/day; mean FCR=2.6±0.1 kg/kg; mean MORT=6.8±2.0%; and mean CSW=117.7±3.6 kg). Multiple correspondence analysis was used to identify factors associated with the level of technical performance. Infections with PRRSV and PCV2 were associated with the cluster having the lowest performance values. This cluster also featured farrow-to-finish type herds, a short interval between successive batches of pigs (≤3 weeks) and mixing of pigs from different batches in the growing or/and finishing steps. Inconsistency between nursery and fattening building management was another factor associated with the low-performance cluster. The odds of a herd showing low growing-finishing performance was significantly increased when infected by PRRSV in the growing-finishing steps (OR=8.8, 95% confidence interval [95% CI]: 1.8-41.7) and belonging to a farrow-to-finish type herd (OR=5.1, 95% CI=1.1-23.8). Herd management and viral infections significantly influenced the performance levels of the swine herds included in this study.

Assessing the effect of dietary inulin and resistant starch on gastrointestinal fermentation in pigs

B.U. Metzler-Zebeli[1], L. Montagne[2], N. Canibe[3], J. Freire[4], P. Bosi[5], J.A.M. Prates[6], S. Tanghe[7] and P. Trevisi[8]
[1]University of Veterinary Medicine Vienna, Veterinaerplatz, 1210 Vienna, Austria, [2]INRA – Agrocampus Ouest, 65 Rue de Saint-Brieuc, 35000 Rennes, France, [3]Aarhus University, Blichers Allé, 8830 Tjele, Denmark, [4]University of Lisbon, Tapada da Ajuda, 1349-017 Lisbon, Portugal, [5]University of Bologna, Viale Giuseppe Fanin, 40127 Bologna, Italy, [6]University of Lisbon, Avenida da Universidade Técnica, 1300-477 Lisbon, Portugal, [7]Nutrition Sciences N.V., Booiebos, 9031 Drongen, Belgium, [8]University of Bologna, Viale Giuseppe Fanin, 40127 Bologna, Italy; barbara.metzler@vetmeduni.ac.at

Prebiotics may support gut health in pigs. In conducting two meta-analyses, we aimed to evaluate the capability of inulin (IN) and resistant starch (RS) type 2 to modify intestinal short-chain fatty acids (SCFA), pH and gut health-related bacteria in pigs. Prediction models were computed, accounting for inter- and intra-study variability. Backward elimination was used to test the effects of pig's BW and experimental time on response variables. Dietary IN levels ranged from 0.1 to 25.8%, whereas RS levels ranged from 0 to 78.0%. A negative relationship existed between dietary IN and gastric pH in weaned pigs (P<0.01); gastric pH decreased by 0.12 log units with 3% dietary IN. Increasing dietary IN linearly decreased colonic enterobacteria and faecal lactobacilli (P<0.01). Similarly, negative linear relationships between RS and hindgut pH existed, with the effect being stronger in the colon and faeces (P<0.01), where 15% RS lower the pH by 0.4 to 0.6 log units. Increasing RS enhanced the proportion of propionate in mid-colon and reduced those of acetate and butyrate in mid-colon (P<0.05). In faeces, increasing RS levels promoted lactobacilli and bifidobacteria (P<0.01), with a minimum of 10% RS for a 0.5 log unit-increase in abundance. Best-fit equations indicated pig's age-related changes in the IN and RS effects and different short- and long-term effects on parameters (P<0.05). In conclusion, meta-regressions support that dietary IN and RS may have certain beneficial abilities to support gut homeostasis in pigs, with IN being more efficient to decrease gastric pH, while RS may be used to lower hindgut pH and increase faecal lactic acid-producing bacteria.

Session 70 Poster 14

Sensitivity analysis identifies candidate genes for regulation of immune responses in chicken

A. Polewko-Klim[1], W. Lesiński[1], A. Golińska[1], M. Siwek[2], K. Mnich[3] and W.R. Rudnicki[1,3,4]

[1]University of Bialystok, Institute of Informatics, Konstantego Ciołkowskiego 1M, 15, 245 Bialystok, Poland, [2]UTP University of Science and Technology, Animal Biochemistry and Biotechnology, Mazowiecka 28, 85-084 Bydgoszcz, Poland, [3]University of Bialystok, Computational Centre, Konstantego Ciołkowskiego 1M, 15, 245 Bialystok, Poland, [4]University of Warsaw, Interdisciplinary Centre for Mathematical and Computational Modelling, Pawińskiego 5A, 02-106 Warszawa, Poland; siwek@utp.edu.pl

The goal of the study was to identify candidate genes and mutations related to the immune responses, both innate and adaptive, in chicken using machine learning. Sensitivity analysis of machine learning models was performed for entire genes and SNPs located in candidate genes defined in QTL regions. The adaptive immunity was represented by the specific antibody response toward keyhole lymphet heamocyanin (KLH), whereas the innate immunity was represented by natural antibodies toward lipopolysaccharide (LPS) and lipoteichoic acid (LTA). Machine learning regression models were built applying Random Forest algorithm, using SNPs as descriptive variables and level of response as a decision variable. The importance of the SNP was estimated based on the permutation importance measure. The analysis consisted of three basic steps. First an identification of candidate SNPs by means of feature selection was performed, with the help of Boruta and MDFs algorithms. Then optimization of the feature set was completed using recursive feature elimination, by elimination of features with lowest importance scores. Finally the gene-level sensitivity analysis was done and final models were selected. The best results were obtained for adaptive immune response towards KLH. The predictive model based on five genes explains 14.9% of variance for KLH adaptive response. The models obtained for innate immunity towards LPS and LTA utilize more genes: 9 genes for LPS and 7 genes for LTA. The predictive power of these models is lower. Current study shows that machine learning methods applied to systems with complex interaction network can discover phenotype/genotype associations with much higher sensitivity than traditional statistical models. The research was supported by the National Science Center (UMO-2014/13/B/NZ9/02123, UMO-2013/09/B/ST6/01550).

Session 70 Poster 15

Accumulation of an emergent contaminant, α-hexabromocyclododecane (α-HBCDD), in tissues of pigs

E. Royer[1], R. Cariou[2], E. Baéza[3], A. Travel[4] and C. Jondreville[5]

[1]Ifip-institut du porc, 34 bd de la gare, 31500 Toulouse, France, [2]ONIRIS, INRA, Université Bretagne Loire, Laberca, CS 50707, 44307 Nantes, France, [3]INRA, URA, Centre de Tours, 37380 Nouzilly, France, [4]ITAVI, Centre INRA de Tours, 37380 Nouzilly, France, [5]INRA, Université de Lorraine, AFPA, BP 20163, 54500 Vandoeuvre-lès-Nancy, France; eric.royer@ifip.asso.fr

Hexabromocyclododecane (HBCDD) is a brominated flame retardant used into materials such as polystyrenes. Its presence in some food samples from animal origin may contribute to human exposure to this endocrine disruptor. A total of 56 fattening barrows (LW×Ld) × Piétrain (27.7 kg) were individually exposed to controls or diets containing 3.2 or 29.7 µg α-HBCDD per kg, during a growing period of 49 days followed by a finishing period of 63 days. Decontamination parameters were assessed in pigs successively exposed during 49 days and depurated during 63 days. Serial slaughtering of 3 pigs per exposed or decontaminated group was performed at days 19, 49, 70, 91, and 112 to collect liver, dorsal fat and semimembranosus muscle samples. Performance was not affected by α-HBCDD exposure. No isomerization of α- to β- or γ-HBCDD forms occurred, while OH-HBCDD was identified as a minor product of α-HBCDD metabolism. With the 29.7 µg/kg diet, α-HBCDD concentration in the muscle quickly increased up to 116 ng/g lipid at day 49. It decreased down to 31 ng/g after 63 d of depuration, partly because of the body weight gain. HBCDD concentration on a lipid weight basis was higher in the adipose tissue than in the muscle and was lower in the liver. The overall elimination half-lives were 27 and 32 d in semimembranosus muscle and dorsal fat but, without the dilution due to the growth and the associated lipid deposition, these half-lives would have been increased by a factor 2 and 5, respectively. It is concluded that HBCDD accumulation in pig tissues is possible, justifying a monitoring of exposure sources.

Session 70 Poster 16

Evaluation of micropollutants and chemical residues in organic and conventional pig meat

G. Dervilly-Pinel[1], T. Guérin[2], E. Royer[3], B. Minvielle[3], E. Dubreil[4], S. Mompelat[4], F. Hommet[2], M. Nicolas[2], V. Hort[2], C. Inthavong[2], M. Saint-Hilaire[2], C. Chafey[2], J. Parinet[2], R. Cariou[1], P. Marchand[1], B. Le Bizec[1], E. Verdon[4] and E. Engel[5]

[1]ONIRIS, Laberca, CS 50707, Nantes, France, [2]ANSES, rue Curie, Maisons-Alfort, France, [3]Ifip-institut du porc, bd de la gare, Toulouse, France, [4]ANSES, Laboratoire Fougères, Javené, France, [5]INRA, QuaPA, Site Theix, St-Genès-Champanelle, France; eric.royer@ifip.asso.fr

Even if there is no clear evidence that organic food products are healthier than conventional ones, the presumed absence of chemical contaminants is reported as main driver for organic consumers.. To provide occurrence data in a context of chronic exposure, samples of liver and meat (*psoas major* muscle) were collected in 2014 in six French slaughterhouses representing 70 pig farms, including 30 organic, 12 Label Rouge and 28 conventional. Each sample corresponded to a pool of tissues of three carcasses. Environmental contaminants (17 polychlorinated dibenzodioxins/dibenzofurans (Dioxins), 18 polychlorinated biphenyls (PCBs), 3 hexabromocyclododecane (HBCD) isomers, 6 mycotoxins, 6 trace metal elements) and residues from production inputs (75 antimicrobials and 121 pesticides) were investigated using the most sensitive methods. Contamination levels were measured below regulatory limits in all the samples. However, some differences were observed between types of farming. Dioxins, PCBs and HBCD concentrations were thus observed as significantly higher in organic meat samples. Cu, Zn and As were measured at slightly higher levels in organic meat without differences between organic and Label Rouge. Liver samples from conventional and Label Rouge farms exhibited higher contents in Zn and Cd than the organic ones. Ochratoxin A was the only mycotoxin quantified in 25 samples (36%) and detected in another 22 samples (31%) of the livers analysed, without significant differences between farming systems. A correlation could be observed between mycotoxins concentrations in meat and liver. All meat samples exhibited pesticides levels below the detection limits, whereas only 3 conventional or organic samples (overall: 3.5%) displayed residual concentrations of authorized veterinary antimicrobials, but with concentrations far below the regulatory limits.

Session 70 Poster 17

Air quality in nurseries: a room for swine environmental conditions improvements?

C. Fablet, F. Bidan, V. Dorenlor, F. Eono, E. Eveno and N. Rose

Anses, Ploufragan/Plouzané Laboratory, Unité de Recherche en Epidémiologie et Bien-Etre du Porc, B.P.53, 22440 Ploufragan, France; christelle.fablet@anses.fr

A study was carried out in 128 swine herds to firstly describe the air quality in nursery rooms by five parameters (temperature, humidity, respirable dust concentration, CO_2 and NH_3 levels) and secondly, identify and quantify the effect of factors related on the one hand to building design, management practices and internal equipment and on the other hand, to external climatic conditions on air quality in the nursery rooms. Temperature, relative humidity, CO_2 concentration and respirable dust levels were continuously monitored over a 20-hour period in 128 nursery rooms. Ammonia concentrations were measured with a one-spot electro-chemical device. A questionnaire was filled in to collect data on hygiene measures, management, feeding practices and housing conditions in the nursery and finishing rooms selected for measuring climatic conditions. Relationships between the climatic parameters were investigated by principal component analysis and air quality levels were defined by hierarchical clustering. Multifactorial analyses were used to identify factors associated with sub-optimal air quality in 94 of the 128 nursery houses. The overall mean air temperature, relative humidity, CO_2, NH_3 and respirable dust levels were 25.7 °C, 63.3%, 2,162 ppm, 5.2 ppm and 0.13 mg/m respectively in the nursery rooms. Three levels of air quality were identified by hierarchical clustering. Eleven factors were associated with sub-optimal air quality in nurseries. They may be grouped into three main areas related to external climatic conditions; building design and engineering; and building management practices. These findings should be considered when designing and developing strategies to enhance environmental conditions in pig buildings.

Estimation of heritability of feeding behaviour traits in Finnish Yorkshire

A.T. Kavlak, A. Riihimäki and P. Uimari
University of Helsinki, Koetilantie 5, 00014 University of Helsinki, Finland; alper.kavlak@helsinki.fi

A major proportion of the costs of pork production relates to feeding. Feed conversion rate or residual feed intake is thus commonly included in the breeding programs. Feeding behaviour traits do not directly have economic importance but if correlated with residual feed intake, growth rate or carcass traits can be used as early indicators of these economically important traits. The aim of this study was to estimate the heritability of feeding behaviour, production and carcass traits and their genetic correlations in Finnish Yorkshire breed. Data were available from 3,627 purebred Yorkshire animals (2,622 boars, 536 gilts, and 469 castrates). In this first part of the study, feeding behaviour was measured as a number of visits per day, daily feed intake and time spent in feeding, mean feed intake per visit, time spent per visit and mean feed intake rate. Behaviour traits were separately calculated for five test periods: from the beginning of the test to 20 days in the test, 21-40, 41-60, 61-80, and 81-93 days in the test. The statistical model included sex, herd and year as fixed effects and batch×pen and animal as random effects for all the traits. Estimates of heritability were relatively high for all behaviour traits varying from 0.4 to 0.6. Genetic correlations between the different time categories of the same trait were mostly over 0.8 and between the different traits within the same time category varied from 0.1 to 0.9. Genetic correlations between the behaviour traits and production and carcass traits will be estimated next. The results will be used in breeding value estimation and give insight if the early feeding behaviour relates to economically important traits such as residual feed intake, growth traits and carcass and meat quality.

An analysis of miR-208b polymorphism in relation to growth and carcass traits in pigs

D. Polasik[1], M. Tyra[2], D. Zagrobelny[1], G. Żak[2] and A. Terman[1]
[1]West Pomeranian University of Technology in Szczecin, AL. Piastów 17, 70-310 Szczecin, Poland, [2]National Research Institute of Animal Production in Krakow, Ul. Sarego 2, 31-047 Kraków, Poland; grzegorz.zak@izoo.krakow.pl

The class of microRNAs expressed abundantly in skeletal and heart muscles, called myomiRs plays an important role in myogenesis, embryonic muscle growth and cardiac function. Kim *et al.* detected polymorphic site in pig genome, located 105 bp downstream of pre-miR-208b (g.17104G>A), which is correlated with expression of miR-208b and the host gene – MYH7 (myosin heavy chain 7). It was also proved that MYH7 gene shows differential expression between pigs characterized by extremely high and low growth rate. Taking into consideration these facts, the aim of this study was to evaluate miR-208b SNP in relation to six growth and nine carcass traits in pigs reared in Poland. Investigations were conducted on 582 pigs belonging to five breeds: Landrace, Large White, Puławska, Pietrain, Duroc and Hampshire. miR-208b variants were determined by of means PCR-RFLP method. Statistical analysis (GLM procedure) was performed for first three breeds separately and all animals together. Obtained results showed that AA genotype was positively associated with test daily gain, average daily gain, and negatively with age at slaughter and number of days on test ($P \leq 0.05$ or $P \leq 0.01$) in Large White, Puławska and whole group. AA genotype in Landrace breed was excluded from analysis because of low frequency. Among carcass traits, miR-208b variants were associated only with mean backfat thickness from 5 measurements ($P \leq 0.05$) in the same groups. Highest values of this parameters were observed for AA genotype. We also noticed some associations for slaughter efficiency, weight of loin without backfat and skin, loin eye area and meat percentage ($P \leq 0.05$ or $P \leq 0.01$) but results were not consistent among breeds and whole group. Obtained results indicate that miR-208b SNP could be considered as a genetic marker for improve some growth traits and backfat thickness in pigs.

Analysis of rs81286101 polymorphism related to intramuscular fat content in different breeds of pigs

D. Zagrobelny[1], D. Polasik[1], M. Tyra[2], G. Żak[2] and A. Terman[1]
[1]West Pomeranian University of Technology in Szczecin, AL. Piastów 17, 70-310 Szczecin, Poland, [2]National Research Institute of Animal Production in Krakow, Ul. Sarego 2, 31-047 Kraków, Poland; grzegorz.zak@izoo.krakow.pl

Numerous studies have shown that the intramuscular fat content (IMF) is correlated with the sensory quality of meat. Luo *et al.* have performed genome-wide association analysis of meat quality traits and indicated SNP MARC0017000 (rs81286101) as the most significant for the IMF in pigs. It is located on SSC12, 28.6 kbp from the PIRT gene. The aim of this study was to design PCR based test for the rs81286101 polymorphism analysis and to determine frequency of genotypes and alleles in different pig breeds. The analysis included 425 individuals belonging to five breeds of pigs: Duroc, Pietrain, Polish Landrace, Puławska and Polish Large White. Genotypes were determined by means of ACRS-PCR method. Population parameters of breeds were calculated using PowerMarker v3.25 software. The analysis have shown the presence of three genotypes: AA, AG, GG. AA genotype has appeared in three breeds (Polish Landrace, Polish Large White, Puławska) and its frequency was low. The highest frequency of GG genotype was observed in Duroc, however lowest in Puławska breed. AA genotype was most frequent in Puławska breed. Obtained results were compared with the data published by Polish Union of Pig Breeders and Producers. Pig breeds, which have meat content higher than 60%, and mean backfat thickness below 10 mm were characterized by high frequency of GG genotype and G allele. In Puławska breed, which is marked by meat content less than 60% and mean backfat thickness about 15 mm, we recorded highest frequency of AA and AG genotype as well as G allele. Puławska is polish native breed, under protection program, known from tasty meat and high intramuscular fat content. Based on the above results, it can be assumed that the AA and AG genotypes may be associated with higher intramuscular fat content, however GG genotype with lower value of this parameter. In the future IMF will be determined in collected samples and association analysis will be performed.

Estimates of genetic parameters of semen characteristics in Japanese Duroc boars

K. Ishii[1], M. Kimata[2] and O. Sasaki[1]
[1]NARO Institute of Livestock and Grassland Science, Ikenodai 2, Tsukuba city, Ibaraki, 305-0901, Japan, [2] CIMCO Co., Ltd, Kameido 2-3-13, Koto ward, Tokyo, 136-0071, Japan; kazishi@affrc.go.jp

Semen characteristics are important traits in Duroc boars, because artificial insemination is becoming main method recently. Our aim of this study was to investigate genetic parameters of semen characteristics. Data on characteristics of 28,670 ejaculates from 686 Duroc boars were obtained from 4 farms between 2001 and 2015. We estimated the heritabilities of boar semen traits by using the REML procedure applied to multi-trait repeatability animal models. The heritability estimates of semen volume, sperm concentration, percentage of normal sperm, motility, total number of sperms, and functional number of sperms were 0.17±0.03, 0.22±0.02, 0.23±0.04, 0.17±0.03, 0.20±0.04, and 0.16±0.04, respectively. The genetic correlation between semen volume and sperm concentration was high and negative -0.56±0.06. The genetic correlations of percentage of normal sperm with motility and total number of sperms with functional number of sperm were high and positive at 0.84±0.05 and 0.95±0.01, respectively. Permanent-environmental ratios to phenotypic variances of semen volume, sperm concentration, percentage of normal sperm, motility, total number of sperms and functional number of sperms were 0.22±0.02, 0.16±0.02, 0.36±0.04, 0.12±0.02, 0.24±0.03 and 0.27±0.03, respectively.

Effects of arginine supplementation during late gestation on reproductive performance and piglet uni

J.S. Hong, Y.G. Han, L.H. Fang, J.H. Jeong and Y.Y. Kim
Seoul National University, 1 Gwanak-ro, Gwanak-gu, Seoul, 08826, Korea, South; jinsu13@snu.ac.kr

This study was conducted to evaluate the effects of arginine supplementation levels during late gestation on reproductive performance and piglet uniformity in sows. A total of 40 F1 multiparous sows (Yorkshire × Landrace), with an average body weight of 246.1 kg, were allotted to one of four treatment groups in a completely randomized design (CRD). The dietary treatments were divided by the supplementation level of arginine during the late-gestation period (70-115 d), as follows: (1) Arg0.72%: corn-SBM based diet + L-Arg 0% (Arg 0.72%), (2) Arg1.0%: basal diet + L-Arg 0.28% (Arg 1.0%), (3) Arg1.5%: basal diet + L-Arg 0.79% (Arg 1.5%), and (4) Arg2.0%: basal diet + L-Arg 1.35% (Arg 2.0%). The same lactation diet was provided *ad libitum* during the lactation period. All collected data were analysed by least squares mean comparisons and were evaluated with the GLM procedure of SAS. Orthogonal polynomial contrasts were used to determine linear and quadratic effects by increasing arginine supplementation level. There were no significant differences in body weight and backfat thickness in sows during gestation and lactation. In addition, dietary arginine had no significant effects on the number of total born, stillborn, total born alive, and piglet growth during lactation. Increasing arginine supplemented levels improved total litter weight (linear, P=0.08) and alive litter weight (quadratic, P=0.07). However, additional arginine did not showed the piglet uniformity in piglet weight at birth and at day 21 of lactation. Although there was no significant difference in blood profiles in gestating sows, the level of blood urea nitrogen of lactating sows was increased as dietary arginine level was higher (linear, P<0.05). Additional arginine supplementation had no influence on casein, lactose, or protein contents of colostrum and milk. Consequently, dietary arginine up to 1.5% in a late-gestation diet improved total litter weight and alive litter weight at farrowing, but it did not affect piglet uniformity.

Comparison of three different farrowing systems with special regard to air quality

E. Lühken, J. Schulz and N. Kemper
University of Veterinary Medicine Hannover, Foundation, Institute for Animal Hygiene, Animal Welfare and Farm Animal Behaviour, Bischofsholer Damm 15, 30173 Hannover, Germany; eyke.luehken@tiho-hannover.de

Because of animal welfare efforts, new farrowing systems have been invented to replace the conventional crate-system, in which sows are fixated during the farrowing period. This study compared the air quality of three compartments with different farrowing systems (loose-housing-system (LHS), group-system (GS) and crate-pen-system (CS)). Eight batches with six sows per system were examined from August 2016 to August 2017. Indoor and outdoor temperature, carbon dioxide- and ammonia concentrations were measured, and dust-samples were taken on day 5, 19 and 33 of each 5-week farrowing-period. Statistical analysis was carried out by using the procedure GLM (multifactorial-ANOVA) of SAS. Over all systems, mean values of dust load varied between 0.383 and 1.824 mg/m3, and gaseous ammonia varied between 10.8 and 15.8 ppm. There was a significant increase of dust load from day 5 to 19 and to 33 in all systems (P<0.02). Concentration of dust was significantly higher in GS on day 19 (P<0.03), and concentration of gaseous ammonia was significantly higher in GS on day 5 (P<0.04), compared to the other systems. The higher dust-concentration in GS can possibly be related to an increased activity of the animals, whereas the higher ammonia content on day 5 was possibly due to worse cleaning-results. According to visual assessment, the GS was more difficult to clean than the other systems. The increase of dust load over the lactation period can be ascribed to the longer occupation time and the increased activity of the growing piglets. The new farrowing-systems did not affect the air-quality adversely in general. The observed negative effects on air quality in GS can probably be solved by modifications of the floor.

Session 70 Poster 24

Effect of dietary rapeseed oil and humus-containing mineral preparation on cooked ham quality

A.M. Salejda[1], M. Sczygiol[1], A. Zwyrzykowska-Wodzińska[2] and G. Krasnowska[1]
[1]Wrocław University of Environmental and Life Sciences, Department of Animal Products Technology and Quality Management, Chełmońskiego 37, 51-630 Wrocław, Poland, [2]Wrocław University of Environmental and Life Sciences, Department of Environment Hygiene and Animal Welfare, Chełmońskiego 38 c, 51-630 Wrocław, Poland; amsalejda@gmail.com

The concentration of omega-3 fatty acids and minerals is generally too low in the everyday life diet. But as the nutrient composition of pork meat and adipose tissue is influenced by the feed given to the animals, the product can be changed to support nutrient demands. Rapeseed oil is used as a carrier of polyunsaturated fatty acids, in particular n-3 fatty acids and vitamins E of antioxidant properties. Humus-containing mineral preparation increases the biological value of feed, showing bactericidal and fungicidal properties, also used in pig nutrition positive influence on digestive (prevents diarrhoea), stimulate immunity of animals, as well as have a positive influence the quality of animal raw materials The aim of the study was to assess the influence of dietary rapeseed oil and humus-containing mineral preparation on selected quality factors of cooked ham from finishing pigs. Obtained results confirmed the possibility of modifying the chemical composition of meat and fat of pigs in the direction of improving the dietary value, while maintaining good technological parameters.

Session 70 Poster 25

Conjugated linoleic acid and betaine affect gene expression in adipose tissue of Iberian pigs

I. Fernandez-Figares[1], J. Die[2], B. Rabanal[3] and M. Lachica[1]
[1]Estacion Experimental del Zaidin. CSIC, Animal Nutrition, Profesor Albareda 1, 18008 Granada, Spain, [2]Universidad de Cordoba, Genetica, Edificio Gregor Mendel. Campus de Rabanales, 14071 Córdoba, Spain, [3]Universidad de Leon, Laboratorio de Tecnicas Instrumentales, Campus de Vegazana s/n, 24071 Leon, Spain; ifigares@eez.csic.es

To investigate possible mechanisms explaining energy partitioning in pigs fed betaine or CLA, the expression of genes involved in cellular energy metabolism were evaluated: peroxisome proliferator activated receptor (PPAR) γ and genes involved in AMPK signalling pathway (AMPKα1, AMPKα2, PPAR coactivated 1α (PGC1-α), and sirtuin). Sixteen Iberian barrows (19 kg BW) were randomly assigned to one of four dietary treatments: control, 5 g/kg betaine, 10 g/kg CLA, or betaine + CLA. At 61 kg BW pigs were slaughtered and samples from adipose tissue obtained from each pig, snapped frozen in liquid N and stored at -80 °C until analysis. Gene expression was evaluated using RT-qPCR and values were normalized using cyclophilin A and β-actin as reference genes. Data were analysed using a mixed-model ANOVA that included the fixed effect of treatments and the random effect of each pig. PPARγ was downregulated by CLA ($P<0.05$) compared to control treatment. CLA also tended to downregulate PGC1-α ($P=0.011$) and AMPKα2 ($P=0.080$), compared to control. CLA did not affect AMPKα1 and sirtuin expression. On the other hand, betaine and betaine in combination with CLA had no effect on the expression of the genes evaluated ($P>0.15$). Downregulation of AMPKα2 and PGC1-α in adipose tissue may indicate a lipolytic effect and decreased fatty acid oxidation in Iberian pigs fed CLA supplemented diets. Reduced PPARγ plays a role in downregulating the lipogenic enzymes, thus leading to reduced lipid accumulation in pigs fed CLA supplemented diets. Furthermore, by antagonizing PPARγ activity, CLA may block adipogenesis, lipogenesis, and glucose uptake in adipocytes. Additionally, reduced PPARγ agrees with decreased PGC1-α and may decrease mitochondrial biogenesis and substrate oxidation which may indicate a lower energy needs in adipose tissue of pigs fed CLA supplemented diets. Funded by Ministerio de Economía y Competitividad (AGL2016-80231).

Effect of genotype and leucine levels on the amino acids concentration in tissues of weaned pigs

M. Bertocchi[1], P. Bosi[1], D. Luise[1], V. Motta[1], C. Salvarani[1], A. Ribani[1], A. Simongiovanni[2], B. Makoto[2], E. Corrent[2], L. Fontanesi[1], T. Chalvondemersay[2] and P. Trevisi[1]
[1]University of Bologna, Viale Fanin 50, 40127 Bologna, Italy, [2]Ajinomoto Eurolysine S.A.S., 153 Rue de Courcelles, 75017 Paris, France; micol.bertocchi2@unibo.it

Proper essential amino acids (AA) requirement in pig diet formulation can guarantee feed efficiency and performance. Blood AA values are important indicators of the AA requirements. Dose-response data to dietary leucine (Leu) in weaned pigs are scarce but needed, also for its metabolic properties; furthermore, the effect of genotype was generally not considered. Different dietary Leu levels effects on AA metabolism were assessed in a 3-weeks dose-response trial with a 2 (genotype) × 5 (diets) factorial arrangement on one-hundred pigs (9 to 20 kg). Pigs were selected at weaning (24d old) after being genotyped for α-aminoadipate d-semialdehyde synthase (AASS) gene polymorphism, coding for a lysine (Lys) metabolism enzyme, to have the two balanced homozygous genotypes equally distributed in the litters. After 1 adaptation week, pigs were fed experimental diets (d7 to d28) obtained by variation of the standardized ileal digestible (SID) Leu:Lys ratios, that were 70, 85, 100, 115 and 13%. At d7, 21 and 27 blood was sampled and plasma was collected; saliva was collected at d21. At d28 animals were sacrificed and liver, muscle and urine were sampled. All samples were analysed for AA concentration. Statistical analysis was carried out by one factor ANOVA (Leu level) and multiple comparison with adjustment via Bonferroni. AASS genotype did not affect the AA levels. At d7 AA plasma levels were similar within groups, while a strong effect was recorded at d21 and 28: at 70% Leu:Lys, levels of AA such as Asparagine, Glutamic acid, Glycine, Histidine, Isoleucine, Lysine, Serine, Tyrosine and Valine increased, due to the deep Leu deficiency which blocked protein synthesis. Moreover, the increase to 115 and 130% Leu:Lys highly decreased Valine and Isoleucine in plasma, due to the Leu-dependent system of their catabolism. In liver, muscle and urine AA concentrations followed the same trend, but slightly less evident. No effect was seen in saliva.

The detection of ochratoxin A in the tissues of pigs fed mildly contaminated feed

C.A. Moran[1], A. Yiannikouris[2], J.D. Keegan[1], S. Kainulainen[3] and J. Apajalahti[3]
[1]Alltech SARL, Rue Charles Amand, 14500 Vire, France, [2]Alltech Inc., 3031 Catnip Hill Road, Nicholasville, KY40356, USA, [3]Alimetrics Ltd, Koskelontie 19 B, Espoo, 02920, Finland; cmoran@alltech.com

The objective of this pilot study was to provide pigs with a diet artificially contaminated with Ochratoxin A (OTA), in order to detect the subsequent deposition of the mycotoxin in various swine tissues. Six pigs (25-27 kg) were individually housed and fed a commercial feed for an acclimatisation period of 7 days. Feed and water was provided *ad libitum*. On day 7, the pigs were randomly assigned to either a control group (n=3) receiving no OTA, or a treatment group (n=3) receiving OTA at a level of 50 μg/kg, which is the upper limit of OTA allowed in pig feed in the EU. After 21 days receiving the experimental diets, the pigs were sacrificed using the captive bolt technique. Liver, kidney, pancreas, blood and thigh muscle samples were taken immediately post-mortem and frozen at -20 °C until analysis for OTA content. In order to determine the OTA content, each sample was first subjected to grinding, after which the OTA was extracted from the ground tissue and then concentrated by immunoaffinity chromatography before finally being analysed quantitatively by HPLC. All tissues recovered from the pigs receiving the OTA diet had detectable levels of the mycotoxin. The highest concentration of OTA was found in the blood of the OTA group (114 ng/kg). This was followed by the kidneys, liver and pancreas with OTA levels of 96, 40 and 37 ng/kg, respectively. The lowest amount was detected in the muscle tissue at 14 ng/kg. For the control animals OTA was below the limit of detection in blood, pancreas and muscle, while the level of OTA detected in liver and kidney samples was below 1.4 ng/kg in each case. This pilot study demonstrates that OTA concentrations in swine tissues can be accurately detected when present in low concentrations in feed and the suitability of this experimental model as a means to investigate potential detoxification strategies.

Session 70 Poster 28

Effect of increasing phosphorus levels on growth performance and minerals status in weaned piglets

P. Schlegel
Agroscope, Tioleyre 4, 1725 Posieux, Switzerland; patrick.schlegel@agroscope.admin.ch

Phosphorus (P) and calcium (Ca) are essential minerals for adequate bone development in pigs. The 41-day study evaluated the effect of apparent total tract digestible P levels (dP=2.5, 3.5, 4.5 and 5.5 g/kg diet containing 14 MJ digestible energy/kg diet) on growth performance and minerals status of 56 weaned piglets (7.6±1.2 kg body weight). Monocalcium phosphate was used as supplemental P source and the dietary Ca to digestible P ratio was fixed at 2.8. Measured individual parameters were weekly feed intake and body weight, serum analytes, metacarpal III and IV mineral concentration and tibia breaking strength and mineral density by dual-energy X-ray absorptiometry (DXA). Feed intake (561±117 g/d) and body weight gain (345±85 g/d) were independent of dP levels, but feed conversion ratio was impaired ($P<0.01$) in the 4.5 and 5.5 compared to the other dP levels. Blood serum Ca (2.96±0.20 mmol/l), P (3.03±0.39 mmol/l) and alkaline phosphatase activity (286±51 U/l) were not affected by dP levels. Metacarpal ash content was 5% lower in the 2.5 than in the 3.5 dP group and 5% lower ($P<0.01$) in the 3.5 than in the 5.5 dP group. Metacarpal ash increased linearly ($P<0.05$) with increasing daily dP intake. Metacarpal Zn content was not affected by increasing dietary Ca and P. Tibia breaking strength, DXA mineral contents and DXA mineral density were improved ($P<0.001$) with increasing dP levels. With increasing daily dP intake, tibia breaking strength and DXA tibia mineral content and mineral density increased linearly ($P<0.001$) and DXA tibia mineral density increased quadratically ($P<0.01$). Tibia breaking strength and DXA mineral content and mineral density were positively correlated ($R^2=0.89$ and 0.78 respectively, $P<0.001$). A dP level of 3.5 g/kg allowed a maximal growth efficiency while greater levels of dP further improved bone mineralization and physical bone parameters. Finally, bone mineral density measured by DXA may present an alternative measurement for bone characteristics to the labour intensive breaking strength measurements.

Session 70 Poster 29

Rapeseed meal and enzyme supplementation on growth performance and nutrient digestibility in pigs

A.D.B. Melo[1,2], B. Villca[2], E. Esteve-García[2] and R. Lizardo[2]
[1]PUCPR, Rua Imaculada Conceição, 1155, 80215-901 Prado Velho, Curitiba, PR, Brazil, [2]IRTA, Centre Mas de Bover, Crta Reus-El Morell km. 3.8, 43120 Constantí, Spain; rosil.lizardo@irta.es

The EU animal production relies on protein imports to satisfy the demand of the animal feed industry. However, these imports are subject of increasing concerns and the use of alternative EU protein sources is recommended. Rapeseed meal (RSM), a by-product of oil industry, contains large amounts of protein but also contains antinutritional factors (ANF) such as glucosinolates or a high fibre content that limit its use. Processing of RSM to eliminate ANF might increase its protein content resulting more attractive for animal feeding. The aim of this study is to evaluate the influence of a high-protein RSM in combination with enzyme supplementation of diets for growing pigs. A 2×3 factorial arrangement of diets containing conventional (35% CP) or high-protein (40% CP) RSM in combination with supplementation of protease or NSPase and protease enzymes was used. One hundred forty-four pigs were allocated into 12 blocks of body weight and sex, and housed by 2 in 72 pens. The trial lasted 6 weeks and faeces samples were collected during the 5th week and oven-dried for 72 h at 60 °C before lab analysis. Productive performance and major nutrient digestibility parameters were used as response criteria. Data were analysed using the GLM procedure of SAS. Body weight gain was not affected by the dietary treatments (NS). However, feed conversion ratio was improved ($P<0.01$) on pigs fed the high-protein RSM diets due to a reduction of feed intake ($P<0.01$). On the contrary, enzymes inclusion into diets did not affect productive performances ($P>0.05$). Digestibility of dry matter, nitrogen or energy were not affected by the diet. Only fat digestibility was improved ($P<0.01$) due to high-protein RSM utilization. In conclusion, high-protein RSM improves feed efficiency of growing pigs and can positively contributes to reduce EU protein imports for animal feeding.

Session 70 Poster 30

The criteria for the purchase of crossbred gilts required for the sow herd replacement

M. Sviben[1], Z. Jiang[2] and N. Jančo[3]

[1]Freelance consultant, Siget 22B, 10020 Zagreb, Croatia, [2]Washington State University, Dpt. of Animal Sciences, P.O. Box 646310, Pullman, WA 99164-6310, USA, [3]OPG Goran Jančo, Ulica Matije Gupca 19, 31424 Punitovci, Croatia; marijan.sviben@zg.t-com.hr

Recently, at the 11th WCGALP it was shown that crossbred gilts express the prolificacy depending on the selection differentials at generations of purebred mothers. After 1990 the world largest farms in Croatia and in neighbour states were destroyed but the data on the pigs' traits were saved. It was possible to do the research using the data on the prolificacy expressed during the 1987 at the 1st farrowings of 1,712 BOSTOP (Improved Swedish Landrace) sows and at the 1st litters of their 2,411 daughters used at The Pig Farm in Nova Topola, Bosnia and Herzegovina. The calculations were made respecting the equations $\bar{X}_{DSM} = \bar{X}_{DAM} + SE$, $SE = SD \times h^2$, $SD = \bar{X}_{SM} - \bar{X}_{AM}$, $h^2_r = SE/SD$ (DSM – Daughters of Selected Mothers, DAM – Daughters of All Mothers, SM – Selected Mothers, AM – All Mothers, SE – Selection Effect, SD – Selection Differential). The prolificacy was expressed above $\bar{X}_{DAM}$ according to the magnitudes of SE which were: 0.171 piglets/litter when 996 daughters originated from 650 mothers which bore at least 10 piglets, 0.217 p/l when 534 daughters originated from 349 mothers which gave at least 11 piglets, 0.349 p/l when 251 daughters originated from 157 mothers which had at least 12 piglets, 0.477 p/l when 82 daughters originated from 48 mothers which farrowed at least 13 piglets. Corresponding SDs were 2.521, 3.054, 3.599 and 4.198 piglets/litters respectively. The magnitudes of h^2_r were 6.78, 7.10, 9.70 and 10.65.The mean numbers were $\bar{X}_{AM}$ 9.314 and $\bar{X}_{DAM}$ 9.504 piglets/litter. Nowadays the average of born piglets in the 1st litters is above 13. It is recommendable to buy crossbred gilts, required for the sow herd replacement, originated from pure-bred mothers which at the 1st farrowings bore at least 14 piglets – alive at the parturition and at the weaning (no stillborn, no died piglets during the sucking period!).

Session 70 Poster 31

Study of faecal microbiota in pigs exposed to high-protein and high-carotene diets

R. González-Prendes[1], R.N. Pena[1], E. Solé[1], A.R. Seradj[1], J. Estany[1] and Y. Ramayo-Caldas[2]

[1]Universitat de Lleida-Agrotecnio Center, Departament de Ciència Animal, Av. Alcalde Rovira Roure, 191, 25198 Lleida, Spain, [2]IRTA, Torre Marimon, 08140, Caldes de Montbui, Spain; rayner.prendes@gmail.com

In this study we investigated the faecal microbial composition and its change during the last month of fattening in purebred Duroc pigs. From 70 to 165 d of age, 32 pigs were divided into two groups fed either a high-protein (HP) or a regular-protein (RP) diet. In the last month of fattening, from 165 to 195 d, all 32 pigs received a different RP diet, either carotene-enriched (CE) (8 pigs from HP and 8 pigs from RP previous diets) or not (NE) (8 pigs from HP and 8 pigs from RP previous diets). Faecal samples were collected at 165 d (T1) and at 193 d (T2). Faecal microbiota composition was studied using high-throughput 16S ribosomal RNA sequencing in Illumina MiSeq platform. In each sample, the amplicon sequence variant (ASV) abundance, taxonomy classification and diversity (α-diversity, β-diversity and richness indexes) were evaluated. After the sequence quality control, we identified a total 3,142 distinct ASVs. At taxonomic level, 16 phyla and 65 genera were detected. *Firmicutes*, *Bacteroidetes* and *Proteobacterius* were the most abundant phyla at both T1 and T2, with the most abundant genera including *Prevotella* and *Oscillospira*. Differences were observed between T1 and T2 for microbiota composition at ASVs, phylum and genus levels. At T1, the composition of microbiota was affected by the HP diet for ASVs and phylum classification. In contrast, although 160 ASVs were differentially abundant across CE and NE at T2, the carotene treatment did not impact any of the diversity indexes at the end of the fattening period. Cluster analysis evidenced two enterotypes with different stability across T1 and T2 time-points. In summary, a strong effect of the diet was observed between T1 and T2 on the microbiota composition, which may explain the weak enterotype stability. We are currently evaluating the functional composition of the detected ASVs.

Session 70 Poster 32

Effects of dietary Hy-D® on gestating and lactating sows on reproductive performance and its progeny

Y.G. Han, J.H. Jeong, L.H. Fang, W.L. Chung, J.S. Hong and Y.Y. Kim
Seoul National University, 1 Gwanak-ro, Gwanak-gu, Seoul, 08826, Korea, South; hanyounggur@naver.com

This experiment was conducted to evaluate the Influence of dietary Hy-D® on gestating and lactating sows on reproductive performance and its progeny growth. A total of 30 gilts (Yorkshire×Landrace) with 150.03±11.00 kg of body weight (BW) were allotted to 3 dietary treatments by body weight and backfat thickness in a completely randomized design (CRD) with 10 replicates. All experimental diets for gestating gilts were based on corn-soybean meal and Hy-D® supplemented by levels (0, 0.05 and 0.10%, respectively). Reproductive performance, lactating performance, growth performance of their progeny and blood profiles were evaluated by supplementation of Hy-D®. Data were analysed by ANOVA with a completely randomized design using the GLM procedure in SAS. Orthogonal polynomial contrasts were used to determine the linear and quadratic effects by increasing the Hy-D® levels in gestation for all measurements of sows and piglets. When the significance was declared, fisher's least significance difference (LSD) method was used to separate the means. In growth performance of gilts, backfat thickness at 24 h postpartum was highly significantly increased when dietary Hy-D® level was increased (linear, $P<0.01$). Backfat thickness of sows fed 0.10% Hy-D® was significantly increased ($P<0.01$). In litter performance, the number of total born and born alive piglet tended to be increased as increasing dietary Hy-D® level (linear, $P=0.081$, $P=0.091$, respectively). Also, litter weight at weaning tended to be improved when dietary Hy-D® level was increased (linear, $P=0.067$). In blood profiles, the concentration of 25-OH-D3 of sows fed Hy-D® was significantly increased in 35 d, 70 d of gestation period and 0 day of lactation period. Also, when sows were provided Hy-D®, the concentration of 25-OH-D3 of their piglets was increased. In piglet growth, body weight of weaning pig at 5 weeks was linearly increased when Hy-D® was provided to sows (linear, $P=0.041$). This experiment demonstrated that 0.1% inclusion of Hy-D® in the diet of gestation sow improved reproductive performance of sows and their progeny. Consequently, 0.1% inclusion of Hy-D® in the diet of gestating sow is recommended.

Session 70 Poster 33

Comparison of classical, ancestral, partial and genomic inbreeding to achieve genetic diversity

J. Schäler[1], B. Krüger[2], G. Thaller[1] and D. Hinrichs[3]
[1]Christian-Albrechts-University of Kiel, Institute of Animal Breeding and Husbandry, Hermann-Rodewald-Straße 6, 24098 Kiel, Germany, [2]Christian-Albrechts-University of Kiel, Institute of Zoology, Am Botanischen Garten 1-9, 24118 Kiel, Germany, [3]University of Kassel, Department of Animal Science, Nordbahnhofstraße 1a, 37213 Witzenhausen, Germany; jschaeler@tierzucht.uni-kiel.de

The avoidance of inbreeding is a primary goal for the management of small populations and very important for the design of breeding programs in order to create productive progeny. The objective of this study was the investigation and comparison of different pedigree and genomic inbreeding coefficients for 76 selection candidates from a local pig breed. The data set included the pedigree of 1,273 individuals born between 1980 and 2015 and genotypes of selection candidates born between 2004 and 2014. Pedigree based classical, ancestral, and partial inbreeding coefficients were calculated with the GRAIN and VANRAD programs within the FORTRAN 77 software PEDIG. Genomic coefficients were calculated with the GCTA software by using three approaches: (1) variance of additive genetic values, (2) SNP homozygosity, and (3) uniting gametes. The lowest and highest pedigree inbreeding were estimated within partial and ancestral concepts. Genomic inbreeding showed an obvious difference among all genomic coefficients. Comparison between pedigree and genomic inbreeding showed all significant differences ($P<0.01$). Correlations between pedigree and genomic inbreeding coefficients ranged from -0.44 to 0.45. By applying a threshold <0.05 for inbreeding coefficients, 17 inbred selection candidates were identified and removed from parental generation. Ballou's concept of ancestral inbreeding was moderate correlated with genomic inbreeding measurements according to homozygosity and uniting gametes. Kalinowski's concept of ancestral inbreeding was negatively correlated with genomic inbreeding estimates regarding the variance of additive genetic values and uniting gametes. Correlations of partial inbreeding vary between 0.25 and 0.41, where Kalinowski's concept of 'new' inbreeding was correlated with all genomic inbreeding coefficients. However, Lacy's concept of partial inbreeding only correlated with genomic inbreeding regarding homozygosity.

Session 70 Poster 34

Effects of feeding level at early gestation on body condition and reproductive performances in sows

S. Seoane[1], J.M. Lorenzo[2], P. González[2], L. Pérez-Ciria[3] and M.A. Latorre[3]

[1]COREN, Av. da Habana, 32003 Ourense, Spain, [2]Cent. Tecnol. da Carne, Rúa Galicia 4, Parq. Tec. de Galicia, 32900 San Cibrao das Viñas, Ourense, Spain, [3]IA2-Universidad de Zaragoza, C/ Miguel Servet 177, 50013 Zaragoza, Spain; leticiapcgm@gmail.com

A total of 36 hyperprolific Danbred second- (n=12), third- (n=12) and fourth-parity (n=12) sows were used to evaluate the effect of feeding level at early gestation on the sow body weight (BW), backfat and loin muscle measures and reproductive performances. During the pregnancy, all the animals received a standard gestation diet. From day 1 to 30 after the first insemination, sows were given increasing feeding levels: 2.5 (control, n=12), 3.0 (n=12) or 3.5 kg/d (n=12). From day 31 to 90, all of them were fed 2.5 kg/d and from day 91 to the farrowing 3.0 kg/d. The BW of all sows from the previous weaning to the trial was recorded. Before and after farrowing, sows given at least 3.0 kg/d during the first 30 d post-insemination were heavier than those given 2.5 kg/d ($P<0.05$). Also, sow BW gain was affected by the diet and by the number of parity. So, groups were fed higher levels of feed (3.0 and 3.5 kg/d) gained approx. 44.0 kg from the previous weaning to the farrowing, whereas group that was fed the lowest level (2.5 kg/d) gained 40.0 kg ($P=0.05$). In addition, the BW gain from the previous weaning to the subsequent one decreased as the number of parity increased ($P=0.005$). On the other hand, backfat and loin muscle depths at farrowing were lower with 2.5 than with 3.5 kg/d ($P=0.08$ and $P=0.001$, respectively) and the fatness also increased with the number of parity ($P=0.03$). Similar results were observed in backfat and muscle gains from the previous weaning to the farrowing. No effect of parity number was detected on reproductive performances at birth but litter size tended to be lower with increasing levels of feed from 2.5 to 3.5 kg/d (+3.6 total piglets born; $P=0.06$ and +2.9 piglets born alive; $P=0.07$). However, sows given the lowest feed level had the lightest newborn piglets ($P=0.02$) and their weights tended to be more heterogeneous ($P=0.08$). It can be concluded that increasing feeding level during the first 30 d of pregnancy from 2.5 to 3.5 kg/d increased sow BW and backfat and muscle gain and carried out smaller litter sizes with heavier piglets.

Session 71 Theatre 1

Challenging selection for consistency of the rank in endurance competitions

I. Cervantes[1], L. Bodin[2] and J.P. Gutiérrez[1]

[1]Departamento de Producción Animal, Universidad Complutense de Madrid, Avda. Puerta del Hierro s/n, 28040 Madrid, Spain, [2]GenPhySE, INRA, 31320 Castanet-Tolosan, France; icervantes@vet.ucm.es

Horse performance in endurance competitions is finally defined by the position the horse reaches. The challenge of carrying out a genetic evaluation in such a trait has been usually faced by using atypical complex models, but they are not useful when both, good positioning and consistency of positions across races are aspired. The objective of this work was to estimate genetic parameters useful to select for decreasing the mean of the positions and reducing the variability of these positions of a given horse in several endurance races, as well as the genetic correlation between that variability and the position. An appropriate transformation of the trait was tested, and different models were tested including for both traits, the mean and its variability, the sex, the age and the competition as fixed effects and the rider, the rider-horse or the permanent environmental effect as an additional random effect besides the additive genetic. The total data set consisted of 2,863 ranking records from 621 horses (254 males, 253 females and 114 geldings) aged between 5 and 24 years. The pedigree contained 9,527 animals. The rank trait was transformed from its uniform distribution to a normalized variable to counteract skewness. A Double Hierarchical Generalized Linear model was applied using the ASReml program. The model including the rider-horse effect as additional random effect for both the mean trait and its variability was the only one providing reliable estimates, with a heritability for the mean of 0.09 and an additive genetic variance for the environmental variability of 0.08. These results were in the range of values estimated for other traits and species. A null genetic correlation between position and its variability suggests the possibility of independently selecting to reduce the variability of the position in endurance competitions.

Session 71 Theatre 2

Genome-wide association studies for performance traits in endurance horses

M. Chassier[1], C. Robert[2], C. Morgenthaler[1], J. Rivière[1], N. Mach[1], M. Vidament[3,4], X. Mata[1], L. Schibler[1], A. Ricard[1,4] and E. Barrey[1]

[1]INRA Jouy-en-Josas, UMR 1313 Génétique Animale et Biologie Intégrative, Allée de Vilvert, 78352 Jouy-en-Josas, France, [2]Ecole Nationale Vétérinaire d'Alfort, 7 Avenue du Général de Gaulle, 94700 Maisons-Alfort, France, [3]INRA Nouzilly, UMR 85 Physiologie de la Reproduction et des Comportements, INRA, 37380 Nouzilly, France, [4]Institut Français du Cheval et de l'Equitation, Pôle développement, Innovation et Recherche, La Jumenterie, 61310 Exmes, France; marjorie.chassier@inra.fr

The objective of the study was to identify single nucleotide polymorphisms (SNPs) associated with gaits (walk and trot), morphometric, cardiac and behaviour traits in endurance horses by performing genome-wide association studies (GWAS). For different subsets of 1,013 Arabian horses, information on in total 63 traits was available: conformation (n=1,012), morphometry (n=716), walk (n=673) and trot characteristics (n=668), cardiac dimensions (n=326) and behaviour traits (n=370). Genotypes were available for 1,309 horses, including those 1,013 horses with phenotypes and of which 86% were genotyped for 54k SNPs (EquineSNP50 BeadChip, Illumina) and 14% for 670k SNPs (Axiom Equine genotyping array, Affymetrix). An imputation from medium to high density allowed merging all the genotypic data. To identify associations between genotypes and the various phenotypes, we used a single-marker mixed model that included a genotype effect and a polygenic effect to take into account familial structures. For the skin thickness (subcutaneous fat marker), the GWAS revealed three significant quantitative trait loci (QTL) (10^{-5}<P-values<10^{-13}) corresponding to 70 SNPs distributed on chromosomes 4, 7 and 30. At the walk, GWAS revealed that the forward propulsion and dorsoventral power were significantly (10^{-5}<P-values<10^{-8}) associated to 30 and 7 SNPs, respectively, distributed on 15 chromosomes. For others traits relating to gaits (walk and trot), conformation, morphometric, cardiac and behaviour traits, GWAS revealed some significant SNPs (10^{-5}<P-values<10^{-7}), allocated on different chromosomes. This study indicated the polygenic determinism of aspects relating to endurance ability, and candidate genes will be investigated in the main QTL regions.

Session 71 Theatre 3

Genetic analysis of data from Swedish stallion performance test

Å. Viklund, L. Granberg and S. Eriksson

Swedish University of Agricultural Sciences, Animal Breeding and Genetics, P.O. Box 7023, 75007 Uppsala, Sweden; asa.viklund@slu.se

The stallion performance test (SPT) is the most important step in the breeding program for Swedish warmblood horses (SWB). In 2002, SPT was shortened from nine to seven days, and two major changes were introduced. The SPT for four- and five-year-old stallions were divided into two phases, where all tested stallions participate in the first phase, and the most promising stallions continue to the specialized second phase as either dressage or jumping stallions. Three-year-old stallions can now participate and be temporarily approved. The aim of this study was to estimate genetic parameters for SPT traits and genetic correlations between SPT and competition traits. SPT data consisted of 660 test results from 569 stallions tested in 2002-2017 which originated from 20 different studbooks. Competition data consisted of 50,943 horses born in 1980-2012. Genetic parameters were estimated with an animal model. Heritabilities ranged for conformation and gaits at hand from 0.04 (correctness of legs) to 0.81 (trot) and for gaits under rider from 0.44 (walk) to 0.62 (trot). For jumping (free and under rider), the heritability estimates were 0.55 (technique and ability) and 0.50 (temperament). Genetic correlations with dressage competition ranged from 0.73 (walk) to 0.79 (canter) for gait traits at SPT. Between jumping traits at SPT and show jumping the genetic correlations were 0.68 (temperament) and 0.74 (technique and ability). We conclude that SPT is a good tool for selection of warmblood stallions. The estimated average relationship between dressage and jumping stallions was much lower than between stallions within the same discipline, which can have influenced the estimates. Specialization within SWB has increased over time, and it would be interesting to investigate consequences of this further. Due to strong preselection of stallions for discipline, information on SPT traits is limited and should be combined with competition or young horse test data in genetic analyses.

Session 71 Theatre 4

Accelerometry for genetic improvement of gait of jumping horse

A. Ricard[1,2], B. Dumont Saint Priest[1], M. Chassier[2], E. Barrey[2] and S. Danvy[1]
[1]Institut français du Cheval et de l'Equitation, Pôle développement, Innovation et Recherche, 61310 Exmes, France, [2]Institut National de la Recherche Agronomique, Génétique Animale et Biologie Intégrative, 78350 Jouy en Josas, France; anne.ricard@inra.fr

Gaits of 1,477 young jumping horses (4 and 5 years old) were recorded using accelerometer device Equimetrix during 27 sport events in 2015 and 2016. From the raw data of acceleration recorded at 100 Hertz in the 3 dimensions for 10 seconds, 8 parameters were calculated for each gait. Genetic analysis used mixed model methodology with an animal model involving 10,907 ancestors for the animal effect and velocity, sex, age, event for the fixed effects. Genetic correlations with jumping were calculated using all performances in competition since birth year 1998, i.e. 232,952 horses, 406,750 ancestors and 458,269 annual performances from 2002 to 2016. Genome-wide analysis was performed including 541,175 SNP after quality control. Results showed a high heritability for dorsoventral displacement and stride frequency (>0.41), moderate for longitudinal activity (>0.19) and low for lateral activity (<0.07). The analogous characteristics of trot and canter were genetically correlated (>0.56). Heritabilities for walk were lower. Genetic correlation with jumping performance was null except one negative correlation (-0.22) for longitudinal activity at canter. GWAS revealed the importance of withers height for gaits and motivated including this trait as covariate in subsequent analyses to investigate gaits at constant height and velocity. Jumping ability and gaits characteristics may be selected mostly as independent traits. The only exception was longitudinal activity at the canter corrected for speed and height, so perhaps a sign of loss of energy.

Session 71 Theatre 5

Genome wide recognition and analysis of long non-coding RNAs in the transcriptome of Arabian horses

K. Żukowski[1], M. Stefaniuk-Szmukier[2], K. Ropka Molik[1] and M. Bugno-Poniewierska[1]
[1]National Research Institute of Animal Production, Krakowska 1, 32-083 Balice, Poland, [2]University of Agriculture in Cracow, Department of Horse Breeding, Institute of Animal Science, Adama Mickiewicza 21, 31-121 Cracow, Poland; kacper.zukowski@izoo.krakow.pl

Long noncoding RNAs (lncRNAs) were described as transcripts, usually larger than 200 bp without any protein-coding potential. However, easy to use lncRNAs definition does not parallel the computational reconstruction and recognition of transcripts within RNASeq experiments for both genome-guided analysis and de novo transcript assembly. Moreover, the role of non-coding fraction of transcriptome in regulation of gene expression and control of translation has not been clearly established, especially considering horses subjected to a certain training regime. The aim of the study was to recognize and analyse previously identified and novel lncRNAs in Arabian horses across different genome releases and tools. As material we used horse transcriptomes acquired from blood and muscles of 37 Arabian horses. The RNASeq analysis based on STAR alignment and Cufflinks transcript reconstruction models against two reference genomes EquCab2.0 and EquCab3.0. In both cases, we compared database annotated lncRNAs with those identified by analytic tools like FEELnc. The identified lncRNAs were filtered and functionally enriched. Due to early stage of EquCab3.0 genome assembly, our preliminary study results were based on EquCab2.0 release. However, the results of genome comparisons showed that the total number of lncRNAs in the database increased from 4,369 to 10,850 according to EquCab2.0 and EquCab3.0, respectively. The results of our de novo transcript assembly revealed more than 20,000 lncRNAs. The study was funded by the National Science Centre in Cracow, Poland (project no. 2014/15/D/NZ9/05256).

ROH as hint of selection in the genome of a modern sport horse breed

M. Ablondi[1,2], Å. Viklund[1], C.J. Rubin[3], G. Lindgren[1], S. Eriksson[1] and S. Mikko[1]
[1]Swedish University of Agricultural Sciences, Animal Breeding and Genetics, Almas Allé, 8, 7023 Uppsala, Sweden, [2]Università di Parma, Dipartimento Scienze Medico-Veterinarie, Via del Taglio, 10, 43126 Parma, Italy, [3] Uppsala University, Medical Biochemistry and Microbiology, Husarg, 3, 75236 Uppsala, Sweden; michela.ablondi@studenti.unipr.it

The horse breeding sector has to meet the growing demand of horses for high level competitions, while guaranteeing good health standards. Knowledge of the genetic background of sport-related traits could benefit the selection process. Recent advances in genome mapping are paving the way to further explore effects of selection. Strong selective pressure reduces genetic diversity which results in increased homozygosity at genome level. In this project, Runs of Homozygosity (ROH) were studied to find genomic regions under selective pressure of an equestrian sport horse breed. To identify differences in selective pressure between sport horses and a pony breed with low selective pressure on sport performance, data from 380 Swedish Warmblood horses (SWB), genotyped with a high density SNP array (670K) were analysed and compared to Exmoor pony genotypes. ROH were detected using a sliding windows approach in PLINK v1.90: short ROH reflect population history, whereas long ROH are indicators of recent selection. Length and location of ROH differed between the two breeds in agreement with the known breed history, with ROH more evenly distributed in the genome of the inbred Exmoor ponies. Long ROH were the rarest, although in both breeds they covered the largest proportion of the genome. Specific patterns in the location of long shared ROH (85% of the population) were found in SWB horses but not in Exmoor ponies. In SWB horses, enrichment analyses of the 65 shared ROH pointed out genomic regions in ECA4, ECA6 and ECA7, harbouring genes involved in muscle function, excitatory synaptic transmission, and development of central nervous system. Altogether, this indicates that movements, cognitive functions and personality related traits represent important targets of selection in the SWB breed, which has become a breed shaped for sport purposes.

Genetic structure and analysis of connectivity in the Pura Raza Español horse meta-population

M. Solé[1], M. Valera[2] and J. Fernández[3]
[1]Universidad de Córdoba, Genética, Ctra. Madrid-Cadiz km 396a, 14071 Córdoba, Spain, [2]Universidad de Sevilla, Ciencias Agroforestales, Ctra. Utrera km 1, 41013 Sevilla, Spain, [3]INIA, Mejora Genética Animal, Ctra. Coruña km 7.5, 28040 Madrid, Spain; jmj@inia.es

The Pura Raza Español (PRE) is an autochthonous Spanish horse population which is bred throughout 65 countries. This study analysed the genetic structure in this meta-population (MP) of 281,052 animals (on average 9.2 complete generations). Only those countries with at least 80 active breeding animals were considered: 27 countries comprising 77% of the complete pedigree. Additionally, genotypes from active animals (59% of complete pedigree) were available. Results indicated that F rate decreased in the last two decades (1990-2013) for the MP implying that there may be an explicit management against the rise of inbreeding. In general, unbalanced contributions of founders were found reflecting the high loss of genetic diversity over generations with N_{ef} being as low as 32 for the whole MP. Of the current MP, 73.6% was originated by females belonging to the Carthusian strain. Despite this differential contribution of some founders, the proportional contribution of each country to the global diversity was similar. The highest coancestry within country was found for Cuba (0.1509), being the only country with highly inbred individuals (F above 12%), and the lowest value was obtained for Spain (0.0574). This must be taken into account to avoid further decline in genetic variability and considerably increase in F, especially in the case of smaller countries like Cuba. Only 9 different countries presented private founders (founders with descendants only in a single population), which is a signal of the common origin for all the countries and/or a substantial exchange of genetic material within MP. These results encourage to go on with a coordinated management strategy for populations from every country, promoting the exchange of genetic material to increase the effective population size, and thus helping to maintain genetic diversity in the PRE population.

Session 71 Theatre 8

Genomic and genealogical coancestries within and between the Norwegian and the Swedish Fjord horses

S. Tenhunen, H. Fjerdingby Olsen, N.I. Dolvik, D.I. Våge and G. Klemetsdal
Norwegian University of Life Sciences, P.O. Box 5003, 1432 Aas, Norway; saija.r.tenhunen@gmail.com

One of the challenges in conservation of native breeds, like the Norwegian Fjord Horse, is how to evaluate relatedness between animals in different countries. Traditionally, pedigree data have been used to calculate relationships between animals, but pedigree information is often not comparable between different countries and breeding organisations. Therefore, methods based on genomic relationships are a good alternative. This study was based on 413 genotyped samples from Fjord Horses, of which 311 were Norwegian and 102 were Swedish. Coancestry was evaluated between genotyped animals based on pedigree (PED), molecular homozygosity (HOM) and shared homozygous segments (SEG) of different segment sizes. The smaller segment sizes (100 and 500 kb) seemed to detect more ancient inbreeding, whereas the larger sizes (1.5 and 2 Mb) seemed to detect more recent inbreeding. The fitting was worse for the larger than for the smaller segment sizes, when comparing to the other methods, and the average coancestry based on HOM showed higher correlations to coancestry from PED than from SEG. The complete generation equivalent for Norwegian genotyped horses was 13.7 and for Swedish 12.4. The effective population size (N_e) was calculated from the increase in coancestry (Δf) per generation, based on the Norwegian, the Swedish and the mixed population. Figures for N_e were larger with PED and HOM than with SEG. These results suggest that the HOM method may be more accurate than the SEG method to evaluate relationships between animals in different populations. Still, other studies have shown that use of SEG could be more beneficial when using selection tools like optimal contribution selection with commercial livestock. The breeding schemes in horses differ quite much from the ones in commercial livestock breeding, implying the need of further research on genomic relationship between horse populations by use of shared homozygous segments.

Session 71 Theatre 9

Genetic diversity of draft horses in the Netherlands

A. Schurink[1,2], S.J. Hiemstra[1], J.K. Oldenbroek[1], A. De Wit[1,2], S. Janssens[3], B.J. Ducro[2] and J.J. Windig[1,2]
[1]Wageningen University & Research, Centre for Genetic Resources, the Netherlands, P.O. Box 338, 6700 AH Wageningen, the Netherlands, [2]Wageningen University & Research, Animal Breeding and Genomics, P.O. Box 338, 6700 AH Wageningen, the Netherlands, [3]KU Leuven, Division Animal and Human Health Engineering, P.O. Box 2456, 3001 Leuven, Belgium; anouk.schurink@wur.nl

The draft horse population is composed of rare horse breeds in the Netherlands and closely related to the population in Belgium. The number of draft horses decreased drastically after mechanisation in agriculture, when their draft power was no longer needed. The aim of this study was to investigate genetic diversity in the Dutch draft horse population based on pedigree data. Pedigree data contained 25,672 horses. Inbreeding and kinship were calculated. The number of foals born per year decreased from 649 in 2009 to 320 in 2016 as a result of a decreased number of mares and stallions used for breeding. Average generation interval was 7.3 years. Effective population size was 130, and deltaF from 1960 to 2017 was 0.62% per generation. Kinship between parents increased during the last decade, which resulted in a deltaF of >1% per generation in that period. Parameters indicated that the genetic management of the draft horse population needs to be adapted in an attempt to reduce the risk of problems due to inbreeding. Extension of collaboration and exchange of genetic material between the Dutch and Belgian populations, which face similar problems, could be beneficial for both populations. A combined analysis of pedigree or DNA information from both populations will increase our understanding of exchange and relationship between these populations.

Cupidon, a tool of the French Society of Working Equids to support mating planning in small breeds

C. Bonnin[1], M. Sabbagh[2], O. Lecampion[3] and S. Danvy[2]
[1]Société Française des Equidés de Travail, 83-85 Boulevard Vincent Auriol, 75013 Paris, France, [2]Institut Français du Cheval et de l'Equitation, pôle DIR, La jumenterie du Pin, 61310 Exmes, France, [3]Association Nationale des Races Mulassières du Poitou, 2 rue du Port Brouillac, 79510 Coulon, France; cleme.bonnin@wanadoo.fr

How to properly reason crossbreeding and maintain genetic variability within breed when the choice of breeding mating partners is limited. In order to support their optimization efforts, the French Society of Working Equids (SFET) has decided to set up a common tool bringing together various relevant indicators selected from among the most relevant ones studied. At first, the tool was finalized and tested for seven breeds of donkeys, then made available on internet to breeders. The same indicators are calculated and disseminated regardless of breed. However, genealogies are not known with uniform depth and certainty and each studbook had slightly different challenges and expectations. Accordingly, definitions of thresholds of advice were breed-specific. For each of its females, a breeder will have individual advice on all available males, including basic data like genealogy, type of covering, place, number of progeny. Indicators of genetic variability such as the coefficient of consanguinity of a potential progeny and the coefficient of originality (IOG) of the donkey in his studbook are included. The IOG is calculated for each donkey as follows: IOG = ΣAM (ΦBAM.cAM) with: AM a major ancestor, ΦBAM the kinship coefficient between the individual considered and the ancestor AM, and cAM the marginal genetic contribution of the AM ancestor in the breed. In order to be easily usable, these indicators are given numerically and imaged with stickers of colours of green (mating advised) to black (mating disadvised). Launched for the 2017 covering season on the SFET website, this tool has been very well received by breeders. The first year of use has been such a success that this interactive tool will be developed and put online for the 18 other working equine breeds of the SFET (draft horses and territory horses) for the 2018 breeding season.

Genetic and statistical analysis of the coat colour roan in Icelandic horses

K. Voß, D. Becker, J.L. Tetens and G. Thaller
Institute of Animal Breeding and Husbandry, Kiel University, Hermann-Rodewald-Straße 6, 24118 Kiel, Germany; kvoss@tierzucht.uni-kiel.de

The white patterning coat colour roan is a mixture of white and coloured hairs in the body while the head, lower legs, mane and tail remain fully coloured. Roan is known to be embryonically lethal in Belgian draft horses when homozygous. Furthermore, roan is dominantly inherited and shows an association to the *KIT* gene. This study aimed to identify the causative genetic variant responsible for the patterning and examine if roan in Icelandic horses is lethal in a homozygous condition. Therefore, all coding exons of *KIT* were screened in 30 cases and 23 controls, and an association analysis was performed. Additionally, a statistical analysis of mating information available through the database Worldfengur was conducted. Worldfengur contains 460,000 registered Icelandic horses, of which 2,166 (0.45%) are roan. Matings of roan and non-roan horses resulted in a ratio of 1 roan to 1 non-roan offspring according to the expectations for a dominantly inherited trait. The observed offspring ratio for matings of horses that were both roan was 3 roan to 1 non-roan, implying that roan is not lethal in a homozygous condition in Icelandic horses. Sequencing of the coding exons of the *KIT* gene identified 16 polymorphisms, of which five were located in exons. None of the polymorphisms are causative for the roan phenotype. However, an association analysis of the polymorphisms showed a highly significant association ($P<0.001$) to the coat colour roan. Further genome analyses will be performed to elucidate the causative genetic variant coding for the coat colour roan.

Session 71 Theatre 12

Validation of a molecular screening tool for the detection of chromosomal abnormalities in donkey

J. Poyato-Bonilla[1], G. Anaya[2], J. Dorado[3] and S. Demyda-Peyrás[4]
[1]ETSIA, Universidad de Sevilla, Departamento de Ciencias Agroforestales, Ctra. Utrera, km 1, 41013 Sevilla, Spain, [2]Laboratorio de Diagnóstico Genético Veterinario, Grupo Investigación MERAGEM, Universidad de Córdoba, Departamento de Genética, CN IV km 396, Edificio Gregor Mendel, Campus Rabanales, 14071 Córdoba, Spain, [3]Veterinary Reproduction Group, Universidad de Córdoba, Department of Animal Medicine and Surgery, CN IV km 396, Campus Rabanales, 14071 Córdoba, Spain, [4]IGEVET-Instituto de Genética Veterinaria. Ing. Fernando N. Dulout, UNLP-CONICET LA PLATA, Facultad de Ciencias Veterinarias UNLP, 1900 La Plata, Buenos Aires, Argentina; juliapb92@gmail.com

Chromosomal abnormalities are one of the main causes of infertility and reproductive problems in horses. Nowadays, the detection of individuals showing this type of aberrations is rising due to the use of new diagnostic tools based on molecular markers located along the autosomal and sexual chromosomes. In contrast, despite its great similarities with the horse, there is only one recent report of sterility associated with chromosomal abnormalities in the domestic donkey (*Equus asinus*), a scarcely studied species in spite of its importance for the human being and the endangered status of certain breeds. In the present study, we analysed the possibility of applying an STR (Single-Tandem-Repeat)-based molecular method developed for horses as a diagnostic tool for these abnormalities in donkeys. The frequencies of five X-linked (*LEX003*, *LEX026*, *TKY38*, *TKY270* and *UCEDQ502*) molecular markers and one Y-linked gene (Sex-Determining Region Y, *SRY*) were determined in 121 donkeys of two different Spanish breeds (Andaluza and Encartaciones) and 58 donkeys from north Africa (Moruna). Taking as reference the analysed population, sensitivity and specificity of the diagnostic tool were determined based on expected profiles of chromosomal abnormalities and results of heterozygosity of the molecular markers used. The molecular panel showed 100% sensitivity and 98.78% specificity. Hence, its use complementarily with other cytogenetic techniques constitutes a highly specific, rapid and low cost detection tool for chromosomal abnormalities and their characterization in domestic donkey.

Session 71 Theatre 13

Impact of alternative RNA splicing on β-casein transcripts in mare's mammary gland

A. Cividini[1], P. Jamnik[2], M. Narat[1] and P. Dovč[1]
[1]University of Ljubljana, Biotechnical faculty, Department of Animal Science, Jamnikarjeva 101, 1000 Ljubljana, Slovenia, [2]University of Ljubljana, Biotechnical faculty, Department of Food Science and Technology, Jamnikarjeva 101, 1000 Ljubljana, Slovenia; angela.cividini@bf.uni-lj.si

The result of alternative exon splicing and posttranslational modifications of mare milk β-casein (β-CN) gene transcripts are different isoforms of mRNA for β-CN. The β-CN mRNA showed three different splicing patterns due to differential splicing of weak exons 5 and 8, and due to an alternative donor splice site in exon 7. The known three splicing patterns differed in the presence and relative quantities of the four transcripts, representing the full-length variant (1), variable splicing patterns (2) and short variant of β-CN (3). The aim of this study was to investigate whether correlation between alternative RNA splicing patterns and isoforms of mare β-CN can be established. In 19 lactating Lipizzan mares we confirmed three different splicing patterns of weak exons and expected isoforms of β-CN that reflect differential splicing events. To determine the inclusion of weak exons from the β-CN gene, RT-PCR method was performed. Amplified fragments were analysed electrophoretically and quantified using capillary electrophoresis. The isolated casein fractions from mare's milk were separated by two-dimensional gel electrophoresis (2-DE) coupling IEF in the first and SDS-PAGE in the second dimension. 2-DE protein gel images were analysed using image processing 2-D Dymension programme. To confirm the identification of the separated caseins, the N-terminal amino acid sequencing was performed. In Lipizzan mare's milk prevailed full-length β-CN. The full-length β-CN and β-CN lacking exon 5 were located in the area of 32 kDa and 30 kDa, respectively. Indeed, 2-DE profile of pure equine β-CN is difficult to interpret because of numerous phosphorylated variants. However, the results showed, that the differences between β-CN isoforms on 2-DE profile could be identified due to spot normalized volume of the isoforms. We found, that the percentage of full-length β-CN and β-CN lacking exon 5 were correlated with splicing events (1, 2 and 3) affecting weak exons. Further investigations should be performed to determine phosphorylated variants on 2-DE profile.

Session 71 Poster 14

New phenotypes based on morphology to benefit from genetic correlation with gait

F. Bussiman[1], B. Abreu Silva[1], J. Eler[1], J.B. Ferraz[1], E. Mattos[1] and J.C. Balieiro[2]
[1]College of Animal Science and Food Engineering, Veterinary Medicine, Rua Duque de Caxias Norte, 225, 13635-900, Brazil, [2]College of Veterinary Medicine and Animal Science, Department of Animal Nutrition and Production, Rua Duque de Caxias Norte, 225, 13635900, Brazil; jbferraz@usp.br

In Campolina horse, all registered animals are evaluated at around 36 months of age, considering 16 morphometrics traits and eight gait scores. In this study, it is proposed to use combined morphological measurements in order to benefit from genetic correlations between morphology and gait. Around 41,125 phenotypes and 107,951 pedigree information were used to perform a multi-trait analysis, considering the fixed effects of contemporary group (birth year and season), herd, year of registration and age as covariate, and for gait trait additionally the technician as uncorrelated random effect. Traits were: difference between height at withers (HW) and height at croup (HC), overall harmony (1 when HW=HC=BL; BL= body length, else 0), difference between head length and neck length (HLNL), difference between head length and shoulder length, and gait scores (1 to 5 classes). The multi-trait model was implemented in THRGIBBS1F90 with 800,000 samples, a burn-in period of 200,000 and a thinning interval of 100 samples. Heritability estimates ranged from low (0.08 for HLNL) to very high (0.88 for harmony). Technician effect accounted for 73.4% of the phenotypic variance for gait scores, showing the importance of new strategies of phenotyping for gait in this breed. Genetic correlations between gait scores and harmony-related morphometric traits ranged from -0.94 (HWHC) to 0.00 (HLNL). According to our results, gait had no relationship with one single measurement but with sets of combined measurements. Genetic correlation between HWHC and Gait was favourable for selection purposes, with the gait class increasing with shorter distance between HW and HC. Gait in Campolina horse breed has a complex genetic relationship with morphology, and more studies and more phenotypes, also for gait, are necessary to determine better approaches to select for this trait.

Session 71 Poster 15

Genetic parameters in the Bardigiano horse breed population

M. Ablondi[1], V. Beretti[1], M. Vasini[2], P. Superchi[1] and A. Sabbioni[1]
[1]Università di Parma, Dipartimento Scienze Medico-Veterinarie, Via del Taglio, 10, 43126 Parma, Italy, [2]Libro Genealogico Cavallo Bardigiano, Associazione Regionale Allevatori dell'Emilia Romagna, Strada dei Mercati, 17, 43126 Parma, Italy; michela.ablondi@studenti.unipr.it

The Bardigiano is an Italian native horse breed with excellent resilience, well adapted to mountain areas. In 1977, the Bardigiano studbook was founded to improve the use of this breed for riding purposes while maintaining its distinctive features. To facilitate the conversion of the Bardigiano to an equestrian horse type, one Arabian stallion was included in the breeding program in 1992. Body measurements (height at withers, chest girth, cannon bone circumference, shoulder length) have been recorded since 1977. In 1987, conformation, attitude and gait-related traits, evaluated as 10 grading traits and 10 linear scores, were also introduced in the assessment of horses to improve rideability. Therefore, for nearly 3,000 out of 5,135 horses with body measurements, additional traits were also available. The 10 grading traits, measured on a scale from 1 (extremely undesirable) to 10 (excellent), are summed and used to calculate a total score, which is used to approve stallions for breeding. This study aimed at assessing the suitability of body measurements and grading traits for genetic evaluation. The heritabilities and genetic correlations were estimated in univariate and bivariate animal models, using a statistical model that accounted for: gender, date of birth, age at evaluation, and percentage of Arabian blood. The body measurements showed heritabilities ranging from 0.23 (length of shoulder) to 0.64 (height at withers). The conformation and attitude traits displayed heritabilities from 0.13 to 0.31, whereas for gait-related traits the heritability was 0.10. The genetic correlations between the total score and body measurements were moderate to high, ranging from 0.30 to 0.87. The highest genetic correlations were found between body measurements and the grading trait 'development' (0.55 to 0.97). We conclude that body measurements can be used effectively as indicators of conformation and attitude traits in the Bardigiano breed. Further studies combining body measurements in morphometric indices are suggested to optimise the breeding strategy.

Session 71 — Poster 16

The expression profile of VAV3 gene in whole blood of Arabian horses during competing at race track

M. Stefaniuk-Szmukier[1], K. Ropka-Molik[2], K. Piórkowska[2] and M. Bugno-Poniewierska[2]
[1]University of Agriculture in Kraków, Department of Horse Breeding, Aleje Mickiewicza 21, 21-120 Kraków, Poland, [2]National Research Institute of Animal Production, Krakowska 1, 32-083 Balice, Poland; katarzyna.ropka@izoo.krakow.pl

Arabian horses are competing at the race track when 3 years old. The most successful horses start in more than one racing season. During training schedule, the skeletal reinforcement occurs, but assessment of skeletal condition is difficult. Failure of proper bone maintenance during conditioning can result in lameness in performing horses, which is one of the most important reasons for loss of training days and poor performance, and by that generating substantial economic loss. The objective of this study was to evaluate the expression patterns of VAV3 gene, recognized as major factor in the organisation of the cytoskeleton of osteoclasts, in whole blood of Arabian horses during 3 racing seasons. A total of 53 Arabian horses were introduced to race track training when they were 2.5 years old and started participating in races. According to training schedule, horses (samples) have been divided into 10 groups. The expression profile of the analysed gene was assessed using real-time PCR method. Analysis was performed using EvaGreen® qPCR Mix Plus (ROX) in three technical replicates for each sample for the analysed gene and the endogenous control (GAPDH). For determining significant differences in estimated transcript abundance between the investigated periods, the one-way ANOVA with Duncan's post hoc test was used. Our results indicated that expression pattern of VAV3 is influenced by exercise. We observed significant decrease of transcript abundance ($P \leq 0.05$) in horses after phases containing intensive canter workload compared to untrained horses and horses after light training. These results suggest that intensive workload could downregulate osteoclastogenesis through VAV3 interactions, which indicates new possibilities in the field of searching new biomarkers of bone activity in race horses. Founded by the National Science Centre (NCN) (project no. 2014/15/D/NZ9/05256).

Session 71 — Poster 17

Identification of polymorphisms within ACTN3 locus in Equus caballus

K. Ropka-Molik[1], A.D. Musiał[2], M. Stefaniuk-Szmukier[3], T. Szmatoła[1] and M. Bugno-Poniewierska[1]
[1]National Research Institute of Animal Production, Krakowska 1, 32-083 Balice, Poland, [2]Jagiellonian University, Gołębia 24, 31-007 Kraków, Poland, [3]University of Agriculture in Cracow, Aleje Mickiewicza 21, 31-120 Kraków, Poland; katarzyna.ropka@izoo.krakow.pl

ACTN3 gene codes for α-actinin-3 (alpha-actinin skeletal muscle isoform 3), a protein localized in the Z-line of skeletal muscle. The actinin 3 is critical to anchor the myofibrillar actin filaments and plays a key role during muscle contraction. Due to the localization at Z-line, ACTN3 cross-linked glycogen phosphorylase is the key enzyme catalysing glycogen metabolism. The aim of this study was to identify polymorphisms in ACTN3 gene in Arabian horses, which can be related to racing performance traits. Sanger sequencing was performed on 72 pure breed Arabian horses and allowed to detect 17 polymorphisms localized in promoter region (7 SNPs), exons (4 SNPs), introns (5 SNPs) and 3'UTR (one SNP). From all synonymous variants identified in exons (8, 13, 16, 19), the ENSECAT00000021149.1:c. 2353C>T mutation has been selected for further analysis and genotyped in 440 Arabian horses using PCR-RFLP method. The novel c. 2353C>T SNP, localized in exon 19, is considered as splice region variant with potential effect on slicing process. Genotype analysis performed on a larger horse population (n=440) showed that horses with TT genotype were most numerous (72%), while the opposite homozygotes accounted only for 2%. The genotyping of 5 SNPs detected in intron 15 (A>G change, rs1144429650; rs1144535704; rs1138678833; rs1140963666; rs396707901) using Sanger sequencing showed the significant strong linkage disequilibrium of rs1144429650 - rs1144535704 (D' – 1.0) and rs1140963666 - rs396707901 (D' – 1.0) SNPs, which created two LD blocks. The detected polymorphisms might affect splicing process of ACTN3 gene and as a result transcript abundance of actinin 3 gene. Future research is needed to establish the potential association of identified polymorphisms and racing performance traits in horses. Founded by the National Science Centre (NCN) (project no. 2014/15/D/NZ9/05256).

Genetic and phenotype variation of Croatian Posavina horse

A. Ivanković[1], J. Ramljak[1], I. Šubek[2], H. Grabić[2] and M. Glasnović[3]
[1]University of Zagreb Faculty of Agriculture, Department of Animal Science and Technology, Svetošimunska cesta 25, 10000 Zagreb, Croatia, [2]Breeders Association of Croatian Posavina horse, Martinska Ves Desna 67, 44201 Martinska Ves, Croatia, [3]Public Institution for the Management of Protected Areas of Nature in the Zagreb County 'Green ring', 151.samoborske brigade HV 1, 10430 Samobor, Croatia; aivankovic@agr.hr

Croatian Posavina Horse is one of the autochthonous horse breeds in Croatia that has been systematically preserved for more than two decades. Conservation program is based on the balance between sire and dam lines with the main goal to preserve observed genetic diversity. Breeding is regularly monitored through population trends by analysing condition, phenotype and genotype variability in Posavina Horse populations. In collaboration with the breeding association, 50 reproductively active stallions were phenotyped and genotyped (for 15 microsatellites). Obtained measures of withers height, chest circumference, and cannon bone circumference were 141.9±4.34 cm, 194.4±8.82 cm, and 22.2±0.95 cm, respectively. Comparison of these results with previous findings revealed decrease in withers height and chest circumference and constancy of cannon bone circumference over time. Comparison of dispersion parameters revealed decrease in phenotypic variability in all traits examined, which may indicate consolidation of conformation traits. Mean number of alleles of 7.3 determined in microsatellite analysis across all the examined loci was similar to previous findings. The observed heterozygosity was higher than expected (0.743 vs 0.705), and F_{IS} value (-0.043) revealed slight excess of heterozygous individuals. By taking into account the results of both phenotypic and genotypic analysis, it can be stated that phenotypic consolidation of the breed may not have caused major loss of initial genetic variability. This indirectly justifies the selected strategy for in *in situ* conservation. Monitoring of the breed including the phenotypic and genotypic level, and meticulously planned mating should be continued in order to protect existence of the breed. To also ensure long term viability, stronger efforts should be made in promotion and economic affirmation of the Croatian Posavina Horse breed.

Contribution of male lines to biodiversity conservation of the Silesian horse population

I. Tomczyk-Wrona and A. Chełmińska
National Research Institute of Animal Production, Department of Horse Breeding, Sarego 2, 31-047 Kraków, Poland; agnieszka.chelminska@izoo.krakow.pl

The goal of the conservation programme of the Silesian horse is to maintain its old type that represents a specific breeding standard and meets specific pedigree conditions related to historical determinants of breed creation. The Silesian horse is largely derived from the Oldenburg breed. With regard to biodiversity conservation, the relationship of approved Silesian horses with established male lines was analysed. To implement the conservation programme for the 2016/2017 season, 299 Silesian stallions were approved for mating to 915 mares in the programme. Analysis of the Silesian horse population showed representatives of six male lines, which were created based on imported pedigree Oldenburg horses. Five of the six lines derived from the Anglo-Norman stallion NORMANN (Introuvable AN – Seduisante SF) born 1868 in France. In 1871 to 1887 this stallion produced 12 stallions and 71 brood mares, giving rise to a strong line that still dominates the Oldenburg and Silesian breeds today. Most credit for propagating the NORMANN stallion line belonged to his son RUBICO, born in 1877. He produced two outstanding sons who consolidated the line by giving two parallel branches. These were RUTHARD and WITTELSBACHER. The subsequent branches of the line were established by their sons and grandsons. The ERBGRAF stallion line, well known to Oldenburg breeders, is currently not found in Silesian horse breeding. The present-day lines are EDELMANN (Ed), RUTER (Ru), ROLAND (Ro), GIDO (Gi), GAMBO (Ga). The sixth line was established by the Anglo-Norman stallion CONDOR (Fondroyant II xx – Sėduisante AN) of German breeding. The stallion CONDOR (Co) founded the Anglo-Norman line in Oldenburg breeding, which is also abundant in Silesian horse breeding. Most of the stallions originate from two main lines EDELMANN and GAMBO, which form more than 60% of the entire approved population of Silesian stallions. The lines most at risk are RUTHER and ROLAND, which together constitute less than 10% of the stallion population. The loss of these lines would considerably impoverish the biodiversity of the conserved population of Silesian horses and should be avoided.

Crypto-Tobiano horses as a breeding problem in the Hucul population in Poland

M. Pasternak, A. Gurgul and J. Krupiński
National Research Institute of Animal Production, Krakowska 1, 32-083 Balice, Poland; marta.pasternak@izoo.krakow.pl

Knowledge of coat colour genetics is especially important in Hucul horse bred in Poland, where there has long been a problem in distinguishing white spotting pattern (Tobiano) from white markings. According to the provisions of the breeding standard published by the Polish Horse Breeders Association, white markings disqualifies a Hucul horse from breeding, whereas Tobiano spotting is accepted. Horses with white markings, especially stallions, are not accepted for breeding and cannot be entered into the registry. However, a horse with apparent white spots may be crypto-Tobiano and have, for example, only white spots on legs. The Tobiano spotting pattern is associated with chromosome 3 inversion, with the proximal end located between the *ADH1C* and *PDLIM5* genes (*ECA3q13*), and the distal end between the *KDR* and *KIT* genes (*ECA3q21*). This study aimed to determine the scale of the problem of crypto-Tobiano pattern occurrence in Hucul horses in Poland, using a PCR method described by Brooks *et al*. The test material used was blood sampled from 96 piebald Hucul horses and 55 horses described in the passport as single-coloured with white markings. DNA isolated from the blood was used to identify the inversion on chromosome 3. The results showed the presence of chromosome 3 inversion in all of the 96 piebald horses (100%), thus confirming the presence of the Tobiano gene. The inversion was also located in 10 out the 55 horses described in the passport as single-coloured with white markings, so 18% of the animals were misclassified for colour. In fact, these horses have a crypto-Tobiano pattern, expressed mainly as white spots on the legs. These results demonstrate the essential need for DNA tests to verify coat colour in doubtful cases, in order to avoid valuable horses being eliminated from breeding.

Author index

C

D

L

M

N

O

P

S

T

Z

Printed in the United States
by Baker & Taylor Publisher Services